THE THEORY OF MEASURES AND INTEGRATION

THE THEORY OF MEASURES AND INTEGRATION

Eric M. Vestrup

WILEY-INTERSCIENCE

A JOHN WILEY & SONS, INC., PUBLICATION

This text is printed on acid-free paper. ♾

Published by John Wiley & Sons, Inc., Hoboken, New Jersey.
Published simultaneously in Canada.

Library of Congress Cataloging-in-Publication is available.

Vestrup, Eric M., 1971–
The theory of measures and integration / Eric M. Vestrup.
p. cm. — (Wiley series in probability and statistics)
Includes bibliographical references and index.
ISBN 0-471-24977-7 (acid-free paper)
1. Measure theory. 2. Integrals, Generalized. I. Title. II. Series.
QA325.V47 2004
515'.42—dc21

2003053475

Printed in the United States of America.

10 9 8 7 6 5 4 3 2 1

To my parents,
Raymond R. Vestrup
and
Linda M. Vestrup

Contents

Preface

The goal of this book is to present the major results of classical measure and integration theory in as clear, rigorous, and detailed a fashion as possible. This book assumes that the reader has studied advanced calculus and elementary analysis. For example, the first eight chapters of Walter Rudin's *Principles of Mathematical Analysis* should provide a strong preparatory framework for the material in this book.

Measure Theory and its sister Integration Theory are viewed by some as special topics to be subsumed under the larger heading of *Analysis*, and many analysis books cover these topics in varying degrees of detail. However, these topics are rich enough that they can stand alone as deep and fascinating topics. Modern probability theory and mathematical statistics rely heavily on measure and integration theory, and to understand the former topics, one must be competent in the latter topics.

In the writing of proofs and in my choice of notation, I have tried to be as explicit as possible. Elegance or a sense of aesthetics have without exception deferred to clarity, although one can certainly be both clear and elegant in certain situations. If anything, I have erred on the side of perhaps showing too many details, although I believe most readers will at some time be grateful for the details.

Anybody who knows the area well and purviews this book will see the influence of the texts of Walter Rudin, Kai Lai Chung, and Paul Halmos. One will also see, in tone and in the style of presentation, the influence of those magisterial and stupendous works *Real and Abstract Analysis* by Edwin Hewitt and Karl Stromberg, and *Probability and Measure* by Patrick Billingsley. All of these men and their works were my teachers in this area, and any apparent profundity or excellence on my part is merely a reflection of their mastery and presentation. I make absolutely no claim to originality in this book, as all of the material within is quite classical. My contribution comes [so I hope] in making certain difficult areas of the classical material very clear and accessible to the reader.

There are exercises found at the end of almost every section. These exercises range from being trivialities to being quite substantial developments in and of themselves. Hints or outlines are provided for many of the more substantive claims made in the exercises. As Hewitt and Stromberg say in their introduction, the more heroic readers may ignore the hints and outlines, but the hope is that every reader will be grateful for some of them. Certain exercises are marked with an asterisk; these exercises are results that are used implicitly or explicitly in later work. I believe that some of the exercises, besides providing useful extensions or side comments pertaining to the results in the text, are also interesting on their own merits. Exercises are quite often stated as assertions. For example, if *P has property X* appears in the exercise, this should be interpreted as *Give a proof of the claim that P has property X.* At any rate, the reader shouldn't have any problem figuring out exactly what is asked of him in the exercises. As with the material in the text, the exercises are again of a classical nature; I again make no claim of originality in this regard.

I believe that this book could serve as a primary or supplementary text for (1) a semester-long or year-long real analysis sequence that deals heavily with integration and measure theory, (2) a complementary background text for graduate courses in analysis, probability, and mathematical statistics, and, of course, (3) self study by those with mathematical maturity and the requisite background. This book is solidly "graduate-level" if we average over all of the contents. [There are some areas that solid and uncommonly mature undergraduates would find accessible, and there are some areas that require set theoretic considerations that are usually not found until later in graduate school, if it all.]

It may be helpful to briefly discuss the material in this book, so a relatively short description of the chapters will now be given.

Chapter 1. Sections 1-5 are required, and Sections 6 and 7 are optional. The first five sections are straightforward, and shouldn't pose much of a problem. I have tried to give the reader as explicit a description of Borel sets as possible without being too pedantic. Sections 6 and 7 are fairly deep and involved, but the advanced reader may find them worth the study. With only a few very tangential exceptions, nothing from Sections 6 and 7 will be used in later work.

Chapter 2. Those who merely want to study measures in the abstract can possibly get by with only the first two sections. However, this would require missing out on the construction of Lebesgue measure, which surely is an important item. Therefore, Sections 3 and 4 are also highly recommended. The first two sections are used everywhere in the text. Sections 3 and 4 are used primarily in Chapters 3, 4, and 7 when dealing with Euclidean space. The instructor should at least cover Sections 1-3 thoroughly.

Chapter 3. This is where we obtain our first collection of deep and foundational results. A minimalist approach covers the first six sections, although really all ten sections should be covered in detail. This chapter is far deeper than the first two chapters, representing a collection of some of the most important results in measure theory.

Chapter 4. Sections 1-4 are required, as they discuss our rigorous formulation of length, area, volume, and so on. These sections serve as a sort of apologetic for all of the work done in Chapters 2 and 3 in the sense that we show that Lebesgue measure does everything [or almost everything] that we expect length, area, volume, [and so on] to do. Sections 5-9 are optional, but in my opinion they are fascinating on their own merits. It is my hope that those who read these latter sections find the results of interest.

Chapter 5. Sections 1-7 are required, while Section 8 is optional. Like Chapters 1 and 2, this section is really nothing more than a collection of definitions and some basic results that may not be seen as anything more than necessary evils for the more interesting things to come. This chapter in essence is the grab-bag of the results needed to construct the Lebesgue integral in Chapter 6.

Chapter 6. This chapter is the very soul of the book, and nothing can be safely skipped by the reader. In this chapter, we painstakingly construct the abstract Lebesgue integral, proving many properties concerning it. This chapter will be heavily used in all that follows.

Chapter 7. This fairly short chapter demonstrates that Lebesgue and Riemann integrals coincide in a large class of situations. Those pressed for time may skip this chapter and take its results for granted. However, there is nothing particularly hard about the theory. Section 5 may seem unimportant at first, but it will be used in the [optional] discussion of convolutions in Chapter 10. I would have liked to include a careful buildup to a general change of variables result for multiple integrals, but space considerations prevent such a buildup. The same space considerations prevent a detailed discussion of the famous theorems of Lebesgue that relate derivatives to Lebesgue integrals.

Chapter 8. This and the following three chapters present the main avenues through which one takes the theory of Chapter 6. For this sizable chapter, the first seven sections are absolutely required, as these results and discussions are foundational for both analysis and measure theory. Sections 8 and 9 may be thought of as optional, especially in light of the fact that the main result of these two sections will be obtained [in a slightly weaker form] using the techniques of Chapter 9.

Chapter 9. Section 1 is essential, presenting the famous and important Radon-Nikodym Theorem. Section 2 is a fairly deep section, and some may justifiably find that the generalization to the Radon-Nikodym Theorem obtained in that section is not worth all of the extra work. It is there for those who are interested, and those who are not interested [or are pressed for time] need not worry that this section will be needed for later work. Section 3 presents a second proof of the Riesz Representation Theorem as originally given in Section 8.9. Readers who skipped Sections 8.8 and 8.9 might want to study this section so that the Riesz Representation Theorem may be seen at least once. Section 4 is a long and involved section that may be safely skipped by those who are not planning to study Chapter 11, or who want to study the Strong Laws in Chapter 11 without spending so much time on the constructions given there. [Section 4 in no way requires Sections 2 or 3.] For those who will want all of the details behind the Strong Laws, this section is absolutely essential, forming a key component for the convergence results given in Section 3 of Chapter 11.

Chapter 10. This is the longest chapter in the book, and presents the basic theory regarding product measures and integrals over product spaces. Despite the length of this chapter, those who are in a hurry can get by with merely covering the first two sections, where a very complete and lucid presentation of the standard theory is given. The remaining four sections present interesting applications of the Fubini Theorems and the general theory of product measures, and are there for those who want to see "Fubini in action" and the like. However, these four sections are not used in Chapter 11, so the lecturer may pretty much do as he pleases with these sections. It is hoped that some readers will appreciate the discussion of convolutions and the famous Hardy-Littlewood Maximal Theorem.

Chapter 11. Chapter 11 discusses the product of two measure spaces, while this chapter discusses arbitrary products of measure spaces. There are five sections in this chapter. The first two sections carefully [some might say tediously] construct the requisite items, showing that certain objects actually exist and make sense. Depending on one's tastes and desires, the work produced by these two foundational sections may be viewed as interesting or an annoyance. Those who are under a time constraint might be able to get away with covering the notation and definitions, while presenting the basic ideas and results without spending a lengthy amount of time on their proofs. However, part of measure theory and mathematics in general is a quest for rigor and details, so I would encourage everybody to study the first two sections, painful as they might be. Section 3 of this chapter may be thought of as a first application of the theory of Sections 1 and 2, or it may be regarded as more machinery needed to state and prove the Strong Laws of Large Numbers. The same comments regarding Sections 1 and 2 apply to Section 3. Sections 4–6 represent the pinnacle of the chapter [and perhaps of the entire book], where careful proofs of the famous Strong Laws are presented. It would have been of interest to present some further topics regarding infinite-dimensional product spaces, such as Martingale Theory, Kakutani's Dichotomy, Brownian Motion, and a general treatment of stochastic processes, but limitations of space [and the author's energy!] and a general desire to not wander too far from the beaten path prevent their inclusion.

My hope has been to write well enough and clearly enough that even the most advanced concepts appear accessible to general readers pursuing a Ph.D. or graduate degree in mathematics, statistics, or a field that requires a rigorous background in integration and measure theory. Those who read the book will of course be able to make up their own minds on whether I have succeeded in this regard.

ERIC VESTRUP

Downers Grove
Illinois

Acknowledgments

Many people have directly or indirectly contributed to this book. I would like to acknowledge the great amounts of help given by Professors Alan Fenech, Wesley Johnson, Prabir Burman, Ed Mack, Jessica Utts, Rahman Azari, Robert Shumway, Rich Levine, P. K. Bhattacharya, and George Roussas in matters mathematical and nonmathematical during my times as a graduate student and then as a professor. Special mention must be made of Professor Francisco Samaniego, under whom I studied and wrote my dissertation, and with whom I still have the pleasure of working.

Outside of UC Davis, I would next like to thank Professors Jonathan Cohen, Jeffrey Bergen, Jerry Goldman, Roger Jones, Stefan Catoiu, Constantine Georgakis, Gang Wang, Ayse Sahin, Effat Moussa, and David Webb for helpful assistance in mathematical matters and such.

Also, the following professors at Sacramento State University [my undergraduate school] are acknowledged: Geetha and Doraiswamy Ramachandran, Tom Schulte, Earl Kymala, and Don Taranto.

I next wish to thank Cathy Hertz, whose editing saved many errors from seeing the light of day.

Finally, the acknowledgement that is supreme must be made: I thank the Triune God of the Old and New Testaments for allowing me to finish this project in one piece and on time.

1

Set Systems

This chapter talks about the various collections of sets associated with measure theory. The need for these many set structures defined and discussed within comes from the desire to rigorously define the framework wherein measure-theoretic concepts reside. Of the seven sections in this chapter, the first five are essential. The last two sections are optional and show the profundity of certain seemingly simple concepts.

Preliminary Notation. We will use the standard symbols $\mathbb{N}$, $\mathbb{Q}$, $\mathbb{Z}$, and $\mathbb{R}$ to denote the sets of positive integers, rational numbers, integers, and real numbers respectively. Note that $\pm\infty \notin \mathbb{R}$. There are 9 types of intervals in $\mathbb{R}$, and we use standard notation for them:

$$(-\infty,\infty),\ (-\infty,x],\ (-\infty,x),\ [x,\infty),\ (x,\infty),\ (a,b),\ (a,b],\ [a,b),\ [a,b],$$

where $x, a, b \in \mathbb{R}$. In particular, intervals of the form $(a, b]$ are called *right semiclosed* [*rsc*]; intervals of the form $[a, b)$ are called *left semiclosed intervals* [*lsc*].

For $k \in \mathbb{N}$, the set $\mathbb{R}^k$ is called *k-dimensional Euclidean space*, and $\mathbb{R}^k$ is defined as the k-fold Cartesian product $\mathbb{R} \times \cdots \times \mathbb{R}$. Given $\boldsymbol{x} \in \mathbb{R}^k$, we will write $\boldsymbol{x} = (x_1, \cdots, x_k)$, where $x_1, \cdots, x_k \in \mathbb{R}$. The nine types of intervals above have their k-dimensional analogs, denoted by

$$(-\boldsymbol{\infty},\boldsymbol{\infty}),\ (-\boldsymbol{\infty},\boldsymbol{x}],\ (-\boldsymbol{\infty},\boldsymbol{x}),\ [\boldsymbol{x},\boldsymbol{\infty}),\ (\boldsymbol{x},\boldsymbol{\infty}),\ (\boldsymbol{a},\boldsymbol{b}),\ (\boldsymbol{a},\boldsymbol{b}],\ [\boldsymbol{a},\boldsymbol{b}),\ [\boldsymbol{a},\boldsymbol{b}],$$

where $\boldsymbol{x}, \boldsymbol{a}, \boldsymbol{b} \in \mathbb{R}^k$. The interpretations of these k-dimensional rectangles is similar to the one-dimensional case: $(-\boldsymbol{\infty}, \boldsymbol{x}]$ denotes the set of $\boldsymbol{y} \in \mathbb{R}^k$

with $-\infty < y_i \le x_i$ for $i = 1, \cdots, k$; $(\boldsymbol{a}, \boldsymbol{b}]$ denotes the set of $\boldsymbol{x} \in \mathbb{R}^k$ with $a_i < x_i \le b_i$ for $i = 1, \cdots, k$, and so on. We will write $\boldsymbol{x} \le \boldsymbol{y}$ [or $\boldsymbol{x} < \boldsymbol{y}$] to mean that $x_i \le y_i$ [or $x_i < y_i$] for $i = 1, \cdots, k$. Thus, we may write $[\boldsymbol{x}, \boldsymbol{\infty}) = \{\boldsymbol{y} \in \mathbb{R}^k : \boldsymbol{y} \ge \boldsymbol{x}\}$, $(\boldsymbol{a}, \boldsymbol{b}) = \{\boldsymbol{x} \in \mathbb{R}^k : \boldsymbol{a} < \boldsymbol{x} < \boldsymbol{b}\}$, etc. We will refer to $(\boldsymbol{a}, \boldsymbol{b}]$ as a *rsc rectangle*; $[\boldsymbol{a}, \boldsymbol{b})$ is a *lsc rectangle.*

The set $\bar{\mathbb{R}}$ of *extended real numbers* is defined as $\mathbb{R} \cup \{-\infty, +\infty\}$. While $\pm\infty \notin \mathbb{R}$, we do have $\pm\infty \in \bar{\mathbb{R}}$. By definition, $-\infty < +\infty$ and $-\infty < x < +\infty$ for all $x \in \mathbb{R}$. Subsets of $\bar{\mathbb{R}}$ having the form A, $A \cup (x, \infty]$, $A \cup [-\infty, x)$, or $A \cup [-\infty, x) \cup (y, \infty]$, where $A \subseteq \mathbb{R}$ is open and $x, y \in \mathbb{R}$, are declared open. In particular, both $\varnothing$ and $\bar{\mathbb{R}}$ are open subsets of $\bar{\mathbb{R}}$.

For $k \in \mathbb{N}$, the set $\bar{\mathbb{R}}^k$ is called *k-dimensional extended Euclidean space,* and is the k-fold Cartesian product of $\bar{\mathbb{R}}$. Therefore, an element $\boldsymbol{x} \in \bar{\mathbb{R}}^k$ has some, none, or possibly all of its coordinates equal to $+\infty$ and/or $-\infty$, and all other coordinates are real numbers. We write $+\boldsymbol{\infty} = (+\infty, \cdots, +\infty)$ and $-\boldsymbol{\infty} = (-\infty, \cdots, -\infty)$.

Given a set Ω, we denote its cardinal number by card(Ω). We will write $\mathfrak{c} = \text{card}(\mathbb{R})$, $\aleph_0 = \text{card}(\mathbb{N})$, and $\mathfrak{2} = \text{card}(\{0, 1\})$. The set of all subsets of Ω [or the *power set of* Ω] is denoted by 2^Ω; we write $\mathfrak{2}^{\text{card}(\Omega)}$ for card(2^Ω). A set Ω is *at most countable* [amc] iff Ω is finite or $\text{card}(\Omega) = \aleph_0$, that is, if $\text{card}(\Omega) \le \aleph_0$. Ω will be called *uncountable* iff it is not amc.

1.1 π-SYSTEMS, λ-SYSTEMS, AND SEMIRINGS

This section deals with three systems of sets used in measure theory, and is broken up into three subsections, each with exercises.

Definition. A nonvoid collection $\mathcal{P}$ of subsets of a nonvoid set Ω is called a *π-system [of subsets of Ω]* iff $A, B \in \mathcal{P}$ implies $A \cap B \in \mathcal{P}$, that is, $\mathcal{P}$ is *closed under intersection.* We will also say that *$\mathcal{P}$ is a π-system on/over* Ω, or, if no confusion can arise, *$\mathcal{P}$ is a π-system.*

Example 1. Let (Ω, ρ) denote a metric space. Recall that if $A, B \subseteq \Omega$ and both A and B are open with respect to ρ, then $A \cap B$ is open. Thus, the collection of open subsets of Ω forms a π-system of subsets of Ω. Next, recall that the intersection of two closed subsets of Ω is itself a closed subset of Ω, and hence the collection of closed subsets of Ω is a π-system as well.

Example 2. Let $\Omega = \mathbb{R}$, and let $\mathcal{P}$ consist of $\varnothing$ and the rsc intervals. It may be painlessly verified that $\mathcal{P}$ is a π-system on $\mathbb{R}$. Next, let $\Omega = \mathbb{R}^k$, and let $\mathcal{P}_k$ consist of $\varnothing$ and the k-dimensional rsc rectangles. *We claim that $\mathcal{P}_k$ is a π-system.* To see this, let $A, B \in \mathcal{P}_k$. If A or B is empty, then $A \cap B = \varnothing \in \mathcal{P}_k$. If $A = (\boldsymbol{a}, \boldsymbol{b}]$ and $B = (\mathbf{c}, \boldsymbol{d}]$, then $A \cap B = C_1 \times \cdots \times C_k$, where $C_i = (a_i, b_i] \cap (c_i, d_i]$, $i = 1, \cdots k$. Since $\mathcal{P}$ is a π-system, we have $C_1, \cdots, C_k \in \mathcal{P}$ by what has been shown in the case $k = 1$. Since a k-fold Cartesian product of $\mathcal{P}$-sets is a $\mathcal{P}_k$-set, we have $A \cap B \in \mathcal{P}_k$, and thus $\mathcal{P}_k$ is closed under intersection.

In general, it is automatic that a nonempty collection $\mathcal{P}$ is a π-system on a nonempty set Ω iff for every finite collection $A_1, \cdots, A_n$ of $\mathcal{P}$-sets, we have $\bigcap_{i=1}^n A_i \in \mathcal{P}$ *[closure under finite intersections]*. At least two π-systems of subsets of Ω always exist: 2^Ω and the collection $\{\varnothing, \Omega\}$. Thus, the concept of a π-system is never logically vacuous.

Exercises.

1*. Let $\Omega = (\alpha, \beta]$. Let $\mathcal{P}$ consist of $\varnothing$ along with the rsc subintervals of Ω. $\mathcal{P}$ is a π-system of subsets of $(\alpha, \beta]$.

2. Must $\varnothing$ be in every π-system?

3. List all π-systems consisting of at least two subsets of $\{\omega_1, \omega_2, \omega_3\}$.

4*. If $\mathcal{P}$ consists of the empty set and the k-dimensional rectangles of any one form, then $\mathcal{P}$ is a π-system of subsets of $\mathbb{R}^k$.

5. Let $\mathcal{P}$ consist of $\varnothing$ and all subsets of $\mathbb{R}^k$ that are neither open nor closed. Then $\mathcal{P}$ is not a π-system of subsets of $\mathbb{R}^k$.

6*. For each α in a nonempty index set A, let $\mathcal{P}_\alpha$ be a π-system over Ω.
(a) The collection $\bigcap_{\alpha \in A} \mathcal{P}_\alpha$ is a π-system on Ω.
(b) Let $\mathcal{A} \subseteq 2^\Omega$. Suppose that $\{\mathcal{P}_\alpha : \alpha \in A\}$ is the "exhaustive list" of all the π-systems that contain $\mathcal{A}$. In other words, each $\mathcal{P}_\alpha \supseteq \mathcal{A}$, and any π-system that contains $\mathcal{A}$ coincides with some $\mathcal{P}_\alpha$. Then $\bigcap_{\alpha \in A} \mathcal{P}_\alpha$ is a π-system that contains $\mathcal{A}$. If $\mathcal{Q}$ is a π-system containing $\mathcal{A}$, then $\bigcap_{\alpha \in A} \mathcal{P}_\alpha \subseteq \mathcal{Q}$. [The collection $\bigcap_{\alpha \in A} \mathcal{P}_\alpha$ is called the *[minimal] π-system [on Ω] generated by $\mathcal{A}$*.] The minimal π-system generated by $\mathcal{A}$ always exists.
(c) Suppose that $\mathcal{P}$ is a π-system with $\mathcal{P} \supseteq \mathcal{A}$, and suppose that $\mathcal{P}$ is contained in any other π-system that contains $\mathcal{A}$. Then $\mathcal{P} = \bigcap_{\alpha \in A} \mathcal{P}_\alpha$, with notation as in (b). The minimal π-system containing $\mathcal{A}$ [which always exists] is also unique.

It will turn out that π-systems play a role in an important uniqueness question in Chapter 3. At this stage, with the definition and the exercises, there is not much else to say. The next definition will give us a collection of sets that has a stronger set of closure properties.

Definition. A nonempty collection $\mathcal{L}$ of subsets of a nonempty set Ω is called a *λ-system [on Ω]* iff
(λ_1) $\Omega \in \mathcal{L}$,
(λ_2) $A \in \mathcal{L}$ implies $A^c \in \mathcal{L}$, and
(λ_3) For every disjoint sequence $\{A_n\}_{n=1}^\infty$ of $\mathcal{L}$-sets, we have $\bigcup_{n=1}^\infty A_n \in \mathcal{L}$.

Properties (λ_2) and (λ_3) are called *closure under complementation* and *closure under countable disjoint unions*, respectively. By (λ_1) and (λ_2), it is automatic that $\varnothing$ is in every λ-system. Next, observe that the collections $\{\varnothing, \Omega\}$ and 2^Ω are λ-systems, hence a λ-system over Ω always exists.

Example 3. For any $k \in \mathbb{N}$, if Ω consists of $2k$ elements, then the collection that consists of Ω, $\varnothing$, and all k-element subsets of Ω is a λ-system of subsets of Ω.

The additional closure properties listed below will be used in the exercises:

(λ_2') For all $A, B \in \mathcal{L}$, $A \subseteq B$ implies $B - A \in \mathcal{L}$;
(λ_4) For all $A, B \in \mathcal{L}$, $A \cap B = \varnothing$ implies $A \cup B \in \mathcal{L}$;
(λ_5) For any nondecreasing sequence $\{A_n\}_{n=1}^{\infty}$ of $\mathcal{L}$-sets, $\bigcup_{n=1}^{\infty} A_n \in \mathcal{L}$;
(λ_6) For any nonincreasing sequence $\{A_n\}_{n=1}^{\infty}$ of $\mathcal{L}$-sets, $\bigcap_{n=1}^{\infty} A_n \in \mathcal{L}$.

These properties are called *closure under proper differences*, *closure under disjoint unions*, *closure under nondecreasing countable unions*, and *closure under nonincreasing countable intersections*. We use the term *nondecreasing* relative to $\{A_n\}_{n=1}^{\infty}$ to mean that $A_1 \subseteq A_2 \subseteq \cdots$, where any of the containment relations may in fact be equality. We say that $A_1, A_2, \cdots$ are *[strictly] increasing* if each of the containments is proper: $A_1 \subsetneq A_2 \subsetneq \cdots$. Similar comments apply to the terms *nonincreasing* $[A_1 \supseteq A_2 \supseteq \cdots]$ and *[strictly] decreasing* $[A_1 \supsetneq A_2 \supsetneq \cdots]$.

Exercises.

7*. This exercise explores some equivalent definitions of a λ-system.
(a) $\mathcal{L}$ is a λ-system iff $\mathcal{L}$ satisfies (λ_1), (λ_2'), and (λ_3).
(b) Every λ-system additionally satisfies (λ_4), (λ_5), and (λ_6).
(c) $\mathcal{L}$ is a λ-system iff $\mathcal{L}$ satisfies (λ_1), (λ_2'), and (λ_5).
(d) If a collection $\mathcal{L}$ is nonempty and satisfies (λ_2) and (λ_3), then $\mathcal{L}$ is a λ-system.

8*. If $\mathcal{L}$ is a λ-system and a π-system, then $\bigcup_{n=1}^{\infty} A_n \in \mathcal{L}$ whenever $A_n \in \mathcal{L}$ for all $n \in \mathbb{N}$. That is, $\mathcal{L}$ is *closed under countable unions*.

9. A λ-system is not necessarily a π-system.

10. Find all λ-systems over $\{\omega_1, \omega_2, \omega_3, \omega_4\}$ with at least three elements.

11. The collection consisting of $\varnothing$ and the rsc intervals is not a λ-system on $\mathbb{R}$.

12*. Suppose that for each α in a nonempty index set A, $\mathcal{L}_\alpha$ is a λ-system over Ω.
(a) The collection $\bigcap_{\alpha \in A} \mathcal{L}_\alpha$ is a λ-system on Ω.
(b) Suppose that $\mathcal{A} \subseteq 2^\Omega$ is such that $\mathcal{A}$ is contained in each $\mathcal{L}_\alpha$, and suppose that $\{\mathcal{L}_\alpha : \alpha \in A\}$ is the "exhaustive list" of all the λ-systems that contain $\mathcal{A}$. Then $\bigcap_{\alpha \in A} \mathcal{L}_\alpha$ is a λ-system that contains $\mathcal{A}$. If $\mathcal{Q}$ is a λ-system on Ω that contains $\mathcal{A}$, then $\bigcap_{\alpha \in A} \mathcal{L}_\alpha \subseteq \mathcal{Q}$. [The collection $\bigcap_{\alpha \in A} \mathcal{L}_\alpha$ is also called the *[minimal] λ-system [over Ω] generated by $\mathcal{A}$*.] The minimal λ-system generated by $\mathcal{A}$ always exists.
(c) Let $\mathcal{L}$ denote a λ-system over Ω with $\mathcal{L} \supseteq \mathcal{A}$ and where $\mathcal{L}$ is contained in any other λ-system also containing $\mathcal{A}$. Then $\mathcal{L} = \bigcap_{\alpha \in A} \mathcal{L}_\alpha$, with notation as in (b). Therefore, the λ-system generated by $\mathcal{A}$ always exists and is unique.

The third type of collection of sets in this section will now be presented.

Definition. A *semiring on/over* Ω is a collection $\mathcal{A}$ of subsets of Ω satisfying
(SR1) $\varnothing \in \mathcal{A}$,
(SR2) $\mathcal{A}$ is a π-system, and
(SR3) If $A, B \in \mathcal{A}$ with $A \subseteq B$, then there exist disjoint $C_1, \cdots, C_k \in \mathcal{A}$ with $B - A = C_1 \cup \cdots \cup C_k$. [Equivalently, we have $B = A \cup C_1 \cup \cdots \cup C_k$.]

Example 4. Both 2^Ω and $\{\varnothing, \Omega\}$ are [trivial] semirings of subsets of Ω.

Example 5. Let $\Omega = \mathbb{R}$, and let $\mathcal{A}$ consist of $\varnothing$ and all rsc intervals. Property (SR1) is automatic, and Example 2 gives (SR2). To verify (SR3), pick $A \subseteq B$

with $A, B \in \mathcal{A}$. If $A = \varnothing$, then $B - A = B$ and (SR3) trivially holds. Next, let $A = (a, b]$ and $B = (c, d]$ with $c \le a < b \le d$, and consider the cases (i) $a = c$ and $b = d$, (ii) $a = c$ and $b < d$, (iii) $c < a$ and $b = d$, and (iv) $c < a$ and $b < d$. For these cases, $B - A$ is $\varnothing$, $(b, d]$, $(c, a]$, and $(c, a] \cup (b, d]$, respectively. In every case (SR3) holds, hence $\mathcal{A}$ is a semiring on $\mathbb{R}$. If we instead used lsc intervals, $\mathcal{A}$ would still be a semiring, but if we used intervals of any other form $\mathcal{A}$ would fail to be a semiring. This consideration is why we will concentrate on rsc intervals when working in subsequent chapters.

Example 6. Let $\Omega = \mathbb{R}^k$, and let $\mathcal{A}_k$ consist of $\varnothing$ and the k-dimensional rsc rectangles. *We claim that $\mathcal{A}_k$ is a semiring.* (SR1) is satisfied by definition, and Example 2 shows that (SR2) holds. We now turn to verifying (SR3). Let $A, B \in \mathcal{A}_k$ with $A \subseteq B$. If $A = \varnothing$, then $B - A = B$, and (SR3) trivially holds in this case. Otherwise, suppose that $A \neq \varnothing$ and write $A = I_1 \times \cdots \times I_k$ and $B = J_1 \times \cdots \times J_k$, where each of $I_1, \cdots, I_k, J_1, \cdots, J_k$ is a rsc interval. Since $A \subseteq B$, we have $I_i \subseteq J_i$, $i = 1, \cdots, k$. By Example 5, $J_i - I_i$ is a disjoint union $A_i \cup B_i$, where A_i and B_i freely denote rsc intervals or $\varnothing$, $i = 1, \cdots, k$. Consider the 3^k k-dimensional sets $C_1 \times \cdots \times C_k$, where C_i denotes either I_i, A_i, or B_i, $i = 1, \cdots k$; the sets $C_1 \times \cdots \times C_k$ are disjoint [some might be empty]. One of the sets $C_1 \times \cdots \times C_k$ is A, and $B - A$ is the disjoint union of the remaining sets $C_1 \times \cdots \times C_k$. Since each set of the form $C_1 \times \cdots \times C_k$ is either empty or a k-dimensional rsc rectangle, (SR3) holds.

The role of semirings will become apparent in discussing the results relating to the extension and uniqueness of measures, found in Chapter 3.

Exercises.

13*. Is $\{\varnothing\} \cup \{(0, x] : 0 < x \le 1\}$ a semiring over $(0, 1]$?

14*. This exercise explores some alternative definitions of a semiring.
(a) Some define $\mathcal{A}$ to be a semiring iff $\mathcal{A}$ is a nonempty π-system such that both $E, F \in \mathcal{A}$ and $E \subseteq F$ imply the existence of a finite collection $C_0, C_1, \cdots, C_n \in \mathcal{A}$ with $E = C_0 \subseteq C_1 \subseteq \cdots \subseteq C_n \subseteq F$ and $C_i - C_{i-1} \in \mathcal{A}$ for $i = 1, \cdots, n$. This definition of a semiring is equivalent to our definition of a semiring.
(b) Some define $\mathcal{A}$ to be a semiring by stipulating (SR1), (SR2), and the following property: $A, B \in \mathcal{A}$ implies the existence of disjoint $\mathcal{A}$-sets $C_0, C_1, \cdots, C_n$ with $B - A = \bigcup_{i=0}^n C_i$. Note that here $B - A$ is not necessarily a *proper* difference. If $\mathcal{A}$ is a semiring by this definition, then $\mathcal{A}$ is a semiring by our definition, but the converse is not necessarily true.

15*. Let $\mathcal{A}$ consist of $\varnothing$ as well as all rsc rectangles $(\boldsymbol{a}, \boldsymbol{b}]$. The collection of all finite disjoint unions of $\mathcal{A}$-sets is a semiring over $\mathbb{R}^k$.

16. An arbitrary intersection of semirings on Ω is not necessarily a semiring on Ω.

17. If $\mathcal{A}$ is a semiring over Ω, must $\Omega \in \mathcal{A}$?

18*. Let $\mathcal{A}$ denote a semiring. Pick $n \in \mathbb{N}$, and let $A, A_1, \cdots, A_n \in \mathcal{A}$. Then there exists a finite collection $\{C_1, \cdots, C_m\}$ of disjoint $\mathcal{A}$-sets with $A - \bigcup_{i=1}^n A_i = \bigcup_{j=1}^m C_j$. [When $n = 1$, write $A - A_1 = A - (A \cap A_1)$ and invoke (SR3). Assume the result is true for a fixed n and show that $A - \bigcup_{i=1}^{n+1} A_i = \bigcup_{j=1}^m (C_j - A_{n+1})$. Apply the $n = 1$ case to each of the $C_j - A_{n+1}$ terms.]

19. Other books deal with a system called a *ring*. We will not deal with rings of sets in this text, but since the reader might refer to other books that deal with rings, it is worthy to discuss the concept. A collection $\mathcal{R}$ of subsets of a nonempty set Ω is called a *ring* of subsets of Ω iff (R1) $\mathcal{R} \neq \varnothing$, (R2) $A, B \in \mathcal{R}$ implies $A \cup B \in \mathcal{R}$, and (R3) $A, B \in \mathcal{R}$ implies $A - B \in \mathcal{R}$. That is, a ring is a nonempty collection of subsets closed under unions and differences.
(a) $\varnothing$ is in every ring.
(b) $\mathcal{R}$ is a ring iff $\mathcal{R}$ satisfies (R1), (R2), and (R4): $A, B \in \mathcal{R}$ with $A \subseteq B$ implies $B - A \in \mathcal{R}$. [Use the identity $B - A = (B \cup A) - A$.]
(c) Every ring satisfies (R5): $A, B \in \mathcal{R}$ implies $A \Delta B \in \mathcal{R}$. [$A \Delta B$ is the usual *symmetric difference of A and B*, the set of elements that are in exactly one of A or B. That is, $A \Delta B = (A - B) \cup (B - A)$.]
(d) Every ring is a π-system. [Verify first that $A \cap B = (A \cup B) - (A \Delta B)$.]
(e) Every ring is closed under finite unions and finite intersections.
(f) $\mathcal{R}$ is a ring iff $\mathcal{R}$ is a nonempty π-system that satisfies (R4) along with (R6): $A, B \in \mathcal{R}$ and $A \cap B = \varnothing$ imply $A \cup B \in \mathcal{R}$. [First, show that $A \cup B$ coincides with $[A - (A \cap B)] \cup [B - (A \cap B)] \cup (A \cap B)$.]
(g) $\mathcal{R}$ is a ring iff $\mathcal{R}$ is a nonempty π-system that satisfies (R5). [First, verify the identities $A \cup B = (A \Delta B) \Delta (A \cap B)$ and $A - B = A \Delta (A \cap B)$.]
(h) Suppose that $\{\mathcal{R}_\alpha : \alpha \in A\}$ is the "exhaustive list" of all rings that contain $\mathcal{A}$. Then $\bigcap_{\alpha \in A} \mathcal{R}_\alpha$ is a ring that contains $\mathcal{A}$, and $\bigcap_{\alpha \in A} \mathcal{R}_\alpha$ is contained in any ring that contains $\mathcal{A}$. [The collection $\bigcap_{\alpha \in A} \mathcal{R}_\alpha$ is called the *[minimal] ring generated by* $\mathcal{A}$.] The minimal ring containing $\mathcal{A}$ always exists and is unique.
(i) The collection of finite unions of rsc intervals is a ring on $\mathbb{R}$.
(j) Let Ω be uncountable. The collection of all amc subsets of Ω is a ring on Ω.

20. This problem explores the relationship between semirings and rings.
(a) Every ring is a semiring. However, not every semiring is a ring.
(b) Let $\mathcal{A}$ denote a semiring on Ω, and let $\mathcal{R}$ consist of the finite disjoint unions of $\mathcal{A}$-sets. Then $\mathcal{R}$ is closed under finite intersections and disjoint unions.
(c) If $A, B \in \mathcal{A}$ and $A \subseteq B$, then $B - A \in \mathcal{R}$.
(d) $A \in \mathcal{A}$, $B \in \mathcal{R}$, and $A \subseteq B$ imply $B - A \in \mathcal{R}$.
(e) $A, B \in \mathcal{R}$ and $A \subseteq B$ imply $B - A \in \mathcal{R}$.
(f) $\mathcal{R}$ is the minimal ring generated by $\mathcal{A}$. [See Exercise 19(h).]
(g) A semiring that satisfies (R2) is a ring.

21. Let Ω be infinite, and let $\mathcal{A} \subseteq 2^\Omega$ have cardinality $\aleph_0$. We will show that the ring generated by $\mathcal{A}$ has cardinality $\aleph_0$.
(a) Given $\mathcal{C} \subseteq 2^\Omega$, let $\mathcal{C}^*$ denote the collection of all finite unions of differences of $\mathcal{C}$-sets. If $\text{card}(\mathcal{C}) = \aleph_0$, then $\text{card}(\mathcal{C}^*) = \aleph_0$. Also, $\varnothing \in \mathcal{C}$ implies $\mathcal{C} \subseteq \mathcal{C}^*$.
(b) Let $\mathcal{A}_0 = \mathcal{A}$, and define $\mathcal{A}_n = \mathcal{A}_{n-1}^*$ for $n \geq 1$. Then $\mathcal{A} \subseteq \bigcup_{n=0}^\infty \mathcal{A}_n \subseteq R(\mathcal{A})$, where $R(\mathcal{A})$ is the minimal ring generated by $\mathcal{A}$ and where [without loss of generality] $\varnothing \in \mathcal{A}$. Also, $\text{card}(\bigcup_{n=0}^\infty \mathcal{A}_n) = \aleph_0$.
(c) $\bigcup_{n=0}^\infty \mathcal{A}_n$ is a ring on Ω, and from the fact that $R(\mathcal{A})$ is the minimal ring containing $\mathcal{A}$, we have $\bigcup_{n=0}^\infty \mathcal{A}_n = R(\mathcal{A})$, and thus $\text{card}(R(\mathcal{A})) = \aleph_0$.
(d) We may generalize: if $\mathcal{A}$ is infinite, then $\text{card}(\mathcal{A}) = \text{card}(R(\mathcal{A}))$.

1.2 FIELDS

This section deals with a new system of sets that is more stringent in its requirements than are the systems of the previous section. We will call this new system a *field* of sets. As before, Ω denotes a nonempty set.

Definition. A *field on* Ω is a collection $\mathcal{F} \subseteq 2^{\Omega}$ such that
(F1) $\Omega \in \mathcal{F}$,
(F2) $A \in \mathcal{F}$ implies $A^c \in \mathcal{F}$, and
(F3) If $A, B \in \mathcal{F}$, then $A \cup B \in \mathcal{F}$.

Using induction, it is easy to see that every field is closed under arbitrary finite unions and intersections: if $A_1, \cdots, A_n \in \mathcal{F}$ then $\bigcup_{i=1}^{n} A_i$, $\bigcap_{i=1}^{n} A_i \in \mathcal{F}$. It is also easily verified that every field is a π-system and a semiring; however, a field is not necessarily a λ-system. Also, $\mathcal{F}$ is a field iff $\mathcal{F}$ satisfies (F1), (F2), and is a π-system; this latter characterization is often useful.

Example 1. Both 2^{Ω} and $\{\varnothing, \Omega\}$ are trivial fields of subsets of Ω.

Example 2. Let $\Omega = (0, 1]$ and let $\mathcal{F}$ consist of $\varnothing$ and all rsc subintervals $(a, b]$ of $(0, 1]$. Note immediately that $\Omega \in \mathcal{F}$ and that $\mathcal{F}$ is a π-system, but requirement (F2) might fail: any $A = (a, b]$ with $0 < a < b < 1$ will be such that A^c is neither $\varnothing$ nor a rsc subinterval, and hence $A^c \notin \mathcal{F}$. Accordingly, $\mathcal{F}$ is not a field over Ω.

Example 3. Keep $\Omega = (0, 1]$ as in the previous example, but amend $\mathcal{F}$ to consist of $\varnothing$ and the *finite disjoint unions* of rsc subintervals of $(0, 1]$. That is, a non-empty $A \in \mathcal{F}$ iff for some $n \in \mathbb{N}$ we have $A = \bigcup_{i=1}^{n} (a_i, b_i]$, where $(a_i, b_i] \subseteq (0, 1]$ for $i = 1, \cdots, n$ and the union is disjoint. Note that this definition of $\mathcal{F}$ includes the sets in the collection from the above example.

The verification that $\mathcal{F}$ is a field is a direct one. Clearly $\Omega \in \mathcal{F}$, so (F1) holds. To verify (F2), pick $A \in \mathcal{F}$. If $A = \varnothing$, then $A^c = (0, 1] = \Omega \in \mathcal{F}$. If $A = (0, 1] = \Omega$, then $A^c = \varnothing \in \mathcal{F}$. For the nontrivial case, let $A = (a_1, b_1] \cup \cdots \cup (a_n, b_n]$ with the constituent $(a_i, b_i]$'s denoting disjoint proper rsc subintervals of $(0, 1]$. Then $A^c = (0, a_1] \cup (b_1, a_2] \cup \cdots \cup (b_n, 1]$, which is a finite disjoint union of rsc subintervals of Ω. [We will have $(0, a_1] = \varnothing$ and $(b_n, 1] = \varnothing$ when $a_1 = 0$ and $b_n = 1$.] Thus (F2) holds in all cases.

We now show that $\mathcal{F}$ is a π-system, which in conjunction with the above will show that $\mathcal{F}$ is a field. Let A and B be $\mathcal{F}$-sets. If A or B is empty, then $A \cap B = \varnothing \in \mathcal{F}$. We now deal with the nontrivial case, writing $A \in \mathcal{F}$ and $B \in \mathcal{F}$ as the disjoint unions $A = \bigcup_{i=1}^{n} (a_i, b_i]$ and $B = \bigcup_{j=1}^{m} (c_j, d_j]$. Thus $A \cap B = \bigcup_{i=1}^{n} \bigcup_{j=1}^{m} C_{ij}$, where each $C_{ij} = (a_i, b_i] \cap (c_j, d_j]$ and where both unions are disjoint. Each C_{ij} will either be a rsc subinterval of Ω or $\varnothing$, hence $A \cap B$ is either $\varnothing$ [hence in $\mathcal{F}$] or $A \cap B$ is a finite disjoint union of rsc subintervals of Ω [hence in $\mathcal{F}$]. Thus for every case we have $A \cap B \in \mathcal{F}$ whenever $A, B \in \mathcal{F}$, and $\mathcal{F}$ is accordingly a π-system.

Were Ω any other rsc interval, then the collection consisting of $\varnothing$ and the finite disjoint unions of rsc subintervals of Ω would still be a field.

Example 4. Let $\Omega = \mathbb{R}$, and let $\mathcal{F}$ denote the collection consisting of $\varnothing$ and the finite disjoint unions of rsc intervals. $\mathcal{F}$ is not a field, for (F1) and (F2) fail.

Exercises.

1. The collection $\{A \subseteq \Omega : \ A \text{ is finite or } A^c \text{ is finite}\}$ is a field on Ω.

2. Let $\mathcal{F} \subseteq 2^\Omega$ be such that $\Omega \in \mathcal{F}$ and $A - B \in \mathcal{F}$ whenever $A, B \in \mathcal{F}$. Then $\mathcal{F}$ is a field on Ω.

3. Every λ-system that is closed under *arbitrary* differences is a field.

4. Let $\mathcal{F} \subseteq 2^\Omega$ satisfy (F1) and (F2), and suppose that $\mathcal{F}$ is closed under finite *disjoint* unions. Then $\mathcal{F}$ is not necessarily a field.

5. Suppose that $\mathcal{F}_1 \subseteq \mathcal{F}_2 \subseteq \mathcal{F}_3 \subseteq \cdots$, where $\mathcal{F}_n$ is a field on Ω for each $n \in \mathbb{N}$. Then $\bigcup_{n=1}^\infty \mathcal{F}_n$ is a field on Ω.

6. The collection consisting of $\mathbb{R}^k$, $\varnothing$, and all k-dimensional rectangles of all forms fails to be a field on $\mathbb{R}^k$.

7. The collection consisting of $\varnothing$ and the finite disjoint unions of k-dimensional rsc subrectangles of the given k-dimensional rsc rectangle $(\boldsymbol{a}, \boldsymbol{b}]$ is a field on Ω.

8*. An arbitrary intersection of fields on Ω is a field on Ω.

9*. Let Ω be arbitrary, and let $\mathcal{A} \subseteq 2^\Omega$. There exists a unique field $\mathcal{F}$ on Ω with the properties that (i) $\mathcal{A} \subseteq \mathcal{F}$, and (ii) if $\mathcal{G}$ is a field with $\mathcal{A} \subseteq \mathcal{G}$, then $\mathcal{F} \subseteq \mathcal{G}$. This field $\mathcal{F}$ is called the *[minimal] field [on Ω] generated by* $\mathcal{A}$. [2^Ω is a field on Ω that contains $\mathcal{A}$, hence a field containing $\mathcal{A}$ always exists. Let $\mathcal{F}$ denote the intersection of all fields on Ω that contain $\mathcal{A}$.]

10. Let $A_1, \cdots, A_n \subsetneq \Omega$ be disjoint. What does a typical element in the minimal field generated by $\{A_1, \cdots, A_n\}$ look like? [See Exercise 9.]

11. Let S be finite, and let Ω denote the set of sequences of elements of S. For each $\omega \in \Omega$, write $\omega = (z_1(\omega), z_2(\omega), \cdots)$, so that $z_k(\omega)$ denotes the kth term of ω for all $k \in \mathbb{N}$. For $n \in \mathbb{N}$ and $H \subseteq S^n$, let $C_n(H) = \{\omega \in \Omega : \ (z_1(\omega), \cdots, z_n(\omega)) \in H\}$. Let $\mathcal{F} = \{C_n(H) : \ n \in \mathbb{N}, \ H \subseteq S^n\}$. Then $\mathcal{F}$ is a field of subsets of S^∞. [The sets $C_n(H)$ are called *cylinders of rank n*, and $\mathcal{F}$ is collection of all cylinders of all ranks.]

12*. Suppose that $\mathcal{A}$ is a semiring on Ω with $\Omega \in \mathcal{A}$. The collection of finite disjoint unions of $\mathcal{A}$-sets is a field on Ω. [Compare with Example 3 and Exercise 7.]

13*. Let $f : \Omega \to \Omega'$. Given $\mathcal{A}' \subseteq 2^{\Omega'}$, let $f^{-1}(\mathcal{A}') = \{f^{-1}(A') : \ A' \in \mathcal{A}'\}$, where $f^{-1}(A')$ is the usual inverse image of A' under f.
(a) If $\mathcal{A}'$ is a field on Ω', then $f^{-1}(\mathcal{A}')$ is a field on Ω.
(b) $f(\mathcal{A})$ may not be a field over Ω' even if $\mathcal{A}$ is a field on Ω.

14. Let Ω be infinite, and let $\mathcal{A} \subseteq 2^\Omega$ have cardinality $\aleph_0$. Let $f(\mathcal{A})$ denote the minimal field generated by $\mathcal{A}$ [Exercise 9]. We will show that $\text{card}(f(\mathcal{A})) = \aleph_0$.
(a) Given a collection $\mathcal{C}$, let $\mathcal{C}^*$ denote the collection of (i) finite unions of $\mathcal{C}$-sets, (ii) finite unions of differences of $\mathcal{C}$-sets, and (iii) finite unions of complements of $\mathcal{C}$-sets. If $\varnothing \in \mathcal{C}$, then $\mathcal{C} \subseteq \mathcal{C}^*$. If $\text{card}(\mathcal{C}) = \aleph_0$, then $\text{card}(\mathcal{C}^*) = \aleph_0$.
(b) Define $\mathcal{A}_0 = \mathcal{A}$ and $\mathcal{A}_n = \mathcal{A}_{n-1}^*$ for $n \in \mathbb{N}$. Without any loss of generality we may assume that $\varnothing \in \mathcal{A}_0$ when considering the collection $\bigcup_{n=0}^\infty \mathcal{A}_n$.
(c) $\mathcal{A} \subseteq \bigcup_{n=0}^\infty \mathcal{A}_n \subseteq f(\mathcal{A})$, and $\text{card}(\bigcup_{n=0}^\infty \mathcal{A}_n) = \aleph_0$.
(d) $\bigcup_{n=0}^\infty \mathcal{A}_n$ is a field on Ω that contains $\mathcal{A}$.
(e) From (c) and (d), we have $f(\mathcal{A}) = \bigcup_{n=0}^\infty \mathcal{A}_n$, hence $\text{card}(f(\mathcal{A})) = \aleph_0$.
(f) We may generalize: if $\mathcal{A} \subseteq 2^\Omega$ is infinite, then $\text{card}(\mathcal{A}) = \text{card}(f(\mathcal{A}))$.

15. Some books work with a system of sets called an *algebra*. An *algebra on* Ω is a nonempty collection of subsets of Ω that satisfies (F2) and (F3).
(a) $\mathcal{F}$ is an algebra on Ω iff $\mathcal{F}$ is a ring on Ω with $\Omega \in \mathcal{F}$.
(b) $\mathcal{F}$ is an algebra iff $\mathcal{F}$ is a field. Thus *algebra* and *field* are synonymous.

1.3 σ-FIELDS

The most important set system is called a *σ-field.* Other authors might equivalently use the term *σ-algebra.* This set system will eventually be found in most of our work. A σ-field is a field with one extra assumption of a rather strong nature.

Definition. A collection $\mathcal{F}$ is a *σ-field on* Ω iff
(S1) $\Omega \in \mathcal{F}$,
(S2) $A \in \mathcal{F}$ implies $A^c \in \mathcal{F}$, and
(S3) $\bigcup_{n=1}^{\infty} A_n \in \mathcal{F}$ whenever $A_1, A_2, \cdots \in \mathcal{F}$.

Closure properties (S1) and (S2) are exactly those for a field, but (S3) deals with closure under *countably infinite* unions, whereas property (F3) for fields dealt merely with closure under *finite* unions. From (S1) and (S2) it is automatic that $\varnothing$ is in every σ-field. From (S2) and (S3), a σ-field $\mathcal{F}$ is closed under countable intersections of $\mathcal{F}$-sets. Also, the collections $\{\varnothing, \Omega\}$ and 2^{Ω} are σ-fields, hence the concept of a σ-field is never vacuous.

Every σ-field is a field. To see this, let $\mathcal{F}$ denote a σ-field with $A, B \in \mathcal{F}$. Since $\varnothing \in \mathcal{F}$, we have $A \cup B = A \cup B \cup \varnothing \cup \varnothing \cup \cdots \in \mathcal{F}$ by (S3). Thus every σ-field is closed under finite unions as well. Similarly, every σ-field is also closed under finite intersections. Also, every finite field is a σ-field.

The main thing about σ-fields is that they are closed under the application of countably many of the standard set manipulations. The standard set operations are union, intersection, complementation, difference, and symmetric difference, and all of these can be expressed in terms of unions and complements. Thus, when one works with a collection of sets in a σ-field, one will never by using at most countably many set operations on these sets produce a set outside the σ-field.

Example 1. Let Ω be infinite, and let $\mathcal{F} = \{A \subseteq \Omega : A \text{ is finite or } A^c \text{ is finite}\}$. It is trivial that $\mathcal{F}$ satisfies (S1) and (S2). To see that $\mathcal{F}$ is closed under union, let $A, B \in \mathcal{F}$. There are two cases: (i) both A and B finite, and (ii) at least one of A^c or B^c is finite. For (i) we have that $A \cup B$ is finite, whence $A \cup B \in \mathcal{F}$. For (ii), assume that B^c is finite. We have $(A \cup B)^c = A^c \cap B^c \subseteq B^c$, and thus $(A \cup B)^c$ is finite, so that again $A \cup B \in \mathcal{F}$. Therefore, $\mathcal{F}$ is a field.

However, $\mathcal{F}$ fails to be a σ-field. To see this, pick a countably infinite subset $\{\omega_1, \omega_2, \cdots\}$ of Ω. Define $A_n = \{\omega_{2n}\}$ for $n \in \mathbb{N}$, hence $A_n \in \mathcal{F}$ for all $n \in \mathbb{N}$. We have $\bigcup_{n=1}^{\infty} A_n = \{\omega_2, \omega_4, \omega_6 \cdots\}$, and $(\bigcup_{n=1}^{\infty} A_n)^c$ contains $\{\omega_1, \omega_3, \omega_5, \cdots\}$, which means that both $\bigcup_{n=1}^{\infty} A_n$ and $(\bigcup_{n=1}^{\infty} A_n)^c$ are infinite, giving $\bigcup_{n=1}^{\infty} A_n \notin \mathcal{F}$. Thus, $\mathcal{F}$ fails to satisfy (S3). This gives a nontrivial example of a field that is not a σ-field.

Example 2. Let Ω be infinite, and let $\mathcal{F} = \{A \subseteq \Omega : A \text{ is amc or } A^c \text{ is amc}\}$. Again, (S1) and (S2) are trivially satisfied. To see (S3), let $\{A_n\}_{n=1}^{\infty}$ denote a sequence of $\mathcal{F}$-sets, and consider two cases: (i) each A_n is amc, and (ii) at least one A_n is such that A_n^c is amc. Since the countable union of amc sets is itself amc, (S3) holds in case (i). In case (ii), we assume without any loss of generality that A_1^c is amc, and we have that $(\bigcup_{n=1}^{\infty} A_n)^c = \bigcap_{n=1}^{\infty} A_n^c \subseteq A_1^c$, so that $(\bigcup_{n=1}^{\infty} A_n)^c$ is amc. It follows that $\bigcup_{n=1}^{\infty} A_n \in \mathcal{F}$ for case (ii) as well. Therefore, $\bigcup_{n=1}^{\infty} A_n \in \mathcal{F}$ in both cases, so $\mathcal{F}$ is closed under countable unions, hence $\mathcal{F}$ is a σ-field.

Observe that if $\Omega = [0, 1]$ and $U = [0, \frac{1}{2}]$, then both U and U^c are uncountable, whence $U \notin \mathcal{F}$. This shows that *a σ-field does not necessarily contain every subset of Ω*. Also, if we consider all singletons $\{x\}$ where $x \in U$, then for each $x \in U$ we have $\{x\} \in \mathcal{F}$ but $\bigcup_{x \in U} \{x\} = U \notin \mathcal{F}$, so that *a σ-field is not necessarily closed under completely arbitrary [uncountable] unions.*

The following claim is mimicked by Exercises 1.6, 1.12, 1.19(h), 2.8, and 2.9, and is of great importance for later devlopments. A simple fact used in the proof is this: *every arbitrary intersection of σ-fields on some common set Ω is itself a σ-field on Ω.* [The reader may easily verify this.]

Claim 1. *Let $\varnothing \neq \mathcal{A} \subseteq 2^{\Omega}$. There exists a unique σ-field $\sigma(\mathcal{A})$ on Ω such that (i) $\mathcal{A} \subseteq \sigma(\mathcal{A})$, and (ii) any σ-field $\mathcal{G}$ on Ω with $\mathcal{A} \subseteq \mathcal{G}$ is such that $\sigma(\mathcal{A}) \subseteq \mathcal{G}$.*

Proof: Let $\Sigma(\mathcal{A})$ denote the family of σ-fields on Ω that contain $\mathcal{A}$, and observe that $2^{\Omega} \in \Sigma(\mathcal{A})$, hence $\Sigma(\mathcal{A}) \neq \varnothing$. Let $\sigma(\mathcal{A}) = \bigcap_{\mathcal{F} \in \Sigma(\mathcal{A})} \mathcal{F}$. We have that $\sigma(\mathcal{A})$ is a σ-field since it is the intersection of σ-fields, and it is clear that $\mathcal{A} \subseteq \sigma(\mathcal{A})$. Therefore (i) holds. To verify (ii), if $\mathcal{G}$ is a σ-field on Ω that contains $\mathcal{A}$, then $\mathcal{G} \in \Sigma(\mathcal{A})$, hence $\sigma(\mathcal{A}) = \bigcap_{\mathcal{F} \in \Sigma(\mathcal{A})} \mathcal{F} \subseteq \mathcal{G}$.

All that remains is to show that $\sigma(\mathcal{A})$ is the unique σ-field on Ω with properties (i) and (ii). Suppose that $\mathcal{H}$ is a σ-field on Ω with (i) $\mathcal{A} \subseteq \mathcal{H}$ and (ii) any σ-field $\mathcal{G}$ with $\mathcal{A} \subseteq \mathcal{G}$ is such that $\mathcal{H} \subseteq \mathcal{G}$. By property (i) for $\sigma(\mathcal{A})$ and property (ii) for $\mathcal{H}$, we have that $\sigma(\mathcal{A}) \subseteq \mathcal{H}$. By property (ii) for $\sigma(\mathcal{A})$, we have $\mathcal{H} \subseteq \sigma(\mathcal{A})$. Therefore $\mathcal{H} = \sigma(\mathcal{A})$, and the entire proof is complete. ∎

Definition. We will call $\sigma(\mathcal{A})$ the *[minimal] σ-field [on Ω] generated by $\mathcal{A}$.*

Some properties are notable: $\varnothing \subsetneqq \mathcal{A} \subseteq \mathcal{B} \subseteq 2^{\Omega}$ *implies* $\sigma(\mathcal{A}) \subseteq \sigma(\mathcal{B})$. To see this, observe that any σ-field over Ω that contains $\mathcal{B}$ must also contain $\mathcal{A}$, so $\Sigma(\mathcal{B}) \subseteq \Sigma(\mathcal{A})$. It follows that $\sigma(\mathcal{A}) = \bigcap_{\mathcal{F} \in \Sigma(\mathcal{A})} \mathcal{F} \subseteq \bigcap_{\mathcal{F} \in \Sigma(\mathcal{B})} \mathcal{F} = \sigma(\mathcal{B})$.

Next, *$\mathcal{A}$ is a σ-field iff $\mathcal{A} = \sigma(\mathcal{A})$.* To see this, let $\mathcal{A}$ denote a σ-field. By definition, $\mathcal{A} \subseteq \sigma(\mathcal{A})$; however, part (ii) of Claim 1 forces $\sigma(\mathcal{A}) \subseteq \mathcal{A}$, and hence $\mathcal{A} = \sigma(\mathcal{A})$. Conversely, if $\mathcal{A} = \sigma(\mathcal{A})$, then $\mathcal{A}$ is a σ-field since $\sigma(\mathcal{A})$ is.

Letting $\varnothing \subsetneq \mathcal{C} \subseteq 2^{\Omega}$, we take $\mathcal{A} = \sigma(\mathcal{C})$ in the previous paragraph. $\mathcal{A}$ is a σ-field over Ω, hence $\sigma(\mathcal{A}) = \mathcal{A}$, that is, $\sigma(\sigma(\mathcal{C})) = \sigma(\mathcal{C})$. Therefore, *$\sigma(\sigma(\mathcal{C})) = \sigma(\mathcal{C})$ for any nonempty collection $\mathcal{C}$ of subsets of Ω.*

The concept of a minimal σ-field on Ω generated by a collection $\mathcal{A}$ may be extended to π-systems, λ-systems, and fields by the same reasoning exhibited in Claim 1. If $\varnothing \neq \mathcal{A} \subseteq 2^{\Omega}$, we let $\Pi(\mathcal{A})$, $\Lambda(\mathcal{A})$, and $F(\mathcal{A})$ denote the collections of all π-systems, λ-systems, and fields of subsets of Ω that contain $\mathcal{A}$, respectively. With this, we may define $\pi(\mathcal{A}) = \bigcap_{\mathcal{P}\in\Pi(\mathcal{A})} \mathcal{P}$, $\lambda(\mathcal{A}) = \bigcap_{\mathcal{L}\in\Lambda(\mathcal{A})} \mathcal{L}$, and $f(\mathcal{A}) = \bigcap_{\mathcal{F}\in F(\mathcal{A})} \mathcal{F}$, and we may show that these quantities are π-systems, λ-systems, and fields that contain $\mathcal{A}$. We may also show that if $\mathcal{G}$ is a π-system, λ-system, or field that contains $\mathcal{A}$, then $\pi(\mathcal{A})$, $\lambda(\mathcal{A})$, and $f(\mathcal{A})$ are contained in $\mathcal{G}$. Furthermore, each of $\pi(\mathcal{A})$, $\lambda(\mathcal{A})$, and $f(\mathcal{A})$ are the unique π-systems, λ-systems, and fields with such properties. Such demonstrations were the points of Exercises 1.6, 1.12, 2.8, and 2.9.

Finally, $\pi(\mathcal{A}) \subseteq f(\mathcal{A}) \subseteq \sigma(\mathcal{A})$ *and* $\pi(\mathcal{A}) \subseteq \lambda(\mathcal{A}) \subseteq \sigma(\mathcal{A})$. To see this, observe that any σ-field containing $\mathcal{A}$ is a field containing $\mathcal{A}$, and any field containing $\mathcal{A}$ is a π-system containing $\mathcal{A}$, whence $\Sigma(\mathcal{A}) \subseteq F(\mathcal{A}) \subseteq \Pi(\mathcal{A})$, hence $\bigcap_{\mathcal{F}\in\Pi(\mathcal{A})} \mathcal{F} \subseteq \bigcap_{\mathcal{F}\in F(\mathcal{A})} \mathcal{F} \subseteq \bigcap_{\mathcal{F}\in\Sigma(\mathcal{A})} \mathcal{F}$, so $\pi(\mathcal{A}) \subseteq f(\mathcal{A}) \subseteq \sigma(\mathcal{A})$. The other containment claim follows similarly and is left as an exercise.

The following result is important, and will be used later to characterize a certain collection of sets.

Claim 2. *Let $X : \Omega \to \Omega'$ be given with $\mathcal{A}' \subseteq 2^{\Omega'}$. Then (i) $X^{-1}(\sigma(\mathcal{A}'))$ is a σ-field on Ω, and (ii) we have $\sigma(X^{-1}(\mathcal{A}')) = X^{-1}(\sigma(\mathcal{A}'))$.*

Proof: We first show (i). Since $\Omega = X^{-1}(\Omega')$ and $\Omega' \in \sigma(\mathcal{A}')$, we have that $\Omega \in X^{-1}(\sigma(\mathcal{A}'))$. Next, if $A \in X^{-1}(\sigma(\mathcal{A}'))$, then $A = X^{-1}(B)$ for some $B \in \sigma(\mathcal{A}')$. Therefore, $A^c = X^{-1}(B)^c = X^{-1}(B^c)$, and $B^c \in \sigma(\mathcal{A}')$ since $\sigma(\mathcal{A}')$ is a σ-field. It follows that $A^c \in X^{-1}(\sigma(\mathcal{A}'))$, so that $X^{-1}(\sigma(\mathcal{A}'))$ is closed under complementation. To see that $X^{-1}(\sigma(\mathcal{A}'))$ is closed under countable unions, let $\{A_n\}_{n=1}^{\infty}$ denote a sequence of sets in $X^{-1}(\sigma(\mathcal{A}'))$. For each $n \in \mathbb{N}$, there is $B_n \in \sigma(\mathcal{A}')$ with $A_n = X^{-1}(B_n)$. Therefore $\bigcup_{n=1}^{\infty} A_n = \bigcup_{n=1}^{\infty} X^{-1}(B_n) = X^{-1}(\bigcup_{n=1}^{\infty} B_n)$, and $\bigcup_{n=1}^{\infty} B_n \in \sigma(\mathcal{A}')$ since $\sigma(\mathcal{A}')$ is a σ-field. It follows that $\bigcup_{n=1}^{\infty} A_n \in X^{-1}(\sigma(\mathcal{A}'))$, and thus $X^{-1}(\sigma(\mathcal{A}'))$ is closed under countable unions. Thus (i) stands proven.

We now turn to (ii). Since $\mathcal{A}' \subseteq \sigma(\mathcal{A}')$, we have $X^{-1}(\mathcal{A}') \subseteq X^{-1}(\sigma(\mathcal{A}'))$, and hence $\sigma(X^{-1}(\mathcal{A}')) \subseteq \sigma(X^{-1}(\sigma(\mathcal{A}'))) = X^{-1}(\sigma(\mathcal{A}'))$, where the equality follows from (i). This gives one containment relation. For the reverse containment relation, let $\mathcal{A}^* = \{A' \subseteq \Omega' : X^{-1}(A') \in \sigma(X^{-1}(\mathcal{A}'))\}$. It is direct to verify that (a) $\mathcal{A}^*$ contains $\mathcal{A}'$ and (b) $\mathcal{A}^*$ is a σ-field on Ω'. Since $\sigma(\mathcal{A}')$ is the smallest σ-field containing $\mathcal{A}'$, (a) and (b) yield $\sigma(\mathcal{A}') \subseteq \mathcal{A}^*$, so $X^{-1}(\sigma(\mathcal{A}')) \subseteq X^{-1}(\mathcal{A}^*) \subseteq \sigma(X^{-1}(\mathcal{A}'))$, where the last containment follows by the definition of $\mathcal{A}^*$. This gives the reverse containment relation, and the proof of (ii) is complete. ∎

Exercises.

1*. A collection $\mathcal{F}$ of sets is called a *monotone class* iff (MC1) for every nondecreasing sequence $\{A_n\}_{n=1}^{\infty}$ of $\mathcal{F}$-sets we have $\bigcup_{n=1}^{\infty} A_n \in \mathcal{F}$, and (MC2) for every nonincreasing sequence $\{A_n\}_{n=1}^{\infty}$ of $\mathcal{F}$-sets we have $\bigcap_{n=1}^{\infty} A_n \in \mathcal{F}$.
(a) If $\mathcal{F}$ is both a field and a monotone class, then $\mathcal{F}$ is a σ-field.
(b) A field is a monotone class if and only if it is a σ-field.

Monotone classes will be a main tool in discussing product measures and Fubini's Theorem in later chapters.

2*. This problem discusses some equivalent formulations of a σ-field.
(a) $\mathcal{F}$ satisfies (SF1), (SF2), and closure under amc intersections iff $\mathcal{F}$ is a σ-field.
(b) Every field that is closed under countable disjoint unions is a σ-field.
(c) If $\mathcal{F}$ satisfies (S1), closure under differences, and closure under countable unions or closure under countable intersections, then $\mathcal{F}$ is a σ-field.

3. Prove the following claims.
(a) A finite union of σ-fields on Ω is not necessarily a field on Ω.
(b) If a finite union of σ-fields on Ω is a field, then it is a σ-field as well.
(c) Given σ-fields $\mathcal{F}_1 \subsetneqq \mathcal{F}_2 \subsetneqq \cdots$ on Ω, it is not necessarily the case that $\bigcup_{n=1}^{\infty} \mathcal{F}_n$ is a σ-field. [Let $\Omega = \mathbb{N}$ and for all $n \in \mathbb{N}$ let $\mathcal{F}_n = \sigma(\{\{1\}, \cdots, \{n\}\})$.]

4*. Does Exercise 2.11 hold when *field* is replaced with *σ-field*?

5. A subset $A \subseteq \mathbb{R}$ is called *nowhere dense* iff every open interval I contains an open interval J such that $J \cap A = \varnothing$. Clearly $\varnothing$ and all subsets of a nowhere dense set are nowhere dense. A subset $A \subseteq \mathbb{R}$ is called a *set of the first category* iff A is a countable union of nowhere dense sets.
(a) An amc union of sets of the first category is of the first category.
(b) Let $\mathcal{F} = \{A \subseteq \mathbb{R} : A \text{ or } A^c \text{ is a set of the first category}\}$. Then $\mathcal{F}$ is a σ-field of subsets of $\mathbb{R}$. [To verify closure under countable unions, let $\{A_n\}_{n=1}^{\infty}$ denote a sequence of $\mathcal{F}$-sets. To show that $\bigcup_{n=1}^{\infty} A_n \in \mathcal{F}$, consider two cases: (i) each A_n is of the first category, and (ii) some A_n^c is of the first category.]

6. A *σ-ring of subsets of* Ω is a nonempty collection of subsets of Ω that is closed under differences as well as countable unions.
(a) Every σ-ring is closed under finite unions and amc intersections.
(b) $\mathcal{F}$ is a σ-field iff $\mathcal{F}$ is a σ-ring with $\Omega \in \mathcal{F}$.
(c) State and prove an existence and uniqueness result regarding the *[minimal] σ-ring generated by a collection $\mathcal{A}$ of subsets of Ω.*

7. This exercise continues Exercise 6. Let $\mathcal{A}$ denote a collection of subsets of some set Ω, and let $S(\mathcal{A})$ denote the minimal σ-ring that contains $\mathcal{A}$.
(a) For any collection $\mathcal{C}$ of sets, let $\mathcal{C}^*$ denote the collection of all countable unions of differences of $\mathcal{C}$-sets. Define $\mathcal{A}_0 = \mathcal{A}$, and for any ordinal $\alpha \succ 0$, define $\mathcal{A}_\alpha = (\bigcup_{0 \preceq \beta \prec \alpha} \mathcal{A}_\beta)^*$. Then $0 \preceq \alpha \prec \beta$ implies $\mathcal{A} \subseteq \mathcal{A}_\alpha \subseteq \mathcal{A}_\beta \subseteq S(\mathcal{A})$.
(b) Letting ω denote the first uncountable ordinal, show that $S(\mathcal{A}) = \bigcup_{0 \preceq \alpha \prec \omega} \mathcal{A}_\alpha$. [Hint: for any sequence of ordinals $\{\alpha_n\}_{n=1}^{\infty}$ with each $\alpha_n \prec \omega$, there exists an ordinal $\gamma \prec \omega$ with $\alpha_n \prec \gamma$ for all $n \in \mathbb{N}$.]
(c) If $\text{card}(\mathcal{A}) \leq \mathfrak{c}$, then $\text{card}(S(\mathcal{A})) \leq \mathfrak{c}$. [Use the fact that a union of continuum-many sets that are amc must have cardinality no greater than $\mathfrak{c}$.]

8. We will show that no σ-field can ever be countably infinite.
(a) If Ω is finite, then any σ-field on Ω is finite.
(b) For the remainder, assume that Ω is infinite and that there is a countably infinite σ-field $\mathcal{F} = \{A_1, A_2, \cdots\}$ on Ω. For each $\omega \in \Omega$, let $B_\omega = \bigcap\{A_n : \omega \in A_n\}$. For any distinct $\omega, \omega' \in \Omega$, either $B_\omega = B_{\omega'}$ or $B_\omega \cap B_{\omega'} = \varnothing$.
(c) From (b), there exists a disjoint collection $\{C_1, C_2, \cdots\}$ of nonempty $\mathcal{F}$-sets with $\bigcup_{n=1}^{\infty} C_n = \Omega$.

(d) For each sequence $e = (e_1, e_2, \cdots)$ of 0's and 1's, let $D_e = \bigcup\{C_n : e_n = 1\}$. Then $e \neq e'$ forces $D_e \neq D_{e'}$, and a contradiction ensues. Hence, there cannot exist a countably infinite σ-field on our infinite Ω.

9*. This problem is meant to give practice for some simple claims.
(a) If $\mathcal{A} \subseteq \mathcal{A}' \subseteq \sigma(\mathcal{A})$, then $\sigma(\mathcal{A}') = \sigma(\mathcal{A})$.
(b) For any collection $\varnothing \neq \mathcal{A} \subseteq 2^\Omega$, $\pi(\mathcal{A}) \subseteq \lambda(\mathcal{A}) \subseteq \sigma(\mathcal{A})$.
(c) If the nonempty collection $\mathcal{A}$ is finite, then $\sigma(\mathcal{A}) = f(\mathcal{A})$.
(d) For arbitrary collections $\mathcal{A}$, we have $\sigma(\mathcal{A}) = \sigma(f(\mathcal{A}))$.
(e) For arbitrary collections $\mathcal{A}$, we have $f(\sigma(\mathcal{A})) = \sigma(f(\mathcal{A}))$.

10. We will show that $f(\mathcal{A})$ is the class of sets of the form $\bigcup_{i=1}^m \bigcap_{j=1}^{n_i} A_{ij}$, where for each (i,j) pair either A_{ij} or A_{ij}^c is in $\mathcal{A}$, and where $\bigcap_{j=1}^{n_1} A_{1j}, \cdots, \bigcap_{j=1}^{n_m} A_{mj}$ are disjoint.
(a) Let $\mathcal{C}$ denote the class of sets of the stated form. Then $\Omega \in \mathcal{C}$ and $\mathcal{C}$ is closed under intersection.
(b) Show that $\mathcal{C}$ is closed under complementation by first showing that

$$\left(\bigcup_{i=1}^m \bigcap_{j=1}^{n_i} A_{ij}\right)^c = \bigcap_{i=1}^m \left(A_{i1}^c \cup \bigcup_{j=2}^{n_i} \left(A_{ij}^c \cap \bigcap_{k=1}^{j-1} A_{ik}\right)\right),$$

where if $n_i < 2$, $\bigcup_{j=2}^{n_i} \left(A_{ij}^c \cap \bigcap_{k=1}^{j-1} A_{ik}\right)$ is defined as $\varnothing$, $i = 1, \cdots, m$.
(c) For $i = 1, \cdots, m$, $\bigcup_{j=2}^{n_i} \left(A_{ij}^c \cap \bigcap_{k=1}^{j-1} A_{ik}\right) \in \mathcal{C}$. Also, by closure under intersection, we have $\left(\bigcup_{i=1}^m \bigcap_{j=1}^{n_i} A_{ij}\right)^c \in \mathcal{C}$.
(d) $\mathcal{C}$ is a field that contains $\mathcal{A}$, and thus $f(\mathcal{A}) \subseteq \mathcal{C}$.
(e) Every $\mathcal{C}$-set must be in any field containing $\mathcal{A}$, hence $\mathcal{C} \subseteq f(\mathcal{A})$.

11. If $\text{card}(\mathcal{A}) = \aleph_0$, then $\text{card}(f(\mathcal{A})) = \aleph_0$. [Use Exercise 10.]

12. When $\mathcal{A} = \{A_1, \cdots, A_n\}$, $\sigma(\mathcal{A})$ has at most 2^{2^n} sets. The bound is achieved if $A_1, \cdots, A_n$ are disjoint.

13. Let $\mathcal{A} = \{\{\omega\} : \omega \in \Omega\}$, where Ω as usual denotes a nonempty set.
(a) $f(\mathcal{A}) = \{A \subseteq \Omega : A \text{ is finite or } A^c \text{ is finite}\}$.
(b) When Ω is amc, $\sigma(\mathcal{A}) = 2^\Omega$.
(c) For general Ω, $\sigma(\mathcal{A}) = \{A \subseteq \otimes : A \text{ or } A^c \text{ is amc}\}$.

14. Suppose that $\{\mathcal{F}_n\}_{n=1}^\infty$ is a sequence of σ-fields on Ω with the property that each $\mathcal{F}_n$ is generated by a countable collection of subsets of Ω. Then the minimal σ-field containing each of $\mathcal{F}_1, \mathcal{F}_2, \cdots$, that is, the intersection of all σ-fields on Ω that contain each of $\mathcal{F}_1, \mathcal{F}_2, \cdots$, is generated by an amc collection of subsets of Ω.

15. Let $\mathcal{F}$ denote any σ-field on Ω other than 2^Ω, and let $H \subsetneqq \Omega$ be such that $H \notin \mathcal{F}$. Then $\sigma(\mathcal{F} \cup \{H\}) = \{(H \cap A) \cup (H^c \cap B) : A, B \in \mathcal{F}\}$. [Let $\mathcal{C}$ denote the sets of the form $(H \cap A) \cup (H^c \cap B)$, where A and B range over $\mathcal{F}$. One handily obtains $\sigma(\mathcal{F} \cup \{H\}) \supseteq \mathcal{C}$. The reverse inclusion is more difficult. Argue that $\mathcal{C}$ is a σ-field of subsets. To show (S2), argue that $(A \cup B)^c \subseteq (H \cap A^c) \cup (H^c \cap B^c)$. Show that $\mathcal{F} \cup \{H\} \subseteq \mathcal{C}$, and invoke minimality.]

16. Let $\mathcal{F} = \sigma(\mathcal{A})$, where $\varnothing \neq \mathcal{A} \subseteq 2^\Omega$. For each $B \in \mathcal{F}$ there exists a countable subcollection $\mathcal{A}_B \subseteq \mathcal{A}$ with $B \in \sigma(\mathcal{A}_B)$. [Let $\mathcal{B}$ to be the set of all $B \in \mathcal{F}$ with this property. Thus $\mathcal{B} \subseteq \mathcal{F}$. Next, show that $\mathcal{B}$ is itself a σ-field containing each $\mathcal{A}$-set, so that $\mathcal{F} \subseteq \mathcal{B}$.]

17. The following parts are of a similar flavor.
(a) Suppose that $\mathcal{A}$ is such that $\sigma(\mathcal{A}) = 2^\Omega$. For each distinct pair $\omega, \omega' \in \Omega$ there exists $A \in \mathcal{A}$ with $\omega \in A$ but $\omega' \notin A$.
(b) Let Ω be countable, and suppose that for each distinct pair $\omega, \omega' \in \Omega$, there exists a set $A \in \mathcal{A}$ such that $\omega \in A$ but $\omega' \notin A$. Then $\sigma(\mathcal{A}) = 2^\Omega$.

18. Given $\varnothing \neq \mathcal{A} \subseteq 2^\Omega$ and $\varnothing \neq B \subseteq \Omega$, let $\mathcal{A} \cap B = \{A \cap B : A \in \mathcal{A}\}$ and let $\sigma(\mathcal{A}) \cap B = \{A \cap B : A \in \sigma(\mathcal{A})\}$.
(a) $\sigma(\mathcal{A}) \cap B$ is a σ-field on B.
(b) Next, define $\sigma_B(\mathcal{A} \cap B)$ to be the minimal σ-field over B generated by the class $\mathcal{A} \cap B$. Then $\sigma_B(\mathcal{A} \cap B) = \sigma(\mathcal{A}) \cap B$. [Use (a) to argue that $\sigma_B(\mathcal{A} \cap B) \subseteq \sigma(\mathcal{A}) \cap B$. Next, let $\mathcal{C} = \{A \in \sigma(\mathcal{A}) : A \cap B \in \sigma_B(\mathcal{A} \cap B)\} \subseteq \sigma(\mathcal{A})$. Verify that $\mathcal{C}$ is a σ-field that contains $\mathcal{A}$, hence $\sigma(\mathcal{A}) = \mathcal{C}$. This will give $\sigma(\mathcal{A}) \cap B \subseteq \sigma_B(\mathcal{A} \cap B)$.]

19*. Suppose that $\mathcal{A} = \{A_1, A_2, \cdots\}$ is a disjoint sequence of subsets of Ω with $\bigcup_{n=1}^\infty A_n = \Omega$. Then each $\sigma(\mathcal{A})$-set is the union of an at most countable subcollection of $A_1, A_2, \cdots$. [Define $\mathcal{C}$ as the class of $A \in \sigma(\mathcal{A})$ such that A is an at most countable union of $\mathcal{A}$-sets. Show that $\mathcal{C}$ is a σ-field of subsets of Ω with $\mathcal{C} \supseteq \mathcal{A}$, which gives $\mathcal{C} = \sigma(\mathcal{A})$.]

20*. Let $\mathcal{P}$ denote a π-system on Ω, and let $\mathcal{L}$ denote a λ-system on Ω with $\mathcal{P} \subseteq \mathcal{L}$. We will show that $\sigma(\mathcal{P}) \subseteq \mathcal{L}$. Let $\lambda(\mathcal{P})$ denote the λ-system generated by $\mathcal{P}$, and for each subset $A \subseteq \Omega$ we define $\mathcal{G}_A = \{C \subseteq \Omega : A \cap C \in \lambda(\mathcal{P})\}$.
(a) $\mathcal{P} \subseteq \lambda(\mathcal{P}) \subseteq \mathcal{L}$.
(b) If $\lambda(\mathcal{P})$ is shown to be a π-system, then $\sigma(\mathcal{P}) \subseteq \mathcal{L}$.
(c) For any $A \in \mathcal{P}$, $\mathcal{G}_A$ is a λ-system containing $\mathcal{P}$. If $A \in \lambda(\mathcal{P})$, then $\mathcal{G}_A$ is a λ-system.
(d) For all $A \in \mathcal{P}$, $\lambda(\mathcal{P}) \subseteq \mathcal{G}_A$.
(e) For all $A \in \mathcal{P}$ and $B \in \lambda(\mathcal{P})$ we have $A \in \mathcal{G}_B$.
(f) For all $B \in \lambda(\mathcal{P})$, we have $\lambda(\mathcal{P}) \subseteq \mathcal{G}_B$.
(g) $\lambda(\mathcal{P})$ is a π-system, hence $\sigma(\mathcal{P}) \subseteq \mathcal{L}$.

Comment: This result is called *Dynkin's π-λ Theorem*, and it is a tool used in proving a uniqueness question regarding measures in Chapter 3. All parts of this exercise should be straightforward.

21*. Let $\mathcal{F}$ denote a field on Ω, and let $\mathcal{M}$ denote a monotone class on Ω [See Exercise 1]. We will show that $\mathcal{F} \subseteq \mathcal{M}$ implies $\sigma(\mathcal{F}) \subseteq \mathcal{M}$. Let $m(\mathcal{F})$ denote the *minimal monotone class on Ω generated by $\mathcal{F}$*. That is, $m(\mathcal{F})$ is the intersection of all monotone classes on Ω containing the collection $\mathcal{F}$.
(a) To prove the claim, it is sufficient to show that $\sigma(\mathcal{F}) \subseteq m(\mathcal{F})$.
(b) If $m(\mathcal{F})$ is a field, then $\sigma(\mathcal{F}) \subseteq m(\mathcal{F})$.
(c) $\Omega \in m(\mathcal{F})$.
(d) Let $\mathcal{G} = \{A \subseteq \Omega : A^c \in m(\mathcal{F})\}$. $\mathcal{G}$ is a monotone class on Ω and $m(\mathcal{F}) \subseteq \mathcal{G}$.
(e) $m(\mathcal{F})$ is indeed closed under complementation.
(f) Let $\mathcal{G}_1 = \{A \subseteq \Omega : A \cup B \in m(\mathcal{F}) \text{ for all } B \in \mathcal{F}\}$. Then $\mathcal{G}_1$ is a monotone class such that $\mathcal{F} \subseteq \mathcal{G}_1$ and $m(\mathcal{F}) \subseteq \mathcal{G}_1$.
(g) Let $\mathcal{G}_2 = \{B \subseteq \Omega : A \cup B \in m(\mathcal{F}) \text{ for all } A \in m(\mathcal{F})\}$. Then $\mathcal{G}_2$ is a monotone class such that $\mathcal{F} \subseteq \mathcal{G}_2$, and $m(\mathcal{F}) \subseteq \mathcal{G}_2$.
(h) $m(\mathcal{F})$ is closed under finite unions, and hence is a field.

Comment: This result is called *Halmos' Monotone Class Theorem*, and is similar to the π-λ *Theorem* above.

1.4 THE BOREL σ-FIELD

This and the next section deal with the two very important special cases of the abstract theory regarding minimal σ-fields presented in the previous section. The definition and discussions follow.

Definition. The *Borel σ-field [on $\mathbb{R}$]* is denoted by $\mathcal{B}$ and is defined as $\sigma(\{(a,b] : -\infty < a < b < +\infty\})$. The sets in $\mathcal{B}$ are called *[linear, one-dimensional] Borel sets.*

The concepts that make $\mathcal{B}$ worth careful study are not at all obvious, and there seems to be nothing so intrinsically special about the rsc subintervals of $\mathbb{R}$ that the σ-field on $\mathbb{R}$ generated by them is noteworthy. An explanation of the significance of $\mathcal{B}$ is to be found in Chapter 3 and more so in Chapter 4, when Lebesgue measure is discussed.

Notation. We will use the following notation again and again in this section, and it is best to define everything in one place. Here, $x, a, b \in \mathbb{R}$ with $a < b$.

$\mathcal{A}_1$ = intervals of the form $(-\infty, x]$	$\mathcal{A}_2$ = intervals of the form $(-\infty, x)$
$\mathcal{A}_3$ = intervals of the form $[x, \infty)$	$\mathcal{A}_4$ = intervals of the form (x, ∞)
$\mathcal{A}_5$ = intervals of the form $[a, b]$	$\mathcal{A}_6$ = intervals of the form $(a, b]$
$\mathcal{A}_7$ = intervals of the form $[a, b)$	$\mathcal{A}_8$ = intervals of the form (a, b)
$\mathcal{A}_9$ = open subsets of $\mathbb{R}$	$\mathcal{A}_{10}$ = closed subsets of $\mathbb{R}$.

For $n = 1, \cdots, 8$, $\mathcal{A}_n^*$ will denote the collection of intervals having the same form as those in $\mathcal{A}_n$, except that the endpoints are rational. For example, $\mathcal{A}_6^*$ denotes the set of intervals of the form $(a, b]$ with $a, b \in \mathbb{Q}$, $\mathcal{A}_8^*$ denotes the set of open intervals (a, b) with $a, b \in \mathbb{Q}$, etc.

In the definition above, $\mathcal{B}$ was defined as $\sigma(\mathcal{A}_6)$, and it might be wondered just why $\sigma(\mathcal{A}_6)$ is of importance instead of [say] $\sigma(\mathcal{A}_3)$ or any of the other σ-fields generated by $\mathcal{A}_n$, $n \neq 3, 6$. It will be seen below that the choice of $\mathcal{A}_6$ for a generating class as compared to [say] $\mathcal{A}_3$ or [say] $\mathcal{A}_8^*$ is wholly arbitrary, as any of $\mathcal{A}_1, \cdots, \mathcal{A}_{10}, \mathcal{A}_1^*, \cdots, \mathcal{A}_8^*$ generates $\mathcal{B}$. This result is important to know because various books define $\mathcal{B}$ differently. Some define $\mathcal{B}$ as $\sigma(\mathcal{A}_6)$, while some write $\mathcal{B} = \sigma(\mathcal{A}_9)$. Halmos' text on measure theory writes $\mathcal{B} = \sigma(\mathcal{A}_7)$, etc. We will show that all of these definitions are equivalent, so that these seemingly different definitions are all ultimately saying the same thing.

Claim 1. *Let $\mathcal{C}, \mathcal{D} \subseteq 2^\Omega$ each be nonempty. If each $\mathcal{C}$-set may be written in terms of complements of $\mathcal{D}$-sets, amc unions of $\mathcal{D}$-sets, and amc intersections of $\mathcal{D}$-sets, then $\sigma(\mathcal{C}) \subseteq \sigma(\mathcal{D})$. If in addition each $\mathcal{D}$-set may be written similarly in terms of $\mathcal{C}$-sets, then $\sigma(\mathcal{C}) = \sigma(\mathcal{D})$.*

Proof: The hypotheses imply that each $\mathcal{C}$-set will be in every σ-field containing $\mathcal{D}$, for σ-fields are closed under complements, amc unions, amc intersections, and any amc sequences of such operations. Therefore, each $\mathcal{C}$-set is in the intersection of each σ-field containing $\mathcal{D}$: $\mathcal{C} \subseteq \sigma(\mathcal{D})$. Thus $\sigma(\mathcal{C}) \subseteq \sigma(\sigma(\mathcal{D})) = \sigma(\mathcal{D})$. Under the additional assumptions of the claim, the reverse inclusion $\sigma(\mathcal{D}) \subseteq \sigma(\mathcal{C})$ follows in the exact same fashion. ■

Claim 2. $\mathcal{B} = \sigma(\mathcal{A}_1) = \cdots = \sigma(\mathcal{A}_{10}) = \sigma(\mathcal{A}_1^*) = \cdots = \sigma(\mathcal{A}_8^*)$.

Proof: Let $x, a, b \in \mathbb{R}$ with $a < b$. Letting the unions and intersections below be taken over all $n \in \mathbb{N}$, it may be verified that

$$(-\infty, x) = \bigcup(-\infty, x - n^{-1}], \qquad [x, +\infty) = (-\infty, x)^c,$$

$$(x, +\infty) = \bigcup[x + n^{-1}, +\infty), \qquad [a, b] = (b, +\infty)^c - \left[\bigcup(a - n^{-1}, +\infty)\right]^c,$$

$$(a, b] = \bigcup[a + n^{-1}, b], \qquad [a, b) = \left((a, b] \cup \bigcap(a - n^{-1}, a]\right) - \bigcap(b - n^{-1}, b],$$

$$\text{and} \quad (a, b) = \bigcup[a + n^{-1}, b).$$

Recall next that every open subset of $\mathbb{R}$ is an amc union of sets of the form (a, b), every closed subset of $\mathbb{R}$ is the complement of an open subset of $\mathbb{R}$, and $(a, b] = [(-\infty, a] \cup \bigcup[b + n^{-1}, +\infty)]^c$, which is the complement of a countable union of closed sets. Hence, for $n = 2, \cdots, 10$, we have written a typical element of $\mathcal{A}_n$ in terms of amc unions, amc intersections, and complements of sets in $\mathcal{A}_{n-1}$, and $\mathcal{A}_1$ has been similarly expressed in terms of $\mathcal{A}_{10}$-sets. We may therefore invoke Claim 1 to obtain $\sigma(\mathcal{A}_1) \subseteq \cdots \subseteq \sigma(\mathcal{A}_{10}) \subseteq \sigma(\mathcal{A}_1)$, and thus these inclusions are actually equalities.

Next, taking $x, a, b \in \mathbb{Q}$ above, noting that $(-\infty, x] = (\bigcup(x, b + n))^c$, and using Claim 1 yields $\sigma(\mathcal{A}_1^*) \subseteq \cdots \subseteq \sigma(\mathcal{A}_8^*) \subseteq \sigma(\mathcal{A}_1^*)$, and hence these inclusions are actually equalities. We thus have $\sigma(\mathcal{A}_1) = \cdots = \sigma(\mathcal{A}_{10})$ and $\sigma(\mathcal{A}_1^*) = \cdots = \sigma(\mathcal{A}_8^*)$.

Finally, observe that for any $x \in \mathbb{R}$ we may find an increasing rational sequence $\{r_n\}_{n=1}^{\infty}$ with limit x. Since $(-\infty, x) = \bigcup(-\infty, r_n)$, elements of $\mathcal{A}_2$ can be expressed in terms of amc unions of $\mathcal{A}_2^*$-sets, whence $\sigma(\mathcal{A}_2) \subseteq \sigma(\mathcal{A}_2^*)$. But $\mathcal{A}_2^* \subseteq \mathcal{A}_2$, hence $\sigma(\mathcal{A}_2^*) \subseteq \sigma(\mathcal{A}_2)$, ergo $\sigma(\mathcal{A}_2) = \sigma(\mathcal{A}_2^*)$, which, with the other equalites, finishes the proof. ■

Since $\mathcal{A} \subseteq \sigma(\mathcal{A})$ for arbitrary $\mathcal{A}$, the above theorem says that $\mathcal{B}$ contains $\mathcal{A}_1, \cdots, \mathcal{A}_{10}$. That is, $\mathcal{B}$ contains every interval of every form, every open set, and every closed set. Furthermore, $\mathcal{B}$ contains any set that can be obtained from the sets in these classes via amc combinations of amc unions, amc intersections, and complements. It will eventually turn out that $\mathcal{B}$ will contain just about every subset of $\mathbb{R}$ that one can imagine, although it will also turn out that (i) there are *many* subsets of $\mathbb{R}$ that are *not* in $\mathcal{B}$, and (ii) there are many sets in $\mathcal{B}$ that are inexpressible in terms of amc combinations of amc unions, amc intersections, and complements of sets in any and all of the 10 aforementioned classes. Assertions (i) and (ii) are given rigorous proofs in Sections 1.6 and 1.7.

A few extra questions of a somewhat courageous nature might have suggested themselves by now. These questions are now listed in no special order.

(Q1) *What is* card($\mathcal{B}$)*? That is, "how many" Borel sets are there?*

(Q2) *What does a generic Borel set "look like?"*

(Q3) *Is there some type of "constructive algorithm" for generating $\mathcal{B}$ from the sets in some or all of the $\mathcal{A}_1, \cdots, \mathcal{A}_{10}, \mathcal{A}_1^*, \cdots, \mathcal{A}_8^*$?*

(Q4) *Why look at the minimal σ-field generated by any of the classes in Claim 2 when one could have the largest σ-field simply by dealing with $2^{\mathbb{R}}$?*

Such questions actually invoke some of the very interesting and foundational items from modern set theory and analysis, such as the Axiom of Choice[1] (AC) and the various forms of the Continuum Hypothesis[2] (CH). We will take both of these items as true in this text, and shall attempt loose sketches of the answers to questions (Q1)-(Q4) as posed.

Since card($\mathcal{A}_6$) $= \mathfrak{c}$ and $\mathcal{A}_6 \subseteq \mathcal{B}$, we have card($\mathcal{B}$) $\geq \mathfrak{c}$. Since $\mathcal{B} \subseteq 2^{\mathbb{R}}$ and card($2^{\mathbb{R}}$) $= 2^{\mathfrak{c}}$, we have card($\mathcal{B}$) $\leq 2^{\mathfrak{c}}$. Therefore $\mathfrak{c} \leq$ card($\mathcal{B}$) $\leq 2^{\mathfrak{c}}$. If we adopt the Generalized Continuum Hypothesis[3] (GCH), this fact forces either card($\mathcal{B}$) $= \mathfrak{c}$ or card($\mathcal{B}$) $= 2^{\mathfrak{c}}$. It is not obvious which possibility is true, for both seem plausible. In Section 1.6 we rigorously show that card($\mathcal{B}$) $= \mathfrak{c}$. If we accept this as true, then card($2^{\mathbb{R}}$) $= 2^{\mathfrak{c}} > \mathfrak{c} =$ card($\mathcal{B}$), so that there are subsets of $\mathbb{R}$ outside $\mathcal{B}$. In fact, since $2^{\mathfrak{c}} - \mathfrak{c} = 2^{\mathfrak{c}}$, "most" subsets of $\mathbb{R}$ are not Borel sets. In Chapter 4, we will use (AC) to exhibit as explicitly as possible some sets outside $\mathcal{B}$.

The answers to (Q2) and (Q3) require substantial work. For now, the answer to (Q2) is that there is no "nice" generic representation formula for an arbitrary $B \in \mathcal{B}$. One might very well think that a typical $B \in \mathcal{B}$ might have the form $B = A_1 \star A_2 \star A_3 \star \cdots$, where each $\star$ can freely denote $\cup$ or $\cap$, and where each A_n freely denotes an interval of any form, a singleton, a finite set, an open set, a closed set, a countable set, or a complement of any one of these types of "nice" sets. However, this scheme, while accounting for a subset of $\mathcal{B}$ with cardinality $\mathfrak{c}$, will *not* be sufficient to account for *all* Borel sets. In fact,

[1] One form of (AC) is as follows: Given any collection of nonempty sets $\mathcal{A} = \{A_i : i \in I\}$, there exists a function ϕ with domain $\mathcal{A}$ and the property that $\phi(A_i) \in A_i$ for each $i \in I$. More loosely put, given an arbitrary family of nonempty sets, we can form a new set consisting of one element from each set in the family.

[2] One form of (CH) is this: There exists no cardinal number strictly between $\aleph_0$ and $\mathfrak{c}$. In advanced set theory, it is shown that (i) both (CH) and (AC) are independent of the usual Zermelo-Fraenkel (ZF) axioms, and (ii) in the presence of (ZF), (CH) and (AC) are independent.

[3] (GCH) asserts that for any infinite cardinal $\mathfrak{u}$ there is no cardinal $\mathfrak{v}$ with $\mathfrak{u} < \mathfrak{v} < 2^{\mathfrak{u}}$. In advanced set theory, it is shown that (GCH) implies both (AC) and (CH), so we are forced to adopt (GCH) to pursue this discussion.

the collection of all sets generated by this scheme will be a "small" proper subset of $\mathcal{B}$. One might think that sets of the form $B = B_1 \star B_2 \star B_3 \star \cdots$, where each B_n is a set like $A_1 \star A_2 \star \cdots$ in the previous formula, would account for all of the Borel sets. Again, $\mathfrak{c}$ more Borel sets will be accounted for by this more comprehensive representation, but many will be excluded. One might iterate this process countably many times, obtaining $\aleph_0 \mathfrak{c} = \mathfrak{c}$ Borel sets, but surprisingly there will still be $\mathfrak{c}$ Borel sets that are excluded. [See Section 1.7.] Thus, in proving things about $\mathcal{B}$, we cannot take a generic $B \in \mathcal{B}$ and invoke a "nice" representation of B, using this hoped-for representation to prove a property about our generic B and hence about all Borel sets. Any argument involving $\mathcal{B}$ will be "nonconstructive" in the same sense that many of the exercises in the previous section were.

The answer to (Q3) is in the negative if we want an algorithm that is "practical" or not too esoteric. In imprecise language, it will be shown that we cannot arrive at all of $\mathcal{B}$ by taking all intervals of all forms and employing a recursive scheme that uses countably many operations "in a given order" at each stage, no matter how many stages we let the scheme run. On the other hand, there *is* a scheme that uses countably many operations "not performed in a simple sequence" at each stage that "after countably many stages" will produce $\mathcal{B}$; the discussion of such a scheme requires using ordinal numbers. These utterly imprecise answers would make a professional set theorist or logician chafe, and it might be wondered just what all of these loosely worded answers really mean. Perhaps it is best to just say that (Q3) has a negative answer, and a rigorous formulation of (Q3) is found in Section 1.6.

The answer to (Q4) will be completely answered in a later chapter. It suffices to say for the present that $\mathcal{B}$ is, in a certain vague sense, the "largest σ-field containing all of the 'ordinary' sets from analysis that will not cause mathematical and logical problems with (AC) and our upcoming definition of what will be called Lebesgue measure." This very nebulous statement will be completely demystified in the proofs of the two "impossibility theorems" found at the end of Chapter 4.

The final topic of this section concerns the *extended Borel σ-field on $\bar{\mathbb{R}}$*, and it is a simple extension of what has been discussed regarding $\mathcal{B}$ and $\mathbb{R}$.

Definition. The *extended Borel σ-field on* $\bar{\mathbb{R}}$ is denoted by $\bar{\mathcal{B}}$, and is defined by $\bar{\mathcal{B}} \equiv \sigma(\{(a, b] : a, b \in \bar{\mathbb{R}}, a < b\})$. Sets in $\bar{\mathcal{B}}$ are called *extended Borel sets.*

Just as different authors use different definitions for $\mathcal{B}$, so too $\bar{\mathcal{B}}$ has various definitions. Some other definitions are (1) the class of sets that are $\mathcal{B}$-sets or are $\mathcal{B}$-sets enlarged by one or both of $\pm\infty$, (2) the σ-field on $\bar{\mathbb{R}}$ generated by the class of intervals of the form (a, b), $[-\infty, a)$, and $(b, +\infty]$, where $a, b \in \mathbb{R}$, (3) the σ-field on $\bar{\mathbb{R}}$ generated by intervals of the form $[-\infty, x)$, $x \in \mathbb{R}$, (4) $\sigma(\mathcal{B} \cup \{\{-\infty\}, \{+\infty\}\})$, and (5) the σ-field on $\bar{\mathbb{R}}$ generated by the open subsets of $\bar{\mathbb{R}}$, sets of the form A, $A \cup (x, \infty]$, $A \cup [-\infty, x)$, and $A \cup [-\infty, x) \cup (y, +\infty]$, where A is an open subset of $\mathbb{R}$ and $x, y \in \mathbb{R}$. The reader may rest assured that all of these definitions are equivalent to ours. In fact, (1) may be the most useful working characterization of $\bar{\mathcal{B}}$-sets. It is straightforward to verify the equivalences; see Exercise 6.

There are some final points that should be stated. First, *every Borel set is an extended Borel set.* Second, *the definition of $\bar{\mathcal{B}}$ implies that* $\{-\infty, +\infty\}$, $\{+\infty\}$, *and* $\{-\infty\}$ *are extended Borel sets.* These facts are obtained by using (1): If every $\bar{\mathcal{B}}$-set is a $\mathcal{B}$-set possibly augmented by one or both points $\pm\infty$, then, taking $\varnothing \in \mathcal{B}$, it follows that $\varnothing \cup \{-\infty, +\infty\}$, $\varnothing \cup \{-\infty\}$, and $\varnothing \cup \{+\infty\}$ are $\bar{\mathcal{B}}$-sets. Third, the answers to questions (Q1)-(Q4) are not changed in any way by now allowing consideration of $\pm\infty$. Finally, the natural extension of Claim 2 holds in the extended Borel setting.

Exercises.

1*. Show directly that $\sigma(\mathcal{A}_3) = \sigma(\mathcal{A}_3^*)$, $\sigma(\mathcal{A}_4) = \sigma(\mathcal{A}_7)$, and $\sigma(\mathcal{A}_4^*) = \sigma(\mathcal{A}_{10})$.

2*. All amc subsets of $\mathbb{R}$ are Borel sets. All subsets of $\mathbb{R}$ that differ from a Borel set by at most countably many points are Borel sets. That is, if the symmetric difference $C \triangle B$ is amc and $B \in \mathcal{B}$, then $C \in \mathcal{B}$.

3*. The *Borel σ-field on* $(0, 1]$ is denoted by $\mathcal{B}_{(0,1]}$ and is defined as the σ-field on $(0, 1]$ generated by the rsc subintervals of $(0, 1]$. $\mathcal{B}_{(0,1]}$ may be equivalently defined by $\{B \cap (0, 1] : B \in \mathcal{B}\}$.

4*. $\mathcal{B}$ is generated by the compact subsets of $\mathbb{R}$.

5. $\mathcal{B}$ is not generated by the following:
(a) Any finite collection of subsets of $\mathbb{R}$
(b) The collection of real singletons.
(c) The collection of all finite subsets of $\mathbb{R}$.
(d) The collection of all amc subsets of $\mathbb{R}$.

6*. The representations (1)-(5) give equivalent formulations of $\bar{\mathcal{B}}$. [Let $\mathcal{C}$ denote the generating class for $\bar{\mathcal{B}}$ given in the definition, and let $\mathcal{D}$ denote the collection of sets that are in $\mathcal{B}$ or are $\mathcal{B}$-sets possibly augmented by one or both of $-\infty$ and $+\infty$. It is nearly automatic that $\mathcal{D}$ is a σ-field on $\bar{\mathbb{R}}$. Clearly an interval of the form $(a, b]$ where $-\infty \leq a < b \leq +\infty$ is in $\mathcal{D}$, so $\mathcal{C} \subseteq \mathcal{D}$, hence $\sigma(\mathcal{C}) \subseteq \sigma(\mathcal{D}) = \mathcal{D}$. The reverse inclusion $\mathcal{D} \subseteq \sigma(\mathcal{C})$ follows from the facts that (i) $\mathcal{D}$ contains $\mathcal{C}$ and (ii) $\mathcal{D}$ is clearly the smallest such σ-field containing $\mathcal{C}$. This gives $\sigma(\mathcal{C}) = \mathcal{D}$, and therefore the definition of $\bar{\mathcal{B}}$ is equivalent to the formulation given in (1). The equivalence between the definition and (2)-(5) follows in the same fashion.]

1.5 THE k-DIMENSIONAL BOREL σ-FIELD

This section generalizes the last section to k-dimensional Euclidean space for arbitrary $k \in \mathbb{N}$. There are some interesting questions that arise in higher dimensions; these cannot be asked when interest is restricted to one dimension.

Definition. Let $k \in \mathbb{N}$. The *k-dimensional Borel σ-field [on $\mathbb{R}^k$]* is denoted by $\mathcal{B}^k$, and is $\sigma(\{(\boldsymbol{a},\boldsymbol{b}] : \boldsymbol{a} < \boldsymbol{b},\ \boldsymbol{a},\boldsymbol{b} \in \mathbb{R}^k\})$. The sets in $\mathcal{B}^k$ are called *[k-dimensional] Borel sets.*

The symbol $\mathcal{B}^k$ is *not* to be construed as the k-fold Cartesian product $\mathcal{B}\times\cdots\times\mathcal{B}$ even though the notation suggests this; see the following discussion for further details.

Notation. We desire to adapt the notation of the previous section to $\mathbb{R}^k$ for any $k \in \mathbb{N}$. To that end, we will let $\mathcal{A}_1^k$ denote $\{(-\infty,\boldsymbol{x}] : \boldsymbol{x} \in \mathbb{R}^k\}$, $\mathcal{A}_1^{k*}$ will denote $\{(-\infty,\boldsymbol{x}] : \boldsymbol{x} \in \mathbb{Q}^k\}$, with $\mathcal{A}_2^k$, $\mathcal{A}_2^{k*}$, $\cdots$, $\mathcal{A}_8^k$, $\mathcal{A}_8^{k*}$ defined in analogous fashion; the classes $\mathcal{A}_9^k$ and $\mathcal{A}_{10}^k$ will denote the open subsets of $\mathbb{R}^k$ and the closed subsets of $\mathbb{R}^k$, respectively. When $k = 1$, these definitions reduce exactly to the definitions given in the last section.

According to the definition of $\mathcal{B}^k$, we have $\mathcal{B}^k = \sigma(\mathcal{A}_6^k)$. As in the previous section, various authors define $\mathcal{B}^k$ as the σ-field of subsets of $\mathbb{R}^k$ generated by $\mathcal{A}_1^{k*}$, $\mathcal{A}_7^{k*}$, and $\mathcal{A}_9^{k*}$, among other things. The following result says that any of the eighteen classes listed will generate $\mathcal{B}^k$.

Claim 1. $\mathcal{B}^k = \sigma(\mathcal{A}_1^k) = \cdots = \sigma(\mathcal{A}_{10}^k) = \sigma(\mathcal{A}_1^{k*}) = \cdots = \sigma(\mathcal{A}_8^{k*})$.

The proof of Claim 1 proceeds along the same tedious lines as does Claim 2 from the previous section; we merely need to make the natural modifications for k dimensions.

By Claim 1, the following types of subsets of $\mathbb{R}^k$ are k-dimensional Borel sets: all rectangles of all forms, all singletons, all finite subsets, all countably infinite sets, all open sets, all closed sets, and all sets that differ from a $\mathcal{B}^k$-set by at most countably many points. $\mathcal{B}^k$ will contain many more sets than these aforementioned types of sets, and, just as when $k = 1$, $\mathcal{B}^k$ will in the end be seen to include just about every set that one encounters in doing analysis on $\mathbb{R}^k$. Also, the questions (Q1)-(Q4) from the previous section, modified for k dimensions, have the same answers for all $k \geq 1$, with the proofs and careful discussions being postponed until later. In other words, we may state in colloquial terms that (i) there are $\mathfrak{c}$ sets in $\mathcal{B}^k$, (ii) there is no "nice" representation of a generic $\mathcal{B}^k$-set, (iii) there is a transfinite algorithm for obtaining all of $\mathcal{B}^k$ from countably infinite set operations performed on k-dimensional rectangles of all forms, and (iv) certain logical problems with the Axiom of Choice will arise in conjunction with future developments if we consider every subset of $\mathbb{R}^k$ as compared to merely some of them.

Notation. $\mathbb{R}^k$ is the k-fold Cartesian product $\mathbb{R} \times \cdots \times \mathbb{R}$. For example, $\mathbb{R}^3 = \mathbb{R} \times \mathbb{R} \times \mathbb{R}$, and the elements of $\mathbb{R}^3$ are ordered triples (x_1, x_2, x_3), where $x_1, x_2, x_3 \in \mathbb{R}$. Strictly speaking, $(\mathbb{R} \times \mathbb{R}) \times \mathbb{R}$, $\mathbb{R} \times (\mathbb{R} \times \mathbb{R})$, and $\mathbb{R} \times \mathbb{R} \times \mathbb{R}$ are distinct; this is because the elements of $(\mathbb{R} \times \mathbb{R}) \times \mathbb{R}$ are ordered pairs of the form $((x_1, x_2), x_3)$, whereas the elements of $\mathbb{R} \times (\mathbb{R} \times \mathbb{R})$ are ordered pairs of the form $(x_1, (x_2, x_3))$. Clearly $((x_1, x_2), x_3) \neq (x_1, (x_2, x_3))$, and none of these equal (x_1, x_2, x_3). However, there is an obvious one-to-one correspondence between the elements of $(\mathbb{R} \times \mathbb{R}) \times \mathbb{R}$ and $\mathbb{R} \times (\mathbb{R} \times \mathbb{R})$, namely, $(x_1, (x_2, x_3)) \leftrightarrow ((x_1, x_2), x_3)$. We will accordingly view these forms as identical, and we assert that this identification will not cause any logical or set-theoretic problems for us. For example, we will treat the element $(((x_1, x_2), x_3), ((x_4, x_5), (x_6, x_7)))$ of $((\mathbb{R}\times\mathbb{R})\times\mathbb{R})\times((\mathbb{R}\times\mathbb{R})\times(\mathbb{R}\times\mathbb{R}))$ as equal to the element $(x_1, x_2, x_3, x_4, x_5, x_6, x_7)$ of $\mathbb{R} \times \mathbb{R} \times \mathbb{R} \times \mathbb{R} \times \mathbb{R} \times \mathbb{R} \times \mathbb{R} = \mathbb{R}^7$. These items are not formally the same, but there is an obvious one-to-one correspondence between the elements of $((\mathbb{R} \times \mathbb{R}) \times \mathbb{R}) \times ((\mathbb{R} \times \mathbb{R}) \times (\mathbb{R} \times \mathbb{R}))$ and the elements of $\mathbb{R}^7$, namely

$$(((x_1, x_2), x_3), ((x_4, x_5), (x_6, x_7))) \leftrightarrow (x_1, x_2, x_3, x_4, x_5, x_6, x_7),$$

hence we will not run aground in carrying through this convention of identification. All other cases are handled similarly.

More Notation and the Structure of $\mathcal{B}^k$. Just as the symbol $\mathbb{R}^k$ denotes the k-fold Cartesian product of $\mathbb{R}$ with itself, the symbol $\mathcal{B}^k$ suggests that $\mathcal{B}^k$ is the k-fold Cartesian product $\mathcal{B} \times \cdots \times \mathcal{B}$. One might also understandably think that $\mathcal{B}^{n+m} = \mathcal{B}^n \times \mathcal{B}^m$ is a notational truism for each $m, n \in \mathbb{N}$; after all, we identify $\mathbb{R}^{n+m}$ with $\mathbb{R}^n \times \mathbb{R}^m$ for each $m, n \in \mathbb{N}$. *While the notation regarding the Borel σ-field suggests this, such identities regarding the Borel σ-field are not true.* To see a simple counterexample to the claim that [say] $\mathcal{B} \times \mathcal{B} = \mathcal{B}^2$, consider the open unit disk $D = \{\boldsymbol{x} \in \mathbb{R}^2 : x_1^2 + x_2^2 < 1\}$. D is an open subset of $\mathbb{R}^2$, hence in $\mathcal{B}^2$, but there do not exist $B_1, B_2 \subseteq \mathbb{R}$, let alone $B_1, B_2 \in \mathcal{B}$, such that $B_1 \times B_2 = D$. Thus $D \in \mathcal{B}^2$, but $D \notin \mathcal{B} \times \mathcal{B}$. Therefore, *the superscript k in the symbol $\mathcal{B}^k$ is merely part of the symbol – it does not denote any k-fold Cartesian product of $\mathcal{B}$ with itself.* This is of course not what the notation suggests, but it is the standard notation. We may think of $\mathbb{R}^2$ as $\mathbb{R} \times \mathbb{R}$, but $\mathcal{B}^2$ may *not* be thought of as $\mathcal{B} \times \mathcal{B}$. We may think of $\mathbb{R}^4$ as $\mathbb{R} \times \mathbb{R} \times \mathbb{R} \times \mathbb{R}$ or $(\mathbb{R} \times \mathbb{R} \times \mathbb{R}) \times \mathbb{R}$ or $(\mathbb{R} \times \mathbb{R}) \times (\mathbb{R} \times \mathbb{R})$ and so on, but we may *not* think of $\mathcal{B}^4$ as $\mathcal{B} \times \mathcal{B} \times \mathcal{B} \times \mathcal{B}$ or $(\mathcal{B} \times \mathcal{B} \times \mathcal{B}) \times \mathcal{B}$ or $(\mathcal{B} \times \mathcal{B}) \times (\mathcal{B} \times \mathcal{B})$, etc. However, we do have the following containment.

Claim 2. *For all $n, m \in \mathbb{N}$ we have $\mathcal{B}^n \times \mathcal{B}^m \subsetneq \mathcal{B}^{n+m}$.*

Proof: Denote a typical element of $\mathbb{R}^{n+m}$ by $(x_1, \cdots, x_n, y_1, \cdots, y_m)$, and using the notational convention discussed above, we will write this generic element as $(\boldsymbol{x}, \boldsymbol{y})$, where $\boldsymbol{x} \in \mathbb{R}^n$ and $\boldsymbol{y} \in \mathbb{R}^m$. We first show that

$$\{B \times \mathbb{R}^m : B \in \mathcal{B}^n\} \subseteq \mathcal{B}^{n+m}. \tag{1.1}$$

Define the mapping $f : \mathbb{R}^{n+m} \to \mathbb{R}^n$ for all $(\boldsymbol{x}, \boldsymbol{y}) \in \mathbb{R}^{n+m}$ by $f((\boldsymbol{x}, \boldsymbol{y})) = \boldsymbol{x}$. This f is continuous, and hence the inverse f-image of an open set in $\mathbb{R}^n$ is an open set in $\mathbb{R}^{n+m}$. Therefore, $f^{-1}(\mathcal{A}_9^n) \subseteq \mathcal{A}_9^{n+m}$. It follows from Claim 1 of this section and Claim 2 of Section 1.3 that

$$f^{-1}(\mathcal{B}^n) = f^{-1}(\sigma(\mathcal{A}_9^n)) = \sigma(f^{-1}(\mathcal{A}_9^n)) \subseteq \sigma(\mathcal{A}_9^{n+m}) = \mathcal{B}^{n+m}.$$

Since $B \times \mathbb{R}^m = \{(\boldsymbol{x}, \boldsymbol{y}) \in \mathbb{R}^{n+m} : f((\boldsymbol{x}, \boldsymbol{y})) \in B\} = f^{-1}(B)$ for each $B \in \mathcal{B}^n$, (1.1) is established. The exact same type of argument gives

$$\{\mathbb{R}^n \times C : C \in \mathcal{B}^m\} \subseteq \mathcal{B}^{n+m}. \tag{1.2}$$

Putting (1.1) and (1.2) together, we have

$$\mathcal{B}^n \times \mathcal{B}^m = \{B \times \mathbb{R}^m : B \in \mathcal{B}^n\} \cap \{\mathbb{R}^n \times C : C \in \mathcal{B}^m\} \subseteq \mathcal{B}^{n+m}.$$

To see that the containment is strict, observe that the open unit ball D in $\mathbb{R}^{n+m}$ cannot be written as $A_1 \times A_2$ with $A_1 \subseteq \mathbb{R}^n$ and $A_2 \subseteq \mathbb{R}^m$, let alone with $A_1 \in \mathcal{B}^n$ and $A_2 \in \mathcal{B}^m$. Therefore $D \in \mathcal{B}^{n+m}$ but $D \notin \mathcal{B}^n \times \mathcal{B}^m$. This completes the proof. ■

Not only does $\mathcal{B}^n \times \mathcal{B}^m$ fail to equal $\mathcal{B}^{n+m}$, but $\mathcal{B}^n \times \mathcal{B}^m$ is not even a σ-field on $\mathbb{R}^{n+m}$. To see this, observe that $\mathcal{A}_6^n \subseteq \mathcal{B}^n$ and $\mathcal{A}_6^m \subseteq \mathcal{B}^m$, hence

$$\mathcal{A}_6^{n+m} = \mathcal{A}_6^n \times \mathcal{A}_6^m \subseteq \mathcal{B}^n \times \mathcal{B}^m \subsetneq \mathcal{B}^{n+m}.$$

If $\mathcal{B}^n \times \mathcal{B}^m$ is a σ-field on $\mathbb{R}^{n+m}$, then

$$\mathcal{B}^{n+m} = \sigma(\mathcal{A}_6^{n+m}) \subseteq \sigma(\mathcal{B}^n \times \mathcal{B}^m) = \mathcal{B}^n \times \mathcal{B}^m,$$

contradicting Claim 2. Therefore, $\mathcal{B}^n \times \mathcal{B}^m$ fails to be a σ-field on $\mathbb{R}^{n+m}$.

Definition. Let $n, m \in \mathbb{N}$. The *product σ-field of $\mathcal{B}^n$ and $\mathcal{B}^m$* is denoted by the symbol $\mathcal{B}^n \otimes \mathcal{B}^m$, and is defined as $\sigma(\mathcal{B}^n \times \mathcal{B}^m)$.

Clearly $\mathcal{B}^n \otimes \mathcal{B}^m$ is a σ-field on $\mathbb{R}^{n+m}$, and $\mathcal{B}^n \times \mathcal{B}^m \subseteq \mathcal{B}^n \otimes \mathcal{B}^m$. While authors use the symbol $\times$ for *both* the ordinary Cartesian product and our symbol $\otimes$, we use $\otimes$ for notational clarity.

Claim 3. *For every $n, m \in \mathbb{N}$ we have $\mathcal{B}^{n+m} = \mathcal{B}^n \otimes \mathcal{B}^m$.*

Proof: Claim 2 gives $\mathcal{B}^n \times \mathcal{B}^m \subsetneq \mathcal{B}^{n+m}$, thus

$$\mathcal{B}^n \otimes \mathcal{B}^m = \sigma(\mathcal{B}^n \times \mathcal{B}^m) \subseteq \sigma(\mathcal{B}^{n+m}) = \mathcal{B}^{n+m}.$$

Next, we have $\mathcal{A}_1^{n+m} = \mathcal{A}_1^n \times \mathcal{A}_1^m \subseteq \mathcal{B}^n \times \mathcal{B}^m$, hence

$$\mathcal{B}^{n+m} = \sigma(\mathcal{A}_1^{n+m}) \subseteq \sigma(\mathcal{B}^n \times \mathcal{B}^m) = \mathcal{B}^n \otimes \mathcal{B}^m,$$

completing the proof. ■

We now turn to a k-dimensional version of the extended Borel field mentioned in the previous section. This definition is a mere extension of the concepts in the last section, and moving from one to k dimensions creates no real difficulty.

Definition. The *extended k-dimensional Borel σ-field [on $\bar{\mathbb{R}}^k$]* will be denoted by $\bar{\mathcal{B}}^k$, and is defined as $\sigma(\{(\boldsymbol{a}, \boldsymbol{b}] : -\infty \leq \boldsymbol{a} < \boldsymbol{b} \leq +\infty\})$.

The various equivalent formulations of $\bar{\mathcal{B}}$ given at the end of the previous section generalize to k dimensions with no problems, and the formulations modified for k dimensions are still equivalent; see Exercise 7.

Exercises.

1*. For each $k \in \mathbb{N}$ and $j = 1, \cdots, k$, every $(k-j)$-dimensional hyperplane is in $\mathcal{B}^k$.

2. Consider the circumference of the unit circle in $\mathbb{R}^2$, the set of points (x, y) with $x^2 + y^2 = 1$. The set of points on the circumference of the unit circle having at least one irrational coordinate is a two-dimensional Borel set.

3. The set of points in $\mathbb{R}^k$ having all transcendental coordinates is in $\mathcal{B}^k$.

4. For any rectangle [of any form] $A \subseteq \mathbb{R}^k$, we define the *k-dimensional Borel σ-field over A* to be the σ-field generated by all k-dimensional rsc subrectangles of A; call this class $\mathcal{B}^k_A$. $\mathcal{B}^k_A$ may be equivalently defined as $\{B \cap A : B \in \mathcal{B}^k\}$.

5*. $\mathcal{B}^k$ is generated by a countable class.

6. Show that $\mathcal{B}^k$ is not generated by the singletons of $\mathbb{R}^k$, nor by the finite subsets of $\mathbb{R}^k$, nor by the amc subsets of $\mathbb{R}^k$.

7. Here are five extra definitions of $\bar{\mathcal{B}}^k$ that are found elsewhere. These five definitions are equivalent to our definition of $\bar{\mathcal{B}}^k$.
(a) The class of subsets of $\bar{\mathbb{R}}^k$ that are $\mathcal{B}^k$-sets or are $\mathcal{B}^k$ sets augmented by points that have $\pm\infty$ in at least one coordinate.
(b) The σ-field of subsets of $\bar{\mathbb{R}}^k$ generated by the class of open rectangles in $\bar{\mathbb{R}}^k$ of the form $(\boldsymbol{a}, \boldsymbol{b})$, $[-\infty, \boldsymbol{a})$, and $(\boldsymbol{a}, +\infty]$, where $\boldsymbol{a}, \boldsymbol{b} \in \mathbb{R}^k$.
(c) The σ-field on $\bar{\mathbb{R}}^k$ generated by $\{(-\infty, \boldsymbol{x}] : \boldsymbol{x} \in \bar{\mathbb{R}}^k\}$.
(d) $\sigma(\mathcal{B}^k \cup \{\{\boldsymbol{x}\} :$ each x_i freely equals $+\infty$ or $-\infty\})$.
(e) The σ-field on $\bar{\mathbb{R}}^k$ generated by the open subsets of $\bar{\mathbb{R}}^k$, that is, the sets of the form A, $A \cup (\boldsymbol{x}, +\infty]$, $A \cup [-\infty, \boldsymbol{x})$, and $A \cup [-\infty, \boldsymbol{x}) \cup (\boldsymbol{y}, +\infty]$, where A is an open subset of $\mathbb{R}^k$ and $\boldsymbol{x}, \boldsymbol{y} \in \mathbb{R}^k$.

8. Let Ω and Ω' be sets, and let $\mathcal{F}$ and $\mathcal{F}'$ denote fields over Ω and Ω', respectively. Let $\mathcal{F} * \mathcal{F}'$ denote the collection of all finite disjoint unions of sets of the form $A \times B$ where $A \in \mathcal{F}$ and $B \in \mathcal{F}'$.
(a) $\mathcal{F} * \mathcal{F}'$ is a field over $\Omega \times \Omega'$. [Showing $\Omega \times \Omega' \in \mathcal{F} * \mathcal{F}'$ is trivial. Verify the identities $(A_1 \times B_1) \cap (A_2 \times B_2) = (A_1 \cap A_2) \times (B_1 \cap B_2)$ and

$$(A_1 \times B_1) - (A_2 \times B_2) = [(A_1 - A_2) \times B_1] \cup [(A_1 \cap A_2) \times (B_1 - B_2)]$$

to show that $\mathcal{F} * \mathcal{F}'$ is closed under intersection and differences.]
(b) $\sigma(\mathcal{F}) \otimes \sigma(\mathcal{F}') = \sigma(\mathcal{F} * \mathcal{F}')$.

1.6 σ-FIELDS: CONSTRUCTION AND CARDINALITY

In the previous two sections, we have claimed that (i) card$(\mathcal{B}^k) = \mathfrak{c}$ for all $k \in \mathbb{N}$ and (ii) there is an "algorithm" for producing all of the Borel sets from a simple class via set-theoretic manipulations, although such an algorithm does not shed much if any light onto the structure of $\mathcal{B}^k$. In this section, we give rigorous discussions and proofs of these claims, assuming the Axiom of Choice (AC) and the Continuum Hypothesis (CH) throughout.

Cardinal Arithmetic. We assume the following standard results of cardinal arithmetic, where $\mathfrak{a}$ denotes a cardinal number with $\mathfrak{a} \geq 2$, and where $\aleph_0$ and $\mathfrak{c}$ denote the cardinality of $\mathbb{N}$ and $\mathbb{R}$, respectively.

(C1) ${\aleph_0}^{\aleph_0} = \mathfrak{c}$,
(C2) $\mathfrak{a} \geq \aleph_0$ implies $2\mathfrak{a} = \mathfrak{a}$,
(C3) $\mathfrak{a}^{\aleph_0}\aleph_0 = \mathfrak{a}^{\aleph_0}$,
(C4) $(\mathfrak{a}^{\aleph_0})^{\aleph_0} = \mathfrak{a}^{\aleph_0}$, and
(C5) $\mathfrak{a}^{\aleph_0}\mathfrak{c} = \mathfrak{a}^{\aleph_0}$.

For example, (C1) states that the set of functions $\mathbb{N} \to \mathbb{N}$ may be placed into a one-to-one correspondence with the elements of $\mathbb{R}$; (C2) states for any infinite A that $\{0, 1\} \times A$ can be placed into a one-to-one correspondence with A. The others statements are similarly interpreted.

Ordinal Numbers. In this section, lowercase Greek letters such as α, β, and ω will denote ordinal numbers. Given a set A, we will write ord(A) for the ordinal number of A. The symbols 0 and 1 will denote ord$(\varnothing)$ and ord$(\{1\})$, respectively.[4] The usual order relations among ordinal numbers will be denoted by $\prec$ and $\preceq$, with equality denoted by $\approx$. For each ordinal α, define $P_\alpha = \{\beta : 0 \preceq \beta \prec \alpha\}$, so that P_α is the set of ordinal predecessors of α. We will repeatedly use the following facts for ordinal numbers[5].

There exists a smallest ordinal number ω such that
(O1) P_ω is a well-ordered set.
(O2) For all $\alpha \in P_\omega$, the set P_α is amc.
(O3) P_ω is uncountable.
(O4) If $C \subseteq P_\omega$ is amc, then there is $\beta \in P_\omega$ with $\alpha \preceq \beta$ for each $\alpha \in C$.

The ordinal ω above is often called the *first* [or *smallest*] *uncountable ordinal*, and the usual symbol for card(P_ω) is $\aleph_1$. Under (CH), $\aleph_1 = \mathfrak{c}$. [Without (CH), we cannot prove that card$(P_\omega) = \mathfrak{c}$.]

Definition. Given any set Ω and any nonempty $\mathcal{A} \subseteq 2^\Omega$, let

$$\mathcal{A}^* = \left\{ \bigcup_{n=1}^{\infty} A_n : A_n \in \mathcal{A} \text{ or } A_n^c \in \mathcal{A} \text{ for each } n \in \mathbb{N} \right\}.$$

Claim 1. *We have the following: (i) if $\varnothing \in \mathcal{A}$, then $\mathcal{A} \subseteq \mathcal{A}^*$, and (ii) $\mathcal{C} \subseteq \mathcal{D}$ implies $\mathcal{C}^* \subseteq \mathcal{D}^*$.*

[4] In the expression $1 = \text{ord}(\{1\})$, the lefthand 1 is an ordinal, the righthand 1 is in $\mathbb{N}$. This notational solecism will not cause any confusion. Also, 1 is the smallest ordinal α [relative to the ordinal order relation $\succeq$] such that $\alpha \succ 0$.
[5] Hewitt and Stromberg, p. 29.

Proof: To see (i), let $A \in \mathcal{A}$ and suppose that $\varnothing \in \mathcal{A}$. Define $A_1 = A$ and $A_j = \varnothing$ for all $j \geq 2$. Then $A_n \in \mathcal{A}$ for all $n \in \mathbb{N}$, hence $A = \bigcup_{n=1}^{\infty} A_n \in \mathcal{A}^*$, giving (i). For (ii), let $A \in \mathcal{C}^*$. Then $A = \bigcup_{n=1}^{\infty} B_n$, where for each $n \in \mathbb{N}$ we have $B_n \in \mathcal{C}$ or $B_n^c \in \mathcal{C}$. Since $\mathcal{C} \subseteq \mathcal{D}$, we have for each $n \in \mathbb{N}$ that $B_n \in \mathcal{D}$ or $B_n^c \in \mathcal{D}$, hence $A \in \mathcal{D}^*$, and the proof is complete. ■

Definition. Let $\mathcal{A} \subseteq 2^{\Omega}$ be nonempty. We obtain the family $\{\mathcal{A}_\alpha : \alpha \in P_\omega\}$ by transfinite induction: $\mathcal{A}_0$ is defined as $\mathcal{A}$, and, for any $\alpha \in P_\omega$, if $\mathcal{A}_\beta$ has been defined for each $\beta \in P_\alpha$, then $\mathcal{A}_\alpha$ is defined as $(\bigcup_{\beta \in P_\alpha} \mathcal{A}_\beta)^*$.

Claim 2. *Let $\mathcal{A} \subseteq 2^{\Omega}$ be such that $\varnothing \in \mathcal{A}$. Then*
(i) $\beta \preceq \alpha$ implies $\mathcal{A}_\beta \subseteq \mathcal{A}_\alpha$.
(ii) If $0 \preceq \beta \prec \delta \prec \omega$, then $\mathcal{A}_\beta^* \subseteq \mathcal{A}_\delta$.
(iii) For any $\beta \in P_\omega$, we have $\mathcal{A}_\beta^* \subseteq \bigcup_{\alpha \in P_\omega} \mathcal{A}_\alpha$.

Proof: We first do (i). If $\beta \preceq \alpha$, then $P_\beta \subseteq P_\alpha$, hence $\bigcup_{\gamma \in P_\beta} \mathcal{A}_\gamma \subseteq \bigcup_{\gamma \in P_\alpha} \mathcal{A}_\gamma$. By (ii) of Claim 1, $(\bigcup_{\gamma \in P_\beta} \mathcal{A}_\gamma)^* \subseteq (\bigcup_{\gamma \in P_\alpha} \mathcal{A}_\gamma)^*$, hence $\mathcal{A}_\beta \subseteq \mathcal{A}_\alpha$.

For (ii), let $0 \preceq \beta \prec \delta \prec \omega$. Then $\mathcal{A}_\beta \subseteq \bigcup_{\gamma \in P_\delta} \mathcal{A}_\gamma$. By (ii) of Claim 1, $\mathcal{A}_\beta^* \subseteq (\bigcup_{\gamma \in P_\delta} \mathcal{A}_\gamma)^* = \mathcal{A}_\delta$.

For (iii), let $\beta \in P_\omega$. By (O2) P_β is amc, and P_ω is uncountable by (O3). Thus, there exist uncountably many $\delta \in P_\omega$ with $\delta \succeq \beta$. By (O1), any such δ not equal to β is necessarily such that $\delta \succ \beta$; pick such a δ. By (ii), $\mathcal{A}_\beta^* \subseteq \mathcal{A}_\delta$, and clearly $\mathcal{A}_\delta \subseteq \bigcup_{\alpha \in P_\omega} \mathcal{A}_\alpha$. This completes (iii). ■

The upcoming theorem describes somewhat the structure of the minimal σ-field generated by an arbitrary collection $\mathcal{A}$. The representation of $\sigma(\mathcal{A})$ as the union given below is nonconstructive, but it will allow the cardinality claim of the beginning of this section to be demonstrated.

Claim 3. *Let $\mathcal{A} \subseteq 2^{\Omega}$ with $\varnothing \in \mathcal{A}$. Then $\sigma(\mathcal{A}) = \bigcup_{\alpha \in P_\omega} \mathcal{A}_\alpha$.*

Proof: We first obtain $\sigma(\mathcal{A}) \subseteq \bigcup_{\alpha \in P_\omega} \mathcal{A}_\alpha$. Since $\mathcal{A} = \mathcal{A}_0 \subseteq \bigcup_{\alpha \in P_\omega} \mathcal{A}_\alpha$, it is sufficient to show that $\bigcup_{\alpha \in P_\omega} \mathcal{A}_\alpha$ is a σ-field, for $\sigma(\mathcal{A})$ is contained in every σ-field containing $\mathcal{A}$. Clearly $\varnothing \in \bigcap_{\alpha \in P_\omega} \mathcal{A}_\alpha \subseteq \bigcup_{\alpha \in P_\omega} \mathcal{A}_\alpha$.

We now show that $\bigcup_{\alpha \in P_\omega} \mathcal{A}_\alpha$ is closed under complementation. Pick an element $A \in \bigcup_{\alpha \in P_\omega} \mathcal{A}_\alpha$. Then there exists $\beta \in P_\omega$ with $A \in \mathcal{A}_\beta$. Since $A^c = A^c \cup \varnothing \cup \varnothing \cup \cdots$, we have $A^c \in \mathcal{A}_\beta^*$. By (iii) of Claim 2, it follows that $\mathcal{A}_\beta^* \subseteq \bigcup_{\alpha \in P_\omega} \mathcal{A}_\alpha$, so $A^c \in \bigcup_{\alpha \in P_\omega} \mathcal{A}_\alpha$.

We now show that $\bigcup_{\alpha \in P_\omega} \mathcal{A}_\alpha$ is closed under countable unions. Pick a sequence $\{A_n\}_{n=1}^{\infty}$ of sets in $\bigcup_{\alpha \in P_\omega} \mathcal{A}_\alpha$. Thus, for every $n \in \mathbb{N}$ there exists $\beta_n \in P_\omega$ such that $A_n \in \mathcal{A}_{\beta_n}$. The set $\{\beta_1, \beta_2, \cdots\}$ is an amc subset of P_ω, and (O4) forces the existence of $\beta \in P_\omega$ such that $\beta_n \preceq \beta$ for every $n \in \mathbb{N}$. By (i) of Claim 2, $A_n \in \mathcal{A}_\beta$ for all $n \in \mathbb{N}$, whence $\bigcup_{n=1}^{\infty} A_n \in \mathcal{A}_\beta^* \subseteq \bigcup_{\alpha \in P_\omega} \mathcal{A}_\alpha$, where the containment follows from (iii) of Claim 2.

These facts show that $\bigcup_{\alpha \in P_\omega} \mathcal{A}_\alpha$ is a σ-field on Ω, hence we have shown that $\sigma(\mathcal{A}) \subseteq \bigcup_{\alpha \in P_\omega} \mathcal{A}_\alpha$.

We now demonstrate the reverse inclusion $\bigcup_{\alpha \in P_\omega} \mathcal{A}_\alpha \subseteq \sigma(\mathcal{A})$ by using transfinite induction to show that $\mathcal{A}_\alpha \subseteq \sigma(\mathcal{A})$ for each $\alpha \in P_\omega$. Clearly, we have $\mathcal{A}_0 = \mathcal{A} \subseteq \sigma(\mathcal{A})$. Next, let $\alpha \in P_\omega$, and suppose that $\mathcal{A}_\beta \subseteq \sigma(\mathcal{A})$ for all $\beta \prec \alpha$. Let $A \in \mathcal{A}_\alpha$, so that $A = \bigcup_{n=1}^\infty B_n$, where for each $n \in \mathbb{N}$ there exists $\beta_n \prec \alpha$ with $B_n \in \mathcal{A}_{\beta_n}$ or $B_n^c \in \mathcal{A}_{\beta_n}$. By the induction hypothesis, $B_n \in \bigcup_{m=1}^\infty \mathcal{A}_{\beta_m} \subseteq \sigma(\mathcal{A})$ for all $n \in \mathbb{N}$, and thus $\bigcup_{n=1}^\infty B_n \in \sigma(\mathcal{A})$. This gives $A \in \sigma(\mathcal{A})$ and hence $\mathcal{A}_\alpha \subseteq \sigma(\mathcal{A})$, whence the entire proof is complete. ■

Example. As an application of Claim 3, we prove a slightly modified form of Claim 2 from Section 3. *If $X : \Omega \to \Omega'$ is given, and if $\varnothing \in \mathcal{C} \subseteq 2^{\Omega'}$, then we have $X^{-1}(\sigma(\mathcal{C})) = \sigma(X^{-1}(\mathcal{C}))$.* Using Claim 3 gives

$$X^{-1}(\sigma(\mathcal{C})) = X^{-1}\left(\bigcup_{\alpha \in P_\omega} \mathcal{C}_\alpha\right) = \bigcup_{\alpha \in P_\omega} X^{-1}(\mathcal{C}_\alpha), \tag{1.3}$$

where $\mathcal{C}_0 = \mathcal{C}$ and $\mathcal{C}_\alpha = (\bigcup_{\gamma \in P_\alpha} \mathcal{C}_\gamma)^*$ for each $\alpha \in P_\omega$. If we can show for each $0 \prec \alpha \prec \omega$ that $X^{-1}(\mathcal{C}_\alpha) = (\bigcup_{\beta \in P_\alpha} X^{-1}(\mathcal{C}_\beta))^*$, then by Claim 3 we will have that $\bigcup_{\alpha \in P_\omega} X^{-1}(\mathcal{C}_\alpha) = \sigma(X^{-1}(\mathcal{C}))$, which by (1.3) proves the claim.

Pick $\alpha \in P_\omega$ and consider an arbitrary $A \in X^{-1}(\mathcal{C}_\alpha)$, so that $A = X^{-1}(C)$ for some $C \in \mathcal{C}_\alpha$. Since $C \in \mathcal{C}_\alpha$, we have that $C = \bigcup_{n=1}^\infty C_n$, where for each $n \in \mathbb{N}$ there exists $\beta_n \prec \alpha$ such that $C_n \in \mathcal{C}_{\beta_n}$ or $C_n^c \in \mathcal{C}_{\beta_n}$. Therefore, for all $n \in \mathbb{N}$, $X^{-1}(C_n) \in X^{-1}(\mathcal{C}_{\beta_n})$ or $X^{-1}(C_n^c) = X^{-1}(C_n)^c \in X^{-1}(\mathcal{C}_{\beta_n})$. Since we have $A = X^{-1}(C) = X^{-1}\left(\bigcup_{n=1}^\infty C_n\right) = \bigcup_{n=1}^\infty X^{-1}(C_n)$, we see that our $A \in X^{-1}(\mathcal{C}_\alpha)$ is a countable union of sets, each set having the property that it or its complement is from $\bigcup_{n=1}^\infty X^{-1}(\mathcal{C}_{\beta_n}) \subseteq \bigcup_{\beta \in P_\alpha} X^{-1}(\mathcal{C}_\beta)$. Therefore, our $A \in X^{-1}(\mathcal{C}_\alpha)$ is necessarily in $(\bigcup_{\beta \in P_\alpha} X^{-1}(\mathcal{C}_\beta))^*$, whence $X^{-1}(\mathcal{C}_\alpha) \subseteq (\bigcup_{\beta \in P_\alpha} X^{-1}(\mathcal{C}_\beta))^*$.

For the reverse containment, let $A \in (\bigcup_{\beta \in P_\alpha} X^{-1}(\mathcal{C}_\beta))^*$. Thus $A = \bigcup_{n=1}^\infty D_n$, where each D_n is such that (i) there exists $\beta_n \prec \alpha$ and $E_n \in \mathcal{C}_{\beta_n}$ such that $D_n = X^{-1}(E_n)$, or (ii) there exists $\beta_n \prec \alpha$ and $E_n \in \mathcal{C}_{\beta_n}$ such that $D_n^c = X^{-1}(E_n)$, that is, $D_n = X^{-1}(E_n^c)$. Therefore, for each $n \in \mathbb{N}$, we have that the equality $D_n = X^{-1}(B_n)$ holds for some $B_n \in \mathcal{C}_{\beta_n}^* \subseteq \mathcal{C}_\alpha$, the containment following by (ii) of Claim 2. Since $A = \bigcup_{n=1}^\infty D_n = \bigcup_{n=1}^\infty X^{-1}(B_n) = X^{-1}(\bigcup_{n=1}^\infty B_n)$, and since $\bigcup_{n=1}^\infty B_n \in \bigcup_{n=1}^\infty \mathcal{C}_{\beta_n} \subseteq \mathcal{C}_\alpha$, it follows that $A \in X^{-1}(\mathcal{C}_\alpha)$. This shows that $(\bigcup_{\beta \in P_\alpha} X^{-1}(\mathcal{C}_\beta))^* \subseteq X^{-1}(\mathcal{C}_\alpha)$, and the entire proof is complete.

If $\mathcal{A}$ denotes the collection consisting of $\varnothing$ and all k-dimensional rsc rectangles, then $\sigma(\mathcal{A}) = \mathcal{B}^k$, and the previous claim shows that $\mathcal{B}^k = \bigcup_{\alpha \in P_\omega} \mathcal{A}_\alpha$, where $\mathcal{A}_0 = \mathcal{A}$ and the $\mathcal{A}_\alpha$'s are defined inductively in the same transfinite fashion as before. However, this algorithm does not allow us to represent what a generic $\mathcal{B}^k$ set "looks like" unless we are content to say that a $\mathcal{B}^k$-set is a set that is in $\mathcal{A}_\alpha$ for some $\alpha \in P_\omega$, which is not too satisfying. In fact, if α is the ordinal number corresponding to a well-ordered countably infinite set, we still can't even write out explicitly what the elements of $\mathcal{A}_\alpha$ are. In other words, the claim works out in a formal sense, but this construction does not lend itself to intuition, nor does it specify what *sequence* of operations on what $\mathcal{A}$-sets should be performed in order to arrive at a given $\sigma(\mathcal{A})$-set.

We now prove that $\text{card}(\mathcal{B}^k) = \mathfrak{c}$ for each $k \in \mathbb{N}$. This demonstration uses the representation found in the previous claim. It should also be noted that this construction does not involve (GCH), but only (AC) and (CH).

Claim 4. *Let $\mathcal{A}$ denote a collection of subsets of Ω such that $\varnothing \in \mathcal{A}$ and $\mathfrak{a} = \text{card}(\mathcal{A}) \geq \aleph_0$. Then* $\text{card}(\sigma(\mathcal{A})) \leq \mathfrak{a}^{\aleph_0}$.

Claim 5. *For each $k \in \mathbb{N}$,* $\text{card}(\mathcal{B}^k) = \mathfrak{c}$.

Proof of Claim 5. Let $\mathcal{A}$ consist of $\varnothing$ and the rsc rectangles $(\boldsymbol{a}, \boldsymbol{b}]$ with $\boldsymbol{a}, \boldsymbol{b} \in \mathbb{Q}^k$, hence $\mathcal{B}^k = \sigma(\mathcal{A})$. Since $\text{card}(\mathcal{A}) = \aleph_0$, Claim 4 and (C1) give $\text{card}(\mathcal{B}^k) \leq {\aleph_0}^{\aleph_0} = \mathfrak{c}$. Clearly $\text{card}(\mathcal{B}^k) \geq \mathfrak{c}$, and hence $\text{card}(\mathcal{B}^k) = \mathfrak{c}$ [this deduction requires (AC)]. This completes the proof. ∎

Proof of Claim 4. Suppose that it is shown that $\text{card}(\mathcal{A}_\alpha) \leq \mathfrak{a}^{\aleph_0}$ for each ordinal $\alpha \in P_\omega$. Then, invoking [in the following order] Claim 3, the fact that $\text{card}(\mathcal{A}_\alpha) \leq a^{\aleph_0}$ for each $\alpha \in P_\omega$, the definition of cardinal multiplication, (O3), (CH), and (C5), we will obtain

$$\text{card}(\sigma(\mathcal{A})) = \text{card}\left(\bigcup\{\mathcal{A}_\alpha : \alpha \in P_\omega\}\right) \leq \mathfrak{a}^{\aleph_0}\,\text{card}(P_\omega) = \mathfrak{a}^{\aleph_0}\mathfrak{c} = \mathfrak{a}^{\aleph_0}.$$

We will use transfinite induction to show that $\text{card}(\mathcal{A}_\alpha) \leq \mathfrak{a}^{\aleph_0}$ for all $\alpha \in P_\omega$.

Note that $\text{card}(\mathcal{A}_0) = \mathfrak{a} \leq \mathfrak{a}^{\aleph_0}$. Next, let $\mathcal{D}_1$ denote the set of $C \subseteq \Omega$ such that $C \in \mathcal{A}_0$ or C is the complement of a set in $\mathcal{A}_0$. It follows that $\text{card}(\mathcal{D}_1) \leq 2\mathfrak{a} = \mathfrak{a}$, the last equality following from (C2). A generic element $A \in \mathcal{A}_1$ is a countable union of $\mathcal{D}_1$-sets, and is obtained by specifying a function $\phi : \mathcal{D}_1 \to \mathbb{N} \cup \{0\}$, where for each n we define $\phi(C) = n$ iff C is the nth term in the union of $\mathcal{D}_1$-sets defining A, and $\phi(C) = 0$ otherwise.[6] It follows that $\text{card}(\mathcal{A}_1)$ is no more than the cardinal number of the set of such functions ϕ; this latter cardinal is $\text{card}(\mathcal{D}_1)^{\text{card}(\mathbb{N}\cup\{0\})} \leq \mathfrak{a}^{\aleph_0}$, hence we have shown that the assumption $\text{card}(\mathcal{A}_0) \leq \mathfrak{a}^{\aleph_0}$ forces $\text{card}(\mathcal{A}_1) \leq \mathfrak{a}^{\aleph_0}$.

Next, let $\alpha \in P_\omega$ and assume that $\text{card}(\mathcal{A}_\beta) \leq \mathfrak{a}^{\aleph_0}$ for all $\beta \in P_\alpha$. We desire to show that $\text{card}(\mathcal{A}_\alpha) \leq \mathfrak{a}^{\aleph_0}$. Using the definition of cardinal multiplication, (O2), and (C3), we have $\text{card}\left(\bigcup_{\beta \in P_\alpha} \mathcal{A}_\beta\right) \leq \mathfrak{a}^{\aleph_0}\,\text{card}(P_\alpha) = \mathfrak{a}^{\aleph_0}\aleph_0 = \mathfrak{a}^{\aleph_0}$. Next, let $\mathcal{D}_\alpha$ denote the set of $C \subseteq \Omega$ such that $C \in \bigcup_{\beta \in P_\alpha} \mathcal{A}_\beta$ or C is the complement of a set in $\bigcup_{\beta \in P_\alpha} \mathcal{A}_\beta$, so that the induction hypothesis and (C2) give $\text{card}(\mathcal{D}_\alpha) \leq 2\,\text{card}(\bigcup_{\beta \in P_\alpha} \mathcal{A}_\beta) \leq 2\mathfrak{a}^{\aleph_0} = \mathfrak{a}^{\aleph_0}$. A typical element of $\mathcal{A}_\alpha$ is obtained by forming a countable union of $\mathcal{D}_\alpha$-sets. Thus, each element of $\mathcal{A}_\alpha$ is obtained by specifying a function $\phi : \mathcal{D}_\alpha \to \mathbb{N} \cup \{0\}$, where $\phi(C) = n$ iff C is the nth term in the union of $\mathcal{D}_\alpha$-sets that defines our typical $\mathcal{A}_\alpha$-set, and $\phi(C) = 0$ otherwise.[7] The cardinality of the set of such functions ϕ is no more than $\text{card}(\mathcal{D}_\alpha)^{\text{card}(\mathbb{N}\cup\{0\})} = (\mathfrak{a}^{\aleph_0})^{\aleph_0} = \mathfrak{a}^{\aleph_0}$; the last equality follows by (C4). Thus, $\text{card}(\mathcal{A}_\alpha) \leq \mathfrak{a}^{\aleph_0}$. By transfinite induction, $\text{card}(\mathcal{A}_\alpha) \leq \mathfrak{a}^{\aleph_0}$ for all $\alpha \in P_\omega$, and the entire proof is now complete. ∎

[6] If a set $D \in \mathcal{D}_1$ appears more than once in the union defining A, the appearances beyond the first appearance are redundant. Therefore, without any loss of generality, we may stipulate that any $D \in \mathcal{D}_1$ that appears in the union that defines A occurs exactly once. Therefore, $\phi(D)$ is not ambiguous, and hence ϕ is well-defined by this convention.

[7] The same comments in the previous footnote apply to ensure that ϕ is well-defined.

1.7 A CLASS OF ETHEREAL BOREL SETS

The main point of this section is to actually construct a Borel set so ethereal that it cannot be arrived at by any amc sequence of set operations applied to the elements of a generating class, each operation consisting of amc unions, amc intersections, and complements. The construction is based on that found in Section 2 of Billingsley's most excellent *Probability and Measure.* The presentation consists of many small lemmas, and will be summarized in the form of a final result which is immediate from these lemmas.

Notation, Part I. We let $\Omega = (0,1]$, and we write $\mathcal{B}_{(0,1]}$ to denote the Borel sets of $(0,1]$. From Exercise 4.3, $\mathcal{B}_{(0,1]}$ consists exactly of those Borel sets contained in $(0,1]$. Next, for all that follows, we pick a double sequence $\{m_{ij} : i,j \in \mathbb{N}\}$ of positive intergers such that every positive integer appears exactly once in the sequence. For arbitrary sets $A_1, A_2, \cdots$ we define

$$\Psi(A_1, A_2, \cdots) = A_1^c \cup A_2 \cup A_3 \cup \cdots, \qquad \Phi_0(A_1, A_2, \cdots) = A_1,$$

and for each $n \in \mathbb{N}$ we inductively define

$$\Phi_n(A_1, A_2, \cdots) = \Psi(\Phi_{n-1}(A_{m_{11}}, A_{m_{12}}, \cdots), \Phi_{n-1}(A_{m_{21}}, A_{m_{22}}, \cdots), \cdots);$$

Finally, we define

$$\Phi(A_1, A_2, \cdots) \equiv \bigcup_{n=1}^{\infty} \Phi_n(A_{m_{n1}}, A_{m_{n2}}, \cdots).$$

For any collection $\mathcal{A}$ of subsets of $(0,1]$, we will let $\mathcal{A}^*$ to denote the collection consisting of the sets in $\mathcal{A}$, complements of $\mathcal{A}$-sets, and amc unions of $\mathcal{A}$-sets. We will let $\mathcal{I}_0$ consist of $\varnothing$ and all rsc subintervals $(a,b] \subseteq (0,1]$ with $a, b \in \mathbb{Q}$, hence $\mathcal{B}_{(0,1]} = \sigma(\mathcal{I}_0)$. For each $n \in \mathbb{N}$, we will inductively write $\mathcal{I}_n \equiv \mathcal{I}_{n-1}^*$, so that $\mathcal{I}_n$ consists of all $\mathcal{I}_{n-1}$-sets, complements of $\mathcal{I}_{n-1}$-sets, and amc unions of $\mathcal{I}_{n-1}$-sets. Clearly, $\mathcal{I}_0 \subseteq \mathcal{I}_1 \subseteq \cdots$.

Before starting, consider the following. Clearly $\mathcal{I}_0 \subsetneqq \mathcal{B}_{(0,1]}$. One might augment $\mathcal{I}_0$ by adding to it the complements of sets in $\mathcal{I}_0$ as well as amc unions of sets in $\mathcal{I}_0$; this new larger class is $\mathcal{I}_0^* = \mathcal{I}_1$. It is easily verified that there exist $\mathcal{B}_{(0,1]}$-sets outside $\mathcal{I}_1$, hence $\mathcal{I}_0 \subsetneqq \mathcal{I}_1 \subsetneqq \mathcal{B}$. One takes the sets in $\mathcal{I}_1$, and augments these sets by adding the complements and the amc unions of sets in $\mathcal{I}_1$. This new larger class is $\mathcal{I}_1^* = \mathcal{I}_2$, and it is easily shown that there are $\mathcal{B}_{(0,1]}$-sets outside $\mathcal{I}_2$, hence $\mathcal{I}_0 \subsetneqq \mathcal{I}_1 \subsetneqq \mathcal{I}_2 \subsetneqq \mathcal{B}$. At this stage, one augments the sets in $\mathcal{I}_2$ as before, obtaining $\mathcal{I}_2^* = \mathcal{I}_3$, and we will end up having $\mathcal{I}_0 \subsetneqq \mathcal{I}_1 \subsetneqq \mathcal{I}_2 \subsetneqq \mathcal{I}_3 \subsetneqq \mathcal{B}_{(0,1]}$. We proceed ad infinitum with this process and are curious about what happens with $\mathcal{I}_n$ as $n \to +\infty$.

Observe that even when n is quite modest in size, it is nontrivial to find a set in $\mathcal{B}_{(0,1]} - \mathcal{I}_n$, if indeed such a set exists. One might think that for some large value of n it will be the case that $\mathcal{I}_n = \mathcal{B}_{(0,1]}$. However, *it will turn out that* $\bigcup_{n=0}^{\infty} \mathcal{I}_n \subsetneq \mathcal{B}_{(0,1]}$. This seemingly comprehensive iterative procedure applied to the initial class $\mathcal{I}_0$ will never at any stage nor "in the limit" have produced all of $\mathcal{B}_{(0,1]} = \sigma(\mathcal{I}_0)$.

Claim. *There exists a set in $\mathcal{B}_{(0,1]}$ that fails to be in $\mathcal{I}_n$ for any $n \in \mathbb{N}$.*

The proof of the claim is accomplished by a series of lemmas that in turn require some extra notation. The lemmas aren't too difficult conceptually, but the notation may seem a bit forbidding.

Lemma 1. *Let $n \geq 0$. We can find $A_1, A_2, \cdots \in \mathcal{I}_0$ with $\Phi_n(A_1, A_2, \cdots) = (0, 1]$, and we can find $A_1, A_2, \cdots \in \mathcal{I}_0$ with $\Phi_n(A_1, A_2, \cdots) = \varnothing$.*

Proof: Let $n = 0$. Note that $\Phi_0((0,1], \varnothing, \cdots) = (0,1]$ and $\Phi_0(\varnothing, \varnothing, \cdots) = \varnothing$, and $\varnothing, (0,1] \in \mathcal{I}_0$. Thus the lemma holds when $n = 0$. Suppose the lemma holds for a fixed n. Then for each $i \in \mathbb{N}$ there exist $C_{i1}, C_{i2}, \cdots \in \mathcal{I}_0$ such that $\Phi_n(C_{i1}, C_{i2}, \cdots) = \varnothing$. Define $A_{m_{ij}} = C_{ij}$ for each $i, j \in \mathbb{N}$. Then $A_1, A_2, \cdots, \in \mathcal{I}_0$ and $\Phi_{n+1}(A_1, A_2, \cdots) = \varnothing^c \cup \varnothing \cup \varnothing \cup \cdots = (0, 1]$. Also, there are $D_{11}, D_{21}, \cdots \in \mathcal{I}_0$ such that $\Phi_n(D_{11}, D_{12}, \cdots) = (0,1]$. If $A_{1j} = D_{1j}$ and $A_{ij} = C_{ij}$ for $i \geq 2$ and $j \in \mathbb{N}$, then $\Phi_{n+1}(A_1, A_2, \cdots) = (0,1]^c \cup \varnothing \cup \varnothing \cup \cdots = \varnothing$. This proves that the claim holds in the $n+1$ case, and the lemma is complete. ■

Lemma 2. *Let $n \geq 0$. Then each $\mathcal{I}_n$-set has the form $\Phi_n(A_1, A_2, \cdots)$ for some appropriately chosen $A_1, A_2, \cdots \in \mathcal{I}_0$.*

Proof: The $n = 0$ case is trivially true. Assume that the lemma holds for some $n \in \mathbb{N}$, and let $A \in \mathcal{I}_{n+1}$. There are four cases: (i) $A \in \mathcal{I}_n$, (ii) $A = B^c$ for some $B \in \mathcal{I}_n$, (iii) $A = \bigcup_{i=1}^{k} B_i$ for some $B_1, \cdots, B_k \in \mathcal{I}_n$, and (iv) $A = \bigcup_{i=1}^{\infty} B_i$ for some sequence $\{B_k\}_{k=1}^{\infty}$ of $\mathcal{I}_n$-sets. Case (i) is automatic by the induction hypothesis.

For (ii), use the induction hypothesis to obtain sets $B_{ij} \in \mathcal{I}_0$, $i \geq 2$ and $j \in \mathbb{N}$, such that $\Phi_n(B_{i1}, B_{i2}, \cdots) = \varnothing$ for all $i \geq 2$, and again use the induction hypothesis to find sets $B_{11}, B_{12}, \cdots \in \mathcal{I}_0$ such that $\Phi_n(B_{11}, B_{12}, \cdots) = B$. If $A_1, A_2, \cdots$ are such that $A_{m_{ij}} = B_{ij}$ for all $i, j \in \mathbb{N}$, then $\Phi_{n+1}(A_1, A_2, \cdots) = B^c \cup \varnothing \cup \varnothing \cup \cdots = B^c = A$, and the proof of (ii) is complete.

For (iii), use the induction hypothesis to obtain sets $B_{ij} \in \mathcal{I}_0$, $i = 2, \cdots, k+1$ and $j \in \mathbb{N}$, such that $\Phi_n(B_{i1}, B_{i2}, \cdots) = B_i$. Again use the induction hypothesis to obtain sets $B_{ij} \in \mathcal{I}_0$, $i \in \{1, k+2, k+3, \cdots\}$ and $j \in \mathbb{N}$, such that $\Phi_n(B_{11}, B_{12}, \cdots) = (0,1]$ and $\Phi_n(B_{i1}, B_{i2}, \cdots) = \varnothing$ for $i \geq k+2$. Let $A_{m_{ij}} = B_{ij}$ for all $i, j \in \mathbb{N}$. Then $\Phi_{n+1}(A_1, A_2, \cdots) = (0,1]^c \cup B_1 \cup \cdots \cup B_k \cup \varnothing \cup \cdots = \bigcup_{i=1}^{k} B_i$, finishing (iii). The proof of (iv) is similar to that of (iii), and hence the lemma complete. ■

Lemma 3. *For every $A \in \bigcup_{n=0}^{\infty} \mathcal{I}_n$, we have $A = \Phi(A_1, A_2, \cdots)$ for appropriately chosen $A_1, A_2, \cdots \in \mathcal{I}_0$.*

Proof: Let $A \in \bigcup_{n=0}^{\infty} \mathcal{I}_n$ so that there exists $N \in \mathbb{N}$ with $A \in \mathcal{I}_N$. By Lemma 2, $A = \Phi_N(B_1, B_2, \cdots)$ for some sequence $B_1, B_2, \cdots \in \mathcal{I}_0$. By Lemma 1, we may choose $C_1, C_2, \cdots \in \mathcal{I}_0$ such that $\Phi_n(C_{m_{i1}}, C_{m_{i2}}, \cdots) = \varnothing$ for all $n \neq N$. Next, for all $i \neq N$ and all $j \geq 1$ define $A_{m_{ij}} = C_{m_{ij}}$; for all $j \geq 1$ define $A_{m_{Nj}} = B_j$, so that

$A_1, A_2, \cdots \in \mathcal{I}_0$. Since

$$\begin{aligned}
\Phi(A_1, A_2, \cdots) &= \bigcup_{k=1}^{\infty} \Phi_k(A_{m_{k1}}, A_{m_{k2}}, \cdots) \\
&= \bigcup_{k \neq N} \Phi_k(A_{m_{k1}}, A_{m_{k2}}, \cdots) \ \cup \ \Phi_N(A_{m_{N1}}, A_{m_{N2}}, \cdots) \\
&= \varnothing \cup \Phi_N(B_1, B_2, \cdots) \\
&= A,
\end{aligned}$$

the proof is complete. ∎

Lemma 4. *If* $A_1, A_2, \cdots \in \mathcal{B}_{(0,1]}$, *then* $\Psi(A_1, A_2, \cdots) \in \mathcal{B}_{(0,1]}$. *[Trivial.]*

Lemma 5. $A_1, A_2, \cdots \in \mathcal{B}_{(0,1]}$ *implies* $\Phi_n(A_1, A_2, \cdots) \in \mathcal{B}_{(0,1]}$ *for all* $n \geq 0$.

Proof: Let $A_1, A_2, \cdots \in \mathcal{B}_{(0,1]}$. For $n = 0$, the claim is immediate. Suppose now that the lemma is true for a fixed $n \geq 0$, and recall that

$$\Phi_{n+1}(A_1, A_2, \cdots) = \Psi(\Phi_n(A_{m_{11}}, A_{m_{12}}, \cdots), \Phi_n(A_{m_{21}}, A_{m_{22}}, \cdots), \cdots).$$

By the induction hypothesis, $\Phi_n(A_{m_{k1}}, A_{m_{k2}}, \cdots) \in \mathcal{B}_{(0,1]}$ for all $k \geq 1$. By Lemma 4, it follows that $\Phi_{n+1}(A_1, A_2, \cdots) \in \mathcal{B}_{(0,1]}$. ∎

Lemma 6. *If* $A_1, A_2, \cdots \in \mathcal{B}_{(0,1]}$, *then* $\Phi(A_1, A_2, \cdots) \in \mathcal{B}_{(0,1]}$.

Proof: Let $A_1, A_2, \cdots \in \mathcal{B}_{(0,1]}$. By Lemma 5, $\Phi_n(A_{m_{n1}}, A_{m_{n2}}, \cdots) \in \mathcal{B}_{(0,1]}$ for each $n \in \mathbb{N}$, hence $\Phi(A_1, A_2, \cdots) = \bigcup_{n=1}^{\infty} \Phi_n(A_{m_{n1}}, A_{m_{n2}}, \cdots) \in \mathcal{B}_{(0,1)}$. ∎

Notation, Part II. For each fixed $\omega \in (0, 1]$, we consider the dyadic [base 2] expansion of ω *such that the expansion has a nonterminating sequence of 1's from some point on.* Thus, we let $0.0\bar{1}$ be the expansion of $\frac{1}{2}$ instead of $0.1\bar{0}$. Given $\omega \in (0, 1]$, let ω_1 denote the position of the first 1 in the expansion of ω. For each $k \geq 2$, inductively define ω_k by equating $\sum_{j=1}^{k} \omega_j$ with the position of the kth 1 in the expansion of ω. For example, let $\omega = 0.00101000100000010\bar{1}$; then $\omega_1 = 3$, $\omega_2 = 2$, $\omega_3 = 4$, $\omega_4 = 7$, $\omega_5 = 2$, and $\omega_k = 1$ for all $k \geq 6$.

Lemma 7. *For each* $\omega \in (0, 1]$, *there exists a unique sequence* $\{\omega_k\}_{k=1}^{\infty}$ *and vice-versa, where the values* $\omega_1, \omega_2, \cdots$ *are given by the above. [Trivial.]*

Notation, Part III. We let $I_1, I_2, \cdots$ denote a fixed listing of the countably many sets constituting $\mathcal{I}_0$. Define a function $\varphi : (0, 1] \to 2^{(0,1]}$ by writing $\varphi(\omega) = \Phi(I_{\omega_1}, I_{\omega_2}, \cdots)$ whenever $\omega \leftrightarrow \{\omega_k\}_{k=1}^{\infty}$. By Lemma 7, each $\omega \in (0, 1]$ corresponds with exactly one sequence $\{\omega_k\}_{k=1}^{\infty}$, so φ is well-defined.

Next, let $B = \{\omega \in (0, 1] : \omega \notin \varphi(\omega)\}$[8], so that $B \neq \varphi(\omega)$ for all $\omega \in (0, 1]$[9]. *This B is going to turn out to be the Borel set that lies outside* $\mathcal{I}_n$ *for each*

[8]To see that $B \neq \varnothing$, use Lemma 3 [taking $A = \varnothing$ there] to find the $\omega \in (0, 1]$ such that $\varphi(\omega) = \varnothing$. For this ω we have $\omega \notin \varphi(\omega)$, hence this ω is in B.

[9]Suppose to the contrary that there is $\omega^* \in (0, 1]$ such that $B = \varphi(\omega^*)$. There are two cases: (i) if $\omega^* \in B = \varphi(\omega^*)$, then $\omega^* \notin \varphi(\omega^*)$; (ii) if $\omega^* \notin B = \varphi(\omega^*)$, then $\omega^* \in \varphi(\omega^*)$. We deduce a contradiction from both (i) and (ii).

$n \in \mathbb{N}$. Next, for all $k, n \in \mathbb{N}$ define $D_k = \{\omega \in (0,1] : \omega \in I_{\omega_k}\}$ and $L_{k,n} = \{\omega \in (0,1] : \omega_1 + \cdots + \omega_k = n\}$.

Lemma 8. *The set B as defined above lies outside $\bigcup_{n=0}^{\infty} \mathcal{I}_n$.*

Proof: By Lemma 3, each element of $\bigcup_{n=0}^{\infty} \mathcal{I}_n$ has the form $\Phi(A_1, A_2, \cdots)$ for some $A_1, A_2, \cdots \in \mathcal{I}_0$. Since $I_1, I_2, \cdots$ constitutes a listing of all of the elements of $\mathcal{I}_0$, we have that each element of $\bigcup_{n=0}^{\infty} \mathcal{I}_n$ has the form $\Phi(I_{\omega_1}, I_{\omega_2}, \cdots)$ for some sequence of positive integers $\{\omega_k\}_{k=1}^{\infty}$. By Lemma 7, there exists $\omega \in (0,1]$ with $\omega \leftrightarrow \{\omega_k\}_{k=1}^{\infty}$. Therefore, by the definition of φ, it follows that each element of $\bigcup_{n=0}^{\infty} \mathcal{I}_n$ has the form $\varphi(\omega)$ for some $\omega \in (0,1]$. Since $\varphi(\omega) \neq B$ for all $\omega \in (0,1]$, B cannot possibly be in $\bigcup_{n=0}^{\infty} \mathcal{I}_n$. ∎

With Lemma 8 in tow, all that remains is to show that the set B as defined above is in $\mathcal{B}_{(0,1]}$. The following lemmas build up to showing this fact.

Lemma 9. *For all $k, n \in \mathbb{N}$ we have $L_{k,n} \in \mathcal{B}_{(0,1]}$.*

Proof: Let $k, n \in \mathbb{N}$. If $k > n$, then $L_{k,n} = \varnothing \in \mathcal{B}_{(0,1]}$. If $k \leq n$, then $L_{k,n}$ is a finite disjoint union of subintervals of $(0,1]$, and hence is in $\mathcal{B}_{(0,1]}$. ∎

Lemma 10. $\{\omega \in (0,1] : \omega_k = n\} \in \mathcal{B}_{(0,1]}$ *for all $k, n \in \mathbb{N}$.*

Proof: Let $n \in \mathbb{N}$. If $k = 1$, then $\{\omega \in (0,1] : \omega_1 = n\} = L_{1,n} \in \mathcal{B}_{(0,1]}$ by Lemma 9. If $k \geq 2$, we have $\{\omega \in (0,1] : \omega_k = n\} = \bigcup_{m=1}^{\infty} (L_{k-1,m} \cap L_{k,m+n}) \in \mathcal{B}_{(0,1]}$ for every $k, n \in \mathbb{N}$, where the membership follows by another appeal to Lemma 9. ∎

Lemma 11. $D_k \in \mathcal{B}_{(0,1]}$ *for all $k \in \mathbb{N}$.*

Proof: Let $k \in \mathbb{N}$. Observe that

$$D_k = \{\omega \in (0,1] : \omega \in I_{\omega_k}\} = \bigcup_{n=1}^{\infty} (\{\omega \in (0,1] : \omega_k = n\} \cap I_n).$$

By Lemma 10, $\{\omega \in (0,1] : \omega_k = n\} \in \mathcal{B}_{(0,1]}$ for all $n, k \in \mathbb{N}$; by definition $I_n \in \mathcal{B}_{(0,1]}$ for all $n \in \mathbb{N}$. Thus D_k is the countable union $\mathcal{B}_{(0,1]}$-sets, hence $D_k \in \mathcal{B}_{(0,1]}$. ∎

Lemma 12. *For all $n \geq 0$ and for all sequences $\{u_k\}_{k=1}^{\infty}$ of positive integers,*

$$\{\omega \in (0,1] : \omega \in \Phi_n(I_{\omega_{u_1}}, I_{\omega_{u_2}}, \cdots)\} = \Phi_n(D_{u_1}, D_{u_2}, \cdots).$$

Proof: Let $n = 0$, and pick a sequence of positive integers $\{u_k\}_{k=1}^{\infty}$. By the definitions of Φ_0 and of the D_k's, we obtain

$$\{\omega : \omega \in \Phi_0(I_{\omega_{u_1}}, I_{\omega_{u_2}}, \cdots)\} = \{\omega : \omega \in I_{\omega_{u_1}}\} = D_{u_1} = \Phi_0(D_{u_1}, D_{u_2}, \cdots).$$

Next, pick $n \in \mathbb{N}$ and assume that the lemma holds for this n. By the induction hypothesis and the definitions of Ψ and Φ_{n+1}, we obtain for all $\omega \in (0,1]$ that

$$\begin{aligned}
&\omega \in \Phi_{n+1}(I_{\omega_{u_1}}, I_{\omega_{u_2}}, \cdots) \\
\text{iff}\quad &\omega \in \Psi(\Phi_n(I_{\omega_{u_{m_{11}}}}, I_{\omega_{u_{m_{12}}}}, \cdots), \Phi_n(I_{\omega_{u_{m_{21}}}}, I_{\omega_{u_{m_{22}}}}, \cdots), \cdots) \\
\text{iff}\quad &\omega \in \Phi_n(I_{\omega_{u_{m_{11}}}}, I_{\omega_{u_{m_{12}}}}, \cdots)^c \cup \bigcup_{i=2}^{\infty} \Phi_n(I_{\omega_{u_{m_{i1}}}}, I_{\omega_{u_{m_{i2}}}}, \cdots) \\
\text{iff}\quad &\omega \in \Phi_n(D_{u_{m_{11}}}, D_{u_{m_{12}}}, \cdots)^c \cup \bigcup_{i=2}^{\infty} \Phi_n(D_{u_{m_{i1}}}, D_{u_{m_{i2}}}, \cdots) \\
\text{iff}\quad &\omega \in \Psi(\Phi_n(D_{u_{m_{11}}}, D_{u_{m_{12}}}, \cdots), \Phi_n(D_{u_{m_{21}}}, D_{u_{m_{22}}}, \cdots), \cdots) \\
\text{iff}\quad &\omega \in \Phi_{n+1}(D_{u_1}, D_{u_2}, \cdots),
\end{aligned}$$

hence $\Phi_{n+1}(I_{\omega_{u_1}}, I_{\omega_{u_2}}, \cdots) = \Phi_{n+1}(D_{u_1}, D_{u_2}, \cdots)$, finishing the proof. ■

Lemma 13. $B \in \mathcal{B}_{(0,1]}$.

Proof: For each $\omega \in (0,1]$, we have that $\omega \in B^c$ iff $\omega \in \varphi(\omega)$ iff $\omega \in \Phi(I_{\omega_1}, I_{\omega_2}, \cdots)$. However, Lemma 12 gives

$$\begin{aligned}
\omega \in \Phi(I_{\omega_1}, I_{\omega_2}, \cdots) \quad \text{iff}\quad &\omega \in \bigcup_{n=1}^{\infty} \Phi_n(I_{\omega_{m_{n1}}}, I_{\omega_{m_{n2}}}, \cdots) \\
\text{iff}\quad &\omega \in \bigcup_{n=1}^{\infty} \Phi_n(D_{m_{n1}}, D_{m_{n2}}, \cdots) \\
\text{iff}\quad &\omega \in \Phi(D_1, D_2, \cdots),
\end{aligned}$$

so that $B^c = \Phi(D_1, D_2, \cdots)$. By Lemma 11, each $D_j \in \mathcal{B}_{(0,1]}$; by Lemma 6 we have $\Phi(D_1, D_2, \cdots) \in \mathcal{B}_{(0,1]}$. This gives $B^c \in \mathcal{B}_{(0,1]}$, whence $B \in \mathcal{B}_{(0,1]}$ and the proof of this lemma is complete. ■

The set B is not visualizable; it exists as a well-defined symbol but it can't be grasped as more than such. The simple nonconstructive definition of $\mathcal{B}$ as given in Sections 4 and 5 "is what it is"; there is no "practical" way of describing $\mathcal{B}$; we must content ourselves with nonconstructively describing $\mathcal{B}$ as the intersection of all σ-fields that contain [say] the rsc subintervals of $\mathbb{R}$. Every set in $\bigcup_{n=0}^{\infty} \mathcal{I}_n$ is a Borel set, but there are $\mathcal{B}_{(0,1]}$-sets not in $\mathcal{I}_n$ for any $n \in \mathbb{N}$. The exercises contain some other assertions and constructions related to this interesting development. In particular, even more "ethereal" Borel sets exist than the one constructed here.

Exercises.

1. Find the smallest n so that (a) $\{\{x\} : 0 < x \leq 1\} \subseteq \mathcal{I}_n$, (b) each finite subset of $(0,1]$ is in $\mathcal{I}_n$, and (c) each countable subset of $(0,1]$ is in $\mathcal{I}_n$.

2. Directly prove that $\mathcal{I}_0 \subsetneqq \mathcal{I}_1 \subsetneqq \mathcal{I}_2 \subsetneqq \mathcal{I}_3 \subsetneqq \mathcal{I}_4$.

3. $\mathcal{I}_4$ contains *every* subinterval of $(0,1]$ of every form.

4. If $\mathcal{I}_n \subsetneqq \mathcal{I}_{n+1}$ for all $n \geq 0$, then $\mathcal{I}_N \subsetneqq \mathcal{B}_{(0,1]}$ for all $N \in \mathbb{N}$.

5. Let $\Omega \neq \varnothing$ be generic and let $\mathcal{A} = \{A_1, A_2, \cdots\}$ be such that the A_n's are disjoint and nonempty with $\bigcup_{n=1}^{\infty} A_n = \Omega$. Show that $\mathcal{A}^* = \sigma(\mathcal{A})$. Therefore, in this very special case, sets in a generated σ-field can be conveniently described in terms of sets in the generating class.

6. Instantly construct $\mathfrak{c}$ other sets in $\mathcal{B}_{(0,1]} - \bigcup_{n=0}^{\infty} \mathcal{I}_n$. [Hint: recall that B as constructed depended on the particular fixed double sequence $\{m_{ij} : i, j \in \mathbb{N}\}$.]

7. card$(\mathcal{I}_n) = \mathfrak{c}$ for all $n \in \mathbb{N}$, card$(\bigcup_{n=0}^{\infty} \mathcal{I}_n) = \mathfrak{c}$, and card$(\mathcal{B}_{(0,1]} - \bigcup_{n=0}^{\infty} \mathcal{I}_n) = \mathfrak{c}$. The Borel sets constructible by the seemingly comprehensive procedure given at the beginning of this section form but a "negligible" subcollection of $\mathcal{B}_{(0,1]}$. That is, "most" elements of $\mathcal{B}_{(0,1]}$ are not describable by the construction given at the beginning of this section.

8. Let $\Omega = \mathbb{R}$, and let $\mathcal{I}_0$ denote the class consisting of $\varnothing$ and all rsc subintervals $(a, b]$ of $\mathbb{R}$ with $a, b \in \mathbb{Q}$, so that $\mathcal{B} = \sigma(\mathcal{I}_0)$. Construct a set in $\mathcal{B} - \bigcup_{n=0}^{\infty} \mathcal{I}_n$ by modifying where appropriate this section's developments.

9. For each $\omega \in (0,1]$, write $\omega = 0.d_1(\omega)d_2(\omega)d_3(\omega)\cdots$, where each $d_j(\omega)$ represents the jth digit in the dyadic expansion of ω and thus is either 0 or 1. For definiteness, we restrict ourselves to take the expansion that has a non-terminating sequence of 1's: if $\omega = \frac{1}{2}$ [say] we write $\omega = 0.0\bar{1}$ instead of $\omega = 0.1\bar{0}$. For each $n \in \mathbb{N}$, define $s_n(\omega) = \sum_{j=1}^{n} d_j(\omega)$, so that $s_n(\omega)$ denotes the number of 1's in the first n digits of the dyadic expansion of ω. The famous set of *normal numbers* as studied by Borel is $N = \left\{\omega \in (0,1] : \lim_{n\to\infty} \frac{s_n(\omega)}{n} \text{ exists and equals } \frac{1}{2}\right\}$. Demonstrate that $N \in \mathcal{B}_{(0,1]}$, and show that $N \in \mathcal{I}_6$.

10. An even more extreme set in $\mathcal{B}_{(0,1]}$ can be constructed. This exercise will provide an outline for constructing this set. We will use the notation for ordinals as given in Section 1.6, except here we will use ξ to denote the first uncountable ordinal number. For each $\alpha \in P_\xi$, find a sequence $\{\beta_\alpha(n)\}_{n=1}^{\infty}$ of ordinals such that (i) $\beta_\alpha(n) \prec \alpha$ for all $n \in \mathbb{N}$ and (ii) the set $\{\beta_\alpha(n) : n \text{ is such that } \beta_\alpha(n) = \gamma\}$ is infinite for all $\gamma \prec \alpha$. Let $\mathcal{I}_0$ be as before, except the subscript 0 is now the ordinal number 0. For each $\alpha \in P_\xi$, with $\mathcal{I}_\beta$ defined for each $0 \preceq \beta \prec \alpha$, define $\mathcal{I}_\alpha = (\bigcup_{\beta \in P_\alpha} \mathcal{I}_\beta)^*$. This is a transfinitely inductive definition that produces the nondecreasing class $\{\mathcal{I}_\alpha : \alpha \in P_\xi\}$. Next, given sets $A_1, A_2, \cdots$, define $\Psi(A_1, A_2, \cdots) = A_1^c \cup A_2 \cup A_3 \cup \cdots$ as before. Define $\Phi_0(A_1, A_2, \cdots) = A_1$. We define $\Phi_\alpha(A_1, A_2, \cdots)$ for all $\alpha \in P_\xi$ as follows: if $\Phi_\beta(A_1, A_2, \cdots)$ is defined for all $0 \preceq \beta \prec \alpha$, write

$$\Phi_\alpha(A_1, A_2, \cdots) = \Psi\left(\Phi_{\beta_\alpha(1)}(A_{m_{11}}, A_{m_{12}}, \cdots), \Phi_{\beta_\alpha(2)}(A_{m_{21}}, A_{m_{22}}, \cdots), \cdots\right),$$

where $\{m_{ij} : i, j \in \mathbb{N}\}$ is as given in this section.

(a) Let $\alpha \in P_\xi$. There exist $A_1, A_2, \cdots \in \mathcal{I}_0$ such that $\Phi_\alpha(A_1, A_2, \cdots) = (0,1]$. There exist $A_1, A_2, \cdots \in \mathcal{I}_0$ such that $\Phi_\alpha(A_1, A_2, \cdots) = \varnothing$.
(b) Let $\alpha \in P_\xi$. Each $\mathcal{I}_\alpha$-set has the form $\Phi_\alpha(A_1, A_2, \cdots)$ for some appropriately chosen $A_1, A_2, \cdots \in \mathcal{I}_0$.

Parts (a) and (b) correspond with Lemmas 1 and 2. We do not need an analog of Lemma 3. We will need Lemma 4, whose statement and proof remain the same.

(c) Let $\alpha \in P_\xi$. If $A_1, A_2, \cdots \in \mathcal{B}_{(0,1]}$, then $\Phi_\alpha(A_1, A_2, \cdots) \in \mathcal{B}_{(0,1]}$.

Part (c) corresponds to Lemma 5. We do not need an analog of Lemma 6. The notation preceding Lemma 7 and Lemma 7 itself remain unchanged. We again let $I_1, I_2, \cdots$ denote a fixed listing of the countable many sets comprising $\mathcal{I}_0$. For each $\alpha \in P_\xi$, define the function $\varphi_\alpha : (0,1] \to 2^{(0,1]}$ by writing $\varphi_\alpha(\omega) = \Phi_\alpha(I_{\omega_1}, I_{\omega_2}, \cdots)$ whenever $\omega \leftrightarrow \{\omega_k\}_{k=1}^{\infty}$. Lemma 7 ensures that φ_α is well-defined for each $\alpha \in P_\xi$. Next, for each $\alpha \in P_\xi$, let B_α denote the set $\{\omega \in (0,1] : \omega \notin \varphi_\alpha(\omega)\}$. The next part is the analog of Lemma 8.

(d) $B_\alpha \notin \mathcal{I}_\alpha$ for all $\alpha \in P_\xi$.

The sets $L_{k,n}$ and D_k remain the same for all $k, n \in \mathbb{N}$. Lemmas 9, 10, and 11 remain unchanged. Lemma 12 is modified in the next part.

(e) For all $\alpha \in P_\xi$ and sequences $\{u_k\}_{k=1}^{\infty}$ of positive integers,

$$\{\omega \in (0,1] : \omega \in \Phi_\alpha(I_{\omega_{u_1}}, I_{\omega_{u_2}}, \cdots)\} = \Phi_\alpha(D_{u_1}, D_{u_2}, \cdots).$$

Finally, Lemma 13 is modified to the following part.

(f) For each $\alpha \in P_\xi$, $B_\alpha \in \mathcal{B}_{(0,1]}$.

Therefore, for each $0 \preceq \alpha \prec \xi$, there exists a set $B_\alpha \in \mathcal{B}_{(0,1]} - \mathcal{I}_\alpha$. While we have that $\mathcal{B}_{(0,1]} = \bigcup_{0 \preceq \alpha \prec \xi} \mathcal{I}_\alpha$ by an argument similar to Claim 3 of Section 6, we can still find sets in $\mathcal{B}_{(0,1]}$ outside of $\bigcup_{0 \preceq \beta \prec \alpha} \mathcal{I}_\beta$ even for *infinite* ordinal numbers α. Such sets B_α are of course completely inaccessible to human visualization, but they are interesting in that they show the existence of very ethereal entities whose claim to reality is just as strong as those mathematical entities we consider "intuitive."

2

Measures

When we talk of the "length" of a subset of $\mathbb{R}$, the "area" of a subset of $\mathbb{R}^2$, the "volume" of a subset of $\mathbb{R}^3$, and so on, we are assigning a number to a set. The idea of assigning a number to a set can and will be generalized beyond these examples, but consider first the idea of "finding the length" of a given $A \subseteq \mathbb{R}$. What properties are intrinsic to "length?" First, "negative length" makes no logical sense, hence "length" is nonnegative; if we allow discussions of the "length" of unbounded intervals — and it seems natural to allow such discussions — we would be forced to allow $+\infty$ as a "length." Second, if $A \subseteq \mathbb{R}$ can be written as $A = \bigcup_n A_n$ where the A_n's are disjoint nonempty subintervals of A, we would think that the "length" of A should equal the sum of the "lengths" of the component A_n's, regardless of whether there are finitely many A_n's or countably many A_n's. We do not have to consider the case of when there are uncountably many disjoint nonempty subintervals, since such a case is an impossibility.[1] If one is allowed to talk about "the length of $\varnothing$," the logical choice is zero. Of course, we also have to discuss which $A \subseteq \mathbb{R}$ even *have* a well-defined "length".

These properties of "length" will be abstracted and studied on their own without reliance on a Euclidean framework. This chapter discusses the abstraction of the situation above by defining what are called *measures* and by proving certain important facts about them. The reader has already dealt with measures without calling them as such when doing such things as inte-

[1] If A can be written as an uncountable disjoint union of nonempty subintervals, then, since each subinterval contains a distinct rational, we have $\operatorname{card}(\mathbb{Q}) = \mathfrak{c}$, a contradiction.

grating nonnegative functions or computing probabilities. Thus the concept is not completely new – it shall merely be elucidated in a general framework.

Before starting, we must discuss the role of $\pm\infty$ in addition and subtraction. For addition, we write $\pm\infty + x = x \pm \infty = \pm\infty$ for all $x \in \mathbb{R}$, *but we leave the expressions* $\infty - \infty$ *and* $-\infty + \infty$ *undefined — proofs must avoid these two expressions.* Finally, we write $\infty + \infty = \infty$ and $-\infty - \infty = -\infty$.

The final item to consider is that of an *indicator function.* Consider a set Ω, and let $A \subseteq \Omega$. We define the function I_A on Ω by writing $I_A(\omega) = 1$ if $\omega \in A$ and $I_A(\omega) = 0$ if $\omega \notin A$. Clearly the function I_A is well-defined for each $A \subseteq \Omega$. We call I_A the *indicator function of A.*

2.1 MEASURES

For this section, Ω is a generic nonempty set, and $\varnothing \in \mathcal{A} \subseteq 2^\Omega$. We define the main object of this chapter, giving some examples of it, and deriving some basic properties. Note that $\mathcal{A}$ is not assumed to be a σ-field.

Definition. A *[countably additive] measure [on $\mathcal{A}$]* is a function $\mu : \mathcal{A} \to \bar{\mathbb{R}}$ with

(M1) $\mu(A) \geq 0$ for each $A \in \mathcal{A}$,

(M2) $\mu(\varnothing) = 0$, and

(M3) For any sequence of disjoint sets $A_1, A_2, \dots \in \mathcal{A}$ with $\bigcup_{n=1}^\infty A_n \in \mathcal{A}$, the equality $\mu\left(\bigcup_{n=1}^\infty A_n\right) = \sum_{n=1}^\infty \mu(A_n)$ obtains.

When $A \in \mathcal{A}$, we will refer to $\mu(A)$ as *the μ-measure of A* or simply *the measure of A* when no confusion can arise.

Property (M1) is called *nonnegativity,* and it should be noted that $+\infty$ is allowed as a possible value for the measure of a set $A \in \mathcal{A}$.

Property (M3) is called *countable additivity* and is the key property as far as the definition is concerned. We equivalently say that *μ is countably additive.* The equality in (M3) should be construed in one of three ways: (i) both $\mu(\bigcup_{n=1}^\infty A_n)$ and $\sum_{n=1}^\infty \mu(A_n)$ are finite, where the convergence of the series $\sum_{n=1}^\infty \mu(A_n)$ is convergence in the usual calculus sense;[2] (ii) both $\mu(\bigcup_{n=1}^\infty A_n)$ and $\sum_{n=1}^\infty \mu(A_n)$ are $+\infty$ with $\mu(A_n) < +\infty$ for each $n \in \mathbb{N}$, that is, $\sum_{n=1}^\infty \mu(A_n)$ diverges in the usual calculus sense;[3] and (iii) both $\mu(\bigcup_{n=1}^\infty A_n)$ and $\sum_{n=1}^\infty \mu(A_n)$ are $+\infty$, where $\mu(A_N) = +\infty$ for some $N \in \mathbb{N}$. In all three cases we are dealing with an infinite sum of nonnegative terms; a standard result from analysis tells us that convergence or divergence is not affected in any way by summing in a different order[4].

[2]That is, the limit of the sequence of partial sums $\{\sum_{j=1}^n \mu(A_j)\}_{n=1}^\infty$ exists and is finite.
[3]That is, given any $M > 0$ there is $N_M \in \mathbb{N}$ such that $\sum_{n=1}^N \mu(A_n) > M$ for all $N \geq N_M$.
[4]A more formal statement goes as follows: if j is any one-to-one mapping of $\mathbb{N}$ onto $\mathbb{N}$, then the nonnegativity of each $\mu(A_n)$ forces $\sum_{n=1}^\infty \mu(A_n) = \sum_{n=1}^\infty \mu(A_{j(n)})$, so that both series either have a common finite sum or both series have an infinite sum.

Definition. Let μ denote any function from $\mathcal{A}$ into $\bar{\mathbb{R}}$. If there exist $\mathcal{A}$-sets $A_1, A_2, \cdots$ with $\mu(A_n) < +\infty$ for all $n \in \mathbb{N}$ and $\Omega = \bigcup_{n=1}^{\infty} A_n$, we say that μ is *σ-finite [on $\mathcal{A}$]*. If μ is a measure with $\Omega \in \mathcal{A}$ and $\mu(\Omega) < +\infty$, we will say that μ is a *finite measure [on $\mathcal{A}$]*.

Every finite measure is σ-finite: take $A_1 = \Omega$ and $A_2 = A_3 = \cdots = \varnothing$ in the definition above. We shall drop the words *on $\mathcal{A}$* in the terms above when no confusion about the underlying space can arise.

Example 1. Fix a given $\omega_0 \in \Omega$. Let $\mathcal{A} = 2^\Omega$, so that $\mathcal{A}$ is a σ-field. Define $\mu : \mathcal{A} \to \bar{\mathbb{R}}$ by writing $\mu(A) = I_A(\omega_0)$ for all $A \subseteq \Omega$. Since $\mu(A)$ is 0 or 1, (M1) is satisfied. Since $\mu(\varnothing) = I_\varnothing(\omega_0) = 0$, (M2) is satisfied. To demonstrate (M3), let $A_1, A_2, \cdots \subseteq \Omega$ be disjoint. Then it is either the case that (a) ω_0 is in exactly one of $A_1, A_2, \cdots$, or (b) $\omega_0 \notin A_n$ for all $n \in \mathbb{N}$. In situation (a), both $\mu(\bigcup_{n=1}^{\infty} A_n)$ and $\sum_{n=1}^{\infty} \mu(A_n)$ equal 1. In situation (b), both $\mu(\bigcup_{n=1}^{\infty} A_n)$ and $\sum_{n=1}^{\infty} \mu(A_n)$ equal 0. (M3) is satisfied for both cases, and thus μ is a measure on $\mathcal{A}$. Also, μ is a finite measure, hence μ is σ-finite. This particular μ is called a *unit mass at ω_0*.

Example 2. Let $\Omega = \{\omega_1, \omega_2, \cdots\}$, a countably infinite set. Let $\mathcal{A} = 2^\Omega$, and let $\{p_n\}_{n=1}^{\infty}$ denote a sequence of nonnegative elements of $\bar{\mathbb{R}}$. We define $\mu : \mathcal{A} \to \bar{\mathbb{R}}$ by writing $\mu(\varnothing) = 0$ and by writing $\mu(A) = \sum\{p_j : j \text{ is such that } \omega_j \in A\}$ for each nonempty $A \in \mathcal{A}$. Since $p_k \geq 0$ for all $k \in \mathbb{N}$, the sum is invariant to the order in which the appropriate p_j's are summed. Accordingly, μ is well-defined; there is no $A \subseteq \Omega$ for which $\mu(A)$ is ambiguous.

We now show that μ is a meausure. Clearly (M1) and (M2) hold. We now verify the countable additivity of μ on $\mathcal{A}$. Let $A_1, A_2, \cdots \in \mathcal{A}$ be disjoint. Then

$$\begin{aligned} \mu\left(\bigcup_{n=1}^{\infty} A_n\right) &= \sum\{p_k : k \text{ is such that } \omega_k \in A_n \text{ for some } n \in \mathbb{N}\} \\ &= \sum_{n=1}^{\infty} \sum\{p_k : k \text{ is such that } \omega_k \in A_n\} \\ &= \sum_{n=1}^{\infty} \mu(A_n), \end{aligned}$$

where the second equality follows by the disjointness of $A_1, A_2, \cdots$ and the same reasoning used to show that μ is well-defined. This proves (M3), hence μ is a measure on $\mathcal{A}$. If $\sum_{n=1}^{\infty} p_n < +\infty$ then μ is finite. If $\sum_{k=1}^{\infty} p_k = +\infty$ and each p_k is finite, then μ is σ-finite: take $A_n = \{\omega_n\}$ for each $n \in \mathbb{N}$ in the definition. On the other hand, if $p_k = +\infty$ for at least one k, then μ is not σ-finite. The case $\sum_{k=1}^{\infty} p_k = 1$ is the framework for discrete probability.

With the basic definition illustrated, we now derive some basic properties of measures that are conclusions from the definitional properties (M1)-(M3).

(M4) *For any finite disjoint collection of sets $A_1, \cdots, A_n \in \mathcal{A}$ such that $\bigcup_{k=1}^{n} A_k \in \mathcal{A}$, we have $\mu\left(\bigcup_{k=1}^{n} A_k\right) = \sum_{k=1}^{n} \mu(A_k)$.*

Proof: Pick disjoint $A_1, \cdots, A_n \in \mathcal{A}$ and suppose that $\bigcup_{k=1}^{n} A_k \in \mathcal{A}$. Define $A_N = \varnothing$ for all $N \geq n+1$, so that $A_N \in \mathcal{A}$ and $\mu(A_N) = 0$ [by (M2)] for all $N \geq n+1$. The sequence $\{A_n\}_{n=1}^{\infty}$ is thus a disjoint $\mathcal{A}$-sequence with $\bigcup_{k=1}^{\infty} A_k = \bigcup_{k=1}^{n} A_k \in \mathcal{A}$. Applying the countable additivity (M3) of μ gives $\mu\left(\bigcup_{k=1}^{n} A_k\right) = \mu\left(\bigcup_{k=1}^{\infty} A_k\right) = \sum_{k=1}^{\infty} \mu(A_k) = \sum_{k=1}^{n} \mu(A_k)$. ■

Property (M4) is called *finite additivity*, and it shows that every measure is *finitely additive*. The derivation of (M4) shows that countable additivity of a measure implies finite additivity of a measure. Exercise 7 will show that the conjunction of finite additivity along with (M1) and (M2) does not necessarily yield countable additivity, so that countable additivity is a stronger condition than is finite additivity.

Definition. Let $\varnothing \in \mathcal{A} \subseteq 2^{\Omega}$. A function $\mu : \mathcal{A} \to \bar{\mathbb{R}}$ that satisfies (M1), (M2), and (M4) is called a *finitely additive measure.*

The next two properties are valid for finitely additive measures [and hence for measures], as nowhere in their proofs is (M3) invoked.

(M5) *Suppose $A \subseteq B$ with $A, B, B-A \in \mathcal{A}$. Then $\mu(A) \leq \mu(B)$. Furthermore, if $\mu(A) < +\infty$ or $\mu(B) < +\infty$, then $\mu(B-A) = \mu(B) - \mu(A)$.*

Proof: Assume the hypotheses, and observe that B is the disjoint union of A and $B-A$. (M4) gives $\mu(B) = \mu(A \cup (B-A)) = \mu(A) + \mu(B-A) \geq \mu(A)$, where the inequality follows by (M1). This proves the first claim.

For the second claim, consider the equality $\mu(B) = \mu(A) + \mu(B-A)$ from the previous paragraph. If $\mu(A) < +\infty$, we may meaningfully subtract $\mu(A)$ from both sides to prove the result; the form $+\infty - \infty$ is avoided. If $\mu(B) < +\infty$, then the previous paragraph gives $\mu(A) \leq \mu(B) < +\infty$, and hence we may again meaningfully subtract $\mu(A)$ from both sides to get the result. ■

The first property in (M5) is called *monotonicity*, and we equivalently state that *μ is monotone.* This is something that makes sense intuitively. The second property in (M5) also makes intuitive sense, although one must have some care in using this, so that the undefined expression $+\infty - \infty$ is avoided; the equality is not true [or even meaningful] when $\mu(A) = +\infty$.

An immediate consequence of (M5) and (M1) is the following: *if $A, B \in \mathcal{A}$ with $\mu(B) = 0$, then $\mu(A) = 0$.* We shall often use this very simple fact.

(M6) *Let $n \in \mathbb{N}$, and let $A_1, \cdots, A_n \in \mathcal{A}$, where $\mathcal{A}$ is a [σ-]field. Let μ denote a measure on $\mathcal{A}$, and suppose that $\mu(A_k) < +\infty$ for $k = 1, \cdots, n$. Then*

$$\mu\left(\bigcup_{k=1}^{n} A_k\right) = \sum_{k=1}^{n} \sum \left\{(-1)^{k-1} \mu\left(\bigcap_{j=1}^{k} A_{i_j}\right) : 1 \leq i_1 < \cdots < i_k \leq n\right\}. \tag{2.1}$$

When $n = 1$, (2.1) reduces to asserting that $\mu(A_1) = \mu(A_1)$. When $n = 2$, (2.1) reduces to asserting that $\mu(A_1 \cup A_2) = \mu(A_1) + \mu(A_2) - \mu(A_1 \cap A_2)$.

When $n = 3$, (2.1) asserts that

$$\mu\left(\bigcup_{k=1}^{3} A_i\right) = \sum_{i=1}^{3} \mu(A_i) - \mu(A_1 \cap A_2) - \mu(A_1 \cap A_3) - \mu(A_2 \cap A_3) + \mu\left(\bigcap_{k=1}^{3} A_i\right).$$

Property (M6) is called the *inclusion-exclusion rule.*

Proof: We first prove (2.1) when $n = 2$. Observe that

$$A_1 \cup A_2 = [A_1 - (A_1 \cap A_2)] \cup [A_2 - (A_1 \cap A_2)] \cup (A_1 \cap A_2).$$

Since the unions on the right side are disjoint, and since $\mu(A_1), \mu(A_2) < +\infty$ implies $\mu(A_1 \cap A_2) < +\infty$ by the first part of (M5), we invoke (M4) and the second part of (M5) to obtain

$$\mu(A_1 \cup A_2) = [\mu(A_1) - \mu(A_1 \cap A_2)] + [\mu(A_2) - \mu(A_1 \cap A_2)] + \mu(A_1 \cap A_2),$$

which is (2.1) when $n = 2$.

Next, let $n \in \mathbb{N}$ and assume that (2.1) holds for $1, 2, \cdots, n$. For each $k = 1, \cdots, n$, we will let $\sum$ denote summation over all k-tuples $(i_1, \cdots, i_k)$ with $1 \leq i_1 < \cdots < i_k \leq n$. Observe that

$$\begin{aligned}
\mu\left(\bigcup_{k=1}^{n+1} A_k\right) &= \mu\left(\bigcup_{k=1}^{n} A_k \cup A_{n+1}\right) \\
&= \mu\left(\bigcup_{k=1}^{n} A_k\right) + \mu(A_{n+1}) - \mu\left(\bigcup_{k=1}^{n}(A_k \cap A_{n+1})\right) \quad \text{[result for } n = 2\text{]} \\
&= \sum_{k=1}^{n}\sum(-1)^{k-1}\mu\left(\bigcap_{j=1}^{k} A_{i_j}\right) + \mu(A_{n+1}) \\
&\quad -\sum_{k=1}^{n}\sum(-1)^{k-1}\mu\left(\bigcap_{j=1}^{k} A_{i_j} \cap A_{n+1}\right) \\
&= \sum_{k=1}^{n+1}\sum\left\{(-1)^{k-1}\mu\left(\bigcap_{j=1}^{k} A_{i_j}\right) : 1 \leq i_1 < \cdots < i_k \leq n+1\right\},
\end{aligned}$$

where the third equality follows by twice applying the induction hypothesis, and the fourth equality follows upon expanding the terms in the third equality and simplifying. [This is not hard, merely tedious.] This shows that (2.1) holds in the $n + 1$ case, and the entire proof is complete. ∎

Property (M6) will play an important role in a certain uniqueness question to be addressed in Chapter 3. The $n = 2$ and $n = 3$ cases are widely used in probability calculations.

The two inequalities in (M7) below are called *countable subadditivity* and *finite subadditivity*, respectively, and we will say that *μ is countably subadditive* and *μ is finitely subadditive*, respectively. The first inequality below is also known as *Boole's Inequality.* Although the bound is not sharp by any means, it is quite useful.

(M7) *Let $\mathcal{A}$ denote a σ-field, and let $A_1, A_2, \cdots \in \mathcal{A}$. Then we have the inequality $\mu(\bigcup_{k=1}^{\infty} A_k) \leq \sum_{k=1}^{\infty} \mu(A_k)$. Furthermore, for each $n \in \mathbb{N}$ we have $\mu(\bigcup_{k=1}^{n} A_k) \leq \sum_{k=1}^{n} \mu(A_k)$.*

Proof: Let $B_1 = A_1$, and for $k \geq 2$, let $B_k = A_k \cap \bigcap_{j=1}^{k-1} A_j^c$. Then for each $k \in \mathbb{N}$ we have that $B_k \in \mathcal{A}$ and $B_k \subseteq A_k$, so that $\mu(B_k)$ is defined with $\mu(B_k) \leq \mu(A_k)$ [by (M5)]. Observe next that $\bigcup_{k=1}^{\infty} B_k = \bigcup_{k=1}^{\infty} A_k \in \mathcal{A}$. Since $B_1, B_2, \cdots$ are disjoint we apply (M3) and (M5) to obtain

$$\mu\left(\bigcup_{k=1}^{\infty} A_k\right) = \mu\left(\bigcup_{k=1}^{\infty} B_k\right) = \sum_{k=1}^{\infty} \mu(B_k) \leq \sum_{k=1}^{\infty} \mu(A_k),$$

giving countable subadditivity. To prove the finite subadditivity claim for any $n \in \mathbb{N}$, simply let $A_{n+1} = A_{n+2} = \varnothing$ above and apply (M2). ∎

As a consequence of (M7), we have the following for the situation where $\mathcal{A}$ is a σ-field: *an amc union of $\mathcal{A}$-sets of measure zero is a set in $\mathcal{A}$ of measure zero.* To see this, let $A_1, A_2, \cdots \in \mathcal{A}$ be such that $\mu(A_n) = 0$ for each n. [Here n is finite or n ranges over all of $\mathbb{N}$.] Then using (M1) and (M7) yields $0 \leq \mu(\bigcup_n A_n) \leq \sum_n \mu(A_n) = \sum_n 0 = 0$, giving $\mu(\bigcup_n A_n) = 0$. We shall use this property several times in what follows.

Definition. Let Ω be given, let $\mathcal{A}$ denote a σ-field on Ω, and let μ denote a measure on $\mathcal{A}$. We call the triple $(\Omega, \mathcal{A}, \mu)$ a *measure space*. The pair $(\Omega, \mathcal{A})$ is called a *measurable space*. If μ is a measure on $\mathcal{A}$ but $\mathcal{A}$ is not a σ-field on Ω, then we will not call $(\Omega, \mathcal{A}, \mu)$ a measure space. Therefore, proving that $(\Omega, \mathcal{A}, \mu)$ is a measure space means showing that $\mathcal{A}$ is a σ-field on Ω and μ is a measure with domain $\mathcal{A}$.

The properties (M1)-(M7) listed in this section are of one type, namely, they are direct consequences of the definition. The next section will give additional important properties of measures with regard to limits, but some machinery will need to be developed first.

Exercises.

1*. This problem deals with some other variants of properties (M1)-(M3).
(a) Some define a *probability measure P on a σ-field $\mathcal{A}$ of subsets of Ω* by stipulating that (i) $0 \leq P(A) \leq 1$ for all $A \in \mathcal{A}$, (ii) $P(\Omega) = 1$, and (iii) P is countably additive. This is a special case of our definition of a measure.
(b) If (M1) and (M3) hold for a set function μ defined on a field $\mathcal{A}$ with $\mu(A) < +\infty$ for some $A \in \mathcal{A}$, then μ is a measure on $\mathcal{A}$.

2*. Let $\Omega = \{\omega_1, \cdots, \omega_n\}$, and let $p_1, \cdots, p_n \in [0, +\infty]$. Define μ on 2^{Ω} as in Example 2. Then $(\Omega, 2^{\Omega}, \mu)$ is a measure space, and μ is σ-finite iff $p_n < +\infty$ for each $n \in \mathbb{N}$.

3. Let $\mathcal{A} = \{\varnothing, \Omega\}$, $\mu(\varnothing) = 0$, and $\mu(\Omega) = +\infty$. Then $(\Omega, \mathcal{A}, \mu)$ is a measure space, but μ fails to be σ-finite.

4*. Let Ω be uncountable. Let $\mathcal{A} = \{A \subseteq \Omega : A \text{ is amc or } A^c \text{ is amc}\}$. Write $\mu(A) = 0$ if A is amc and $\mu(A) = +\infty$ if A^c is amc. Then $(\Omega, \mathcal{A}, \mu)$ is a measure space, and μ is not σ-finite.

5. Let Ω be arbitrary, and let $\mathcal{A} = \{A \subseteq \Omega : A \text{ is amc or } A^c \text{ is amc}\}$. Define μ over $\mathcal{A}$ by stating that $\mu(A) = 0$ if A is amc, and $\mu(A) = 1$ if A^c is amc.
(a) μ is not well-defined if Ω is amc, but μ is well-defined if Ω is uncountable.
(b) μ is a σ-finite measure on the σ-field $\mathcal{A}$ when Ω is uncountable.

6. Suppose that $\mathcal{A}$ is a *finite* σ-field on Ω. Suppose that μ is defined on $\mathcal{A}$ such that (M1), (M2), and (M4) hold. Then $(\Omega, \mathcal{A}, \mu)$ is a measure space.

7. Let $\mathcal{A} = \{A \subseteq \Omega : A \text{ finite or } A^c \text{ finite}\}$. Define μ on $\mathcal{A}$ by $\mu(A) = 0$ if A is finite, and $\mu(A) = 1$ if A^c is finite.
(a) μ fails to be well-defined when Ω is finite.
(b) If Ω is infinite, then μ satisfies (M1), (M2), and (M4).
(c) Let $\text{card}(\Omega) = \aleph_0$. Then μ is finitely additive but not countably subadditive.
(d) When Ω is uncountable, μ is a measure. Is μ a σ-finite?

8. Let $\text{card}(\Omega) = \aleph_0$ and $\mathcal{A} = 2^{\Omega}$. Let $\mu(A) = 0$ if A is finite and let $\mu(A) = +\infty$ if A is infinite. Then μ is well-defined, μ satisfies (M1), (M2), and (M4), and that (M3) fails. Also, μ is σ-finite.

9. (M5) is not true if the hypothesis $\mu(A) < +\infty$ is omitted.

10*. Let μ denote a measure on a σ-field $\mathcal{A}$, and let $A, A_1, A_2, \cdots \in \mathcal{A}$.
(a) $\mu(A) = \sum_{k=1}^{\infty} \mu(A \cap A_k)$ when the A_k's are disjoint with $\bigcup_{k=1}^{\infty} A_k = \Omega$.
(b) $\mu(A_1 \Delta A_2) = 0$ iff $\mu(A_1) = \mu(A_2) = \mu(A_1 \cap A_2)$. [Consider two cases: (i) both $\mu(A_1), \mu(A_2) < +\infty$, and (ii) at least one of $\mu(A_1), \mu(A_2) = +\infty$.]
(c) $\mu(A_2) = 0$ forces both $\mu(A_1 \cup A_2) = \mu(A_1)$ and $\mu(A_1 \Delta A_2) = 0$.
(d) $\mu(A_2) = 0$ forces $\mu(A_1 - A_2) = \mu(A_1)$.
[All parts should be direct. These facts will be used in discussing completeness in Chapter 3.]

11. Let $(\Omega, \mathcal{A}, \mu)$ be a measure space such that there is $B \in \mathcal{A}$ with $0 < \mu(B) < +\infty$. Fix such a B, and define $\mu_B : \mathcal{A} \to \bar{\mathbb{R}}$ by the formula $\mu_B(A) = \mu(A \cap B)/\mu(B)$.
(a) $(\Omega, \mathcal{A}, \mu_B)$ is a measure space.
(b) Suppose in addition that Ω is the disjoint union of an amc collection of sets $B_n \in \mathcal{A}$ such that each B_n has finite measure, and suppose that μ is finite. Then for all $A \in \mathcal{A}$ we have $\mu(A) = \sum_n \mu_{B_n}(A)\mu(B_n)$. Also, for each $i \in \mathbb{N}$ we have

$$\mu_{B_i}(A) = \frac{\mu_A(B_i)\mu(A)}{\sum_n \mu_A(B_n)\mu(A)}.$$

This formula is known as *Bayes' Rule.*

12. Let $S = \{s_1, \cdots, s_m\}$, and let $\{p_u : u \in S\}$ denote a collection of nonnegative numbers with $\sum_{u \in S} p_u = 1$. Let Ω denote the set of sequences of S. For each $\omega \in \Omega$, write $\omega = (z_1(\omega), z_2(\omega), \cdots)$. Given $n \in \mathbb{N}$ and $H \subseteq S^n$, let

$$C_n(H) = \{\omega \in \Omega : (z_1(\omega), \cdots, z_n(\omega)) \in H\}.$$

Such a set is called a *cylinder of rank n*. Let $\mathcal{F} = \{C_n(H) : n \in \mathbb{N}, H \subseteq S^n\}$, so that $\mathcal{F}$ consists of all cylinders of all ranks. Define $\mu : \mathcal{F} \to \bar{\mathbb{R}}$ as follows: if for some $n \in \mathbb{N}$ and $H \subseteq S^n$ we have $A = \{\omega \in \Omega : (z_1(\omega), \cdots, z_n(\omega)) \in H\}$, write $\mu(A) = \sum\{p_{i_1} \cdots p_{i_n} : (i_1, \cdots, i_n) \in H\}$.

(a) μ is well-defined.

(b) Show that μ satisfies (M1) and (M2), and is finitely additive.

Accept without proof or recall a special case of Tychonoff's Theorem: *Let $A_1, A_2, \cdots$ denote a sequence of cyliders with $A_1 \supseteq A_2 \supseteq \cdots$. If $\bigcap_{n=1}^{\infty} A_n = \varnothing$, then $A_N = \varnothing$ for some $n \in \mathbb{N}$.*

(c) Any function ν on $\mathcal{F}$ that satisfies (M1), (M2), and (M4) also satisfies (M3). Therefore, μ is a measure, often referred to as *product measure.*

2.2 CONTINUITY OF MEASURES

Section 1 defined a measure and discussed some basic results. This section will state and prove some additional important properties of measures. To do this, some additional notation and terminology must first be presented.

Definition. Let $\{A_n\}_{n=1}^{\infty}$ denote a sequence of subsets of some nonempty set Ω. The sets $\overline{\lim}\, A_n$ and $\underline{\lim}\, A_n$ are given by

$$\overline{\lim}\, A_n = \bigcap_{k=1}^{\infty} \bigcup_{n=k}^{\infty} A_n \quad \text{and} \quad \underline{\lim}\, A_n = \bigcup_{k=1}^{\infty} \bigcap_{n=k}^{\infty} A_n.$$

If $\underline{\lim}\, A_n = \overline{\lim}\, A_n$, we define $\lim A_n$ to be the common set to which $\underline{\lim}\, A_n$ and $\overline{\lim}\, A_n$ are equal. If $\underline{\lim}\, A_n \neq \overline{\lim}\, A_n$, then $\lim A_n$ will be left undefined[5].

The sets $\overline{\lim}\, A_n$ and $\underline{\lim}\, A_n$ are referred to as the *limit superior* and *limit inferior* of the sequence $\{A_n\}_{n=1}^{\infty}$. Alternative names are *superior [inferior] limit* and *upper [lower] limit.* Notational variants for $\overline{\lim}\, A_n$ and $\underline{\lim}\, A_n$ are $\limsup A_n$ and $\liminf A_n$, respectively.

The subscript n in the symbols $\underline{\lim}\, A_n$, $\overline{\lim}\, A_n$, and $\lim A_n$ is really best viewed as a dummy variable, much as x in $\int_a^b f(x)\, dx$ is viewed as a dummy variable. The limit inferior, limit superior, and limit [when it exists] are derived from the *entire sequence* $\{A_n\}_{n=1}^{\infty}$, and we could equally write [say] $\underline{\lim}\, A_k$ or $\underline{\lim}\, A_m$ for $\underline{\lim}\, A_n$, with similar comments for the superior limit and limit. There will be situations where the subscript is important. For example, if $\{B_n\}_{n=1}^{\infty}$ is another sequence of subsets of Ω, then [say] $\underline{\lim}(A_n \cap B_k)$ is ambiguous, whereas $\underline{\lim}_n(A_n \cap B_k)$ and $\underline{\lim}_k(A_n \cap B_k)$ denote quite different things; in the former k is fixed and we are looking for the inferior limit of the sequence $\{A_n \cap B_k\}_{n=1}^{\infty}$, and in the latter n is fixed and we are looking for the inferior limit of the sequence $\{A_n \cap B_k\}_{k=1}^{\infty}$. On the other hand, we may write [say] $\underline{\lim}(A_n \cap B_n)$ or $\underline{\lim}_n(A_n \cap B_n)$ without any worry. Other situations are

[5]The convention above is the same convention as used for sequences $\{x_n\}$, where $\lim x_n$ exists iff $\underline{\lim}\, x_n = \overline{\lim}\, x_n$, otherwise $\lim x_n$ is undefined.

handled analogously and it is trusted that no confusion will arise with any of the notation.

The basic properties of inferior and superior limits are embodied in the following claim. It will be observed that these properties mimic those of inferior and superior limits for sequences of numbers.

Claim 1. *Let $\{A_n\}_{n=1}^{\infty}$ denote a sequence of subsets of Ω. Then we have the following properties:*
(a) $\underline{\lim} A_n = \{\omega \in \Omega :$ ω is in all but finitely many of $A_1, A_2, \cdots\}$.
(b) $\overline{\lim} A_n = \{\omega \in \Omega :$ ω is in infinitely many of $A_1, A_2, \cdots\}$.
(c)[6] $\underline{\lim} A_n \subseteq \overline{\lim} A_n$.
(d)[7] *If $\{A_n\}_{n=1}^{\infty}$ is nondecreasing, then* $\lim A_n$ *exists and equals* $\bigcup_{n=1}^{\infty} A_n$.
(e) If $\{A_n\}_{n=1}^{\infty}$ is nonincreasing, then $\lim A_n$ *exists and equals* $\bigcap_{n=1}^{\infty} A_n$.
(f) If $A_1, A_2, \cdots$ are disjoint, then $\lim A_n$ *exists and equals* $\varnothing$.

Proof: We first do (a). Assume first that $\underline{\lim} A_n = \varnothing$. Then for all $k \in \mathbb{N}$ we have that $\bigcap_{n=k}^{\infty} A_n = \varnothing$. But then no $\omega \in \Omega$ can possibly be in all but finitely many of $A_1, A_2, \cdots$, and hence (a) holds if $\underline{\lim} A_n = \varnothing$. Next, assume that $\underline{\lim} A_n \neq \varnothing$, and let $\omega \in \underline{\lim} A_n$. Then for some $k \in \mathbb{N}$, ω is in each of $A_k, A_{k+1}, \cdots$. Hence ω is in all but finitely many of $A_1, A_2, \cdots$. Conversely, if ω is in all but finitely many of $A_1, A_2, \cdots$, then there is some $k \in \mathbb{N}$ such that $\omega \in \bigcap_{n=k}^{\infty} A_k$. Therefore $\omega \in \bigcup_{k=1}^{\infty} \bigcap_{n=k}^{\infty} A_k$, hence $\omega \in \underline{\lim} A_k$. This finishes (a) in the case where $\underline{\lim} A_n \neq \varnothing$, and (a) is proven in all cases. The proof of (b) is done in the same fashion, and is left to the reader.

For (c), the claim is trivial if $\underline{\lim} A_n = \varnothing$. If $\underline{\lim} A_n \neq \varnothing$, we may pick $\omega \in \underline{\lim} A_n$. This ω is in all but finitely many of $A_1, A_2, \cdots$, and hence is in infinitely many of $A_1, A_2, \cdots$. Therefore, $\omega \in \overline{\lim} A_n$, and (c) holds.

For (d), the claim is trivial if $A_n = \varnothing$ for each $n \in \mathbb{N}$. We therefore assume that $A_N \neq \varnothing$ for some $N \in \mathbb{N}$. Since $A_1 \subseteq A_2 \subseteq \cdots$, any ω is in infinitely many of $A_1, A_2, \cdots$ iff ω is in all but finitely many of $A_1, A_2, \cdots$, and hence $\underline{\lim} A_n = \overline{\lim} A_n$ by parts (a) and (b). Thus $\lim A_n$ exists and equals $\underline{\lim} A_n = \bigcup_{k=1}^{\infty} \bigcap_{n=k}^{\infty} A_k = \bigcup_{k=1}^{\infty} A_k$. This gives (d); (e) holds similarly.

To do (f), merely observe that if $A_1, A_2, \cdots$ are disjoint, then no $\omega \in \Omega$ can possibly be in infinitely many of $A_1, A_2, \cdots$, hence $\overline{\lim} A_n = \varnothing$. By (c), $\underline{\lim} A_n = \varnothing$. It follows that $\lim A_n$ exists and is empty. ∎

Notation. When $\{A_n\}_{n=1}^{\infty}$ is a sequence of sets such that $\lim A_n$ exists and equals some set A, we will equivalently write $A_n \to A$ or $\lim A_n = A$.[8]

As an example, if $\{A_n\}_{n=1}^{\infty}$ is a monotone sequence, then we may write $A_n \to \bigcup_{m=1}^{\infty} A_m$ or $A_n \to \bigcap_{m=1}^{\infty} A_m$ as $A_1, A_2, \cdots$ are nondecreasing or

[6]This mimics the corresponding inequality $\underline{\lim}\, x_n \leq \overline{\lim}\, x_n$ for sequences $\{x_n\}_{n=1}^{\infty}$.
[7]Compare this and (e) with the corresponding result for sequences: if $\{x_n\}_{n=1}^{\infty}$ is a monotone sequence, then $\overline{\lim}\, x_n = \underline{\lim}\, x_n$, hence $\lim x_n$ exists.
[8]This is similar to writing $x_n \to x$ for sequences, where x is nothing other than the common value of $\underline{\lim}\, x_n$ and $\overline{\lim}\, x_n$.

nonincreasing. If $A_1, A_2, \cdots$ are disjoint we may write $A_n \to \varnothing$ or $\lim A_n = \varnothing$. These are restatements of (d)-(f) in Claim 1. In general the set A in the equivalent expressions $A_n \to A$ and $\lim A_n = A$ is nothing other than the set to which both $\overline{\lim}\, A_n$ and $\underline{\lim}\, A_n$ are equal.

Exercises.

Throughout, $\{A_n\}_{n=1}^{\infty}$ and $\{B_n\}_{n=1}^{\infty}$ are generic sequences of subsets of a generic Ω. It is hoped that the reader will work through most or all of them, being sure to work Exercise 9, which will be used to prove an important and general convergence result later in this section. Analogies between sequences of numbers and sequences of sets are given in the footnotes.

1. $\underline{\lim}\, I_{A_n} = I_{\underline{\lim}\, A_n}$ and $\overline{\lim}\, I_{A_n} = I_{\overline{\lim}\, A_n}$.

2. Show that $\underline{\lim}\, A_n \subseteq \overline{\lim}\, A_n$ without using the representations of $\underline{\lim}\, A_n$ and $\overline{\lim}\, A_n$ given in parts (a) and (b) of Claim 1.

3*[9]**.** $(\underline{\lim}\, A_n)^c = \overline{\lim}\, A_n^c$ and $(\overline{\lim}\, A_n)^c = \underline{\lim}\, A_n^c$.

4*[10]**.** If $B_n \neq A_n$ for at most finitely many $n \in \mathbb{N}$, then $\underline{\lim}\, A_n = \underline{\lim}\, B_n$ and $\overline{\lim}\, A_n = \overline{\lim}\, B_n$. In other words, $\underline{\lim}\, A_n$, $\overline{\lim}\, A_n$ [and $\lim A_n$] are not changed if a finite number of A_k's are altered.

5*. We have the following relations:
(a)[11] $\overline{\lim}(A_n \cup B_n) = \overline{\lim}\, A_n \cup \overline{\lim}\, B_n$ and $\underline{\lim}(A_n \cap B_n) = \underline{\lim}\, A_n \cap \underline{\lim}\, B_n$.
(b)[12] $\overline{\lim}(A_n \cap B_n) \subseteq \overline{\lim}\, A_n \cap \overline{\lim}\, B_n$ and $\underline{\lim}(A_n \cup B_n) \supseteq \underline{\lim}\, A_n \cup \underline{\lim}\, B_n$. Both containment relations can be strict.
(c) $I_{\overline{\lim}(A_n \cup B_n)} = \max\left(I_{\overline{\lim}\, A_n}, I_{\overline{\lim}\, B_n}\right)$ and $I_{\underline{\lim}(A_n \cap B_n)} = \min\left(I_{\underline{\lim}\, A_n}, I_{\underline{\lim}\, B_n}\right)$.

6*. If $A_n \to A$ and $B_n \to B$, then $A_n \cup B_n \to A \cup B$, $A_n^c \to A^c$, $A_n \cap B_n \to A \cap B$, $A_n - B_n \to A - B$, and $A_n \Delta B_n \to A \Delta B$.

7. If A_n is B or C as n is even or odd, then $\underline{\lim}\, A_n = B \cap C$ and $\overline{\lim}\, A_n = B \cup C$.

8*. $\overline{\lim}\, A_n - \underline{\lim}\, A_n = \overline{\lim}_n(A_n \cap A_{n+1}^c) = \overline{\lim}_n(A_n^c \cap A_{n+1})$.

9*. The following results will be used to verify a very important continuity result regarding measures.
(a) $\overline{\lim}_n\, \underline{\lim}_k(A_n \cap A_k^c) = \varnothing$. [Use Exercise 5(a), the fact that $\underline{\lim}_k\, A_n = A_n$ for all $n \in \mathbb{N}$, Exericse 3, and then Exercise 5(b).]
(b) $A - \overline{\lim}_k\, A_k = \underline{\lim}_k(A - A_k)$. [Argue that $\underline{\lim}_k(A - A_k) = \underline{\lim}_k\, A \cap \underline{\lim}_k\, A_k^c = A \cap (\overline{\lim}_k\, A_k)^c$ by again using Exercises 5 and 3.]
(c) $\overline{\lim}_n(\underline{\lim}_k\, A_k - A_n) = \varnothing$. [Argue that

$$\overline{\lim_n}(\underline{\lim_k}\, A_k - A_n) \subseteq \overline{\lim_n}\,\underline{\lim_k}\, A_k \cap \overline{\lim_n}\, A_n^c = \underline{\lim_k}\, A_k \cap (\underline{\lim_n}\, A_n)^c,$$

again by using Exercises 5 and 3.]
(d) $\overline{\lim}_n(A - A_n) = A - \underline{\lim}_n\, A_n$ and $\overline{\lim}_n(A_n - A) = \overline{\lim}_n\, A_n - A$.

[9]Compare with $-\underline{\lim}\, x_n = \overline{\lim}(-x_n)$ and $-\overline{\lim}\, x_n = \underline{\lim}(-x_n)$.
[10]Analogously, the lower and upper limit of a sequence $\{x_n\}_{n=1}^{\infty}$ is unchanged if finitely many x_n's are altered.
[11]$\overline{\lim} \max\{x_n, y_n\} = \max\{\overline{\lim}\, x_n, \overline{\lim}\, y_n\}$, $\underline{\lim} \min\{x_n, y_n\} = \min\{\underline{\lim}\, x_n, \underline{\lim}\, y_n\}$.
[12]$\overline{\lim} \min\{x_n, y_n\} \leq \min\{\overline{\lim}\, x_n, \overline{\lim}\, y_n\}$, $\underline{\lim} \max\{x_n, y_n\} \geq \max\{\underline{\lim}\, x_n, \underline{\lim}\, y_n\}$.

(e) $\overline{\lim}_n(A \triangle A_n) = (A - \underline{\lim}_n A_n) \cup \overline{\lim}_n(A_n - A)$. [Write out $A \triangle A_n$, use Exercise 5(a), and then apply (d).]
(f) $A_n \to A$ implies that $\overline{\lim}_n(A \triangle A_n) = \overline{\lim}_n(A_n - A)$.
(g) For arbitrary sets E, F, G, and H we have $(E \triangle F) \triangle (G \triangle H) = (E \triangle G) \triangle (F \triangle H)$. We also have for any set A that

$$\begin{aligned} \overline{\lim_k} A_k - \underline{\lim_k} A_k &= \underline{\lim_k} A_k \triangle \overline{\lim_k} A_k \\ &= (\underline{\lim_k} A_k \triangle \overline{\lim_k} A_k) \triangle (A \triangle A) \\ &= (\underline{\lim_k} A_k \triangle A) \triangle (\overline{\lim_k} A_k \triangle A). \end{aligned}$$

We now come to an important property of measures, which property is called *continuity*. Just as [roughly] $x_n \to x$ implies $f(x_n) \to f(x)$ for continuous functions f, so too under certain conditions we will have [roughly] that $A_n \to A$ implies $\mu(A_n) \to \mu(A)$ for measures μ.

We will assume that $(\Omega, \mathcal{A}, \mu)$ is a measure space in all that follows.

In order to avoid repetition, observe first that because $\mathcal{A}$ is a σ-field, given any sequence $A_1, A_2, \cdots \in \mathcal{A}$, we will always have $\underline{\lim} A_n \in \mathcal{A}$ and $\overline{\lim} A_n \in \mathcal{A}$; if $\lim A_n$ exists, then it will be in $\mathcal{A}$ as well. *In the proofs of the following properties, it is left to the reader to trivially verify that each set to which μ is applied is in the domain of μ, namely, the σ-field $\mathcal{A}$. Also, unless otherwise stated, a set is assumed to be in $\mathcal{A}$.*

Properties (M8) and (M9) are the simplest continuity results on measures that we have. Property (M8) is called *continuity from below*, and we equivalently state that μ is *continuous from below*. Property (M9) is called *continuity from above*, and we equivalently state that μ is *continuous from above*.

(M8) *Let $A_1 \subseteq A_2 \subseteq \cdots$. Then $\mu(A_1) \le \mu(A_2) \le \cdots$, and we have that $\mu(\lim A_n) = \lim \mu(A_n) = \mu(\bigcup_{n=1}^{\infty} A_n)$.*

Proof: Assume the hypotheses of the claim. By (M5), $\{\mu(A_k)\}_{k=1}^{\infty}$ is a nondecreasing sequence. Next, define $B_1 = A_1$ and $B_k = A_k - A_{k-1}$ for all $k \ge 2$; it follows that $B_1, B_2, \cdots$ are disjoint and $\bigcup_{k=1}^{\infty} B_k = \bigcup_{k=1}^{\infty} A_k$. By (M3) and the fact that $\lim A_n$ exists and equals $\bigcup_{n=1}^{\infty} A_n$,

$$\mu(\lim A_n) = \mu\left(\bigcup_{k=1}^{\infty} A_k\right) = \mu\left(\bigcup_{k=1}^{\infty} B_k\right) = \sum_{k=1}^{\infty} \mu(B_k) = \lim \sum_{k=1}^{n} \mu(B_k).$$

Invoking (M4) and using the fact that $\bigcup_{k=1}^{n} B_k = A_n$ for all $n \in \mathbb{N}$, we obtain

$$\lim \sum_{k=1}^{n} \mu(B_k) = \lim \mu\left(\bigcup_{k=1}^{n} B_k\right) = \lim \mu(A_n),$$

which completes the proof. ∎

(M9) *Let* $A_1 \supseteq A_2 \supseteq \cdots$ *be such that* $\mu(A_N) < +\infty$ *for some* N. *Then* $\mu(A_1) \geq \mu(A_2) \geq \cdots$ *and* $\mu\left(\bigcap_{k=1}^{\infty} A_k\right) = \mu(\lim A_n) = \lim \mu(A_n)$.

Proof: Assume the hypotheses and let $B_k = A_{N-1+k}$ for all $k \geq 1$, so that $B_1 \supseteq B_2 \supseteq \cdots$; (M5) gives $\mu(B_k) \leq \mu(B_1) = \mu(A_N) < +\infty$ for all $k \in \mathbb{N}$. Define $C_k = B_1 - B_k$ for all $k \in \mathbb{N}$, so that $C_1 \subseteq C_2 \subseteq \cdots$. Since $\mu(B_1) < +\infty$, we have $\mu(C_k) = \mu(B_1) - \mu(B_k)$ for all $k \in \mathbb{N}$ by (M5). Also, $\bigcup_{k=1}^{\infty} C_k = \bigcup_{k=1}^{\infty}[B_1 - B_k] = B_1 - \bigcap_{k=1}^{\infty} B_k$. Applying (M8) to the increasing sequence $C_1 \subseteq C_2 \subseteq \cdots$, we obtain

$$\mu\left(\bigcup_{k=1}^{\infty} C_k\right) = \lim \mu(C_k) = \lim_k [\mu(B_1) - \mu(B_k)] = \mu(B_1) - \lim \mu(B_k).$$

[Note how the finiteness of $\mu(B_1)$ keeps the last equality from possibly being of the undefined form $+\infty - \infty$.] However, (M5) again gives

$$\mu\left(\bigcup_{k=1}^{\infty} C_k\right) = \mu\left(B_1 - \bigcap_{k=1}^{\infty} B_k\right) = \mu(B_1) - \mu\left(\bigcap_{k=1}^{\infty} B_k\right).$$

[Since $\mu(B_1) < +\infty$, the last expression is always defined.] Therefore,

$$\mu(B_1) - \lim \mu(B_k) = \mu(B_1) - \mu\left(\bigcap_{k=1}^{\infty} B_k\right).$$

Since $\mu(B_1) < +\infty$, we obtain $\mu\left(\bigcap_{k=1}^{\infty} B_k\right) = \lim \mu(B_k)$. Since we have $\bigcap_{k=1}^{\infty} B_k = \bigcap_{k=N}^{\infty} A_k = \bigcap_{k=1}^{\infty} A_k$ and $\lim \mu(B_k) = \lim \mu(A_k)$, it follows that $\mu(\bigcap_{k=1}^{\infty} A_k) = \lim \mu(A_k)$, whence the proof is complete. ∎

Observe that (M9) has an extra condition when compared to (M8). This extra condition is not superfluous; the result may not hold in cases when $\mu(A_n) = +\infty$ for each $n \in \mathbb{N}$.

The next property is a useful one for establishing certain convergence results. It combines the concepts of inferior and superior limits for sets with inferior and superior limits for sequences of numbers.

(M10) *We have that*

$$\mu(\underline{\lim}\, A_n) \leq \underline{\lim}\, \mu(A_n) \leq \overline{\lim}\, \mu(A_n).$$

If $\mu(\bigcup_{k=N}^{\infty} A_k) < +\infty$ *for some* N, *then* $\overline{\lim}\, \mu(A_n) \leq \mu(\overline{\lim}\, A_n)$ *as well.*

Proof: Define $B_k = \bigcap_{n=k}^{\infty} A_n$ for all $k \in \mathbb{N}$, so that $\{B_k\}_{k=1}^{\infty}$ is nondecreasing with $B_k \subseteq A_k$ for all $k \in \mathbb{N}$. Thus $\underline{\lim}\, A_n = \bigcup_{k=1}^{\infty} \bigcap_{n=k}^{\infty} A_n = \bigcup_{k=1}^{\infty} B_k$, and hence (M8) yields

$$\mu(\underline{\lim}\, A_n) = \mu\left(\bigcup_{k=1}^{\infty} B_k\right) = \lim \mu(B_n) = \underline{\lim}\, \mu(B_n) \leq \underline{\lim}\, \mu(A_n),$$

where the inequality follows from observing that $\mu(B_n) \leq \mu(A_n)$ for all $n \in \mathbb{N}$ by (M5). Next, the inequality $\underline{\lim}\, \mu(A_n) \leq \overline{\lim}\, \mu(A_n)$ is automatic. This shows that $\mu(\underline{\lim}\, A_n) \leq \underline{\lim}\, \mu(A_n) \leq \overline{\lim}\, \mu(A_n)$.

To complete affairs, assume that $\mu(\bigcup_{k=N}^{\infty} A_k) < +\infty$ for some $N \in \mathbb{N}$. Let $C_n = \bigcup_{k=n}^{\infty} A_k$ for all $n \in \mathbb{N}$, so that for some $N \in \mathbb{N}$ we have $\mu(C_N) < +\infty$. Then $\{C_n\}_{n=1}^{\infty}$ is a nonincreasing sequence with $C_n \supseteq A_n$ for all $n \in \mathbb{N}$, hence $\mu(C_n) \geq \mu(A_n)$ for all $n \in \mathbb{N}$ by (M5). Observe next that $\lim \mu(C_n)$ exists since $\{\mu(C_n)\}_{n=1}^{\infty}$ is monotone. Furthermore,

$$\lim C_n = \bigcap_{n=1}^{\infty} C_n = \bigcap_{n=1}^{\infty} \bigcup_{k=n}^{\infty} A_k = \overline{\lim}\, A_n.$$

An application of (M9) yields

$$\overline{\lim}\, \mu(A_n) \leq \overline{\lim}\, \mu(C_n) = \lim \mu(C_n) = \mu(\lim C_n) = \mu(\overline{\lim}\, A_n),$$

which completes the proof. ■

The following property is the culmination of our continuity results. Unlike (M8) and (M9), we do not require that the sequence $\{A_n\}_{n=1}^{\infty}$ be monotone.

(M11) *If $\mu(A \triangle \overline{\lim}\, A_n) = 0$, $\mu(A \triangle \underline{\lim}\, A_n) = 0$, and $\mu(\bigcup_{j=N}^{\infty}(A \triangle A_j)) < +\infty$ for some $N \in \mathbb{N}$, then $\lim_n \mu(A \triangle A_n)$ exists and equals zero. If in addition $\mu(A) < +\infty$, and if there is $N \in \mathbb{N}$ such that $\mu(A_n) < +\infty$ for all $n \geq N$, then $\lim \mu(A_n)$ exists and equals $\mu(A)$.*

Proof: We apply (M10) to the sequence $\{A \triangle A_j\}_{j=1}^{\infty}$ and obtain

$$\begin{aligned}
\overline{\lim}\, \mu(A \triangle A_n) &\leq \mu\left(\overline{\lim}(A \triangle A_n)\right) \\
&= \mu\left((A - \underline{\lim}\, A_n) \cup \overline{\lim}(A_n - A)\right) \qquad \text{(Exercise 9(e))} \\
&\leq \mu\left(A - \underline{\lim}\, A_n\right) + \mu\left(\overline{\lim}(A_n - A)\right) \qquad \text{(by (M7))} \\
&= \mu\left(\overline{\lim}(A_n - A)\right) \qquad \text{(assumption and (M5))} \\
&= \mu(\overline{\lim}\, A_n - A) \qquad \text{(Exercise 9(d))} \\
&= 0 \qquad \text{(assumption and (M5))},
\end{aligned}$$

hence $\overline{\lim}\, \mu(A \triangle A_n)$ exists and equals zero. By (M1), $\underline{\lim}\, \mu(A \triangle A_n) = 0$, hence $\lim \mu(A \triangle A_n)$ exists and equals zero. This proves the first claim.

For the proof of the second claim, assume that $\mu(A) < +\infty$ and that there is $N_1 \in \mathbb{N}$ such that $\mu(A_n) < +\infty$ for all $n \geq N_1$. Given $\epsilon > 0$, use the first claim to find N_2 so that $n \geq N_2$ implies $\mu(A \triangle A_n) < \epsilon$. That is, if $n \geq N_2$, then we have $\mu((A \cup A_n) - (A \cap A_n)) < \epsilon$. Let $N = \max\{N_1, N_2\}$, and thus for any $n \geq N$ we have that both $\mu(A \cup A_n)$ and $\mu(A \cap A_n)$ are finite. By (M5), $n \geq N$ implies that $\mu(A \cup A_n) - \mu(A \cap A_n) < \epsilon$, which gives $\mu(A \cup A_n) < \mu(A \cap A_n) + \epsilon$. Since by (M5) we have $\mu(A_n) \leq \mu(A \cup A_n)$ and $\mu(A \cap A_n) \leq \mu(A)$, we have that $n \geq N$ implies $\mu(A_n) \leq \mu(A) + \epsilon$. Similarly, we have $\mu(A) \leq \mu(A \cup A_n)$ and $\mu(A \cap A_n) \leq \mu(A_n)$, so $n \geq N$ implies that $\mu(A) \leq \mu(A_n) + \epsilon$ for $n \geq N$. Since $\mu(A) < +\infty$ and $\mu(A_n) < +\infty$ for $n \geq N$, we have that $|\mu(A_n) - \mu(A)| \leq \epsilon$ for all $n \geq N$. Since $\epsilon > 0$ was arbitrary, we conclude that $\lim \mu(A_n)$ exists and equals $\mu(A)$. ■

(M12) *Let μ denote a finite measure. If $A_1, A_2, \cdots$ are such that* $\lim A_n$ *exists, then* $\mu(\lim A_n) = \lim \mu(A_n)$.

Proof: This immediately follows by taking A to be $\lim A_n$ in (M11). ■

(M13) *If $A_n \to A$ and $\mu(A) = +\infty$, then* $\lim \mu(A_n)$ *exists and equals* $+\infty$.

Proof: By (M10) we have $\mu(A) = \mu(\underline{\lim}\, A_n) \leq \underline{\lim}\, \mu(A_n) \leq \overline{\lim}\, \mu(A_n)$. Since $\mu(A) = +\infty$, we have that $\underline{\lim}\, \mu(A_n) = +\infty$, hence $\overline{\lim}\, \mu(A_n) = +\infty$, whence $\lim \mu(A_n)$ exists and equals $+\infty$. ■

In summary, if μ is a measure on a σ-field $\mathcal{A}$ and if $A_1, A_2, \cdots \in \mathcal{A}$ are such that $\lim A_n$ exists, we will always have $\lim \mu(A_n) = \mu(\lim A_n)$ when $\mu(\lim A_n) = +\infty$ by virtue of (M13). However, if $\mu(\lim A_n) < +\infty$, care must be taken before asserting that $\lim \mu(A_n) = \mu(\lim A_n)$, for (M11) asserts that one must verify that (i) $\mu(\bigcup_{j=N}^{\infty}(\lim A_n \,\triangle\, A_j)) < +\infty$ for some $N \in \mathbb{N}$, and that (ii) there exists $M \in \mathbb{N}$ such that $\mu(A_n) < +\infty$ for each $n \geq M$. Of course, if μ is a finite measure, all of these conditions are met and the limit interchange is justifed.

Example 1. Let $\Omega = \mathbb{N}$, let $\mathcal{A} = 2^{\Omega}$, and let $\mu(A)$ denote the number of points in A when A is finite, with $\mu(A) = +\infty$ otherwise. Then $(\Omega, \mathcal{A}, \mu)$ is a measure space. If we let $A_n = \{n, n+1, \cdots\}$ for each $n \in \mathbb{N}$, then $\lim A_n$ exists and is empty, hence $\mu(\lim A_n) = 0$. However, $\mu(A_n) = +\infty$ for each $n \in \mathbb{N}$, so that $\lim \mu(A_n) = +\infty \neq 0 = \mu(\lim A_n)$.

Example 2. Let Ω, $\mathcal{A}$, and μ be the same as in Example 1. Let $A_n = \{n\}$ for each $n \in \mathbb{N}$. The collection $\{A_1, A_2, \cdots\}$ is disjoint, hence $\lim A_n$ exists and is empty, whence $\mu(\lim A_n) = 0$. However, $\mu(A_n) = 1$ for all $n \in \mathbb{N}$, so that $\lim \mu(A_n) \neq \mu(\lim A_n)$. Here, $\mu(\bigcup_{j=N}^{\infty}(\lim A_n \,\triangle\, A_j)) = \mu(\{N, N+1, \cdots\}) = +\infty$ for all $N \in \mathbb{N}$. This shows that even if $A_n \to A$, $\mu(A) < +\infty$, and $\mu(A_n) < +\infty$ for all n, we may still have $\lim \mu(A_n) \neq \mu(\lim A_n)$. In other words, even with all of these assumptions, the condition $\mu(\bigcup_{j=N}^{\infty}(\lim A_n \,\triangle\, A_j)) < +\infty$ for some $N \in \mathbb{N}$ is not a redundant one.

Exercises.

10*. Let $\Omega = \mathbb{N}$, and let $\mathcal{A} = 2^{\Omega}$. Define μ on $\mathcal{A}$ by $\mu(A) =$ number of points in A if A is finite; define $\mu(A) = +\infty$ if A is infinite.
(a) μ is a measure on $\mathcal{A}$. This measure is called the *counting measure.*
(b) There exists a nonincreasing sequence $\{A_n\}_{n=1}^{\infty}$ of $\mathcal{A}$-sets with $\mu(A_n) = +\infty$ for all $n \in \mathbb{N}$ but $\mu(\lim_n A_n) = 0$, thus (M9) accordingly fails to hold, hence the assumption that some A_k must have finite measure cannot be dropped.

11*. Let $(\Omega, \mathcal{A}, \mu)$ denote a measure space. Let $\{A_x : x \in \mathbb{R},\ x > 0\}$ denote a collection of $\mathcal{A}$-sets.

(a) Suppose that $0 < x < y$ implies $A_x \subseteq A_y$. Then (i) $\bigcup_{x>0} A_x \in \mathcal{A}$, (ii) $x_1 < x_2$ implies $\mu(A_{x_1}) \leq \mu(A_{x_2})$, and (iii) $\mu(A_x) \to \mu(\bigcup_{y>0} A_y)$ as $x \to +\infty$.
(b) Suppose that $0 < x < y$ implies $A_x \supseteq A_y$. Also, further assume that $\mu(A_z) < \infty$ for some $z > 0$. Then (i) $\bigcap_{x>0} A_x \in \mathcal{A}$, (ii) $x_1 < x_2$ implies $\mu(A_{x_1}) \geq \mu(A_{x_2})$, and (iii) $\mu(A_x) \to \mu\left(\bigcap_{y>0} A_y\right)$ as $x \to +\infty$. The assumption $\mu(A_z) < +\infty$ for some $z > 0$ cannot be dropped. This and (a) generalize (M8) and (M9) from monotone sequences of sets to monotone [uncountable] collections of sets.

12*. Let $(\Omega, \mathcal{A}, \mu)$ denote a measure space. This important exercise gives a few easy-to-verify claims.
(a) μ is σ-finite iff there is a nondecreasing sequence $A_1 \subseteq A_2 \subseteq \cdots$ of $\mathcal{A}$-sets with $\mu(A_n) < +\infty$ for all $n \in \mathbb{N}$ and $\bigcup_{n=1}^{\infty} A_n = \Omega$.
(b) μ is σ-finite iff there is a disjoint sequence $A_1, A_2, \cdots$ of $\mathcal{A}$-sets with $\mu(A_n) < +\infty$ for all $n \in \mathbb{N}$ and $\bigcup_{n=1}^{\infty} A_n = \Omega$.
(c) Let $\mu_1, \cdots, \mu_n$ denote σ-finite measures on $\mathcal{A}$. Then there exists a sequence $\{A_m\}_{m=1}^{\infty}$ of $\mathcal{A}$-sets such that (i) $\mu_i(A_j) < +\infty$ for all $i = 1, \cdots, m$ and $j \in \mathbb{N}$ and (ii) $\Omega = \bigcup_{m=1}^{\infty} A_m$. These sets may be chosen to be nondecreasing or disjoint. [This fact will be called on in later chapters.]
(d) Does (c) hold if we have countably many σ-finite measures on $\mathcal{A}$ as compared to finitely many σ-finite measures on $\mathcal{A}$?

13. Let μ denote a measure on a σ-field $\mathcal{A}$, and let $A_1, A_2, \cdots \in \mathcal{A}$ be such that $\mu(\bigcup_{j=N}^{\infty} \underline{\lim}_k (A_j \cap A_k^c)) < +\infty$ for some $N \in \mathbb{N}$. Use (M10) and parts (a)-(c) of Exercise 9 to show the following claims:
(a) $\lim_n \mu(\underline{\lim}_k (A_n \cap A_k^c))$ exists and equals zero.
(b) $\lim_n \mu(A_n - \overline{\lim}_k A_k)$ exists and equals zero.
(c) $\lim_n \mu((\underline{\lim}_k A_k) - A_n)$ exists and equals zero.

14*. Let $(\Omega, \mathcal{A}, \mu)$ denote a σ-finite measure space, where μ is not the measure that is identically zero. We will show that any collection of disjoint $\mathcal{A}$-sets having positive measure is amc.
(a) For any $A \in \mathcal{A}$ with $\mu(A) < +\infty$ and for any $\epsilon > 0$, it is the case that $\{\eta \in H : \mu(A \cap C_\eta) \geq \epsilon\}$ is finite. [Suppose that for some $A \in \mathcal{A}$ with $\mu(A) < +\infty$ there is $\epsilon > 0$ such that $\mu(A \cap C_{\eta_k}) \geq \epsilon$ for some sequence $\{\eta_k\}_{k=1}^{\infty}$ in H. From $A \supseteq \bigcup_{k=1}^{\infty}(A \cap C_{\eta_k})$, obtain $\mu(A) \geq \sum_{k=1}^{\infty} \mu(A \cap C_{\eta_k}) = +\infty$, a contradiction.]
(b) Prove that $\{\eta \in H : \mu(A \cap C_\eta) > 0\}$ is amc for any $A \in \mathcal{A}$ with $\mu(A) < +\infty$. [Argue that $\{\eta \in H : \mu(A \cap C_\eta) > 0\} = \bigcup_{\epsilon>0, \epsilon\in\mathbb{Q}}\{\eta \in H : \mu(A \cap C_\eta) \geq \epsilon\}$.]
(c) Write Ω as an amc union $\bigcup_k A_k$ where $\mu(A_k) < +\infty$ for all $k \in \mathbb{N}$. Define $H_k = \{\eta \in H : \mu(A_k \cap C_\eta) > 0\}$ for all $k \in \mathbb{N}$. Then H_k is amc for all $k \in \mathbb{N}$.
(d) $\bigcup_k H_k = H$, hence H is amc. [Each $H_k \subseteq H$, hence $\bigcup_k H_k \subseteq H$. To obtain $H \subseteq \bigcup_k H_k$, suppose that there is $\eta \in H$ with $\mu(A_k \cap C_\eta) = 0$ for all k. Since $C_\eta = \bigcup_k (A_k \cap C_\eta)$, we have $\mu(C_\eta) \leq \sum_k \mu(A_k \cap C_\eta) = 0$, a contradiction.]

15. Let $(\Omega, \mathcal{A}, \mu)$ be a measure space. Suppose that whenever $A \in \mathcal{A}$ is such that $\mu(A) = +\infty$, there exists $B \subsetneq A$ with $0 < \mu(B) < +\infty$. Furthermore, suppose that there does not exist an uncountable disjoint collection of $\mathcal{A}$-sets, each of positive measure. Then μ is σ-finite. [One approach is to use Zorn's Lemma.]

16*. Let $\mathcal{A}$ be a field on Ω, and suppose that $\mu : \mathcal{A} \to \bar{\mathbb{R}}$ satisfies (M1) with $\mu(\Omega) < +\infty$, (M2), (M4) [and hence (M5)], and in addition is continuous from above at $\varnothing$: if $A_1 \supseteq A_2 \supseteq \cdots$ is a sequence of $\mathcal{A}$-sets with $\bigcap_{k=1}^{\infty} A_k = \varnothing$, then

$\mu(A_k) \to 0$. Then μ is a measure. [Let $\{B_n\}_{n=1}^{\infty}$ denote a disjoint sequence of $\mathcal{A}$-sets with $\bigcup_{n=1}^{\infty} B_n \in \mathcal{A}$. For each $n \geq 2$ define $C_n = \bigcup_{k=n}^{\infty} B_k = \bigcup_{k=1}^{\infty} B_k - \bigcup_{k=1}^{n-1} B_k$. Then $C_1 \supseteq C_2 \supseteq \cdots$ and $\lim C_n = \varnothing$. Thus $\lim \mu(C_n) = 0$, and application of (M4) and (M5) yields $\mu(\bigcup_{k=1}^{\infty} B_k) - \lim_n \sum_{k=1}^{n-1} \mu(B_k) = 0$, which gives (M3).]

17. Let $\Omega = (0,1]$, and let $\mathcal{F}$ consist of $\varnothing$ and the finite disjoint unions of rsc subintervals of $(0,1]$. Then $\mathcal{F}$ is a field. Define μ on $\mathcal{F}$ as follows: $\mu(A) = 1$ if there exists $\epsilon_A > 0$ with $(\frac{1}{2}, \frac{1}{2} + \epsilon] \subseteq A$ and $\mu(A) = 0$ otherwise. Then μ is well-defined and satisfies (M1), (M2), and (M4), but μ is not countably additive.

18. Let $(\Omega, \mathcal{A}, \mu)$ be a measure space. Suppose that μ is *nonatomic*: $A \in \mathcal{A}$ and $\mu(A) > 0$ imply that there exists $B \subseteq A$ with $B \in \mathcal{A}$ with $0 < \mu(B) < \mu(A)$.
(a) The measure μ of Example 2 in Section 1 is atomic.
(b) Suppose $A \in \mathcal{A}$ is such that $\mu(A) > 0$, and let $\epsilon > 0$ be given. Then there exists $B \in \mathcal{A}$ with $B \subseteq A$ and $0 < \mu(B) < \epsilon$. [Hint: Find $B \in \mathcal{A}$ with $B \subseteq A$ and $0 < \mu(B) < \mu(A)$. Show that this inequality implies that (i) $0 < \mu(B) = \mu(B \cap A) \leq \frac{1}{2}\mu(A)$ or (ii) $0 < \mu(B \cap A^c) \leq \frac{1}{2}\mu(A)$. If (i), find $C \in \mathcal{A}$ with $C \subseteq B$ and $0 < \mu(C) < \mu(B)$; argue that $0 < \mu(C) = \mu(C \cap B) < \frac{1}{2}\mu(B)$ or $0 < \mu(C \cap B^c) < \frac{1}{2}\mu(B)$. If (ii), Find $C \in \mathcal{A}$ with $C \subseteq B \cap A^c$ and $0 < \mu(C) < \mu(B \cap A^c)$; argue that $0 < \mu(C \cap B \cap A^c) = \mu(C) < \frac{1}{2}\mu(B \cap A^c)$ or $0 < \mu(C \cap (B \cap A^c)^c) < \frac{1}{2}\mu(B \cap A^c)$. Argue that this process may be iterated as many times as needed until the desired set is obtained.]
(c) Let $A \in \mathcal{A}$ be such that $\mu(A) > 0$. Given any $0 \leq \alpha \leq \mu(A)$ there exists a set $B \in \mathcal{A}$ with $B \subseteq A$ and $\mu(B) = \alpha$. [Let $\mathfrak{C}$ denote the family of collections $\mathcal{D}$ of disjoint $\mathcal{A}$-sets contained in A such that the sum of the measures of the $\mathcal{A}$-sets in $\mathcal{D}$ is $\leq \alpha$. For $\mathcal{D}, \mathcal{E} \in \mathfrak{C}$, write $\mathcal{D} \preceq \mathcal{E}$ iff (i) the sum of the measures of the sets in $\mathcal{D}$ is no more than the sum of the measures of the sets in $\mathcal{E}$ and (ii) $\mathcal{D} \subseteq \mathcal{E}$. Show that $\preceq$ is a partial ordering on $\mathfrak{C}$. Next, show that every totally ordered subset of $\mathfrak{C}$ has an upper bound with respect to $\preceq$. Invoke Zorn's Lemma to obtain a maximal element $\mathcal{F} \in \mathfrak{C}$, then show that the disjoint union of the sets in $\mathcal{F}$ has measure equal to α.]

19*. Let $(\Omega, \mathcal{A}, \mu)$ be a measure space. Let $B \in \mathcal{A}$ and $\mathcal{A}_B = \{A \in \mathcal{A} : A \subseteq B\}$. Then $\mathcal{A}_B$ is a σ-field on B, and the restriction of μ to $\mathcal{A}_B$ is a measure on $\mathcal{A}_B$. [This exercise, while very straightforward, is also important for later work.]

2.3 A CLASS OF MEASURES

Some examples of measures have been given in the previous two sections, but nothing has been presented that mimics the examples of "length", "area", and "volume" as given in the introduction to this chapter. This deficiency shall be partially rectified, although mathematical problems will present themselves; these problems will be resolved in Chapters 3 and 4. For this section, our background space Ω will be $\mathbb{R}^k$, where $k \in \mathbb{N}$ is arbitrary. We will also use $\mathcal{A}_k$ to denote $\{\varnothing\} \cup \{(\boldsymbol{a}, \boldsymbol{b}] : -\infty < \boldsymbol{a} < \boldsymbol{b} < +\infty\}$; $\mathcal{A}_k$ is a semiring on $\mathbb{R}^k$ by Example 6 of Section 1.1.

Definition. A function $F : \mathbb{R}^k \to \mathbb{R}$ is called *nondecreasing* iff $\boldsymbol{x} \leq \boldsymbol{y}$ implies $F(\boldsymbol{x}) \leq F(\boldsymbol{y})$. We will also say that F is *continuous from above [cfa] at* $\boldsymbol{x}$ iff $F(\boldsymbol{x}_n) \to F(\boldsymbol{x})$ whenever $\{\boldsymbol{x}_n\}_{n=1}^{\infty}$ is a nondecreasing sequence with $\boldsymbol{x}_n \geq \boldsymbol{x}$

for all $n \in \mathbb{N}$ and $\boldsymbol{x}_n \to \boldsymbol{x}$. Finally, we will say that F is *continuous from above [cfa]* iff F is cfa at $\boldsymbol{x}$ for every $\boldsymbol{x} \in \mathbb{R}^k$.

Definition. Given $A = (\boldsymbol{a}, \boldsymbol{b}] \in \mathcal{A}_k$, the set $\{a_1, b_1\} \times \cdots \times \{a_k, b_k\}$ is called *the vertex set of* A and is denoted by $V(A)$. We also define $V(\varnothing) = \varnothing$. The set $V(A)$ consists of 2^k elements.

Definition. Let $A = (\boldsymbol{a}, \boldsymbol{b}] \in \mathcal{A}_k$. The *signum function of* A is the function $s_A : V(A) \to \{-1, 1\}$ such that

$$s_A(\boldsymbol{x}) = \begin{cases} +1 & \text{the number of } i\text{'s such that } x_i = a_i \text{ is even} \\ -1 & \text{the number of } i\text{'s such that } x_i = a_i \text{ is odd.} \end{cases}$$

The following properties, valid for all $A = (\boldsymbol{a}, \boldsymbol{b}] \in \mathcal{A}_k$, are directly shown:
(i) $s_A(\boldsymbol{x}) = 1$ for 2^{k-1} elements $\boldsymbol{x} \in V(A)$ and $s_A(\boldsymbol{x}) = -1$ for 2^{k-1} elements $\boldsymbol{x} \in V(A)$.
(ii) If $\boldsymbol{x}, \boldsymbol{y} \in V(A)$ and if $\boldsymbol{y}$ differs from $\boldsymbol{x}$ in exactly m coordinates, then $s_A(\boldsymbol{x}) = (-1)^m s_A(\boldsymbol{y})$.

Definition. For any $F : \mathbb{R}^k \to \mathbb{R}$, we define the function $\Delta_F : \mathcal{A}_k \to \mathbb{R}$ by

$$\Delta_F(A) = \begin{cases} \sum_{\boldsymbol{x} \in V(A)} s_A(\boldsymbol{x}) F(\boldsymbol{x}) & A = (\boldsymbol{a}, \boldsymbol{b}] \\ 0 & A = \varnothing. \end{cases}$$

Example 1. Let $k = 1$ and let $A = (a, b] \in \mathcal{A}_1$. Here, $V(A) = \{a, b\}$, $s_A(a) = -1$, and $s_A(b) = 1$. It follows that

$$\Delta_F(A) = \sum_{x \in \{a,b\}} s_A(x) F(x) = s(a)F(a) + s(b)F(b) = F(b) - F(a).$$

In the case $k = 2$, where $A = (\boldsymbol{a}, \boldsymbol{b}] = (a_1, b_1] \times (a_2, b_2] \in \mathcal{A}_2$, we have

$$V(A) = \{(a_1, a_2), (a_1, b_2), (b_1, a_2), (b_1, b_2)\},$$

$s_A(a_1, a_2) = s_A(b_1, b_2) = +1$, and $s_A(a_1, b_2) = s_A(b_1, a_2) = -1$. We thus have

$$\begin{aligned} \Delta_F(A) &= s_A(b_1, b_2)F(b_1, b_2) + s_A(a_1, b_2)F(a_1, b_2) + s_A(b_1, a_2)F(b_1, a_2) \\ &\quad + s_A(a_1, a_2)F(a_1, a_2) \\ &= F(b_1, b_2) - F(a_1, b_2) - F(b_1, a_2) + F(a_1, a_2). \end{aligned}$$

One can write out $\Delta_F((\boldsymbol{a}, \boldsymbol{b}])$ for each $k \geq 3$ in similar but tedious fashion.

Definition. A function $F : \mathbb{R}^k \to \mathbb{R}$ that is cfa with $\Delta_F(A) \geq 0$ for each $A \in \mathcal{A}_k$ will be called a *[k-dimensional] Stieltjes function.* We will let $\mathfrak{S}_k$ denote the collection of k-dimensional Stieltjes functions. When $k = 1$, we will write $\mathfrak{S}$ for $\mathfrak{S}_1$.

Note that when $k = 1$, $F \in \mathfrak{S}$ iff F is right continuous and nondecreasing. However, when $k \geq 2$, it is not necessarily the case that if F is cfa and nondecreasing, then $F \in \mathfrak{S}_k$. For a nontrivial counterexample, let $F : \mathbb{R}^2 \to \mathbb{R}$

be given by $F(\boldsymbol{x}) = 1$ when $x_1 + x_2 \geq 0$, and let $F(\boldsymbol{x}) = 0$ otherwise. This F is both continuous from above and nondecreasing. Yet, if $A = (-1, 0] \times (0, 1]$, then a routine computation gives $\Delta_F(A) = -1 < 0$, hence $F \notin \mathfrak{S}_2$.

The following theorem has an arduous proof that is given in the next section; we shall assume the theorem as true for the remainder of this section.

Stieltjes Theorem. *For any $k \in \mathbb{N}$ and $F \in \mathfrak{S}_k$, $\Delta_F : \mathcal{A}_k \to \mathbb{R}$ is a measure that assigns finite measure to each $\mathcal{A}_k$-set [and hence Δ_F is σ-finite on $\mathcal{A}_k$].*

Example 2. Let $k = 1$, and let $F : \mathbb{R} \to \mathbb{R}$ denote the identity function: $F(x) = x$ for all $x \in \mathbb{R}$. Observe that F is continuous and increasing, so $F \in \mathfrak{S}$, whence Δ_F is a measure on $\mathcal{A}_1$ by the theorem. Since $\Delta_F((a, b]) = F(b) - F(a) = b - a$ for each $(a, b] \in \mathcal{A}_1$, the measure Δ_F corresponds to our intuitive notion of "length," at least for rsc intervals. However, Δ_F can be applied only to the empty set and to bounded intervals of the form $(a, b]$; $\Delta_F((a, b))$, $\Delta_F([a, b))$, and $\Delta_F([a, b])$ are all undefined, even though one might think that the "length" of these intervals is also $b - a$. Similarly, $\Delta_F(\{x\})$ is undefined, although one will naturally think that the "length of a single point" is zero. Thus, $\Delta_F(A)$ corresponds to our intuitive notion of "the length of A" only in the restrictive case where $A = \varnothing$ or $A = (a, b]$.

Example 3. We revisit Example 2, but let $k \in \mathbb{N}$ be arbitrary. Define $F_k : \mathbb{R}^k \to \mathbb{R}$ by writing $F_k(\boldsymbol{x}) = \prod_{i=1}^k x_i$ for all $\boldsymbol{x} \in \mathbb{R}^k$; it is automatic that F_k is cfa for each $k \in \mathbb{N}$. We now proceed to show that $\Delta_{F_k}(A) \geq 0$ for all $A \in \mathcal{A}_k$. The case $k = 1$ with $F_1(x) = x$ for all $x \in \mathbb{R}$ is given in Example 2. Next, suppose that $\Delta_{F_k}(A) \geq 0$ for all $A \in \mathcal{A}_k$ when $F_k(\boldsymbol{x}) = x_1 \cdots x_k$ for all $\boldsymbol{x} \in \mathbb{R}^k$. We will show that this forces $\Delta_{F_{k+1}}(A) \geq 0$ for all $A \in \mathcal{A}_{k+1}$. Pick and fix arbitrary $A = (\boldsymbol{a}, \boldsymbol{b}] \in \mathcal{A}_{k+1}$. Letting $\boldsymbol{x} \in \mathbb{R}^{k+1}$ be arbitrary, write $A^* = (a_1, b_1] \times \cdots \times (a_k, b_k]$, $\boldsymbol{x}^* = (x_1, \cdots, x_k)$, $V_1(A) = \{(\boldsymbol{x}^*, a_{k+1}) : \boldsymbol{x}^* \in V(A^*)\}$, and $V_2(A) = \{(\boldsymbol{x}^*, b_{k+1}) : \boldsymbol{x}^* \in V(A^*)\}$. Observing that $V(A)$ is the disjoint union of $V_1(A)$ and $V_2(A)$, we have

$$
\begin{aligned}
& \Delta_{F_{k+1}}(A) \\
= \; & \sum_{\boldsymbol{x} \in V(A)} s_A(\boldsymbol{x}) F_{k+1}(\boldsymbol{x}) \\
= \; & \sum_{\boldsymbol{x} \in V_1(A)} s_A(\boldsymbol{x}) F_{k+1}(\boldsymbol{x}) + \sum_{\boldsymbol{x} \in V_2(A)} s_A(\boldsymbol{x}) F_{k+1}(\boldsymbol{x}) \\
= \; & \sum_{\boldsymbol{x}^* \in V(A^*)} s_A(\boldsymbol{x}^*, a_{k+1}) F_{k+1}(\boldsymbol{x}^*, a_{k+1}) + \sum_{\boldsymbol{x}^* \in V(A^*)} s_A(\boldsymbol{x}^*, b_{k+1}) F_{k+1}(\boldsymbol{x}^*, b_{k+1}) \\
\stackrel{1}{=} \; & \sum_{\boldsymbol{x}^* \in V(A^*)} (-1) s_{A^*}(\boldsymbol{x}^*) F_{k+1}(\boldsymbol{x}^*, a_{k+1}) + \sum_{\boldsymbol{x}^* \in V(A^*)} s_{A^*}(\boldsymbol{x}^*) F_{k+1}(\boldsymbol{x}^*, b_{k+1}) \\
\stackrel{2}{=} \; & \sum_{\boldsymbol{x}^* \in V(A^*)} s_{A^*}(\boldsymbol{x}^*)(b_{k+1} - a_{k+1}) F_k(\boldsymbol{x}^*) \\
\stackrel{3}{=} \; & (b_{k+1} - a_{k+1}) \sum_{\boldsymbol{x}^* \in V(A^*)} s_{A^*}(\boldsymbol{x}^*) F_k(\boldsymbol{x}^*) \\
\geq \; & 0
\end{aligned}
$$

where equality 1 follows by property (ii) given for the signum function, equality 2 follows by the definitions of F_k and F_{k+1} [this is the only place in the above where we use the specific forms of F_{k+1} and F_k], and the inequality follows by the induction hypothesis. This shows that $\Delta_{F_{k+1}}(A) \geq 0$ for the generic $A \in \mathcal{A}_{k+1}$. By induction we have that $F_k \in \mathfrak{S}_k$ for each $k \in \mathbb{N}$. It follows from the theorem that Δ_{F_k} is a measure on $\mathcal{A}_k$ for all $k \in \mathbb{N}$.

We now claim for all $k \in \mathbb{N}$ that $\Delta_{F_k}((\boldsymbol{a}, \boldsymbol{b}]) = \prod_{i=1}^{k}(b_i - a_i)$ for every element $(\boldsymbol{a}, \boldsymbol{b}] \in \mathcal{A}_k$. If $k = 1$, then $\Delta_{F_1}((a, b]) = F_1(b) - F_1(a) = b - a$. We now assume that the result is true for k dimensions, and we desire to show that the equality $\Delta_{F_{k+1}}((\boldsymbol{a}, \boldsymbol{b}]) = \prod_{i=1}^{k+1}(b_i - a_i)$ holds for all $(\boldsymbol{a}, \boldsymbol{b}] \in \mathcal{A}_{k+1}$. Utilizing equality 3 and the induction hypothesis, we have for all $(\boldsymbol{a}, \boldsymbol{b}] \in \mathcal{A}_{k+1}$ that

$$\begin{aligned}\Delta_{F_{k+1}}((\boldsymbol{a}, \boldsymbol{b}]) \quad = \quad (b_{k+1} - a_{k+1})\Delta_{F_k}((\boldsymbol{a}, \boldsymbol{b}]^*) \quad &= \quad (b_{k+1} - a_{k+1})\prod_{i=1}^{k}(b_i - a_i) \\ &= \quad \prod_{i=1}^{k+1}(b_i - a_i).\end{aligned}$$

It follows by induction that $\Delta_{F_k}((\boldsymbol{a}, \boldsymbol{b}]) = \prod_{i=1}^{k}(b_i - a_i)$ for all $k \in \mathbb{N}$ and $(\boldsymbol{a}, \boldsymbol{b}] \in \mathcal{A}_k$, which is what we wanted to show.

If $k = 1$, $\Delta_{F_1}((a, b]) = b - a$, which corresponds to the "length of an interval," which is what Example 2 covered. If $k = 2$, the measure Δ_{F_2} corresponds to the "area of the rectangle" $(a_1, b_1] \times (a_2, b_2]$:

$$\Delta_{F_2}((a_1, b_1] \times (a_2, b_2]) = (b_1 - a_1)(b_2 - a_2).$$

If $k = 3$, the measure corresponds to the "volume of a box":

$$\Delta_{F_3}((a_1, b_1] \times (a_2, b_2] \times (a_3, b_3]) = (b_1 - a_1)(b_2 - a_2)(b_3 - a_3).$$

For general $k \in \mathbb{N}$, $\Delta_F((a_1, b_1] \times \cdots \times (a_k, b_k])$ corresponds to the "k-dimensional volume" $\prod_{i=1}^{k}(b_i - a_i)$ of the k-dimensional rsc rectangle $(\boldsymbol{a}, \boldsymbol{b}]$.

Just as a point was made in Example 2 to show how limited the connection was between Δ_{F_1} and our intuitive notion of "length", so too the connection between Δ_{F_k} and our intuitive notion of "k-dimensional volume" is weak. For example, taking $k = 2$, the sets $(0, 1] \times (0, 1]$, $[0, 1] \times [0, 1]$, $[0, 1) \times (0, 1)$, etc., should all have "area" equal to one. However, "area" in the sense of the measure Δ_{F_2} applies only to $(0, 1] \times (0, 1]$; Δ_{F_2} cannot be applied to $[0, 1] \times [0, 1]$ nor $[0, 1) \times (0, 1)$, say. This deficiency will be overcome in Chapters 3 and 4.

Example 4. Let $k = 1$, and let $f : \mathbb{R} \to [0, +\infty)$ denote a continuous function with $\int_{-\infty}^{+\infty} f(t)\, dt < +\infty$, and define $F : \mathbb{R} \to \mathbb{R}$ by writing $F(x) = \int_{-\infty}^{x} f(t)\, dt$ for all $x \in \mathbb{R}$, where the integral is the usual calculus-based improper Riemann integral. Invoking the First Fundamental Theorem of the Calculus, we have that F is continuous. Also, F is nondecreasing. Therefore $F \in \mathfrak{S}$, and Δ_F is a measure on $\mathcal{A}_1$ by the theorem. Finally, observe for each $A = (a, b] \in \mathcal{A}_1$ that

$$\Delta_F(A) = F(b) - F(a) = \int_{-\infty}^{b} f(t)\, dt - \int_{-\infty}^{a} f(t)\, dt = \int_{a}^{b} f(t)\, dt.$$

This is the justification for the earlier claim in the introduction to this chapter that when one integrates a nonnegative function, one is working with a measure.

It might be noticed that all the examples in this section have dealt with measures whose domain is the semiring $\mathcal{A}_k$. No measure over a σ-field of subsets of $\mathbb{R}^k$ has been exhibited. In particular, the reader will notice that no examples of measures on the σ-field $\mathcal{B}^k$ have been given.

Ultimately, we desire to show that there exists a measure λ_k on $\mathcal{B}^k$ such that [among many other things]

$$\lambda_k((\boldsymbol{a},\boldsymbol{b}]) = \lambda_k((\boldsymbol{a},\boldsymbol{b})) = \lambda_k([\boldsymbol{a},\boldsymbol{b}]) = \lambda_k([\boldsymbol{a},\boldsymbol{b})) = \prod_{i=1}^{k}(b_i - a_i),$$

$$\lambda_k((-\infty,\boldsymbol{x}]) = \lambda_k((-\infty,\boldsymbol{x})) = \lambda_k((\boldsymbol{x},+\infty)) = \lambda_k([\boldsymbol{x},+\infty)) = +\infty,$$

$\lambda_k(\{\boldsymbol{x}\}) = 0$ for each $\boldsymbol{x} \in \mathbb{R}^k$, and $\lambda_k(\{\boldsymbol{x}_1,\cdots,\boldsymbol{x}_n\}) = 0$. The desire for *this* particular measure comes from the fact such a measure has been tacitly assumed to exist in treatments of advanced calculus, yet we have no a priori guarantee that such a measure exists. If such a measure fails to exist, we have serious foundational problems in analysis; if such a measure exists, we have as of yet to construct such an object and verify that it does not conflict with the logical foundations of mathematics. It will turn out that we *can* rigorously construct such a measure that mimics our idea of "length," "area," and so on, but it will take careful work to do so. Chapter 3 will do the requisite careful work in total generality, and Chapter 4 will specialize to the situation of this section.

Exercises.

1. Let $p_1,\cdots,p_n > 0$. Fix n real numbers $x_1 < x_2 < \cdots < x_n$, and define $F(x) = 0$ for $x < x_1$, $F(x) = p_1 + \cdots + p_j$ if there exists $1 \le j \le n$ such that $x_j \le x < x_{j+1}$, and $F(x) = 1$ if $x \ge x_n$. Δ_F is a measure on $\mathcal{A}_1$, and

$$\Delta_F((a,b]) = \begin{cases} \sum\{p_j : j \text{ is such that } a < x_j \le b\} & (a,b] \cap \{x_1,\cdots,x_n\} \neq \varnothing \\ 0 & \text{otherwise} \end{cases}.$$

This is an important framework in discrete probability theory.

2. This problem generalizes Example 2. Let $f : \mathbb{R}^k \to \mathbb{R}$ be such that f is continuous and nonnegative. Further suppose that $\int_{-\infty}^{+\infty} \cdots \int_{-\infty}^{+\infty} f(\boldsymbol{t})\, d\boldsymbol{t} < +\infty$. Define a function $F : \mathbb{R}^k \to \mathbb{R}$ by

$$F(\boldsymbol{x}) = \int_{-\infty}^{x_1} \cdots \int_{-\infty}^{x_k} f(\boldsymbol{t})\, d\boldsymbol{t}, \qquad \boldsymbol{x} \in \mathbb{R}^k.$$

[This is the k-dimensional Riemann integral from calculus.] Then Δ_F is a measure on the semiring $\mathcal{A}_k$, and for all $(\boldsymbol{a},\boldsymbol{b}] \in \mathcal{A}_k$ we have

$$\Delta_F((\boldsymbol{a},\boldsymbol{b}]) = \int_{a_1}^{b_1} \cdots \int_{a_k}^{b_k} f(t_1,\cdots,t_k)\, dt_k \cdots dt_1.$$

3*. Let $F_1,\cdots,F_k \in \mathfrak{S}$. For each $\boldsymbol{x} \in \mathbb{R}^k$, write $F(\boldsymbol{x}) = \prod_{i=1}^{k} F_i(x_i)$.
(a) $F \in \mathfrak{S}_k$, hence Δ_F is a measure on $\mathcal{A}_k$.

(b) $\Delta_F((\boldsymbol{a},\boldsymbol{b}]) = \prod_{i=1}^{k}(F_i(b_i) - F_i(a_i))$ for all $(\boldsymbol{a},\boldsymbol{b}] \in \mathcal{A}_k$.
(c) Fit Example 3 into this framework.
(d) Suppose that in Exercise 2 we replace $f(\boldsymbol{t})$ everywhere with $f_1(t_1)\cdots f_k(t_k)$, where each f_i is nonnegative and continuous with $\int_{-\infty}^{\infty} f_i(t_i)\, dt_i < +\infty$. Fit this into the framework of this exercise.

4. Suppose that $F_i \in \mathfrak{S}_{k_i}$ for $i = 1, \cdots, n$. Suppose that $F : \mathbb{R}^{\sum_{i=1}^{n} k_i} \to \mathbb{R}$ is such that $F(\boldsymbol{x}^{(1)}, \cdots, \boldsymbol{x}^{(n)}) = \prod_{i=1}^{n} F_i(\boldsymbol{x}^{(i)})$ for each $\boldsymbol{x}^{(1)} \in \mathbb{R}^{k_1}$, $\cdots$, and $\boldsymbol{x}^{(n)} \in \mathbb{R}^{k_n}$. Then Δ_F is a measure on $\mathcal{A}_{\sum_{i=1}^{n} k_i}$. Also, $\Delta_F(A_1 \times \cdots \times A_n) = \prod_{i=1}^{n} \Delta_{F_i}(A_i)$ for each $A_1 \in \mathcal{A}_{k_1}$, $\cdots$, and $A_n \in \mathcal{A}_{k_n}$.

5. For each $k \in \mathbb{N}$ we have the following: $\Delta_F(A) = 0$ for all $A \in \mathcal{A}_k$ iff F is a finite sum of functions, each function being a function of only at most $k-1$ of the coordinates. [If $k = 1$, then a function of at most $k-1$ coordinates is a constant function.]

2.4 APPENDIX: PROOF OF THE STIELTJES THEOREM

The theorem given in Section 2.3 was merely stated but not proved so that time could be spent on the applications of the Stieltjes Theorem; this section exists solely to prove the theorem. All notation from Section 2.3 carries over to this section. The key to proving the theorem lies in the following two claims.

Claim 1. *Let $n \in \mathbb{N}$, and let $A, A_1, \cdots, A_n \in \mathcal{A}$, where $\mathcal{A}$ is a semiring. Then there exist disjoint $C_1, \cdots, C_m \in \mathcal{A}$ such that $A \cap \bigcap_{k=1}^{n} A_k^c = \bigcup_{j=1}^{m} C_j$.*

Proof: Let $n = 1$. Then $A \cap A_1^c = A - (A \cap A_1)$, which by the definition of a semiring is a finite disjoint union of $\mathcal{A}$-sets. Assume that the result now holds for an arbitrary but fixed $n \in \mathbb{N}$: given $A, A_1, \cdots, A_n \in \mathcal{A}$ there are disjoint $\mathcal{A}$-sets $C_1, \cdots, C_m$ with $A \cap \bigcap_{k=1}^{n} A_k^c = \bigcup_{j=1}^{m} C_j$. Next, let $A_{n+1} \in \mathcal{A}$; we desire to show that the result holds for the case $n+1$. We have that

$$A \cap \bigcap_{k=1}^{n+1} A_k^c = \left[A \cap \bigcap_{k=1}^{n} A_k^c\right] \cap A_{n+1}^c = \left[\bigcup_{j=1}^{m} C_j\right] \cap A_{n+1}^c = \bigcup_{j=1}^{m} (C_j \cap A_{n+1}^c).$$

From the $n = 1$ case, for each $j = 1, \cdots, m$ we have that there exist disjoint $D_{j1}, \cdots, D_{jm_j} \in \mathcal{A}$ with $C_j \cap A_{n+1}^c = \bigcup_{i=1}^{m_j} D_{ji}$; furthermore, the collection $\{D_{ji} : j = 1, \cdots, m,\ i = 1, \cdots, m_j\}$ is disjoint. Therefore

$$A \cap \bigcap_{k=1}^{n+1} A_k^c = \bigcup_{j=1}^{m} \bigcup_{i=1}^{m_j} D_{ji}, \quad \text{a finite disjoint union of } \mathcal{A} \text{ sets.}$$

This establishes the $n+1$ case, and the result follows by induction. ■

Claim 2. *Let $\mathcal{A}$ denote a semiring on $\mathbb{R}^k$, and let $\mu : \mathcal{A} \to \mathbb{R}$ have properties (M1), (M2), and (M4). Let $A, A_1, A_2, \cdots \in \mathcal{A}$. Then for all $n \in \mathbb{N}$ we have:*

(i) If $\bigcup_{i=1}^n A_i \subseteq A$ *and* $A_1, \cdots, A_n$ *are disjoint, then* $\sum_{i=1}^n \mu(A_i) \leq \mu(A)$.
(ii) If $\bigcup_{i=1}^\infty A_i \subseteq A$ *and* $A_1, A_2, \cdots$ *are disjoint, then* $\sum_{i=1}^\infty \mu(A_i) \leq \mu(A)$.
(iii) If $A \subseteq \bigcup_{i=1}^n A_i$*, then* $\mu(A) \leq \sum_{i=1}^n \mu(A_i)$.
(iv) μ *is countably subadditive*[13] *iff* μ *is countably additive*[14].

Proof: Let $n \in \mathbb{N}$ and assume the hypotheses of the claim. By Claim 1, there are disjoint $\mathcal{A}$-sets $C_1, \cdots, C_m$ such that $A - \bigcup_{i=1}^n A_i = \bigcup_{j=1}^m C_j$. Since $\bigcup_{i=1}^n A_i \subseteq A$, we have that $A = \bigcup_{i=1}^n A_i \cup \bigcup_{j=1}^m C_j$, so that A is the finite disjoint union of $A_1, \cdots, A_n, C_1, \cdots, C_m$. By (M4) and (M1), we obtain

$$\mu(A) = \sum_{i=1}^n \mu(A_i) + \sum_{j=1}^m \mu(C_j) \geq \sum_{i=1}^n \mu(A_i),$$

finishing (i).

For (ii), let $\bigcup_{i=1}^\infty A_i$ denote a disjoint union of $\mathcal{A}$-sets with $\bigcup_{i=1}^\infty A_i \subseteq A$. It follows for all $n \in \mathbb{N}$ that $\bigcup_{i=1}^n A_i \subseteq A$ and $A_1, \cdots, A_n \in \mathcal{A}$ are disjoint. By (i), $\sum_{i=1}^n \mu(A_i) \leq \mu(A)$ for each $n \in \mathbb{N}$, hence $\sum_{i=1}^\infty \mu(A_i) \leq \mu(A)$.

For (iii), let $n \in \mathbb{N}$ be arbitrary, and suppose that $A \subseteq \bigcup_{i=1}^n A_i$, with $A, A_1, \cdots, A_n \in \mathcal{A}$. Let $B_1 = A \cap A_1$ and $B_i = A \cap A_i \cap \bigcap_{j=1}^{i-1} A_j^c$ for $i = 2, \cdots, n$. Observe that $B_1, \cdots, B_n$ are disjoint subsets of A with

$$\bigcup_{i=1}^n B_i = A \cap \left[A_1 \cup \bigcup_{i=2}^n (A_i \cap A_1^c \cap \cdots \cap A_{i-1}^c)\right] = A \cap \bigcup_{i=1}^n A_i = A.$$

Apply Claim 1 to each of $B_2, \cdots, B_n$: for each $i = 2, \cdots, n$ there is a disjoint collection $C_{i1}, \cdots, C_{im_i}$ of $\mathcal{A}$-sets with $B_i = \bigcup_{j=1}^{m_i} C_{ij}$. For $i = 1$, we have $B_1 \in \mathcal{A}$, defining $m_1 = 1$ and $C_{1m_1} = C_{11} = B_1$. Therefore, $\bigcup_{j=1}^{m_i} C_{ij} \subseteq A_i$ for $i = 1, \cdots, n$. By (i), it follows that $\sum_{j=1}^{m_i} \mu(C_{ij}) \leq \mu(A_i)$. Since A is the disjoint union of $C_{11}, \cdots, C_{nm_n}$, the finite additivity of μ gives

$$\mu(A) = \sum_{i=1}^n \sum_{j=1}^{m_i} \mu(C_{ij}) \leq \sum_{i=1}^n \mu(A_i),$$

giving (iii).

For (iv), assume that μ is countably subadditive; we desire to show that μ is countably additive. Let $A_1, A_2, \cdots \in \mathcal{A}$ be disjoint with $\bigcup_{n=1}^\infty A_n \in \mathcal{A}$. Then $\mu(\bigcup_{n=1}^\infty A_n) \leq \sum_{n=1}^\infty \mu(A_n)$ by assumption. Taking $A = \bigcup_{n=1}^\infty A_n$ in (ii) gives $\sum_{n=1}^\infty \mu(A_n) \leq \mu(\bigcup_{n=1}^\infty A_n)$. Therefore, if $\{A_n\}_{n=1}^\infty$ is a disjoint sequence of $\mathcal{A}$ sets with $\bigcup_{n=1}^\infty A_n \in \mathcal{A}$, then $\mu(\bigcup_{n=1}^\infty A_n) = \sum_{n=1}^\infty \mu(A_n)$, so μ is countably additive on $\mathcal{A}$. The proof of the converse is automatic, and the entire proof is now complete. ∎

[13]That is, if $A_1, A_2, \cdots \in \mathcal{A}$ with $\bigcup_{n=1}^\infty A_n \in \mathcal{A}$, then $\mu(\bigcup_{n=1}^\infty A_n) \leq \sum_{n=1}^\infty \mu(A_n)$.
[14]That is, $\mu(\bigcup_{n=1}^\infty A_n) = \sum_{n=1}^\infty \mu(A_n)$ whenever $A_1, A_2, \cdots \in \mathcal{A}$ are disjoint with $\bigcup_{n=1}^\infty A_n \in \mathcal{A}$.

Stieltjes Theorem. *For any $k \in \mathbb{N}$ and $F \in \mathfrak{S}_k$, $\Delta_F : \mathcal{A}_k \to \mathbb{R}$ is a measure that assigns finite measure to each $\mathcal{A}_k$ set [and hence Δ_F is σ-finite on $\mathcal{A}_k$].*

Proof: Let $k \in \mathbb{N}$ and $F \in \mathfrak{S}_k$. Observe that Δ_F trivially satisfies (M1) and (M2). We will show that Δ_F is finitely additive and countably subadditive on $\mathcal{A}_k$. By part (iv) of Claim 2, this will show that Δ_F is a measure on $\mathcal{A}_k$. We will first prove something needed to establish finite additivity.

Pick $A = (\boldsymbol{a}, \boldsymbol{b}] \in \mathcal{A}_k$, and write $A = I_1 \times \cdots \times I_k$, where $I_i = (a_i, b_i]$ for all $i = 1, \cdots, k$. Suppose that for each $i = 1, \cdots, k$, we let $n_i \geq 1$ and pick $n_i + 1$ points $t_{i0}, \cdots, t_{in_i}$ such that $a_i = t_{i0} < t_{i1} < \cdots < t_{in_i} = b_i$. We also consider the disjoint intervals $J_{i1}, \cdots, J_{in_i}$, where $J_{ij} = (t_{i,j-1}, t_{ij}]$ for each $j = 1, \cdots, n_i$. Observe that $I_i = (a_i, b_i] = \bigcup_{j=1}^{n_i} J_{ij}$ for $i = 1, \cdots, k$, the union being disjoint. For all values $j_1, \cdots, j_k$ with $1 \leq j_i \leq n_i$ for $i = 1, \cdots, k$, write $B_{j_1, \cdots, j_k} = J_{1j_1} \times \cdots \times J_{kj_k}$. Next, define

$$R(A) = \{J_{1j_1} \times \cdots \times J_{kj_k} : 1 \leq j_i \leq n_i \text{ for } i = 1, \cdots, k\},$$

so that $R(A)$ consists of the disjoint rsc subrectangles of A produced by the partitioning, that is, $A = \bigcup\{B_{j_1, \cdots, j_k} : B_{j_1, \cdots, j_k} \in R(A)\}$, a disjoint union. As a final piece of notation, let U denote the set of $\boldsymbol{x}$ such that $\boldsymbol{x}$ is a vertex of at least one $B_{j_1, \cdots, j_k}$ but not a vertex of A, and let W denote the set of $\boldsymbol{x}$ such that $\boldsymbol{x}$ is a vertex of at least one $B_{j_1, \cdots, j_k}$ and a vertex of A. It is immediate that U and W are disjoint, $W = V(A)$, and $U \cup W$ is the set of $\boldsymbol{x}$ such that $\boldsymbol{x}$ is a vertex of at least one $B_{j_1, \cdots, j_k}$.

Before proving finite additivity (M4), we will need to show the following:

Claim. *We have*

$$\Delta_F(A) = \sum\{\Delta_F(B_{j_1, \cdots, j_k}) : B_{j_1, \cdots, j_k} \in R(A)\}. \tag{2.2}$$

Proof of Claim: Consider now the following chain of equalities:

$$\sum_{B_{j_1, \cdots, j_k} \in R(A)} \Delta_F(B_{j_1, \cdots, j_k})$$

$$= \sum_{B_{j_1, \cdots, j_k} \in R(A)} \sum_{\boldsymbol{x} \in V(B_{j_1, \cdots, j_k})} s_{B_{j_1, \cdots, j_k}}(\boldsymbol{x}) F(\boldsymbol{x}) \tag{2.3}$$

$$= \sum_{\boldsymbol{x} \in U \cup W} F(\boldsymbol{x}) \sum_{B_{j_1, \cdots, j_k}} s_{B_{j_1, \cdots, j_k}}(\boldsymbol{x}) \tag{2.4}$$

$$= \sum_{\boldsymbol{x} \in U} F(\boldsymbol{x}) \sum_{B_{j_1, \cdots, j_k}} s_{B_{j_1, \cdots, j_k}}(\boldsymbol{x}) + \sum_{\boldsymbol{x} \in W} F(\boldsymbol{x}) \sum_{B_{j_1, \cdots, j_k}} s_{B_{j_1, \cdots, j_k}}(\boldsymbol{x}). \tag{2.5}$$

Equality (2.3) follows by definition. Equality (2.4) follows by interchanging the summations: whereas in (2.3) the outer sum was over the $B_{j_1, \cdots, j_k} \in R(A)$, and for each fixed $B_{j_1, \cdots, j_k} \in R(A)$ the inner sum was over the vertices of $B_{j_1, \cdots, j_k}$, we have switched vantage points, so that we now have the outer sum over all vertices of all $B_{j_1, \cdots, j_k}$'s, and for each fixed vertex $\boldsymbol{x}$ we sum over the $B_{j_1,\cdots,j_k}$'s that claim this particular $\boldsymbol{x}$ as a vertex. Equality (2.5) follows because U and W are disjoint.

We now examine the two double sums in (2.5). Suppose that $\boldsymbol{x} \in U$, that is, $\boldsymbol{x}$ is a vertex for at least one $B_{j_1,\cdots,j_k}$ but is not a vertex of A. Then, for at least one $1 \leq i \leq k$, say $i = 1$, we have that x_1 is neither a_1 nor b_1; x_1 instead equals t_{1j} for some $1 \leq j \leq n_1 - 1$. Suppose that $j_2, \cdots, j_k$ are such that $\boldsymbol{x} \in V(B_{j,j_2,\cdots,j_k})$. For this $\boldsymbol{x}$, we have the following facts:

(a) $\boldsymbol{x}$ is a vertex of $B_{j+1,j_2,\cdots,j_k}$;

(b) $s_{B_{j,j_2,\cdots,j_k}}(\boldsymbol{x}) = -s_{B_{j+1,j_2,\cdots,j_k}}(\boldsymbol{x})$;

(c) letting $m(\boldsymbol{x})$ denote the number of $i = 1, \cdots, k$ such that $x_i \notin \{a_i, b_i\}$, we have that $\boldsymbol{x}$ is a vertex of $2^{m(\boldsymbol{x})}$ rsc subrectangles in $R(A)$;

(d) $2^{m(\boldsymbol{x})-1}$ of the rsc subrectangles for which $\boldsymbol{x}$ is a vertex have a signum of $+1$, and the other $2^{m(\boldsymbol{x})-1}$ of the rsc rectangles have a signum of -1;

(e) For this $\boldsymbol{x}$ that is a vertex of the $2^{m(\boldsymbol{x})}$ rsc rectangles mentioned in (c),

$$\sum \{s_{B_{j_1,\cdots,j_k}}(\boldsymbol{x}) : B_{j_1,\cdots,j_k} \text{ claims } \boldsymbol{x} \text{ as a vertex}\} = 2^{m(\boldsymbol{x})-1} - 2^{m(\boldsymbol{x})-1} = 0.$$

From (e) we have for every $\boldsymbol{x} \in U$ that

$$\sum \{s_{B_{j_1,\cdots,j_k}}(\boldsymbol{x}) : B_{j_1,\cdots,j_k} \text{ claims } \boldsymbol{x} \text{ as a vertex}\} = 0,$$

and hence

$$\sum_{\boldsymbol{x} \in U} F(\boldsymbol{x}) \sum_{B_{j_1,\cdots,j_k}} s_{B_{j_1,\cdots,j_k}}(\boldsymbol{x}) = 0.$$

Thus, (2.5) gives

$$\sum_{B_{j_1,\cdots,j_k} \in R(A)} \Delta_F(B_{j_1,\cdots,j_k}) = \sum_{\boldsymbol{x} \in W} F(\boldsymbol{x}) \sum_{B_{j_1,\cdots,j_k}} s_{B_{j_1,\cdots,j_k}}(\boldsymbol{x}). \tag{2.6}$$

We will now show that the right side of (2.6) is $\Delta_F(A)$. If $\boldsymbol{x} \in W$, that is, if $\boldsymbol{x}$ is a vertex of at least one rsc subrectangle $B_{j_1,\cdots,j_k}$ as well as a vertex of A, then for all $1 \leq i \leq k$ we have $x_i = a_i = t_{i0}$ or $x_i = b_i = t_{in_i}$. Therefore, $\boldsymbol{x} \in W$ can be a vertex of one and only one $B_{j_1,\cdots,j_k} \in R(A)$, namely, the $B_{j_1,\cdots,j_k} \in R(A)$ with $j_i = 1$ or n_i as $x_i = a_i$ or b_i, where $1 \leq i \leq k$. For this unique $B_{j_1,\cdots,j_k} \in R(A)$ that claims our fixed $\boldsymbol{x} \in W$ as a vertex, we have $s_{B_{j_1,\cdots,j_k}}(\boldsymbol{x}) = s_A(\boldsymbol{x})$, whence

$$\sum \{s_{B_{j_1,\cdots,j_k}}(\boldsymbol{x}) : B_{j_1,\cdots,j_k} \text{ claims } \boldsymbol{x} \text{ as a vertex}\} = s_A(\boldsymbol{x}).$$

Since $\boldsymbol{x} \in W$ was generic, we obtain

$$\sum_{\boldsymbol{x} \in W} F(\boldsymbol{x}) \sum_{B_{j_1,\cdots,j_k}} s_{B_{j_1,\cdots,j_k}}(\boldsymbol{x}) = \sum_{\boldsymbol{x} \in W} F(\boldsymbol{x}) s_A(\boldsymbol{x}) = \Delta_F(A),$$

where the second equality follows because $W = V(A)$. This and (2.6) give (2.2), completing the proof of our claim. ■

Proof of Stieltjes Theorem [Resumed]: We will first show that Δ_F is finitely additive. Let $A = (\boldsymbol{a}, \boldsymbol{b}] \in \mathcal{A}_k$ be the disjoint union of the $\mathcal{A}_k$-sets $A_1, \cdots, A_n$, and write $A = (\boldsymbol{a}, \boldsymbol{b}] = I_1 \times \cdots \times I_k$ with $A_l = I_{1l} \times \cdots \times I_{kl}$, where $1 \le l \le n$. Note that for each $1 \le i \le k$, we have $\bigcup_{m=1}^{n} I_{im} = I_i$ although the union is not necessarily disjoint. However, the endpoints of $I_{i1}, \cdots, I_{in}$ split I_i into disjoint intervals $J_{i1}, \cdots, J_{in_i}$, where each I_{il} is the disjoint union of certain of the $J_{i1}, \cdots, J_{in_i}$. Let $B_{j_1, \cdots, j_k} = J_{1j_1} \times \cdots \times J_{kj_k}$ as before, where each $1 \le j_i \le n_i$, and continue to write

$$R(A) = \{B_{j_1, \cdots, j_k} : \ 1 \le j_i \le n_i \text{ for each } 1 \le i \le k\}.$$

Observe that each $B_{j_1, \cdots, j_k} \in R(A)$ is in exactly one of the A_l's, and that the elements of $R(A)$ are disjoint with a union that equals A. It follows that

$$\begin{aligned} \Delta_F(A) = \sum_{B_{j_1, \cdots, j_k} \in R(A)} \Delta_F(B_{j_1, \cdots, j_k}) &= \sum_{l=1}^{n} \sum_{B_{j_1, \cdots, j_k} \subseteq A_l} \Delta_F(B_{j_1, \cdots, j_k}) \\ &= \sum_{l=1}^{n} \Delta_F(A_l), \end{aligned}$$

where the first equality follows because the bounded subrectangles in $R(A)$ partition A and thus (2.2) of the claim applies, and where the third equality follows from the fact that the bounded subrectangles in $R(A) \cap A_l$ partition A_l and thus (2.2) of the claim again applies. The second equality follows by writing

$$\{B_{j_1, \cdots, j_k} : \ B_{j_1, \cdots, j_k} \in R(A)\} = \bigcup_{l=1}^{n} \{B_{j_1, \cdots, j_k} : \ B_{j_1, \cdots, j_k} \subseteq A_l\},$$

where the union is disjoint. It therefore follows that $\Delta_F(A) = \sum_{l=1}^{n} \Delta_F(A_l)$ when $A \in \mathcal{A}_k$ is the finite disjoint union of the $\mathcal{A}_k$-sets $A_1, \cdots, A_n$. Δ_F is thus finitely additive on $\mathcal{A}_k$.

We now show that Δ_F is countably subadditive, which by (iv) of Claim 2 is sufficient to show that Δ_F is countably additive and hence is a measure. Let $A = (\boldsymbol{a}, \boldsymbol{b}] \in \mathcal{A}_k$, and suppose that $A \subseteq \bigcup_{n=1}^{\infty} A_n$, where $A_n = (\boldsymbol{a}_n, \boldsymbol{b}_n] \in \mathcal{A}_k$ for all $n \in \mathbb{N}$. We desire to show that $\Delta_F(A) \le \sum_{n=1}^{\infty} \Delta_F(A_n)$. This will be accomplished by showing that $\Delta_F(A) \le \sum_{n=1}^{\infty} \Delta_F(A_n) + \epsilon$ for every $\epsilon > 0$. Let $\epsilon > 0$ be arbitrary. Since F is cfa, we have the following facts:

(f) There is $\delta > 0$ such that $\Delta_F((\boldsymbol{a}, \boldsymbol{b}]) - \frac{\epsilon}{2} < \Delta_F((\boldsymbol{a} + \delta \mathbf{1}, \boldsymbol{b}])$, where $\mathbf{1}$ is a k-vector of ones. Writing $B = (\boldsymbol{a} + \delta \mathbf{1}, \boldsymbol{b}]$, we have that $\Delta_F(A) - \frac{\epsilon}{2} < \Delta_F(B)$.

(g) For each $n \in \mathbb{N}$, there is $\delta_n > 0$ so that $\Delta_F(B_n) - \Delta_F(A_n) < \frac{\epsilon}{2^{n+1}}$, where $B_n = (\boldsymbol{a}_n, \boldsymbol{b}_n + \delta_n \mathbf{1}]$.

Letting B^- and B° denote the closure and interior of B, respectively, and proceeding with the $\delta > 0$ and the positive $\delta_1, \delta_2, \cdots$ from above, we have

$$B \subseteq B^- \subseteq A \subseteq \bigcup_{n=1}^{\infty} A_n \subseteq \bigcup_{n=1}^{\infty} B_n^\circ.$$

Therefore, we have that $\{B_n^\circ\}_{n=1}^\infty$ is an open cover of B^-. By the Heine-Borel Theorem, B^- is compact, hence there exists $N \in \mathbb{N}$ such that

$$B \subseteq B^- \subseteq \bigcup_{n=1}^{N} B_n^\circ \subseteq \bigcup_{n=1}^{N} B_n.$$

By part (iii) of Claim 2, we have $\Delta_F(B) \leq \sum_{n=1}^N \Delta_F(B_n)$. Therefore, by (f) and then by (g), we obtain

$$\begin{aligned}\Delta_F(A) - \frac{\epsilon}{2} \ < \ \Delta_F(B) \ &\leq \ \sum_{n=1}^{N} \Delta_F(B_n) \\ &\leq \ \sum_{n=1}^{N} \left[\Delta_F(A_n) + \frac{\epsilon}{2^{n+1}}\right] \\ &\leq \ \sum_{n=1}^{\infty} \Delta_F(A_n) + \frac{\epsilon}{2}.\end{aligned}$$

It therefore follows that $\Delta_F(A) \leq \sum_{n=1}^\infty \Delta_F(A_n) + \epsilon$. Since $\epsilon > 0$ is arbitrary, we have for any $A, A_1, A_2, \cdots \in \mathcal{A}_k$ that $A \subseteq \bigcup_{n=1}^\infty A_n$ implies the inequality $\Delta_F(A) \leq \sum_{n=1}^\infty \Delta_F(A_n)$. In particular, if $A = \bigcup_{n=1}^\infty A_n$ and $A, A_1, A_2, \cdots$ are in $\mathcal{A}_k$, then the just-mentioned inequality shows that Δ_F is countably subadditive on $\mathcal{A}_k$. Therefore, Δ_F is a measure on $\mathcal{A}_k$.

Since $F(\boldsymbol{x}) < +\infty$ for each $\boldsymbol{x} \in \mathbb{R}^k$, we will clearly have $\Delta_F(A) < +\infty$ for each $A \in \mathcal{A}_k$, for $\Delta_F(A)$ is merely a finite sum of finite numbers.

To see that Δ_F is σ-finite, merely write $\mathbb{R}^k = \bigcup_{n=1}^\infty (-\mathbf{1}n, \mathbf{1}n]$ and observe that $\Delta_F((-\mathbf{1}n, \mathbf{1}n]) < +\infty$ for all $n \in \mathbb{N}$ by what has just been stated in the previous paragraph. At long last the entire proof is complete. ∎

The proof of the theorem is quite complicated, and it is the first of the two major steps that we will take towards the goal of making rigorous the notions of "length," "area," and "volume" that we have taken for granted throughout calculus and advanced calculus. If [say] "length" is to satisfy the properties of a measure — and intuition dictates that it should – then by taking F to be the identity function and recalling Example 2 of Section 3 we have shown that there is a measure that does what "length" should do, namely, assign the value $b - a$ to each $(a, b]$. If $F(\boldsymbol{x}) = x_1 x_2$, then we have shown that there exists a measure that does what "area" should do, namely assign the value $(b_1 - a_1)(b_2 - a_2)$ to $(a_1, b_1] \times (a_2, b_2]$, and so on. As mentioned, the shortcoming of these aforementioned measures is that they can be applied only to a class of sets that is far too small in comparison with the class of sets that one encounters in advanced mathematics. The next chapter will deal with "extending" our rigorously defined measure Δ_F to a larger domain, such as $\mathcal{B}^k$.

Exercises.

1*. Verify the simple claims (a)-(g) found in the proof of the Stieltjes Theorem.

2. For each $k \in \mathbb{N}$, let $\mathcal{A}_k^*$ denote the collection of finite disjoint unions of $\mathcal{A}_k$-sets. Let $F \in \mathfrak{S}_k$. Define a new function $\Delta_F^* : \mathcal{A}_k^* \to \mathbb{R}$ by defining $\Delta_F^*(\varnothing) = 0$ and $\Delta_F^*(A) = \sum_{i=1}^n \Delta_F(A_i)$ whenever A is the disjoint union of $A_1, \cdots, A_n \in \mathcal{A}_k$.

(a) Δ_F^* is well-defined. [To do this, suppose that a generic $A \in \mathcal{A}_k^*$ has two different representations as a finite disjoint union of $\mathcal{A}_k$-sets: $A = \bigcup_{i=1}^n A_i$ and $A = \bigcup_{j=1}^m B_j$. We desire the equality $\Delta_F^*(\bigcup_{i=1}^n A_i) = \Delta_F^*(\bigcup_{j=1}^m B_j)$, that is, $\sum_{i=1}^n \Delta_F(A_i) = \sum_{j=1}^m \Delta_F(B_j)$. Write $A_i = \bigcup_{j=1}^m (A_i \cap B_j)$ and $B_j = \bigcup_{i=1}^n (A_i \cap B_j)$, note that these unions are disjoint, and use the finite additivity of Δ_F, obtaining

$$\sum_{i=1}^n \Delta_F(A_i) = \sum_{i=1}^n \sum_{j=1}^m \Delta_F(A_i \cap B_j) = \sum_{j=1}^m \sum_{i=1}^n \Delta_F(A_i \cap B_j) = \sum_{j=1}^m \Delta_F(B_j).]$$

(b) Δ_F^* is a measure on $\mathcal{A}_k^*$. [Let $A_1, A_2, \cdots \in \mathcal{A}_k^*$ be disjoint. Write each nonempty A_k as $A_k = \bigcup_{j=1}^{n_k} I_{kj}$ where each $I_{kj} \in \mathcal{A}_k$ and the I_{kj}'s are disjoint. Observe that for $\Delta_F^*(\bigcup_{k=1}^\infty A_k)$ to be defined, we need $\bigcup_{k=1}^\infty A_k$ to be of the form $\bigcup_{l=1}^m J_l$ where each $J_l \in \mathcal{A}_k$ and the J_l's are disjoint. Verify that

$$\begin{aligned}
\sum_{k=1}^\infty \Delta_F^*(A_k) &= \sum_{k=1}^\infty \Delta_F^*\left(\bigcup_{j=1}^{n_k} I_{kj}\right) = \sum_{k=1}^\infty \sum_{j=1}^{n_k} \Delta_F(I_{kj}) \\
&= \sum_{k=1}^\infty \sum_{j=1}^{n_k} \Delta_F\left(\bigcup_{l=1}^m (J_l \cap I_{kj})\right) = \sum_{k=1}^\infty \sum_{j=1}^{n_k} \sum_{l=1}^m \Delta_F(J_l \cap I_{kj}) \\
&= \sum_{l=1}^m \sum_{k=1}^\infty \sum_{j=1}^{n_k} \Delta_F(J_l \cap I_{kj}) = \sum_{l=1}^m \Delta_F\left(\bigcup_{k=1}^\infty \bigcup_{j=1}^{n_k} (J_l \cap I_{kj})\right) \\
&= \sum_{l=1}^m \Delta_F\left(J_l \cap \bigcup_{k=1}^\infty \bigcup_{j=1}^{n_k} I_{kj}\right) = \sum_{l=1}^m \Delta_F(J_l) \\
&= \Delta_F^*\left(\bigcup_{l=1}^m J_l\right) = \Delta_F^*\left(\bigcup_{k=1}^\infty A_k\right).
\end{aligned}$$

This and the fact that (M1) and (M2) are automatic finishes things.]

(c) Δ_F^* agrees with Δ_F on $\mathcal{A}_k$.

(d) Nothing particular about $\mathcal{A}_k$ nor $\mathcal{A}_k^*$ was used, and hence we have proven the following general result: *Let μ denote a measure defined on a semiring $\mathcal{A}$. Let $\mathcal{A}^*$ consist of the finite disjoint unions of $\mathcal{A}$-sets. Define $\mu^* : \mathcal{A}^* \to \mathbb{R}$ by writing $\mu^*(A) = \sum_{i=1}^k \mu(A_i)$ whenever A is the disjoint union of $A_1, \cdots, A_n \in \mathcal{A}$. Then μ^* is well defined and is a measure on $\mathcal{A}^*$. Moreover, $\mu^*(A) = \mu(A)$ for all $A \in \mathcal{A}$.*

3. Suppose that $F : \mathbb{R}^k \to \mathbb{R}$ is such that $\Delta_F(A) \geq 0$ for all $A \in \mathcal{A}_k$, but suppose that F fails to be cfa at some $\boldsymbol{x} \in \mathbb{R}^k$. Then Δ_F fails to be a measure on $\mathcal{A}_k$.

3

Extensions of Measures

One of our goals is to find measures that mimic what we call "length," "area," and so on. In particular, if $F : \mathbb{R}^k \to \mathbb{R}$ is such that $F(\boldsymbol{x}) = \prod_{i=1}^k x_i$ for each $\boldsymbol{x} \in \mathbb{R}^k$, then $F \in \mathfrak{S}_k$ [see Examples 2 and 3 of Section 2.3]; the Stieltjes Theorem ensures that Δ_F is a measure on the semiring $\mathcal{A}_k$ consisting of $\varnothing$ and all bounded rsc k-dimensional rectangles. Since $\Delta_F((\boldsymbol{a}, \boldsymbol{b}])$ was shown in the aforementioned examples to equal $\prod_{i=1}^k (b_i - a_i)$ for each $(\boldsymbol{a}, \boldsymbol{b}] \in \mathcal{A}_k$, Δ_F does what we think "length," "area," and "volume" should do when $k = 1$, $k = 2$, and $k = 3$, respectively. However, Δ_F has the very limited domain $\mathcal{A}_k$, and we would like to consider "k-dimensional volume" for sets far more complicated than $\mathcal{A}_k$-sets.

Let us take $k = 1$ for specificity, and consider a quest to rigorously define "length." In our framework, we are seeking a measure $\lambda : 2^{\mathbb{R}} \to [0, +\infty]$ that "agrees with intuition": λ should assign measure $b - a$ to any interval with endpoints $-\infty < a < b < +\infty$; λ should also assign $+\infty$ to any unbounded interval, λ should assign the value 0 to any singleton, which by (M3) forces us to believe that λ should assign the value 0 to any amc set. We would also think that for any $A \subseteq \mathbb{R}$ and $x \in \mathbb{R}$ the λ-measures of A and $\{a + x : a \in A\}$ [the set A "shifted x units to the right"] are identical. That is, λ should be *translation invariant.* The attempt to find a measure λ that satisfies these properties [among other things] is called *the problem of measure.*

Early in the twentieth century, mathematicians attempted to solve the problem of measure. However, it turned out that mathematicians were asking for too much, for it was shown that *no such measure on $2^{\mathbb{R}}$ with all of the properties above can simultaneously exist with the Axiom of Choice (AC).*

Since (AC) is accepted as a part of everyday mathematics, the problem of measure is *not* solved by discarding (AC), for it is fundamental to much of what ones does, even if one is not quite aware of its use.[1]

The sad but interesting fact that no such measure on $2^{\mathbb{R}}$ can co-exist with (AC) naturally rained somewhat on the mathematical parade, and there were two major paths that could be taken to attempt a reconciliation of the situation. The first path was to insist that the hoped-for measure would still have domain $2^{\mathbb{R}}$, but some property or properties intuitively associated with "length" would no longer be required. One would jettison the requirements that caused any logical problems with (AC) until there were no longer any inconsistencies with (AC), and one would just learn to be content with this scaled-down measure. The second path was to try to keep all the properties associated with "length" intact while no longer requiring that every set have a "length," hoping that the resulting measure would still have a "large enough" domain while not butting logical heads with (AC). That is, one hopes that by scaling down the domain of λ one can preserve all the intuitive properties that λ should have while still being able to apply λ to a "large enough" collection of subsets of $\mathbb{R}$.

If one opts for the first path, it is not easy to determine which requirement[s] would or should be abandoned. Should translation invariance be sacrificed on the altar of logical necessity, or should the fact that each singleton is to have zero "length" be the odd man out? Why should this or that or some other property be the one to sacrifice? After all, each of the properties above associated with "length" seems quite natural – axiomatic, almost – and it would appear to be a significant breach of concord to throw away one or more of the desired properties and still call the resulting measure "length." Therefore, this first path is abandoned in favor of the second path.

Regarding the second path, it will eventually be shown that a measure corresponding to our intuitive notion of "length" can be consistently defined on a class of subsets of $\mathbb{R}$ "large enough" to include every "ordinary" set from analysis. In particular, the σ-field $\mathcal{B}$ on $\mathbb{R}$ contains just about every set in analysis that one can imagine, and we will show that $\mathcal{B}$ can be included in the domain of such a measure. This chapter and the next are committed to constructing this measure [among others] and showing that this measure indeed has the desired properties.

[1]Some examples: (i) the proof that every infinite set contains a countably infinite subset requires (AC), (ii) if A and B are nonempty subsets with an onto function $f : A \to B$, the proof of the claim that there exists a function $g : B \to A$ with $g(y) \in f^{-1}(\{y\})$ for each $y \in B$ requires a statement that is equivalent to (AC), (iii) the proof that every vector space contains a Hamel basis requires Tukey's Lemma, which lemma is logically equivalent to (AC), and (iv) proving for cardinals $\mathfrak{a}$ and $\mathfrak{b}$ that $\mathfrak{a} \leq \mathfrak{b}$ and $\mathfrak{b} \leq \mathfrak{a}$ forces $\mathfrak{a} = \mathfrak{b}$.

3.1 EXTENSIONS AND RESTRICTIONS

Extensions and *restrictions* aren't particularly difficult concepts, but they are important enough to have their own section. Throughout, Ω will denote an arbitrary nonempty set which will serve as our background space.

Definition. Let $\varnothing \neq \mathcal{G} \subseteq \mathcal{H} \subseteq 2^{\Omega}$. Let $\mu : \mathcal{G} \to \bar{\mathbb{R}}$ and let $\nu : \mathcal{H} \to \bar{\mathbb{R}}$. Furthermore, suppose that $\mu(A) = \nu(A)$ for all $A \in \mathcal{G}$. We say that ν is an *extension of μ [from $\mathcal{G}$] to $\mathcal{H}$*, and we equivalently say that μ is the *restriction of ν [from $\mathcal{H}$] to $\mathcal{G}$*. If ν happens to be a measure [regardless of whether μ is a measure], then ν is called a *measure extension of μ [from $\mathcal{G}$] to $\mathcal{H}$*. If μ happens to be a measure [regardless of whether or not ν is a measure], then μ is called a *measure restriction of ν [from $\mathcal{H}$] to $\mathcal{G}$*.

In the above definition, no particular structure is assumed regarding the collections $\mathcal{G}$ and $\mathcal{H}$, and μ and ν are not necessarily measures. If it seems that no confusion can arise, we will merely say that μ is the restriction of ν and ν is an extension of μ.

Example. Keep notation as in Sections 2.3 and 2.4, and recall the Stieltjes Theorem as well as Examples 2 and 3 of Section 2.3. Let $\Omega = \mathbb{R}$, and let $\mathcal{A}_1$ denote the semiring consisting of $\varnothing$ and the bounded rsc intervals. Taking $F \in \mathfrak{S}$ to be the identity function, $\Delta_F : \mathcal{A}_1 \to \mathbb{R}$ is a measure, and $\Delta_F((a,b]) = b - a$ for each $(a,b] \in \mathcal{A}_1$. In the *problem of measure*, we desire to find a measure λ on the σ-field $2^{\mathbb{R}}$ such that $\lambda = \Delta_F$ on $\mathcal{A}_1$. That is, we want the measure $\lambda : 2^{\mathbb{R}} \to \bar{\mathbb{R}}$ to be such that $\lambda((a,b]) = b - a = \Delta_F((a,b])$ for all $(a,b] \in \mathcal{A}_1$. In the terminology of this section, we are seeking a measure extension of Δ_F from $\mathcal{A}_1$ to $2^{\mathbb{R}}$.

However, it was stated [without proof] in the introduction to this chapter that no such λ exists. That is, Δ_F fails to have a measure extension from $\mathcal{A}_1$ to $2^{\mathbb{R}}$. However, in this chapter we will construct a measure extension of Δ_F from $\mathcal{A}_1$ to the σ-field $\mathcal{B}$. We will also construct a measure extension of Δ_F from $\mathcal{A}_1$ to a σ-field that contains $\mathcal{B}$. These measure extensions will [by definition] agree with Δ_F on $\mathcal{A}_1$, that is, these measure extensions will assign the value $b - a$ to every $(a,b] \in \mathcal{A}_1$.

Now consider $\mathbb{R}^k$ and the semiring $\mathcal{A}_k$ consisting of $\varnothing$ and the bounded rsc rectangles $(\boldsymbol{a}, \boldsymbol{b}]$. Given any $F \in \mathfrak{S}_k$, we will construct a measure extension of Δ_F from the semiring $\mathcal{A}_k$ to $\mathcal{B}^k$. This measure extension by definition will assign the value $\Delta_F(A) = \sum_{\boldsymbol{x} \in V(A)} s_A(\boldsymbol{x}) F(\boldsymbol{x})$ to every $A \in \mathcal{A}_k$. If $F \in \mathfrak{S}_k$ is such that $F(\boldsymbol{x}) = \prod_{i=1}^k x_i$ for all $\boldsymbol{x} \in \mathbb{R}^k$, then this measure extension of the measure Δ_F from $\mathcal{A}_k$ to $\mathcal{B}^k$ will assign $\prod_{i=1}^k (b_i - a_i)$ to every $(\boldsymbol{a}, \boldsymbol{b}] \in \mathcal{A}_k$. Everything in this example will be subsumed by the abstract and detailed presentation of this chapter.

The following two claims are near-trivial consequences of the definitions given so far, and their proofs are left as simples exercise for the reader.

Claim 1. *Assume the notation of the definition. If μ is σ-finite on $\mathcal{G}$, then ν is σ-finite on $\mathcal{G}$ as well.*

Claim 2. *Suppose that $\varnothing \in \mathcal{G}$, and let $\mathcal{G} \subseteq \mathcal{H} \subseteq 2^{\Omega}$. Let $\nu : \mathcal{H} \to \bar{\mathbb{R}}$ denote a measure. Then the restriction of ν to $\mathcal{G}$ is a measure.*

Exercises.

1*. Let $(\Omega, \mathcal{A}, \mu)$ denote a measure space. Pick $E \in \mathcal{A}$ and define $\mathcal{A}_E = \{F \in \mathcal{A} : F \subseteq E\}$. Then $\mathcal{A}_E$ is a σ-field on E, $\mathcal{A}_E = \{A \cap E : A \in \mathcal{A}\}$, and the restriction μ_E of μ [from $\mathcal{A}$] to $\mathcal{A}_E$ is a measure. That is, $(E, \mathcal{A}_E, \mu_E)$ is a measure space and $\mu_E = \mu$ on $\mathcal{A}_E$.

2*. Quickly prove Claims 1 and 2.

3.2 OUTER MEASURES

In this section, we define an *outer measure.* For us, outer measures are not by themselves important, but they are tools that are useful for proving significant results of measure theory, such as the Extension Theorem which is coming up in Section 4. In order to keep that section as uncluttered as possible, we give this concept its own section. At this point, we proceed to the definition, letting Ω denote a nonempty set as usual.

Definition. An *outer measure* is a function $\nu : 2^\Omega \to \bar{\mathbb{R}}$ such that
(O1) $\nu(A) \geq 0$ for every $A \subseteq \Omega$,
(O2) $\nu(\varnothing) = 0$,
(O3) ν is monotone: $A \subseteq B \subseteq \Omega$ implies $\nu(A) \leq \nu(B)$, and
(O4) ν is countably subadditive on 2^Ω: $\nu(\bigcup_{n=1}^\infty A_n) \leq \sum_{n=1}^\infty \nu(A_n)$ for any sequence $\{A_n\}_{n=1}^\infty$ of subsets of Ω.

Every measure with domain 2^Ω is an outer measure. However, a measure is not an outer measure unless its domain is 2^Ω. Also, it follows from (O2) and (O4) that every outer measure is finitely subadditive: for every $n \in \mathbb{N}$ and $A_1, \cdots, A_n \subseteq \Omega$ the inequality $\nu(\bigcup_{k=1}^n A_k) \leq \sum_{k=1}^n \nu(A_k)$ obtains.

Example 1 below is the most important case of an outer measure, and will be extensively referenced throughout this chapter.

Example 1. Let $\mathcal{A} \subseteq 2^\Omega$ with $\varnothing \in \mathcal{A}$. Let $\mu : \mathcal{A} \to \bar{\mathbb{R}}$ such that $\mu(\varnothing) = 0$ and $\mu(A) \geq 0$ for all $A \in \mathcal{A}$. Define a function $\mu^* : 2^\Omega \to \bar{\mathbb{R}}$ as follows: if there exists a sequence $\{A_n\}_{n=1}^\infty$ of $\mathcal{A}$-sets such that $A \subseteq \bigcup_{n=1}^\infty A_n$, then define

$$\mu^*(A) = \inf \left\{ \sum_{n=1}^\infty \mu(A_n) : A_1, A_2, \cdots \in \mathcal{A},\ A \subseteq \textstyle\bigcup_{n=1}^\infty A_n \right\}.$$

If it happens to be the case that there is no sequence of $\mathcal{A}$-sets whose union contains A, we define $\mu^*(A) = +\infty$. For brevity, we will refer to a sequence $\{A_n\}_{n=1}^\infty$ of $\mathcal{A}$-sets with $A \subseteq \bigcup_{n=1}^\infty A_n$ as an *$\mathcal{A}$-covering of A.*

We claim that μ^* is an outer measure. This function μ^* is called *the outer measure induced by μ*, and it is a key ingredient in the coming sections. To see that μ^* deserves the title *outer measure*, we must verify (O1)-(O4). Since (O1) is automatic, we need only verify (O2)-(O4).

To see (O2), observe that the sequence $\{A_n\}_{n=1}^{\infty}$, where $A_n = \varnothing$ for each $n \in \mathbb{N}$, is an $\mathcal{A}$-covering of $\varnothing$. Since $\sum_{n=1}^{\infty} \mu(A_n) = \sum_{n=1}^{\infty} \mu(\varnothing) = 0$, we must have $\mu^*(\varnothing) \leq 0$ by the definition of μ^* as an infimum. By (O1), $\mu^*(\varnothing) = 0$.

To see (O3), we let $A \subseteq B \subseteq \Omega$. Consider the two cases of where (i) no $\mathcal{A}$-covering of B exists and (ii) at least one $\mathcal{A}$-covering of B exists. In (i) we have $\mu^*(B) = +\infty$, hence $\mu^*(A) \leq \mu^*(B)$ in a trivial fashion. For (ii), suppose that an $\mathcal{A}$-covering of B exists. Since $B \supseteq A$, this $\mathcal{A}$-covering of B is necessarily an $\mathcal{A}$-covering of A. Thus $\{\sum_{n=1}^{\infty} \mu(A_n) : \{A_n\}_{n=1}^{\infty}$ is an $\mathcal{A}$-covering of $A\}$ is a superset of $\{\sum_{n=1}^{\infty} \mu(A_n) : \{A_n\}_{n=1}^{\infty}$ is an $\mathcal{A}$-covering of $B\}$; the infimum of the former set is accordingly no more than the infimum of the larger set, which means that $\mu^*(A) \leq \mu^*(B)$ for case (ii), and thus monotonicity is established for all cases.

To see (O4), we let $\{A_n\}_{n=1}^{\infty}$ denote a sequence of subsets of Ω; we desire to show that $\mu^*(\bigcup_{n=1}^{\infty} A_n) \leq \sum_{n=1}^{\infty} \mu^*(A_n)$. There are two cases: (i) $\mu^*(A_n) = +\infty$ for some $n \in \mathbb{N}$, and (ii) $\mu^*(A_n) < +\infty$ for all $n \in \mathbb{N}$. In case (i), $\sum_{n=1}^{\infty} \mu^*(A_n) = +\infty$, and thus we trivially obtain $\mu^*(\bigcup_{n=1}^{\infty} A_n) \leq \sum_{n=1}^{\infty} \mu^*(A_n)$. For case (ii), we pick $\epsilon > 0$ and show that $\mu^*(\bigcup_{n=1}^{\infty} A_n) \leq \sum_{n=1}^{\infty} \mu^*(A_n) + \epsilon$, which will prove the result. Observe that if $\mu^*(A_n) < +\infty$ for all $n \in \mathbb{N}$, then for each $n \in \mathbb{N}$ the set A_n possesses at least one $\mathcal{A}$-covering. In particular, the definition of $\mu^*(A_n)$ as an infimum gives the existence of an $\mathcal{A}$-covering $\{B_{nk}\}_{k=1}^{\infty}$ of A_n with $\mu^*(A_n) + \frac{\epsilon}{2^n} > \sum_{k=1}^{\infty} \mu(B_{nk})$, $n \in \mathbb{N}$. It follows that the collection $\{B_{nk} : n, k \in \mathbb{N}\}$ is an $\mathcal{A}$-covering of $\bigcup_{n=1}^{\infty} A_n$, giving

$$\mu^*\left(\bigcup_{n=1}^{\infty} A_n\right) \leq \sum_{n,k\in\mathbb{N}} \mu(B_{nk}) = \sum_{n=1}^{\infty}\sum_{k=1}^{\infty} \mu(B_{nk}) \leq \sum_{n=1}^{\infty}\left[\mu^*(A_n) + \frac{\epsilon}{2^n}\right] \leq \sum_{n=1}^{\infty} \mu^*(A_n) + \epsilon.$$

[The first inequality follows by the definition of μ^* as an infimum and the first equality follows from the fact that a countably infinte set of nonnegative numbers may be summed in any order.] Since $\epsilon > 0$ is arbitrary, we have established countable subadditivity for case (ii), and thus (O4) is satisfied in all cases. This completes the proof of the claim that μ^* is an outer measure.

We also mention another property concerning μ^* in addition to (O1)-(O4) given above: *for any* $A \in \mathcal{A}$ *we have* $\mu^*(A) \leq \mu(A)$. To see this, let $A_1 = A$ and $A_j = \varnothing$ for all $j \geq 2$. Then $\{A_n\}_{n=1}^{\infty}$ is an $\mathcal{A}$-covering of A, and hence $\mu^*(A)$ cannot exceed $\sum_{n=1}^{\infty} \mu(A_n) = \mu(A)$ by how $\mu^*(A)$ is defined as an infimum.

Example 2. Let us apply the important previous example to the case where $\Omega = \mathbb{R}^k$ and the collection $\mathcal{A}$ is the semiring $\mathcal{A}_k$ consisting of $\varnothing$ and the [bounded] rsc k-dimensional rectangles. Let the function μ denote Δ_F, where $F \in \mathfrak{S}_k$ is given for all $\boldsymbol{x} \in \mathbb{R}^k$ by $F(\boldsymbol{x}) = \prod_{i=1}^{n} x_i$. We already know that $\mu = \Delta_F$ is nonnegative and $\mu(\varnothing) = \Delta_F(\varnothing) = 0$, so we are in the framework of Example 1. Observe that any $A \subseteq \mathbb{R}^k$ has an $\mathcal{A}_k$-covering, namely, $\{(-n\mathbf{1}, n\mathbf{1}]\}_{n=1}^{\infty}$ [the symbol $\mathbf{1}$ denotes $(1, \cdots, 1)$]. Thus, the outer measure $\mu^* = \Delta_F^*$ induced by $\mu = \Delta_F$ is given for all $A \subseteq \mathbb{R}^k$ by the formula

$$\mu^*(A) = \Delta_F^*(A) = \inf\left\{\sum_{n=1}^{\infty} \Delta_F(A_n) : \{A_n\}_{n=1}^{\infty} \text{ is an } \mathcal{A}_k\text{-covering of } A\right\}.$$

Since (i) $\varnothing \in \mathcal{A}_k$, (ii) $\Delta_F(\varnothing) = 0$, (iii) $\varnothing$ contributes nothing to unions, and (iv) $\Delta_F((\boldsymbol{a}, \boldsymbol{b}]) = \prod_{i=1}^k (b_i - a_i)$, the above formula becomes

$$\begin{aligned} \mu^*(A) = \Delta_F^*(A) &= \inf\left\{\sum_n \Delta_F((\boldsymbol{a}_n, \boldsymbol{b}_n]) : A \subseteq \bigcup_n (\boldsymbol{a}_n, \boldsymbol{b}_n]\right\} \\ &= \inf\left\{\sum_n \prod_{i=1}^k (b_{ni} - a_{ni}) : A \subseteq \bigcup_n (\boldsymbol{a}_n, \boldsymbol{b}_n]\right\}, \quad A \subseteq \mathbb{R}^k, \end{aligned}$$

where here $\sum_n$ and $\bigcup_n$ denote amc sums and unions. This formula will eventually be taken to be our definition of k-dimensional volume for certain $A \subseteq \mathbb{R}^k$. For the case $k = 1$, we have

$$\mu^*(A) = \Delta_F^*(A) = \inf\left\{\sum_n (b_n - a_n) : A \subseteq \bigcup_n (a_n, b_n]\right\}, \quad A \subseteq \mathbb{R}.$$

This formula will eventually be our definition of the "length" of certain $A \subseteq \mathbb{R}$.

It is not possible to give a truly intuitive justification of outer measures and their importance, not is it possible to avoid the seeming formalism and over-abstraction of Examples 1 and 2. It merely suffices for now to say that they let us get to where we want in an efficient fashion. Other concepts in this chapter will also have the same formalistic flair, and understanding them in their fully general setting will take a good amount of work.

Exercises.

1. Let $(\Omega, \mathcal{A}, \mu)$ denote an arbitrary measure space. Define ν on 2^Ω by writing $\nu(B) = \inf\{\mu(A) : B \subseteq A, A \in \mathcal{A}\}$ for each $B \subseteq \Omega$. Then ν is an outer measure.

2*. Let $\nu : 2^\Omega \to \bar{\mathbb{R}}$ be an outer measure, and suppose in addition that ν is finitely additive: $\nu(A \cup B) = \nu(A) + \nu(B)$ whenever $A, B \subseteq \Omega$ are disjoint. Then ν is a measure. That is, $(\Omega, 2^\Omega, \nu)$ is a measure space. [Observe for every $N \in \mathbb{N}$ that $\nu(\bigcup_{n=1}^\infty A_n) \geq \nu(\bigcup_{n=1}^N A_n) = \sum_{n=1}^N \nu(A_n)$ by (O3) and the assumption. Let $N \to \infty$ and combine the resulting inequality with (O4).]

3. Suppose that ϕ and ξ are outer measures [relative to some common background set Ω], and suppose that we define a new function $\nu : 2^\Omega \to \bar{\mathbb{R}}$ for all $A \subseteq \Omega$ by writing $\nu(A) = \max\{\phi(A), \xi(A)\}$. Then ν is an outer measure [relative to Ω].

4. Let ν denote an outer measure, and let $A \subseteq \Omega$. Define a new set function ν_A on 2^Ω by writing $\nu_A(B) = \nu(B \cap A)$ for each $B \subseteq \Omega$. Then ν_A is an outer measure relative to Ω.

5. Let $\{\nu_n\}_{n=1}^\infty$ denote a sequence of outer measures [relative to some common Ω], and let $\{a_n\}_{n=1}^\infty$ denote a sequence of nonnegative numbers. For each $A \subseteq \Omega$, let $\nu(A) = \sum_{n=1}^\infty a_n \nu_n(A)$. Then ν is an outer measure relative to Ω.

3.3 CARATHÉODORY'S CRITERION

As stated in the beginning of this chapter, one of our goals is to find [among other things] a measure that mimics "length," and in the introduction we asserted [but did not prove] that this measure could not completely mimic what intuition demands of length and still have domain $2^{\mathbb{R}}$, for (AC) blocks the path. Instead, the domain of this hoped-for measure would have to be smaller than $2^{\mathbb{R}}$, but hopefully the domain would still be sufficiently large.

When this problem was being tackled, attempts were made to find a collection of "good" subsets of $\mathbb{R}$ such that including these "good" sets in the domain of the aspiring "length" measure would cause no consistency problems with (AC). It was hoped that this collection of "good" sets would in the end include $\mathcal{B}$ or at least most of $\mathcal{B}$, for $\mathcal{B}$ contains many of the sets that one encounters in doing analysis. Various criteria for "goodness" were proposed, but the criterion of the German mathematician Constantin Carathéodory (1873-1950) turns out to be the standard criterion for "good" sets – that is, sets that would have a meaningfully defined "length".

The main idea is this: if the measure Δ_F and the semiring $\mathcal{A}_1$ were as stated in the introduction to this chapter,[2] and if Δ_F^* denotes the outer measure on $2^{\mathbb{R}}$ induced by Δ_F [from Examples 1 and 2 of the previous section], a set $A \subseteq \mathbb{R}$ will be considered "good" if it meets Carathéodory's Criterion:

$$A \subseteq \mathbb{R} \text{ is "good" iff } \Delta_F^*(T) = \Delta_F^*(T \cap A) + \Delta_F^*(T \cap A^c) \text{ for all } T \subseteq \mathbb{R}.$$

It is difficult if not impossible to see the motivation for such a criterion, and hence one is not attempted. The genius and justification of his approach is that it not only works — it also yields very powerful results without demanding too much work in exchange.

For the definitions and claims of this section, we will merely have a generic background set Ω and an arbitrary outer measure ν. This is more general than the specific case considered by Carathéodory and the earlier mathematicians, but the generality will make things more clear.

Definition. Given an outer measure ν, we define the collection

$$\mathcal{M}(\nu) = \{A \subseteq \Omega : \ \nu(T) = \nu(T \cap A) + \nu(T \cap A^c) \text{ for all } T \subseteq \Omega\}.$$

Various authors refer to the sets in $\mathcal{M}(\nu)$ by different monikers, such as *measurable sets* and *ν-measurable sets.* We, on the other hand, will merely call such sets *$\mathcal{M}(\nu)$-sets.* It is perhaps an unfortunate choice of terminology to use the moniker *ν-measurable* for the sets in $\mathcal{M}(\nu)$ since one might incorrectly but understandably think that a ν-measurable set is a set to which one can apply ν, but at this point it is standard terminology. We will always write

[2]That is, $F \in \mathfrak{S}$ is the identity function, so that $\Delta_F((a, b]) = b - a$ for all $(a, b] \in \mathcal{A}_1$.

$A \in \mathcal{M}(\nu)$ or the equivalent A *is a* $\mathcal{M}(\nu)$*-set* instead of the equivalent A *is* ν*-measurable* or A *is measurable.*

Outer Measure Theorem. *If ν is an outer measure, then $\mathcal{M}(\nu)$ is a σ-field on Ω, and the restriction $\nu_{\mathcal{M}(\nu)}$ of ν from 2^Ω to $\mathcal{M}(\nu)$ is a measure restriction. That is, $(\Omega, \mathcal{M}(\nu), \nu_{\mathcal{M}(\nu)})$ is a measure space.*

Example. Let $k \in \mathbb{N}$, $F \in \mathfrak{S}_k$, and consider the measure $\Delta_F : \mathcal{A}_k \to \bar{\mathbb{R}}$ of the Stieltjes Theorem. If

$$\Delta_F^*(A) = \inf\{\sum_n \Delta_F(A_n) : A \subseteq \bigcup_n A_n, \text{ each } A_n \text{ is in } \mathcal{A}_k\}$$

for each $A \subseteq \mathbb{R}^k$, [$\sum_n$ and $\bigcup_n$ are amc sums and unions], then $\Delta_F^*(A) : 2^{\mathbb{R}^k} \to \bar{\mathbb{R}}$ is the outer measure induced by Δ_F in accordance with Examples 1 and 2 of the previous section. By the Outer Measure Theorem, the collection $\mathcal{M}(\Delta_F^*)$ of $A \subseteq \mathbb{R}^k$ such that $\Delta_F^*(T) = \Delta_F^*(T \cap A) + \Delta_F^*(T \cap A^c)$ for all $T \subseteq \mathbb{R}^k$ is a σ-field of subsets of $\mathbb{R}^k$, and the restriction of Δ_F^* to $\mathcal{M}(\Delta_F^*)$ is a measure on $\mathcal{M}(\Delta_F^*)$. At this stage, it is not clear which subsets of $\mathbb{R}^k$ [if any] are in $\mathcal{M}(\Delta_F^*)$, and we have no practical way of actually computing what value the restriction of Δ_F^* to $\mathcal{M}(\Delta_F^*)$ assigns to a given set in $\mathcal{M}(\Delta_F^*)$.

The Outer Measure Theorem is the main result of this section, and it is the key to the proofs of the Extension Theorem and the Uniqueness Theorem found shortly after this section. The proof of the theorem will be given by the following series of claims.

Claim 1. $\varnothing, \Omega \in \mathcal{M}(\nu)$. *[Appealing to (O2) makes this immediate.]*

Claim 2. *We have the following:*

$$\mathcal{M}(\nu) = \{A \subseteq \Omega : \nu(T) \geq \nu(T \cap A) + \nu(T \cap A^c) \text{ for all } T \subseteq \Omega \text{ with } \nu(T) < +\infty\}. \tag{3.1}$$

Proof: If $A \in \mathcal{M}(\nu)$, then $\nu(T) = \nu(T \cap A) + \nu(T \cap A^c)$ for every $T \subseteq \Omega$, hence $\nu(T) \geq \nu(T \cap A) + \nu(T \cap A^c)$ for every $T \subseteq \Omega$. In particular, the inequality holds for those $T \subseteq \Omega$ with $\nu(T) < +\infty$. This shows that $\mathcal{M}(\nu)$ is contained in the collection on the right side of (3.1).

Next, let $A \subseteq \Omega$ be such that $\nu(T) \geq \nu(T \cap A) + \nu(T \cap A^c)$ for every $T \subseteq \Omega$ with $\nu(T) < +\infty$. Observing that $T = (T \cap A) \cup (T \cap A^c)$ for every $T \subseteq \Omega$ and invoking the finite subadditivity of ν, we obtain

$$\nu(T) = \nu((T \cap A) \cup (T \cap A^c)) \leq \nu(T \cap A) + \nu(T \cap A^c), \quad T \subseteq \Omega. \tag{3.2}$$

(3.2) and the assumed reverse inequality yield the equality

$$\nu(T) = \nu(T \cap A) + \nu(T \cap A^c), \qquad T \subseteq \Omega \text{ and } \nu(T) < +\infty. \tag{3.3}$$

For those $T \subseteq \Omega$ such that $\nu(T) = +\infty$, (3.2) is actually an equality, hence

$$\nu(T) = \nu(T \cap A) + \nu(T \cap A^c), \qquad T \subseteq \Omega \text{ and } \nu(T) = +\infty. \tag{3.4}$$

Putting (3.3) and (3.4) together yields $\nu(T) = \nu(T \cap A) + \nu(T \cap A^c)$ for all $T \subseteq \Omega$ and hence $A \in \mathcal{M}(\nu)$. It follows that the right side of (3.1) is contained in $\mathcal{M}(\nu)$, and the entire proof is complete. ■

Claim 3. *$\mathcal{M}(\nu)$ is a field on Ω.*

Proof: $\Omega \in \mathcal{M}(\nu)$ by Claim 1, and $\mathcal{M}(\nu)$ is clearly closed under complementation. To complete the proof, it is sufficient to show that $\mathcal{M}(\nu)$ is closed under intersection. By Claim 2, this means showing that $A, B \in \mathcal{M}(\nu)$ implies

$$\nu(T) \geq \nu(T \cap (A \cap B)) + \nu(T \cap (A \cap B)^c) \quad T \subseteq \Omega,\ \nu(T) < +\infty. \tag{3.5}$$

Let $A, B \in \mathcal{M}(\nu)$, and let $T \subseteq \Omega$ be such that $\nu(T) < +\infty$. Since $B \in \mathcal{M}(\nu)$, we have $\nu(T) = \nu(T \cap B) + \nu(T \cap B^c)$. Since $A \in \mathcal{M}(\nu)$, we have

$$\nu(T \cap B) = \nu((T \cap B) \cap A) + \nu((T \cap B) \cap A^c)$$

and

$$\nu(T \cap B^c) = \nu((T \cap B^c) \cap A) + \nu((T \cap B^c) \cap A^c).$$

With these three equalities and finite subadditivity, we obtain

$$\begin{aligned} & \nu(T) \\ = \;& \nu(T \cap B) + \nu(T \cap B^c) \\ = \;& \nu(T \cap B \cap A) + \nu(T \cap B \cap A^c) + \nu(T \cap B^c \cap A) + \nu(T \cap B^c \cap A^c) \\ \geq \;& \nu(T \cap (A \cap B)) + \nu((T \cap B \cap A^c) \cup (T \cap B^c \cap A) \cup (T \cap B^c \cap A^c)) \\ = \;& \nu(T \cap (A \cap B)) + \nu(T \cap (A \cap B)^c), \end{aligned}$$

whence (3.5) holds, and it follows that $A \cap B \in \mathcal{M}(\nu)$. ■

One of our goals is to show that $\mathcal{M}(\nu)$ is a σ-field, which means that we still must show that $\mathcal{M}(\nu)$ is closed under countable unions. We will work up to this by a series of intermediate claims.

Claim 4. *Let $n \in \mathbb{N}$, and let $A_1, \cdots, A_n \in \mathcal{M}(\nu)$ be disjoint. Then*

$$\nu(T) = \sum_{i=1}^{n} \nu(T \cap A_i) + \nu\left(T \cap \left(\bigcup_{i=1}^{n} A_i\right)^c\right), \quad T \subseteq \Omega. \tag{3.6}$$

In particular, if $T = \bigcup_{i=1}^n A_i$, then $\nu(\bigcup_{i=1}^n A_i) = \sum_{i=1}^n \nu(A_i)$, so that the restriction of ν to $\mathcal{M}(\nu)$ is finitely additive.

Proof: If $n = 1$, (3.6) is automatic by the definition of $\mathcal{M}(\nu)$. Suppose now that (3.6) holds for all $m = 1, \cdots, n$ with $n \in \mathbb{N}$ fixed and generic; we show that (3.6) holds in the $n + 1$ case. Let $A_1, \cdots, A_{n+1} \in \mathcal{M}(\nu)$ be disjoint.

Then, for all $T \subseteq \Omega$, we obtain

$$\begin{aligned}
&\nu(T) \\
=\;& \nu(T \cap A_{n+1}) + \nu(T \cap A_{n+1}^c) \\
=\;& \nu(T \cap A_{n+1}) + \sum_{i=1}^{n} \nu((T \cap A_{n+1}^c) \cap A_i) + \nu\left((T \cap A_{n+1}^c) \cap \left(\bigcup_{i=1}^{n} A_i\right)^c\right) \\
=\;& \nu(T \cap A_{n+1}) + \sum_{i=1}^{n} \nu(T \cap A_i) + \nu\left(T \cap \left(\bigcup_{i=1}^{n+1} A_i\right)^c\right) \\
=\;& \sum_{i=1}^{n+1} \nu(T \cap A_i) + \nu\left(T \cap \left(\bigcup_{i=1}^{n+1} A_i\right)^c\right).
\end{aligned}$$

[The first equality follows since $A_{n+1} \in \mathcal{M}(\nu)$. The second equality follows by the induction hypothesis. The third equality follows from the disjointness of $A_1, \cdots, A_{n+1}$.] This shows that (3.6) holds in the $n+1$ case. Therefore, (3.6) holds for all $n \in \mathbb{N}$. The second part of the claim follows by direct substitution. ■

Claim 5. *Let $\{A_m\}_{m=1}^{\infty}$ denote a disjoint sequence of $\mathcal{M}(\nu)$-sets. Then*

$$\nu(T) = \sum_{m=1}^{\infty} \nu(T \cap A_m) + \nu\left(T \cap \left(\bigcup_{m=1}^{\infty} A_m\right)^c\right), \quad T \subseteq \Omega \tag{3.7}$$

Proof: Let $\{A_m\}_{m=1}^{\infty}$ denote a sequence of disjoint $\mathcal{M}(\nu)$-sets. We show the above equality (3.7) by showing first for all $T \subseteq \Omega$ that

$$\nu(T) \le \sum_{m=1}^{\infty} \nu(T \cap A_m) + \nu\left(T \cap \left(\bigcup_{m=1}^{\infty} A_m\right)^c\right), \tag{3.8}$$

and then demonstrating for all $T \subseteq \Omega$ that

$$\nu(T) \ge \sum_{m=1}^{\infty} \nu(T \cap A_m) + \nu\left(T \cap \left(\bigcup_{m=1}^{\infty} A_m\right)^c\right). \tag{3.9}$$

To show (3.8), pick $T \subseteq \Omega$. We therefore have

$$\begin{aligned}
\nu(T) &= \nu\left(\left(T \cap \bigcup_{m=1}^{\infty} A_m\right) \cup \left(T \cap \left(\bigcup_{m=1}^{\infty} A_m\right)^c\right)\right) \\
&\le \nu\left(\bigcup_{m=1}^{\infty} (T \cap A_m)\right) + \nu\left(T \cap \left(\bigcup_{m=1}^{\infty} A_m\right)^c\right) \\
&\le \sum_{m=1}^{\infty} \nu(T \cap A_m) + \nu\left(T \cap \left(\bigcup_{m=1}^{\infty} A_m\right)^c\right),
\end{aligned}$$

where the first inequality follows by finite subadditivity, and the second inequality follows by countable subadditivity. Therefore, (3.8) holds.

To establish the reverse inequality (3.9), pick $T \subseteq \Omega$ and consider the cases where $\nu(T) = +\infty$ and where $\nu(T) < +\infty$. If $T \subseteq \Omega$ is such that $\nu(T) = +\infty$, then (3.9) trivially holds. Therefore, we thus only need show that (3.9) holds for all $T \subseteq \Omega$ with $\nu(T) < +\infty$ in order to complete the proof. We proceed by picking an arbitrary $T \subseteq \Omega$ with $\nu(T) < +\infty$. We recall Claim 4, which asserts that

$$\nu(T) = \sum_{i=1}^{n} \nu(T \cap A_i) + \nu\left(T \cap \left(\bigcup_{i=1}^{n} A_i\right)^c\right)$$

for every $n \in \mathbb{N}$. Next, we observe that

$$\begin{aligned} \nu(T) &= \lim_{n\to\infty}\left(\sum_{m=1}^{n} \nu(T\cap A_m) + \nu\left(T \cap \left(\bigcup_{m=1}^{n} A_m\right)^c\right)\right) \\ &= \lim_{n\to\infty}\sum_{m=1}^{n} \nu(T\cap A_m) + \lim_{n\to\infty} \nu\left(T \cap \left(\bigcup_{m=1}^{n} A_m\right)^c\right) \\ &\geq \sum_{m=1}^{\infty} \nu(T\cap A_m) + \nu\left(T \cap \left(\bigcup_{m=1}^{\infty} A_m\right)^c\right). \end{aligned}$$

[The inequality follows because (i) $\nu(T \cap (\bigcup_{m=1}^{n} A_m)^c) \geq \nu(T \cap (\bigcup_{m=1}^{\infty} A_m)^c)$ for all $n \in \mathbb{N}$, and (ii) the sequence of values $\{\nu(T \cap (\bigcup_{m=1}^{n} A_m)^c)\}_{n=1}^{\infty}$ is nondecreasing and hence has a limit.] This establishes (3.9) for any $T \subseteq \Omega$ such that $\nu(T) < +\infty$, and the proof of this claim is complete. ■

Claim 6. *Let $\{A_n\}_{n=1}^{\infty}$ denote a disjoint sequence of $\mathcal{M}(\nu)$-sets. Then*
(i) $\nu\left(T \cap \bigcup_{n=1}^{\infty} A_n\right) = \sum_{n=1}^{\infty} \nu(T \cap A_n)$ for any $T \subseteq \Omega$, and
(ii) $\nu(\bigcup_{n=1}^{\infty} A_n) = \sum_{n=1}^{\infty} \nu(A_n)$, so that the restriction of ν to $\mathcal{M}(\nu)$ is countably additive.

Proof: For (i), go to Claim 5, replace T by $T \cap \bigcup_{n=1}^{\infty} A_n$, and use (O2). For (ii), replace T by $\bigcup_{n=1}^{\infty} A_n$ in (i). ■

Claim 7. *The field $\mathcal{M}(\nu)$ is in fact a σ-field.*

Proof: We need only to show that $\mathcal{M}(\nu)$ is closed under countable unions. Since any field that is closed under countable disjoint unions is closed under countable unions,[3] it is sufficient to show that $\mathcal{M}(\nu)$ is closed under countable disjoint unions. Let $\{A_k\}_{k=1}^{\infty}$ denote a sequence of disjoint sets in $\mathcal{M}(\nu)$, and

[3]Proof: Let $A_1, A_2, \cdots \in \mathcal{F}$ where $\mathcal{F}$ is a field closed under countable disjoint unions. Let $B_1 = A_1$ and $B_k = A_k \cap \bigcap_{j=1}^{k-1} A_j^c$ for all $k \geq 2$. Then $B_1, B_2, \cdots \in \mathcal{F}$ are disjoint with $\bigcup_{k=1}^{\infty} B_k \in \mathcal{F}$. Since $\bigcup_{k=1}^{\infty} A_k = \bigcup_{k=1}^{\infty} B_k$, we have $\bigcup_{k=1}^{\infty} A_k \in \mathcal{F}$.

pick $T \subseteq \Omega$ with $\nu(T) < +\infty$. Observe that for each $n \in \mathbb{N}$ we have

$$\begin{aligned} \nu(T) &= \nu\left(T \cap \left(\bigcup_{k=1}^{n} A_k\right)\right) + \nu\left(T \cap \left(\bigcup_{k=1}^{n} A_k\right)^c\right) \\ &\geq \sum_{k=1}^{n} \nu(T \cap A_k) + \nu\left(T \cap \left(\bigcup_{k=1}^{\infty} A_k\right)^c\right), \end{aligned} \tag{3.10}$$

where the equality follows from the fact that $\mathcal{M}(\nu)$ is a field and hence $\bigcup_{k=1}^{n} A_k \in \mathcal{M}(\nu)$ for each $n \in \mathbb{N}$, and the inequality follows because (i) $\nu(\bigcup_{k=1}^{n}(T \cap A_k)) = \sum_{k=1}^{n} \nu(T \cap A_k)$ [Claim 4] and (ii) ν is monotone. Since $n \in \mathbb{N}$ is arbitrary, we let $n \to +\infty$ in (3.10), obtaining

$$\begin{aligned} \nu(T) &\geq \sum_{k=1}^{\infty} \nu(T \cap A_k) + \nu\left(T \cap \left(\bigcup_{k=1}^{\infty} A_k\right)^c\right) \\ &= \nu\left(T \cap \bigcup_{k=1}^{\infty} A_k\right) + \nu\left(T \cap \left(\bigcup_{k=1}^{\infty} A_k\right)^c\right), \end{aligned}$$

where the equality follows from Claim 6. By the fact that our choice of $T \subseteq \Omega$ with $\nu(T) < +\infty$ is generic, Claim 2 gives $\bigcup_{k=1}^{\infty} A_k \in \mathcal{M}(\nu)$, and thus $\mathcal{M}(\nu)$ is closed under countable disjoint unions. ■

Claim 8. *The restriction $\nu_{\mathcal{M}(\nu)}$ of ν from 2^{Ω} to $\mathcal{M}(\nu)$ is a measure.*

Proof: It is clear that $\nu_{\mathcal{M}(\nu)}$ is the restriction of a nonnegative function and hence is nonnegative; also note that $\varnothing \in \mathcal{M}(\nu)$ by Claim 1, hence $\nu_{\mathcal{M}(\nu)}(\varnothing)$ is defined and equals $\nu(\varnothing) = 0$. To show that $\nu_{\mathcal{M}(\nu)}$ is countably additive, we merely appeal to Claim 6. ■

By putting Claims 7 and 8 together, the proof of the Outer Measure Theorem is complete. This theorem holds for all outer measures, not just the special outer measure given in Example 1 of Section 2. This theorem will also be the main ingredient for the Extension Theorem presented in the next section. We close this section with the following simple claim used in later work.

Claim 9. *If $A \subseteq \Omega$ and $\nu(A) = 0$, then $A \in \mathcal{M}(\nu)$.*

Proof: Assume the hypotheses, and let $T \subseteq \Omega$. Observe that $T \cap A \subseteq A$, hence monotonicity gives $\nu(T \cap A) \leq \nu(A) = 0$, so $\nu(T \cap A) = 0$. Observe next that $T \supseteq T \cap A^c$, whence $\nu(T) \geq \nu(T \cap A^c)$. Therefore $\nu(T) \geq \nu(T \cap A) + \nu(T \cap A^c)$, hence $A \in \mathcal{M}(\nu)$ by Claim 2. ■

Exercises.

1. Show *directly* that if $A, B \in \mathcal{M}(\nu)$, then $A \cup B, A - B \in \mathcal{M}(\nu)$. [Hint: for unions, first prove the identity $T \cap (A \cup B) = (T \cap A) \cup (T \cap A^c \cap B)$.]

2. Suppose that Ω may be written as $\bigcup_{n=1}^{\infty} A_n$, where $\{A_n\}_{n=1}^{\infty}$ is a nondecreasing sequence of subsets of Ω. If $A \subseteq \Omega$ is such that $A \cap A_k \in \mathcal{M}(\nu)$ for all k exceeding some constant k_A, then $A \in \mathcal{M}(\nu)$.

3. Let ν denote an outer measure such that $\nu(\Omega) < +\infty$, and further suppose that if $A \subseteq \Omega$ with $\nu(A) < +\infty$, then there exists $B \in \mathcal{M}(\nu)$ such that $A \subseteq B$ and $\nu(A) = \nu(B)$. Then $E \in \mathcal{M}(\nu)$ iff $\nu(\Omega) = \nu(E) + \nu(E^c)$.

4. Suppose that ν is an arbitrary outer measure, and let $A, B \subseteq \Omega$ with $A \in \mathcal{M}(\nu)$. Show that $\nu(A \cup B) + \nu(A \cap B) = \nu(A) + \nu(B)$.

5. Let ν denote an outer measure, and let $\{A_n\}_{n=1}^{\infty}$ denote a nondecreasing sequence of $\mathcal{M}(\nu)$-sets. Show that $\nu(\lim(A \cap A_n)) = \lim \nu(A \cap A_n)$ for any $A \subseteq \Omega$. State and prove an analogous result for nonincreasing sequences of $\mathcal{M}(\nu)$-sets. [This will require an additional finiteness assumption.]

6. Let ν denote an outer measure such that the following holds: if $A \subseteq \Omega$ with $\nu(A) < +\infty$, then there is $B \in \mathcal{M}(\nu)$ with $A \subseteq B$ and $\nu(A) = \nu(B)$. Then, for any nondecreasing sequence $\{A_n\}_{n=1}^{\infty}$ of subsets of Ω, we have $\nu(\lim A_n) = \lim \nu(A_n)$.

7. In each of the following parts, (i) describe the outer measure μ^* on 2^Ω induced by the given μ, (ii) describe the collection $\mathcal{M}(\mu^*)$ and determine if $\mathcal{M}(\mu^*)$ is a σ-field, and (iii) check to see whether $\mu^* = \mu$ on the given collection $\mathcal{A}$.
(a) $\Omega = \{1, 2, 3\}$, $\mathcal{A} = \{\varnothing, \{1\}, \{2, 3\}, \Omega\}$, and μ is a measure on $\mathcal{A}$ such that $\mu(\Omega) = 1$ and $\mu(\{1\}) = 0$.
(b) Ω and $\mathcal{A}$ are as in (a), but ν is a measure on $\mathcal{A}$ such that $\nu(\Omega) = 1$ and $\nu(\{2, 3\}) = 0$. Also, show that $\mathcal{M}(\mu^*) \neq \mathcal{M}(\nu^*)$.
(c) Let $\mathcal{A}$ denote a field, let $\omega_0 \in \Omega$ be such that $\{\omega_0\} \in \sigma(\mathcal{A})$, and let μ be defined on $\mathcal{A}$ by $\mu(A) = I_A(\omega_0)$ for all $A \in \mathcal{A}$.
(d) Let Ω be countably infinite, let $\mathcal{A}$ denote the field of sets A such that A is finite or A^c is finite, and write $\mu(A) = 0$ if A is finite and $\mu(A) = 1$ if A^c is finite.
(e) Redo (d) assuming that Ω is uncountable.
(f) Let Ω be uncountable, and let $\mathcal{A}$ denote the field of sets A such that A is amc or A^c is amc. Let $\mu(A) = 0$ if A is amc, and let $\mu(A) = 1$ if A^c is amc.

3.4 EXISTENCE OF EXTENSIONS

The famous theorem of this section is known as *Carathéodory's Extension Theorem.* This theorem is the main tool that allows us to assert the existence of measures with large domains that can mimic "length," "area," and the like. The theorem also justifies our earlier study of seemingly nonintuitive set systems such as semirings.

Extension Theorem. *Let $\mathcal{A}$ denote a semiring on Ω. Let $\mu : \mathcal{A} \to \bar{\mathbb{R}}$ be such that*

(i) $\mu(A) \geq 0$ for all $A \in \mathcal{A}$,
(ii) $\mu(\varnothing) = 0$,
(iii) μ is finitely additive, and
(iv) μ is countably subadditive.
Then there exists a measure extension of μ from $\mathcal{A}$ to $\sigma(\mathcal{A})$.

To be sure, condition (iii) means that if $A_1, \cdots, A_n \in \mathcal{A}$ are disjoint with $\bigcup_{i=1}^n A_i \in \mathcal{A}$, then $\mu(\bigcup_{i=1}^n A_i) = \sum_{i=1}^n \mu(A_i)$. Condition (iv) means that if $\{A_n\}_{n=1}^\infty$ is a sequence of $\mathcal{A}$-sets with $\bigcup_{n=1}^\infty A_n \in \mathcal{A}$, then the inequality $\mu(\bigcup_{n=1}^\infty A_n) \leq \sum_{n=1}^\infty \mu(A_n)$ holds.

The conclusion of the theorem is that there exists a measure $\nu : \sigma(\mathcal{A}) \to \bar{\mathbb{R}}$ such that $\nu(A) = \mu(A)$ for every $A \in \mathcal{A}$. At this stage, we do not know if the measure ν is necessarily unique. That is, there may be some other measure $\xi : \sigma(\mathcal{A}) \to \bar{\mathbb{R}}$ such that $\xi(A) = \mu(A)$ for every $A \in \mathcal{A}$, but $\xi(B) \neq \nu(B)$ for some $B \in \sigma(\mathcal{A}) - \mathcal{A}$. The next section will discuss conditions under which such a measure extension ν from $\mathcal{A}$ to $\sigma(\mathcal{A})$ is unique.

The theorem may appear somewhat abstract at first, but its profundity will hopefully become more clear over time. The theorem will ultimately be used [among other things] to justify those concepts of analysis and calculus that have been taken for granted.

Before proving the theorem, a very important example of the theorem's application will be presented.

Example. Let $k \in \mathbb{N}$ and let $\Omega = \mathbb{R}^k$. Consider the familiar semiring $\mathcal{A}_k$ consisting of $\varnothing$ and the [bounded] k-dimensional rsc rectangles. If $F(\boldsymbol{x}) = \prod_{i=1}^k x_i$ for each $\boldsymbol{x} \in \mathbb{R}^k$, then we have $F \in \mathfrak{S}_k$ and $\Delta_F((\boldsymbol{a}, \boldsymbol{b}]) = \prod_{i=1}^k (b_i - a_i)$ for each $(\boldsymbol{a}, \boldsymbol{b}] \in \mathcal{A}_k$. By the Stieltjes Theorem, Δ_F is a measure, and Δ_F corresponds to our idea of the "k-dimensional volume" of $(\boldsymbol{a}, \boldsymbol{b}] \in \mathcal{A}_k$. [Recall Examples 1 and 2 of Section 2.3.]

By the Extension Theorem, there is at least one measure — call it λ_k — defined on $\sigma(\mathcal{A}_k)$, such that $\lambda_k((\boldsymbol{a}, \boldsymbol{b}]) = \prod_{i=1}^k (b_i - a_i)$ for all $(\boldsymbol{a}, \boldsymbol{b}] \in \mathcal{A}_k$. Since $\sigma(\mathcal{A}_k) = \mathcal{B}^k$, we have a measure defined on $\mathcal{B}^k$ that does what we think "k-dimensional volume" should do, at least for those elements of $\mathcal{A}_k$. The question is whether λ_k does what we think "k-dimensional volume" should do for those sets in $\mathcal{B}^k - \mathcal{A}_k$.

Let us take $k = 1$, denoting λ_1 by λ. Intuition demands that each of the $\mathcal{B}$-sets $(a, b), [a, b], [a, b)$, and $(a, b]$ should have "length" $b - a$ for each $a, b \in \mathbb{R}$ with $a < b$. Intuition also demands that each of the $\mathcal{B}$-sets $(-\infty, x), (-\infty, x], (x, +\infty)$, and $[x, +\infty)$ should have "length" $+\infty$ for each $x \in \mathbb{R}$. The "length" of each $\{x\} \in \mathcal{B}$ should be 0 as well, and so on. *For such a λ [unique or not], our desires are met.*

To prove such a claim, observe that $\{x\} = \lim(x - n^{-1}, x]$ for any $x \in \mathbb{R}$. Since λ is a measure that agrees with Δ_F on $\mathcal{A}_1$, we have by (M9) that

$$\begin{aligned} \lambda(\{x\}) = \lambda(\lim(x - n^{-1}, x]) = \lim \lambda((x - n^{-1}, x]) &= \lim \Delta_F((x - n^{-1}, x]) \\ &= \lim n^{-1} = 0. \end{aligned}$$

It follows by (M3) or (M4) that for any amc set $\{x_1, x_2, \cdots\}$ we have the equation $\lambda(\{x_1, x_2, \cdots\}) = \sum_n \lambda(\{x_n\}) = 0$. Next, by (M5),

$$\lambda((a, b)) = \lambda((a, b] - \{b\}) = \lambda((a, b]) - \lambda(\{b\}) = \Delta_F((a, b]) - 0 = b - a$$

for any $a, b \in \mathbb{R}$ with $a < b$. The other bounded intervals are handled similarly. Next, observe for all $x \in \mathbb{R}$ that (M8) gives

$$\begin{aligned} \lambda((x, +\infty]) = \lambda(\lim(x, x + n]) = \lim \lambda((x, x + n]) &= \lim \Delta_F((x, x + n]) \\ &= \lim n = +\infty. \end{aligned}$$

It is similarly shown that λ assigns $+\infty$ to any other unbounded interval. In this context then, any measure λ guaranteed by the Extension Theorem will mimic our idea of "length" — at least in simple situations — while agreeing with Δ_F on $\mathcal{A}_1$.

The Extension Theorem is why we spent considerable effort in proving the Stieltjes Theorem, a theorem that us to assert the existence of a large class of measures on the semiring $\mathcal{A}_k$. Observe that for any $F \in \mathfrak{S}_k$ the measure Δ_F as given in the Stieltjes Theorem automatically satisfies the requirements of the Extension Theorem. [The previous example deals with the special $F \in \mathfrak{S}_k$ such that $F(\boldsymbol{x}) = \prod_{i=1}^k x_i$ for all $\boldsymbol{x} \in \mathbb{R}^k$.]

Like the Outer Measure Theorem of the previous section, the Extension Theorem's proof will be broken into various claims. Throughout the claims, note the importance of the outer measure μ^* induced by μ as given in Example 1 of Section 2.

Claim 1. *μ is monotone: if $A \subseteq B$ and $A, B \in \mathcal{A}$, then $\mu(A) \leq \mu(B)$.*

Proof: Let $A \subseteq B$ with $A, B \in \mathcal{A}$. Since $\mathcal{A}$ is a semiring, there exists a finite collection of disjoint sets $C_1, \cdots, C_m \in \mathcal{A}$ with $B = A \cup \bigcup_{i=1}^m C_i$, both unions being disjoint. [This follows by Claim 1 of Section 2.4.] By nonnegativity and finite additivity, we have

$$\mu(B) = \mu(A) + \sum_{i=1}^m \mu(C_i) \geq \mu(A),$$

and the proof is complete. ∎

Recall that the outer measure $\mu^* : 2^\Omega \to \bar{\mathbb{R}}$ induced by $\mu : \mathcal{A} \to \bar{\mathbb{R}}$ is defined as follows: if $A \subseteq \Omega$ is such that there exists a sequence of $\mathcal{A}$-sets $\{A_n\}_{n=1}^\infty$ with $A \subseteq \bigcup_{n=1}^\infty A_n$, that is, if $A \subseteq \Omega$ has an $\mathcal{A}$-covering, then we define

$$\mu^*(A) = \inf\left\{\sum_{n=1}^\infty \mu(A_n) : \{A_n\}_{n=1}^\infty \text{ is an } \mathcal{A}\text{-covering of } A\right\}.$$

If no $\mathcal{A}$-covering of A exists, we set $\mu^*(A) = +\infty$. Define $\mathcal{M}(\mu^*)$ as before:

$$\mathcal{M}(\mu^*) = \{A \subseteq \Omega : \mu^*(T) = \mu^*(T \cap A) + \mu^*(T \cap A^c) \text{ for all } T \subseteq \Omega\}.$$

By the Outer Measure Theorem, *$\mathcal{M}(\mu^*)$ is a σ-field on Ω, and the restriction of the outer measure μ^* to the σ-field $\mathcal{M}(\mu^*)$ is a measure.* Let us denote this restriction of μ^* to $\mathcal{M}(\mu^*)$ by the symbol $\mu^*_{\mathcal{M}(\mu^*)}$. We now have three functions: (i) the original $\mu : \mathcal{A} \to \bar{\mathbb{R}}$, (ii) the outer measure $\mu^* : 2^\Omega \to \bar{\mathbb{R}}$ induced by μ, and (iii) $\mu^*_{\mathcal{M}(\mu^*)} : \mathcal{M}(\mu^*) \to \bar{\mathbb{R}}$, which agrees with μ^* on $\mathcal{M}(\mu^*)$ and is a measure on $\mathcal{M}(\mu^*)$.

Claim 2. *$\mathcal{A} \subseteq \mathcal{M}(\mu^*)$.*

Proof: Let $A \in \mathcal{A}$. By Claim 2 of Section 3, we desire to show that

$$\mu^*(T) \geq \mu^*(T \cap A) + \mu^*(T \cap A^c) \quad \text{for all } T \subseteq \Omega \text{ with } \mu^*(T) < +\infty.$$

To show this, we will demonstrate for every $\epsilon > 0$ that

$$\mu^*(T) + \epsilon \geq \mu^*(T \cap A) + \mu^*(T \cap A^c) \quad \text{for all } T \subseteq \Omega \text{ with } \mu^*(T) < +\infty.$$

Let $\epsilon > 0$, and choose $T \subseteq \Omega$ with $\mu^*(T) < +\infty$. By the definition of μ^*, we may find an $\mathcal{A}$-covering $\{A_n\}_{n=1}^{\infty}$ of T with $\sum_{n=1}^{\infty} \mu(A_n) < \mu^*(T) + \epsilon$. For each $n \in \mathbb{N}$, let $B_n = A \cap A_n$, so that we have $B_n \in \mathcal{A}$ and $B_n \subseteq A_n$ for each $n \in \mathbb{N}$. Next,

$$A \cap T \subseteq A \cap \bigcup_{n=1}^{\infty} A_n = \bigcup_{n=1}^{\infty} (A \cap A_n) = \bigcup_{n=1}^{\infty} B_n$$

and

$$\begin{aligned} A^c \cap T \subseteq A^c \cap \bigcup_{n=1}^{\infty} A_n = \bigcup_{n=1}^{\infty} (A_n - A) &= \bigcup_{n=1}^{\infty} (A_n - B_n) \\ &= \bigcup_{n=1}^{\infty} \bigcup_{k=1}^{m_n} C_{nk} \quad \text{[say]}, \end{aligned}$$

where the last equality follows because for each $n \in \mathbb{N}$ we have $A_n, B_n \in \mathcal{A}$ with $B_n \subseteq A_n$, and hence there exists a finite collection $C_{n1}, \cdots, C_{nm_n}$ of disjoint $\mathcal{A}$-sets with $A_n - B_n = \bigcup_{k=1}^{m_n} C_{nk}$. [This is Claim 1 of Section 2.4.] Using these relations, Claim 1, countable subadditivity, and the definition of μ^* as an infimum, we obtain

$$\begin{aligned} \mu^*(T \cap A) + \mu^*(T \cap A^c) &\leq \mu^*\left(\bigcup_{n=1}^{\infty} B_n\right) + \mu^*\left(\bigcup_{n=1}^{\infty} \bigcup_{k=1}^{m_n} C_{nk}\right) \\ &\leq \sum_{n=1}^{\infty} \mu(B_n) + \sum_{n=1}^{\infty} \sum_{k=1}^{m_n} \mu(C_{nk}) \\ &= \sum_{n=1}^{\infty} \left(\mu(B_n) + \sum_{k=1}^{m_n} \mu(C_{nk})\right) \\ &= \sum_{n=1}^{\infty} \mu(A_n) \\ &< \mu^*(T) + \epsilon, \end{aligned}$$

where the final equality holds because we have $A_n = B_n \cup \bigcup_{k=1}^{m_n} C_{nk}$ [a disjoint union] for each $n \in \mathbb{N}$ and because μ is finitely additive. Since $\epsilon > 0$ was arbitrary, this completes the proof. ■

Claim 3. $\mathcal{A} \subseteq \sigma(\mathcal{A}) \subseteq \mathcal{M}(\mu^*) \subseteq 2^{\Omega}$.

Proof: $\mathcal{M}(\mu^*)$ is a σ-field that contains $\mathcal{A}$ by virtue of Claim 2. However, $\sigma(\mathcal{A})$ is the minimal σ-field that contains $\mathcal{A}$, hence $\sigma(\mathcal{A}) \subseteq \mathcal{M}(\mu^*)$. The other inclusions $\mathcal{A} \subseteq \sigma(\mathcal{A})$ and $\mathcal{M}(\mu^*) \subseteq 2^{\Omega}$ are trivial. ■

Claim 4. *We have*
(i) $\mu^* = \mu$ *on* $\mathcal{A}$, *and*
(ii) $\mu^*_{\mathcal{M}(\mu^*)} = \mu$ *on* $\mathcal{A}$.
That is, μ, μ^*, *and* $\mu^*_{\mathcal{M}(\mu^*)}$ *agree on* $\mathcal{A}$.

Proof: In Example 1 of Section 2 it was shown that $\mu^*(A) \leq \mu(A)$ for each $A \in \mathcal{A}$. We thus only need to show that $\mu(A) \leq \mu^*(A)$ for each $A \in \mathcal{A}$. Pick $A \in \mathcal{A}$. If $\mu^*(A) = +\infty$, we are trivially done. We thus assume that $\mu^*(A) < +\infty$, so that at least one $\mathcal{A}$-covering of A exists. Pick a generic $\mathcal{A}$-covering $\{A_n\}_{n=1}^{\infty}$ of A. It follows that $A = \bigcup_{n=1}^{\infty}(A_n \cap A)$ and hence

$$\mu(A) = \mu\left(\bigcup_{n=1}^{\infty}(A_n \cap A)\right) \leq \sum_{n=1}^{\infty}\mu(A_n \cap A) \leq \sum_{n=1}^{\infty}\mu(A_n),$$

where the first inequality follows because of the countable subadditivity of μ; the second inequality follows by noting that $A_n \cap A \subseteq A_n$ for all $n \in \mathbb{N}$ and invoking Claim 1. We have thus shown that $\mu(A) \leq \sum_{n=1}^{\infty}\mu(A_n)$ for an arbitrary $\mathcal{A}$-covering $\{A_n\}_{n=1}^{\infty}$ of A, and thus $\mu(A)$ is no more than the infimum of the set of sums of the form $\sum_{n=1}^{\infty}\mu(A_n)$, where $\{A_n\}_{n=1}^{\infty}$ is an arbitrary $\mathcal{A}$-covering of A. Thus, we have $\mu(A) \leq \mu^*(A)$ for every $A \in \mathcal{A}$, and the proof of (i) is complete.

We now prove (ii). If $A \in \mathcal{A}$, then Claim 3 gives $A \in \mathcal{M}(\mu^*)$. Since $\mu^*_{\mathcal{M}(\mu^*)} = \mu^*$ on $\mathcal{M}(\mu^*)$, we use (i) and obtain $\mu^*_{\mathcal{M}(\mu^*)}(A) = \mu^*(A) = \mu(A)$. ∎

Claim 5. *The restriction of the measure* $\mu^*_{\mathcal{M}(\mu^*)}$ *from the* σ*-field* $\mathcal{M}(\mu^*)$ *to the sub-*σ*-field* $\sigma(\mathcal{A})$ *is a measure that agrees with* μ *on* $\mathcal{A}$. *Equivalently, the restriction of* μ^* *from* 2^{Ω} *to* $\sigma(\mathcal{A})$ *is a measure that agrees with* μ *on* $\mathcal{A}$.

Proof: This follows from Claims 3 and 4 of this section and from Claim 2 of Section 1. ∎

The Extension Theorem's proof is complete. If $\mu^*_{\sigma(\mathcal{A})} : \sigma(\mathcal{A}) \to \bar{\mathbb{R}}$ denotes the above measure extension of μ from $\mathcal{A}$ to $\sigma(\mathcal{A})$, the "explicit" formula for $\mu^*_{\sigma(\mathcal{A})}(A)$ is given for all $A \in \sigma(\mathcal{A})$ by

$$\mu^*_{\sigma(\mathcal{A})}(A) = \inf\left\{\sum_{n=1}^{\infty}\mu(A_n) : \{A_n\}_{n=1}^{\infty} \text{ is an } \mathcal{A}\text{-covering of } A\right\}.$$

Since $\mu^*_{\sigma(\mathcal{A})}$ is an extension of μ from $\mathcal{A}$ to $\sigma(\mathcal{A})$, $\mu^*_{\sigma(\mathcal{A})}(A) = \mu(A)$ if $A \in \mathcal{A}$.

We have actually proved something along the way that is stronger than the Extension Theorem's claim, for we have a measure extension of μ not only to $\sigma(\mathcal{A})$, but we actually have a measure extension of μ to the σ-field $\mathcal{M}(\mu^*)$ that contains $\sigma(\mathcal{A})$, namely, $\mu^*_{\mathcal{M}(\mu^*)}$. Given any $A \in \mathcal{M}(\mu^*)$, the "explicit" formula for $\mu^*_{\mathcal{M}(\mu^*)}(A)$ is given by

$$\mu^*_{\mathcal{M}(\mu^*)}(A) = \inf\left\{\sum_{n=1}^{\infty}\mu(A_n) : \{A_n\}_{n=1}^{\infty} \text{ is an } \mathcal{A}\text{-covering of } A\right\}.$$

By Claim 4, this will reduce to $\mu(A)$ if $A \in \mathcal{A}$.

The question of whether the measure extensions $\mu^*_{\sigma(\mathcal{A})}$ and $\mu^*_{\mathcal{M}(\mu^*)}$ of μ to $\sigma(\mathcal{A})$ and $\mathcal{M}(\mu^*)$ [respectively] are unique will be settled in the next section. There, it is shown that if μ is σ-finite on $\mathcal{A}$, then the measure extension of μ from $\mathcal{A}$ to $\sigma(\mathcal{A})$ is unique.

One of the difficulties in fully comprehending exactly what is going on is that we are dealing with four functions, each function having its own domain. The four functions in our generic framework are

μ with domain $\mathcal{A}$ [a semiring],
$\mu^*_{\sigma(\mathcal{A})}$ with domain $\sigma(\mathcal{A})$ [a σ-field, of course],
$\mu^*_{\mathcal{M}(\mu^*)}$ with domain $\mathcal{M}(\mu^*)$ [also a σ-field], and
μ^* with domain 2^Ω, where
$\mathcal{A} \subseteq \sigma(\mathcal{A}) \subseteq \mathcal{M}(\mu^*) \subseteq 2^\Omega$ by Claim 3.

With the proof all done, we may choose our "vantage point" as we see fit. For example, we may view μ on $\mathcal{A}$ in several different ways: (i) as a given, (ii) as the restriction of $\mu^*_{\sigma(\mathcal{A})}$ to $\mathcal{A}$, (iii) as the restriction of $\mu^*_{\mathcal{M}(\mu^*)}$ to $\mathcal{A}$, or (iv) as the restriction of μ^* to $\mathcal{A}$. Given an $A \in \mathcal{A}$, we have

$$\mu(A) = \mu^*_{\sigma(\mathcal{A})}(A) = \mu^*_{\mathcal{M}(\mu^*)}(A) = \mu^*(A),$$

and hence we may write $\mu(A)$, $\mu^*_{\sigma(\mathcal{A})}(A)$, $\mu^*_{\mathcal{M}(\mu^*)}(A)$, or $\mu^*(A)$ with utter indifference. Similarly, we may view the measure $\mu^*_{\sigma(\mathcal{A})}$ with domain $\sigma(\mathcal{A})$ as (i) the measure extension of μ from $\mathcal{A}$ to $\sigma(\mathcal{A})$, (ii) the restriction of the measure $\mu^*_{\mathcal{M}(\mu^*)}$ from $\mathcal{M}(\mu^*)$ to $\sigma(\mathcal{A})$, or (iii) the restriction of the outer measure μ^* from 2^Ω to $\sigma(\mathcal{A})$, and, given $A \in \sigma(\mathcal{A})$, we may be indifferent as to whether we write $\mu^*_{\sigma(\mathcal{A})}(A)$, $\mu^*_{\mathcal{M}(\mu^*)}(A)$, or $\mu^*(A)$. We shall continue to keep this notation throughout the rest of the chapter, and it is important that it be completely comprehended.

Exercises.

1*. Let $k \in \mathbb{N}$ and refer to the measure $\lambda_k : \mathcal{B}^k \to \bar{\mathbb{R}}$ that assigns the value $\prod_{i=1}^k (b_i - a_i)$ to every $(\boldsymbol{a}, \boldsymbol{b}] \in \mathcal{A}_k$, as given in this section's example. For this measure λ_k, we have the following;
(a) $\lambda_k((\boldsymbol{a}, \boldsymbol{b}]) = \lambda_k((\boldsymbol{a}, \boldsymbol{b})) = \lambda_k([\boldsymbol{a}, \boldsymbol{b})) = \lambda_k([\boldsymbol{a}, \boldsymbol{b}]) = \prod_{i=1}^k (b_i - a_i)$.
(b) λ_k applied to any k-dimensional rectangle that contains a k-dimensional open set and is unbounded in at least one dimension gives $+\infty$.
(c) λ_k applied to any bounded $\mathcal{B}^k$-set yields a finite number.
(d) It might be thought that if $A \subseteq B$ with $B \in \mathcal{B}^k$ with $\lambda_k(B) = 0$, then $\lambda_k(A)$ must exist and equal 0. Show that if there exists an uncountable set $C \in \mathcal{B}^k$ with $\lambda_k(C) = 0$, then $\lambda_k(A)$ need not even exist, let alone equal zero, and thus λ_k violates our intuition in this regard. [Use the facts that (i) $\text{card}(\mathcal{B}^k) = \mathfrak{c}$ and (ii) there are $2^{\mathfrak{c}}$ subsets of a set with cardinal number $\mathfrak{c}$.]

2*. This problem reviews the Extension Theorem.
(a) Where or how is the fact that $\mu(\varnothing) = 0$ used?
(b) What happens if $\mathcal{A} = 2^\Omega$?

3*. Consider the Extension Theorem framework. If we have $\mu(A) < +\infty$ for each $A \in \mathcal{A}$, it might not be the case that the measure extension $\mu^*_{\sigma(\mathcal{A})}$ assigns finite measure to every set in $\sigma(\mathcal{A})$. However, if μ is σ-finite on $\mathcal{A}$, then the measure extension $\mu^*_{\sigma(\mathcal{A})}$ is σ-finite on $\sigma(\mathcal{A})$, and the measure extension $\mu^*_{\mathcal{M}(\mu^*)}$ is σ-finite on $\mathcal{M}(\mu^*)$.

4. There is a measure $\nu : \mathcal{B} \to \bar{\mathbb{R}}$ with $\nu(\mathbb{R}) = 1$ and $\nu((a,b]) = \int_a^b (2\pi)^{-1/2} e^{-z^2/2}\, dz$, where the integral is the familiar Riemann integral from calculus. [One approach is by C. Georgakis, *Mathematics Magazine*, February 1994.]

5*. Consider the Extension Theorem framework again.
(a) If $A \subseteq B \subseteq \Omega$ with $B \in \mathcal{M}(\mu^*)$ and $\mu^*_{\mathcal{M}(\mu^*)}(B) = 0$, then $A \in \mathcal{M}(\mu^*)$ and $\mu^*_{\mathcal{M}(\mu^*)}(A) = 0$.
(b) If in (a) we replace every occurrence of $\mathcal{M}(\mu^*)$ [including instances where it appears as a subscript] with $\sigma(\mathcal{A})$, then the claim is not necessarily true.

6. In other analysis books, it is common to extend a given measure μ defined on a given *field* $\mathcal{F}$ [as compared to a *semiring*] to a measure on $\sigma(\mathcal{F})$. Our more general theorem contains this setup as a special case. However, it is worthwhile practice to walk through this approach. To this end, let $\mathcal{F}$ denote a field on Ω, and let $\mu : \mathcal{F} \to \bar{\mathbb{R}}$ denote a measure. Let $\mu^* : 2^\Omega \to \bar{\mathbb{R}}$ be given by

$$\mu^*(A) = \inf\left\{\sum_{n=1}^{\infty} \mu(A_n) : \{A_n\}_{n=1}^{\infty} \text{ is a } \mathcal{F}\text{-covering of } A\right\}, \qquad A \subseteq \Omega.$$

Then μ^* is an outer measure, and the restriction of μ^* to the σ-field $\mathcal{M}(\mu^*)$ is a measure. With these facts, we have that $\mu^* = \mu$ on $\mathcal{F}$ and $\mathcal{F} \subseteq \mathcal{M}(\mu^*)$. Finally, there exists a measure extension of μ to $\sigma(\mathcal{F})$.

7. Let $\mathcal{F}$ denote a field on Ω. Suppose that $\nu : \mathcal{F} \to \bar{\mathbb{R}}$ is such that (i) $\nu(A) \geq 0$ for all $A \in \mathcal{F}$, (ii) ν is finitely additive, and (iii) if $\{A_n\}_{n=1}^{\infty}$ is a nonincreasing sequence of $\mathcal{F}$-sets with $\lim A_n = \varnothing$, then $\lim \nu(A_n) = 0$. Define $\nu^* : 2^\Omega \to \bar{\mathbb{R}}$ for all $A \subseteq \Omega$ by writing

$$\nu^*(A) = \inf\left\{\sum_{n=1}^{\infty} \nu(A_n) : \{A_n\}_{n=1}^{\infty} \text{ is an } \mathcal{F}\text{-covering of } A\right\}.$$

(a) ν^* is an outer measure.
(b) $\mathcal{M}(\nu^*)$ is a σ-field on Ω.
(c) The restriction of ν^* to $\mathcal{M}(\nu^*)$ is a measure on $\mathcal{M}(\nu^*)$.
(d) $\mathcal{F} \subseteq \mathcal{M}(\nu^*)$.
(e) There exists a measure extension of ν to $\sigma(\mathcal{F})$.

3.5 UNIQUENESS OF MEASURES AND EXTENSIONS

Given a semiring $\mathcal{A}$ on Ω and a nonnegative, finitely additive, countably subadditive function $\mu : \mathcal{A} \to \bar{\mathbb{R}}$ with $\mu(\varnothing) = 0$, the Extension Theorem says that there exists a measure $\mu_{\sigma(\mathcal{A})} : \sigma(\mathcal{A}) \to \bar{\mathbb{R}}$ with $\mu_{\sigma(\mathcal{A})}(A) = \mu(A)$ for each $A \in \mathcal{A}$. Does there exist a measure $\nu : \sigma(\mathcal{A}) \to \bar{\mathbb{R}}$ such that $\nu(A) = \mu(A)$

for all $A \in \mathcal{A}$ but $\nu(B) \neq \mu^*_{\sigma(\mathcal{A})}(B)$ for at least one $B \in \sigma(\mathcal{A}) - \mathcal{A}$? In other words, is the measure extension $\mu^*_{\sigma(\mathcal{A})}$ of μ from $\mathcal{A}$ to $\sigma(\mathcal{A})$ unique? We shall answer the question of whether the measure extension is unique by actually proving a much more general result, which in turn depends on the following technical claim.

Claim 1. [Pi-Lambda Theorem.] *Let $\mathcal{P}$ denote a π-system on Ω, and let $\mathcal{L}$ denote a λ-system on Ω with $\mathcal{L} \supseteq \mathcal{P}$. Then $\mathcal{L} \supseteq \sigma(\mathcal{P})$.*

Proof: Let $\lambda(\mathcal{P})$ denote the λ-system generated by $\mathcal{P}$, that is, let $\lambda(\mathcal{P})$ be the intersection of all λ-systems on Ω that contain $\mathcal{P}$; we therefore have $\mathcal{P} \subseteq \lambda(\mathcal{P}) \subseteq \mathcal{L}$. If $\lambda(\mathcal{P})$ is shown to be a π-system, then, since every λ-system that is a π-system is a σ-field,[4] $\lambda(\mathcal{P})$ will be a σ-field that contains $\mathcal{P}$, and hence we will have $\mathcal{P} \subseteq \sigma(\mathcal{P}) \subseteq \lambda(\mathcal{P}) \subseteq \mathcal{L}$, which forces the claim of this theorem. Accordingly, we will show that $\lambda(\mathcal{P})$ is a π-system.

For each $A \subseteq \Omega$, define $\mathcal{G}_A = \{C \subseteq \Omega : A \cap C \in \lambda(\mathcal{P})\}$. We now make a series of subclaims.

(a) *If $A \in \mathcal{P}$, then $\mathcal{G}_A$ is a λ-system that contains $\mathcal{P}$.*

Proof: Let $A \in \mathcal{P}$. Since $A \cap \Omega = A \in \lambda(\mathcal{P})$, we have (λ_1): $\Omega \in \mathcal{G}_A$. We now show (λ_2'): if $C_1 \subseteq C_2$ with $C_1, C_2 \in \mathcal{G}_A$, then $C_2 - C_1 \in \mathcal{G}_A$. To see this, we have $A \cap C_1 \subseteq A \cap C_2$ and $A \cap C_1, A \cap C_2 \in \lambda(\mathcal{P})$ by assumption. Since every λ-system is closed under proper differences,[5] we have $A \cap (C_2 - C_1) = (A \cap C_2) - (A \cap C_1) \in \lambda(\mathcal{P})$, and hence (λ_2') holds. We now show that $\mathcal{G}_A$ is closed under countable disjoint unions [property (λ_3)]. Let $\{C_n\}_{n=1}^\infty$ denote a disjoint collection of $\mathcal{G}_A$-sets, so that $\{A \cap C_n\}_{n=1}^\infty$ is a disjoint sequence of $\lambda(\mathcal{P})$-sets. Since $\lambda(\mathcal{P})$ is a λ-system and hence satisfies (λ_3), we have $A \cap \bigcup_{n=1}^\infty C_n = \bigcup_{n=1}^\infty (A \cap C_n) \in \lambda(\mathcal{P})$, so that $\bigcup_{n=1}^\infty C_n \in \mathcal{G}_A$, hence (λ_3) holds. Thus, $A \in \mathcal{P}$ implies that $\mathcal{G}_A$ satisfes (λ_1), (λ_2'), and (λ_3), so that $\mathcal{G}_A$ is a λ-system.[6] To see that $\mathcal{G}_A \supseteq \mathcal{P}$ for each $A \in \mathcal{P}$, observe that for any $P \in \mathcal{P}$ we have $A \cap P \in \mathcal{P} \subseteq \lambda(\mathcal{P})$, so that $P \in \mathcal{G}_A$.

(b) *If $A \in \lambda(\mathcal{P})$, then $\mathcal{G}_A$ is a λ-system.*

Proof: The arguments of (a) carry over exactly.

(c) *If $A \in \mathcal{P}$, then $\lambda(\mathcal{P}) \subseteq \mathcal{G}_A$.*

[4]Proof: Let $\mathcal{F}$ denote a λ-system that is a π-system. We have that (i) $\Omega \in \mathcal{F}$ and (ii) $\mathcal{F}$ is closed under complementation from the defining properties of a λ-system. Next, let $A_1, A_2, \dots \in \mathcal{F}$. Let $B_1 = A_1$ and $B_k = A_k \cap \bigcap_{j=1}^{k-1} A_j^c$ for each $k \geq 2$. The B_k's are disjoint, each B_k in $\mathcal{F}$ since $\mathcal{F}$ is a π-system, and thus $\bigcup_{k=1}^\infty A_k = \bigcup_{k=1}^\infty B_k \in \mathcal{F}$ since $\mathcal{F}$ is closed under countable disjoint unions. This shows that $\mathcal{F}$ is closed under countable unions.

[5]Proof: Let $\mathcal{F}$ denote a λ-system, and let $A, B \in \mathcal{F}$ with $A \subseteq B$. Then $B^c \in \mathcal{F}$, hence the disjoint union $A \cup B^c$ is in $\mathcal{F}$, hence $B - A = (A \cup B^c)^c \in \mathcal{F}$.

[6]Let $\mathcal{F}$ denote a collection that satisfies (λ_1), (λ_2'), and (λ_3). If $A \in \mathcal{F}$, then $A^c = \Omega - A \in \mathcal{F}$ by (λ_1) and (λ_2'), so that (λ_2) holds, hence $\mathcal{F}$ is a λ-system.

Proof: Let $A \in \mathcal{P}$. By (a), $\mathcal{G}_A$ is a λ-system containing $\mathcal{P}$, hence $\mathcal{G}_A$ will contain the minimal λ-system that contains $\mathcal{P}$.

(d) *If $D \in \mathcal{P}$ and $E \in \lambda(\mathcal{P})$, then $D \in \mathcal{G}_E$.*

Proof: Let $D \in \mathcal{P}$ and $E \in \lambda(\mathcal{P})$. Since $D \in \mathcal{P}$, (c) gives $\lambda(\mathcal{P}) \subseteq \mathcal{G}_D$. That is, $D \in \mathcal{P}$ implies that the intersection of any $\lambda(\mathcal{P})$-set with D is itself a $\lambda(\mathcal{P})$-set. In particular, $E \cap D \in \lambda(\mathcal{D})$, which by definition gives $D \in \mathcal{G}_E$.

(e) *If $E \in \lambda(\mathcal{P})$, then $\lambda(\mathcal{P}) \subseteq \mathcal{G}_E$.*

Proof: Let $D \in \mathcal{P}$ and $E \in \lambda(\mathcal{P})$. Part (d) gives $D \in \mathcal{G}_E$. Since $D \in \mathcal{P}$ is arbitrary, we have $\mathcal{P} \subseteq \mathcal{G}_E$. Since $E \in \lambda(\mathcal{P})$, (b) shows that $\mathcal{G}_E$ is a λ-system, hence $\mathcal{G}_E$ is a λ-system that contains $\mathcal{P}$. It follows that $\mathcal{G}_E \supseteq \lambda(\mathcal{P})$.

With subclaims (a)-(e) in tow, we are now ready to complete the proof by carrying out the original program of showing that $\lambda(\mathcal{P})$ is a π-system. Let $A, B \in \lambda(\mathcal{P})$; we desire to show that $A \cap B \in \lambda(\mathcal{P})$. Since $A \in \lambda(\mathcal{P})$, (e) gives $\lambda(\mathcal{P}) \subseteq \mathcal{G}_A$. In particular, the $\lambda(\mathcal{P})$-set B must be in $\mathcal{G}_A$, hence $A \cap B \in \lambda(\mathcal{P})$, whence the entire proof is complete. ■

Uniqueness Theorem. *Let μ_1 and μ_2 denote measures with domain $\sigma(\mathcal{P})$, where $\mathcal{P}$ denotes a π-system. If μ_1 is σ-finite with respect to $\mathcal{P}$ and if $\mu_1 = \mu_2$ on $\mathcal{P}$, then μ_2 is σ-finite on $\mathcal{P}$ and $\mu_1 = \mu_2$ on $\sigma(\mathcal{P})$.*

Before proving this theorem, it will be applied to a very important setting.

Example. Let $k \in \mathbb{N}$, $\Omega = \mathbb{R}^k$, and let $\mathcal{A}_k$ denote the usual semiring consisting of $\varnothing$ and the [bounded] k-dimensional rsc rectangles. Since every semiring is a π-system, $\mathcal{A}_k$ may serve as the π-system of the theorem, and to show that two arbitrary measures $\mu_1 : \mathcal{B}^k \to \bar{\mathbb{R}}$ and $\mu_2 : \mathcal{B}^k \to \bar{\mathbb{R}}$ are identical, one would only need to argue that (i) $\mathbb{R}^k$ may be written as an amc union $\bigcup_n (\boldsymbol{a}_n, \boldsymbol{b}_n]$ with $\mu_1((\boldsymbol{a}_n, \boldsymbol{b}_n]) < +\infty$ for each n, and (ii) $\mu_1((\boldsymbol{a}, \boldsymbol{b}]) = \mu_2((\boldsymbol{a}, \boldsymbol{b}])$ for each $(\boldsymbol{a}, \boldsymbol{b}] \in \mathcal{A}_k$. That is, instead of having to argue that $\mu_1(B) = \mu_2(B)$ for every $B \in \mathcal{B}^k$, one can show via (ii) that $\mu_1 = \mu_2$ on a much smaller and more easily describable collection of subsets of $\mathbb{R}^k$. Since most Borel sets can't be described in simple constructive terms [Sections 1.6 and 1.7], this theorem greatly reduces the effort required to show that two measures are equal on $\mathcal{B}^k$.

By the Extension Theorem, we know from the example of Section 4 that there is at least one measure $\lambda_k : \mathcal{B}^k \to \bar{\mathbb{R}}$ such that $\lambda_k((\boldsymbol{a}, \boldsymbol{b}]) = \prod_{i=1}^k (b_i - a_i)$ for each $(\boldsymbol{a}, \boldsymbol{b}] \in \mathcal{A}_k$. *We claim that λ_k is the unique measure with domain $\mathcal{B}^k$ and such that $\lambda_k((\boldsymbol{a}, \boldsymbol{b}]) = \prod_{i=1}^k (b_i - a_i)$ for each $(\boldsymbol{a}, \boldsymbol{b}] \in \mathcal{A}_k$.* In other words, we claim that when $F \in \mathfrak{S}_k$ is the function that assigns $\prod_{i=1}^k x_i$ to each $\boldsymbol{x} \in \mathbb{R}^k$, the measure extension of Δ_F from $\mathcal{A}_k$ to $\mathcal{B}^k$ is unique. To see this, let $\nu : \mathcal{B}^k \to \bar{\mathbb{R}}$ denote a measure such that $\nu((\boldsymbol{a}, \boldsymbol{b}]) = \prod_{i=1}^k (b_i - a_i)$ for all $(\boldsymbol{a}, \boldsymbol{b}] \in \mathcal{A}_k$. Therefore, $\nu = \lambda_k$ on the semiring $\mathcal{A}_k$. Since λ_k is σ-finite on $\mathcal{A}_k$, it follows that $\nu = \lambda_k$ on all of $\mathcal{B}^k$, proving the uniqueness claim. The practical importance of this is that there is but one measure on $\mathcal{B}^k$ that corresponds to our notion of "k-dimensional volume"; we needn't worry about picking among rival measures that mimic "k-dimensional volume."

The proof of the Uniqueness Theorem will require the following piece of technical skullduggery.

Claim 2. *Let μ_1 and μ_2 denote measures each having domain $\sigma(\mathcal{P})$, where $\mathcal{P}$ is a π-system on Ω. Let μ_1 be σ-finite on $\mathcal{P}$, and let $\mu_1 = \mu_2$ on $\mathcal{P}$. Pick $B \in \mathcal{P}$ such that $\mu_1(B) = \mu_2(B) < +\infty$. Define*

$$\mathcal{L}_B = \{A \in \sigma(\mathcal{P}) :\ \mu_1(B \cap A) = \mu_2(B \cap A)\}.$$

Then we have the following:
(i) $\mathcal{L}_B$ is a λ-system on Ω that contains $\mathcal{P}$, and
(ii) $\mathcal{L}_B \supseteq \sigma(\mathcal{P})$.

Proof: The fact that a $B \in \mathcal{P}$ with $\mu_1(B) = \mu_2(B) < +\infty$ may be chosen comes from the facts that (a) μ_1 is σ-finite on $\mathcal{P}$ and (ii) $\mu_1 = \mu_2$ on $\mathcal{P}$. We verify (i) in straightforward fashion. Observe first that $\Omega \in \sigma(\mathcal{P})$ and

$$\mu_1(B \cap \Omega) = \mu_1(B) = \mu_2(B) = \mu_2(B \cap \Omega),$$

so $\Omega \in \mathcal{L}_B$. Next, let $A \in \mathcal{L}_B$. Since (a) $\mu_i(B) = \mu_i(B \cap A) + \mu_i(B \cap A^c)$ for $i = 1, 2$, (b) $\mu_1(B) = \mu_2(B) < +\infty$, and (c) $\mu_1(B \cap A) = \mu_2(B \cap A) < +\infty$, we have that $\mu_1(B \cap A^c) = \mu_2(B \cap A^c)$, whence $A^c \in \mathcal{L}_B$: $\mathcal{L}_B$ is closed under complementation. Finally, let $A_1, A_2, \cdots \in \mathcal{L}_B$ be disjoint; we have $\bigcup_{k=1}^{\infty} A_k \in \sigma(\mathcal{P})$. We also have that $\mu_1(B \cap A_k) = \mu_2(B \cap A_k)$ for all $k \in \mathbb{N}$, and $\{B \cap A_k\}_{k=1}^{\infty}$ is a disjoint sequence of $\sigma(\mathcal{P})$-sets. Since

$$\mu_i\left(B \cap \bigcup_{k=1}^{\infty} A_k\right) = \mu_i\left(\bigcup_{k=1}^{\infty} (B \cap A_k)\right) = \sum_{k=1}^{\infty} \mu_i(B \cap A_k)$$

for $i = 1, 2$, we have that $\mu_1(B \cap \bigcup_{k=1}^{\infty} A_k) = \mu_2(B \cap \bigcup_{k=1}^{\infty} A_k)$, whence $\bigcup_{k=1}^{\infty} A_k \in \mathcal{L}_B$. It follows that $\mathcal{L}_B$ is a λ-system.

To see that $\mathcal{L}_B$ contains $\mathcal{P}$, let $A \in \mathcal{P}$. Since $B \cap A \in \mathcal{P}$ and since $\mu_1 = \mu_2$ on $\mathcal{P}$, we have $\mu_1(B \cap A) = \mu_2(B \cap A)$, hence $A \in \mathcal{L}_B$. This completes the proof of (i). Part (ii) follows directly from the π-λ Theorem. ∎

Proof of the Uniqueness Theorem: Since μ_1 is σ-finite on $\mathcal{P}$, we may write $\Omega = \bigcup_{n=1}^{\infty} B_n$ for some sequence $\{B_n\}_{n=1}^{\infty}$ of $\mathcal{P}$-sets with $\mu_1(B_n) < +\infty$ for each $n \in \mathbb{N}$. Since $\mu_2 = \mu_1$ on $\mathcal{P}$, we have $\mu_2(B_n) < +\infty$ for each $n \in \mathbb{N}$, and hence μ_2 is σ-finite on $\mathcal{P}$ as well.

Next, pick a generic $A \in \sigma(\mathcal{P})$; we desire to show that $\mu_1(A) = \mu_2(A)$. For $i = 1, 2$, observe that

$$\begin{aligned} \mu_i(A) &= \mu_i\left(\bigcup_{k=1}^{\infty} (B_k \cap A)\right) = \mu_i\left(\lim_n \bigcup_{k=1}^{n} (B_k \cap A)\right) \\ &= \lim_n \mu_i\left(\bigcup_{k=1}^{n} (B_k \cap A)\right), \end{aligned}$$

where the last equality follows from (M8). To show $\mu_1(A) = \mu_2(A)$ for our generic $A \in \sigma(\mathcal{P})$, we need to show that

$$\lim_n \mu_1\left(\bigcup_{k=1}^n (B_k \cap A)\right) = \lim_n \mu_2\left(\bigcup_{k=1}^n (B_k \cap A)\right),$$

which in turn will hold if we show that $\mu_1(\bigcup_{k=1}^n (B_k \cap A)) = \mu_2(\bigcup_{k=1}^n (B_k \cap A))$ for each $n \in \mathbb{N}$. To do this, use the inclusion-exclusion formula in measure property (M6) to write

$$\begin{aligned} &\mu_i\left(\bigcup_{k=1}^n (B_k \cap A)\right) \\ = \;&\sum_{k=1}^n \sum \{(-1)^{k-1}\mu_i(B_{i_1} \cap \cdots \cap B_{i_k} \cap A) : 1 \leq i_1 < \cdots < i_k \leq n\} \end{aligned}$$

for each $n \in \mathbb{N}$ and $i = 1, 2$. It is sufficient in turn to show that

$$\mu_1(B_{i_1} \cap \cdots \cap B_{i_k} \cap A) = \mu_2(B_{i_1} \cap \cdots \cap B_{i_k} \cap A) \tag{3.11}$$

for each $n \in \mathbb{N}$, $k = 1, \cdots, n$, and $i_1, \cdots, i_k$ with $1 \leq i_1 < \cdots < i_k \leq n$. Given $n \in \mathbb{N}$, $k = 1, \cdots, n$, and $1 \leq i_1 < \cdots < i_k \leq n$, we have $B_{i_1} \cap \cdots \cap B_{i_k} \in \mathcal{P}$ since $\mathcal{P}$ is a π-system and $B_{i_1}, \cdots, B_{i_k} \in \mathcal{P}$. By monotonicity and the fact that $\mu_1 = \mu_2$ on $\mathcal{P}$, we have that $B_{i_1} \cap \cdots \cap B_{i_k}$ has finite and identical μ_1- and μ_2-measure. Since $A \in \sigma(\mathcal{P})$ and $\sigma(\mathcal{P}) \subseteq \mathcal{L}_{B_{i_1} \cap \cdots \cap B_{i_k}}$ by Claim 1, it follows that $A \in \mathcal{L}_{B_{i_1} \cap \cdots \cap B_{i_k}}$, so that (3.11) holds, completing the proof. ■

With these results, we may now completely settle the uniqueness question in regards to the Extension Theorem given in the previous section.

Claim 3. [Uniqueness of Extensions.] *Consider the conditions and notation of the Extension Theorem. If μ is σ-finite on the semiring $\mathcal{A}$, then the measure extension of μ to $\sigma(\mathcal{A})$ is unique and is σ-finite on $\mathcal{A}$.*

Proof: Claim 1 in Section 1 shows that if μ is σ-finite on $\mathcal{A}$, then any extension of μ to $\sigma(\mathcal{A})$ will be σ-finite on $\mathcal{A}$. For uniqueness, let $\mu^*_{\sigma(\mathcal{A})}$ denote the extension constructed in the Extension Theorem, and let $\nu : \sigma(\mathcal{A}) \to \bar{\mathbb{R}}$ denote a measure such that $\nu = \mu^*_{\sigma(\mathcal{A})}$ on $\mathcal{A}$. Since $\mathcal{A}$ is a π-system, the previous theorem ensures that $\nu = \mu^*_{\sigma(\mathcal{A})}$ on $\sigma(\mathcal{A})$. Therefore, the only measure with domain $\sigma(\mathcal{A})$ that agrees with μ on $\mathcal{A}$ is $\mu^*_{\sigma(\mathcal{A})}$. ■

Another question is this: *if μ is σ-finite [on $\mathcal{A}$], is $\mu^*_{\mathcal{M}(\mu^*)}$ the only measure with domain $\mathcal{M}(\mu^*)$ that agrees with μ on $\mathcal{A}$?* The answer to this question is in the affirmative, but we will need to further elucidate the nature of $\mathcal{M}(\mu^*)$ and $\mu^*_{\mathcal{M}(\mu^*)}$ to show this. This further elucidation takes place in the next section, where the proof of this other uniqueness claim is given after the necessary buildup. The reason for this extra required buildup is because the Uniqueness

Theorem deals with an underlying π-system and the σ-field generated by that π-system, and we cannot fit this latest question into the framework of the Uniqueness Theorem given in this section.

Exercises.

1*. If μ_1 and μ_2 are finite measures with domain $\sigma(\mathcal{P})$ [where $\mathcal{P}$ denotes a π-system on Ω], if Ω can be expressed as an amc union of $\mathcal{P}$-sets, and if $\mu_1 = \mu_2$ on $\mathcal{P}$, then $\mu_1 = \mu_2$ on $\sigma(\mathcal{P})$.

2. Let μ_1 and μ_2 denote finite measures with domain $\sigma(\mathcal{P})$, where $\mathcal{P}$ is a π-system on Ω, and further suppose that $\mu_1 = \mu_2$ on $\mathcal{P}$. Then $\mu_1 = \mu_2$. [This, of course, is a trivial consequence of the theorem of this section. The point though is to give a "first principles" proof without invoking the Uniqueness Theorem. Start by letting $\mathcal{L}$ denote the class of sets $A \in \sigma(\mathcal{P})$ such that $\mu_1(A) = \mu_2(A)$. Show that $\mathcal{L}$ is a λ-system on Ω that contains $\mathcal{P}$ and invoke the π-λ Theorem. The proof should be simpler than the general proof due to the assumed finiteness of μ_1 and μ_2.]

3. Let $\Omega = \{\omega_1, \omega_2, \omega_3, \omega_4\}$, let $\mathcal{A}$ consist of $\varnothing$, $\{\omega_1, \omega_2\}$, $\{\omega_1, \omega_3\}$, $\{\omega_2, \omega_4\}$, $\{\omega_3, \omega_4\}$, and Ω, and let $\mu : \mathcal{A} \to \mathbb{R}$ be defined as follows: $\mu(\Omega) = 6$, $\mu(\varnothing) = 0$, and $\mu(\{\omega_1, \omega_2\}) = \mu(\{\omega_1, \omega_3\}) = \mu(\{\omega_2, \omega_4\}) = \mu(\{\omega_3, \omega_4\}) = 3$.
(a) $\mathcal{A}$ is neither a π-system nor a semiring, and $\sigma(\mathcal{A}) = 2^\Omega$.
(b) μ is a measure.
(c) Define two new distinct measures ν and ξ on the σ-field 2^Ω by the following: $\nu(\{\omega_1\}) = \nu(\{\omega_4\}) = 1$, $\nu(\{\omega_2\}) = \nu(\{\omega_3\}) = 2$, $\xi(\{\omega_2\}) = \xi(\{\omega_3\}) = 1$, and $\xi(\{\omega_1\}) = \xi(\{\omega_4\}) = 2$. Then ν and ξ are distinct measure extensions of μ from $\mathcal{A}$ to $\sigma(\mathcal{A}) = 2^\Omega$.
(d) Let $\mu^* : 2^\Omega \to \mathbb{R}$ denote the outer measure induced by μ. Then $\mu^* = \mu$ on $\mathcal{A}$ and μ^* is a measure.
(e) μ^*, ν, and ξ are distinct measures. This shows that *if $\mathcal{A}$ fails to be a semiring, a measure extension of a function μ from $\mathcal{A}$ to $\sigma(\mathcal{A})$ need not be unique, even if μ satisfies the conditions of the Extension Theorem.*

4. Let $\Omega = \mathbb{Q}$, and let $\mathcal{A}$ consist of the sets $\{\omega \in \mathbb{Q} : a < \omega \leq b\}$, where $a < b$ and $a, b \in \mathbb{Q}$. Also, let $\varnothing \in \mathcal{A}$.
(a) $\mathcal{A}$ is a semiring and $\sigma(\mathcal{A}) = 2^\Omega$.
(b) Define $\mu : \mathcal{A} \to \bar{\mathbb{R}}$ by stipulating that $\mu(A)$ equals the number of points in A if A is finite and $\mu(A) = +\infty$ if A is infinite. Then $\mu(A) = +\infty$ for all $A \in \mathcal{A} - \{\varnothing\}$, hence μ fails to be σ-finite on $\mathcal{A}$.
(c) Define $\nu : 2^\Omega \to \bar{\mathbb{R}}$ by stating that $\nu(A)$ equals the number of points in A if A is finite and $\nu(A) = +\infty$ if A is infinite. Next, define ξ by writing $\xi(A) = 2\nu(A)$ for all $A \subseteq \Omega$. Then ν and ξ are distinct σ-finite measure extensions of μ from $\mathcal{A}$ to $\sigma(\mathcal{A})$. Therefore, *when μ fails to be σ-finite on $\mathcal{A}$, there may be more than one measure extension of μ from $\mathcal{A}$ to $\sigma(\mathcal{A})$.*
(d) The outer measure $\mu^* : 2^\Omega \to \bar{\mathbb{R}}$ induced by μ is such that $\mu^*(A) = +\infty$ for any nonempty $A \subseteq \Omega$, and μ^* is a non-σ-finite measure. Therefore, while we have two distinct σ-finite measure extensions of μ to $\sigma(\mathcal{A})$, we also have a non-σ-finite extension of μ to $\sigma(\mathcal{A})$ as well.
(e) $\mathcal{M}(\mu^*) = 2^\Omega = \sigma(\mathcal{A})$.

5. This and the next exercise build off of Exercises 6 and 7 in Section 4 to provide different proofs of the uniqueness of an extension. We assume the setup of Exercise

6 there. The aim of this exercise is to show that if ν is a measure with domain $\sigma(\mathcal{F})$ such that $\nu = \mu$ on the field $\mathcal{F}$, then ν coincides with the measure extension of μ to $\sigma(\mathcal{F})$ guaranteed by Exercise 6.

(a) If $B \in \sigma(\mathcal{F})$, then $\nu(B) \le \mu^*(B)$. [Let $\mathcal{F}_\sigma$ denote the family of all countable unions of $\mathcal{F}$-sets. Argue that if $A \in \mathcal{F}_\sigma$, then $A = \bigcup_{n=1}^\infty A_n$, where each $A_n \in \mathcal{F}$, and hence A can be written as a countable disjoint union of $\mathcal{F}$-sets, namely $A = A_1 \cup \bigcup_{n=2}^\infty (A_n \cap A_1^c \cap \cdots \cap A_{n-1}^c)$. From this show that $\nu(A) = \mu^*(A)$ by invoking the measure properties of ν and using the fact that $\nu = \mu = \mu^*$ on $\mathcal{F}$. Next, show that $\mu^*(B) = \inf\{\mu^*(A) : B \subseteq A \in \mathcal{F}_\sigma\} = \inf\{\nu(A) : B \subseteq A \in \mathcal{F}_\sigma\} \ge \nu(B)$.]

(b) If $F \in \sigma(\mathcal{F})$ and $\mu^*(F) < +\infty$, then $\nu(F) = \mu^*(F)$. [To get $\mu^*(F) \le \nu(F)$, pick $\epsilon > 0$, and find $A \in \mathcal{F}_\sigma$ such that $\mu^*(A) < \mu^*(F) + \epsilon$. Use part (a) to show that

$$\mu^*(F) \le \mu^*(A) = \nu(A) = \nu(F) + \nu(A - F) \le \nu(F) + \mu^*(A - F) < \nu(F) + \epsilon.]$$

(c) If μ is σ-finite on $\mathcal{F}$, then $\nu(E) = \mu^*(E)$ for all $E \in \sigma(\mathcal{F})$. This gives the uniqueness of the measure extension whose existence is guaranteed by Exercise 6. [Write $\Omega = \bigcup_{n=1}^\infty F_n$, where $F_n \in \mathcal{F}$ and $\mu(F_n) < +\infty$ for each $n \in \mathbb{N}$; the F_n's may be taken as disjoint. Obtain $\nu(E) = \sum_{n=1}^\infty \nu(E \cap F_n) = \sum_{n=1}^\infty \mu^*(E \cap F_n) = \mu^*(E)$ for each $E \in \sigma(\mathcal{F})$ from (b).]

6. A *lattice on* Ω is a collection $\mathcal{L} \subseteq 2^\Omega$ such that (i) $\Omega \in \mathcal{L}$, (ii) $\varnothing \in \mathcal{L}$, (iii) $\mathcal{L}$ is closed under [finite] unions, and (iv) $\mathcal{L}$ is a π-system. We also define the following two collections: $\mathcal{D} = \{B - A : A, B \in \mathcal{L}, A \subseteq B\}$, and $\mathcal{U}$ will denote the collection of all finite disjoint unions of $\mathcal{D}$-sets.

(a) $\mathcal{D}$ is a π-system. [Write $D_i = B_i - A_i$ where $A_i \subseteq B_i$ and $A_i, B_i \in \mathcal{L}$ for $i = 1, 2$; show that $D_1 \cap D_2 = (B_1 \cap B_2) \cap [(A_1 \cap B_2) \cup (A_2 \cap B_1)]^c$.]

(b) $\mathcal{U}$ is a π-system.

(c) $\mathcal{U}$ is closed under complementation. [Pick a generic $U \in \mathcal{U}$, so that U is the finite disjoint union of $D_1, \cdots, D_n \in \mathcal{D}$. Use an induction argument on n.]

(d) $\mathcal{U}$ coincides with the minimal field containing the lattice $\mathcal{L}$. That is, in the notation of Section 1.3, $\mathcal{U} = f(\mathcal{L})$. [Show that $\mathcal{U}$ is a field with $\mathcal{U} \supseteq \mathcal{L}$, and hence $\mathcal{U} \supseteq f(\mathcal{L})$. To show that $f(\mathcal{L}) \supseteq \mathcal{U}$, argue that any field containing $\mathcal{L}$ must contain all of the $\mathcal{U}$-sets.]

(e) Let $\mathcal{A}$ denote a σ-field on Ω that contains $\mathcal{L}$. Suppose that μ and ν are measures with domain $\mathcal{A}$ such that $\mu = \nu$ on $\mathcal{L}$. Furthermore, suppose that $\Omega = \bigcup_{n=1}^\infty A_n$, where $A_n \in \mathcal{L}$ and $\mu(A_n) < +\infty$ for each $n \in \mathbb{N}$. Then $\mu = \nu$ on $\sigma(\mathcal{L})$. [Argue first that $\mu = \nu$ on $\mathcal{U}$.]

7*. Let μ and ν denote two σ-finite measures with domain $\mathcal{B}^k$ such that $\mu = \nu$ on the subcollection of open subsets of $\mathbb{R}^k$.

(a) Show that $\mu = \nu$ by appealing to the Uniqueness Theorem. [This is the important part of the exercise.]

(b) Show that $\mu = \nu$ by alternatively appealing to Exercise 6.

(c) Repeat parts (a) and (b) by replacing the word *open* with *closed.* [What extra work if any does this require?]

(d) Repeat parts (a) and (b) by replacing the word *open* with *compact.*

3.6 THE COMPLETION THEOREM

We begin with an example that will hopefully motivate the mathematics of this section to a certain degree.

Example 1. Let $\mathcal{A}$ denote the semiring consisting of $\varnothing$ and the [bounded] rsc intervals $(a, b]$, and let $\lambda : \mathcal{B} \to \bar{\mathbb{R}}$ denote the unique measure with domain $\mathcal{B}$ such that $\lambda((a, b]) = b - a$ for each $(a, b] \in \mathcal{A}$. [See the examples of Sections 4 and 5.] This measure λ mimics our idea of "length." In Section 1.6 it was shown that $\text{card}(\mathcal{B}) = \mathfrak{c}$. We assume for the moment that there exists an uncountable set $C \in \mathcal{B}$ with $\lambda(C) = 0$ [such sets will be exhibited in Chapter 4]. Since this C has "length zero," we would think that any $A \subseteq C$ would also have "length zero," for it seems natural and even axiomatic that the "length of a subset" should be no more than the "length of its corresponding superset." However, C has $2^{\mathfrak{c}}$ subsets, and therefore $2^{\mathfrak{c}}$ subsets of C will not be in $\mathcal{B}$. For a set $A \subseteq C$ with $A \notin \mathcal{B}$, we cannot write $\lambda(A) \leq \lambda(C) = 0$ to conclude that $\lambda(A) = 0$, for the symbol $\lambda(A)$ is not even defined. Is there a way of enlarging the domain of λ to include [say] all $2^{\mathfrak{c}}$ subsets of C, so that these subsets may be said to "have length zero" as well? Can such an enlargement be performed without causing any logical problems?

The answer to the question in Example 1 is in the affirmative: we *can* enlarge the domain of λ as specified above in a way that causes no logical inconsistencies nor surrenders the current properties of λ that mimic what we consider to be axiomatic or intuitive regarding "length." The fact that we can do this comes from the upcoming Completion Theorem.

For the following definitions, Ω is generic, $\mathcal{F}$ is a collection of subsets of Ω with $\varnothing \in \mathcal{F}$, and $\nu : \mathcal{F} \to \bar{\mathbb{R}}$ is an arbitrary function with $\nu(\varnothing) = 0$.

Definition. A set A will be called *[ν]-null* iff $A \in \mathcal{F}$ and $\nu(A) = 0$.

Definition. We shall let $\mathcal{N}(\nu)$ denote the collection of ν-null sets. That is, $A \in \mathcal{N}(\nu)$ iff $A \in \mathcal{F}$ and $\nu(A) = 0$. Note that $\varnothing \in \mathcal{N}(\nu) \subseteq \mathcal{F}$ by definition.

Definition. We shall let $\mathcal{N}_0(\nu)$ denote the collection of all subsets of sets in $\mathcal{N}(\nu)$: $B \in \mathcal{N}_0(\nu)$ iff there exists $N \in \mathcal{F}$ with $\nu(N) = 0$ and $B \subseteq N$. Observe that $\mathcal{N}(\nu) \subseteq \mathcal{N}_0(\nu)$ by definition.

Definition. We will say that $\mathcal{F}$ is *[ν-] complete* iff $\mathcal{N}_0(\nu) \subseteq \mathcal{F}$. That is, $\mathcal{F}$ will be called ν-complete iff every subset of every ν-null set is in $\mathcal{F}$. Equivalently, $\mathcal{F}$ is ν-complete iff $A \subseteq B$, $B \in \mathcal{F}$, and $\nu(B) = 0$ force $A \in \mathcal{F}$.

For applications, $\mathcal{F}$ will denote a σ-field, and $\nu : \mathcal{F} \to \bar{\mathbb{R}}$ will denote a measure. The Completion Theorem will be stated and proved in this setting after some technical claims are presented. The first claim is a straightforward exercise in set theory, and the proof is left to the reader.

Claim 1. *Let A, B and N denote subsets of Ω with $B \subseteq N$. Then we have:*
(a) $A \cup B = (A - N) \,\triangle\, [N \cap (A \cup B)]$,
(b) $A \,\triangle\, B = (A - N) \cup [N \cap (A \triangle B)]$,

(c) $A \cup B = (A \cup N) - [(N - B) - A]$, *and*
(d) $A - B = (A - N) \cup (A \cap N \cap B^c)$.

Claim 2. *Let* $\mathcal{F}$ *denote a* σ*-field, and let* $\nu : \mathcal{F} \to \bar{\mathbb{R}}$ *denote a measure. Define*
(1) $\bar{\mathcal{F}}_1 = \{A \cup B : A \in \mathcal{F}, B \in \mathcal{N}_0(\nu)\}$,
(2) $\bar{\mathcal{F}}_2 = \{A \triangle B : A \in \mathcal{F}, B \in \mathcal{N}_0(\nu)\}$, *and*
(3) $\bar{\mathcal{F}}_3 = \{A - B : A \in \mathcal{F}, B \in \mathcal{N}_0(\nu)\}$.
Then $\mathcal{F} \subseteq \bar{\mathcal{F}}_1 = \bar{\mathcal{F}}_2 = \bar{\mathcal{F}}_3$.

Proof: Let $C \in \bar{\mathcal{F}}_1$. Then $C = A \cup B$, where $A \in \mathcal{F}$ and B is contained in some $N \in \mathcal{F}$ with $\nu(N) = 0$. By Claim 1(a), $C = (A - N) \triangle [N \cap (A \cup B)]$. Since $\mathcal{F}$ is a σ-field with $A, N \in \mathcal{F}$, we have $A - N \in \mathcal{F}$. Since $N \cap (A \cup B) \subseteq N$ and $\nu(N) = 0$, we have that C is the symmetric difference of an $\mathcal{F}$-set and a $\mathcal{N}_0(\nu)$-set, so that $C \in \bar{\mathcal{F}}_2$. Therefore, $\bar{\mathcal{F}}_1 \subseteq \bar{\mathcal{F}}_2$. For the reverse inclusion, let $D \in \bar{\mathcal{F}}_2$, so that $D = A \triangle B$, where $A \in \mathcal{F}$ and B is contained in some $N \in \mathcal{F}$ with $\nu(N) = 0$. By Claim 1(b), we have that $D = (A - N) \cup [N \cap (A \triangle B)]$. We have $A - N \in \mathcal{F}$ and $N \cap (A \triangle B) \subseteq N$, and hence D is the union of an $\mathcal{F}$-set with a $\mathcal{N}_0(\nu)$-set, hence $D \in \bar{\mathcal{F}}_1$. This gives $\bar{\mathcal{F}}_2 \subseteq \bar{\mathcal{F}}_1$, so $\bar{\mathcal{F}}_1 = \bar{\mathcal{F}}_2$.

We now show that $\bar{\mathcal{F}}_1 \subseteq \bar{\mathcal{F}}_3$. Let $C \in \bar{\mathcal{F}}_1$, so that $C = A \cup B$, where $A \in \mathcal{F}$ and B is contained in some $N \in \mathcal{F}$ with $\nu(N) = 0$. By Claim 1(c), $A \cup B = (A \cup N) - [(N - B) - A]$. Observe that $A \cup N \in \mathcal{F}$ and $(N - B) - A \subseteq N$. Thus C is equal to an $\mathcal{F}$-set minus a $\mathcal{N}_0(\nu)$-set, whence $C \in \bar{\mathcal{F}}_3$, giving $\bar{\mathcal{F}}_1 \subseteq \bar{\mathcal{F}}_3$. We now show that $\bar{\mathcal{F}}_3 \subseteq \bar{\mathcal{F}}_1$, which will force $\bar{\mathcal{F}}_1 = \bar{\mathcal{F}}_2 = \bar{\mathcal{F}}_3$. If $C \in \mathcal{F}_3$, then $C = A - B$, where $A \in \mathcal{F}$ and B is contained in some $N \in \mathcal{F}$ with $\nu(N) = 0$. By Claim 1(d), we have that $C = (A - N) \cup (A \cap N \cap B^c)$. Observe that $A - N \in \mathcal{F}$ and $A \cap N \cap B^c \subseteq N$. Thus, C is the union of an $\mathcal{F}$-set and a $\mathcal{N}_0(\nu)$-set, hence $C \in \bar{\mathcal{F}}_1$. This gives $\bar{\mathcal{F}}_3 \subseteq \bar{\mathcal{F}}_1$, showing that $\bar{\mathcal{F}}_1 = \bar{\mathcal{F}}_2 = \bar{\mathcal{F}}_3$.

We will now complete the proof by showing that $\mathcal{F} \subseteq \bar{\mathcal{F}}_1$. Let $A \in \mathcal{F}$. Then $A = A \cup \varnothing$, and observe that $\varnothing \in \mathcal{N}(\nu) \subseteq \mathcal{N}_0(\nu)$. Therefore, A may be expressed as the union of an $\mathcal{F}$-set with a $\mathcal{N}_0(\nu)$-set, and hence $A \in \bar{\mathcal{F}}_1$. ∎

Before stating and proving the next claim, recall the following facts, where $(\Omega, \mathcal{F}, \mu)$ denotes a measure space. Here, $A, B, N, A_1, A_2, \cdots$ are all in $\mathcal{F}$:
(1) if $\mu(A_n) = 0$ for all $n \in \mathbb{N}$, then $\mu(\bigcup_{n=1}^{\infty} A_n) = 0$,
(2) if $\mu(A \triangle B) = 0$, then $\mu(A) = \mu(B)$, and
(3) if $\mu(N) = 0$, then $\mu(A - N) = \mu(A)$.

Fact (1) follows immediately from the countable subadditivity property: $\mu(\bigcup_{n=1}^{\infty} A_n) \le \sum_{n=1}^{\infty} \mu(A_n) = 0$. Facts (2) and (3) are from Exercise 10 of Section 2.1.

Claim 3. *Assume the framework of Claim 2, and write* $\bar{\mathcal{F}} = \bar{\mathcal{F}}_1 = \bar{\mathcal{F}}_2 = \bar{\mathcal{F}}_3$ *. Define* $\bar{\nu}_1$, $\bar{\nu}_2$, *and* $\bar{\nu}_3$ *on* $\bar{\mathcal{F}}$ *as follows:*
(1) $\bar{\nu}_1(C) = \nu(A)$ *whenever* $C = A \cup B$, *where* $A \in \mathcal{F}$ *and* $B \in \mathcal{N}_0(\nu)$,
(2) $\bar{\nu}_2(C) = \nu(A)$ *whenever* $C = A \triangle B$, *where* $A \in \mathcal{F}$ *and* $B \in \mathcal{N}_0(\nu)$, *and*
(3) $\bar{\nu}_3(C) = \nu(A)$ *whenever* $C = A - B$, *where* $A \in \mathcal{F}$ *and* $B \in \mathcal{N}_0(\nu)$.

Then $\bar{\nu}_1$, $\bar{\nu}_2$, and $\bar{\nu}_3$ are well-defined and $\bar{\nu}_1 = \bar{\nu}_2 = \bar{\nu}_3$ on $\bar{\mathcal{F}}$.

Proof: We first show that $\bar{\nu}_1$ is well-defined. Suppose that $C \in \mathcal{F}$ may be written as $A_1 \cup B_1$ and $A_2 \cup B_2$, where $A_1, A_2 \in \mathcal{F}$ and B_i is contained in some $N_i \in \mathcal{F}$ with $\nu(N_i) = 0$ for $i = 1, 2$. We desire to show that $\bar{\nu}_1$ does not depend on which representation of C that is used, which in turn means showing that $\nu(A_1) = \nu(A_2)$. By Claim 1(a) and the fact that $A_1 \cup B_1 = A_2 \cup B_2 = C$, we have

$$(A_1 - N_1) \,\triangle\, (N_1 \cap C) = (A_2 - N_2) \,\triangle\, (N_2 \cap C).$$

Since $P \triangle Q = R \triangle S$ implies that $P \triangle R = Q \triangle S$, we have that

$$\begin{aligned} (A_1 - N_1) \,\triangle\, (A_2 - N_2) &= (N_1 \cap C) \,\triangle\, (N_2 \cap C) \\ &\subseteq N_1 \cup N_2. \end{aligned}$$

Therefore, $\nu((A_1 - N_1) \triangle (A_2 - N_2)) \leq \nu(N_1 \cup N_2) = 0$, so that we have $\nu((A_1 - N_1) \triangle (A_2 - N_2)) = 0$, whence $\nu(A_1 - N_1) = \nu(A_2 - N_2)$. Since $\nu(N_1) = \nu(N_2) = 0$, we have $\nu(A_1) = \nu(A_2)$, so $\bar{\nu}_1$ is well-defined.

We now show that $\bar{\nu}_2$ is well-defined. Suppose that $C \in \mathcal{F}$ may be written as $A_1 \triangle B_1$ and $A_2 \triangle B_2$, where A_1, A_2, B_1, B_2 are as above; we desire to show that $\bar{\nu}_2(C)$ does not depend on which representation of C is used. This means that we again desire the equality $\nu(A_1) = \nu(A_2)$. Since $A_1 \triangle B_1 = A_2 \triangle B_2$, we again use the symmetric difference result mentioned in the preceding paragraph to obtain $A_1 \triangle A_2 = B_1 \triangle B_2 \subseteq B_1 \cup B_2 \subseteq N_1 \cup N_2 \in \mathcal{F}$, and $\nu(N_1 \cup N_2) = 0$. It follows that $\nu(A_1 \triangle A_2) = 0$, whence $\nu(A_1) = \nu(A_2)$, so $\bar{\nu}_2$ is well-defined.

We now show that $\bar{\nu}_3$ is well-defined. Suppose that $C \in \mathcal{F}$ may be written as $A_1 - B_1$ and $A_2 - B_2$, where A_1, A_2, B_1, B_2 are as above; we desire yet again to show that $\nu(A_1) = \nu(A_2)$. Since $C = (A_i - N_i) \cup (A_i \cap N_i \cap B_i^c)$ for $i = 1, 2$ by Claim 1(d), and since $\bar{\nu}_1$ is well-defined, it follows that $\bar{\nu}_1$ assigns the same value to both $(A_1 - N_1) \cup (A_1 \cap N_1 \cap B_1^c)$ and $(A_2 - N_2) \cup (A_2 \cap N_2 \cap B_2^c)$. But then $\nu(A_1 - N_1) = \nu(A_2 - N_2)$. Since $\nu(N_1) = \nu(N_2) = 0$, we have $\nu(A_1) = \nu(A_2)$. Therefore, $\bar{\nu}_3$ is well-defined.

We now show that $\bar{\nu}_1 = \bar{\nu}_2$ on $\bar{\mathcal{F}}$. Select $C \in \bar{\mathcal{F}} = \bar{\mathcal{F}}_1$, so that $C = A \cup B$, where $A \in \mathcal{F}$ and B is contained in some $N \in \mathcal{F}$ with $\nu(N) = 0$. Then by definition we have $\bar{\nu}_1(C) = \nu(A)$. However, C also may be written as $(A - N) \triangle [N \cap (A \cup B)]$ by Claim 1(a). Since $N \cap (A \cup B) \subseteq N \in \mathcal{F}$ and $\nu(N) = 0$, we have the equation $\bar{\nu}_2(C) = \nu(A - N)$. Since $\nu(N) = 0$, we have $\nu(A - N) = \nu(A)$, and hence $\bar{\nu}_2(C) = \nu(A)$. It follows that $\bar{\nu}_1(C) = \bar{\nu}_2(C)$, the common value being $\nu(A)$. Therefore $\bar{\nu}_1 = \bar{\nu}_2$ on $\bar{\mathcal{F}}$.

Finally, we show that $\bar{\nu}_3 = \bar{\nu}_1$ on $\bar{\mathcal{F}}$. Let $C \in \bar{\mathcal{F}} = \bar{\mathcal{F}}_3$, so that $C = A - B$, where $A \in \mathcal{F}$ and B is contained in some $N \in \mathcal{F}$ with $\nu(N) = 0$. By Claim 1(d), $C = (A - N) \cup (A \cap N \cap B^c)$, so $\bar{\nu}_1(C) = \nu(A - N) = \nu(A) = \bar{\nu}_3(C)$. This completes the entire proof. ∎

In the proof of the following theorem, we will utilize (1) and (3) of Claim 2, and (1) of Claim 3.

Completion Theorem. *Let $(\Omega, \mathcal{F}, \nu)$ be a measure space. Define*

$$\bar{\mathcal{F}} = \{A \cup B : A \in \mathcal{F}, B \in \mathcal{N}_0(\nu)\},$$

and define $\bar{\nu} : \bar{\mathcal{F}} \to \bar{\mathbb{R}}$ by writing

$$\bar{\nu}(C) = \nu(A) \quad \textit{whenever} \quad C = A \cup B \textit{ wtih } A \in \mathcal{F} \textit{ and } B \in \mathcal{N}_0(\nu).$$

Then the following results obtain:
(a) $\mathcal{F} \subseteq \bar{\mathcal{F}}$,
(b) $\bar{\mathcal{F}}$ is a σ-field,
(c) $\bar{\nu}$ is well-defined,
(d) $\bar{\nu}$ is a measure [hence $(\Omega, \bar{\mathcal{F}}, \bar{\nu})$ is a measure space],
(e) $\bar{\nu} = \nu$ on $\mathcal{F}$,
(f) $\bar{\mathcal{F}}$ is $\bar{\nu}$-complete, and
(g) $\bar{\nu}$ is the unique measure with domain $\bar{\mathcal{F}}$ that agrees with ν on $\mathcal{F}$.

Proof: We note that (a) is automatic from Claim 2 and that (c) is automatic from Claim 3, and hence we need only to prove (b), (d), (e), (f), and (g).

We now prove (b). Since $\Omega \in \mathcal{F}$, (a) gives $\Omega \in \bar{\mathcal{F}}$. To show that $\bar{\mathcal{F}}$ is closed under complementation, pick $E \in \bar{\mathcal{F}}$, so that $E = A \cup B$, where $A \in \mathcal{F}$ and B is contained in some $N \in \mathcal{F}$ with $\nu(N) = 0$. We obtain that $E^c = A^c - B$, where $A^c \in \mathcal{F}$, $B \subseteq N \in \mathcal{F}$, and $\nu(N) = 0$. Using (3) in Claim 2, we conclude that $E^c \in \bar{\mathcal{F}}$. To see that $\bar{\mathcal{F}}$ is closed under countable unions, let $\{E_n\}_{n=1}^{\infty}$ denote a sequence of $\bar{\mathcal{F}}$-sets, and for each $n \in \mathbb{N}$ write $E_n = A_n \cup B_n$ where $A_n \in \mathcal{F}$ and B_n is contained in some $N_n \in \mathcal{F}$ with $\nu(N_n) = 0$. By set algebra, we have that $\bigcup_{n=1}^{\infty} E_n = \bigcup_{n=1}^{\infty} A_n \cup \bigcup_{n=1}^{\infty} B_n$, where $\bigcup_{n=1}^{\infty} A_n \in \mathcal{F}$, $\bigcup_{n=1}^{\infty} B_n \subseteq \bigcup_{n=1}^{\infty} N_n \in \mathcal{F}$, and $\nu(\bigcup_{n=1}^{\infty} N_n) = 0$. It follows that $\bigcup_{n=1}^{\infty} E_n \in \bar{\mathcal{F}}$. This concludes the proof of part (b).

We now prove (d). Clearly $\bar{\nu}$ is nonnegative since ν is nonnegative. Next, $\varnothing$ may be written as $\varnothing \cup \varnothing$, where $\varnothing \in \mathcal{F}$ and $\varnothing \in \mathcal{N}_0(\nu)$. By the definition of $\bar{\nu}$ we accordingly have $\bar{\nu}(\varnothing) = \nu(\varnothing) = 0$. To show that $\bar{\nu}$ is countably additive, let $\{C_n\}_{n=1}^{\infty}$ denote a disjoint sequence of $\bar{\mathcal{F}}$-sets. For each $n \in \mathbb{N}$, write $C_n = A_n \cup B_n$, where $A_n \in \mathcal{F}$ and B_n is contained in some $N_n \in \mathcal{F}$ with $\nu(N_n) = 0$. Observe that $\bigcup_{n=1}^{\infty} C_n = \bigcup_{n=1}^{\infty} A_n \cup \bigcup_{n=1}^{\infty} B_n$, where $\bigcup_{n=1}^{\infty} A_n \in \mathcal{F}$, $\bigcup_{n=1}^{\infty} B_n \subseteq \bigcup_{n=1}^{\infty} N_n \in \mathcal{F}$, and $\nu(\bigcup_{n=1}^{\infty} N_n) = 0$. Observing that $A_1, A_2, \cdots$ are disjoint, we have that

$$\bar{\nu}\left(\bigcup_{n=1}^{\infty} C_n\right) = \nu\left(\bigcup_{n=1}^{\infty} A_n\right) = \sum_{n=1}^{\infty} \nu(A_n) = \sum_{n=1}^{\infty} \bar{\nu}(C_n),$$

which completes the proof of (d).

For (e), let $C \in \mathcal{F}$. Then $C = C \cup \varnothing$, where $C \in \mathcal{F}$ and $\varnothing \in \mathcal{N}_0(\nu)$, hence $\bar{\nu}(C) = \nu(C)$ by definition. This completes the proof of (e).

For (f), suppose that A is a generic subset of some $B \in \bar{\mathcal{F}}$ with $\bar{\nu}(B) = 0$. We desire to show that $A \in \bar{\mathcal{F}}$ as well. To see this, write $B = C \cup D$, where $C \in \mathcal{F}$ and D is contained in some $N \in \mathcal{F}$ with $\nu(N) = 0$. Since

$\bar{\nu}(B) = 0$, we have $\nu(C) = 0$. Next, $A \subseteq C \cup N$, and $C \cup N \in \mathcal{F}$ with $\bar{\nu}(C \cup N) \leq \bar{\nu}(C) + \bar{\nu}(N) = \nu(C) + \nu(N) = 0$ by (e). Since $\varnothing \in \mathcal{F}$ and $A = \varnothing \cup A$, it follows that $A \in \bar{\mathcal{F}}$, and hence $\bar{\mathcal{F}}$ is accordingly $\bar{\nu}$-complete.

For (g), let $\xi : \bar{\mathcal{F}} \to \bar{\mathbb{R}}$ denote an arbitrary measure that agrees with ν on $\mathcal{F}$. Let $A \in \bar{\mathcal{F}}$, so that $A = B \cup C$ where $B \in \mathcal{F}$ and C is contained in some $N \in \mathcal{F}$ with $\nu(N) = \xi(N) = 0$. Then $\bar{\nu}(A) = \nu(B) = \xi(B) \leq \xi(A)$ by the definition of $\bar{\nu}$, the fact that $\xi = \nu$ on $\mathcal{F}$, and monotonicity. However, $\xi(A) \leq \xi(B \cup N) \leq \xi(B) + \xi(N) = \xi(B) = \bar{\nu}(A)$, so that $\bar{\nu}(A) = \xi(A)$ for our generic $A \in \bar{\mathcal{F}}$. This completes the entire proof. ∎

The construction above will be called the $(\Omega, \mathcal{F}, \nu)$-*completion* or [equivalently] *the completion of* $(\Omega, \mathcal{F}, \nu)$. That is, given a measure space $(\Omega, \mathcal{F}, \nu)$, the process of completion consists of (i) adding to $\mathcal{F}$ all subsets of the form $A \cup B$, where $A \in \mathcal{F}$ and B is contained in some $N \in \mathcal{F}$ with $\nu(N) = 0$, and (ii) [uniquely] extending the measure ν with domain $\mathcal{F}$ to the measure $\bar{\nu}$ with domain $\bar{\mathcal{F}}$ by any one of the three representations in Claim 3. Upon performing (i) and (ii), we then arrive at the measure space $(\Omega, \bar{\mathcal{F}}, \bar{\nu})$. Therefore, in dealing with an arbitrary measure space $(\Omega, \mathcal{G}, \xi)$, one may assume from the outset that the measure space is complete, for if not, one could without any loss of generality work instead with the completed measure space $(\Omega, \bar{\mathcal{G}}, \bar{\xi})$.

Example 2. Let us continue with the framework of Example 1. The Completion Theorem says that if $\bar{\mathcal{B}} = \{B \cup C : \ B \in \mathcal{B},\ C \in \mathcal{N}_0(\lambda)\}$, then $\bar{\mathcal{B}}$ is a σ-field with $\mathcal{B} \subseteq \bar{\mathcal{B}}$. Furthermore, if $\bar{\lambda} : \bar{\mathcal{B}} \to \bar{\mathbb{R}}$ is defined by writing $\bar{\lambda}(A) = \lambda(B)$ when $A = B \cup C$ with $B \in \mathcal{B}$ and $C \in \mathcal{N}_0(\lambda)$, then $\bar{\lambda}$ is the unique measure with domain $\bar{\mathcal{B}}$ that agrees with λ on $\mathcal{B}$. Because of this $\bar{\lambda}$ mimics "length" exactly as λ does, but $\bar{\lambda}$ has the additional intuitive property that if $B \in \mathcal{B}$ with $\lambda(B) = 0$ [and hence $\bar{\lambda}(B) = 0$ as well], then any $A \subseteq B$ will be in $\bar{\mathcal{B}}$ and $\bar{\lambda}(A)$ will equal zero.

If $C \in \mathcal{B}$ is the uncountable set with $\lambda(C) = 0$ from Example 1, then each of the $2^{\mathfrak{c}}$ subsets of C are in $\bar{\mathcal{B}}$, and hence $\text{card}(\bar{\mathcal{B}}) = 2^{\mathfrak{c}}$. Since $\text{card}(\mathcal{B}) = \mathfrak{c}$, we see that there are many more sets in $\bar{\mathcal{B}}$ than there are in $\mathcal{B}$. The completion of $(\mathbb{R}, \mathcal{B}, \lambda)$ has added $2^{\mathfrak{c}}$ new sets whose "length" we may now rigorously discuss.

Because nothing is lost by considering $(\Omega, \bar{\mathcal{B}}, \bar{\lambda})$ instead of $(\Omega, \mathcal{B}, \lambda)$, we will try at all times to deal with $\bar{\mathcal{B}}$ and $\bar{\lambda}$ instead of $\mathcal{B}$ and λ. Chapter 4 is devoted to discussing the nuances of $\bar{\mathcal{B}}$ and $\bar{\lambda}$ as well as the relationships between $\bar{\mathcal{B}}$ and $\mathcal{B}$, among other things. There, we shall also show that both λ and $\bar{\lambda}$ have additional properties that we would consider "intuitive." Finally, the fact that we are dealing with $\mathbb{R}$ instead of $\mathbb{R}^k$ has absolutely nothing to do with any of the comments here; the same comments would hold for any $k \in \mathbb{N}$.

If in the Completion Theorem we replaced $\bar{\mathcal{F}}$ with either $\bar{\mathcal{F}}_2$ or $\bar{\mathcal{F}}_3$ from Claim 2, and if $\bar{\nu}$ was accordingly replaced by $\bar{\nu}_2$ or $\bar{\nu}_3$, then absolutely nothing would change by virtue of Claim 3.

The final claim of this section says that completion is a process that may be applied at most once; anything more is redundant. Informally, we may say that the completion of an already complete measure space does not introduce any changes.

Claim 4. *$(\Omega, \mathcal{F}, \nu) = (\Omega, \bar{\mathcal{F}}, \bar{\nu})$ if $(\Omega, \mathcal{F}, \nu)$ is already complete.*

Proof: We have $\mathcal{F} \subseteq \bar{\mathcal{F}}$ by the Completion Theorem. Next, let $C \in \bar{\mathcal{F}}$, so that $C = A \cup B$ for some $A \in \mathcal{F}$ and $B \in \mathcal{N}_0(\nu)$. Since $\mathcal{F}$ is ν-complete, we have $\mathcal{N}_0(\nu) \subseteq \mathcal{F}$, hence $B \in \mathcal{F}$. It follows that $A \cup B \in \mathcal{F}$, whence $C \in \mathcal{F}$. This gives $\bar{\mathcal{F}} \subseteq \mathcal{F}$, hence $\bar{\mathcal{F}} = \mathcal{F}$. Since the Completion Theorem asserts that $\bar{\nu}$ is the unique measure on $\bar{\mathcal{F}}$ that agrees with ν on $\mathcal{F}$, and since $\bar{\mathcal{F}} = \mathcal{F}$, we have that $\bar{\nu} = \nu$ on $\bar{\mathcal{F}} = \mathcal{F}$. ∎

The results of this section will be used in the next section to further clarify the relationship between the σ-fields $\sigma(\mathcal{A})$ and $\mathcal{M}(\mu^*)$ in the framework wherein are found our discussions of the Outer Measure Theorem, the Extension Theorem, and the Uniqueness Theorem.

Exercises.

1. Let $k \geq 2$, and let $\lambda_k : \mathcal{B}^k \to \bar{\mathbb{R}}$ denote the unique measure with domain $\mathcal{B}^k$ that assigns the value $\prod_{i=1}^k (b_i - a_i)$ to each k-dimensional rsc rectangle $(\boldsymbol{a}, \boldsymbol{b}]$. Let $A = \{\boldsymbol{x} \in \mathbb{R}^k : x_2 = \cdots = x_k = 0\}$. Then $A \in \mathcal{B}^k$, A is uncountable, and $\lambda_k(A) = 0$, hence $\text{card}(\overline{\mathcal{B}^k}) = 2^{\mathfrak{c}}$.

2*. Let $(\Omega, \mathcal{F}, \nu)$ denote a measure space. If $A, B \in \mathcal{F}$ with $A \subseteq E \subseteq B$ and $\nu(B - A) = 0$, then $E \in \bar{\mathcal{F}}$ and $\bar{\nu}(E) = \nu(A) = \nu(B)$.

3. Let $(\Omega, \mathcal{F}, \nu)$ denote a measure space. Furthermore, let $\mathcal{F}_1$ denote a sub-σ-field of $\mathcal{F}$. Then there exists a minimal σ-field $\mathcal{F}_2$ such that $\mathcal{F}_1 \subseteq \mathcal{F}_2 \subseteq \mathcal{F}$ and $\mathcal{N}(\nu) \subseteq \mathcal{F}_2$. Also, $A \in \mathcal{F}_2$ iff there exists $B \in \mathcal{F}_1$ with $A \triangle B \in \mathcal{N}(\nu)$.

3.7 THE RELATIONSHIP BETWEEN $\sigma(\mathcal{A})$ AND $\mathcal{M}(\mu^*)$

Let us briefly review what has transpired so far. Let $\mathcal{A}$ denote a semiring on Ω, and let $\mu : \mathcal{A} \to \bar{\mathbb{R}}$ be such that μ is nonnegative, $\mu(\varnothing) = 0$, μ is finitely additive, μ is countably subadditive, and μ is σ-finite on $\mathcal{A}$. Define $\mu^* : 2^\Omega \to \bar{\mathbb{R}}$ by requiring that

$$\mu^*(A) = \inf\left\{\sum_{n=1}^{\infty} \mu(A_n) : \{A_n\} \text{ is an } \mathcal{A}\text{-covering of } A\right\}$$

when at least one $\mathcal{A}$-covering of A exists, and set $\mu(A) = +\infty$ when no such $\mathcal{A}$-covering exists; μ^* is the outer measure induced by μ as given in Example 1 of Section 2. If we let

$$\mathcal{M}(\mu^*) = \{A \subseteq \Omega : \mu^*(T) = \mu^*(T \cap A) + \mu^*(T \cap A^c) \text{ for all } T \subseteq \Omega\},$$

then the Outer Measure Theorem states that (i) $\mathcal{M}(\mu^*)$ is a σ-field and (ii) the restriction $\mu^*_{\mathcal{M}(\mu^*)} : \mathcal{M}(\mu^*) \to \bar{\mathbb{R}}$ of μ^* to $\mathcal{M}(\mu^*)$ is a measure. In the proof of the Extension Theorem we showed that $\mathcal{A} \subseteq \sigma(\mathcal{A}) \subseteq \mathcal{M}(\mu^*) \subseteq 2^\Omega$ and $\mu^* = \mu$ on $\mathcal{A}$. The restriction $\mu^*_{\sigma(\mathcal{A})}$ of μ^* from 2^Ω to $\sigma(\mathcal{A})$ [or equivalently the restriction of $\mu^*_{\mathcal{M}(\mu^*)}$ from $\mathcal{M}(\mu^*)$ to $\sigma(\mathcal{A})$] is a measure with the property that $\mu^*_{\sigma(\mathcal{A})}(A) = \mu(A)$ for every $A \in \mathcal{A}$. By virtue of the fact that μ is σ-finite on $\mathcal{A}$, the Uniqueness Theorem and Claim 3 of Section 3.5 imply that $\mu^*_{\sigma(\mathcal{A})}$ is the unique measure with domain $\sigma(\mathcal{A})$ that agrees with μ on $\mathcal{A}$. *We shall assume this setup for the remainder of the section.*

This section aims to explore two relationships. The first is that of $\sigma(\mathcal{A})$ to $\mathcal{M}(\mu^*)$, and the second is that of $\mu^*_{\sigma(\mathcal{A})}$ to $\mu^*_{\mathcal{M}(\mu^*)}$. Under the above framework we already know that $\sigma(\mathcal{A}) \subseteq \mathcal{M}(\mu^*)$, and it would also be of interest to ask if there is a way of expressing $\mathcal{M}(\mu^*)$-sets in terms of $\sigma(\mathcal{A})$-sets. We shall see that such a representation is possible. In the language of the previous section, we shall show that $\mathcal{M}(\mu^*) = \overline{\sigma(\mathcal{A})}$ *and* $\mu^*_{\mathcal{M}(\mu^*)} = \overline{\mu^*_{\sigma(\mathcal{A})}}$, *that is,* $(\Omega, \mathcal{M}(\mu^*), \mu^*_{\mathcal{M}(\mu^*)})$ *is the completion of* $(\Omega, \sigma(\mathcal{A}), \mu^*_{\sigma(\mathcal{A})})$. In other words, this means that (i) every $A \in \mathcal{M}(\mu^*)$ may be written as $B \cup C$ [or $B - C$ or $B \Delta C$], where $B \in \sigma(\mathcal{A})$ and C is a subset of some $N \in \sigma(\mathcal{A})$ with $\mu^*_{\sigma(\mathcal{A})}(N) = 0$, and (ii) $\mu^*_{\mathcal{M}(\mu^*)}(A) = \mu^*_{\sigma(\mathcal{A})}(B)$ when A has the stated representation.

Claim 1. *Given any outer measure* $\nu : 2^\Omega \to \bar{\mathbb{R}}$, *we have* $\mathcal{N}(\nu) \subseteq \mathcal{M}(\nu)$.

Proof: This is a mere restatement of Claim 9 in Section 3 in terms of the notation of the previous section. ■

Claim 2. *Let* $\nu : 2^\Omega \to \bar{\mathbb{R}}$ *denote an outer measure, and let* $\nu_{\mathcal{M}(\nu)}$ *denote the measure restricion of* ν *from* 2^Ω *to the* σ*-field* $\mathcal{M}(\nu)$ *in accordance with the Outer Measure Theorem. Then* $(\Omega, \mathcal{M}(\nu), \nu_{\mathcal{M}(\nu)})$ *is a complete measure space.*

Proof: Let $A \subseteq B \subseteq \Omega$ with $B \in \mathcal{M}(\nu)$ and $\nu_{\mathcal{M}(\nu)}(B) = 0$. We desire to show that $A \in \mathcal{M}(\nu)$. Since $\nu_{\mathcal{M}(\nu)} = \nu$ on $\mathcal{M}(\nu)$, we have that $\nu(B) = 0$. Since $A \subseteq B$, we obtain $\nu(A) = 0$, hence $A \in \mathcal{M}(\nu)$ by Claim 1. ■

The previous claims were wholly general, but the following claim utilizes the specific framework as given in the introduction to this section.

Claim 3. *Assume the notation and framework of the Extension Theorem. Let* $E \subseteq \Omega$. *Then there exists* $A_E \in \sigma(\mathcal{A})$ *with* $E \subseteq A_E$ *and* $\mu^*(E) = \mu^*(A_E)$.

Proof: Let $n \in \mathbb{N}$ and $\mu^*(E) < +\infty$. We may find a sequence of $\mathcal{A}$-sets $\{A_{nm}\}_{m=1}^\infty$ such that $E \subseteq \bigcup_{m=1}^\infty A_{nm}$ and $\sum_{m=1}^\infty \mu(A_{nm}) \le \mu^*(E) + \frac{1}{n}$. Define $A_E = \bigcap_{n=1}^\infty \bigcup_{m=1}^\infty A_{nm}$, so that $A_E \in \sigma(\mathcal{A})$ and $E \subseteq A_E$. Observe that $\mu^*(E) \le \mu^*(A_E)$ by monotonicity. Next, observe for all $n \in \mathbb{N}$ that

$$
\begin{aligned}
\mu^*(A_E) = \mu^*\left(\bigcap_{n=1}^\infty \bigcup_{m=1}^\infty A_{nm}\right) \le \mu^*\left(\bigcup_{m=1}^\infty A_{nm}\right) &\le \sum_{m=1}^\infty \mu^*(A_{nm}) \\
&= \sum_{m=1}^\infty \mu(A_{nm}) \\
&\le \mu^*(E) + \frac{1}{n}.
\end{aligned}
$$

[The first inequality is due to monotonicity, the second inequality follows by countable subadditivity, and the second equality follows because $\mu^* = \mu$ on $\mathcal{A}$ as proven in the Extension Theorem.] It follows that $\mu^*(A_E) \leq \mu^*(E)$, and hence $\mu^*(A_E) = \mu^*(E)$, so the proof is complete for the case $\mu^*(E) < +\infty$. The case $\mu^*(E) = +\infty$ is trivial, for we may take $A_E = \Omega$. ∎

Claim 4. *In the framework of the Extension Theorem, we have the following:*

(a) $(\Omega, \mathcal{M}(\mu^*), \mu^*_{\mathcal{M}(\mu^*)})$ *is the completion of* $(\Omega, \sigma(\mathcal{A}), \mu^*_{\sigma(\mathcal{A})})$.

(b) The measure $\mu^*_{\mathcal{M}(\mu*)}$ *is the unique measure with domain* $\mathcal{M}(\mu^*)$ *that (i) agrees with* $\mu^*_{\sigma(\mathcal{A})}$ *on* $\sigma(\mathcal{A})$ *and (ii) agrees with* μ *on* $\mathcal{A}$.

Proof: We first show that $\overline{\sigma(\mathcal{A})} \subseteq \mathcal{M}(\mu^*)$. Pick $C \in \overline{\sigma(\mathcal{A})}$. Therefore, C has the form $A \cup B$, where $A \in \sigma(\mathcal{A})$ and B is contained in some $N \in \sigma(\mathcal{A})$ with $\mu^*_{\sigma(\mathcal{A})}(N) = \mu^*(N) = 0$. Since $\sigma(\mathcal{A}) \subseteq \mathcal{M}(\mu^*)$, we have that $A \in \mathcal{M}(\mu^*)$. Since $B \subseteq N$ and $\mu^*(N) = 0$, we have $\mu^*(B) = 0$, hence $B \in \mathcal{M}(\mu^*)$ by Claim 1. Since $\mathcal{M}(\mu^*)$ is a σ-field, we have $C = A \cup B \in \mathcal{M}(\mu^*)$, which gives $\overline{\sigma(\mathcal{A})} \subseteq \mathcal{M}(\mu^*)$.

We now show the reverse inclusion $\mathcal{M}(\mu^*) \subseteq \overline{\sigma(\mathcal{A})}$. Pick $E \in \mathcal{M}(\mu^*)$. There are two cases for this E: (i) $\mu^*(E) < +\infty$, and (ii) $\mu^*(E) = +\infty$. We handle case (i) first. By Claim 3, we may find a set $A_E \in \sigma(\mathcal{A})$ and hence in $\mathcal{M}(\mu^*)$ such that $E \subseteq A_E$ and $\mu^*(E) = \mu^*(A_E)$, hence $\mu^*(A_E) < +\infty$. Observe that $A_E - E \in \mathcal{M}(\mu^*)$ and $\mu^*(A_E - E) = \mu^*(A_E) - \mu^*(E) = 0$ by (M5). Apply Claim 3 to $A_E - E$ to find a set $G \in \sigma(\mathcal{A})$ such that $A_E - E \subseteq G$ and $\mu^*(G) = \mu^*(A_E - E)$, so that $\mu^*(G) = 0$. Since $E = (A_E \cap G^c) \cup (E \cap G)$ and $A_E \cap G^c \in \sigma(\mathcal{A})$ with both $E \cap G \subseteq G \in \sigma(\mathcal{A})$ and $\mu^*(G) = 0$, we have $E \in \overline{\sigma(\mathcal{A})}$. This proves case (i).

To handle case (ii), we suppose that $E \in \mathcal{M}(\mu^*)$ is such that $\mu^*(E) = +\infty$. Invoke the σ-finiteness of μ on $\mathcal{A}$ to write $\Omega = \bigcup_{n=1}^{\infty} A_n$ where $A_n \in \mathcal{A}$ and $\mu(A_n) < +\infty$ for each $n \in \mathbb{N}$. Write $E_n = E \cap A_n$ for each $n \in \mathbb{N}$. Since $\mathcal{A} \subseteq \mathcal{M}(\mu^*)$, we have for every $n \in \mathbb{N}$ that $E_n \in \mathcal{M}(\mu^*)$ and $\mu^*(E_n) < +\infty$; it also holds that $\bigcup_{n=1}^{\infty} E_n = E$. We now apply case (i) to assert that $E_n \in \overline{\sigma(\mathcal{A})}$ for each $n \in \mathbb{N}$. Therefore, for every $n \in \mathbb{N}$ we have $E_n = C_n \cup D_n$, where $C_n \in \sigma(\mathcal{A})$ and D_n is contained in some $N_n \in \sigma(\mathcal{A})$ with $\mu^*(N_n) = 0$. It follows that $E = \bigcup_{n=1}^{\infty} E_n = \bigcup_{n=1}^{\infty} C_n \cup \bigcup_{n=1}^{\infty} D_n$, where $\bigcup_{n=1}^{\infty} C_n \in \sigma(\mathcal{A})$, $\bigcup_{n=1}^{\infty} D_n \subseteq \bigcup_{n=1}^{\infty} N_n \in \sigma(\mathcal{A})$, and $\mu^*(\bigcup_{n=1}^{\infty} N_n) = 0$, hence $E \in \overline{\sigma(\mathcal{A})}$. This completes case (ii), and for both cases we have shown that $\mathcal{M}(\mu^*) \subseteq \overline{\sigma(\mathcal{A})}$. We have therefore shown that $\mathcal{M}(\mu^*) = \overline{\sigma(\mathcal{A})}$.

We now turn to showing that $\overline{\mu^*_{\sigma(\mathcal{A})}} = \mu^*_{\mathcal{M}(\mu^*)}$ on $\mathcal{M}(\mu^*) = \overline{\sigma(\mathcal{A})}$. Pick $C \in \mathcal{M}(\mu^*) = \overline{\sigma(\mathcal{A})}$, so that $C = A \cup B$ with $A \in \overline{\sigma(\mathcal{A})} = \mathcal{M}(\mu^*)$ and where B is contained in some $N \in \overline{\sigma(\mathcal{A})} = \mathcal{M}(\mu^*)$ with $\mu^*(N) = 0$. Since Claim 2 shows that $(\Omega, \mathcal{M}(\mu^*), \mu^*_{\mathcal{M}(\mu^*)})$ is complete, we have $B \in \mathcal{M}(\mu^*) = \overline{\sigma(\mathcal{A})}$, and $\mu^*_{\mathcal{M}(\mu^*)}(B) = 0$. It follows that

$$\overline{\mu^*_{\sigma(\mathcal{A})}}(C) = \mu^*_{\sigma(\mathcal{A})}(A) = \mu^*_{\mathcal{M}(\mu^*)}(A) = \mu^*_{\mathcal{M}(\mu^*)}(A \cup B) = \mu^*_{\mathcal{M}(\mu^*)}(C),$$

where the first equality follows by the definition of $\overline{\mu^*_{\sigma(\mathcal{A})}}$, the second equality follows because $A \in \sigma(\mathcal{A})$ and $\mu^*_{\mathcal{M}(\mu^*)} = \mu^*_{\sigma(\mathcal{A})}$ on $\sigma(\mathcal{A})$, and the third equality follows because $\mu^*_{\mathcal{M}(\mu^*)}(B) = 0$. This shows that $\overline{\mu^*_{\sigma(\mathcal{A})}} = \mu^*_{\mathcal{M}(\mu^*)}$ on their common domain $\mathcal{M}(\mu^*) = \overline{\sigma(\mathcal{A})}$. The proof of (a) is now complete.

We now consider (b). By part (g) of the Completion Theorem, $\overline{\mu^*_{\sigma(\mathcal{A})}}$ is the unique measure with domain $\overline{\sigma(\mathcal{A})}$ that agrees with $\mu^*_{\sigma(\mathcal{A})}$ on $\sigma(\mathcal{A})$. Since we have shown that $\overline{\mu^*_{\sigma(\mathcal{A})}} = \mu^*_{\mathcal{M}(\mu^*)}$ and $\overline{\sigma(\mathcal{A})} = \mathcal{M}(\mu^*)$, it follows that $\mu^*_{\mathcal{M}(\mu^*)}$ is the unique measure with domain $\mathcal{M}(\mu^*)$ that agrees with $\mu^*_{\sigma(\mathcal{A})}$ on $\sigma(\mathcal{A})$. This proves the first claim of (b).

For the second claim of (b), we let ξ denote a measure on $\mathcal{M}(\mu^*)$ with $\xi = \mu$ on $\mathcal{A}$; the goal is to show that $\xi = \mu^*_{\mathcal{M}(\mu^*)}$ on $\mathcal{M}(\mu^*)$. We let $\xi_{\sigma(\mathcal{A})}$ denote the [measure] restriction of ξ from $\mathcal{M}(\mu^*)$ to $\sigma(\mathcal{A})$. We have shown that $\mu^*_{\sigma(\mathcal{A})}$ is the unique measure on $\sigma(\mathcal{A})$ that agrees with μ on $\mathcal{A}$, hence we have $\xi_{\sigma(\mathcal{A})} = \mu^*_{\sigma(\mathcal{A})}$ on $\sigma(\mathcal{A})$. Pick $C \in \mathcal{M}(\mu^*)$. Then $C = A \cup B$, where $A \in \sigma(\mathcal{A}) \subseteq \mathcal{M}(\mu^*)$ and B is contained in some $N \in \sigma(\mathcal{A}) \subseteq \mathcal{M}(\mu^*)$ with $\mu^*_{\sigma(\mathcal{A})}(N) = 0$, so that $\xi(N) = \xi_{\sigma(\mathcal{A})}(N) = \mu^*_{\sigma(\mathcal{A})}(N) = 0$. Accordingly, the set B is in $\mathcal{M}(\mu^*)$ and $\xi(B) \leq \xi(N) = 0$, whence $\xi(B) = 0$. It follows that

$$\xi(C) = \xi(A \cup B) = \xi(A) = \xi_{\sigma(\mathcal{A})}(A) = \mu^*_{\sigma(\mathcal{A})}(A) = \mu^*_{\mathcal{M}(\mu^*)}(C).$$

This shows that $\xi = \mu^*_{\mathcal{M}(\mu^*)}$ on $\mathcal{M}(\mu^*)$, completing the entire proof. ■

Exercise.

1. Let Ω be uncountable, let $\mathcal{A}$ denote the σ-field $\{A \subseteq \Omega : A \text{ is amc or } A^c \text{ is amc}\}$, and define $\mu : \mathcal{A} \to \bar{\mathbb{R}}$ by stipulating that $\mu(A)$ denotes the number of points in A if A is finite and $\mu(A) = +\infty$ if A is infinite.

(a) $(\Omega, \mathcal{A}, \mu)$ is a non-σ-finite measure space.

(b) $(\Omega, \mathcal{A}, \mu)$ is complete.

(c) Letting μ^* denote the outer measure induced by μ, the [σ-field] $\mathcal{M}(\mu^*)$ coincides with 2^Ω. [Since $\mathcal{A}$ is a σ-field, we have $\sigma(\mathcal{A}) = \mathcal{A}$. Use a relevant claim to argue that $\bar{\mathcal{A}} = \mathcal{M}(\mu^*)$ and $\overline{\mu^*_{\sigma(\mathcal{A})}} = \mu^*_{\mathcal{M}(\mu^*)}$. However, here $\overline{\sigma(\mathcal{A})} = \bar{\mathcal{A}} = \mathcal{A}$ since $\mathcal{A}$ is already a complete σ-field with respect to μ, and clearly $\mathcal{A} \neq 2^\Omega = \mathcal{M}(\mu^*)$. Thus, in this example where μ fails to be σ-finite, we have $\overline{\sigma(\mathcal{A})} \neq \mathcal{M}(\mu^*)$, which shows that the σ-finiteness of μ with respect to $\mathcal{A}$ is essential to Claim 4.]

3.8 APPROXIMATIONS

In this section, we assume that $\mathcal{S}$ is a semiring, and we let $\nu : \sigma(\mathcal{S}) \to \bar{\mathbb{R}}$ denote a measure that is σ-finite on $\mathcal{S}$. We will let $\nu_{\mathcal{S}}$ denote the restriction of ν from $\sigma(\mathcal{S})$ to $\mathcal{S}$, so that $\nu = \nu_{\mathcal{S}}$ on $\mathcal{S}$. We will let $\nu^* : 2^\Omega \to \bar{\mathbb{R}}$ be defined by stipulating that

$$\begin{aligned}\nu^*(A) &= \inf\left\{\sum_{n=1}^{\infty}\nu(A_n) : \{A_n\}_{n=1}^{\infty} \text{ is an } \mathcal{S}\text{-covering of } A\right\} \\ &= \inf\left\{\sum_{n=1}^{\infty}\nu_{\mathcal{S}}(A_n) : \{A_n\}_{n=1}^{\infty} \text{ is an } \mathcal{S}\text{-covering of } A\right\}\end{aligned}$$

when at least one $\mathcal{S}$-covering of A exists; let $\nu^*(A) = +\infty$ otherwise. We have seen that the restriction of ν^* from 2^Ω to $\sigma(\mathcal{S})$ is a measure, and furthermore it is the unique measure on $\sigma(\mathcal{S})$ that agrees with $\nu_{\mathcal{S}}$ on $\mathcal{S}$. Since ν is a measure on $\sigma(\mathcal{S})$ with $\nu = \nu_{\mathcal{S}}$ on $\mathcal{S}$, it follows that the restriction of ν^* from 2^Ω to $\sigma(\mathcal{S})$ coincides with ν. In other words, $\nu^* = \nu$ on $\sigma(\mathcal{S})$.

Using the above framework and the fact just derived, the idea of this section is to show that the sets in $\sigma(\mathcal{S})$ can be approximated in a certain sense by sets in the smaller collection $\mathcal{S}$, which collection may be easily describable in comparison to $\sigma(\mathcal{S})$. The precise statements are given by the following claims, and they will then be applied to measures on $\mathcal{B}^k$.

Claim 1. *Let $n \in \mathbb{N}$, and let $A, A_1, \cdots, A_n \in \mathcal{S}$. Then there exist disjoint $\mathcal{S}$-sets $C_1, \cdots, C_m$ such that $A \cap \bigcap_{k=1}^n A_k^c = \bigcup_{j=1}^m C_j$.*

Proof: This is a restatement of Claim 1 in Section 2.4. ∎

Claim 2. *Suppose that $B \in \sigma(\mathcal{S})$ with $\epsilon > 0$ given and $\nu(B) < +\infty$. Then there exists a disjoint amc collection $\{A_1, A_2, \cdots\}$ of $\mathcal{S}$-sets with $B \subseteq \bigcup_k A_k$ and $\nu(\bigcup_k A_k - B) < \epsilon$.*

Proof: Let $B \in \sigma(\mathcal{S})$ be such that $\nu(B) = \nu^*(B) < +\infty$, and let $\epsilon > 0$. Since $\nu^*(B) < +\infty$, there are $D_1, D_2, \cdots \in \mathcal{S}$ such that $B \subseteq \bigcup_k D_k$ and $\nu(\bigcup_k D_k) \le \sum_k \nu(D_k) < \nu(B) + \epsilon$. By (M5), this forces $\nu(\bigcup_k D_k - B) < \epsilon$. Let $C_1 = D_1$ and for $k \ge 2$ let $C_k = D_k - (D_1 \cup \cdots \cup D_{k-1})$, so that the C_k's are disjoint with $\bigcup_k C_k = \bigcup_k D_k$. From Claim 1, we may write each C_k as a finite disjoint union $\bigcup_{j=1}^{m_k} E_{kj}$ where $E_{k1}, \cdots, E_{km_k} \in \mathcal{S}$. It follows that $\nu(\bigcup_k \bigcup_{j=1}^{m_k} E_{kj} - B) < \epsilon$. Define the collection $\{A_1, A_2, \cdots\}$ to consist of the sets $E_{11}, \cdots, E_{1m_1}, E_{21}, \cdots, E_{2m_2}, \cdots$. We therefore have an amc disjoint collection $\{A_1, A_2, \cdots\}$ of $\mathcal{S}$-sets with $\nu(\bigcup_k A_k - B) < \epsilon$. ∎

Claim 3. *Suppose that $B \in \sigma(\mathcal{S})$ with $\epsilon > 0$ given and $\nu(B) = +\infty$. Then there exists a disjoint amc collection $\{A_1, A_2, \cdots\}$ of $\mathcal{S}$-sets with $B \subseteq \bigcup_k A_k$ and $\nu(\bigcup_k A_k - B) < \epsilon$.*

Proof: Let $B \in \sigma(\mathcal{S})$ be such that $\nu(B) = +\infty$, and let $\epsilon > 0$. Invoke the σ-finiteness of ν on $\mathcal{S}$ to write Ω as $\bigcup_{m=1}^{\infty} E_m$, where $E_m \in \mathcal{S}$ with $\nu(E_m) < +\infty$ for all $m \in \mathbb{N}$. For each $m \in \mathbb{N}$, we have $\nu(B \cap E_m) < +\infty$, and Claim 2 gives a disjoint amc collection $\{C_{m1}, C_{m2}, \cdots\}$ of $\mathcal{S}$-sets with $B \cap E_m \subseteq \bigcup_k C_{mk}$ and $\nu((\bigcup_k C_{mk}) - (B \cap E_m)) < \epsilon/2^m$. Using the relationship

$\bigcup_m R_m - \bigcup_m S_m \subseteq \bigcup_m (R_m - S_m)$ and countable subadditivity, we have

$$\begin{aligned}
\nu\left(\bigcup_{m=1}^{\infty}\bigcup_k C_{mk} - B\right) &= \nu\left(\bigcup_{m=1}^{\infty}\bigcup_k C_{mk} - \bigcup_{m=1}^{\infty}(B\cap E_m)\right)\\
&\leq \nu\left(\bigcup_{m=1}^{\infty}\left[\bigcup_k C_{mk} - (B\cap E_m)\right]\right)\\
&\leq \sum_{m=1}^{\infty}\nu\left(\bigcup_k C_{mk} - (B\cap E_m)\right)\\
&\leq \sum_{m=1}^{\infty}\frac{\epsilon}{2^m} = \epsilon.
\end{aligned}$$

Let us simplify notation by writing $C_{11}, C_{12}, \cdots, C_{21}, \cdots$ as $D_1, D_2, \cdots$, so that $B \subseteq \bigcup_k D_k$ and $\nu(\bigcup_k D_k - B) < \epsilon$. Let $F_1 = D_1$ and for $k \geq 2$ let $F_k = D_k \cap \bigcap_{j=1}^{k=1} D_j^c$. Then $F_1, F_2, \cdots$ are disjoint with $\bigcup_k F_k = \bigcup_k D_k$, hence $B \subseteq \bigcup_k F_k$ and $\nu(\bigcup_k F_k - B) < \epsilon$. By Claim 1, we may write each F_k as a finite disjoint union $\bigcup_{i=1}^{n_k} G_{ik}$ of $\mathcal{S}$-sets, so that we have $B \subseteq \bigcup_k \bigcup_{i=1}^{n_k} G_{ik}$ and $\nu(\bigcup_k \bigcup_{i=1}^{n_k} G_{ik} - B) < \epsilon$. The collection $\{G_{11}, \cdots, G_{1n_1}, G_{21}, \cdots\}$ is an amc disjoint collection of $\mathcal{S}$-sets. Rewriting this collection as $\{A_1, A_2, \cdots\}$, we have that $\{A_1, A_2, \cdots\}$ is an amc disjoint collection of $\mathcal{S}$-sets such that $B \subseteq \bigcup_k A_k$ and $\nu(\bigcup_k A_k - B) < \epsilon$. ∎

Claim 4. *Let $B \in \sigma(\mathcal{S})$ with $\nu(B) < +\infty$. Then, given $\epsilon > 0$, one can find a finite disjoint collection $A_1, \cdots, A_n \in \mathcal{S}$ such that $\nu(B \triangle \bigcup_{k=1}^n A_k) < \epsilon$.*

Proof: Assume that $B \in \sigma(\mathcal{S})$, $\nu(B) < +\infty$, and $\epsilon > 0$. From Claim 2, there exists a disjoint amc collection $\{A_1, A_2, \cdots\}$ of $\mathcal{S}$-sets with $B \subseteq \bigcup_k A_k$ and $\nu(\bigcup_k A_k - B) < \frac{\epsilon}{2}$. Since $\nu(B) < +\infty$, we have $\nu(\bigcup_k A_k) < +\infty$. Observe that $\{\bigcup_{k>n} A_k\}_{n=1}^{\infty}$ is a decreasing sequence of $\mathcal{S}$-sets with $\lim_n \bigcup_{k>n} A_k = \varnothing$. Since $\nu(\bigcup_k A_k) < +\infty$, (M9) gives $\lim_n \nu(\bigcup_{k>n} A_k) = 0$, so that there is $N \in \mathbb{N}$ such that $n > N$ implies $\nu(\bigcup_{k>n} A_k) < \frac{\epsilon}{2}$. Therefore, for all $n > N$, we have

$$\begin{aligned}
\nu\left(B \triangle \bigcup_{k\leq n} A_k\right) &= \nu\left(\left[B \triangle \bigcup_k A_k\right] \triangle \left[\bigcup_{k\leq n} A_k \triangle \bigcup_k A_k\right]\right)\\
&\leq \nu\left(B \triangle \bigcup_k A_k\right) + \nu\left(\bigcup_{k\leq n} A_k \triangle \bigcup_k A_k\right)\\
&= \nu\left(\bigcup_k A_k - B\right) + \nu\left(\bigcup_{k>n} A_k\right)\\
&< \frac{\epsilon}{2} + \frac{\epsilon}{2} = \epsilon,
\end{aligned}$$

where the first equality follows from the identity $P \triangle Q = (P \triangle R) \triangle (Q \triangle R)$, and the first inequality follows from the inclusion $P \triangle Q \subseteq P \cup Q$ and the monotonicity of ν. This completes the proof. ■

Example. Let $\Omega = \mathbb{R}^k$, and let $\mathcal{A}_k$ denote the usual semiring consisting of $\varnothing$ and the [bounded] k-dimensional rsc rectangles $(\boldsymbol{a}, \boldsymbol{b}]$, and let $\lambda_k : \mathcal{B}^k \to \bar{\mathbb{R}}$ denote the unique σ-finite measure on $\mathcal{B}^k$ such that $\lambda_k((\boldsymbol{a}, \boldsymbol{b}]) = \prod_{i=1}^k (b_i - a_i)$ for each $(\boldsymbol{a}, \boldsymbol{b}] \in \mathcal{A}_k$. Claims 2 and 3 state that no matter how esoteric a $B \in \mathcal{B}^k$ may be, we may find an amc disjoint collection $(\boldsymbol{a}_1, \boldsymbol{b}_1], (\boldsymbol{a}_2, \boldsymbol{b}_2], \cdots \in \mathcal{A}_k$ such that $B \subseteq \bigcup_j (\boldsymbol{a}_j, \boldsymbol{b}_j]$ and $\lambda_k(\bigcup_j (\boldsymbol{a}_j, \boldsymbol{b}_j] - B)$ is as small as desired. If $\lambda_k(B) < +\infty$, this states that, given any $\epsilon > 0$, we may cover $B \in \mathcal{B}^k$ by disjoint $(\boldsymbol{a}_1, \boldsymbol{b}_1], (\boldsymbol{a}_2, \boldsymbol{b}_2], \cdots \in \mathcal{A}_k$ so efficiently that $\lambda_k(B) + \epsilon > \sum_j \prod_{i=1}^k (b_{ji} - a_{ji})$, where for each j we write $\boldsymbol{a}_j = (a_{j1}, \cdots, a_{jk})$ and $\boldsymbol{b}_j = (b_{j1}, \cdots, b_{jk})$. If $\lambda_k(B) < +\infty$, Claim 4 states that we may find a finite disjoint collection $(\boldsymbol{c}_1, \boldsymbol{d}_1], \cdots, (\boldsymbol{c}_n, \boldsymbol{d}_n] \in \mathcal{A}_k$ such that $\lambda_k(\bigcup_{j=1}^n (\boldsymbol{c}_j, \boldsymbol{d}_j] \triangle B)$ is as small as desired. In the sense of this example and section we may accordingly state that any $\mathcal{B}^k$-set can be approximated as closely as desired by an appropriate amc disjoint union of easily describable sets, namely, by [bounded] k-dimensional rsc rectangles. Given the fact that most Borel sets cannot be described in any tangible fashion [as shown in Sections 1.6 and 1.7], this is an interesting development.

Any measure $\nu : \mathcal{B}^k \to \bar{\mathbb{R}}$ that assigns finite measure to every bounded $\mathcal{B}^k$-set is σ-finite on the semiring $\mathcal{A}_k$ consisting of $\varnothing$ and the bounded k-dimensional rsc rectangles $(\boldsymbol{a}, \boldsymbol{b}]$, and therefore the approximation results of the previous three claims apply to such a ν. In particular, the measures with domain $\mathcal{B}^k$ that arise from extending the measure Δ_F $[F \in \mathfrak{S}_k]$ from $\mathcal{A}_k$ to $\mathcal{B}^k$ via the Extension Theorem assign finite values to bounded $\mathcal{B}^k$-sets [via the Stieltjes Theorem], and hence the approximation results may be used for such measures as well.

The following results use the specifics of Euclidean space to discuss approximation in the topological context of open, closed, and compact subsets of $\mathbb{R}^k$.

Claim 5. *Let $\nu : \mathcal{B}^k \to \bar{\mathbb{R}}$ denote a measure that assigns finite measure to bounded $\mathcal{B}^k$-sets.*

(i) For any $A \in \mathcal{B}^k$ and $\epsilon > 0$ we can find a closed set C and an open set G with $C \subseteq A \subseteq G$ and $\nu(G - C) < \epsilon$.

(ii) The equation

$$\nu(A) = \sup\{\nu(C) : C \subseteq A,\ C \text{ closed}\} = \inf\{\nu(G) : A \subseteq G,\ G \text{ open}\}.$$

obtains.

Proof: Let $(\boldsymbol{a}, \boldsymbol{b}] \in \mathcal{A}_k$, and let $G_n = (\boldsymbol{a}, \boldsymbol{b} + n^{-1}\mathbf{1})$ for each $n \in \mathbb{N}$. Therefore, for each $n \in \mathbb{N}$ we have that G_n is an open set, $(\boldsymbol{a}, \boldsymbol{b}] \subseteq G_n$, and $\nu(G_n) < +\infty$ [since ν assigns finite measure to each bounded $\mathcal{B}^k$-set]. Also, observe that $\{G_n\}_{n=1}^\infty$ is a decreasing sequence with $\lim G_n = (\boldsymbol{a}, \boldsymbol{b}]$. By (M9), $\{\nu(G_n)\}_{n=1}^\infty$ is a nonincreasing sequence of finite numbers with $\lim \nu(G_n) = \nu((\boldsymbol{a}, \boldsymbol{b}])$. Therefore, given $\epsilon > 0$, we may find $N \in \mathbb{N}$ such that $n \geq N$ implies that

$\nu(G_n) - \nu((\boldsymbol{a},\boldsymbol{b}]) < \epsilon$, that is, $\nu(G_n - (\boldsymbol{a},\boldsymbol{b}]) < \epsilon$ [by (M5)]. It follows that *every $\mathcal{A}_k$-set can be approximated from above by bounded open k-dimensional rectangles to an arbitrary degree of accuracy.*

We now reset everything and let $A \in \mathcal{B}^k$ and $\epsilon > 0$ be given. By Claim 3 or Claim 4 [as $\nu(A) < +\infty$ or $\nu(A) = +\infty$] there exists an amc disjoint collection $A_1, A_2, \cdots \in \mathcal{A}_k$ with $A \subseteq \bigcup_n A_n$ and $\nu(\bigcup_n A_n - A) < \frac{\epsilon}{4}$. For each n, use the first paragraph to find an open bounded k-dimensional rectangle G_n with $A_n \subseteq G_n$ and $\nu(G_n - A_n) < \frac{\epsilon}{2^{n+2}}$. Observe that $\bigcup_n G_n$ is open and $A \subseteq \bigcup_n A_n \subseteq \bigcup_n G_n$. It follows that

$$\begin{aligned}
\nu\left(\bigcup_n G_n - A\right) &= \nu\left(\bigcup_n G_n - \bigcup_n A_n\right) + \nu\left(\bigcup_n A_n - A\right) \\
&< \nu\left(\bigcup_n (G_n - A_n)\right) + \frac{\epsilon}{4} \\
&\leq \sum_n \nu(G_n - A_n) + \frac{\epsilon}{4} \\
&< \sum_n \frac{\epsilon}{2^{n+2}} + \frac{\epsilon}{4} \leq \frac{\epsilon}{2},
\end{aligned}$$

where the equality follows because (i) $\bigcup_n G_n - A$ is the disjoint union of the sets $\bigcup_n G_n - \bigcup_n A_n$ and $\bigcup_n A_n - A$ and (ii) finite additivity, the first inequality follows since $\bigcup_n G_n - \bigcup_n A_n \subseteq \bigcup_n (G_n - A_n)$, and the second inequality follows by subadditivity. Therefore, we may take $G = \bigcup_n G_n$ to complete the first part of the proof, as $A \subseteq G$, G is open, and $\nu(G - A) < \frac{\epsilon}{2}$.

We now find closed $C \subseteq A$ with $\nu(A - C) < \frac{\epsilon}{2}$. Now $A \in \mathcal{B}^k$ implies that $A^c \in \mathcal{B}^k$, and by what has just been proved we may find an open H with $A^c \subseteq H$ and $\nu(H - A^c) < \frac{\epsilon}{2}$. Let $C = H^c$, so that C is closed and $C \subseteq A$. Since $A - C = H - A^c$, we have that $\nu(A - C) = \nu(H - A^c) < \frac{\epsilon}{2}$.

Finally, $G - C$ is the disjoint union of $G - A$ and $A - C$, and hence we obtain $\nu(G - C) = \nu(G - A) + \nu(A - C) < \frac{\epsilon}{2} + \frac{\epsilon}{2} = \epsilon$. This completes the proof of (i).

Now assume that $\nu(A) < +\infty$, and let $\epsilon > 0$. By what we have done, we may find open G and closed C with $C \subseteq A \subseteq G$ and $\nu(G - C) < \epsilon$. Since $A - C \subseteq G - C$, we have $\nu(A - C) < \epsilon$. Since $\nu(A) < +\infty$, (M5) gives $\nu(A) - \epsilon < \nu(C)$, and hence $\nu(A)$ is the supremum of the numbers $\nu(C)$ where $C \subseteq A$ is closed.

Reset everything, keeping $\nu(A) < +\infty$ and picking $\epsilon > 0$. As before, find open G and closed C with $C \subseteq A \subseteq G$ and $\nu(G - C) < \epsilon$. Since $G - A \subseteq G - C$, we have $\nu(G - A) < \epsilon$, and hence $\nu(G) - \epsilon < \nu(A)$ by (M5). This shows that $\nu(A)$ is the infimum of the numbers $\nu(G)$ where G is an open superset of A. This and the previous paragraph prove (ii) for when $\nu(A) < +\infty$.

We now prove (ii) for the case where $A \in \mathcal{B}^k$ is such that $\nu(A) = +\infty$. Observe that any $G \in \mathcal{B}^k$ with $A \subseteq G$ is such that $\nu(G) = +\infty$, hence in particular we have $\nu(G) = +\infty$ whenever G is an open superset of A. This gives $\nu(A) = +\infty = \inf\{\nu(G) : A \subseteq G, G \text{ open}\}$.

All that remains is to show that $\sup\{\nu(C) : C \subseteq A, C \text{ closed}\} = +\infty$. Let $A_n = [-n\mathbf{1}, n\mathbf{1}]$ for each $n \in \mathbb{N}$, so that $\{A_n\}_{n=1}^{\infty}$ is an increasing sequence of bounded $\mathcal{B}^k$-sets [and hence having finite ν-measure] with limit $\mathbb{R}^k$. Pick $0 < M < +\infty$. Now $\{A \cap A_n\}_{n=1}^{\infty}$ is an increasing sequence of bounded $\mathcal{B}^k$-sets having limit A, and hence by (M8) we have that

$$+\infty = \nu(A) = \nu(\lim(A \cap A_n)) = \lim \nu(A \cap A_n).$$

This means that there is $N \in \mathbb{N}$ such that $M < \nu(A \cap A_N) < +\infty$. From the fact that (ii) holds for sets of finite ν-measure, we may find closed $C \subseteq A \cap A_N$ [and hence $C \subseteq A$] such that $M < \nu(C) \leq \nu(A \cap A_N)$. In other words, given $0 < M < +\infty$, there is a closed subset C of A with $\nu(C) > M$. It follows that $\sup\{\nu(C) : C \subseteq A, C \text{ closed}\} = +\infty$. This and the previous paragraph prove (ii) for the case where $\nu(A) = +\infty$, and hence (ii) is complete in all cases. ■

Claim 6. *Let $\nu : \mathcal{B}^k \to \bar{\mathbb{R}}$ be a measure that assigns finite measure to bounded sets, and let $A \in \mathcal{B}^k$ be such that $\nu(A) < +\infty$. Then for every $\epsilon > 0$ there exists a bounded $B \in \mathcal{B}^k$ with $B \subseteq A$ and $\nu(A - B) < \epsilon$.*

Proof: Let $A \in \mathcal{B}^k$ be such that $\nu(A) < +\infty$, and let $\epsilon > 0$ be given. From Claim 4, there exists a finite disjoint collection $A_1, \cdots, A_n \in \mathcal{A}_k$ with $\nu(A \triangle \bigcup_{j=1}^n A_j) < \epsilon$. Therefore, $\nu(A - \bigcup_{j=1}^n A_j) < \epsilon$. Let $B = A \cap \bigcup_{j=1}^n A_j$, so that $B \subseteq A$ and $B \in \mathcal{B}^k$; we also have that B is bounded since $B \subseteq \bigcup_{j=1}^n A_j$ and each of $A_1, \cdots, A_n$ is bounded. Since $A - B = A - \bigcup_{j=1}^n A_j$, we have that $\nu(A - B) < \epsilon$, which completes the proof. ■

For the following claim, recall that the Heine-Borel Theorem asserts that the compact sets in $\mathbb{R}^k$ are precisely those sets that are closed and bounded. Also, recall that every compact subset of $\mathbb{R}^k$ is in $\mathcal{B}^k$.

Claim 7. *Let $\nu : \mathcal{B}^k \to \bar{\mathbb{R}}$ be a measure that assigns finite measure to bounded sets, and let $A \in \mathcal{B}^k$. Then*

$$\nu(A) = \sup\{\nu(K) : K \subseteq A, K \textit{ compact}\}. \tag{3.12}$$

Proof: Let $A \in \mathcal{B}^k$. Clearly $\nu(K) \leq \nu(A)$ for each compact $K \subseteq A$, hence

$$\nu(A) \geq \sup\{\nu(K) : K \subseteq A, K \text{ compact}\}. \tag{3.13}$$

This holds whether $\nu(A)$ is infinite or finite. We desire the reverse inequality

$$\nu(A) \leq \sup\{\nu(K) : K \subseteq A, K \text{ compact}\}. \tag{3.14}$$

Let $\epsilon > 0$ be given, and let $A \in \mathcal{B}^k$ be such that $\nu(A) < +\infty$. We will find a compact subset $K \subseteq A$ such that $\nu(A) - \epsilon < \nu(K)$, which will yield (3.14). In conjunction with (3.13), this will yield the desired equality (3.12). By Claim 6, find a bounded $B \in \mathcal{B}^k$ with $B \subseteq A$ and $\nu(A - B) < \frac{\epsilon}{2}$. From Claim 5, there exists a closed $K \in \mathcal{B}^k$ with $K \subseteq B$ and $\nu(B - K) < \frac{\epsilon}{2}$. This K is bounded, and hence K is compact by the Heine-Borel Theorem. Since $A - K$ is the disjoint union of $A - B$ and $B - K$, we have that

$$\nu(A - K) = \nu(A - B) + \nu(B - K) < \frac{\epsilon}{2} + \frac{\epsilon}{2} = \epsilon.$$

Since $\nu(K) < +\infty$, we have by (M5) that $\nu(A) < \nu(K) + \epsilon$ for our compact $K \subseteq A$. This completes the proof of (3.14) for when $\nu(A) < +\infty$. The case where $A \in \mathcal{B}^k$ is such that $\nu(A) = +\infty$ is left as an exercise.

For the case where $A \in \mathcal{B}^k$ is such that $\nu(A) = +\infty$, we show (3.12) directly. Let $0 < M < +\infty$; we will find compact $K \subseteq A$ with $\nu(K) > M$. Let $A_n = [-n\mathbf{1}, n\mathbf{1}]$ for each $n \in \mathbb{N}$. Observe that $\{A \cap A_n\}_{n=1}^{\infty}$ is a sequence of bounded $\mathcal{B}^k$-sets [thus having finite ν-measure] that increase to A, hence $\lim \nu(A \cap A_n)$ exists and is $\nu(\lim(A \cap A_n)) = \nu(A) = +\infty$. Therefore, there exists $N \in \mathbb{N}$ with $M < \nu(A \cap A_N) < +\infty$. By what we have proved above for the case of finite measure, $\nu(A \cap A_N)$ is the supremum of the numbers $\nu(K)$, where K ranges over the compact subsets of $A \cap A_N$, hence there is compact $K \subseteq A \cap A_N$ with $M < \nu(K) \leq \nu(A \cap A_n)$. Thus, given the generic $M > 0$, we have found a compact subset of $A \cap A_N$ and hence of A such that $\nu(K) > M$. It follows that

$$\sup\{\nu(K) : \ K \subseteq A, \ K \text{ compact}\} = +\infty = \nu(A),$$

completing the proof of (3.12) in the case where $\nu(A) + \infty$. This establishes (3.12) for every $A \in \mathcal{B}^k$, completing the proof. ∎

Exercises.

1. The assumption $\nu(B) = +\infty$ is not superfluous in Claim 4, and the assumption $\nu(A) < +\infty$ is not superfluous in Claim 6. [Consider $(\mathbb{R}, \mathcal{B}, \lambda)$. For Claim 4, take $B = \mathbb{R}$; for Claim 6, take $A = \mathbb{R}$. Recall that λ is the unique measure on $\mathcal{B}$ such that $\lambda((a, b]) = b - a$ whenever $-\infty < a < b < +\infty$. The semiring of interest here consists of $\varnothing$ and the bounded rsc intervals.]

2. Let $\nu : \mathcal{B}^k \to \bar{\mathbb{R}}$ be nonnegative and finitely additive with $\nu(\mathbb{R}^k) < +\infty$. Suppose that $\nu(A) = \sup\{\nu(K) : \ K \subseteq A, \ K \text{ compact}\}$ for each $A \in \mathcal{B}^k$. Then ν is a finite measure.

3. Let $\nu : \mathcal{B}^k \to \bar{\mathbb{R}}$ be a measure such that $\nu(B) < +\infty$ for every bounded $B \in \mathcal{B}^k$. Let $A \subseteq \mathbb{R}^k$, where A is not necessarily in $\mathcal{B}^k$. There exist sets F and G such that $F \subseteq A \subseteq G$, F is a countable union of closed sets, and G is a countable intersection of open sets, and $\nu(G - F) = 0$.

4. Suppose that $\nu : \mathcal{B}^k \to \mathbb{R}^k$ is a finite measure such that if $A \in \mathcal{B}^k$ and $\nu(A) > 0$, then there exists $B \in \mathcal{B}^k$ with $B \subseteq A$ and $0 < \nu(B) < \nu(A)$. Then there exists an uncountable compact set K such that $K \subseteq A$ and $\nu(K) = 0$.

5. Let $\mathcal{A}$ denote a field on Ω, and let μ and ν denote measures with domain $\sigma(\mathcal{A})$ that are σ-finite on $\mathcal{A}$. Let $E \in \sigma(\mathcal{A})$ be such that $\mu(E), \nu(E) < +\infty$, and let $\epsilon > 0$ be given. Then there exists $E_0 \in \mathcal{A}$ such that $\mu(E \triangle E_0)$, $\nu(E \triangle E_0) < \epsilon$.

3.9 A FURTHER DESCRIPTION OF $\mathcal{M}(\mu^*)$

Assume the usual framework and notation of the Extension and Uniqueness Theorems. That is, $\mathcal{A}$ denotes a semiring on Ω, and $\mu : \mathcal{A} \to \bar{\mathbb{R}}$ is nonnegative, finitely additive, countably subadditive, σ-finite on $\mathcal{A}$, and such that $\mu(\varnothing) = 0$. Where $\mu^* : 2^\Omega \to \bar{\mathbb{R}}$ is the outer measure induced by μ, the measure restrictions $\mu^*_{\mathcal{M}(\mu^*)}$ and $\mu^*_{\sigma(\mathcal{A})}$ of μ^* to $\mathcal{M}(\mu^*)$ and $\sigma(\mathcal{A})$ are the unique measures with the aforementioned domains that agree with μ on $\mathcal{A}$. From our discussion on completeness, we have that $\mathcal{M}(\mu^*)$ consists of the sets $A \cup B$, $A - B$, and $A \triangle B$, where $A \in \sigma(\mathcal{A})$ and B is contained in some $N \in \sigma(\mathcal{A})$ with $\mu^*_{\sigma(\mathcal{A})}(N) = 0$. This section will give a further description of $\mathcal{M}(\mu^*)$, providing a sufficient condition for a set to be in $\mathcal{M}(\mu^*)$ and a necessary condition for a set to be in $\mathcal{M}(\mu^*)$.

Definition. We define $\mu^{**} : 2^\Omega \to \bar{\mathbb{R}}$ by writing

$$\mu^{**}(A) = \inf\{\mu^*_{\sigma(\mathcal{A})}(B) : A \subseteq B, B \in \sigma(\mathcal{A})\}.$$

We are never in the position of taking the infimum of $\varnothing$ since $\mu^*_{\sigma(\mathcal{A})}(\Omega)$ is an element in the above set.

Claim 1. $\mu^* = \mu^{**}$ *[on all of 2^Ω, of course].*

Proof: Pick $A \subseteq \Omega$, and assume that $\mu^*(A) = +\infty$. Let

$$\mathcal{C} = \{\mu^*_{\sigma(\mathcal{A})}(B) : A \subseteq B, B \in \sigma(\mathcal{A})\}.$$

Pick $c \in \mathcal{C}$; c must be of the form $\mu^*_{\sigma(\mathcal{A})}(B)$, where $A \subseteq B \in \sigma(\mathcal{A})$. Since $\mu^*_{\sigma(\mathcal{A})}(B) = \mu^*(B) \geq \mu^*(A) = +\infty$, we have that $c = +\infty$. This forces $\mathcal{C} = \{+\infty\}$, hence $\inf \mathcal{C} = +\infty$, whence $\mu^{**}(A) = +\infty = \mu^*(A)$. This proves the case when $A \subseteq \Omega$ is such that $\mu^*(A) = +\infty$.

We now assume that $A \subseteq \Omega$ is such that $\mu^*(A) < +\infty$. Let

$$\mathcal{D} = \left\{\sum_{n=1}^{\infty} \mu(A_n) : \{A_n\}_{n=1}^{\infty} \text{ is an } \mathcal{A}\text{-covering of } A\right\},$$

so that $\inf \mathcal{D} = \mu^*(A)$. Pick $d \in \mathcal{D}$ [$\mathcal{D}$ is nonempty since $\mu^*(A) < +\infty$]. Then $d = \sum_{n=1}^{\infty} \mu(A_n)$, where we have $A_1, A_2, \cdots \in \mathcal{A}$ and $A \subseteq \bigcup_{n=1}^{\infty} A_n$. If $B = \bigcup_{n=1}^{\infty} A_n$, then $A \subseteq B \in \sigma(\mathcal{A})$, so that $\mu^*_{\sigma(\mathcal{A})}(B) \in \mathcal{C}$. Next, observe that $\mu^*_{\sigma(\mathcal{A})}(B) \leq \sum_{n=1}^{\infty} \mu^*_{\sigma(\mathcal{A})}(A_n) = \sum_{n=1}^{\infty} \mu(A_n)$ by the countable subadditivity of μ^* and the fact that $\mu^*_{\sigma(\mathcal{A})} = \mu$ on $\mathcal{A}$. Thus, for our generic $d \in \mathcal{D}$ we have found a $c \in \mathcal{C}$ with $c \leq d$. It follows that $\inf \mathcal{C} \leq \inf \mathcal{D}$, so that $\mu^{**}(A) \leq \mu^*(A)$.

To show that $\mu^*(A) \leq \mu^{**}(A)$ in the case where $\mu^*(A) < +\infty$, it is sufficient to show that given any $c \in \mathcal{C}$ and $\epsilon > 0$, there is $d \in \mathcal{D}$ with $d < c + \epsilon$, for then $\inf \mathcal{D} \leq \inf \mathcal{C}$. Pick $c \in \mathcal{C}$ and $\epsilon > 0$. Then $c = \mu^*_{\sigma(\mathcal{A})}(B)$ for some $B \in \sigma(\mathcal{A})$ with $A \subseteq B$. By the definition of μ^*, there is an $\mathcal{A}$-covering $\{A_n\}_{n=1}^{\infty}$ of A with $\sum_{n=1}^{\infty} \mu(A_n) < \mu^*(A) + \epsilon$. Since $\mu^*(A) \leq \mu^*(B) = \mu^*_{\sigma(\mathcal{A})}(B)$, we have

$\sum_{n=1}^{\infty} \mu(A_n) < \mu^*_{\sigma(\mathcal{A})}(B) + \epsilon$. If $d = \sum_{n=1}^{\infty} \mu(A_n)$, then $d \in \mathcal{D}$ and $d < c + \epsilon$, and we have successfully shown that $\mu^*(A) \le \mu^{**}(A)$ when $\mu^*(A) < +\infty$. This inequality and the reverse inequality demonstrated in the previous paragraph give $\mu^*(A) = \mu^{**}(A)$ in the case where $\mu^*(A) < +\infty$. Combining this fact with the $\mu^*(A) = +\infty$ case yields the result. ■

We have used the term *outer measure* in our work, and the usage of such a term makes it natural to believe that such a thing as an *inner measure* is used in various capacities. Historically, inner measures were used in the construction and discussion of extensions, but it turned out that outer measures alone were sufficient to obtain everything that was desired. [This is another justification for Carathéodory's approach.] However, the concept of an inner measure will allow us to further describe $\mathcal{M}(\mu^*)$, and this further description will aid us when we desire to show that certain sets are not Borel sets. The function in the following definition is one form of an inner measure.

Definition. Define the function $\mu_* : 2^{\Omega} \to \bar{\mathbb{R}}$ by writing

$$\mu_*(A) = \sup\{\mu^*_{\sigma(\mathcal{A})}(C) : C \subseteq A,\ C \in \sigma(\mathcal{A})\}.$$

The following facts are immediate or nearly so: (i) μ_* is nonnegative, (ii) μ_* is monotone [$A \subseteq B$ implies $\mu_*(A) \le \mu_*(B)$], (iii) $\mu_*(\varnothing) = 0$, and (iv) $\mu_*(A) \le \mu^*(A)$ for every $A \subseteq \Omega$. Using Claim 1, we immediately have the following claim.

Claim 2. *$\mu_*(A) \le \mu^*(A) = \mu^{**}(A)$ for every $A \subseteq \Omega$.*

Claim 3. *We have the following:*
(a) If $A \in \mathcal{M}(\mu^*)$, then $\mu_*(A) = \mu^*(A) = \mu^{**}(A)$, and
(b) If $A \subseteq \Omega$ is such that $\mu_*(A) = \mu^*(A) = \mu^{**}(A) < +\infty$, then $A \in \mathcal{M}(\mu^*)$.

Proof: We first do (a). Let $A \in \mathcal{M}(\mu^*)$. By Claim 2, we have $\mu_*(A) \le \mu^*(A)$. If $\mu_*(A) = +\infty$ then $\mu^*(A) = +\infty = \mu^*(A)$.

Next, assume that $A \in \mathcal{M}(\mu^*)$ is such that $\mu_*(A) < +\infty$, and further assume that $\mu_*(A) < \mu^*(A) = \mu^{**}(A)$. Write $A = E \cup F$, where $E \in \sigma(\mathcal{A})$ and F is contained in some $N \in \sigma(\mathcal{A})$ with $\mu^*_{\sigma(\mathcal{A})}(N) = 0$. Since we have assumed that $\mu_*(A) < \mu^{**}(A)$, then every number of the form $\mu^*_{\sigma(\mathcal{A})}(C)$ with $C \subseteq A$ and $C \in \sigma(\mathcal{A})$ is strictly less than every number of the form $\mu^*_{\sigma(\mathcal{A})}(B)$, where $A \subseteq B \in \sigma(\mathcal{A})$. Take $C = E$ and $B = E \cup N$. Then $C \subseteq A$, $C \in \sigma(\mathcal{A})$, and $A \subseteq B \in \sigma(\mathcal{A})$, hence $\mu^*_{\sigma(\mathcal{A})}(C) < \mu^*_{\sigma(\mathcal{A})}(B)$. However, we also have $\mu^*_{\sigma(\mathcal{A})}(C) = \mu^*_{\sigma(\mathcal{A})}(B)$, for $\mu^*_{\sigma(\mathcal{A})}(N) = 0$. Therefore, the assumption that $\mu_*(A) < \mu^{**}(A)$ entails a contradiction, hence $\mu_*(A) \ge \mu^{**}(A)$. This and Claim 2 show that $\mu_*(A) = \mu^*(A)$ when $\mu_*(A) < +\infty$. Combining the $\mu_*(A) < +\infty$ and $\mu_*(A) = +\infty$ cases completely finishes (a).

We do (b). Let $A \subseteq \Omega$ be such that $\mu_*(A) = \mu^*(A) = \mu^{**}(A) < +\infty$. Then for every $n \in \mathbb{N}$ the definitions of μ_* and μ^* imply the existence of B_n, C_n in $\sigma(\mathcal{A})$ with $B_n \subseteq A \subseteq C_n$ and $\mu^*_{\sigma(\mathcal{A})}(C_n - B_n) < \frac{1}{n}$. Letting $B = \bigcup_{n=1}^{\infty} B_n$

and $C = \bigcap_{n=1}^{\infty} C_n$, we have $B, C \in \sigma(\mathcal{A})$ and $B_n \subseteq B \subseteq A \subseteq C \subseteq C_n$ for every $n \in \mathbb{N}$, hence $\mu^*_{\sigma(\mathcal{A})}(C - B) \leq \mu^*_{\sigma(\mathcal{A})}(C_n - B_n) < \frac{1}{n}$ for every $n \in \mathbb{N}$, so that $\mu^*_{\sigma(\mathcal{A})}(C - B) = 0$. Since $A = B \cup (A - B)$, with $A - B \subseteq C - B \in \sigma(\mathcal{A})$, $\mu^*_{\sigma(\mathcal{A})}(C - B) = 0$, and $B \in \sigma(\mathcal{A})$, we have that $A \in \mathcal{M}(\mu^*)$. This completes the proof of (b), and the entire proof is now complete. ∎

Exercises.

1. If $A_1, A_2, \cdots \subseteq \Omega$ are disjoint, then $\mu_*(\bigcup_{n=1}^{\infty} A_n) \geq \sum_{n=1}^{\infty} \mu_*(A_n)$. This property is called *countable superadditivity*.

2. For any $A \subseteq \Omega$, $\mu^*(A) = \inf\{\mu^*_{\mathcal{M}(\mu^*)}(B) : A \subseteq B \in \mathcal{M}(\mu^*)\} = \mu^{**}(A)$.

3. For any $A \subseteq \Omega$, $\mu_*(A) = \sup\{\mu^*_{\mathcal{M}(\mu^*)}(B) : B \subseteq A,\ B \in \mathcal{M}(\mu^*)\}$.

4. For any $A \subseteq \Omega$, there is $E \in \sigma(\mathcal{A})$ such that $E \subseteq A$ and $\mu^*_{\sigma(\mathcal{A})}(E) = \mu_*(A)$.

5. The infimum that defines $\mu^*(A)$ and the supremum that defines $\mu_*(A)$ are achieved for each $A \in \Omega$. That is, there exist $C, B \in \sigma(\mathcal{A})$ with $C \subseteq A \subseteq B$ and $\mu^*_{\sigma(\mathcal{A})}(C) = \mu_*(A)$ and $\mu^*_{\sigma(\mathcal{A})}(B) = \mu^*(A)$.

6. Let $A \subseteq \Omega$ and let $\{A_n\}$ denote a disjoint sequence of $\mathcal{M}(\mu^*)$-sets. Then we have $\mu_*(\bigcup_{n=1}^{\infty}(A \cap A_n)) = \sum_{n=1}^{\infty} \mu_*(A \cap A_n)$.

7. If $A, B \subseteq \Omega$ are disjoint, then $\mu_*(A \cup B) \leq \mu_*(A) + \mu^*(B) \leq \mu^*(A \cup B)$.

8. If $A \in \mathcal{M}(\mu^*)$ and $B \subseteq \Omega$, then $\mu^*_{\mathcal{M}(\mu^*)}(A) = \mu_*(B \cap A) + \mu^*(B^c \cap A)$.

9. Let $\Omega = (0,1] \times (0,1]$, and let $\mathcal{F} = \{B \times (0,1] : B \in \mathcal{B}, B \subseteq (0,1]\}$. Define $\mu : \mathcal{F} \to \bar{\mathbb{R}}$ by requiring that $\mu(B \times (0,1]) = \nu(B)$ whenever $B \in \mathcal{B}$ and $B \subseteq (0,1]$, where ν is the unique measure on $\{B \in \mathcal{B} : B \subseteq (0,1]\}$ that assigns measure $b - a$ to each $(a, b] \subseteq (0,1]$.
(a) $\mathcal{F}$ is a σ-field.
(b) μ is well-defined and is a measure on $\mathcal{F}$.
(c) Let $C = (0,1] \times \{\frac{1}{2}\}$. Then $C \notin \mathcal{F}$, $\nu_*(C) = 0$ and $\nu^*(C) = 1$.

10. Is the hypothesis of finiteness essential in part (b) of Claim 3?

3.10 A CORRESPONDENCE THEOREM

The goal of this section is to connect the material from Sections 2.3 and 2.4 with the material in this chapter. *We will use the notation of Sections 2.3 and 2.4.* Therefore, $\mathcal{A}_k$ denotes the semiring consisting of $\varnothing$ and the bounded k-dimensional rsc rectangles, and $\mathfrak{S}_k$ denotes the set of functions $F : \mathbb{R}^k \to \mathbb{R}$ that are continuous from above [cfa] with $\Delta_F(A) = \sum_{\boldsymbol{x} \in V(A)} s_A(\boldsymbol{x}) F(\boldsymbol{x}) \geq 0$ for every $A \in \mathcal{A}_k$. We will ultimately show two things:

(i) For every $F \in \mathfrak{S}_k$ there exists a unique measure $\mu_F : \mathcal{B}^k \to \bar{\mathbb{R}}$ that agrees with Δ_F on $\mathcal{A}_k$, and
(ii) Given a measure $\mu : \mathcal{B}^k \to \bar{\mathbb{R}}$ that assigns finite measure to each bounded $\mathcal{B}^k$-set, there exists a unique equivalence class Ξ of elements of $\mathfrak{S}_k$ with the following property: for any $F \in \Xi$, the unique measure extension of Δ_F from $\mathcal{A}_k$ to $\mathcal{B}^k$ coincides with μ.

We first start out with some claims and then proceed to a precise statement of the main result.

Definition. Let $k \in \mathbb{N}$. A function $\phi : \mathbb{R}^k \to \mathbb{R}$ is *k-degenerate* iff ϕ is a function of at most $k-1$ arguments. The collection of k-degenerate functions will be denoted by $\mathfrak{D}_k$.

Given $k \in \mathbb{N}$, any $\phi : \mathbb{R}^k \to \mathbb{R}$ that is constant will be viewed as a function of 0 arguments, and hence such a ϕ will be in $\mathfrak{D}_k$.

Claim 1. *Let $k \in \mathbb{N}$ and $\phi \in \mathfrak{D}_k$. Then $\Delta_\phi(A) = 0$ for every $A \in \mathcal{A}_k$.*

Proof. If ϕ is constant [a function of 0 arguments], say $\phi(\boldsymbol{x}) = c$ for all $\boldsymbol{x} \in \mathbb{R}^k$, the result follows from the equation

$$\Delta_\phi(A) = \sum_{\boldsymbol{x} \in V(A)} s_A(\boldsymbol{x})\phi(\boldsymbol{x}) = c \sum_{\boldsymbol{x} \in V(A)} s_A(\boldsymbol{x}) = c \times 0 = 0.$$

Next, let $1 \leq j \leq k-1$, and assume that ϕ is a function of j arguments; without loss of generality assume that $\phi(x_1, \cdots, x_j, x_{j+1}, \cdots, x_k)$ depends only on $x_1, \cdots, x_j$. Let $A = (\boldsymbol{a}, \boldsymbol{b}] \in \mathcal{A}_k$, and for any $\boldsymbol{x} \in \mathbb{R}^k$ write $\boldsymbol{x}^* = (x_1, \cdots, x_j)$, $\boldsymbol{x}^{**} = (x_{j+1}, \cdots, x_k)$, $A^* = (\boldsymbol{a}^*, \boldsymbol{b}^*]$, and $A^{**} = (\boldsymbol{a}^{**}, \boldsymbol{b}^{**}]$. Therefore we have $\boldsymbol{x} = (\boldsymbol{x}^*, \boldsymbol{x}^{**})$ and $A = (\boldsymbol{a}^*, \boldsymbol{b}^*] \times (\boldsymbol{a}^{**}, \boldsymbol{b}^{**}]$, hence

$$\begin{aligned}
\Delta_\phi(A) = \sum_{\boldsymbol{x} \in V(A)} s_A(\boldsymbol{x})\phi(\boldsymbol{x}) &= \sum_{\boldsymbol{x}^* \in V(A^*)} \sum_{\boldsymbol{x}^{**} \in V(A^{**})} s_A(\boldsymbol{x}^*, \boldsymbol{x}^{**})\phi(\boldsymbol{x}^*, \boldsymbol{x}^{**}) \\
&= \sum_{\boldsymbol{x}^* \in V(A^*)} \phi(\boldsymbol{x}^*, \boldsymbol{x}^{**}) \sum_{\boldsymbol{x}^{**} \in V(A^{**})} s_A(\boldsymbol{x}^*, \boldsymbol{x}^{**}),
\end{aligned}$$

where the last equality follows since $\phi(\boldsymbol{x}^*, \boldsymbol{x}^{**})$ depends only on $\boldsymbol{x}^*$ for all $\boldsymbol{x} \in \mathbb{R}^k$. For each $\boldsymbol{x} \in V(A)$, $\sum_{\boldsymbol{x}^{**} \in V(A^{**})} s_A(\boldsymbol{x}^*, \boldsymbol{x}^{**})$ is the sum of 2^{k-j} terms, of which 2^{k-j-1} terms are $+1$ and the other 2^{k-j-1} terms are -1. It follows that $\sum_{\boldsymbol{x}^{**} \in V(A^{**})} s_A(\boldsymbol{x}^*, \boldsymbol{x}^{**}) = 0$ for all $\boldsymbol{x} \in V(A)$, so $\Delta_\phi(A) = 0$. ∎

Claim 2. Let $H : \mathbb{R}^k \to \mathbb{R}$ be arbitrary. Then $\Delta_H(A) = 0$ for all $A \in \mathcal{A}_k$ iff there is $n \in \mathbb{N}$ such that $H = \phi_1 + \cdots + \phi_n$, for elements $\phi_1, \cdots, \phi_n \in \mathfrak{D}_k$.

Proof: Let $A \in \mathcal{A}_k$, and let $H = \phi_1 + \cdots + \phi_n$ for functions $\phi_1, \cdots, \phi_n \in \mathfrak{D}_k$. By Claim 1 we have $\Delta_{\phi_i}(A) = 0$ for $i = 1, \cdots, n$, and hence

$$\Delta_H(A) = \sum_{\boldsymbol{x} \in V(A)} s_A(\boldsymbol{x})H(\boldsymbol{x}) = \sum_{i=1}^{n} \sum_{\boldsymbol{x} \in V(A)} s_A(\boldsymbol{x})\phi_i(\boldsymbol{x}) = \sum_{i=1}^{n} \Delta_{\phi_i}(A) = 0.$$

We now handle the reverse implication via mathematical induction. Let $k = 1$, and suppose that $H : \mathbb{R} \to \mathbb{R}$ is such that $\Delta_H((a, b]) = 0$ for all rsc intervals $(a, b]$. It follows that $H(b) - H(a) = 0$ for any $a, b \in \mathbb{R}$ with $a < b$, hence H is constant, so that $H \in \mathfrak{D}_1$. Therefore H trivially satisfies the requirements of the claim, completing the $k = 1$ case. Next, assume that for any $k = 1, \cdots, m$ it is the case that $\Delta_\psi(A) = 0$ for all $A \in \mathcal{A}_k$ implies that ψ is a finite sum of functions in $\mathfrak{D}_k$. We now let $H : \mathbb{R}^{m+1} \to \mathbb{R}$ be such that $\Delta_H(A) = 0$ for all $A = (\boldsymbol{a}, \boldsymbol{b}] \in \mathcal{A}_{m+1}$. For each $\boldsymbol{x} \in \mathbb{R}^{m+1}$, let $\boldsymbol{x}^* = (x_1, \cdots, x_m)$, and write $A^* = (\boldsymbol{a}^*, \boldsymbol{b}^*]$, so that $A = A^* \times (a_{m+1}, b_{m+1}]$. With this notation, we have for all $A \in \mathcal{A}_{m+1}$ that

$$\begin{aligned}
& 0 \\
= \quad & \Delta_H(A) \\
= \quad & \sum_{\boldsymbol{x} \in V(A)} s_A(\boldsymbol{x}) H(\boldsymbol{x}) \\
= \quad & \sum_{\boldsymbol{x}^* \in V(A^*)} s_A(\boldsymbol{x}^*, b_{m+1}) H(\boldsymbol{x}^*, b_{m+1}) + \sum_{\boldsymbol{x}^* \in V(A^*)} s_A(\boldsymbol{x}^*, a_{m+1}) H(\boldsymbol{x}^*, a_{m+1}) \\
= \quad & \sum_{\boldsymbol{x}^* \in V(A^*)} s_A(\boldsymbol{x}^*, b_{m+1}) H(\boldsymbol{x}^*, b_{m+1}) - \sum_{\boldsymbol{x}^* \in V(A^*)} s_A(\boldsymbol{x}^*, b_{m+1}) H(\boldsymbol{x}^*, a_{m+1}) \\
= \quad & \sum_{\boldsymbol{x}^* \in V(A^*)} s_A(\boldsymbol{x}^*, b_{m+1}) [H(\boldsymbol{x}^*, b_{m+1}) - H(\boldsymbol{x}^*, a_{m+1})] \\
= \quad & \sum_{\boldsymbol{x}^* \in V(A^*)} s_{A^*}(\boldsymbol{x}^*) [H(\boldsymbol{x}^*, b_{m+1}) - H(\boldsymbol{x}^*, a_{m+1})].
\end{aligned}$$

Therefore, for fixed but arbitrary $a_{m+1} < b_{m+1}$ the induction hypothesis asserts that there exists a finite collection

$$\phi_{1, a_{m+1}, b_{m+1}}, \cdots, \phi_{n_{a_{m+1}, b_{m+1}}, a_{m+1}, b_{m+1}}$$

of functions $\mathbb{R}^m \to \mathbb{R}$ that each depend on at most $m-1$ arguments and such that

$$H(\boldsymbol{x}^*, b_{m+1}) - H(\boldsymbol{x}^*, a_{m+1}) = \sum_{i=1}^{n_{a_{m+1}, b_{m+1}}} \phi_{i, a_{m+1}, b_{m+1}}(\boldsymbol{x}^*), \qquad \boldsymbol{x}^* \in \mathbb{R}^m.$$

Therefore, given $\boldsymbol{y} \in \mathbb{R}^{m+1}$, we have [letting $\boldsymbol{y}^* = (y_1, \cdots, y_m)$]

$$H(\boldsymbol{y}) = \begin{cases} H(\boldsymbol{y}^*, 0) + \sum_{i=1}^{n_{0, y_{m+1}}} \phi_{i, 0, y_{m+1}}(\boldsymbol{y}^*) & \boldsymbol{y} \in \mathbb{R}^{m+1}, \ y_{m+1} > 0 \\ H(\boldsymbol{y}^*, 0) & \boldsymbol{y} \in \mathbb{R}^{m+1}, \ y_{m+1} = 0 \\ H(\boldsymbol{y}^*, 0) + \sum_{i=1}^{n_{y_{m+1}, 0}} -\phi_{i, y_{m+1}, 0}(\boldsymbol{y}^*) & \boldsymbol{y} \in \mathbb{R}^{m+1}, \ y_{m+1} < 0. \end{cases}$$

In other words, for each $\boldsymbol{y} \in \mathbb{R}^{m+1}$ we have that

$$\begin{aligned}
H(\boldsymbol{y}) \quad = \quad & H(\boldsymbol{y}^*, 0) + \sum_{i=1}^{n_{0, y_{m+1}}} \phi_{i, 0, y_{m+1}}(\boldsymbol{y}^*) I_{(0, +\infty)}(y_{m+1}) \\
& + \sum_{i=1}^{n_{y_{m+1}, 0}} -\phi_{i, y_{m+1}, 0}(\boldsymbol{y}^*) I_{(-\infty, 0)}(y_{m+1}).
\end{aligned}$$

Now the mapping $\boldsymbol{y} \mapsto H(\boldsymbol{y}^*, 0)$ clearly depends on at most m arguments. Each $\phi_{i, 0, y_{m+1}}$ depends [possibly] on y_{m+1} and at most $m-1$ of $y_1, \cdots, y_m$, and hence $\phi_{i, 0, y_{m+1}}(\boldsymbol{y}^*) I_{(0, +\infty)}(y_{m+1})$ depends on y_{m+1} and at most $m-1$ of

$y_1, \cdots, y_m$, so that the first sum above is a finite sum of functions on $\mathbb{R}^{m+1}$, each function depending on at most m arguments. The second sum is similarly seen to be a finite sum of functions on $\mathbb{R}^{m+1}$, each function depending on at most m arguments. Therefore, H is a finite sum of functions on $\mathbb{R}^{m+1}$, each function depending on at most m arguments. This finishes the entire proof. ■

Definition. For F and G in $\mathfrak{G}_k$, we will write $F \sim G$ iff $F - G$ may be represented as $\sum_{i=1}^{n} \phi_i$ for some $n \in \mathbb{N}$ and $\phi_1, \cdots, \phi_n \in \mathfrak{D}_k$.

Claim 3. The relation $\sim$ is an equivalence relation. Also, if $F, G \in \mathfrak{G}_k$, then $F \sim G$ iff $\Delta_F = \Delta_G$.

Proof: That $\sim$ is an equivalence relation is nearly automatic. For the other claim, let $A \in \mathcal{A}_k$. Observe that if $F \sim G$, then $F = G + \sum_{i=1}^{n} \phi_i$ for some $n \in \mathbb{N}$ and $\phi_1, \cdots, \phi_n \in \mathcal{D}_k$, hence

$$\begin{aligned} \Delta_F(A) = \sum_{\boldsymbol{x} \in V(A)} s_A(\boldsymbol{x}) F(\boldsymbol{x}) &= \sum_{\boldsymbol{x} \in V(A)} s_A(\boldsymbol{x}) G(\boldsymbol{x}) + \sum_{i=1}^{n} \sum_{\boldsymbol{x} \in V(A)} s_A(\boldsymbol{x}) \phi_i(\boldsymbol{x}) \\ &= \sum_{\boldsymbol{x} \in V(A)} s_A(\boldsymbol{x}) G(\boldsymbol{x}) \\ &= \Delta_G(A), \end{aligned}$$

where the third equality follows from Claim 2. Conversely, if $\Delta_F(A) = \Delta_G(A)$ for all $A \in \mathcal{A}_k$, then

$$0 = \Delta_F(A) - \Delta_G(A) = \sum_{\boldsymbol{x} \in V(A)} s_A(\boldsymbol{x})[F(\boldsymbol{x}) - G(\boldsymbol{x})] = \Delta_{F-G}(A).$$

By Claim 2, $F - G$ has the form $\sum_{i=1}^{n} \phi_i$ for some $n \in \mathbb{N}$, where $\phi_1, \cdots, \phi_n$ are in $\mathfrak{D}_k$, hence $F \sim G$. This completes the proof. ■

Given $k \in \mathbb{N}$, $t \in \mathbb{R}$ and $\boldsymbol{x} \in \mathbb{R}^k$, write

$$I_t = \begin{cases} (t, 0] & t < 0 \\ (0, t] & t > 0 \\ \varnothing & t = 0 \end{cases} \quad \text{and} \quad \phi_k(\boldsymbol{x}) = \begin{cases} +1 & \#\text{ of } x_1, \cdots, x_k < 0 \text{ even} \\ -1 & \#\text{ of } x_1, \cdots, x_k < 0 \text{ odd}, \end{cases}$$

and we also write $A_{\boldsymbol{x}} = I_{x_1} \times \cdots \times I_{x_n}$.

Claim 4. *Let $k \in \mathbb{N}$, and suppose that $\mu : \mathcal{B}^k \to \bar{\mathbb{R}}$ is a measure that assigns finite measure to bounded $\mathcal{B}^k$-sets. If $F_k : \mathbb{R}^k \to \mathbb{R}$ is such that $F_k(\boldsymbol{x}) = \phi_k(\boldsymbol{x})\mu(A_{\boldsymbol{x}})$ for all $\boldsymbol{x} \in \mathbb{R}^k$, then $\Delta_F = \mu$ on $\mathcal{A}_k$.*

Proof: Let $k = 1$. Here, $\Delta_{F_1}((a, b]) = \phi_1(b)\mu(A_b) - \phi_1(a)\mu(A_a)$ for every $(a, b] \in \mathcal{A}_1$. There are five cases to consider: (i) $0 < a < b$, (ii) $a = 0 < b$, (iii) $a < 0 < b$, (iv) $a < b = 0$, and (v) $a < b < 0$. Observing that

$$\phi_1(b) = \begin{cases} +1 & \text{(i),(ii),(iii)} \\ -1 & \text{else}, \end{cases} \qquad \phi_1(a) = \begin{cases} -1 & \text{(i),(ii)} \\ +1 & \text{else}, \end{cases}$$

$$\mu(A_b) = \begin{cases} \mu((0,b]) & \text{(i),(ii),(iii)} \\ \mu(\varnothing) = 0 & \text{(iv)} \\ \mu((b,0]) & \text{(v)}, \end{cases} \quad \text{and} \quad \mu(A_a) = \begin{cases} \mu((0,a]) & \text{(i)} \\ \mu(\emptyset) = 0 & \text{(ii)} \\ \mu((a,0]) & \text{(iii),(iv),(v)}, \end{cases}$$

we obtain that

$$\phi_1(b)\mu(A_b) - \phi_1(a)\mu(A_a) = \begin{cases} \mu((0,b]) - \mu((0,a]) & \text{(i)} \\ \mu((0,b]) - 0 & \text{(ii)} \\ \mu((0,b]) + \mu((a,0]) & \text{(iii)} \\ -0 + \mu((a,0]) & \text{(iv)} \\ -\mu((b,0]) + \mu((a,0]) & \text{(v)} \end{cases} = \mu((a,b]),$$

and thus $\Delta_F((a,b]) = \mu((a,b])$ in every case when $k = 1$.

We now assume that the claim holds for a generic fixed k; we shall show that the claim holds for the $k+1$ case. Let $\mu : \mathcal{B}^{k+1} \to \bar{\mathbb{R}}$ be such that μ assigns finite measure to bounded $\mathcal{B}^{k+1}$-sets, let $F_{k+1} : \mathbb{R}^{k+1} \to \mathbb{R}$ be such that $F_{k+1}(\boldsymbol{x}) = \phi_{k+1}(\boldsymbol{x})\mu(A_{\boldsymbol{x}})$ for each $\boldsymbol{x} \in \mathbb{R}^{k+1}$, and let $A = (\boldsymbol{a}, \boldsymbol{b}] \in \mathcal{A}_{k+1}$. We desire to show that $\Delta_{F_{k+1}}((\boldsymbol{a}, \boldsymbol{b}]) = \mu((\boldsymbol{a}, \boldsymbol{b}])$. Given $\boldsymbol{x} \in \mathbb{R}^{k+1}$, we will write $\boldsymbol{x}^* = (x_1, \cdots, x_k)$, $A^* = (\boldsymbol{a}^*, \boldsymbol{b}^*]$. Next, observe for all $\boldsymbol{x} \in V(A)$ that

$$s_A(\boldsymbol{x}) = \begin{cases} s_{A^*}(\boldsymbol{x}^*) & x_{k+1} = b_{k+1} \\ -s_{A^*}(\boldsymbol{x}^*) & x_{k+1} = a_{k+1}. \end{cases}$$

We again consider five cases: (i) $0 < a_{k+1} < b_{k+1}$, (ii) $a_{k+1} = 0 < b_{k+1}$, (iii) $a_{k+1} < 0 < b_{k+1}$, (iv) $a_{k+1} < b_{k+1} = 0$, and (v) $a_{k+1} < b_{k+1} < 0$. We have for each $\boldsymbol{x} \in V(A)$ that

$$\text{cases (i) and (ii):} \ \phi_{k+1}(\boldsymbol{x}) = \begin{cases} \phi_k(\boldsymbol{x}^*, b_{k+1}) & x_{k+1} = b_{k+1} \\ \phi_k(\boldsymbol{x}^*, a_{k+1}) & x_{k+1} = a_{k+1}, \end{cases}$$

$$\text{cases (iii) and (iv):} \ \phi_{k+1}(\boldsymbol{x}) = \begin{cases} \phi_k(\boldsymbol{x}^*) & x_{k+1} = b_{k+1} \\ -\phi_k(\boldsymbol{x}^*) & x_{k+1} = a_{k+1}, \end{cases}$$

and

$$\text{case (v):} \ \phi_{k+1}(\boldsymbol{x}) = \begin{cases} -\phi_k(\boldsymbol{x}^*) & x_{k+1} = b_{k+1} \\ -\phi_k(\boldsymbol{x}^*) & x_{k+1} = a_{k+1}. \end{cases}$$

Also, observe that for each $\boldsymbol{x} \in V(A)$, we have

$$\text{case (i):} \ \mu(A_{\boldsymbol{x}}) = \mu(I_{x_1} \times \cdots \times I_{x_k} \times (0, x_{k+1}]), \qquad x_{k+1} \in \{a_{k+1}, b_{k+1}\},$$

$$\text{case (ii):} \ \mu(A_{\boldsymbol{x}}) = \begin{cases} \mu(I_{x_1} \times \cdots \times I_{x_k} \times (0, b_{k+1}]) & x_{k+1} = b_{k+1} \\ \mu(I_{x_1} \times \cdots \times I_{x_k} \times \varnothing) = \mu(\varnothing) = 0 & x_{k+1} = a_{k+1}, \end{cases}$$

$$\text{case (iii):} \ \mu(A_{\boldsymbol{x}}) = \begin{cases} \mu(I_{x_1} \times \cdots \times I_{x_k} \times (0, b_{k+1}]) & x_{k+1} = b_{k+1} \\ \mu(I_{x_1} \times \cdots \times I_{x_k} \times (a_{k+1}, 0]) & x_{k+1} = a_{k+1}, \end{cases}$$

$$\text{case (iv): } \mu(A_{\boldsymbol{x}}) = \begin{cases} \mu(I_{x_1} \times \cdots \times I_{x_k} \times \varnothing) = \mu(\varnothing) = 0 & x_{k+1} = b_{k+1} \\ \mu(I_{x_1} \times \cdots \times I_{x_k} \times (a_{k+1}, 0]) & x_{k+1} = a_{k+1}, \end{cases}$$

and

$$\text{case (v): } \mu(A_{\boldsymbol{x}}) = \mu(I_{x_1} \times \cdots \times I_{x_k} \times (x_{k+1}, 0]), \;\; x_{k+1} \in \{a_{k+1}, b_{k+1}\}.$$

For our generic but fixed $(\boldsymbol{a}, \boldsymbol{b}] \in \mathcal{A}_{k+1}$, it may be verified in straightforward fashion that for each of cases (i)-(v) we have

$$\begin{aligned} \Delta_{F_{k+1}}(A) &= \sum_{\boldsymbol{x} \in V(A)} s_A(\boldsymbol{x})\phi_{k+1}(\boldsymbol{x})F_{k+1}(\boldsymbol{x}) \\ &= \sum_{\boldsymbol{x}^* \in V(A^*)} s_{A^*}(\boldsymbol{x}^*)\phi_k(\boldsymbol{x}^*)\mu(I_{x_1} \times \cdots \times I_{x_k} \times (a_{k+1}, b_{k+1}]). \end{aligned}$$

Define the function $\nu : \mathcal{B}^k \to \bar{\mathbb{R}}$ by writing $\nu(B) = \mu(B \times (a_{k+1}, b_{k+1}])$; it may be easily verified that ν is a measure. With this, we obtain that

$$\Delta_{F_{k+1}}(A) = \sum_{\boldsymbol{x}^* \in V(A^*)} s_{A^*}(\boldsymbol{x}^*)\phi_k(\boldsymbol{x}^*)\nu(A_{\boldsymbol{x}^*}) = \nu(A^*) = \mu(A),$$

where the second equality follows by the induction hypothesis applied to ν. It follows that $\Delta_{F_{k+1}} = \mu$ on $\mathcal{A}_{k+1}$, and the entire proof is complete. ■

Claim 5. *Let the notation and hypotheses be as in Claim 4. For each $k \in \mathbb{N}$, we have $F_k \in \mathfrak{S}_k$.*

Proof: Let $k \in \mathbb{N}$. From Claim 4, we have that $\Delta_{F_k} = \mu$ on $\mathcal{A}_k$; since μ is nonnegative on $\mathcal{B}^k$, Δ_{F_k} will certainly be nonnegative on $\mathcal{A}_k$. To complete the proof, we need only show that F_k is continuous from above at every $\boldsymbol{x} \in \mathbb{R}^k$. It is sufficient to show that both ϕ_k and the mapping $\boldsymbol{x} \mapsto \mu(A_{\boldsymbol{x}})$ are cfa, for F_k is the product of these two mappings.

To see this, let $\boldsymbol{x} \in \mathbb{R}^k$ be given, and let $\{\boldsymbol{x}_n\}_{n=1}^\infty = \{(x_{n1}, \cdots, x_{nk})\}_{n=1}^\infty$ denote a sequence in $\mathbb{R}^k$ that decreases to $\boldsymbol{x}$. It may be verifed in a straightforward fashion that $\overline{\lim} A_{\boldsymbol{x}_n} = \underline{\lim} A_{\boldsymbol{x}_n} = A_{\boldsymbol{x}}$, hence $\lim A_{\boldsymbol{x}_n}$ exists and equals $A_{\boldsymbol{x}}$. Observe the following: (i) $\mu(A_{\boldsymbol{x}} \triangle \overline{\lim} A_{\boldsymbol{x}_n}) = \mu(\varnothing) = 0$, (ii) $\mu(A_{\boldsymbol{x}} \triangle \underline{\lim} A_{\boldsymbol{x}_n}) = \mu(\varnothing) = 0$, (iii) $\bigcup_{n=1}^\infty (A_{\boldsymbol{x}} \triangle A_{\boldsymbol{x}_n})$ is bounded and hence $\mu(\bigcup_{n=1}^\infty (A_{\boldsymbol{x}} \triangle A_{\boldsymbol{x}_n})) < +\infty$, (iv) $A_{\boldsymbol{x}}$ is bounded and hence $\mu(A_{\boldsymbol{x}}) < +\infty$, and (v) for each $n \in \mathbb{N}$ we have that $A_{\boldsymbol{x}_n}$ is bounded, hence $\mu(A_{\boldsymbol{x}_n}) < +\infty$. By property (M11) of measures, we must have $\lim \mu(A_{\boldsymbol{x}_n}) = \mu(\lim A_{\boldsymbol{x}_n}) = \mu(A_{\boldsymbol{x}})$. This shows that the mapping $\boldsymbol{x} \mapsto \mu(A_{\boldsymbol{x}})$ is cfa.

We now show that ϕ_k is cfa for each $k \in \mathbb{N}$. Clearly, ϕ_1 is cfa. Next, for $k \geq 2$ observe that for each $\boldsymbol{x} \in \mathbb{R}^k$ we have $\phi_k(\boldsymbol{x}) = \prod_{i=1}^n \phi_1(x_i)$. Since the mapping $\boldsymbol{x} \mapsto \phi_1(x_i)$ is cfa for $i = 1, \cdots, k$, and since products of cfa mappings are cfa, we have that ϕ_k. It follows that ϕ_k is cfa for all $k \in \mathbb{N}$, which completes the proof. ■

With these results now given, the main theorem of this section may now be precisely stated and proved. The point of the theorem is that it connects measures, which are functions whose domains are classes of sets, with certain types of real-valued functions.

Correspondence Theorem. *We have the following correspondences:*

(a) Given $F \in \mathfrak{S}_k$, there is a unique measure $\mu_F : \mathcal{B}^k \to \bar{\mathbb{R}}$ with $\mu_F = \Delta_F$ on $\mathcal{A}_k$. This measure μ_F assigns finite measure to bounded $\mathcal{B}^k$-sets.

(b) Given a measure $\mu : \mathcal{B}^k \to \bar{\mathbb{R}}$ that assigns finite measure to bounded $\mathcal{B}^k$-sets, there exists a unique equivalence class Ξ of functions in $\mathfrak{S}_k$ with the following property: any $F \in \Xi$ is such that μ is the unique measure extension of the measure Δ_F from $\mathcal{A}_k$ to $\mathcal{B}^k$.

Proof: We first do part (a). Let $F \in \mathfrak{S}_k$ be given. By the Stieltjes Theorem, Δ_F is a measure on $\mathcal{A}_k$ that assigns finite measure to each $\mathcal{A}_k$-set, and hence Δ_F is σ-finite. Thus, there exists a measure extension μ_F of Δ_F from $\mathcal{A}_k$ to $\sigma(\mathcal{A}_k) = \mathcal{B}^k$ by the Extension Theorem. This extension is unique by Claim 3 of Section 5. To see that μ_F assigns finite measure to bounded $\mathcal{B}^k$-sets, let $B \in \mathcal{B}^k$ be bounded, so that there is some $A \in \mathcal{A}_k$ with $B \subseteq A$. Since $\mu_F(B) \leq \mu_F(A) = \Delta_F(A) < +\infty$, part (a) is complete.

We next prove (b). If $\mu : \mathcal{B}^k \to \bar{\mathbb{R}}$ assigns finite measure to bounded $\mathcal{B}^k$-sets, and if $F_k : \mathbb{R}^k \to \mathbb{R}$ is given for all $\boldsymbol{x} \in \mathbb{R}^k$ by $F_k(\boldsymbol{x}) = \phi_k(\boldsymbol{x})\mu(A_{\boldsymbol{x}})$, then by Claim 4 we have that $\Delta_{F_k} = \mu$ on $\mathcal{A}_k$, and by Claim 5 we have $F_k \in \mathfrak{S}_k$. By part (a), μ is the unique measure extension of Δ_{F_k} from $\mathcal{A}_k$ to $\mathcal{B}^k$. This shows that at least one $F \in \mathfrak{S}_k$ is such that μ is the unique measure extension of Δ_F from $\mathcal{A}_k$ to $\mathcal{B}^k$. If $G \in \mathfrak{S}_k$ is such that $F \sim G$, then $\Delta_F = \Delta_G$ by Claim 3, and hence part (a) shows that μ is the unique measure extension of $\Delta_F = \Delta_G$ from $\mathcal{A}_k$ to $\mathcal{B}^k$. Therefore, there exists an equivalence class Ξ of functions in $\mathfrak{S}_k$ with the property asserted by (b). To show that the equivalence class Ξ is the unique equivalence class of functions in $\mathfrak{S}_k$ with the property asserted by (b), observe that if $H \in \mathfrak{S}_k$ is such that μ is the unique extension of Δ_H from $\mathcal{A}_k$ to $\mathcal{B}^k$, then $\Delta_H = \mu = \Delta_F$ on $\mathcal{A}_k$, so $F \sim H$ by Claim 3. This completes the proof of (b). ∎

Exercises.

1. Let $\mu : \mathcal{B}^k \to \bar{\mathbb{R}}$ denote a measure that assigns finite measure to all bounded $\mathcal{B}^k$-sets. There does not exist an $F : \mathbb{R}^k \to \mathbb{R}$ such that $F \notin \mathfrak{S}_k$ and $\mu = \Delta_F$ on $\mathcal{A}_k$.

2. Let $\mu : \mathcal{B}^k \to \bar{\mathbb{R}}$ denote a measure such that $\mu(\mathbb{R}^k) = 1$. There is a unique $F \in \mathfrak{S}_k$ that simultaneously has the following properties:
(i) $\lim_{\boldsymbol{x}\to-\infty} F(\boldsymbol{x}) = 0$,
(ii) $\lim_{\boldsymbol{x}\to+\infty} F(\boldsymbol{x}) = 1$, and
(iii) μ is the unique measure extension of Δ_F from $\mathcal{A}_k$ to $\mathcal{B}^k$.
[Such a result is foundational to mathematical statistics and probability.]

3. There exists a set $B \subseteq \mathbb{R}$ that fails to be in $\mathcal{M}(\Delta_F^*)$ for every $F \in \mathfrak{S}$ that is continuous. [Here Δ_F^* is the outer measure induced by Δ_F.] Follow this outline.
(a) The cardinality of the collection of uncountable closed subsets of $\mathbb{R}$ is $\mathfrak{c}$.
(b) Let ω denote the smallest uncountable ordinal number. From (a), we can form a well-ordering $\{F_\eta : \eta \prec \omega\}$ of the uncountable closed subsets of $\mathbb{R}$.

(c) Let x_0 and y_0 denote two distinct points of F_0. Observe that $A_1 = \{x_0, y_0\}$ has cardinal number less than $\mathfrak{c}$, hence $\text{card}(F_1 - A_1) = \mathfrak{c}$. Therefore, we may select distinct x_1 and y_1 from $F_1 - A_1$. Observe that $A_2 = \{x_0, y_0, x_1, y_1\}$ has cardinal number less than $\mathfrak{c}$, hence $\text{card}(F_2 - A_2) = \mathfrak{c}$. Pick two distinct points $x_2, y_2 \in F_2 - A_2$, and continue. Employ the following transfinite recursion scheme: let $\eta \prec \omega$ denote an infinite ordinal, and suppose that x_γ and y_γ have been defined for all $\gamma \prec \eta$. The set $A_\eta = \{x_\gamma : \gamma \prec \eta\} \cup \{y_\gamma : \gamma \prec \eta\}$ has cardinality less than $\mathfrak{c}$ [prove!], and thus $\text{card}(F_\eta - A_\eta) = \mathfrak{c}$, hence we may choose distinct $x_\gamma, y_\gamma \in F_\eta - A_\eta$. Let $B = \{x_\eta : \eta \prec \omega\}$. Then $B \cap F_\eta$ and $B^c \cap F_\eta$ are nonempty for each $\eta \prec \omega$.
(d) Let $F \in \mathfrak{S}$ be continuous and nonconstant. Then $\Delta_F(A) > 0$ for some $A = (a, b]$.
(e) Let $F \in \mathfrak{S}$ be continuous and nonconstant, so that Δ_F and Δ_F^* are nontrivial [that is, they are not identically zero]. If $B \in \mathcal{M}(\Delta_F^*)$, then

$$\Delta^*_{F,\mathcal{M}(\Delta_F^*)}(B) = \sup\{\Delta^*_{F,\mathcal{M}(\Delta_F^*)}(F) : F \subseteq B, F \text{ compact}\},$$

where $\Delta^*_{F,\mathcal{M}(\Delta_F^*)}$ is the restriction of the outer measure Δ_F^* to $\mathcal{M}(\Delta_F^*)$.
(f) Given how B was defined, the only compact subsets of B are amc.
(g) For each $x \in \mathbb{R}$, $\Delta^*_{F,\mathcal{M}(\Delta_F^*)}(\{x\}) = 0$. [$F$ is of course continuous in this and all subsequent parts.]
(h) By (e)-(g), $\Delta^*_{F,\mathcal{M}(\Delta_F^*)}(B) = 0$.
(i) $\Delta^*_{F,\mathcal{M}(\Delta_F^*)}(B^c) = 0$.
(j) The results of (h) and (i) force F to be constant, and hence a contradiction ensues. Conclude that $B \notin \mathcal{M}(\Delta_F^*)$ for any continuous non-constant $F \in \mathfrak{S}$.

4. In the same spirit as the previous exercise, the Continuum Hypothesis (CH) and the Axiom of Choice (AC) imply that there exists a set $S \subseteq \mathbb{R}$ with $\text{card}(S) = \mathfrak{c}$ and $\Delta^*_{F,\mathcal{M}(\Delta_F^*)}(S) = 0$ for every continuous $F \in \mathfrak{S}$. Follow this outline.
(a) There exists an indexing $\{C_\eta : \eta \prec \omega\}$ of the collection of nowhere dense closed subsets of $[0, 1]$, where ω again denotes the smallest uncountable ordinal.
(b) Employ a transfinitely recursive scheme to obtain a set $\{x_\eta : \eta \prec \omega\}$ by the following: let $x_0 \in [0, 1] - C_0$; suppose that for each $\eta \prec \omega$ the number x_γ is defined for each $\gamma \prec \eta$. Pick $x_\eta \in [0, 1] - \bigcup_{\gamma \prec \eta}(C_\gamma \cup \{x_\gamma\})$. This construction requires (AC). [Precisely where?]
(c) Let $S = \{x_\eta : \eta \prec \omega\}$. Then (CH) forces $\text{card}(S) = \mathfrak{c}$.
(d) For every $\eta \prec \omega$, we have that $S \cap C_\eta$ is amc.
(e) $\Delta^*_{F,\mathcal{M}(\Delta_F^*)}(S) = 0$ for every continuous $F \in \mathfrak{S}$.

4 Lebesgue Measure

The theory of the previous chapter may now be put to use in the important context of k-dimensional Euclidean space $\mathbb{R}^k$, where $k \in \mathbb{N}$. More specifically, we at last discuss the rigorous formulation of "k-dimensional volume." For $k = 1$, we will have a rigorous formulation of "length," for $k = 2$ we will have a rigorous formulation of "area," and so on. We will first establish the existence and uniqueness of the measure that will be taken to represent what we call "k-dimensional volume." From this, properties of those sets for which the "k-dimensional volume" exists will be examined, and it will be seen that the constructed measure does what intuition demands. After that, the relationship between "k-dimensional volume" and linear transformations will be discussed in detail. The rest of the chapter talks about some sets that are of mathematical interest with respect to the measure constructed in this chapter.

4.1 LEBESGUE MEASURE: EXISTENCE AND UNIQUENESS

Let $k \in \mathbb{N}$ be given, and consider the semiring $\mathcal{A}_k$ consisting of $\varnothing$ and all k-dimensional bounded rsc rectangles. Let $F_k : \mathbb{R}^k \to \mathbb{R}$ be given for all $\boldsymbol{x} \in \mathbb{R}^k$ by $F(\boldsymbol{x}) = \prod_{i=1}^{k} x_i$. In Example 3 of Section 2.3, it was seen that $F_k \in \mathfrak{S}_k$ and $\Delta_{F_k}((\boldsymbol{a}, \boldsymbol{b}]) = \prod_{i=1}^{k}(b_i - a_i)$ for every $(\boldsymbol{a}, \boldsymbol{b}] \in \mathcal{A}_k$. The Stieltjes Theorem asserts in turn that $\Delta_{F_k} : \mathcal{A}_k \to \mathbb{R}$ is a measure. Let $\Delta^*_{F_k} : 2^{\mathbb{R}^k} \to \bar{\mathbb{R}}$ denote the outer measure induced by Δ_{F_k} [Example 1 of Section 3.2.] By the Extension Theorem and Claim 3 of Section 3.5, the restriction $\Delta^*_{F_k, \mathcal{B}^k}$ of $\Delta^*_{F_k}$

from $2^{\mathbb{R}^k}$ to $\sigma(\mathcal{A}_k) = \mathcal{B}^k$ is the unique measure with domain $\mathcal{B}^k$ such that $\Delta^*_{F_k,\mathcal{B}^k}((\boldsymbol{a},\boldsymbol{b}]) = \prod_{i=1}^k (b_i - a_i)$ for every $(\boldsymbol{a},\boldsymbol{b}] \in \mathcal{A}_k$. Note too that $\Delta^*_{F_k,\mathcal{B}^k}$ assigns finite measure to all bounded $\mathcal{B}^k$-sets. Next, let $\Delta^*_{F_k,\mathcal{M}(\Delta^*_{F_k})}$ denote the measure restriction of $\Delta^*_{F_k}$ from $2^{\mathbb{R}^k}$ to the σ-field $\mathcal{M}(\Delta^*_{F_k})$. By Claim 4 of Section 3.7, the measure space $(\mathbb{R}^k, \mathcal{M}(\Delta^*_{F_k}), \Delta^*_{F_k,\mathcal{M}(\Delta^*_{F_k})})$ is the completion of the measure space $(\mathbb{R}^k, \sigma(\mathcal{A}_k) = \mathcal{B}^k, \Delta^*_{F_k,\mathcal{B}^k})$. This means that the σ-field $\mathcal{M}(\Delta^*_{F_k})$ consists precisely of the sets of the form $B \cup C$, $B - C$, and $B \Delta C$, where $B \in \mathcal{B}^k$ and C is contained in some $N \in \mathcal{B}^k$ with $\Delta^*_{F_k,\mathcal{B}^k}(N) = 0$. Also, Claim 4 of Section 3.7 asserts that $\Delta^*_{F_k,\mathcal{M}(\Delta^*_{F_k})}$ is the unique measure having domain $\mathcal{M}(\Delta^*_{F_k})$ that (i) agrees with Δ_{F_k} on $\mathcal{A}_k$ and (ii) agrees with $\Delta^*_{F_k,\mathcal{B}^k}$ on $\mathcal{B}^k = \sigma(\mathcal{A}_k)$.

Understanding the discussion above hinges on how well the theory of Chapter 3 is understood, for what has just been discussed is merely an instantiation of the very general framework given in the previous chapter. The above discussion forms the basis for the following definitions. These definitions represent a triumph for us, as they could not be made without the buildup of the previous three chapters.

Definition. The measure $\Delta^*_{F_k,\mathcal{B}^k}$ with domain $\mathcal{B}^k$ will be henceforth referred to as *[k-dimensional] Lebesgue measure.* We will use the symbol λ_k in place of $\Delta^*_{F_k,\mathcal{B}^k}$. If $k = 1$, we write λ for λ_1.

By the previous discussion, the measure λ_k is the unique measure with domain $\mathcal{B}^k$ such that $\lambda_k((\boldsymbol{a},\boldsymbol{b}]) = \prod_{i=1}^k (b_i - a_i)$ for every $(\boldsymbol{a},\boldsymbol{b}] \in \mathcal{A}_k$. That is, λ_k is the unique measure extension of Δ_{F_k} from the semiring $\mathcal{A}_k$ to $\mathcal{B}^k$. When we say *the k-dimensional volume of a set* $B \in \mathcal{B}^k$, we mean $\lambda_k(B)$.

From the above definition and the preceding discussion, the measure space $(\mathbb{R}^k, \mathcal{M}(\Delta^*_{F_k}), \Delta^*_{F_k,\mathcal{M}(\Delta^*_{F_k})})$ is the completion of $(\mathbb{R}^k, \mathcal{B}^k, \lambda_k)$. The next definition merely relabels certain objects for notational simplicity and tradition's sake.

Definition. The σ-field $\mathcal{M}(\Delta^*_{F_k})$ will be denoted by $\mathcal{L}^k$, and the elements of this σ-field are called *[k-dimensional] Lebesgue sets.* The measure $\Delta^*_{F_k,\mathcal{M}(\Delta^*_{F_k})}$ with domain $\mathcal{M}(\Delta^*_{F_k}) = \mathcal{L}^k$ will be denoted by the symbol $\bar{\lambda}_k$. The measure $\bar{\lambda}_k$ on $\mathcal{L}^k$ is also referred to as *[k-dimensional] Lebesgue measure.*

From the above notation, $(\mathbb{R}^k, \mathcal{L}^k, \bar{\lambda}_k)$ is the completion of $(\mathbb{R}^k, \mathcal{B}^k, \lambda_k)$: $(\mathbb{R}^k, \mathcal{L}^k, \bar{\lambda}_k) = (\mathbb{R}^k, \overline{\mathcal{B}^k}, \overline{\lambda_k})$. That is, $\mathcal{L}^k$ consists precisely of the sets of the form $B \cup C$, $B - C$, and $B \Delta C$, where $B \in \mathcal{B}^k$ and C is contained in some $N \in \mathcal{B}^k$ with $\lambda_k(N) = 0$. From the above discussion, $\bar{\lambda}_k$ is the unique measure with domain $\mathcal{L}^k$ that (i) agrees with Δ_{F_k} on $\mathcal{A}_k$ and (ii) agrees with λ_k on $\mathcal{B}^k$. We will write $\mathcal{L}$ for $\mathcal{L}^1$ and $\bar{\lambda}$ for $\bar{\lambda}_1$.

The final definition of this section is also nothing more than a renaming of certain items.

Definition. We will write λ_k^* to denote the outer measure $\Delta_{F_k}^* : 2^{\mathbb{R}^k} \to \bar{\mathbb{R}}$ induced by $\Delta_{F_k} : \mathcal{A}_k \to \mathbb{R}$.

Using Examples 1 and 2 of Section 3.2, the explicit formula for λ_k^* is

$$\begin{aligned}\lambda_k^*(A) &= \inf\left\{\sum_n \Delta_{F_k}((\boldsymbol{a}_n, \boldsymbol{b}_n]) : A \subseteq \bigcup_n (\boldsymbol{a}_n, \boldsymbol{b}_n]\right\} \\ &= \inf\left\{\sum_n \prod_{i=1}^k (b_{ni} - a_{ni}) : A \subseteq \bigcup_n (\boldsymbol{a}_n, \boldsymbol{b}_n]\right\}, \quad A \subseteq \mathbb{R}^k,\end{aligned}$$

where $\sum_n$ and $\bigcup_n$ run over $n \in \mathbb{N}$ or n in a set of the form $\{1, 2, \cdots, M\}$ with $M \in \mathbb{N}$. Since $\lambda_k^* = \lambda_k$ on $\mathcal{B}^k$ and $\lambda_k^* = \bar{\lambda}_k$ on $\mathcal{L}^k$, the infima above also give a representation for $\lambda_k(A)$ when $A \in \mathcal{B}^k$ and $\bar{\lambda}_k(A)$ when $A \in \mathcal{L}^k$. With this notation, Claim 1 of Section 3.7 becomes the following: *if $A \subseteq \mathbb{R}^k$ is such that $\lambda_k^*(A) = 0$, then $A \in \mathcal{L}^k$.*

We put all of this notation together and summarize. The inclusion chain

$$\mathcal{A} \subseteq \sigma(\mathcal{A}) \subseteq \mathcal{M}(\mu^*) \subseteq 2^{\Omega}$$

given in the section on the Extension Theorem here in our specific Euclidean setting becomes

$$\mathcal{A}_k \subseteq \mathcal{B}^k \subseteq \mathcal{L}^k \subseteq 2^{\mathbb{R}^k}.$$

We also have that

$$\mu, \mu^*_{\sigma(\mathcal{A})}, \mu^*_{\mathcal{M}(\mu^*)}, \mu^* \quad \text{here become [respectively]} \quad \Delta_{F_k}, \lambda_k, \bar{\lambda}_k, \lambda_k^*.$$

With these definitions and facts, we may now begin to explore the various properties of Lebesgue measure and Lebesgue sets. At this stage, the reader should review the example in Section 3.4, which reviews some properties of Lebesgue measure when $k = 1$.

Exercises.

1*. Let $x \in \mathbb{R}$ and $k \geq 2$. Then $\lambda(\{x\}) = \lambda_k(\{x\} \times \mathbb{R}^{k-1}) = 0$. Next, for any $j \in \{1, \cdots, k-1\}$ and $\boldsymbol{x} \in \mathbb{R}^j$ we have $\lambda_k(\{\boldsymbol{x}\} \times \mathbb{R}^{k-j}) = 0$.

2. Enumerate the rationals in $(0, 1]$ by $\{q_1, q_2, \cdots\}$. Given arbitrarily small $\epsilon > 0$, remove the interval $A_n = (q_n - \epsilon 2^{-n-1}, q_n + \epsilon 2^{-n-1}) \cap (0, 1]$. Let $A = \bigcup_{n=1}^{\infty} A_n$. Then $\bar{\lambda}(A) \leq \epsilon$, despite the fact that A is an open dense subset of $(0, 1]$. Also, we have $\bar{\lambda}((0, 1] - A) \geq 1 - \epsilon$, even though $(0, 1] - A$ is a nowhere dense subset of $(0, 1]$.

3. There cannot exist a closed subset of $(0, 1]$ whose interior is empty, yet has $\bar{\lambda}$-measure of one.

4. λ_k is nonatomic: any $A \in \mathcal{B}^k$ with $\lambda_k(A) > 0$ has a proper subset $B \in \mathcal{B}^k$ with $0 < \lambda_k(B) < \lambda_k(A)$. This forces $\bar{\lambda}_k$ to be nonatomic as well.

5. Let $k \geq 2$. There exists an uncountable set $U \in \mathcal{B}^k$ with $\lambda_k(U) = 0$. [See Exercise 1. We will discuss the $k = 1$ case later.]

6. $\text{card}(\mathcal{L}^k) = 2^{\mathfrak{c}}$ and $\text{card}(\mathcal{B}^k) = \mathfrak{c}$ for each $k \in \mathbb{N}$. [Use Exercise 5 when $k \geq 2$, and for the $k = 1$ case assume gratis that there exists an uncountable set $U \in \mathcal{B}$ with $\lambda(U) = 0$. Since $\mathcal{L}^k$ is $\bar{\lambda}_k$-complete, every subset of U is in $\mathcal{L}^k$.]

7. The following argument is taken from Galambos (1988) who in turn cites Freedman (1985). Assume that $\mathcal{L} = 2^{\mathbb{R}}$; in particular every one of the $2^{\mathfrak{c}}$ subsets of $[0,1]$ is a Lebesgue set. Let $B = \{\overline{\lambda}(A) : A \subseteq [0,1], \overline{\lambda}(A) \notin A\}$. Consideration of the set B [which is in $\mathcal{L}$ by assumption] leads to contradictions. [Is $\overline{\lambda}(B)$ in B?]

8*. Let $k \geq 2$. Every line[1] is in $\mathcal{B}^k$, has λ_k-measure zero, and hence has $\bar{\lambda}_k$-measure zero. [First show that any line *segment* is in $\mathcal{B}^k$ and has λ_k-measure zero. Write a line as an increasing limit of line segments and use the appropriate continuity result for measures. Note the similarity of this exercise to Exercise 1.]

9*. Let $\mathcal{B}_{(0,1]} = \sigma(\{(a,b] : (a,b] \subseteq (0,1]\})$.
(a) $\mathcal{B}_{(0,1]} = \{B \subseteq (0,1] : B \in \mathcal{B}\}$ and $\overline{\mathcal{B}_{(0,1]}} = \{B \subseteq (0,1] : B \in \mathcal{L}\}$.
(b) Construct Lebesgue measure on both $\mathcal{B}_{(0,1]}$ and $\overline{\mathcal{B}_{(0,1]}}$. Call these measures $\lambda_{(0,1]}$ and $\bar{\lambda}_{(0,1]}$, and denote $\overline{\mathcal{B}_{(0,1]}}$ by $\mathcal{L}_{(0,1]}$.
(c) $\lambda_{(0,1]}$ as constructed is the measure restriction of λ from $\mathcal{B}$ to $\mathcal{B}_{(0,1]}$ and $\bar{\lambda}_{(0,1]}$ as constructed is the measure restriction of $\bar{\lambda}$ from $\mathcal{L}$ to $\mathcal{L}_{(0,1]}$.
[Everything in this exercise should be straightforward with not too much difficulty.]

4.2 LEBESGUE SETS

In Sections 1.4 and 1.5 we carefully discussed various properties of the σ-field $\mathcal{B}^k$. This section discusses similar things with regards to sets in $\mathcal{L}^k$. We will make heavy use of the representation of a set $A \in \mathcal{L}^k$ as the union of some $B \in \mathcal{B}^k$ and C, where C is contained in some $N \in \mathcal{B}^k$ with $\lambda_k(N) = 0$.

Claim 1. *Let* $(\boldsymbol{a}, \boldsymbol{b}] \in \mathcal{A}_k$ *and* $(\boldsymbol{c}, \boldsymbol{d}] \in \mathcal{A}_l$. *Then* $(\boldsymbol{a}, \boldsymbol{b}] \times (\boldsymbol{c}, \boldsymbol{d}] \in \mathcal{A}_k \times \mathcal{A}_l$ *and*

$$\lambda_{k+l}\left((\boldsymbol{a}, \boldsymbol{b}] \times (\boldsymbol{c}, \boldsymbol{d}]\right) = \lambda_k\left((\boldsymbol{a}, \boldsymbol{b}]\right) \lambda_l\left((\boldsymbol{c}, \boldsymbol{d}]\right). \tag{4.1}$$

Proof: Trivial, as both sides of (4.1) equal $\prod_{i=1}^k (b_i - a_i) \prod_{j=1}^l (d_j - c_j)$. ∎

Claim 2. *Let* $N \in \mathcal{B}^k$ *be such that* $\lambda_k(N) = 0$. *Then for every* $A \in \mathcal{B}^l$ *we have* $N \times A \in \mathcal{B}^{k+l}$ *and* $\lambda_{k+l}(N \times A) = 0$.

Proof: Let $N \in \mathcal{B}^k$ be such that $\lambda_k(N) = 0$. By Claim 2 of Section 1.5, we have $N \times A \in \mathcal{B}^{k+l}$ for any $A \in \mathcal{B}^l$, giving the first assertion of the claim. For the second assertion, let $\mathcal{F} = \{A \in \mathcal{B}^l : \lambda_{k+l}(N \times A) = 0\}$, so that $\mathcal{F} \subseteq \mathcal{B}^l$. We will show that $\mathcal{F}$ is a σ-field on $\mathbb{R}^l$ that contains the semiring $\mathcal{A}_l$. From this, it will follow that $\mathcal{B}^l \subseteq \mathcal{F}$, giving $\mathcal{F} = \mathcal{B}^l$ and completing the proof.

We show that $\mathcal{A}_l \subseteq \mathcal{F}$. Let $(\boldsymbol{a}, \boldsymbol{b}] \in \mathcal{A}_l$; we want $\lambda_{k+l}(N \times (\boldsymbol{a}, \boldsymbol{b}]) = 0$. It is sufficient to show that $\lambda_{k+l}(N \times (\boldsymbol{a}, \boldsymbol{b}]) < \epsilon$ for any $\epsilon > 0$. Since $N \in \mathcal{B}^k$ and $\lambda_k(N) = \lambda_k^*(N) = 0$, there is for given $\epsilon > 0$ a sequence of $\mathcal{A}_k$-sets $\{A_n\}_{n=1}^\infty$

[1] More formally, we mean the set of points in $\mathbb{R}^k$ that comprise the line.

such that $N \subseteq \bigcup_{n=1}^{\infty} A_n$ and $\sum_{n=1}^{\infty} \lambda_k(A_n) < \epsilon/\lambda_l((\boldsymbol{a}, \boldsymbol{b}])$. We obtain

$$\begin{aligned}\lambda_{k+l}(N \times (\boldsymbol{a}, \boldsymbol{b}]) \leq \lambda_{k+l}\left(\bigcup_{n=1}^{\infty} (A_n \times (\boldsymbol{a}, \boldsymbol{b}])\right) &\leq \sum_{n=1}^{\infty} \lambda_{k+l}(A_n \times (\boldsymbol{a}, \boldsymbol{b}]) \\ &= \sum_{n=1}^{\infty} \lambda_k(A_n)\lambda_l((\boldsymbol{a}, \boldsymbol{b}]) \\ &< \frac{\epsilon}{\lambda_l((\boldsymbol{a}, \boldsymbol{b}])}\lambda_l((\boldsymbol{a}, \boldsymbol{b}]) = \epsilon,\end{aligned}$$

where the first equality follows from Claim 1. This shows that $(\boldsymbol{a}, \boldsymbol{b}] \in \mathcal{F}$. Since $\varnothing$ is obviously in $\mathcal{F}$, we have $\mathcal{A}_l \subseteq \mathcal{F}$.

We now show that $\mathbb{R}^l \in \mathcal{F}$. For each $n \in \mathbb{N}$, let $A_n = (-n\mathbf{1}, n\mathbf{1}] \in \mathcal{A}_l \subseteq \mathcal{F}$. Observe that $\{N \times A_n\}_{n=1}^{\infty}$ is an increasing sequence of $\mathcal{B}^{k+l}$-sets with limit $N \times \mathbb{R}^l$. Since each A_n is in $\mathcal{F}$ we have $\lambda_{k+l}(N \times A_n) = 0$ for all $n \in \mathbb{N}$. By (M8), we obtain

$$\lambda_{k+l}(N \times \mathbb{R}^l) = \lambda_{k+l}\left(\lim_{n\to\infty} (N \times A_n)\right) = \lim_{n\to\infty} \lambda_{k+l}(N \times A_n) = 0,$$

and thus $\mathbb{R}^l \in \mathcal{F}$.

We now show that $\mathcal{F}$ is closed under complementation. Let $A \in \mathcal{F}$. Since $N \times A^c = (N \times \mathbb{R}^l) - (N \times A)$ and $\lambda_k(N \times A) = 0$, we have

$$\lambda_{k+l}(N \times A^c) = \lambda_{k+l}(N \times \mathbb{R}^l) - \lambda_{k+l}(N \times A) = 0 - 0 = 0,$$

where the second equality follows since $\mathbb{R}^l \in \mathcal{F}$. Therefore $A^c \in \mathcal{F}$.

We complete the proof by showing that $\mathcal{F}$ is closed under countable unions. Let $\{A_n\}_{n=1}^{\infty}$ denote a sequence of $\mathcal{F}$-sets, so that $\lambda_{k+l}(N \times A_n) = 0$ for each $n \in \mathbb{N}$. Observe that

$$\lambda_{k+l}\left(N \times \bigcup_{n=1}^{\infty} A_n\right) = \lambda_{k+l}\left(\bigcup_{n=1}^{\infty} (N \times A_n)\right) \leq \sum_{n=1}^{\infty} \lambda_{k+l}(N \times A_n) = 0,$$

and hence $\bigcup_{n=1}^{\infty} A_n \in \mathcal{F}$. This shows that $\mathcal{F}$ is a σ-field, which completes the proof. ■

There is a symmetric version of Claim 2: *for any $N \in \mathcal{B}^l$ with $\lambda_l(N) = 0$, we have for all $A \in \mathcal{B}^k$ that $A \times N \in \mathcal{B}^{k+l}$ and $\lambda_{k+l}(A \times N) = 0$.* The proof of this claim is the same as that given above.

For the upcoming claim, *we assume that for each k that there is a bounded set $A \subseteq \mathbb{R}^k$ with $A \notin \mathcal{L}^k$.* That such a set exists will be shown in Section 5, and the construction which provides such a set does not depend on these immediately upcoming results.

Claim 3. *For every $k, l \in \mathbb{N}$ we have $\mathcal{L}^k \times \mathcal{L}^l \subsetneqq \mathcal{L}^{k+l}$.*

Proof: Let $E \in \mathcal{L}^k$ and $F \in \mathcal{L}^l$. Thus, $E = A_1 \cup B_1$, where $A_1 \in \mathcal{B}^k$ and $B_1 \subseteq N_1 \in \mathcal{B}^k$ with $\lambda_k(N_1) = 0$, and $F = A_2 \cup B_2$, where $A_2 \in \mathcal{B}^l$ and

$B_2 \subseteq N_2 \in \mathcal{B}^l$ with $\lambda_l(N_2) = 0$. Observe that $A_1 \times A_2 \in \mathcal{B}^{k+l}$ by Claim 2 of Section 1.5 and

$$E \times F = (A_1 \times A_2) \cup [(B_1 \times A_2) \cup (A_1 \times B_2) \cup (B_1 \times B_2)].$$

Since $B_1 \subseteq N_1$ and $B_2 \subseteq N_2$, we have

$$(B_1 \times A_2) \cup (A_1 \times B_2) \cup (B_1 \times B_2) \subseteq (N_1 \times A_2) \cup (A_1 \times N_2) \cup (N_1 \times N_2) \in \mathcal{B}^{k+l},$$

where the membership follows since $N_1 \times A_2$, $A_1 \times N_2$, and $N_1 \times N_2$ are in $\mathcal{B}^k \times \mathcal{B}^l \subseteq \mathcal{B}^{k+l}$. By finite subadditivity and Claim 2, we have

$$\begin{aligned} & \lambda_{k+l}((N_1 \times A_2) \cup (A_1 \times N_2) \cup (N_1 \times N_2)) \\ \leq\ & \lambda_{k+l}(N_1 \times A_2) + \lambda_{k+l}(A_1 \times N_2) + \lambda_{k+l}(N_1 \times N_2) \\ =\ & 0 + 0 + 0 = 0, \end{aligned}$$

hence $E \times F$ may be written as the union of a $\mathcal{B}^{k+l}$-set and a subset of a $\mathcal{B}^{k+l}$-set of λ_{k+l}-measure zero, so $E \times F \in \mathcal{L}^{k+l}$. This shows that $\mathcal{L}^k \times \mathcal{L}^l \subseteq \mathcal{L}^{k+l}$.

We now show that the containment is strict. Let $\mathbf{0}$ denote the origin of $\mathbb{R}^l$, hence $\{\mathbf{0}\} \in \mathcal{L}^l$. Pick bounded $A \subseteq \mathbb{R}^k$ such that $A \notin \mathcal{L}^k$. [Refer to the comments preceding this claim.] Then we have $A \times \{\mathbf{0}\} \notin \mathcal{L}^k \times \mathcal{L}^l$, and there exists $m \in \mathbb{N}$ such that $A \subseteq (-m\mathbf{1}, m\mathbf{1}]$, giving $A \times \{\mathbf{0}\} \subseteq (-m\mathbf{1}, m\mathbf{1}] \times \{\mathbf{0}\}$. Employing (M9), Claim 1, and Claim 2, we have

$$\begin{aligned} \lambda_{k+l}((-m\mathbf{1}, m\mathbf{1}] \times \{\mathbf{0}\}) &= \lambda_{k+l}(\lim_n\, (-m\mathbf{1}, m\mathbf{1}] \times (-n^{-1}\mathbf{1}, \mathbf{0}]) \\ &= \lim_n \lambda_{k+l}((-m\mathbf{1}, m\mathbf{1}] \times (-n^{-1}\mathbf{1}, \mathbf{0}]) \\ &= \lim_n \lambda_k((-m\mathbf{1}, m\mathbf{1}])\lambda_l((-n^{-1}\mathbf{1}, \mathbf{0}]) \\ &= \lim_n (2m)^k n^{-l} = 0. \end{aligned}$$

Thus $A \times \{\mathbf{0}\}$ is a subset of a λ_{k+l}-null set, hence $A \times \{\mathbf{0}\} \in \mathcal{L}^{k+l}$. This demonstrates that $\mathcal{L}^k \times \mathcal{L}^l \subsetneq \mathcal{L}^{k+l}$. ■

Claim 4. *Let $E \in \mathcal{L}^k$ and $F \in \mathcal{L}^l$. Then $\bar{\lambda}_{k+l}(E \times F) = \bar{\lambda}_k(E)\bar{\lambda}_l(F)$.*

Proof: We first show that $\bar{\lambda}_{k+l}(E \times F) = \bar{\lambda}_k(E)\bar{\lambda}_l(F)$ for bounded $E \in \mathcal{L}^k$ and bounded $F \in \mathcal{L}^l$, following the approach of Kingman and Taylor (1966). Let E and F denote bounded $\mathcal{L}^k$- and $\mathcal{L}^l$-sets, respectively. Find $J \in \mathcal{A}_k$ such that $E \subseteq J$, and find $K \in \mathcal{A}_l$ such that $F \subseteq K$. Let $\epsilon > 0$, and let $\epsilon^* > 0$ denote the number such that $\epsilon^*(\bar{\lambda}_k(E) + \bar{\lambda}_l(F) + \epsilon^*) = \epsilon$. Find a sequence $\{Q_i\}_{i=1}^{\infty}$ of $\mathcal{A}_k$-sets such that

$$E \subseteq \bigcup_{i=1}^{\infty} Q_i \subseteq J \quad \text{and} \quad \sum_{i=1}^{\infty} \lambda_k(Q_i) < \bar{\lambda}_k(E) + \epsilon^*,$$

and find a sequence $\{R_j\}_{j=1}^{\infty}$ of $\mathcal{A}_l$-sets such that

$$F \subseteq \bigcup_{j=1}^{\infty} R_j \subseteq K \quad \text{and} \quad \sum_{j=1}^{\infty} \lambda_l(R_j) < \bar{\lambda}_l(F) + \epsilon^*.$$

Such sequences exist by the definition of λ_k^* as an infimum[2]. It follows that $E \times F \subseteq \bigcup_{i=1}^{\infty} Q_i \times \bigcup_{j=1}^{\infty} R_j = \bigcup_{i=1}^{\infty} \bigcup_{j=1}^{\infty} (Q_i \times R_j)$, and hence

$$\begin{aligned}
\bar{\lambda}_{k+l}(E \times F) &\leq \bar{\lambda}_{k+l}\left(\bigcup_{i=1}^{\infty}\bigcup_{j=1}^{\infty}(Q_i \times R_j)\right) \\
&\leq \sum_{i=1}^{\infty}\sum_{j=1}^{\infty} \bar{\lambda}_{k+l}(Q_i \times R_j) \\
&= \sum_{i=1}^{\infty}\sum_{j=1}^{\infty} \bar{\lambda}_k(Q_i)\bar{\lambda}_l(R_j) \quad \text{[Claim 1]} \\
&= \sum_{i=1}^{\infty} \bar{\lambda}_k(Q_i) \times \sum_{j=1}^{\infty} \bar{\lambda}_l(R_j) \\
&< \left(\bar{\lambda}_k(E) + \epsilon^*\right)\left(\bar{\lambda}_l(F) + \epsilon^*\right) \\
&= \bar{\lambda}_k(E)\bar{\lambda}_l(F) + \epsilon^*\left(\bar{\lambda}_k(E) + \bar{\lambda}_l(F) + \epsilon^*\right) \\
&= \bar{\lambda}_k(E)\bar{\lambda}_l(F) + \epsilon.
\end{aligned}$$

Therefore, $\bar{\lambda}_{k+l}(E \times F) < \bar{\lambda}_k(E)\bar{\lambda}_l(F) + \epsilon$ for any $\epsilon > 0$, and hence we have $\bar{\lambda}_{k+l}(E \times F) \leq \bar{\lambda}_k(E)\bar{\lambda}_l(F)$. We have shown the following: *for any bounded sets $M \in \mathcal{L}^k$ and $N \in \mathcal{L}^l$ the inequality $\bar{\lambda}_{k+l}(M \times N) \leq \bar{\lambda}_k(M)\bar{\lambda}_l(N)$ obtains.*

We shall use the above inequality to obtain actual equality. We have

$$J \times K = (E \times F) \cup [(J - E) \times F) \cup (E \times (K - F)) \cup ((J - E) \times (K - F)],$$

and it follows by finite subadditivity that

$$\begin{aligned}
&\bar{\lambda}_{k+l}(J \times K) \\
\leq\; &\bar{\lambda}_{k+l}(E \times F) + \bar{\lambda}_{k+l}((J - E) \times F) + \bar{\lambda}_{k+l}(E \times (K - F)) \\
&+\bar{\lambda}_{k+l}((J - E) \times (K - F)) \qquad (4.2) \\
\leq\; &\bar{\lambda}_k(E)\bar{\lambda}_l(F) + \bar{\lambda}_k(J - E)\bar{\lambda}_l(F) + \bar{\lambda}_k(E)\bar{\lambda}_l(K - F) \\
&+\bar{\lambda}_k(J - E)\bar{\lambda}_l(K - F) \qquad (4.3) \\
=\; &\left(\bar{\lambda}_k(E) + \bar{\lambda}_k(J - E)\right)\left(\bar{\lambda}_l(F) + \bar{\lambda}_l(K - F)\right) \\
=\; &\bar{\lambda}_k(J)\bar{\lambda}_l(K) \\
=\; &\bar{\lambda}_{k+l}(J \times K).
\end{aligned}$$

[Inequality (4.3) follows by applying the inequality just proved for bounded Lebesgue sets to each of the four terms in (4.2). The last equality follows

[2]There exists at least one sequence $\{Q_i\}_{i=1}^{\infty}$ of $\mathcal{A}_k$-sets with $E \subseteq \bigcup_{i=1}^{\infty} Q_i \subseteq J$, namely, the sequence with $Q_1 = J$ and $Q_2 = Q_3 = \cdots = \varnothing$. If $\sum_{i=1}^{\infty} \lambda_k(Q_i) = \lambda_k(J) < \bar{\lambda}_k(E) + \epsilon^*$, we have the desired sequence. Otherwise, there is a sequence $\{S_i\}_{i=1}^{\infty}$ of $\mathcal{A}_k$-sets with $E \subseteq \bigcup_{i=1}^{\infty} S_i$ and $\sum_{i=1}^{\infty} \lambda_k(S_i) < \bar{\lambda}_k(E) + \epsilon^*$. The sequence $\{S_i \cap J\}_{i=1}^{\infty}$ may then serve as the desired sequence $\{Q_i\}_{i=1}^{\infty}$.

by Claim 1.] From the fact that each term in (4.2) is no more than the corresponding term of (4.3), it follows that each term of (4.2) equates with the corresponding term of (4.3). In particular, the first terms of (4.2) and (4.3) coincide, giving $\bar{\lambda}_{k+l}(E \times F) = \bar{\lambda}_k(E)\bar{\lambda}_l(F)$ for the bounded sets $E \in \mathcal{L}^k$ and $F \in \mathcal{L}^l$. *We have therefore proved the claim for bounded Lebesgue sets.*

We complete the proof by removing the condition of boundedness. Let $E \in \mathcal{L}^k$ and $F \in \mathcal{L}^l$. For each $n \in \mathbb{N}$ let $E_n = E \cap (-n\mathbf{1}, n\mathbf{1}] \in \mathcal{L}^k$ and let $F_n = F \cap (-n\mathbf{1}, n\mathbf{1}] \in \mathcal{L}^l$. For each $n \in \mathbb{N}$ we have that (i) E_n and F_n are bounded with $E_n \times F_n \in \mathcal{L}^{k+l}$, and (ii) $\bar{\lambda}_{k+l}(E_n \times F_n) = \bar{\lambda}_k(E_n)\bar{\lambda}_l(F_n)$. Observe that $\{E_n \times F_n\}_{n=1}^{\infty}$ is a sequence of bounded $\mathcal{L}^{k+l}$-sets that increases to $E \times F$, and thus $\lim_{n\to\infty} \bar{\lambda}_{k+l}(E_n \times F_n)$ exists and equals $\bar{\lambda}_{k+l}(E \times F)$ by (M8). Putting all of these facts together yields

$$\begin{aligned} \bar{\lambda}_{k+l}(E \times F) = \lim_n \bar{\lambda}_{k+l}(E_n \times F_n) &= \lim_n \bar{\lambda}_k(E_n)\bar{\lambda}_l(F_n) \\ &= \bar{\lambda}_k(E)\bar{\lambda}_k(F), \end{aligned}$$

where the final equality holds by noting that $\{E_n\}_{n=1}^{\infty}$ and $\{F_n\}_{n=1}^{\infty}$ are increasing to E and F, respectively, and then by twice applying (M8). This completes the entire proof. ∎

Since $\mathcal{L}^k \times \mathcal{L}^l \subsetneq \mathcal{L}^{k+l}$, we are now interested in the relationship between $\sigma(\mathcal{L}^k \times \mathcal{L}^l)$ and $\mathcal{L}^{k+l}$: are they equal or do we have $\sigma(\mathcal{L}^k \times \mathcal{L}^l) \subsetneq \mathcal{L}^{k+l}$? It turns out that $\sigma(\mathcal{L}^k \times \mathcal{L}^l) \subsetneq \mathcal{L}^{k+l}$. To prove this assertions, some further machinery is required.

Definition. Let Ω_i denote a nonempty set, and let $\mathcal{F}_i$ denote a σ-field on Ω_i, where $i = 1, 2$. We will write $\mathcal{F}_1 \otimes \mathcal{F}_2 = \sigma(\mathcal{F}_1 \times \mathcal{F}_2)$. For any $E \subseteq \Omega_1 \times \Omega_2$, $\omega_1 \in \Omega_1$, and $\omega_2 \in \Omega_2$, we define

$$E_{\omega_1} = \{\omega_2 \in \Omega_2 : (\omega_1, \omega_2) \in E\} \text{ and } E^{\omega_2} = \{\omega_1 \in \Omega_1 : (\omega_1, \omega_2) \in E\}.$$

If $\Omega_1 = \mathbb{R}^k$, $\Omega_2 = \mathbb{R}^l$, $\mathcal{F}_1 = \mathcal{B}^k$, and $\mathcal{F}_2 = \mathcal{B}^l$, then the above definition of $\mathcal{F}_1 \otimes \mathcal{F}_2$ is consistent with the definition of $\mathcal{B}^k \otimes \mathcal{B}^l$ given in Section 1.5. That is, the above definition is merely a generalization of the definition in Section 1.5.

Claim 5. *If $E \in \mathcal{F}_1 \otimes \mathcal{F}_2$, then for every $\omega_1 \in \Omega_1$ and $\omega_2 \in \Omega_2$ we have $E_{\omega_1} \in \mathcal{F}_2$ and $E^{\omega_2} \in \mathcal{F}_1$.*

Proof: Define the collection

$$\mathcal{C} = \{E \subseteq \Omega_1 \times \Omega_2 : E_{\omega_1} \in \mathcal{F}_2 \text{ and } E^{\omega_2} \in \mathcal{F}_1 \text{ for all } \omega_1 \in \Omega_1,\ \omega_2 \in \Omega_2\}.$$

We first claim that $\mathcal{F}_1 \times \mathcal{F}_2 \subseteq \mathcal{C}$. To see this, let $E_1 \in \mathcal{F}_1$ and $E_2 \in \mathcal{F}_2$. Then $(E_1 \times E_2)_{\omega_1}$ is E_2 or $\varnothing$ as $\omega_1 \in E_1$ or $\omega_1 \notin E_1$. Thus, for each $\omega_1 \in \Omega_1$, $(E_1 \times E_2)_{\omega_1} \in \mathcal{F}_2$. Symmetrically, $(E_1 \times E_2)^{\omega_2}$ is E_1 or $\varnothing$ as $\omega_2 \in E_2$ or $\omega_2 \notin E_2$. Therefore, $(E_1 \times E_2)^{\omega_2} \in \mathcal{F}_1$ for each $\omega_2 \in \Omega$, and thus $E_1 \times E_2 \in \mathcal{C}$. It follows that $\mathcal{F}_1 \times \mathcal{F}_2 \subseteq \mathcal{C}$. In particular, $\Omega_1 \times \Omega_2 \in \mathcal{C}$.

Next, it is easily verified that $(E^c)_{\omega_1} = (E_{\omega_1})^c$ and $(E^c)^{\omega_2} = (E^{\omega_2})^c$ for all $E \subseteq \Omega_1 \times \Omega_2$, $\omega_1 \in \Omega_1$, and $\omega_2 \in \Omega_2$. If $E \in \mathcal{C}$, then $(E^c)_{\omega_1} = (E_{\omega_1})^c \in \mathcal{F}_2$ and $(E^c)^{\omega_2} = (E^{\omega_2})^c \in \mathcal{F}_1$ for each $\omega_1 \in \Omega_1$ and $\omega_2 \in \Omega_2$, hence $E^c \in \mathcal{C}$. Therefore, $\mathcal{C}$ is closed under complementation.

Furthermore, it is easily verified that if $\{E_n\}_{n=1}^{\infty}$ is a sequence of subsets of $\Omega_1 \times \Omega_2$, then $(\bigcup_{n=1}^{\infty} E_n)_{\omega_1} = \bigcup_{n=1}^{\infty}(E_n)_{\omega_1}$ and $(\bigcup_{n=1}^{\infty} E_n)^{\omega_2} = \bigcup_{n=1}^{\infty}(E_n)^{\omega_2}$ for each $\omega_1 \in \Omega_1$ and $\omega_2 \in \Omega_2$. Therefore, if $\{E_n\}_{n=1}^{\infty}$ is a sequence of $\mathcal{C}$-sets, then $(\bigcup_{n=1}^{\infty} E_n)_{\omega_1} = \bigcup_{n=1}^{\infty}(E_n)_{\omega_1} \in \mathcal{F}_2$ and $(\bigcup_{n=1}^{\infty} E_n)^{\omega_2} = \bigcup_{n=1}^{\infty}(E_n)^{\omega_2} \in \mathcal{F}_1$ for each $\omega_1 \in \Omega_1$ and $\omega_2 \in \Omega_2$, hence $\mathcal{C}$ is closed under countable unions.

We have shown that $\mathcal{C}$ is a σ-field on $\Omega_1 \times \Omega_2$ such that $\mathcal{F}_1 \times \mathcal{F}_2 \subseteq \mathcal{C}$. It follows that $\sigma(\mathcal{F}_1 \times \mathcal{F}_2) \subseteq \mathcal{C}$, hence $\mathcal{F}_1 \otimes \mathcal{F}_2 \subseteq \mathcal{C}$. This proves the result. ■

Claim 6. *Under the same notation and hypotheses as in Claim 5, we have that $E_1 \times E_2 \in \mathcal{F}_1 \otimes \mathcal{F}_2$ iff $E_1 \in \mathcal{F}_1$ and $E_2 \in \mathcal{F}_2$.*

Proof: The "if" part follows since $\mathcal{F}_1 \times \mathcal{F}_2 \subseteq \sigma(\mathcal{F}_1 \times \mathcal{F}_2) = \mathcal{F}_1 \otimes \mathcal{F}_2$. We now prove the "only if" claim. From Claim 5, we know that $E_1 \times E_2 \in \mathcal{F}_1 \otimes \mathcal{F}_2$ forces $E_1 \times E_2 \in \mathcal{C}$. This means that for every $\omega_1 \in \Omega_1$ and $\omega_2 \in \Omega_2$ we have $(E_1 \times E_2)_{\omega_1} \in \mathcal{F}_2$ and $(E_1 \times E_2)^{\omega_2} \in \mathcal{F}_1$. Since $(E_1 \times E_2)_{\omega_1}$ is E_2 or $\varnothing$ as $\omega_1 \in E$ or $\omega_1 \in E^c$, we have that $E_2 \in \mathcal{F}_2$. The same reasoning gives $E_1 \in \mathcal{F}_1$. ■

Claim 7. *We have the following facts:*

(a) $\mathcal{L}^k \times \mathcal{L}^l \subsetneqq \mathcal{L}^k \otimes \mathcal{L}^l \subsetneqq \mathcal{L}^{k+l}$.

(b) Let $\bar{\lambda}_k \otimes \bar{\lambda}_l$ denote the measure restriction of $\bar{\lambda}_{k+l}$ from $\mathcal{L}^{k+l}$ to the proper sub-σ-field $\mathcal{L}^k \otimes \mathcal{L}^l$. Then $(\mathbb{R}^{k+l}, \mathcal{L}^k \otimes \mathcal{L}^l, \bar{\lambda}_k \otimes \bar{\lambda}_l)$ fails to be complete.

(c) $(\mathbb{R}^{k+l}, \mathcal{L}^{k+l}, \bar{\lambda}_{k+l})$ is the completion of $(\mathbb{R}^{k+l}, \mathcal{L}^k \otimes \mathcal{L}^l, \bar{\lambda}_k \otimes \bar{\lambda}_l)$.

Proof: We first do (a). We have $\mathcal{L}^k \times \mathcal{L}^l \subsetneqq \mathcal{L}^{k+l}$ by Claim 3. Since $\mathcal{L}^{k+l}$ is a σ-field containing $\mathcal{L}^k \times \mathcal{L}^l$, we have that $\sigma(\mathcal{L}^k \times \mathcal{L}^l) \subseteq \mathcal{L}^{k+l}$, whence $\mathcal{L}^k \otimes \mathcal{L}^l \subseteq \mathcal{L}^{k+l}$. Next, let $A \subseteq \mathbb{R}^k$ be outside $\mathcal{L}^k$ and such that A is bounded. [See the paragraph preceding Claim 3.] In the proof of Claim 3 we showed that $A \times \{\mathbf{0}\} \in \mathcal{L}^{k+l}$, where $\mathbf{0}$ is the origin of $\mathbb{R}^l$. Were $A \times \{\mathbf{0}\} \in \mathcal{L}^k \otimes \mathcal{L}^l$, the fact that $\{\mathbf{0}\} \in \mathcal{L}^l$ and Claim 6 would force $A \in \mathcal{L}_k$, a contradiction. Therefore, $A \times \{\mathbf{0}\} \notin \mathcal{L}^k \otimes \mathcal{L}^l$. This gives $\mathcal{L}^k \otimes \mathcal{L}^l \subsetneqq \mathcal{L}^{k+l}$.

We next show that $\mathcal{L}^k \times \mathcal{L}^l \subsetneqq \mathcal{L}^k \otimes \mathcal{L}^l$. Containment is trivial by the definition of $\mathcal{L}^k \otimes \mathcal{L}^l$. For strict containment, observe that the open unit ball in $\mathbb{R}^{k+l}$ is an element of $\mathcal{B}^{k+l} = \mathcal{B}^k \otimes \mathcal{B}^l \subseteq \mathcal{L}^k \otimes \mathcal{L}^l$ [the equality follows from Claim 3 in Section 1.5]. However, the unit ball cannot be written as a Cartesian product of a subset of $\mathbb{R}^k$ with a subset of $\mathbb{R}^l$, let alone as an $\mathcal{L}^k \times \mathcal{L}^l$-set. Thus, the unit ball is not an element of $\mathcal{L}^k \times \mathcal{L}^l$. This gives (a).

For (b), we showed in Claim 3 that $A \times \{\mathbf{0}\}$ was a subset of a $\mathcal{B}^{k+l}$-set of λ_{k+l}-measure zero. Since $\mathcal{B}^{k+l} = \mathcal{B}^k \otimes \mathcal{B}^l \subseteq \mathcal{L}^k \otimes \mathcal{L}^l$ and $\bar{\lambda}_k \otimes \bar{\lambda}_l$ agrees with λ_{k+l} on $\mathcal{B}_{k+l}$, we have that $A \times \{\mathbf{0}\}$ is a subset of a $\mathcal{L}^k \otimes \mathcal{L}^l$ set of $\bar{\lambda}_k \otimes \bar{\lambda}_l$-measure zero. However, $A \times \{\mathbf{0}\} \notin \mathcal{L}^k \otimes \mathcal{L}^l$, so not every subset of the $\bar{\lambda}_k \otimes \bar{\lambda}_l$-null sets is in $\mathcal{L}^k \otimes \mathcal{L}^l$. This gives (b).

For (c), let $E \in \overline{\mathcal{L}^k \otimes \mathcal{L}^l}$. Then $E = A \cup B$, where $A \in \mathcal{L}^k \otimes \mathcal{L}^l$ and B is a subset of some $\bar{\lambda}_k \otimes \bar{\lambda}_l$-null set $N \in \mathcal{L}^k \otimes \mathcal{L}^l$. Since $\mathcal{L}^k \otimes \mathcal{L}^l \subseteq \mathcal{L}^{k+l}$, we have that $A \in \mathcal{L}^{k+l}$ and $N \in \mathcal{L}^{k+l}$ with $\bar{\lambda}_{k+l}(N) = 0$. Since $(\mathbb{R}^{k+l}, \mathcal{L}^{k+l}, \bar{\lambda}_{k+l})$ is complete, we have that $B \in \mathcal{L}^{k+l}$, so $A \cup B \in \mathcal{L}^{k+l}$, whence $E \in \mathcal{L}^{k+l}$. This shows that $\overline{\mathcal{L}^k \otimes \mathcal{L}^l} \subseteq \mathcal{L}^{k+l}$. At the same time,

$$\mathcal{L}^{k+l} = \overline{\mathcal{B}^{k+l}} = \overline{\mathcal{B}^k \otimes \mathcal{B}^l} \subseteq \overline{\mathcal{L}^k \otimes \mathcal{L}^l}$$

by Claim 3 of Section 1.5, hence $\overline{\mathcal{L}^k \otimes \mathcal{L}^l} = \mathcal{L}^{k+l}$. Because of this and the fact that $\overline{\bar{\lambda}_k \otimes \bar{\lambda}_l} = \bar{\lambda}_k \otimes \bar{\lambda}_l = \bar{\lambda}_{k+l}$ on $\mathcal{L}^k \otimes \mathcal{L}^l$, we have that $\overline{\bar{\lambda}_k \otimes \bar{\lambda}_l} = \bar{\lambda}_{k+l}$ on $\mathcal{L}^{k+l}$ by Claim 4 of Section 3.7, and the entire proof is now complete. ∎

In Section 8 of Chapter 3 we proved a number of approximation results regarding $\mathcal{B}^k$-sets. Since $\mathcal{L}^k$-sets are [roughly] $\mathcal{B}^k$-sets "modified on a set of $\bar{\lambda}_k$-measure zero," one would expect the same approximation results to hold for $\mathcal{L}^k$-sets, and in fact they do. While we have usually written a set $A \in \mathcal{L}^k$ as $B \cup C$ where $B \in \mathcal{B}^k$ and C is a subset of some λ_k-null set N, recall that we may also represent such an A as a set of the form $B - D$ or $B \triangle E$, where $B \in \mathcal{B}^k$ and D and E are subsets of λ_k-null sets.

The following theorem is an amalgamation of the various claims in Section 3.8. There, we will take the generic semiring $\mathcal{S}$ to be the semiring $\mathcal{A}_k$, and the measure ν there with domain $\sigma(\mathcal{S})$ will be taken as λ_k with domain $\mathcal{B}^k$.

Approximation Theorem for Lebesgue Measure. *We have the following results:*

(a) Given any $A \in \mathcal{L}^k$ and $\epsilon > 0$, there exists a disjoint amc collection $\{A_1, A_2, \cdots\}$ of $\mathcal{A}_k$-sets such that $A \subseteq \bigcup_n A_n$ and $\bar{\lambda}_k(\bigcup_n A_n - A) < \epsilon$.

(b) Given any $A \in \mathcal{L}^k$ with $\bar{\lambda}_k(A) < +\infty$ and $\epsilon > 0$, there exists a finite disjoint collection $A_1, \cdots, A_n \in \mathcal{A}_k$ with $\bar{\lambda}_k(A \triangle \bigcup_{m=1}^n A_m) < \epsilon$.

(c) For any $A \in \mathcal{L}^k$ we can find a closed set C and an open set G such that $C \subseteq A \subseteq G$ and $\lambda_k(G - C) < \epsilon$. We also have

$$\bar{\lambda}_k(A) = \sup\{\lambda_k(C) : C \text{ closed}, C \subseteq A\} = \inf\{\lambda_k(G) : G \text{ open}, A \subseteq G\}.$$

(d) For any $A \in \mathcal{L}^k$, we have $\bar{\lambda}_k(A) = \sup\{\lambda_k(K) : K \text{ compact}, K \subseteq A\}$.

(e) If $A \in \mathcal{L}^k$, then

$$\lambda_k^*(A) = \inf\{\lambda_k(B) : A \subseteq B \in \mathcal{B}^k\} = \sup\{\lambda_k(C) : C \subseteq A, C \in \mathcal{B}^k\} = \lambda_{k*}(A),$$

the common value being $\bar{\lambda}_k(A)$. [The λ_{k*} notation mimics the μ_* notation from Section 3.9.]

(f) If $A \subseteq \mathbb{R}^k$ is such that $\lambda_{k*}(A) = \lambda_k^*(A) < +\infty$, then $A \in \mathcal{L}^k$.

Proof: We first do (a). Let $\epsilon > 0$ and $A \in \mathcal{L}^k$, so that $A = B - C$, where $B \in \mathcal{B}^k$ and C is contained in some λ_k-null set N. From Claim 2 or Claim 3 in Section 3.8 [as $\lambda_k(B)$ is finite or infinite], there exists a disjoint amc collection

$A_1, A_2, \cdots \in \mathcal{A}_k$ such that $B \subseteq \bigcup_n A_n$ and $\lambda_k(\bigcup_n A_n - B) < \epsilon$. Since $A \subseteq B$, we have $A \subseteq \bigcup_n A_n$ and

$$\bar{\lambda}_k\left(\bigcup_n A_n - A\right) = \bar{\lambda}_k\left(\bigcup_n A_n - B\right) + \bar{\lambda}_k(B - A) < \epsilon + \bar{\lambda}_k(N) = \epsilon,$$

where the inequality follows because $B - A \subseteq N$. This completes (a).

We now prove (b). Let $\epsilon > 0$ and let $A \in \mathcal{L}^k$ be such that $\bar{\lambda}_k(A) < +\infty$. Write $A = B \triangle C$, where $B \in \mathcal{B}^k$ and C is a subset of some λ_k-null set N. From Claim 4 of Section 3.8, there exists a finite disjoint collection $A_1, \cdots, A_n \in \mathcal{A}_k$ such that $\lambda_k(B \triangle \bigcup_{m=1}^n A_m) < \epsilon$. Therefore,

$$\begin{aligned} \bar{\lambda}_k\left((B\triangle C) \triangle \bigcup_{m=1}^n A_m\right) &= \bar{\lambda}_k\left(C\triangle \left[B \triangle \bigcup_{m=1}^n A_m\right]\right) \\ &\leq \bar{\lambda}_k(C) + \bar{\lambda}_k\left(B \triangle \bigcup_{m=1}^n A_m\right) \\ &< 0 + \epsilon = \epsilon, \end{aligned}$$

and thus $A \triangle \bigcup_{m=1}^n A_m$ has $\bar{\lambda}_k$ measure less than ϵ, giving (b).

We now deal with part (c). Let $A \in \mathcal{L}^k$, and let $\epsilon > 0$. Write $A = B \cup C$, where $B \in \mathcal{B}^k$ and C is contained in some λ_k-null set N. From Claim 5 in Section 3.8, there is a closed set C and open set G_1 with $C \subseteq B \subseteq G_1$ and $\lambda_k(G_1 - C) < \frac{\epsilon}{2}$, and hence $\lambda_k(B - C) < \frac{\epsilon}{2}$. Therefore, $C \subseteq A$ and

$$\bar{\lambda}_k(A - C) = \bar{\lambda}_k(A - B) + \bar{\lambda}_k(B - C) \leq \bar{\lambda}_k(N) + \frac{\epsilon}{2} = 0 + \frac{\epsilon}{2} = \frac{\epsilon}{2}.$$

Next, rewrite A as $D - E$, where $D \in \mathcal{B}^k$ and E is a subset of some λ_k-null set M. From Claim 5 in Section 3.8, there is an open set G and a closed set C_1 with $C_1 \subseteq D \subseteq G$ and $\lambda_k(G) - \lambda_k(C_1) < \frac{\epsilon}{2}$, giving $\lambda_k(G - D) < \frac{\epsilon}{2}$. Therefore, $A \subseteq G$ and

$$\bar{\lambda}_k(G - A) = \bar{\lambda}_k(G - D) + \bar{\lambda}_k(D - A) < \frac{\epsilon}{2} + \bar{\lambda}_k(M) = \frac{\epsilon}{2} + 0 = \frac{\epsilon}{2}.$$

It follows that $C \subseteq A \subseteq G$ and $\bar{\lambda}_k(G - C) = \bar{\lambda}_k(G - A) + \bar{\lambda}_k(A - C) < \epsilon$, and the proof of the first part of (c) is complete. This gives the first part of (c).

For the second part of (c), write $A = B \cup C$, where B and C are as before. Then

$$\begin{aligned} \bar{\lambda}_k(A) = \lambda_k(B) &= \sup\{\lambda_k(D) : D \text{ closed}, D \subseteq B\} \\ &\leq \sup\{\lambda_k(D) : D \text{ closed}, D \subseteq A\}, \end{aligned}$$

where the second equality follows by Claim 5 of Section 3.8. At the same time, we have $\lambda_k(D) = \bar{\lambda}_k(D) \leq \bar{\lambda}_k(A)$ whenever D is a closed subset of A, giving the reverse inequality

$$\bar{\lambda}_k(A) \geq \sup\{\lambda_k(D) : D \text{ closed}, D \subseteq A\}.$$

This gives $\bar{\lambda}_k(A) = \sup\{\lambda_k(D) : \ D \text{ closed}, D \subseteq A\}$. The other desired equation $\bar{\lambda}_k(A) = \inf\{\lambda_k(G) : \ G \text{ open}, G \subseteq A\}$ follows similarly, giving (c).

We now prove (d). Write $A \in \mathcal{L}^k$ as $B \cup C$, where $B \in \mathcal{B}^k$ and C is contained in some λ_k-null set N. By Claim 7 of Section 3.8, the fact that $B \subseteq A$, and monotonicity, we obtain

$$\begin{aligned}\bar{\lambda}_k(A) = \lambda_k(B) &= \sup\{\lambda_k(K) : \ K \text{ compact}, K \subseteq B\} \\ &\le \sup\{\lambda_k(K) : \ K \text{ compact}, K \subseteq A\} \\ &\le \bar{\lambda}_k(A).\end{aligned}$$

[The last inequality follows because $\lambda_k(K) = \bar{\lambda}_k(K) \le \bar{\lambda}_k(A)$ for every compact subset K of A.] Therefore, the inequalities are equalities, proving (d).

Parts (e) and (f) are merely the results of Claims 1 and 3 from Section 3.9 applied to our situation. This completes the proof. ∎

4.3 TRANSLATION INVARIANCE OF LEBESGUE MEASURE

Regarding length and k-dimensional volume in general, intuition demands that the location of an interval or a k-dimensional rectangle has no effect on the measure. When $k = 1$, an interval with endpoints a and b should have exactly the same length as the interval with endpoints $a + x$ and $b + x$, where $x \in \mathbb{R}$ is arbitrary. Similarly, the area of any planar region should remain unaffected if we move that region any distance in any direction "while preserving the shape and directional orientation of the region." We shall show that k-dimensional Lebesgue measure satisfies this important property for every $k \in \mathbb{N}$, and we shall also make rigorous the concepts just stated.

Definition. Let $A \subseteq \mathbb{R}^k$ and $\boldsymbol{x} \in \mathbb{R}^k$. The *translate [or translation] of* A *by* $\boldsymbol{x}$ is $A \oplus \boldsymbol{x} = \{\boldsymbol{a} + \boldsymbol{x} : \ \boldsymbol{a} \in A\}$. We will write $A \ominus \boldsymbol{x} = A \oplus (-\boldsymbol{x})$.

Example. Let $A = (\boldsymbol{a}, \boldsymbol{b}] \in \mathcal{A}_k$, where as usual $\mathcal{A}_k$ consists of $\varnothing$ and the bounded k-dimensional rsc rectangles. Given any $\boldsymbol{x} \in \mathbb{R}^k$, we have that

$$A \oplus \boldsymbol{x} = (\boldsymbol{a}, \boldsymbol{b}] \oplus \boldsymbol{x} = \{\boldsymbol{y} + \boldsymbol{x} \in \mathbb{R}^k : \ \boldsymbol{a} < \boldsymbol{y} \le \boldsymbol{b}\} = (\boldsymbol{a} + \boldsymbol{x}, \boldsymbol{b} + \boldsymbol{x}] \in \mathcal{A}_k.$$

Therefore $\lambda_k(A \oplus \boldsymbol{x})$ is defined and

$$\lambda_k(A \oplus \boldsymbol{x}) = \prod_{i=1}^{k} [(b_i + x_i) - (a_i + x_i)] = \prod_{i=1}^{k} (b_i - a_i) = \lambda_k(A).$$

We also have for all $\boldsymbol{x} \in \mathbb{R}^k$ that $\varnothing \oplus \boldsymbol{x} = \varnothing \in \mathcal{A}_k$ and $\lambda_k(\varnothing \oplus \boldsymbol{x}) = \lambda_k(\varnothing) = 0$. Putting everything together, we have shown that if $A \in \mathcal{A}_k$ and $\boldsymbol{x} \in \mathbb{R}^k$, then $A \oplus \boldsymbol{x} \in \mathcal{A}_k$ and $\bar{\lambda}_k(A \oplus \boldsymbol{x}) = \bar{\lambda}_k(A)$. It follows that (i) $\mathcal{A}_k$ is *closed under translations* and (ii) the restriction of $\bar{\lambda}_k$ to $\mathcal{A}_k$ is *translation invariant.*

Claim 1. *For any* $A \in \mathcal{B}^k$ *and* $\boldsymbol{x} \in \mathbb{R}^k$, *we have* $A \oplus \boldsymbol{x} \in \mathcal{B}^k$. *In other words,* $\mathcal{B}^k$ *is closed under translations.*

Proof: Define $\mathcal{C} = \{A \in \mathcal{B}^k : A \oplus \boldsymbol{x} \in \mathcal{B}^k \text{ for all } \boldsymbol{x} \in \mathbb{R}^k\}$, so that $\mathcal{C} \subseteq \mathcal{B}^k$. The example shows that $\mathcal{A}_k \subseteq \mathcal{C}$. If $\mathcal{C}$ is a σ-field on $\mathbb{R}^k$, then minimality will give $\mathcal{B}^k = \sigma(\mathcal{A}_k) \subseteq \mathcal{C}$, giving $\mathcal{B}^k = \mathcal{C}$, completing the proof.

We verify that $\mathcal{C}$ is a σ-field. Since $\mathbb{R}^k \oplus \boldsymbol{x} = \mathbb{R}^k \in \mathcal{B}^k$ for each $\boldsymbol{x} \in \mathbb{R}^k$, we have $\mathbb{R}^k \in \mathcal{C}$. If $A \in \mathcal{C}$, then $A^c \oplus \boldsymbol{x} = (A \oplus \boldsymbol{x})^c \in \mathcal{B}^k$ for each $\boldsymbol{x} \in \mathbb{R}^k$, hence $A^c \in \mathcal{C}$, so $\mathcal{C}$ is closed under complementation. Next, let $A_1, A_2, \cdots \in \mathcal{C}$. Since $(\bigcup_{n=1}^\infty A_n) \oplus \boldsymbol{x} = \bigcup_{n=1}^\infty (A_n \oplus \boldsymbol{x}) \in \mathcal{B}^k$ for each $\boldsymbol{x} \in \mathbb{R}^k$, we have $\bigcup_{n=1}^\infty A_n \in \mathcal{C}$, so $\mathcal{C}$ is closed under countable unions and hence is a σ-field. ∎

Claim 2. *For every $B \in \mathcal{B}^k$ and $\boldsymbol{x} \in \mathbb{R}^k$ we have $\lambda_k(B \oplus \boldsymbol{x}) = \lambda_k(B)$. In other words, λ_k is translation invariant.*

Proof: Pick and fix $\boldsymbol{x} \in \mathbb{R}^k$. Define $\nu : \mathcal{B}^k \to \bar{\mathbb{R}}$ by writing $\nu(B) = \lambda_k(B \oplus \boldsymbol{x})$ for every $B \in \mathcal{B}^k$. [$\nu(B)$ is defined for each $B \in \mathcal{B}^k$ by Claim 1.] Clearly ν is nonnegative and $\nu(\varnothing) = 0$. If $B_1, B_2, \cdots \in \mathcal{B}^k$ are disjoint, then so are $B_1 \oplus \boldsymbol{x}, B_2 \oplus \boldsymbol{x}, \cdots$, hence

$$\begin{aligned}\nu\left(\bigcup_{n=1}^\infty B_n\right) = \lambda_k\left(\bigcup_{n=1}^\infty B_n \oplus \boldsymbol{x}\right) &= \lambda_k\left(\bigcup_{n=1}^\infty (B_n \oplus \boldsymbol{x})\right) \\ &= \sum_{n=1}^\infty \lambda_k(B_n \oplus \boldsymbol{x}) \\ &= \sum_{n=1}^\infty \nu(B_n),\end{aligned}$$

whence ν is a measure with domain $\mathcal{B}^k$. It was shown in the example that $\nu = \lambda_k$ on the semiring $\mathcal{A}_k$. Since λ_k is σ-finite on $\mathcal{A}_k$ and $\sigma(\mathcal{A}_k) = \mathcal{B}^k$, the Uniqueness Theorem in Section 3.5 implies that $\nu = \lambda_k$ on $\mathcal{B}^k$: $B \in \mathcal{B}^k$ implies $\lambda_k(B \oplus \boldsymbol{x}) = \lambda_k(B)$. This completes the proof. ∎

We may now put the previous two claims together to prove the following theorem on translation invariance. This theorem will be used in our construction of a set outside $\mathcal{L}^k$ to be found in Section 5.

Translation Invariance of Lebesgue Measure. *For every $A \in \mathcal{L}^k$ and $\boldsymbol{x} \in \mathbb{R}^k$, we have that $A \oplus \boldsymbol{x} \in \mathcal{L}^k$ and $\bar{\lambda}_k(A \oplus \boldsymbol{x}) = \bar{\lambda}_k(A)$. In other words, $\mathcal{L}^k$ is closed under translations, and $\bar{\lambda}_k$ is translation invariant.*

Proof: Fix $A \in \mathcal{L}^k$ and $\boldsymbol{x} \in \mathbb{R}^k$. Then $A = B \cup C$, where $B \in \mathcal{B}^k$ and C is contained in some λ_k-null set N. Observe that

$$A \oplus \boldsymbol{x} = (B \cup C) \oplus \boldsymbol{x} = (B \oplus \boldsymbol{x}) \cup (C \oplus \boldsymbol{x}),$$

where $B \oplus \boldsymbol{x} \in \mathcal{B}^k$ by Claim 1. Also observe that $C \oplus \boldsymbol{x} \subseteq N \oplus \boldsymbol{x} \in \mathcal{B}^k$ by Claim 1, and $\lambda_k(N \oplus \boldsymbol{x}) = \lambda_k(N) = 0$ by Claim 2. It therefore follows that $A \oplus \boldsymbol{x}$ is the union of a $\mathcal{B}^k$-set and a subset of λ_k-null set, hence $A \oplus \boldsymbol{x} \in \mathcal{L}^k$. We may now apply $\bar{\lambda}_k$ to $A \oplus \boldsymbol{x}$ and invoke Claim 2 to obtain

$$\bar{\lambda}_k(A \oplus \boldsymbol{x}) = \lambda_k(B \oplus \boldsymbol{x}) = \lambda_k(B) = \bar{\lambda}_k(A),$$

which completes the proof. ■

The mathematics in this brief section is at a close, and Lebesgue measure [either λ_k or $\bar{\lambda}_k$] does exactly what intuition demands. Since the mathematics of this section is not particularly difficult, it may be somewhat surprising that this property is what will be the sticking point in trying to construct a measure defined for *every* subset of $\mathbb{R}^k$ that mimics length [and thus is translation invariant]. It will be shown that the property of translation invariance cannot coexist with the Axiom of Choice (AC); the work required to prove such an assertion will be decidedly involved.

Exercises.

1. Let $\mu : \mathcal{B}^k \to \bar{\mathbb{R}}$ denote a measure that assigns finite measure to bounded $\mathcal{B}^k$-sets. Furthermore, suppose that μ is translation invariant: $\mu(B \oplus \boldsymbol{x}) = \mu(B)$ for all $\boldsymbol{x} \in \mathbb{R}^k$. Then there is $c > 0$ such that $\mu(B) = c\lambda_k(B)$ for all $B \in \mathcal{B}^k$.

Exercises 2-8 give the outline of a subset of $\mathbb{R}$ that fails to be in $\mathcal{L}$, and hence fails to be in $\mathcal{B}$ as well. Note the explicit use of the Axiom of Choice (AC) and the concepts of translation invariance in the construction. We will construct such a set in $\mathbb{R}^k$ later this chapter. The following outline explicitly follows Halmos' development.

2. Let $A \in \mathcal{L}$ be such that $\bar{\lambda}(A) > 0$, and let $c \in [0, 1)$. There exists an open interval U such that $\bar{\lambda}(A \cap U) \geq c\lambda(U)$. [Recall that $\bar{\lambda}(A)$ is the infimum of the numbers $\lambda(G)$, where $A \subseteq G$ and G is open. Argue that this implies the existence of an open G containing A with $c\lambda(G) \leq \bar{\lambda}(A)$. Write G as a disjoint union $\bigcup_{n=1}^{\infty} G_n$ of open intervals, and write $c\lambda(G) = c\sum_{n=1}^{\infty} \lambda(G_n) \leq \bar{\lambda}(A) = \sum_{n=1}^{\infty} \bar{\lambda}(A \cap G_n)$. For some n we must have $c\lambda(G_n) \leq \bar{\lambda}(A \cap G_n)$; let U denote such a G_n.]

3. Let $A \in \mathcal{L}$ contain an open interval. Then there exists $a > 0$ such that $(-a, a)$ is contained in $D(A) = \{x - y : x, y \in A\}$. [This is obvious geometrically.]

4. Let $A \in \mathcal{L}$ be such that $\bar{\lambda}(A) > 0$. Then there exists $a > 0$ such that $(-a, a)$ is contained in $D(A) = \{x - y : x, y \in A\}$. [Use Exercise 2 to find an open interval U with $\bar{\lambda}(A \cap U) \geq \frac{3}{4}\lambda(U)$. From this, argue that a may be taken as $\frac{1}{2}\lambda(U)$ as follows. (i) If $|x| < \frac{1}{2}\lambda(U)$, demonstrate that $(A \cap U) \cup [(A \cap U) \oplus x]$ is contained in the open interval $U \cup (U \oplus x)$ and $\lambda(U \cup (U \oplus x)) < \frac{3}{2}\lambda(U)$. (ii) Show that if $A \cap U$ and $(A \cap U) \oplus x$ are disjoint, then we have by translation invariance that $\lambda((A \cap U) \cup [(A \cap U) \oplus x]) = 2\lambda(A \cap U) \geq \frac{3}{2}\lambda(U)$, which contradicts the fact that $\bar{\lambda}((A \cap U) \cup [(A \cap U) \oplus x]) \leq \bar{\lambda}(U \cup (U \oplus x)) < \frac{3}{2}\lambda(U)$. Conclude that $A \cap U$ and $(A \cap U) \oplus x$ are not disjoint. (iii) Show that $x \in D(A)$ by using (ii) to find y in both $A \cap U$ and $(A \cap U) \oplus x$, so that $y = z + x$ for some $z \in A \cap U$. Thus $x = y - z$, where $y, z \in A$. Thus, this x with $|x| < \frac{1}{2}\lambda(U)$ is in $D(A)$.]

5. Let A be a dense subset of $\mathbb{R}$. Then $cA = \{ca : a \in A\}$ is dense for any $c \neq 0$.

6. Let ξ be an irrational number.
(a) Let $A = \{n + m\xi : n, m \in \mathbb{Z}\}$. Then A is a dense subset of $\mathbb{R}$.
(b) Let $B = \{n + m\xi : n, m \in \mathbb{Z},\ n \text{ even}\}$. Then B is a dense subset of $\mathbb{R}$.
(c) Let $C = \{n + m\xi : n, m \in \mathbb{Z},\ n \text{ odd}\}$. Then C is a dense subset of $\mathbb{R}$.
[Part (a) is a standard analysis result. To prove (b), observe that $\frac{\xi}{2}$ is also irrational, and thus $\{n + m\frac{\xi}{2} : n, m \in \mathbb{Z}\}$, is dense. Therefore, $\{2(n + m\frac{\xi}{2}) : n, m \in \mathbb{Z}\}$ is dense by the previous exercse, so B is dense. Finally, $C = B \oplus 1$, and translates of dense sets are obviously dense.]

7. For $x, y \in \mathbb{R}$ write $x \sim y$ iff $x - y \in A = \{n + m\xi : n, m \in \mathbb{Z}\}$, where ξ is a fixed irrational number as in the previous exercise. Then $\sim$ is an equivalence relation, and hence $\mathbb{R}$ may be partitioned into disjoint equivalence classes.

We now invoke (AC) to form a set E_0 consisting of exactly one element from each of the equivalence classes in the previous exercise.

8. We will now show that $E_0 \notin \mathcal{L}$.
(a) There exist Borel subsets of E_0. [Recall that singletons are Borel sets.]
(b) Let $F \subseteq E_0$ be a Borel set. Then $D(F)$ cannot contain any nonzero elements of A, where A is the set in Exercise 6.
(c) By (b), there cannot exist an open interval containing the origin that is contained in $D(F)$, hence $\lambda(F) = 0$. [Use Exercise 4.]
(d) From (c), we have $\lambda_*(E_0) = 0$. [Recall the appropriate part of the Approximation Theorem of this section.] Therefore, if $E_0 \in \mathcal{L}$ it would follow that $\bar{\lambda}(E_0) = 0$.
(e) If a_1 and a_2 are distinct elements of $A = \{n + m\xi : n, m \in \mathbb{Z}\}$, then $E_0 \oplus a_1$ and $E_0 \oplus a_2$ are disjoint. [Argue by contradiction. If $z = x_1 + a_1$ and $z = x_2 + a_2$ where $x_1, x_2 \in E_0$, then $x_1 - x_2 = a_1 - a_2 \in A$, so $x_1 \sim x_2$ and $x_1 \neq x_2$. But E_0 is constructed by choosing exactly one point from each equivalence class.]
(f) $\mathbb{R} = \bigcup\{E_0 \oplus a : a \in A\}$, the countable union being disjoint. [Let $x \in \mathbb{R}$. Thus, x is in some equivalence class, hence $x \sim y$ for some $y \in E_0$. Therefore, $x - y = n + m\xi$ for some $n, m \in \mathbb{Z}$, hence $x = y + n + m\xi \in E_0 \oplus (n + m\xi)$.]
(g) If $E_0 \in \mathcal{L}$, then $\bar{\lambda}(E_0 + a) = 0$ for each $a \in A$, hence $\bar{\lambda}(\mathbb{R}) = 0$. Therefore, since the assumption that $E_0 \in \mathcal{L}$ leads to an absurdity, it must be the case that $E_0 \notin \mathcal{L}$, and hence there exists a subset of $\mathbb{R}$ that fails to be a Lebesgue set.

Comment: The above construction, as taken from Halmos' development, is originally due to Vitali in 1905. The set $E_0 \notin \mathcal{L}$ exists as a formal entity and is in no way "visualizable" due to the fact that (AC) was required for the construction. It may be conjectured that there is a simpler way of constructing a non-Lebesgue set, but a metamathematical theorem due to Solovay shows that every explicit exhibition of a non-Lebesgue set requires the use of (AC). See Section 5 for more details. Exercise 10 shows that under (AC), a large class of non-Lebesgue sets may be exhibited.

9. In conjunction with the previous exercises, it is not possible to have Lebesgue measure defined for every subset of $\mathbb{R}$. That is, a measure extension of Δ_F [where $F \in \mathfrak{S}$ is the identity function] from the semiring $\mathcal{A}$ consisting of $\varnothing$ and the bounded rsc intervals $(a, b]$ to $2^{\mathbb{R}}$ fails to exist. [Argue by contradiction for a quick answer: if such a measure extension μ to $2^{\mathbb{R}}$ exists then the previous work would force $\mu(\mathbb{R}) = 0$, contradicting what we expect of length.]

10. Let $B \in \mathcal{L}$ be such that $\bar{\lambda}(B) > 0$. We will show that there is $E_1 \subseteq B$ with $E_1 \notin \mathcal{L}$ by following the following outline. The set A appearing below is the set A from Exercise 7, consisting of the numbers $n + m\xi$, where $n, m \in \mathbb{Z}$ and ξ is a fixed irrational number.
(a) λ_* is monotone and translation invariant.
(b) $B = \bigcup\{(E_0 \oplus a) \cap B : a \in A\}$, the union being disjoint.
(c) There exists $a^* \in A$ such that $\lambda^*((E_0 \oplus a^*) \cap B) > 0$. [By (b), we have

$$0 < \bar{\lambda}(B) = \lambda^*(B) \leq \sum \{\lambda^*((E_0 \oplus a) \cap B) : a \in A\}.]$$

(d) Let $E_1 = (E_0 \oplus a^*) \cap B$. Then $E_1 \subseteq B$ and $E_1 \notin \mathcal{L}$. [We have

$$\lambda_*(E_1) = \lambda_*((E_0 \oplus a^*) \cap B) \leq \lambda_*(E_0 \oplus a^*) = \lambda_*(E_0) = 0$$

by part (a) and Exercise 8(d). But at the same time we have $\lambda^*(E_1) > 0$. Invoke the relevant part of the Approximation Theorem.]
(e) We may actually generalize further and state that *any subset* $B \subseteq \mathbb{R}$ *with* $\lambda^*(B) > 0$ *[B not necessarily in* $\mathcal{L}$*] contains a subset outside of* $\mathcal{L}$.

11. Let $A \in \mathcal{L}$, let B be a dense subset of $\mathbb{R}$, and suppose that $\bar{\lambda}(A \triangle (A \oplus x)) = 0$ for every $x \in B$. Then either $\bar{\lambda}(A) = 0$ or $\bar{\lambda}(A^c) = 0$. [This exercise is independent of the previous exercises.]

4.4 LINEAR TRANSFORMATIONS

There are two main goals of this section: (i) to show that nonsingular linear transformations of Borel and Lebesgue sets are Borel and Lebesgue sets, respectively, and (ii) to discuss how linear transformations affect the Lebesgue measure of Borel and Lebesgue sets. Everything is done in k-dimensional Euclidean space, and the reader is expected to be familiar with some basic results from linear algebra.

Definition. A mapping $\boldsymbol{T} : \mathbb{R}^k \to \mathbb{R}^k$ is a *linear transformation* iff for every $\boldsymbol{x}, \boldsymbol{y} \in \mathbb{R}^k$ and $\alpha, \beta \in \mathbb{R}$ we have $\boldsymbol{T}(\alpha\boldsymbol{x} + \beta\boldsymbol{y}) = \alpha\boldsymbol{T}(\boldsymbol{x}) + \beta\boldsymbol{T}(\boldsymbol{y})$. The linear transformation $\boldsymbol{T}$ will be called *nonsingular* iff for every $\boldsymbol{y} \in \mathbb{R}^k$ there is $\boldsymbol{x} \in \mathbb{R}^k$ such that $\boldsymbol{T}(\boldsymbol{x}) = \boldsymbol{y}$. [That is, $\boldsymbol{T}$ is onto $\mathbb{R}^k$.]

Recall that for every linear transformation $\boldsymbol{T}$, there exists a unique $k \times k$ matrix $\boldsymbol{M}$ such that $\boldsymbol{T}(\boldsymbol{x}) = \boldsymbol{M}\boldsymbol{x}$ for every $\boldsymbol{x} \in \mathbb{R}^k$. Furthermore, $\boldsymbol{T}$ is nonsingular [as a linear transformation] iff $\boldsymbol{M}$ is nonsingular [as a matrix]. To keep notation simple, we will use the same symbol to denote both the linear transformation as well as the unique matrix corresponding to the linear transformation, hence we are indifferent in writing $\boldsymbol{T}(\boldsymbol{x})$ [the value of the linear transformation $\boldsymbol{T}$ at $\boldsymbol{x}$] or $\boldsymbol{T}\boldsymbol{x}$ [the matrix $\boldsymbol{T}$ times the vector $\boldsymbol{x}$]. We shall view $\boldsymbol{T}$ in this dual fashion, and no confusion will arise.

Nonsingular linear transformations $\boldsymbol{T}$ have the following properties: (i) $\boldsymbol{T}\boldsymbol{0} = \boldsymbol{0}$, (ii) $\boldsymbol{T}\boldsymbol{x} = \boldsymbol{0}$ iff $\boldsymbol{x} = \boldsymbol{0}$, and (iii) $\boldsymbol{T}$ is one-to-one. For (i), write $\boldsymbol{T}(\boldsymbol{0}) = \boldsymbol{T}(\boldsymbol{0} + \boldsymbol{0}) = \boldsymbol{T}(\boldsymbol{0}) + \boldsymbol{T}(\boldsymbol{0})$ to obtain $\boldsymbol{T}\boldsymbol{0} = \boldsymbol{0}$. Property (ii) is a well-known fact concerning nonsingular matrices. For (iii), the assumption $\boldsymbol{T}\boldsymbol{x}_1 = \boldsymbol{T}\boldsymbol{x}_2$ forces $\boldsymbol{0} = \boldsymbol{T}(\boldsymbol{x}_1 - \boldsymbol{x}_2)$, hence $\boldsymbol{x}_1 - \boldsymbol{x}_2 = \boldsymbol{0}$ by (ii), whence $\boldsymbol{x}_1 = \boldsymbol{x}_2$.

Definition. Given any mapping $\boldsymbol{T} : \mathbb{R}^k \to \mathbb{R}^k$ and $A \subseteq \mathbb{R}^k$, we will write $\boldsymbol{T}A = \{\boldsymbol{T}\boldsymbol{x} : \boldsymbol{x} \in A\}$ for the image of A under $\boldsymbol{T}$.

Using this notation, a linear transformation $\boldsymbol{T}$ is nonsingular iff $\boldsymbol{T}\mathbb{R}^k = \mathbb{R}^k$. Therefore, *every nonsingular linear transformation is one-to-one and onto* $\mathbb{R}^k$.

Also note that if $\boldsymbol{T}$ is nonsingular, then

$$\boldsymbol{T}(A^c) = \boldsymbol{T}(\mathbb{R}^k - A) = \boldsymbol{T}\mathbb{R}^k - \boldsymbol{T}A = \mathbb{R}^k - \boldsymbol{T}A = (\boldsymbol{T}A)^c,$$

where the second and third equalities are consequences of nonsingularity. Finally, recall that for any mapping $\boldsymbol{T}$ [nonsingular or not] and any collection $\{A_\gamma : \gamma \in \Gamma\}$ of subsets of $\mathbb{R}^k$ we will have $\boldsymbol{T}\bigcup_{\gamma\in\Gamma} A_\gamma = \bigcup_{\gamma\in\Gamma} \boldsymbol{T}A_\gamma$.

Our first goal will be to show that $\mathcal{B}^k$ and $\mathcal{L}^k$ are closed under nonsingular linear transformations.

Claim 1. *If $\boldsymbol{T}$ is a nonsingular $k \times k$ matrix, then there exists an $\eta > 0$ such that for all $\boldsymbol{x} \in \mathbb{R}^k$ we have $||\boldsymbol{T}\boldsymbol{x}|| \geq \eta||\boldsymbol{x}||$.*

Proof: Let $\gamma_1 \geq \cdots \geq \gamma_k$ denote the eigenvalues corresponding to the symmetric matrix $\boldsymbol{T}'\boldsymbol{T}$, respectively. Since every eigenvalue corresponding to a nonsingular matrix is positive, and since the nonsingularity of $\boldsymbol{T}$ forces the nonsingularity of $\boldsymbol{T}'\boldsymbol{T}$, we have $\gamma_i > 0$, $i = 1, \cdots, k$. Also, let $\boldsymbol{e}_1, \cdots, \boldsymbol{e}_k$ denote the corresponding orthonormal eigenvectors of $\boldsymbol{T}'\boldsymbol{T}$. Write $\boldsymbol{T}'\boldsymbol{T} = \boldsymbol{O}\boldsymbol{D}\boldsymbol{O}'$, where $\boldsymbol{D}$ is the diagonal matrix with (j, j)th element γ_j, and where $\boldsymbol{O}$ is the orthogonal matrix whose jth column is $\boldsymbol{e}_j$, $j = 1, \cdots, k$. Letting $\boldsymbol{x} \in \mathbb{R}^k$ and writing $\boldsymbol{y} = \boldsymbol{O}'\boldsymbol{x}$, we have $||\boldsymbol{y}||^2 = \boldsymbol{x}'\boldsymbol{O}\boldsymbol{O}'\boldsymbol{x} = \boldsymbol{x}'\boldsymbol{I}\boldsymbol{x} = ||\boldsymbol{x}||^2$ and

$$||\boldsymbol{T}\boldsymbol{x}||^2 = \boldsymbol{x}'\boldsymbol{T}'\boldsymbol{T}\boldsymbol{x} = \boldsymbol{y}'\boldsymbol{D}\boldsymbol{y} = \sum_{j=1}^{k} \gamma_j y_j^2 \geq \gamma_k \sum_{j=1}^{k} y_j^2 = \gamma_k||\boldsymbol{y}||^2 = \gamma_k||\boldsymbol{x}||^2.$$

Therefore, we may take $\eta > 0$ to be $\sqrt{\gamma_k}$, the square root of the smallest eigenvalue γ_k for the matrix $\boldsymbol{T}'\boldsymbol{T}$, and the proof is complete. ∎

The next claim is actually a special case of a result for Banach spaces that follows from the famous Open Mapping Theorem. We give a direct proof in the framework of k-dimensional Euclidean space.

Claim 2. *Let $\boldsymbol{T}$ denote an arbitrary nonsingular $k \times k$ matrix. Then for any open subset $A \subseteq \mathbb{R}^k$ we have that $\boldsymbol{T}A$ is an open subset of $\mathbb{R}^k$.*

Proof: Let $A \subseteq \mathbb{R}^k$, and let $\boldsymbol{T}$ denote a nonsingular $k \times k$ matrix. If $A = \varnothing$, then A is open and of course $\boldsymbol{T}A = \boldsymbol{T}\varnothing = \varnothing$ is open. We may therefore assume that $A \neq \varnothing$. We proceed by contraposition, assuming that $\boldsymbol{T}A$ is not open and hoping to show that A is not open. If $\boldsymbol{T}A$ is not open, there is $\boldsymbol{y} \in \boldsymbol{T}A$ with the following property: for every $\epsilon > 0$ there is $\boldsymbol{y}_\epsilon$ with $||\boldsymbol{y} - \boldsymbol{y}_\epsilon|| < \epsilon$ and $\boldsymbol{y}_\epsilon \notin \boldsymbol{T}A$. By the assumed nonsingularity of $\boldsymbol{T}$, $\boldsymbol{y} = \boldsymbol{T}\boldsymbol{x}$ for some unique $\boldsymbol{x} \in A$ [namely, $\boldsymbol{x} = \boldsymbol{T}^{-1}\boldsymbol{y}$], and for any $\epsilon > 0$ there will exist a unique $\boldsymbol{x}_\epsilon$ such that $\boldsymbol{y}_\epsilon = \boldsymbol{T}\boldsymbol{x}_\epsilon$ [namely, $\boldsymbol{x}_\epsilon = \boldsymbol{T}^{-1}\boldsymbol{y}_\epsilon$]. Furthermore, the nonsingularity of $\boldsymbol{T}$ as well as the condition that no $\boldsymbol{y}_\epsilon$ is in $\boldsymbol{T}A$ forces each $\boldsymbol{x}_\epsilon \notin A$. We therefore have the following: *for every $\epsilon > 0$ there is $\boldsymbol{x}_\epsilon \notin A$ with*

$||\boldsymbol{T}\boldsymbol{x} - \boldsymbol{T}\boldsymbol{x}_\epsilon|| < \epsilon$. Observe that for any $\epsilon > 0$ we have

$$\begin{aligned} ||\boldsymbol{T}\boldsymbol{x} - \boldsymbol{T}\boldsymbol{x}_\epsilon|| < \epsilon \text{ and } \boldsymbol{x}_\epsilon \notin A &\leftrightarrow ||\boldsymbol{T}(\boldsymbol{x} - \boldsymbol{x}_\epsilon)|| < \epsilon \text{ and } \boldsymbol{x}_\epsilon \notin A \\ &\rightarrow \eta||\boldsymbol{x} - \boldsymbol{x}_\epsilon|| < \epsilon \text{ and } \boldsymbol{x}_\epsilon \notin A \\ &\leftrightarrow ||\boldsymbol{x} - \boldsymbol{x}_\epsilon|| < \frac{\epsilon}{\eta} \text{ and } \boldsymbol{x}_\epsilon \notin A, \end{aligned}$$

where in the second implication we have invoked the result and notation of Claim 1. Since $\boldsymbol{x} \in A$ and since $\epsilon > 0$ is generic, we have that any open ball with center $\boldsymbol{x}$ contains a point outside A. Therefore, $\boldsymbol{x}$ fails to be an interior point of A. In turn, this means that not every point of A is an interior point of A, hence A cannot possibly be open. This completes the proof by contraposition. ∎

Claim 3. *Let $\boldsymbol{T}$ denote an arbitrary nonsingular $k \times k$ matrix. Then for any $A \in \mathcal{B}^k$ we have $\boldsymbol{T}A \in \mathcal{B}^k$. That is, $\mathcal{B}^k$ is closed under nonsingular linear transformations.*

Proof: Let $\mathcal{C} = \{A \in \mathcal{B}^k : \boldsymbol{T}A \in \mathcal{B}^k\}$, so that $\mathcal{C} \subseteq \mathcal{B}^k$. By Claim 2 and the fact that $\mathcal{B}^k$ contains all of the open subsets of $\mathbb{R}^k$, it follows that every open subset of $\mathbb{R}^k$ is in $\mathcal{C}$. We will now show that $\mathcal{C}$ is a σ-field on $\mathbb{R}^k$, which by minimality forces $\mathcal{B}^k \subseteq \mathcal{C}$, giving $\mathcal{C} = \mathcal{B}^k$ and proving the claim.

Since $\boldsymbol{T}$ is nonsingular, we have $\boldsymbol{T}\mathbb{R}^k = \mathbb{R}^k \in \mathcal{B}^k$, hence $\mathbb{R}^k \in \mathcal{C}$. To see that $\mathcal{C}$ is closed under complementation, let $A \in \mathcal{C}$, so that $A \in \mathcal{B}^k$ and $\boldsymbol{T}A \in \mathcal{B}^k$. Observe that $A^c \in \mathcal{B}^k$, and the nonsingularity of $\boldsymbol{T}$ forces $\boldsymbol{T}A^c = (\boldsymbol{T}A)^c \in \mathcal{B}^k$, hence $A^c \in \mathcal{C}$. To show that $\mathcal{C}$ is closed under countable unions, let $\{A_n\}_{n=1}^\infty$ denote a sequence of $\mathcal{C}$-sets, so that $A_n \in \mathcal{B}^k$ and $\boldsymbol{T}A_n \in \mathcal{B}^k$ for every $n \in \mathbb{N}$. Since $\bigcup_{n=1}^\infty A_n \in \mathcal{B}^k$ and $\boldsymbol{T}\bigcup_{n=1}^\infty A_n = \bigcup_{n=1}^\infty \boldsymbol{T}A_n \in \mathcal{B}^k$, it follows that $\bigcup_{n=1}^\infty A_n \in \mathcal{C}$. Therefore, $\mathcal{C}$ is σ-field on $\mathbb{R}^k$. ∎

Definition. A one-to-one mapping π of $\{1, \cdots, k\}$ onto $\{1, \cdots, k\}$ will be called a *permutation mapping.*

If $\boldsymbol{T}$ is the mapping corresponding to the permutation π, that is, if $\boldsymbol{T}$ maps $\boldsymbol{x} = (x_1, \cdots, x_k)'$ to $(x_{\pi(1)}, \cdots, x_{\pi(k)})'$, then the jth column of $\boldsymbol{T}$ has a 1 in the $\pi(j)$th slot and 0's elsewhere, $j = 1, \cdots, k$. Observe that $|\det \boldsymbol{T}| = 1$, hence $\boldsymbol{T}$ is a nonsingular linear transformation.

Claim 4. *Let $\boldsymbol{T}$ denote a $k \times k$ permutation matrix. Then $A \in \mathcal{B}^k$ implies $\boldsymbol{T}A \in \mathcal{B}^k$, and for any $c > 0$ we have $\lambda_k(\boldsymbol{T}(\boldsymbol{0}, c\boldsymbol{1}]) = \lambda_k((\boldsymbol{0}, c\boldsymbol{1}]) = c^k$.*

Proof: The first part follows by Claim 3, and the second part is obvious since $\boldsymbol{T}(\boldsymbol{0}, c\boldsymbol{1}] = (\boldsymbol{0}, c\boldsymbol{1}]$ for all $c > 0$. ∎

Claim 5. *Let $p \neq 0$, and let $\boldsymbol{T}(\boldsymbol{x}) = (px_1, x_2, \cdots, x_k)'$ for all $\boldsymbol{x} \in \mathbb{R}^k$. Then $\boldsymbol{T}$ is nonsingular, [so that $\boldsymbol{T}A \in \mathcal{B}^k$ for any $A \in \mathcal{B}^k$ by Claim 3], and for any $c > 0$ we have $\lambda_k(\boldsymbol{T}(\boldsymbol{0}, c\boldsymbol{1}]) = |p|c^k = |p|\lambda_k((\boldsymbol{0}, c\boldsymbol{1}])$.*

Proof: For any $p \neq 0$, the matrix $\boldsymbol{T}$ is a $k \times k$ diagonal matrix with first diagonal element p and all other diagonal elements of 1. Since p is nonzero,

$\det \boldsymbol{T} = p \neq 0$, and hence $\boldsymbol{T}$ is nonsingular. We consider the cases $p > 0$ and $p < 0$ separately in order to prove the other parts of the claim. For $p > 0$, we have $\boldsymbol{T}(\mathbf{0}, c\mathbf{1}] = (0, pc] \times (0, c]^{k-1}$, and hence

$$\lambda_k\left(\boldsymbol{T}(\mathbf{0}, c\mathbf{1}]\right) = \lambda_k\left((0, pc] \times (0, c]^{k-1}\right) = |p|c^k = |p|\lambda_k\left((\mathbf{0}, c\mathbf{1}]\right),$$

[the second and third equalities are direct calculations], and thus the claim is proved for $p > 0$. For $p < 0$, we have $\boldsymbol{T}(\mathbf{0}, c\mathbf{1}] = [pc, 0) \times (0, c]^{k-1}$, so

$$\lambda_k\left(\boldsymbol{T}(\mathbf{0}, c\mathbf{1}]\right) = \lambda_k\left([pc, 0) \times (0, c]^{k-1}\right) = -pc^k = |p|c^k = |p|\lambda_k\left((\mathbf{0}, c\mathbf{1}]\right),$$

and the proof is complete. ■

Claim 6. *Let $k \geq 2$, let $\boldsymbol{T}(\boldsymbol{x}) = (x_1 + x_2, x_2, \cdots, x_k)'$ for each $\boldsymbol{x} \in \mathbb{R}^k$, and let $c > 0$. Then $\boldsymbol{T}$ is a nonsingular linear transformation, [so that $\boldsymbol{T}A \in \mathcal{B}^k$ whenever $A \in \mathcal{B}^k$ by Claim 3], and $\lambda_k(\boldsymbol{T}(\mathbf{0}, c\mathbf{1}]) = c^k = \lambda_k((\mathbf{0}, c\mathbf{1}])$.*

Proof: Take $k \geq 2$ and $c > 0$, and define the sets

$$B = \begin{cases} \{\boldsymbol{x} \in \mathbb{R}^k : 0 < x_i \leq c \text{ for } i = 3, \cdots, k\} & \text{if } k \geq 3 \\ \mathbb{R}^2 & \text{if } k = 2, \end{cases}$$

$$B_1 = \{\boldsymbol{x} \in \mathbb{R}^k : 0 < x_1 \leq x_2 \leq c\} \cap B,$$

$$B_2 = \{\boldsymbol{x} \in \mathbb{R}^k : 0 < x_2 < x_1 \leq c\} \cap B,$$

and

$$B_3 = \{\boldsymbol{x} \in \mathbb{R}^k : c < x_1 \leq c + x_2,\ 0 < x_2 \leq c\} \cap B.$$

Since $\boldsymbol{T}$ has $t_{ii} = 1$ for $i = 1, \cdots, k$, $t_{12} = 1$, and $t_{ij} = 0$ otherwise, it follows that $\det \boldsymbol{T} = 1$, hence $\boldsymbol{T}$ is nonsingular. Observe that B_1, B_2, B_3 are disjoint bounded Borel sets with $B_1 \oplus (c, 0, \cdots, 0) = B_3$, so that $\lambda_k(B_1) = \lambda_k(B_3)$ by the translation invariance of Lebesgue measure. Also, observe that we have $(\mathbf{0}, c\mathbf{1}] = B_1 \cup B_2$. Finally, observe that $\boldsymbol{T}(\mathbf{0}, c\mathbf{1}] = B_2 \cup B_3$, and thus

$$\begin{aligned} \lambda_k(\boldsymbol{T}(\mathbf{0}, c\mathbf{1}]) = \lambda_k(B_2) + \lambda_k(B_3) &= \lambda_k(B_2) + \lambda_k(B_1 \oplus c) \\ &= \lambda_k(B_2) + \lambda_k(B_1) \\ &= \lambda_k((\mathbf{0}, c\mathbf{1}]), \end{aligned}$$

where the third equality follows by translation invariance. ■

Claim 7. *Let $k \geq 2$, let $\boldsymbol{T}(\boldsymbol{x}) = (x_1 - x_2, x_2, \cdots, x_k)'$ for each $\boldsymbol{x} \in \mathbb{R}^k$, and let $c > 0$. Then $\boldsymbol{T}$ is a nonsingular linear transformation, [so that $\boldsymbol{T}A \in \mathcal{B}^k$ whenever $A \in \mathcal{B}^k$ by Claim 3], and $\lambda_k(\boldsymbol{T}(\mathbf{0}, c\mathbf{1}]) = c^k = \lambda_k((\mathbf{0}, c\mathbf{1}])$.*

Proof: The proof is similar to that of Claim 6, and is left as an exercise. ■

Claim 8. *Let $k \in \mathbb{N}$, and let $(\boldsymbol{a}, \boldsymbol{b}] \in \mathcal{A}_k$ be such that $\boldsymbol{a}, \boldsymbol{b} \in \mathbb{Q}^k$. Then $(\boldsymbol{a}, \boldsymbol{b}]$ may be written as a finite disjoint union of k-dimensional hypercubes, each cube having sides parallel to the coordinate axes of $\mathbb{R}^k$.*

Proof: When $k = 1$, we have that $b - a = \frac{n}{m}$ for some positive integers m and n, and thus

$$(a, b] = \bigcup_{l=1}^{n} \left(a + \frac{(l-1)}{m}, a + \frac{l}{m} \right],$$

which is a disjoint union of n intervals of equal length $\frac{1}{m}$. This proves the $k = 1$ claim, and now we let $k \geq 2$. Since each a_i and b_i are rational for $i = 1, \cdots, k$, it follows that there exist positive integers n_i and m_i such that $b_i - a_i = n_i/m_i$. Since this is true for $i = 1, \cdots, k$ we may write

$$b_i - a_i = \frac{n_i}{m_i} = \frac{n_i \prod_{l \neq i} m_l}{m_i \prod_{l \neq i} m_l} = \frac{n_i \prod_{l \neq i} m_l}{\prod_{l=1}^{k} m_l} = \frac{n_i M_i}{M},$$

where $M = \prod_{l=1}^{k} m_l$ and $M_i = \prod_{l \neq i} m_l$ for $i = 1, \cdots, k$. Therefore $(\boldsymbol{a}, \boldsymbol{b}]$ may be written as the finite disjoint union

$$\bigcup_{l_1=1}^{n_1 M_1} \cdots \bigcup_{l_k=1}^{n_k M_k} \left(a_1 + \frac{l_1 - 1}{M}, a_1 + \frac{l_1}{M} \right] \times \cdots \times \left(a_k + \frac{l_k - 1}{M}, a_k + \frac{l_k}{M} \right],$$

of $\prod_{l=1}^{k} n_l M_l$ sub-cubes, each cube having sides of length[3] $\frac{1}{M}$. This completes the proof. ∎

We almost have the needed machinery to prove the section's main theorem, but there are a few more preliminary matters that require our attention.

Definition. Let $k \geq 2$. A linear transformation $\boldsymbol{T} : \mathbb{R}^k \to \mathbb{R}^k$ is *special*[4] iff $\boldsymbol{T}$ is a mapping as given in any of Claims 4 through 7. We will use the term *special matrices* for the matrices associated with such transformations.

Claim 9. *Let $\boldsymbol{T}$ denote a special linear transformation. Then for any $A \in \mathcal{B}^k$ we have $\boldsymbol{T}A \in \mathcal{B}^k$, and for every $c > 0$ we have*

$$\lambda_k(\boldsymbol{T}(\boldsymbol{0}, c\boldsymbol{1}]) = |\det \boldsymbol{T}| \lambda_k((\boldsymbol{0}, c\boldsymbol{1}]) = |\det \boldsymbol{T}| c^k.$$

Proof: If $\boldsymbol{T}$ corresponds to any special mapping, then $\boldsymbol{T}$ is nonsingular by Claims 4-7, so $\boldsymbol{T}A \in \mathcal{B}^k$ for any $A \in \mathcal{B}^k$ by Claim 3. The formula follows by again appealing to the formulas established in Claims 4-7 and noting that (i) $|\det \boldsymbol{T}|$ is 1 for the matrices corresponding to the special mappings in Claims 4, 6, and 7, and (ii) $|\det \boldsymbol{T}|$ is $|p|$ for the matrices corresponding to the special mappings of Claim 5. This completes the proof of the claim. ∎

[3]If we wanted to be formal, we would have instead stated something such as *the $\bar{\lambda}$-measure of the projection of the cube onto the ith coordinate axis is* $M^{-1} = (\prod_{i=1}^{k} m_i)^{-1}$, but it seems better to state things as we did in the main text, since no rigor is lost.

[4]This is a makeshift term used only in this section.

The following claim is a standard and nontrivial fact of linear algebra, and is accordingly stated without proof.

Claim 10. *Let $k \in \mathbb{N}$. Any nonsingular $k \times k$ matrix $\boldsymbol{T}$ may be written as a finite product $\boldsymbol{T}_N \cdots \boldsymbol{T}_2\boldsymbol{T}_1$ of nonsingular $k \times k$ matrices, where $\boldsymbol{T}_j$ is nonsingular and freely denotes the matrix corresponding to any of the special linear transformations given in Claims 4-7, $j = 1, \cdots, n$. Conversely, every finite product $\boldsymbol{T}_N \cdots \boldsymbol{T}_2\boldsymbol{T}_1$ of $k \times k$ special matrices is nonsingular.*

With 10 claims behind us, a fair amount of work has been done. At last, we are now ready to state and prove the main theorem.

Claim 11. *Let $k \in \mathbb{N}$, and let $\boldsymbol{T}$ denote a nonsingular transformation.*

(i) For every $A \in \mathcal{B}^k$ we have $\boldsymbol{T}A \in \mathcal{B}^k$ and

$$\lambda_k(\boldsymbol{T}A) = |\det \boldsymbol{T}|\lambda_k(A). \tag{4.4}$$

(ii) For every $A \in \mathcal{L}^k$, we have $\boldsymbol{T}A \in \mathcal{L}^k$ and

$$\bar{\lambda}_k(\boldsymbol{T}A) = |\det \boldsymbol{T}|\bar{\lambda}_k(A). \tag{4.5}$$

Proof: We first prove (i) for $k \geq 2$, leaving the proof of (4.4) when $k = 1$ as a simple exercise. By Claim 10, $\boldsymbol{T} = \boldsymbol{T}_M \cdots \boldsymbol{T}_1$ for some $M \in \mathbb{N}$, where $\boldsymbol{T}_j$ is a special $k \times k$ matrix, $j = 1, \cdots, M$. Let us define the following proposition $H(N)$ for all $N \in \mathbb{N}$:

> *If each of $\boldsymbol{T}_1, \cdots, \boldsymbol{T}_N$ freely denote $k \times k$ special matrices, and if $A \in \mathcal{B}^k$, then $\boldsymbol{T}_N \cdots \boldsymbol{T}_1 A \in \mathcal{B}^k$ and $\lambda_k(\boldsymbol{T}_N \cdots \boldsymbol{T}_1 A) = |\det(\boldsymbol{T}_N \cdots \boldsymbol{T}_1)|\lambda_k(A)$.*

If $H(N)$ is true for all $N \in \mathbb{N}$, then (4.4) holds by taking $N = M$. In verifying $H(N)$ for each N, we needn't bother with checking to see whether $A \in \mathcal{B}^k$ implies $\boldsymbol{T}_N \cdots \boldsymbol{T}_1 A \in \mathcal{B}^k$, for we merely use Claim 3 N times in succession. We thus only need to verify the equation of $H(N)$ for each $N \in \mathbb{N}$.

We show that $H(1)$ is true. Let $\mu_1 : \mathcal{B}^k \to \bar{\mathbb{R}}$ be such that $\mu_1(A) = \lambda_k(\boldsymbol{T}_1 A)$ for all $A \in \mathcal{B}^k$, and let $\mu_2 : \mathcal{B}^k \to \bar{\mathbb{R}}$ be such that $\mu_2(A) = |\det \boldsymbol{T}_1|\lambda_k(A)$. It is easily verified that both μ_1 and μ_2 are measures. Furthermore, both are σ-finite on the semiring $\mathcal{A}_k^*$ consisting of the empty set and the k-dimensional rsc rectangles $(\boldsymbol{a}, \boldsymbol{b}]$ with $\boldsymbol{a}, \boldsymbol{b} \in \mathbb{Q}^k$. Also, recall that $\mathcal{B}^k = \sigma(\mathcal{A}_k^*)$. By the Uniqueness Theorem from Section 3.5, we need only show that $\mu_1(A) = \mu_2(A)$ for all $A \in \mathcal{A}_k^*$ to obtain that $\mu_1 = \mu_2$, which will prove $H(1)$. The case $A = \varnothing$ is trivially true; we need only show that $\mu_1((\boldsymbol{a}, \boldsymbol{b}]) = \mu_2((\boldsymbol{a}, \boldsymbol{b}])$ for all $(\boldsymbol{a}, \boldsymbol{b}] \in \mathcal{A}_k^*$. By Claim 8, we may write any generic $(\boldsymbol{a}, \boldsymbol{b}] \in \mathcal{A}_k^*$ as a finite disjoint union of n cubes $\bigcup_{i=1}^n C_i$, where each cube has sides of some common length c that are parallel to the coordinate axes of $\mathbb{R}^k$. From this, it follows that there is $\boldsymbol{x}_i \in \mathbb{R}^k$ such that $C_i = (\boldsymbol{0}, c\boldsymbol{1}] \oplus \boldsymbol{x}_i$, $i = 1, \cdots, n$. Next, observe

that

$$\begin{aligned}\mu_1((\boldsymbol{a},\boldsymbol{b}]) = \lambda_k\left(\boldsymbol{T}_1(\boldsymbol{a},\boldsymbol{b}]\right) = \lambda_k\left(\boldsymbol{T}_1\bigcup_{i=1}^{n} C_i\right) &= \lambda_k\left(\bigcup_{i=1}^{n}\boldsymbol{T}_1 C_i\right)\\ &= \sum_{i=1}^{n}\lambda_k\left(\boldsymbol{T}_1 C_i\right)\\ &= \sum_{i=1}^{n}\lambda_k\left(\boldsymbol{T}_1\left((\mathbf{0},c\mathbf{1}]\oplus\boldsymbol{x}_i\right)\right)\\ &= \sum_{i=1}^{n}\lambda_k\left(\boldsymbol{T}_1(\mathbf{0},c\mathbf{1}]\oplus\boldsymbol{T}_1\boldsymbol{x}_i\right)\\ &= \sum_{i=1}^{n}\lambda_k\left(\boldsymbol{T}_1(\mathbf{0},c\mathbf{1}]\right)\\ &= n\lambda_k\left(\boldsymbol{T}_1(\mathbf{0},c\mathbf{1}]\right),\end{aligned}$$

where the fourth equality follows because the disjointness of $C_1, \cdots, C_n$ and the nonsingularity of $\boldsymbol{T}$ force the disjointness of $\boldsymbol{T}_1C_1, \cdots, \boldsymbol{T}_1C_n$, and the seventh equality follows because λ_k is translation invariant. We next observe that

$$\begin{aligned}\mu_2((\boldsymbol{a},\boldsymbol{b}]) = |\det\boldsymbol{T}_1|\lambda_k((\boldsymbol{a},\boldsymbol{b}]) &= |\det\boldsymbol{T}_1|\lambda_k\left(\bigcup_{i=1}^{n} C_i\right)\\ &= |\det\boldsymbol{T}_1|\sum_{i=1}^{n}\lambda_k(C_i)\\ &= |\det\boldsymbol{T}_1|\sum_{i=1}^{n}c^k\\ &= |\det\boldsymbol{T}_1|nc^k.\end{aligned}$$

Because n is finite and $\lambda_k(\boldsymbol{T}_1(0,c]^k) = |\det(\boldsymbol{T}_1)|c^k$ [Claim 9], it follows that $\mu_1((\boldsymbol{a},\boldsymbol{b}]) = \mu_2((\boldsymbol{a},\boldsymbol{b}])$ for each $(\boldsymbol{a},\boldsymbol{b}] \in \mathcal{A}_k^*$, and thus $H(1)$ is established.

We now show that $H(N)$ is true for any positive integer $N \geq 2$. Suppose for a fixed $N \geq 2$ that $H(1), \cdots, H(N)$ are true. Pick arbitrary $A \in \mathcal{B}^k$ and let $A^* = \boldsymbol{T}_N\cdots\boldsymbol{T}_1A$, so that $A^* \in \mathcal{B}^k$ by $H(N)$. We have

$$\begin{aligned}\lambda_k(\boldsymbol{T}_{N+1}\cdots\boldsymbol{T}_1A) = \lambda_k(\boldsymbol{T}_{N+1}A^*) &= |\det\boldsymbol{T}_{N+1}|\lambda_k(A^*)\\ &= |\det\boldsymbol{T}_{N+1}|\lambda_k(\boldsymbol{T}_N\cdots\boldsymbol{T}_1A)\\ &= |\det\boldsymbol{T}_{N+1}||\det(\boldsymbol{T}_N\cdots\boldsymbol{T}_1)|\lambda_k(A)\\ &= |\det(\boldsymbol{T}_{N+1}\boldsymbol{T}_N\cdots\boldsymbol{T}_1)|\lambda_k(A),\end{aligned}$$

where the second equality follows from $H(1)$ and the fourth equality follows from $H(N)$. This chain of equalities shows that if $H(1), \cdots, H(N)$ are true, then $H(N+1)$ is true. Therefore, $H(N)$ is true for all $N \in \mathbb{N}$, and thus (4.4) follows, proving part (i) of the claim when $k \geq 2$.

Using (i), we now prove part (ii) for general $k \in \mathbb{N}$. Let $A \in \mathcal{L}^k$. Then $A = B \cup C$, where $B \in \mathcal{B}^k$ and C is contained in some $N \in \mathcal{B}^k$ with $\lambda_k(N) = 0$. It follows that $\boldsymbol{T}A = \boldsymbol{T}(B \cup C) = \boldsymbol{T}B \cup \boldsymbol{T}C$, where $\boldsymbol{T}B \in \mathcal{B}^k$ and $\boldsymbol{T}C \subseteq \boldsymbol{T}N$ with $\lambda_k(\boldsymbol{T}N) = |\det \boldsymbol{T}|\lambda_k(N) = |\det \boldsymbol{T}| \times 0 = 0$ (by (4.4)). It follows that $\boldsymbol{T}A$ is the union of a $\mathcal{B}^k$-set and a subset of a $\mathcal{B}^k$-set with λ_k-measure zero, and so $\boldsymbol{T}A \in \mathcal{L}^k$. Next, we have

$$\bar{\lambda}_k(\boldsymbol{T}A) = \lambda_k(\boldsymbol{T}B) = |\det \boldsymbol{T}|\lambda_k(B) = |\det \boldsymbol{T}|\bar{\lambda}_k(A),$$

where the first and third equalities follow by the definition of $\bar{\lambda}_k$, and the second equality follows by (4.4). Formula (4.5) is therefore established, and the proof of (ii) is complete. This completes the entire proof. ■

Example 1. A *rigid motion* consists of a rotation about the origin and a translation. The previous claim may be used to show that Lebesgue measure is invariant under rigid motions. To see this, let $\boldsymbol{T} : \mathbb{R}^k \to \mathbb{R}^k$ denote a rigid motion. Therefore, there exists some $\boldsymbol{x}_0 \in \mathbb{R}^k$ and an orthogonal [hence nonsingular] matrix $\boldsymbol{O}$ such that $\boldsymbol{T}\boldsymbol{x} = \boldsymbol{O}\boldsymbol{x} + \boldsymbol{x}_0$ for every $\boldsymbol{x} \in \mathbb{R}^k$. For any $A \in \mathcal{B}^k$, we have that $\boldsymbol{O}A$ and hence $\boldsymbol{O}A \oplus \boldsymbol{x}_0 \in \mathcal{B}^k$, and thus $\lambda_k(\boldsymbol{O}A \oplus \boldsymbol{x}_0) = \lambda_k(\boldsymbol{O}A) = |\det \boldsymbol{O}|\lambda_k(A) = \lambda_k(A)$, where the second equality follows by (i) of the previous claim and the final equality follows since orthogonal matrices have determinant ± 1. The fact that this result holds true accords perfectly with intuition, and the analogous result for Lebesgue sets holds by using (ii) of the previous claim and a similar argument.

Example 2. Reflections of Borel sets about $(k-1)$-dimensional hyperplanes that are perpendicular to an axis of $\mathbb{R}^k$ are Borel sets and Lebesgue measure is preserved under such reflections. Reflections of Borel sets about arbitrary $(k-1)$-dimensional hyperplanes are Borel sets, and Lebesgue measure is preserved under such reflections. This latter claim is true because a reflection about an arbitrary $(k-1)$-dimensional hyperplane is obtained by rotating about the origin, reflecting about a $(k-1)$-dimensional plane perpendicular to an axis of $\mathbb{R}^k$, translating, and then rotating again about the origin.[5]

We would now like to discuss hyperplanes of dimension less than k. When $k = 1$, the only hyperplanes are the points on the real line. When $k = 2$, the hyperplanes are the lines and the points in the plane. When $k = 3$, the hyperplanes are the two-dimensional planes, all lines, and all points in space. We have the following definition.

Definition. Let $k \in \mathbb{N}$ and let $j \in \{0, \cdots, k-1\}$. A subset $A \subseteq \mathbb{R}^k$ is a hyperplane of dimension j iff there exists a singular $k \times k$ matrix $\boldsymbol{T}$ such that (i) the dimension of the column space of $\boldsymbol{T}$ is j, and (ii) $\boldsymbol{T}\mathbb{R}^k = A \oplus \boldsymbol{x}$ for some $\boldsymbol{x} \in \mathbb{R}^k$.

[5]Not rotating at all is considered a rotation, and not translating at all is considered a translation by $\mathbf{0}$.

Since hyperplanes are easily visualizable, intuition demands that hyperplanes are Borel sets. Furthermore, hyperplanes should have zero Lebesgue measure. We shall verify both of these conjectures.

Claim 12. *Let $k \in \mathbb{N}$. Then $\mathbb{R}^j \times \{\mathbf{0}\} \in \mathcal{B}^k$ and $\lambda_k(\mathbb{R}^j \times \{\mathbf{0}\}) = 0$ for each $j = 1, \cdots, k-1$, where $\mathbf{0}$ is the origin of $\mathbb{R}^{k-j}$.*

Proof: Let $A = \mathbb{R}^j \times \{\mathbf{0}\}$. Then $A \in \mathcal{B}^j \times \mathcal{B}^{k-j} \subseteq \mathcal{B}^k$ by Claim 2 of Section 1.5. Next, let $A_n = (-n, n)^j \times \{\mathbf{0}\}$ for each $n \in \mathbb{N}$. Therefore, we have for every $n \in \mathbb{N}$ that $A_n \in \mathcal{B}^j \times \mathcal{B}^{k-j} \subseteq \mathcal{B}^k$ and

$$\lambda_k(A_n) = \lambda_k((-n, n)^j \times \{\mathbf{0}\}) = \lambda_j((-n, n)^j)\lambda_{k-j}(\{\mathbf{0}\}) = (2n)^j \times 0 = 0,$$

where the second equality follows from Claim 4 of Section 4.2. Since the A_n's increase to A, (M8) gives $\lim_n \lambda_k(A_n) = \lambda_k(A)$, so that $\lambda_k(A) = 0$. ■

Claim 13. *Let $k \in \mathbb{N}$ and $j \in \{0, \cdots, k-1\}$. Every j-dimensional hyperplane H is in $\mathcal{B}^k$, and $\lambda_k(H) = 0$.*

Proof: The case of $k = 1$ and $j = 0$ is trivial. Next, let $k \geq 2$ and pick arbitrary $j \in \{0, \cdots, k-1\}$. Let A denote a j-dimensional hyperplane passing through the origin, and find a $k \times k$ matrix $\boldsymbol{T}$ such that (i) the columns of $\boldsymbol{T}$ span A and (ii) $\boldsymbol{T}\mathbb{R}^k = A$. Let $\{\boldsymbol{e}_1, \cdots, \boldsymbol{e}_k\}$ denote an orthonormal basis of $\mathbb{R}^k$ such that $\{\boldsymbol{e}_1, \cdots, \boldsymbol{e}_j\}$ is an orthonormal basis for A. Let $\boldsymbol{S} = (\boldsymbol{e}_1, \cdots, \boldsymbol{e}_k)$, so that $\boldsymbol{S}$ is the $k \times k$ orthonormal matrix [hence nonsingular] whose columns consist of the basis vectors $\boldsymbol{e}_1, \cdots, \boldsymbol{e}_k$. By construction, $\boldsymbol{S}(\mathbb{R}^j \times \{\mathbf{0}\}) = A$, where $\mathbf{0}$ is the origin of $\mathbb{R}^{k-j}$. By Claims 11 and 12, $\mathbb{R}^j \times \{\mathbf{0}\} \in \mathcal{B}^k$; since $\boldsymbol{S}$ is nonsingular we have that $\boldsymbol{S}(\mathbb{R}^j \times \{\mathbf{0}\}) \in \mathcal{B}^k$, hence $A \in \mathcal{B}^k$. Using Claim 11, we have

$$\lambda_k(A) = \lambda_k(\boldsymbol{S}(\mathbb{R}^j \times \{\mathbf{0}\})) = |\det \boldsymbol{S}|\lambda_k(\mathbb{R}^j \times \{\mathbf{0}\}) = 1 \times 0 = 0,$$

which proves the claim for hyperplanes passing through the origin.

We now deal with general hyperplanes. A general hyperplane H can be represented as $A \oplus \boldsymbol{x}$ for some hyperplane A passing through the origin and some $\boldsymbol{x} \in \mathbb{R}^k$, so that the previous paragraph gives $H \in \mathcal{B}^k$. By translation invariance and the previous paragraph, we have $\lambda_k(H) = \lambda_k(A) = 0$. This completes the proof. ■

In stating and proving that $\text{card}(\mathcal{L}^k) = 2^c$ for each $k \in \mathbb{N}$, we invoked the existence of an uncountable $\mathcal{B}^k$-set with λ_k-measure of zero. The previous claims justify this invocation when $k \geq 2$. For the case $k = 1$, the only hyperplanes are the singletons, and while every singleton has λ-measure of zero, it is far from uncountable. The $k = 1$ accordingly takes more work, being handled in our presentation of the Cantor Set found in Section 6.

To summarize this entire section, we may say after a substantial amount of toil that Lebesgue measure as rigorously constructed accords with our intuition. The measure whose existence we have tacitly assumed from our earliest days on through analysis does in fact exist, and has a rigorous logical foundation.

Exercises.

1. The inverse of a special matrix is a special matrix.

2. Any nonsingular $k \times k$ matrix $\boldsymbol{T}$ may be written as a finite product $\boldsymbol{T}_N \cdots \boldsymbol{T}_2 \boldsymbol{T}_1$ of nonsingular $k \times k$ matrices, each matrix $\boldsymbol{T}_j$ being nonsingular and freely denoting the matrix corresponding to any one of the special linear transformations from $\mathbb{R}^k$ to $\mathbb{R}^k$ described earlier. [Suggestion: find a finite sequence $\boldsymbol{T}_1, \cdots, \boldsymbol{T}_{N_1}$ of special matrices such that

$$\boldsymbol{T}_{N_1} \cdots \boldsymbol{T}_1 \boldsymbol{T} = \begin{pmatrix} 1 & -- \\ -- & -- \end{pmatrix}.$$

Then find another finite sequence of matrices $\boldsymbol{T}_{N_1+1}, \cdots, \boldsymbol{T}_{N_1+N_2}$ such that

$$\boldsymbol{T}_{N_1+N_2} \cdots \boldsymbol{T}_1 \boldsymbol{T} = \begin{pmatrix} 1 & ----- \\ 0 & ----- \\ \vdots & \\ 0 & ----- \end{pmatrix}.$$

Then, find another finite sequence of matrices $\boldsymbol{T}_{N_2+1}, \cdots, \boldsymbol{T}_{N_2+N_3}$ such that

$$\boldsymbol{T}_{N_1+N_2} \cdots \boldsymbol{T}_1 \boldsymbol{T} \boldsymbol{T}_{N_2+1} \cdots \boldsymbol{T}_{N_2+N_3} = \begin{pmatrix} 1 & 0 & \cdots & 0 \\ 0 & & & \\ \vdots & & & \\ 0 & & & \end{pmatrix}.$$

Argue that the above matrix contains a $(k-1) \times (k-1)$ *invertible* submatrix [the blank spot in the above matrix] to which we can repeat this entire procedure *without disturbing the 1's and 0's of the above matrix* in order to obtain a matrix of the form

$$\begin{pmatrix} 1 & 0 & 0 & \cdots & 0 \\ 0 & 1 & 0 & \cdots & 0 \\ 0 & 0 & & & \\ \vdots & \vdots & & & \\ 0 & 0 & & & \end{pmatrix}.$$

Argue that we may carry on this procedure until we have [written symbolically] a sequence $\boldsymbol{T}_1, \cdots, \boldsymbol{T}_{\alpha+\beta}$ of special matrices such that

$$\boldsymbol{T}_\alpha \cdots \boldsymbol{T}_1 \boldsymbol{T} \boldsymbol{T}_{\alpha+1} \cdots \boldsymbol{T}_{\alpha+\beta} = \boldsymbol{I}_k.$$

Therefore,

$$\boldsymbol{T} = \boldsymbol{T}_1^{-1} \cdots \boldsymbol{T}_\alpha^{-1} \boldsymbol{T}_{\alpha+\beta}^{-1} \cdots \boldsymbol{T}_{\alpha+1}^{-1}.$$

Since the inverse of a special matrix is a special matrix by Exercise 1, the proof is complete. This gives an outline of the proof of Claim 10.]

3. A mapping $F : \mathbb{R}^k \to \mathbb{R}^k$ is called a *rigid motion [on $\mathbb{R}^k$]* iff there exists $\boldsymbol{x}_0 \in \mathbb{R}^k$ and an orthonormal matrix $\boldsymbol{O}$ such that $F(\boldsymbol{x}) = \boldsymbol{O}\boldsymbol{x} + \boldsymbol{x}_0$ for all $\boldsymbol{x} \in \mathbb{R}^k$.
(a) Translations, rotations, permutations, and reflections are rigid motions.
(b) Let I denote the identity mapping [$I(\boldsymbol{x}) = \boldsymbol{x}$ for all $\boldsymbol{x} \in \mathbb{R}^k$], and let F, G, and H denote rigid motions. We have:

(i) $F \circ G$ is a rigid motion.

(ii) $(F \circ G) \circ H = F \circ (G \circ H)$.

(iii) $F \circ I = I \circ F = F$.

(iv) There exists a rigid motion J such that $F \circ J = J \circ F = I$.

Thus, the set of rigid motions on $\mathbb{R}^k$ forms a group with respect to the binary operation of composition.

4. Let $\boldsymbol{a}_1, \cdots, \boldsymbol{a}_k \in \mathbb{R}^k$, and define

$$A = \{t_1\boldsymbol{a}_1 + \cdots + t_k\boldsymbol{a}_k : \ 0 \leq t_i \leq 1 \text{ for all } 1 \leq i \leq k\}.$$

Use the main result of this section to show that $\lambda_k(A)$ is the absolute value of the determinant of the matrix whose columns are $\boldsymbol{a}_1, \cdots, \boldsymbol{a}_k$.

5. Consider a parallelogram with vertices $(0,0)$, $(a_1, 0)$, (h, a_2), and $(a_1 + h, a_2)$, where $a_1, a_2 > 0$ and $0 \leq h \leq a_1$. Find the matrix $\boldsymbol{T}$ that maps the unit square onto this parallelogram, and obtain the formula for the area of a parallelogram by using this section's main result.

6. Use Exercise 14 of Section 2.2 to give a quick proof that the Lebesgue measure of any hyperplane must be zero. [Any disjoint collection of $\mathcal{B}^k$-sets having positive λ_k-measure must be at most countable by Exercise 14 in Section 2.2.]

7*. If $\boldsymbol{T} : \mathbb{R}^k \to \mathbb{R}^k$ is a singular linear transformation, then we do not necessarily have $\boldsymbol{T}A^c = (\boldsymbol{T}A)^c$ for every $A \subseteq \mathbb{R}^k$. Also, we do not necessarily have that $\{\boldsymbol{T}A_\gamma : \ \gamma \in \Gamma\}$ is a disjoint collection of subsets of $\mathbb{R}^k$ whenever $\{A_\gamma : \ \gamma \in \Gamma\}$ is a disjoint collection of subsets of $\mathbb{R}^k$.

8. Let $\boldsymbol{T}$ denote a nonsingular $k \times k$ matrix, and let $\gamma_1 \geq \cdots \geq \gamma_k$ denote the [positive] eigenvalues of $\boldsymbol{T}'\boldsymbol{T}$. Then $\gamma_k\|\boldsymbol{x}\| \leq \|\boldsymbol{T}\boldsymbol{x}\| \leq \gamma_1\|\boldsymbol{x}\|$ for every $\boldsymbol{x} \in \mathbb{R}^k$. This generalizes Claim 1.

9*. Prove Claim 7 by adapting the proof of Claim 6 accordingly.

10*. Prove Claim 11(i) when $k = 1$. [Let $T(x) = px$ for some $p \neq 0$. By Claim 3, $A \in \mathcal{B}$ forces $pA = \{pa : \ a \in A\} \in \mathcal{B}$. Let $\nu(A) = |p|\lambda(A)$. Show that ν is a measure with domain $\mathcal{B}$, then show that $\nu(A) = \lambda(pA)$ whenever $A \in \mathcal{A}_k$. Appeal to the Uniqueness Theorem in Section 3.5 to finish things.]

4.5 THE EXISTENCE OF NON-LEBESGUE SETS

In the discussion of the structure of $\mathcal{L}^k$ found in Section 2, we assumed the existence of a bounded subset of $\mathbb{R}^k$ that was not in $\mathcal{L}^k$. This section demonstrates the existence of such sets in k-dimensional Euclidean space.[6] In addition, we show that every $A \subseteq \mathbb{R}^k$ with $\lambda_k^*(A) > 0$ contains a subset that is not in $\mathcal{L}^k$. The point of this is that sets outside of $\mathcal{L}^k$, as hard as they are to visualize and comprehend, are still ubiquitous.

[6]The exercises in Section 3.10 and Section 4.3 provided outlines for constructing sets not in $\mathcal{L}$.

Recall that the existence of sets outside $\mathcal{L}^k$ is the price we must pay in the *problem of measure* outlined in the beginning of Chapter 3. We can't have a measure that (a) accords with intuition concerning k-dimensional volume and (b) is applicable to every subset of $\mathbb{R}^k$, and so we agreed to discuss only the Lebesgue measures of certain sets. The words *certain sets* ended up meaning those sets in $\mathcal{B}^k$ and $\mathcal{L}^k$.

The construction will be given by a series of modest claims, and will rely on the Approximation Theorem in Section 4.2 as well as the results in Section 4.3 concerning the translation invariance of Lebesgue measure.

Claim 1. *For each $\boldsymbol{x}$ and $\boldsymbol{y}$ in $\mathbb{R}^k$, write $\boldsymbol{x} \sim \boldsymbol{y}$ iff there is $\boldsymbol{r} \in \mathbb{Q}^k$ such that $\boldsymbol{x} + \boldsymbol{r} = \boldsymbol{y}$. Then $\sim$ is an equivalence relation.*

The proof of Claim 1 is a routine exercise and is left to the reader. The relation $\sim$ determines an infinite number of equivalence classes, each class itself being infinite.

We now invoke the Axiom of Choice (AC) to form a set $E \subseteq \mathbb{R}^k$ consisting of exactly one element from each equivalence class. We now consider the collection $\{E \oplus \boldsymbol{r} : \boldsymbol{r} \in \mathbb{Q}^k\}$ of the translates of E by the elements of $\mathbb{Q}^k$.

Claim 2. *If $\boldsymbol{r}_1, \boldsymbol{r}_2 \in \mathbb{Q}^k$ are distinct, then $E \oplus \boldsymbol{r}_1$ and $E \oplus \boldsymbol{r}_2$ are disjoint.*

Proof: Let $\boldsymbol{r}_1, \boldsymbol{r}_2 \in \mathbb{Q}^k$ with $\boldsymbol{r}_1 \neq \boldsymbol{r}_2$, and suppose to the contrary that there is $\boldsymbol{y}$ in both $E \oplus \boldsymbol{r}_1$ and $E \oplus \boldsymbol{r}_2$. Then there exist $\boldsymbol{x}_1, \boldsymbol{x}_2 \in E$ such that $\boldsymbol{y} = \boldsymbol{x}_1 + \boldsymbol{r}_1$ and $\boldsymbol{y} = \boldsymbol{x}_2 + \boldsymbol{r}_2$. It follows that $\boldsymbol{x}_1 + (\boldsymbol{r}_1 - \boldsymbol{r}_2) = \boldsymbol{x}_2$. Since $\boldsymbol{r}_1 - \boldsymbol{r}_2 \in \mathbb{Q}^k$, we have $\boldsymbol{x}_1 \sim \boldsymbol{x}_2$, so that $\boldsymbol{x}_1$ and $\boldsymbol{x}_2$ are in the same equivalence class. Since $\boldsymbol{x}_1, \boldsymbol{x}_2 \in E$, and since E consists of *exactly one* member of each equivalence class, we must have $\boldsymbol{x}_1 = \boldsymbol{x}_2$, hence $\boldsymbol{r}_1 = \boldsymbol{r}_2$, a contradiction. ∎

Claim 3. $\mathbb{R}^k = \bigcup\{(E \oplus \boldsymbol{r}) : \boldsymbol{r} \in \mathbb{Q}^k\}$, *the union being disjoint.*

Proof: The disjointness of the union follows from Claim 2. Let $\boldsymbol{y} \in \mathbb{R}^k$, so that $\boldsymbol{y}$ is in some equivalence class. Since E contains one member from every equivalence class, there is $\boldsymbol{x} \in E$ such that $\boldsymbol{y} \sim \boldsymbol{x}$. Since $\boldsymbol{y} \sim \boldsymbol{x}$, there exists $\boldsymbol{r}^* \in \mathbb{Q}^k$ with $\boldsymbol{y} = \boldsymbol{x} + \boldsymbol{r}^*$, hence $\boldsymbol{y} \in E \oplus \boldsymbol{r}^*$, which gives $\boldsymbol{y} \in \bigcup_{\boldsymbol{r} \in \mathbb{Q}^k}(E \oplus \boldsymbol{r})$. This gives $\mathbb{R}^k \subseteq \bigcup_{\boldsymbol{r} \in \mathbb{Q}^k}(E \oplus \boldsymbol{r})$. The reverse inclusion is trivial, and so the proof is complete. ∎

Claim 4. *For the set E as given, $\lambda_k^*(E) > 0$.*

Proof: Suppose to the contrary that $\lambda_k^*(E) = 0$. Then by Claim 1 of Section 3.7 [or see the paragraph following the definition of λ_k^* given in Section 4.1] we have $E \in \mathcal{L}^k$, and $\bar{\lambda}_k(E) = \lambda_k^*(E) = 0$. By our results on translation invariance, we obtain that $E \oplus \boldsymbol{r} \in \mathcal{L}^k$ and $\bar{\lambda}_k(E \oplus \boldsymbol{r}) = \bar{\lambda}_k(E) = 0$ for each $\boldsymbol{r} \in \mathbb{Q}^k$. Using these facts and Claim 3, we obtain

$$\bar{\lambda}_k(\mathbb{R}^k) = \bar{\lambda}_k\left(\bigcup_{\boldsymbol{r} \in \mathbb{Q}^k}(E \oplus \boldsymbol{r})\right) = \sum_{\boldsymbol{r} \in \mathbb{Q}^k} \bar{\lambda}_k(E \oplus \boldsymbol{r}) = \sum_{\boldsymbol{r} \in \mathbb{Q}^k} \bar{\lambda}_k(E) = 0,$$

a contradiction. Accordingly, we must have $\lambda_k^*(E) > 0$. ■

Claim 5. *For the set E as given,* $\sup\{\lambda_k(K) : K \text{ compact}, K \subseteq E\} = 0$.

Proof: Let $K \subseteq E$ be compact,[7] and let $A = \{\boldsymbol{r} \in \mathbb{Q}^k : \|\boldsymbol{r}\| < 1\}$ [a countable set]. Finally, write $K \oplus A = \{\boldsymbol{x} + \boldsymbol{r} : \boldsymbol{x} \in K, \boldsymbol{r} \in A\}$. Since compact sets are bounded, the boundedness of A forces $K \oplus A$ to be bounded. Also, $K \oplus A = \bigcup_{\boldsymbol{r} \in A}(K \oplus \boldsymbol{r})$, where the union is disjoint since $K \oplus \boldsymbol{r} \subseteq E \oplus \boldsymbol{r}$ and the collection $\{E \oplus \boldsymbol{r} : \boldsymbol{r} \in \mathbb{Q}^k\}$ is disjoint [Claim 2]. Finally, for each $\boldsymbol{r} \in A$ we have $K \oplus \boldsymbol{r} \in \mathcal{B}^k$ since $K \in \mathcal{B}^k$, whence $K \oplus A \in \mathcal{B}^k$. It follows that

$$
\begin{aligned}
+\infty > \lambda_k(K \oplus A) = \lambda_k\left(\bigcup_{\boldsymbol{r} \in A}(K \oplus \boldsymbol{r})\right) &= \sum_{\boldsymbol{r} \in A} \lambda_k(K \oplus \boldsymbol{r}) \\
&= \sum_{\boldsymbol{r} \in A} \lambda_k(K).
\end{aligned}
$$

[The inequality holds since $K \oplus A$ is bounded.] Since $\sum_{\boldsymbol{r} \in A} \lambda_k(K) < +\infty$, we must have $\lambda_k(K) = 0$. Since K is a generic compact subset of E, the proof is complete. ■

Claim 6. *The set E as given is not in* $\mathcal{L}^k$.

Proof: Suppose to the contrary that $E \in \mathcal{L}^k$. By the Approximation Theorem in Section 4.2 and Claim 5 we have $\bar{\lambda}_k(E) = 0$. However, $E \in \mathcal{L}^k$ implies that $\bar{\lambda}_k(E) = \lambda^*(E) > 0$ by Claim 4. Assuming that $E \in \mathcal{L}^k$ yields a contradiction; the only conclusion is that $E \notin \mathcal{L}^k$. ■

This construction is a standard variant of the classical construction given by Vitali in 1905, and the proof as presented owes its structure to the most excellent presentation of Jones. Most discussions of the existence of a non-$\mathcal{L}^k$-set are variants of this presentation[8].

We had to explicitly resort to (AC) to construct such a set, and constructions involving (AC) invite the question of whether there is a more explicit presentation of a set not in $\mathcal{L}^k$. It turns out that one cannot construct "bad" sets – sets that are so pathological that [say] they fail to be outside of $\mathcal{L}^k$ – without using (AC) or one of its equivalents. In 1970, Solovay published a metamathematical paper[9] that [among other things] shows that (AC) is required to construct any set outside of $\mathcal{L}^k$.

A natural question is this: is such a set $E \notin \mathcal{L}^k$ a "rarity," or are there "many other" such sets? This question is not trivial, for both $\mathcal{L}^k$ and $2^{\mathbb{R}^k}$ have the identical cardinal number $2^{\mathfrak{c}}$. [See Exercise 6 of Section 4.1.] The following claim asserts that sets outside $\mathcal{L}^k$ are rather common.

[7]There exist compact subsets of E, namely, the singletons.
[8]Compare this approach with that similarly given by the exercises in Section 4.3. For a different approach — but one that still requires (AC) — see the exercises for Section 3.10.
[9]Solovay, R. M. (1970) A Model of Set Theory In Which Every Set of Reals is Lebesgue Measurable, *Ann. Math.* **92**, 1-56.

Claim 7. *Let $A \subseteq \mathbb{R}^k$ be such that $\lambda_k^*(A) > 0$. Then there exists a subset of A that fails to be in $\mathcal{L}^k$.*

Proof: For each $D \in \mathcal{L}^k$, let $\hat{\lambda}_k(D) = \sup\{\lambda_k(K) : K \subseteq D, K \text{ compact}\}$. It is straightforward to verify that $\hat{\lambda}$ is monotone and translation invariant. Let $A \subseteq \mathbb{R}^k$ be such that $\lambda_k^*(A) > 0$. By Claim 3, we have

$$A = \mathbb{R}^k \cap A = \bigcup_{\boldsymbol{r} \in \mathbb{Q}^k} (E \oplus \boldsymbol{r}) \cap A = \bigcup_{\boldsymbol{r} \in \mathbb{Q}^k} ((E \oplus \boldsymbol{r}) \cap A),$$

where the unions are disjoint and $E \notin \mathcal{L}^k$ is the set constructed earlier. Next, observe that

$$0 < \lambda_k^*(A) = \lambda_k^* \left(\bigcup_{\boldsymbol{r} \in \mathbb{Q}^k} ((E \oplus \boldsymbol{r}) \cap A) \right) \leq \sum_{\boldsymbol{r} \in \mathbb{Q}^k} \lambda_k^*((E \oplus \boldsymbol{r}) \cap A).$$

It follows that $\lambda_k^*((E \oplus \boldsymbol{r}^*) \cap A) > 0$ for some $\boldsymbol{r}^* \in \mathbb{Q}^k$. Next, let us write $B = (E \oplus \boldsymbol{r}^*) \cap A$, so that $B \subseteq A$ and $\lambda_k^*(B) > 0$. By the monotonicity and translation invariance of $\hat{\lambda}_k$, we have

$$\hat{\lambda}_k(B) = \hat{\lambda}_k((E \oplus \boldsymbol{r}^*) \cap A) \leq \hat{\lambda}_k(E \oplus \boldsymbol{r}^*) = \hat{\lambda}_k(E) = 0,$$

where the last equality follows from the proof of Claim 5. Were $B \in \mathcal{L}^k$, then we would have $\bar{\lambda}_k(B) = \hat{\lambda}_k(B) = 0$ [Approximation Theorem]. But at the same time we have $\bar{\lambda}_k(B) = \lambda_k^*(B) > 0$, a contradiction. We conclude that $B \notin \mathcal{L}^k$, completing the proof. ∎

The previous claim shows that the assumption of a bounded $A \subseteq \mathbb{R}^k$ with $A \notin \mathcal{L}^k$ [as made in Section 4.2] was not a gratuitous assumption, for any bounded k-dimensional rectangle has positive λ_k^*-measure, and hence there is a subset of this rectangle that is outside of $\mathcal{L}^k$; this subset is obviously bounded.

These are the standard results concerning the existence of non-Lebesgue sets. From this construction, one who adopts the Continuum Hypothesis (CH) may vary which point is chosen to represent each of the $\mathfrak{c}$ equivalence classes[10] formed by Claim 1, and one can obtain $2^{\mathfrak{c}}$ additional sets outside of $\mathcal{L}^k$. One can then obtain $2^{\mathfrak{c}}$ more sets outside $\mathcal{L}^k$ by considering the $2^{\mathfrak{c}}$ subsets of $\mathbb{R}^k$ with positive outer measure and the construction therein. Therefore, if one permits (AC) and (CH) in the construction, non-Lebesgue sets are hardly rare subsets of $\mathbb{R}^k$. Since $\mathcal{B}^k \subseteq \mathcal{L}^k$, we have that non-Borel sets are also a large constituency of the collection of all subsets of $\mathbb{R}^k$.

[10]The set $E \notin \mathcal{L}^k$ is formed by choosing exactly one point from each equivalence class. If there are $\aleph_0$ equivalence classes, then $\text{card}(E) = \aleph_0$, so $E \in \mathcal{B}^k$, a contradiction. Under the Continuum Hypothesis, $\text{card}(E) > \aleph_0$ forces $\text{card}(E) = \mathfrak{c}$, hence there are $\mathfrak{c}$ equivalence classes.

Exercises.

1. Let $E \notin \mathcal{L}^k$ denote the set constructed in this section, and let $A \subseteq E$ be such that $A \in \mathcal{L}^k$. Then $\bar{\lambda}_k(A) = 0$. [This and the next exercise can be answered handily if the construction given in this section is understood.]

2. There exists no translation invariant measure μ defined for all subsets of $\mathbb{R}^k$ such that $\mu((\boldsymbol{a}, \boldsymbol{b}]) = \prod_{i=1}^k (b_i - a_i)$ for all bounded k-dimensional rsc rectangles $(\boldsymbol{a}, \boldsymbol{b}]$.

3. Here is an alternative but quite similar construction of a set outside $\mathcal{L}^k$. We shall consider $(0, 1]$ as our space. Let $\dot{+}$ denote addition modulo 1: for $x, y \in (0, 1]$,

$$x \dot{+} y = \begin{cases} x + y & \text{if } x + y \leq 1 \\ x + y - 1 & \text{if } x + y > 1. \end{cases}$$

For each $A \subseteq (0, 1]$ and $x \in (0, 1]$, we shall [for this exercise only] write $A \oplus x$ as translation modulo 1: $A \oplus x = \{a \dot{+} x : a \in A\}$. Finally, [for this exercise only] we shall let $\mathbb{Q}$ denote the rational numbers of $(0, 1]$.
(a) For $x, y \in (0, 1]$ write $x \sim y$ iff there is $r \in \mathbb{Q}$ such that $x \dot{+} r = y$. Then $\sim$ is an equivalence relation.
(b) Let F denote the set formed by choosing exactly one element from each equivalence class [use (AC)]. Show that $(F \oplus r_1) \cap (F \oplus r_2)$ is empty if r_1 and r_2 are distinct elements of $\mathbb{Q}$.
(c) $(0, 1] = \bigcup_{r \in \mathbb{Q}} (F \oplus r)$, where the union is disjoint.
(d) $F \notin \mathcal{L}$ and $0 < \lambda^*(F) \leq 1$. [Argue by contradiction: if $F \in \mathcal{L}$, then

$$1 = \bar{\lambda}((0, 1]) = \sum_{r \in \mathbb{Q}} \bar{\lambda}(F \oplus r) = \sum_{r \in \mathbb{Q}} \bar{\lambda}(F).$$

If $\bar{\lambda}(F) = 0$, we obtain $1 = 0$. If $\bar{\lambda}(F) > 0$, then we obtain $1 = +\infty$. Conclude that $\bar{\lambda}(F)$ cannot exist, which means that $F \notin \mathcal{L}$. For the other part, since $F \subseteq (0, 1]$, clearly $\lambda^*(F) \leq 1$. To show that $\lambda^*(F) > 0$, write

$$1 = \lambda^*((0, 1]) = \lambda^* \left(\bigcup_{r \in \mathbb{Q}} (F \oplus r) \right) \leq \sum_{r \in \mathbb{Q}} \lambda^*(F \oplus r) = \sum_{r \in \mathbb{Q}} \lambda^*(F).]$$

(e) If there exists a translation invariant measure μ defined on a σ-field on $(0, 1]$ that includes F and is closed under translations, then $\mu((0, 1])$ is either 0 or $+\infty$.
(f) Show that there can exist no translation invariant measure defined on all subsets of $(0, 1]$ that agrees with $\bar{\lambda}$ on the Lebesgue sets of $(0, 1]$.

4. Prove the following assertions.
(a) There exists a sequence $\{E_n\}_{n=1}^\infty$ of disjoint sets [not necessarily in $\mathcal{L}$] such that $\lambda^*(\bigcup_{n=1}^\infty E_n) < \sum_{n=1}^\infty \lambda^*(E_n)$. [First, show that λ^* is translation invariant. Let $\{r_n\}_{n=1}^\infty$ denote the rationals in $(0, 1]$, and let $E_n = F \oplus r_n$, where F is the set from the previous exercise and $\oplus$ denotes translation modulo 1.]
(b) There exists a nonincreasing sequence $\{E_n\}_{n=1}^\infty$ of sets [not necessarily in $\mathcal{L}$] with $\lambda^*(E_n) < +\infty$ for each $n \in \mathbb{N}$ such that $\lambda^*(\bigcap_{n=1}^\infty E_n) < \lim \lambda^*(E_n)$.

5. Let $A \in \mathcal{L}^k$ be such that $\bar{\lambda}_k(A) > 0$. There exist $\boldsymbol{x}, \boldsymbol{y} \in A$ so that $\boldsymbol{x} - \boldsymbol{y} \in \mathbb{Q}^k$.

6. Let $\boldsymbol{T} : \mathbb{R}^2 \to \mathbb{R}$ be such that $\boldsymbol{T}\boldsymbol{x} = x_1$ for all $\boldsymbol{x} \in \mathbb{R}^2$. That is, $\boldsymbol{T}$ projects points [perpendicularly] onto the x_1-axis.

(a) Such a mapping is linear and singular.
(b) There exists $A \in \mathcal{B}^2$ such that (i) $\boldsymbol{T}A$, viewed as a subset of $\mathbb{R}$, is *not* in $\mathcal{B}$, while (ii) $\boldsymbol{T}A$, viewed as a subset of $\mathbb{R}^2$, is in $\mathcal{B}^2$. [Consider the set $E \times \{0\}$, where E is the one-dimensional form of the set constructed in this section.]

7. For any linear mapping $\boldsymbol{T}$ [singular or nonsingular] defined on $\mathbb{R}^k$, $A \in \mathcal{L}^k$ forces $\boldsymbol{T}A \in \mathcal{L}^k$. [The result has been established for nonsingular $\boldsymbol{T}$. For singular $\boldsymbol{T}$, observe that $\boldsymbol{T}A \subseteq \boldsymbol{T}\mathbb{R}^k$, which is a hyperplane, hence a Borel set of λ_k-measure zero.]

Comment: One might think that $A \in \mathcal{B}^k$ forces $\boldsymbol{T}A \in \mathcal{B}^k$ for arbitrary linear transformations $\boldsymbol{T}$ defined on $\mathbb{R}^k$. We have shown in Section 4.4 that this result holds if $\boldsymbol{T}$ is nonsingular. We have not shown the nonsingular case because *the assertion is not true in this case.* For the situation in Exercise 6(b), there exists a set $A \in \mathcal{B}^2$ such that the projection of A perpendicularly onto the x_1-axis — a very simple transformation that happens to be singular — fails to be a $\mathcal{B}^2$-set. Such a result is quite counterintuitive, and the construction of such a set $A \in \mathcal{B}^2$ is quite involved and beyond the scope of this book.

4.6 THE CANTOR SET AND THE LEBESGUE FUNCTION

In this section, we shall construct an uncountable subset of $[0,1]$ with λ-measure zero. We shall then use this set to obtain a nondecreasing continuous function $F : [0,1] \to \mathbb{R}$ with $F(0) = 0$, $F(1) = 1$,

$$\{x \in [0,1] : \ F'(x) \text{ exists and equals } 0\} \in \mathcal{L},$$

and

$$\bar{\lambda}\left(\{x \in [0,1] : \ F'(x) \text{ exists and equals } 0\}\right) = 1.$$

This function will be used in the next section to exhibit a set that is in $\mathcal{L}$ but not in $\mathcal{B}$. We shall now construct the classic example of an uncountable subset of $\mathbb{R}$ with λ-measure zero, which is known as the *Cantor Set.*

We have claimed for each $k \in \mathbb{N}$ that there exists an uncountable $\mathcal{B}^k$-set of λ_k-measure zero. The case $k \geq 2$ was handled at the end of Section 4.4, and the results of this section will give the $k = 1$ case.

Definition. The set of *dyadic rationals of (0,1)* is

$$D = \left\{d \in (0,1) : \ d = \frac{m}{2^n}, \ n \in \mathbb{N}, \ m = 1, \cdots, 2^n - 1\right\},$$

where we agree to write each $d \in D$ in lowest terms.

Example 1. Examples of elements of D include $\frac{1}{2}$, $\frac{3}{8}$, and $\frac{37}{512}$. The numbers $\frac{1}{3}$ and $\frac{6}{7}$ are not in D. [Observe too that $0, 1 \notin D$.] The words *in lowest terms* from the above definition prevent ambiguity in writing out an element of D. For example, the midpoint of $[0,1]$ is represented by $\frac{1}{2}$, not $\frac{2}{4}$ nor $\frac{12}{24}$. As another example, we write $\frac{3}{8}$ and not $\frac{12}{32}$.

We now construct the *Cantor Set*, which will be the main example of an uncountable $\mathcal{B}$-set of zero Lebesgue measure. We start at "Stage 0" with $[0, 1]$.

Construction, Stage 1. Remove the open interval $(\frac{1}{3}, \frac{2}{3})$, which is the "middle third" of $[0, 1]$. Denote this interval by $J_{\frac{1}{2}}$. What is left is the disjoint union of $[0, \frac{1}{3}]$ and $[\frac{2}{3}, 1]$, two sets of Lebesgue measure $\frac{1}{3}$ each.

Construction, Stage 2. From $[0, \frac{1}{3}]$, remove the "middle third" $(\frac{1}{9}, \frac{2}{9})$, and from $[\frac{2}{3}, 1]$ remove the "middle third" $(\frac{7}{9}, \frac{8}{9})$. Denote these two removed intervals by the symbols $J_{\frac{1}{4}}$ and $J_{\frac{3}{4}}$, respectively. What remains is the union of the four disjoint closed intervals $[0, \frac{1}{9}]$, $[\frac{2}{9}, \frac{3}{9}]$, $[\frac{6}{9}, \frac{7}{9}]$, and $[\frac{8}{9}, 1]$, each interval having Lebesgue measure $\frac{1}{9}$.

Construction, Stage 3. From each of the 4 intervals $[0, \frac{1}{9}]$, $[\frac{2}{9}, \frac{3}{9}]$, $[\frac{6}{9}, \frac{7}{9}]$, and $[\frac{8}{9}, 1]$, we remove the "middle third". This means removing $(\frac{1}{27}, \frac{2}{27})$, $(\frac{7}{27}, \frac{8}{27})$, $(\frac{19}{27}, \frac{20}{27})$, and $(\frac{25}{27}, \frac{26}{27})$. Denote these four removed intervals by the symbols $J_{\frac{1}{8}}$, $J_{\frac{3}{8}}$, $J_{\frac{5}{8}}$, and $J_{\frac{7}{8}}$. What remains is the disjoint union of the 8 closed intervals $[0, \frac{1}{27}]$, $[\frac{2}{27}, \frac{3}{27}]$, $[\frac{6}{27}, \frac{7}{27}]$, $[\frac{8}{27}, \frac{9}{27}]$, $[\frac{18}{27}, \frac{19}{27}]$, $[\frac{20}{27}, \frac{21}{27}]$, $[\frac{24}{27}, \frac{25}{27}]$, and $[\frac{26}{27}, 1]$. Each of these eight disjoint closed intervals has Lebesgue measure $\frac{1}{27}$.

The pattern here should be clear.

Construction, Stage n [$n \in \mathbb{N}$ arbitrary]. At the beginning, there are 2^{n-1} disjoint closed intervals, each of length $(\frac{1}{3})^{n-1}$. From each of these 2^{n-1} closed intervals, remove the "middle third". Denote these 2^{n-1} removed open intervals by $J_{\frac{1}{2^n}}, J_{\frac{3}{2^n}}, \cdots$, and $J_{\frac{2^n-1}{2^n}}$. This leaves 2^n disjoint closed intervals, each having Lebesgue measure $(\frac{1}{3})^n$.

One carries on this construction for each $n \in \mathbb{N}$ according to the scheme above. Having done this, one will have removed the intervals $\{J_d : d \in D\}$ from $[0, 1]$. The points of $[0, 1]$ that remain constitute the Cantor Set:

Definition. The Cantor Set is $C = [0, 1] - \bigcup_{d \in D} J_d$.

Claim 1. *The Cantor Set C has the following properties.*
(a) C is closed [and hence compact].

(b) $x \in C$ iff x has a unique ternary expansion [terminating or nonterminating] consisting exclusively of 0's and 2's.

(c) C is uncountable.

(d) $C \in \mathcal{B}$ and $\lambda(C) = 0$.

(e) C contains no interior points [and by (a) C is thus nowhere dense].

Proof: We first deal with (a). Observe that J_d is open for each $d \in D$, hence $\bigcup_{d \in D} J_d$ is open, whence $(\bigcup_{d \in D} J_d)^c$ is closed. Since $C = [0, 1] \cap \left(\bigcup_{d \in D} J_D\right)^c$, C is the intersection of two closed sets and therefore is closed.

We now prove (b). Pick $x \in C$. After stage 1, x is in $[0, \frac{1}{3}]$ or $[\frac{2}{3}, 1]$, so that the first digit in the ternary expansion of x is 0 or 2. If x is in $[0, \frac{1}{3}]$, then x is in $[0, \frac{1}{9}]$ or $[\frac{2}{9}, \frac{3}{9}]$ so that the ternary expansion of x has 0 for the first digit and either 0 or 2 for the second digit. On the other hand, if x is in $[\frac{2}{3}, 1]$, then x is in $[\frac{6}{9}, \frac{7}{9}]$ or $[\frac{8}{9}, 1]$, so that the ternary expansion of x has 2 for the first digit and 0 or 2 for the second digit. We continue on through the rest of the stages in this fashion, and it is clear that x will have only one ternary expansion consisting exclusively of zeros and twos.[11]

For the converse, suppose that x has a unique ternary expansion consisting of 0's and 2's. Since no digits of this expansion equal 1, x is not in the middle third of any remaining closed interval at any stage of the Cantor Set's construction. Therefore, x cannot be removed at any stage, which means that $x \in C$. This proves (b).

The proof of (c) is simple, for by (b) there is a one-to-one correspondence between the points of C and the collection of sequences $\{a_n\}_{n=1}^{\infty}$ consisting exclusively of 0's and 2's, namely $0.a_1a_2\cdots$ [base 3] $\leftrightarrow \{a_n\}_{n=1}^{\infty}$, and the collection of such sequences has cardinality $\mathfrak{c}$.

We now deal with (d). By (a) and the fact that closed sets are Borel sets, we have $C \in \mathcal{B}$, so that we may meaningfully discuss the number $\lambda(C)$. Next, since the collection $\{J_d : d \in D\}$ is disjoint, we have

$$\begin{aligned}\lambda(C) = \lambda([0,1]) - \lambda\left(\bigcup_{d\in D} J_d\right) = 1 - \sum_{d\in D}\lambda(J_d) &= 1 - \frac{1}{3}\sum_{n=0}^{\infty}\left(\frac{2}{3}\right)^n \\ &= 1 - 1 = 0.\end{aligned}$$

[The third equality follows because the $\lambda(J_d)$'s are nonnegative numbers and can be summed in any order. In particular we may write

$$\begin{aligned}&\sum_{d\in D}\lambda(J_d) \\ = \ &\lambda(J_{\frac{1}{2}}) + \lambda(J_{\frac{1}{4}}) + \lambda(J_{\frac{3}{4}}) + \lambda(J_{\frac{1}{8}}) + \lambda(J_{\frac{3}{8}}) + \lambda(J_{\frac{5}{8}}) + \lambda(J_{\frac{7}{8}}) + \cdots \\ = \ &\frac{1}{3} + \frac{1}{9} + \frac{1}{9} + \frac{1}{27} + \frac{1}{27} + \frac{1}{27} + \frac{1}{27} + \frac{1}{81} + \cdots \\ = \ &\frac{1}{3}\sum_{n=0}^{\infty}\left(\frac{2}{3}\right)^n .]\end{aligned}$$

In light of (d), the proof of (e) is simple. If C contains an interior point, then C contains an open interval [a $\mathcal{B}$-set of positive λ-measure], and hence $\lambda(C) > 0$, contradicting the fact that $\lambda(C) = 0$. Therefore, C cannot contain an interior point. This finishes the entire proof. ∎

The Cantor Set C is interesting because the set is "large" with respect to cardinality, but it is "insignificant" with respect to Lebesgue measure. It is

[11] Some clarification is in order. Consider the point $1/3 \in C$, which has two different ternary expansions, 0.1 and $0.0\bar{2}$. What we are saying is that if we want to write the ternary expansion of $1/3$ using only 0's and 2's, then the only choice is $0.0\bar{2}$ – there is no other ternary expansion of $1/3$ using exclusively 0's and 2's.

hard to visualize an uncountable subset of $\mathbb{R}$ with "zero length," but such a set has just been constructed.

We have proved that $\text{card}(\mathcal{L}^k) = 2^c$ for each $k \geq 2$; the set C allows us to show that $\text{card}(\mathcal{L}) = 2^c$ in the exact same fashion: since $C \in \mathcal{L}$ and $\bar{\lambda}(C) = 0$, each of the 2^c subsets of C must be in $\mathcal{L}$ as well since $(\mathbb{R}, \mathcal{L}, \bar{\lambda})$ is complete.

The first half of this section is at a close, and some exercises follow. Of particular interest for the second half is Exercise 8, which is a review of some well-known facts from advanced calculus that will be needed for the second half of this section.

Exercises.

1. If $x \in [0,1]$ has a finite ternary expansion that ends in two or more 1's, then $x \notin C$.

2*. Let $x \in C$.
(a) x has exactly two different ternary expansions iff there exists $d \in D$ such that x is an endpoint of the removed interval J_d.
(b) For each $d \in D$, either $J_d = (0.1, 0.2)$ [base 3], or for some $n \in \mathbb{N}$ the interval J_d has the form $(0.a_1 \cdots a_n 1, 0.a_1 \cdots a_n 2)$, where $a_1, \cdots, a_n \in \{0, 2\}$.

3. $\frac{1}{4} \in C$, even though $\frac{1}{4}$ fails to be an endpoint of any removed interval J_d. Also, $0, 1 \in C$.

4*. There is a certain symmetry concerning C.
(a) C is symmetric about $\frac{1}{2}$: $x \in C$ iff $1 - x \in C$.
(b) Let $d \in D$, and write $J_d = (\alpha, \beta)$. Then $1 - d \in D$ and $J_{1-d} = (1-\beta, 1-\alpha)$.

5. $\{x - y : x, y \in C\} = [-1, 1]$ and $\{x + y : x, y \in C\} = [0, 2]$, so that

$$\lambda(\{x - y : x, y \in C\}) = \lambda(\{x + y : x, y \in C\}) = 2.$$

That is, while $\lambda(C) = 0$, both $C \oplus C$ and $C \ominus C$ have positive Lebesgue measure.

6. C is *perfect*: each $x \in C$ is the limit of a sequence of *distinct* points of C.

7. Let I_C be the indicator function for C: $I_C(x)$ is 1 or 0 as $x \in C$ or $x \notin C$. Then I_C is continuous on $[0,1] - C$, and I_C is discontinuous at each $x \in C$.

8*. This problem reviews some basic analysis facts that will be used in the next part of this section.
(a) Every collection of disjoint nonempty open intervals is amc. [Each such interval contains a distinct rational, and there are only countably many rationals.]
(b) Let $f : (a, b) \to \mathbb{R}$ be monotone. Then the set of points of discontinuity of f is amc, and the only discontinuities are jump discontinuities. [Recall that f has a jump discontinuity at x if and only if both $f(x-)$ and $f(x+)$ exist, where at least one of $f(x-)$ or $f(x+)$ differs from $f(x)$.]
(c) If $f : [a, b] \to [c, d]$ is monotone and onto, then f is continuous. [Use (b).]

We now use the Cantor Set C to carry out the second part of the program stated at the beginning of this section. Let $f : C \to \mathbb{R}$ be defined as follows: if $x = 0.a_1 a_2 \cdots$ is the unique ternary expansion of $x \in C$ consisting exclusively of 0's and 2's, then

$$f(x) = \sum_{n=1}^{\infty} \frac{a_n/2}{2^n} = 0.\frac{a_1}{2}\frac{a_2}{2}\cdots \quad \text{[base 2]}.$$

We may equivalently say that if $x = 0.(2b_1)(2b_2)\cdots$ [base 3] where each $b_n \in \{0,1\}$, then $f(x) = \sum_{n=1}^{\infty} \frac{b_n}{2^n} = 0.b_1b_2\cdots$ [base 2]. Since each $x \in C$ has a unique ternary expansion consisting exclusively of 0's and 2's, the function f is well defined.

Example 2. Let $x = \frac{1}{3}$. Since $x = 0.0\bar{2}$ [base 3], we have $x = \frac{1}{3} \in C$. It follows that $f(x) = 0.0\bar{1}$ [base 2], so that $f(\frac{1}{3}) = \frac{1}{2}$. Now let $x = \frac{2}{3} = 0.2\bar{0}$ [base 3], so that $x \in C$. Since $x = 0.2$ [base 3], $f(x) = 0.1$ [base 2], so that $f(\frac{2}{3}) = \frac{1}{2} = f(\frac{1}{3})$.

As some other examples, both $\frac{1}{9}, \frac{2}{9}$ are in C, with ternary expansions respectively given by $0.00\bar{2}$ and 0.02. We obtain that $f(\frac{1}{9}) = 0.00\bar{1} = 0.01 = f(\frac{2}{9})$ [base 2], hence $f(\frac{1}{9}) = f(\frac{2}{9}) = \frac{1}{4}$.

Next, observe that $\frac{7}{9}, \frac{8}{9} \in C$, with respective ternary expansions given by $0.20\bar{2}$ and 0.22. We have $f(\frac{7}{9}) = 0.10\bar{1} = 0.11 = f(\frac{8}{9})$ [base 2], so that $f(\frac{7}{9}) = f(\frac{8}{9}) = \frac{3}{4}$.

We have shown so far that (i) f assigns the value $\frac{1}{2}$ to the endpoints of $J_{\frac{1}{2}}$, (ii) f assigns the value $\frac{1}{4}$ to the endpoints of $J_{\frac{1}{4}}$, and (iii) f assigns the value $\frac{3}{4}$ to the endpoints of $J_{\frac{3}{4}}$. This is no coincidence, as it turns out that *for each $n \in \mathbb{N}$, f assigns the same value to $x = 0.a_1\cdots a_n1$ and $y = 0.a_1\cdots a_n2$, where $a_1,\cdots,a_n \in \{0,2\}$.* [Since $0.a_1\cdots a_n1 = 0.a_1\cdots a_n0\bar{2} \in C$, and since y is clearly in C, both $f(x)$ and $f(y)$ exist.] To see this, observe that

$$f(x) = 0.\frac{a_1}{2}\cdots\frac{a_n}{2}0\bar{1} = 0.\frac{a_1}{2}\cdots\frac{a_n}{2}1\bar{0} = f(0.a_1\cdots a_n2\bar{0}) = f(y),$$

where the expansions are in base 2.

The converse holds as well: *if $x, y \in C$ with $x < y$ and $f(x) = f(y)$, then for some $n \in \mathbb{N}$ we have that (i) x and y have the respective ternary expansions $0.a_1\cdots a_n1$ and $0.a_1\cdots a_n2$, where $a_1,\cdots,a_n \in \{0,2\}$, or (ii) $x = \frac{1}{3}$ and $y = \frac{2}{3}$.* To see this, let us suppose that $x, y \in C$ with $x < y$ and

$$f(x) = 0.b_1b_2\cdots = 0.b_1'b_2'\cdots = f(y) \qquad \text{[base 2]}.$$

Assume first that $b_1 \neq b_1'$, hence $b_1 = 0$ and $b_1' = 1$. This forces $b_2 = b_3 = \cdots = 1$ and $b_2' = b_3' = \cdots = 0$. It follows that $x = 0.0\bar{2} = 0.1$ [base 3] and $y = 0.2$ [base 3], so that $x = \frac{1}{3}$ and $y = \frac{2}{3}$. Next, suppose that there exists $n \in \mathbb{N}$ such that $b_1 = b_1', \cdots, b_n = b_n'$, but $b_{n+1} \neq b_{n+1}'$. Then $b_{n+1} = 0$ and $b_{n+1}' = 1$. Therefore, we have

$$f(x) = 0.b_1\cdots b_n0b_{n+2}\cdots = 0.b_1'\cdots b_n'1b_{n+2}'\cdots = f(y),$$

where both expansions are in base 2. Since $b_i = b_i'$ for $i = 1,\cdots,n$, we must have $b_{n+2} = b_{n+3} = \cdots = 1$ and $b_{n+2}' = b_{n+3}' = \cdots = 0$. It follows that

$$x = 0.(2b_1)\cdots(2b_n)0\bar{2} = 0.(2b_1)\cdots(2b_n)1\bar{0}$$

and

$$y = 0.(2b_1)\cdots(2b_n)2 \text{ [recall that } b_i = b_i' \text{ for } i = 1,\cdots,n],$$

where both expansions are in base 3. Letting $a_i = 2b_i$ for $i = 1, \cdots, n$ shows that in this case $f(x) = f(y)$ implies $x = 0.a_1 \cdots a_n 1$ and $y = 0.a_1 \cdots a_n 2$. This completes the proof of the converse claim.

Putting everything together, we have the following: *f(x)=f(y) if and only if one of the three conditions holds: (i) $x = y$, (ii) $x = \frac{1}{3}$ and $y = \frac{2}{3}$, or (iii) for some $n \in \mathbb{N}$, $x = 0.a_1 \cdots a_n 1$ and $y = 0.a_1 \cdots a_n 2$, where $a_1, \cdots, a_n \in \{0, 2\}$.* Condition (ii) is equivalent to the assertion that x and y are the endpoints of $J_{\frac{1}{2}}$. Condition (iii) is equivalent to saying that x and y are endpoints of some J_d with $d \in D - \{\frac{1}{2}\}$.

Claim 2. *The function f as given is onto $[0, 1]$ and nondecreasing. We also have $f(x) = \sup\{f(y) : y \in C, y \leq x\}$ for every $x \in C$.*

Proof: Pick $y \in [0, 1]$, and let $0.b_1 b_2 \cdots$ denote the binary expansion of y. Let $a_n = 2b_n$ for all $n \in \mathbb{N}$. Since each $b_n \in \{0, 1\}$, we have that each $a_n \in \{0, 2\}$, giving $0.a_1 a_2 \cdots \in C$ and $f(0.a_1 a_2 \cdots) = y$. Therefore, f is onto $[0, 1]$.

We now deal with the other claims. If f is shown to be nondecreasing, then we trivially have the asserted equation, and thus we only need to show that f is nondecreasing. Pick $x, y \in C$ with $x < y$. There are two cases to consider: (i) $x = 0.0a_2 a_3 \cdots$ and $y = 0.2b_2 b_3 \cdots$ [base 3], and (ii) there is $n \in \mathbb{N}$ such that x and y have the ternary expansions $0.a_1 \cdots a_n 0 a_{n+2} \cdots$ and $0.a_1 \cdots a_n 2 b_{n+2} \cdots$, respectively.

We do case (i). Suppose that $x = 0.0a_2 a_3 \cdots$ and $y = 0.2b_2 b_3 \cdots$, where both expansions are ternary. Then

$$f(x) = 0.0\frac{a_2}{2}\frac{a_3}{2} \cdots \leq 0.0\bar{1} = 0.1 \leq 0.1\frac{b_2}{2}\frac{b_3}{2} \cdots = f(y),$$

where the expansions above are binary. This gives case (i).

We now do case (ii). Assume for some $n \in \mathbb{N}$ that x and y have the ternary expansions $0.a_1 \cdots a_n 0 a_{n+2} \cdots$ and $0.a_1 \cdots a_n 2 b_{n+2} \cdots$, respectively. It follows that

$$\begin{aligned} f(x) = 0.\frac{a_1}{2} \cdots \frac{a_n}{2} 0 \frac{a_{n+2}}{2} \cdots \quad &\leq \quad 0.\frac{a_1}{2} \cdots \frac{a_n}{2} 0\bar{1} \\ &= \quad 0.\frac{a_1}{2} \cdots \frac{a_n}{2} 1 \\ &\leq \quad 0.\frac{a_1}{2} \cdots \frac{a_n}{2} 1 \frac{b_{n+2}}{2} \frac{b_{n+3}}{2} \cdots \\ &= \quad f(y), \end{aligned}$$

where the expansions above are in base 2. We have shown that in both cases that if $x, y \in C$ and $x < y$, then $f(x) \leq f(y)$. This shows that f is nondecreasing and completes the entire proof. ∎

Observe that we could not strengthen the word *nondecreasing* above to *strictly increasing*, for in Example 2 we saw that if $x = \frac{1}{3}$ and $y = \frac{2}{3}$ then $f(x) = \frac{1}{2} = f(y)$.

Definition. The *Lebesgue [Singular] Function* $F : [0,1] \to R$ is defined by writing $F(x) = \sup\{f(y) : y \in C, \ y \leq x\}$ for every $x \in [0,1]$, where f and C are as defined.

The function F by its very definition is well-defined. This function will be used in the next section to produce a set that is in $\mathcal{L}$ but not in $\mathcal{B}$.

Claim 3. *The Lebesgue Function F has the following properties:*
(a) $F = f$ on C. In particular, $F(0) = 0$ and $F(1) = 1$.

(b) F is onto $[0,1]$.

(c) F is nondecreasing.

(d) F is continuous [and since F is defined on the compact set $[0,1]$, F is in fact uniformly continuous].

(e) For each $d \in D$, $F(x) = d$ for each $x \in J_d$. [Therefore $F'(x)$ exists and equals 0 on each J_d.]

Proof: The first part of (a) is trivial, and the second part follows after easily verifying that $0, 1 \in C$, $f(0) = 0$, and $f(1) = 1$. Claim (b) is automatic by (a) and the fact that f has been shown to be onto [0,1].

We now deal with (c). We take $x_1, x_2 \in [0,1]$ with $x_1 < x_2$, and we desire to show $F(x_1) \leq F(x_2)$. There are four cases: (i) $x_1, x_2 \in C$, (ii) $x_1 \in C$ and $x_2 \notin C$, (iii) $x_1 \notin C$ and $x_2 \in C$, and (iv) $x_1, x_2 \notin C$. In (i) we have $F(x_1) = f(x_1) \leq f(x_2) = F(x_2)$. For (ii), we have

$$F(x_1) = f(x_1) \leq \sup\{f(y) : y \in C, \ y \leq x_2\} = F(x_2).$$

For (iii), we argue by contradiction. If $F(x_1) > F(x_2)$, then $F(x_1) > f(x_2)$. By the definition of $F(x_1)$, there is $x^* \in C$ with $x^* < x_1$ and $f(x^*) > f(x_2)$. But $x^* < x_2$, and since f is nondecreasing on C we obtain $f(x^*) \leq f(x_2)$, which is a contradiction. It follows that $F(x_1) \leq F(x_2)$, and hence (iii) holds. For (iv), we return to direct argumentation, observing that

$$\{f(y) : y \in C, \ y \leq x_1\} \subseteq \{f(y) : y \in C, \ y \leq x_2\},$$

so that

$$\sup\{f(y) : y \in C, \ y \leq x_1\} \leq \sup\{f(y) : y \in C, \ y \leq x_2\},$$

forcing $F(x_1) \leq F(x_2)$. All four cases are complete, and thus F is nondecreasing on [0,1]. This completes the proof of (c).

For (d), we know that F is monotone and onto [0,1]. By Exercise 8, it automatically follows that F is continuous.

We now deal with (e). First, consider $J_{\frac{1}{2}} = (\frac{1}{3}, \frac{2}{3})$. We have already seen that $f(\frac{1}{3}) = f(\frac{2}{3}) = \frac{1}{2}$, and because $F = f$ on C and F is nondecreasing, it follows that $F(x) = \frac{1}{2}$ for each $x \in J_{\frac{1}{2}}$. Next, consider $J_{\frac{1}{4}} = (\frac{1}{9}, \frac{2}{9})$. We have seen that $f(\frac{1}{9}) = f(\frac{2}{9}) = \frac{1}{4}$. Since $F = f$ on C and F is nondecreasing, it follows that $F(x) = \frac{1}{4}$ for each $x \in J_{\frac{1}{4}}$. Similarly, $F(x) = \frac{3}{4}$ for each

$x \in J_{\frac{3}{4}}$. This type of reasoning carries on for all $d \in D$, and the proof of (e) is complete. ■

By Claim 3, we obtain

$$\{x \in [0,1] : F'(x) \text{ exists and equals } 0\} \supseteq \bigcup_{d \in D} J_D,$$

and $\lambda(\bigcup_{d \in D} J_d) = 1$. Since (i) $C \in \mathcal{B}$ with $\lambda(C) = 0$, (ii) $(\mathbb{R}, \mathcal{L}, \bar{\lambda})$ is complete, and (iii)

$$\{x \in [0,1] : F'(x) \text{ does not exist or } F'(x) \neq 0\} \subseteq C,$$

it follows that

$$\{x \in [0,1] : F'(x) \text{ does not exist or } F'(x) \neq 0\} \in \mathcal{L},$$

and therefore

$$\bar{\lambda}(\{x \in [0,1] : F'(x) \text{ does not exist or } F'(x) \neq 0\}) = 0$$

by the fact that $\lambda(C) = 0$. In turn, it follows that

$$\bar{\lambda}(\{x \in [0,1] : F'(x) \text{ exists and equals zero}\}) = 1.$$

It is hard to imagine a *continuous* function g defined on $[0,1]$ with $g(0) = 0$, $g(1) = 1$, is constant "almost everywhere" and yet is *onto* $[0,1]$, but F is such a function.

Exercises.

9. We actually have $\{x \in [0,1] : F'(x) \text{ exists and equals zero}\} \in \mathcal{B}$. [Show that this set actually equals the Borel set $\bigcup_{d \in D} J_d$.]

10. Let $n \in \mathbb{N}$.
(a) If $x, y \in C$ and $|x - y| < 3^{-n}$, then $|f(x) - f(y)| \leq 2^{-n}$.
(b) If $x, y \in [0,1]$ and $|x - y| < 3^{-n}$, then $|F(x) - F(y)| \leq 2^{-n}$. This constitutes another proof that F is uniformly continuous.

11. There is an alternative way of constructing the Lebesgue function. Define $g : \bigcup_{d \in D} J_d \to [0,1]$ by writing $g(x) = d$ for all $x \in J_d$, $d \in D$.
(a) g is nondecreasing.
(b) g is uniformly continuous.
(c) So far g is defined only on $\bigcup_{d \in D} J_d$. Define a new function G with domain $[0,1]$ by writing $G(x) = \lim_n g(x_n)$, where $\{x_n\}_{n=1}^{\infty}$ is a sequence of points in $\bigcup_{d \in D} J_d$ such that $x_n \to x$. For each $x \in [0,1]$ the value of $G(x)$ does not depend on the sequence $\{x_n\}_{n=1}^{\infty}$ of points in $\bigcup_{d \in D} J_d$ that converges to x, so G is well-defined.
(d) $G = g$ on $\bigcup_{d \in D} J_d$, and part (b) forces G to be continuous on $[0,1]$.
(e) $G = F$, where F is the Lebesgue Singular Function as we have constructed it.

12. Let D_α denote a subset of $[0,1]$ constructed in the same fashion as the Cantor Set C, except for the fact that [for each $n \in \mathbb{N}$] at the nth step, instead of removing the "middle third" of the 2^{n-1} remaining closed intervals, we instead remove the middle portion of length $\alpha 3^{-n}$, where $\alpha \in [0,1]$. [The set C corresponds to D_1, and $[0,1]$ corresponds to D_0.] Then D_α is closed, hence in $\mathcal{B}$. Also, D_α^c is a dense subset of $[0,1]$, and $\lambda(D_\alpha) = 1 - \alpha$.

Comment: Such sets D_α for $0 < \alpha < 1$ are called *fat Cantor Sets* because they are constructed in a fashion similar to that by which C was constructed, but unlike C, they have positive Lebesgue measure.

4.7 A NON-BOREL LEBESGUE SET

We have considered the σ-fields $\mathcal{B}^k$ and $\mathcal{L}^k$, and we have shown that $\mathcal{B}^k$ is a proper sub-σ-field of $\mathcal{L}^k$ because card$(\mathcal{B}^k) = \mathfrak{c} < 2^{\mathfrak{c}} =$ card$(\mathcal{L}^k)$. Therefore, for all $k \in \mathbb{N}$, there are sets in $\mathcal{L}^k - \mathcal{B}^k$. This is a "crude cardinality argument," and no actual set in $\mathcal{L}^k - \mathcal{B}^k$ is exhibited by such a proof.

The main point of this section is to exhibit as explicitly as possible a set in $\mathcal{L}^k - \mathcal{B}^k$. For simplicity, we shall take $k = 1$. The construction below hinges on two facts: (i) a set outside $\mathcal{L}$ exists, and (ii) every Lebesgue set of positive $\bar{\lambda}$-measure contains a proper subset not in $\mathcal{L}$. These facts were proven in Section 5 of this chapter. Recall that (AC) is required for the proof of (i), and the proof of (ii) explicitly uses (i). The set constructed in this section will therefore require (AC), at least implicitly.

The demonstration is fairly brief and follows classical lines, using the Cantor Set $C = [0,1] - \bigcup_{d \in D} J_d$ and the Lebesgue Singular Function F, both discussed in the last section. We shall also use the notation of the previous section.

Definition. Let X and Y denote sets, and let $\phi : X \to Y$ denote a function that is (a) one-to one, (b) onto, (c) continuous, and (d) in possession of a continuous inverse. We call ϕ a *homeomorphism from X to Y*, or simply a *homeomorphism* if no confusion can result.

The mathematical buildup now begins to exhibit what has been promised. We begin with a standard analysis fact:

> *Let $\phi : X \to Y$ be continuous, where $X, Y \subseteq \mathbb{R}$. Then for every open $B \subseteq \mathbb{R}$, there is an open subset $A \subseteq \mathbb{R}$ with $\phi^{-1}(B) = X \cap A$.*

If [in the notation of the above fact] X is open, then so is $X \cap A$, and hence this fact becomes the standard fact that *the inverse image of an open set is open if the function is continuous.*

Claim 1. *Let $X, Y \in \mathcal{B}$, and let $\phi : X \to Y$ be continuous. For any $B \in \mathcal{B}$, we have $\phi^{-1}(B) \in \mathcal{B}$.*

Proof: Let ϕ, X, and Y be as stated, and define the collection

$$\mathcal{C} = \{B \subseteq \mathbb{R} : \ B \in \mathcal{B},\ \phi^{-1}(B) \in \mathcal{B}\} \qquad [\text{so } \mathcal{C} \subseteq \mathcal{B}].$$

Let $H \subseteq \mathbb{R}$ be open, so that $H \in \mathcal{B}$. By the analysis fact cited above, there exists an open subset $G \subseteq \mathbb{R}$ [hence $G \in \mathcal{B}$] so that $\phi^{-1}(H) = X \cap G \in \mathcal{B}$. It follows that $H \in \mathcal{C}$, so that $\mathcal{C}$ contains the open subsets of $\mathbb{R}$. If it is shown that $\mathcal{C}$ is a σ-field, then by minimality we will have $\mathcal{B} \subseteq \mathcal{C}$, and it will then follow that $\mathcal{C} = \mathcal{B}$, completing the proof.

We now show that $\mathcal{C}$ is a σ-field. First, $\mathbb{R} \in \mathcal{B}$ and $\phi^{-1}(\mathbb{R}) = X \in \mathcal{B}$, so $\mathbb{R} \in \mathcal{C}$. Second, if $B \in \mathcal{C}$, so that $B \in \mathcal{B}$ and $\phi^{-1}(B) \in \mathcal{B}$, then $B^c \in \mathcal{B}$ and $\phi^{-1}(B^c) = \phi^{-1}(B)^c \in \mathcal{B}$, whence $B^c \in \mathcal{C}$. This shows that $\mathcal{C}$ is closed under complementation. Next, let $\{B_n\}_{n=1}^{\infty}$ denote a sequence of $\mathcal{C}$-sets, so that $B_n \in \mathcal{B}$ and $\phi^{-1}(B_n) \in \mathcal{B}$ for each $n \in \mathbb{N}$. It follows that $\bigcup_{n=1}^{\infty} B_n \in \mathcal{B}$ and $\phi^{-1}(\bigcup_{n=1}^{\infty} B_n) = \bigcup_{n=1}^{\infty} \phi^{-1}(B_n) \in \mathcal{B}$, hence $\bigcup_{n=1}^{\infty} B_n \in \mathcal{C}$, so $\mathcal{C}$ is closed under countable unions. $\mathcal{C}$ is therefore a σ-field, and the proof is complete. ∎

Claim 2. *Let $X, Y \in \mathcal{B}$, let $\phi : X \to Y$ be a homeomorphism, and let $A \subseteq X$. Then $A \in \mathcal{B}$ iff $\phi(A) \in \mathcal{B}$.*

Proof: Let $A \subseteq X$, and suppose that $\phi(A) \in \mathcal{B}$. Since ϕ is a homeomorphism, we have that ϕ is continuous, and the last result implies that $\phi^{-1}(\phi(A)) \in \mathcal{B}$. Since ϕ is one-to-one, we have $\phi^{-1}(\phi(A)) = A$, so that $A \in \mathcal{B}$. Therefore, $\phi(A) \in \mathcal{B}$ implies $A \in \mathcal{B}$.

We now do the converse. Let $A \in \mathcal{B}$, where $A \subseteq X$. Since ϕ is a homeomorphism from X onto Y, it follows that ϕ^{-1} is a homeomorphism from Y onto X. Since $\phi(A) = (\phi^{-1})^{-1}(A)$, and since $(\phi^{-1})^{-1}(A)$ is the inverse image of the $\mathcal{B}$-set A under the continuous mapping ϕ^{-1}, Claim 1 implies that $(\phi^{-1})^{-1}(A) \in \mathcal{B}$, which means that $\phi(A) \in \mathcal{B}$. Therefore, $A \in \mathcal{B}$ with $A \subseteq X$ implies that $\phi(A) \in \mathcal{B}$. The proof of this claim is accordingly complete. ∎

Definition. Define $\phi : [0,1] \to \mathbb{R}$ by writing $\phi(x) = x + F(x)$ for every x in $[0,1]$, where F is the Lebesgue Singular Function from the previous section.

Claim 3. *ϕ as defined is a homeomorphism from $[0,1]$ to $[0,2]$.*

Proof: Since (a) $\phi(0) = 0+F(0) = 0+0 = 0$ and $\phi(1) = 1+F(1) = 1+1 = 2$, (b) ϕ is the sum of two continuous functions [the identity function and F] and hence is continuous, and (c) ϕ is the sum of a strictly increasing function and a nondecreasing function and is therefore strictly increasing, it follows that ϕ is onto $[0,2]$ and one-to-one. Therefore, the inverse ϕ^{-1} of ϕ exists, and inverses of continuous functions [when they exist] are themselves continuous. These facts prove the claim. ∎

Construction: Pick an arbitrary $d \in D$ [the set of dyadic rationals from last section], and consider the open interval J_d as given in the construction of the Cantor Set C, writing $J_d = (\alpha, \beta)$. Since $F(x) = d$ for all $x \in J_d$, we have that

$$\begin{aligned}
\phi(J_d) &= \{y \in [0,2] : \text{ there is } x \in (\alpha,\beta) \text{ such that } y = \phi(x)\} \\
&= \{y \in [0,2] : \text{ there is } x \in (\alpha,\beta) \text{ such that } y = x + d\} \\
&= (\alpha + d, \beta + d),
\end{aligned}$$

hence $\lambda(J_d) = \lambda(\phi(J_d))$ *for every* $d \in D$, the common value being $\beta - \alpha$.

We next claim that *the collection* $\{\phi(J_d) : d \in D\}$ *is a disjoint collection.* This follows because ϕ is one-to-one and the J_d's are disjoint.

Next, observe that because $\phi([0,1]) = [0,2]$, we have

$$\phi(C) = \phi\left([0,1] - \bigcup_{d\in D} J_d\right) = [0,2] - \phi\left(\bigcup_{d\in D} J_d\right) = [0,2] - \bigcup_{d\in D} \phi(J_d) \in \mathcal{B},$$

and it therefore follows that

$$\begin{aligned}\lambda(\phi(C)) = \lambda\left([0,2] - \bigcup_{d\in D} \phi(J_d)\right) &= \lambda([0,2]) - \lambda\left(\bigcup_{d\in D} \phi(J_d)\right) \\ &= 2 - \sum_{d\in D} \lambda(\phi(J_d)) \\ &= 2 - \sum_{d\in D} \lambda(J_d) \\ &= 2 - 1 = 1,\end{aligned}$$

where the fact that $\sum_{d\in D} \lambda(J_d) = 1$ was shown in the last section. It follows that $\phi(C)$ *has positive Lebesgue measure.*

We now recall that every $\mathcal{L}$-set of positive Lebesgue measure contains a subset outside of $\mathcal{L}$. [This was shown in Section 5.] *Let B denote a subset of $\phi(C)$ such that $B \notin \mathcal{L}$.* We claim that the inverse image of this set under ϕ, while in $\mathcal{L}$, is not in $\mathcal{B}$.

Claim 4. $\phi^{-1}(B) \in \mathcal{L} - \mathcal{B}$.

Proof: Since $B \subseteq \phi(C)$, we have $\phi^{-1}(B) \subseteq \phi^{-1}(\phi(C)) = C$, so that by the facts (i) $C \in \mathcal{B}$, (ii) $\lambda(C) = 0$ and (iii) $(\mathbb{R}, \mathcal{L}, \bar{\lambda})$ is complete, we have $\phi^{-1}(B) \in \mathcal{L}$. We now show that this set is outside of $\mathcal{B}$ by contradiction. If to the contrary $\phi^{-1}(B) \in \mathcal{B}$, Claim 2 entails that $B = \phi(\phi^{-1}(B)) \in \mathcal{B}$. But $\mathcal{B} \subseteq \mathcal{L}$ and $B \notin \mathcal{L}$ by construction, so $B \notin \mathcal{B}$. Therefore, assuming that $\phi^{-1}(B) \in \mathcal{B}$ leads to the absurd conclusion that $B \in \mathcal{B}$ and $B \notin \mathcal{B}$, hence $\phi^{-1}(B) \notin \mathcal{B}$. It follows that $\phi^{-1}(B) \in \mathcal{L} - \mathcal{B}$, completing the proof. ■

We can try to be more specific about the form of $\phi^{-1}(B)$ if we use the notation of Section 5 and recall the construction of the set $E \notin \mathcal{L}$ given there. In the subsequent proof that every Lebesgue set A of positive Lebesgue measure contains a subset outside of $\mathcal{L}$, the subset of A outside of $\mathcal{L}$ that was exhibited was $(E \oplus r^*) \cap A$ for an "appropriate rational number r^*." Thus, since we are taking the A of that result to be $\phi(C)$ here, we might represent the set B given above by $B = (E \oplus r^*) \cap \phi(C)$ for an "appropriate rational" r^*, so that

$$\begin{aligned}\phi^{-1}(B) = \phi^{-1}((E \oplus r^*) \cap \phi(C)) &= \phi^{-1}(E \oplus r^*) \cap \phi^{-1}(\phi(C)) \\ &= \phi^{-1}(E \oplus r^*) \cap C.\end{aligned}$$

This is interesting, but it doesn't help matters much, for we have no constructive idea on just how to choose the "appropriate rational number" r^*, and E is quite nonconstructive as well.

At the same time though, we have a recipe of sorts for obtaining $2^{\mathfrak{c}}$ subsets of C that, while being in $\mathcal{L}$, fail to be in $\mathcal{B}$. To see this, recall that the non-Lebesgue set E constructed in Section 5 was formed by choosing exactly one element from each of $\mathfrak{c}$ equivalence classes, where each equivalence class had infinitely many elements. There are $2^{\mathfrak{c}}$ ways of choosing exactly one element from each of $\mathfrak{c}$ equivalence classes.[12] hence there are $2^{\mathfrak{c}}$ such "E's." Finding the "appropriate" rational r^* for each of the $2^{\mathfrak{c}}$ E's allows us to assert that $\phi^{-1}(E \oplus r^*) \cap C$ is in $\mathcal{L} - \mathcal{B}$ for each of the $2^{\mathfrak{c}}$ E's.

Exercises.

1. Construct a set in $\mathcal{L}^k - \mathcal{B}^k$ for each $k \geq 2$. [When $k \geq 2$, there are very simple uncountable Borel sets of Lebesgue measure zero, namely, the hyperplanes. Use these and an appropriate homeomorphism to carry out the construction. Alternatively, one can find the answer given somewhat implicitly in Section 2.]

2. The homeomorphic image of a Borel set is a Borel set.

3. The homeomorphic image of a Lebesgue set is not necessarily a Lebesgue set. Therefore, the Lebesgue and Borel sets behave differently in this regard. [From the construction, $\phi^{-1}(B) \in \mathcal{L}$. The image of $\phi^{-1}(B)$ under the homeomorphism ϕ is $B \notin \mathcal{L}$.]

4.8 THE IMPOSSIBILITY THEOREM

At the beginning of Chapter 3, the *problem of measure* was discussed. The main point of that discussion was that k-dimensional Lebesgue measure could not be consistently defined for all subsets of $\mathbb{R}^k$ if one wanted to allow the Axiom of Choice (AC) as a working tool [and we do]. In Section 5, we exhibited a set $E \notin \mathcal{L}^k$, so that $\mathcal{L}^k$, even though it has cardinality $2^{\mathfrak{c}} = \text{card}(2^{\mathbb{R}^k})$, cannot consist of every subset of $\mathbb{R}^k$. More generally, the results of Section 5 show that we cannot have a measure with domain $2^{\mathbb{R}^k}$ that agrees with Lebesgue measure on $\mathcal{L}^k$. This result is one of the "impossibility theorems" alluded to earlier in the book.

This section aims to prove a stronger result that (a) implicitly requires the Well Ordering Principle[13] [an equivalent from of (AC)] and (b) explicitly requires the Continuum Hypothesis (CH). The result we shall prove is this.

[12] A more precise statement is as follows: the cardinality of the set of choice functions is that of the set of all subsets of $\mathbb{R}$. We should also note that we really shouldn't talk about *the* set $E \notin \mathcal{L}$ constructed in Section 5, but instead we should say something like *a particular representative of the construction process in Section 4.5*. However, being colloquial doesn't cause any confusion in this instance.

[13] The Well Ordering Principle asserts that every nonempty set can be well-ordered.

Impossibility Theorem. *There can exist no measure* $\mu : 2^{\mathbb{R}} \to \bar{\mathbb{R}}$ *with* $\mu(\mathbb{R}) > 0$ *that simultaneously*
(a) assigns finite measure to bounded sets, and
(b) assigns zero measure to each singleton.
In particular, Lebesgue measure cannot be consistently defined for every subset of $\mathbb{R}$.

The proof of this theorem will be performed by contradiction and relies on some claims involving ordinal numbers and (CH) as given in Section 1.6.

Notation. For this section, the following notation will be employed.

Sequences. We will let $\boldsymbol{S}$ denote the set of all sequences of positive integers, and elements of $\boldsymbol{S}$ will be denoted by bold lowercase letters such as $\boldsymbol{s}, \boldsymbol{t}$, and so on. We will write $\boldsymbol{s} = \{s_n\}_{n=1}^{\infty}$, $\boldsymbol{t} = \{t_n\}_{n=1}^{\infty}$, and so on.

Ordinals. We will use lowercase Greek letters such as α, β, and ω for ordinal numbers. We will use $\prec$ and $\preceq$ for the usual ordering relation between ordinal numbers. We will let ω denote the smallest uncountable ordinal, and for each ordinal $\alpha \preceq \omega$ we will let $P_\alpha = \{\beta : \beta \prec \alpha\}$. In particular, P_ω, the set of ordinal numbers preceding the first uncountable ordinal, is well ordered with respect to $\preceq$, and we denote its cardinal number by $\aleph_1$.

Order Relations. We have the following order relations: $\leq$, $\preceq$, $\trianglerighteq$, and $\leq_w$.

The order relation $\leq$ is the usual order relation between real numbers.

The order relation $\preceq$ is the usual order relation for ordinal numbers.

The order relation $\trianglerighteq$ will apply to members of $\boldsymbol{S}$ only, and we shall write $\boldsymbol{s} \trianglerighteq \boldsymbol{t}$ iff $s_n \leq t_n$ for all $n \in \mathbb{N}$. This ordering is a partial ordering; it is one of three orderings that we will define for elements of $\boldsymbol{S}$.

The order relation $\leq_w$ is the second order relation defined for elements of $\boldsymbol{S}$, but it needs some explanation. We have that $\text{card}(\boldsymbol{S}) = \mathfrak{c}$ and $\text{card}(P_\omega) = \aleph_1$. One form of (CH) asserts that $\mathfrak{c} = \aleph_1$, so under (CH) there exists a one-to-one correspondence between the elements $\boldsymbol{s} \in \boldsymbol{S}$ and the elements $\alpha \in P_\omega$. For elements $\boldsymbol{s}, \boldsymbol{t} \in \boldsymbol{S}$ with corresponding ordinals $\alpha_{\boldsymbol{s}}$ and $\alpha_{\boldsymbol{t}}$, we shall write $\boldsymbol{s} \leq_w \boldsymbol{t}$ [resp. $\boldsymbol{s} <_w \boldsymbol{t}$] iff $\alpha_{\boldsymbol{s}} \preceq \alpha_{\boldsymbol{t}}$ [resp. $\alpha_{\boldsymbol{s}} \prec \alpha_{\boldsymbol{t}}$], hence $\leq_w$ is a total ordering and a well ordering on $\boldsymbol{S}$.

The third order relation on $\boldsymbol{S}$ is called *rejection.* We shall say that $\boldsymbol{t}$ *rejects* $\boldsymbol{s}$ and will write $\boldsymbol{t} \operatorname{rej} \boldsymbol{s}$ iff $\boldsymbol{t} <_w \boldsymbol{s}$ and $\boldsymbol{s} \trianglerighteq \boldsymbol{t}$.

Claim 1. *For each* $\boldsymbol{s} \in \boldsymbol{S}$ *we have that* $\{\boldsymbol{t} \in \boldsymbol{S} : \boldsymbol{t} \leq_w \boldsymbol{s}\}$ *is amc.*

Proof: Pick $\boldsymbol{s} \in \boldsymbol{S}$, and let $\alpha \in P_\omega$ denote the ordinal corresponding to $\boldsymbol{s}$. Since P_α is amc and $\leq_w$ is defined according to the one-to-one correspondence that exists between $\boldsymbol{S}$ and P_ω, the result immediately follows. ∎

Claim 2. *Each of the relations $\leq$, $\preceq$, $\trianglerighteq$, $\leq_w$, and* rej *are transitive relations.*

Proof: The first three relations are automatically transitive. The relation $\leq_w$ is transitive because $\preceq$ is transitive. Finally, suppose that $\boldsymbol{r} \operatorname{rej} \boldsymbol{s}$ and $\boldsymbol{s} \operatorname{rej} \boldsymbol{t}$. Then $\boldsymbol{r} <_w \boldsymbol{s}$ and $\boldsymbol{s} \trianglerighteq \boldsymbol{r}$; we also have $\boldsymbol{s} <_w \boldsymbol{t}$ and $\boldsymbol{t} \trianglerighteq \boldsymbol{s}$. By the transitivity of both $\leq_w$ and $\trianglerighteq$ we obtain $\boldsymbol{r} <_w \boldsymbol{t}$ and $\boldsymbol{t} \trianglerighteq \boldsymbol{r}$, so that $\boldsymbol{r} \operatorname{rej} \boldsymbol{t}$. This shows that rej is transitive, which completes the proof. ∎

Claim 3. *Let $\boldsymbol{T} = \{\boldsymbol{s} \in \boldsymbol{S} :$ there is no $\boldsymbol{t} \in \boldsymbol{S}$ with $\boldsymbol{t} \operatorname{rej} \boldsymbol{s}\}$ denote the set of unrejected elements of $\boldsymbol{S}$. For all $\boldsymbol{s} \notin \boldsymbol{T}$, $\boldsymbol{t} \operatorname{rej} \boldsymbol{s}$ implies $\boldsymbol{t} \in \boldsymbol{T}$.*

Proof: Let $\boldsymbol{s} \notin \boldsymbol{T}$, so that the set of elements in $\boldsymbol{S}$ that reject $\boldsymbol{s}$ is nonempty. Since $\leq_w$ is a well ordering on $\boldsymbol{S}$, every nonempty subset of $\boldsymbol{S}$ has a smallest element with respect to $\leq_w$. In particular, there is a smallest element $\boldsymbol{t} \in \boldsymbol{S}$ [with respect to $\leq_w$] such that $\boldsymbol{t} \operatorname{rej} \boldsymbol{s}$. We desire to show that $\boldsymbol{t} \in \boldsymbol{T}$. Suppose to the contrary that $\boldsymbol{t} \notin \boldsymbol{T}$, so that there is $\boldsymbol{u} \in \boldsymbol{S}$ with $\boldsymbol{u} \operatorname{rej} \boldsymbol{t}$, hence $\boldsymbol{u} <_w \boldsymbol{t}$. Since $\boldsymbol{t} \operatorname{rej} \boldsymbol{s}$ and rej is transitive, we obtain $\boldsymbol{u} \operatorname{rej} \boldsymbol{s}$. This contradicts the fact that $\boldsymbol{t}$ is the smallest element of $\boldsymbol{S}$ with respect to $<_w$ that rejects $\boldsymbol{s}$. It follows that $\boldsymbol{t} \in \boldsymbol{T}$. ∎

Claim 4. $\operatorname{card}(\boldsymbol{T}) = \mathfrak{c}$.

Proof: Since $\boldsymbol{T} \subseteq \boldsymbol{S}$, we have $\operatorname{card}(\boldsymbol{T}) \leq \operatorname{card}(\boldsymbol{S}) = \mathfrak{c}$. By (CH), it follows that $\boldsymbol{T}$ is either finite, countably infinite, or has cardinality c.

Suppose that $\operatorname{card}(\boldsymbol{T}) = \aleph_0$; in particular, write $\boldsymbol{T} = \{\boldsymbol{t}^1, \boldsymbol{t}^2, \cdots\}$. Consider the element $\boldsymbol{t}^* \in \boldsymbol{S}$ with the property that $t_k^* = t_k^k + 1$ for all $k \in \mathbb{N}$. Therefore, it is not possible for any $\boldsymbol{t}^i$ to have the property that $t_j^* \leq t_j^i$ for all $j \in \mathbb{N}$. This means that for no $i \in \mathbb{N}$ is it possible to have $\boldsymbol{t}^* \trianglerighteq \boldsymbol{t}^i$. In turn, this implies that no $\boldsymbol{t}^i$ can reject $\boldsymbol{t}^*$. By Claim 3, if $\boldsymbol{t}^*$ is to be rejected, then it is to be rejected by an element of $\boldsymbol{T}$. But $\boldsymbol{t}^*$ is not rejected by any element of $\boldsymbol{T}$, hence $\boldsymbol{t}^*$ is an unrejected element, ergo $\boldsymbol{t}^* \in \boldsymbol{T}$. However, $\boldsymbol{t}^*$ is distinct from each $\boldsymbol{t}^i$, and thus $\{\boldsymbol{t}^1, \boldsymbol{t}^2, \cdots\}$, which was our assumed complete enumeration of $\boldsymbol{T}$, is at the same time incomplete, a contradiction. Therefore, $\boldsymbol{T}$ is either finite or has cardinal number $\mathfrak{c}$.

If $\boldsymbol{T}$ is assumed finite, say $\boldsymbol{T} = \{\boldsymbol{t}^1, \cdots, \boldsymbol{t}^l\}$ for some $l \in \mathbb{N}$, then the same argument used in the above paragraph works. [The main thing is that we can list out the elements of $\boldsymbol{T}$, not that the list is finite or countably infinite.] Thus $\boldsymbol{T}$ cannot be finite. The only remaining option is that $\operatorname{card}(\boldsymbol{T}) = \mathfrak{c}$. ∎

Claim 5. *Let $m \in (0, +\infty)$ and let $\Omega = (l, l+1]$ for some $l \in \mathbb{Z}$. Suppose that μ is a measure [on some σ-field over Ω] such that $\mu(\Omega) = m$. Furthermore, suppose that $\mu(C_n) \geq m[1 - (\frac{1}{2})^{n+1}]$ for all $n \in \mathbb{N}$. Then $\mu(\bigcap_{n=1}^\infty C_n) > 0$.*

Proof: Observing that $\mu(C_n^c) = m - \mu(C_n) \leq m(\frac{1}{2})^{n+1}$ for all $n \in \mathbb{N}$, we have

$$\mu\left(\bigcap_{n=1}^\infty C_n\right) = m - \mu\left(\bigcup_{n=1}^\infty C_n^c\right) \geq m - \sum_{n=1}^\infty \mu(C_n^c) \geq m - \sum_{n=1}^\infty m\left(\frac{1}{2}\right)^{n+1} = m - \frac{m}{2} > 0,$$

completing the proof. ∎

Proof of the Impossibility Theorem: Suppose to the contrary that there exists a measure μ with domain $2^{\mathbb{R}}$ and such that (i) $\mu(\mathbb{R}) > 0$ [possibly infinite], (ii) μ assigns finite measure to all bounded sets, and (iii) $\mu(\{x\}) = 0$ for every $x \in \mathbb{R}$. Observe that

$$0 < \mu(\mathbb{R}) = \mu\left(\bigcup_{l \in \mathbb{Z}} (l, l+1]\right) = \sum_{l \in \mathbb{Z}} \mu((l, l+1]),$$

hence $\mu((l, l+1]) > 0$ for some $l \in \mathbb{Z}$. Pick such an $l \in \mathbb{Z}$, and suppose that $\mu((l, l+1]) = m$ for some $0 < m < +\infty$. Now both $(l, l+1]$ and $\boldsymbol{T}$ have cardinal number $\mathfrak{c}$ [Claim 4], hence there exists a one-to-one and onto function $\psi : \boldsymbol{T} \to (l, l+1]$.

Next, for each $i, k \in \mathbb{N}$ let $A_k^i = \{\psi(\boldsymbol{t}) : \boldsymbol{t} \in \boldsymbol{T},\ t_i = k\}$. Since for every $\boldsymbol{t} \in \boldsymbol{S}$ and $i \in \mathbb{N}$ there is $k \in \mathbb{N}$ with $t_i = k$, it follows for each $i \in \mathbb{N}$ that

$$\begin{aligned} \bigcup_{k=1}^{\infty} A_k^i = \{\psi(\boldsymbol{t}) : \boldsymbol{t} \in \boldsymbol{T},\ t_i \in \mathbb{N}\} = \{\psi(\boldsymbol{t}) : \boldsymbol{t} \in \boldsymbol{T}\} &= \psi(\boldsymbol{T}) \\ &= (l, l+1], \end{aligned}$$

and hence $\mu\left(\bigcup_{k=1}^{\infty} A_k^i\right) = \mu((l, l+1]) = m$ for each $i \in \mathbb{N}$. By property (M8), we have for each $i \in \mathbb{N}$ that there exists a minimal $u_i \in \mathbb{N}$ such that

$$\mu\left(\bigcup_{k=1}^{u_i} A_k^i\right) \geq m\left(1 - \left(\frac{1}{2}\right)^{i+1}\right). \tag{4.6}$$

We will let $\boldsymbol{u}$ denote the element of $\boldsymbol{S}$ whose terms are $u_1, u_2, \cdots$ as given just above.

We now define

$$A = \bigcap_{i=1}^{\infty} \bigcup_{k=1}^{u_i} A_k^i.$$

Since for each $i \in \mathbb{N}$ we have

$$\bigcup_{k=1}^{u_i} A_k^i = \bigcup_{k=1}^{u_i} \{\psi(\boldsymbol{t}) : \boldsymbol{t} \in \boldsymbol{T},\ t_i = k\} = \{\psi(\boldsymbol{t}) : \boldsymbol{t} \in \boldsymbol{T},\ t_i \leq u_i\},$$

we thus have

$$A = \bigcap_{i=1}^{\infty} \bigcup_{k=1}^{u_i} A_k^i = \bigcap_{i=1}^{\infty} \{\psi(\boldsymbol{t}) : \boldsymbol{t} \in \boldsymbol{T},\ t_i \leq u_i\} = \{\psi(\boldsymbol{t}) : \boldsymbol{t} \in \boldsymbol{T},\ \boldsymbol{t} \trianglerighteq \boldsymbol{u}\}. \tag{4.7}$$

Since μ has domain $2^{\mathbb{R}}$, $\mu(A)$ exists. By (4.6) and Claim 5 it follows that $\mu(A) > 0$.

We now claim that there exists $\boldsymbol{t}_0 \in \boldsymbol{T}$ such that $\boldsymbol{u} \trianglerighteq \boldsymbol{t}_0$. To see this, consider the cases where $\boldsymbol{u} \in \boldsymbol{T}$ and $\boldsymbol{u} \notin \boldsymbol{T}$. If $\boldsymbol{u} \in \boldsymbol{T}$ we may take $\boldsymbol{t}_0 = \boldsymbol{u}$. If $\boldsymbol{u} \notin \boldsymbol{T}$ then by the definition of $\boldsymbol{T}$ there must exist some $\boldsymbol{t}_0 \in \boldsymbol{S}$ that rejects $\boldsymbol{u}$ and hence $\boldsymbol{u} \trianglerighteq \boldsymbol{t}_0$. By Claim 3, this $\boldsymbol{t}_0$ that rejects $\boldsymbol{u}$ is necessarily in $\boldsymbol{T}$. Therefore, in either case there exists $\boldsymbol{t}_0 \in \boldsymbol{T}$ such that $\boldsymbol{u} \trianglerighteq \boldsymbol{t}_0$.

We next claim that $\{\psi(\boldsymbol{t}) : \boldsymbol{t} \in \boldsymbol{T},\ \boldsymbol{t} \trianglerighteq \boldsymbol{u}\}$ is an amc subset of $(l, l+1]$. Since ψ is a one-to-one function from $\boldsymbol{T}$ onto $(l, l+1]$, we may equivalently show that $\{\boldsymbol{t} \in \boldsymbol{T} :\ \boldsymbol{t} \trianglerighteq \boldsymbol{u}\}$ is an amc subset of $\boldsymbol{T}$. To show this, let $\boldsymbol{t} \in \boldsymbol{T}$ be such that $\boldsymbol{t} \trianglerighteq \boldsymbol{u}$. Invoke the previous paragraph to find $\boldsymbol{t}_0 \in \boldsymbol{T}$ such that $\boldsymbol{u} \trianglerighteq \boldsymbol{t}_0$. Since $\trianglerighteq$ is transitive, we have $\boldsymbol{t} \trianglerighteq \boldsymbol{t}_0$, where both $\boldsymbol{t}$ and $\boldsymbol{t}_0$ are in $\boldsymbol{T}$. There are two options with regard to the order relation $\leq_w$: either $\boldsymbol{t}_0 <_w \boldsymbol{t}$, or $\boldsymbol{t} \leq_w \boldsymbol{t}_0$. If $\boldsymbol{t}_0 <_w \boldsymbol{t}$ holds, then in conjunction with the fact that $\boldsymbol{t} \trianglerighteq \boldsymbol{t}_0$ we obtain $\boldsymbol{t}_0 \operatorname{rej} \boldsymbol{t}$, hence $\boldsymbol{t} \notin \boldsymbol{T}$, a contradiction. Therefore, it must be the case that $\boldsymbol{t} \leq_w \boldsymbol{t}_0$. We have therefore shown that

$$\{\boldsymbol{t} \in \boldsymbol{T} :\ \boldsymbol{t} \trianglerighteq \boldsymbol{u}\} \subseteq \{\boldsymbol{t} \in \boldsymbol{T} :\ \boldsymbol{t} \leq_w \boldsymbol{t}_0\}.$$

From the fact that $\boldsymbol{T} \subseteq \boldsymbol{S}$, we obtain

$$\{\boldsymbol{t} \in \boldsymbol{T} :\ \boldsymbol{t} \trianglerighteq \boldsymbol{u}\} \subseteq \{\boldsymbol{t} \in \boldsymbol{S} :\ \boldsymbol{t} \leq_w \boldsymbol{t}_0\}.$$

By Claim 1, $\{\boldsymbol{t} \in \boldsymbol{S} :\ \boldsymbol{t} \leq_w \boldsymbol{t}_0\}$ is amc. Therefore, $\{\boldsymbol{t} \in \boldsymbol{T} :\ \boldsymbol{t} \trianglerighteq \boldsymbol{u}\}$ is itself amc, so that A is amc by (4.7). Since $\mu(\{x\}) = 0$ for all $x \in \mathbb{R}$, it follows that $\mu(A) = 0$, which contradicts the earlier fact that $\mu(A) > 0$. The assumption that there exists such a measure $\mu : 2^{\mathbb{R}} \to \bar{\mathbb{R}}$ as stated leads to the contradiction of asserting the existence of a subset $A \subseteq \mathbb{R}$ such that $\mu(A)$ is both zero and positive. The only logical alternative then is that no such μ can exist. ∎

If we allow $\mu(\mathbb{R}) = 0$ as a possibility in the statement of the previous theorem, then the theorem is false, for the measure μ that is identically zero for all subsets serves as a counterexample. However, measures that assign the value 0 to the whole space are completely uninteresting trivialities.

It should be noted that this proof [and indeed even the notation] is adapted from the presentation of Billingsley's *Probability and Measure.*

Exercise.

1. There can exist no measure $\mu : 2^{\mathbb{R}^k} \to \bar{\mathbb{R}}$ such that (i) $0 < \mu(\mathbb{R}^k) \leq +\infty$, (ii) μ assigns a finite nonnegative number to each bounded subset of $\mathbb{R}^k$, and (iii) $\mu(\{\boldsymbol{x}\}) = 0$ for each $\boldsymbol{x} \in \mathbb{R}^k$. This is the k-dimensional version of the Impossibility Theorem.

4.9 EXCURSUS: "EXTREMELY NONMEASURABLE SETS"

In Section 5, we construced a set $E \subseteq \mathbb{R}^k$ with $E \notin \mathcal{L}^k$. This set, of course, does not have a Lebesgue measure; the symbols $\lambda_k(E)$ and $\bar{\lambda}_k(E)$ are undefined. This is a specific exhibition of a set not in $\mathcal{M}(\mu^*)$ for the Extension

Theorem framework of Chapter 3. It would be interesting to produce measure spaces other than $(\mathbb{R}^k, \mathcal{L}^k, \bar{\lambda}_k)$ and some subsets of these spaces that fail to be in the corresponding σ-field $\mathcal{M}(\mu^*)$ or the sub-σ-field $\sigma(\mathcal{A})$. Such sets would be "nonmeasurable" in the sense that one could not meaningfully apply $\mu^*_{\mathcal{M}(\mu^*)}$ or $\mu^*_{\sigma(\mathcal{A})}$ to them.

The general problem of finding such "nonmeasurable" sets is a very difficult one, as no "recipe" for creating such sets exists for general measure spaces. However, there are interesting constructions for certain specific settings. This section produces a collection of subsets of $\mathbb{R}^k$ that fail to have a meaningfully defined measure for any measure in a very large class of measures.

We use the framework and notation of Sections 2.3, 2.4, and 3.10, and we follow the construction of Hewitt and Stromberg (1969). Let $\mathcal{A}_k$ denote the semiring consisting of $\varnothing$ and the bounded k-dimensional rsc rectangles $(\boldsymbol{a}, \boldsymbol{b}]$, and let $F \in \mathfrak{S}_k$. We have shown that $\Delta_F : \mathcal{A}_k \to \mathbb{R}$ is a measure, and if $\Delta^*_F : 2^{\mathbb{R}^k} \to \bar{\mathbb{R}}$ denotes the outer measure induced by Δ_F, then the restriction $\Delta^*_{F,\mathcal{M}(\Delta^*_F)}$ of Δ^*_F to $\mathcal{M}(\Delta^*_F)$ is a measure, and the restriction $\Delta^*_{F,\mathcal{B}^k}$ of Δ^*_F to $\mathcal{B}^k$ is a measure as well. *We will construct a subset of $\mathbb{R}^k$ outside $\mathcal{M}(\Delta^*_F)$ for any continuous $F \in \mathfrak{S}_k$ that is not a finite sum of k-degenerate functions.*

Claim 1. *For any $A \in \mathcal{M}(\Delta^*_F)$ [with $F \in \mathfrak{S}_k$], we have*

$$\Delta^*_{F,\mathcal{M}(\Delta^*_F)}(A) = \sup\{\Delta^*_{F,\mathcal{B}^k}(K) : \ K \subseteq A, \ K \textit{ compact}\}.$$

Proof: For any $B \in \mathcal{B}^k$, we have

$$\Delta^*_{F,\mathcal{B}^k}(B) = \sup\{\Delta^*_{F,\mathcal{B}^k}(K) : \ K \subseteq B, \ K \text{ compact}\}$$

by the fact that $\Delta^*_{F,\mathcal{B}^k}$ assigns finite measure to bounded sets and Claim 7 of Section 3.8. If $A \in \mathcal{M}(\Delta^*_F)$, then $A = B \cup C$, where $B \in \mathcal{B}^k$ and C is contained in some $\Delta^*_{F,\mathcal{B}^k}$-null set N. Since

$$\begin{aligned} \Delta^*_{F,\mathcal{M}(\Delta^*_F)}(A) = \Delta^*_{F,\mathcal{B}^k}(B) &= \sup\{\Delta^*_{F,\mathcal{B}^k}(K) : \ K \subseteq B, \ K \text{ compact}\} \\ &\leq \sup\{\Delta^*_{F,\mathcal{B}^k}(K) : \ K \subseteq A, \ K \text{ compact}\} \\ &\leq \Delta^*_{F,\mathcal{M}(\Delta^*_F)}(A). \end{aligned}$$

[The first equality is by definition and the first inequality holds since $B \subseteq A$.] The inequalities above are therefore equalities and the proof is complete. ∎

Claim 2. *The collection of open subsets of $\mathbb{R}^k$ has cardinal number $\mathfrak{c}$.*

Proof: Recall that the countable collection $\mathcal{O} = \{O_1, O_2, \cdots\}$ of open k-dimensional rectangles with rational endpoints forms a base for the usual topology in $\mathbb{R}^k$ induced by the Euclidean metric. That is, every open $A \subseteq \mathbb{R}^k$ is an amc union of $\mathcal{O}$-sets, where without any loss of generality we state that (i) every $\mathcal{O}$-set appears at most once in the union that defines A, and (ii) the elements of $\mathcal{O}$ whose union is A are written in the order of increasing indices.

Therefore, specifying an open $A \subseteq \mathbb{R}^k$ is equivalent to specifying a function $\phi_A : \mathcal{O} \to \mathbb{N} \cup \{0\}$, where $\phi_A(O_i) = n$ iff O_i is the nth $\mathcal{O}$-set that appears in the amc union that defines A, and $\phi_A(O_i) = 0$ if O_i does not appear in the amc union that defines A. [ϕ_A is well defined by stipulations (i) and (ii).] We have that

$$\begin{aligned} \text{card}(\{\phi_A : A \subseteq \mathbb{R} \text{ open}, \phi_A \text{ as above}\}) \le \text{card}(\mathcal{O})^{\text{card}(\mathbb{N}\cup\{0\})} &= {\aleph_0}^{\aleph_0} \\ &= \mathfrak{c}. \end{aligned}$$

Therefore, there are at most $\mathfrak{c}$ open subsets of $\mathbb{R}^k$. However, there are also clearly at least $\mathfrak{c}$ open subsets of $\mathbb{R}^k$ [consider the rectangles $(-x\mathbf{1}, x\mathbf{1})$ where $x \in \mathbb{R}$], and hence there are $\mathfrak{c}$ open subsets of $\mathbb{R}^k$. ■

Claim 3. *The collection of uncountable closed subsets of $\mathbb{R}^k$ has cardinal number $\mathfrak{c}$.*

Proof: Since a subset of $\mathbb{R}^k$ is closed iff it is open, Claim 2 shows that there are $\mathfrak{c}$ closed subsets of $\mathbb{R}^k$, and hence the number of uncountable closed subsets of $\mathbb{R}^k$ is no more than $\mathfrak{c}$. At the same time, each of the closed rectangles $[-x\mathbf{1}, x\mathbf{1}]$ with $x \in \mathbb{R}$ is an uncountable closed subset of $\mathbb{R}^k$, and there are at least $\mathfrak{c}$ such subsets. This proves the claim. ■

For each ordinal α, let $P_\alpha = \{\gamma : \gamma \prec \alpha\}$ denote the set of predecessors of α. We will let ω denote the smallest uncountable ordinal. We will now begin the construction of the promised set.

Construction. Let $\{F_\eta : \eta \in P_\omega\}$ denote a well ordering of the $\mathfrak{c}$ uncountable closed subsets of $\mathbb{R}^k$; by Claim 3 and (AC) this is possible. We employ the following scheme, courtesy of transfinite induction:

Stage 0. Let $\boldsymbol{x}_0$ and $\boldsymbol{y}_0$ be distinct points of F_0.

Stage 1. The set $A_1 = \{\boldsymbol{x}_0, \boldsymbol{y}_0\}$ clearly has cardinal number less than $\mathfrak{c}$, hence $\text{card}(F_1 - A_1) = \mathfrak{c}$. Pick distinct $\boldsymbol{x}_1, \boldsymbol{y}_1 \in F_1 - A_1$.

Stage 2. The set $A_2 = \{\boldsymbol{x}_0, \boldsymbol{x}_1, \boldsymbol{y}_0, \boldsymbol{y}_1\}$ clearly has cardinal number less than $\mathfrak{c}$, hence $\text{card}(F_2 - A_2) = \mathfrak{c}$. Pick distinct $\boldsymbol{x}_2, \boldsymbol{y}_2 \in F_2 - A_2$

$$\vdots$$

Stage η. Let $\eta \in P_\omega$. If $A_\eta = \{\boldsymbol{x}_\gamma : \gamma \prec \eta\} \cup \{\boldsymbol{y}_\gamma : \gamma \prec \eta\}$, then

$$\begin{aligned} \text{card}(A_\eta) &= \text{card}(\{\boldsymbol{x}_\gamma : \gamma \prec \eta\}) + \text{card}(\{\boldsymbol{y}_\gamma : \gamma \prec \eta\}) \\ &= \text{card}(P_\eta) + \text{card}(P_\eta) \\ &< \mathfrak{c} + \mathfrak{c} = \mathfrak{c}. \end{aligned}$$

[The inequality holds because $\eta \prec \omega$ and ω is the smallest ordinal such that $\text{card}(P_\omega) = \mathfrak{c}$.] Therefore, we may pick distinct $\boldsymbol{x}_\eta, \boldsymbol{y}_\eta \in F_\eta - A_\eta$.

Carrying out the transfinitely inductive scheme for all $\gamma \prec \omega$, We will define $B = \{\boldsymbol{x}_\eta : \eta \prec \omega\}$, where such a set exists by invocation of (AC). This set B will end up being the promised set.

Claim 4. *For any $\eta \prec \omega$, we have $B \cap F_\eta \neq \varnothing$ and $B^c \cap F_\eta \neq \varnothing$. [The complement is of course with respect to $\mathbb{R}^k$.]*

Proof: Let $\eta \prec \omega$. Observe that by construction we have $\boldsymbol{x}_\eta \in B$ and $\boldsymbol{x}_\eta \in F_\eta$, hence $B \cap F_\eta \neq \varnothing$. To see that $B^c \cap F_\eta \neq \varnothing$, observe that while $\boldsymbol{y}_\eta \in F_\eta$, the definition of B forces $\boldsymbol{y}_\eta \notin B$. ∎

The following claim is a mere consequence of definitions, making no use of anything done in this section. The proof is a straightforward exercise.

Claim 5. *If $F \in \mathfrak{S}_k$ is continuous, then $\Delta^*_{F,\mathcal{B}^k}(\{\boldsymbol{x}\}) = 0$ for every $\boldsymbol{x} \in \mathbb{R}^k$.*

Claim 6. *Compact subsets of B and B^c are necessarily amc.*

Proof: Let $K \subseteq B$ be compact, so that K is closed among other things. The set K is either uncountable or amc. If K is uncountable, then $K = F_\eta$ for some $\eta \prec \omega$. By Claim 4, K contains elements not in B. This contradicts the assumption $K \subseteq B$, and hence K must be amc. Since Claim 4 is symmetric in B and B^c, the same argument works for B^c as well. ∎

Claim 7. *If $\Delta^*_{F,\mathcal{B}^k}(\mathbb{R}^k) = 0$, then F is a finite sum of k-degenerate functions.*

Proof: If $\Delta^*_{F,\mathcal{B}^k}(\mathbb{R}^k) = 0$, then $\Delta_F(A) = 0$ for every $A \in \mathcal{A}_k$. By Claim 2 of Section 3.10, the result follows. ∎

Claim 7, like Claim 5, depends nowhere on anything specific to this section, but is merely a standard consequence of the material from Section 3.10.

The next and final claim culminates the development of this section, showing that the set B constructed earlier is so pathological that we can't meaningfully discuss its measure for a large collection of measures.

Claim 8. *$B \notin \mathcal{M}(\Delta^*_F)$ for any continuous $F \in \mathfrak{S}_k$ that is not a finite sum of k-degenerate functions.*

Proof: Suppose to the contrary that $B \in \mathcal{M}(\Delta^*_F)$ for some continuous F in $\mathfrak{S}_k$ that is not a finite sum of k-degenerate functions. By Claim 1, we have

$$\Delta^*_{F,\mathcal{M}(\Delta^*_F)}(B) = \sup\{\Delta^*_{F,\mathcal{B}^k}(K) : K \subseteq B, \ K \text{ compact}\}.$$

and

$$\Delta^*_{F,\mathcal{M}(\Delta^*_F)}(B^c) = \sup\{\Delta^*_{F,\mathcal{B}^k}(K) : K \subseteq B^c, \ K \text{ compact}\}.$$

However, a typical compact subset K_1 of B is amc, and a typical compact subset K_2 of B^c is amc, hence $\Delta^*_{F,\mathcal{B}^k}(K_1) = \Delta^*_{F,\mathcal{B}^k}(K_2) = 0$ by Claim 5 and countable additivity. Therefore, we have that $\Delta^*_{F,\mathcal{M}(\Delta^*_F)}(B) = \Delta^*_{F,\mathcal{M}(\Delta^*_F)}(B^c) = 0$, hence $\Delta^*_{F,\mathcal{B}^k}(\mathbb{R}^k) = 0$. Claim 7 forces F to be a finite sum of k-degenerate functions, a contradiction. This shows that $B \notin \mathcal{M}(\Delta^*_F)$ for any continuous $F \in \mathfrak{S}_k$ that is not a finite sum of k-degenerate functions. ∎

In particular, if $F \in \mathfrak{S}_k$ is such that $F(\boldsymbol{x}) = \prod_{i=1}^k x_i$ [so that F is continuous and is not a finite sum of k-degenerate functions], then the set B as constructed is not in $\mathcal{L}^k$ and hence is not in $\mathcal{B}^k$.

Exercises. [Assume the notation of this section.]

1*. $\Delta^*_{F,\mathcal{B}^k}(\{\boldsymbol{x}\}) = 0$ for all $\boldsymbol{x} \in \mathbb{R}^k$ iff $F \in \mathfrak{S}_k$ is continuous. [The "if" direction is Claim 5.]

2. Let $F \in \mathfrak{S}$ be continuous. For each $\epsilon > 0$ there exists a nowhere dense perfect set $A \subseteq [0,1]$ such that $A \in \mathcal{M}(\Delta^*_F)$ and $\Delta^*_{F,\mathcal{M}(\Delta^*_F)}(A) > \Delta^*_{F,\mathcal{B}}([0,1]) - \epsilon$. [Let $q_1, q_2, \cdots$ denote the rational numbers in $[0,1]$. For each $n \in \mathbb{N}$, find ϵ_n such that $B_n = (q_n - \epsilon_n, q_n + \epsilon_n)$ has $\Delta^*_{F,\mathcal{B}}$-measure less than 2^{-n}, which is possible by the continuity of F. Let $A = [0,1] - \bigcup_{n=1}^{\infty} B_n$. A is closed and the interior of A is empty by construction.] In the case of Lebesgue measure, this says [roughly] that there is a subset of $[0,1]$ with Lebesgue measure ["length"] arbitrarily close to 1, yet this set contains no open subinterval of $[0,1]$.

3. Let $F \in \mathfrak{S}$ be continuous. There exists a set $B \subseteq [0,1]$ with (i) $B \in \mathcal{M}(\Delta^*_F)$, (ii) B is a countable union of nowhere dense sets, and (iii) $\Delta^*_{F,\mathcal{M}(\Delta^*_F)}(B) = \Delta^*_{F,\mathcal{B}}([0,1])$. [For each $n \in \mathbb{N}$, use Exercise 2 to find a nowhere dense perfect set $A_n \subseteq [0,1]$ with $\Delta^*_{F,\mathcal{M}(\Delta^*_F)}(A_n) > \Delta^*_{F,\mathcal{B}}([0,1]) - n^{-1}$. Let $B = \bigcup_{n=1}^{\infty} A_n$.]

4. Let $F \in \mathfrak{S}$ be continuous. There exists a set $C \subseteq [0,1]$ such that (i) C is the countable intersection of open subsets of $[0,1]$, (ii) C is not a countable union of nowhere dense sets, and (iii) the $\Delta^*_{F,\mathcal{M}(\Delta^*_F)}$-measure of C is 0. [Use Exercise 3.]

5. The Continuum Hypothesis (CH) and the Axiom of Choice (AC) imply that there exists a set $S \subseteq \mathbb{R}$ with $\text{card}(S) = \mathfrak{c}$, $S \in \mathcal{M}(\Delta^*_F)$, and $\Delta^*_{F,\mathcal{M}(\Delta^*_F)}(S) = 0$ for every continuous $F \in \mathfrak{S}$. Follow this outline.
(a) There exists a well-ordering $\{C_\eta : \eta \prec \omega\}$ of the collection of nowhere dense closed subsets of $[0,1]$, where ω again denotes the smallest uncountable ordinal.
(b) Employ a transfinitely recursive scheme to obtain a set $\{x_\eta : \eta \prec \omega\}$ by the following: let $x_0 \in [0,1] - C_0$; suppose that for each $\eta \prec \omega$ the number x_γ is defined for each $\gamma \prec \eta$. Pick $x_\eta \in [0,1] - \bigcup_{\gamma \prec \eta}(C_\gamma \cup \{x_\gamma\})$. This construction requires (AC). [Precisely where?]
(c) Let $S = \{x_\eta : \eta \prec \omega\}$. Then (CH) forces $\text{card}(S) = \mathfrak{c}$.
(d) For every $\eta \prec \omega$, we have that $S \cap C_\eta$ is amc.
(e) $S \in \mathcal{M}(\Delta^*_F)$ and $\Delta^*_{F,\mathcal{M}(\Delta^*_F)}(S) = 0$ for every continuous $F \in \mathfrak{S}$.
(f) S cannot contain a nonempty perfect subset.

Comment: The uncountable set S contained in the separable and complete metric space $[0,1]$ cannot possibly be a Borel set [even though the exercise asks you to show that S is in the larger collection $\mathcal{M}(\Delta^*_F)$]. This is due to the following theorem, found as Theorem 120 in Sierpiński (1956): *An uncountable Borel subset of a complete separable metric space cannot contain a nonempty perfect subset.* [This theorem is stated without proof.] Taking F to be the identity function, we therefore have yet another example of a set in $\mathcal{L} - \mathcal{B}$.

5

Measurable Functions

After completion of the deep theorems in Chapters 2-4 regarding the existence, extension, and uniqueness of measures, the next major thing to discuss is the abstract Lebesgue integral. It is possible at this stage to directly proceed to such a concept. However, the technical results that one would prove along the way would, although fairly straightforward, clutter the presentation of the main constructions and results to a certain degree. Therefore, to keep the development of the Lebesgue integral as clean as possible, we will use this chapter to discuss these aforementioned technical results, which results involve the notion of a *measurable function.*

Before proceeding to discuss this concept, a short review of arithmetic is in order. We will be working with the extended real line $\bar{\mathbb{R}} = \mathbb{R} \cup \{-\infty, +\infty\}$ with the usual ordering: $-\infty < +\infty$ and $-\infty < x < +\infty$ for all $x \in \mathbb{R}$. We define $-(-\infty) = +\infty$ and $-(+\infty) = -\infty$. We shall often write ∞ for $+\infty$. Special care must be taken with the four basic arithmetic operations in $\bar{\mathbb{R}}$:

Addition. Addition is defined as follows: $x + \pm\infty = \pm\infty + x = \pm\infty$ for each $x \in \mathbb{R}$. We also have $+\infty + \infty = +\infty$ and $-\infty + (-\infty) = -\infty$. *The two expressions* $+\infty + (-\infty)$ *and* $-\infty + \infty$ *are left undefined.*[1]

Subtraction. Subtraction is defined as follows: $x - (\pm\infty) = \mp\infty$ for each $x \in \mathbb{R}$. We also have $-\infty - (+\infty) = -\infty$ and $+\infty - (-\infty) = +\infty$. *The expressions* $+\infty - (+\infty)$ *and* $-\infty - (-\infty)$ *are left undefined.*

[1]Therefore, not every element of $\bar{\mathbb{R}}$ has an additive inverse, hence $\bar{\mathbb{R}}$ fails to be a group under the operation $+$. This fact will not cause us any difficulty.

Multiplication. For multiplication, we have $(\pm\infty)x = \pm\infty$ for $0 < x \leq +\infty$ and $(\pm\infty)x = \mp\infty$ for $-\infty \leq x < 0$. *We adopt the convention of writing* $(\pm\infty)(0) = (0)(\pm\infty) = 0$. There are no undefined expressions involving multiplication as there are for addition and subtraction.

Division. For division, we have $\frac{\pm\infty}{x} = \pm\infty$ for $0 < x < +\infty$ and $\frac{\pm\infty}{x} = \mp\infty$ when $-\infty < x < 0$. *The expression* $\frac{\pm\infty}{x}$ *is undefined for* $x \in \{-\infty, 0, +\infty\}$. Also, for every $x \in \mathbb{R}$ we have $\frac{x}{\pm\infty} = 0$. *The expression* $\frac{x}{\pm\infty}$ *is undefined if* $x \in \{-\infty, +\infty\}$.

Therefore, in dealing with arguments that involve arithmetic in $\bar{\mathbb{R}}$ [such as many of the claims and proofs in this chapter], we must at all times avoid the undefined expressions given above.

Topology. We use the standard topology of open sets for $\bar{\mathbb{R}}$. That is, the open subsets of $\bar{\mathbb{R}}$ are sets of the form

$$A, \quad A \cup [-\infty, x), \quad A \cup (x, +\infty], \quad \text{and} \quad A \cup [-\infty, x) \cup (y, +\infty],$$

where $A \subseteq \mathbb{R}$ is open in $\mathbb{R}$ and $x, y \in \mathbb{R}$.

Notation for Borel Sets. It will be noted that in Chapter 1 we used $\bar{\mathcal{B}}$ to denote the extended Borel σ-field on $\bar{\mathbb{R}}$, whereas in Chapter 3 the symbol $\bar{\mathcal{B}}$ denoted the completion of $\mathcal{B}$ with respect to Lebesgue measure λ, that is, $\bar{\mathcal{B}} = \mathcal{L}$. No confusion arose from this even though the Borel σ-field on $\bar{\mathbb{R}}$ is not the same thing as the σ-field of Lebesgue subsets of $\mathbb{R}$, but we must now reform our notation. *From this point forward we will use* $\mathcal{B}^*$ *to denote the extended Borel* σ*-field on* $\bar{\mathbb{R}}$, *and we will use the symbol* $\mathcal{B}^{k*}$ *to denote the extended Borel* σ*-field on* $\bar{\mathbb{R}}^k$. The symbols $\mathcal{L}$ and $\mathcal{L}^k$ retain their original meaning, and a horizontal bar will denote the completion process described in Chapters 3 and 4.

What we are about to present in this chapter would be much simpler if we considered only elements of $\mathbb{R}$. However, we will be dealing with suprema, infima, superior limits, inferior limits, and limits [when they exist], and it is quite common for such things to be infinite. If we want a comprehensive treatment of these cases, we are forced to admit the values $\pm\infty$ into our discussions. This will create some extra work, but in the end we will have a rigorous theory that includes the infinite and not only the finite cases.

Notation Review. This introduction closes with a review of some standard notation. Let Ω_1 and Ω_2 be nonempty sets, let $\mathcal{A}_1 \subseteq 2^{\Omega_1}$ and $\mathcal{A}_2 \subseteq 2^{\Omega_2}$, and let f denote a function from Ω_1 into Ω_2. Given any subset A_2 of Ω_2, $f^{-1}(A_2) = \{\omega_1 \in \Omega_1 : f(\omega_1) \in A_2\}$ denotes the usual inverse image of the set A_2. Given $A_1 \subseteq \Omega_1$, $f(A_1) = \{f(\omega_1) : \omega_1 \in A_1\}$ is the usual image of A_1. We will also write $f^{-1}(\mathcal{A}_2) = \{f^{-1}(A_2) : A_2 \in \mathcal{A}_2\}$, so that $f^{-1}(\mathcal{A}_2)$ is the collection of inverse images of $\mathcal{A}_2$-sets. Similarly, we define the collection $f(\mathcal{A}_1) = \{f(A_1) : A_1 \in \mathcal{A}_1\}$, which consists of the images of $\mathcal{A}_1$-sets.

5.1 MEASURABILITY

Throughout this section, $(\Omega, \mathcal{F})$ will denote a measurable space: Ω will denote a nonempty background set, and $\mathcal{F}$ will denote a σ-field on Ω. We proceed immediately to the main definition.

Definition. Let $A \in \mathcal{F}$ be nonempty, and let $f : A \to \bar{\mathbb{R}}$ denote a function. We will say that f is $\mathcal{F}/\mathcal{B}^*$*-measurable* iff $f^{-1}(\mathcal{B}) \subseteq \mathcal{F}$ and both $f^{-1}(\{-\infty\})$ and $f^{-1}(\{+\infty\})$ are in $\mathcal{F}$.

If the range of f is contained in $\mathbb{R}$, the above definition states that f is $\mathcal{F}/\mathcal{B}^*$-measurable iff $f^{-1}(B) \in \mathcal{F}$ for each $B \in \mathcal{B}$, since in this case we have $f^{-1}(\{-\infty\}) = f^{-1}(\{+\infty\}) = \varnothing \in \mathcal{F}$. That is, if f is real-valued, then the condition for $\mathcal{F}/\mathcal{B}^*$-measurability becomes $f^{-1}(\mathcal{B}) \subseteq \mathcal{F}$.

There are various definitions of $\mathcal{F}/\mathcal{B}^*$-measurable functions found in various treatments of measure theory, given by (a)-(f) below. The first order of business is to show that these definitions are in fact equivalent.

(MF1) *Let D denote a dense subset of $\mathbb{R}$. The following requirements are all logically equivalent:*

(a) $f^{-1}((x, +\infty]) \in \mathcal{F}$ for each $x \in D$.

(b) $f^{-1}([x, +\infty]) \in \mathcal{F}$ for each $x \in D$.

(c) $f^{-1}([-\infty, x)) \in \mathcal{F}$ for each $x \in D$.

(d) $f^{-1}([-\infty, x]) \in \mathcal{F}$ for each $x \in D$.

(e) $f^{-1}(B) \in \mathcal{F}$ for every $B \in \mathcal{B}^*$. [That is, $f^{-1}(\mathcal{B}^*) \subseteq \mathcal{F}$.]

(f) $f^{-1}(B) \in \mathcal{F}$ for every $B \in \mathcal{B}$ [that is, $f^{-1}(\mathcal{B}) \subseteq \mathcal{F}$] with both $f^{-1}(\{-\infty\})$ and $f^{-1}(\{+\infty\})$ in $\mathcal{F}$. [This is the original definition.]

Consequently, f is $\mathcal{F}/\mathcal{B}^*$-measurable iff any of (a)-(f) hold.

Proof: We shall show (a) $\to$ (b) $\to$ (c) $\to$ (d) $\to$ (a), and we shall then show that (a) $\leftrightarrow$ (e) $\leftrightarrow$ (f).

Suppose that f satisfies (a), and let $x \in D$. Since D is dense, there exists a sequence $x_1 < x_2 < \cdots < x$ with $x_n \to x$ and $x_n \in D$ for each $n \in \mathbb{N}$. We have $f^{-1}([x, +\infty]) = f^{-1}(\bigcap_{n=1}^{\infty}(x_n, +\infty]) = \bigcap_{n=1}^{\infty} f^{-1}((x_n, +\infty]) \in \mathcal{F}$ by the properties of inverse images and by the fact that $f^{-1}((x_n, +\infty]) \in \mathcal{F}$ for each $n \in \mathbb{N}$. Therefore (a) implies (b).

Now assume that f satisfies (b), and let $x \in D$. Observe that we have $f^{-1}([-\infty, x)) = f^{-1}([x, +\infty]^c) = f^{-1}([x, +\infty])^c$, so $f^{-1}([-\infty, x))$ is the complement of an $\mathcal{F}$-set and hence is in $\mathcal{F}$. Since $x \in D$ was generic, we have shown that (b) implies (c).

To see that (c) implies (d), let $x \in D$, and let $\{x_n\}_{n=1}^{\infty}$ denote a sequence of elements of D such that $x_1 > x_2 > \cdots > x$ and $x_n \to x$. Since the assumption of (c) forces $f^{-1}([-\infty, x_n)) \in \mathcal{F}$ for each $n \in \mathbb{N}$, it follows that we have $f^{-1}([-\infty, x]) = f^{-1}(\bigcap_{n=1}^{\infty}[-\infty, x_n)) = \bigcap_{n=1}^{\infty} f^{-1}([-\infty, x_n)) \in \mathcal{F}$, verifying (d). That is, (c) implies (d).

Next, suppose that f satisfies (d). Then $f^{-1}([-\infty, x]) \in \mathcal{F}$ for each $x \in D$, hence $f^{-1}([-\infty, x])^c \in \mathcal{F}$ for each $x \in D$, so $f^{-1}([-\infty, x]^c) \in \mathcal{F}$ for each $x \in D$. It follows that $f^{-1}((x, +\infty]) \in \mathcal{F}$ for each $x \in D$, whence f satisfies (a). We have shown that (d) implies (a), and therefore (a)-(d) are equivalent.

We now connect (a) and (e). Let $\mathcal{A} = \{(x, +\infty] : x \in D\}$. It follows that[2] $\sigma(\mathcal{A}) = \mathcal{B}^*$. By (a), we have $f^{-1}(\mathcal{A}) \subseteq \mathcal{F}$, so that $\sigma(f^{-1}(\mathcal{A})) \subseteq \sigma(\mathcal{F}) = \mathcal{F}$. By Claim 2 of Section 1.3, we have $f^{-1}(\mathcal{B}^*) = f^{-1}(\sigma(\mathcal{A})) = \sigma(f^{-1}(\mathcal{A})) \subseteq \mathcal{F}$, hence f satisfies (e). Therefore (a) implies (e). Conversely, if f satisfies (e), then f trivially satisfies (a) since $(x, +\infty] \in \mathcal{B}^*$ for each $x \in D$. Therefore (e) implies (a), so that (a) and (e) are equivalent.

We complete the proof by showing that (e) and (f) are equivalent. That (e) implies (f) trivially follows from the facts that $\mathcal{B} \subseteq \mathcal{B}^*$ and $\{-\infty\}, \{+\infty\} \in \mathcal{B}^*$. To see that (f) implies (e), let $A \in \mathcal{B}^*$; we desire to show that $f^{-1}(A) \in \mathcal{F}$. Since $A \in \mathcal{B}^*$, we have $A = B \cup C$, where $B \in \mathcal{B}$ and C is a subset of $\{-\infty, +\infty\}$. Since $B \in \mathcal{B}$, $f^{-1}(B) \in \mathcal{F}$ by assumption. Next, $f^{-1}(C)$ is exactly one of $f^{-1}(\{-\infty\})$, $f^{-1}(\{+\infty\})$, $f^{-1}(\varnothing)$, or $f^{-1}(\{-\infty, +\infty\})$, each set being in $\mathcal{F}$. Therefore, $f^{-1}(A) = f^{-1}(B \cup C) = f^{-1}(B) \cup f^{-1}(C) \in \mathcal{F}$, so (e) follows from (f). This completes the proof. ■

Given a function f for which we desire to establish $\mathcal{F}/\mathcal{B}^*$-measurability, we may choose which of (a)-(f) is most convenient for the situation at hand. For certain claims (e) or (f) will be the most appropriate condition with which to work. For other claims, one of (a)-(d) will be the most appropriate.

The next property says that the restrictions of $\mathcal{F}/\mathcal{B}^*$-measurable functions to nonempty $\mathcal{F}$-sets are themselves $\mathcal{F}/\mathcal{B}^*$-measurable.

(MF2) *Let $A \in \mathcal{F}$ be nonempty, let $f : A \to \bar{\mathbb{R}}$ be $\mathcal{F}/\mathcal{B}^*$-measurable, and let B denote a nonempty $\mathcal{F}$-subset of A. If f_B denotes the restriction of f from A to B, then f_B is $\mathcal{F}/\mathcal{B}^*$-measurable.*

Proof: We have that $f_B^{-1}(C) = B \cap f^{-1}(C)$ for any $C \subseteq \bar{\mathbb{R}}$. By assumption, $B \in \mathcal{F}$. If $C \in \mathcal{B}^*$, then $f^{-1}(C) \in \mathcal{F}$ by (e) of (MF1). Therefore, $C \in \mathcal{B}^*$ implies that $f_B^{-1}(C)$ is the intersection of two $\mathcal{F}$-sets and hence is in $\mathcal{F}$. It follows by (e) of (MF1) that f_B is $\mathcal{F}/\mathcal{B}^*$-measurable. ■

Definition. Let $A \in \mathcal{B}$ be nonempty, and let $f : A \to \bar{\mathbb{R}}$ be $\mathcal{B}/\mathcal{B}^*$-measurable. We then say that f is *Borel measurable.* If $A \in \mathcal{L}$ is nonempty and $f : A \to \bar{\mathbb{R}}$ is $\mathcal{L}/\mathcal{B}^*$ measurable, we then say that f is *Lebesgue measurable.*

The above definition merely gives synonyms for the term $\mathcal{F}/\mathcal{B}^*$*-measurable* in the special cases of where the measurable space $(\Omega, \mathcal{F})$ is $(\mathbb{R}, \mathcal{B})$ or $(\mathbb{R}, \mathcal{L})$.

[2]Let $\mathcal{C} = \{(x, +\infty] : x \in \mathbb{R}\}$. We know that $\sigma(\mathcal{C}) = \mathcal{B}^*$. Given $(x, +\infty]$ with $x \in \mathbb{R}$, we may write $(x, +\infty] = \bigcup_{n=1}^{\infty}(x_n, +\infty]$, where $\{x_n\}_{n=1}^{\infty}$ is a nonincreasing sequence of elements of D having limit x. Therefore $(x, +\infty]$ may be written in terms of amc unions of elements of $\mathcal{C}$. By Claim 1 of Section 1.4, we have $\sigma(\mathcal{A}) = \sigma(\mathcal{C}) = \mathcal{B}^*$.

From (i) the definition above, (ii) the fact that $\mathcal{B} \subseteq \mathcal{L}$, and (iii) (e) of (MF1), we may conclude that every Borel measurable function is also Lebesgue measurable.

The use of the word *measurable* above is somewhat incongruous at this stage, since nothing given so far in this section explicitly deals with a given measure. Rather, the definition of $\mathcal{F}/\mathcal{B}^*$-measurability is set-theoretic, dealing only with inverse images, membership, and σ-fields.

Let $A \subseteq \Omega$ be arbitrary. Recall that the *indicator function of* A, denoted by I_A, is the function with domain Ω such that $I_A(\omega) = 1$ if $\omega \in A$ and $I_A(\omega) = 0$ if $\omega \notin A$. The following claim, while simple, is important.

(MF3) *The function I_A is $\mathcal{F}/\mathcal{B}^*$-measurable iff $A \in \mathcal{F}$.*

Proof: For any $B \subseteq \bar{\mathbb{R}}$, it is easy to verify that

$$(I_A)^{-1}(B) = \begin{cases} \Omega & 0, 1 \in B \\ \varnothing & 0 \notin B,\ 1 \notin B \\ A & 0 \notin B,\ 1 \in B \\ A^c & 0 \in B,\ 1 \notin B. \end{cases}$$

If I_A is $\mathcal{F}/\mathcal{B}^*$-measurable, then $(I_A)^{-1}(B) \in \mathcal{F}$ for every $B \in \mathcal{B}^*$. In particular, $(I_A)^{-1}(B) \in \mathcal{F}$ for those $B \in \mathcal{B}^*$ that contain 1 but not 0, so that $A \in \mathcal{F}$. Conversely, if $A \in \mathcal{F}$, then $A^c \in \mathcal{F}$ as well. Since any $B \in \mathcal{B}^*$ will be such that $(I_A)^{-1}(B)$ is one of A, A^c, $\varnothing$, or Ω, [all in $\mathcal{F}$], it follows that $(I_A)^{-1}(B) \in \mathcal{F}$, whence I_A is $\mathcal{F}/\mathcal{B}^*$-measurable by (e) of (MF1). ∎

For the next result, we will take our background space Ω to be $\mathbb{R}^k$, where $k \in \mathbb{N}$ is arbitrary. Here, it is the case that $\mathcal{F}/\mathcal{B}^*$-measurable functions exist in profusion.

(MF4) *Let $\Omega = \mathbb{R}^k$, and let $\mathcal{F}$ denote an arbitrary σ-field on $\mathbb{R}^k$ that contains the open subsets of $\mathbb{R}^k$. Then for any nonempty $A \in \mathcal{F}$ and every continuous $f : A \to \bar{\mathbb{R}}$, f is $\mathcal{F}/\mathcal{B}^*$-measurable.*

Proof: Let $\mathcal{O}$ denote the open subsets of $\bar{\mathbb{R}}$, so that $\sigma(\mathcal{O}) = \mathcal{B}^*$, and let $f : A \to \bar{\mathbb{R}}$ be continuous, where $A \in \mathcal{F} - \{\varnothing\}$. Since f is continuous, $f^{-1}(\mathcal{O}) \subseteq \mathcal{F}$. Therefore, $f^{-1}(\mathcal{B}^*) = f^{-1}(\sigma(\mathcal{O})) = \sigma(f^{-1}(\mathcal{O})) \subseteq \sigma(\mathcal{F}) = \mathcal{F}$, where the second equality follows by Claim 2 of Section 1.3. By (e) of (MF1), f is $\mathcal{F}/\mathcal{B}^*$-measurable. ∎

In the above claim nothing is special about Euclidean space, as Ω can denote any metric space [or even a topological space], and $\mathcal{F}$ can denote any σ-field on Ω that contains the open subsets of Ω.

It should be noted that (MF4) holds if $\Omega = \bar{\mathbb{R}}^k$ and $\mathcal{F}$ is an arbitrary σ-field on $\bar{\mathbb{R}}^k$ that contains the open subsets of $\bar{\mathbb{R}}^k$. The proof of this claim is the same as before.

The following example is a direct application of (MF4).

Example 1. Let $\Omega = \mathbb{R}$ and $\mathcal{F} = \mathcal{B}$. Then any polynomial of any degree is continuous, hence is $\mathcal{B}/\mathcal{B}^*$-measurable [that is, Borel measurable]. Any rational function whose denominator is nonzero is continuous, hence such a function is $\mathcal{B}/\mathcal{B}^*$-measurable [that is, Borel measurable]. For further examples, the exponential functions [with any base] and the sine and cosine functions are Borel measurable. All of these functions are $\mathcal{L}/\mathcal{B}^*$-measurable [that is, Lebesgue measurable] since Borel measurability implies Lebesgue measurability.

Example 2. The usual natural logarithm function $\log : (0, +\infty) \to \mathbb{R}$ is $\mathcal{F}/\mathcal{B}^*$-measurable by (MF4). We may also give a simple direct proof by using (a) of (MF1). To see this, observe for any $x \in \mathbb{R}$ that $(\log)^{-1}((x, +\infty]) = (e^x, +\infty) \in \mathcal{B}^*$.

Related to (MF4) are the following properties, which assert [roughly] that $\mathcal{B}^*/\mathcal{B}^*$-measurable and/or continuous functions of $\mathcal{F}/\mathcal{B}^*$-measurable functions are themselves $\mathcal{F}/\mathcal{B}^*$-measurable. The background space Ω is once again generic. It is left as an exercise [see Exercise 14] to show that we may replace *continuous* with either of *lower semicontinuous* or *upper semicontinuous* in the statement of (MF4).

(MF5) *Let $A \in \mathcal{F}$ be nonempty, let $f : A \to \bar{\mathbb{R}}$ be $\mathcal{F}/\mathcal{B}^*$-measurable, and let $\phi : \bar{\mathbb{R}} \to \bar{\mathbb{R}}$ be $\mathcal{B}^*/\mathcal{B}^*$-measurable. Then $\phi \circ f$ is $\mathcal{F}/\mathcal{B}^*$-measurable. In particular, if $\phi : \bar{\mathbb{R}} \to \bar{\mathbb{R}}$ is continuous, then $(\phi \circ f)$ is $\mathcal{F}/\mathcal{B}^*$-measurable.*

Proof: Let $B \in \mathcal{B}^*$. Then $\phi^{-1}(B) \in \mathcal{B}^*$ by the assumption that ϕ is $\mathcal{B}^*/\mathcal{B}^*$-measurable. Since $\phi^{-1}(B) \in \mathcal{B}^*$ and f is $\mathcal{F}/\mathcal{B}^*$-measurable, we have $f^{-1}(\phi^{-1}(B)) \in \mathcal{F}$. Since $(\phi \circ f)^{-1}(B) = f^{-1}(\phi^{-1}(B))$, it follows that $(\phi \circ f)^{-1}(B) \in \mathcal{F}$. Therefore, $\phi \circ f$ is $\mathcal{F}/\mathcal{B}^*$-measurable by (e) of (MF1).

For the other claim, if ϕ is continuous, then ϕ is $\mathcal{B}^*/\mathcal{B}^*$-measurable by (MF4) [or the comments following (MF4)], so the conclusion follows directly from the first claim. ■

Example 3. Let $f : \Omega \to \bar{\mathbb{R}}$ be $\mathcal{F}/\mathcal{B}^*$-measurable. Let $\phi(x) = x^2$ for every $x \in \bar{\mathbb{R}}$, so that ϕ is continuous. [Recall that $(\pm\infty)^2 = +\infty$ by our conventions, so there are no logical problems with the squaring function as a function on $\bar{\mathbb{R}}$.] By (MF5), $(\phi \circ f) = f^2$ is $\mathcal{F}/\mathcal{B}^*$-measurable. If we define $e^{+\infty} = +\infty$ and $e^{-\infty} = 0$, then the exponential function [now viewed as a function on $\bar{\mathbb{R}}$] is continuous, and $\exp(f)$ is $\mathcal{F}/\mathcal{B}^*$-measurable by again appealing to (MF5).

Example 4. Define $\phi : \bar{\mathbb{R}} \to \bar{\mathbb{R}}$ by writing

$$\phi(x) = \begin{cases} \frac{1}{x} & x \in \bar{\mathbb{R}} - \{0\} \\ 0 & x = 0. \end{cases}$$

[Recall that $\frac{1}{\pm\infty} = 0$ with our conventions, so ϕ is defined at $\pm\infty$.] *We claim that ϕ is $\mathcal{B}^*/\mathcal{B}^*$-measurable.* We will show that $\phi^{-1}((t, +\infty]) \in \mathcal{B}^*$ for each real t, which by (a) of (MF1) suffices to prove the claim. Now

$$\phi^{-1}((t, +\infty]) = \phi^{-1}((t, +\infty)) \cup \phi^{-1}(\{+\infty\}) = \phi^{-1}((t, +\infty)) \cup \varnothing = \phi^{-1}((t, +\infty)).$$

Therefore, it suffices to show that $\phi^{-1}((t,+\infty)) \in \mathcal{B}^*$ for each real t. We will consider two cases: $t \geq 0$ and $t < 0$. If $t \geq 0$, then $\phi^{-1}((t,+\infty)) = (0, \frac{1}{t}) \in \mathcal{B}^*$. If $t < 0$, then

$$\phi^{-1}((t,+\infty)) = \phi^{-1}((t,0)) \cup \phi^{-1}(\{0\}) \cup \phi^{-1}((0,+\infty)).$$

We have that $\phi^{-1}((0,+\infty)) \in \mathcal{B}^*$ by what we have just shown in the $t \geq 0$ case, and $\phi^{-1}(\{0\}) = \{-\infty, 0, +\infty\} \in \mathcal{B}^*$. Since $\phi^{-1}((t,0)) = (-\infty, \frac{1}{t}) \in \mathcal{B}^*$, we have that $\phi^{-1}((t,+\infty)) \in \mathcal{B}^*$ for all $t < 0$. Combining the $t \geq 0$ and the $t < 0$ cases gives $\phi^{-1}((t,+\infty)) \in \mathcal{B}^*$ for all $t \in \mathbb{R}$. It follows that $\phi^{-1}((t,+\infty]) \in \mathcal{B}^*$ for all $t \in \mathbb{R}$, whence ϕ is $\mathcal{B}^*/\mathcal{B}^*$-measurable.

We may now apply (MF5) to say the following: if $A \in \mathcal{F} - \{\varnothing\}$ and if $f : A \to \bar{\mathbb{R}}$ is $\mathcal{F}/\mathcal{B}^*$-measurable, then $\phi \circ f : A \to \bar{\mathbb{R}}$ is $\mathcal{F}/\mathcal{B}^*$-measurable, where for each $\omega \in A$ we have

$$(\phi \circ f)(\omega) = \begin{cases} \frac{1}{f(\omega)} & \text{if } \omega \in A \text{ is such that } f(\omega) \neq 0 \\ 0 & \text{if } \omega \in A \text{ is such that } f(\omega) = 0 \end{cases}.$$

It should be noted that the choice of 0 for the value of $\phi(0)$ is wholly arbitrary. If we had instead defined $\phi(0) = c$ [where $c \in \mathbb{R}$], say, then very minor modifications of the above argument would still show that $\phi \circ f$ is $\mathcal{F}/\mathcal{B}^*$-measurable, with

$$(\phi \circ f)(\omega) = \begin{cases} \frac{1}{f(\omega)} & \text{if } \omega \in A \text{ is such that } f(\omega) \neq 0 \\ c & \text{if } \omega \in A \text{ is such that } f(\omega) = 0 \end{cases}$$

for each $\omega \in A$.

The next result says roughly that a function defined piecewise is $\mathcal{F}/\mathcal{B}^*$-measurable if the pieces are in $\mathcal{F}$ and the component functions are $\mathcal{F}/\mathcal{B}^*$-measurable.

(MF6) *Let $A \in \mathcal{F} - \{\varnothing\}$, and let $A_1, \cdots, A_n$ denote nonempty disjoint $\mathcal{F}$-sets with union Ω. Suppose that $f_i : A_i \to \bar{\mathbb{R}}$ is $\mathcal{F}/\mathcal{B}^*$-measurable for $i = 1, \cdots, n$. If $f : A \to \bar{\mathbb{R}}$ is such that $f(\omega) = f_i(\omega)$ iff $\omega \in A_i$, then f is $\mathcal{F}/\mathcal{B}^*$-measurable.*

Proof: Let $B \in \mathcal{B}^*$, so that $f_i^{-1}(B) \in \mathcal{F}$ for $i = 1, \cdots, n$, hence $\bigcup_{i=1}^n f_i^{-1}(B)$ is in $\mathcal{F}$. Since $f^{-1}(B) = \bigcup_{i=1}^n f_i^{-1}(B)$, we have that $f^{-1}(B) \in \mathcal{F}$. By (e) of (MF1), it follows that f is indeed $\mathcal{F}/\mathcal{B}^*$-measurable. ∎

The above claim also holds if we have a countable collection $A_1, A_2, \cdots$ of nonempty disjoint $\mathcal{F}$-sets with union Ω as compared to a finite collection $A_1, \cdots, A_n$ of nonempty disjoint $\mathcal{F}$-sets with union Ω; the proof is exactly the same.

Example 5. Let $\Omega = \bar{\mathbb{R}}$ and $\mathcal{F} = \mathcal{B}^*$. Let $c > 0$, pick two arbitrary numbers y_- and y_+ in $\bar{\mathbb{R}}$, and define $\phi : \bar{\mathbb{R}} \to \bar{\mathbb{R}}$ by

$$\phi(\omega) = \begin{cases} |\omega|^c & \text{if } \omega \in \mathbb{R} \\ y_+ & \text{if } \omega = +\infty \\ y_- & \text{if } \omega = -\infty. \end{cases}$$

[If $y_+ = y_- = +\infty$, then ϕ is continuous, hence $\mathcal{F}/\mathcal{B}^*$-measurable by (MF4).] In (MF6), let $A = \bar{\mathbb{R}}$, $A_1 = \mathbb{R}$, $A_2 = \{+\infty\}$, and $A_3 = \{-\infty\}$. Define $f_1 : A_1 \to \bar{\mathbb{R}}$ by writing $f_1(\omega) = |\omega|^c$, so that f_1 is continuous and hence $\mathcal{F}/\mathcal{B}^*$-measurable by (MF4). Define $f_2 : \{+\infty\} \to \bar{\mathbb{R}}$ by writing $f_2(+\infty) = y_+$. Then, for any $B \in \mathcal{B}^*$, $f_2^{-1}(B)$ equals $\{+\infty\}$ or $\varnothing$ as $y_+ \in B$ or $y_+ \notin B$. In any case, $f_2^{-1}(B) \in \mathcal{F}$. Since $B \in \mathcal{B}^*$ is generic, (e) of (MF1) establishes that f_2 is $\mathcal{F}/\mathcal{B}^*$-measurable. Next, let $f_3 : \{-\infty\} \to \bar{\mathbb{R}}$ be given by $f_3(-\infty) = y_-$. By the same arguments as just given, f_3 is $\mathcal{F}/\mathcal{B}^*$-measurable. Applying (MF6) to this situation shows that ϕ is $\mathcal{F}/\mathcal{B}^*$-measurable.

Even though we have spent time on the concept of $\mathcal{F}/\mathcal{B}^*$-measurability, it has not yet been stated exactly why the definition of $\mathcal{F}/\mathcal{B}^*$-measurability is an important or useful one in measure theory. There are two preliminary and very informal answers that we can give with the hope that they will satisfy the reader for now or at least seem plausible.

One answer for now is that we will want to discuss certain properties of functions in finer detail than to merely say that [for example] *Property P holds everywhere* or *P holds on such and such a set in the domain of the function.* We will want to discuss "how large" the sets are where P holds, using a notion of "largeness" other than cardinality. We will want to discuss "how large" the sets are where P holds by applying a given measure to them. However, measures cannot usually be applied to any subset of the space, so if we want to apply a measure to that subset of the domain possessing property P, we have to make sure that the subset under discussion is in fact in the domain of the measure under discussion. The theory we are developing gives a rigorous foundation for making such discussion possible.

The second answer is somewhat more vague. Our development of the abstract Lebesgue integral of a function to be found in the next chapter requires four stages, and these stages depend ultimately on the function being $\mathcal{F}/\mathcal{B}^*$-measurable for the entire development to make sense. While this answer is somewhat circular, it is hoped that by the end of the next chapter the connections of $\mathcal{F}/\mathcal{B}^*$-measurability and the integral will be more clear.

Exercises.

1. Let $f : \Omega \to \bar{\mathbb{R}}$ be $\mathcal{F}/\mathcal{B}^*$-measurable. Let $y \in \bar{\mathbb{R}}$, and let $h : \Omega \to \bar{\mathbb{R}}$ be such that $h(\omega) = \sqrt{f(\omega)}$ for those $\omega \in \Omega$ such that $f(\omega) \geq 0$, and $h(\omega) = y$ for those $\omega \in \Omega$ such that $f(\omega) < 0$. Then h is $\mathcal{F}/\mathcal{B}^*$-measurable.

Comment on Exercises 2-4: We are tacitly assuming the Axiom of Choice in these exercises, as (AC) was used in the sections from which these exercises draw their impetus.

2. There exists a continuous function $f : \mathbb{R} \to \mathbb{R}$ and a subset $A \subseteq \mathbb{R}$ such that $A \in \mathcal{L}$ but $f^{-1}(A) \notin \mathcal{L}$. [Let $f(x) = 0$, $\phi^{-1}(x)$, and 1 as $x < 0$, $0 \leq x \leq 2$, and $x > 2$, where ϕ is the homeomorphism from $[0, 1]$ to $[0, 2]$ constructed in Section 4.7. Then f is continuous. Let B denote the non-Lebesgue subset of $\phi(C)$ constructed therein, where C is the Cantor Set. We showed that $\phi^{-1}(B) \in \mathcal{L}$. Let $A = \phi^{-1}(B)$, so that $A \in \mathcal{L}$. Thus $f^{-1}(A) = \phi(\phi^{-1}(B)) = B \notin \mathcal{L}$.]

Comment: A situation as above cannot happen for Borel sets. That is, if the function $f : \mathbb{R} \to \mathbb{R}$ is continuous, then $f^{-1}(B) \in \mathcal{B}$ for every $B \in \mathcal{B}$. The fact that the above situation can occur for even such stringently well-behaved functions such as continuous functions is why the definition of measurability is what it is. If we instead declared a function $f : \Omega \to \mathbb{R}$ to be "measurable" iff $f^{-1}(B) \in \mathcal{F}$ for every $B \in \mathcal{L}$, say, then we would not be able to state and prove an analog of (MF4), which asserts [roughly] that every continuous function is "measurable."

3. Using the notation and framework of Section 4.7 and the previous exercise, we have $I_B = I_{\phi^{-1}(B)} \circ \phi^{-1}$ and $I_{\phi^{-1}(B)}$ is $\mathcal{L}/\mathcal{B}^*$-measurable, while I_B is not $\mathcal{L}/\mathcal{B}^*$-measurable. Therefore, an $\mathcal{L}/\mathcal{B}^*$-measurable function of a homeomorphism is not necessarily $\mathcal{L}/\mathcal{B}^*$-measurable.

Comment: On the other hand, a homeomorphism of an $\mathcal{F}/\mathcal{B}^*$-measurable function is itself $\mathcal{F}/\mathcal{B}^*$-measurable by noting that homeomorphisms are continuous and invoking (MF5).

4. A Lebesgue measurable function is not necessarily Borel measurable. [There exists a set $E \in \mathcal{L} - \mathcal{B}$; consider the Lebesgue measurable function I_E.]

5*. Suppose that $f : \Omega \to \bar{\mathbb{R}}$ is $\mathcal{F}/\mathcal{B}^*$-measurable.
(a) If $\mathcal{F} = 2^\Omega$, f can be any function from Ω into $\bar{\mathbb{R}}$.
(b) If $\mathcal{F} = \{\varnothing, \Omega\}$, then f must be constant.
(c) If $\mathcal{F} = \sigma(\{A_1, \cdots, A_n\})$, where $A_1, \cdots, A_n$ are disjoint subsets of Ω such that $\Omega = \bigcup_{i=1}^n A_i$, then f must have the form $f = \sum_{i=1}^n c_i I_{A_i}$, where $c_1, \cdots, c_n \in \bar{\mathbb{R}}$.

6. If $f : \Omega \to \bar{\mathbb{R}}$ is such that $f^{-1}(\{x\}) \in \mathcal{F}$ for every $x \in \bar{\mathbb{R}}$, then f is not necessarily $\mathcal{F}/\mathcal{B}^*$-measurable. [Take $\Omega = \mathbb{R}$ and $\mathcal{F} = \mathcal{L}$. Let $A \notin \mathcal{L}$, and let $f : \mathbb{R} \to \mathbb{R}$ be given by $f(x) = x$ or $-x$ as $x \in A$ or $x \notin A$. Then f fails to be $\mathcal{L}/\mathcal{B}^*$-measurable, yet $f^{-1}(\{x\}) \in \mathcal{F}$.]

7. If $A \subseteq \mathbb{R}$ is any type of interval and $f : A \to \bar{\mathbb{R}}$ is monotone, then f is both Borel and Lebesgue measurable.

8. Let $f : \Omega \to \bar{\mathbb{R}}^k$ be given by the rule $f(\omega) = (f_1(\omega), \cdots, f_k(\omega))$, where we have that $f_i : \Omega \to \bar{\mathbb{R}}$ is a function, $i = 1, \cdots, k$. Then $f^{-1}(B) \in \mathcal{F}$ for every $B \in \mathcal{B}^{k*}$ iff f_i is $\mathcal{F}/\mathcal{B}^*$-measurable for $i = 1, \cdots, k$. [If $f_1, \cdots, f_k$ are $\mathcal{F}/\mathcal{B}^*$-measurable, then $f_i^{-1}([-\infty, x]) \in \mathcal{F}$ for each $x \in \mathbb{R}$, so $f^{-1}([-\infty, \boldsymbol{x}]) = \bigcap_{i=1}^k f_i^{-1}([-\infty, x_i]) \in \mathcal{F}$. Let $\mathcal{A} = \{[-\infty, \boldsymbol{x}] : \boldsymbol{x} \in \mathbb{R}^k\}$. Since $f^{-1}(\mathcal{A}) \subseteq \mathcal{F}$, Claim 2 of Section 1.3 gives $\sigma(f^{-1}(\mathcal{A})) = f^{-1}(\sigma(\mathcal{A})) = f^{-1}(\mathcal{B}^{k*}) \subseteq \mathcal{F}$. On the other hand, if $f^{-1}(B) \in \mathcal{F}$ for every $B \in \mathcal{B}^{k*}$, then, fixing $i = 1$ and letting $\mathcal{A}$ now denote the collection $\{[-\infty, x] : x \in \mathbb{R}\}$, we have for each $x \in \mathbb{R}$ that $[-\infty, x] \times \bar{\mathbb{R}} \times \cdots \times \bar{\mathbb{R}} \in \mathcal{B}^{k*}$, hence $f^{-1}([-\infty, x] \times \bar{\mathbb{R}} \times \cdots \times \bar{\mathbb{R}}) \in \mathcal{F}$ for each $x \in \mathbb{R}$. But this latter set is $f_1^{-1}([-\infty, x])$, so f_1 is $\mathcal{F}/\mathcal{B}^*$-measurable. The same type of argument works for $f_2, \cdots, f_k$.]

9*. $\mathcal{B}^k$ is the minimal σ-field $\mathcal{F}$ on $\mathbb{R}^k$ such that every continuous $f : \mathbb{R}^k \to \bar{\mathbb{R}}$ is $\mathcal{F}/\mathcal{B}^*$-measurable. [Recall that if f is continuous, then $f^{-1}(\mathcal{O})$ is a subcollection of the open subsets in $\mathbb{R}^k$, where $\mathcal{O}$ denotes the open subsets of $\bar{\mathbb{R}}$. Use Claim 2, Section 1.3.]

10. Let $f : \Omega \to \bar{\mathbb{R}}$ be $\sigma(\mathcal{A})/\mathcal{B}^*$-measurable, where $\mathcal{A} \subseteq 2^\Omega$. There exists an amc subcollection $\mathcal{A}_f$ of $\mathcal{A}$ such that f is $\sigma(\mathcal{A}_f)/\mathcal{B}^*$-measurable.

11. Let $f : \Omega \to \bar{\mathbb{R}}$, and suppose that $\Omega = \bigcup_{n=1}^\infty A_n$, where $A_1, A_2, \cdots$ are disjoint $\mathcal{F}$-sets [$\mathcal{F}$ is a σ-field on Ω]. Let $\mathcal{F}_n = \{A \in \mathcal{F} : A \subseteq A_n\}$ for each $n \in \mathbb{N}$. Then $\mathcal{F}_n$

is a σ-field for each $n \in \mathbb{N}$. Let f_n denote the restriction of f from Ω to A_n, $n \in \mathbb{N}$. Then f is $\mathcal{F}/\mathcal{B}^*$-measurable iff f_n is $\mathcal{F}_n/\mathcal{B}^*$-measurable for each $n \in \mathbb{N}$.

12. Show that the function $\phi : \bar{\mathbb{R}} \to \bar{\mathbb{R}}$ given in Example 4 is $\mathcal{B}^*/\mathcal{B}^*$-measurable by suitably appealing to (MF6).

13. The minimal σ-field $\mathcal{F}$ on Ω such that $f : \Omega \to \bar{\mathbb{R}}$ is $\mathcal{F}/\mathcal{B}^*$-measurable is $f^{-1}(\mathcal{B}^*)$. [Show that $f^{-1}(\mathcal{B}^*)$ is a σ-field on Ω and apply the definition of $\mathcal{F}/\mathcal{B}^*$-measurability.] The σ-field $f^{-1}(\mathcal{B}^*)$ is often denoted by $\sigma(f)$ and is referred to as *the σ-field generated by f*.

14*. The word *continuous* in (MF4) may be replaced by either of *lower semicontinuous* and *upper semicontinuous*. [Recall or show that if f is lower semicontinuous, then $f^{-1}((t, +\infty])$ is open. For the upper semicontinuous case use or show the following fact: if f is upper semicontinuous, then $-f$ is lower semicontinuous. With these facts the result follows nearly automatically.]

5.2 COMBINING MEASURABLE FUNCTIONS

The main goal of this section is to show that sums, differences, products, and quotients of $\mathcal{F}/\mathcal{B}^*$-measurable functions are themselves $\mathcal{F}/\mathcal{B}^*$-measurable. This statement is crude and will be made precise in the language of the upcoming claims. We will again let Ω denote an arbitrary nonempty background set, and $\mathcal{F}$ will denote a σ-field on Ω.

(MF7) *Let $A \in \mathcal{F}$ be nonempty, and let $f : A \to \bar{\mathbb{R}}$ be $\mathcal{F}/\mathcal{B}^*$-measurable. The following statements hold:*

(a) $f + c$ is $\mathcal{F}/\mathcal{B}^*$-measurable for any $c \in \mathbb{R}$.

(b) cf is $\mathcal{F}/\mathcal{B}^*$-measurable for any $c \in \mathbb{R}$. In particular, $-f$ is $\mathcal{F}/\mathcal{B}^*$-measurable.

(c) Let $c > 0$ and $y_-, y_+ \in \bar{\mathbb{R}}$. For each $\omega \in A$, define

$$g(\omega) = \begin{cases} |f(\omega)|^c & \omega \in A \text{ is such that } |f(\omega)| < +\infty \\ y_- & \omega \in A \text{ is such that } f(\omega) = -\infty \\ y_+ & \omega \in A \text{ is such that } f(\omega) = +\infty. \end{cases}$$

Then $g : A \to \bar{\mathbb{R}}$ is $\mathcal{F}/\mathcal{B}^*$-measurable.

(d) Define $g : A \to \bar{\mathbb{R}}$ for all $\omega \in A$ by writing

$$g(\omega) = \begin{cases} \frac{1}{f(\omega)} & \text{if } \omega \in A \text{ is such that } f(\omega) \neq 0 \\ 0 & \text{if } \omega \in A \text{ is such that } f(\omega) = 0. \end{cases}$$

Then g is $\mathcal{F}/\mathcal{B}^*$-measurable.

Proof: Let $c \in \mathbb{R}$ and $t \in \mathbb{R}$. Then $(f+c)^{-1}((t, +\infty]) = f^{-1}((t-c, +\infty]) \in \mathcal{F}$, so $f + c$ is $\mathcal{F}/\mathcal{B}^*$-measurable by (a) of (MF1). This gives (a).

We now do (b). First consider the case $c = 0$. Then $cf = 0$ on A and for every $t \in \mathbb{R}$ we have that $(cf)^{-1}((t, +\infty])$ is A or $\varnothing$ as $t < 0$ or $t \geq 0$. That is, $(cf)^{-1}((t, +\infty]) \in \mathcal{F}$ for each real t, and by (a) of (MF1) we therefore have that cf is $\mathcal{F}/\mathcal{B}^*$-measurable when $c = 0$. We now consider the cases $c > 0$ and $c < 0$. For any $t \in \mathbb{R}$, observe that

$$(cf)^{-1}((t, +\infty]) = \begin{cases} f^{-1}((\frac{t}{c}, +\infty]) & c > 0 \\ f^{-1}([-\infty, \frac{t}{c})) & c < 0. \end{cases}$$

By (a) and (c) of (MF1), we have $(cf)^{-1}((t, +\infty]) \in \mathcal{F}$ for every $t \in \mathbb{R}$. By (a) of (MF1), we have that cf is $\mathcal{F}/\mathcal{B}^*$-measurable for the cases $c < 0$ and $c > 0$. Therefore, (b) has been established for all real c.

For (c), let $c > 0$ and $y_-, y_+ \in \bar{\mathbb{R}}$, and for each $x \in \bar{\mathbb{R}}$ let

$$\phi(x) = \begin{cases} x^c & x \in \mathbb{R} \\ y_+ & x = +\infty \\ y_- & x = -\infty. \end{cases}$$

The function ϕ was shown to be $\mathcal{B}^*/\mathcal{B}^*$-measurable in Example 5 of Section 1, so $g = \phi \circ f$ is $\mathcal{F}/\mathcal{B}^*$-measurable by appealing to (MF5).

For (d), we merely appeal to Example 4 of Section 1, and the proof is accordingly complete. ∎

To show that sums, differences, products, and quotients of functions that are $\mathcal{F}/\mathcal{B}^*$-measurable produce functions that are $\mathcal{F}/\mathcal{B}^*$-measurable, we will need the following little technical property, whose proof is fairly direct.

(MF8) *Let $A \in \mathcal{F}$ be nonempty, and let $f_1 : A \to \bar{\mathbb{R}}$ and $f_2 : A \to \bar{\mathbb{R}}$ be $\mathcal{F}/\mathcal{B}^*$-measurable. We have the following:*

(a) $\{\omega \in B : f_1(\omega) > f_2(\omega)\} \in \mathcal{F}$ for any $B \in \mathcal{F}$ with $B \subseteq A$.

(b) $\{\omega \in B : f_1(\omega) \geq f_2(\omega)\} \in \mathcal{F}$ for any $B \in \mathcal{F}$ with $B \subseteq A$.

(c) $\{\omega \in B : f_1(\omega) = f_2(\omega)\} \in \mathcal{F}$ for any $B \in \mathcal{F}$ with $B \subseteq A$

Proof: Let $B \in \mathcal{F}$ with $B \subseteq A$. Observe that

$$\{\omega \in B : f_1(\omega) > f_2(\omega)\} = B \cap \bigcup_{q \in \mathbb{Q}} \left(f_1^{-1}((q, +\infty]) \cap f_2^{-1}([-\infty, q))\right),$$

which is in $\mathcal{F}$ since $f_1^{-1}((q, +\infty]), f_2^{-1}([-\infty, q)) \in \mathcal{F}$ for every $q \in \mathbb{Q}$ by the assumption that f_1 and f_2 are $\mathcal{F}/\mathcal{B}^*$-measurable. This gives (a). Next,

$$\{\omega \in B : f_1(\omega) \geq f_2(\omega)\} = \{\omega \in B : f_2(\omega) > f_1(\omega)\}^c,$$

and $\{\omega \in B : f_2(\omega) > f_1(\omega)\} \in \mathcal{F}$ by permuting f_1 with f_2 in (a), hence $\{\omega \in B : f_1(\omega) \geq f_2(\omega)\} \in \mathcal{F}$, giving (b). For (c), observe that

$$\{\omega \in B : f_1(\omega) = f_2(\omega)\} = \{\omega \in B : f_1(\omega) \leq f_2(\omega)\} \cap \{\omega \in B : f_2(\omega) \leq f_1(\omega)\},$$

and both sets in the intersection are $\mathcal{F}$-sets by an appeal to (b). It follows that $\{\omega \in B : f_1(\omega) = f_2(\omega)\} \in \mathcal{F}$, giving (c) and completing the proof. ∎

We are now ready to show that sums and products of $\mathcal{F}/\mathcal{B}^*$-measurable functions are $\mathcal{F}/\mathcal{B}^*$-measurable. Some care must be exercised in dealing with sums of functions that possibly allow infinite values, so that the undefined expressions $+\infty - \infty$ and $-\infty + \infty$ are avoided. Given two functions f and g with common domain A, saying that $f(\omega) + g(\omega)$ *is defined* means merely that it is not the case that $\omega \in A$ is such that (i) $f(\omega) = +\infty$ and $g(\omega) = -\infty$ or (ii) $f(\omega) = -\infty$ and $g(\omega) = +\infty$. Saying that $f(\omega) + g(\omega)$ *is defined* is taken to be the negation of the statement $f(\omega) + g(\omega)$ *is undefined*, which means that both the negations of (i) and (ii) hold. Analogous comments hold for the statements $f(\omega) - g(\omega)$ *is defined* and $f(\omega) - g(\omega)$ *is undefined.*

(MF9) *Let $y \in \bar{\mathbb{R}}$, let $A \in \mathcal{F}$ be nonempty, and let $f : A \to \bar{\mathbb{R}}$ and $g : A \to \bar{\mathbb{R}}$ be $\mathcal{F}/\mathcal{B}^*$-measurable.*

(a) Let $h : A \to \bar{\mathbb{R}}$ be such that

$$h(\omega) = \begin{cases} f(\omega) + g(\omega) & \text{if } \omega \in A \text{ is such that } f(\omega) + g(\omega) \text{ is defined} \\ y & \text{if } \omega \in A \text{ is such that } f(\omega) + g(\omega) \text{ is undefined} \end{cases}$$

for every $\omega \in A$. Then $h : A \to \bar{\mathbb{R}}$ is $\mathcal{F}/\mathcal{B}^*$-measurable.

(b) Let $h : A \to \bar{\mathbb{R}}$ be such that

$$h(\omega) = \begin{cases} f(\omega) - g(\omega) & \text{if } \omega \in A \text{ is such that } f(\omega) - g(\omega) \text{ is defined} \\ y & \text{if } \omega \in A \text{ is such that } f(\omega) - g(\omega) \text{ is undefined} \end{cases}$$

for every $\omega \in A$. Then $h : A \to \bar{\mathbb{R}}$ is $\mathcal{F}/\mathcal{B}^*$-measurable.

Proof: We do (a). Let $y \in \bar{\mathbb{R}}$, and let $x \in \mathbb{R}$. Define

$$A_y = \begin{cases} [f^{-1}(\{+\infty\}) \cap g^{-1}(\{-\infty\})] \cup [f^{-1}(\{-\infty\}) \cap g^{-1}(\{+\infty\})] & y > x \\ \varnothing & y \le x. \end{cases}$$

Observe that (i) $A_y \subseteq A$, and (ii) the assumption of $\mathcal{F}/\mathcal{B}^*$-measurability for f and g forces $A_y \in \mathcal{F}$ [and hence $A - A_y \in \mathcal{F}$]. Next,

$$\begin{aligned} h^{-1}((x, +\infty]) &= \{\omega \in A : h(\omega) > x\} \\ &= \{\omega \in A - A_y : h(\omega) > x\} \cup \{\omega \in A_y : h(\omega) > x\} \\ &= \{\omega \in A - A_y : f(\omega) + g(\omega) > x\} \cup A_y \\ &= \{\omega \in A - A_y : f(\omega) > x - g(\omega)\} \cup A_y. \end{aligned}$$

The third equality follows because (i) $h(\omega) = f(\omega) + g(\omega)$ for each $\omega \in A - A_y$ and (ii) $\{\omega \in A_y : h(\omega) > x\}$ equals A_y for both of the cases $y > x$ and $y \le x$. By (a) and (b) of (MF7), the $\mathcal{F}/\mathcal{B}^*$-measurability of g forces $x - g$

to be $\mathcal{F}/\mathcal{B}^*$-measurable, hence $\{\omega \in A - A_y : f(\omega) > x - g(\omega)\} \in \mathcal{F}$ by (MF8). It follows that $h^{-1}((x, +\infty]) \in \mathcal{F}$ for the generic $x \in \mathbb{R}$, hence h is $\mathcal{F}/\mathcal{B}^*$-measurable by (a) of (MF1), completing (a).

Part (b) follows by appealing to (a) and (MF7), or it can be done by trivially modifying the proof of (a). This completes the proof. ∎

The preceding claim says that when we have two $\mathcal{F}/\mathcal{B}^*$ measurable functions f and g with domain $A \in \mathcal{F}$, we can then form an $\mathcal{F}/\mathcal{B}^*$-measurable function h such that $h = f + g$ on the set of $\omega \in A$ where $f(\omega) + g(\omega)$ won't be of the undefined form $-\infty + \infty$ or $+\infty - \infty$, and we can define h to be constant on the set of $\omega \in A$ such that $f(\omega) + g(\omega)$ is undefined. In the next chapter on integration, we shall take h to be 0 on the set of $\omega \in A$ such that $f(\omega) + g(\omega)$ is undefined. Analogous comments hold when sums are replaced by differences.

In the exercises for this section, the reader is asked to use induction on part (a) of (MF9) to obtain the following result for arbitrary finite sums: *if $f_1, \cdots, f_n$ are $\mathcal{F}/\mathcal{B}^*$-measurable functions with common domain $A \in \mathcal{F}$, if $y \in \bar{\mathbb{R}}$, and if h is defined for all $\omega \in A$ by the rule*

$$h(\omega) = \begin{cases} f_1(\omega) + \cdots + f_n(\omega) & \text{if } f_1(\omega) + \cdots + f_n(\omega) \text{ is defined} \\ y & \text{if } f_1(\omega) + \cdots + f_n(\omega) \text{ is undefined,} \end{cases}$$

then h is $\mathcal{F}/\mathcal{B}^*$-measurable.

For the upcoming claim, recall the identity $ab = \frac{1}{4}[(a+b)^2 - (a-b)^2]$ for $a, b \in \mathbb{R}$. Thus, for $a, b, x \in \mathbb{R}$, we have $ab > x$ iff $(a+b)^2 - (a-b)^2 > 4x$. Also, if f and g are functions from A into $\bar{\mathbb{R}}$, there are no problems with fg, since multiplication in $\bar{\mathbb{R}}$ causes no possible problems regarding undefined expressions, in contrast to the situation for sums, differences, and quotients. However, it will be seen that a rigorous proof is probably much more complicated than one would expect.

(MF10) *Let $A \in \mathcal{F} - \{\varnothing\}$, and let $f : A \to \bar{\mathbb{R}}$ and $g : A :\to \bar{\mathbb{R}}$ denote $\mathcal{F}/\mathcal{B}^*$-measurable functions. The following hold:*

(a) fg is $\mathcal{F}/\mathcal{B}^*$-measurable.

(b) Let $y \in \bar{\mathbb{R}}$ and $h : A \to \bar{\mathbb{R}}$ be such that

$$h(\omega) = \begin{cases} \frac{f(\omega)}{g(\omega)} & \text{if } \omega \in A \text{ is such that } \frac{f(\omega)}{g(\omega)} \text{ is defined} \\ y & \text{otherwise} \end{cases}$$

for each $\omega \in A$. Then h is $\mathcal{F}/\mathcal{B}^*$-measurable.

Proof: Let $h = fg$, and let $x \in \mathbb{R}$. We will show that $h^{-1}((x, +\infty]) \in \mathcal{F}$ for every $x \in \mathbb{R}$, which by (a) of (MF1) establishes part (a). We consider two cases: $x \geq 0$ and $x < 0$.

We first do the $x \geq 0$ case. Here, $h^{-1}((x, +\infty])$ is the union of the five sets

$$f^{-1}(\{+\infty\}) \cap g^{-1}((0, +\infty]), \tag{5.1}$$

$$f^{-1}((0,+\infty]) \cap g^{-1}(\{+\infty\}), \tag{5.2}$$

$$f^{-1}([-\infty,0)) \cap g^{-1}(\{-\infty\}), \tag{5.3}$$

$$f^{-1}(\{-\infty\}) \cap g^{-1}([-\infty,0)), \tag{5.4}$$

and [using the identity stated before the proof]

$$\{\omega \in A \cap f^{-1}(\mathbb{R}) \cap g^{-1}(\mathbb{R}) : (f(\omega)+g(\omega))^2 - (f(\omega)-g(\omega))^2 > 4x\}. \tag{5.5}$$

[The union of (5.1)-(5.4) is $h^{-1}(\{+\infty\})$, and (5.5) is $h^{-1}((x,+\infty))$, and thus $h^{-1}((x,+\infty])$ is the union of the sets (5.1)-(5.5).] Since f and g are $\mathcal{F}/\mathcal{B}^*$-measurable, each of (5.1)-(5.4) is in $\mathcal{F}$. Therefore, we will have shown that $h^{-1}((x,+\infty])$ is an $\mathcal{F}$-set in the $x \geq 0$ case when we show that (5.5) is an $\mathcal{F}$-set for each $x \geq 0$.

We thus turn to showing that (5.5) is an $\mathcal{F}$-set when $x \geq 0$. There are two cases. If $A \cap f^{-1}(\mathbb{R}) \cap g^{-1}(\mathbb{R}) = \varnothing$, then (5.5) is empty, hence in $\mathcal{F}$. The second case is where $A \cap f^{-1}(\mathbb{R}) \cap g^{-1}(\mathbb{R})$ is nonempty. Since f and g are $\mathcal{F}/\mathcal{B}^*$-measurable and $A, f^{-1}(\mathbb{R}), g^{-1}(\mathbb{R}) \in \mathcal{F}$, we have $A \cap f^{-1}(\mathbb{R}) \cap g^{-1}(\mathbb{R}) \in \mathcal{F}$. The restrictions f_0 and g_0 of f and g from A to $A \cap f^{-1}(\mathbb{R}) \cap g^{-1}(\mathbb{R})$ are $\mathcal{F}/\mathcal{B}^*$-measurable by (MF2) [and both f_0 and g_0 are finite-valued]. By (MF9) and (MF7), we have that $(f_0+g_0)^2 - (f_0-g_0)^2$ is $\mathcal{F}/\mathcal{B}^*$-measurable, hence the set (5.5) may be written as

$$\{\omega \in A \cap f^{-1}(\mathbb{R}) \cap g^{-1}(\mathbb{R}) : (f_0(\omega)+g_0(\omega))^2 - (f_0(\omega)-g_0(\omega))^2 > 4x\},$$

which is the inverse image of the $\mathcal{B}^*$-set $(4x,+\infty)$ under $(f_0+g_0)^2-(f_0-g_0)^2$, hence (5.5) is in $\mathcal{F}$ by (e) of (MF1).

We summarize: it has been shown that $h^{-1}((x,+\infty]) \in \mathcal{F}$ for each real $x \geq 0$. The $\mathcal{F}/\mathcal{B}^*$-measurability of h will be established when we show that $h^{-1}((x,+\infty]) \in \mathcal{F}$ for all $x < 0$.

To do this, let $x < 0$ be real and observe that

$$h^{-1}((x,+\infty]) = h^{-1}((x,0)) \cup h^{-1}(\{0\}) \cup h^{-1}((0,+\infty]).$$

Furthermore, we have (i) $h^{-1}((0,+\infty]) \in \mathcal{F}$ by what we have just shown, and (ii) $h^{-1}(\{0\}) = f^{-1}(\{0\}) \cup g^{-1}(\{0\}) \in \mathcal{F}$ by (e) of (MF1). Therefore, $h^{-1}((x,+\infty])$ will have been shown to be an $\mathcal{F}$-set in the $x < 0$ case when it is shown that $h^{-1}((x,0)) \in \mathcal{F}$ for $x < 0$. As in the $x \geq 0$ case, there are two cases to consider. The first case is where $h^{-1}((x,0)) = \varnothing$, and the second case is where $h^{-1}((x,0)) \neq \varnothing$. In the first case, $h^{-1}((x,0)) \in \mathcal{F}$ automatically. In the second case, let $q_0 = (f_0+g_0)^2 - (f_0-g_0)^2$, so that q_0 has domain $A \cap f^{-1}(\mathbb{R}) \cap g^{-1}(\mathbb{R}) \in \mathcal{F}$ and is $\mathcal{F}/\mathcal{B}^*$-measurable. Observe that [again using the identity stated earlier]

$$\begin{aligned} & h^{-1}((x,0)) \\ = \; & \{\omega \in A \cap f^{-1}(\mathbb{R}) \cap g^{-1}(\mathbb{R}) : 4x < q_0(\omega) < 0\} = q_0^{-1}((4x,0)) \in \mathcal{F} \end{aligned}$$

by invoking (e) of (MF1). We have therefore shown that $h^{-1}((x, +\infty]) \in \mathcal{F}$ for all $x < 0$. Putting the $x \geq 0$ and $x < 0$ cases together, we have that $h^{-1}((x, +\infty]) \in \mathcal{F}$ for each $x \in \mathbb{R}$, which by (a) of (MF1) completes the proof of part (a).

The proof of (b) may be handled similarly, or one may put together some previous results, being careful to avoid the expressions that are undefined under division. This completes the proof. ∎

It should be noted that all of the above proofs simplify substantially if the functions have range contained in $\mathbb{R}$, that is, if the values $\pm\infty$ are not allowed as possible output values for functions. In this regard, Exercises 4 and 5 are of interest.

Exercises.

1*. If $f : \Omega \to \bar{\mathbb{R}}$ is $\mathcal{F}/\mathcal{B}^*$-measurable, then $|f|$ is $\mathcal{F}/\mathcal{B}^*$-measurable. However, if $|f|$ is $\mathcal{F}/\mathcal{B}^*$-measurable, then f is not necessarily $\mathcal{F}/\mathcal{B}^*$-measurable. [This simple but fundamental result is used in our study of integration.]

2*. Let $n \in \mathbb{N}$, and let $f_1, \cdots, f_n$ denote $\mathcal{F}/\mathcal{B}^*$-measurable functions with common domain $A \in \mathcal{F}$.
(a) Both $\max\{f_1, f_2\}$ and $\min\{f_1, f_2\}$ are $\mathcal{F}/\mathcal{B}^*$-measurable functions. [Let $g = \max\{f_1, f_2\}$. Observe for $t \in \mathbb{R}$ that $g(\omega) \leq t$ iff both $f_1(\omega) \leq t$ and $f_2(\omega) \leq t$. Invoke (d) of (MF1) to complete matters. If $h = \min\{f_1, f_2\}$, observe for $t \in \mathbb{R}$ that $h(\omega) > t$ iff both $f_1(\omega) > t$ and $f_2(\omega) > t$. Invoke (a) of (MF1) to complete things.]
(b) Both $\max\{f_1, \cdots, f_n\}$ and $\min\{f_1, \cdots, f_n\}$ are $\mathcal{F}/\mathcal{B}^*$-measurable functions. [Use (a) and induction.]

3. Let $(\Omega, \mathcal{F}, \mu)$ denote a measure space, and let $f : \Omega \to \bar{\mathbb{R}}$ denote a $\mathcal{F}/\mathcal{B}^*$-measurable mapping. Let $\nu : \mathcal{B}^* \to \bar{\mathbb{R}}$ be such that $\nu(B) = \mu(f^{-1}(B))$ for every $B \in \mathcal{B}^*$. That is, $\nu = \mu \circ f^{-1}$. Then $(\bar{\mathbb{R}}, \mathcal{B}^*, \nu)$ is a measure space. The measure $\nu = \mu \circ f^{-1}$ is often referred to as *the measure induced by f*. Furthermore, even if μ is σ-finite, ν is not necessarily σ-finite.

4*. This exercise concerns itself with (MF9).
(a) Prove part (b) of (MF9) by suitably adapting the proof of (a).
(b) Prove part (b) of (MF9) by using (a) [of (MF9)] and (MF7).
(c) It was claimed that if the functions f and g were real valued instead of extended real valued, then the proof of (MF9) would greatly simplify. Show that

$$\{\omega \in A : f(\omega) + g(\omega) < x\} = \bigcup_{\substack{r_1, r_2 \in \mathbb{Q} \\ r_1 + r_2 < x}} \{\omega \in A : f(\omega) < r_1\} \cap \{\omega \in A : g(\omega) < r_2\}$$

for every $x \in \mathbb{R}$ and hence conclude that if f and g are real valued $\mathcal{F}/\mathcal{B}^*$-measurable functions, then so is $f + g$. [Observe how much more direct the proof is.]
(d) Repeat part (c) for $f - g$ by proving an analogous identity.
(e) Let us now generalize (MF9) to an arbitrary finite number of functions. Let $y \in \bar{\mathbb{R}}$, $n \geq 2$, and for $i = 1, \cdots, n$, let $f_i : A \to \bar{\mathbb{R}}$ denote a $\mathcal{F}/\mathcal{B}^*$-measurable function. [The set A is, of course, in $\mathcal{F} - \{\varnothing\}$.] Let $h : A \to \bar{\mathbb{R}}$ be defined for all

$\omega \in A$ by the rule

$$h(\omega) = \begin{cases} f_1(\omega) + \cdots + f_n(\omega) & \text{if } f_1(\omega) + \cdots + f_n(\omega) \text{ is defined} \\ y & \text{if } f_1(\omega) + \cdots + f_n(\omega) \text{ is undefined.} \end{cases}$$

Show that h is $\mathcal{F}/\mathcal{B}^*$-measurable. [Outline: Pick $x \in \mathbb{R}$ and assume that the result is valid for $n = 1, \cdots, m$. Consider the case $n = m + 1$. If $y > x$, set A_y to be the set of $\omega \in A$ such that $f_1(\omega) + \cdots + f_{m+1}(\omega)$ is undefined. If $y \leq x$, set $A_y = \varnothing$. Show that $A_y \in \mathcal{F}$ in all cases. Next, show that

$$h^{-1}((x, +\infty]) = \{\omega \in A - A_y : f_1(\omega) + \cdots + f_{m+1}(\omega) > x\} \cup A_y$$

by the same reasoning as in (M9). Since $A_y \in \mathcal{F}$, the proof reduces to showing that

$$\{\omega \in A - A_y : (f_1(\omega) + \cdots + f_m(\omega)) + f_{m+1}(\omega) > x\} \in \mathcal{F}.$$

Use the induction hypothesis and do a little work to show that the preceding set is indeed in $\mathcal{F}$.]

5*. This exercise concerns itself with (MF10).

(a) Directly prove (a) of (MF10) assuming that f and g are real valued instead of extended real valued. Observe that the proof is much simpler than that given in the text which handles the most general case. [Observe for $x \in \mathbb{R}$ that

$$(fg)^{-1}((x, +\infty]) = \{\omega \in A : (f(\omega) + g(\omega))^2 - (f(\omega) - g(\omega))^2 > 4x\},$$

which is easily seen to be in $\mathcal{F}$ by (MF7) and (MF9).]

(b) Prove (b) of (MF10) assuming that f and g are real valued and that g is nonzero on A. [Write h as $f \times \frac{1}{g}$ and then use (a).]

(c) Using the previous part, now prove (b) of (MF10) in full generality. [Pick $x \in \mathbb{R}$. If $y > x$, define A_y to consist of the $\omega \in A$ such that $\frac{f(\omega)}{g(\omega)}$ is undefined. If $y \leq x$, set $A_y = \varnothing$. Show that $A_y \in \mathcal{F}$ in all cases. Next, show that

$$h^{-1}((x, +\infty]) = \left\{\omega \in A - A_y : x < \frac{f(\omega)}{g(\omega)} \leq +\infty\right\} \cup A_y.$$

By considering cases [each of which is straightforward], show that the first set on the right side above is in $\mathcal{F}$.]

5.3 SEQUENCES OF MEASURABLE FUNCTIONS

The previous section showed that [roughly] basic arithmetical combinations of $\mathcal{F}/\mathcal{B}^*$-measurable functions defined on a common set A are themselves $\mathcal{F}/\mathcal{B}^*$-measurable. In this section, we shall consider operations defined for sequences of $\mathcal{F}/\mathcal{B}^*$-measurable functions having a common domain. More specifically stated, we desire to show that $\mathcal{F}/\mathcal{B}^*$-measurability is preserved under the formations of suprema, infima, inferior limits, superior limits, and limits. Again, Ω denotes a nonempty set and $\mathcal{F}$ denotes a σ-field on Ω.

Definition. Let $\{f_n\}_{n=1}^{\infty}$ denote a sequence of functions from some common set A into $\bar{\mathbb{R}}$. We define the functions $\sup_n f_n$, $\inf_n f_n$, $\overline{\lim}_n f_n$, and $\underline{\lim}_n f_n$ on A by writing

$$\left(\sup_n f_n\right)(\omega) = \sup\{f_1(\omega), f_2(\omega), \cdots\},$$

$$\left(\inf_n f_n\right)(\omega) = \inf\{f_1(\omega), f_2(\omega), \cdots\},$$

$$(\overline{\lim_n} f_n)(\omega) = \inf_{k\in\mathbb{N}} \sup\{f_k(\omega), f_{k+1}(\omega), \cdots\},$$

and

$$(\underline{\lim_n} f_n)(\omega) = \sup_{k\in\mathbb{N}} \inf\{f_k(\omega), f_{k+1}(\omega), \cdots\}$$

for every $\omega \in A$. If $\lim_n f_n(\omega)$ exists for each $\omega \in A$, then we define $\lim_n f_n$ on A by writing $(\lim_n f_n)(\omega) = \lim_n f_n(\omega)$ for every $\omega \in A$.

The above definitions are the usual pointwise definitions for the supremum, infimum, inferior limit, and superior limit of a sequence of functions. With this definition, we wish to show that if $\{f_n\}_{n=1}^{\infty}$ is a sequence of $\mathcal{F}/\mathcal{B}^*$-measurable functions having common domain $A \in \mathcal{F}$, then $\sup_n f_n$, $\inf_n f_n$, $\underline{\lim}_n f_n$, and $\overline{\lim}_n f_n$ are themselves $\mathcal{F}/\mathcal{B}^*$-measurable.

(MF11) *Let $A \in \mathcal{F}$ be nonempty, and let $f_n : A \to \bar{\mathbb{R}}$ denote an $\mathcal{F}/\mathcal{B}^*$-measurable function for each $n \in \mathbb{N}$. Then the following hold:*

(a) $\inf_n f_n$, $\sup_n f_n$, $\underline{\lim}_n f_n$, and $\overline{\lim}_n f_n$ are $\mathcal{F}/\mathcal{B}^*$-measurable.

(b) $\{\omega \in B : (\lim_n f_n)(\omega)$ exists$\} \in \mathcal{F}$ for any $B \in \mathcal{F}$ with $B \subseteq A$.

(c) If $\lim_n f_n$ exists on A, then $\lim_n f_n$ is $\mathcal{F}/\mathcal{B}^*$-measurable.

(d) Let $y \in \bar{\mathbb{R}}$, and define $h : A \to \bar{\mathbb{R}}$ by writing

$$h(\omega) = \begin{cases} \lim_n f_n(\omega) & \text{if } \lim_n f_n(\omega) \text{ exists} \\ y & \text{otherwise.} \end{cases}$$

Then h is $\mathcal{F}/\mathcal{B}^*$-measurable.

Proof: We first handle (a). Let $x \in \mathbb{R}$. Since f_n is assumed to be $\mathcal{F}/\mathcal{B}^*$-measurable for each $n \in \mathbb{N}$, we have by (a) of (MF1) that $f_n((x, +\infty]) \in \mathcal{F}$ for each $n \in \mathbb{N}$. Since

$$\{\omega \in A : \inf_n f_n(\omega) > x\} = \bigcap_{n=1}^{\infty} f_n^{-1}((x, +\infty]),$$

we have that $(\inf_n f_n)^{-1}((x, +\infty]) \in \mathcal{F}$ for each $x \in \mathbb{R}$, so $\inf_n f_n$ is $\mathcal{F}/\mathcal{B}^*$-measurable by (a) of (MF1).

Next, use (MF7) and what has just been proved regarding infima: $-f_n$ is $\mathcal{F}/\mathcal{B}^*$-measurable for every $n \in \mathbb{N}$, hence $\inf(-f_n)$ is $\mathcal{F}/\mathcal{B}^*$-measurable, so $-\inf(-f_n)$ is $\mathcal{F}/\mathcal{B}^*$-measurable by a second application of (MF7). Since $\sup f_n = -\inf(-f_n)$, it follows that $\sup f_n$ is $\mathcal{F}/\mathcal{B}^*$-measurable.

Next, the hypotheses imply that $\{f_n\}_{n=k}^{\infty}$ is a sequence of $\mathcal{F}/\mathcal{B}^*$-measurable functions for each $k \in \mathbb{N}$, hence the infimum of this sequence of functions, $\inf_{n \geq k} f_n$, is $\mathcal{F}/\mathcal{B}^*$-measurable for each $k \in \mathbb{N}$. Since $\{\inf_{n \geq k} f_n\}_{k=1}^{\infty}$ is a sequence of $\mathcal{F}/\mathcal{B}^*$-measurable functions, its supremum is $\mathcal{F}/\mathcal{B}^*$-measurable as well. That is, $\sup_{k \in \mathbb{N}} \inf_{n \geq k} f_n$ is $\mathcal{F}/\mathcal{B}^*$-measurable. Since by definition we have $\underline{\lim}_n f_n = \sup_{k \in \mathbb{N}} \inf_{n \geq k} f_n$, it follows that $\underline{\lim}_n f_n$ is $\mathcal{F}/\mathcal{B}^*$-measurable. That $\overline{\lim}_n f_n$ is $\mathcal{F}/\mathcal{B}^*$-measurable follows by the same type of reasoning, and hence (a) is complete.

The proof of (b) is direct. Let $B \subseteq A$ with $B \in \mathcal{F}$. The claim is trivial if $B = \varnothing$, so we may assume that $B \neq \varnothing$. Observe that both $\overline{\lim}_n f_n$ and $\underline{\lim}_n f_n$ are $\mathcal{F}/\mathcal{B}^*$-measurable by (a) and everywhere defined on B, hence (MF8) and the definition of a limit show that

$$\left\{\omega \in B : \left(\lim_n f_n\right)(\omega) \text{ exists}\right\} = \left\{\omega \in B : \left(\overline{\lim_n} f_n\right)(\omega) = \left(\underline{\lim_n} f_n\right)(\omega)\right\}$$

is in $\mathcal{F}$, completing (b).

Part (c) is equally direct. If $\lim_n f_n$ exists on A, then $\overline{\lim}_n f_n$ and $\underline{\lim}_n f_n$ are identical on A and $\lim_n f_n$ is equal to both. Since both $\overline{\lim} f_n$ and $\underline{\lim} f_n$ are $\mathcal{F}/\mathcal{B}^*$-measurable by (a), it follows that $\lim_n f_n$ is $\mathcal{F}/\mathcal{B}^*$-measurable when $\lim_n f_n$ exists on A. This completes the proof of (c).

We now turn to (d). Let $B = \{\omega \in A : \lim_n f_n(\omega) \text{ exists}\}$, and let g_0 be the constant function on $A - B$ that assigns ω to y. If $B = \varnothing$, then $h = g_0$, so h is $\mathcal{F}/\mathcal{B}^*$-measurable. Assume next that $B \neq \varnothing$. We have by (b) that $B \in \mathcal{F}$, and if we define g_n to denote the restriction of f_n from A to B for each $n \in \mathbb{N}$, (MF2) shows that $\{g_n\}_{n=1}^{\infty}$ is a sequence of $\mathcal{F}/\mathcal{B}^*$-measurable functions with domain B, and hence (c) implies that the function $\lim_n g_n$ with domain B is $\mathcal{F}/\mathcal{B}^*$-measurable. Therefore, h is g_0 on $A - B$ and $\lim_n g_n$ on B. The function h is a piecewise defined function whose component functions g_0 and $\lim_n g_n$ are $\mathcal{F}/\mathcal{B}^*$-measurable and whose component sets $A - B$ and B are in $\mathcal{F}$, hence h is $\mathcal{F}/\mathcal{B}^*$-measurable by (MF6). This completes (d) and the entire proof. ■

What follows now is a simple consequence of the work just done, but it could also be proved directly from first principles or from the work given in the last section.

(MF12) *Let $n \in \mathbb{N}$, let $A \in \mathcal{F}$ be nonempty, and let $f_j : A \to \bar{\mathbb{R}}$ be $\mathcal{F}/\mathcal{B}^*$-measurable for $j = 1, \cdots, n$. Then the functions $\max\{f_1, \cdots, f_n\}$ and $\min\{f_1, \cdots, f_n\}$ are $\mathcal{F}/\mathcal{B}^*$-measurable.*

Proof: Define a sequence $\{g_n\}_{n=1}^{\infty}$ of functions with domain A by requiring that $g_1 = f_1, \cdots, g_{n-1} = f_{n-1}$ and $g_m = f_n$ for each $m \in \{n, n+1, n+2, \cdots\}$.

Then $\{g_n\}_{n=1}^{\infty}$ is a sequence of $\mathcal{F}/\mathcal{B}^*$-measurable functions, hence $\sup_k g_k$ and $\inf_k g_k$ are $\mathcal{F}/\mathcal{B}^*$-measurable by (MF11). The proof is complete upon observing that $\max\{f_1, \cdots, f_n\} = \sup_k g_k$ and $\min\{f_1, \cdots, f_n\} = \inf_k g_k$. ■

We close this brief section with an important definition that will be used in our discussion of integration. As with (MF12), these items could be proven directly from first principles, but we will quickly prove them using the work done up to the present.

Definition. Let $f : A \to \bar{\mathbb{R}}$ denote a function, where $A \in \mathcal{F}$ is nonempty. We define the functions f^+ and f^- on A by the rules

$$f^+(\omega) = \max\{f(\omega), 0\} \qquad \text{and} \qquad f^-(\omega) = -\min\{f(\omega), 0\}.$$

The symbol 0 used in the definition denotes the function on A that is identically zero. The function f^+ is called the *positive part of* f, and f^- is called the *negative part of* f.

By considering the cases $f(\omega) \geq 0$ and $f(\omega) < 0$, we may quickly prove the following:

(i) f^+ and f^- are everywhere nonnegative,

(ii) For every $\omega \in A$ we have that at least one of $f^+(\omega)$ and $f^-(\omega)$ is zero, and

(iii) $f(\omega) = f^+(\omega) - f^-(\omega)$ for all $\omega \in A$, hence we may write $f = f^+ - f^-$.

The following property states some other basic properties regarding positive and negative parts.

(MF13) *Let $f : A \to \bar{\mathbb{R}}$, $g : A \to \bar{\mathbb{R}}$, and $h : A \to \bar{\mathbb{R}}$ be arbitrary. Then f is $\mathcal{F}/\mathcal{B}^*$-measurable iff both f^+ and f^- are $\mathcal{F}/\mathcal{B}^*$-measurable. In addition, we also have the following:*

(a) $|f| = f^+ + f^-$. In particular, $f \geq 0$ iff $f = f^+$ and $f^- = 0$. Also, $f < 0$ iff $f^- = -f$ and $f^+ = 0$.

(b) Let $c \in \mathbb{R}$. Then

$$(cf)^+ = \begin{cases} cf^+ & c \geq 0 \\ -cf^- & c < 0 \end{cases} \quad \textit{and} \quad (cf)^- = \begin{cases} cf^- & c \geq 0 \\ -cf^+ & c < 0. \end{cases}$$

(c) We do not necessarily have $(f+g)^+ = f^+ + g^+$, nor do we necessarily have $(f+g)^- = f^- + g^-$, but we do have

$$0 \leq (f+g)^+ \leq f^+ + g^+ \quad \textit{and} \quad 0 \leq (f+g)^- \leq f^- + g^-.$$

(d) If $|g| \leq f$ then $0 \leq g^+, g^- \leq f$.

(e) If $f = g - h$ and $g, h \geq 0$, then $f^+ \leq g$ and $f^- \leq h$.

Proof: We shall concentrate on the initial measurability claim preceding (a)-(e), and shall leave the proofs of (a)-(e) as very simple exercises for the reader.

Assume that f is $\mathcal{F}/\mathcal{B}^*$-measurable. Then, since 0 is $\mathcal{F}/\mathcal{B}^*$-measurable, it follows by (MF12) that $f^+ = \max\{f, 0\}$ is $\mathcal{F}/\mathcal{B}^*$-measurable. Also, (MF7) and (MF12) show that $f^- = -\min\{f, 0\}$ is $\mathcal{F}/\mathcal{B}^*$-measurable. Conversely, if both f^+ and f^- are $\mathcal{F}/\mathcal{B}^*$-measurable, then since $f = f^+ - f^-$, (MF9) gives the $\mathcal{F}/\mathcal{B}^*$-measurability of f. ■

Exercises.

1*. Prove (a)-(e) of (MF13).
2. The class of $\mathcal{F}/\mathcal{B}^*$-measurable functions is not necessarily closed under uncountable suprema and infima. The following outline gives a simple instantiation of this claim. Let $\Omega = \mathbb{R}$ and $\mathcal{F} = \mathcal{B}$.
(a) Let E denote a non-Borel set as constructed in Section 4.5. Argue that E cannot be at most countable.
(b) For each $x \in E$, define $f_x : \mathbb{R} \to \bar{\mathbb{R}}$ by writing $f_x(\omega) = I_{\{x\}}(\omega)$ for each $\omega \in \mathbb{R}$. Then f_x is $\mathcal{F}/\mathcal{B}^*$-measurable for each $x \in E$, but $\sup_{x\in E} f_x = I_E$, hence $\sup_{x\in E} f_x$ is not $\mathcal{F}/\mathcal{B}^*$-measurable.

5.4 ALMOST EVERYWHERE

Here, we discuss the concept of "almost everywhere." The reason that this concept is important is because we will often deal with properties that are true not for every point ω in a given $\mathcal{F}$-set A, but for a large enough $\mathcal{F}$-subset of A that the set of $\omega \in A$ where such a property fails to hold may be considered "negligible" or "unessential." This section makes precise what has just been stated.

Definition. Let $(\Omega, \mathcal{F}, \mu)$ denote a measure space, let $A \in \mathcal{F}$, and let P denote a property such that $P(\omega)$ has a truth value for each $\omega \in A$. The property P is said to hold *μ-almost everywhere on A* iff $\{\omega \in A: P \text{ fails to hold for } \omega\}$ is in $\mathcal{F}$ and $\mu(\{\omega \in A: P \text{ fails to hold for } \omega\}) = 0$. The colloquialism *$P$ holds for almost all* $\omega \in A$ is also employed. We use *μ-a.e.* as notation for *μ-almost everywhere.* Finally, we will at times equivalently write *P holds for μ-a.e.* $\omega \in A$ as an equivalent expression.

If both A and $\{\omega \in A: P \text{ fails to hold for } \omega\}$ are in $\mathcal{F}$, then

$$\{\omega \in A: P \text{ holds for } \omega\} = A - \{\omega \in A: P \text{ fails to hold for } \omega\}$$

is also in $\mathcal{F}$ since $\mathcal{F}$ is closed under differences. Therefore, we actually have that the statement *P holds μ-a.e. on* A is logically equivalent to stating that
(i) $\{\omega \in A: P \text{ fails to hold for } \omega\} \in \mathcal{F}$ and is μ-null, and
(ii) $\{\omega \in A: P \text{ holds for } \omega\} \in \mathcal{F}$ and has μ-measure equal to $\mu(A)$.[3]

[3] $\mu(\{\omega \in A: P \text{ holds for } \omega\}) = \mu(A - \{\omega \in A: P \text{ fails to hold for } \omega\}) = \mu(A) - 0 = \mu(A)$ by measure property (M5).

It should be observed that nothing precludes the case where $A \in \mathcal{F}$ is such that $\mu(A) = 0$. In this trivial case, P holds μ-a.e. on A for any well-defined property P.

Example 1. Consider the measure space $(\mathbb{R}, \mathcal{B}, \lambda)$. Then, taking $A = \mathbb{R}$ and noting that $\mathbb{Q} \in \mathcal{B}$ and $\lambda(\mathbb{Q}) = 0$, we may say that almost every $\omega \in \mathbb{R}$ is irrational. Equivalently, we could say that the function $I_{\mathbb{Q}^c} = 1$ λ-a.e. [on $A = \mathbb{R}$].

Example 2. Let $(\Omega, \mathcal{F}, \mu)$ be an arbitrary measure space. If f is a function defined μ-a.e. on Ω, this means that (i) $(\operatorname{dom} f)^c, \operatorname{dom} f \in \mathcal{F}$, (ii) $\mu((\operatorname{dom} f)^c) = 0$, and (iii) $\mu(\operatorname{dom} f) = \mu(A)$. If f is a function defined μ-a.e. on some nonempty $A \in \mathcal{F}$, then (i) $\operatorname{dom} f \subseteq A$, (ii) $(\operatorname{dom} f)^c, \operatorname{dom} f \in \mathcal{F}$, (iii) $\mu(A - \operatorname{dom} f) = 0$, and (iv) $\mu(\operatorname{dom} f) = \mu(A)$.

Example 3. Continue as in Example 2, but further assume that $\{f_n\}_{n=1}^{\infty}$ is a sequence of functions, each function defined μ-a.e. on A. This means that for each $n \in \mathbb{N}$ we have (i) $(\operatorname{dom} f_n)^c, \operatorname{dom} f_n \in \mathcal{F}$, (ii) $\mu(A - \operatorname{dom} f_n) = 0$, and (iii) $\mu(\operatorname{dom} f_n) = \mu(A)$.

Under this setup, if we say that $f_n \to f$ μ-a.e. on A, this merely means that

$$\{\omega \in A : \lim f_n(\omega) \text{ fails to exist or exists and differs from } f(\omega)\} \in \mathcal{F}$$

and is μ-null, and hence its complement $\{\omega \in A : \lim f_n(\omega) = f(\omega)\}$[4] is in $\mathcal{F}$ and has μ-measure equal to $\mu(A)$.

In the next claim, we will let $(\Omega, \mathcal{F}, \mu)$ denote a *complete* measure space. That is, every subset of a μ-null set is itself an $\mathcal{F}$-set, and hence is itself μ-null.

(MF14) *Let $(\Omega, \mathcal{F}, \mu)$ denote a complete measure space. Pick nonempty A in $\mathcal{F}$, and let f be defined μ-a.e. on A and $\mathcal{F}/\mathcal{B}^*$-measurable.*

(a) If g is defined μ-a.e. on A and $f = g$ μ-a.e. on A, then g is $\mathcal{F}/\mathcal{B}^*$-measurable.

(b) If g is defined everywhere on A and $f = g$ μ-a.e. on A, then g is $\mathcal{F}/\mathcal{B}^*$-measurable.

Proof: If (a) is shown, then (b) is trivial since being defined everywhere on A is a special case of being defined μ-a.e. on A. We thus need only show (a). Let $B = \{\omega \in \operatorname{dom} f \cap \operatorname{dom} g : f(\omega) = g(\omega)\}$. The assumption that $f = g$ μ-a.e. on A forces $B \in \mathcal{F}$ and $\mu(A - B) = 0$. Pick $x \in \mathbb{R}$, and observe that

$$\begin{aligned} g^{-1}((x, +\infty]) &= [g^{-1}((x, +\infty]) \cap B] \cup [g^{-1}((x, +\infty]) \cap (A - B)] \\ &= [f^{-1}((x, +\infty]) \cap B] \cup [g^{-1}((x, +\infty]) \cap (A - B)]. \end{aligned}$$

Since $f^{-1}((x, +\infty]) \in \mathcal{F}$ and $B \in \mathcal{F}$, we have that $f^{-1}((x, +\infty]) \cap B \in \mathcal{F}$. Next, $g^{-1}((x, +\infty]) \cap (A - B)$ is a subset of $A - B$, and $\mu(A - B) = 0$.

[4]We use the notation "$\lim f_n(\omega) = f(\omega)$" as a shorthand for the more cumbersome "$\lim f_n(\omega)$ exists and equals $f(\omega)$." Similarly, we will write "$\lim f_n(\omega) \neq f(\omega)$" for the more cumbersome "$\lim f_n(\omega)$ fails to exist or exists but is not equal to $f(\omega)$."

Since $(\Omega, \mathcal{F}, \mu)$ is complete, we have $g^{-1}((x, +\infty]) \cap (A - B) \in \mathcal{F}$. Therefore, $g^{-1}((x, +\infty])$ is the union of two $\mathcal{F}$-sets and hence is in $\mathcal{F}$. Since $x \in \mathbb{R}$ was arbitrary, g is $\mathcal{F}/\mathcal{B}^*$-measurable by (a) of (MF1). ■

The next result will be of use to us in our discussion of integration to be found in the next chapter. In this result, we will twice call on (MF14).

(MF15) *Let $(\Omega, \mathcal{F}, \mu)$ denote a complete measure space, and let $A \in \mathcal{F} - \{\varnothing\}$. Let $\{f_n\}_{n=1}^{\infty}$ denote a sequence of $\mathcal{F}/\mathcal{B}^*$-measurable functions, each of which is defined μ-a.e. on A. Suppose that f is defined μ-a.e. on A and that $\lim f_n(\omega) = f(\omega)$ μ-a.e. on A. Then f is $\mathcal{F}/\mathcal{B}^*$-measurable.*

Proof: By hypothesis, $\operatorname{dom} f, \operatorname{dom} f_1, \operatorname{dom} f_2, \cdots \in \mathcal{F}$; the hypotheses also imply that $\{\omega \in A : \lim f_n(\omega) = f(\omega)\} \in \mathcal{F}$. Let

$$B = (\operatorname{dom} f) \cap \left(\bigcap_{n=1}^{\infty} \operatorname{dom} f_n \right) \cap \{\omega \in A : \lim_n f_n(\omega) = f(\omega)\},$$

so that $B \in \mathcal{F}$. Also, the $\mathcal{F}$-sets $A - (\operatorname{dom} f)$, $A - (\operatorname{dom} f_1), A - (\operatorname{dom} f_2), \cdots$ and $\{\omega \in A : \lim f_n(\omega) \neq f(\omega)\}$ have μ-measure zero by the hypotheses. Since

$$A - B = (A - \operatorname{dom} f) \cup \left(\bigcup_{n=1}^{\infty} (A - \operatorname{dom} f_n) \right) \cup \{\omega \in A : \lim_n f_n(\omega) \neq f(\omega)\},$$

we have $\mu(A - B) = 0$. For each $n \in \mathbb{N}$, define $g_n : A \to \bar{\mathbb{R}}$ and $g : A \to \bar{\mathbb{R}}$ by writing

$$g_n(\omega) = \begin{cases} f_n(\omega) & \omega \in B \\ 0 & \omega \in A - B \end{cases} \quad \text{and} \quad g(\omega) = \begin{cases} f(\omega) & \omega \in B \\ 0 & \omega \in A - B \end{cases}$$

for each $\omega \in A$. Since $\mu(A - B) = 0$, we have that $g_n = f_n$ μ-a.e. on A for each $n \in \mathbb{N}$, hence g_n is $\mathcal{F}/\mathcal{B}^*$-measurable for each $n \in \mathbb{N}$ by (MF14). Also, $g_n(\omega) \to g(\omega)$ for each $\omega \in A$; since each g_n is $\mathcal{F}/\mathcal{B}^*$-measurable, (MF11) forces g to be $\mathcal{F}/\mathcal{B}^*$-measurable. Since $f = g$ μ-a.e. on A, we appeal to (MF14) yet again to obtain the $\mathcal{F}/\mathcal{B}^*$-measurability of f. This completes the proof. ■

Exercises.

1. In (MF15), the completeness of $(\Omega, \mathcal{F}, \mu)$ is not a redundant assumption. [Give a counterexample to (MF15) in the case where $(\Omega, \mathcal{F}, \mu)$ fails to be complete.]

2. Suppose that $f : \mathbb{R} \to \mathbb{R}$ is a differentiable function. Then f' [the derivative of f] is a Borel measurable function. [Let $f_n(x) = f(x+n^{-1})$ for all $x \in \mathbb{R}$ and $n \in \mathbb{N}$. Then f_n is a Borel measurable function for each $n \in \mathbb{N}$ since f is continuous. Therefore, $n(f_n - f)$ is a Borel measurable function for each $n \in \mathbb{N}$. Thus $f' = \lim_n n(f_n - f)$, which exists by assumption, is Borel measurable.]

3. Let $f : \mathbb{R} \to \bar{\mathbb{R}}$. The statement *$f$ is continuous at almost every $x \in \mathbb{R}$* neither implies nor is implied by the statement *there exists continuous $g : \mathbb{R} \to \bar{\mathbb{R}}$ such that $f = g$ μ-a.e. on $\mathbb{R}$.*

5.5 SIMPLE FUNCTIONS

The next chapter's development of the abstract Lebesgue integral will first be defined in terms of an operation applied to a very easily describable class of functions, namely, those functions that can assume only a finite number of possible values. These functions, appropriately enough, are called *simple functions*, and are the topic of this section.

Definition. A function is called *simple* iff it assumes at most finitely many different values.

Definition. Given any $A \subseteq B \subseteq \Omega$, we will define $I_{A;B} : B \to \bar{\mathbb{R}}$ by writing

$$I_{A;B}(\omega) = \begin{cases} 1 & \text{if } \omega \in A \\ 0 & \text{if } \omega \notin A \end{cases}$$

for every $\omega \in B$. That is, $I_{A;B}$ is the restriction of $I_A : \Omega \to \bar{\mathbb{R}}$ to the set B. If $B = \Omega$, we write I_A [as usual] for $I_{A;\Omega}$.

By (MF3), I_A is $\mathcal{F}/\mathcal{B}^*$-measurable iff $A \in \mathcal{F}$. If $A \subseteq B$ and $A, B \in \mathcal{F}$, then $I_{A;B}$, being the restriction of I_A to an $\mathcal{F}$-set, is $\mathcal{F}/\mathcal{B}^*$-measurable by (MF2).

The following result is a representation result for simple functions that is straightforward.

(MF16) *Let $s : A \to \bar{\mathbb{R}}$ denote a simple function [where $A \in \mathcal{F} - \{\varnothing\}$] with range $\{c_1, \cdots, c_n\}$, where $c_1, \cdots, c_n$ are distinct. If $A_j = \{\omega \in A : s(\omega) = c_j\}$ for $j = 1, \cdots, n$, then $s = \sum_{j=1}^{n} c_j I_{A_j;A}$. Furthermore, s is $\mathcal{F}/\mathcal{B}^*$-measurable iff $A_1, \cdots, A_n \in \mathcal{F}$.*

Proof: The first claim is trivial. For the second claim, suppose that s is $\mathcal{F}/\mathcal{B}^*$-measurable. Then, since $\{c_j\} \in \mathcal{B}^*$ for $j = 1, \cdots, n$, we have $s^{-1}(\{c_j\}) \in \mathcal{F}$ for $j = 1, \cdots, n$. But $s^{-1}(\{c_j\}) = A_j$ for $j = 1, \cdots, n$, hence $A_j \in \mathcal{F}$ for $j = 1, \cdots, n$. Conversely, if $A_1, \cdots, A_n \in \mathcal{F}$, then $I_{A_1;A}, \cdots, I_{A_n;A}$ are $\mathcal{F}/\mathcal{B}^*$-measurable. By (MF7), $c_j I_{A_j;A}$ is $\mathcal{F}/\mathcal{B}^*$-measurable for $j = 1, \cdots, n$. By (MF9), we have that $\sum_{j=1}^{n} c_j I_{A_j;A}$ is $\mathcal{F}/\mathcal{B}^*$-measurable, which completes the proof. ■

(MF17) *Finite linear combinations of simple functions with common domain A are themselves simple functions with domain A. [Trivial.]*

The immediately preceding claim will be used in our proof of Lusin's Theorem to be found in Section 7.

The following result is of the greatest importance in our discussion of integration.

(MF18) *Let $A \in \mathcal{F}$ be nonempty, and let $f : A \to \bar{\mathbb{R}}$ denote a nonnegative $\mathcal{F}/\mathcal{B}^*$-measurable function. Then there exists a nondecreasing sequence*

$\{s_n\}_{n=1}^{\infty}$ of nonnegative $\mathcal{F}/\mathcal{B}^*$-measurable and finite-valued simple functions on A such that $s_n \to f$ on A.

Proof: For every $n \in \mathbb{N}$ and $k = 1, \cdots, n2^n$, define

$$A_{n,k} = \left\{\omega \in A : \frac{k-1}{2^n} \leq f(\omega) < \frac{k}{2^n}\right\} \quad \text{and} \quad B_n = \{\omega \in A : f(\omega) \geq n\}.$$

[Observe that $A = \bigcup_{k=1}^{n2^n} A_{n,k} \cup B_n$ for each $n \in \mathbb{N}$, the union being disjoint.] For each $n \in \mathbb{N}$, define $s_n : A \to \mathbb{R}$ by writing

$$s_n(\omega) = \sum_{k=1}^{n2^n} \frac{k-1}{2^n} I_{A_{n,k};A}(\omega) + nI_{B_n;A}(\omega), \qquad \omega \in A.$$

We claim that this sequence satisfies the assertions of the theorem. First, it is clear that $0 \leq s_n < +\infty$ for each $n \in \mathbb{N}$. Second, $A_{n,k} \in \mathcal{F}$ and $B_n \in \mathcal{F}$ for every $n \in \mathbb{N}$ and $k = 1, \cdots, n2^n$ since f is assumed $\mathcal{F}/\mathcal{B}^*$-measurable, hence s_n is $\mathcal{F}/\mathcal{B}^*$-measurable for each $n \in \mathbb{N}$ by (MF16). Demonstrating that $\{s_n\}_{n=1}^{\infty}$ is nondecreasing and converges to f on A takes a bit more work.

We now show that $s_n(\omega) \to f(\omega)$ for each $\omega \in A$. To see this, let $\omega \in A$, and consider the cases (i) $f(\omega) < +\infty$ and (ii) $f(\omega) = +\infty$. We first handle (i), picking arbitrary $\epsilon > 0$. Since $f(\omega) < +\infty$, there is $N \in \mathbb{N}$ such that both $f(\omega) < N$ and $2^{-N} < \epsilon$ hold. Next, let $n > N$. For this n, we have $\omega \in A_{n,k}$ for some $k = 1, \cdots, n2^n$, hence $\frac{k-1}{2^n} \leq f(\omega) < \frac{k}{2^n}$ and $s_n(\omega) = \frac{k-1}{2^n}$ for this particular k, whence $|f(\omega) - s_n(\omega)| < 2^{-n} < 2^{-N} < \epsilon$. We have shown that for every $\epsilon > 0$ there is $N \in \mathbb{N}$ such that $n > N$ implies $|f(\omega) - s_n(\omega)| < \epsilon$, and hence $s_n(\omega) \to f(\omega)$ for all $\omega \in A$ such that $f(\omega) < +\infty$, giving case (i). For (ii), if $\omega \in A$ is such that $f(\omega) = +\infty$, then $\omega \in B_n$ for every $n \in \mathbb{N}$, hence $s_n(\omega) = n$ for every $n \in \mathbb{N}$, whence $s_n(\omega) \to +\infty = f(\omega)$, giving (ii). We therefore have that $s_n(\omega) \to f(\omega)$ for each $\omega \in A$.

We complete the proof by showing that $\{s_n\}_{n=1}^{\infty}$ is a nondecreasing sequence. Pick $n \in \mathbb{N}$ and let $\omega \in A$; we must show that $s_n(\omega) \leq s_{n+1}(\omega)$. There are two cases: (a) $\omega \in A_{n,k}$ for some $k = 1, \cdots, n2^n$, and (b) $\omega \in B_n$. For (a), we have (i) $\frac{k-1}{2^n} \leq f(\omega) < \frac{k-1}{2^n} + \frac{1}{2^{n+1}}$ or (ii) $\frac{k-1}{2^n} + \frac{1}{2^{n+1}} \leq f(\omega) < \frac{k}{2^n}$. [Case (i) corresponds to where $f(\omega)$ is in the left half of $A_{n,k}$, and (ii) corresponds to where $f(\omega)$ is in the right half of $A_{n,k}$.] By algebra, we have in (i) that $\frac{2k-2}{2^{n+1}} \leq f(\omega) < \frac{2k-1}{2^{n+1}}$; in (ii) we have that $\frac{2k-1}{2^{n+1}} \leq f(\omega) < \frac{2k}{2^{n+1}}$. That is, for (a) we either have (i) $\omega \in A_{n+1,2k-1}$ or (ii) $\omega \in A_{n+1,2k}$. It follows that in case (i) we have $s_{n+1}(\omega) = \frac{2k-2}{2^{n+1}} = \frac{k-1}{2^n} = s_n(\omega)$ and in case (ii) we have $s_{n+1}(\omega) = \frac{2k-1}{2^{n+1}} > \frac{k-1}{2^n} = s_n(\omega)$, so in case (a) we have $s_{n+1}(\omega) \geq s_n(\omega)$. For (b), [where $s_n(\omega) = n$], there are again two cases: (i) $n \leq f(\omega) < n+1$ and (ii) $f(\omega) \geq n+1$. For (i), we have $\omega \in A_{n+1,k}$ for some $k = n2^{n+1} + 1, \cdots, (n+1)2^{n+1}$. Therefore,

$$s_{n+1}(\omega) = \frac{k-1}{2^{n+1}} \geq \frac{(n2^{n+1}+1)-1}{2^{n+1}} = n = s_n(\omega).$$

For (ii), we have $\omega \in B_{n+1}$, hence $s_{n+1}(\omega) = n+1 > n = s_n(\omega)$, hence (b) holds in both subcases. Putting both (a) and (b) together proves that $\{s_n\}_{n=1}^{\infty}$ is a nondecreasing sequence, and the entire proof is now complete. ∎

The above proof is an instance where [arguably] drawing a picture is more helpful than the analytic proof given, especially in the proof that the sequence $\{s_n\}_{n=1}^{\infty}$ is nondecreasing. It is important to note that each s_n is finite-valued by the construction process given. See Exercise 1 for some very important extensions of (MF18).

Exercises.

1*. Refer to (MF18).
(a) If $f : \Omega \to \bar{\mathbb{R}}$ is a general $\mathcal{F}/\mathcal{B}^*$-measurable function, then there exists a sequence $\{s_n\}_{n=1}^{\infty}$ of $\mathcal{F}/\mathcal{B}^*$-measurable and finite valued simple functions such that $s_n \to f$, having the additional property that $0 \le |s_1| \le |s_2| \le \cdots$. [Write $f = f^+ - f^-$ and apply (MF18) separately to the nonnegative functions f^+ and f^-.]
(b) If in (a) the function f is also bounded, then $s_n \to f$ uniformly on A.
(c) The following converse to (MF18) holds: if $f : A \to \bar{\mathbb{R}}$ is such that there exists a sequence $\{s_n\}_{n=1}^{\infty}$ of simple $\mathcal{F}/\mathcal{B}^*$-measurable functions with $s_n \to f$, then f is $\mathcal{F}/\mathcal{B}^*$-measurable.

2*. Consider the measure space $(\mathbb{R}^k, \mathcal{L}^k, \bar{\lambda}_k)$. Let $f : \mathbb{R}^k \to \bar{\mathbb{R}}$ denote a Lebesgue measurable function. We will show that there exists a Borel measurable function $g : \mathbb{R}^k \to \bar{\mathbb{R}}$ with $|g| \le |f|$ and $f = g$ $\bar{\lambda}_k$-a.e.
(a) Let $f \ge 0$. There exists a sequence $\{s_n\}_{n=1}^{\infty}$ of nonnegative finite-valued Lebesgue measurable simple functions such that $0 \le s_1 \le s_2 \le \cdots \le f$ and $s_n \to f$.

Pick $m \in \mathbb{N}$, and write $s_m = \sum_{j=1}^{n_m} c_{mj} I_{A_{mj}}$, where $0 \le c_{m1}, \cdots, c_{mn_m} < +\infty$ and $A_{m1}, \cdots, A_{mn_m}$ are disjoint $\mathcal{L}^k$-sets with $\bigcup_{j=1}^{n_m} A_{mj} = \mathbb{R}^k$. Write the set A_{mj} as $B_{mj} \cup C_{mj}$, where $B_{mj} \in \mathcal{B}^k$ and C_{mj} is contained in some λ_k-null set N_{mj}. Define $s_m^* = \sum_{j=1}^{n_m} c_{mj} I_{B_{mj}}$.

(b) For each $m \in \mathbb{N}$, s_m^* is a Borel measurable simple function such that $0 \le s_m^* \le s_m$ and $s_m^* = s_m$ $\bar{\lambda}_k$-a.e.

Define $N = \bigcup_{m=1}^{\infty} N_m$, where $N_m = \{\boldsymbol{x} \in \mathbb{R}^n : s_m(\boldsymbol{x}) \ne s_m^*(\boldsymbol{x})\}$ for each $m \in \mathbb{N}$, and let $g = \sup_{m \in \mathbb{N}} s_m^*$.

(c) N is $\bar{\lambda}_k$-null, $0 \le g \le f$, and $g = f$ $\bar{\lambda}_k$-a.e.
(d) g is Borel measurable, hence the proof is complete in the nonnegative case. [Observe that g is the supremum of a sequence of Borel measurable functions.]
(e) The claim holds when $f : \mathbb{R}^k \to \bar{\mathbb{R}}$ is an arbitrary Lebesgue measurable function. [Write $f = f^+ - f^-$. Both f^+ and f^- are nonnegative Lebesgue measurable functions. Use the work done in the nonnegative case to find a Borel measurable g_1 with $0 \le g_1 \le f$ and $g_1 = f_1$ $\bar{\lambda}_k$-a.e. Find a Borel measurable g_2 with $0 \le g_2 \le f$ and $g_2 = f_2$ $\bar{\lambda}_k$-a.e. Let $g = g_1 - g_2$, so that g is Borel measurable. Showing that $g = f$ $\bar{\lambda}_k$-a.e. and $|g| \le |f|$ should be straightforward.]

3*. In Exercise 2, show that the Borel measurable function $g : \mathbb{R}^k \to \bar{\mathbb{R}}$ may be chosen such that $|g| \ge |f|$ and $f = g$ $\bar{\lambda}_k$-a.e. [Follow the outline of Exercise 2, but this time write $A_{mj} = B_{mj} - C_{mj}$.]

Comment: Putting the previous two exercises together proves the following theorem: *If $f : \mathbb{R}^k \to \bar{\mathbb{R}}$ is Lebesgue measurable, then there exist Borel measurable functions $g : \mathbb{R}^k \to \bar{\mathbb{R}}$ and $h : \mathbb{R}^k \to \bar{\mathbb{R}}$ such that $|g| \le |f| \le |h|$ and $g = f = h$ $\bar{\lambda}_k$-a.e.*

4. Let $A \in \mathcal{F} - \{\varnothing\}$, and let $f : A \to \bar{\mathbb{R}}$ denote a nonnegative $\mathcal{F}/\mathcal{B}^*$-measurable function. Then there exists a nondecreasing sequence $\{s_n\}_{n=1}^{\infty}$ of nonnegative $\mathcal{F}/\mathcal{B}^*$-measurable functions such that each s_n assumes at most countably many values, all values are finite, and $s_n \to f$ uniformly. [Such functions s_n are called *elementary functions.* Examine the proof of (MF18) and make a simple modification.]

5.6 SOME CONVERGENCE CONCEPTS

Let $(\Omega, \mathcal{F}, \mu)$ denote a measure space, and let $A \in \mathcal{F}$ be nonempty. For each $n \in \mathbb{N}$, let $f_n : A \to \mathbb{R}$ denote an $\mathcal{F}/\mathcal{B}^*$-measurable function. We also let $f : A \to \mathbb{R}$ denote an $\mathcal{F}/\mathcal{B}^*$-measurable function. Two modes of convergence familiar to the reader are: (i) *pointwise convergence* and (ii) *uniform convergence.* Pointwise convergence is just the idea that we write $f_n \to f$ *on* A to mean that $f_n(\omega) \to f(\omega)$ for each $\omega \in A$. This version of convergence has nothing to do with measure-theoretic considerations and has everything to do with topological considerations. Uniform convergence is a special case of pointwise convergence, and hence is also free of measure-theoretic considerations on first glance.

This section deals with two additional modes of convergence, *convergence μ-a.e.* and *convergence in μ-measure*, the main idea being that we now bring the underlying measure μ into the mix.

Definition. In the framework above, we say that $f_n \to f$ *μ-a.e. [on A]* iff

$$\mu(\{\omega \in A : \lim f_n(\omega) \neq f(\omega)\}) = 0.$$

[Writing "$\lim f_n(\omega) \neq f(\omega)$" means precisely that either (i) $\lim f_n(\omega)$ fails to exist, or (ii) $\lim f_n(\omega)$ exists but fails to be equal to $f(\omega)$.]

Observe in general that $\lim f_n(\omega) = f(\omega)$ iff for every $m \in \mathbb{N}$ there is $N \in \mathbb{N}$ such that $n \ge N$ implies $|f_n(\omega) - f(\omega)| \le \frac{1}{m}$. Translating this definition into set notation gives

$$\{\omega \in A : \lim f_n(\omega) = f(\omega)\} = \bigcap_{m=1}^{\infty} \bigcup_{N=1}^{\infty} \bigcap_{n=N}^{\infty} \{\omega \in A : |f_n(\omega) - f(\omega)| \le m^{-1}\}. \tag{5.6}$$

Since each of $f, f_1, f_2, \cdots$ are $\mathcal{F}/\mathcal{B}^*$-measurable, it follows that $|f_n - f|$ is $\mathcal{F}/\mathcal{B}^*$-measurable for each $n \in \mathbb{N}$, hence for each $n, m \in \mathbb{N}$ we have that $\{\omega \in A : |f_n(\omega) - f(\omega)| \le m^{-1}\} \in \mathcal{F}$. Since σ-fields are closed under amc unions and intersections, it follows that the set (5.6) is in $\mathcal{F}$. Therefore, since σ-fields are closed under complementation, it makes sense to discuss the μ-measure of $\{\omega \in A : \lim f_n(\omega) \neq f(\omega)\}$. In other words, the definition has no logical problems. We also note that the functions [so far] are assumed to have range in $\mathbb{R}$ instead of $\bar{\mathbb{R}}$ so as to avoid any complications caused by allowing the undefined expressions $+\infty - \infty$ and $-\infty + \infty$ to enter the fray.

We now introduce the second new mode of convergence, which is most easily distinguished by the first definition by the fact that the limit is on the outside of the measure.

Definition. Given the framework above, we say that $f_n \to f$ *in μ-measure [on A]* iff for every $\epsilon > 0$ we have

$$\lim_n \mu(\{\omega \in A : |f_n(\omega) - f(\omega)| > \epsilon\}) = 0.$$

If no confusion about the underlying measure μ can arise, we will merely say that $f_n \to f$ *in measure [on A]*.

By the same comments as above, the $\mathcal{F}/\mathcal{B}^*$-measurability of the functions $f, f_1, f_2, \cdots$ forces $\{\omega \in A : |f_n(\omega) - f(\omega)| > \epsilon\} \in \mathcal{F}$ for every $n \in \mathbb{N}$ and $\epsilon > 0$, hence for every $\epsilon > 0$ it makes sense to talk about the sequence of numbers

$$\{\mu(\{\omega \in A : |f_n(\omega) - f(\omega)| > \epsilon\})\}_{n=1}^{\infty}.$$

In particular, we may ask if the sequence above converges to 0 [if at all].

These definitions and the desire to discuss convergence exemplify why the definition of $\mathcal{F}/\mathcal{B}^*$-measurability is important. To say that $f_n \to f$ μ-a.e. carries a fair amount of baggage because we want to ensure that the μ-measure of the set given in the definition even *exists*, let alone is equal to a value such as zero. Without the assumption that $f, f_1, f_2, \cdots$ are $\mathcal{F}/\mathcal{B}^*$-measurable, and without the assumption that their common domain A is in $\mathcal{F}$, we cannot guarantee that the two definitions above make any sense. The underlying concept of $\mathcal{F}/\mathcal{B}^*$-measurability is what gives these convergence concepts their logical coherency.

We will now discuss some properties related to these definitions and to uniform convergence. The following result is attributed to F. Riesz.

Claim 1. *Let $(\Omega, \mathcal{F}, \mu)$ denote a measure space, and let $A \in \mathcal{F} - \{\varnothing\}$. Let $f, f_1, f_2, \cdots$ be $\mathcal{F}/\mathcal{B}^*$-measurable functions with domain A and range in $\mathbb{R}$ such that $f_n \to f$ in μ-measure. Then there exists a subsequence $\{f_{n_k}\}_{k=1}^{\infty}$ of $\{f_n\}_{n=1}^{\infty}$ such that $f_{n_k} \to f$ μ-a.e. on A [as $k \to +\infty$].*

Proof: Pick $n_1 \in \mathbb{N}$ such that $\mu(\{\omega \in A : |f_{n_1}(\omega) - f(\omega)| \geq 1\}) < \frac{1}{2}$. Such an $n_1 \in \mathbb{N}$ certainly exists since $f_n \to f$ in μ-measure. Next, we know that there is $N_2 \in \mathbb{N}$ such that $n \geq N_2$ implies $\mu(\{\omega \in A : |f_{n_2}(\omega) - f(\omega)| \geq \frac{1}{2}\}) < \frac{1}{4}$. Let $n_2 = \max\{n_1 + 1, N_2\}$, so that $n_1 < n_2$. Next, we know that there is $N_3 \in \mathbb{N}$ such that $n \geq N_3$ implies that $\mu(\{\omega \in A : |f_{n_3}(\omega) - f(\omega)| \geq \frac{1}{3}\}) < \frac{1}{8}$. Define $n_3 = \max\{n_2 + 1, N_3\}$, so that $n_1 < n_2 < n_3$. In general, given $n_1 < \cdots < n_k$, there is $N_{k+1} \in \mathbb{N}$ with $\mu(\{\omega \in A : |f_{n_{k+1}}(\omega) - f(\omega)| \geq \frac{1}{k+1}\}) < \frac{1}{2^{k+1}}$ whenever $n \geq N_{k+1}$. Let $n_{k+1} = \max\{n_k, N_{k+1}\}$, so that $n_1 < \cdots < n_k < n_{k+1}$. Carry through this process for all $k \in \mathbb{N}$. What arises is a strictly increasing sequence

$n_1 < n_2 < \cdots$ of indices such that $\mu(\{\omega \in A : |f_{n_k}(\omega) - f(\omega)| \geq \frac{1}{k}\}) < \frac{1}{2^k}$ for every $k \in \mathbb{N}$. We claim that the subsequence $\{f_{n_k}\}_{k=1}^{\infty}$, with $n_1 < n_2 < \cdots$ as above, satisfies the requirements of the theorem. That is, we claim that $f_{n_k} \to f$ μ-a.e. on A.

Let $B_m = \bigcup_{j=m}^{\infty}\{\omega \in A : |f_{n_j}(\omega) - f(\omega)| \geq \frac{1}{j}\}$ for each $m \in \mathbb{N}$. Observe that $\{B_m\}_{m=1}^{\infty}$ is a nonincreasing sequence of $\mathcal{F}$-sets, hence $\lim_m B_m$ exists, equals $\bigcap_{m=1}^{\infty} B_m$, and is in $\mathcal{F}$. Since

$$\begin{aligned} \mu(B_1) &= \mu\left(\bigcup_{j=1}^{\infty}\left\{\omega \in A : |f_{n_j}(\omega) - f(\omega)| \geq \frac{1}{j}\right\}\right) \\ &\leq \sum_{j=1}^{\infty} \mu\left(\left\{\omega \in A : |f_{n_j}(\omega) - f(\omega)| \geq \frac{1}{j}\right\}\right) \\ &< \sum_{j=1}^{\infty} 2^{-j} \quad [< +\infty], \end{aligned}$$

we may apply measure property (M9) and countable subadditivity to obtain

$$\begin{aligned} \mu\left(\lim_m B_m\right) = \lim_m \mu(B_m) &\leq \lim_m \sum_{j=m}^{\infty} \mu\left(\left\{\omega \in A : |f_{n_j}(\omega) - f(\omega)| \geq \frac{1}{j}\right\}\right) \\ &\leq \lim_m \sum_{j=m}^{\infty} \left(\frac{1}{2}\right)^j = \lim_m \left(\frac{1}{2}\right)^{m-1} = 0. \end{aligned}$$

Therefore, $\mu(\lim_m B_m) = 0$. Writing C for $\lim_m B_m$, we now claim that $f_{n_k}(\omega) \to f(\omega)$ for all $\omega \in A - C = \bigcup_{m=1}^{\infty}(A - B_m)$. Let $\omega \in A - C$. Then there is $m \in \mathbb{N}$ such that $\omega \in A - B_m$, hence for every $j \geq m$ we have $|f_{n_j}(\omega) - f(\omega)| < \frac{1}{j}$. Let $\epsilon > 0$ be given, and choose $j_0 > \max\{m, \frac{1}{\epsilon}\}$. It follows that whenever $k \geq j_0$ we have $|f_{n_k}(\omega) - f(\omega)| < \frac{1}{k} \leq \frac{1}{j_0} < \epsilon$. Therefore, it is the case that $f_{n_k} \to f$ on $A - C$.

We now summarize. We have $f_{n_k} \to f$ *everywhere* on $A - C$, and $\mu(C) = 0$. The set $\{\omega \in A : \lim f_{n_k}(\omega) \neq f(\omega)\}$, which is in $\mathcal{F}$ by the discussion following the definition of μ-a.e. convergence, must therefore be a subset of C. Since $\mu(C) = 0$, we must have $\mu(\{\omega \in A : \lim f_{n_k}(\omega) \neq f(\omega)\}) = 0$. It follows that $f_{n_k} \to f$ μ-a.e. on A, and the proof is complete. ■

Claim 2. *Let $(\Omega, \mathcal{F}, \mu)$ denote a measure space with $\mu(\Omega) < +\infty$. Pick nonempty $A \in \mathcal{F}$. Suppose that (i) each of $f, f_1, f_2, \cdots$ are $\mathcal{F}/\mathcal{B}^*$-measurable functions with domain A and range contained in $\mathbb{R}$, and (ii) $f_n \to f$ μ-a.e. on A. Then, given any $\epsilon > 0$ and $\delta > 0$ we may find a set $B \in \mathcal{F}$ and $n_0 \in \mathbb{N}$ such that $\mu(B^c) < \epsilon$ and $|f_n(\omega) - f(\omega)| < \delta$ for every $\omega \in B$ and $n \geq n_0$.*

Proof: Let $\epsilon > 0$ and $\delta > 0$, and let $C = \{\omega \in A : \lim f_n(\omega) = f(\omega)\}$. By the hypotheses, $C, C^c \in \mathcal{F}$ and $\mu(C^c) = 0$. For every $m \in \mathbb{N}$, let

$$C_m = \{\omega \in C : \text{for every } n \geq m \text{ we have } |f_n(\omega) - f(\omega)| < \delta\}.$$

Observe the following facts:

(a) $\{C_m\}_{m=1}^{\infty}$ is a nondecreasing sequence of $\mathcal{F}$-sets [obvious], and

(b) $\lim_m C_m$ exists and equals C.

To see (b), we have by (a) that $\lim_m C_m = \bigcup_{m=1}^{\infty} C_m$. Since $C_m \subseteq C$ for each $m \in \mathbb{N}$, we have $\lim_m C_m \subseteq C$. On the other hand, if $\omega \in C$, then $\lim f_n(\omega) = f(\omega)$, so there is $M \in \mathbb{N}$ such that $n \geq M$ implies that $|f_n(\omega) - f(\omega)| < \delta$ whenever $n \geq M$, hence $\omega \in C_M \subseteq \lim_m C_m$. It follows that $C \subseteq \lim_m C_m$, giving $\lim_m C_m = C$.

From this information, it follows that $\{C_m^c\}_{m=1}^{\infty}$ is a nonincreasing sequence of $\mathcal{F}$-sets with limit C^c. Since $\mu(C_1^c) \leq \mu(\Omega) < +\infty$ by assumption, we invoke measure property (M9) to obtain $\lim_m \mu(C_m^c) = \mu(C^c) = 0$. Choose $n_0 \in \mathbb{N}$ such that $n \geq n_0$ yields $\mu(C_n^c) < \epsilon$, and set $B = C_{n_0}$. Then we indeed have $\mu(B^c) = \mu(C_{n_0}^c) < \epsilon$ and $|f_n(\omega) - f(\omega)| < \delta$ for every $\omega \in B$ and $n \geq n_0$, completing the proof. ■

The next theorem uses the preceding result and is credited to H. Lebesgue. Roughly speaking, it shows that μ-a.e. convergence is a stronger condition than convergence in μ-measure when $\mu(\Omega) < +\infty$.

Claim 3. *Let $(\Omega, \mathcal{F}, \mu)$ denote a measure space with $\mu(\Omega) < +\infty$, and let $\varnothing \subsetneq A \in \mathcal{F}$. Suppose that each of $f, f_1, f_2, \cdots$ are $\mathcal{F}/\mathcal{B}^*$-measurable functions with domain A and range contained in $\mathbb{R}$ such that $f_n \to f$ μ-a.e. on A. Then $f_n \to f$ in μ-measure on A.*

Proof: Let $\epsilon > 0$ and $\delta > 0$, and, utilizing the notation of the previous theorem, let $C = \{\omega \in A : \lim f_n(\omega) = f(\omega)\}$. Use Claim 2 to find $B \in \mathcal{F}$ and $n_0 \in \mathbb{N}$ such that $\mu(B^c) < \epsilon$ and $|f_n(\omega) - f(\omega)| < \delta$ for every $\omega \in B$ and $n \geq n_0$. Therefore, for every $n \geq n_0$ we have

$$\{\omega \in C : |f_n(\omega) - f(\omega)| \geq \delta\} \subseteq B^c,$$

hence for every $n \geq n_0$ we have

$$\mu(\{\omega \in C : |f_n(\omega) - f(\omega) \geq \delta\}) \leq \mu(B^c) < \epsilon.$$

Since $\epsilon, \delta > 0$ are arbitrary, we have $\lim_n \mu(\{\omega \in C : |f_n(\omega) - f(\omega)| \geq \delta\}) = 0$ for every $\delta > 0$. It follows that

$$\begin{aligned}
& \lim_n \mu(\{\omega \in A : |f_n(\omega) - f(\omega)| \geq \delta\}) \\
= \;& \lim_n [\mu(\{\omega \in C : |f_n(\omega) - f(\omega)| \geq \delta\}) + \mu(\{\omega \in A - C : |f_n(\omega) - f(\omega)| \geq \delta\})] \\
\leq \;& \lim_n \mu(\{\omega \in C : |f_n(\omega) - f(\omega)| \geq \delta\}) + \mu(A - C) \\
\leq \;& 0 + 0 \;=\; 0,
\end{aligned}$$

where the equation $\mu(A - C) = 0$ follows because $f_n \to f$ μ-a.e. on A. It follows that

$$\lim_n \mu(\{\omega \in A : |f_n(\omega) - f(\omega)| \geq \delta\}) = 0$$

for the generic $\delta > 0$. From this, we have $f_n \to f$ in μ-measure. ∎

The hypothesis $\mu(\Omega) < +\infty$ is essential in the previous claim; an example of where the conclusion of the theorem fails when $\mu(\Omega) = +\infty$ is given in the exercises.

The following result is called *Egorov's Theorem*, which roughly says that if $f_n \to f$ μ-a.e., then $f_n \to f$ uniformly except on a set of arbitrarily small μ-measure. That is, $f_n \to f$ "uniformly almost everywhere." The statement of the theorem will clarify these colloquial remarks.

Claim 4. *Let $(\Omega, \mathcal{F}, \mu)$ denote a measure space with $\mu(\Omega) < +\infty$, and let $A \in \mathcal{F}$ be nonempty. Let each of $f, f_1, f_2, \cdots$ be $\mathcal{F}/\mathcal{B}^*$-measurable functions with domain A and range contained in $\mathbb{R}$ such that $f_n \to f$ μ-a.e. on A. Then for every $\epsilon > 0$ there exists $B \in \mathcal{F}$ such that $\mu(B^c) < \epsilon$ and $f_n \to f$ uniformly on B. [The set B, of course, depends on ϵ.]*

Proof: Let $\epsilon > 0$. For each $m \in \mathbb{N}$, use Claim 2 to find $B_m \in \mathcal{F}$ and $n_m \in \mathbb{N}$ such that $\mu(B_m^c) < \frac{\epsilon}{2^m}$ and $|f_n(\omega) - f(\omega)| < \frac{1}{m}$ for every $\omega \in B_m$ and $n \geq n_m$. We claim that the desired set B is none other than $\bigcap_{m=1}^{\infty} B_m$. To see this, observe that $B^c = \bigcup_{m=1}^{\infty} B_m^c$, and

$$\mu(B^c) = \mu\left(\bigcup_{m=1}^{\infty} B_m^c\right) \leq \sum_{m=1}^{\infty} \mu(B_m^c) < \sum_{m=1}^{\infty} \frac{\epsilon}{2^m} = \epsilon.$$

We need only show that $f_n \to f$ uniformly on $B = \bigcap_{m=1}^{\infty} B_m$. To see this, let $m \in \mathbb{N}$, and observe that if $n \geq n_m$ then

$$\sup_{\omega \in B} |f_n(\omega) - f(\omega)| \leq \sup_{\omega \in B_m} |f_n(\omega) - f(\omega)| \leq \frac{1}{m}.$$

This shows that $f_n \to f$ uniformly on B, and thus completes the proof. ∎

The preceding results are the major results concerning μ-a.e. convergence and convergence in μ-measure. The following exercises, which state some subsidiary results concerning the modes of convergence mentioned, are important and should be worked carefully. Following the exercises is a discussion that generalizes the definitions and results of this section.

Exercises.

These exercises discuss some basic properties of convergence in μ-measure and convergence μ-a.e. We shall have a generic measure space $(\Omega, \mathcal{F}, \mu)$, and all functions are defined on some nonempty $A \in \mathcal{F}$ and have range contained in $\mathbb{R}$. We furthermore assume that all functions mentioned are $\mathcal{F}/\mathcal{B}^*$-measurable.

1. This result concerns uniqueness.
(a) If $f_n \to f$ μ-a.e. and $f_n \to g$ μ-a.e., then $f = g$ μ-a.e.
(b) If $f_n \to f$ in μ-measure and $f_n \to g$ in μ-measure, then $f = g$ μ-a.e.

Comment: Whereas limits of sequences of numbers are unique, the limits of sequences of functions are not unique, but are "unique up to a set of μ-measure zero".

For (a), write $\{\omega \in A : f(\omega) \neq g(\omega)\}$ as the union of four $\mathcal{F}$-sets:

$$A_1 = \{\omega \in A : f(\omega) \neq g(\omega),\ f_n(\omega) \to f(\omega),\ g_n(\omega) \to g(\omega)\},$$

$$A_2 = \{\omega \in A : f(\omega) \neq g(\omega),\ f_n(\omega) \to f(\omega),\ g_n(\omega) \not\to g(\omega)\},$$

$$A_3 = \{\omega \in A : f(\omega) \neq g(\omega),\ f_n(\omega) \not\to f(\omega),\ g_n(\omega) \to g(\omega)\},$$

and

$$A_4 = \{\omega \in A : f(\omega) \neq g(\omega),\ f_n(\omega) \not\to f(\omega),\ g_n(\omega) \not\to g(\omega)\}.$$

Since limits of sequences of numbers are unique, $A_1 = \varnothing$. Observe that each of A_2, A_3, and A_4 are contained in μ-null sets; for example, $A_2 \subseteq \{\omega \in A : g_n(\omega) \not\to g(\omega)\}$. Conclude that $\mu(\{\omega \in A : f(\omega) \neq g(\omega)\}) = 0$. For (b), show first that for any $m \in \mathbb{N}$ we have

$$\begin{aligned} &\{\omega \in A : |f(\omega) - g(\omega)| > 1/m\} \\ \subseteq\ &\{\omega \in A : |f_n(\omega) - f(\omega)| \geq 1/2m\} \cup \{\omega \in A : |f_n(\omega) - g(\omega)| > 1/2m\}, \end{aligned}$$

hence for any $m \in \mathbb{N}$ it follows that $\mu(\{\omega \in A : |f(\omega) - g(\omega)| > 1/m\})$ can be made as small as desired, say less than $\epsilon/2^m$. Since

$$\mu(\{\omega \in A : f(\omega) \neq g(\omega)\}) \leq \sum_{m=1}^{\infty} \mu(\{\omega \in A : |f(\omega) - g(\omega)| > 1/m\}) < \epsilon,$$

the conclusion follows.

2. Suppose that $f_n \to f$ and $g_n \to g$, both in μ-measure.
(a) $f_n - f \to 0$ in μ-measure and $|f_n| \to |f|$ in μ-measure.
(b) If $a, b \in \mathbb{R}$, then $af_n + b \to af + b$ in μ-measure.
(c) If $a, b \in \mathbb{R}$, then $af_n + bg_n \to af + bg$ in μ-measure.
(d) $f_n^+ \to f^+$ and $f_n^- \to f^-$, both in μ-measure.
(e) If $\mu(A) < +\infty$ and $\eta > 0$, there is $M > 0$ with $\mu(\{\omega \in A : |g(\omega)| > M\}) < \eta$.
(f) If $\mu(A) < +\infty$, then $f_n g \to fg$ in μ-measure.
(g) If $\mu(A) < +\infty$, then $f_n g_n \to fg$ in μ-measure.
(h) If $\mu(A) = +\infty$, then $f_n g_n$ does not necessarily converge to fg in μ-measure.
(i) It is not necessarily the case that $\frac{f_n}{g_n} \to \frac{f}{g}$ in μ-measure when $\mu(A) = +\infty$, even if $g(\omega) \neq 0$ and $g_n(\omega) \neq 0$ for every $\omega \in A$ and $n \in \mathbb{N}$. However, if $\mu(A) < +\infty$, the result follows.

Comment: The first claim of (a) is trivial. For the second claim, observe that

$$\{\omega \in A : |\,|f_n(\omega)| - |f(\omega)|\,| > \delta\} \subseteq \{\omega \in A : |f_n(\omega) - f(\omega)| > \delta\}$$

for all $\delta > 0$, $n \in \mathbb{N}$, and $\omega \in A$. Part (b) is nearly trivial. For (c), one really needs to only prove that $f_n + g_n \to f + g$. For any $\delta > 0$ and $n \in \mathbb{N}$, observe that

$$\begin{aligned} &\{\omega \in A : |(f_n(\omega) + g_n(\omega)) - (f(\omega) - g(\omega))| > \delta\} \\ \subseteq\ &\{\omega \in A : |f_n(\omega) - f(\omega)| > \delta/2\} \cup \{\omega \in A : |g_n(\omega) - g(\omega)| > \delta/2\}. \end{aligned}$$

For (d), observe for any $\delta > 0$ and $n \in \mathbb{N}$ that

$$\{\omega \in A : |\max\{f_n(\omega), 0\} - \max\{f(\omega), 0\}| > \delta\} \subseteq \{\omega \in A : |f_n(\omega) - f(\omega)| > \delta\}.$$

The other claim of (d) is handled similarly. For (e), let $A_m = \{\omega \in A : |g(\omega)| > m\}$ for all $m \in \mathbb{N}$. Observe that $\{A_m\}_{m=1}^{\infty}$ decreases to $\varnothing$ and that $\mu(A_1) \leq \mu(A) < +\infty$, hence $\mu(A_m)$ decreases to 0 as $m \to \infty$. Pick M to be any m such that $\mu(A_m) < \eta$. For (f), observe for all $\delta > 0$, $M > 0$, and $n \in \mathbb{N}$ that

$$\begin{aligned} & \{\omega \in A : |g(\omega)||f_n(\omega) - f(\omega)| > \delta\} \\ \subseteq\ & \{\omega \in A : |g(\omega)| > M,\ |g(\omega)||f_n(\omega) - f(\omega)| > \delta\} \\ & \cup \{\omega \in A : |g(\omega)| \leq M,\ |g(\omega)||f_n(\omega) - f(\omega)| > \delta\} \\ \subseteq\ & \{\omega \in A : |g(\omega)| > M\} \cup \{\omega \in A : |f_n(\omega) - f(\omega)| > \delta/M\}. \end{aligned}$$

By assumption and part (e), given $\epsilon > 0$, we can choose $M > 0$ so large that $\mu(\{\omega \in A : |g(\omega)||f_n(\omega) - f(\omega)| > \delta\}) < \epsilon$. For (g), we have that

$$\{\omega \in A : |f_n(\omega)g_n(\omega) - f(\omega)g(\omega)| > \delta\}$$

is contained in

$$\begin{aligned} & \{\omega \in A : |f_n(\omega) - f(\omega)| > \sqrt{\delta/3}\} \cup \{\omega \in A : |g_n(\omega) - g(\omega)| > \sqrt{\delta/3}\} \\ & \cup \{\omega \in A : |g(\omega)||f_n(\omega) - f(\omega)| > \delta/3\} \cup \{\omega \in A : |f(\omega)||g_n(\omega) - g(\omega)| > \delta/3\} \end{aligned}$$

for every $n \in \mathbb{N}$ and $\delta > 0$. Use the assumptions and (f) to make the measure of these pieces arbitrarily small. For (i), one really needs only show that if $\mu(A) < +\infty$, then $\frac{1}{g_n} \to \frac{1}{g}$.

3. This exercise, which deals with μ-a.e. convergence, is the dual of the previous exercise which dealt with convergence in μ-measure. Suppose that $f_n \to f$ μ-a.e. and $g_n \to g$ μ-a.e.
(a) $f_n - f \to 0$ μ-a.e. and $|f_n| \to |f|$ μ-a.e.
(b) If $a, b \in \mathbb{R}$, then $af_n + b \to af + b$ μ-a.e.
(c) If $a, b \in \mathbb{R}$, then $af_n + bg_n \to af + bg$ μ-a.e.
(d) $f_n^+ \to f^+$ μ-a.e. and $f_n^- \to f^-$ μ-a.e.
(e) $f_n g_n \to fg$ in μ-measure.
(f) If $g, g_n \neq 0$ μ-a.e. for each $n \in \mathbb{N}$, then $\frac{f_n}{g_n} \to \frac{f}{g}$ μ-a.e.
Comment: Part (a) is nearly automatic. For (b), the case $a = 0$ is trivial, and for $a \neq 0$, observe that

$$\{\omega \in A : af_n(\omega) + b \not\to af(\omega) + b\} = \{\omega \in A : f_n(\omega) \not\to f(\omega)\}.$$

For (c), (e), and (f), show that

$$\{\omega \in A : af_n(\omega) + bg_n(\omega) \not\to af(\omega) + bg(\omega)\},$$

$$\{\omega \in A : f_n(\omega)g_n(\omega) \not\to f(\omega)g(\omega)\},$$

and

$$\{\omega \in A : f_n(\omega)/g_n(\omega) \not\to f(\omega)/g(\omega)\}$$

are all contained in

$$\{\omega \in A : f_n(\omega) \not\to f(\omega)\} \cup \{\omega \in A : g_n(\omega) \not\to g(\omega)\}.$$

4. Suppose for each $n \in \mathbb{N}$ that $f_n = g_n$ μ-a.e.

(a) If $f_n \to f$ μ-a.e., then $g_n \to f$ μ-a.e.
(b) If $f_n \to f$ in μ-measure, then $g_n \to f$ in μ-measure.
(c) If $f_n \to f$ in μ-measure and $g_n \to g$ in μ-measure, then $f = g$ μ-a.e.
(d) If $f_n \to f$ μ-a.e. and $g_n \to g$ μ-a.e., then $f = g$ μ-a.e.
(e) If $f_n \to f$ μ-a.e. and $f_n \to g$ μ-a.e., then $f = g$ μ-a.e.

Comment: These can be handled similarly to the previous exercises, and it doesn't seem necessary to give detailed hints here.

5. Prove the following statements connecting convergence in μ-measure with convergence μ-a.e.
(a) $f_n \to f$ in μ-measure iff $f_{n_j} \to f$ in μ-measure [as $j \to \infty$] for every subsequence $\{n_j\}_{j=1}^{\infty}$ of the positive integers.
(b) $f_n \to f$ in μ-measure iff each subsequence of $\{f_n\}_{n=1}^{\infty}$ has a sub-subsequence that converges to f μ-a.e.

Comment: One direction of (a) is trivial while the other direction is straightforward. For (b), suppose that it is not the case that $f_n \to f$ in μ-measure. Then there exists $\epsilon > 0$ and $\delta > 0$ such that $\mu(\{\omega \in A : |f_n(\omega) - f(\omega)| > \delta\}) > \epsilon$ for infinitely many choices of $n \in \mathbb{N}$, say $n_1 < n_2 < n_3 < \cdots$. It follows that no subsequence of $\{f_{n_k}\}_{k=1}^{\infty}$ can possibly converge in μ-measure to f, hence no subsequence of $\{f_{n_k}\}_{k=1}^{\infty}$ can possibly converge μ-a.e. to f. Conversely, suppose that $f_n \to f$ in μ-measure. Let $\{n_j\}_{j=1}^{\infty}$ denote a subsequence of $1, 2, \cdots$. By (a), $f_{n_j} \to f$ in μ-measure. By Claim 1, there exists a subsequence $\{n'_j\}_{j=1}^{\infty}$ of the original subsequence $\{n_j\}_{j=1}^{\infty}$ such that $f_{n'_j} \to f$ μ-a.e.

6. Suppose that $\phi : \mathbb{R} \to \mathbb{R}$ is continuous.
(a) If $f_n \to f$ μ-a.e., then $\phi \circ f_n \to \phi \circ f$ μ-a.e.
(b) If $f_n \to f$ in μ-measure, then $\phi \circ f_n \to \phi \circ f$ in μ-measure.

Comment: For (a), observe that if $\omega \in A$ is such that $\phi(f_n(\omega)) \not\to \phi(f(\omega))$, then $f_n(\omega) \not\to f(\omega)$. For (b), use the previous exercise. Let $\{n_j\}_{j=1}^{\infty}$ denote a subsequence of $1, 2, \cdots$. Since $f_n \to f$ in μ-measure, there exists a subsequence $\{n'_j\}_{j=1}^{\infty}$ of $\{n_j\}_{j=1}^{\infty}$ such that $f_{n'_j} \to f$ μ-a.e. By (a), $\phi \circ f_{n'_j} \to \phi \circ f$ μ-a.e. Since $\{n_j\}_{j=1}^{\infty}$ was an arbitrary subsequence of $1, 2, \cdots$, we have $\phi \circ f_n \to \phi \circ f$ in μ-measure.

7. Let $\Omega = [0, 1]$, let $\mathcal{F} = \{B \subseteq [0, 1] : B \in \mathcal{L}\}$, and let μ denote the restriction of Lebesgue measure $\bar{\lambda}$ to $\mathcal{F}$. Let $A_1 = [0, 1]$, $A_2 = [0, \frac{1}{2}]$, $A_3 = [\frac{1}{2}, 1]$, $A_4 = [0, \frac{1}{4}]$, $A_5 = [\frac{1}{4}, \frac{1}{2}]$, $A_6 = [\frac{1}{2}, \frac{3}{4}]$, $A_7 = [\frac{3}{4}, 1]$, $A_8 = [0, \frac{1}{8}]$, etc. On Ω, define $f_n = I_{A_n}$ for each $n \in \mathbb{N}$.
(a) $f_n \to 0$ in μ-measure.
(b) For each $\omega \in [0, 1]$, $\underline{\lim}\, f_n(\omega) = 0 < 1 = \overline{\lim}\, f_n(\omega)$, hence $\lim f_n(\omega)$ fails to exist for any $\omega \in [0, 1]$. Therefore, the sequence $\{f_n\}_{n=1}^{\infty}$ converges to no function with domain $[0, 1]$. In particular, $\{f_n\}$ fails to converge μ-a.e. to 0, so that convergence in μ-measure does not imply convergence μ-a.e., even if $\mu(\Omega) < +\infty$.

8. Consider the measure space $(\mathbb{R}, \mathcal{L}, \bar{\lambda})$, and let $f_n(\omega) = I_{[n,n+1]}(\omega)$ for each $n \in \mathbb{N}$ and $\omega \in \mathbb{R}$.
(a) $f_n \to 0$ $\bar{\lambda}$-a.e.
(b) $\{f_n\}_{n=1}^{\infty}$ fails to converge in $\bar{\lambda}$-measure to 0, hence the hypothesis $\mu(\Omega) < +\infty$ in Claim 3 is not redundant.
(c) If $B \in \mathcal{L}$ and $\bar{\lambda}(B^c) < 1$, then $\{f_n\}_{n=1}^{\infty}$ cannot converge uniformly to 0 on B. This shows that the hypothesis $\mu(\Omega) < +\infty$ is not redundant in Claim 4.

9. If $\mu(\{\omega \in A : |f_n(\omega) - f(\omega)| > \epsilon \text{ for infinitely many } n\}) = 0$ for each $\epsilon > 0$, then $f_n \to f$ μ-a.e.

We have assumed throughout that the functions $f, f_1, f_2, \cdots$ have domain A and are finite-valued. We may generalize the previous definitions and discussions to the case where each of $f, f_1, f_2, \cdots$ are defined μ-a.e. on A and each of $f, f_1, f_2, \cdots$ are finite μ-a.e. on A. Let B denote the set of $\omega \in A$ such that

(i) $f(\omega)$ is defined and finite, [this set is $\operatorname{dom} f \cap f^{-1}(\mathbb{R})$] and

(ii) $f_n(\omega)$ is defined and finite for each $n \in \mathbb{N}$. [This set is

$$\bigcap_{n=1}^{\infty} \operatorname{dom} f_n \cap \bigcap_{n=1}^{\infty} f_n^{-1}(\mathbb{R}).]$$

Therefore, we have

$$B = \operatorname{dom} f \cap \bigcap_{n=1}^{\infty} \operatorname{dom} f_n \cap f^{-1}(\mathbb{R}) \cap \bigcap_{n=1}^{\infty} f_n^{-1}(\mathbb{R}),$$

and $A - B$ is the same set as

$$(A - \operatorname{dom} f) \cup \left(\bigcup_{n=1}^{\infty} (A - \operatorname{dom} f_n) \right) \cup f^{-1}(\{-\infty, +\infty\}) \cup \bigcup_{n=1}^{\infty} f_n^{-1}(\{-\infty, +\infty\}),$$

which is a countable union of μ-null sets, so that $\mu(A - B) = 0$. Accordingly, we may decompose A into the two disjoint $\mathcal{F}$-sets B and $A - B$, and since we have $\mu(A - B) = 0$, we may in the context of convergence μ-a.e. and convergence in μ-measure equivalently think of having $g, g_1, g_2, \cdots$ denoting $\mathcal{F}/\mathcal{B}^*$-measurable functions with domain B and range contained in $\mathbb{R}$, where $g, g_1, g_2, \cdots$ denote the restrictions of $f, f_1, f_2, \cdots$ [respectively] from their respective domains to B. In other words, we break A into two $\mathcal{F}$-sets, of which one is μ-null while the other is "almost all of A." In the context of convergence μ-a.e. we are not interested in what happens on the μ-null set $A - B$, but we are interested in what happens on B.

In this more general situation, we have the following definitions.

Definition. Under the more general framework just given, we will say that $f_n \to f$ μ-a.e. on A iff $\mu(\{\omega \in B : g_n(\omega) \not\to g(\omega)\}) = 0$, where B, g, and $\{g_n\}_{n=1}^{\infty}$ are as defined above, and where the notation $g_n(\omega) \not\to g(\omega)$ means that either $\lim g_n(\omega)$ fails to exist or $\lim g_n(\omega)$ does exist but does not equal $g(\omega)$. That is, we will say that $f_n \to f$ μ-a.e. on A iff $g_n \to g$ μ-a.e. on B.

Observe that if $f, f_1, f_2, \cdots$ are defined everywhere on A and are finite-valued [as in all the material preceding the exercises], then $B = A$ and $g = f$ and $g_n = f_n$ for all $n \in \mathbb{N}$, so the case considered at the beginning of the section is truly a special case of the more general situation.

Definition. Under the more general framework and notation just given, we will say that $f_n \to f$ in μ-measure on A iff $g_n \to g$ in μ-measure on B. That is, we will write $f_n \to f$ in μ-measure on A iff for every $\epsilon > 0$ and $\delta > 0$ we have $\lim_n \mu(\{\omega \in B : |g_n(\omega) - g(\omega)| > \delta\}) = 0$.

The theorems and the exercises carry over to this more general situation with ease. For example, let us revisit Claim 1 in a more general setting.

Generalized Claim 1. *Let $(\Omega, \mathcal{F}, \mu)$ denote a measure space, and let $A \in \mathcal{F} - \{\varnothing\}$. Let $f, f_1, f_2, \cdots$ be $\mathcal{F}/\mathcal{B}^*$-measurable functions defined and finite μ-a.e. on A such that $f_n \to f$ in μ-measure. Then there exists a subsequence $\{f_{n_k}\}_{k=1}^{\infty}$ of $\{f_n\}_{n=1}^{\infty}$ such that $f_{n_k} \to f$ μ-a.e. on A.*

In this restatement, we have removed the assumption that the domain of each of $f, f_1, f_2, \cdots$ is equal to A and that the functions have ranges contained in $\mathbb{R}$, and instead have generalized matters as stated earlier. A proof of this generalized theorem might look like this.

Proof: Let B, g, and $\{g_n\}_{n=1}^{\infty}$ be as in the preceding discussion. By the original Claim 1, there exists a subsequence $\{g_{n_k}\}_{k=1}^{\infty}$ of $\{g_n\}_{n=1}^{\infty}$ such that $g_{n_k} \to g$ μ-a.e. on B. Since $g_{n_k} \to g$ μ-a.e. on B, it follows by definition that $f_{n_k} \to f$ μ-a.e. on A. ∎

One can generalize the other theorems to this setting, and one may prove them in the same easy fashion as above. In particular, the claims of the exercises carry over with the same ease. In other words, no truly new work is required in restating and reproving the theorems and claims in the exercises for the more general case. This similar approach will be followed in our discussions of integration to be found in subsequent work, where we will state and prove results for functions defined everywhere [and finite everywhere] on some $\mathcal{F}$-set A, and then it will be seen that nothing is really changed by allowing functions to be defined μ-a.e. on A and possibly infinite for some points in their domain, so long as the subset of the domain where the function assumes an infinite value is "negligible" in the sense of having μ-measure zero.

Exercises.

10. State and prove Claim 2 in this more general setting. [Mimic the spirit of example of Claim 1 given earlier.]

11. Do the same for Claim 3.

12. Do the same for Claim 4.

5.7 CONTINUITY AND MEASURABILITY

Consider the background space $\Omega = \mathbb{R}^k$, where $k \in \mathbb{N}$ is arbitrary. By (MF4), any continuous $f : \mathbb{R}^k \to \bar{\mathbb{R}}$ is Borel measurable. The converse to this claim is false. [Indicator functions of Borel subsets of $\mathbb{R}^k$ are not always continuous but are always Borel measurable.] However, there is an interesting result that may be viewed as a sort-of-converse. This result is credited to Lusin,[5] and it says [roughly] that $\mathcal{F}/\mathcal{B}^*$-measurable functions may be approximated to any degree of accuracy by continuous functions. This statement is very loose, since we have not specified just what *approximation* means. The precise statement will follow upon completion of a few minor claims.

Definition. Let $A \subseteq \mathbb{R}^k$ be nonempty. The *[Euclidean] distance function for* A is denoted by d_A and given for all $\boldsymbol{x} \in \mathbb{R}^k$ by the rule

$$d_A(\boldsymbol{x}) = \inf\{||\boldsymbol{x} - \boldsymbol{y}|| : \boldsymbol{y} \in A\}.$$

In this context, $||\boldsymbol{x} - \boldsymbol{y}||$ denotes the usual k-dimensional Euclidean distance $\sqrt{\sum_{i=1}^k (x_i - y_i)^2}$ between $\boldsymbol{x} = (x_1, \cdots, x_k)$ and $\boldsymbol{y} = (y_1, \cdots, y_k)$.

The above definition is merely a special case of the distance function given in the elementary study of metric spaces. In that elementary study, two important facts regarding d_A were established: (i) d_A *is continuous*, and (ii) $d_A(\boldsymbol{x}) = 0$ *iff* $\boldsymbol{x} \in \bar{A}$, where $\bar{A}$ denotes the closure of A. An equivalent form of (ii) is $d_A(\boldsymbol{x}) > 0$ *iff* $\boldsymbol{x} \in \bar{A}^c$.

Claim 1. *For any nonempty* $A \subseteq \mathbb{R}^k$ *and* $\boldsymbol{x} \in \mathbb{R}^k$ *we have the following:*
(a) d_A *is continuous.*
(b) $d_A(\boldsymbol{x}) = 0$ *iff* $\boldsymbol{x} \in \bar{A}$.
(c) $d_A(\boldsymbol{x}) > 0$ *iff* $\boldsymbol{x} \in \bar{A}^c$.

Proof: Let $\varnothing \subsetneq A \subseteq \mathbb{R}^k$ and $\boldsymbol{x} \in \mathbb{R}^k$. We first establish (a), picking $\boldsymbol{x}' \in \mathbb{R}^k$ and $\boldsymbol{y} \in A$. Observe that by the definition and the triangle inequality we have

$$d_A(\boldsymbol{x}) = \inf\{||\boldsymbol{x} - \boldsymbol{y}'|| : \boldsymbol{y}' \in A\} \leq ||\boldsymbol{x} - \boldsymbol{y}|| \leq ||\boldsymbol{x} - \boldsymbol{x}'|| + ||\boldsymbol{x}' - \boldsymbol{y}||,$$

and hence $d_A(\boldsymbol{x}) - ||\boldsymbol{x} - \boldsymbol{x}'|| \leq ||\boldsymbol{x}' - \boldsymbol{y}||$. Since $y \in A$ is generic, we have $d_A(\boldsymbol{x}) - ||\boldsymbol{x} - \boldsymbol{x}'|| \leq d_A(\boldsymbol{x}')$. This gives $d_A(\boldsymbol{x}) - d_A(\boldsymbol{x}') \leq ||\boldsymbol{x} - \boldsymbol{x}'||$. By the exact same reasoning, we may reverse the roles of $\boldsymbol{x}$ and $\boldsymbol{x}'$ to obtain $d_A(\boldsymbol{x}') - d_A(\boldsymbol{x}) \leq ||\boldsymbol{x} - \boldsymbol{x}'||$. Therefore $|d_A(\boldsymbol{x}) - d_A(\boldsymbol{x}')| \leq ||\boldsymbol{x} - \boldsymbol{x}'||$ for our generic $\boldsymbol{x}$ and $\boldsymbol{x}'$ in $\mathbb{R}^k$.

With this inequality established, it is easy to show (a): let $\boldsymbol{x} \in \mathbb{R}^k$, $\epsilon > 0$, and take $\delta = \epsilon$. Then, for any $\boldsymbol{x} \in \mathbb{R}^k$ with $||\boldsymbol{x} - \boldsymbol{x}'|| < \delta = \epsilon$, we have that $|d_A(\boldsymbol{x}) - d_A(\boldsymbol{x}')| < ||\boldsymbol{x} - \boldsymbol{x}'|| < \delta = \epsilon$, so d_A is continuous at the generic $\boldsymbol{x} \in \mathbb{R}^k$, giving (a).

For (b), pick $\boldsymbol{x} \in \mathbb{R}^k$ and observe that the following statements are logically equivalent: (1) $d_A(\boldsymbol{x}) = 0$, (2) for every $\epsilon > 0$ there exists $\boldsymbol{y} \in A$ with $||\boldsymbol{x} - \boldsymbol{y}|| < \epsilon$, (3) we may find an element of A arbitrarily close to $\boldsymbol{x}$, and (4) $\boldsymbol{x} \in \bar{A}$. This gives (b). Part (c) follows immediately from (b), and hence the proof is complete. ∎

[5]Lusin is also spelled as Luzin by various authors.

The following claim is a special case of what is known as *Urysohn's Lemma.* We will use the three results of Claim 1 repeatedly in the proof.

Claim 2. *Let $\varnothing \subsetneq F \subseteq G \subseteq \mathbb{R}^k$, where F is closed and G is open. Then there exists a continuous function $\phi : \mathbb{R}^k \to \mathbb{R}$ such that*
(i) $0 \le \phi(\boldsymbol{x}) \le 1$ for each $\boldsymbol{x} \in \mathbb{R}^k$,
(ii) $\phi(\boldsymbol{x}) = 1$ for each $\boldsymbol{x} \in F$,
(iii) $\phi(\boldsymbol{x}) = 0$ for each $\boldsymbol{x} \in G^c$, and
(iv) $0 < \phi(\boldsymbol{x}) < 1$ for each $\boldsymbol{x} \in G - F$.

Proof: For each $\boldsymbol{x} \in \mathbb{R}^k$, define

$$\phi(\boldsymbol{x}) = \frac{d_{G^c}(\boldsymbol{x})}{d_{G^c}(\boldsymbol{x}) + d_F(\boldsymbol{x})}.$$

We first claim that $\phi(\boldsymbol{x})$ is meaningfully defined for each $\boldsymbol{x} \in \mathbb{R}^k$. This means that we must verify for every $\boldsymbol{x} \in \mathbb{R}^k$ that at least one of $d_{G^c}(\boldsymbol{x})$ and $d_F(\boldsymbol{x})$ is positive [or else the denominator of $\phi(\boldsymbol{x})$ is zero]. There are three cases:
(a) If $\boldsymbol{x} \in F$, then $d_F(\boldsymbol{x}) = 0$ and $\boldsymbol{x} \notin G^c = \overline{G^c}$, so $d_{G^c}(\boldsymbol{x}) > 0$.
(b) If $\boldsymbol{x} \in G - F$, then $\boldsymbol{x} \notin F = \bar{F}$ and $\boldsymbol{x} \notin G^c = \overline{G^c}$, hence $d_F(\boldsymbol{x}) > 0$ and $d_{G^c}(\boldsymbol{x}) > 0$.
(c) If $\boldsymbol{x} \in G^c$, then $d_{G^c}(\boldsymbol{x}) = 0$ and $\boldsymbol{x} \notin F = \bar{F}$, hence $d_F(\boldsymbol{x}) > 0$.
Putting these three cases together shows that $\phi(\boldsymbol{x})$ is meaningfully defined on $\mathbb{R}^k$. Since (1) the denominator of ϕ is nonzero, (2) d_{G^c} is continuous, and (3) d_F is continuous, it follows that ϕ itself is continuous. All that remains is to verify that ϕ satisfies (i)-(iv). Requirement (i) is automatically fulfilled. For (ii), let $\boldsymbol{x} \in F$, and refer to (a) above to see in this case that $\phi(\boldsymbol{x}) = \frac{d_{G^c}(\boldsymbol{x})}{d_{G^c}(\boldsymbol{x})} = 1$. For (iii), let $\boldsymbol{x} \in G^c$, and refer to (c) above to see that in this case we have $\phi(\boldsymbol{x}) = \frac{0}{d_F(\boldsymbol{x})} = 0$. For (iv), let $\boldsymbol{x} \in G - F$, and refer to part (b) to see that both $d_F(\boldsymbol{x}), d_{G^c}(\boldsymbol{x}) > 0$, giving $0 < \phi(\boldsymbol{x}) < 1$. This completes the proof. ∎

The important requirement in Claim 2 is the continuity requirement. While it is easy to construct noncontinuous $\phi : \mathbb{R}^k \to \mathbb{R}$ satisfying (i)-(iv) above, it is not obvious that there is a continuous $\phi : \mathbb{R}^k \to \mathbb{R}$ satisfying (i)-(iv) above.

Claim 3. *Let $K \subseteq \mathbb{R}^k$ be compact. Then there exists an open $G \subseteq \mathbb{R}^k$ such that $K \subseteq G$ and $\bar{G}$ is compact.*

Proof: For any $\boldsymbol{x} \in \mathbb{R}^k$, let $B(\boldsymbol{x}) = \{\boldsymbol{y} \in \mathbb{R}^k : \|\boldsymbol{y} - \boldsymbol{x}\| < 1\}$, the open unit ball with center $\boldsymbol{x}$. Observe that $\{B(\boldsymbol{x}) : \boldsymbol{x} \in K\}$ is an open covering of K. Since K is compact, there exists a finite subcovering $\{B(\boldsymbol{x}_1), \cdots, B(\boldsymbol{x}_n)\}$ of K, where $\boldsymbol{x}_1, \cdots, \boldsymbol{x}_n \in K$. Let $G = \bigcup_{i=1}^n B(\boldsymbol{x}_i)$. Then G is open and $K \subseteq G$. Next, observe that $\bar{G} = \overline{\bigcup_{i=1}^n B(\boldsymbol{x}_i)} = \bigcup_{i=1}^n \overline{B(\boldsymbol{x}_i)}$ is a finite union of closed and bounded sets, so $\bar{G}$ is closed and bounded, whence $\bar{G}$ is compact by the Heine-Borel Theorem. Therefore, $G = \bigcup_{i=1}^n B(\boldsymbol{x}_i)$ satisfies the requirements of the claim. ∎

Before proceeding to Lusin's Theorem, let $A \subseteq \mathbb{R}^k$ be compact, and consider an $\mathcal{F}/\mathcal{B}^*$-measurable function $f : \mathbb{R}^k \to [0,1)$ with $f(\omega) = 0$ for every $\omega \notin A$. For $n \in \mathbb{N}$ and $j = 1, \cdots, n2^n$, let $A_{n,j} = \{\omega \in A : \frac{j-1}{2^n} \leq f(\omega) < \frac{j}{2^n}\}$ and $B_n = \{\omega \in A : f(\omega) \geq n\}$, and write

$$s_n(\omega) = \sum_{j=1}^{n2^n} \frac{j-1}{2^n} I_{A_{n,j}}(\omega) + I_{B_n}(\omega), \qquad \omega \in \mathbb{R}^k. \tag{5.7}$$

This is the setup of (MF18), hence $\{s_n\}_{n=1}^{\infty}$ is a nonnegative and nondecreasing sequence of $\mathcal{F}/\mathcal{B}^*$-measurable simple functions with $s_n(\omega) \to f(\omega)$ for each $\omega \in \mathbb{R}^k$. However, since $0 \leq f < 1$, we have that $B_n = A_{n,j} = \varnothing$ for each $n \in \mathbb{N}$ and $j = 2^n + 1, 2^n + 2, \cdots, n2^n$, hence (5.7) is the same thing as

$$s_n(\omega) = \sum_{j=1}^{2^n} \frac{j-1}{2^n} I_{A_{n,j}}(\omega), \qquad \omega \in \mathbb{R}^k. \tag{5.8}$$

Observe that for any $n \geq 2$ and $\omega \in \mathbb{R}^k$, $s_n(\omega) - s_{n-1}(\omega)$ will be either 0 or 2^{-n} as $f(\omega)$ is in the left half of some $A_{n-1,j}$ or $f(\omega)$ is in the right half of some $A_{n-1,j}$. Therefore, for $n \geq 2$, $2^n(s_n - s_{n-1})$ is the indicator function of $\bigcup_{j=1}^{2^n} f^{-1}(\text{right half of } A_{n-1}, j)$, and hence $2^n(s_n - s_{n-1})$ is $\mathcal{F}/\mathcal{B}^*$-measurable. Denoting the just-mentioned union by T_n, we have for every $n \geq 2$ that $2^n(s_n - s_{n-1}) = I_{T_n}$.

For $n = 1$, let $T_1 = f^{-1}((\frac{1}{2}, 1)) \in \mathcal{F}$, so that $T_1 \subseteq A$ and $2s_1 = I_{T_1}$. The notation of this and previous paragraphs will be used in the proof of the first stage of Lusin's Theorem.

We require only a final definition.

Definition. Let $f : \mathbb{R}^k \to \bar{\mathbb{R}}$ be an arbitrary function. The *support of* f is the set $\text{supp}(f) = \overline{\{\omega \in \mathbb{R}^k : f(\omega) \neq 0\}}$. If this closed set is bounded and hence compact, we say that f *has compact support.*

Lusin's Theorem. *Consider the space* $(\mathbb{R}^k, \mathcal{F}, \mu)$, *where* $\mathcal{F}$ *contains* $\mathcal{B}^k$, *and where* μ *is such that*

(i) $\mu(A) = \inf\{\mu(G) : A \subseteq G,\ G \text{ open}\}$ *for every* $A \in \mathcal{F}$, *and*

(ii) $\mu(A) = \sup\{\mu(K) : K \subseteq A,\ K \text{ compact}\}$ *for every* $A \in \mathcal{F}$.

Let $A \in \mathcal{F}$ *be such that* $\mu(A) < +\infty$, *and suppose that* $f : \mathbb{R}^k \to \mathbb{R}$ *is* $\mathcal{F}/\mathcal{B}^*$-*measurable with* $f(\omega) = 0$ *for each* $\omega \notin A$. *Then for any* $\epsilon > 0$ *there exists a function* $g : \mathbb{R}^k \to \mathbb{R}$ *with the following properties:*

(a) g *is continuous and has compact support*

(b) $\mu(\{\omega \in \mathbb{R}^k : f(\omega) \neq g(\omega)\}) < \epsilon$

(c) $\sup_{\omega \in \mathbb{R}^k} |g(\omega)| \leq \sup_{\omega \in \mathbb{R}^k} |f(\omega)|$.

Proof: We shall prove the theorem in increasing cases of generality. We find a g satisfying (a) and (b) in stages of increasing generality:

Stage 1: $0 \leq f < 1$ and A is compact.

Stage 2: f is nonnegative and bounded, while A is still compact.
Stage 3: f is bounded [not necessarily nonnegative] and A is compact.
Stage 4: f is bounded and A is such that $\mu(A) < +\infty$.
Stage 5: f and A are as assumed in the statement of the theorem.
From this, we will establish (c), completing the proof.

Stage 1: Assume that A is compact and $0 \leq f < 1$, and recall the notation found after Claim 3. Use Claim 3 to find an open V such that $A \subseteq V$ and $\bar{V}$ is compact. Consider the functions $0 \leq s_1 \leq s_2 \leq \cdots$ given by (5.8). Let $t_1 = s_1$ and $t_n = s_n - s_{n-1}$ for all $n \geq 2$. Since $\sum_{i=1}^{n} t_i = s_n$ for every $n \in \mathbb{N}$ and $\lim s_n = f$, we have $\sum_{i=1}^{\infty} t_i = f$. We have for all $n \in \mathbb{N}$ that $2^n t_n = I_{T_n}$, $\mu(T_n) < +\infty$, and both $2^n t_n$ and t_n are $\mathcal{F}/\mathcal{B}^*$-measurable. For each $n \in \mathbb{N}$, use (i) and (ii) to find compact K_n and open V_n with $K_n \subseteq T_n \subseteq V_n \subseteq V$ and $\mu(V_n - K_n) < \frac{\epsilon}{2^n}$. For each $n \in \mathbb{N}$, invoke Claim 2 to find a continuous function $\phi_n : \mathbb{R}^k \to [0, 1]$ such that

$$\phi_n(\omega) = \begin{cases} 1 & \text{if } \omega \in K_n \\ 0 & \text{if } \omega \in V_n^c \end{cases}$$

with $0 < \phi_n(\omega) < 1$ if $\omega \in V_n - K_n$. This forces $\{\omega \in \mathbb{R}^k : \phi_n(\omega) \neq 0\} = V_n$ for each $n \in \mathbb{N}$, hence $\operatorname{supp}(\phi_n) = \overline{V_n}$ for each $n \in \mathbb{N}$. Let $g = \sum_{n=1}^{\infty} 2^{-n}\phi_n$. We claim that g satisfies (a) and (b).

To see that g is continuous, let $h_n = \sum_{i=1}^{n} 2^{-i}\phi_i$ for each $n \in \mathbb{N}$. For each $n \in \mathbb{N}$, h_n is a sum of continuous functions and hence is continuous. We claim that $h_n \to g$ uniformly. To see this, let $\epsilon > 0$ and $\omega \in \mathbb{R}^k$. Let N be such that $\sum_{i=N+1}^{\infty} 2^{-i} < \epsilon$, so that N does not depend on ω. If $n \geq N$, we have

$$|h_n(\omega) - g(\omega)| = \sum_{i=n+1}^{\infty} 2^{-i}\phi_i(\omega) \leq \sum_{i=n+1}^{\infty} 2^{-i} \leq \sum_{i=N+1}^{\infty} 2^{-i} < \epsilon.$$

This shows that $h_n \to g$ uniformly, and we have therefore shown that g is indeed continuous.[6]

We now show that g has compact support. Observe that if $\omega \in \mathbb{R}^k$ is such that $g(\omega) \neq 0$, then for some $n \in \mathbb{N}$ it must be the case that $\phi_n(\omega) \neq 0$ and conversely, hence

$$\{\omega \in \mathbb{R}^k : g(\omega) \neq 0\} = \bigcup_{n=1}^{\infty} \{\omega \in \mathbb{R}^k : \phi_n(\omega) \neq 0\},$$

so that

$$\operatorname{supp}(g) = \overline{\bigcup_{n=1}^{\infty} \{\omega \in \mathbb{R}^k : \phi_n(\omega) \neq 0\}} \subseteq \bigcup_{n=1}^{\infty} \operatorname{supp}(\phi_n) = \bigcup_{n=1}^{\infty} \bar{V}_n \subseteq \bar{V}.$$

[6]We are using the following result from advanced calculus: if $\{f_n\}_{n=1}^{\infty}$ is a sequence of continuous functions with domain X, and if $f_n \to f$ uniformly, where f is a function with domain X, then f is continuous.

Since $\bar{V}$ is compact and hence bounded, the closed set supp(g) is bounded, hence g has compact support. This g accordingly satisfies (a).

We now show that g satisfies (b), which will complete the proof of Stage 1. Observe for each $n \in \mathbb{N}$ that $\omega \in K_n$ forces $2^{-n}\phi_n(\omega) = 2^{-n} = t_n(\omega)$. Next, observe that $\omega \in V_n^c$ forces $2^{-n}\phi_n(\omega) = 0 = t_n(\omega)$. Therefore, for each $n \in \mathbb{N}$ we have that $2^{-n}\phi_n = t_n$ on $V_n^c \cup K_n$. For each $n \in \mathbb{N}$ and $\omega \in (V_n^c \cup K_n)^c = V_n - K_n$, we have $0 < 2^{-n}\phi_n(\omega) < 2^{-n}$; since $t_n(\omega)$ is 0 or 2^{-n}, we have $2^{-n}\phi_n(\omega) \neq t_n(\omega)$, and hence $2^{-n}\phi_n \neq t_n$ on $V_n - K_n$. Recalling that $f = \sum_{n=1}^{\infty} t_n$ and $g = \sum_{n=1}^{\infty} 2^{-n}\phi_n$, it follows that

$$\{\omega \in \mathbb{R}^k : f(\omega) \neq g(\omega)\} = \bigcup_{n=1}^{\infty} \{\omega \in \mathbb{R}^k : \; 2^{-n}\phi_n(\omega) \neq t_n(\omega)\} = \bigcup_{n=1}^{\infty} (V_n - K_n).$$

Therefore, by countable subadditivity, we obtain

$$\mu\left(\{\omega \in \mathbb{R}^k : \; f(\omega) \neq g(\omega)\}\right) \leq \sum_{n=1}^{\infty} \mu(V_n - K_n) < \sum_{n=1}^{\infty} \frac{\epsilon}{2^n} = \epsilon,$$

so it follows that $\mu(\{\omega \in \mathbb{R}^k : \; f(\omega) \neq g(\omega)\}) < \epsilon$, whence g satisfies (b) as well. This completes the proof of (a) and (b) of the theorem for when $0 \leq f < 1$ and A is compact.

Stage 2: Keep A compact, let $\epsilon > 0$, but assume instead that f is nonnegative and bounded. Therefore, there is $0 < M < +\infty$ such that $0 \leq f(\omega) < M$ for each $\omega \in \mathbb{R}^k$. Then $0 \leq \frac{f}{M} < 1$. From Stage 1, there exists a continuous $g : \mathbb{R}^k \to \mathbb{R}$ with compact support and $\mu(\{\omega \in \mathbb{R}^k : \frac{f(\omega)}{M} \neq g(\omega)\}) < \epsilon$. Therefore, $\mu(\{\omega \in \mathbb{R}^k : \; f(\omega) \neq Mg(\omega)\}) < \epsilon$, and Mg is continuous with compact support, hence Mg satisfies (a) and (b) in this case. This proves (a) and (b) for when f is nonnegative and bounded with A compact.

Stage 3: Keep A compact and take $\epsilon > 0$, but assume now that f is merely bounded, say $|f| \leq M$ for some $0 < M < +\infty$. By (MF13) [or by inspection] we have that $0 \leq f^+, f^- < M$. Apply the result of Stage 2 to f^+ to obtain a continuous $g_1 : \mathbb{R}^k \to \mathbb{R}$ with compact support such that

$$\mu(\{\omega \in \mathbb{R}^k : \; f^+(\omega) \neq g_1(\omega)\}) < \frac{\epsilon}{2}.$$

Apply Stage 2 to f^- to obtain a continuous $g_2 : \mathbb{R}^k \to \mathbb{R}$ with compact support such that

$$\mu(\{\omega \in \mathbb{R}^k : \; f^-(\omega) \neq g_2(\omega)\}) < \frac{\epsilon}{2}.$$

Let $g = g_1 - g_2$. Then g is the difference of two continuous functions, hence g is continuous. Next, observe that if $\omega \in \mathbb{R}^k$ is such that $g_1(\omega) = 0 = g_2(\omega)$, then $g(\omega) = 0$, so that

$$\{\omega \in \mathbb{R}^k : \; g(\omega) \neq 0\} \subseteq \{\omega \in \mathbb{R}^k : \; g_1(\omega) \neq 0\} \cup \{\omega \in \mathbb{R}^k : \; g_2(\omega) \neq 0\}.$$

It follows that $\text{supp}(g) \subseteq \text{supp}(g_1) \cup \text{supp}(g_2)$. Since the union of a finite number of compact sets is compact, we have that the closed set $\text{supp}(g)$ is a subset of a compact set and hence is compact. Thus, this g satisfies (a). To see that g satisfies (b), observe that if $\omega \in \mathbb{R}^k$ is such that $f^+(\omega) = g_1(\omega)$ and $f^-(\omega) = g_2(\omega)$, then

$$f(\omega) = f^+(\omega) - f^-(\omega) = g_1(\omega) - g_2(\omega) = g(\omega),$$

so that

$$\begin{aligned} & \{\omega \in \mathbb{R}^k : f(\omega) \neq g(\omega)\} \\ \subseteq\ & \{\omega \in \mathbb{R}^k : f^+(\omega) \neq g_1(\omega)\} \cup \{\omega \in \mathbb{R}^k : f^-(\omega) \neq g_2(\omega)\}. \end{aligned}$$

From this and subadditivity, it follows that

$$\begin{aligned} & \mu(\{\omega \in \mathbb{R}^k : f(\omega) \neq g(\omega)\} \\ \leq\ & \mu(\{\omega \in \mathbb{R}^k : f^+(\omega) \neq g_1(\omega)\}) + \mu(\{\omega \in \mathbb{R}^k : f^-(\omega) \neq g_2(\omega)\} \\ <\ & \frac{\epsilon}{2} + \frac{\epsilon}{2} = \epsilon. \end{aligned}$$

The function g therefore satisfies (a) and (b) in Stage 3.

Stage 4: Assume that $\mu(A) < +\infty$ and keep f bounded as before. By (ii), there is compact $K \subseteq A$ with $\mu(A) < \mu(K) + \frac{\epsilon}{3}$. Since $\mu(A) < +\infty$, this is equivalent to $\mu(A - K) < \frac{\epsilon}{3}$. Let $f_K : \mathbb{R}^k \to \bar{\mathbb{R}}$ be such that $f_K(\omega) = f(\omega)$ if $\omega \in K$, and let $f_K(\omega) = 0$ for $\omega \notin K$. Since f is $\mathcal{F}/\mathcal{B}^*$-measurable and $K \in \mathcal{F}$, f_K is $\mathcal{F}/\mathcal{B}^*$-measurable by (MF6). Since f is bounded, f_K is bounded. From Stage 3, there is a continuous g with compact support such that $\mu(\{\omega \in \mathbb{R}^k : f_K(\omega) \neq g(\omega)\}) < \frac{\epsilon}{3}$, hence

$$\mu(\{\omega \in K : f_K(\omega) \neq g(\omega)\}),\ \mu(\{\omega \in A^c : f_K(\omega) \neq g(\omega)\}) < \frac{\epsilon}{3}.$$

Therefore, recalling that $f(\omega) = 0 = f_K(\omega)$ for each $\omega \in A^c$, we have that

$$\{\omega \in A^c : f(\omega) \neq g(\omega)\} = \{\omega \in A^c : f_K(\omega) \neq g(\omega)\},$$

hence

$$\begin{aligned} & \{\omega \in \mathbb{R}^k : f(\omega) \neq g(\omega)\} \\ =\ & \{\omega \in K : f(\omega) \neq g(\omega)\} \cup \{\omega \in K - A : f(\omega) \neq g(\omega)\} \\ & \cup \{\omega \in A^c : f(\omega) \neq g(\omega)\} \\ \subseteq\ & \{\omega \in K : f_K(\omega) \neq g(\omega)\} \cup (K - A) \cup \{\omega \in A^c : f_K(\omega) \neq g(\omega)\}, \end{aligned}$$

so finite subadditivity implies that

$$\begin{aligned} & \mu(\{\omega \in \mathbb{R}^k : f(\omega) \neq g(\omega)\}) \\ \leq\ & \mu(\{\omega \in K : f_K(\omega) \neq g(\omega)\}) + \mu(K - A) + \mu(\{\omega \in A^c : f_K(\omega) \neq g(\omega)\}) \\ <\ & \frac{\epsilon}{3} + \frac{\epsilon}{3} + \frac{\epsilon}{3} = \epsilon, \end{aligned}$$

hence the g that satisfies (a) and (b) for f_K satisfies (a) and (b) for f as well. This completes Stage 4.

Stage 5: Let $A \in \mathcal{F}$ be such that $\mu(A) < +\infty$, and remove the boundedness of f, only assuming that $f : \mathbb{R}^k \to \bar{\mathbb{R}}$ is an $\mathcal{F}/\mathcal{B}^*$-measurable function such that $f(\omega) = 0$ for each $\omega \in A^c$. Let $B_n = \{\omega \in \mathbb{R}^k : |f(\omega)| \geq n\}$ for each $n \in \mathbb{N}$, so that $B_n \in \mathcal{F}$ for every $n \in \mathbb{N}$. Since $f = 0$ on A^c it follows that $B_n \subseteq A$ for each $n \in \mathbb{N}$, hence $\mu(B_n) \leq \mu(A) < +\infty$ for each $n \in \mathbb{N}$. Next, the sequence $\{B_n\}_{n=1}^{\infty}$ is decreasing to $\{\omega \in \mathbb{R}^k : |f(\omega)| = +\infty\} = \varnothing$, so that $\mu(B_n) \to \mu(\varnothing) = 0$ by measure property (M9). It follows that there is $N \in \mathbb{N}$ such that $n \geq N$ implies $\mu(B_n) < \frac{\epsilon}{2}$. Next, for each $n \in \mathbb{N}$ let $f_n = I_{B_n^c} f$, so that $f = f_n$ on B_n^c. Since each f_n is a bounded $\mathcal{F}/\mathcal{B}^*$-measurable function that is zero on B_n [where $\mu(B_n) < +\infty$], we invoke Stage 4 to assert that there exists a continuous g_n with compact support such that $\mu(\{\omega \in \mathbb{R}^k : f_n(\omega) \neq g_n(\omega)\}) < \frac{\epsilon}{2}$. In particular, we have

$$\begin{aligned} & \mu(\{\omega \in \mathbb{R}^k : f(\omega) \neq g_N(\omega)\}) \\ = \; & \mu(\{\omega \in B_N^c : f(\omega) \neq g_N(\omega)\}) + \mu(\{\omega \in B_N : f(\omega) \neq g_N(\omega)\}) \\ \leq \; & \mu(\{\omega \in B_N^c : f_N(\omega) \neq g_N(\omega)\}) + \mu(B_N) \\ < \; & \frac{\epsilon}{2} + \frac{\epsilon}{2} = \epsilon, \end{aligned}$$

so that g_N satisfies (a) and (b) of the theorem, whence Stage 5 is complete.

Proof of (c): Given $\epsilon > 0$, we have shown that a g satisfying both (a) and (b) exists under the assumptions given, but nothing yet guarantees that $\sup_{\omega \in \mathbb{R}^k} |g(\omega)| \leq \sup_{\omega \in \mathbb{R}^k} |f(\omega)|$. There are three cases: (1) the case where (c) in fact holds for g [in which case we're through], (2) $\sup_{\omega \in \mathbb{R}^k} |f(\omega)| = +\infty$, [so (c) trivially holds], and (3) $\sup_{\omega \in \mathbb{R}^k} |f(\omega)| < \sup_{\omega \in \mathbb{R}^k} |g(\omega)| \leq +\infty$. For (3), we tinker with g to make (c) hold. Let $S = \{x \in \mathbb{R} : |x| \leq \sup_{\omega' \in \mathbb{R}^k} |f(\omega')|\}$ and $T = \{x \in \mathbb{R} : |x| \leq \sup_{\omega' \in \mathbb{R}^k} |g(\omega')|\}$, so that $S \subseteq T$. Define $\phi : T \to S$ by writing

$$\phi(x) = \begin{cases} x & \text{if } x \in S \\ \sup_{\omega' \in \mathbb{R}^k} |f(\omega')| & \text{if } x \in T - S. \end{cases}$$

We have that ϕ is onto S and $\sup_{x \in T} |\phi(x)| = \sup_{\omega \in \mathbb{R}^k} |f(\omega)|$. Furthermore, ϕ is continuous. Let $h = \phi \circ g$; h is the composition of continuous functions and hence is continuous. Also, if $\omega \in \mathbb{R}^k$ is such that $g(\omega) = 0$, then $h(\omega) = \phi(g(\omega)) = \phi(0) = 0$, whence

$$\{\omega \in \mathbb{R}^k : \; h(\omega) \neq 0\} \subseteq \{\omega \in \mathbb{R}^k : \; g(\omega) \neq 0\},$$

so that $\operatorname{supp}(h) \subseteq \operatorname{supp}(g)$. Since $\operatorname{supp}(h)$ is a closed subset of the compact set $\operatorname{supp}(g)$, it follows that h has compact support. This h that has resulted from tampering with g so far satisfies (a). The function h is also seen to satisfy (c), as $\sup_{\omega \in \mathbb{R}^k} |h(\omega)| = \sup_{\omega \in \mathbb{R}^k} |f(\omega)|$. We will show that h satisfies (b), which will complete the proof. If $\omega \in \mathbb{R}^k$ is such that $f(\omega) = g(\omega)$, then

$\phi(f(\omega)) = \phi(g(\omega))$. By the definitions of ϕ and h, this forces $f(\omega) = h(\omega)$. It follows that

$$\{\omega \in \mathbb{R}^k : f(\omega) \neq h(\omega)\} \subseteq \{\omega \in \mathbb{R}^k : f(\omega) \neq g(\omega)\}.$$

Since the latter set has μ-measure less than ϵ, so does the former set. Thus h satisfies (a)-(c) of the theorem, and the entire proof is finally complete. ∎

All of the claims and Lusin's Theorem actually in a much more general setting. We may replace $\mathbb{R}^k$ with a locally compact Hausdorff space and $\mathcal{F}$ with any σ-field that contains the sets open under whatever topology one is using. The proofs given in this section — Urysohn's Lemma excepted[7] — work with fairly simple modifications if one wants to prove the claim under these more general conditions.

Exercises.

1*. In Lusin's Theorem, we may replace the condition that f has range contained in $\mathbb{R}$ with the more general assumption that f is finite μ-a.e. on A.

2. If $|f| \leq 1$ in Lusin's Theorem, there exists a sequence $\{g_n\}_{n=1}^{\infty}$ of $\mathcal{F}/\mathcal{B}^*$-measurable functions such that (a) $|g_n| \leq 1$ for all $n \in \mathbb{N}$, (b) g_n is continuous and has compact support for each $n \in \mathbb{N}$, and (c) $\lim g_n = f$ μ-a.e.

3. Let $F \subseteq G \subseteq \mathbb{R}^k$, where F is closed and G is open. Without using Urysohn's Lemma [Claim 2], quickly construct a lower semicontinuous $\phi : \mathbb{R}^k \to [0,1]$ such that $\phi(\boldsymbol{x}) = 1$ if $\boldsymbol{x} \in F$ and $\phi(\boldsymbol{x}) = 0$ if $\boldsymbol{x} \notin G$. Then quickly construct an upper semicontinuous ϕ with the exact same properties. [Take $\phi = I_F$ and $\phi = I_G$. The point of Urysohn's Lemma is that we can state something much stronger, namely, that there exists a *continuous* ϕ with these properties.]

4. We defined the support of a function $g : \mathbb{R}^k \to \bar{\mathbb{R}}$ to be the closure of the set $\{\boldsymbol{x} \in \mathbb{R}^k : g(\boldsymbol{x}) \neq 0\}$. This is equivalent to defining the support of g to be the complement of the interior of the set $\{\boldsymbol{x} \in \mathbb{R}^k : g(\boldsymbol{x}) = 0\}$.

5. Is it the case that every compact subset of $\mathbb{R}^k$ is the support of some continuous function f defined on $\mathbb{R}^k$?

6. Consider the measure space $(\mathbb{R}^k, \mathcal{F}, \mu)$, where μ assigns finite measure to bounded $\mathcal{B}^k$-sets and $\mathcal{F} \supseteq \mathcal{B}^k$. Let $f : \mathbb{R}^k \to \bar{\mathbb{R}}$ be given. If for every $\epsilon > 0$ there exists a closed set K such that (i) $\mu(K^c) < \epsilon$ and (ii) the restriction of f to K is continuous, then f is $\mathcal{F}/\mathcal{B}^*$-measurable.

7. Consider the measure space in the previous exercise. Let $A \in \mathcal{F} - \{\varnothing\}$ be such that $\mu(A) < +\infty$. Let $\{f_n\}_{n=1}^{\infty}$ denote a sequence of $\mathcal{F}/\mathcal{B}^*$-measurable functions that are finite μ-a.e., and suppose that there exists an $\mathcal{F}/\mathcal{B}^*$-measurable function $f : A \to \mathbb{R}$ with $\lim f_n = f$ μ-a.e. Then, for any $\epsilon > 0$, there exists $B \in \mathcal{F} - \{\varnothing\}$ with $B \subseteq A$, $\mu(A - B) < \epsilon$, and $f_n \to f$ uniformly on B.

[7] A general proof of this lemma under the more general setting would itself require a few more claims. The "metric space" proof as given is disproportionately simple when compared with the proof for the "topological space" setting.

5.8 A GENERALIZED DEFINITION OF MEASURABILITY

Our setting thus far in the chapter has been to have a generic measure space $(\Omega, \mathcal{F}, \mu)$, a nonempty set $A \in \mathcal{F}$, and a function $f : A \to \bar{\mathbb{R}}$. We have stated that the function f is $\mathcal{F}/\mathcal{B}^*$-measurable if and only if $f^{-1}(B) \in \mathcal{F}$ for every $B \in \mathcal{B}^*$. That is, f is $\mathcal{F}/\mathcal{B}^*$-measurable iff $f^{-1}(\mathcal{B}^*) \subseteq \mathcal{F}$. In this very brief section, we discuss the case where f still has domain A, but is not necessarily real valued nor extended real-valued. In other words, we shall let the output values of f be elements of a generic space.

Notation. We will let $(\Omega, \mathcal{F}, \mu)$ denote a generic measure space. We will let Ω' denote a nonempty set, with $\mathcal{F}'$ denoting a σ-field of subsets of Ω'. The symbol A will, as usual, be reserved exclusively for a nonempty set in $\mathcal{F}$.

In the previous parts of this chapter, $\Omega' = \bar{\mathbb{R}}$ or $\mathbb{R}$ and $\mathcal{F}'$ has denoted $\mathcal{B}^*$ [or $\mathcal{B}$]. Most of our subsequent work will deal with this setting, but it is good to deal with things in total abstraction for a little while.

Definition. A function $f : A \to \Omega'$ will be called *$\mathcal{F}/\mathcal{F}'$-measurable* iff it is the case that $f^{-1}(A') \in \mathcal{F}$ for every $A' \in \mathcal{F}'$.

We now present the following properties. These properties are analogs to properties (MF1) and (MF5), respectively.

(MF1′) *If $\mathcal{C}' \subseteq 2^{\Omega'}$ is such that $\sigma(\mathcal{C}') = \mathcal{F}'$, and if $f^{-1}(C') \in \mathcal{F}$ for every $C' \in \mathcal{C}'$, then $f : A \to \Omega'$ is $\mathcal{F}/\mathcal{F}'$-measurable.*

(MF5′) *Let Ω'' denote a nonempty set and let $\mathcal{F}''$ denote a σ-field on Ω''. Furthermore, assume that $f : A \to \Omega'$ is $\mathcal{F}/\mathcal{F}'$-measurable and $\phi : \Omega' \to \Omega''$ is $\mathcal{F}'/\mathcal{F}''$-measurable. Then $\phi \circ f$ is $\mathcal{F}/\mathcal{F}''$-measurable.*

Proof of (MF1′): By assumption, we have that $f^{-1}(\mathcal{C}') \subseteq \mathcal{F}$. Therefore, we have $\sigma(f^{-1}(\mathcal{C}')) \subseteq \sigma(\mathcal{F})$. By Claim 2 of Section 1.3 and the fact that $\mathcal{F}$ is a σ-field, we have that $f^{-1}(\sigma(\mathcal{C}')) \subseteq \mathcal{F}$. Since $\sigma(\mathcal{C}') = \mathcal{F}'$ by assumption, we have $f^{-1}(\mathcal{F}') \subseteq \mathcal{F}$, so that f is indeed $\mathcal{F}/\mathcal{F}'$-measurable.

Proof of (MF5′): Let $A'' \in \mathcal{F}''$. Then $(\phi \circ f)^{-1}(A'') = f^{-1}(\phi^{-1}(A''))$. Since ϕ is $\mathcal{F}'/\mathcal{F}''$-measurable, we have $\phi^{-1}(A'') \in \mathcal{F}'$. Since f is $\mathcal{F}/\mathcal{F}'$-measurable and $\phi^{-1}(A'') \in \mathcal{F}'$, we have $f^{-1}(\phi^{-1}(A'')) \in \mathcal{F}$. This shows that $(\phi \circ f)^{-1}(A'') \in \mathcal{F}$ for the generic $A'' \in \mathcal{F}''$, so $\phi \circ f$ is $\mathcal{F}/\mathcal{F}''$-measurable. ∎

As a rough rule, those properties that involve purely set-theoretic concerns and nothing that invokes the topological or metrical properties of $\bar{\mathbb{R}}$ carry over with the appropriate modification. For example, (MF2) carries over and may be written as follows.

(MF2′) *Let $f : A \to \Omega'$ be $\mathcal{F}/\mathcal{F}'$-measurable, and let $B \in \mathcal{F} - \{\varnothing\}$ be such that $B \subseteq A$. Then the restriction of f from A to B is also $\mathcal{F}/\mathcal{F}'$-measurable.*

It seems unnecessary to give a proof, as the proof of (MF2) carries over with only the most minor modifications: replace $\bar{\mathbb{R}}$ by Ω', and replace each occurrence of $\mathcal{B}^*$ with $\mathcal{F}'$. Observe that nothing involves anything specific about $\bar{\mathbb{R}}$ or the structure of the extended Borel σ-field. Similarly, (MF6) carries over and may be modified to this more general setting without any substantial work.

However, properties such as [say] (MF3), (MF4), and the "continuity" part of (MF5) sneak in at least one special property of $\bar{\mathbb{R}}$, and hence cannot really be generalized to this more abstract setting. Of course, those properties in Sections 2 and 3 which deal with arithmetic, suprema, infima, and limits involve special order properties of $\bar{\mathbb{R}}$ and hence cannot really be generalized either. The fact that only a few properties can be generalized is not troubling, as virtually every [if not every] application for us [at least in this book] deals with mappings into $\bar{\mathbb{R}}$.

We now deal with the subject of measures induced by measurable mappings.

Definition. Keeping the notation the same as before and letting $f : A \to \Omega'$ denote a function that is $\mathcal{F}/\mathcal{F}'$-measurable, consider the function $\nu : \mathcal{F}' \to \bar{\mathbb{R}}$ given for all $A' \in \mathcal{F}'$ by

$$\nu(A') = \mu(\{\omega \in A : \ f(\omega) \in A'\}) = \mu(f^{-1}(A')).$$

We call ν *the measure induced by* f. We also write the formalism $\mu \circ f^{-1}$ as an alternative notation for the measure induced by f.

That ν really is a measure is nearly automatic and is left as a simple exercise. The easy exercises below contain some other simple properties about ν. This concept will be used when we discuss one form of the change of variables theorem in our study of integration.

Exercises. [Very, very easy.]

1*. Regarding the measure ν:

(a) ν is really a meausre.

(b) If μ is finite [that is, if $\mu(\Omega) < +\infty$], then so is ν.

(c) If μ is σ-finite, ν need not be σ-finite.

2*. Modify (MF14) and prove it in the more general setting of this section.

6

The Lebesgue Integral

In this chapter, the Lebesgue integral is defined and discussed. The Lebesgue integral is the standard generalization of the calculus-based Riemann integral, and it will be seen to enjoy the properties of the Riemann integral while at the same time being able to handle a large class of functions.

From the middle of the nineteenth century on, it became apparent that certain results in the study of Fourier series ultimately depended on equations of the form

$$\lim_{n\to\infty} \int_a^b f_n(x)\, dx = \int_a^b \lim_{n\to\infty} f_n(x)\, dx.$$

The problem of interchanging the operations of limit and Riemann integral can be handled fairly smoothly, provided that the limit is *uniform*, that is, if [roughly] $f_n \to f$ uniformly. However, if the limit is merely *pointwise*, then the Riemann integral is a stubborn object with which to work if one wants to interchange limit and integral. During the change from the nineteenth to the twentieth centuries, various new approaches to integration that subsumed the traditional Riemann integral were put forth by Jordan, Borel, and Lebesgue [among others]. Ultimately, the pride of place has gone to Lebesgue's 1902 construction which dealt with [what we have called] Lebesgue measure on the real line. Our construction takes place in a much more abstract setting, but will ultimately subsume the Riemann integral material learned from calculus on through advanced calculus.

Our approach to constructing the abstract Lebesgue integral is to first construct a functional defined on a certain small class of functions, and to then extend this functional's domain to a larger class of functions, eventually arriving at an extended functional that will operate on a very large of functions. This chapter may seem somewhat tedious because the same result will often have to be proven for each new stage of the construction. However, the chapter is not too difficult conceptually.

All throughout this chapter, we will be dealing with a completely abstract measure space $(\Omega, \mathcal{F}, \mu)$, and this will not change at all throughout the construction process. Our construction of the abstract Lebesgue integral requires dealing with four stages of increasing generality: (i) nonnegative finite-valued simple $\mathcal{F}/\mathcal{B}^*$-measurable functions defined on Ω [as defined in the previous chapter], (ii) nonnegative $\mathcal{F}/\mathcal{B}^*$-measurable functions defined on Ω, (iii) general $\mathcal{F}/\mathcal{B}^*$-measurable functions defined on Ω, and, finally, (iv) general $\mathcal{F}/\mathcal{B}^*$-measurable functions defined μ-a.e. on Ω. Each section deals with a separate stage, and the measurability results of Chapter 5 are used extensively, whether implicitly or explicitly.

This chapter and the next form a conceptual unit. For this chapter, we construct the integral and discuss many of its properties. The next chapter justifies the assertion that the new constructed integral subsumes the Riemann integral.

6.1 STAGE ONE: SIMPLE FUNCTIONS

This section commences the first of the four stages that lead to the abstract Lebesgue integral. Without further ado, we proceed to the definitions and notation, all relative to a general measure space $(\Omega, \mathcal{F}, \mu)$.

Definition. We define $\mathfrak{S}$ to denote the collection of finite-valued nonnegative $\mathcal{F}/\mathcal{B}^*$-measurable simple functions with domain Ω. If any confusion about the underlying space can arise, we will write $\mathfrak{S}(\Omega, \mathcal{F}, \mu)$ or $\mathfrak{S}(\Omega, \mathcal{F})$ in place of $\mathfrak{S}$.

By (MF16) in Chapter 5, every $s \in \mathfrak{S}$ has a representation $s = \sum_{i=1}^{n} c_i I_{A_i}$ for some $n \in \mathbb{N}$, where $0 \leq c_1, \cdots, c_n < +\infty$, $A_1, \cdots, A_n$ are disjoint $\mathcal{F}$-sets whose union is Ω, and where $I_{A_1}, \cdots, I_{A_n}$ are the indicator functions [with domain Ω] of $A_1, \cdots, A_n$. Usually, if not always, an element of $\mathfrak{S}$ has infinitely many representations in this form. The upcoming example illustrates this.

Example 1. Consider the measure space $(\mathbb{R}, \mathcal{B}, \lambda)$. Consider the element of $\mathfrak{S}$ that assigns 1 to any $\omega \in [0, 1]$, 0 to any $\omega \notin [0, 1]$. This element may be written as $1I_{[0,1]} + 0I_{[0,1]^c}$, or as $1I_{[0,\frac{1}{2})} + 1I_{[\frac{1}{2},1)} + 1I_{\{1\}} + 0I_{[0,1]^c}$, say. These are two different representations [out of the infinitely many] of this element of $\mathfrak{S}$. For any $n \in \mathbb{N}$ and $A_1, \cdots, A_n \subseteq [0, 1]$ that are disjoint with $\bigcup_{i=1}^{n} A_i = [0, 1]$, $\sum_{i=1}^{n} 1I_{A_i} + 0I_{[0,1]^c}$ is a representation of the aforementioned element of $\mathfrak{S}$.

We will now define the main item of this section, which represents the first of the four phases of the program.

Definition. Given $E \in \mathcal{F}$, we define $\mathcal{I}_E$ on $\mathfrak{S}$ by

$$\mathcal{I}_E(s) = \sum_{i=1}^{n} c_i \mu(A_i \cap E) \quad \text{whenever } s \text{ is of the form } \textstyle\sum_{i=1}^{n} c_i I_{A_i} \text{ as above.}$$

We will equivalently write $\mathcal{I}_E(s;\mu)$ for $\mathcal{I}_E(s)$ if there is more than one measure with domain $\mathcal{F}$ and clarity demands it. If $E = \Omega$, we will write $\mathcal{I}$ for $\mathcal{I}_\Omega$.

Since $A_1, \cdots, A_n, E \in \mathcal{F}$, we have $A_1 \cap E, \cdots, A_n \cap E \in \mathcal{F}$, hence the numbers $\mu(A_1 \cap E), \cdots, \mu(A_n \cap E)$ are defined. The components in the expression that defines $\mathcal{I}_E(s)$ therefore make sense. However, it is not immediately a priori obvious that the number $\mathcal{I}_E(s)$ is itself well-defined. In other words, if t and s are different representations of the same element of $\mathfrak{S}$, will $\mathcal{I}_E(t)$ be the same number as $\mathcal{I}_E(s)$? That the answer is affirmative will be shown.

Example 2. Revisit Example 1, letting $s = I_{[0,\frac{1}{2})} + I_{[\frac{1}{2},1)} + I_{\{1\}} + 0 I_{[0,1]^c}$ and $t = I_{[0,1]} + 0 I_{[0,1]^c}$. Then s and t are equivalent representations of the same element of $\mathfrak{S}$. Furthermore, the definition gives

$$\mathcal{I}(s) = (1 \times \lambda([0,1/2))) + (1 \times \lambda([1/2,1))) + (1 \times \lambda(\{1\})) + (0 \times \lambda([0,1]^c)) = 1$$

[recall that $0 \times +\infty = 0$ by our conventions], and

$$\mathcal{I}(t) = (1 \times \lambda([0,1])) + (0 \times \lambda([0,1]^c)) = 1,$$

hence $\mathcal{I}$ assigns the same value to the equivalent representations s and t. The question is whether $\mathcal{I}(r) = 1$ for every other equivalent representation $r \in \mathfrak{S}$.

We now list the important properties of $\mathcal{I}_E$ below. Most of the properties are simple, and they mimic to a large extent the behavior of the Riemann integral.

In all that follows, $(\Omega, \mathcal{F}, \mu)$ is a generic measure space, $s, t \in \mathfrak{S}$ with $s = \sum_{i=1}^{n} c_i I_{A_i}$, $t = \sum_{j=1}^{m} d_j I_{B_j}$. Also, assume that $E, F \in \mathcal{F}$.

(S1) *If s and t are equivalent representations of the same element of $\mathfrak{S}$, then $\mathcal{I}_E(s) = \mathcal{I}_E(t)$, hence $\mathcal{I}_E$ is well-defined.*

Proof: For each $i = 1, \cdots, n$ and $j = 1, \cdots, m$, write $A_i = \bigcup_{j=1}^{m}(A_i \cap B_j)$, [a disjoint union], $B_j = \bigcup_{i=1}^{n}(A_i \cap B_j)$ [also a disjoint union], $e_{ij} = c_i$, and $e'_{ij} = d_j$. Then, using basic properties of indicator functions, we have

$$s = \sum_{i=1}^{n} c_i I_{\bigcup_{j=1}^{m}(A_i \cap B_j)} = \sum_{i=1}^{n} c_i \sum_{j=1}^{m} I_{A_i \cap B_j} = \sum_{i=1}^{n}\sum_{j=1}^{m} e_{ij} I_{A_i \cap B_j}. \tag{6.1}$$

By the same reasoning, we also have

$$t = \sum_{j=1}^{m} d_j I_{\bigcup_{i=1}^{n}(A_i \cap B_j)} = \sum_{j=1}^{m} d_j \sum_{i=1}^{n} I_{A_i \cap B_j} = \sum_{i=1}^{n}\sum_{j=1}^{m} e'_{ij} I_{A_i \cap B_j}. \tag{6.2}$$

Since s and t are equivalent representations of the same element of $\mathfrak{S}$, it follows that $e_{ij} = e'_{ij}$ for every $i = 1, \cdots, n$ and $j = 1, \cdots, m$ such that $A_i \cap B_j \neq \varnothing$. For those i and j with $A_i \cap B_j = \varnothing$, we have

$$e_{ij}\mu(A_i \cap B_j \cap E) = 0 = e'_{ij}\mu(A_i \cap B_j \cap E).$$

Therefore, the definition of $\mathcal{I}_E$, (6.1), (6.2), and the previous display yield

$$\mathcal{I}_E(s) = \sum_{i=1}^{n}\sum_{j=1}^{m} e_{ij}\mu(A_i \cap B_j \cap E) = \sum_{i=1}^{n}\sum_{j=1}^{m} e'_{ij}\mu(A_i \cap B_j \cap E) = \mathcal{I}_E(t),$$

which proves the claim. ∎

Now that $\mathcal{I}_E$ has been shown to be well-defined, we may actually start proving some properties regarding it. Observe first that $\mathcal{I}_E$ applied to the element of $\mathfrak{S}$ that is identically c [where $0 \leq c < +\infty$] yields the number $c\mu(E)$. In symbols, $\mathcal{I}_E(c) = c\mu(E)$. This follows immediately from the definition. Also, it is immediate that $\mathcal{I}_E$ maps $\mathfrak{S}$ into $[0, +\infty]$.

(S2) *We have the following facts:*

(a) If $\mu(E) = 0$, then $\mathcal{I}_F(s) = 0$ whenever $F \subseteq E$.

(b) $\mathcal{I}_F(sI_E) = \mathcal{I}_{E\cap F}(s)$.

(c) $\mathcal{I}_F(I_E) = \mu(E \cap F)$.

Proof: If $\mu(E) = 0$ and $F \subseteq E$, then

$$0 \leq \mathcal{I}_F(s) = \sum_{i=1}^{n} c_i\mu(A_i \cap F) \leq \sum_{i=1}^{n} c_i\mu(A_i \cap E) \leq \sum_{i=1}^{n} c_i\mu(E) = 0,$$

and hence $\mathcal{I}_F(s) = 0$, giving (a). For (b), observe that $sI_E = \sum_{i=1}^{n} c_i I_{A_i\cap E}$, hence $\mathcal{I}_F(sI_E) = \sum_{i=1}^{n} c_i\mu(A_i \cap E \cap F) = \mathcal{I}_{E\cap F}(s)$. Part (c) follows by taking $s = 1$ in (b). ∎

The next three properties are also simple and they illustrate the ordering on $\mathfrak{S}$ that $\mathcal{I}_E$ induces.

(S3) *If $s = t$ on E, then $\mathcal{I}_F(s) = \mathcal{I}_F(t)$ for every $F \subseteq E$.*

Proof: By hypothesis we have $sI_E = tI_E$, hence $\mathcal{I}_F(sI_E) = \mathcal{I}_F(tI_E)$ by (S1), whence $\mathcal{I}_{F\cap E}(s) = \mathcal{I}_{F\cap E}(t)$ by (S2). Since $F \subseteq E$, we have $\mathcal{I}_F(s) = \mathcal{I}_F(t)$. ∎

(S4) *$s \leq t$ [on Ω] implies $\mathcal{I}_E(s) \leq \mathcal{I}_E(t)$. More generally, if $s \leq t$ on E, then $\mathcal{I}_F(s) \leq \mathcal{I}_F(t)$ for every $F \subseteq E$.*

Proof: Refer back to the proof of (S1) and observe that $s \leq t$ iff $e_{ij} \leq e'_{ij}$ for every $i = 1, \cdots, n$ and $j = 1, \cdots, m$ such that $A_i \cap B_j \neq \varnothing$. For those i and

j with $A_i \cap B_j = \varnothing$, we have $e_{ij}\mu(A_i \cap B_j \cap E) = 0 = e'_{ij}\mu(A_i \cap B_j \cap E)$. By (6.1) and (6.2) we have

$$\mathcal{I}_E(s) = \sum_{i=1}^{n}\sum_{j=1}^{m} e_{ij}\mu(A_i \cap B_j \cap E) \le \sum_{i=1}^{n}\sum_{j=1}^{m} e'_{ij}\mu(A_i \cap B_j \cap E) = \mathcal{I}_E(t).$$

For the second claim, let $F \subseteq E$ and assume that $s \le t$ on E. Then $sI_E \le tI_E$, hence $\mathcal{I}_F(sI_E) \le \mathcal{I}_F(tI_E)$. By (S2), we have $\mathcal{I}_{F\cap E}(s) \le \mathcal{I}_{F\cap E}(t)$. Since $F \subseteq E$, we have $\mathcal{I}_F(s) \le \mathcal{I}_F(t)$. ∎

(S5) *If $E \subseteq F$, then $\mathcal{I}_E(s) \le \mathcal{I}_F(s)$.*

Proof: If $E \subseteq F$, then $sI_E \le sI_F$ [on Ω], hence $\mathcal{I}(sI_E) \le \mathcal{I}(sI_F)$ by (S4), whence $\mathcal{I}_E(s) \le \mathcal{I}_F(s)$ by (S2). ∎

The next result says that finite summations and constants may be interchanged with $\mathcal{I}_E$; this foreshadows our upcoming claim that the Lebesgue integral is a linear operator. In the proof, we use the simple fact that any finite linear combination of elements of $\mathfrak{S}$ is itself an element of $\mathfrak{S}$.

(S6) *For any $n \in \mathbb{N}$ and constants $0 \le c_1, \cdots, c_n < +\infty$, we have the equation $\mathcal{I}_E\left(\sum_{i=1}^n c_i s_i\right) = \sum_{i=1}^n c_i \mathcal{I}_E(s_i)$.*

Proof: Let $0 \le c < +\infty$ and write $s = \sum_{i=1}^n c_i I_{A_i}$ as before. Then we have $cs = \sum_{i=1}^n cc_i I_{A_i}$, which gives

$$\mathcal{I}_E(cs) = \sum_{i=1}^{n} cc_i\mu(A_i \cap E) = c\sum_{i=1}^{n} c_i\mu(A_i \cap E) = c\mathcal{I}_E(s),$$

proving the claim for the case $n = 1$.

Next, let $n = 2$ and $0 \le a, b < +\infty$. Representations (6.1) and (6.2) give $as + bt = \sum_{i=1}^n \sum_{j=1}^m (ae_{ij} + be'_{ij}) I_{A_i \cap B_j}$, so that

$$\begin{aligned}\mathcal{I}_E(as+bt) &= \sum_{i=1}^{n}\sum_{j=1}^{m}(ae_{ij}+be'_{ij})\mu(A_i \cap B_j \cap E) \\ &= a\sum_{i=1}^{n}\sum_{j=1}^{m} e_{ij}\mu(A_i \cap B_j \cap E) + b\sum_{i=1}^{n}\sum_{j=1}^{m} e'_{ij}\mu(A_i \cap B_j \cap E) \\ &= a\mathcal{I}_E(s) + b\mathcal{I}_E(t),\end{aligned}$$

where the third equality again follows from (6.1) and (6.2). This gives the case where $n = 2$. The proof for general $n \in \mathbb{N}$ now follows by induction. ∎

The next and final result of this section shows that our functional $\mathcal{I}_E$ behaves like a measure.

(S7) *Fix $s \in \mathfrak{S}$, with $s = \sum_{i=1}^n c_i I_{A_i}$ as before. Define $\nu_s : \mathcal{F} \to \bar{\mathbb{R}}$ by writing $\nu_s(B) = \mathcal{I}_B(s)$ for each $B \in \mathcal{F}$. Then ν_s is a measure.*

Proof: Clearly ν_s is nonnegative and (S2) gives $\nu_s(\varnothing) = 0$. We need only show that ν_s is countably additive. Let $\{B_m\}_{m=1}^\infty$ denote a disjoint sequence of $\mathcal{F}$-sets. Then we have that $\{A_i \cap B_m\}_{m=1}^\infty$ is a disjoint sequence of $\mathcal{F}$-sets and $A_i \cap \bigcup_{m=1}^\infty B_m = \bigcup_{m=1}^\infty (A_i \cap B_m)$ for each $i = 1, \cdots, n$, hence

$$\begin{aligned}\mathcal{I}_{\bigcup_{m=1}^\infty B_m}(s) = \sum_{i=1}^n c_i \mu\left(\bigcup_{m=1}^\infty (A_i \cap B_m)\right) &= \sum_{i=1}^n c_i \sum_{m=1}^\infty \mu(A_i \cap B_m) \\ &= \sum_{m=1}^\infty \sum_{i=1}^n c_i \mu(A_i \cap B_m) \\ &= \sum_{m=1}^\infty \mathcal{I}_{B_m}(s),\end{aligned}$$

where the interchange of summations follows because all terms are nonnegative. This shows that $\nu_s(\bigcup_{m=1}^\infty B_m) = \sum_{m=1}^\infty \nu_s(B_m)$, hence ν_s is countably additive. It follows that ν_s is a measure as claimed. ∎

Example 3. We list some important consequences of (S7). First and foremost, we have that if $E \in \mathcal{F}$ is the disjoint union of the $\mathcal{F}$-sets $E_1, E_2, \cdots$, then we may write $\mathcal{I}_E(s) = \sum_{n=1}^\infty \mathcal{I}_{E_n}(s)$. By finite additivity, this result holds in the similar case of where E is the finite disjoint union of $E_1, \cdots, E_n \in \mathcal{F}$.[1] We shall use this fact quite often in the next section.

Another useful consequence of (S7) is this: if $\{E_n\}_{n=1}^\infty$ denotes a nondecreasing sequence of $\mathcal{F}$-sets [so that $\lim E_n = \bigcup_{n=1}^\infty E_n$], then $\nu_s(\lim E_n) = \lim \nu_s(E_n)$ by measure property (M8), hence we may write $\mathcal{I}_{\lim E_n}(s) = \lim \mathcal{I}_{E_n}(s)$.

There is also the interpretation of (M9) in this context. Using (S7), we have that if $\{E_n\}_{n=1}^\infty$ is a nonincreasing sequence of $\mathcal{F}$-sets such that $\mathcal{I}_{E_N}(s) < +\infty$ for some N, then $\mathcal{I}_{\lim E_n}(s) = \lim \mathcal{I}_{E_n}(s)$. [Recall that here $\lim E_n = \bigcap_{n=1}^\infty E_n$.] The other properties involving the interchange of limit and measure may be transcribed in terms of our operator in the same simple fashion.

The properties in this section are not ends in and of themselves, but instead are merely means to an end, which is the development of the Lebesgue integral. In the next section, we shall generalize $\mathcal{I}_E$ to a functional that operates on a class of functions that includes the elements of $\mathfrak{S}$. The generalization will be done in a fashion that uses the buildup of this section, so it is good to be thoroughly familiar with the properties presented in this section.

Exercises.

These exercises are all intended to be straightforward "finger exercises" designed more for familiarity with the notation and the properties proved than for a rigorous challenge, and the starred problems should all be worked, as some of these results will be called upon explicitly or implicitly in later developments. We will continue to work in a generic measure space $(\Omega, \mathcal{F}, \mu)$.

[1] This is the abstract version of that property of the Riemann integral that lets us write $\int_a^b f(x)\,dx = \int_a^c f(x)\,dx + \int_c^b f(x)\,dx$, where $a \le c \le b$, say.

1*. Let $E \in \mathcal{F}$.
(a) $\mu(E) = 0$ iff $\mathcal{I}_E(s) = 0$ for every $s \in \mathfrak{S}$.
(b) For any $c \geq 0$, $\mathcal{I}_E(c) = c\mu(E)$. [In $\mathcal{I}_E(c)$, c denotes the element of $\mathfrak{S}$ that maps each $\omega \in \Omega$ to the number c. The c on the right, of course, is the number c.]

2. Let $t, s, s_1, s_2, \cdots \in \mathfrak{S}$. Why can't we say that $\mathcal{I}_E(as + bt) = a\mathcal{I}_E(s) + b\mathcal{I}_E(t)$ for every $a, b \in \mathbb{R}$, as compared to saying that the result holds for every $0 \leq a, b < +\infty$? Also, why can't we necessarily write $\mathcal{I}_E(\sum_{i=1}^{\infty} c_i s_i) = \sum_{i=1}^{\infty} c_i \mathcal{I}_E(s_i)$? [What is the domain of $\mathcal{I}_E$?]

3*. Let $E \in \mathcal{F}$ be such that $\mu(E) > 0$. Then $\mathcal{I}_E(s) = 0$ iff $s = 0$ μ-a.e. on E. In particular, $\mathcal{I}(s) = 0$ iff $s = 0$ μ-a.e. [on Ω].

4*. Let $E \in \mathcal{F}$ and $s \in \mathfrak{S}$.
(a) If $\mu(E) < +\infty$, then $\mathcal{I}_E(s) < +\infty$, but the converse is not necessarily true.
(b) If $\mu(E) = +\infty$ and $\mathcal{I}_E(s) = +\infty$, then $\mu(\{\omega \in E : s(\omega) > 0\}) > 0$, but the converse is not necessarily true.
(c) Let $\mu(E) = +\infty$. Then $\mathcal{I}_E(s) < +\infty$ iff $\mu(\{\omega \in E : s(\omega) > 0\}) < +\infty$.

5*. Suppose that $\{E_n\}_{n=1}^{\infty}$ is a nonincreasing sequence of $\mathcal{F}$-sets [so that $\lim E_n$ exists and equals $\bigcap_{n=1}^{\infty} E_n$]. Also, let $s \in \mathfrak{S}$.
(a) It is not necessarily the case that $\lim_n \mathcal{I}_{E_n}(s) = \mathcal{I}_{\lim_n E_n}(s)$.
(b) If $\mathcal{I}_{E_n}(s) < +\infty$ for some $n \in \mathbb{N}$, then $\lim_n \mathcal{I}_{E_n}(s) = \mathcal{I}_{\lim_n E_n}(s)$.
(c) If $\{E_n\}_{n=1}^{\infty}$ is no longer nonincreasing but is still such that $\lim_n E_n$ exists, state conditions under which $\lim_n \mathcal{I}_{E_n}(s) = \mathcal{I}_{\lim_n E_n}(s)$.

6*. Let $s \in \mathfrak{S}$ and recall (S7), where ν_s is the measure on $\mathcal{F}$ with $\nu_s(B) = \mathcal{I}_B(s)$ for every $B \in \mathcal{F}$. Then, for any $E \in \mathcal{F}$ and $t \in \mathfrak{S}$, we have $\mathcal{I}_E(t; \nu_s) = \mathcal{I}_E(ts; \mu)$.

7*. Let $A \in \mathcal{F}$ be nonempty, and let $s \in \mathfrak{S}$. Let $\mathcal{F}_A = \{E \subseteq A : E \in \mathcal{F}\}$, and let μ_A denote the restriction of μ from $\mathcal{F}$ to $\mathcal{F}_A$. Finally, let s_A denote the restriction of s from Ω to A.
(a) $(A, \mathcal{F}_A, \mu_A)$ is a measure space.
(b) If $\mathcal{J}$ denotes this section's functional *relative to* $(A, \mathcal{F}_A, \mu_A)$, then we have $\mathcal{I}_E(s; \mu) = \mathcal{J}_E(s_A; \mu_A)$ for every $E \in \mathcal{F}_A$. [Given any $C \subseteq \Omega$, let $I_{C;A}$ denote the restriction of I_C to A. If $s = \sum_{i=1}^{n} c_i I_{A_i}$, then $s_A = \sum_{i=1}^{n} c_i I_{A_i \cap A; A}$. Next, we have

$$\mathcal{J}_E(s_A; \mu_A) = \sum_{i=1}^{n} c_i \mu_A(A_i \cap A \cap E) = \sum_{i=1}^{n} c_i \mu(A_i \cap E) = \mathcal{I}_E(s; \mu)$$

since $\mu_A = \mu$ on $\mathcal{F}_A$ and $E \subseteq A$.]

8*. Let $A \in \mathcal{F} - \{\varnothing\}$, and suppose that $s : A \to [0, +\infty)$ is simple and $\mathcal{F}/\mathcal{B}^*$-measurable. Let $E \in \mathcal{F}_A$, where $\mathcal{F}_A$ is defined in the previous exercise. Consider two programs:
Program 1: Extend s from A to Ω as follows: Let $s^* \in \mathfrak{S}$ be such that $s^* = s$ on A and $s^* = 0$ on A^c.
Program 2: Do not extend s from A to Ω as in Program 1, but instead view s as a function defined everywhere relative to the measure space $(A, \mathcal{F}_A, \mu_A)$, where the notation is as in the previous exercise.

These two programs are equivalent in the sense that $\mathcal{I}_E(s^*; \mu) = \mathcal{J}_E(s; \mu_A)$, where $\mathcal{I}_E$ is this section's functional relative to the measure space $(\Omega, \mathcal{F}, \mu)$, and $\mathcal{J}_E$ is this section's functional relative to the measure space $(A, \mathcal{F}_A, \mu_A)$.

Comment: This and Exercise 7, while simple, are important with respect to logical consistency. Our definitions and results in the section exposition assumed that functions were everywhere defined on Ω, and such an approach may seem somewhat restrictive, since it may not be possible for some naturally occurring functions to be defined everywhere on Ω. These two exercises show that such functions are not excluded from consideration nor left out of the scope of the results of this section, since we are justified in "switching our perspective" from $(\Omega, \mathcal{F}, \mu)$ to an appropriate "sub-measure-space." Or, if we wish, we may equivalently just extend the domain of our function by defining it to be zero everywhere outside of the original domain, and this extended function now has domain Ω and the theory relative to $(\Omega, \mathcal{F}, \mu)$ applies. In either case, the corresponding functionals give the same numbers.

9*. Quickly prove the following "almost everywhere" modifications of (S3) and (S4). As usual, $s, t \in \mathfrak{S}$ and all sets are in $\mathcal{F}$.
(a) If $s = t$ μ-a.e. on E, then $\mathcal{I}_F(s) = \mathcal{I}_F(t)$ for every $F \subseteq E$.
(b) If $s \leq t$ μ-a.e. on E, then $\mathcal{I}_F(s) \leq \mathcal{I}_F(t)$ for every $F \subseteq E$.

6.2 STAGE TWO: NONNEGATIVE FUNCTIONS

In the introduction to this chapter, it was stated that the construction of the Lebesgue integral would require four stages. In the previous section, we discussed the first stage, defining for each $E \in \mathcal{F}$ the functional $\mathcal{I}_E$ on the collection $\mathfrak{S}$ of nonnegative finite-valued simple $\mathcal{F}/\mathcal{B}^*$-measurable functions with domain Ω. In this section, we will define for each $E \in \mathcal{F}$ an extension of $\mathcal{I}_E$ to the larger class of nonnegative $\mathcal{F}/\mathcal{B}^*$-measurable functions defined on Ω, and we will derive various useful properties of this extended functional. This will constitute the second and most arduous stage of the program. However, the length comes from having to prove many simple properties, as compared to having a smaller number of very complex and intricate properties. As usual, everything will be done in the context of a generic measure space $(\Omega, \mathcal{F}, \mu)$.

Definition. We will let $\mathfrak{N}$ denote the set of nonnegative $\mathcal{F}/\mathcal{B}^*$-measurable functions on Ω. If any confusion about the underlying measure space can arise, we will write $\mathfrak{N}(\Omega, \mathcal{F}, \mu)$ or $\mathfrak{N}(\Omega, \mathcal{F})$ for $\mathfrak{N}$. Given $f \in \mathfrak{N}$, we define

$$\mathfrak{S}_f = \{s \in \mathfrak{S} : 0 \leq s \leq f\},$$

and we write $\mathfrak{S}_f(\Omega, \mathcal{F}, \mu)$ for $\mathfrak{S}_f$ if any confusion is possible.

We have $0 \in \mathfrak{S}_f \subseteq \mathfrak{S} \subseteq \mathfrak{N}$ for every $f \in \mathfrak{N}$, where 0 denotes the zero function. Also, *nonnegative* means that $+\infty$ is allowed, as usual.

We now define for every $E \in \mathcal{F}$ the extension of $\mathcal{I}_E$ from $\mathfrak{S}$ to $\mathfrak{N}$.

Definition. Let $E \in \mathcal{F}$. We define $\bar{\mathcal{I}}_E$ on $\mathfrak{N}$ by writing

$$\bar{\mathcal{I}}_E(f) = \sup_{s \in \mathfrak{S}_f} \mathcal{I}_E(s) = \sup\{\mathcal{I}_E(s) : \ s \in \mathfrak{S},\ 0 \leq s \leq f\}$$

for every $f \in \mathfrak{N}$. We will equivalently write $\bar{\mathcal{I}}_E(f; \mu)$ for $\bar{\mathcal{I}}_E(f)$ if extra clarity regarding the measure μ is required. We will also write $\bar{\mathcal{I}}(f)$ for $\bar{\mathcal{I}}_\Omega(f)$.

It is immediately observed that there are no problems with the definition and $0 \le \bar{\mathcal{I}}_E(f) \le +\infty$ for each $E \in \mathcal{F}$ and $f \in \mathfrak{N}$. It is also nearly immediate from the definition and (S2) that $\mu(E) = 0$ forces $\bar{\mathcal{I}}_E(f) = 0$ for each $f \in \mathfrak{N}$.

The new functional $\bar{\mathcal{I}}_E$ is intimately connected with the functional $\mathcal{I}_E$ from the previous section, and many proofs in this section will be based upon the properties for $\mathcal{I}_E$ given in the previous section.

As in the previous section, everything that follows is done in the context of a generic measure space $(\Omega, \mathcal{F}, \mu)$. We will use f and g to denote generic functions in $\mathfrak{N}$, and s and t will denote generic simple functions in $\mathfrak{S}$. All sets mentioned in the statements of the claims are assumed to be in $\mathcal{F}$.

(N1) *$\bar{\mathcal{I}}_E$ consistently extends $\mathcal{I}_E$ from $\mathfrak{S}$ to $\mathfrak{N}$: $\bar{\mathcal{I}}_E(s) = \mathcal{I}_E(s)$ for each $s \in \mathfrak{S}$.*

Proof: Let $s \in \mathfrak{S}$. Then $s \in \mathfrak{S}_s$, so $\mathcal{I}_E(s) \le \sup_{t \in \mathfrak{S}_s} \mathcal{I}_E(t) = \bar{\mathcal{I}}_E(s)$. Next, by (S4), any $t \in \mathfrak{S}_s$ is such that $\mathcal{I}_E(t) \le \mathcal{I}_E(s)$, hence we have the reverse inequality $\bar{\mathcal{I}}_E(s) = \sup_{t \in \mathfrak{S}_s} \mathcal{I}_E(t) \le \mathcal{I}_E(s)$, giving $\bar{\mathcal{I}}_E(s) = \mathcal{I}_E(s)$. ■

By (N1), the properties from the previous section carry over to $\bar{\mathcal{I}}_E$ whenever they are applied to elements of $\mathfrak{S}$. For example, we have $\bar{\mathcal{I}}_E(sI_F) = \bar{\mathcal{I}}_{E\cap F}(s)$ for each $s \in \mathfrak{S}$.[2] As another example, we have $\bar{\mathcal{I}}_E(c) = c\mu(E)$ for each $c \ge 0$,[3] where c denotes the function in $\mathfrak{S}$ that is identically equal to c. Many of the upcoming proofs will use (N1) implicitly or explicitly, and it will become tedious to cite every instance of its use, hence it is left to the reader's good judgment to know when it is being used.

(N2) $\bar{\mathcal{I}}_F(fI_E) = \bar{\mathcal{I}}_{F\cap E}(f)$.

Proof: We must show that $\sup_{t \in \mathfrak{S}_{fI_E}} \mathcal{I}_F(t) = \sup_{s \in \mathfrak{S}_f} \mathcal{I}_{F\cap E}(s)$. By (S2), this is equivalent to showing that

$$\sup\{\mathcal{I}_F(t) : t \in \mathfrak{S},\ 0 \le t \le fI_E\} = \sup\{\mathcal{I}_F(sI_E) : s \in \mathfrak{S},\ 0 \le s \le f\}. \tag{6.3}$$

Let A and B denote the sets on the left side and right side of (6.3), respectively. We will show that $\sup A = \sup B$ by showing that $A = B$.

We first show that $B \subseteq A$. Let $x \in B$, so that $x = \mathcal{I}_F(sI_E)$ for some $s \in \mathfrak{S}$ with $0 \le s \le f$. If $t = sI_E$, then $x = \mathcal{I}_F(t)$ with $t \in \mathfrak{S}$ and $0 \le t \le fI_E$, so that $x \in A$. This shows that $B \subseteq A$.

We now complete the proof by showing that $A \subseteq B$. Let $x \in A$, so that $x = \mathcal{I}_F(t)$ for some $t \in \mathfrak{S}$ with $0 \le t \le fI_E$. Since $F = (F \cap E) \cup (F \cap E^c)$ [a disjoint union], (S7) and (S2) yield

$$x = \mathcal{I}_F(t) = \mathcal{I}_{F\cap E}(t) + \mathcal{I}_{F\cap E^c}(t) = \mathcal{I}_F(tI_E) + \mathcal{I}_F(tI_{E^c}). \tag{6.4}$$

[2]Since $\bar{\mathcal{I}}_E(sI_F) = \mathcal{I}_E(sI_F) = \mathcal{I}_{E\cap F}(s) = \bar{\mathcal{I}}_{E\cap F}(s)$, where the first and third equalities are courtesy of (N1), and the second is by (S2).

[3]Since $\bar{\mathcal{I}}_E(c) = \mathcal{I}_E(c) = c\mu(E)$.

Since $0 \le t \le fI_E$ and $fI_E \le f$, we have $0 \le t \le f$. Furthermore, $t = 0$ on E^c, so that $tI_{E^c} = 0$, hence $\mathcal{I}_F(tI_{E^c}) = 0$. Substitution of these facts back into (6.4) yields $x = \mathcal{I}_F(tI_E)$, where $t \in \mathfrak{S}$ and $0 \le t \le f$. It follows that $x \in B$, whence $A \subseteq B$. This shows $A = B$, and the proof is complete. ■

(N3) *If $f = g$ on E, then $\bar{\mathcal{I}}_F(f) = \bar{\mathcal{I}}_F(g)$ for any $F \subseteq E$.*

Proof: If $f = g$ on E, then $fI_E = gI_E$ on Ω, so $\bar{\mathcal{I}}_F(fI_E) = \bar{\mathcal{I}}_F(gI_E)$, hence (N2) forces $\bar{\mathcal{I}}_{F\cap E}(f) = \bar{\mathcal{I}}_{F\cap E}(g)$. Since $F \subseteq E$, this gives $\bar{\mathcal{I}}_F(f) = \bar{\mathcal{I}}_F(g)$. ■

(N4) *If $f \le g$ on E, then $\bar{\mathcal{I}}_F(f) \le \bar{\mathcal{I}}_F(g)$ for any $F \subseteq E$.*

Proof: If $f \le g$ on Ω, then $\mathfrak{S}_f \subseteq \mathfrak{S}_g$, hence $\sup_{s\in\mathfrak{S}_f} \mathcal{I}_E(s) \le \sup_{s\in\mathfrak{S}_g} \mathcal{I}_E(s)$, whence $\bar{\mathcal{I}}_E(f) \le \bar{\mathcal{I}}_E(g)$. Therefore, $f \le g$ on Ω forces $\bar{\mathcal{I}}_E(f) \le \bar{\mathcal{I}}_E(g)$ for every $E \in \mathcal{F}$.

Applying this, we may prove the claim itself. If $f \le g$ on E and $F \subseteq E$, then $fI_E \le gI_E$ on Ω, hence $\bar{\mathcal{I}}_F(fI_E) \le \bar{\mathcal{I}}_F(gI_E)$ by what has just been shown. By (N2) and the fact that $F \subseteq E$ forces $F \cap E = F$, we obtain the inequality $\bar{\mathcal{I}}_F(f) \le \bar{\mathcal{I}}_F(g)$. ■

If $f \in \mathfrak{N}$ and $0 \le c <\le +\infty$, then it is clear that $cf \in \mathfrak{N}$, so that we may discuss $\bar{\mathcal{I}}_E(cf)$. This fact is used implicitly in the proof of the next property.

(N5) *For any $0 \le c < +\infty$ we have $\bar{\mathcal{I}}_E(cf) = c\bar{\mathcal{I}}_E(f)$.*

Proof: The case $c = 0$ is trivial,[4] so we consider the case where $0 < c < +\infty$. If $0 < c < +\infty$, we first claim that

$$\{\mathcal{I}_E(s') : s' \in \mathfrak{S}_{cf}\} = \{\mathcal{I}_E(cs) : s \in \mathfrak{S}_f\}. \tag{6.5}$$

If $x = \mathcal{I}_E(s')$ for some $s' \in \mathfrak{S}_{cf}$, then $x = \mathcal{I}_E(c\frac{s'}{c})$ and $\frac{s'}{c} \in \mathfrak{S}_f$, hence x is in the right side of (6.5). This gives $\{\mathcal{I}_E(s') : s' \in \mathfrak{S}_{cf}\} \subseteq \{\mathcal{I}_E(cs) : s \in \mathfrak{S}_f\}$. Next, if $y = \mathcal{I}_E(cs)$ for some $s \in \mathfrak{S}_f$, then $y = \mathcal{I}_E(s')$ for some $s' \in \mathfrak{S}_{cf}$, namely, $s' = cs$. Thus $\{\mathcal{I}_E(s') : s' \in \mathfrak{S}_{cf}\} \supseteq \{\mathcal{I}_E(cs) : s \in \mathfrak{S}_f\}$, and (6.5) follows. Using (6.5) and (S6), we have

$$\begin{aligned}\bar{\mathcal{I}}_E(cf) = \sup_{s'\in\mathfrak{S}_{cf}} \mathcal{I}_E(s') = \sup_{s\in\mathfrak{S}_f} \mathcal{I}_E(cs) = \sup_{s\in\mathfrak{S}_f} c\mathcal{I}_E(s) &= c \sup_{s\in\mathfrak{S}_f} \mathcal{I}_E(s) \\ &= c\bar{\mathcal{I}}_E(s),\end{aligned}$$

hence the proof is complete. ■

We can not yet prove that $\bar{\mathcal{I}}_E(f+g) = \bar{\mathcal{I}}_E(f) + \bar{\mathcal{I}}_E(g)$ for general $f, g \in \mathfrak{N}$, so we can not yet strengthen the claim of (N5) to assert that $\bar{\mathcal{I}}_E$ behaves in

[4]If f assumes the value $+\infty$ or if $\bar{\mathcal{I}}_E(f) = +\infty$, our convention that $+\infty(0) = 0(+\infty) = 0$ saves the day.

a linear fashion. It will turn out that this property does indeed hold, but the proof of such a property will require some more development.

(N6) *If* $E \subseteq F$, *then* $\bar{\mathcal{I}}_E(f) \leq \bar{\mathcal{I}}_F(f)$.

Proof: If $E \subseteq F$, then $fI_E \leq fI_F$, hence $\bar{\mathcal{I}}(fI_E) \leq \bar{\mathcal{I}}(fI_F)$ by (N4). By (N2), this gives $\bar{\mathcal{I}}_E(f) \leq \bar{\mathcal{I}}_F(f)$. ∎

The following property (N7) is one of the most important results if not the most important result in measure theory. It is known as *[Lebesgue's] Monotone Convergence Theorem*, and its value will be made clear through seeing how often it will be called in subsequent work. We will denote this result by MCT. In the statement and proof of MCT, we use (MF11) to assert the following: *if* $\{f_n\}_{n=1}^{\infty}$ *denotes a nondecreasing sequence of elements of* $\mathfrak{N}$, *then* $\lim f_n \in \mathfrak{N}$.

(N7) *For any nondecreasing sequence* $\{f_n\}_{n=1}^{\infty}$ *of functions in* $\mathfrak{N}$, *we have* $\lim_n \bar{\mathcal{I}}_E(f_n) = \bar{\mathcal{I}}_E(\lim_n f_n)$.

Proof: By the preceding comments, it makes sense to discuss $\bar{\mathcal{I}}_E(\lim_n f_n)$. Also, by (N4), the sequence $\{\bar{\mathcal{I}}_E(f_n)\}_{n=1}^{\infty}$ is nondecreasing, hence $\lim_n \bar{\mathcal{I}}_E(f_n)$ exists. The quantities $\lim_n \bar{\mathcal{I}}_E(f_n)$ and $\bar{\mathcal{I}}_E(\lim_n f_n)$ accordingly exist; we now show equality by showing that each is no more than the other.

We first claim that $\lim_m \bar{\mathcal{I}}_E(f_m) \leq \bar{\mathcal{I}}_E(\lim_n f_n)$. To see this, observe that $f_m \leq \lim_n f_n$ for every $m \in \mathbb{N}$, hence $\bar{\mathcal{I}}_E(f_m) \leq \bar{\mathcal{I}}_E(\lim_n f_n)$ for each $m \in \mathbb{N}$ by (N4). Since $\lim_m \bar{\mathcal{I}}_E(f_m)$ exists, we have $\lim_m \bar{\mathcal{I}}_E(f_m) \leq \bar{\mathcal{I}}_E(\lim_n f_n)$ as claimed. Therefore, to prove the theorem, we must demonstrate the reverse inequality $\bar{\mathcal{I}}_E(\lim_n f_n) \leq \lim_m \bar{\mathcal{I}}_E(f_m)$.

To prove the stated reverse inequality, pick $s \in \mathfrak{S}_{\lim_n f_n}$ and $0 < c < 1$, and for each $n \in \mathbb{N}$ write $A_n = \{\omega \in \Omega : cs(\omega) \leq f_n(\omega)\}$. Since s is $\mathcal{F}/\mathcal{B}^*$-measurable along with each of $f_1, f_2, \cdots$, we have $A_n \in \mathcal{F}$ for each $n \in \mathbb{N}$. Furthermore, the hypothesis $f_1 \leq f_2 \leq \cdots$ shows that $\{A_n\}_{n=1}^{\infty}$ is a nondecreasing sequence of $\mathcal{F}$-sets, hence $\lim_n A_n$ exists and equals $\bigcup_{n=1}^{\infty} A_n$.

Before proceeding further, we claim that $\lim_n A_n = \Omega$. To see this, let $\omega \in \Omega$. There are two cases for this fixed generic $\omega \in \Omega$: $s(\omega) = 0$ and $s(\omega) > 0$.

(i) $s(\omega) = 0$: Here, $\omega \in A_n$ for all $n \in \mathbb{N}$, hence $\omega \in \lim_n A_n$.

(ii) $s(\omega) > 0$: Here, $cs(\omega) < s(\omega) \leq \lim_n f_n(\omega)$. If $cs(\omega) > f_n(\omega)$ for all $n \in \mathbb{N}$, then $cs(\omega) \geq \lim_n f_n(\omega)$, a contradiction to what has just been stated. Therefore, there is $n \in \mathbb{N}$ such that $cs(\omega) \leq f_n(\omega)$, so $\omega \in \lim_n A_n$.

We accordingly have $\lim_n A_n = \Omega$ as claimed.

Using (N6), the fact that $f_n \geq cs$ on $A_n \cap E$ for each $n \in \mathbb{N}$, (N4), (N5), and (N1), we have

$$\bar{\mathcal{I}}_E(f_n) \geq \bar{\mathcal{I}}_{A_n \cap E}(f_n) \geq \bar{\mathcal{I}}_{A_n \cap E}(cs) = c\mathcal{I}_{A_n \cap E}(s) \qquad \text{for all } n \in \mathbb{N}. \tag{6.6}$$

Observe by (S7) that since the sequence $\{A_n \cap E\}_{n=1}^{\infty}$ of $\mathcal{F}$-sets is nondecreasing with limit $\bigcup_{n=1}^{\infty}(A_n \cap E) = E$, we have that $\lim_n \mathcal{I}_{A_n \cap E}(s)$ exists and

$$\lim_n \mathcal{I}_{A_n \cap E}(s) = \mathcal{I}_{\lim_n A_n \cap E}(s) = \mathcal{I}_E(s),$$

hence letting $n \to +\infty$ in (6.6) gives

$$\lim_n \bar{\mathcal{I}}_E(f_n) \geq \lim_n c\mathcal{I}_{A_n \cap E}(s) = c \lim_n \mathcal{I}_{A_n \cap E}(s) = c\mathcal{I}_E(s).$$

Since $c \in (0,1)$ is generic, we have $\lim_n \bar{\mathcal{I}}_E(f_n) \geq \mathcal{I}_E(s)$. Since the function $s \in \mathfrak{S}_{\lim_n f_n}$ is generic, it follows that

$$\lim_n \bar{\mathcal{I}}_E(f_n) \geq \sup_{s \in \mathfrak{S}_{\lim_n f_n}} \mathcal{I}_E(s) = \bar{\mathcal{I}}_E\left(\lim_n f_n\right),$$

which is the desired reverse inequality. This completes the proof of the Monotone Convergence Theorem. ∎

The preceding statement gives the most common form of MCT. However, we can actually weaken the assumptions somewhat for the same result to hold. We now state and prove this simple extension.

(N7′) *For any sequence $\{f_n\}_{n=1}^{\infty}$ of functions in $\mathfrak{N}$ such that $f_1 \leq f_2 \leq \cdots$ on E, we have $\lim_n \bar{\mathcal{I}}_F(f_n) = \bar{\mathcal{I}}_F(g)$ for every $F \subseteq E$ and $g \in \mathfrak{N}$ such that $g = \lim_n f_n$ on E.*

Proof: Let $g \in \mathfrak{N}$ be such that $g = \lim_n f_n$ on E. For each $n \in \mathbb{N}$, let $g_n = f_n I_E$. Then $\{g_n\}_{n=1}^{\infty}$ is a sequence in $\mathfrak{N}$ that is nondecreasing on Ω, hence $\lim_n \bar{\mathcal{I}}_E(g_n) = \bar{\mathcal{I}}_E(\lim_n g_n)$ by (N7). Since $g_n = f_n$ on E for each $n \in \mathbb{N}$, and since $\lim_n g_n = g$ on E, we have $\lim_n \bar{\mathcal{I}}_F(f_n) = \bar{\mathcal{I}}_F(g)$ for every $F \subseteq E$ by (N3). ∎

We now turn to the analog of property (S6).

(N8) *We have the following for functions $g, f, f_1, f_2, \cdots \in \mathfrak{N}$:*

(a) $\bar{\mathcal{I}}_E(\sum_{i=1}^n c_i f_i) = \sum_{i=1}^n c_i \bar{\mathcal{I}}_E(f_i)$ for every $n \in \mathbb{N}$ and choice of $c_1, \cdots, c_n \in [0, +\infty)$, so that $\bar{\mathcal{I}}_E$ is linear in the same sense as is $\mathcal{I}_E$.

(b) If $f \leq g$, $g - f$ is defined on Ω, and $\bar{\mathcal{I}}_E(f) < +\infty$, then we have that $\bar{\mathcal{I}}_E(g - f) = \bar{\mathcal{I}}_E(g) - \bar{\mathcal{I}}_E(f)$.

Proof: For (a), the $n = 1$ case is given by (N5). It clearly suffices to show the result for $n = 2$, since the case for general $n \geq 2$ follows by induction. For the $n = 2$ case, we need only show that $\bar{\mathcal{I}}_E(f + g) = \bar{\mathcal{I}}_E(f) + \bar{\mathcal{I}}_E(g)$ for all $f, g \in \mathfrak{N}$, courtesy again of (N5). Twice applying (MF18), there exist nondecreasing sequences $\{s_n\}_{n=1}^{\infty}$ and $\{t_n\}_{n=1}^{\infty}$ of functions in $\mathfrak{S}_f$ and $\mathfrak{S}_g$ [respectively] such that $\lim_n s_n = f$ and $\lim_n t_n = g$. We now make the following three observations:

(1) $\bar{\mathcal{I}}_E(f) = \bar{\mathcal{I}}_E(\lim_n s_n) = \lim_n \bar{\mathcal{I}}_E(s_n) = \lim_n \mathcal{I}_E(s_n)$, where the second equality follows by MCT and the final equality follows by (N1).

(2) $\bar{\mathcal{I}}_E(g) = \lim_n \mathcal{I}_E(t_n)$ by the same reasoning as in (1).

(3) Since $\{s_n\}_{n=1}^{\infty}$ is nondecreasing with limit f and $\{t_n\}_{n=1}^{\infty}$ is nondecreasing with limit g, we have that $\{s_n + t_n\}_{n=1}^{\infty}$ is nondecreasing with limit $f + g$.

Therefore, these facts along with MCT, (N1), and (S6) give

$$\begin{aligned}\bar{\mathcal{I}}_E(f+g) = \bar{\mathcal{I}}_E\left(\lim_n (s_n + t_n)\right) &= \lim_n \bar{\mathcal{I}}_E(s_n + t_n) \\ &= \lim_n \mathcal{I}_E(s_n + t_n) \\ &= \lim_n \left(\mathcal{I}_E(s_n) + \mathcal{I}_E(t_n)\right) \\ &= \lim_n \mathcal{I}_E(s_n) + \lim_n \mathcal{I}_E(t_n) \\ &= \bar{\mathcal{I}}_E(f) + \bar{\mathcal{I}}_E(g),\end{aligned}$$

proving (a).

With (a) in tow, (b) is simple. Let $f \le g$ be such that $g - f$ is everywhere defined [that is, it is not the case that we have $f(\omega) = g(\omega) = +\infty$ for some $\omega \in \Omega$] and $\bar{\mathcal{I}}_E(f) < +\infty$. Then $g - f \in \mathfrak{N}$, and using (a) with $n = 2$ gives

$$\bar{\mathcal{I}}_E(g) = \bar{\mathcal{I}}_E((g-f)+f) = \bar{\mathcal{I}}_E(g-f) + \bar{\mathcal{I}}_E(f).$$

Since $\bar{\mathcal{I}}_E(f) < +\infty$, we may meaningfully subtract it from both sides of the quantity shown above, giving (b) and completing the entire proof. ■

The upcoming property mimics (S7) from the previous section, and is important.

(N9) *Define $\nu_f : \mathcal{F} \to \bar{\mathbb{R}}$ by writing $\nu_f(B) = \bar{\mathcal{I}}_B(f)$ for every $B \in \mathcal{F}$. Then ν_f is a measure.*

Proof: The nonnegativity of ν_f is obvious, and we have $\nu_f(\varnothing) = 0$ since $\mu(\varnothing) = 0$. To demonstrate that ν_f is countably additive, we first show that ν_f is finitely additive. Let E and F denote disjoint $\mathcal{F}$-sets. Then

$$\begin{aligned}\nu_f(E \cup F) = \bar{\mathcal{I}}_{E\cup F}(f) = \bar{\mathcal{I}}(fI_{E\cup F}) = \bar{\mathcal{I}}(fI_E + fI_F) &= \bar{\mathcal{I}}(fI_E) + \bar{\mathcal{I}}(fI_F) \\ &= \bar{\mathcal{I}}_E(f) + \bar{\mathcal{I}}_F(f) \\ &= \nu_f(E) + \nu_f(F)\end{aligned}$$

via properties (N8) and (N2), so ν_f is indeed finitely additive.

We next show that $\nu_f(\lim_n A_n) = \lim_n \nu_f(A_n)$ whenever $\{A_n\}_{n=1}^{\infty}$ is a nondecreasing sequence of $\mathcal{F}$-sets. That is, we now claim that ν_f is continuous from below. MCT and (N2) make this possible:

$$\begin{aligned}\nu_f\left(\lim_n A_n\right) = \bar{\mathcal{I}}_{\lim_n A_n}(f) = \bar{\mathcal{I}}(fI_{\lim_n A_n}) = \bar{\mathcal{I}}\left(\lim_n fI_{A_n}\right) &= \lim_n \bar{\mathcal{I}}(fI_{A_n}) \\ &= \lim_n \bar{\mathcal{I}}_{A_n}(f) \\ &= \lim_n \nu_f(A_n).\end{aligned}$$

[The fourth equality follows because $\{fI_{A_n}\}_{n=1}^{\infty}$ is a nondecreasing sequence of elements of $\mathfrak{N}$ with limit $fI_{\lim_n A_n}$.]

We are now in a position to show that ν_f is countably additive. Let $\{B_n\}_{n=1}^{\infty}$ denote a disjoint sequence of $\mathcal{F}$-sets. Then $\{\bigcup_{m=1}^{n} B_m\}_{n=1}^{\infty}$ is a nondecreasing sequence of $\mathcal{F}$-sets with limit $\bigcup_{n=1}^{\infty} B_n$, hence

$$\begin{aligned}\nu_f\left(\bigcup_{n=1}^{\infty} B_n\right) = \nu_f\left(\lim_n \bigcup_{m=1}^{n} B_m\right) &= \lim_n \nu_f\left(\bigcup_{m=1}^{n} B_m\right)\\ &= \lim_n \sum_{m=1}^{n} \nu_f(B_m)\\ &= \sum_{n=1}^{\infty} \nu_f(B_n)\end{aligned}$$

by the fact that ν_f has been shown to be finitely additive and continuous from below. This shows that ν_f is countably additive and completes the proof. ■

Property (N9) is one of those "fundamental" properties of $\bar{\mathcal{I}}_E$. It allows us to write $\sum_{n=1}^{\infty} \bar{\mathcal{I}}_{E_n}(f) = \bar{\mathcal{I}}_E(f)$ when E is the union of the disjoint sequence $\{E_n\}_{n=1}^{\infty}$. If $n \in \mathbb{N}$ is given, if $E_{n+1} = E_{n+2} = \cdots = \varnothing$, and if E is the disjoint union of $E_1, \cdots, E_n$, then (N9) allows us to write $\sum_{i=1}^{n} \bar{\mathcal{I}}_{E_i}(f) = \bar{\mathcal{I}}_E(f)$. Finally, if $E = E_1 \cup E_2$ [a disjoint union], and if $\mu(E_2) = 0$, we have

$$\bar{\mathcal{I}}_E(f) = \bar{\mathcal{I}}_{E_1}(f) + \bar{\mathcal{I}}_{E_2}(f) = \bar{\mathcal{I}}_{E_1}(f) + 0 = \bar{\mathcal{I}}_{E_1}(f).$$

That is, in discussing $\bar{\mathcal{I}}_E(f)$, we may consider instead $\bar{\mathcal{I}}_F(f)$, where F is any subset of E with $\mu(E - F) = 0$. That is, we may "ignore" what happens on a μ-null subset of E. For example, we may generalize properties (N3) and (N4) on an "almost everywhere" basis due to the just-stated comment:

(N3′) *If $f = g$ μ-a.e. on E, then $\bar{\mathcal{I}}_E(f) = \bar{\mathcal{I}}_E(g)$.*

(N4′) *If $f \leq g$ μ-a.e. on E, then $\bar{\mathcal{I}}_E(f) \leq \bar{\mathcal{I}}_E(g)$.*

The proof of, say, (N3′) is straightforward: write E as the disjoint union of E_1 and E_2, where E_1 denotes the set of $\omega \in E$ such that $f(\omega) = g(\omega)$ and $E_2 = E - E_1$. Since $f = g$ μ-a.e. on E, we have $\mu(E_2) = 0$, hence $\bar{\mathcal{I}}_{E_2}(f) = \bar{\mathcal{I}}_{E_2}(g) = 0$; we also have $\bar{\mathcal{I}}_{E_1}(f) = \bar{\mathcal{I}}_{E_1}(g)$ by applying (N3). It follows that $\bar{\mathcal{I}}_E(f) = \bar{\mathcal{I}}_{E_1}(f) = \bar{\mathcal{I}}_{E_1}(g) = \bar{\mathcal{I}}_E(g)$, completing the proof of (N3′). The proof of (N4′) is quite similar, and is left to the reader as an easy exercise. We shall refer to both properties (N3) and (N3′) by (N3). Similarly, we shall refer to properties (N4) and (N4′) by (N4).

Properties (N10) and (N11) below further characterize the behavior of the functional $\bar{\mathcal{I}}_E$. These two properties are very important despite their humble trappings.

(N10) $\bar{\mathcal{I}}_E(f) = 0$ *iff $f = 0$ μ-a.e. on E.*

Proof: If $f = 0$ μ-a.e. on E, then (N3) gives $\bar{\mathcal{I}}_E(f) = 0$. Conversely, assume that $\bar{\mathcal{I}}_E(f) = 0$. Let $A_n = \{\omega \in E : f(\omega) \geq \frac{1}{n}\}$ for each $n \in \mathbb{N}$. Then, for any $n \in \mathbb{N}$, we have $A_n \in \mathcal{F}$ and $\frac{1}{n} I_{A_n} \leq f$ on E, hence $\bar{\mathcal{I}}_E(\frac{1}{n} I_{A_n}) \leq \bar{\mathcal{I}}_E(f)$ by (N4), whence $\frac{1}{n}\mu(A_n) \leq 0$ by (N5). It follows that $\mu(A_n) = 0$ for all $n \in \mathbb{N}$. Since $\{\omega \in E : f(\omega) > 0\} = \bigcup_{n=1}^{\infty} A_n$ [a countable union of μ-null sets], we have $\mu(\{\omega \in E : f(\omega) > 0\}) = 0$, hence $f = 0$ μ-a.e. on E. Therefore, $\bar{\mathcal{I}}_E(f) = 0$ forces $f = 0$ μ-a.e. on E, proving the converse. ■

(N11) *If $\bar{\mathcal{I}}_E(f) < +\infty$, then f is finite μ-a.e. on E.*

Proof: Let $E_1 = \{\omega \in E : f(\omega) = +\infty\}$. For each $n \in \mathbb{N}$ we have $nI_{E_1} \leq f$ on E, hence $\bar{\mathcal{I}}_E(nI_{E_1}) \leq \bar{\mathcal{I}}_E(f)$ by (N4). This gives $n\mu(E_1) \leq \bar{\mathcal{I}}_E(f)$ for all $n \in \mathbb{N}$. Since $\bar{\mathcal{I}}_E(f)$ is finite, it follows that $\mu(E_1) = 0$. By definition, it now follows that f is finite μ-a.e. on E. ■

We now extend part (a) of (N8) to countable sums. To do this, we need a preliminary fact. If $\{f_n\}_{n=1}^{\infty}$ denotes a sequence of elements of $\mathfrak{N}$, then the sequence $\{\sum_{i=1}^{n} f_i\}_{n=1}^{\infty}$ is a nondecreasing sequence of elements of $\mathfrak{N}$, and its limit $\sum_{i=1}^{\infty} f_i$ is nonnegative and $\mathcal{F}/\mathcal{B}^*$-measurable by (MF11) from Chapter 5, hence $\sum_{n=1}^{\infty} f_n \in \mathfrak{N}$. We may therefore discuss the quantity $\bar{\mathcal{I}}_E(\sum_{n=1}^{\infty} f_n)$.

(N12) *Let $\{f_n\}_{n=1}^{\infty}$ denote a sequence of elements of $\mathfrak{N}$. Then the equation $\bar{\mathcal{I}}_E(\sum_{n=1}^{\infty} f_n) = \sum_{n=1}^{\infty} \bar{\mathcal{I}}_E(f_n)$ holds.*

Proof: Let $g_n = \sum_{i=1}^{n} f_i$ for each $n \in \mathbb{N}$. Observe that $\{g_n\}_{n=1}^{\infty}$ is a nondecreasing sequence of elements of $\mathfrak{N}$. By MCT and (N8), we have

$$\bar{\mathcal{I}}_E\left(\sum_{n=1}^{\infty} f_n\right) = \bar{\mathcal{I}}_E\left(\lim_n g_n\right) = \lim_n \bar{\mathcal{I}}_E(g_n) = \lim_n \sum_{m=1}^{n} \bar{\mathcal{I}}_E(f_m) = \sum_{n=1}^{\infty} \bar{\mathcal{I}}_E(f_n),$$

whence the proof is complete. ■

As we generalized properties (N3) and (N4), we may generalize (N12) to the following result; the proof is easy and is left to the reader.

(N12′) *Let $\{f_n\}_{n=1}^{\infty}$ denote a sequence of functions in $\mathfrak{N}$. Let $g \in \mathfrak{N}$ be such that $g = \sum_{n=1}^{\infty} f_n$ μ-a.e. on E. Then $\bar{\mathcal{I}}_E(g) = \sum_{n=1}^{\infty} \bar{\mathcal{I}}_E(f_n)$.*

The following property is known as *Fatou's Lemma*, and it is the main piece in the *Dominated Convergence Theorem* to be found in the next section.

(N13) *Let $\{f_n\}_{n=1}^{\infty}$ denote a sequence of elements of $\mathfrak{N}$. Then the inequality $\bar{\mathcal{I}}_E(\underline{\lim}_n f_n) \leq \underline{\lim}_n \bar{\mathcal{I}}_E(f_n)$ obtains.*

Proof: By (MF11) we have $\underline{\lim}_n f_n \in \mathfrak{N}$, hence we may discuss $\bar{\mathcal{I}}_E(\underline{\lim}_n f_n)$. Let $g_n = \inf\{f_n, f_{n+1}, \cdots\}$ for every $n \in \mathbb{N}$, so that $\{g_n\}_{n=1}^{\infty}$ is a nondecreasing sequence of functions in $\mathfrak{N}$ with $\lim_n g_n = \underline{\lim}_n f_n$ and $g_n \le f_n$ for all $n \in \mathbb{N}$, hence $\bar{\mathcal{I}}_E(g_n) \le \bar{\mathcal{I}}_E(f_n)$ for all $n \in \mathbb{N}$ by (N4). Invoking MCT, we have

$$\begin{aligned}\bar{\mathcal{I}}_E\left(\underline{\lim}_n f_n\right) = \bar{\mathcal{I}}_E\left(\lim_n g_n\right) = \lim_n \bar{\mathcal{I}}_E(g_n) &= \underline{\lim}_n\ \bar{\mathcal{I}}_E(g_n) \\ &\le \underline{\lim}_n\ \bar{\mathcal{I}}_E(f_n).\end{aligned}$$

The third equality follows because $\{\bar{\mathcal{I}}_E(g_n)\}_{n=1}^{\infty}$ is nondecreasing and hence has a limit [which is equal to the inferior limit]. This completes the proof. ∎

As with (N12), we may generalize Fatou's Lemma: *Under the hypothesis of (N13) above, we have* $\bar{\mathcal{I}}_E(g) \le \underline{\lim}_n \bar{\mathcal{I}}_E(f_n)$, *where* $g \in \mathfrak{N}$ *is such that* $g = \underline{\lim}_n f_n$ μ*-a.e. on* E. The proof of this claim follows from (N13) in the same way that (N12′) follows from (N12).

We now turn to some "miscellaneous" properties involving our functional that will serve us at one time or another. They are presented in no special order, and are generally straightforward.

(N14) *Let* ν_f *be as in (N9). Then* $\bar{\mathcal{I}}_E(g; \nu_f) = \bar{\mathcal{I}}_E(fg, \mu)$.

Proof: We will first prove the theorem when $g \in \mathfrak{S}$, and then we will invoke MCT to prove the theorem for the more general case when $g \in \mathfrak{N}$. To start things, let $g \in \mathfrak{S}$, writing $g = \sum_{i=1}^n c_i I_{B_i}$, where $B_1, \cdots, B_n \in \mathcal{F}$ are disjoint with union Ω. Then by (N1), (N2), and (N8), we have

$$\begin{aligned}\bar{\mathcal{I}}_E(g; \nu_f) = \sum_{i=1}^n c_i \nu_f(B_i \cap E) = \sum_{i=1}^n c_i \bar{\mathcal{I}}_{B_i \cap E}(f; \mu) &= \sum_{i=1}^n c_i \bar{\mathcal{I}}_E(f I_{B_i}; \mu) \\ &= \bar{\mathcal{I}}_E\left(f \sum_{i=1}^n c_i I_{B_i}; \mu\right) \\ &= \bar{\mathcal{I}}_E(fg; \mu).\end{aligned}$$

This proves the theorem when $g \in \mathfrak{S}$.

For general $g \in \mathfrak{N}$, let $\{s_n\}_{n=1}^{\infty}$ denote a nondecreasing sequence of functions in $\mathfrak{S}_g$ with $\lim_n s_n = g$ in accordance with (MF18). Then, using MCT twice and using what has just been shown, we have

$$\begin{aligned}\bar{\mathcal{I}}_E(g; \nu_f) = \bar{\mathcal{I}}_E\left(\lim_n s_n; \nu_f\right) = \lim_n \bar{\mathcal{I}}_E(s_n; \nu_f) &= \lim_n \bar{\mathcal{I}}_E(f s_n; \mu) \\ &= \bar{\mathcal{I}}_E\left(\lim_n f s_n; \mu\right) \\ &= \bar{\mathcal{I}}_E(fg; \mu),\end{aligned}$$

which finishes the proof. ∎

The function f above is often denoted by the formalism $\frac{d\nu_f}{d\mu}$, or one writes the formalism $d\nu_f = f\, d\mu$. There is no independent meaning ascribed to either of the symbols $d\nu_f$ or $d\mu$. This notation merely means that f is such that the conclusion of the previous property holds. Such notational formalisms and their motivation will be thoroughly discussed in our treatment of the Radon-Nikodym Theorem in Chapter 9.

The preceding results are consequences of MCT, and it is hoped that the earlier statement that MCT is perhaps the most profound result in the book is not seen as too much of an overstatement.

The following result shows that $\bar{\mathcal{I}}_E$ is continuous in a certain sense.

(N15) *Suppose that $\bar{\mathcal{I}}_E(f) < +\infty$ and $\epsilon > 0$. Then there exists $\delta > 0$ with the following property: any $F \subseteq E$ with $\mu(F) < \delta$ is such that $\bar{\mathcal{I}}_F(f) < \epsilon$.*

Proof: Let $\epsilon > 0$. For each $n \in \mathbb{N}$, define $\phi_n : \Omega \to \bar{\mathbb{R}}$ by writing

$$\phi_n(\omega) = \begin{cases} f(\omega) & \text{if } \omega \in \Omega \text{ is such that } f(\omega) \le n \\ n & \text{if } \omega \in \Omega \text{ is such that } f(\omega) > n \end{cases}$$

for every $\omega \in \Omega$. Observe by (MF6) that $\{\phi_n\}_{n=1}^{\infty}$ is a nondecreasing sequence of functions in $\mathfrak{N}$ such that $\lim_n \phi_n$ exists and equals f. Also, $\phi_n \le f$ for each $n \in \mathbb{N}$, hence $\bar{\mathcal{I}}_E(\phi_n) \le \bar{\mathcal{I}}_E(f)$ for each $n \in \mathbb{N}$ by (N4). By MCT, it follows that

$$\lim_n \bar{\mathcal{I}}_E(\phi_n) = \bar{\mathcal{I}}_E\left(\lim_n \phi_n\right) = \bar{\mathcal{I}}_E(f).$$

Since $\bar{\mathcal{I}}_E(f) < +\infty$ by assumption, the above says that there is $n \in \mathbb{N}$ such that $\bar{\mathcal{I}}_E(f) - \bar{\mathcal{I}}_E(\phi_n) < \frac{\epsilon}{2}$, which forces $\bar{\mathcal{I}}_E(f - \phi_n) < \frac{\epsilon}{2}$ by (N8). Take $\delta = \frac{\epsilon}{2n}$; we claim that any $F \subseteq E$ with $F \in \mathcal{F}$ and $\mu(F) < \delta$ is such that $\bar{\mathcal{I}}_F(f) < \epsilon$. To see this, let $F \subseteq E$ with $F \in \mathcal{F}$ and $\mu(F) < \delta = \frac{\epsilon}{2n}$. Observing that $\phi_n \le n$ [the function on Ω that is identically equal to n], we have

$$\bar{\mathcal{I}}_F(\phi_n) \le \bar{\mathcal{I}}_F(n) = n\mu(F) < n\frac{\epsilon}{2n} = \frac{\epsilon}{2}$$

by (N4). Therefore, (N8) and (N6) imply

$$\bar{\mathcal{I}}_F(f) = \bar{\mathcal{I}}_F((f - \phi_n) + \phi_n) = \bar{\mathcal{I}}_F(f - \phi_n) + \bar{\mathcal{I}}_F(\phi_n) < \frac{\epsilon}{2} + \frac{\epsilon}{2} = \epsilon.$$

This shows that our choice of δ works, and the proof is complete. ∎

The upcoming result says that in a certain sense the values of the functional $\bar{\mathcal{I}}_E(f)$ over all $E \in \mathcal{F}$ [while $f \in \mathfrak{N}$ is fixed] "determine" f up to a μ-null set.

(N16) *Suppose that $(\Omega, \mathcal{F}, \mu)$ is a σ-finite measure space, and further suppose that $\bar{\mathcal{I}}_E(f) = \bar{\mathcal{I}}_E(g)$ for every $E \in \mathcal{F}$. Then $f = g$ μ-a.e. on Ω.*

Proof: Pick and fix arbitrary $f, g \in \mathfrak{N}$ but merely assume that $\bar{\mathcal{I}}_E(f) \le \bar{\mathcal{I}}_E(g)$ for each $E \in \mathcal{F}$. We will show that

$$\mu(\{\omega \in \Omega : 0 \le g(\omega) < f(\omega),\ g(\omega) < +\infty\}) = 0. \tag{6.7}$$

Noting that the set in (6.7) coincides with $\{\omega \in \Omega : 0 \leq g(\omega) < f(\omega)\}$, it will follow that $f \leq g$ μ-a.e. on Ω. By then reversing the roles of f and g, we will obtain $g \leq f$ μ-a.e. on Ω, and hence we will have $f = g$ μ-a.e. on Ω.

Therefore, everything hinges on assuming that $\bar{\mathcal{I}}_E(f) \leq \bar{\mathcal{I}}_E(g)$ for every $E \in \mathcal{F}$ and obtaining (6.7) as a consequence. Since $(\Omega, \mathcal{F}, \mu)$ is a σ-finite measure space, we may write Ω as the union of a nondecreasing sequence $\{A_n\}_{n=1}^{\infty}$ of $\mathcal{F}$-sets such that $\mu(A_n) < +\infty$ for each $n \in \mathbb{N}$. For each $n \in \mathbb{N}$, define

$$B_n = \{\omega \in \Omega : 0 \leq g(\omega) < f(\omega),\ g(\omega) \leq n\}.$$

Observe for each $n \in \mathbb{N}$ that $\mu(A_n \cap B_n) \leq \mu(A_n) < +\infty$ and $g \leq n$ on all of $A_n \cap B_n$, hence

$$\bar{\mathcal{I}}_{A_n \cap B_n}(f) \leq \bar{\mathcal{I}}_{A_n \cap B_n}(g) \leq \bar{\mathcal{I}}_{A_n \cap B_n}(n) = n\mu(A_n \cap B_n) < +\infty,$$

where the first inequality follows by hypothesis and the second follows by (N4). Therefore, for each $n \in \mathbb{N}$, the subtraction $\bar{\mathcal{I}}_{A_n \cap B_n}(f) - \bar{\mathcal{I}}_{A_n \cap B_n}(g)$ makes sense. Next, (N2), (N8), and the fact that $I_{A_n \cap B_n}(f - g) \in \mathfrak{N}$ yield

$$\begin{aligned} \bar{\mathcal{I}}_{A_n \cap B_n}(f) - \bar{\mathcal{I}}_{A_n \cap B_n}(g) &= \bar{\mathcal{I}}(I_{A_n \cap B_n} f) - \bar{\mathcal{I}}(I_{A_n \cap B_n} g) \\ &= \bar{\mathcal{I}}(I_{A_n \cap B_n}(f - g)). \end{aligned} \tag{6.8}$$

The left side of (6.8) is nonpositive, while the right side of (6.8) is nonnegative. It follows that $\bar{\mathcal{I}}(I_{A_n \cap B_n}(f - g)) = 0$ for every $n \in \mathbb{N}$. By (N10), we have for each $n \in \mathbb{N}$ that $I_{A_n \cap B_n}(f - g) = 0$ μ-a.e. on Ω. However, $f - g > 0$ on $A_n \cap B_n$ for each $n \in \mathbb{N}$, so it must be the case that $I_{A_n \cap B_n} = 0$ μ-a.e. on Ω for each $n \in \mathbb{N}$, hence $\mu(A_n \cap B_n) = \bar{\mathcal{I}}(I_{A_n \cap B_n}) = 0$ for each $n \in \mathbb{N}$ [by (N10)]. Since the sequence $\{A_n \cap B_n\}_{n=1}^{\infty}$ is increasing with limit equal to the set in (6.7), and since $\mu(A_n \cap B_n) = 0$ for each $n \in \mathbb{N}$, measure property (M8) gives the equation (6.7), which proves the result. ■

For the next property, again let $(\Omega, \mathcal{F}, \mu)$ denote a measure space. Let $(\Omega', \mathcal{F}')$ denote a measurable space, and suppose that $T : \Omega \to \Omega'$ is $\mathcal{F}/\mathcal{F}'$-measurable, with $\mu \circ T^{-1}$ denoting the measure on $\mathcal{F}'$ induced by μ. [See Section 5.8 for the relevant details.] In a descriptive and symbolic form whose only goal is to clarify, we might write the following:

$$(\Omega, \mathcal{F}, \mu) \xrightarrow{T} (\Omega', \mathcal{F}', \mu \circ T^{-1}) \xrightarrow{f} (\bar{\mathbb{R}}, \mathcal{B}^*).$$

In the picture above, we can consider two mappings into $\bar{\mathbb{R}}$: (i) the mapping f defined on Ω', and (ii) the mapping $f \circ T$ defined on Ω. When we write the symbol $\bar{\mathcal{I}}(f \circ T; \mu)$, we are talking about this section's functional relative to $(\Omega, \mathcal{F}, \mu)$. When we write $\bar{\mathcal{I}}(f; \mu \circ T^{-1})$, we are talking about this section's functional relative to $(\Omega', \mathcal{F}', \mu \circ T^{-1})$. The following result is known in some quarters as the *Change of Variable Theorem*, not to be confused with another theorem of the same name used in calculus to make multivariate substitutions in multiple integrals. The main point is that $\bar{\mathcal{I}}(f \circ T; \mu)$ and $\bar{\mathcal{I}}(f; \mu \circ T^{-1})$ coincide.

(N17) *Let f denote a nonnegative $\mathcal{F}'/\mathcal{B}^*$-measurable function. We have the two equalities*

$$\bar{\mathcal{I}}(f \circ T; \mu) = \bar{\mathcal{I}}(f; \mu \circ T^{-1}) \tag{6.9}$$

and

$$\bar{\mathcal{I}}_{T^{-1}(A')}(f \circ T; \mu) = \bar{\mathcal{I}}_{A'}(f; \mu \circ T^{-1}) \quad \textit{for each } A' \in \mathcal{F}'. \tag{6.10}$$

Proof: We give a sketch of the proof and trust the reader to fill in the straightforward details as an exercise. Let $f = I_{A'}$ where $A' \in \mathcal{F}$. Then $f \circ T = I_{T^{-1}(A')}$, and both sides of (6.9) are equal to $\mu(T^{-1}(A'))$. The reader may employ (N8) to show that (6.9) holds when $f \in \mathfrak{S}(\Omega', \mathcal{F}', \mu \circ T^{-1})$. Next, let $f \in \mathfrak{N}(\Omega', \mathcal{F}', \mu \circ T^{-1})$. Then, letting $\{s_n\}_{n=1}^{\infty}$ denote a nondecreasing sequence in $\mathfrak{S}_f(\Omega', \mathcal{F}', \mu \circ T^{-1})$ with $s_n \to f$ [courtesy of (MF18)], we have that $\{s_n \circ T\}_{n=1}^{\infty}$ is a nondecreasing sequence of simple functions with limit $f \circ T$. Applying MCT yields the result (6.9). The proof of (6.10) follows by replacing f by $f I_{A'}$ in (6.9). ■

The proof above has a structure that is common for proofs regarding our functional here and in later sections: show the claim is true for simple functions, then use MCT and (MF18) to show that the claim is true for general nonnegative functions.

We have proved many properties regarding the functional $\bar{\mathcal{I}}_E$ with domain $\mathfrak{N}$, and much of the work of this chapter is now done. In the next section, we will extend $\bar{\mathcal{I}}_E$ to handle general $\mathcal{F}/\mathcal{B}^*$-measurable f defined on Ω such that $\bar{\mathcal{I}}_E(f^+) < +\infty$ or $\bar{\mathcal{I}}_E(f^-) < +\infty$, where f^+ and f^- denote the positive and negative parts of f as described in Section 3 of Chapter 5. Most [but not all] of what follows comes from applying properties (N1)-(N17) separately to the nonnegative functions f^+ and f^-.

Exercises.

As with the exercises in the previous section, let $(\Omega, \mathcal{F}, \mu)$ denote a generic measure space. As with the last section, most if not all of the exercises should be worked for maximal understanding, as some of these results will be invoked in later work. All functions are assumed to be in $\mathfrak{N}$, and all sets are assumed to be in $\mathcal{F}$. Exercises 6 and 7 should be worked.

1*. Some of the properties discussed in this section may be modified on an "almost everywhere" basis in a simple fashion. Using the original properties in any way desired, quickly prove the following modified properties, where the notation is the same as in the section's exposition.

(a) (N3′) If $f = g$ μ-a.e. on E, then $\bar{\mathcal{I}}_F(f) = \bar{\mathcal{I}}_F(g)$ for any $F \subseteq E$.
(b) (N4′) If $f \le g$ μ-a.e. on E, then $\bar{\mathcal{I}}_F(f) \le \bar{\mathcal{I}}_F(g)$ for any $F \subseteq E$.
(c) (N7′) If $f, f_1, f_2, \cdots \in \mathfrak{N}$ and $f = \lim_n f_n$ μ-a.e. on E, then $\bar{\mathcal{I}}_E(f) = \lim_n \bar{\mathcal{I}}_E(f_n)$.
(d) If $f \le M$ μ-a.e. on E for some $M \in [0, +\infty]$, then $\bar{\mathcal{I}}_E(f) \le M\mu(E)$.

2*. It was claimed in (N5) that $\bar{\mathcal{I}}_E(cf) = c\bar{\mathcal{I}}_E(f)$ for every $c \in [0, +\infty)$. This result in fact holds if $c = +\infty$ as well: $\bar{\mathcal{I}}_E(+\infty f) = +\infty\bar{\mathcal{I}}_E(f)$. Therefore, (N5) holds for all $c \in [0, +\infty]$. Similarly, we may allow the numbers $c_1, \cdots, c_n$ to be in $[0, +\infty]$ in the statement (N8).

3. This exercise concerns Fatou's Lemma.
(a) Let $\{A_n\}_{n=1}^{\infty}$ denote a sequence of $\mathcal{F}$-sets. Show that $\mu(\underline{\lim}\, A_n) \le \underline{\lim}\, \mu(A_n)$ by using Fatou's Lemma applied to the sequence of indicator functions $\{I_{A_n}\}_{n=1}^{\infty}$. [You will need to recall or show that $I_{\underline{\lim}\, A_n} = \underline{\lim}\, I_{A_n}$.]
(b) Consider $(\mathbb{R}, \mathcal{B}, \lambda)$. If $s_n = n^2 I_{[0,\frac{1}{n}]}$ for each $n \in \mathbb{N}$, then $\bar{\mathcal{I}}(\underline{\lim}\, s_n) = 0$ while $\underline{\lim}\, \bar{\mathcal{I}}(s_n) = +\infty$, hence strict inequality may hold in Fatou's Lemma.
(c) In (b), with $E \in \mathcal{B}$, the sequence $I_E, 1 - I_E, I_E, 1 - I_E, \cdots$ provides another example where strict inequality holds in Fatou's Lemma.
(d) It is not necessarily the case that $\overline{\lim}\, \bar{\mathcal{I}}_E(f_n) \le \bar{\mathcal{I}}_E(\overline{\lim}\, f_n)$ if $\mu(E) = +\infty$. However, if $\mu(E) < +\infty$, the inequality holds, hence we have

$$\bar{\mathcal{I}}_E(\underline{\lim}\, f_n) \le \underline{\lim}\, \bar{\mathcal{I}}_E(f_n) \le \overline{\lim}\, \bar{\mathcal{I}}_E(f_n) \le \bar{\mathcal{I}}_E(\overline{\lim}\, f_n)$$

by putting everything together.

4. Let Ω denote an arbitrary nonempty set, and fix attention upon a particular $\omega_0 \in \Omega$. Let $\mathcal{F} = 2^{\Omega}$, and define $\mu : \mathcal{F} \to \bar{\mathbb{R}}$ by writing $\mu(A) = 1$ if $\omega_0 \in A$ and $\mu(A) = 0$ if $\omega_0 \notin A$.
(a) $(\Omega, \mathcal{F}, \mu)$ is a measure space.
(b) Every $f : \Omega \to \mathbb{R}$ is $\mathcal{F}/\mathcal{B}^*$-measurable.
(c) Let $E \in \mathcal{F}$ and $f \in \mathfrak{N}$. Then $\bar{\mathcal{I}}_E(f) = f(\omega_0) I_E(\omega_0)$.

5. Let Ω denote an uncountable set, and let $\mathcal{F} = \{A \subseteq \Omega :\ A \text{ is amc or } A^c \text{ is amc}\}$. Define $\mu : \mathcal{F} \to \bar{\mathbb{R}}$ for all $A \in \mathcal{F}$ by writing $\mu(A) = 0$ if A is amc and $\mu(A) = 1$ if A^c is amc. [μ is well-defined since Ω is uncountable.]
(a) $(\Omega, \mathcal{F}, \mu)$ is a measure space.
(b) $f : \Omega \to \bar{\mathbb{R}}$ is $\mathcal{F}/\mathcal{B}^*$-measurable iff there is $c \in \bar{\mathbb{R}}$ [depending on f] with $f = c$ μ-a.e. on Ω.

Outline: It should be straightforward to show that if $f = c$ μ-a.e. on Ω, then f is $\mathcal{F}/\mathcal{B}^*$-measurable. To show the more difficult implication, suppose that f is $\mathcal{F}/\mathcal{B}^*$-measurable. Let $C = \{t \in \bar{\mathbb{R}} : f^{-1}([-\infty, t)) \text{ is amc}\}$. We have $-\infty \in C$, and if $-\infty < b < a$ with $a \in C$, then $b \in C$ as well. Therefore, $C = \{-\infty\}$ or C is some type of unbounded interval containing $-\infty$. If $C = \{-\infty\}$, then $f^{-1}([-\infty, t))$ is not amc for any $t > -\infty$, hence $f^{-1}([t, +\infty])$, which is in $\mathcal{F}$ since f is $\mathcal{F}/\mathcal{B}^*$-measurable, must be amc for each $t > -\infty$. Hence $\mu(\{\omega \in A : f(\omega) \neq -\infty\}) = 0$, so $f = -\infty$ μ-a.e. on Ω. If C is an interval containing $-\infty$, let $c = \sup C$, so that $-\infty < c \le +\infty$. Let $t_1 \le t_2 \le \cdots$ be such that $t_n \to c$ and $t_n < c$ for each $n \in \mathbb{N}$. Then $f^{-1}([-\infty, t_n))$ is amc for each $n \in \mathbb{N}$; from this deduce that $f^{-1}([-\infty, c))$ is amc, hence $c \in C$. If $c = +\infty$, then A itself is amc, hence $\mu(A) = 0$ and the conclusion is trivially satisfied. Therefore, let $c < +\infty$. Since $c = \sup C$, any $d > c$ is such that $f^{-1}([-\infty, d))$ is not amc. Since f is $\mathcal{F}/\mathcal{B}^*$-measurable, we have $f^{-1}([d, +\infty])$ is amc. Now let $t_1 \ge t_2 \ge \cdots$ be such that $t_n \to c$ and $t_n > c$ for each $n \in \mathbb{N}$. Then $f^{-1}([t_n, +\infty])$ is amc for each $n \in \mathbb{N}$; deduce that $f^{-1}((c, +\infty])$ is amc. Therefore, since $f^{-1}([-\infty, c))$ is amc, we have $\mu(\{\omega \in A : f(\omega) \neq c\}) = 0$.
(c) Let $E \in \mathcal{F}$ and $f \in \mathfrak{N}$, then $\bar{\mathcal{I}}_E(f) = c\mu(E)$, where c is the constant such that $f = c$ μ-a.e. on Ω [as given in (b)].

6*. This exercise is necessary for the next exercise. Let Ω denote an arbitrary nonempty set, let $A \subseteq \Omega$, and let $f : \Omega \to [0, +\infty]$ be given. Write

$$\sum_{\omega \in A}^{*} f(\omega) = \sup \left\{ \sum_{\omega \in F} f(\omega) : \ F \subseteq A, \ F \text{ finite} \right\}.$$

In other words, the left side above is defined by the expression on the right side.
(a) Suppose $A = \{\omega_1, \cdots, \omega_n\}$ [a finite set]. Then $\sum_{\omega \in A}^{*} f(\omega) = \sum_{i=1}^{n} f(\omega_i)$, hence the definition above is consistent with what we're used to in the finite case.
(b) Suppose $A = \{\omega_1, \omega_2, \cdots\}$ [a countable set]. Then $\sum_{\omega \in A}^{*} f(\omega) = \sum_{i=1}^{\infty} f(\omega_i)$, where $\sum_{i=1}^{\infty} f(\omega_i)$ denotes summation in the sense that we're already used to [and $+\infty$ is allowed as sum]. Therefore, the definition is consistent with what we're used to in the at most countable case.

Comment: The definition above lets us extend the idea of sums from amc sets to uncountable sets in a way that is consistent with what we're used to in the at most countable situation. The need for this extension comes in discussing integration with respect to counting measure in the next exercise.

7*. Let Ω denote a nonempty set, let $\mathcal{F} = 2^{\Omega}$, and let $\mu : \mathcal{F} \to \bar{\mathbb{R}}$ be such that $\mu(A) =$ the number of points in A when A is finite, and $\mu(A) = +\infty$ otherwise.
(a) $(\Omega, \mathcal{F}, \mu)$ is a measure space. The measure μ is called the *counting measure* since μ "counts" the number of points in each $\mathcal{F}$-set.
(b) Every $f : \Omega \to \bar{\mathbb{R}}$ is $\mathcal{F}/\mathcal{B}^*$-measurable.
(c) Given any $E \in \mathcal{F}$ and $f \in \mathfrak{N}$, we have $\bar{\mathcal{I}}_E(f) = \sum_{\omega \in E}^{*} f(\omega)$.

Outline for (c): First show that if $f \in \mathfrak{S}$ and $f > 0$ only on a finite subset of E, then $\mathcal{I}_E(f) = \sum_{\omega \in E}^{*} f(\omega)$. Now let $f \in \mathfrak{N}$ and let $F \subseteq E$ be finite. Then $f I_F \in \mathfrak{S}$ and $f > 0$ only on a finite subset of F, hence

$$\bar{\mathcal{I}}_E(f) \geq \bar{\mathcal{I}}_E(f I_F) = \bar{\mathcal{I}}_F(f) = \sum_{\omega \in F}^{*} f(\omega) = \sum_{\omega \in F} f(\omega)$$

by what was shown at the beginning and the previous exercise. Since F was an arbitrary finite subset of E, we have $\bar{\mathcal{I}}_E(f) \geq \sum_{\omega \in E}^{*} f(\omega)$. This gives one inequality, so one must prove that $\bar{\mathcal{I}}_E(f) \leq \sum_{\omega \in E}^{*} f(\omega)$. If $\sum_{\omega \in E}^{*} f(\omega) = +\infty$, the previous inequality forces the result, so we may assume that $\sum_{\omega \in E}^{*} f(\omega) < +\infty$. For any $s \in \mathfrak{S}_f$ we have $\sum_{\omega \in E}^{*} s(\omega) \leq \sum_{\omega \in E}^{*} f(\omega) < +\infty$; from this deduce that $s > 0$ only on a finite subset of E, hence, we may find finite $F \subseteq E$ with $s = 0$ on $E - F$. Verify that

$$\mathcal{I}_E(s) = \sum_{\omega \in E}^{*} s(\omega) = \sum_{\omega \in F}^{*} s(\omega) \leq \sum_{\omega \in F}^{*} f(\omega) \leq \sum_{\omega \in E}^{*} f(\omega).$$

Since the above holds for any $s \in \mathfrak{S}_f$, it follows that $\bar{\mathcal{I}}_E(f) \leq \sum_{\omega \in E}^{*} f(\omega)$, which gives the reverse inequality.

Comment: Usually, the symbol $\sum_{\omega \in E}^{*} f(\omega)$ is written as $\sum_{\omega \in E} f(\omega)$. This is perfectly fine if E is at most countable, but a slight solecism is committed when E is uncountable. Nonetheless, the convention is to omit the extra * superscript.

8. The previous exercise regarding counting measure has some interesting applications to statements regarding infinite series.

(a) If $a_{ij} \geq 0$ for each $i, j \in \mathbb{N}$, then

$$\sum_{i=1}^{\infty}\sum_{j=1}^{\infty} a_{ij} = \sum_{j=1}^{\infty}\sum_{i=1}^{\infty} a_{ij} = \sum_{(i,j)\in\mathbb{N}^2} a_{ij},$$

where the last expression above is a standard unordered sum from advanced calculus. [This is a result that is demonstrated in advanced calculus courses independently of any measure or integration theory, and hence our usage of it before this exercise is not in any way circular. Here, the idea is to prove this by constructing the appropriate space and functions and using the results of the previous exercise.]
(b) Let $\{a_{nm} : n, m \in \mathbb{N}\}$ denote a double sequence of nonnegative numbers. Show that $\sum_{n=1}^{\infty} \underline{\lim}_m a_{nm} \leq \underline{\lim}_m \sum_{n=1}^{\infty} a_{nm}$.
Comment: For (a), let $\Omega = \{\omega_{ij} : i, j \in \mathbb{N}\}$, $\mathcal{F} = 2^{\Omega}$, and define $f : \Omega \to [0, +\infty]$ by writing $f(\omega_{ij}) = a_{ij}$ for each $i, j \in \mathbb{N}$. Then $f \in \mathfrak{N}$ and

$$\bar{\mathcal{I}}(f) = \sum_{\omega_{ij}\in\Omega} f(\omega_{ij}) = \sum_{i,j\in\mathbb{N}} a_{ij}.$$

Let $A_i = \{a_{i1}, a_{i2}, \cdots\}$ for each $i \in \mathbb{N}$, so that $A_1, A_2, \cdots \in \mathcal{F}$ are disjoint with $\bigcup_{i=1}^{\infty} A_i = \Omega$. Observe that

$$\bar{\mathcal{I}}(f) = \bar{\mathcal{I}}\left(\sum_{i=1}^{\infty} f I_{A_i}\right) = \sum_{i=1}^{\infty} \bar{\mathcal{I}}_{A_i}(f) = \sum_{i=1}^{\infty} \sum_{\omega_{ij}\in A_i} f(\omega_{ij}) = \sum_{i=1}^{\infty}\sum_{j=1}^{\infty} a_{ij}.$$

Similar reasoning gives $\bar{\mathcal{I}}(f) = \sum_{j=1}^{\infty}\sum_{i=1}^{\infty} a_{ij}$.

Comment for the next three exercises: The following three exercises are important in that they explore equivalent definitions of the operator $\bar{\mathcal{I}}$ as given in this section. Different books on analysis and the Lebesgue integral define the Lebesgue integral differently, and it is useful to see that these alternative definitions are in fact equivalent definitions.

9. Consider the measure space $(\Omega, \mathcal{F}, \mu)$, and let $E \in \mathcal{F}$. For each $f \in \mathcal{N}$, define

$$\bar{\mathcal{J}}_E(f) = \sup \sum_{i=1}^{n} \inf\{f(\omega) : \omega \in A_i \cap E\}\mu(A_i \cap E),$$

where the supremum is taken over all finite disjoint collections $\{A_1, \cdots, A_n\}$ of $\mathcal{F}$-sets with $\bigcup_{i=1}^{n} A_i = \Omega$. Then $\bar{\mathcal{J}}_E(f) = \bar{\mathcal{I}}_E(f)$, so that this is an equivalent formulation of what we have done in this section. [Recall the arithmetical conventions $+\infty \times 0 = 0 \times +\infty = 0$ and $\inf \varnothing = +\infty$ for the case when some $A_i = \varnothing$.]

10. Consider the measure space $(\Omega, \mathcal{F}, \mu)$, and let $E \in \mathcal{F}$ and $f \in \mathfrak{N}$. Define $\bar{\mathcal{J}}_E(f) = \lim_n \mathcal{I}_E(s_n)$, where $\{s_n\}_{n=1}^{\infty}$ is a nondecreasing sequence of functions in $\mathfrak{S}_f$ such that $s_n \to f$. We will show that $\bar{\mathcal{J}}_E(f) = \bar{\mathcal{I}}_E(f)$ *without using MCT*, so that this definition could have instead served as our definition.
(a) We first want to show that $\bar{\mathcal{J}}_E(f)$ is well-defined, since there are possibly many nondecreasing sequences $\{s_n\}_{n=1}^{\infty}$ of elements in $\mathfrak{S}_f$ with $s_n \to f$. Let $t \in \mathfrak{S}_f$, so that $t \leq \lim_n s_n = f$. If $\mathcal{I}_E(t) = +\infty$, then $\lim_n \mathcal{I}_E(s_n) = +\infty$.

(b) If $\mathcal{I}_E(t) < +\infty$, then $\lim_n \mathcal{I}_E(s_n) \geq \mathcal{I}_E(t)$. This with (a) gives the inequality $\lim_n \mathcal{I}_E(s_n) \geq \mathcal{I}_E(t)$ in any case.
(c) Now we may show that this definition is well-defined. Let $\{s_n\}_{n=1}^\infty$ and $\{t_n\}_{n=1}^\infty$ denote two nondecreasing sequences in $\mathfrak{S}_f$ with $\lim_n s_n = \lim_n t_n = f$. Show that $\lim_n \mathcal{I}_E(s_n) = \lim_n \mathcal{I}_E(t_n)$ by using the results of (a) and (b).
(d) Finally, show that $\bar{\mathcal{J}}_E(f) = \bar{\mathcal{I}}_E(f)$, which shows that that this definition is indeed truly equivalent to our definition. Observe that this alternative definition is defined to directly make MCT automatically true, so that MCT follows as a definition, but the price to be paid is in showing (a)-(c), that is, in showing that this alternative definition is well-defined.

Outline: For (a), write $t = \sum_{i=1}^k c_i I_{E_i}$. If $\mathcal{I}_E(t) = \sum_{i=1}^k c_i \mu(E_i \cap E) = +\infty$, then there exists $i^* \in \{1, \cdots, k\}$ such that $c_{i^*} > 0$ and $\mu(E_{i^*} \cap E) = +\infty$. Choose $\epsilon \in (0, c_{i^*})$, and for each $n \in \mathbb{N}$ write $A_n = \{\omega \in E : s_n(\omega) + \epsilon > t(\omega)\}$. Argue that $\{A_n \cap E_{i^*}\}_{n=1}^\infty$ is nondecreasing with limit $E_{i^*} \cap E$, hence $\lim_n \mu(A_n \cap E_{i^*})$ exists and equals $+\infty$. From this, show for all $n \in \mathbb{N}$ that

$$\mathcal{I}_E(s_n) \geq \mathcal{I}_E(s_n I_{A_n \cap E_{i^*}}) \geq (c_{i^*} - \epsilon)\mu(A_n \cap E_{i^*}),$$

hence $\lim_n \mathcal{I}_E(s_n) = +\infty = \mathcal{I}_E(t)$, giving (a).

For (b), assume that $\mathcal{I}_E(t) < +\infty$. Let $c = \min\{c_i : c_i > 0\}$ and

$$A = \{\omega \in E : t(\omega) > 0\} = \bigcup\{E_i : i \text{ such that } c_i > 0\}.$$

Argue that $t \in \mathfrak{S}$ forces $c > 0$ and $\mu(A) < +\infty$. Pick $\epsilon \in (0, c)$, and again write $A_n = \{\omega \in E : s_n(\omega) + \epsilon > t(\omega)\}$ for each $n \in \mathbb{N}$. Verify the following for each $n \in \mathbb{N}$:

$$\begin{aligned} \mathcal{I}_E(s_n) \geq \mathcal{I}_E(s_n I_{A_n \cap A}) &\geq \mathcal{I}_E((t - \epsilon) I_{A_n \cap A}) \\ &= \mathcal{I}_E(t I_{A_n \cap A}) - \epsilon\mu(A_n \cap A) \\ &\geq \mathcal{I}_E(t I_{A_n \cap A}) - \epsilon\mu(A), \end{aligned}$$

and then show that we have

$$\mathcal{I}_E(t I_{A_n \cap A}) - \epsilon\mu(A) \geq \mathcal{I}_E(t I_A) - \epsilon - \epsilon\mu(A)$$

for sufficiently large enough n by appealing to the fact that for each i with $c_i > 0$ we have $\lim_n \mu(A_n \cap E_i) = \mu(E_i)$. Recalling the definition of A, and recalling that $\epsilon \in (0, c)$ was arbitrary, argue that this forces $\mathcal{I}_E(s_n) \geq \mathcal{I}_E(t)$ for sufficiently large n, whence $\lim_n \mathcal{I}_E(s_n) \geq \mathcal{I}_E(t)$.

For (c), let $s_1 \leq s_2 \leq \cdots \to f$ and $t_1 \leq t_2 \leq \cdots \to f$. For each $m \in \mathbb{N}$ we have that $f = \lim_n s_n \geq t_m$, hence what has been shown in (a) and (b) forces $\lim_n \mathcal{I}_E(s_n) \geq \mathcal{I}_E(t_m)$ for each $m \in \mathbb{N}$, hence $\lim_n \mathcal{I}_E(s_n) \geq \lim_m \mathcal{I}_E(t_m)$ follows upon letting $m \to +\infty$. The symmetry of the problem in the sequences $\{s_n\}_{n=1}^\infty$ and $\{t_m\}_{m=1}^\infty$ instantly yields the reverse inequality as well, giving (c).

11. Let $(\Omega, \mathcal{F}, \mu)$ denote a measure space with $E \in \mathcal{F}$ such that $\mu(E) < +\infty$, and let $f \in \mathfrak{N}$. Lebesgue's original version of our operator is

$$\bar{\mathcal{J}}_E(f) = \lim_n \sum_{i=1}^{n2^n} \frac{i-1}{2^n} \mu\left(\left\{\omega \in E : \frac{i-1}{2^n} \leq f(\omega) < \frac{i}{2^n}\right\}\right).$$

(a) The preceding limit exists, hence the preceding definition makes sense.
(b) This definition is equivalent to our definition of $\bar{\mathcal{I}}_E(f)$. [Compare with (MF18) in Chapter 5.]

12*. Let Ω denote a nonempty set, and let $\mathcal{F}$ denote a σ-field on Ω. For each $n \in \mathbb{N}$, let μ_n denote a measure with domain $\mathcal{F}$. For each $n \in \mathbb{N}$, let $\sum_{i=1}^{n} \mu_i$ denote the measure that assigns the value $\sum_{i=1}^{n} \mu_i(A)$ to each $A \in \mathcal{F}$.
(a) Let $s \in \mathfrak{S}$ and $n \in \mathbb{N}$. Then $\mathcal{I}_E(s; \sum_{i=1}^{n} \mu_i) = \sum_{i=1}^{n} \mathcal{I}_E(s; \mu_i)$.
(b) Let $f \in \mathfrak{N}$ and $n \in \mathbb{N}$. Then $\bar{\mathcal{I}}_E(f; \sum_{i=1}^{n} \mu_i) = \sum_{i=1}^{n} \bar{\mathcal{I}}_E(f; \mu_i)$.

13. Let $A \in \mathcal{F}$ be nonempty, and suppose that $f : A \to [0, +\infty]$ is $\mathcal{F}/\mathcal{B}^*$-measurable. Let $E \in \mathcal{F}_A$, where $\mathcal{F}_A$ is the set of $\mathcal{F}$-subsets of A. Consider two programs:
Program 1: Extend f from A to Ω as follows: Let $f^* \in \mathfrak{N}$ be such that $f^* = f$ on A and $f^* = 0$ on A^c.
Program 2: Do not extend f from A to Ω as in Program 1, but instead view f as a function defined everywhere relative to the measure space $(A, \mathcal{F}_A, \mu_A)$, where μ_A is the restriction of μ from $\mathcal{F}$ to $\mathcal{F}_A$.

These two programs are equivalent in the sense that $\bar{\mathcal{I}}_E(f^*; \mu) = \bar{\mathcal{J}}_E(f; \mu_A)$, where $\bar{\mathcal{I}}_E$ is this section's functional relative to the measure space $(\Omega, \mathcal{F}, \mu)$, and $\bar{\mathcal{J}}_E$ is this section's functional relative to the measure space $(A, \mathcal{F}_A, \mu_A)$.

Comment: This exercise, while simple, is important with respect to logical consistency. Our definitions and results in the section exposition assumed that functions were everywhere defined on Ω, and such an approach may seem somewhat restrictive, since it may not be possible for some naturally occurring functions to be defined everywhere on Ω. These two exercises show that such functions are not excluded from consideration nor left out of the scope of the results of this section, since we are justified in "switching our perspective" from $(\Omega, \mathcal{F}, \mu)$ to an appropriate "sub-measure-space." Or, if we wish, we may equivalently just extend the domain of our function by defining it to be zero everywhere outside of the original domain, and this extended function now has domain Ω and the theory relative to $(\Omega, \mathcal{F}, \mu)$ applies. In either case, the corresponding functionals give the same numbers.

14*. Keep $(\Omega, \mathcal{F}, \mu)$ general. Suppose that $\mathcal{F}_0 \subseteq \mathcal{F}$ is a σ-field on Ω, and let μ_0 denote the restriction of μ to μ_0. [Then $(\Omega, \mathcal{F}_0, \mu_0)$ is clearly a measure space.] If a nonnegative $\mathcal{F}/\mathcal{B}^*$-measurable function f defined on Ω also happens to be $\mathcal{F}_0/\mathcal{B}^*$-measurable in addition, then $\bar{\mathcal{I}}_E(f; \mu) = \bar{\mathcal{I}}_E(f; \mu_0)$ for every $E \in \mathcal{F}_0$.

Comment: This exercise is important and should be worked. First, consider the case of $f = I_E$ where $E \in \mathcal{F}_0$. Then $\bar{\mathcal{I}}_E(f; \mu) = \mu(E) = \mu_0(E) = \bar{\mathcal{I}}_E(f; \mu_0)$. Next, consider the case of where f is a simple finite-valued nonnegative $\mathcal{F}_0/\mathcal{B}^*$-measurable function, and obtain the desired equality by using the previous result along with linearity. For the general case, let $\{s_n\}_{n=1}^{\infty}$ denote a nondecreasing sequence of simple finite-valued nonnegative $\mathcal{F}_0/\mathcal{B}^*$-measurable functions with limit f [use (MF18) to ensure the existence of such a sequence]. Use MCT and the result for simple functions to obtain the desired equality.

15*. We have the following claims related to (N16):
(a) If $\bar{\mathcal{I}}(f)$ and $\bar{\mathcal{I}}(g)$ are finite, and if $\bar{\mathcal{I}}_E(f) = \bar{\mathcal{I}}_E(g)$ for every $E \in \mathcal{F}$, then $f = g$ μ-a.e. on Ω. [Observe that here we don't need σ-finiteness because of the assumed finiteness. A simple proof is this: if $\bar{\mathcal{I}}_E(f) \leq \bar{\mathcal{I}}_E(g)$ for each $E \in \mathcal{F}$, then

$\bar{\mathcal{I}}(I_A(f-g)) = 0$, where $A = \{\omega \in \Omega : g(\omega) < f(\omega)\}$. Conclude that $\mu(A) = 0$. Reverse the roles of f and g for the result.]

(b) Suppose that $\mathcal{P}$ is a π-system that generates $\mathcal{F}$: $\sigma(\mathcal{P}) = \mathcal{F}$. If $\bar{\mathcal{I}}(f)$ and $\bar{\mathcal{I}}(g)$ are finite, and if $\bar{\mathcal{I}}_E(f) = \bar{\mathcal{I}}_E(g)$ for every $E \in \mathcal{P}$, then $f = g$ μ-a.e. on Ω.

16*. Fill in the straightforward details for the proof of (N17). [This should be completely mundane.]

6.3 STAGE THREE: GENERAL MEASURABLE FUNCTIONS

Consider a generic measure space $(\Omega, \mathcal{F}, \mu)$ with $E \in \mathcal{F}$. In Section 1, we defined the functional $\mathcal{I}_E$ on the class $\mathfrak{S}$ of simple finite-valued nonnegative $\mathcal{F}/\mathcal{B}^*$-measurable functions with domain Ω. In Section 2, we extended the functional $\mathcal{I}_E$ to $\bar{\mathcal{I}}_E$, and this extended functional had the larger domain $\mathfrak{N}$, the class of nonnegative $\mathcal{F}/\mathcal{B}^*$-measurable functions with domain Ω. The third stage in the construction consists of removing the restriction of nonnegativity in a fashion that is consistent with what has been done.

Definition. We will let $\mathfrak{M}$ denote the collection of $\mathcal{F}/\mathcal{B}^*$-measurable functions $f : \Omega \to \bar{\mathbb{R}}$. Given $E \in \mathcal{F}$, we will let $\mathfrak{M}(E)$ denote the collection of those $f \in \mathfrak{M}$ such that at least one of the nonnegative numbers $\bar{\mathcal{I}}_E(f^+)$ and $\bar{\mathcal{I}}_E(f^-)$ is finite. If confusion is possible, we will write $\mathfrak{M}(\Omega, \mathcal{F}, \mu)$ for $\mathfrak{M}$, and we will write $\mathfrak{M}(E, (\Omega, \mathcal{F}, \mu))$ for $\mathfrak{M}(E)$.

Observe that we now have $\mathfrak{S} \subseteq \mathfrak{N} \subseteq \mathfrak{M}$.

Recall that $f^+ = \max\{f, 0\} \geq 0$ and $f^- = -\min\{f, 0\} \geq 0$. By (MF13), f is $\mathcal{F}/\mathcal{B}^*$-measurable iff both f^+ and f^- are $\mathcal{F}/\mathcal{B}^*$-measurable. Therefore, $f \in \mathfrak{M}$ iff $f^+, f^- \in \mathfrak{N}$, so it makes sense to discuss the quantities $\bar{\mathcal{I}}_E(f^+)$ and $\bar{\mathcal{I}}_E(f^-)$ in the definition above.

We desire to define a suitable extension of $\bar{\mathcal{I}}_E$ from $\mathfrak{N}$ to the larger domain $\mathfrak{M}$. Care will have to be exercised because the upcoming definition involves subtraction of nonnegative numbers, and we must watch out for undefined differences of the form $+\infty - (+\infty)$.

Definition. For any $E \in \mathcal{F}$, we define $\bar{\bar{\mathcal{I}}}_E : \mathfrak{M}(E) \to \bar{\mathbb{R}}$ by writing

$$\bar{\bar{\mathcal{I}}}_E(f) = \bar{\mathcal{I}}_E(f^+) - \bar{\mathcal{I}}_E(f^-) \quad \text{for every } f \in \mathfrak{M}(E).$$

If clarity demands it, we will write $\bar{\bar{\mathcal{I}}}_E(f; \mu)$ for $\bar{\bar{\mathcal{I}}}_E(f)$. Also, we will write $\bar{\bar{\mathcal{I}}}(f)$ for $\bar{\bar{\mathcal{I}}}_\Omega(f)$.

Observe that the expression $\bar{\bar{\mathcal{I}}}_E(f)$ makes perfectly good logical sense for $f \in \mathfrak{M}(E)$ since we will never have $\bar{\mathcal{I}}_E(f^+) = \bar{\mathcal{I}}_E(f^-) = +\infty$, hence we will never have the undefined expression $+\infty - (+\infty)$ on the right side. That is, $\bar{\bar{\mathcal{I}}}_E(f)$ exists for every $f \in \mathfrak{M}(E)$ in the sense that $\bar{\bar{\mathcal{I}}}_E(f)$ is a well-defined element of $[-\infty, +\infty]$. *We therefore have that the following statements are logically equivalent:*

(1) $\bar{\bar{\mathcal{I}}}_E(f)$ exists [as an element of $\bar{\mathbb{R}}$],
(2) $f \in \mathfrak{M}(E)$, and
(3) at least one of $\bar{\mathcal{I}}_E(f^+)$ and $\bar{\mathcal{I}}_E(f^-)$ is finite.

Given $f \in \mathfrak{M}(E)$ [so that $\bar{\bar{\mathcal{I}}}_E(f)$ exists], we have that $|\bar{\bar{\mathcal{I}}}_E(f)| < +\infty$ iff both $\bar{\mathcal{I}}_E(f^+)$ and $\bar{\mathcal{I}}_E(f^-)$ are finite nonnegative numbers.

Notation: We will use the acronym *eaif* to denote *exists and is finite.* This is taken to mean *exists as a well-defined element of $\bar{\mathbb{R}}$ and in fact is actually in $\mathbb{R}$.* We will be indifferent between writing $\bar{\bar{\mathcal{I}}}_E(f)$ eaif and writing $|\bar{\bar{\mathcal{I}}}_E(f)| < +\infty$.

As with the previous two sections, we aim to establish a variety of properties concerning the defined functional. Some properties will be very mundane, while others will take some work. Many of the properties for $\bar{\bar{\mathcal{I}}}_E$ are proved by recalling the corresponding results established for $\bar{\mathcal{I}}_E$. Care will have to be taken in avoiding undefined expressions such as $\infty - \infty$.

> *In all that follows, $(\Omega, \mathcal{F}, \mu)$ denotes a measure space, and all sets are assumed to be $\mathcal{F}$-sets. Furthermore, all functions are assumed to be elements of $\mathfrak{M}$.*

(G1) *We have the following:*

(a) $\bar{\bar{\mathcal{I}}}_E(f)$ exists and equals $\bar{\mathcal{I}}_E(f)$ for each $f \in \mathfrak{N}$, hence $\bar{\bar{\mathcal{I}}}_E$ consistently extends $\bar{\mathcal{I}}_E$.

(b) If $f \in \mathfrak{M}(E)$, then $f \in \mathfrak{M}(F)$ for every $F \subseteq E$. Furthermore, if $|\bar{\bar{\mathcal{I}}}_E(f)| < +\infty$, then $|\bar{\bar{\mathcal{I}}}_F(f)| < +\infty$ for every $F \subseteq E$.

(c) If $\mu(E) = 0$, then $\bar{\bar{\mathcal{I}}}_E(f) = 0$ for every $f \in \mathfrak{M}$.

Proof: Let $f \in \mathfrak{N}$, so that $f^+ = f$ and $f^- = 0$. Then $\bar{\mathcal{I}}_E(f^-) = 0$, so that at least one of $\bar{\mathcal{I}}_E(f^+)$ and $\bar{\mathcal{I}}_E(f^-)$ is finite, hence $\bar{\bar{\mathcal{I}}}_E(f)$ exists. We thus have

$$\bar{\bar{\mathcal{I}}}_E(f) = \bar{\mathcal{I}}_E(f^+) - \bar{\mathcal{I}}_E(f^-) = \bar{\mathcal{I}}_E(f) - 0 = \bar{\mathcal{I}}_E(f).$$

This proves (a).

We now prove (b). If $f \in \mathfrak{M}(E)$, then at least one of $\bar{\mathcal{I}}_E(f^+)$ and $\bar{\mathcal{I}}_E(f^-)$ is finite. If $F \subseteq E$, then

$$\bar{\mathcal{I}}_F(f^+) \le \bar{\mathcal{I}}_E(f^+) \qquad \text{and} \qquad \bar{\mathcal{I}}_F(f^-) \le \bar{\mathcal{I}}_E(f^-) \tag{6.11}$$

by (N6), and hence at least one of $\bar{\mathcal{I}}_F(f^+)$ and $\bar{\mathcal{I}}_F(f^-)$ is finite. Therefore, $f \in \mathfrak{M}(F)$. If $|\bar{\bar{\mathcal{I}}}_E(f)| < +\infty$ in addition, then both $\bar{\mathcal{I}}_E(f^+)$ and $\bar{\mathcal{I}}_E(f^-)$ are finite. By (6.11) we have that both $\bar{\mathcal{I}}_F(f^+)$ and $\bar{\mathcal{I}}_F(f^-)$ are finite, hence $\bar{\bar{\mathcal{I}}}_F(f)$ is finite as well. This proves (b).

We now do (c). If $\mu(E) = 0$, then $\bar{\mathcal{I}}_E(f^+) = 0 = \bar{\mathcal{I}}_E(f^-)$, hence we have $\bar{\bar{\mathcal{I}}}_E(f) = 0 - 0 = 0$. This completes the proof. ■

By part (a) of the property above, we may replace $\bar{\mathcal{I}}_E$ by $\bar{\bar{\mathcal{I}}}_E$ everywhere it occurs in the previous section. Part (a) also implies that $\bar{\bar{\mathcal{I}}}_E(f)$ always exists [but may not necessarily be finite] for $f \in \mathfrak{N}$.

(G2) *If* $\bar{\bar{\mathcal{I}}}_E(f)$ *eaif and* $\{E_n\}_{n=1}^{\infty}$ *is a sequence of disjoint* $\mathcal{F}$*-subsets of* E*, then*

(a) $\bar{\bar{\mathcal{I}}}_{\bigcup_{n=1}^{\infty} E_n}(f)$ *eaif and* $\bar{\bar{\mathcal{I}}}_{E_n}(f)$ *eaif for each* $n \in \mathbb{N}$*,*

(b) $\bar{\bar{\mathcal{I}}}_{\bigcup_{n=1}^{\infty} E_n}(f) = \sum_{n=1}^{\infty} \bar{\bar{\mathcal{I}}}_{E_n}(f)$*, and*

(c) $\bar{\bar{\mathcal{I}}}_{\bigcup_{k=1}^{n} E_k}(f)$ *eaif and equals* $\sum_{k=1}^{n} \bar{\bar{\mathcal{I}}}_{E_k}(f)$*.*

Proof: Assume that $\bar{\bar{\mathcal{I}}}_E(f)$ eaif. Then, since $\bigcup_{n=1}^{\infty} E_n$ and $E_1, E_2, \cdots$ are $\mathcal{F}$-subsets of E, (G1) shows that each of $\bar{\bar{\mathcal{I}}}_{\bigcup_{n=1}^{\infty} E_n}(f), \bar{\bar{\mathcal{I}}}_{E_1}(f), \bar{\bar{\mathcal{I}}}_{E_2}(f), \cdots$ exist and are finite, giving (a). We may now concentrate on (b). By (a), $\bar{\mathcal{I}}_{\bigcup_{n=1}^{\infty} E_n}(f^+)$ and $\bar{\mathcal{I}}_{\bigcup_{n=1}^{\infty} E_n}(f^-)$ are finite. Similarly, for each $n \in \mathbb{N}$, both $\bar{\mathcal{I}}_{E_n}(f^+)$ and $\bar{\mathcal{I}}_{E_n}(f^-)$ are finite. We have

$$\begin{aligned}\bar{\bar{\mathcal{I}}}_{\bigcup_{n=1}^{\infty} E_n}(f) &= \bar{\mathcal{I}}_{\bigcup_{n=1}^{\infty} E_n}(f^+) - \bar{\mathcal{I}}_{\bigcup_{n=1}^{\infty} E_n}(f^-) \\ &= \sum_{n=1}^{\infty} \bar{\mathcal{I}}_{E_n}(f^+) - \sum_{n=1}^{\infty} \bar{\mathcal{I}}_{E_n}(f^-) \\ &= \sum_{n=1}^{\infty} \left[\bar{\mathcal{I}}_{E_n}(f^+) - \bar{\mathcal{I}}_{E_n}(f^-)\right] \\ &= \sum_{n=1}^{\infty} \bar{\bar{\mathcal{I}}}_{E_n}(f).\end{aligned}$$

Let us be careful in verifying this seemingly innocent chain of equalities. The first equality follows by definition and the fact that $\bar{\bar{\mathcal{I}}}_{\bigcup_{n=1}^{\infty} E_n}(f)$ exists; note that both terms in the difference are finite. The second equality follows by (N9); observe that both $\sum_{n=1}^{\infty} \bar{\mathcal{I}}_{E_n}(f^+)$ and $\sum_{n=1}^{\infty} \bar{\mathcal{I}}_{E_n}(f^-)$ are finite [in particular, the series are series of finite nonnegative numbers that converge in the usual calculus sense], hence the third equality follows in basic calculus fashion. The fourth equality follows because $\bar{\bar{\mathcal{I}}}_{E_n}(f)$ eaif for each $n \in \mathbb{N}$, namely, because both $\bar{\mathcal{I}}_{E_n}(f^+)$ and $\bar{\mathcal{I}}_{E_n}(f^-)$ are finite for each $n \in \mathbb{N}$. This completes the proof of (b).

For (c), the claims of existence and finiteness follow almost trivially from the previous parts. The identity $\bar{\bar{\mathcal{I}}}_{\bigcup_{k=1}^{n} E_k}(f) = \sum_{k=1}^{n} \bar{\bar{\mathcal{I}}}_{E_k}(f)$ follows from (b) by taking $E_{n+1} = E_{n+2} = \cdots = \varnothing$ and using (G1). ■

(G3) *If* $\bar{\bar{\mathcal{I}}}_E(f)$ *eaif, then* f *is finite* μ*-a.e. on* E*.*

Proof: By hypothesis, f is such that both of $\bar{\mathcal{I}}_E(f^+)$ and $\bar{\mathcal{I}}_E(f^-)$ are finite. By (N11), f^+ is finite μ-a.e. on E and f^- is finite μ-a.e. on E. Since

$$\begin{aligned}&\{\omega \in E : f(\omega) = \pm\infty\} \\ = \ &\{\omega \in E : f^+(\omega) = +\infty\} \cup \{\omega \in E : f^-(\omega) = +\infty\}\end{aligned}$$

for such an f, and since these latter two sets are μ-null, it follows that we have $\mu(\{\omega \in E : f(\omega) = \pm\infty\}) = 0$. That is, f is finite μ-a.e. on E. ■

(G4) *If one of $\bar{\bar{\mathcal{I}}}_F(fI_E)$ and $\bar{\bar{\mathcal{I}}}_{F\cap E}(f)$ exists, then so does the other and both terms are equal.*

Proof: Suppose that $\bar{\bar{\mathcal{I}}}_F(fI_E)$ exists, so that at least one of $\bar{\mathcal{I}}_F((fI_E)^+)$ and $\bar{\mathcal{I}}_F((fI_E)^-)$ are finite. Since $(fI_E)^+ = f^+I_E$ and $(fI_E)^- = f^-I_E$, it follows that at least one of $\bar{\mathcal{I}}_F(f^+I_E)$ and $\bar{\mathcal{I}}_F(f^-I_E)$ is finite. By (N2), we have that at least one of $\bar{\mathcal{I}}_{F\cap E}(f^+)$ and $\bar{\mathcal{I}}_{F\cap E}(f^-)$ is finite. But then $\bar{\bar{\mathcal{I}}}_{F\cap E}(f)$ exists. We have shown that the existence of $\bar{\bar{\mathcal{I}}}_F(fI_E)$ implies the existence of $\bar{\bar{\mathcal{I}}}_{F\cap E}(f)$; the fact that the existence of $\bar{\bar{\mathcal{I}}}_{F\cap E}(f)$ implies that of $\bar{\bar{\mathcal{I}}}_F(fI_E)$ follows by reversing the steps. To see that $\bar{\bar{\mathcal{I}}}_F(fI_E) = \bar{\bar{\mathcal{I}}}_{F\cap E}(f)$, put what has been stated together:

$$\begin{aligned}\bar{\bar{\mathcal{I}}}_F(fI_E) = \bar{\mathcal{I}}_F((fI_E)^+) - \bar{\mathcal{I}}_F((fI_E)^-) &= \bar{\mathcal{I}}_F(f^+I_E) - \bar{\mathcal{I}}_F(f^-I_E)\\ &= \bar{\mathcal{I}}_{F\cap E}(f^+) - \bar{\mathcal{I}}_{F\cap E}(f^-)\\ &= \bar{\bar{\mathcal{I}}}_{F\cap E}(f).\end{aligned}$$

This completes the proof. ■

(G5) *If $f = g$ μ-a.e. on E and $f \in \mathfrak{M}(E)$, then we have both $g \in \mathfrak{M}(E)$ and $\bar{\bar{\mathcal{I}}}_E(f) = \bar{\bar{\mathcal{I}}}_E(g)$.*

Proof: Assume that $f = g$ μ-a.e. on E and $f \in \mathfrak{M}(E)$. Define the $\mathcal{F}$-sets $E_1 = \{\omega \in E : f(\omega) = g(\omega)\}$ and $E_2 = \{\omega \in E : f(\omega) \neq g(\omega)\}$, so that $\mu(E_2) = 0$. We first desire to show that $g \in \mathfrak{M}(E)$. If $f \in \mathfrak{M}(E)$, then we have that at least one of $\bar{\mathcal{I}}_{E_1}(f^+)$ and $\bar{\mathcal{I}}_{E_1}(f^-)$ is finite by (G1); we also have

$$\bar{\mathcal{I}}_{E_2}(g^+) = \bar{\mathcal{I}}_{E_2}(g^-) = \bar{\mathcal{I}}_{E_2}(f^+) = \bar{\mathcal{I}}_{E_2}(f^-) = 0$$

by the fact that $\mu(E_2) = 0$. By the fact that $f = g$ on E_1 and (N3), we have that $\bar{\mathcal{I}}_{E_1}(g^+) = \bar{\mathcal{I}}_{E_1}(f^+)$ and $\bar{\mathcal{I}}_{E_1}(g^-) = \bar{\mathcal{I}}_{E_2}(f^-)$, hence at least one of $\bar{\mathcal{I}}_{E_1}(g^+)$ and $\bar{\mathcal{I}}_{E_1}(g^-)$ is finite. By (N9) or the comments immediately following it, we have

$$\bar{\mathcal{I}}_E(g^+) = \bar{\mathcal{I}}_{E_1}(g^+) + \bar{\mathcal{I}}_{E_2}(g^+) = \bar{\mathcal{I}}_{E_1}(g^+)$$

and

$$\bar{\mathcal{I}}_E(g^-) = \bar{\mathcal{I}}_{E_1}(g^-) + \bar{\mathcal{I}}_{E_2}(g^-) = \bar{\mathcal{I}}_{E_1}(g^-),$$

hence at least one of $\bar{\mathcal{I}}_E(g^+)$ and $\bar{\mathcal{I}}_E(g^-)$ is finite. It follows that $g \in \mathfrak{M}(E)$. Using all of these quantities whose existence has been shown, we have

$$\begin{aligned}\bar{\bar{\mathcal{I}}}_E(g) = \bar{\mathcal{I}}_E(g^+) - \bar{\mathcal{I}}_E(g^-) &= \bar{\mathcal{I}}_{E_1}(g^+) - \bar{\mathcal{I}}_{E_1}(g^-)\\ &= \bar{\mathcal{I}}_{E_1}(f^+) - \bar{\mathcal{I}}_{E_1}(f^-)\\ &= \bar{\mathcal{I}}_E(f^+) - \bar{\mathcal{I}}_E(f^-)\\ &= \bar{\bar{\mathcal{I}}}_E(f),\end{aligned}$$

which completes the proof. ∎

(G6) *If* $f \le g$ *on* μ*-a.e. on* E *and* $f, g \in \mathfrak{M}(E)$*, then* $\bar{\bar{\mathcal{I}}}_E(f) \le \bar{\bar{\mathcal{I}}}_E(g)$.

Proof: We do the proof for when $f \le g$ everywhere on E, and leave the μ-a.e. generalization as a simple exercise for the reader. [Merely follow the same type of procedure given in the proof of (G5).] If $f \le g$ on E, then we have $f^+ \le g^+$ on E and $g^- \le f^-$ on E. By (N4), we have that $\bar{\mathcal{I}}_E(f^+) \le \bar{\mathcal{I}}_E(g^+)$ and $\bar{\mathcal{I}}_E(g^-) \le \bar{\mathcal{I}}_E(f^-)$. Therefore, we have $\bar{\mathcal{I}}_E(f^+) - \bar{\mathcal{I}}_E(f^-) \le \bar{\mathcal{I}}_E(g^+) - \bar{\mathcal{I}}_E(g^-)$. [We do not encounter the undefined form $\infty - \infty$ here because of the assumptions $f \in \mathfrak{M}(E)$ and $g \in \mathfrak{M}(E)$.] Therefore, $\bar{\bar{\mathcal{I}}}_E(f) \le \bar{\bar{\mathcal{I}}}_E(g)$. ∎

The proofs of the previous properties have more or less the same flavor: show that it makes sense to discuss the subtraction $\bar{\mathcal{I}}_E(f^+) - \bar{\mathcal{I}}_E(f^-)$ and then obtain a desired equality or inequality by calling on the appropriate result from the previous section. There really isn't much to this, and no real difficulties present themselves up to this point.

The next property is very important. In particular, part (b) greatly generalizes the triangle inequality.

(G7) *We have the following:*

(a) $\bar{\bar{\mathcal{I}}}_E(f)$ *eaif iff* $\bar{\bar{\mathcal{I}}}_E(|f|)$ *eaif.*

(b) If $\bar{\bar{\mathcal{I}}}_E(f)$ *exists [finite or not], then* $|\bar{\bar{\mathcal{I}}}_E(f)| \le \bar{\bar{\mathcal{I}}}_E(|f|)$.

Proof: We do (a). If $\bar{\bar{\mathcal{I}}}_E(f)$ eaif, then both $\bar{\mathcal{I}}_E(f^+)$ and $\bar{\mathcal{I}}_E(f^-)$ are finite. Therefore, $\bar{\mathcal{I}}_E(f^+) + \bar{\mathcal{I}}_E(f^-)$ is finite. By (N8), $\bar{\mathcal{I}}_E(f^+ + f^-)$ is finite. Since $|f| = f^+ + f^-$, it follows that $\bar{\mathcal{I}}_E(|f|)$ is finite. Since $|f|$ is nonnegative, (G1) shows that $\bar{\bar{\mathcal{I}}}_E(|f|)$ exists and equals $\bar{\mathcal{I}}_E(|f|)$, so that $\bar{\bar{\mathcal{I}}}_E(|f|)$ exists and is finite.

Conversely, suppose that $\bar{\bar{\mathcal{I}}}_E(|f|)$ eaif. Since $\bar{\bar{\mathcal{I}}}_E$ consistently extends $\bar{\mathcal{I}}_E$, it follows that $\bar{\mathcal{I}}_E(|f|) < +\infty$. Since $|f| = f^+ + f^-$, (N8) gives the inequality $\bar{\mathcal{I}}_E(f^+) + \bar{\mathcal{I}}_E(f^-) < +\infty$. It follows that $\bar{\mathcal{I}}_E(f^+)$ and $\bar{\mathcal{I}}_E(f^-)$ are finite, so $\bar{\bar{\mathcal{I}}}_E(f)$ exists and is finite. This gives (a).

For (b), assume that $\bar{\bar{\mathcal{I}}}_E(f)$ exists, so that the difference $\bar{\mathcal{I}}_E(f^+) - \bar{\mathcal{I}}_E(f^-)$ is not of the undefined form $\infty - \infty$. By (N8) and (G1),

$$\begin{aligned}|\bar{\bar{\mathcal{I}}}_E(f)| = |\bar{\mathcal{I}}_E(f^+) - \bar{\mathcal{I}}_E(f^-)| \le \bar{\mathcal{I}}_E(f^+) + \bar{\mathcal{I}}_E(f^-) &= \bar{\mathcal{I}}_E(f^+ + f^-) \\ &= \bar{\mathcal{I}}_E(|f|) \\ &= \bar{\bar{\mathcal{I}}}_E(|f|),\end{aligned}$$

hence (b) and the entire claim stand complete. ∎

Before proving the next property, which shows [roughly] that $\bar{\bar{\mathcal{I}}}_E$ is linear, we wish to make some terminology clear. Given $f : \Omega \to \bar{\mathbb{R}}$ and $g : \Omega \to \bar{\mathbb{R}}$, define $h : \Omega \to \bar{\mathbb{R}}$ for all $\omega \in \Omega$ by

$$h(\omega) = \begin{cases} f(\omega) + g(\omega) & \text{if } f(\omega) + g(\omega) \text{ is defined} \\ 0 & \text{if } f(\omega) + g(\omega) \text{ is of an undefined form.}\end{cases}$$

If f and g are such that $f(\omega) + g(\omega)$ is always defined for each $\omega \in \Omega$, then $h(\omega) = f(\omega) + g(\omega)$ for each $\omega \in \Omega$, hence we may write $h = f + g$, which reduces to the familiar definition in the finite case. Regardless of the situation, if f and g are in $\mathfrak{M}$, then h is in $\mathfrak{M}$ by (MF9).

The proof of the following very important property is longer than what one might think; this is due to the fact that we are conscious of showing that certain quantities [which may possibly not exist] do in fact exist.

(G8) *Let $\bar{\bar{\mathcal{I}}}_E(f)$ and $\bar{\bar{\mathcal{I}}}_E(g)$ both exist and be finite. We have the following:*

(a) For every $c \in \mathbb{R}$, $\bar{\bar{\mathcal{I}}}_E(cf)$ eaif. Moreover, $\bar{\bar{\mathcal{I}}}_E(cf) = c\bar{\bar{\mathcal{I}}}_E(f)$.

(b) $\bar{\bar{\mathcal{I}}}_E(h)$ eaif, and $\bar{\bar{\mathcal{I}}}_E(h) = \bar{\bar{\mathcal{I}}}_E(f) + \bar{\bar{\mathcal{I}}}_E(g)$, where the definition of h is as given immediately prior to this statement.

Proof: Suppose that $\bar{\bar{\mathcal{I}}}_E(f)$ eaif, and let $c \in \mathbb{R}$. There are three cases to consider: (i) $c = 0$, (ii) $c > 0$, and (iii) $c < 0$. We will show that $\bar{\bar{\mathcal{I}}}_E(cf)$ exists and equals $c\bar{\bar{\mathcal{I}}}_E(f)$ in each of the three cases.

If $c = 0$, then $(cf)^+ = (cf)^- = 0$, hence $\bar{\mathcal{I}}_E((cf)^+) = \bar{\mathcal{I}}_E((cf)^-) = 0$, hence $\bar{\bar{\mathcal{I}}}_E(cf)$ exists and equals $c\bar{\bar{\mathcal{I}}}_E(f)$, the common value being zero.

If $c > 0$, observe that $(cf)^+ = cf^+$ and $(cf)^- = cf^-$ by (MF13). Since $\bar{\bar{\mathcal{I}}}_E(f)$ eaif, it follows that both $\bar{\mathcal{I}}_E(f^+)$ and $\bar{\mathcal{I}}_E(f^-)$ are finite, hence both $c\bar{\mathcal{I}}_E(f^+)$ and $c\bar{\mathcal{I}}_E(f^-)$ are finite. (N8) implies that $\bar{\mathcal{I}}_E(cf^+)$ and $\bar{\mathcal{I}}_E(cf^-)$ are finite, so that $\bar{\mathcal{I}}_E((cf)^+)$ and $\bar{\mathcal{I}}_E((cf)^-)$ are finite. This shows that $\bar{\bar{\mathcal{I}}}_E(cf)$ exists and is finite. Now we may show the identity by (N8):

$$\begin{aligned}\bar{\bar{\mathcal{I}}}_E(cf) = \bar{\mathcal{I}}_E((cf)^+) - \bar{\mathcal{I}}_E((cf)^-) &= \bar{\mathcal{I}}_E(cf^+) - \bar{\mathcal{I}}_E(cf^-) \\ &= c\bar{\mathcal{I}}_E(f^+) - c\bar{\mathcal{I}}_E(f^-) \\ &= c\bar{\bar{\mathcal{I}}}_E(f).\end{aligned}$$

This completes the $c > 0$ case.

For the $c < 0$ case, observe that $(cf)^+ = -cf^-$ and $(cf)^- = -cf^+$ by (MF13). The same type of argument used in the $c > 0$ case to establish that $\bar{\bar{\mathcal{I}}}_E(cf)$ eaif may be used in this case. Using the existence and finiteness of all quantities involved, we have

$$\begin{aligned}\bar{\bar{\mathcal{I}}}_E(cf) = \bar{\mathcal{I}}_E((cf)^+) - \bar{\mathcal{I}}_E((cf)^-) &= \bar{\mathcal{I}}_E(-cf^-) - \bar{\mathcal{I}}_E(-cf^+) \\ &= -c\bar{\mathcal{I}}_E(f^-) - (-c)\bar{\mathcal{I}}_E(f^+) \\ &= c[\bar{\mathcal{I}}_E(f^+) - \bar{\mathcal{I}}_E(f^-)] \\ &= c\bar{\bar{\mathcal{I}}}_E(f),\end{aligned}$$

which completes the proof in the $c < 0$ case and establishes (a).

For the second part of the program, suppose that $\bar{\bar{\mathcal{I}}}_E(f)$ and $\bar{\bar{\mathcal{I}}}_E(g)$ exist and are finite, and additionally assume that f and g are everywhere finite. [We will remove this assumption shortly.] We want to first show that $\bar{\bar{\mathcal{I}}}_E(f+g)$

eaif, and then to show that $\bar{\bar{\mathcal{I}}}_E(f+g) = \bar{\bar{\mathcal{I}}}_E(f) + \bar{\bar{\mathcal{I}}}_E(g)$. First, observe that the finiteness of f and g on Ω means that $f+g$ is defined everywhere on Ω. That is, for no $\omega \in E$ will $f(\omega) + g(\omega)$ be of the undefined form $\infty - \infty$ or $-\infty + \infty$. To show that $\bar{\bar{\mathcal{I}}}_E(f+g)$ eaif, we must show that both $\bar{\mathcal{I}}_E((f+g)^+)$ and $\bar{\mathcal{I}}_E((f+g)^-)$ are finite. Since $(f+g)^+ \le f^+ + g^+$ and $(f+g)^- \le f^- + g^-$ by (MF13), and since the existence and finiteness of $\bar{\bar{\mathcal{I}}}_E(f)$ and $\bar{\bar{\mathcal{I}}}_E(g)$ forces the finiteness of each of $\bar{\mathcal{I}}_E(f^+)$, $\bar{\mathcal{I}}_E(f^-)$, $\bar{\mathcal{I}}_E(g^+)$, and $\bar{\mathcal{I}}_E(g^-)$, we have

$$\bar{\mathcal{I}}_E((f+g)^+) \le \bar{\mathcal{I}}_E(f^+ + g^+) = \bar{\mathcal{I}}_E(f^+) + \bar{\mathcal{I}}_E(g^+) < +\infty \tag{6.12}$$

and

$$\bar{\mathcal{I}}_E((f+g)^-) \le \bar{\mathcal{I}}_E(f^- + g^-) = \bar{\mathcal{I}}_E(f^-) + \bar{\mathcal{I}}_E(g^-) < +\infty \tag{6.13}$$

by (N8), so it is the case that $\bar{\bar{\mathcal{I}}}_E(f+g)$ exists and is finite. We may now establish the identity. From the equations $f+g = (f+g)^+ - (f+g)^-$ and $f+g = f^+ - f^- + g^+ - g^-$, we have

$$(f+g)^+ + f^- + g^- = (f+g)^- + f^+ + g^+,$$

hence

$$\bar{\mathcal{I}}_E\left((f+g)^+ + f^- + g^-\right) = \bar{\mathcal{I}}_E\left((f+g)^- + f^+ + g^+\right).$$

By (N8), this becomes

$$\bar{\mathcal{I}}_E((f+g)^+) + \bar{\mathcal{I}}_E(f^-) + \bar{\mathcal{I}}_E(g^-) = \bar{\mathcal{I}}_E((f+g)^-) + \bar{\mathcal{I}}_E(f^+) + \bar{\mathcal{I}}_E(g^+).$$

All quantities above are finite numbers, so we have

$$\bar{\mathcal{I}}_E((f+g)^+) - \bar{\mathcal{I}}_E((f+g)^-) = \bar{\mathcal{I}}_E(f^+) - \bar{\mathcal{I}}_E(f^-) + \bar{\mathcal{I}}_E(g^+) - \bar{\mathcal{I}}_E(g^-),$$

which means that $\bar{\bar{\mathcal{I}}}_E(f+g) = \bar{\bar{\mathcal{I}}}_E(f) + \bar{\bar{\mathcal{I}}}_E(g)$ as claimed. This completes the proof of (b) for the case of when both f and g are finite on Ω.

We now wish to remove the assumption that f and g are finite everywhere on Ω. Therefore, we reset everything and merely assume that $\bar{\bar{\mathcal{I}}}_E(f)$ and $\bar{\bar{\mathcal{I}}}_E(g)$ exist and are finite. By (G3), f is finite μ-a.e. on E and g is finite μ-a.e. on E. We must first show that $\bar{\bar{\mathcal{I}}}_E(h)$ eaif. To see this, observe that we still have $h^+ \le f^+ + g^+$ and $h^- \le f^- + g^-$ *everywhere on* Ω, hence the quantities $\bar{\mathcal{I}}_E(h^+)$ and $\bar{\mathcal{I}}_E(h^-)$ may be shown to be finite by repeating the same argument that leads to (6.12) and (6.13). Therefore, $\bar{\bar{\mathcal{I}}}_E(h)$ exists.

We now establish the identity $\bar{\bar{\mathcal{I}}}_E(h) = \bar{\bar{\mathcal{I}}}_E(f) + \bar{\bar{\mathcal{I}}}_E(g)$. Write E as a disjoint union $E_1 \cup E_2$ of $\mathcal{F}$-sets, where $E_1 = \{\omega \in E : f(\omega), g(\omega) \text{ are finite}\}$ and $E_2 = \{\omega \in E : f(\omega) + g(\omega) \text{ is infinite or undefined}\}$. Since the $\mathcal{F}$-set E_2 equals $\{\omega \in E : f(\omega) = \pm\infty\} \cup \{\omega \in E : g(\omega) = \pm\infty\}$ [a union of μ-null sets], it follows that $\mu(E_2) = 0$, hence $\bar{\bar{\mathcal{I}}}_{E_2}(h) = \bar{\bar{\mathcal{I}}}_{E_2}(f) = \bar{\bar{\mathcal{I}}}_{E_2}(g) = 0$. By (G2) we only need $\bar{\bar{\mathcal{I}}}_{E_1}(h) = \bar{\bar{\mathcal{I}}}_{E_1}(f) + \bar{\bar{\mathcal{I}}}_{E_1}(g)$, that is, $\bar{\bar{\mathcal{I}}}(hI_{E_1}) = \bar{\bar{\mathcal{I}}}(fI_{E_1}) + \bar{\bar{\mathcal{I}}}(gI_{E_1})$ [by (G4)]. We have that both fI_{E_1} and gI_{E_1} are finite everywhere on Ω and

$hI_{E_1} = fI_{E_1} + gI_{E_1}$ on Ω, so the "finite everywhere" case established a little earlier gives

$$\bar{\bar{\mathcal{I}}}(hI_{E_1}) = \bar{\bar{\mathcal{I}}}(fI_{E_1} + gI_{E_1}) = \bar{\bar{\mathcal{I}}}(fI_{E_1}) + \bar{\bar{\mathcal{I}}}(gI_{E_1}),$$

completing the proof. ■

Putting the two claims in (G8) together lets us say that (i) the subset of functions f in $\mathfrak{M}(E)$ with $|\bar{\bar{\mathcal{I}}}_E(f)| < \infty$ is a vector [sub]space of $\mathfrak{M}(E)$, and (ii) the restriction of $\bar{\bar{\mathcal{I}}}_E$ to this vector subspace is a linear functional on the subspace. Often part (b) is often written as $\bar{\bar{\mathcal{I}}}_E(f+g) = \bar{\bar{\mathcal{I}}}_E(f) + \bar{\bar{\mathcal{I}}}_E(g)$, with the tacit understanding that what is really meant is the precise statement as we have given it in (b).

The upcoming property gives a quick way of showing that $\bar{\bar{\mathcal{I}}}_E(g)$ exists and is finite for a given $g \in \mathfrak{M}$.

(G9) *If $\bar{\bar{\mathcal{I}}}_E(f)$ eaif and g is such that $|g| \le f$ μ-a.e. on E, then $\bar{\bar{\mathcal{I}}}_E(g)$ eaif. As a consequence, any $f \in \mathfrak{M}$ that is bounded μ-a.e. on a set E of finite measure and zero μ-a.e. on E^c is such that $\bar{\bar{\mathcal{I}}}_F(f)$ eaif for every $F \in \mathcal{F}$.*

Proof: If $|g| \le f$ on E, then $f \ge 0$ and $g^+ + g^- \le f$ on E, hence $g^+ \le f$ and $g^- \le f$ on E. It follows that $\bar{\mathcal{I}}_E(g^+) \le \bar{\mathcal{I}}_E(f) = \bar{\bar{\mathcal{I}}}_E(f) < +\infty$ by (N4), and $\bar{\mathcal{I}}_E(g^-) < +\infty$ by the same reasoning. Therefore both $\bar{\mathcal{I}}_E(g^+)$ and $\bar{\mathcal{I}}_E(g^-)$ are finite, hence $\bar{\bar{\mathcal{I}}}_E(g)$ eaif. The extension of this proof to the case where $|g| \le f$ μ-a.e. on E is left to the reader and is straightforward.

For the second claim, we again do the "everywhere" proof, leaving the minor complications of the "almost everywhere" case for the reader as a very straightforward exercise. Suppose that $E \in \mathcal{F}$ and $0 \le M < +\infty$ are such that (i) $|f| \le M$ on E, (ii) $\mu(E) < +\infty$, and (iii) $f = 0$ on E^c. Observe that the simple function MI_E is such that $\bar{\bar{\mathcal{I}}}(MI_E)$ exists and is finite, for $\bar{\mathcal{I}}((MI_E)^+) = \bar{\mathcal{I}}(MI_E) = M\bar{\mathcal{I}}(I_E) = M\mu(E) < +\infty$ by (N8), and $\bar{\mathcal{I}}((MI_E)^-) = \bar{\mathcal{I}}(0) = 0$. Furthermore, observe that $|f| \le MI_E$ on Ω, so the first claim shows that $|\bar{\bar{\mathcal{I}}}(f)| < +\infty$. Property (G1) shows that this forces $|\bar{\bar{\mathcal{I}}}_F(f)| < +\infty$ for every $F \in \mathcal{F}$. ■

Before proceeding, it is worth stating again that since $\bar{\bar{\mathcal{I}}}_E = \bar{\mathcal{I}}_E$ on $\mathfrak{N}$, all of the properties of the previous section carry over to this section and its notation. For example, the Monotone Convergence Theorem (N7) becomes the following: *Let $\{f_n\}_{n=1}^\infty$ denote a nondecreasing sequence of nonnegative functions in $\mathfrak{M}$. Then $\bar{\bar{\mathcal{I}}}_E(f_n)$ exists for each $n \in \mathbb{N}$, both $\lim_n \bar{\bar{\mathcal{I}}}_E(f_n)$ and $\bar{\bar{\mathcal{I}}}_E(\lim_n f_n)$ exist, and we have $\lim_n \bar{\bar{\mathcal{I}}}_E(f_n) = \bar{\bar{\mathcal{I}}}_E(\lim_n f_n)$.*

As another example, Fatou's Lemma (N13) states the following under the trivial modification: *Let $\{f_n\}_{n=1}^\infty$ denote a sequence of nonnegative functions in $\mathfrak{M}$. Then $\bar{\bar{\mathcal{I}}}_E(f_n)$ exists for each $n \in \mathbb{N}$, $\bar{\bar{\mathcal{I}}}_E(\underline{\lim}_n f_n)$ exists, and we have $\bar{\bar{\mathcal{I}}}_E(\underline{\lim}_n f_n) \le \underline{\lim}_n \bar{\bar{\mathcal{I}}}_E(f_n)$.* The other statements of the previous section are similarly modified.

The following property, like MCT, occupies a central place in analysis and measure theory. It is known as *Lebesgue's Dominated Convergence Theorem*, and we will denote it by DCT.

(G10) *Let $\{f_n\}_{n=1}^{\infty}$ denote a sequence of functions in $\mathfrak{M}$, and let $g \in \mathfrak{M}$ be nonnegative and such that $\bar{\bar{\mathcal{I}}}_E(g)$ eaif. Assume that* $\lim_n f_n$ *exists on* Ω *and $|f_n| \leq g$ on E for each $n \in \mathbb{N}$. Then*

(a) $\bar{\bar{\mathcal{I}}}_E(\lim_n f_n)$ *eaif,*

(b) $\bar{\bar{\mathcal{I}}}_E(f_n)$ *eaif for each* $n \in \mathbb{N}$*,*

(c) $\lim_n \bar{\bar{\mathcal{I}}}_E(f_n)$ *eaif, and*

(d) $\bar{\bar{\mathcal{I}}}_E(\lim_n f_n) = \lim_n \bar{\bar{\mathcal{I}}}_E(f_n)$.

Proof: Since $|f_n| \leq g$ on E for each $n \in \mathbb{N}$ and $\bar{\bar{\mathcal{I}}}_E(g)$ eaif, it follows from (G9) that $\bar{\bar{\mathcal{I}}}_E(f_n)$ eaif for each $n \in \mathbb{N}$, which gives (b). Next, $\lim_n f_n$ exists by assumption [and is in $\mathfrak{M}$ by (MF11)], and the assumption that $|f_n| \leq g$ on E for each $n \in \mathbb{N}$ forces $|\lim_n f_n| \leq g$ on E. Again, the assumption that $\bar{\bar{\mathcal{I}}}_E(g)$ eaif allows us to invoke (G9) to claim that $\bar{\bar{\mathcal{I}}}_E(\lim_n f_n)$ eaif, giving (a). Parts (a) and (b) accordingly stand complete.

We now turn to proving (c) and (d). Before proceeding, we make some preliminary comments. First, we will write f for $\lim_n f_n$. Second, because the value of $\bar{\bar{\mathcal{I}}}_E(h)$ is not affected by the behavior of h on E^c, we may assume without any loss of generality that each of $f, g, f_1, f_2, \cdots$ are zero on E^c. Third, the assumption that $\bar{\bar{\mathcal{I}}}_E(g)$ exists and is finite implies that g is finite μ-a.e. on E. Since the values of $\bar{\bar{\mathcal{I}}}_E$ are not affected by what happens on μ-null sets, we may assume without any loss of generality that g is finite everywhere on E. Putting everything in this paragraph together, we may without loss of generality make the following assumptions: (1) g is zero on E^c and finite on all of E, and (2) each of $f, f_1, f_2, \cdots$ is zero on E^c. Since $|f_n| \leq g$ on E for each $n \in \mathbb{N}$ and $|f| \leq g$ on E, we additionally have (3) f_n is finite on E for each $n \in \mathbb{N}$ and f is finite on E. Therefore, $g + f_n$ and $g - f_n$ are defined on Ω for each $n \in \mathbb{N}$, and $g + f$ and $g - f$ are defined on Ω. With these facts, we will simultaneously obtain (c) and (d).

First, observe that $\{g + f_n\}_{n=1}^{\infty}$ is a nonnegative sequence of functions on Ω such that $\lim_n (g + f_n)$ exists and equals $g + f$, hence $g + f = \underline{\lim}_n (g + f_n)$. Therefore, (G8) and Fatou's Lemma (N13) give

$$\begin{aligned}
\bar{\bar{\mathcal{I}}}_E(g) + \bar{\bar{\mathcal{I}}}_E(f) = \bar{\bar{\mathcal{I}}}_E(g + f) &= \bar{\bar{\mathcal{I}}}_E(\underline{\lim}_n (g + f_n)) \\
&\leq \underline{\lim}_n \bar{\bar{\mathcal{I}}}_E(g + f_n) \\
&= \underline{\lim}_n \left(\bar{\bar{\mathcal{I}}}_E(g) + \bar{\bar{\mathcal{I}}}_E(f_n)\right) \\
&= \bar{\bar{\mathcal{I}}}_E(g) + \underline{\lim}_n \bar{\bar{\mathcal{I}}}_E(f_n).
\end{aligned}$$

Since $\bar{\bar{\mathcal{I}}}_E(g)$ is finite, we may meaningfully subtract it from both sides of the presented chain, obtaining

$$\bar{\bar{\mathcal{I}}}_E(f) \leq \underline{\lim}_n \bar{\bar{\mathcal{I}}}_E(f_n). \tag{6.14}$$

[Observe that we couldn't obtain (6.14) directly from Fatou's Lemma, since the f_n's and f are not necessarily nonnegative; we did what was presented above so as to obtain a sequence of nonnegative functions to which we could apply Fatou's Lemma.]

Next, observe that $\{g - f_n\}_{n=1}^{\infty}$ is a nonnegative sequence of functions such that $\lim_n (g - f_n)$ exists and equals $g - f$, hence $g - f = \underline{\lim}_n (g - f_n)$. Therefore, (G8) and Fatou's Lemma (N13) give

$$\begin{aligned} \bar{\bar{\mathcal{I}}}_E(g) - \bar{\bar{\mathcal{I}}}_E(f) = \bar{\bar{\mathcal{I}}}_E(g - f) &= \bar{\bar{\mathcal{I}}}_E(\underline{\lim}_n (g - f_n)) \\ &\leq \underline{\lim}_n \bar{\bar{\mathcal{I}}}_E(g - f_n) \\ &= \underline{\lim}_n \left(\bar{\bar{\mathcal{I}}}_E(g) - \bar{\bar{\mathcal{I}}}_E(f_n) \right) \\ &= \bar{\bar{\mathcal{I}}}_E(g) + \underline{\lim}_n \left(-\bar{\bar{\mathcal{I}}}_E(f_n) \right). \end{aligned}$$

Since $\bar{\bar{\mathcal{I}}}_E(g)$ is finite, we may meaningfully subtract it from both sides of the above chain, obtaining $-\bar{\bar{\mathcal{I}}}_E(f) \leq \underline{\lim}_n \left(-\bar{\bar{\mathcal{I}}}_E(f_n) \right)$, which is equivalent to

$$\bar{\bar{\mathcal{I}}}_E(f) \geq \overline{\lim}_n \bar{\bar{\mathcal{I}}}_E(f_n). \tag{6.15}$$

Putting (6.14) and (6.15) together yields

$$\overline{\lim}_n \bar{\bar{\mathcal{I}}}_E(f_n) \leq \bar{\bar{\mathcal{I}}}_E(f) \leq \underline{\lim}_n \bar{\bar{\mathcal{I}}}_E(f_n).$$

Since $\bar{\bar{\mathcal{I}}}_E(f)$ is finite, it follows that $\overline{\lim}_n \bar{\bar{\mathcal{I}}}_E(f_n)$ is finite, hence the inequality above simultaneously says that $\lim_n \bar{\bar{\mathcal{I}}}_E(f_n)$ exists and equals the finite number $\bar{\bar{\mathcal{I}}}_E(f)$. This gives (c) and (d) at once, hence the proof of the Dominated Convergence Theorem is complete. ■

MCT, Fatou's Lemma, and the immediately preceding DCT may be generalized to "almost everywhere" statements. For example, DCT may be written as follows under this slightly more general setting:

(G10′) *Let* $\{f_n\}_{n=1}^{\infty}$ *denote a sequence of functions in* $\mathfrak{M}$, *and let* $g \in \mathfrak{M}$ *be nonnegative and such that* $\bar{\bar{\mathcal{I}}}_E(g)$ *eaif. Assume that* $\lim_n f_n$ *exists* μ*-a.e. on* E *and let* $h : \Omega \to \bar{\mathbb{R}}$ *be defined for all* $\omega \in \Omega$ *by*

$$h(\omega) = \begin{cases} \lim_n f_n(\omega) & \textit{if } \lim_n f_n(\omega) \textit{ exists} \\ 0 & \textit{for all other } \omega \in \Omega. \end{cases}$$

If $|f_n| \le g$ μ*-a.e. on* E *for each* $n \in \mathbb{N}$*, then*

(a) $\bar{\bar{\mathcal{I}}}_E(h)$ *eaif,*

(b) $\bar{\bar{\mathcal{I}}}_E(f_n)$ *eaif for each* $n \in \mathbb{N}$*,*

(c) $\lim_n \bar{\bar{\mathcal{I}}}_E(f_n)$ *eaif, and*

(d) $\bar{\bar{\mathcal{I}}}_E(h) = \lim_n \bar{\bar{\mathcal{I}}}_E(f_n)$*.*

There really isn't too much to prove, as all that is required is for E to be broken up into disjoint sets E_1 and E_2, where E_1 is the set where (i) $\lim_n f_n$ exists and (ii) $|f_n| \le g$ for every $n \in \mathbb{N}$. Because of the "almost everywhere" assumptions, we will have $\mu(E_2) = 0$ after a simple calculation. The original DCT may then be applied over E_1. The uninteresting details may be filled in by the reader.

(G11) *Let* $f \in \mathfrak{M}$ *be nonnegative, and let* ν_f *be as in (N9).*

(a) If $g \in \mathfrak{M}$ *is nonnegative, then* $\bar{\bar{\mathcal{I}}}_E(g;\nu_f) = \bar{\bar{\mathcal{I}}}_E(fg;\mu)$*.*

(b) For every $g \in \mathfrak{M}$ *we have that* $\bar{\bar{\mathcal{I}}}_E(g;\nu_f)$ *eaif iff* $\bar{\bar{\mathcal{I}}}_E(fg;\mu)$ *eaif, in which case we have* $\bar{\bar{\mathcal{I}}}_E(g;\nu_f) = \bar{\bar{\mathcal{I}}}_E(fg;\mu)$*.*

Proof: We first consider (a). By (G1), both of the expressions $\bar{\bar{\mathcal{I}}}_E(g;\nu_f)$ and $\bar{\bar{\mathcal{I}}}_E(fg;\mu)$ exist. The alleged equality follows by (N14), and hence (a) is complete.

We now turn to (b). Let $f \in \mathfrak{M}$ be nonnegative, and let $g \in \mathfrak{M}$. Suppose first that $\bar{\bar{\mathcal{I}}}_E(g;\nu_f)$ eaif. Then both $\bar{\mathcal{I}}_E(g^+;\nu_f)$ and $\bar{\mathcal{I}}_E(g^-;\nu_f)$ are finite. By (N14), we have

$$\bar{\mathcal{I}}_E(g^+;\nu_f) = \bar{\mathcal{I}}_E(fg^+;\mu) \quad \text{and} \quad \bar{\mathcal{I}}_E(g^-;\nu_f) = \bar{\mathcal{I}}_E(fg^-;\mu).$$

Since $f \ge 0$, we have

$$\bar{\mathcal{I}}_E(fg^+;\mu) = \bar{\mathcal{I}}_E((fg)^+;\mu) \quad \text{and} \quad \bar{\mathcal{I}}_E(fg^-;\mu) = \bar{\mathcal{I}}_E((fg)^-;\mu),$$

and thus both $\bar{\mathcal{I}}_E((fg)^+;\mu)$ and $\bar{\mathcal{I}}_E((fg)^-;\mu)$ are finite, hence $\bar{\bar{\mathcal{I}}}_E(fg;\mu)$ eaif. The steps of this argument may be reversed to show that if $\bar{\bar{\mathcal{I}}}_E(fg;\mu)$ eaif, then $\bar{\bar{\mathcal{I}}}_E(g;\nu_f)$ eaif. Now that existence claim has been demonstrated, we may establish the identity; putting things together with (N14) gives

$$\begin{aligned}\bar{\bar{\mathcal{I}}}_E(g;\nu_f) = \bar{\mathcal{I}}_E(g^+;\nu_f) - \bar{\mathcal{I}}_E(g^-;\nu_f) &= \bar{\mathcal{I}}_E(fg^+;\mu) - \bar{\mathcal{I}}_E(fg^-;\mu) \\ &= \bar{\mathcal{I}}_E((fg)^+;\mu) - \bar{\mathcal{I}}_E((fg)^-;\mu) \\ &= \bar{\bar{\mathcal{I}}}_E(fg;\mu).\end{aligned}$$

[Since all terms have been shown to be finite, the undefined form $\infty - \infty$ does not occur.] This completes the proof. ■

Property (N15) of the previous section may be generalized to show that $\bar{\bar{\mathcal{I}}}_E$ is continuous in a certain sense.

(G12) *Let $f \in \mathfrak{M}$ be such that $\bar{\bar{\mathcal{I}}}_E(f)$ eaif. Given any $\epsilon > 0$, there exists $\delta > 0$ with the property that any $F \subseteq E$ with $\mu(F) < \delta$ is such that $\bar{\bar{\mathcal{I}}}_F(f)$ exists and $|\bar{\bar{\mathcal{I}}}_F(f)| < \epsilon$.*

Proof: Let $\epsilon > 0$. Since $\bar{\bar{\mathcal{I}}}_E(f)$ eaif, we have that both $\bar{\mathcal{I}}_E(f^+)$ and $\bar{\mathcal{I}}_E(f^-)$ exist and are finite. We apply (N15) to f^+, finding $\delta_1 > 0$ such that $F \subseteq E$ and $\mu(F) < \delta_1$ imply $\bar{\mathcal{I}}_F(f^+) < \epsilon/2$. We then apply to (N15) to f^-, finding $\delta_2 > 0$ such that $F \subseteq E$ and $\mu(F) < \delta_2$ imply $\bar{\mathcal{I}}_F(f^-) < \epsilon/2$. Let δ denote the minimum of δ_1 and δ_2. Then $F \subseteq E$ and $\mu(F) < \delta$ imply

$$|\bar{\bar{\mathcal{I}}}_F(f)| = |\bar{\mathcal{I}}_F(f^+) - \bar{\mathcal{I}}_F(f^-)| \le \bar{\mathcal{I}}_F(f^+) + \bar{\mathcal{I}}_F(f^-) < \frac{\epsilon}{2} + \frac{\epsilon}{2} = \epsilon,$$

which completes the proof. ■

For the following property, we will be using the more general setup and definitions found in Section 5.8. We have the following:
(i) A measure space $(\Omega, \mathcal{F}, \mu)$ [as usual],
(ii) A nonempty set Ω' and a σ-field $\mathcal{F}'$ of subsets of Ω',
(iii) A function $f : \Omega \to \Omega'$ that is $\mathcal{F}/\mathcal{F}'$-measurable, and
(iv) A function $\phi : \Omega' \to \bar{\mathbb{R}}$ that is $\mathcal{F}'/\mathcal{B}^*$-measurable.

By (MF5′) found in that section, the composition $\phi \circ f : \Omega \to \bar{\mathbb{R}}$ is $\mathcal{F}/\mathcal{B}^*$-measurable. Also, recall the measure $\mu \circ f^{-1}$ with domain $\mathcal{F}'$ which is given by the rule $(\mu \circ f^{-1})(A') = \mu(f^{-1}(A'))$ for each $A' \in \mathcal{F}'$.

(G13) *Assume the immediately preceding notation and setup. We have the following, where $A' \in \mathcal{F}'$:*

(a) Let ϕ be nonnegative. Then both $\bar{\bar{\mathcal{I}}}_{f^{-1}(A')}(\phi \circ f; \mu)$ and $\bar{\bar{\mathcal{I}}}_{A'}(\phi; \mu \circ f^{-1})$ exist and $\bar{\bar{\mathcal{I}}}_{f^{-1}(A')}(\phi \circ f; \mu) = \bar{\bar{\mathcal{I}}}_{A'}(\phi; \mu \circ f^{-1})$.

(b) For general ϕ, $\bar{\bar{\mathcal{I}}}_{f^{-1}(A')}(\phi \circ f; \mu)$ exists iff $\bar{\bar{\mathcal{I}}}_{A'}(\phi; \mu \circ f^{-1})$ exists, in which case the equation $\bar{\bar{\mathcal{I}}}_{f^{-1}(A')}(\phi \circ f; \mu) = \bar{\bar{\mathcal{I}}}_{A'}(\phi; \mu \circ f^{-1})$ again holds.

Proof: We first handle (a). If ϕ is nonnegative, then $\phi \circ f$ is nonnegative, hence both $\bar{\bar{\mathcal{I}}}_{f^{-1}(A')}(\phi \circ f; \mu)$ and $\bar{\bar{\mathcal{I}}}_{A'}(\phi; \mu \circ f^{-1})$ exist. Next, take $\phi \in \mathfrak{S}(\Omega', \mathcal{F}', \mu \circ f^{-1})$, say, $\phi = \sum_{i=1}^n c_i I_{A'_i}$. By (G1), (G4), and (G8), we have

$$\begin{aligned}
\bar{\bar{\mathcal{I}}}_{A'}(\phi; \mu \circ f^{-1}) = \bar{\bar{\mathcal{I}}}(\phi I_{A'}; \mu \circ f^{-1}) &= \bar{\bar{\mathcal{I}}}\left(\sum_{i=1}^n c_i I_{A' \cap A'_i}; \mu \circ f^{-1}\right) \\
&= \sum_{i=1}^n c_i \bar{\bar{\mathcal{I}}}(I_{A' \cap A'_i}; \mu \circ f^{-1}) \\
&= \sum_{i=1}^n c_i (\mu \circ f^{-1})(A' \cap A'_i).
\end{aligned}$$

Next, observe that $\phi \circ f = \sum_{i=1}^{n} c_i I_{f^{-1}(A'_i)}$, hence

$$\begin{aligned}
\bar{\bar{\mathcal{I}}}_{f^{-1}(A')}(\phi \circ f;\mu) = \bar{\bar{\mathcal{I}}}((\phi \circ f) I_{f^{-1}(A')};\mu) &= \bar{\bar{\mathcal{I}}}\left(\sum_{i=1}^{n} c_i I_{f^{-1}(A') \cap f^{-1}(A'_i)};\mu\right) \\
&= \sum_{i=1}^{n} c_i \bar{\bar{\mathcal{I}}}(I_{f^{-1}(A' \cap A'_i)};\mu) \\
&= \sum_{i=1}^{n} c_i \mu(f^{-1}(A' \cap A'_i)) \\
&= \sum_{i=1}^{n} c_i (\mu \circ f^{-1})(A' \cap A'_i).
\end{aligned}$$

Comparing the identical expressions for $\bar{\bar{\mathcal{I}}}_{f^{-1}(A')}(\phi \circ f;\mu)$ and $\bar{\bar{\mathcal{I}}}_{A'}(\phi;\mu \circ f^{-1})$, we see that (a) holds whenever $\phi \in \mathfrak{S}(\Omega', \mathcal{F}', \mu \circ f^{-1})$.

To complete (a), let $\phi \in \mathfrak{N}(\Omega', \mathcal{F}', \mu \circ f^{-1})$. By (MF18), there exists a nondecreasing sequence $\{\phi_n\}_{n=1}^{\infty}$ in $\mathfrak{S}(\Omega', \mathcal{F}', \mu \circ f^{-1})$ with $\lim_n \phi_n = \phi$. It follows that $\{\phi_n \circ f\}_{n=1}^{\infty}$ is a nondecreasing sequence in $\mathfrak{S}(\Omega, \mathcal{F}, \mu)$ with $\lim_n \phi_n \circ f = \phi \circ f$. From this and the result for simple functions, we apply MCT twice to obtain

$$\begin{aligned}
\bar{\bar{\mathcal{I}}}_{f^{-1}(A')}(\phi \circ f;\mu) = \bar{\bar{\mathcal{I}}}_{f^{-1}(A')}\left(\lim_n \phi_n \circ f;\mu\right) &= \lim_n \bar{\bar{\mathcal{I}}}_{f^{-1}(A')}(\phi_n \circ f;\mu) \\
&= \lim_n \bar{\bar{\mathcal{I}}}_{A'}(\phi_n;\mu \circ f^{-1}) \\
&= \bar{\bar{\mathcal{I}}}_{A'}\left(\lim_n \phi_n;\mu \circ f^{-1}\right) \\
&= \bar{\bar{\mathcal{I}}}_{A'}(\phi;\mu \circ f^{-1}).
\end{aligned}$$

This gives (a).

We now focus on (b). For the existence claim, first observe that

$$(\phi \circ f)^+(\omega) = \max\{\phi(f(\omega)), 0\} = \phi^+(f(\omega)) = (\phi^+ \circ f)(\omega)$$

for each $\omega \in \Omega$, hence $(\phi \circ f)^+ = \phi^+ \circ f$; similarly, $(\phi \circ f)^- = \phi^- \circ f$. Using this and part (a), we have that

$$\begin{aligned}
&\bar{\bar{\mathcal{I}}}_{A'}(\phi;\mu \circ f^{-1}) \text{ exists} \\
\text{iff } &\bar{\mathcal{I}}_{A'}(\phi^+;\mu \circ f^{-1}) < +\infty \text{ or } \bar{\mathcal{I}}_{A'}(\phi^-;\mu \circ f^{-1}) < +\infty \\
\text{iff } &\bar{\mathcal{I}}_{f^{-1}(A')}(\phi^+ \circ f;\mu) < +\infty \text{ or } \bar{\mathcal{I}}_{f^{-1}(A')}(\phi^- \circ f;\mu) < +\infty \\
\text{iff } &\bar{\bar{\mathcal{I}}}_{f^{-1}(A')}(\phi \circ f;\mu) \text{ exists.}
\end{aligned}$$

Next, under the existence hypothesis, invoking (a) twice gives

$$\begin{aligned}\bar{\bar{\mathcal{I}}}_{A'}(\phi;\mu\circ f^{-1}) &= \bar{\mathcal{I}}_{A'}(\phi^+;\mu\circ f^{-1}) - \bar{\mathcal{I}}_{A'}(\phi^-;\mu\circ f^{-1})\\ &= \bar{\mathcal{I}}_{f^{-1}(A')}(\phi^+\circ f;\mu) - \bar{\mathcal{I}}_{f^{-1}(A')}(\phi^-\circ f;\mu)\\ &= \bar{\mathcal{I}}_{f^{-1}(A')}((\phi\circ f)^+;\mu) - \bar{\mathcal{I}}_{f^{-1}(A')}((\phi\circ f)^-;\mu)\\ &= \bar{\bar{\mathcal{I}}}_{f^{-1}(A')}(\phi\circ f;\mu).\end{aligned}$$

This gives (b) and thus completes the entire proof. ∎

We have discussed in a complete and thorough fashion the behavior and the properties concerning the functional $\bar{\bar{\mathcal{I}}}_E$. Much, but not all, of the work in the properties given within this section were extensions of the corresponding results for $\bar{\mathcal{I}}_E$ found in the previous section. In turn, this section's results will form the backbone for many of the results to be found in the next section, where the Lebesgue integral is finally defined as a suitable extension of $\bar{\bar{\mathcal{I}}}_E$ to the larger domain of μ-a.e. defined $\mathcal{F}/\mathcal{B}^*$-measurable functions.

At this stage, almost all of the hard work is behind us. The path has been somewhat arduous and possibly tedious, but the prize awaits in the next section. The reader is urged to be thoroughly familiar with the properties in this section for what follows in the next section, as well as for what is in the remainder of this book.

Exercises.

In these exercises, be careful to demonstrate existence of any quantities in question, as did the various proofs of the properties found in the text. Be sure to work the starred problems; some of the results in these problems will be called on at a future time.

1*. Let $E \in \mathcal{F}$ be such that $\mu(E) < +\infty$, and let $f \in \mathfrak{M}$ be such that $f = 0$ on E^c and $m \le f \le M$ on E, where $m, M \in \mathbb{R}$. Then $\bar{\bar{\mathcal{I}}}(f)$ exists and is finite. Furthermore, we have $m\mu(E) \le \bar{\bar{\mathcal{I}}}(f) \le M\mu(E)$. [This should be straightforward.]

2*. Let $E \in \mathcal{F}$, $\epsilon > 0$ and let $f \in \mathfrak{M}$ be such that $\bar{\bar{\mathcal{I}}}_E(f)$ exists and is finite. There exists a subset $F \subseteq E$ such that $\mu(F) < +\infty$ and $|\bar{\bar{\mathcal{I}}}_E(f) - \bar{\bar{\mathcal{I}}}_F(f)| < \epsilon$. [Follow the following outline, verifying each step in straightforward fashion:
(i) It suffices to prove the result for nonnegative finite f.
(ii) Let $f \ge 0$. For every $x > 0$, $\mu(\{\omega \in E : f(\omega) > x\}) < +\infty$. [Argue by contradiction.]
(iii) For each $n \in \mathbb{N}$, let $E_n = \{\omega \in E : f(\omega) \ge \frac{1}{n}\}$. Then the E_n's are increasing with limit $\{\omega \in E : f(\omega) > 0\}$, and $\mu(E_n) < +\infty$ for each $n \in \mathbb{N}$.
(iv) We have $\bar{\bar{\mathcal{I}}}_E(f) = \lim_n \bar{\bar{\mathcal{I}}}_{E_n}(f)$, hence we may take the F of the statement to be some E_n for sufficiently large enough n.

3. f is such that $\bar{\bar{\mathcal{I}}}_E(f)$ exists and is finite iff for any $\epsilon > 0$ there are functions g and h in $\mathfrak{M}$ such that $h \le f \le g$ on E and $\bar{\bar{\mathcal{I}}}_E(g-h) < \epsilon$.

4. Let $f, f_1, f_2, \cdots$ denote a sequence of nonnegative functions in $\mathfrak{M}$. For each $n \in \mathbb{N}$ and $E \in \mathcal{F}$, define $\nu_n(E) = \bar{\bar{\mathcal{I}}}_E(f_n;\mu)$ and $\nu(E) = \bar{\bar{\mathcal{I}}}_E(f;\mu)$. Furthermore, assume that $\nu(\Omega), \nu_1(\Omega), \nu_2(\Omega), \cdots$ are finite and $f_n \to f$ μ-a.e. on Ω. Then

$$\sup\{|\nu(E) - \nu_n(E)| : E \in \mathcal{F}\} \le \bar{\bar{\mathcal{I}}}(|f_n - f|;\mu) \to 0 \quad \text{as } n \to +\infty.$$

This result is known as *Scheffe's [Useful Convergence] Theorem.*

5. Let $(\Omega, \mathcal{F}, \mu) = (\mathbb{R}^k, \mathcal{B}^k, \lambda_k)$ with $E \in \mathcal{F}$, and let $f \in \mathfrak{M}$ be such that $\bar{\bar{\mathcal{I}}}(f)$ exists and is finite.

(a) Suppose that $\{f_n\}_{n=1}^{\infty}$ is a sequence of functions in $\mathfrak{M}$ such that

$$f_n(\boldsymbol{x}) = \begin{cases} f(\boldsymbol{x}) & \text{if } \boldsymbol{x} \text{ is such that } |f(\boldsymbol{x})| \le n \text{ and } \|\boldsymbol{x}\| \le n \\ 0 & \text{otherwise.} \end{cases}$$

Then $\lim_n \bar{\bar{\mathcal{I}}}_E(f_n)$ exists and equals $\bar{\bar{\mathcal{I}}}_E(\lim_n f_n)$.

(b) Reset everything in (a), and this time let $f_n(\boldsymbol{x}) = f(\boldsymbol{x}) \exp(-\|\boldsymbol{x}\|^2/n)$ for each $\boldsymbol{x} \in \mathbb{R}^k$ and $n \in \mathbb{N}$. Then $\lim_n \bar{\bar{\mathcal{I}}}_E(f_n) = \bar{\bar{\mathcal{I}}}_E(f)$.

6*. Suppose that $\mu(\Omega) < +\infty$, $\{f_n\}_{n=1}^{\infty}$ is a sequence of functions in $\mathfrak{M}$ such that there exists $M \in \mathbb{R}$ with $|f_n| \le M$ for each $n \in \mathbb{N}$, and let $f \in \mathfrak{M}$ be such that $f_n \to f$ uniformly on E. Then $\lim_n \bar{\bar{\mathcal{I}}}_E(f_n) = \bar{\bar{\mathcal{I}}}_E(f)$.

7*. Let $\{f_n\}_{n=1}^{\infty}$ denote a nondecreasing sequence of functions in $\mathfrak{M}$ such that $\bar{\bar{\mathcal{I}}}_E(f_n)$ exists and is finite for each $n \in \mathbb{N}$ and $\sup_{n \in \mathbb{N}} \bar{\bar{\mathcal{I}}}_E(f_n) < +\infty$. Then $\bar{\bar{\mathcal{I}}}_E(\lim_n f_n)$ exists, is finite, and equals $\lim_n \bar{\bar{\mathcal{I}}}_E(f_n)$. This is one form of *Beppo Levi's Theorem.*

8. Let $\{f_n\}_{n=1}^{\infty}$, $\{g_n\}_{n=1}^{\infty}$, and $\{h_n\}_{n=1}^{\infty}$ denote sequences of functions in $\mathfrak{M}$ such that $f_n \to f$ μ-a.e., $g_n \to g$ μ-a.e., and $h_n \to h$ μ-a.e. for some functions $f, g, h \in \mathfrak{M}$. Suppose for any $p \in \{g, g_1, g_2, \cdots, h, h_1, h_2, \cdots\}$ that $\bar{\bar{\mathcal{I}}}_E(p)$ exists and is finite. Furthermore, suppose that $\lim_n \bar{\bar{\mathcal{I}}}_E(g_n) = \bar{\bar{\mathcal{I}}}_E(g)$ and $\lim_n \bar{\bar{\mathcal{I}}}_E(h_n) = \bar{\bar{\mathcal{I}}}_E(h)$. Also, assume that $g_n \le f_n \le h_n$ for every $n \in \mathbb{N}$.

(a) $\bar{\bar{\mathcal{I}}}_E(f_n)$ exists and is finite for all $n \in \mathbb{N}$, $\bar{\bar{\mathcal{I}}}_E(f)$ exists and is finite, and $\lim_n \bar{\bar{\mathcal{I}}}_E(f_n)$ equals $\bar{\bar{\mathcal{I}}}_E(f)$.

(b) DCT may be obtained from (a).

9*. In properties (G6) and (G9), we gave a proof for the "everywhere" case. Prove the "almost everywhere" case for these two properties. [No problems should arise.]

10. Suppose that for each $t \in \mathbb{R}$ we have $f_t \in \mathfrak{M}$, and suppose that $\lim_{t \to +\infty} f_t(\omega)$ exists for each $\omega \in \Omega$. Next, assume that there exists nonnegative $g \in \mathfrak{M}$ such that $\bar{\bar{\mathcal{I}}}_E(g)$ exists and is finite; further assume that $|f_t| \le g$ for each $t \in \mathbb{R}$.

(a) $\bar{\bar{\mathcal{I}}}_E(f_t)$ exists and is finite.

(b) $\bar{\bar{\mathcal{I}}}_E(\lim_{t \to +\infty} f_t)$ exists and is finite. [Show first that $\lim_{t \to +\infty} f_t \in \mathfrak{M}$.]

(c) $\lim_{t \to +\infty} \bar{\bar{\mathcal{I}}}_E(f_t)$ exists and is finite.

(d) $\lim_{t \to +\infty} \bar{\bar{\mathcal{I}}}_E(f_t) = \bar{\bar{\mathcal{I}}}_E(\lim_{t \to +\infty} f_t)$. [This is a more general form of DCT.]

(e) Suppose that instead of t ranging over $\mathbb{R}$, we have t ranging over some interval (a, b), where $-\infty \le a < b \le +\infty$, and suppose that limits are now as $t \to a$ from the right [or $t \to b$ from the left]. Then (a)-(d) hold in these two modified scenarios.

11. Suppose that $\{f_n\}_{n=1}^{\infty}$ is a sequence of functions in $\mathfrak{M}$ such that $\lim_n f_n$ exists on Ω. Next, suppose that $\{g_n\}_{n=1}^{\infty}$ is a sequence of nonnegative functions in $\mathfrak{M}$ such that for all $n \in \mathbb{N}$ we have (i) $\bar{\bar{\mathcal{I}}}_E(g_n)$ exists and is finite and (ii) $|f_n| \le g_n$. Finally, suppose that (iii) $\lim_n g_n$ exists on Ω, (iv) $\bar{\bar{\mathcal{I}}}_E(\lim_n g_n)$ exists and is finite, and (v) $\bar{\bar{\mathcal{I}}}_E(\lim_n g_n) = \lim_n \bar{\bar{\mathcal{I}}}_E(g_n)$. Then $\bar{\bar{\mathcal{I}}}_E(\lim_n f_n) = \lim_n \bar{\bar{\mathcal{I}}}_E(f_n)$. [This is yet another more general form of DCT. As with the previous exercise, remember to show that all quantities in fact exist and make sense before showing that they're equal.]

12*. Suppose that $\bar{\bar{\mathcal{I}}}_{\bigcup_{n=1}^{\infty} E_n}(f)$ exists and is finite, where $\{E_n\}_{n=1}^{\infty}$ is a disjoint sequence of $\mathcal{F}$-sets. Then $\bar{\bar{\mathcal{I}}}_{\bigcup_{n=1}^{\infty} E_n}(f) = \sum_{n=1}^{\infty} \bar{\bar{\mathcal{I}}}_{E_n}(f)$, and the convergence of the series is absolute.

13. Assume for each $n \in \mathbb{N}$ that (i) $\bar{\bar{\mathcal{I}}}(f_n)$ exists and is finite and (ii) f_n is finite. Assume also that $\sum_{n=1}^{\infty} \bar{\bar{\mathcal{I}}}(f_n) < +\infty$.
(a) For μ-a.e. $\omega \in \Omega$, $\sum_{n=1}^{\infty} f_n(\omega)$ exists.
(b) Define g on Ω by writing

$$g(\omega) = \begin{cases} \sum_{n=1}^{\infty} f_n(\omega) & \text{if } \sum_{n=1}^{\infty} f_n(\omega) \text{ exists} \\ 0 & \text{otherwise.} \end{cases}$$

Then $\bar{\bar{\mathcal{I}}}(g) = \sum_{n=1}^{\infty} \bar{\bar{\mathcal{I}}}(f_n)$.
(c) The assumption that each f_n is finite may be relaxed to saying that each f_n is finite μ-a.e. on Ω.

14. Suppose that $\{f_n\}_{n=1}^{\infty}$ is a sequence of elements of $\mathfrak{N}$ converging to some f [which is, of course, in $\mathfrak{N}$]. Furthermore, assume that there is $0 \le M < +\infty$ such that $\bar{\bar{\mathcal{I}}}(f_n) \le M$ for each $n \in \mathbb{N}$. Then $\bar{\bar{\mathcal{I}}}(f)$ exists, is finite, and is no more than M.

6.4 STAGE FOUR: ALMOST EVERYWHERE DEFINED FUNCTIONS

We may now complete the process that leads us to a rigorous definition of the abstract Lebesgue integral relative to a generic measure space $(\Omega, \mathcal{F}, \mu)$. Most of the work for this section has been done in the previous three sections. In the previous section, we had a functional $\bar{\bar{\mathcal{I}}}_E$ that admitted functions $f \in \mathfrak{M}(E)$ as arguments. Recall that functions f in $\mathfrak{M}(E)$ are [by definition] $\mathcal{F}/\mathcal{B}^*$-measurable functions with domain Ω satisfying the requirement that $\bar{\mathcal{I}}_E(f^+) < +\infty$ or $\bar{\mathcal{I}}_E(f^-) < +\infty$. In this section, we shall define an extension of $\bar{\bar{\mathcal{I}}}_E$ that allows us to handle situations where functions are defined only μ-a.e. on Ω. We immediately proceed to the definition.

Definition. Let f denote an $\mathcal{F}/\mathcal{B}^*$-measurable function defined μ-a.e. on Ω. We define f^* on all of Ω by writing

$$f^*(\omega) = \begin{cases} f(\omega) & \omega \in \operatorname{dom} f \\ 0 & \text{otherwise.} \end{cases}$$

The function f^* above has domain Ω and is $\mathcal{F}/\mathcal{B}^*$-measurable by (MF6). All that has been done is to take the function f and to define it to be zero over where it was originally undefined.

Definition. Let $E \in \mathcal{F}$, and let f denote an $\mathcal{F}/\mathcal{B}^*$-measurable function defined μ-a.e. on Ω. If $f^* \in \mathfrak{M}(E)$, that is, if $\bar{\bar{\mathcal{I}}}_E(f^*)$ exists as an element of $[-\infty, +\infty]$, we define the *Lebesgue integral of f over E* by writing

$$\int_E f = \bar{\bar{\mathcal{I}}}_E(f^*).$$

If $\bar{\bar{\mathcal{I}}}_E(f^*)$ fails to exist, that is, if both $\bar{\mathcal{I}}_E(f^{*+})$ and $\bar{\mathcal{I}}_E(f^{*-})$ equal $+\infty$, we leave $\int_E f$ undefined. In other words, $\int_E f$ is undefined iff $f^* \notin \mathfrak{M}(E)$.

Notation. Some points regarding notation should be discussed.

(1) Often the symbol $\int_E f$ is written as $\int_E f\,d\mu$. The symbol $d\mu$ is merely a part of the entire symbol $\int_E f\,d\mu$, and it has no independent meaning other than to specify that μ is the measure under consideration [as opposed to some other measure]. When no confusion can arise as to which measure is being considered, we will write $\int_E f$ for $\int_E f\,d\mu$. This is similar to the notation we adopted for the definitions of $\mathcal{I}_E$, $\bar{\mathcal{I}}_E$, and $\bar{\bar{\mathcal{I}}}_E$, where, for example, we would write $\bar{\bar{\mathcal{I}}}_E(f)$ instead of $\bar{\bar{\mathcal{I}}}_E(f;\mu)$ if no confusion could arise.

(2) The symbol $\int_E f\,d\mu$ is intended to notationally mimic the familiar Riemann integral notation from calculus, and this notational mimicry is justified by the fact that the Lebesgue integral behaves quite similarly to the Riemann integral. [This fact is shown in Chapter 7.]

(3) If $E = \Omega$, the usual convention is to write $\int f$ for $\int_\Omega f$, although of course there is nothing wrong with the latter notation. We have used this convention in the previous three sections, writing $\mathcal{I}$ for $\mathcal{I}_\Omega$, $\bar{\mathcal{I}}$ for $\bar{\mathcal{I}}_\Omega$, and $\bar{\bar{\mathcal{I}}}$ for $\bar{\bar{\mathcal{I}}}_\Omega$.

(4) The notations $\int_E f(\omega)\,d\mu(\omega)$ and $\int_E f(\omega)\,\mu(d\omega)$ are alternative expressions for $\int_E f\,d\mu$ or $\int_E f$. [It is purely a notational choice whether to write $d\mu(\omega)$ instead of $\mu(d\omega)$.] Even though the dummy variable ω seems redundant, it may be helpful at times to use this notation. This is the convention used for the Riemann integral. For example, we write $\int_0^1 x^2\,dx$ [say] instead of the more pedantic expression

$$\int_0^1 f, \quad \text{where } f(x) = x^2 \text{ for each } 0 \le x \le 1.$$

Even though the Lebesgue integral is a different item than the Riemann integral, we will follow the same notational conventions that are allowed with the Riemann integral.

It is immediate that no logical problems arise with our definition of the Lebesgue integral. It is also immediate that the functional $\int_E$ consistently extends $\bar{\bar{\mathcal{I}}}_E$, for if $f \in \mathfrak{M}(E)$, then $f^* = f$ on Ω, giving $f^* \in \mathfrak{M}(E)$, whence

$$\int_E f\,d\mu \text{ exists} \quad \text{and} \quad \int_E f\,d\mu = \bar{\bar{\mathcal{I}}}_E(f^*) = \bar{\bar{\mathcal{I}}}_E(f).$$

From the fact that $\int_E$ consistently extends $\bar{\bar{\mathcal{I}}}_E$, the statements of the previous section hold with the symbol $\int_E$ everywhere replacing the symbol $\bar{\bar{\mathcal{I}}}_E$. Since $\bar{\bar{\mathcal{I}}}_E$ extends $\bar{\mathcal{I}}_E$, all statements in Section 2 hold with the symbol $\int_E$ everywhere replacing $\bar{\mathcal{I}}_E$. Since $\bar{\mathcal{I}}_E$ extends $\mathcal{I}_E$, all statements in Section 1 hold with the symbol $\int_E$ everywhere replacing $\mathcal{I}_E$. We shall keep these facts in mind throughout.

The following definition that follows is given in a formal version and a "workaday" version. The workaday version is all that is needed for this section, and hence it is the only one stated. The formal version will be given in Chapter 8 and is really needed only when we discuss bounded linear functionals in later work. [And, in fact, it is really used only to make one point there.] There will be no logical gaps or circularities in using the workaday definition of $L^1(E)$ given below.

Definition. Let $E \in \mathcal{F}$. We will write $L^1(E)$ for the collection of $\mathcal{F}/\mathcal{B}^*$-measurable functions defined μ-a.e. on Ω such that $\int_E f$ exists and is finite.[5] A function $f \in L^1(E)$ is also equivalently said to be *integrable*[6] *[with respect to μ] over E.*

For the final time, we present a long list of properties. Many of these properties are immediate [or nearly so] by recalling the appropriate result from the previous section.

> *In the statements below, all functions are assumed to be $\mathcal{F}/\mathcal{B}^*$-measurable and defined μ-a.e. on Ω, and, as usual, all sets are assumed to be $\mathcal{F}$-sets.*

(L1) *If $\int_E f$ exists and $F \subseteq E$, then $\int_F f$ exists. If $f \in L^1(E)$ and $F \subseteq E$, then $f \in L^1(F)$.*

Proof: By hypothesis, we have that $\bar{\bar{\mathcal{I}}}_E(f^*)$ exists. By (G1), we have that $\bar{\bar{\mathcal{I}}}_F(f^*)$ exists, hence $\int_F f$ exists. This gives the first claim. For the second claim, assume that $\int_E f$ exists and is finite. Therefore, $\bar{\bar{\mathcal{I}}}_E(f^*)$ exists and is finite. By another appeal to (G1), we obtain that $\bar{\bar{\mathcal{I}}}_F(f^*)$ exists and is finite. It follows that $\int_F f$ exists and is finite, giving $f \in L^1(F)$. This completes the entire proof. ■

(L2) *If $\mu(E) = 0$, then $\int_E f = 0$.*

Proof: If $\mu(E) = 0$, then $\bar{\bar{\mathcal{I}}}_E(f^*)$ exists and equals zero by appealing to (G1). Therefore, $\int_E f$ exists and $\int_E f = \bar{\bar{\mathcal{I}}}_E(f^*) = 0$. ■

(L3) *If $f \geq 0$ on its domain and $\int_E f = 0$, then $f = 0$ μ-a.e. on E.*

Proof: By assumption, we have $\bar{\bar{\mathcal{I}}}_E(f^*) = 0$. Since $f^* \geq 0$ on Ω, (N10) forces $f^* = 0$ μ-a.e. on E. Since $f = f^*$ μ-a.e. on Ω, we have $f = 0$ μ-a.e. on E. ■

(L4) *If $f \in L^1(E)$, then f is finite μ-a.e. on E.*

[5] To be sure, if $\int_E f$ is $+\infty$ or $-\infty$, then we may say that $\int_E f$ exists, but we may not say that $f \in L^1(E)$.

[6] Therefore, *integrable* does not mean that "we may integrate f over E and obtain a number in $[-\infty, +\infty]$" even though this is what the word might suggest; it instead means the stronger statement "not only does the integral of f over E exist as an extended real number, but it is actually finite."

Proof: $\bar{\bar{\mathcal{I}}}_E(f^*)$ exists and is finite by assumption, hence (G3) implies that f^* is finite μ-a.e. on E. Since $f = f^*$ μ-a.e. on Ω, we have that f is finite μ-a.e. on E. ∎

Convention. Let f be defined μ-a.e. on Ω, and let I_E denote the usual indicator function of the set E. [I_E has domain Ω.] We define the function fI_E on Ω by writing

$$(fI_E)(\omega) = \begin{cases} f(\omega) & \omega \in E \cap \operatorname{dom} f \\ 0 & \text{otherwise} \end{cases}$$

for each $\omega \in \Omega$. If $E \in \mathcal{F}$ and f is $\mathcal{F}/\mathcal{B}^*$-measurable, then fI_E is $\mathcal{F}/\mathcal{B}^*$-measurable by (MF6). If f is defined on all of Ω, this definition reduces to the familiar one.

(L5) *If one of $\int_F fI_E$ and $\int_{E\cap F} f$ exists, then so does the other and both integrals are equal.*

Proof: Consider four cases for an element $\omega \in \Omega$: (i) $\omega \in E \cap \operatorname{dom} f$, (ii) $\omega \in E \cap (\operatorname{dom} f)^c$, (iii) $\omega \in E^c \cap \operatorname{dom} f$, and (iv) $\omega \in E^c \cap (\operatorname{dom} f)^c$. We have $(fI_E)^* = fI_E$ on Ω since fI_E is defined on Ω, and verifying the equation $(fI_E)(\omega) = (f^*I_E)(\omega)$ for ω in cases (i)-(iv) shows that $fI_E = f^*I_E$ on Ω. Therefore, we obtain that $(fI_E)^* = f^*I_E$ on Ω.

We now use the just-obtained identity: $\int_F fI_E$ exists iff $\bar{\bar{\mathcal{I}}}_F((fI_E)^*)$ exists iff $\bar{\bar{\mathcal{I}}}_F(f^*I_E)$ exists iff $\bar{\bar{\mathcal{I}}}_{E\cap F}(f^*)$ exists iff $\int_{E\cap F} f$ exists. [We have used (G4) and (G5).] This establishes the existence claim. Next, since all quantities involved exist, we have by (G4) and (G5) that

$$\int_F fI_E = \bar{\bar{\mathcal{I}}}_F((fI_E)^*) = \bar{\bar{\mathcal{I}}}_F(f^*I_E) = \bar{\bar{\mathcal{I}}}_{E\cap F}(f^*) = \int_{E\cap F} f,$$

which establishes the claim of equality. This completes the proof. ∎

(L6) *Suppose that $E = \bigcup_{n=1}^{\infty} E_n$, where the union is disjoint. If $f \geq 0$ on its domain or if $f \in L^1(E)$, then $\sum_{n=1}^{\infty} \int_{E_n} f = \int_E f$.*

Proof: If $f \geq 0$ on its domain or if $f \in L^1(E)$, then each of $\int_E f$, $\int_{E_1} f$, $\int_{E_2} f, \cdots$ exist by (L1) and equal $\bar{\bar{\mathcal{I}}}_E(f^*)$, $\bar{\bar{\mathcal{I}}}_{E_1}(f^*)$, $\bar{\bar{\mathcal{I}}}_{E_2}(f^*), \cdots$, respectively. Consider the purported equation

$$\sum_{n=1}^{\infty} \int_{E_n} f = \sum_{n=1}^{\infty} \bar{\bar{\mathcal{I}}}_{E_n}(f^*) = \bar{\bar{\mathcal{I}}}_E(f^*) = \int_E f.$$

The first and third equalities are by definition. If $f \geq 0$ on its domain, then the second equality holds by the comments following (N9). If $f \in L^1(E)$, the second equality follows by (G2). ∎

Property (L6) above also holds when E is the finite disjoint union of $E_1, \cdots, E_n$ [and when $f \geq 0$ on its domain or $f \in L^1(E)$]. To see this, take $E_{n+1} = E_{n+1} = \cdots = \varnothing$ and use (L2).

The next property is important. It generalizes the results of (G5) and (G6) from $\overset{\equiv}{\mathcal{I}}_E$ to $\int_E$.

(L7) *We have the following:*

(a) If $f = g$ μ-a.e. on E and $\int_E f$ exists or $\int_E g$ exists, then both $\int_E f$ and $\int_E g$ exist and $\int_E f = \int_E g$.

(b) If $f \leq g$ μ-a.e. on E and both $\int_E f$ and $\int_E g$ exist, then $\int_E f \leq \int_E g$.

Proof: For (a), we have that $f^* = g^*$ μ-a.e. on E. By (G5), the existence of at least one of $\overset{\equiv}{\mathcal{I}}_E(f^*)$ and $\overset{\equiv}{\mathcal{I}}_E(g^*)$ forces the existence of both of them, and both quantities are equal. Therefore, both $\int_E f$ and $\int_E g$ exist, and

$$\int_E f = \overset{\equiv}{\mathcal{I}}_E(f^*) = \overset{\equiv}{\mathcal{I}}_E(g^*) = \int_E g,$$

giving (a). For (b), the assumptions and (G6) imply that $f^* \leq g^*$ μ-a.e. on E, $\overset{\equiv}{\mathcal{I}}_E(f^*)$ and $\overset{\equiv}{\mathcal{I}}_E(g^*)$ exist, and $\overset{\equiv}{\mathcal{I}}_E(f^*) \leq \overset{\equiv}{\mathcal{I}}_E(g^*)$. Putting everything together gives

$$\int_E f = \overset{\equiv}{\mathcal{I}}_E(f^*) \leq \overset{\equiv}{\mathcal{I}}_E(g^*) = \int_E g,$$

completing the proof of (b). ∎

(L8) $f \in L^1(E)$ *iff* $|f| \in L^1(E)$. *Also, $\int_E f$ exists iff $\int_E |f|$ exists, in which case we have $|\int_E f| \leq \int_E |f|$.*

Proof: Before doing anything, observe first that $|f^*| = |f|^*$ on Ω. Using this, we have that $f \in L^1(E)$ iff $\overset{\equiv}{\mathcal{I}}_E(f^*)$ exists and is finite, which by (G7) is equivalent to saying that $\overset{\equiv}{\mathcal{I}}_E(|f^*|)$ exists and is finite. This in turn holds iff $\overset{\equiv}{\mathcal{I}}_E(|f|^*)$ exists and is finite, which is equivalent to asserting that $|f| \in L^1(E)$. This gives the first assertion.

The second assertion is similar: $\int_E f$ exists iff $\overset{\equiv}{\mathcal{I}}_E(f^*)$ exists iff $\overset{\equiv}{\mathcal{I}}_E(|f^*|)$ exists [(G7) again] iff $\overset{\equiv}{\mathcal{I}}_E(|f|^*)$ exists iff $\int_E |f|$ exists. We next have

$$\left|\int_E f\right| = \left|\overset{\equiv}{\mathcal{I}}_E(f^*)\right| \leq \overset{\equiv}{\mathcal{I}}_E(|f^*|) = \overset{\equiv}{\mathcal{I}}_E(|f|^*) = \int_E |f|$$

by (G7), completing the proof. ∎

The next property is important, but a little bit of notation must be discussed first. Let f and g denote functions defined μ-a.e. on Ω. Then, various problems may arise in writing $f(\omega) + g(\omega)$: (i) ω might not be in the domain of f or the domain of g, and (ii) ω might be in both the domain of f

and the domain of g, but one of the output values is $-\infty$ while the other is $+\infty$. Therefore, even if both $f(\omega)$ and $g(\omega)$ are defined, we may have that $f(\omega)+g(\omega)$ is undefined. Noting this, we will make the following definitions.

Definition. Let $g, f, f_1, f_2, \cdots$ denote functions defined μ-a.e. on Ω. We define $f \oplus g$ on all of Ω by writing

$$(f \oplus g)(\omega) = \begin{cases} f(\omega)+g(\omega) & \text{if } f(\omega)+g(\omega) \text{ is defined} \\ 0 & \text{otherwise.} \end{cases}$$

If $n \in \mathbb{N}$, we define $\bigoplus_{i=1}^n f_i$ on Ω by writing

$$\left(\bigoplus_{i=1}^n f_i\right)(\omega) = \begin{cases} f_1(\omega)+\cdots+f_n(\omega) & \text{if } f_1(\omega)+\cdots+f_n(\omega) \text{ is defined} \\ 0 & \text{otherwise} \end{cases}$$

for every $\omega \in \Omega$. For "subtraction", we will define $f \ominus g$ as $f \oplus (-g)$. That is, $f \ominus g$ is defined on all of Ω by the rule

$$(f \ominus g)(\omega) = \begin{cases} f(\omega)-g(\omega) & \text{if } f(\omega)-g(\omega) \text{ is defined} \\ 0 & \text{otherwise.} \end{cases}$$

By appealing to (MF6), we have that $\bigoplus_{i=1}^n f_i$ is $\mathcal{F}/\mathcal{B}^*$-measurable if each of $f_1, \cdots, f_n$ is $\mathcal{F}/\mathcal{B}^*$-measurable.

The next property is extremely important, and gives the analogue of (G8) [and the comments following that property] for the Lebesgue integral.

(L9) *Let $c \in \mathbb{R}$ and $f, g \in L^1(E)$. We have the following:*
(a) $cf \in L^1(E)$ and $\int_E cf = c\int_E f$.
(b) $f \oplus g \in L^1(E)$ and $\int_E (f \oplus g) = \int_E f + \int_E g$.
(c) $f \ominus g \in L^1(E)$ and $\int_E (f \ominus g) = \int_E f - \int_E g$.

Proof: We prove (a), first taking $E=\Omega$. We have $f \in L^1(\Omega)$, hence $\bar{\bar{\mathcal{I}}}(f^*)$ exists and is finite. By (G8) we have that $\bar{\bar{\mathcal{I}}}(cf^*)$ exists, is finite, and equals $c\bar{\bar{\mathcal{I}}}(f^*)$. Since $cf^* = (cf)^*$ on Ω, we have that $\bar{\bar{\mathcal{I}}}((cf)^*)$ exists, is finite, and equals $\bar{\bar{\mathcal{I}}}(cf^*)$. We accordingly have that

$$\bar{\bar{\mathcal{I}}}((cf)^*) = \bar{\bar{\mathcal{I}}}(cf^*) = c\bar{\bar{\mathcal{I}}}(f^*), \tag{6.16}$$

and all quantities are finite. The left side of (6.16) is $\int cf$ and is finite, hence $cf \in L^1(\Omega)$. Since the right side of (6.16) is $c\int f$, we have established (a) for the case $E=\Omega$.

Using the previous paragraph, we prove (a) when $E \in \mathcal{F}$ is generic. Suppose that $f \in L^1(E)$. Then $\int_E f$ exists and is finite, so that by (L5) $\int fI_E$ exists, is finite, and equals $\int_E f$. That is, we have $f \in L^1(E)$ iff $fI_E \in L^1(\Omega)$. Therefore, if c is real, we have

$$f \in L^1(E) \leftrightarrow fI_E \in L^1(\Omega) \leftrightarrow cfI_E \in L^1(\Omega) \leftrightarrow cf \in L^1(E),$$

where in the second equivalence we have used the fact that (a) has been proved for $E = \Omega$. [The third equivalence is proved in similar fashion as is the first equivalence.] Therefore, $f \in L^1(E)$ forces $cf \in L^1(E)$. That is, the fact that $\int_E f$ exists and is finite forces $\int_E cf$ to exist and be finite. Using this, we have

$$\int_E cf = \int cfI_E = c\int fI_E = c\int_E f,$$

where the second equality follows from the $E = \Omega$ case and the first and third equalities follow from (L5). This completely establishes part (a).

We now turn to (b), first considering the case where $E = \Omega$, so that both f and g are in $L^1(\Omega)$. Besides having the functions f^* and g^* defined on all of Ω, we define $\hat{f}$ and $\hat{g}$ on all of Ω by

$$\hat{f}(\omega) = \begin{cases} f(\omega) & f(\omega) \text{ defined, finite} \\ 0 & \text{otherwise} \end{cases} \quad \text{and} \quad \hat{g}(\omega) = \begin{cases} g(\omega) & g(\omega) \text{ defined, finite} \\ 0 & \text{otherwise.} \end{cases}$$

Appealing to (MF6) shows that $\hat{f}$ and $\hat{g}$ are $\mathcal{F}/\mathcal{B}^*$-measurable. We desire to show that $f \oplus g \in L^1(\Omega)$. To see this, we have the following chain of statements:

(1) $\bar{\bar{\mathcal{I}}}(f^*)$, $\bar{\bar{\mathcal{I}}}(g^*)$ exist and are finite.

Proof: This holds by the assumption that both f and g are in $L^1(\Omega)$.

(2) $\bar{\bar{\mathcal{I}}}(\hat{f})$, $\bar{\bar{\mathcal{I}}}(\hat{g})$ exist and are finite; furthermore, we have $\bar{\bar{\mathcal{I}}}(f^*) = \bar{\bar{\mathcal{I}}}(\hat{f})$ and $\bar{\bar{\mathcal{I}}}(g^*) = \bar{\bar{\mathcal{I}}}(\hat{g})$.

Proof: Since $f \in L^1(\Omega)$, f is finite μ-a.e. on Ω by (L4). Since f is finite μ-a.e. on Ω and defined μ-a.e. on Ω, it follows that $f^* = \hat{f}$ μ-a.e. on Ω. The same argument shows that $g^* = \hat{g}$ μ-a.e. on Ω. By (G5), we have that $\bar{\bar{\mathcal{I}}}(\hat{f})$ exists, is finite, and equals $\bar{\bar{\mathcal{I}}}(f^*)$. The same argument shows that $\bar{\bar{\mathcal{I}}}(\hat{g})$ exists, is finite, and equals $\bar{\bar{\mathcal{I}}}(g^*)$.

(3) $\bar{\bar{\mathcal{I}}}(\hat{f} + \hat{g})$ exists and is finite, and $\bar{\bar{\mathcal{I}}}(\hat{f} + \hat{g}) = \bar{\bar{\mathcal{I}}}(\hat{f}) + \bar{\bar{\mathcal{I}}}(\hat{g})$.

Proof: Both $\bar{\bar{\mathcal{I}}}(\hat{f})$ and $\bar{\bar{\mathcal{I}}}(\hat{g})$ exist and are finite by (2). Invoking (G8) completes matters.

(4) $\bar{\bar{\mathcal{I}}}(f \oplus g)$ exists, is finite, and equals $\bar{\bar{\mathcal{I}}}(\hat{f}) + \bar{\bar{\mathcal{I}}}(\hat{g})$.

Proof: We have that $\hat{f} + \hat{g} = f \oplus g$ μ-a.e. on Ω. By (3), $\bar{\bar{\mathcal{I}}}(\hat{f} + \hat{g})$ exists and is finite, so (G5) asserts that $\bar{\bar{\mathcal{I}}}(f \oplus g)$ exists, is finite, and equals $\bar{\bar{\mathcal{I}}}(\hat{f} + \hat{g})$. By the equation in (3), we have that $\bar{\bar{\mathcal{I}}}(f \oplus g)$ exists, is finite, and equals $\bar{\bar{\mathcal{I}}}(\hat{f}) + \bar{\bar{\mathcal{I}}}(\hat{g})$.

(5) $\bar{\bar{\mathcal{I}}}((f \oplus g)^*)$ exists, is finite, and equals $\bar{\bar{\mathcal{I}}}(f^*) + \bar{\bar{\mathcal{I}}}(g^*)$.

Proof: This holds by observing that $f \oplus g = (f \oplus g)^*$ on Ω and then using (4) and (2) [in that order].

From (5) we conclude that $\int f \oplus g$ exists and is finite, whence $f \oplus g \in L^1(\Omega)$. Furthermore, (5) gives the equation $\int f \oplus g = \int f + \int g$. We have accordingly proved part (b) for the case where $E = \Omega$.

We now prove (b), letting $E \in \mathcal{F}$ be generic. Since $f, g \in L^1(E)$, we have that $fI_E, gI_E \in L^1(\Omega)$, hence $fI_E \oplus gI_E \in L^1(\Omega)$ by appealing to the result of the previous paragraph. Since $fI_E \oplus gI_E = (f \oplus g)I_E$ μ-a.e. on Ω, we have $(f \oplus g)I_E \in L^1(\Omega)$ by (L7). But then $f \oplus g \in L^1(E)$, so that $\int_E f \oplus g$ exists and is finite. We next have

$$\int_E f \oplus g = \int (f \oplus g)I_E = \int fI_E \oplus gI_E = \int fI_E + \int gI_E = \int_E f + \int_E g;$$

we have used (L5) in the first and final equalities, (L7) has been used for the second equality, and the $E = \Omega$ case has been used in the third equality. This completes the proof of (b). Part (c) is immediate from taking $c = -1$ in (a) and then applying (b) to the functions f and $-g$ in $L^1(E)$. This completes the entire proof. ■

The proof of the preceding claim is rather long, but it is fully rigorous. Most authors write $f + g$ for what we have written as $f \oplus g$. After seeing the rigorous proof of the claim, we can dispose of the $f \oplus g$ notation if desired and write $f + g$ in its place, so long as we realize that by $f + g$ we really mean $f \oplus g$ as defined earlier.[7] Most authors write $\int_E (f \pm g) = \int_E f \pm \int_E g$ in place of our more precise [but pedantic] statements in (b) and (c). Therefore, in extending (c) by induction to a formula such as $\int_E \sum_{i=1}^n f_i = \sum_{i=1}^n \int_E f_i$, [where each of $f_1, \cdots, f_n$ is in $L^1(E)$ of course], merely recognize that when we write $\sum_{i=1}^n f_i$, we mean $\bigoplus_{i=1}^n f_i$.

(L10) *If $f \in L^1(E)$ and if $|g| \le f$ μ-a.e. on E, then $g \in L^1(E)$. Also, any f that is bounded μ-a.e. on a set E with $\mu(E) < +\infty$ and is zero μ-a.e. on E^c is in $L^1(E)$.*

The proof of this claim is simple, following from (G9) on some trivial modifications. The reader should have no problems providing a proof.

We now turn to the major results regarding passage of the limit operator inside and outside the integral. We want to state the Monotone Convergence Theorem, Fatou's Lemma, and the Dominated Convergence Theorem in this setting.

(L11) **MCT.** *Let $\{f_n\}_{n=1}^\infty$ denote a sequence of nonnegative functions such that $f_1 \le f_2 \le \cdots$ μ-a.e. on Ω. Then*

$$\lim_n \int_E f_n = \int_E \lim_n f_n.$$

[7] The same comments hold for $f \ominus g$.

Also, for any sequence $\{f_n\}_{n=1}^{\infty}$ of nonnegative functions, we have

$$\sum_{n=1}^{\infty} \int_E f_n = \int_E \sum_{n=1}^{\infty} f_n.$$

Before proving MCT, the notation within must be discussed. First, since each of $f_1, f_2, \cdots$ are nonnegative, $\mathcal{F}/\mathcal{B}^*$-measurable, and defined μ-a.e. on Ω, the integrals $\int_E f_1, \int_E f_2, \cdots$ exist as elements of $[0, +\infty]$. By invoking (L7), we have that $\int_E f_1 \leq \int_E f_2 \leq \cdots$, and hence $\lim_n \int_E f_n$ exists as an element of $[0, +\infty]$.

Second, by the assumption that $f_1 \leq f_2 \leq \cdots$ μ-a.e. on Ω, we have that $\lim_n f_n(\omega)$ exists for μ-a.e. $\omega \in \Omega$. The function $\lim_n f_n$ appearing in the expression $\int_E \lim_n f_n$ is the function defined on the $\mathcal{F}$-set

$$A = \bigcap_{n=1}^{\infty} \operatorname{dom} f_n \ \cap \ \{\omega \in \Omega : \ f_1(\omega) \leq f_2(\omega) \leq \cdots\}$$

that assigns each $\omega \in A$ to the number $\lim_n f_n(\omega)$. The function $\lim_n f_n$ will be taken as undefined on A^c. Since A^c is a countable union of μ-null sets, we have $\mu(A^c) = 0$, so that the mapping $\lim_n f_n$ is defined μ-a.e. on Ω. By (MF6) and (MF11), we have that $\lim_n f_n$ is $\mathcal{F}/\mathcal{B}^*$-measurable. Summarizing, we have that $\lim_n f_n$ as described is $\mathcal{F}/\mathcal{B}^*$-measurable, nonnegative, and defined μ-a.e. on Ω. We may therefore discuss the quantity $\int_E \lim_n f_n$.

For the second assertion of the MCT, the function $\sum_{n=1}^{\infty} f_n$ appearing in the symbol $\int_E \sum_{n=1}^{\infty} f_n$ is the function defined on $\bigcap_{n=1}^{\infty} \operatorname{dom} f_n$ that assigns each $\omega \in \bigcap_{n=1}^{\infty} \operatorname{dom} f_n$ to the value $\sum_{n=1}^{\infty} f_n(\omega)$. We leave $\sum_{n=1}^{\infty} f_n$ undefined on $(\bigcap_{n=1}^{\infty} \operatorname{dom} f_n)^c$. Therefore, we have that $\sum_{n=1}^{\infty} f_n$ is defined μ-a.e. on Ω, nonnegative, and $\mathcal{F}/\mathcal{B}^*$-measurable,[8] so that we may meaningfully discuss the quantity $\int_E \sum_{n=1}^{\infty} f_n$.

The preceding discussion, while pedantic, makes completely rigorous what we mean by the two statements in the Monotone Convergence Theorem. Before showing that two quantities are equal, one must first show that the two quantities are actually meaningfully defined if there is any doubt as to whether the said quantities exist.

Proof of MCT: Let $A = \bigcap_{n=1}^{\infty} \operatorname{dom} f_n \ \cap \ \{\omega \in \Omega : \ f_1(\omega) \leq f_2(\omega) \leq \cdots\}$, so that $A \in \mathcal{F}$ and $\mu(A^c) = 0$. For each $n \in \mathbb{N}$, define $\hat{f}_n$ on all of Ω by the rule

$$\hat{f}_n(\omega) = \begin{cases} f_n(\omega) & \text{if } \omega \in A \\ 0 & \text{if } \omega \notin A. \end{cases}$$

[8] For each $n \in \mathbb{N}$, let g_n denote f_n restricted to $\bigcap_{m=1}^{\infty} \operatorname{dom} f_m$, so that each g_n is $\mathcal{F}/\mathcal{B}^*$-measurable by (MF2). Then for each $n \in \mathbb{N}$, $\sum_{i=1}^{n} g_i$ is $\mathcal{F}/\mathcal{B}^*$-measurable by (MF9) or the comments following (MF9). Now $\lim_n \sum_{i=1}^{n} g_i = \sum_{n=1}^{\infty} f_n$ on $\bigcap_{m=1}^{\infty} \operatorname{dom} f_m$, and hence (MF11) shows that $\sum_{n=1}^{\infty} f_n$ is $\mathcal{F}/\mathcal{B}^*$-measurable.

Then $\{\hat{f}_n\}_{n=1}^\infty$ is a nondecreasing sequence of functions in $\mathfrak{N}$ such that (i) $f_n = f_n^* = \hat{f}_n$ μ-a.e. on Ω for every $n \in \mathbb{N}$, (ii) $\lim_n \hat{f}_n$ is nonnegative and defined on Ω, and (iii) $\lim_n \hat{f}_n = (\lim_n f_n)^*$ μ-a.e. on Ω. Putting all this together and using (N3′) gives

$$\begin{aligned}\int_E \lim_n f_n = \bar{\mathcal{I}}_E\left(\left(\lim_n f_n\right)^*\right) = \bar{\mathcal{I}}_E\left(\lim_n \hat{f}_n\right) = \lim_n \bar{\mathcal{I}}_E(\hat{f}_n) &= \lim_n \bar{\mathcal{I}}_E(f_n^*) \\ &= \lim_n \int_E f_n.\end{aligned}$$

[The third equality follows by the form of MCT (N7) given in Section 2.] This proves the first part of MCT.

We now consider the second claim, resetting everything and letting $\{f_n\}_{n=1}^\infty$ denote a sequence of nonnegative $\mathcal{F}/\mathcal{B}^*$-measurable functions defined μ-a.e. on Ω. Observe that $\sum_{n=1}^\infty f_n^* = (\sum_{n=1}^\infty f_n)^*$ μ-a.e. on Ω, and thus

$$\sum_{n=1}^\infty \int_E f_n = \sum_{n=1}^\infty \bar{\mathcal{I}}_E(f_n^*) = \bar{\mathcal{I}}_E\left(\sum_{n=1}^\infty f_n^*\right) = \bar{\mathcal{I}}_E\left(\left(\sum_{n=1}^\infty f_n\right)^*\right) = \int_E \sum_{n=1}^\infty f_n,$$

where the second equality follows by (N12) and the third equality follows by (N3′). This completes the proof. ∎

The next property is Fatou's Lemma. This result was proven in Section 2 under the guise of (N13). Like MCT given just earlier, a bit of notation must be discussed. Let $\{f_n\}_{n=1}^\infty$ denote a sequence of $\mathcal{F}/\mathcal{B}^*$-measurable nonnegative functions, each defined μ-a.e. on Ω. The function $\underline{\lim}_n f_n$ is taken to be the function with domain $\bigcap_{n=1}^\infty \operatorname{dom} f_n$ that assigns each $\omega \in \bigcap_{n=1}^\infty \operatorname{dom} f_n$ to the nonnegative number $\underline{\lim}_n f_n(\omega)$. The function $\underline{\lim}_n f_n$ is left undefined on $(\bigcap_{n=1}^\infty \operatorname{dom} f_n)^c$. Since $\mu((\bigcap_{n=1}^\infty \operatorname{dom} f_n)^c) = 0$, we have that $\underline{\lim}_n f_n$ is defined μ-a.e. on Ω. Since it is straightforward to show that $\underline{\lim}_n f_n$ is $\mathcal{F}/\mathcal{B}^*$-measurable, we may discuss the number $\int_E \underline{\lim}_n f_n$.

(L12) **Fatou's Lemma.** *Let $\{f_n\}_{n=1}^\infty$ denote a sequence of nonnegative functions. Then $\int_E \underline{\lim}_n f_n \le \underline{\lim}_n \int_E f_n$.*

Proof: The key fact here is that $(\underline{\lim}_n f_n)^* = \underline{\lim}_n f_n^*$ μ-a.e. on Ω. Using this fact, the form of Fatou's Lemma from Section 2, and (N3′), we obtain

$$\int_E \underline{\lim}_n f_n = \bar{\mathcal{I}}_E((\underline{\lim}_n f_n)^*) = \bar{\mathcal{I}}_E(\underline{\lim}_n f_n^*) \le \underline{\lim}_n \bar{\mathcal{I}}_E(f_n^*) = \underline{\lim}_n \int_E f_n,$$

which completes the proof. ∎

(L13) **DCT.** *Let $\{f_n\}_{n=1}^\infty$ denote a sequence of functions such that there is $g \in L^1(E)$ with $|f_n| \le g$ μ-a.e. on E for every $n \in \mathbb{N}$. Furthermore, assume that $\lim_n f_n(\omega)$ exists for μ-a.e. $\omega \in \Omega$.*

(a) $f_n \in L^1(E)$ for every $n \in \mathbb{N}$.

(b) $\lim_n f_n \in L^1(E)$.

(c) $\lim_n \int_E f_n$ *exists and is finite.*

(d) $\lim_n \int_E f_n = \int_E \lim_n f_n$.

Proof: Part (a) is immediate from (L10). By the assumption that $|f_n| \leq g$ μ-a.e. on E for every $n \in \mathbb{N}$, we have that $|\lim_n f_n| \leq g$ μ-a.e. on E, hence a second appeal to (L10) shows that (b) holds.

We now turn to proving (c) and (d). We will first prove these parts by assuming that $E = \Omega$. That is, we assume that $\{f_n\}_{n=1}^{\infty}$ is a sequence of functions such that there is $g \in L^1(\Omega)$ with $|f_n| \leq g$ μ-a.e. on Ω for every $n \in \mathbb{N}$, and we assume that $\lim_n f_n(\omega)$ exists for μ-a.e. $\omega \in \Omega$. Parts (a) and (b) show that $f_n \in L^1(\Omega)$ for each $n \in \mathbb{N}$ and $\lim_n f_n \in L^1(\Omega)$. Our goal is to show that $\lim_n \int f_n$ exists and equals the finite number $\int \lim_n f_n$. We now define the following sets:

$$A = \{\omega \in \Omega : \ f_n(\omega) = \pm\infty \text{ or } |f_n(\omega)| > g(\omega) \text{ for some } n \in \mathbb{N}\},$$

$$B = \{\omega \in \Omega : \ f_n(\omega) \text{ is undefined for some } n \in \mathbb{N}\},$$

$$C = \{\omega \in \Omega : \ g(\omega) = \pm\infty \text{ or } g(\omega) \text{ is undefined}\},$$

and

$$D = \{\omega \in \Omega : \ \lim_n f_n(\omega) \text{ fails to exist}\}.$$

Since $f_n \in L^1(\Omega)$ for each $n \in \mathbb{N}$, (L4) shows that f_n is finite μ-a.e. on Ω for each $n \in \mathbb{N}$, hence $\mu(A) = 0$. It is clear that $\mu(B) = 0$. Since $g \in L^1(\Omega)$, g is finite μ-a.e. on Ω by (L4), and hence $\mu(C) = 0$. It is clear that $\mu(D) = 0$. We therefore have $\mu(A) = \mu(B) = \mu(C) = \mu(D) = 0$, hence $\mu(A \cup B \cup C \cup D) = 0$. For each $n \in \mathbb{N}$, we define $\hat{f}_n$ on Ω by writing

$$\hat{f}_n(\omega) = \begin{cases} f_n(\omega) & \omega \in (A \cup B \cup C \cup D)^c \\ 0 & \omega \in (A \cup B \cup C \cup D) \end{cases}$$

for every $\omega \in \Omega$. It is clear for all $n \in \mathbb{N}$ that $\hat{f}_n = f_n^* = f_n$ μ-a.e. on Ω and $\hat{f}_n$ is $\mathcal{F}/\mathcal{B}^*$-measurable. Next, define $\hat{g}$ on Ω by writing

$$\hat{g}(\omega) = \begin{cases} g(\omega) & \omega \in (A \cup B \cup C \cup D)^c \\ 0 & \omega \in (A \cup B \cup C \cup D) \end{cases}$$

for every $\omega \in \Omega$. The function $\hat{g}$ is $\mathcal{F}/\mathcal{B}^*$-measurable, and for each $n \in \mathbb{N}$ we have $|\hat{f}_n| \leq \hat{g}$ on Ω. Observe too that $\hat{g} = g$ μ-a.e. on Ω. Since $g \in L^1(\Omega)$, (L7) shows that $\hat{g} \in L^1(\Omega)$. Finally, define $\hat{h}$ on Ω by writing

$$\hat{h}(\omega) = \begin{cases} \lim_n f_n(\omega) & \omega \in (A \cup B \cup C \cup D)^c \\ 0 & \omega \in (A \cup B \cup C \cup D) \end{cases}$$

for every $\omega \in \Omega$. The function $\hat{h}$ is $\mathcal{F}/\mathcal{B}^*$-measurable and equals $\lim_n f_n$ μ-a.e. on Ω. Also, $\hat{h} = \lim_n \hat{f}_n$ on all of Ω. Using all of this, we have the following: $\{\hat{f}_n\}_{n=1}^{\infty}$ is a sequence of functions in $\mathfrak{M}$ such that there is $\hat{g} \in \mathfrak{M}(\Omega)$ with $|\hat{f}_n| \leq \hat{g}$ on Ω, and $\hat{h} = \lim_n \hat{f}_n$ exists on Ω. This is precisely the framework of the Dominated Convergence Theorem (G10) as given in the previous section, and from it we obtain that

$$\lim_n \bar{\bar{\mathcal{I}}}(\hat{f}_n) \text{ exists, is finite, and equals } \bar{\bar{\mathcal{I}}}(\hat{h}).$$

Now for every $n \in \mathbb{N}$ we have $\bar{\bar{\mathcal{I}}}(\hat{f}_n) = \bar{\bar{\mathcal{I}}}(f_n^*) = \int f_n$ by (G5), hence

$$\lim_n \int f_n \text{ exists, is finite, and equals } \bar{\bar{\mathcal{I}}}(\hat{h}).$$

Next, we have $\bar{\bar{\mathcal{I}}}(\hat{h}) = \bar{\bar{\mathcal{I}}}((\lim_n f_n)^*) = \int \lim_n f_n$ since $\hat{h} = (\lim_n f_n)^*$ μ-a.e. on Ω. Therefore, we have that

$$\lim_n \int f_n \text{ exists, is finite, and equals } \int \lim_n f_n.$$

This gives (c) and (d) for the case $E = \Omega$.

We now prove (c) and (d) for general $E \in \mathcal{F}$. By the assumptions, we have that $\{f_n I_E\}_{n=1}^{\infty}$ is a sequence of $\mathcal{F}/\mathcal{B}^*$-measurable functions defined on Ω, $|f_n I_E| \leq g I_E$ μ-a.e. on Ω for each $n \in \mathbb{N}$, $g I_E \in L^1(\Omega)$, and $\lim_n f_n I_E$ exists μ-a.e. on Ω. By what was shown in the $E = \Omega$ case, we have that

$$\lim_n \textstyle\int f_n I_E \text{ exists, is finite, and equals } \int \lim_n f_n I_E.$$

By (L5), the fact that $\lim_n f_n I_E = \lim_n f_n$ μ-a.e. on E, and (L7), the above display is equivalent is to asserting that

$$\lim_n \textstyle\int_E f_n \text{ exists, is finite, and equals } \int_E \lim_n f_n.$$

This proves (c) and (d) for general $E \in \mathcal{F}$, and the entire proof is now complete. ■

The next property restates (G11), (N14), and (N9).

(L14) *Let $f \geq 0$ on its domain, and define ν on $\mathcal{F}$ by writing*

$$\nu(E) = \int_E f \, d\mu \quad \textit{for every } E \in \mathcal{F}.$$

We have the following results:

(a) ν is a measure on $\mathcal{F}$.

(b) If g is a nonnegative $\mathcal{F}/\mathcal{B}^*$-measurable function defined μ-a.e. on Ω, then $\int_E g \, d\nu = \int_E fg \, d\mu$.

(c) For $\mathcal{F}/\mathcal{B}^*$-measurable g defined μ-a.e. on Ω, we have that $\int_E g\, d\nu$ exists and is finite iff $\int_E fg\, d\mu$ exists and is finite; in this case, we have $\int_E g\, d\nu = \int_E fg\, d\mu$.

Proof: We first handle (a). Since f is nonnegative, $\nu(E)$ is defined and nonnegative for every $E \in \mathcal{F}$. By (L2), we have $\nu(\varnothing) = \int_\varnothing f = 0$. All that remains for (a) is countable additivity. Let $\{E_n\}_{n=1}^\infty$ denote a disjoint sequence of $\mathcal{F}$-sets. Then

$$\nu\left(\bigcup_{n=1}^\infty E_n\right) = \int_{\bigcup_{n=1}^\infty E_n} f = \sum_{n=1}^\infty \int_{E_n} f = \sum_{n=1}^\infty \nu(E_n),$$

the second equality following by (L6). This shows that ν is countably additive and completes (a).

We now turn to (b). Since g is defined μ-a.e. on Ω, we have $\mu((\operatorname{dom} g)^c) = 0$. By the definition of ν and (L2), we have $\nu((\operatorname{dom} g)^c) = 0$, and hence g is defined ν-a.e. on Ω. Using this and the nonnegativity of g and fg, we see that we may discuss $\int_E g\, d\nu$ and $\int_E fg\, d\mu$. Consider the following chain:

$$\int_E g\, d\nu = \bar{\bar{\mathcal{I}}}_E(g^*; \nu) = \bar{\bar{\mathcal{I}}}_E(f^* g^*; \mu) = \bar{\bar{\mathcal{I}}}_E((fg)^*; \mu) = \int_E fg\, d\mu.$$

The first and fourth equalities follow by the definition of the Lebesgue integral, and the third equality holds by noting that $f^* g^* = (fg)^*$ μ-a.e. on E and invoking (G5). The second equality holds by observing that $\nu(E) = \bar{\bar{\mathcal{I}}}_E(f^*)$ for each $E \in \mathcal{F}$ and invoking (G11). This gives (b).

We now turn to (c). Observe that $g^{*+} f^* = (fg)^{*+}$ and $g^{*-} f^* = (fg)^{*-}$ μ-a.e. on Ω. With this fact, (N3′), and (N14), it is easily seen that the following statements are equivalent: (1) $\int_E g\, d\nu$ exists and is finite, (2) $\bar{\bar{\mathcal{I}}}_E(g^*)$ exists and is finite, (3) both $\bar{\mathcal{I}}_E(g^{*+}; \nu)$ and $\bar{\mathcal{I}}_E(g^{*-}; \nu)$ are finite, (4) both $\bar{\mathcal{I}}_E((g^{*+} f^*; \mu)$ and $\bar{\mathcal{I}}_E(g^{*-} f^*; \mu)$ are finite [since $\nu(E) = \bar{\bar{\mathcal{I}}}_E(f^*; \mu)$ by the nonnegativity of f], (5) $\bar{\mathcal{I}}_E((fg)^{*+}; \mu)$ and $\bar{\mathcal{I}}_E((fg)^{*-}; \mu)$ are finite, (6) $\bar{\bar{\mathcal{I}}}_E((fg)^*; \mu)$ exists and is finite, and (7) $\int_E fg\, d\mu$ exists and is finite. This proves the existence claim for (c). For the equation, we assemble (1)-(7), obtaining

$$\begin{aligned}
\int_E g\, d\nu = \bar{\bar{\mathcal{I}}}_E(g^*) &= \bar{\mathcal{I}}_E((g^*)^+; \nu) - \bar{\mathcal{I}}_E((g^*)^-; \nu) \\
&= \bar{\mathcal{I}}_E(g^{*+} f^*; \mu) - \bar{\mathcal{I}}_E(g^{*-} f^*; \mu) \\
&= \bar{\mathcal{I}}_E((fg)^{*+}; \mu) - \bar{\mathcal{I}}_E((fg)^{*-}; \mu) \\
&= \bar{\bar{\mathcal{I}}}_E((fg)^*; \mu) \\
&= \int_E fg\, d\mu,
\end{aligned}$$

which completes (c) and the entire proof. ■

The function f is often called *the density of ν with respect to μ*; often the formalisms $f = \frac{d\nu}{d\mu}$ and $d\nu = f\ d\mu$ are employed to express the situation where $\nu(E) = \int_E f\ d\mu$ for every $E \in \mathcal{F}$. The symbols $\frac{d\nu}{d\mu}$, $d\nu$, and $d\mu$ have no independent meaning.

The next result generalizes (G12), which is a "continuity result" for the Lebesgue integral. It says [quite roughly] that $\int_E f\ d\mu \to 0$ as $\mu(E) \to 0$.

(L15) *Let $f \in L^1(E)$ and $\epsilon > 0$. Then there exists $\delta > 0$ such that any $F \subseteq E$ with $\mu(F) < \delta$ satisfies $|\int_F f\ d\mu| < \epsilon$.*

Proof: If $f \in L^1(E)$, then $\bar{\bar{\mathcal{I}}}_E(f^*)$ exists and is finite. By (G12), for any $\epsilon > 0$ there is $\delta > 0$ with the property that any $F \subseteq E$ with $\mu(F) < \delta$ satisfies $|\bar{\bar{\mathcal{I}}}_F(f^*)| < \epsilon$. But then any $F \subseteq E$ with $\mu(F) < \delta$ satisfies $|\int_F f\, d\mu| < \epsilon$, completing the proof. ∎

For the next property, refer to the discussion immediately preceding property (G13); we may now only require that $\phi \circ f$ and ϕ are defined μ-a.e. on Ω and $\mu \circ f^{-1}$-a.e. on Ω', respectively. The proof of this property is straightforward in the same fashion as have been the previous proofs, and is left to the reader as an exercise.

(L16) *We have the following, where $A' \in \mathcal{F}'$:*

(a) If $\phi \geq 0$, then $\int_{f^{-1}(A')} \phi \circ f\ d\mu = \int_{A'} \phi\ d(\mu \circ f^{-1})$.

(b) For general ϕ, $\int_{f^{-1}(A')} \phi \circ f\ d\mu$ exists and is finite iff $\int_{A'} \phi\ d(\mu \circ f^{-1})$ exists and is finite, and in this case equality obtains [as in (a)].

(L17) *We have the following:*

(a) Suppose that $f, g \geq 0$ and $\int_E f = \int_E g$ for every $E \in \mathcal{F}$. Furthermore, suppose that μ is σ-finite. Then $f = g$ μ-a.e. on Ω.

(b) If $f, g \in L^1(\Omega)$ and $\int_E f = \int_E g$ for every $E \in \mathcal{F}$, then $f = g$ μ-a.e. on Ω. [Nothing about σ-finiteness assumed here.]

Proof: As stated, part (a) follows nearly immediately from (N16). We merely note that the assumptions give $\bar{\mathcal{I}}_E(f^*) = \bar{\mathcal{I}}_E(g^*)$ for every $E \in \mathcal{F}$, so the nonnegativity of f^* and g^* forces $f^* = g^*$ μ-a.e. on Ω. Since $f = f^*$ μ-a.e. on Ω and $g = g^*$ μ-a.e. on Ω, we have that $f = g$ μ-a.e. on Ω as claimed.

For part (b), pick $f, g \in L^1(\Omega)$ and assume that $\int_E f \leq \int_E g$ for every E in $\mathcal{F}$. In particular, letting $A = \{\omega \in \Omega :\ g(\omega) < f(\omega)\}$, we have $\int_A f \leq \int_A g$. Therefore, subtracting $\int_A g$ from both sides [which may be done because all quantities are assumed finite] yields $\int_A f - \int_A g \leq 0$, so $\int_A (f - g) \leq 0$ by (L9). Using (L5), this forces $\int (f - g) I_A \leq 0$. At the same time, $(f - g) I_A \geq 0$ on Ω, so we must have $\int (f - g) I_A \geq 0$. It follows that $\int (f - g) I_A = 0$. By (L3), $(f - g) I_A = 0$ μ-a.e. on Ω. It follows that $I_A = 0$ μ-a.e. on Ω, hence $\mu(A) = 0$. We have shown the following: if $\int_E f \leq \int_E g$ for every $E \in \mathcal{F}$, then $\mu(\{\omega \in \Omega :\ g(\omega) < f(\omega)\}) = 0$. Switching the roles of f and g gives the following: if $\int_E g \leq \int_E f$ for each $E \in \mathcal{F}$, then $\mu(\{\omega \in \Omega : f(\omega) < g(\omega)\}) = 0$. Combining these two facts gives the following: if $\int_E f = \int_E g$ for every $E \in \mathcal{F}$, then $\mu(\{\omega \in \Omega : f(\omega) \neq g(\omega)\}) = 0$, giving (b). ■

(L18) *Define μ_E on $\mathcal{F}$ by writing $\mu_E(A) = \mu(A \cap E)$ for every $A \in \mathcal{F}$. Then μ_E is a measure and $\int_A f\, d\mu_E = \int_{A\cap E} f\, d\mu$ in the usual sense that if one integral exists, then so does the other and both are equal.*

Proof: The reader may quickly show [or recall] that μ_E is in fact a measure. For the main claim, first assume that $f = I_B$ [where $B \in \mathcal{F}$]. Then

$$\int_A f\, d\mu_E = \mu(A \cap B \cap E) = \int_{A\cap E} f\, d\mu.$$

The claim is therefore true for indicator functions. Next, let $f \in \mathfrak{S}(\Omega, \mathcal{F}, \mu)$; write $f = \sum_{i=1}^n c_i I_{B_i}$. We have

$$\begin{aligned}\int_A f\, d\mu_E = \int_A \sum_{i=1}^n c_i I_{B_i}\, d\mu_E = \sum_{i=1}^n c_i \int_A I_{B_i}\, d\mu_E &= \sum_{i=1}^n c_i \int_{A\cap E} I_{B_i}\, d\mu \\ &= \int_{A\cap E} \sum_{i=1}^n c_i I_{B_i}\, d\mu \\ &= \int_{A\cap E} f\, d\mu\end{aligned}$$

by (L9) and the result just shown for indicator functions. The claim therefore holds for all elements of $\mathfrak{S}(\Omega, \mathcal{F}, \mu)$. Next, assume that $f \in \mathfrak{N}(\Omega, \mathcal{F}, \mu)$. Let $\{s_n\}_{n=1}^\infty$ denote a nondecreasing sequence of elements of $\mathfrak{S}(\Omega, \mathcal{F}, \mu)$ with limit f. By MCT and the result for simple functions, we have

$$\int_A f\, d\mu_E = \lim_n \int_A s_n\, d\mu_E = \lim_n \int_{A\cap E} s_n\, d\mu = \int_{A\cap E} f\, d\mu.$$

The result accordingly holds for every $f \in \mathfrak{N}(\Omega, \mathcal{F}, \mu)$.

Next, let $f \in \mathfrak{M}(\Omega, \mathcal{F}, \mu)$, so that f is $\mathcal{F}/\mathcal{B}^*$-measurable and defined on Ω. Suppose that $\int_A f\, d\mu_E$ exists. Then at least one of $\int_A f^+\, d\mu_E$ and $\int_A f^-\, d\mu_E$ is finite. By the result for nonnegative functions we have that at least one of $\int_{A\cap E} f^+\, d\mu$ and $\int_{A\cap E} f^-\, d\mu$ is finite. But then $\int_{A\cap E} f\, d\mu$ exists. The steps of this argument reverse, hence the existence of $\int_{A\cap E} f\, d\mu$ forces that of $\int_A f\, d\mu_E$. For the equality, we have

$$\begin{aligned}\int_A f\, d\mu_E = \int_A f^+\, d\mu_E - \int_A f^-\, d\mu_E &= \int_{A\cap E} f^+\, d\mu - \int_{A\cap E} f^-\, d\mu \\ &= \int_{A\cap E} f\, d\mu.\end{aligned}$$

This proves the claim for general $f \in \mathfrak{M}(\Omega, \mathcal{F}, \mu)$ such that at least one of $\int_A f\, d\mu_E$ and $\int_{A\cap E} f\, d\mu$ exists. The case for general $\mathcal{F}/\mathcal{B}^*$-measurable f

defined μ-a.e. on Ω and such that at least one of $\int_A f\,d\mu_E$ and $\int_{A\cap E} f\,d\mu$ exists is now nearly immediate. If f is defined μ-a.e. on Ω and $\int_A f\,d\mu_E$ or $\int_{A\cap E} f\,d\mu$ exists, then $\int_A f^*\,d\mu_E$ or $\int_{A\cap E} f^*\,d\mu$ exists. Since $f^* \in \mathfrak{M}(\Omega, \mathcal{F}, \mu)$, the case just proved forces both $\int_A f^*\,d\mu_E$ and $\int_{A\cap E} f^*\,d\mu$ to exist and be equal. This proves the existence claim for the general case. For the equation, we have

$$\int_A f\,d\mu_E = \int_A f^*\,d\mu_E = \int_{A\cap E} f^*\,d\mu = \int_{A\cap E} f\,d\mu,$$

where the second equality follows from what has just been shown. This gives the entire result, and the proof is complete. ∎

With these properties listed [and some more properties are given in the exercises], the construction of the Lebesgue integral is now complete. It has been a tedious process, but this somewhat dull chore is a necessary evil to discuss the major results of measure theory, such as the Radon-Nikodym Theorem, Fubini's Theorem, the completeness of certain spaces of functions, and the construction of product measures. With this work finally at a close, one can move on to more interesting things, hopefully taking most or all of this material for granted.

Exercises. Unless otherwise stated, assume that all functions are defined μ-a.e. on Ω and are $\mathcal{F}/\mathcal{B}^*$-measurable; assume too that all sets are $\mathcal{F}$-sets.

1*. Prove properties (L10) and (L16). [This should be easy.]

2. Suppose that $\sum_{n=1}^{\infty} f_n$ converges μ-a.e. on Ω and that for every $n \in \mathbb{N}$ we have $|\sum_{k=1}^{n} f_k| \le g$ μ-a.e. on Ω, where $g \in L^1(\Omega)$.
(a) For every $n \in \mathbb{N}$ we have $f_n \in L^1(\Omega)$, and $\sum_{n=1}^{\infty} f_n \in L^1(\Omega)$ as well.
(b) The equality $\int \sum_{n=1}^{\infty} f_n = \sum_{n=1}^{\infty} \int f_n$ obtains.
(c) Prove the following corollary: if $\sum_{n=1}^{\infty} \int |f_n|$ converges, then $\sum_{n=1}^{\infty} f_n$ converges absolutely μ-a.e. on Ω and again the equality $\int \sum_{n=1}^{\infty} f_n = \sum_{n=1}^{\infty} \int f_n$ holds.

3. Suppose that $f \in L^1(\Omega)$.
(a) Let $A_n = \{\omega \in \Omega : |f(\omega)| \ge n\}$. Then $\lim_n |f_n(\omega)| I_{A_n}(\omega) = 0$ for μ-a.e. $\omega \in \Omega$.
(b) Use DCT to show that $\lim_n \int_{A_n} |f| = 0$.

4. This exercise deals with the concept of *uniform integrability.* A sequence $\{f_n\}_{n=1}^{\infty}$ is called *uniformly integrable* iff $\lim_m \sup_n \int_{A_m} |f_n| = 0$, where the sets $A_1, A_2, \cdots$ are as given in the previous exercise.
(a) Suppose that $\mu(\Omega) < +\infty$ and $\{f_n\}_{n=1}^{\infty}$ is uniformly integrable. Then each f_n is in $L^1(\Omega)$.
(b) If $\{f_n\}_{n=1}^{\infty}$ and $\{g_n\}_{n=1}^{\infty}$ are uniformly integrable, then so is $\{f_n + g_n\}_{n=1}^{\infty}$.
(c) Suppose that $\mu(\Omega) < +\infty$, $f_n \to f$ μ-a.e. on Ω, and $\{f_n\}_{n=1}^{\infty}$ is uniformly integrable. Then $f \in L^1(\Omega)$ and $\lim \int f_n = \int f$.
(d) Suppose that $\mu(\Omega) < +\infty$ and that $f_n \to f$ μ-a.e. on Ω, where each of $f, f_1, f_2, \cdots$ is nonnegative and in $L^1(\Omega)$. If $\lim \int f_n = \int f$, then $\{f_n\}_{n=1}^{\infty}$ is uniformly integrable.
(e) Suppose that $\mu(\Omega) < +\infty$, $f, f_1, f_2, \cdots \in L^1(\Omega)$, and $f_n \to f$ μ-a.e. on Ω. Then the following statements are logically equivalent:
(i) $\{f_n\}_{n=1}^{\infty}$ is uniformly integrable.
(ii) $\lim \int |f_n - f| = 0$.

(iii) $\lim \int |f_n| \to \int |f|$.
(f) Suppose that $\mu(\Omega) < +\infty$. Then $\{f_n\}_{n=1}^{\infty}$ is uniformly integrable iff (i) $\int |f_n|$ is bounded and (ii) for every $\epsilon > 0$ there is $\delta > 0$ such that $\mu(E) < \delta$ implies $\int_E |f_n| < \epsilon$ for every $n \in \mathbb{N}$.

Comment: The reader may consult Billingsley's *Probability and Measure*, Section 16, for the solutions to (a)-(e).

5. This problem deals with parametric integrals. Here, we consider a measure space $(\Omega, \mathcal{F}, \mu)$ as well as a nondegenerate interval $\Theta \subseteq \mathbb{R}$ of any form. Suppose that for every $\theta \in \Theta$ there is a function f_θ in $L^1(\Omega, \mathcal{F}, \mu)$ with domain Ω. That is, for every $\theta \in \Theta$ we may consider the real number $\int_\Omega f_\theta(\omega)\, d\mu(\omega)$. For each $\theta \in \Theta$, write $F(\theta)$ for $\int_\Omega f_\theta(\omega)\, d\mu(\omega)$. Hold this notation and these assumptions fixed throughout.
(a) Make the following additional assumptions:
(1) There exists $g \in L^1(\Omega, \mathcal{F}, \mu)$ with domain Ω such that $|f_\theta(\omega)| \leq g(\omega)$ for every $\omega \in \Omega$ and $\theta \in \Theta$.
(2) The element $\theta_0 \in \Theta$ is such that $\lim_{\theta \to \theta_0} f_\theta(\omega) = f_{\theta_0}(\omega)$ for μ-a.e. $\omega \in \Omega$.
Show that $\lim_{\theta \to \theta_0} F(\theta) = F(\theta_0)$, so that F is continuous at θ_0.
(b) Forget about the assumptions from (a), and instead make the following new set of assumptions:
(1) For μ-a.e. $\omega \in \Omega$ we have that $f_\theta(\omega)$ is differentiable at each $\theta \in \Theta$.
(2) There exists $g \in L^1(\Omega, \mathcal{F}, \mu)$ with domain Ω such that $|\frac{\partial}{\partial\theta} f_\theta(\omega)| \leq g(\omega)$ for μ-a.e. $\omega \in \Omega$ and all $\theta \in \Theta$.
Show that (i) F is differentiable on Θ, and (ii) $\frac{\partial}{\partial\theta} F(\theta) = \int_\Omega \frac{\partial}{\partial\theta} f_\theta(\omega)\, d\mu(\omega)$.

Suggestion: For (a), let $A = \{\omega \in \Omega : \lim_{\theta \to \theta_0} f_\theta(\omega) = f_{\theta_0}(\omega)\}$, so that $\mu(A^c) = 0$. Let $\{\theta_n\}_{n=1}^{\infty}$ denote a sequence in Θ with limit θ. Use DCT to obtain

$$\lim_{\theta_n \to \theta_0} \int_A f_{\theta_n}(\omega)\, d\mu(\omega) = \int_A \lim_{\theta_n \to \theta_0} f_\theta(\omega)\, d\mu(\omega) = \int_A f_{\theta_0}(\omega)\, d\mu(\omega)$$

where the limit interchange is justified by (1) and the last equality follows for all $\omega \in A$ by (2). Since $\{\theta_n\}_{n=1}^{\infty}$ was an arbitrary sequence of elements of Θ converging to θ_0, and since $\mu(A^c) = 0$ [so that integrals over Ω are equivalent to integrals over A], the above means that $\lim_{\theta \to \theta_0} F(\theta) = F(\theta_0)$. In other words, the claim for (a) follows almost immediately from assumptions (1) and (2) [as well as the assumptions in the prolog of the exercise].

For (b), let $A = \{\omega \in \Omega :\ f_\theta(\omega) \text{ is differentiable for each } \theta \in \Theta\}$, so that (1) gives $\mu(A^c) = 0$. Let $\{\delta_n\}_{n=1}^{\infty}$ be a sequence of numbers tending to zero. We have

$$\begin{aligned}
\lim_n \frac{F(\theta + \delta_n) - F(\theta)}{\delta_n} &= \lim_n \int_A \frac{f_{\theta+\delta_n}(\omega) - f_\theta(\omega)}{\delta_n}\, d\mu(\omega) \\
&= \int_A \lim_n \frac{f_{\theta+\delta_n}(\omega) - f_\theta(\omega)}{\delta_n}\, d\mu(\omega) \\
&= \int_A \frac{\partial}{\partial\theta} f_\theta(\omega)\, d\mu(\omega).
\end{aligned}$$

The first equality is by direct computation and the fact that $\mu(A^c) = 0$; the last equality follows by the definition of A and (1). The limit interchange is the important step, and requires (2) and the Mean Value Theorm. To see this, observe for each

$n \in \mathbb{N}$ that we have

$$\left|\frac{f_{\theta+\delta_n}(\omega) - f_\theta(\omega)}{\delta_n}\right| \leq \left|\frac{\partial}{\partial\theta} f_\theta(\omega)\right|\bigg|_{\theta=\eta_n} \leq g(\omega) \text{ for } \mu\text{-a.e. } \omega \in \Omega,$$

where η_n denotes the number between θ and $\theta + \delta_n$ ensured by the Mean Value Theorem. This allows DCT to be invoked, justifying the limit interchange above. Since $\{\delta_n\}_{n=1}^{\infty}$ is an arbitrary sequence tending to zero, the result stands complete.

6. Assume that f is an $\mathcal{F}/\mathcal{B}^*$-measurable function defined μ-a.e. on Ω such that there are real numbers m and M with $m \leq f \leq M$ μ-a.e. on Ω. Also, assume that $g \in L^1(\Omega, \mathcal{F}, \mu)$. Then there is $\alpha \in [m, M]$ such that $\int f|g| = \alpha \int |g|$.

7. Let $\{f_n\}_{n=1}^{\infty}$ denote a sequence in $\mathfrak{N}$ with $f_1 \geq f_2 \geq \cdots \geq 0$, $f_n \to f$ on Ω, and $f_1 \in L^1(\Omega, \mathcal{F}, \mu)$. Then $\lim_n \int f_n = \int f$, and the requirement that $f_1 \in L^1(\Omega, \mathcal{F}, \mu)$ is essential. [This mimics MCT somewhat.]

8. Suppose that $\mu(\Omega) < +\infty$, and let $\{f_n\}_{n=1}^{\infty}$ denote a uniformly bounded sequence of functions on Ω with $f_n \to f$ *uniformly* on Ω. Then $\lim_n \int f_n = \int f$.

9. Suppose that $\{f_n\}_{n=1}^{\infty}$ is a sequence of functions in $L^1(\Omega, \mathcal{F}, \mu)$ such that the series $\sum_{n=1}^{\infty} \int |f_n|$ converges. Then $\sum_{n=1}^{\infty} f_n$ converges μ-a.e. on Ω to a function $f \in L^1(\Omega, \mathcal{F}, \mu)$ and $\int f = \sum_{n=1}^{\infty} \int f_n$.

10. Let $(\Omega, \mathcal{F}, \mu)$ denote a generic measure space, and let $(\Omega, \bar{\mathcal{F}}, \bar{\mu})$ denote the completion of $(\Omega, \mathcal{F}, \mu)$. Let $\bar{f}$ denote an $\bar{\mathcal{F}}/\mathcal{B}^*$-measurable function on Ω. [The bar on f is just part of the notation; it denotes no operation.]
(a) There exists f on Ω that is $\mathcal{F}/\mathcal{B}^*$-measurable with $f = \bar{f}$ $\bar{\mu}$-a.e. [Here, look back at the outline of Exercise 2 in Section 5.5, where there $\mathcal{F}$ consisted of the Borel sets and $\bar{\mathcal{F}}$ consisted of the Lebesgue sets. Observe that nothing specific to Euclidean space was used in that outline — only the general results concerning completion — hence the same outline will work here.]
(b) If one of $\int \bar{f}\, d\bar{\mu}$ and $\int f\, d\mu$ exists, then so does the other and both integrals coincide. Therefore, integrals of $\bar{\mathcal{F}}/\mathcal{B}^*$-measurable functions with respect to $\bar{\mu}$ are completely determined by the integrals of $\mathcal{F}/\mathcal{B}^*$-measurable functions with respect to μ. [It is perhaps best to use the usual approach in proving results about integrals: show that the claim holds for simple, then nonnegative, then measurable functions.]

7

Integrals Relative to Lebesgue Measure

It was stated in the previous chapter that the Lebesgue integral is a generalization of the Riemann integral. The Lebesgue integral was painstakingly constructed, but it has not as of yet been shown to generalize the Riemann integral. In fact, what we mean by *generalize* has not yet been rigorously stated. This chapter shall define in what sense the word *generalize* is to be taken, and we prove some other useful properties about Lebesgue integrals when the underlying measure is Lebesgue measure.

7.1 SEMICONTINUITY

In this rudimentary section we review in a more or less complete fashion the principles of semicontinuous functions. The point of doing this is that semicontinuity will play a major role in our proof that the Lebesgue integral generalizes the Riemann integral, and such a review allows us to establish some notation while getting some technical points out of the way.

For this section, our background space will be $\mathbb{R}^k$, where $k \in \mathbb{N}$ is generic. *All functions discussed in definitions, claims, or anything else will be assumed to have domain* $\mathbb{R}^k$ *and range contained in* $[-\infty, +\infty]$ *unless explicitly stated to the contrary.* The definitions follow.

Definition. We will say that a function f is *lower semicontinuous at* $\boldsymbol{x} \in \mathbb{R}^k$ iff for every $t < f(\boldsymbol{x})$ there is $\delta > 0$ [depending on t and $\boldsymbol{x}$] such that $f(\boldsymbol{y}) > t$ for every $\boldsymbol{y} \in B(\boldsymbol{x}, \delta)$. [Here $B(\boldsymbol{x}, \delta)$ stands for the open ball with center $\boldsymbol{x}$ and radius δ, where distance is the usual k-dimensional Euclidean distance.] If

f is lower semicontinuous at every $\boldsymbol{x} \in \mathbb{R}^k$, we will merely say that *f is lower semicontinuous.* We use the abbreviation LSC for *lower semicontinuous.*

All we mean when we say that *f is lower semicontinuous at $\boldsymbol{x}$* is that whenever $f(\boldsymbol{x})$ exceeds some real number, call it t, it follows that $f(\boldsymbol{y})$ exceeds the same t for every $\boldsymbol{y}$ "sufficiently close" to $\boldsymbol{x}$. That is, t is a lower bound for f on some neighborhood of $\boldsymbol{x}$, which is from where the adjective *lower* in *lower semicontinuity* comes.

Definition. We will say that a function f is *upper semicontinuous at* $\boldsymbol{x} \in \mathbb{R}^k$ iff for every $t > f(\boldsymbol{x})$ there is $\delta > 0$ [depending on t and $\boldsymbol{x}$] such that $f(\boldsymbol{y}) < t$ for every $\boldsymbol{y} \in B(\boldsymbol{x}, \delta)$. If f is upper semicontinuous at every $\boldsymbol{x} \in \mathbb{R}^k$, we will merely say that *f is upper semicontinuous.* We use the abbreviation USC for *upper semicontinous.*

When we say that *f is upper semicontinuous at $\boldsymbol{x}$*, we merely mean that whenever $f(\boldsymbol{x})$ is less than some real number, call it t, it follows that $f(\boldsymbol{y})$ is less than the same t for every $\boldsymbol{y}$ "sufficiently close" to $\boldsymbol{x}$. That is, t is an upper bound for f on some neighborhood of $\boldsymbol{x}$, which is from where the adjective *upper* in *upper semicontinuity* comes.

Definition. Given a function f, we define the functions $\underline{f}$ and $\bar{f}$ by the following rules for every $\boldsymbol{x} \in \mathbb{R}^k$:

$$\underline{f}(\boldsymbol{x}) = \liminf_{\boldsymbol{y}\to\boldsymbol{x}} f(\boldsymbol{y}) = \sup_{\delta>0} \inf_{\boldsymbol{y}\in B(\boldsymbol{x},\delta)} f(\boldsymbol{y}), \quad \text{and}$$

$$\bar{f}(\boldsymbol{x}) = \limsup_{\boldsymbol{y}\to\boldsymbol{x}} f(\boldsymbol{y}) = \inf_{\delta>0} \sup_{\boldsymbol{y}\in B(\boldsymbol{x},\delta)} f(\boldsymbol{y}).$$

[In both displays, the second equality is the actual definition; the first equality merely identifies a notational variant used by other authors.]

Observe that for each fixed $\boldsymbol{x}$ the quantity $\inf_{\boldsymbol{y}\in B(\boldsymbol{x},\delta)} f(\boldsymbol{y})$ is monotone nondecreasing as δ decreases, hence we may also write

$$\underline{f}(\boldsymbol{x}) = \lim_{\delta\to 0+} \inf_{\boldsymbol{y}\in B(\boldsymbol{x},\delta)} f(\boldsymbol{y}),$$

a "lim inf," instead of a "sup inf." This may help explain why the notation $\liminf_{\boldsymbol{y}\to\boldsymbol{x}} f(\boldsymbol{y})$ is used by some as a notational variant for $\underline{f}(\boldsymbol{x})$. Analogous comments apply to $\bar{f}(\boldsymbol{x})$. For each fixed $\boldsymbol{x}$ the quantity $\sup_{\boldsymbol{y}\in B(\boldsymbol{x},\delta)} f(\boldsymbol{y})$ is monotone nonincreasing as δ decreases, hence we may write [in addition to the definition]

$$\bar{f}(\boldsymbol{x}) = \lim_{\delta\to 0+} \sup_{\boldsymbol{y}\in B(\boldsymbol{x},\delta)} f(\boldsymbol{y}),$$

a "lim sup," instead of an "inf sup." This should help explain why the notation $\limsup_{\boldsymbol{y}\to\boldsymbol{x}} f(\boldsymbol{y})$ is used by some as a notation variant for $\bar{f}(\boldsymbol{x})$.

We now state the basic facts regarding semicontinuous functions. The proofs are more or less direct, and provide no real difficulty.

(SC1) *f is USC at $\boldsymbol{x}$ iff $-f$ is LSC at $\boldsymbol{x}$. Also, f is LSC at $\boldsymbol{x}$ iff $-f$ is USC at $\boldsymbol{x}$.*

Proof: Suppose that f is not USC at $\boldsymbol{x}$. Then there is $t > f(\boldsymbol{x})$ such that for every $\delta > 0$ there is $\boldsymbol{y} \in B(\boldsymbol{x}, \delta)$ with $t \leq f(\boldsymbol{y})$. It follows that there is $r < -f(\boldsymbol{x})$ such that for every $\delta > 0$ there is $\boldsymbol{y} \in B(\boldsymbol{x}, \delta)$ with $r \geq -f(\boldsymbol{y})$. [Merely take $r = -t$.] But then $-f$ is not LSC at $\boldsymbol{x}$. Therefore, if f is not USC at $\boldsymbol{x}$, then $-f$ is not LSC at $\boldsymbol{x}$. The steps used obtaining the just-given implication may be reversed, allowing us to state that if $-f$ is not LSC at $\boldsymbol{x}$, then f is not USC at $\boldsymbol{x}$. Putting the two implications together shows that f is not USC at $\boldsymbol{x}$ iff $-f$ is not LSC at $\boldsymbol{x}$. This is equivalent to the first desired equivalence.

The second part of the claim is now a direct consequence of the first claim: f is LSC at $\boldsymbol{x}$ iff $-(-f)$ is LSC at $\boldsymbol{x}$ iff $-f$ is USC at $\boldsymbol{x}$. The proof is accordingly complete. ■

(SC2) $\underline{f} \leq f \leq \bar{f}$.

Proof: We have

$$\inf_{\boldsymbol{y} \in B(\boldsymbol{x}, \delta)} f(\boldsymbol{y}) \leq f(\boldsymbol{x}) \leq \sup_{\boldsymbol{y} \in B(\boldsymbol{x}, \delta)} f(\boldsymbol{y})$$

for every $\boldsymbol{x} \in \mathbb{R}^k$ and $\delta > 0$. Therefore,

$$\sup_{\delta > 0} \inf_{\boldsymbol{y} \in B(\boldsymbol{x}, \delta)} f(\boldsymbol{y}) \leq f(\boldsymbol{x}) \leq \inf_{\delta > 0} \sup_{\boldsymbol{y} \in B(\boldsymbol{x}, \delta)} f(\boldsymbol{y})$$

for every $\boldsymbol{x} \in \mathbb{R}^k$, which by definition gives the result. ■

(SC3) *If $f(\boldsymbol{x}) = -\infty$, then f is LSC at $\boldsymbol{x}$. If $f(\boldsymbol{x}) = +\infty$, then f is USC at $\boldsymbol{x}$. [Vacuously true.]*

(SC4) *We have the following:*

(a) If $f(\boldsymbol{x}) = +\infty$, then f is LSC at $\boldsymbol{x}$ iff $\lim_{\boldsymbol{y} \to \boldsymbol{x}} f(\boldsymbol{y}) = +\infty$.

(b) If $f(\boldsymbol{x}) = -\infty$, then f is USC at $\boldsymbol{x}$ iff $\lim_{\boldsymbol{y} \to \boldsymbol{x}} f(\boldsymbol{y}) = -\infty$.

Proof: We do (a), first assuming that f is LSC at $\boldsymbol{x}$. Pick $t < f(\boldsymbol{x}) = +\infty$. Since f is LSC at $\boldsymbol{x}$, there is $\delta > 0$ such that $f(\boldsymbol{y}) > t$ for each $\boldsymbol{y} \in B(\boldsymbol{x}, \delta)$. Since t is arbitrary, we have $\lim_{\boldsymbol{y} \to \boldsymbol{x}} f(\boldsymbol{y}) = +\infty$. Conversely, assume that $\lim_{\boldsymbol{y} \to \boldsymbol{x}} f(\boldsymbol{y}) = +\infty$, and pick $t < f(\boldsymbol{x}) = +\infty$. Since $\lim_{\boldsymbol{y} \to \boldsymbol{x}} f(\boldsymbol{y}) = +\infty$, there is $\delta > 0$ such that $\boldsymbol{y} \in B(\boldsymbol{x}, \delta)$ implies $f(\boldsymbol{y}) > t$. Since t is generic, f is LSC at $\boldsymbol{x}$ by definition. This gives (a). Part (b) follows from (a) and (SC1), and is left as an easy exercise. ■

(SC5) *f is continuous at $\boldsymbol{x}$ iff f is both LSC and USC at $\boldsymbol{x}$.*

Proof: We first assume that f is continuous at $\boldsymbol{x}$. There are three cases to consider in showing that f is both LSC and USC at $\boldsymbol{x}$: (i) $f(\boldsymbol{x}) = -\infty$, (ii) $f(\boldsymbol{x}) = +\infty$, and (iii) $-\infty < f(\boldsymbol{x}) < +\infty$.

We handle (i) first. If $f(\boldsymbol{x}) = -\infty$, then f is LSC at $\boldsymbol{x}$ by (SC3). We must show that f is USC at $\boldsymbol{x}$ to complete case (i). To do this, pick $t > -\infty = f(\boldsymbol{x})$. By continuity, there is $\delta > 0$ such that for every $\boldsymbol{y} \in B(\boldsymbol{x}, \delta)$ we have $t > f(\boldsymbol{y})$. Since t is an arbitrary number greater than $f(\boldsymbol{x})$, we have by definition that f is also USC at $\boldsymbol{x}$, finishing case (i).

For (ii), f is automatically USC at $\boldsymbol{x}$ by (SC3). We need to show that f is LSC at $\boldsymbol{x}$ to complete this case. Now f is USC at $\boldsymbol{x}$, hence $-f$ is LSC at $\boldsymbol{x}$ by (SC1) and $-f(\boldsymbol{x}) = -\infty$. By noting that $-f$ is continuous at $\boldsymbol{x}$ and applying case (i) to $-f$, it follows that $-f$ is USC at $\boldsymbol{x}$. By another appeal to (SC1), f is LSC at $\boldsymbol{x}$, giving case (ii).

For (iii), let $t < f(\boldsymbol{x})$ [or $t > f(\boldsymbol{x})$]. Since f is continuous at $\boldsymbol{x}$, there is $\delta > 0$ so that $t < f(\boldsymbol{y})$ [or $t > f(\boldsymbol{y})$] for every $\boldsymbol{y} \in B(\boldsymbol{x}, \delta)$. It follows that f is LSC [or USC] at $\boldsymbol{x}$. This gives (iii), and we have shown in all cases that continuity at $\boldsymbol{x}$ implies upper and lower semicontinuity at $\boldsymbol{x}$.

For the converse, assume that f is both LSC and USC at $\boldsymbol{x}$. Again we must deal with cases (i)-(iii) above. For (i), we require only that f be USC at $\boldsymbol{x}$. To see this, let $t > f(\boldsymbol{x}) = -\infty$. Since f is USC at $\boldsymbol{x}$, there is $\delta > 0$ such that $f(\boldsymbol{y}) < t$ for each $\boldsymbol{y} \in B(\boldsymbol{x}, \delta)$. Since t is arbitrary, we have $\lim_{\boldsymbol{y} \to \boldsymbol{x}} f(\boldsymbol{y}) = -\infty = f(\boldsymbol{x})$, so that f is continuous at $\boldsymbol{x}$. Part (ii) is handled similarly; there we only require the assumption that f is LSC at $\boldsymbol{x}$. We now do (iii), picking $\epsilon > 0$. In the definition of LSC, take $t = f(\boldsymbol{x}) - \epsilon$. For this t, there is $\delta_1 > 0$ such that for all $\boldsymbol{y} \in B(\boldsymbol{x}, \delta_1)$ we have $f(\boldsymbol{x}) - \epsilon < f(\boldsymbol{y})$. In the definition of USC, take $t = f(\boldsymbol{x}) + \epsilon$ and find $\delta_2 > 0$ such that for all $\boldsymbol{y} \in B(\boldsymbol{x}, \delta_2)$ we have $f(\boldsymbol{y}) < f(\boldsymbol{x}) + \epsilon$. Letting $\delta = \min\{\delta_1, \delta_2\} > 0$, we have for every $\boldsymbol{y} \in B(\boldsymbol{x}, \delta)$ that $|f(\boldsymbol{y}) - f(\boldsymbol{x})| < \epsilon$, so that $\lim_{\boldsymbol{y} \to \boldsymbol{x}} f(\boldsymbol{y}) = f(\boldsymbol{x})$, hence f is continuous at $\boldsymbol{x}$. This gives (iii) and finishes the proof. ∎

(SC6) *We have the following:*

(a) f is LSC at $\boldsymbol{x}$ iff $f(\boldsymbol{x}) = \underline{f}(\boldsymbol{x})$,

(b) f is USC at $\boldsymbol{x}$ iff $f(\boldsymbol{x}) = \bar{f}(\boldsymbol{x})$, and

(c) f is continuous at $\boldsymbol{x}$ iff $\underline{f}(\boldsymbol{x}) = f(\boldsymbol{x}) = \bar{f}(\boldsymbol{x})$.

Proof: Suppose that (a) holds. To see (b), we have that f is USC at $\boldsymbol{x}$ iff $-f$ is LSC at $\boldsymbol{x}$, which holds iff $-f(\boldsymbol{x}) = \underline{-f}(\boldsymbol{x})$ [by (a)]. But

$$\underline{-f}(\boldsymbol{x}) = \sup_{\delta>0} \inf_{\boldsymbol{y} \in B(\boldsymbol{x},\delta)} (-f)(\boldsymbol{y}) = -\inf_{\delta>0} \sup_{\boldsymbol{y} \in B(\boldsymbol{x},\delta)} f(\boldsymbol{y}) = -\bar{f}(\boldsymbol{x}),$$

hence f is USC at $\boldsymbol{x}$ iff $f(\boldsymbol{x}) = \bar{f}(\boldsymbol{x})$, giving (b). Part (c) follows from (a), (b), and (SC5). Everything therefore hinges on establishing (a).

To do this, first assume that f is LSC at $\boldsymbol{x}$. By (SC2) we already have $\underline{f}(\boldsymbol{x}) \leq f(\boldsymbol{x})$ [without any assumptions]. We thus need only to show that

$\underline{f}(\boldsymbol{x}) \geq f(\boldsymbol{x})$. There are two cases: (i) $f(\boldsymbol{x}) = -\infty$ and (ii) $f(\boldsymbol{x}) > -\infty$. Case (i) is trivial, hence we assume that $f(\boldsymbol{x}) > -\infty$. Pick $t < f(\boldsymbol{x})$. Since f is LSC at $\boldsymbol{x}$, there is $\delta > 0$ such that $f(\boldsymbol{y}) > t$ for every $\boldsymbol{y} \in B(\boldsymbol{x}, \delta)$. For this $\delta > 0$, we thus have that $\inf_{\boldsymbol{y} \in B(\boldsymbol{x},\delta)} f(\boldsymbol{y}) \geq t$. Since for each $\boldsymbol{x}$ the quantity $\inf_{\boldsymbol{y} \in B(\boldsymbol{x},\delta)} f(\boldsymbol{y})$ is monotone nondecreasing as δ decreases, we have that $\lim_{\delta \to 0^+} \inf_{\boldsymbol{y} \in B(\boldsymbol{x},\delta)} f(\boldsymbol{y}) \geq t$, which gives $\underline{f}(\boldsymbol{x}) \geq t$. Since t is an arbitrary number less than $f(\boldsymbol{x})$, we have $\underline{f}(\boldsymbol{x}) \geq f(\boldsymbol{x})$, giving (ii). One direction of (a) has been established, namely, we have shown that if f is LSC at $\boldsymbol{x}$, then $f(\boldsymbol{x}) = \underline{f}(\boldsymbol{x})$.

For the converse, suppose that $f(\boldsymbol{x}) = \underline{f}(\boldsymbol{x})$; we want f to be LSC at $\boldsymbol{x}$. If $f(\boldsymbol{x}) = -\infty$, then (SC3) gives the result; we may therefore assume that $f(\boldsymbol{x}) > -\infty$. Pick $t < f(\boldsymbol{x}) = \underline{f}(x) = \sup_{\delta>0} \inf_{\boldsymbol{y} \in B(\boldsymbol{x},\delta)} f(\boldsymbol{y})$. There exists $\delta > 0$ such that $t < \inf_{\boldsymbol{y} \in B(\boldsymbol{x},\delta)} f(\boldsymbol{y})$. For this $\delta > 0$, it follows that $t < f(\boldsymbol{y})$ whenever $\boldsymbol{y} \in B(\boldsymbol{x}, \delta)$. But t is a generic number less than $f(\boldsymbol{x})$, hence f is accordingly LSC at $\boldsymbol{x}$. This gives the other direction of (a) and completes the entire proof. ■

(SC7) *We have the following:*

(a) $\underline{f}$ *is LSC, and is the maximal LSC function* $\leq f$.

(b) $\bar{f}$ *is USC, and is the minimal USC function* $\geq f$.

Some clarification is perhaps useful. To say that $\underline{f}$ is the maximal LSC function $\leq f$ means the following: if h is LSC and such that $h(\boldsymbol{x}) \leq f(\boldsymbol{x})$ for all $\boldsymbol{x} \in \mathbb{R}^k$, then $h(\boldsymbol{x}) \leq \underline{f}(\boldsymbol{x})$ for every $\boldsymbol{x} \in \mathbb{R}^k$. Similar comments hold for the statement in (b).

Proof: Let $\boldsymbol{x} \in \mathbb{R}^k$; we will show that $\underline{f}$ is LSC at $\boldsymbol{x}$. If $\underline{f}(\boldsymbol{x}) = -\infty$, then $\underline{f}$ is automatically LSC at $\boldsymbol{x}$ by (SC3), hence we may assume that $\underline{f}(\boldsymbol{x}) > -\infty$. Let $t < \underline{f}(\boldsymbol{x}) = \sup_{\delta>0} \inf_{\boldsymbol{y} \in B(\boldsymbol{x},\delta)} f(\boldsymbol{y})$. It follows that there is $\delta > 0$ such that $t < f(\boldsymbol{y})$ for every $\boldsymbol{y} \in B(\boldsymbol{x}, \delta)$. Since t is a generic number less than $\underline{f}(\boldsymbol{x})$, we have by definition that $\underline{f}$ is LSC at $\boldsymbol{x}$. This gives the first part of (a).

For the second part of (a), let h denote a LSC function with $h \leq f$. Then $\underline{h} \leq \underline{f}$. Since h is LSC, we have $h = \underline{h}$ by (SC6), whence $h \leq \underline{f}$. This establishes maximality and completes the proof of (a). Part (b) follows from (SC1) and (a); it is left as an easy exercise for the reader. ■

Let $f : \mathbb{R}^k \to \bar{\mathbb{R}}$ be given. Recall that f is *Borel measurable* iff f is $\mathcal{B}^k/\mathcal{B}^*$-measurable, and recall that f is *Lebesgue measurable* iff f is $\mathcal{L}^k/\mathcal{B}^*$-measurable. Finally, recall that every Borel measurable function is Lebesgue measurable.

(SC8) *If f is LSC or USC, then f is Borel [hence Lebesgue] measurable. That is, if f is LSC or USC, then f is $\mathcal{B}^k/\mathcal{B}^*$-measurable and hence is $\mathcal{L}^k/\mathcal{B}^*$-measurable. Therefore, in conjunction with (SC7), the functions $\underline{f}$ and $\bar{f}$ are Borel measurable.*

Proof: Suppose that f is LSC. We shall show that f is $\mathcal{B}^k/\mathcal{B}^*$-measurable by showing that $f^{-1}((t,+\infty]) \in \mathcal{B}^k$ for every $t \in \mathbb{R}$. In fact, we shall show something stronger, namely, that $f^{-1}((t,+\infty])$ is an open subset of $\mathbb{R}^k$ for every $t \in \mathbb{R}$. Picking $t \in \mathbb{R}$, the result is trivial if $f^{-1}((t,+\infty]) = \varnothing$, so we may assume that $f^{-1}((t,+\infty])$ is nonempty, say $\boldsymbol{x}^* \in f^{-1}((t,+\infty])$. We have $t < f(\boldsymbol{x}^*) \leq +\infty$, and the assumption that f is LSC ensures the existence of some $\delta > 0$ such that $t < f(\boldsymbol{y})$ for every $\boldsymbol{y} \in B(\boldsymbol{x}^*, \delta)$. For this $\delta > 0$, we have $\boldsymbol{y} \in f^{-1}((t,+\infty])$ for each $\boldsymbol{y} \in B(\boldsymbol{x}^*, \delta)$. This gives $B(\boldsymbol{x}^*, \delta) \subseteq f^{-1}((t,+\infty])$, and hence $\boldsymbol{x}^*$ is an interior point of $f^{-1}((t,+\infty])$. Since $\boldsymbol{x}^*$ is an arbitrary point of $f^{-1}((t,+\infty])$, it follows that $f^{-1}((t,+\infty])$ is an open subset of $\mathbb{R}^k$ in the nonempty case. Putting the empty and the nonempty cases together shows that $f^{-1}((t,+\infty])$ is an open subset of $\mathbb{R}^k$ whenever f is LSC. Since $t \in \mathbb{R}$ is generic, we have established that lower semicontinuity implies Borel [and hence Lebesgue] measurability.

To see that upper semicontinuity implies Borel measurability, we have the following: f is USC iff $-f$ is LSC, which is equivalent to stating that $-f$ is Borel measurable [using what we have just proved], which in turn is equivalent to stating that $-(-f) = f$ is Borel measurable. This completes the proof. ■

(SC9) *Let A denote a generic nonempty index set. For each $\alpha \in A$, suppose that f_α is a function from $\mathbb{R}^k$ into $\bar{\mathbb{R}}$. We have the following:*

(a) If f_α is LSC for each $\alpha \in A$, then $\sup_{\alpha \in A} f_\alpha$ is LSC.

(b) If f_α is USC for each $\alpha \in A$, then $\inf_{\alpha \in A} f_\alpha$ is USC.

The proof of this claim is a straightforward consequence of the definitions and will not pose any problems. The reader may easily provide a short proof.

We have proved all of the semicontinuity properties for functions defined from $\mathbb{R}^k$ into $\bar{\mathbb{R}}$. All of the results hold if the domain, instead of being $\mathbb{R}^k$, is some k-dimensional rectangle of any form.

Exercises. The reader should work at least Exercises 1 and 6. Exercise 6 will be used in the next section, and is straightforward.

1*. The following items were left as simple exercises for the reader:
(a) Part (b) of (SC4).
(b) Part (b) of (SC7).
(c) All of (SC9).

2. Let $E \subset \mathbb{R}^k$.
(a) E is open iff I_E is LSC.
(b) E is closed iff I_E is USC.
(c) We have $\underline{I_E} = I_{E^\circ}$ and $\overline{I_E} = I_{\bar{E}}$, where E° denotes the interior of E and $\bar{E}$ denotes the closure of E.

3. If f_1 and f_2 are LSC [or USC] and finite on $\mathbb{R}^k$, then $f_1 + f_2$ is LSC [or USC]. If $\{f_n\}_{n=1}^\infty$ is a sequence of nonnegative LSC [or USC] functions, is $\sum_{n=1}^\infty f_n$ necessarily LSC [or USC]?

4. Let K denote a compact subset of $\mathbb{R}^k$.

(a) If f is a real-valued function that is USC on K, then f obtains a local maximum on K.
(b) If f is a real-valued function that is LSC on K, then f obtains a local minimum on K.

5. Let A denote a generic nonempty index set, and suppose for each $\alpha \in A$ that f_α is continuous.
(a) It is not necessarily the case that $\sup_{\alpha \in A} f_\alpha$ is continuous.
(b) $\sup_{\alpha \in A} f_\alpha$ is Borel measurable. [Use (SC5), (SC8), and (SC9).]
(c) It is not necessarily the case that $\inf_{\alpha \in A} f_\alpha$ is continuous, but at any rate it is still Borel measurable.

6*. For each $\boldsymbol{x} \in \mathbb{R}^k$ and $\delta > 0$, define $\phi(\boldsymbol{x}) = \inf_{\delta>0} \sup_{\boldsymbol{s},\boldsymbol{t} \in B(\boldsymbol{x},\delta)} |f(\boldsymbol{s}) - f(\boldsymbol{t})|$.
(a) ϕ is USC, hence Borel measurable.
(b) f is continuous at $\boldsymbol{x}$ iff $\phi(\boldsymbol{x}) = 0$.
(c) $\{\boldsymbol{x} \in \mathbb{R}^k : f \text{ is continuous at } \boldsymbol{x}\}$ is a Borel set. This shows that *for any f, the continuity set of f [as well as the set of points where f is discontinuous] is a Borel set.*

7.2 STEP FUNCTIONS IN EUCLIDEAN SPACE

In the previous section, we considered functions with domain $\mathbb{R}^k$ and range contained in $\bar{\mathbb{R}} = [-\infty, +\infty]$. In this section, we will again keep k arbitrary, focusing attention on functions defined on a given bounded k-dimensional closed rectangle

$$I = [\boldsymbol{a}, \boldsymbol{b}] = [a_1, b_1] \times \cdots \times [a_k, b_k],$$

where $-\infty < a_i < b_i < +\infty$ for $i = 1, \cdots, k$. We shall keep this notation fixed throughout the section. Also, we note that the semicontinuity results and the notation from the previous section may be applied to functions with domain I. [See the comments at the end of the preceding section.]

Definition. A finite collection $\{I_1, \cdots, I_n\}$ of closed subrectangles of I is called a *partition of I* iff the open rectangles $I_1^\circ, \cdots, I_n^\circ$ are disjoint and $\bigcup_{i=1}^n I_n = I$. [Recall that I_i° is the interior of I_i, $i = 1, \cdots, n$.]

The point in the definition is that while the boundaries of the rectangles in the partition of I may overlap, the interiors are nonoverlapping.

It is clear on a moment's thought that

$$\lambda_k(I) = \lambda_k\left(\bigcup_{i=1}^n I_i^\circ\right) = \prod_{i=1}^n (b_i - a_i),$$

and hence

$$\lambda_k\left(I - \bigcup_{i=1}^n I_i^\circ\right) = \lambda_k(\{\boldsymbol{x} \in \mathbb{R}^k : \boldsymbol{x} \text{ is a boundary point of some } I_i\}) = 0.$$

[The same comments hold with $\bar{\lambda}_k$ everywhere replacing λ_k.]

Definition. A function $s : I \to \mathbb{R}$ will be called a *step function on I* iff there exists a partition $\{I_1, \cdots, I_n\}$ of I such that
(a) s is constant on I_i° for $i = 1, \cdots, n$, and
(b) s is constant on $I - \bigcup_{i=1}^n I_i^\circ$, the set of points that are on the boundary of at least one of $I_1, \cdots, I_n$.

We now present a list of properties of step functions that will be useful in discussing just how the Lebesgue integral generalizes the Riemann integral.

(SF1) *Every step function s on I is $\mathcal{B}^k/\mathcal{B}^*$-measurable. That is, every step function is Borel [and hence Lebesgue] measurable.*

Proof: Let $\{I_1, \cdots, I_n\}$ be the partition of I associated with s. Each of $I_1^\circ, \cdots, I_n^\circ$ is an open set and hence is in $\mathcal{B}^k$; we also have $I - \bigcup_{i=1}^n I_i^\circ \in \mathcal{B}^k$. The indicator functions[1] of each of $I_1^\circ, \cdots, I_n^\circ$ as well as $I - \bigcup_{i=1}^n I_i^\circ$ are therefore $\mathcal{B}^k/\mathcal{B}^*$-measurable. Since s is a finite linear combination of these indicator functions, s is accordingly $\mathcal{B}^k/\mathcal{B}^*$-measurable. ∎

(SF2) *Every step function s on I is continuous λ_k-a.e. [and $\bar{\lambda}_k$-a.e.] on I. That is, $\{x \in I : s$ is discontinuous at $x\}$ is λ_k-null [and $\bar{\lambda}_k$-null].*

Proof: Let $\{I_1, \cdots, I_n\}$ denote the partition of I associated with s. Clearly s is continuous on $\bigcup_{i=1}^n I_i^\circ$ since s is constant on each of $I_1^\circ, \cdots, I_n^\circ$. It is also clear on a moment's reflection that any discontinuity points of s are restricted to be in $I - \bigcup_{i=1}^n I_i^\circ$. Since the set of discontinuity points of s is a $\mathcal{B}^k$-subset[2] of $I - \bigcup_{i=1}^n I_i^\circ$ and the latter set is λ_k-null, the claim follows. ∎

(SF3) *Let $f : I \to \mathbb{R}$ be continuous λ_k-a.e. and bounded, say $m \le f \le M$ for finite m and M. There exists a sequence $\{s_n\}_{n=1}^\infty$ of step functions on I with the following three properties:*

(a) $s_1 \le s_2 \le \cdots$ λ_k-a.e. on I,

(b) $s_n \le f$ on I for each $n \in \mathbb{N}$, and

(c) $\lim_n s_n \ge f$ λ_k-a.e. on I.

[The entire claim holds if λ_k is replaced everywhere with $\bar{\lambda}_k$.]

Proof: We directly construct such a sequence satisfying (a)-(c). For the moment, fix n. Partition I into 2^{kn} congruent closed rectangles $I_1^{(n)}, \cdots, I_{2^{kn}}^{(n)}$. Define $s_n : I \to \mathbb{R}$ by writing

$$s_n(x) = \begin{cases} \inf_{y \in I_j^{(n)\circ}} f(y) & \text{if } x \text{ is in some } I_j^{(n)\circ} \\ m & \text{otherwise} \end{cases}$$

[1]Just to be sure, these indicator functions have common domain I.
[2]See Exercise 6 of the previous section.

for every $x \in I$. Doing this for each $n \in \mathbb{N}$ produces a sequence $\{s_n\}$ of step functions on I [and a corresponding sequence of partitions] that automatically satisfies (b). We now verify that this sequence $\{s_n\}_{n=1}^{\infty}$ also satisfies requirements (a) and (c).

For (a), pick $n \in \mathbb{N}$ and let $x \in I_j^{(n)}$, where $j \in \{1, \cdots, 2^{kn}\}$. This x will be an element of at least one of the 2^k congruent subrectangles of $I_j^{(n)}$. We consider two cases: (α) x is in fact on the boundary of $I_j^{(n)}$, and (β) x is interior to $I_j^{(n)}$. For (α), we automatically have $s_n(x) \leq s_{n+1}(x)$. For case (β), there are in turn two subcases: (i) x is a boundary point of one of the 2^k congruent subrectangles of $I_j^{(n)}$, and (ii) x is an interior point of one of the 2^k congruent subrectangles of $I_j^{(n)}$. In (i) we have the inequality

$$s_n(x) = \inf_{y \in I_j^{(n)\circ}} f(y) \geq m = s_{n+1}(x).$$

In (ii), writing $I_l^{(n+1)}$ for the subrectangle of $I_j^{(n)}$ for which x is an interior point [to both of $I_l^{(n+1)}$ and $I_j^{(n)}$], we have

$$s_n(x) = \inf_{y \in I_j^{(n)\circ}} f(y) \leq \inf_{y \in I_l^{(n+1)\circ}} f(y) = s_{n+1}(x).$$

Therefore, we have $s_n(x) \leq s_{n+1}(x)$ in case (α) and subcase (ii) of case (β). The only way that we can possibly have $s_n(x) > s_{n+1}(x)$ is when x is interior to $I_j^{(n)}$ and a boundary point of some subrectangle $I_l^{(n+1)}$. However, the set of such x's is λ_k-null,[3] and hence we have $s_n \leq s_{n+1}$ λ_k-a.e. on $I_j^{(n)}$. From the fact that $j \in \{1, \cdots, 2^{kn}\}$ is generic, it follows that $s_n \leq s_{n+1}$ λ_k-a.e. on $I = \bigcup_{j=1}^{2^{kn}} I_j^{(n)}$. In turn, since $n \in \mathbb{N}$ is generic, it follows that we have $s_1 \leq s_2 \leq \cdots$ λ_k-a.e. on I, giving (a).

We now do (c), first observing that part (a) implies the existence of $\lim_n s_n$ λ_k-a.e. on I. Let $x \in I$, and suppose that x is not a boundary point of any congruent [sub]rectangle of I at any stage n. If $\underline{f}(x) > t$, then there is some neighborhood of x such that $f > t$ on this neighborhood. For sufficiently large n, the [sub]rectangle that (i) contains x and (ii) is used in the partition for s_n is contained in this neighborhood, and hence $s_n(x) \geq t$ for large enough n. In turn, we have $\lim_n s_n(x) \geq t$. [Observe that $\lim_n s_n(x)$ exists for all x that that are not boundary points of any subrectangle of I at any stage n.] Since t is a generic number less than $\underline{f}(x)$, we have $\lim_n s_n(x) \geq \underline{f}(x)$. We have shown the following: *if x is not a boundary point of any congruent [sub]rectangle of I at any stage n, then* $\lim_n s_n(x) \geq \underline{f}(x)$. We therefore have that

$$\{x \in I : \text{ it is not the case that } \lim_n s_n(x) \geq \underline{f}(x)\}$$

[3]Again, this is easy to see but somewhat awkward to formally write out.

is a Borel subset of

$$\{\boldsymbol{x} \in I : \boldsymbol{x} \text{ is a boundary point of some subrectangle of } I \text{ at some stage}\}.$$

This latter set is seen on a moment's reflection to be λ_k-null, and hence we have demonstrated (c). ■

There is a symmetric analog to (SF3), and we state it now.

(SF4) *Assume the hypotheses of (SF3). There exists a sequence $\{t_n\}_{n=1}^{\infty}$ of step functions on I with the following three properties:*

(a) $t_1 \geq t_2 \geq \cdots$ λ_k-a.e. on I,

(b) $t_n \geq f$ on I for every $n \in \mathbb{N}$, and

(c) $\lim_n t_n \leq \bar{f}$ λ_k-a.e. on I.

[The entire claim holds if λ_k is everywhere replaced by $\bar{\lambda}_k$.]

Proof: For each $n \in \mathbb{N}$, define

$$t_n(\boldsymbol{x}) = \begin{cases} \sup_{\boldsymbol{y} \in I_j^{(n)\circ}} f(\boldsymbol{y}) & \text{if } \boldsymbol{x} \text{ is in some } I_j^{(n)\circ} \\ M & \text{otherwise} \end{cases}$$

for every $\boldsymbol{x} \in I$. Then the sequence $\{t_n\}_{n=1}^{\infty}$ automatically satisfies (b). Requirements (a) and (c) may be demonstrated in the same fashion as (a) and (c) of (SC3) were demonstrated. ■

(SF5) *Suppose that f is continuous λ_k-a.e. on I, and let $\{s_n\}_{n=1}^{\infty}$ and $\{t_n\}_{n=1}^{\infty}$ denote sequences of step functions satisfying (a)-(c) of (SF3) and (SF4), respectively. Then*

$$\underline{f} = \lim_n s_n = f = \lim_n t_n = \bar{f} \quad \lambda_k\text{-a.e. on } I.$$

[We may everywhere replace λ_k with $\bar{\lambda}_k$ and obtain the same result.]

Proof: By part (c) of (SC6), we have

$$\underline{f} = f = \bar{f} \quad \lambda_k\text{-a.e. on } I.$$

Using (b) and (c) from both (SF3) and (SF4) yields

$$\underline{f} \leq \lim_n s_n \leq f \leq \lim_n t_n \leq \bar{f} \quad \lambda_k\text{-a.e. on } I.$$

Putting these facts together gives the result ■

The preceding property is the crux of the section, and it shall be called on explicitly in the discussion relating the Riemann integral to the Lebesgue integral.

Exercises.

1*. Complete the proof of (SC4).

2. Give a formal writeup of the proof of the following [intuitively obvious] assertion: *The set of $\boldsymbol{x} \in I$ such that $\boldsymbol{x}$ is a boundary point of some subrectangle of I at some stage is λ_k-null.* This assertion was made at the end of the proof of (SC3).

7.3 THE RIEMANN INTEGRAL, PART ONE

In this section, we consider the k-dimensional rectangle

$$I = [\boldsymbol{a}, \boldsymbol{b}] = [a_1, b_1] \times \cdots \times [a_k, b_k],$$

where $-\infty < a_i < b_i < +\infty$ for $i = 1, \cdots, k$. There are two goals for this short section: (i) to recall the definition of the k-dimensional Riemann integral from calculus, and (ii) to express Riemann integrability as a certain property regarding step functions.

Definition. We will say that $f : I \to [m, M]$ is *Riemann integrable [on I]* iff for every $\epsilon > 0$ there exists a partition $\{I_1, \cdots, I_n\}$ of I such that

$$\sum_{j=1}^{n} \sup_{\boldsymbol{x} \in I_j} f(\boldsymbol{x}) \text{vol}(I_j) - \sum_{j=1}^{n} \inf_{\boldsymbol{x} \in I_j} f(\boldsymbol{x}) \text{vol}(I_j) < \epsilon. \tag{7.1}$$

Here m and M are finite numbers with $m \leq M$, and $\text{vol}(I_j)$ stands for the k-dimensional volume of I_j. We will write RI for *Riemann integrable.*

The definition above is the standard definition of Riemann integrability. The symbol $\text{vol}(I_j)$ is the symbol used in advanced calculus for what we are calling $\lambda_k(I_j)$ [or $\bar{\lambda}_k(I_j)$]. However, Lebesgue measure isn't rigorously developed in the calculus, so most books use something like the vol notation.[4] The vol notation has been used merely to make the definition above notationally seem as "calculuslike" as possible.

The following result connects the familiar calculus definition with the concepts and results concerning step functions as given the previous section. The reader should recall the construction of the Lebesgue integral on a generic measure space $(\Omega, \mathcal{F}, \mu)$. Here, the underlying background set Ω is I, the underlying σ-field $\mathcal{F}$ is the collection $\mathcal{B}_I^k = \{B \cap I : B \in \mathcal{B}^k\}$ of Borel subsets of I, and the underlying measure μ is the restriction of k-dimensional Lebesgue measure to the Borel subsets of I, i.e. $\lambda_k|_{\mathcal{B}_I^k}$. We slightly abuse notation and write λ_k for $\lambda_k|_{\mathcal{B}_I^k}$.

Claim 1. *The function $f : I \to [m, M]$ is RI iff for every $\epsilon > 0$ there exist step functions s and t on I with $s \leq f \leq t$ and $\int_I (t - s)\, d\lambda_k < \epsilon$, where this latter integral is the Lebesgue integral constructed in the last chapter.*

Proof: Suppose first that $f : I \to [m, M]$ is RI, and let $\epsilon > 0$. By definition, there exists a partition $\{I_1, \cdots, I_n\}$ of I such that (7.1) holds. Define step

[4]The development of the Riemann integral in advanced calculus is not circular even though it borrows elements of Lebesgue measure. There, one needs only to have something that mimics k-dimensional volume on the closed bounded k-dimensional rectangles. In measure theory, we have a formal measure that mimics k-dimensional volume on a much larger collection of sets, namely $\mathcal{B}^k$ [or $\mathcal{L}^k$]. In calculus, Riemann integrals are defined over compact k-dimensional rectangles, while here we have the machinery to define integrals over sets far more general.

functions s and t on I by writing

$$s(\boldsymbol{x}) = \begin{cases} \inf_{\boldsymbol{y}\in I_j} f(\boldsymbol{y}) & \text{if } \boldsymbol{x} \text{ is in some } I_j^\circ \\ m & \text{otherwise} \end{cases}$$

and

$$t(\boldsymbol{x}) = \begin{cases} \sup_{\boldsymbol{y}\in I_j} f(\boldsymbol{y}) & \text{if } \boldsymbol{x} \text{ is in some } I_j^\circ \\ M & \text{otherwise} \end{cases}$$

for every $\boldsymbol{x} \in I$. We automatically have that $s \leq f \leq t$, and by the definition of the Lebesgue integral we have

$$\begin{aligned} &\int_I s\, d\lambda_k \\ &= \int_{\bigcup_{j=1}^n I_j^\circ} s\, d\lambda_k + \int_{I-\bigcup_{j=1}^n I_j^\circ} s\, d\lambda_k \\ &= \int_{\bigcup_{j=1}^n I_j^\circ} s\, d\lambda_k \quad [I - \textstyle\bigcup_{j=1}^n I_j^\circ \text{ is } \lambda_k\text{-null}] \\ &= \sum_{j=1}^n \int_{I_j^\circ} s\, d\lambda_k \quad [I_1^\circ, \cdots, I_n^\circ \text{ disjoint}] \\ &= \sum_{j=1}^n \inf_{\boldsymbol{x}\in I_j} f(\boldsymbol{x})\lambda_k(I_j^\circ) \quad [s(\boldsymbol{x}) = \inf_{\boldsymbol{y}\in I_j} f(\boldsymbol{y}) \text{ for all } \boldsymbol{x} \in I_j^\circ] \\ &= \sum_{j=1}^n \inf_{\boldsymbol{x}\in I_j} f(\boldsymbol{x})\text{vol}(I_j) \quad [\text{vol}(I_j) = \lambda_k(I_j)]. \end{aligned}$$

By a similar argument, we have

$$\int_I t\, d\lambda_k = \sum_{j=1}^n \sup_{\boldsymbol{x}\in I_j} f(\boldsymbol{x})\text{vol}(I_j).$$

Since (7.1) holds by assumption, we substitute use the last two equations to obtain the desired inequality

$$\int_I (t-s)\, d\lambda_k = \int_I t\, d\lambda_k - \int_I s\, d\lambda_k < \epsilon,$$

giving one implication of the claim.

For the converse implication, completely reset everything, assuming that for every $\epsilon > 0$ there exist step functions s and t on I such that

$$s \leq f \leq t \quad \text{and} \quad \int_I (t-s)\, d\lambda_k < \epsilon.$$

We introduce some notation. Let $\{I_1', \cdots, I_m'\}$ denote the partition of I associated with s; write $\{I_1'', \cdots, I_n''\}$ for the partition of I associated with t. Define the partition $\{I_1, \cdots, I_p\}$ of I to be the common refinement of $\{I'_1, \cdots, I'_m\}$ and $\{I''_1, \cdots, I''_n\}$. [That is, each I_l is of the form $I'_i \cap I''_j$, where $i \in \{1, \cdots, m\}$ and $j \in \{1, \cdots, n\}$.] For $l = 1, \cdots, p$, write s_l for the value of s on I_l°, and similarly write t_l for the value of t on I_l°. We have

$$\begin{aligned}
& \sum_{l=1}^{p} \sup_{\boldsymbol{x} \in I_l} f(\boldsymbol{x})\text{vol}(I_l) - \sum_{l=1}^{p} \inf_{\boldsymbol{x} \in I_l} f(\boldsymbol{x})\text{vol}(I_l) \\
\leq \; & \sum_{l=1}^{p} t_l \text{vol}(I_l) - \sum_{l=1}^{p} s_l \text{vol}(I_l) \\
= \; & \sum_{l=1}^{p} (t_l - s_l)\lambda_k(I_l) \\
= \; & \sum_{l=1}^{p} (t_l - s_l)\lambda_k(I_l^\circ) \\
= \; & \sum_{l=1}^{p} \int_{I_l^\circ} (t - s) \, d\lambda_k \\
= \; & \int_{\bigcup_{l=1}^{p} I_l^\circ} (t - s) \, d\lambda_k \\
= \; & \int_I (t - s) \, d\lambda_k \\
< \; & \epsilon.
\end{aligned}$$

[For each $l = 1, \cdots, p$, we have $f(\boldsymbol{x}) \leq t(\boldsymbol{x})$ for each $\boldsymbol{x} \in I_l$, hence $f(\boldsymbol{x}) \leq t_l$ for each $\boldsymbol{x} \in I_l$, hence $\sup_{\boldsymbol{x} \in I_l} f(\boldsymbol{x}) \leq t_l$. Similarly, for each $l = 1, \cdots, p$ we have $\inf_{\boldsymbol{x} \in I_l} f(\boldsymbol{x}) \geq s_l$. This gives the first inequality above. The second equality follows because $t - s = t_l - s_l$ on I_l° and $\lambda_k(I_l - I_l^\circ) = 0$. The final equality follows because $I - \bigcup_{l=1}^{p} I_l^\circ$ is λ_k-null, hence $\int_{I - \bigcup_{l=1}^{p} I_l^\circ} (t - s) \, d\lambda_k = 0$.] Thus (7.1) holds for the partition $\{I_1, \cdots, I_p\}$, giving the reverse implication. This finishes the entire proof. ∎

We have discussed instances where a function f on I is RI, but we have not talked about what the Riemann integral of f on I actually is in the situation where f is indeed RI on I. The following claim will be used to give the actual definition of the Riemann integral of f on I.

Claim 2. *If $f : I \to [m, M]$ is RI, then*

$$\begin{aligned}
& \sup \left\{ \int_I s \, d\lambda_k : \; s \textit{ is a step function on } I \textit{ and } s \leq f \right\} \\
= \; & \inf \left\{ \int_I t \, d\lambda_k : \; t \textit{ is a step function on } I \textit{ and } t \geq f \right\}.
\end{aligned}$$

Proof: Let A denote the set in the supremum, and let B denote the set in the infimum. For any step functions s and t on I with $s \leq f \leq t$, we have the inequality $\int_I s\, d\lambda_k \leq \int_I t\, d\lambda_k$, and hence every element of A is less than or equal to every element of B, giving $\sup A \leq \inf B$. We desire to strengthen this to equality. Suppose that $\sup A < \inf B$, and let $\epsilon = \inf B - \sup A > 0$. Since f is RI, we use Claim 1 to assert the existence of step functions s and t on I such that

$$s \leq f \leq t \quad \text{and} \quad \int_I t\, d\lambda_k - \int_I s\, d\lambda_k < \epsilon = \inf B - \sup A.$$

But

$$\int_I s\, d\lambda_k \leq \sup A < \inf B \leq \int_I t\, d\lambda_k$$

by appealing to the definitions of A and B, hence at the same time we have the inequality

$$\int_I t\, d\lambda_k - \int_I s\, d\lambda_k \geq \inf B - \sup A = \epsilon,$$

a contradiction. The assumption that $\sup A < \inf B$ yields a contradiction, so it must be the case that $\sup A \geq \inf B$. Combining this with the previously established reverse inequality $\sup A \leq \inf B$ gives $\sup A = \inf B$, completing the proof. ∎

With the previous claim, we may now meaningfully define the Riemann integral of a RI function f on I.

Definition. *If $f : I \to [m, M]$ is RI, then the Riemann integral of f over I, denoted by $\int_I f(x)\, dx$, is defined by the joint equalities*

$$\begin{aligned} \int_I f(x)\, dx &= \sup\left\{ \int_I s\, d\lambda_k : s \textit{ is a step function on } I,\ s \leq f \right\} \\ &= \inf\left\{ \int_I t\, d\lambda_k : t \textit{ is a step function on } I,\ t \geq f \right\}. \end{aligned}$$

It should be noted that everything in this section may be trivially adapted if our measure space is $(I, \mathcal{L}_I^k, \bar{\lambda}_k)$ instead of $(I, \mathcal{B}_I^k, \lambda_k)$. [Here $\mathcal{L}_I^k$ denotes collection of Lebesgue subsets of I and $\bar{\lambda}_k$ really denotes $\bar{\lambda}_k|_{\mathcal{L}_I^k}$ by a similar slight abuse of notation as in the case of λ_k.]

We have now expressed the traditional calculus-based Riemann integral in terms of our measure theory, which means that we have a common measure-theoretic ground on which to discuss and compare the Riemann integral with the Lebesgue integral. This comparison is done in the next section.

Exercise.

1*. $\mathcal{B}_I^k$ is the σ-field on I generated by the k-dimensional right semiclosed subrectangles $(\mathbf{c}, \mathbf{d}]$ of I, and $\mathcal{L}_I^k = \overline{\mathcal{B}_I^k}$, where the completion is with respect to $\lambda_k|_{\mathcal{B}_I^k}$. This exercise justifies using the notation we have used.

7.4 THE RIEMANN INTEGRAL, PART TWO

We are now in the position to show that the Lebesgue integral with respect to Lebesgue measure indeed generalizes the traditional Riemann integral from calculus. We assume all notation and results from the previous section.

The following major result shows just how the Lebesgue integral generalizes the Riemann integral. The underlying measure space is $(I, \mathcal{B}_I^k, \lambda_k)$, where as before λ_k really denotes $\lambda_k|_{\mathcal{B}_I^k}$, the restriction of Lebesgue measure to $\mathcal{B}_I^k$.

Claim 1. *Let $f : I \to [m, M]$ [with $m, M \in \mathbb{R}$ and $m \leq M$] be Borel measurable. We have the following results:*
(a) f is RI iff f is continuous λ_k-a.e. on I.
(b) If f is RI, then the Riemann integral of f over I, denoted by $\int_I f(\boldsymbol{x})\, d\boldsymbol{x}$, coincides with the Lebesgue integral of f over I with respect to λ_k, denoted [as usual] by $\int_I f\, d\lambda_k$:

$$\int_I f(\boldsymbol{x})\, d\boldsymbol{x} = \int_I f\, d\lambda_k.$$

Before beginning the proof, observe that since f is Borel measurable and bounded, the Lebesgue integral $\int_I f\, d\lambda_k$ exists as a finite real number. Therefore, in (b) above, we are justified in discussing $\int_I f\, d\lambda_k$ without worrying about its existence.

Proof: We do (a), assuming first that f is RI. By Claim 1 in the previous section, we have for every $n \in \mathbb{N}$ that there exist step functions s_n and t_n on I with

$$s_n \leq f \leq t_n \quad \text{and} \quad \int_I (t_n - s_n)\, d\lambda_k < \frac{1}{n}.$$

We now observe some facts:

(1) $s_n \leq \underline{f}$ λ_k-a.e. on I for each $n \in \mathbb{N}$.

Proof: For each $n \in \mathbb{N}$, s_n is continuous λ_k-a.e. on I by (SF2), hence $s_n = \underline{s_n}$ λ_k-a.e. on I for each $n \in \mathbb{N}$ by (SC6). Since $s_n \leq f$ on I for each $n \in \mathbb{N}$, we have $\underline{s_n} \leq \underline{f}$ on I for each $n \in \mathbb{N}$. It follows that $s_n \leq \underline{f}$ λ_k-a.e. on I for each $n \in \mathbb{N}$.

(2) $\underline{f} \leq \bar{f}$ on I. [Immediate by (SC2).]

(3) $\bar{f} \leq t_n$ λ_k-a.e. on I for each $n \in \mathbb{N}$.

Proof: For each $n \in \mathbb{N}$, t_n is continuous λ_k-a.e. on I by (SF2), hence $t_n = \overline{t_n}$ λ_k-a.e. on I for each $n \in \mathbb{N}$ by (SC6). Since $t_n \geq f$ on I for each $n \in \mathbb{N}$, we have $\overline{t_n} \geq \bar{f}$ on I for each $n \in \mathbb{N}$. It follows that $t_n \geq \bar{f}$ λ_k-a.e. on I for each $n \in \mathbb{N}$.

Putting the results (1)-(3) together gives

$$s_n \leq \underline{f} \leq \bar{f} \leq t_n \quad \lambda_k\text{-a.e. on } I \quad \text{for every } n \in \mathbb{N}.$$

Next, define $s = \sup_n s_n$ and $t = \inf_n t_n$. Since $s_n \le f \le t_n$ on I for each $n \in \mathbb{N}$, we have

$$s_n \le s \le f \le t \le t_n \quad \text{on } I \text{ for every } n \in \mathbb{N}.$$

Since the s_n's and t_n's are Borel measurable, it follows that s and t are Borel measurable. Next, observe that since $s_n \le \underline{f}$ λ_k-a.e. on I for each $n \in \mathbb{N}$, it follows that $s \le \underline{f}$ λ_k-a.e. on I. Since $\bar{f} \le t_n$ λ_k-a.e. on I for each $n \in \mathbb{N}$, we have $\bar{f} \le t$ λ_k-a.e. on I. We now summarize: for every $n \in \mathbb{N}$ we have

$$s \le \underline{f} \le f \le \bar{f} \le t \;\; \lambda_k\text{-a.e. on } I \qquad \text{and} \qquad t - s \le t_n - s_n \;\text{ on } I.$$

It follows that for every $n \in \mathbb{N}$ we have

$$\int_I (t-s)\, d\lambda_k \le \int_I (t_n - s_n)\, d\lambda_k < \frac{1}{n},$$

and hence $\int_I (t-s)\, d\lambda_k = 0$. Since $t - s \ge 0$ on I, we have $t = s$ λ_k-a.e. by appealing to (L3) in Section 6.4. It follows that $\underline{f} = f = \bar{f}$ λ_k-a.e. on I. By (SC6), we have that f is continuous at λ_k-a.e. $\boldsymbol{x} \in I$. This gives one direction of (a).

Before proving the converse claim in (a), we may now demonstrate (b). To do this, we assume that f is RI, and we keep the notation from the previous pargraphs. Since $m \le f \le M$ and $\lambda_k(I) < +\infty$, we have that $\int_I f\, d\lambda_k$ exists and is finite by (L10). Twice using the definition of the Riemann integral in the previous section, we have for each $n \in \mathbb{N}$ that

$$\int_I f(\boldsymbol{x})\, d\boldsymbol{x} \le \int_I t_n\, d\lambda_k < \int_I s_n\, d\lambda_k + \frac{1}{n} \le \int_I f\, d\lambda_k + \frac{1}{n},$$

and after letting $n \to +\infty$ we have

$$\int_I f(\boldsymbol{x})\, d\boldsymbol{x} \le \int_I f\, d\lambda_k.$$

Again using the definition of the Riemann integral, we have for each $n \in \mathbb{N}$ that

$$\int_I f(\boldsymbol{x})\, d\boldsymbol{x} \ge \int_I s_n\, d\lambda_k > \int_I t_n\, d\lambda_k - \frac{1}{n} \ge \int_I f\, d\lambda_k - \frac{1}{n},$$

and letting $n \to +\infty$ gives

$$\int_I f(\boldsymbol{x})\, d\boldsymbol{x} \ge \int_I f\, d\lambda_k.$$

It follows that

$$\int_I f(\boldsymbol{x})\, d\boldsymbol{x} = \int_I f\, d\lambda_k,$$

which completes the proof of (b).

We now prove the converse of (a), assuming that f is continuous λ_k-a.e. on I and resetting the notation. Let $\{s_n\}_{n=1}^{\infty}$ and $\{t_n\}_{n=1}^{\infty}$ denote the sequences of step functions given in the proofs of (SF3) and (SF4), so that $\{s_n\}_{n=1}^{\infty}$ and $\{t_n\}_{n=1}^{\infty}$ satisfy (a)-(c) of (SF3) and (SF4), respectively. In particular, we have $s_n \leq f \leq t_n$ on I for each $n \in \mathbb{N}$. By (SF5) we have

$$\underline{f} = \lim_n s_n = f = \lim_n t_n = \bar{f} \quad \lambda_k\text{-a.e. on } I. \tag{7.2}$$

It follows that

$$\int_I \lim_n s_n \, d\lambda_k = \int_I \lim_n t_n \, d\lambda_k = \int_I f \, d\lambda_k.$$

[Recall from the comments preceding the proof that $\int_I f \, d\lambda_k$ exists and is finite. Since $\lim_n s_n = f$ λ_k-a.e. on I, we have by (L7) that $\int_I \lim_n s_n \, d\lambda_k$ exists and is finite. Similarly, $\int_I \lim_n t_n \, d\lambda_k$ exists and is finite. Therefore, all three quantities above exist.] Next, for all $n \in \mathbb{N}$ we have $|s_n| \leq \max\{|m|, |M|\}$ on I. [To see this, recall how the function s_n was defined at the beginning of the proof of (SF3).] Since

$$\int_I \max\{|m|, |M|\} \, d\lambda_k = \max\{|m|, |M|\}\lambda_k(I) < +\infty,$$

the Dominated Convergence Theorem (L13) gives

$$\int_I \lim_n s_n \, d\lambda_k = \lim_n \int_I s_n \, d\lambda_k.$$

Next, recall how the step functions $t_1, t_2, \cdots$ were defined at the beginning of the proof of (SF4) to see that $|t_n| \leq \max\{|m|, |M|\}$ on I for each $n \in \mathbb{N}$. A second application of DCT gives

$$\int_I \lim_n t_n \, d\lambda_k = \lim_n \int_I t_n \, d\lambda_k.$$

Therefore, we have

$$\lim_n \int_I s_n \, d\lambda_k = \int_I f \, d\lambda_k = \lim_n \int_I t_n \, d\lambda_k. \tag{7.3}$$

Since all members of (7.3) are finite, we have that $\lim_n \int_I (t_n - s_n) \, d\lambda_k = 0$. In other words, for every $\epsilon > 0$ we may take N so large that $\int_I (t_N - s_N) \, d\lambda_k < \epsilon$. We also have $s_N \leq f \leq t_N$ on I. By Claim 1 of the previous section, f is RI. This gives the converse of (a) as desired, and all of (a) is thus established. The entire proof is now complete. ■

The next result is the previous result stated relative to $(I, \mathcal{L}_I^k, \bar{\lambda}_k)$. [Here $\mathcal{L}_I^k$ denotes the collection of Lebesgue sets of I: $\mathcal{L}_I^k = \{A \in \mathcal{L}^k : A \subseteq I\}$.] In the previous claim, we assumed that the bounded function f on I is Borel measurable. In the next claim, we merely assume that f is bounded, and deduce Lebesgue measurability as a consequence.

Claim 2. *Let $f : I \to [m, M]$ [with m, M finite] be given. We have:*
(a) f is RI iff f is continuous $\bar{\lambda}_k$-a.e. on I.
(b) If f is RI, then f is Lebesgue measurable, and the Riemann integral of f over I, denoted by $\int_I f(\boldsymbol{x})\, d\boldsymbol{x}$, coincides with the Lebesgue integral of f over I with respect to $\bar{\lambda}_k$, denoted by $\int_I f\, d\bar{\lambda}_k$:

$$\int_I f(\boldsymbol{x})\, d\boldsymbol{x} = \int_I f\, d\bar{\lambda}_k.$$

Proof: The proof that the Riemann integrability of f forces f to be continuous $\bar{\lambda}_k$-a.e. on I is exactly the same as the proof of (a) for the previous claim with λ_k everywhere replaced by $\bar{\lambda}_k$. For (b), we again refer to the proof of part (a) and observe that (i) $s = f = t$ $\bar{\lambda}_k$-a.e. on I and (ii) both s and t are Lebesgue measurable. It follows that f is Lebesgue measurable by (MF14) in Section 5.4. The rest of the proof of (b) for $\bar{\lambda}_k$ is obtained by again restating the proof of (b) in the previous claim, replacing λ_k by $\bar{\lambda}_k$ everywhere in the proof. The proof of the converse implication in (a) goes through with the same modification. ∎

The main point of Claim 2 is that whenever the Riemann integral of a bounded function f over a bounded interval $[a, b]$ exists, then we may be indifferent in writing $\int_a^b f(x)\, dx$ [the Riemann integral] or writing any of

$$\int_{(a,b)} f\, d\bar{\lambda}, \quad \int_{(a,b]} f\, d\bar{\lambda}, \quad \int_{[a,b)} f\, d\bar{\lambda}, \quad \text{or} \quad \int_{[a,b]} f\, d\bar{\lambda}.$$

[Recall that all four integrals above are equal since $\bar{\lambda}(\{a\}) = \bar{\lambda}(\{b\}) = 0$.] Often, authors will write $\int_a^b f(x)\, dx$ [the Riemann notation] even if they actually mean one of the four Lebesgue integrals above. If in addition f is Borel measurable, then we may equivalently write any of the above four Lebesgue integrals with λ replacing $\bar{\lambda}$ for $\int_a^b f(x)\, dx$.

Another implication of the results of this section is that familiar computational results from calculus such as the Second Fundamental Theorem can be used without fear. For example, if $f(x) = x^2$ for all real x [say], then we may obtain [say] $\int_{(0,1)} f\, d\lambda = \frac{1}{3}$, since both quantities are equal to $\int_0^1 x^2\, dx = \frac{x^3}{3}\Big|_0^1$. By the comments above, we could replace $(0, 1)$ with any of the other three intervals having 0 and 1 as endpoints, and $\bar{\lambda}$ could replace λ if so desired.

Exercises.

1. Focus attention on a closed and bounded k-dimensional rectangle $I = [\boldsymbol{a}, \boldsymbol{b}]$. Suppose that f and g are RI on I. Using the main results of this section, quickly show that $af + bg$ [where $a, b \in \mathbb{R}$] and fg are also RI. [The idea is to not regurgitate the calculus-based proofs. Observe by Claim 1 or Claim 2 that $af + bg$ and fg are continuous almost everywhere on I if f and g are RI.]

2. Here, take $I = [0,1]$, and let f denote the indicator function of the set of rational numbers in $[0,1]$.
(a) f is not RI on $[0,1]$.
(b) We have that $\int_{[0,1]} f \, d\lambda$ exists and equals 0, so that this is an example of where the Lebesgue integral of f exists while the Riemann integral doesn't exist.

3. Work the following parts:
(a) $\lim_{n\to+\infty} \int_0^n \frac{\sin x}{x} \, dx$ exists and equals $\pi/2$. [Show that $\int_{(n-1)\pi}^{n\pi} \frac{\sin x}{x} \, dx$ alternates in sign and has absolute value decreasing to zero.] In calculus, this limit would be written as an improper Riemann integral: $\int_0^{+\infty} \frac{\sin x}{x} \, dx = \frac{\pi}{2}$.
(b) The Lebesgue integral $\int_{(0,+\infty)} \frac{\sin x}{x} \, d\lambda(x)$ does not exist, as both the Lebesgue integrals $\int_{(0,+\infty)} (\sin x/x)^+ \, d\lambda(x)$ and $\int_{(0,+\infty)} (\sin x/x)^- \, d\lambda(x)$ equal $+\infty$.

4. Obtain $\int_{(0,1)} \frac{1}{\sqrt{x}} \, d\lambda(x)$ if it exists, otherwise show that it fails to exist. Observe that $\frac{1}{\sqrt{x}}$ is not bounded on $(0,1)$, so that one can not immediately appeal to Claim 1 or Claim 2.

5. $\int_{\mathbb{R}} \frac{1}{1+x^2} \, d\lambda(x)$ exists and equals π. [Be more careful here than merely stating that the improper Riemann integral $\int_{-\infty}^{+\infty} \frac{1}{1+x^2} \, dx$ exists and equals π.]

6. Is a RI function necessarily Borel measurable?

7. Show that $\int_1^{+\infty} \frac{1}{x} \, d\lambda(x) = +\infty$. [Be careful here.]

8. Let $f : \mathbb{R} \to \bar{\mathbb{R}}$ be such that $\int_{\mathbb{R}} f \, d\bar{\lambda}$ exists, is finite. Futher suppose that $\int_{(a,b)} f \, d\bar{\lambda} = 0$ for every $-\infty < a < b < +\infty$. Then $f = 0$ $\bar{\lambda}$-a.e. on $\mathbb{R}$. [Work from indicators to simple functions, then to nonnegative functions, then to general integrable functions.]

9. Assume the result of Exercise 8. Show that there cannot exist a set $E \in \mathcal{B}$ such that $\lambda(E \cap [a,b]) = \frac{1}{2}(b-a)$ for all real $a < b$. [Write the left side as $\int_{[a,b]} I_E \, d\lambda$, and write the right side as $\int_{[a,b]} \frac{1}{2} \, d\lambda$. Use the previous exercise, working by contradiction.]

10. There exists a real valued f on $[0,1]$ such that f is continuous on $(0,1]$, $\lim_{\delta\to 0^+} \int_{[\delta,1]} f \, d\bar{\lambda}$ exists and is finite, but $\int_{[0,1]} f \, d\bar{\lambda}$ is undefined.

11. Suppose that f is improperly Riemann integrable over the real line, so that

$$\int_{-\infty}^{+\infty} f(x) \, dx = \lim_{a\to-\infty,\, b\to+\infty} \int_a^b f(x) \, dx$$

exists as a real number. [Observe that this implies that f is RI over each interval $[a,b]$.] Give sufficient conditions for f so that $\int_{\mathbb{R}} f \, d\lambda = \int_{-\infty}^{+\infty} f(x) \, dx$.

7.5 CHANGE OF VARIABLES IN THE LINEAR CASE

In this section, we will work in the measure space $(\mathbb{R}^k, \mathcal{L}^k, \bar{\lambda}_k)$. We shall require the results of Sections 4.2, 4.3, and 4.4 in proving certain assertions. Notationally, we will write $\int$ for $\int_{\mathbb{R}^k}$ in this section.

Claim 1. *Suppose that $\boldsymbol{T}$ denotes a nonsingular transformation from $\mathbb{R}^k$ to $\mathbb{R}^k$. Let $f : \mathbb{R}^k \to \bar{\mathbb{R}}$ denote a Lebesgue measurable mapping. Then*

(a) $f \circ \boldsymbol{T}$ is Lebesgue measurable.

(b) If f is nonnegative, then

$$\int f \, d\bar{\lambda}_k = |\det \boldsymbol{T}| \int f \circ \boldsymbol{T} \, d\bar{\lambda}_k.$$

(c) If $f \in L^1(\mathbb{R}^k, \mathcal{L}^k, \bar{\lambda}_k)$, then $f \circ \boldsymbol{T} \in L^1(\mathbb{R}^k, \mathcal{L}^k, \bar{\lambda}_k)$, and again we have the integral equation

$$\int f \, d\bar{\lambda}_k = |\det \boldsymbol{T}| \int f \circ \boldsymbol{T} \, d\bar{\lambda}_k.$$

Proof: We first do (a). Let $f = I_A$, where $A \in \mathcal{L}^k$. Then $f \circ \boldsymbol{T} = I_{\boldsymbol{T}^{-1}A}$, and $\boldsymbol{T}^{-1}A \in \mathcal{L}$ [apply Claim 11 of Section 4.4 to the nonsingular linear transformation $\boldsymbol{T}^{-1}$], so that $f \circ \boldsymbol{T}$ is Lebesgue measurable by (MF3). Part (a) accordingly holds for indicator functions of Lebesgue sets.

Now reset everything, and assume that f is of the form $\sum_{i=1}^n c_i I_{A_i}$, where $0 \le c_1, \cdots, c_n < +\infty$ and $A_1, \cdots, A_n$ are disjoint $\mathcal{L}^k$-sets with union $\mathbb{R}^k$. [That is, $f \in \mathfrak{S}(\mathbb{R}^k, \mathcal{L}^k, \bar{\lambda}_k)$.] Then $f \circ \boldsymbol{T} = \sum_{i=1}^n c_i I_{\boldsymbol{T}^{-1}A_i}$, which by the previous paragraph is a finite linear combination of the Lebesgue measurable functions $I_{\boldsymbol{T}^{-1}A_1}, \cdots, I_{\boldsymbol{T}^{-1}A_n}$, and hence $f \circ \boldsymbol{T}$ is Lebesgue measurable when $f \in \mathfrak{S}(\mathbb{R}^k, \mathcal{L}^k, \bar{\lambda}_k)$.

Again reset everything, and let $f \in \mathfrak{N}(\mathbb{R}^k, \mathcal{L}^k, \bar{\lambda}_k)$. Then there exists a nondecreasing sequence $\{s_n\}_{n=1}^\infty$ of functions in $\mathfrak{S}(\mathbb{R}^k, \mathcal{L}^k, \bar{\lambda}_k)$ with $\lim s_n = f$. [See (MF18).] Now $f \circ \boldsymbol{T} = (\lim s_n) \circ \boldsymbol{T} = \lim(s_n \circ \boldsymbol{T})$, and $s_n \circ \boldsymbol{T}$ is Lebesgue measurable for each $n \in \mathbb{N}$ by the previous paragraph. Since $f \circ \boldsymbol{T}$ is the limit of a sequence of Lebesgue measurable functions, $f \circ \boldsymbol{T}$ itself is Lebesgue measurable by (MF11). This proves (a) when $f \in \mathfrak{N}(\mathbb{R}^k, \mathcal{L}^k, \bar{\lambda}_k)$.

Finally, if f is a general Lebesgue measurable function, then so are the nonnegative functions f^+ and f^-. Since $f^+ \circ \boldsymbol{T}$ and $f^- \circ \boldsymbol{T}$ are Lebesgue measurable by the previous paragraph, and since

$$f \circ \boldsymbol{T} = (f^+ - f^-) \circ \boldsymbol{T} = (f^+ \circ \boldsymbol{T}) - (f^- \circ \boldsymbol{T}),$$

we have that $f \circ \boldsymbol{T}$ is Lebesgue measurable. This completes the proof of (a).

We now do (b). Let $f \in \mathfrak{S}(\mathbb{R}^k, \mathcal{L}^k, \bar{\lambda}_k)$ have the form $\sum_{i=1}^n c_i I_{A_i}$ as described just above. Then using Claim 11 of Section 4.4, the definition of the

Lebesgue integral, and the fact that $f \circ \boldsymbol{T} = \sum_{i=1}^{n} c_i I_{\boldsymbol{T}^{-1}A_i}$, we have that

$$\begin{aligned}\int f\, d\bar{\lambda}_k = \sum_{i=1}^{n} c_i \bar{\lambda}_k(A_i) &= |\det \boldsymbol{T}| \sum_{i=1}^{n} c_i |\det \boldsymbol{T}^{-1}| \bar{\lambda}_k(A_i) \\ &= |\det \boldsymbol{T}| \sum_{i=1}^{n} c_i \bar{\lambda}_k(\boldsymbol{T}^{-1}A_i) \\ &= |\det \boldsymbol{T}| \int f \circ \boldsymbol{T}\, d\bar{\lambda}_k.\end{aligned}$$

This gives (b) when $f \in \mathfrak{S}(\mathbb{R}^k, \mathcal{L}^k, \bar{\lambda}_k)$.

Next, let $f \in \mathfrak{N}(\mathbb{R}^k, \mathcal{L}^k, \bar{\lambda}_k)$, and let $\{s_n\}_{n=1}^{\infty}$ denote a nondecreasing sequence of functions in $\mathfrak{S}(\mathbb{R}^k, \mathcal{L}^k, \bar{\lambda}_k)$ with $\lim s_n = f$. Using MCT, the previous paragraph, and the fact that $\{s_n \circ \boldsymbol{T}\}_{n=1}^{\infty}$ is a nondecreasing sequence of nonnegative functions with limit $f \circ \boldsymbol{T}$, we have that

$$\int f\, d\bar{\lambda}_k = \lim_n \int s_n\, d\bar{\lambda}_k = \lim_n |\det \boldsymbol{T}| \int s_n \circ \boldsymbol{T}\, d\bar{\lambda}_k = |\det \boldsymbol{T}| \int f \circ \boldsymbol{T}\, d\bar{\lambda}_k.$$

This gives (b) of the theorem.

For (c), we let f denote a function in $L^1(\mathbb{R}^k, \mathcal{L}^k, \bar{\lambda}_k)$. Then both $\int f^+\, d\bar{\lambda}_k$ and $\int f^-\, d\bar{\lambda}_k$ are finite. Since f^+ and f^- are nonnegative and Lebesgue measurable, the equation in (b) asserts that $\int f^+ \circ \boldsymbol{T}\, d\bar{\lambda}_k$ and $\int f^- \circ \boldsymbol{T}\, d\bar{\lambda}_k$ are finite [being equal to $|\det \boldsymbol{T}|^{-1} \int f^+\, d\bar{\lambda}_k$ and $|\det \boldsymbol{T}|^{-1} \int f^-\, d\bar{\lambda}_k$, respectively]. Since $(f^+ \circ \boldsymbol{T}) = (f \circ \boldsymbol{T})^+$ and $(f^- \circ \boldsymbol{T}) = (f \circ \boldsymbol{T})^-$, we have that both of $\int (f \circ \boldsymbol{T})^+\, d\bar{\lambda}_k$ and $\int (f \circ \boldsymbol{T})^-\, d\bar{\lambda}_k$ are finite, hence $f \circ \boldsymbol{T} \in L^1(\mathbb{R}^k, \mathcal{L}^k, \bar{\lambda}_k)$ as claimed. Next, using the equation in (b):

$$\begin{aligned}\int f\, d\bar{\lambda}_k &= \int f^+\, d\bar{\lambda}_k - \int f^-\, d\bar{\lambda}_k \\ &= |\det \boldsymbol{T}| \int f^+ \circ \boldsymbol{T}\, d\bar{\lambda}_k - |\det \boldsymbol{T}| \int f^- \circ \boldsymbol{T}\, d\bar{\lambda}_k \\ &= |\det \boldsymbol{T}| \int (f \circ \boldsymbol{T})^+\, d\bar{\lambda}_k - |\det \boldsymbol{T}| \int (f \circ \boldsymbol{T})^-\, d\bar{\lambda}_k \\ &= |\det \boldsymbol{T}| \int f \circ \boldsymbol{T}\, d\bar{\lambda}_k.\end{aligned}$$

This completes (c) and hence the entire proof. ∎

Of course, we can consider $(\mathbb{R}^k, \mathcal{B}^k, \lambda_k)$ in place of $(\mathbb{R}^k, \mathcal{L}^k, \bar{\lambda}_k)$ and again the theorem will hold true.

A useful corollary to the preceding theorem comes by taking $\boldsymbol{T} = c\boldsymbol{I}$, where $\boldsymbol{I}$ is the identity matrix of order k and $c \in \mathbb{R}$. Noting that $|\det \boldsymbol{T}| = |c|^k$, the equation in parts (b) and (c) of the preceding theorem becomes

$$\int f(\boldsymbol{x})\, d\bar{\lambda}_k(\boldsymbol{x}) = |c|^k \int f(c\boldsymbol{x})\, d\bar{\lambda}_k(\boldsymbol{x}).$$

Related to Claim 1 is the following result concerning translations, which recalls the translation invariance of Lebesgue measure established in Section 4.3.

Claim 2. *Let $f : \mathbb{R}^k \to \bar{\mathbb{R}}$ be Lebesgue measurable. Let $\tau_{\boldsymbol{y}}$ denote translation by some $\boldsymbol{y} \in \mathbb{R}^k$:*

$$\tau_{\boldsymbol{y}} f(\boldsymbol{x}) = f(\boldsymbol{x} + \boldsymbol{y}) \quad \textit{for every } \boldsymbol{x} \in \mathbb{R}^k.$$

We have the following:

(a) $\tau_{\boldsymbol{y}} f$ is Lebesgue measurable.

(b) If f is nonnegative, then

$$\int f \, d\bar{\lambda}_k = \int \tau_{\boldsymbol{y}} f \, d\bar{\lambda}_k.$$

(c) If $f \in L^1(\mathbb{R}^k, \mathcal{L}^k, \bar{\lambda}_k)$, then $\tau_{\boldsymbol{y}} f \in L^1(\mathbb{R}^k, \mathcal{L}^k, \bar{\lambda}_k)$, and again we have the equation $\int f \, d\bar{\lambda}_k = \int \tau_{\boldsymbol{y}} f \, d\bar{\lambda}_k$.

Proof: Let $f \in \mathfrak{S}(\mathbb{R}^k, \mathcal{L}^k, \bar{\lambda}_k)$ have the form $\sum_{i=1}^n c_i I_{A_i}$, where as usual we have that $0 \le c_1, \cdots, c_n < +\infty$ and $A_1, \cdots, A_n \in \mathcal{L}^k$ are disjoint with union $\mathbb{R}^k$. Then $\tau_{\boldsymbol{y}} f = \sum_{i=1}^n c_i I_{-\boldsymbol{y} \oplus A_i}$. Observe that since $A_i \in \mathcal{L}$ for $i = 1, \cdots, n$, translation invariance for Lebesgue measure gives $-\boldsymbol{y} \oplus A_i \in \mathcal{L}^k$ for $i = 1, \cdots, n$. It follows that $\tau_{\boldsymbol{y}} f$ is a linear combination of Lebesgue measurable indicator functions and hence is Lebesgue measurable. Furthermore, translation invariance and the definition of the integral give

$$\int f \, d\bar{\lambda}_k = \sum_{i=1}^n c_i \bar{\lambda}_k(A_i) = \sum_{i=1}^n c_i \bar{\lambda}_k(-\boldsymbol{y} \oplus A_i) = \int \tau_{\boldsymbol{y}} f \, d\bar{\lambda}_k.$$

This proves (a) and (b) whenever $f \in \mathfrak{S}(\mathbb{R}^k, \mathcal{L}^k, \bar{\lambda}_k)$.

Next, let $f \in \mathfrak{N}(\mathbb{R}^k, \mathcal{L}^k, \bar{\lambda}_k)$. Let $\{s_n\}_{n=1}^\infty$ denote a nondecreasing sequence of functions in $\mathfrak{S}(\mathbb{R}^k, \mathcal{L}^k, \bar{\lambda}_k)$ such that $\lim s_n = f$. By the previous paragraph, both (a) and (b) hold for each $\tau_{\boldsymbol{y}} s_n$. Observe that $\{\tau_{\boldsymbol{y}} s_n\}_{n=1}^\infty$ is a nondecreasing sequence of functions in $\mathfrak{S}(\mathbb{R}^k, \mathcal{L}^k, \bar{\lambda}_k)$ with limit $\tau_{\boldsymbol{y}} f$, and hence $\tau_{\boldsymbol{y}} f$ is Lebesgue measurable upon appeal to (MF11). This gives (a) in the nonnegative situation. Using MCT and the fact that $\int s_n \, d\bar{\lambda}_k = \int \tau_{\boldsymbol{y}} s_n \, d\bar{\lambda}_k$ for each $n \in \mathbb{N}$ gives

$$\int f \, d\bar{\lambda}_k = \lim_n \int s_n \, d\bar{\lambda}_k = \lim_n \int \tau_{\boldsymbol{y}} s_n \, d\bar{\lambda}_k = \int \tau_{\boldsymbol{y}} f \, d\bar{\lambda}_k,$$

and hence (b) is completely finished.

We now complete (a). Let f be Lebesgue measurable, so that both of the nonnegative functions f^+ and f^- are Lebesgue measurable. We apply the nonnegative version of (a) to obtain that both $\tau_{\boldsymbol{y}} f^+$ and $\tau_{\boldsymbol{y}} f^-$ are Lebesgue measurable and hence $\tau_{\mathbf{y}}f = \tau_{\mathbf{y}}f^+ - \tau_{\mathbf{y}}f^-$ is Lebesgue measurable. This completely proves (a).

We finish matters by handling (c). Let $f \in L^1(\mathbb{R}^k, \mathcal{L}^k, \bar{\lambda}_k)$, so that both $\int f^+ \, d\bar{\lambda}_k$ and $\int f^- \, d\bar{\lambda}_k$ are finite. From (b), we have that

$$\int f^+ \, d\bar{\lambda}_k = \int \tau_{\mathbf{y}} f^+ \, d\bar{\lambda}_k \quad \text{and} \quad \int f^- \, d\bar{\lambda}_k = \int \tau_{\mathbf{y}} f^- \, d\bar{\lambda}_k.$$

Since $\tau_{\mathbf{y}}f^+ = (\tau_{\mathbf{y}}f)^+$ and $\tau_{\mathbf{y}}f^- = (\tau_{\mathbf{y}}f)^-$, we have that both $\int (\tau_{\mathbf{y}}f)^+ \, d\bar{\lambda}_k$ and $\int (\tau_{\mathbf{y}}f)^- \, d\bar{\lambda}_k$ are finite, hence $\tau_{\mathbf{y}}f \in L^1(\mathbb{R}^k, \mathcal{L}^k, \bar{\lambda}_k)$. These equalities may now be put together to obtain

$$\begin{aligned} \int f \, d\bar{\lambda}_k = \int f^+ \, d\bar{\lambda}_k - \int f^- \, d\bar{\lambda}_k &= \int (\tau_{\mathbf{y}}f)^+ \, d\bar{\lambda}_k - \int (\tau_{\mathbf{y}}f)^- \, d\bar{\lambda}_k \\ &= \int \tau_{\mathbf{y}} f \, d\bar{\lambda}_k, \end{aligned}$$

completing the entire proof. ∎

Despite the length of the proofs, the claims went through without much of a problem. These straightforward results will be invoked in our discussion of convolutions in Chapter 10. There, we shall combine Claims 1 and 2 in a few proofs. The reader is urged to see just how important the Euclidean background involving Lebesgue measure is in these proofs.

Exercises. The following exercises are meant to be worked together as a set. Everything will be done in the context of the measure space $(\mathbb{R}^k, \mathcal{L}^k, \bar{\lambda}_k)$. We will write L^1 for $L^1(\mathbb{R}^k, \mathcal{L}^k, \bar{\lambda}_k)$ in what follows. We will also write $\int$ for $\int_{\mathbb{R}^k}$ as we have done earlier.

1. For general elements $f \in L^1$, define $\|f\|_1 = \int |f| \, d\bar{\lambda}_k$. Then
(a) $0 \leq \|f\|_1 < +\infty$.
(b) $\|f\|_1 = 0$ iff $f = 0$ $\bar{\lambda}_k$-a.e.
(c) $\|cf\|_1 = |c|\|f\|_1$ for every real c.
(d) $\|f + g\|_1 \leq \|f\|_1 + \|g\|_1$ [where $g \in L^1$ as well].
[Parts (a)-(c) are nearly trivial. Use the triangle inequality in (d).]

2. For elements $f, g \in L^1$ write $f \sim g$ iff $f = g$ $\bar{\lambda}_k$-a.e. Then $\sim$ is an equivalence relation.

Comment: We can't formally say that $\|\cdot\|_1$ is a norm on L^1 because it is not the case that $\|f\|_1 = 0$ forces f to be the zero element — the function identically zero — of L^1. However, if we agree by fiat to identify elements of L^1 that are equal $\bar{\lambda}_k$-a.e., and if we view $\|\cdot\|_1$ as being defined on the collection of equivalence classes $L^1/\sim$ of L^1 induced by $\sim$, then $\|\cdot\|_1$ is a norm on $L^1/\sim$, where the "zero element" is the equivalence class of functions that are equal to zero $\bar{\lambda}_k$-a.e. We ignore this distinction, and instead talk about "the norm $\|\cdot\|_1$ defined on L^1" or "the L^1-norm $\|\cdot\|_1$", applying $\|\cdot\|_1$ to functions instead of equivalence classes. That is, we write $\|f\|_1$ [which is slightly incorrect] in place of the much more cumbersome [but completely correct] $\|\{g \in L^1 : g \sim f\}\|_1$.

3. For $f, g \in L^1$, define the L^1*-distance between* f *and* g to be $\|f-g\|_1$. Show that distance function is a metric if we agree to identify functions that are equal almost everywhere.

4. Here we will show that the collection C_c of continuous functions with compact support [see Section 5.7 for the definition] is dense in L^1, where the notion of distance is as given in the previous exercise. That is, given $f \in L^1$ and $\epsilon > 0$, there is $g \in C_c$ such that $\|f-g\|_1 < \epsilon$. Work in the following stages.
(a) $f = I_A$, where A is a bounded Lebesgue subset of $\mathbb{R}^k$. [Find compact K and bounded open G with $K \subseteq A \subseteq G$ and $\bar{\lambda}_k(G-K) < \epsilon$; use the approximation theorem in Section 4.2. Use Urysohn's Lemma in Section 5.7 to find g on $\mathbb{R}^k$ that is continuous with $0 \le g \le 1$, $g = 1$ on K, and $g = 0$ on G^c. Argue that $g - I_A = 0$ for all $\boldsymbol{x} \in K \cup G^c$ and $|g - I_A| \le 1$ on $G - K$. Show that this gives $\|g - I_A\|_1 < \epsilon$.]
(b) f is a nonnegative finite-valued Lebesgue measurable simple function that is zero outside a bounded set, say $f = \sum_{i=1}^n c_i I_{A_i}$, where $0 \le c_1, \cdots, c_n < +\infty$ and $A_1, \cdots, A_n$ are disjoint bounded $\mathcal{L}^k$-sets. [This follows from (a); approximate each I_{A_i} by some $g_i \in C_c$. Show that $\sum_{i=1}^n c_i g_i \in C_c$ approximates $\sum_{i=1}^n c_i I_{A_i}$.]
(c) f is a nonnegative measurable function that is zero outside a bounded set. [Find a nonnegative finite-valued nondecreasing sequence $\{s_n\}_{n=1}^\infty$ of Lebesgue measurable simple functions in (b) that are zero outside of some bounded set with $s_n \to f$. Show that $\lim_n \|s_n - f\|_1 = 0$ by using DCT. Therefore, f may be approximated by the types of functions given in (b).]
(d) f is a measurable function that is zero outside a bounded set. [Write f as $f^+ - f^-$; use (c). This part shows that f may be approximated by functions as given in (c).]
(e) f a completely general element of L^1. [Use DCT to show the limit relation $\lim_n \|f I_{B(\mathbf{0},k)} - f\|_1 = 0$. Thus f may be approximated by functions of the type given in (d). Here $B(\mathbf{0}, k)$ is the open ball centered at the origin with radius k.]
(f) Work backward from (e) to (a) to prove the claim given at the beginning of this problem.

Comment: This claim shall be greatly generalized in the next chapter.

5. We know that if $f \in L^1$, then $\tau_{\boldsymbol{y}} f \in L^1$ for every $\boldsymbol{y} \in \mathbb{R}^k$. [$\tau_{\boldsymbol{y}}$ is the translation operator given in Claim 2.] Show that $\lim_{\boldsymbol{y}\to\mathbf{0}} \|\tau_{\boldsymbol{y}} f - f\|_1 = 0$. [Pick $\epsilon > 0$. Use the previous exercise to find $g \in C_c$ such that $\|f - g\|_1 < \frac{\epsilon}{3}$. Use Claim 2 to show that we also have $\|\tau_{\boldsymbol{y}} f - \tau_{\boldsymbol{y}} g\|_1 < \frac{\epsilon}{3}$. There is $r > 0$ such that $g = 0$ on $B(\mathbf{0}, r)^c$, hence g is uniformly continuous. Find $\delta > 0$ such that

$$\|\boldsymbol{y}\| \le \delta \quad \text{implies} \quad |g(\boldsymbol{x}+\boldsymbol{y}) - g(\boldsymbol{x})| < \frac{\epsilon}{3\bar{\lambda}_k(B(\mathbf{0}, r+1))}$$

for every $\boldsymbol{x} \in \mathbb{R}^k$. Argue that $\tau_{\boldsymbol{y}} g - g = 0$ on $B(\mathbf{0}, r+1)^c$; obtain $\|\tau_{\boldsymbol{y}} g - g\|_1 < \frac{\epsilon}{3}$. Thus $\|y\| \le \delta$ implies $\|\tau_{\boldsymbol{y}} f - f\|_1 \le \|\tau_{\boldsymbol{y}} f - \tau_{\boldsymbol{y}} g\|_1 + \|\tau_{\boldsymbol{y}} g - g\|_1 + \|g - f\|_1 < \epsilon$.]

8

The L^p Spaces

This chapter develops the standard theory for L^p spaces. Major results given here include the Riesz-Fischer Theorem and the Riesz Representation Theorem for bounded linear functionals on L^p, which serves as an exhibition of the Riesz-Fischer Theorem and some very classical analysis. In this sense, the chapter is more "analysis" than "measure theory," but the underlying rigorous structure on which the analysis is performed obtains its impetus from the rigorous discussions of measures and integration given in the earlier chapters.

Throughout this chapter, we consider a generic measure space $(\Omega, \mathcal{F}, \mu)$. For $\mathcal{F}/\mathcal{B}^*$-measurable functions f and g defined μ-a.e. on Ω, we write $f \sim g$ iff $f = g$ μ-a.e. on Ω. Then $\sim$ is an equivalence relation on the collection of $\mathcal{F}/\mathcal{B}^*$-measurable functions defined μ-a.e. on Ω, and hence $\sim$ partitions this collection of functions into disjoint equivalence classes. If $\mathcal{E}$ denotes the collection of $\mathcal{F}/\mathcal{B}^*$-measurable functions defined μ-a.e. on Ω, we use the standard notation $\mathcal{E}/\sim$ to denote the collection of equivalence classes into which $\sim$ partitions $\mathcal{E}$. With this notation, we give the formal definition of L^p space, which by necessity is somewhat clumsy.

Formal Definition. Consider a generic measure space $(\Omega, \mathcal{F}, \mu)$, and let $0 < p < +\infty$. We define L^p by writing

$$L^p = \left\{\mathfrak{A} \in \mathcal{E}/\sim: \text{ there is } f \in \mathfrak{A} \text{ such that } \int |f|^p < +\infty\right\}.$$

Also, given a set $E \in \mathcal{F}$, we define $L^p(E)$ by

$$L^p(E) = \left\{\mathfrak{A} \in \mathcal{E}/\sim: \text{ there is } f \in \mathfrak{A} \text{ such that } \int_E |f|^p < +\infty\right\}.$$

If confusion is possible regarding the underlying measure space, we will write $L^p(\Omega, \mathcal{F}, \mu)$ for L^p, and we will similarly write $L^p(E, (\Omega, \mathcal{F}, \mu))$ for $L^p(E)$.

Formally speaking, L^p is not a subcollection of functions in $\mathcal{E}$. Rather, L^p is a subcollection of equivalence classes in $\mathcal{E}/\sim$. Also, if $f, g \in \mathcal{E}$ are equal μ-a.e. on Ω, then $\int |f|^p = \int |g|^p$. Therefore, an equivalence class $\mathfrak{A}$ is in L^p iff every $f \in \mathfrak{A}$ is such that $\int |f|^p < +\infty$. Similarly, an equivalence class $\mathfrak{A}$ is in $L^p(E)$ iff every $f \in \mathfrak{A}$ is such that $\int_E |f|^p < +\infty$, which in turn holds iff every $f \in \mathfrak{A}$ is such that $f I_E \in L^p$.

Definition. Let $\mathfrak{A}, \mathfrak{B} \in L^p$ and $c \in \mathbb{R}$. We define

$$\mathfrak{A} + \mathfrak{B} = \{f + g : \ f \in \mathfrak{A}, \ g \in \mathfrak{B}\} \quad \text{and} \quad c\mathfrak{A} = \{cf : \ f \in \mathfrak{A}\}.$$

We also define $\mathfrak{o}$ to be $\{f \in \mathcal{E} : \ f = 0 \ \mu\text{-a.e. on } \Omega\}$.

The definition lets us discuss "addition" and "scalar multiplication" in the context of L^p, and the element $\mathfrak{o}$ defined above serves as the "zero element" of L^p. [It is clear that $\mathfrak{o} \in L^p$ since $f \in \mathfrak{o}$ means that $|f|^p = 0$ μ-a.e. on Ω, whence $\int |f|^p = 0 < +\infty$.]

We claim that L^p is a vector space for $0 < p < +\infty$. That is, we claim that (i) $c\mathfrak{A} \in L^p$ whenever $c \in \mathbb{R}$ and $\mathfrak{A} \in L^p$, and (ii) $\mathfrak{A} + \mathfrak{B} \in L^p$ whenever $\mathfrak{A}, \mathfrak{B} \in L^p$. To see (i), pick $c \in \mathbb{R}$ and $\mathfrak{A} \in L^p$. Pick $h \in c\mathfrak{A}$, so that $h = cf$ for some $f \in \mathfrak{A}$. Since $\int |f|^p < +\infty$, we have

$$\int |h|^p = \int |cf|^p = |c|^p \int |f|^p < +\infty,$$

giving $c\mathfrak{A} \in L^p$. The proof of (ii) takes more work, and we accept it for the moment as true.

Having defined L^p space, we now define a very special function on L^p. Again, the formal definition, while correct, is somewhat clumsy to write out.

Formal Definition. Consider a generic measure space $(\Omega, \mathcal{F}, \mu)$, and let $0 < p < +\infty$. We define the function $\| \ \|_p$ on L^p as follows:

$$\|\mathfrak{A}\|_p = \left(\int |f|^p \right)^{\frac{1}{p}} \quad \text{where } f \text{ is any element of } \mathfrak{A}.$$

The function $\| \ \|_p$ is referred to as *the L^p norm.*

No consistency problems arise with the above definition, for if f and g are elements of some $\mathfrak{A} \in L^p$, then $f = g$ μ-a.e. on Ω, hence $\int |f|^p = \int |g|^p$, so that $(\int |f|^p)^{\frac{1}{p}} = (\int |g|^p)^{\frac{1}{p}}$. Observe that $0 \leq \|\mathfrak{A}\|_p < +\infty$ for every $\mathfrak{A} \in L^p$.

We have referred to $\| \ \|_p$ as a norm, and this suggests that $\| \ \|_p$ satisfies the requirements of a norm. This will be seen to be the case provided that

$1 \leq p \leq +\infty$.[1] When we claim that $|| \; ||_p$ is a norm for $1 \leq p < +\infty$, we are saying that $|| \; ||_p$ satisfies four properties:

(1) $0 \leq ||\mathfrak{A}||_p < +\infty$ for every $\mathfrak{A} \in L^p$,
(2) $||\mathfrak{A}||_p = 0$ iff $\mathfrak{A} = \mathfrak{o}$,
(3) $||c\mathfrak{A}||_p = |c|||\mathfrak{A}||_p$ for every $c \in \mathbb{R}$ and $\mathfrak{A} \in L^p$, and
(4) $||\mathfrak{A} + \mathfrak{B}||_p \leq ||\mathfrak{A}||_p + ||\mathfrak{B}||_p$ for every $\mathfrak{A}, \mathfrak{B} \in L^p$.

Property (1) is obvious and has been discussed.

For (2), assume that $\mathfrak{A} = \mathfrak{o}$. Then, if $f \in \mathfrak{A}$, we have that $f = 0$ μ-a.e. on Ω, and so $\int |f|^p = 0$, whence $||\mathfrak{A}||_p = 0$. Conversely, if $||\mathfrak{A}||_p = 0$, then, given f in $\mathfrak{A}$, we have that $(\int |f|^p)^{\frac{1}{p}} = 0$, and hence $\int |f|^p = 0$. It follows that $|f|^p = 0$ μ-a.e. on Ω, hence $f = 0$ μ-a.e. on Ω, so that $f \in \mathfrak{o}$. Since equivalence classes are disjoint, it follows that $\mathfrak{A}$ and $\mathfrak{o}$ coincide. This proves (2). [Note that we can discuss $||\mathfrak{o}||_p$ since $\mathfrak{o} \in L^p$.]

For (3), let $c \in \mathbb{R}$ and $\mathfrak{A} \in L^p$. Pick $h \in c\mathfrak{A}$, so that $h = cf$ for some $f \in \mathfrak{A}$. We obviously have that $(\int |h|^p)^{\frac{1}{p}} = (\int |cf|^p)^{\frac{1}{p}} = |c|(\int |f|^p)^{\frac{1}{p}}$, and hence $||c\mathfrak{A}||_p = |c|||\mathfrak{A}||_p$. [Note that since $\mathfrak{A} \in L^p$ implies $c\mathfrak{A} \in L^p$, we may discuss $||c\mathfrak{A}||_p$.]

Observe that properties (1)-(3) are seen to hold for every $0 < p < +\infty$.

Property (4) is the property that takes some work. For it, we pick elements $\mathfrak{A}$ and $\mathfrak{B}$ in L^p, and we then pick elements $f \in \mathfrak{A}$, $g \in \mathfrak{B}$, and $h \in \mathfrak{A} + \mathfrak{B}$. [Recall that for the moment we are accepting as true the fact that $\mathfrak{A} + \mathfrak{B} \in L^p$ whenever $\mathfrak{A}, \mathfrak{B} \in L^p$, and hence we may discuss the quantity $||\mathfrak{A} + \mathfrak{B}||_p$.] To show that $||\mathfrak{A} + \mathfrak{B}||_p \leq ||\mathfrak{A}||_p + ||\mathfrak{B}||_p$, we must show that

$$\left(\int |h|^p\right)^{\frac{1}{p}} \leq \left(\int |f|^p\right)^{\frac{1}{p}} + \left(\int |g|^p\right)^{\frac{1}{p}}.$$

Observe first that h is of the form $f_0 + g_0$, where $f_0 \in \mathfrak{A}$ and $g_0 \in \mathfrak{B}$. Now $f_0 \in \mathfrak{A}$ implies that $f_0 = f$ μ-a.e. on Ω, and $g_0 \in \mathfrak{B}$ implies that $g_0 = g$ μ-a.e. on Ω. It follows that $h = f_0 + g_0 = f + g$ μ-a.e. on Ω. Since $h = f + g$ μ-a.e. on Ω, proving the inequality displayed above is equivalent to demonstrating the integral inequality

$$\left(\int |f+g|^p\right)^{\frac{1}{p}} \leq \left(\int |f|^p\right)^{\frac{1}{p}} + \left(\int |g|^p\right)^{\frac{1}{p}}.$$

This inequality is a simple consequence of the fact that $|f + g| \leq |f| + |g|$ when $p = 1$, but the fact that the inequality is true when $1 < p < +\infty$ is hardly obvious. [The inequality is *not* true for $0 < p < 1$.] The inequality we are discussing is famous enough to have a name: *Minkowski's Inequality.*

[1]When $0 < p < 1$, $|| \; ||_p$ will turn out *not* to be a norm, although we abuse terminology somewhat and still refer to it as *the L^p norm.* The case $1 \leq p < +\infty$ is considered in the next section. The case $p = +\infty$ will be considered in a later section, as there we give a separate definition of L^∞ and $|| \; ||_\infty$.

An appreciable deal of formalism has been exhibited, and it is good to see things as they really are. However, in workaday matters we agree to make several slight abuses of notation and terminology. The following definition will in fact be the one that we commonly use.

Informal Definition. Let $(\Omega, \mathcal{F}, \mu)$ be given, and let $0 < p < +\infty$. We define

$$L^p = \left\{ f \in \mathcal{E} : \int |f|^p < +\infty \right\}.$$

For functions $f \in L^p$, we write

$$\|f\|_p = \left(\int |f|^p \right)^{\frac{1}{p}}.$$

The solecism committed in the informal definition is that we confuse an equivalence class of functions with the actual elements of the equivalence class. A function f in an equivalence class $\mathfrak{A}$ is not the same as $\mathfrak{A}$ and vice-versa. L^p is *in reality* a collection of equivalence classes of functions, *not* a collection of functions. The operator $\|\ \|_p$ *in reality* accepts equivalence classes as an argument, *not* functions. However, we can usually ignore these fine points and talk about L^p and $\|\ \|_p$ in the informal sense, provided that in certain instances we keep in mind the formal and correct definitions.[2] *We will work in the informal sense unless explicitly stated to the contrary.*

We now discuss a convention that will be used to simplify things somewhat. This convention will be used throughout the chapter unless explicitly stated otherwise. Any function $f \in L^p$ is (a) defined μ-a.e. on Ω and (b) finite μ-a.e. on Ω [among other things]. Since (i) the set of $\omega \in \Omega$ where $f(\omega)$ is undefined or defined and infinite is μ-null, (ii) we will not have to distinguish between functions that are equal μ-a.e. on Ω, and (iii) μ-null sets do not affect Lebesgue integration, we may without any loss of generality view any f in L^p as defined on all of Ω and finite on all of Ω.

Convention. We view all elements f in L^p as functions that are defined and finite everywhere on Ω.

This introduction has been somewhat long, but the formalism is worth seeing once or twice [and probably no more than once or twice]. With these definitions and the discussion above, we may now proceed to some actual mathematics.

[2]This is especially the case in property (2). We would like to say that $\|f\|_p = 0$ iff $f = 0$, where 0 denotes the function *identically* zero on Ω. If $f = 0$ on Ω, then clearly $\|f\|_p = 0$. However, if $\|f\|_p = 0$, then we can only deduce that $f = 0$ μ-a.e. on Ω. Therefore, it is not the case that $\|\ \|_p$ assigns the value zero exclusively to the function 0, and hence $\|\ \|_p$ fails to be a norm when $\|\ \|_p$ is viewed in the informal sense. This fact won't hinder us in anything that we desire to do.

8.1 L^p SPACE: THE CASE $1 \le p < +\infty$

Here, in the context of a general measure space $(\Omega, \mathcal{F}, \mu)$, we consider L^p space in the case where $1 \le p < +\infty$. The cases $0 < p < 1$ and $p = +\infty$ will be considered in Sections 3 and 4, respectively. The main point of this section is to prove two very famous and important inequalities, namely, those of Hölder and Minkowski.

Definition: Given $1 < p < +\infty$, we write $p' = \frac{p}{p-1}$. Note that $1 < p' < +\infty$ and $\frac{1}{p} + \frac{1}{p'} = 1$. We call p' the *Hölder conjugate of* p.

Calculus: For $0 \le a, b < +\infty$ and $1 < p < +\infty$, write $f(a,b) = ab - \frac{1}{p}a^p$. Then for every $0 \le b < +\infty$ we have that $\frac{\partial}{\partial a} f(a,b) \ge 0$ iff $b - a^{p-1} \ge 0$, which in turn holds iff $a \le b^{\frac{1}{p-1}}$. [This last inequality uses the fact that $1 < p < +\infty$.] Therefore, for every $0 \le a, b < +\infty$ we have

$$ab - \frac{1}{p}a^p = f(a,b) \le f\left(b^{\frac{1}{p-1}}, b\right) = b^{\frac{1}{p-1}} b - \frac{1}{p}\left(b^{\frac{1}{p-1}}\right)^p = \frac{1}{p'} b^{p'}.$$

Therefore, whenever $1 < p < +\infty$ and $0 \le a, b < +\infty$, we have

$$ab \le \frac{1}{p}a^p + \frac{1}{p'}b^{p'}.$$

With this inequality, we may now state and prove Hölder's Inequality.

Hölder's Inequality. *Let $f \in L^p$ and $g \in L^{p'}$, where $1 < p < +\infty$. Then $fg \in L^1$ and $|\int fg| \le ||fg||_1 \le ||f||_p ||g||_{p'}$.*[3]

Proof: Suppose that $f \in L^p$ is such that $||f||_p = 0$ or $g \in L^{p'}$ is such that $||g||_{p'} = 0$. Then $f = 0$ μ-a.e. on Ω or $g = 0$ μ-a.e. on Ω. But then $fg = 0$ μ-a.e. on Ω, hence $fg \in L^1$ and the desired inequality in this case is actually the trivial equation 0=0=0.

We next work in the special case of where $f \in L^p$ is such that $||f||_p = 1$ and $g \in L^{p'}$ is such that $||g||_{p'} = 1$. Observe that

$$|f(\omega)g(\omega)| = |f(\omega)||g(\omega)| \le \frac{1}{p}|f(\omega)|^p + \frac{1}{p'}|g(\omega)|^{p'} \quad \text{for all } \omega \in \Omega, \qquad (8.1)$$

[3] Also note that if f and g are nonnegative $\mathcal{F}/\mathcal{B}^*$-measurable functions on Ω such that $\int f^p = +\infty$ or $\int g^{p'} = +\infty$, then we trivially have $\int fg \le (\int f^p)^{\frac{1}{p}} (\int g^{p'})^{\frac{1}{p'}}$.

where the equality is trivial and the inequality follows by the calculus above. We now integrate the left and right sides of (8.1), obtaining

$$\begin{aligned}\int |fg| \le \frac{1}{p}\int |f|^p + \frac{1}{p'}\int |g|^{p'} = \frac{1}{p}\|f\|_p^p + \frac{1}{p'}\|g\|_{p'}^{p'} &= \frac{1}{p}1^p + \frac{1}{p'}1^{p'} \\ &= \frac{1}{p} + \frac{1}{p'} \\ &= 1.\end{aligned}$$

This shows that $|fg| \in L^1$, which is equivalent to asserting that $fg \in L^1$. We have accordingly shown the following: *if $f \in L^p$ and $g \in L^{p'}$ are such that $\|f\|_p = \|g\|_{p'} = 1$, then $fg \in L^1$ and $\int |fg| \le 1$.*

For the general case, we let $f \in L^p$ be such that $\|f\|_p > 0$, and we pick $g \in L^{p'}$ such that $\|g\|_{p'} > 0$. Next, write $f^* = \frac{f}{\|f\|_p}$ and $g^* = \frac{g}{\|g\|_{p'}}$. We have

$$\left(\int |f^*|^p\right)^{\frac{1}{p}} = \left(\int \left|\frac{f}{\|f\|_p}\right|^p\right)^{\frac{1}{p}} = \left(\frac{1}{\|f\|_p^p}\int |f|^p\right)^{\frac{1}{p}} = \frac{\|f\|_p}{\|f\|_p} = 1 < +\infty,$$

which shows that $f^* \in L^p$ and $\|f^*\|_p = 1$. It may similarly be shown that $g^* \in L^{p'}$ and $\|g^*\|_{p'} = 1$. Using what has just been shown gives $f^*g^* \in L^1$ and $\int |f^*g^*| \le 1$. Since $fg = \|f\|_p\|g\|_{p'}f^*g^*$, we have

$$\int |fg| = \int \big|\|f\|_p\|g\|_{p'}f^*g^*\big| = \|f\|_p\|g\|_{p'}\int |f^*g^*| \le \|f\|_p\|g\|_{p'} < +\infty,$$

giving $fg \in L^1$. The integral $\int fg$ accordingly exists, and the previous display gives

$$\left|\int fg\right| \le \int |fg| = \|fg\|_1 \le \|f\|_p\|g\|_{p'},$$

which proves the result. ■

Observe that Hölder's Inequality is given for $1 < p < +\infty$ instead of for $1 \le p < \infty$, as when $p = 1$ the conjugate $1'$ of 1 as given is undefined. In a later section, we will give a special definition for $1'$, the Hölder conjugate of 1. [We will also later give a special definition for $+\infty'$, the Hölder conjugate of $+\infty$.]

Minkowski's Inequality is the other major result of this section, and is given [in this section] for $1 \le p < +\infty$. The proof in the case where $p = 1$ will be seen to be direct, and the case $1 < p < +\infty$ will require Hölder's Inequality.

Minkowski's Inequality. *Let $1 \le p < +\infty$ and $f_1, f_2, \cdots \in L^p$. Then:*

(a) For every $n \in \mathbb{N}$ we have $f_1 + \cdots + f_n \in L^p$ and $\|\sum_{i=1}^n f_i\|_p \le \sum_{i=1}^n \|f_i\|_p$.

(b) If f_n is nonnegative for each $n \in \mathbb{N}$, then $[\int(\sum_{i=1}^\infty f_i)^p]^{\frac{1}{p}} \le \sum_{i=1}^\infty \|f_i\|_p$.

Proof: We consider the cases $p = 1$ and $1 < p < +\infty$ separately. We first let $p = 1$. If $f_1, \cdots, f_n \in L^1$, then integration property (L9) in Section 6.4 [there taking $E = \Omega$] gives $f_1 + \cdots + f_n \in L^1$, and it is clear that

$$\|f_1+\cdots+f_n\|_1 = \int |f_1+\cdots+f_n| \leq \int |f_1|+\cdots+\int |f_n| = \|f_1\|_1+\cdots+\|f_n\|_1.$$

This proves (a) when $p = 1$. For (b), we have

$$\int \sum_{i=1}^{\infty} f_i = \sum_{i=1}^{\infty} \int f_i = \sum_{i=1}^{\infty} \|f_i\|_1$$

by integration property (L11), which shows that the desired inequality for (b) is actually an equality when $p = 1$. This completes the proof of the entire claim when $p = 1$.

We now prove the claim when $1 < p < +\infty$. For (a), we first take $n = 2$, picking $f_1, f_2 \in L^p$. Observe that

$$\begin{aligned} |f_1 + f_2|^p \leq (|f_1| + |f_2|)^p \leq (2\max\{|f_1|, |f_2|\})^p &= 2^p \max\{|f_1|^p, |f_2|^p\} \\ &\leq 2^p(|f_1|^p + |f_2|^p), \end{aligned}$$

and hence

$$\int |f_1 + f_2|^p \leq \int 2^p(|f_1|^p + |f_2|^p) = 2^p(\|f_1\|_p^p + \|f_2\|_p^p) < +\infty.$$

This shows that $f_1 + f_2 \in L^p$. We now wish to demonstrate the inequality $\|f_1+f_2\|_p \leq \|f_1\|_p+\|f_2\|_p$. This inequality is trivial if f_1 and f_2 are such that $\|f_1+f_2\|_p = 0$, so we may assume that f_1 and f_2 are such that $\|f_1+f_2\|_p > 0$. Observe that $f_1, f_2 \in L^p$ forces $|f_1|, |f_2| \in L^p$, and since

$$\left(\int [|f_1 + f_2|^{p-1}]^{p'}\right)^{\frac{1}{p'}} = \left(\int |f_1 + f_2|^p\right)^{\frac{p-1}{p}} = \|f_1 + f_2\|_p^{p-1} < +\infty,$$

we have $|f_1 + f_2|^{p-1} \in L^{p'}$ and $\||f_1 + f_2|^{p-1}\|_{p'} = \|f_1 + f_2\|_p^{p-1}$. Using this information, we have

$$\begin{aligned} \|f_1 + f_2\|_p^p &= \int |f_1 + f_2|^p \\ &= \int |f_1 + f_2||f_1 + f_2|^{p-1} \\ &\leq \int |f_1||f_1 + f_2|^{p-1} + \int |f_2||f_1 + f_2|^{p-1} \\ &\leq \||f_1|\|_p\||f_1 + f_2|^{p-1}\|_{p'} + \||f_2|\|_p\||f_1 + f_2|^{p-1}\|_{p'} \\ &= (\|f_1\|_p + \|f_2\|_p)\|f_1 + f_2\|_p^{p-1}. \end{aligned}$$

[The first two equalities are trivial. The first inequality holds because of the inequality $|f_1 + f_2| \leq |f_1| + |f_2|$. The second inequality follows by applying

Hölder's Inequality to (i) $|f_1| \in L^p$ and $|f_1 + f_2|^{p-1} \in L^{p'}$ and (ii) $|f_2| \in L^p$ and $|f_1 + f_2|^{p-1} \in L^{p'}$. The third equality uses the equation mentioned just before the display above.] Since $\|f_1 + f_2\|_p > 0$, we may divide both sides of the display above by $\|f_1 + f_2\|_p^{p-1}$, obtaining $\|f_1 + f_2\|_p \le \|f_1\|_p + \|f_2\|_p$, which proves (a) in the case where $n = 2$. We may now assert that (a) holds for all $n \in \mathbb{N}$ by using mathematical induction.

We now turn to (b). For each $n \in \mathbb{N}$, write $g_n = \sum_{i=1}^{n} f_n \ge 0$. By (a), we have for all $n \in \mathbb{N}$ that (i) $g_n \in L^p$ and (ii) $\|g_n\|_p \le \sum_{i=1}^{n} \|f_i\|_p \le \sum_{i=1}^{\infty} \|f_i\|_p$. Also, we have that $\{\|g_n\|_p\}_{n=1}^{\infty}$ is nondecreasing, hence $\lim_{n\to\infty} \|g_n\|_p$ exists. We therefore have

$$
\begin{aligned}
\lim_{n\to\infty} \|g_n\|_p = \lim_{n\to\infty} \left(\int g_n^p\right)^{\frac{1}{p}} &= \left(\lim_{n\to\infty} \int g_n^p\right)^{\frac{1}{p}} \\
&= \left(\int \lim_{n\to\infty} g_n^p\right)^{\frac{1}{p}} \\
&= \left(\int \left[\lim_{n\to\infty} g_n\right]^p\right)^{\frac{1}{p}} \\
&= \left(\int \left(\sum_{i=1}^{\infty} f_i\right)^p\right)^{\frac{1}{p}},
\end{aligned}
$$

where the key third equality follows by applying MCT to the nondecreasing sequence $\{g_n^p\}_{n=1}^{\infty}$ of nonnegative functions. Since $\|g_n\|_p \le \sum_{i=1}^{\infty} \|f_i\|_p$ for each $n \in \mathbb{N}$, we have that $\lim_{n\to\infty} \|g_n\|_p \le \sum_{i=1}^{\infty} \|f_i\|_p$, and the previous displayed chain of equalities shows that the conclusion of (b) indeed holds. This completes the entire proof. ■

Even if $f_1, f_2, \cdots \in L^p$, it may not necessarily be the case that $\sum_{i=1}^{\infty} f_i$ is in L^p. Since $\| \|_p$ has domain L^p and $\sum_{i=1}^{\infty} f_i$ may not be in L^p, we cannot in general write $\| \sum_{i=1}^{\infty} f_i\|_p$ for $(\int [\sum_{i=1}^{\infty} f_i]^p)^{\frac{1}{p}}$. In turn, this is why we couldn't write the inequality in (b) as $\| \sum_{i=1}^{\infty} f_i\|_p \le \sum_{i=1}^{\infty} \|f_i\|_p$.

We now return to the world of formalism given in the introduction. Using Minkowski's Inequality, it is now a simple exercise to show that $\mathfrak{A}, \mathfrak{B} \in L^p$ forces $\mathfrak{A} + \mathfrak{B} \in L^p$ and $\|\mathfrak{A} + \mathfrak{B}\|_p \le \|\mathfrak{A}\|_p + \|\mathfrak{B}\|_p$. [See Exercise 1.] Therefore, $\| \|_p$ is truly a norm for $1 \le p < +\infty$.

The ultimate conclusion of this section is as follows: *For $1 \le p < +\infty$, L^p is a normed vector [or linear] space under the definitions of addition and scalar multiplication given in the introduction: $\mathfrak{A}, \mathfrak{B} \in L^p$ and $c \in \mathbb{R}$ imply (i) $c\mathfrak{A} \in L^p$ and (ii) $\mathfrak{A} + \mathfrak{B} \in L^p$. Furthermore, the norm $\| \|_p$ defined on L^p satisfies the following four properties, where $\mathfrak{A}, \mathfrak{B} \in L^p$ and $c \in \mathbb{R}$:*

(1) $0 \le \|\mathfrak{A}\|_p < +\infty$,
(2) $\|\mathfrak{A}\|_p = 0$ *iff* $\mathfrak{A} = o$,
(3) $\|c\mathfrak{A}\|_p = |c|\|\mathfrak{A}\|_p$, *and*
(4) $\|\mathfrak{A} + \mathfrak{B}\|_p \le \|\mathfrak{A}\|_p + \|\mathfrak{B}\|_p$.

Exercises.

1*. Pick $\mathfrak{A}, \mathfrak{B} \in L^p$. Then $\mathfrak{A} + \mathfrak{B} \in L^p$ and $\|\mathfrak{A} + \mathfrak{B}\|_p \leq \|\mathfrak{A}\|_p + \|\mathfrak{B}\|_p$. [We're taking $1 \leq p < +\infty$.] This is property (4) and finishes the proof that $\| \; \|_p$ is truly a norm on L^p for $1 \leq p < +\infty$. [Let $h \in \mathfrak{A} + \mathfrak{B}$, so that $h = f + g$ for some $f \in \mathfrak{A}$ and $g \in \mathfrak{B}$. Now

$$\int |h|^p = \left(\int |f+g|^p\right)^{\frac{p}{p}} \leq \left[\left(\int |f|^p\right)^{\frac{1}{p}} + \left(\int |g|^p\right)^{\frac{1}{p}}\right]^p < +\infty$$

by Minkowski's Inequality, and hence $\mathfrak{A} + \mathfrak{B} \in L^p$. The above also forces the inequality $\|\mathfrak{A} + \mathfrak{B}\|_p^p \leq (\|\mathfrak{A}\|_p + \|\mathfrak{B}\|_p)^p$; taking pth roots gives the result.]

2. Prove the Cauchy-Schwarz inequality $|\sum_{k=1}^n a_k b_k| \leq \sqrt{\sum_{k=1}^n a_k^2}\sqrt{\sum_{k=1}^n b_k^2}$ by using Hölder's Inequality with $p = 2$ and a "counting measure" framework. [Here, the a_k and b_k terms are real. Let $\Omega = \{\omega_1, \cdots, \omega_n\}$ and $\mathcal{F} = 2^\Omega$. Let μ denote counting measure on $\mathcal{F}$. Let f and g be defined on Ω by the rules $f(\omega_i) = a_i$ and $g(\omega_i) = b_i$, $i = 1, \cdots, n$. Recall Exercise 7 in Section 6.2.]

3. Let $1 < p_1, \cdots, p_n < +\infty$ be such that $\frac{1}{p_1} + \cdots + \frac{1}{p_n} = 1$, and pick functions $f_1 \in L^{p_1}, \cdots, f_n \in L^{p_n}$. We wish to generalize Hölder's Inequality by showing that $\prod_{i=1}^n f_i \in L^1$ and $|\int \prod_{i=1}^n f_i| \leq \prod_{i=1}^n \|f_i\|_{p_i}$.
(a) Show first that $a_1 \cdots a_n \leq \frac{a_1^{p_1}}{p_1} + \cdots + \frac{a_n^{p_n}}{p_n}$ by generalizing the calculus result given in the section. [Here $0 \leq a_1, \cdots, a_n < +\infty$.]
(b) If $\|f_1\|_{p_1} = 0$ or $\cdots$ or $\|f_n\|_{p_n} = 0$, the claim is trivial.
(c) Use (a) to prove the claim when $\|f_1\|_{p_1} = \cdots = \|f_n\|_{p_n} = 1$.
(d) Prove the claim when $\|f_1\|_{p_1}, \cdots, \|f_n\|_{p_n}$ are positive. [Write $f_i^* = \frac{f_i}{\|f_i\|_{p_i}}$ and apply (c) to $f_1^*, \cdots, f_n^*$.]

4. We have equality in Hölder's Inequality iff there are nonnegative numbers A and B, not both zero, with $A|f|^p = B|g|^{p'}$ μ-a.e. on Ω. [Explore the calculus and the proof of Hölder's inequality. Show that equality obtains iff $\frac{|f||g|}{\|f\|_p \|g\|_{p'}} = \frac{1}{p}\frac{|f|^p}{\|f\|_p^p} + \frac{1}{p'}\frac{|g|^{p'}}{\|g\|_{p'}^{p'}}$ μ-a.e. on Ω, which in turn holds iff $\frac{|f|^p}{\|f\|_p^p} = \frac{|g|^{p'}}{\|g\|_{p'}^{p'}}$ μ-a.e. on Ω.]

5. Given $f \in L^p$ $[1 < p < +\infty]$, there is $g \in L^{p'}$ with $\|g\|_{p'} = 1$ and $\int fg = \|f\|_p$. [Try $(\frac{f}{\|f\|_p})^{p-1}$.]

6. We now explore conditions for equality in Minkowski's Inequality. Let $f, g \in L^p$.
(a) When $p = 1$, $\|f + g\|_p = \|f\|_p + \|g\|_p$ iff there exists positive $\mathcal{F}/\mathcal{B}^*$-measurable $h > 0$ defined on Ω with $fh = g$ μ-a.e. on $\{\omega \in \Omega : f(\omega)g(\omega) \neq 0\}$.
(b) For $1 < p < +\infty$, equality obtains iff there are nonnegative real numbers A and B, not both zero, such that $Af = Bg$ μ-a.e. on Ω.

7. Let $1 \leq p, q, r < +\infty$ be such that $\frac{1}{r} = \frac{1}{p} + \frac{1}{q}$. Let $f \in L^p$ and $g \in L^q$. Then $fg \in L^r$ and $\|fg\|_r \leq \|f\|_p \|g\|_q$. [Let $p' = \frac{p}{r}$ and $q' = \frac{q}{r}$. Then $\frac{1}{p'} + \frac{1}{q'} = 1$. Apply Hölder's Inequality to $|f|^r$ and $|g|^r$.]

8*. If $1 \leq p < +\infty$, Minkowski's Inequality gives $|\|f\|_p - \|g\|_p| \leq \|f - g\|_p$ for every $f, g \in L^p$. [Write $f = (f - g) + g$.]

9. Let ϕ be a continuous and strictly increasing function defined on $[0, +\infty)$ with $\phi(0) = 0$ and $\lim_{x \to +\infty} \phi(x) = +\infty$. Then for each $0 \leq a, b < +\infty$ we have

$ab \leq \int_0^a \phi(u)\, du + \int_0^b \phi^{-1}(v)\, dv$, with equality iff $\phi(a) = b$. This is known as *Young's Inequality*. [Suggestion: first show that $\int_0^c \phi(u)\, du + \int_0^{\phi(c)} \phi^{-1}(v)\, dv = c\phi(c)$ for every $0 \leq c < +\infty$. A picture might be helpful.] The calculus result in the section may be obtained by setting $\phi(u) = u^{p-1}$ and $\phi^{-1}(v) = v^{\frac{1}{p-1}}$.

10. Prove the following extension of part (c) from Claim 2 of Section 7.5, where $1 \leq p < +\infty$ and $\boldsymbol{y} \in \mathbb{R}^k$: *If* $f \in L^p(\mathbb{R}^k, \mathcal{L}^k, \bar{\lambda}_k)$, *then* $\tau_{\boldsymbol{y}} f \in L^p(\mathbb{R}^k, \mathcal{L}^k, \bar{\lambda}_k)$ *and* $\|\tau_{\boldsymbol{y}} f\|_p = \|f\|_p$. [Let $g = f^p \in L^1(\mathbb{R}^k, \mathcal{L}^k, \bar{\lambda}_k)$. Applying Claim 2 from Section 7.5 to g gives $\tau_{\boldsymbol{y}} g \in L^1(\mathbb{R}^k, \mathcal{L}^k, \bar{\lambda}_k)$ (hence $(\tau_{\boldsymbol{y}} f)^p$ is in $L^1(\mathbb{R}^k, \mathcal{L}^k, \bar{\lambda}_k)$, so that $\tau_{\boldsymbol{y}} f \in L^p(\mathbb{R}^k, \mathcal{L}^k, \bar{\lambda}_k)$) and $\|\tau_{\boldsymbol{y}} g\|_1 = \|g\|_1$. Verify next that

$$\begin{aligned}
\|\tau_{\boldsymbol{y}} g\|_1 = \int_{\mathbb{R}^k} |\tau_{\boldsymbol{y}} g(\boldsymbol{x})|\, d\bar{\lambda}(\boldsymbol{x}) &= \int_{\mathbb{R}^k} |g(\boldsymbol{x}+\boldsymbol{y})|\, d\bar{\lambda}(\boldsymbol{x}) \\
&= \int_{\mathbb{R}^k} |f(\boldsymbol{x}+\boldsymbol{y})|^p\, d\bar{\lambda}(\boldsymbol{x}) \\
&= \int_{\mathbb{R}^k} |\tau_{\boldsymbol{y}} f(\boldsymbol{x})|^p\, d\bar{\lambda}(\boldsymbol{x}) \\
&= \|\tau_{\boldsymbol{y}} f\|_p^p.
\end{aligned}$$

Next, we have $\|g\|_1 = \|f\|_p^p$. Take pth roots for the result.]

8.2 THE RIESZ–FISCHER THEOREM

As with the previous section, we consider L^p space in the context of a generic measure space $(\Omega, \mathcal{F}, \mu)$ with $1 \leq p < +\infty$. From the previous section, we know that (i) L^p is a vector space and (ii) $\|\ \|_p$ is a norm. In this section, we view L^p as a metric space with respect to a certain yet-to-be-defined metric, showing that L^p is in fact complete with respect to this metric.

Formal Definition. Let $\mathfrak{A}, \mathfrak{B} \in L^p$. We define $\mathfrak{A} - \mathfrak{B}$ by writing

$$\mathfrak{A} - \mathfrak{B} = \{f - g : f \in \mathfrak{A},\ g \in \mathfrak{B}\} = \mathfrak{A} + (-\mathfrak{B}).$$

[Observe that $\mathfrak{A} - \mathfrak{B} \in L^p$.] We next define *the distance between* $\mathfrak{A}$ *and* $\mathfrak{B}$, denoted by $\rho(\mathfrak{A}, \mathfrak{B})$, by writing

$$\rho(\mathfrak{A}, \mathfrak{B}) = \|\mathfrak{A} - \mathfrak{B}\|_p.$$

Claim 1. *The function ρ defined above is a metric on L^p. That is, for $\mathfrak{A}$, $\mathfrak{B}$, and $\mathfrak{C}$ in L^p with $1 \leq p < +\infty$, we have*

(a) $0 \leq \rho(\mathfrak{A}, \mathfrak{B}) < +\infty$,

(b) $\rho(\mathfrak{A}, \mathfrak{A}) = 0$,

(c) $\rho(\mathfrak{A}, \mathfrak{B}) = 0$ *iff* $\mathfrak{A} = \mathfrak{B}$,

(d) $\rho(\mathfrak{A}, \mathfrak{B}) = \rho(\mathfrak{B}, \mathfrak{A})$, *and*

(e) $\rho(\mathfrak{A}, \mathfrak{C}) \leq \rho(\mathfrak{A}, \mathfrak{B}) + \rho(\mathfrak{B}, \mathfrak{C})$.

Proof: Part (a) immediately follows from the fact that $\|\ \|_p$ is a norm on L^p.

We now do (b). Pick $h \in \mathfrak{A} - \mathfrak{A}$, so that $h = f - g$ for some $f, g \in \mathfrak{A}$. Since $f, g \in \mathfrak{A}$, we have $f = g$ μ-a.e. on Ω, hence $h = f - g = 0$ μ-a.e. on Ω, whence

$$\rho(\mathfrak{A}, \mathfrak{A}) = \|\mathfrak{A} - \mathfrak{A}\|_p = \int |h|^p = \int 0 = 0.$$

We now turn to (c). Assume first that $\mathfrak{A} = \mathfrak{B}$, and pick $h \in \mathfrak{A} - \mathfrak{B}$. Then $h = f - g$, where $f \in \mathfrak{A}$ and $g \in \mathfrak{B} = \mathfrak{A}$. Since $f, g \in \mathfrak{A}$, we have $f = g$ μ-a.e. on Ω, so that $h = f - g = 0$ μ-a.e. on Ω. It follows that

$$\rho(\mathfrak{A}, \mathfrak{B}) = \|\mathfrak{A} - \mathfrak{B}\|_p = \int |h|^p = \int 0 = 0,$$

so one direction of (c) is complete. Conversely, if $\rho(\mathfrak{A}, \mathfrak{B}) = \|\mathfrak{A} - \mathfrak{B}\|_p = 0$, then $\int |h|^p = 0$ for every $h \in \mathfrak{A} - \mathfrak{B}$. Pick and fix $h \in \mathfrak{A} - \mathfrak{B}$; write $h = f - g$ for appropriate $f \in \mathfrak{A}$ and $g \in \mathfrak{B}$. From the equation $\int |h|^p = 0$, we have $\int |f - g|^p = 0$, so that $|f - g|^p = 0$ μ-a.e. on Ω, which forces $f = g$ μ-a.e. on Ω. It follows that f and g are in the same equivalence class, so that $f, g \in \mathfrak{A} \cap \mathfrak{B}$. Since any two equivalence classes are disjoint or identical, it follows that $\mathfrak{A} = \mathfrak{B}$, and all of (c) is complete.

We now prove (d). Let $h_1 \in \mathfrak{A} - \mathfrak{B}$; write $h_1 = f_1 - g_1$ for appropriate $f_1 \in \mathfrak{A}$ and $g_1 \in \mathfrak{B}$. Next, let $h_2 \in \mathfrak{B} - \mathfrak{A}$; write $h_2 = g_2 - f_2$ for appropriate $g_2 \in \mathfrak{B}$ and $f_2 \in \mathfrak{A}$. Now $f_1 = f_2$ μ-a.e. on Ω and $g_1 = g_2$ μ-a.e. on Ω, hence $g_1 - f_1 = g_2 - f_2$ μ-a.e. on Ω. Next, observe that

$$\begin{aligned}
\rho(\mathfrak{A}, \mathfrak{B}) = \|\mathfrak{A} - \mathfrak{B}\|_p = \left(\int |h_1|^p\right)^{\frac{1}{p}} &= \left(\int |f_1 - g_1|^p\right)^{\frac{1}{p}} \\
&= \left(\int |g_1 - f_1|^p\right)^{\frac{1}{p}} \\
&= \left(\int |g_2 - f_2|^p\right)^{\frac{1}{p}} \\
&= \left(\int |h_2|^p\right)^{\frac{1}{p}} \\
&= \|\mathfrak{B} - \mathfrak{A}\|_p \\
&= \rho(\mathfrak{B}, \mathfrak{A}).
\end{aligned}$$

This proves (d).

For (e), suppose that $\mathfrak{A} - \mathfrak{C} = (\mathfrak{A} - \mathfrak{B}) + (\mathfrak{B} - \mathfrak{C})$. Then

$$\|\mathfrak{A} - \mathfrak{C}\|_p = \|(\mathfrak{A} - \mathfrak{B}) + (\mathfrak{B} - \mathfrak{C})\|_p \leq \|\mathfrak{A} - \mathfrak{B}\|_p + \|\mathfrak{B} - \mathfrak{C}\|_p$$

since $\|\ \|_p$ is a norm [see property (4) for $\|\ \|_p$ from the previous section], giving the desired inequality $\rho(\mathfrak{A}, \mathfrak{C}) \leq \rho(\mathfrak{A}, \mathfrak{B}) + \rho(\mathfrak{B}, \mathfrak{C})$. Therefore, everything hinges on establishing the equality $\mathfrak{A} - \mathfrak{C} = (\mathfrak{A} - \mathfrak{B}) + (\mathfrak{B} + \mathfrak{C})$.

We now establish the aforementioned equality. If $h \in \mathfrak{A} - \mathfrak{C}$, then $h = f - g$ for appropriate $f \in \mathfrak{A}$ and $g \in \mathfrak{C}$. For any element $h_0 \in \mathfrak{B}$, we have that $h = (f - h_0) + (h_0 - g)$. Since $f - h_0 \in \mathfrak{A} - \mathfrak{B}$ and $h_0 - g \in \mathfrak{B} - \mathfrak{C}$, we have that $h \in (\mathfrak{A} - \mathfrak{B}) + (\mathfrak{B} - \mathfrak{C})$. This gives one containment. For the reverse containment, reset everything and let $h \in (\mathfrak{A} - \mathfrak{B}) + (\mathfrak{B} - \mathfrak{C})$. It follows that $h = (f - h_1) + (h_2 - g)$, where $f \in \mathfrak{A}$, $h_1, h_2 \in \mathfrak{B}$, and $g \in \mathfrak{C}$. Now $h_1, h_2 \in \mathfrak{B}$ forces $h_1 = h_2$ μ-a.e. on Ω, and hence $h = f - g$ μ-a.e. on Ω. Since $f - g \in \mathfrak{A} - \mathfrak{C}$, we have that $h \in \mathfrak{A} - \mathfrak{C}$. This gives the reverse containment, and the proof of (e) is now complete. This completes the entire proof. ■

We have defined everything in formal terms, but for workaday matters we use the informal setting that has been discussed. The discussion relative to the informal setting now follows.

Informal Definition. Given $f, g \in L^p$, we define the *L^p distance between f and g* to be $\|f - g\|_p$. [We are taking $1 \leq p < +\infty$.]

There are no problems with the definition. If $f, g \in L^p$, then, since L^p is a vector space, we have $f - g \in L^p$, and hence it is meaningful to discuss the number $\|f - g\|_p$. It is also clear that we have a bona fide distance function in the usual metric space sense provided that we treat functions that are equal μ-a.e. on Ω as one and the same:

(1) $0 \leq \|f - g\|_p < +\infty$ for every $f, g \in L^p$,
(2) $\|f - f\|_p = 0$ for each $f \in L^p$, and $\|f - g\|_p = 0$ forces $f = g$ μ-a.e. on Ω,[4]
(3) $\|f - g\|_p = \|g - f\|_p$ for every $f, g \in L^p$, and
(4) $\|f - h\|_p \leq \|f - g\|_p + \|g - h\|_p$ for every $f, g, h \in L^p$.

If we view L^p as a metric space and use the notion of distance as just given, we may discuss the notion of convergence. The following definition makes this precise.

Definition. Let $f, f_1, f_2, \cdots \in L^p$. We will say that *$\{f_n\}_{n=1}^{\infty}$ converges to f in L^p [or in "pth mean"]* iff $\lim_n \|f_n - f\|_p = 0$. We also use the notation $f_n \xrightarrow{p} f$ to express convergence in L^p. Some authors write $f_n \to f$ *in* L^p to equivalently express this convergence.

We now review some general metric space terminology, letting (X, ρ) denote a metric space. Recall that (X, ρ) is *complete* iff it is the case that every Cauchy sequence of elements of X converges to some element of X. That is, (X, ρ) is complete iff for every Cauchy sequence $\{x_n\}_{n=1}^{\infty}$ of elements of X we have that there is some $x \in X$ with $\lim_n \rho(x_n, x) = 0$. The word *complete* in this sense of course means something entirely different than how the word was used relative to measure spaces.

[4]Property (2) is where we need the proviso that functions equal almost everywhere are identified. With this proviso, the conclusion that $f = g$ μ-a.e. on Ω stated above means that we don't distinguish between f and g.

The famous Riesz-Fischer Theorem, first given nearly a century ago, asserts that L^p, when viewed as a metric space with distance given by the preceding definition, is complete in the general metric space sense. That is, if $\{f_n\}_{n=1}^{\infty}$ denotes a Cauchy sequence of functions in L^p, then there exists a function $f \in L^p$ such that $\lim_n \|f_n - f\|_p = 0$, that is, $f_n \xrightarrow{p} f$. This result, one of the great theorems of analysis, is a key component in the proof of the Riesz Representation Theorem found at the end of this chapter. Before proving the Riesz-Fischer Theorem, some details require attention.

Lemma for the Riesz-Fischer Theorem. *Let $\{f_n\}_{n=1}^{\infty}$ denote a Cauchy sequence in L^p. We have the following:*
(a) If $\{f_n\}_{n=1}^{\infty}$ has a subsequence that converges in L^p to some $f \in L^p$, then $f_n \xrightarrow{p} f$.
(b) There exists a subsequence $\{f_{n_k}\}_{k=1}^{\infty}$ of $\{f_n\}_{n=1}^{\infty}$ with the property that $\|f_{n_{k+1}} - f_{n_k}\|_p \leq (\frac{1}{2})^k$ for every $k \in \mathbb{N}$.

Proof: We first handle part (a). Let $\{f_{n_k}\}_{k=1}^{\infty}$ denote a subsequence of $\{f_n\}_{n=1}^{\infty}$ such that $\{f_{n_k}\}_{k=1}^{\infty}$ converges in pth mean to some $f \in L^p$, and pick arbitrary $\epsilon > 0$. Since $\{f_n\}_{n=1}^{\infty}$ is a Cauchy sequence in L^p, we have that

$$\text{there is } N_1 \in \mathbb{N} \text{ such that } i, j \geq N_1 \text{ implies } \|f_i - f_j\|_p < \tfrac{\epsilon}{2}. \tag{8.2}$$

By our assumptions on $\{f_{n_k}\}_{k=1}^{\infty}$ we have that

$$\text{there is } N_2 \in \mathbb{N} \text{ such that } k \geq N_2 \text{ implies } \|f_{n_k} - f\|_p < \tfrac{\epsilon}{2}. \tag{8.3}$$

Choose k sufficiently large so that $n_k \geq N_1$ and $k \geq N_2$. Next, if $j \geq N_1$, we have $\|f_j - f_{n_k}\|_p < \frac{\epsilon}{2}$ by (8.2) and $\|f_{n_k} - f\|_p < \frac{\epsilon}{2}$ by (8.3). Therefore, $j \geq N_1$ implies that

$$\|f_j - f\|_p = \|(f_j - f_{n_k}) - (f - f_{n_k})\|_p \leq \|f_j - f_{n_k}\|_p + \|f_{n_k} - f\|_p < \frac{\epsilon}{2} + \frac{\epsilon}{2} = \epsilon,$$

which shows that $f_n \xrightarrow{p} f$. This completes (a).

We now prove (b), assuming only that $\{f_n\}_{n=1}^{\infty}$ is a Cauchy sequence in L^p. First, find $N_1 \in \mathbb{N}$ so that $m, n \geq N_1$ forces $\|f_m - f_n\|_p < \frac{1}{2}$. Define $n_1 = N_1$. Next, find $N_2 \in \mathbb{N}$ so that $m, n \geq N_2$ forces $\|f_m - f_n\|_p < \frac{1}{2^2}$. Define $n_2 = \max\{N_2, n_1 + 1\}$, and observe that $n_1 < n_2$. Next, find $N_3 \in \mathbb{N}$ so that $m, n \geq N_3$ forces $\|f_m - f_n\|_p < \frac{1}{2^3}$. Define $n_3 = \max\{N_3, n_2 + 1\}$, and observe that we now have $n_1 < n_2 < n_3$. Continue this process indefinitely. This gives a strictly increasing sequence $\{n_k\}_{k=1}^{\infty}$ with $\lim_k n_k = +\infty$. We also have for every $k \geq \mathbb{N}$ that $n_{k+1} > n_k \geq N_k$, whence $\|f_{n_{k+1}} - f_{n_k}\|_p < \frac{1}{2^k}$. This gives (b) and completes the entire proof. ∎

We may now state and prove the main result of this section. Stated in another form, the upcoming theorem says that if $\{f_n\}_{n=1}^{\infty}$ denotes a Cauchy sequence in L^p with $1 \leq p < +\infty$, then there is $f \in L^p$ such that $f_n \xrightarrow{p} f$.

Riesz-Fischer Theorem. *L^p is a complete metric space. [$1 \leq p < +\infty$.]*

Proof: Let $\{f_n\}_{n=1}^\infty$ denote a Cauchy sequence in L^p. By (a) of the previous lemma, it is sufficient to find a subsequence of $\{f_n\}_{n=1}^\infty$ that converges in L^p to some $f \in L^p$. Next, we use (b) of the previous lemma to find a subsequence $\{f_{n_k}\}_{k=1}^\infty$ of $\{f_n\}_{n=1}^\infty$ such that the inequality $\|f_{n_{k+1}} - f_{n_k}\|_p < (\frac{1}{2})^k$ obtains for every $k \in \mathbb{N}$. For notational convenience we write $g_k = f_{n_k}$ for every $k \in \mathbb{N}$, so that $\|g_{k+1} - g_k\|_p < (\frac{1}{2})^k$ for every $k \in \mathbb{N}$.

With this, we define F on Ω by writing

$$F = |g_1| + \sum_{i=1}^{\infty} |g_{i+1} - g_i|.$$

Then F is $\mathcal{F}/\mathcal{B}^*$-measurable since

$$F = \lim_n \left(|g_1| + \sum_{i=1}^{n} |g_{i+1} - g_i| \right)$$

and $|g_1| + \sum_{i=1}^n |g_{i+1} - g_i|$ is $\mathcal{F}/\mathcal{B}^*$-measurable for each $n \in \mathbb{N}$. Next,

$$\begin{aligned} \left(\int |F|^p \right)^{\frac{1}{p}} \leq \|g_1\|_p + \sum_{i=1}^{\infty} \|g_{i+1} - g_i\|_p &< \|g_1\|_p + \sum_{i=1}^{\infty} \left(\frac{1}{2}\right)^i \\ &= \|g_1\|_p + 1 \\ &< +\infty, \end{aligned}$$

where the first inequality follows by part (b) of the Minkowski Inequality, and the last inequality follows because $g_1 \in L^p$. Since $(\int |F|^p)^{\frac{1}{p}} < +\infty$, we have that F is finite μ-a.e. on Ω. Therefore, there is $N \in \mathcal{F}$ such that F is finite on N^c and $\mu(N) = 0$. For every $\omega \in N^c$, we may discuss the absolutely convergent series $g_1(\omega) + \sum_{i=1}^\infty (g_{i+1}(\omega) - g_i(\omega))$. By algebra, we have

$$g_k(\omega) = g_1(\omega) + \sum_{i=1}^{k-1} (g_{i+1}(\omega) - g_i(\omega)) \quad \text{for every } k \geq 2 \text{ and } \omega \in \Omega,$$

and

$$\lim_k g_k(\omega) = g_1(\omega) + \sum_{i=1}^{\infty} (g_{i+1}(\omega) - g_i(\omega)) \;\; [< +\infty] \quad \text{for every } \omega \in N^c.$$

This allows us to define a new function f on Ω by the rule

$$f(\omega) = \begin{cases} \lim_k g_k(\omega) = g_1(\omega) + \sum_{i=1}^\infty (g_{i+1}(\omega) - g_i(\omega)) & \omega \in N^c \\ 0 & \omega \in N. \end{cases}$$

Observe that f is $\mathcal{F}/\mathcal{B}^*$-measurable since the pieces that define f are themselves $\mathcal{F}/\mathcal{B}^*$-measurable. Next, recalling how f is defined on N^c, we have

$$\begin{aligned} & f(\omega) - g_k(\omega) \\ = & \left(g_1(\omega) + \sum_{i=1}^{\infty}[g_{i+1}(\omega) - g_i(\omega)]\right) - \left(g_1(\omega) + \sum_{i=1}^{k-1}[g_{i+1}(\omega) - g_i(\omega)]\right) \\ = & \sum_{i=k}^{\infty}(g_{i+1}(\omega) - g_i(\omega)) \end{aligned}$$

for every $k \in \mathbb{N}$ and $\omega \in N^c$, hence $|f - g_k| \le \sum_{i=k}^{\infty} |g_{i+1} - g_i|$ on N^c for every $k \in \mathbb{N}$. Using Minkowski's Inequality and the fact that $\mu(N) = 0$ gives

$$\begin{aligned} \left(\int |f - g_k|^p\right)^{\frac{1}{p}} = \left(\int_{N^c} |f - g_k|^p\right)^{\frac{1}{p}} & \le \left(\int_{N^c} \left[\sum_{i=k}^{\infty} |g_{i+1} - g_i|\right]^p\right)^{\frac{1}{p}} \\ & = \left(\int \left[\sum_{i=k}^{\infty} |g_{i+1} - g_i|\right]^p\right)^{\frac{1}{p}} \\ & \le \sum_{i=k}^{\infty} \|g_{i+1} - g_i\|_p \\ & < \sum_{i=k}^{\infty} \left(\frac{1}{2}\right)^i \\ & = \left(\frac{1}{2}\right)^{k-1} \quad [< +\infty] \quad \text{for all } k \in \mathbb{N}. \end{aligned}$$

Therefore, we have that $f - g_k \in L^p$ for every $k \in \mathbb{N}$, and we may rewrite the inequality above as $\|f - g_k\|_p < (\frac{1}{2})^{k-1}$, so that $\lim_k \|f - g_k\|_p = 0$. Since $g_1 \in L^p$, we have that $f = (f - g_1) + g_1 \in L^p$ by the fact that L^p is closed under addition. Therefore, the subsequence $\{g_k\}_{k=1}^{\infty} = \{f_{n_k}\}_{k=1}^{\infty}$ of $\{f_n\}_{n=1}^{\infty}$ converges in L^p to the element $f \in L^p$. This completes the proof. ∎

Corollary to the Riesz-Fischer Theorem. *Suppose that $\{f_n\}_{n=1}^{\infty}$ is a sequence of functions in L^p, and let $f \in L^p$ be such that $f_n \xrightarrow{p} f$. Then there exists a subsequence $\{f_{n_k}\}_{k=1}^{\infty}$ of $\{f_n\}_{n=1}^{\infty}$ such that $f_{n_k} \to f$ μ-a.e. on Ω.*

Proof: Since $\{f_n\}_{n=1}^{\infty}$ is a convergent sequence in L^p, it follows that $\{f_n\}_{n=1}^{\infty}$ is a Cauchy sequence in L^p. The subsequence $\{f_{n_k}\}_{k=1}^{\infty}$ constructed in the proof of the Riesz-Fischer Theorem above is seen to satisfy the requirements for this corollary. ∎

For $1 \le p < +\infty$ we have that (i) L^p is a vector space, (ii) $\| \ \|_p$ is a norm, and (iii) L^p is complete with respect to the metric induced by the norm $\| \ \|_p$. A normed linear space that is complete in the metric induced by its norm is often referred to as a *Banach space*. In this parlance, we may summarize the first two sections of this chapter by the following statement: *L^p is a Banach space for* $1 \leq p < +\infty$.

Exercises.

1*. Return to the formal definition of L^p.
(a) Write out the formal definition of convergence in L^p.
(b) State and prove the formal version of the Riesz-Fischer Theorem. [All parts of this exercise should be quick and easy.]

2. Let (X, ρ) denote a generic metric space. Let $\mathcal{C}$ denote the collection of continuous real-valued bounded functions on X. For $f \in \mathcal{C}$, write $||f|| = \sup_{x \in X} |f(t)|$, the usual supremum norm. Then $\mathcal{C}$ is a Banach space.

3. A function f on $\mathbb{R}^k$ is said to *vanish at infinity* iff $f(\boldsymbol{x}) \to 0$ as $||\boldsymbol{x}|| \to +\infty$. Show that the collection of continuous functions on $\mathbb{R}^k$ that vanish at infinity is a Banach space relative to the supremum norm given in the previous exercise.

4. Let $\mathcal{C}$ denote the collection of continuous functions on $\mathbb{R}^k$ with compact support, and again consider the supremum norm. This collection is dense in the collection in the previous exercise, but it fails to be a Banach space.

5. A sequence $\{f_n\}_{n=1}^{\infty}$ in L^p may converge in pth mean to some $f \in L^p$ but at the same time fail to converge pointwise to f at *any* point in Ω. Therefore, convergence in L^p does not in general imply convergence μ-a.e. [Consider the space $[0, 1]$ equipped with the Borel subsets of $[0, 1]$ and Lebesgue measure restricted to the Borel subsets of $[0, 1]$. Define $f_1, f_2, \cdots$ as the indicator functions of $[0, \frac{1}{2}]$, $[\frac{1}{2}, 1]$, $[0, \frac{1}{4}]$, $[\frac{1}{4}, \frac{1}{2}]$, $[\frac{1}{2}, \frac{3}{4}]$, $[\frac{3}{4}, 1]$, $[0, \frac{1}{8}]$, $\cdots$. Let $f = 0$.]

8.3 L^p SPACE: THE CASE $0 < p < 1$

We have discussed some properties regarding L^p space when $1 \leq p < +\infty$, and in this situation it has been shown that L^p is a Banach space. We have not yet discussed the situation where $0 < p < 1$ or $p = +\infty$, and we partially rectify this deficiency by considering the case where $0 < p < 1$. Unlike the $1 \leq p < +\infty$ case, the Hölder and Minkowski Inequalities do *not* hold when $0 < p < 1$. Because the Minkowski Inequality will not hold in this case, it will follow that $|| \; ||_p$ is *not* a norm when $0 < p < 1$.[5]

In this brief section, we will assume that $0 < p < 1$, and we will prove Hölder-like and Minkowski-like inequalities. As with the first section [where we had $1 \leq p < +\infty$], we will write $p' = \frac{p}{p-1}$ for the Hölder conjugate of p. In the case $0 < p < 1$, we will have $-\infty < p' < 0$. Therefore, in raising things to the p'th power when $0 < p < 1$, we are actually dividing by something, and we must make sure that the something by which we are dividing is nonzero.

[5]Usually, $|| \; ||_p$ is still referred to as *the L^p norm* in the case where $0 < p < 1$, even though $|| \; ||_p$ really isn't a norm. This convention causes no problems.

Observe that we still have $\frac{1}{p} + \frac{1}{p'} = 1$ in the $0 < p < 1$ case [as we did in the $1 \leq p < +\infty$ case].

Hölder-like Inequality. *Let $0 < p < 1$. Let $f \in L^p$ be nonnegative, and let g denote a nonnegative $\mathcal{F}/\mathcal{B}^*$-measurable function such that $\int g^{p'}$ exists and $0 < \int g^{p'} < +\infty$.*[6] *Then $\int fg \geq \|f\|_p (\int |g|^{p'})^{\frac{1}{p'}}$.*

Before proving this claim, note the differences between this and the original Hölder Inequality: (i) the functions f and g here must be nonnegative, and (ii) we do not have the conclusion that $fg \in L^1$. The integral $\int fg$, which exists since $fg \geq 0$, may in fact be $+\infty$ [and hence fg may not be in L^1].

Proof: If f and g are such that $\int fg = +\infty$, the result trivially holds. Therefore, we may assume that $0 \leq \int fg < +\infty$ for the rest of the proof.

Since $0 < \int g^{p'} < +\infty$ and $g \geq 0$ on Ω, we have that $0 < g \leq +\infty$ μ-a.e. on Ω. Without any loss of generality, we may assume that $0 < g \leq +\infty$ on all of Ω. Define $q = \frac{1}{p}$, so that $1 < q < +\infty$ and $1 < q' = \frac{1}{1-p} < +\infty$. On Ω, define the functions $\phi = g^{-\frac{1}{q}}$ and $\psi = f^{\frac{1}{q}} g^{\frac{1}{q}}$. We have that $\phi^{q'} = g^{-\frac{q'}{q}} = g^{p'}$, and since $0 < \int g^{p'} < +\infty$, we have $0 < \int \phi^{q'} < +\infty$, giving $\phi \in L^{q'}$. We are assuming that $0 \leq \int fg < +\infty$, which by the definition of ψ gives $0 \leq \int \psi^q < +\infty$, whence $\psi \in L^q$. Since $1 < q < +\infty$, $\psi \in L^q$, and $\phi \in L^{q'}$, Hölder's Inequality gives

$$\begin{aligned}\int f^p = \int f^{\frac{1}{q}} = \int \phi\psi &\leq \left(\int \psi^q\right)^{\frac{1}{q}} \left(\int \phi^{q'}\right)^{\frac{1}{q'}} \\ &= \left(\int fg\right)^p \left(\int g^{p'}\right)^{\frac{1}{q'}}.\end{aligned}$$

Dividing both sides of the above inequality by $(\int g^{p'})^{\frac{1}{q'}}$, raising everything to the power $\frac{1}{p}$, and noting that $-\frac{1}{q'p} = \frac{1}{p'}$ gives

$$\int fg \geq \|f\|_p \left(\int g^{p'}\right)^{-\frac{1}{q'p}} = \|f\|_p \left(\int g^{p'}\right)^{\frac{1}{p'}},$$

which completes the proof. ∎

We now turn to a Minkowski-like inequality. Note that the direction of the inequality in the case $0 < p < 1$ is the opposite of that in the case $1 \leq p < +\infty$.

Minkowski-like Inequality. *Let $0 < p < 1$, and let $f, g \in L^p$ be nonnegative. Then $f + g \in L^p$ and $\|f + g\|_p \geq \|f\|_p + \|g\|_p$.*

[6] If $\int g^{p'} = 0$, then $(\int g^{p'})^{\frac{1}{p'}}$ is undefined by the fact that $p' < 0$. If $\int g^{p'} = +\infty$, then $(\int g^{p'})^{\frac{1}{p'}} = 0$, and the inequality we are claiming is trivial.

Proof: In the proof of the original Minkowski inequality for $1 \le p < +\infty$, we showed that $|f+g|^p \le 2^p(f^p+g^p)$ on Ω. This inequality did not specifically require $1 \le p < +\infty$; only the fact that $0 < p < +\infty$ was tacitly used. Therefore, this just-mentioned inequality holds in this particular situation. It follows that $\int |f+g|^p \le 2^p(\int f^p + \int g^p) < +\infty$, and hence $f+g \in L^p$. To complete things, we must establish the inequality $\|f+g\|_p \ge \|f\|_p + \|g\|_p$ for our nonnegative functions $f, g \in L^p$.

To establish the desired inequality, first observe that if $\|f+g\|_p = 0$, then $f+g=0$ μ-a.e. on Ω. Since f and g are assumed to be nonnegative, it follows that $f=0$ μ-a.e. on Ω and $g=0$ μ-a.e. on Ω, and hence $\|f\|_p = 0 = \|g\|_p$, showing that the desired inequality is actually the trivial equality $0 = 0+0$. We may therefore proceed under the assumption that $\|f+g\|_p > 0$, which forces $0 < \int (f+g)^p < +\infty$ since $f+g \in L^p$. Let

$$E = \{\omega \in \Omega : f(\omega) > 0\} \cup \{\omega \in \Omega : g(\omega) > 0\},$$

and observe that the inequality $\int (f+g)^p > 0$ forces $\mu(E) > 0$. Since we have $f = g = f+g = 0$ on E^c, we have $\int f^p = \int_E f^p$, $\int g^p = \int_E g^p$, and $\int (f+g)^p = \int_E (f+g)^p$. Observe that

$$\|f+g\|_p^p = \int (f+g)^p = \int_E (f+g)^p \tag{8.4}$$

and

$$\int_E (f+g)^p = \int_E (f+g)(f+g)^{p-1} = \int_E f(f+g)^{p-1} + \int_E g(f+g)^{p-1}. \tag{8.5}$$

Consider the measure space $(E, \mathcal{F}_E, \mu_E)$, where $\mathcal{F}_E = \{A \in \mathcal{F} : A \subseteq E\}$ and μ_E denotes the restriction of μ to $\mathcal{F}_E$. Let f_E and g_E denote the restrictions of f and g to E. Observe that f_E and g_E are nonnegative functions in $L^1(E, \mathcal{F}_E, \mu_E)$. Observe next that $\int_E (f_E + g_E)^{(p-1)\times p'}\, d\mu_E$ exists since $f_E + g_E > 0$ on E; we furthermore have that

$$\begin{aligned} 0 < \int_E (f_E+g_E)^{(p-1)\times p'}\, d\mu_E = \int_E (f_E+g_E)^p\, d\mu_E &= \int_E (f+g)^p\, d\mu \\ &= \int (f+g)^p\, d\mu \\ &< +\infty. \end{aligned}$$

[The first inequality holds since $f_E + g_E > 0$ on E and $\mu_E(E) = \mu(E) > 0$. The first equality follows by algebra. The second equality holds because $f = f_E$ on E, $g = g_E$ on E, and integration property (L18). The third equality holds since $f+g=0$ on E^c. The last inequality follows by the fact that $f+g \in L^p$.] The conditions for the Hölder-like Inequality are satisfied [relative to $(E, \mathcal{F}_E, \mu_E)$] for the functions f_E and $(f_E+g_E)^{p-1}$, and hence we have

$$\int_E f_E(f_E+g_E)^{p-1}\, d\mu_E \ge \left(\int_E f_E^p\, d\mu_E\right)^{\frac{1}{p}} \left(\int_E (f_E+g_E)^{(p-1)\times p'}\, d\mu_E\right)^{\frac{1}{p'}}.$$

Using (L18), this last inequality forces

$$\int_E f(f+g)^{p-1}\,d\mu \geq \left(\int_E f^p\,d\mu\right)^{\frac{1}{p}} \left(\int_E (f+g)^p\,d\mu\right)^{\frac{1}{p'}}.$$

Since $f = f + g = 0$ on E^c, the previous inequality yields

$$\int_E f(f+g)^{p-1}\,d\mu \geq \left(\int f^p\,d\mu\right)^{\frac{1}{p}} \left(\int (f+g)^p\,d\mu\right)^{\frac{1}{p'}},$$

that is,

$$\int_E f(f+g)^{p-1}\,d\mu \geq \|f\|_p \|f+g\|_p^{p-1}, \tag{8.6}$$

where the norm $\|\ \|_p$ is with respect to $(\Omega, \mathcal{F}, \mu)$. Similar reasoning gives

$$\int_E g(f+g)^{p-1}\,d\mu \geq \|g\|_p \|f+g\|_p^{p-1}. \tag{8.7}$$

Putting (8.4)-(8.7) together yields

$$\|f+g\|_p^p \geq \|f\|_p\|f+g\|_p^{p-1} + \|g\|_p\|f+g\|_p^{p-1} = (\|f\|_p + \|g\|_p)\|f+g\|_p^{p-1},$$

and dividing both sides by the finite positive quantity $\|f+g\|_p^{p-1}$ gives the desired inequality for the case where $\|f+g\|_p > 0$. Combining the $\|f+g\|_p = 0$ and $\|f+g\|_p > 0$ cases gives the entire result. ■

The results here for the case $0 < p < 1$ are presented primarily for completeness, although they will be used in proving the Clarkson Inequality found in Section 8 of this chapter. In some presentations of classical analysis and measure theory, the case $0 < p < 1$ is relegated to a lesser status than the case $1 \leq p < +\infty$ because the triangle inequality for $\|\ \|_p$ fails.

Exercises.

1. Let $f, g \in L^p$, where $0 < p < 1$. We know that $f + g \in L^p$ by the Minkowski-like result given earlier.
(a) We have $(a+b)^p \leq a^p + b^p$ for every $0 < a, b < +\infty$.
(b) From (a), we have $\int |f+g|^p \leq \int |f|^p + \int |g|^p$.
(c) If we write $\|f-g\|_p^p$ for the distance between f and g, then this distance function is truly a metric, if we identify functions equal μ-a.e. on Ω.
(d) Writing $\|f-g\|_p$ for the distance between f and g does *not* define a metric on L^p.

2. Consider the space $\Omega = (0,1)$ and let $0 < p < 1$. Write $\mathcal{B}$ for the Borel subsets of $(0,1)$, and write λ for Lebesgue measure restricted to $\mathcal{B}$. We will show that there exists no norm $\|\ \|$ on L^p such that $\lim_k \|f_k\|_p = 0$ forces $\lim_k \|f_k\| = 0$.
(a) Suppose that such a norm $\|\ \|$ exists. Then there is $C \geq 0$ such that $\|f\| \leq C\|f\|_p$ for each $f \in L^p$. [Suppose that for every $C \geq 0$ there is $f \in L^p$ with $\|f\| > C\|f\|_p$. Then for every $k \in \mathbb{N}$ there is $f_k \in L^p$ with $\|f_k\| > k\|f_k\|_p$. Argue further that each f_k may be chosen so that $\|f_k\| = 1$. Then for every $k \in \mathbb{N}$ we have $\|f_k\|_p < \frac{1}{k}$, hence

$\lim_k \|f_k\|_p = 0$, so $\lim_k \|f_k\| = 0$ by the assumption that such a norm exists. But $\lim_k \|f_k\| = 1$ by assumption.]

Pick the minimal such C from (a).

(b) There is $0 < c < 1$ with $\int_0^c |f|^p = \int_c^1 |f|^p = \frac{1}{2}\int_0^1 |f|^p$. [Use a limit interchange result or a continuity result.]

(c) Let $g = fI_{(0,c]}$ and $h = fI_{(c,1)}$, so that $f = g+h$. Then $\|g\|_p = \|h\|_p = 2^{-\frac{1}{p}}\|f\|_p$ and $\|f\| \le \|g\| + \|h\| \le C\|g\|_p + C\|h\|_p = C \times 2^{1-\frac{1}{p}}\|f\|_p$. Use the minimality of C to obtain $C \le C2^{1-\frac{1}{p}}$, and deduce $C = 0$.

(d) Conclude that $\|f\| = 0$ for every $f \in L^p$, and hence the assumption in (a) entails a contradiction.

3. Let $0 < p_0 < +\infty$ and let $f \in L^{p_0}$ be nonnegative. Let $E_1 = \{\omega \in \Omega : f(\omega) = 0\}$, $E_2 = \{\omega \in \Omega : 0 < f(\omega) \le 1\}$, and $E_3 = \{\omega \in \Omega : f(\omega) > 1\}$.

(a) $\lim_{p\to 0+} \int_{E_2} |f|^p = \mu(E_2)$. [Use MCT.]

(b) $\lim_{p\to 0+} \int_{E_3} |f|^p = \mu(E_3)$. [Use DCT.]

(c) $\lim_{p\to 0+} \int |f|^p = \mu(\{\omega \in \Omega : f(\omega) \ne 0\})$. [Let $p \to 0^+$ through a countable set. We have

$$\underline{\lim}_p \int |f|^p \ge \underline{\lim}_p \int_{E_2} |f|^p + \underline{\lim}_p \int_{E_3} |f|^p = \mu(E_2) + \mu(E_3)$$

and

$$\overline{\lim}_p \int |f|^p \le \overline{\lim}_p \int_{E_2} |f|^p + \overline{\lim}_p \int_{E_3} |f|^p = \mu(E_2) + \mu(E_3).]$$

4. Say $\mu(\Omega) = 1$, and let $f \in L^1$ be nonnegative. Write $\log 0 = -\infty$.

(a) $\int \log f \le \log \int f$ if $\log f \in L^1$. [If $\|f\|_1 = 0$, then $\log \int f = -\infty$ and $f = 0$ μ-a.e. on Ω, hence $\log f = -\infty$ μ-a.e., whence $\int \log f = -\infty$. Assume now that $\|f\|_1 > 0$. Since $\log x \le x - 1$ for $0 \le x \le +\infty$ we have $\int \log \frac{f}{\|f\|_1} \le (\int \frac{f}{\|f\|_1}) - 1 = 0$, hence $\int \log f \le \log \|f\|_1$. Note that $\int \log f$ is defined if $\log f \in L^1$, similarly for $\int \log \frac{f}{\|f\|_1}$.]

(b) If $\log f \notin L^1$, then $\int (\log f)^+ < +\infty$ under the assumption $f \in L^1$, so it must be the case that $\int (\log f)^- = +\infty$. Conclude that even if $\log f \notin L^1$, $\int \log f$ still exists and equals $-\infty$, giving the inequality in (a).

(c) $\frac{f^r - 1}{r}$ decreases to $\log f$ as $r \to 0^+$, hence $\lim_{r\to 0+} \frac{1}{r}[\int f^r - 1] = \int \log f$. [Use MCT.]

(d) Verify the inequalities

$$\frac{1}{r}\left[\int f^r - 1\right] \ge \frac{1}{r}\log \int f^r \ge \frac{1}{r}\int \log f^r = \int \log f.$$

(e) Conclude that $\lim_{r\to 0+} \|f\|_r$ exists and equals $\exp(\int \log f)$. If $\log f \notin L^1$, this is interpreted as $\lim_{r\to 0+} \|f\|_r = 0$.

5. Find conditions for equality in the Hölder-like Inequality.

6. Find conditions for equality in the Minkowski-like Inequality.

7. For every $a, b > 0$ and $0 < p < 1$, we have $ab \ge \frac{a^p}{p} + \frac{b^{p'}}{p'}$. Use this to give a calculus-based proof of the Hölder-like Inequality that is similar to the proof of the original Hölder Inequality. [Let $f(a,b) = ab - \frac{a^p}{p}$ for $a, b > 0$,[7] where $0 < p < 1$ is fixed. For each $b \geq 0$, we have that $\frac{\partial}{\partial a} f(a,b) = b - a^{p-1} \geq 0$ iff $a \geq b^{\frac{1}{p-1}}$ [here we use the fact that $p - 1 < 0$]. Therefore, we have that $f(a,b) \geq f(b^{\frac{1}{p-1}}, b)$ for every $b > 0$, which gives $ab - \frac{a^p}{p} \geq \frac{p-1}{p} b^{\frac{p}{p-1}} = \frac{1}{p'} b^{p'}$ for every $b > 0$.]

8. Does the Riesz-Fischer Theorem necessarily hold in the case $0 < p < 1$? [Hint: note the relevance of the original Minkowski inequality in the proof of the Riesz-Fischer Theorem. When $0 < p < 1$, the direction of the inequality is reversed.]

8.4 L^p SPACE: THE CASE $p = +\infty$

We have considered L^p space first for $1 \leq p < +\infty$, and then for $0 < p < 1$. We have discussed forms of the Hölder Inequality for $1 < p < +\infty$ and $0 < p < 1$, but we have not considered such an inequality if $p = 1$ or $p = +\infty$. We have discussed forms of the Minkowski Inequality for $1 \leq p < +\infty$ and $0 < p < 1$, that is, for $0 < p < +\infty$. The Riesz-Fischer Theorem regarding the completeness of L^p was given for $1 \leq p < +\infty$, but again the case $p = +\infty$ has not been considered. In this section, we handle the case $p = +\infty$ for the Hölder Inequality, Minkowski Inequality, and the Riesz-Fischer Theorem. This will require giving a definition of L^∞ and $\| \, \|_\infty$.

Definition. Let f be $\mathcal{F}/\mathcal{B}^*$-measurable and defined on Ω.[8] We define the *essential supremum of f on Ω* by writing

$$\operatorname{ess\,sup} f = \inf \{\alpha \geq 0 : \ \mu(\{\omega \in \Omega : \ |f(\omega)| > \alpha\}) = 0\}.$$

An equivalent definition of $\operatorname{ess\,sup} f$ is

$$\operatorname{ess\,sup} f = \inf\{M \geq 0 : \ |f(\omega)| \leq M \text{ for } \mu\text{-a.e. } \omega \in \Omega\}.$$

This is seen to be true by noting that both sets in the definitions of $\operatorname{ess\,sup} f$ coincide and hence have the same infimum. [Note that $+\infty$ is in both sets, hence the infima above are never taken over the empty set.]

We now turn to a definition of L^p when $p = +\infty$. When $0 < p < +\infty$, we had that $f \in L^p$ iff $\int |f|^p < +\infty$, but something different will have to be given when $p = +\infty$.

Definition. We will write L^∞ to denote the collection of $\mathcal{F}/\mathcal{B}^*$-measurable functions f on Ω with $\operatorname{ess\,sup} f < +\infty$. If confusion is possible, we write $L^\infty(\Omega, \mathcal{F}, \mu)$ for L^∞. It is common to see a function $f \in L^\infty$ referred to as *essentially bounded.*

[7] Unlike the $1 \leq p < +\infty$ case, we do not allow $a = 0$ or $b = 0$ in our calculus discussion.
[8] We really only need f to be defined μ-a.e. on Ω, but we are following the convention of viewing functions as defined everywhere on Ω. Once again, this entails no loss of generality.

Claim 1. *Let $f, g \in L^\infty$ and $c \in \mathbb{R}$.*

(a) $cf \in L^\infty$ and $\operatorname{ess\,sup} cf = |c| \operatorname{ess\,sup} f$.

(b) $|f| \leq \operatorname{ess\,sup} f$ μ-a.e. on Ω.

(c) $f + g \in L^\infty$ and $\operatorname{ess\,sup}(f+g) \leq \operatorname{ess\,sup} f + \operatorname{ess\,sup} g$.

L^∞ is accordingly a vector space.

Proof: We first show (a). There are two cases, $c = 0$ and $c \neq 0$. If $c = 0$, then $cf = 0$ on Ω, hence $\operatorname{ess\,sup} cf = 0 = |c| \operatorname{ess\,sup} f$. If $c \neq 0$, we have that

$$\begin{aligned}
\operatorname{ess\,sup} cf &= \inf\{M \geq 0 : \ |cf| \leq M \ \mu\text{-a.e. on } \Omega\} \\
&= \inf\left\{M \geq 0 : \ |f| \leq \frac{M}{|c|} \ \mu\text{-a.e. on } \Omega\right\} \\
&= \inf\left\{|c|K \geq 0 : \ |f| \leq K \ \mu\text{-a.e. on } \Omega\right\} \\
&= |c| \inf\left\{K \geq 0 : \ |f| \leq K \ \mu\text{-a.e. on } \Omega\right\} \\
&= |c| \operatorname{ess\,sup} f.
\end{aligned}$$

[The third equality follows by making the formal substitution $K = \frac{M}{|c|}$.] This simultaneously shows that (i) if $\operatorname{ess\,sup} f < +\infty$ then $\operatorname{ess\,sup} cf < +\infty$, and (ii) $\operatorname{ess\,sup} cf = |c| \operatorname{ess\,sup} f$. That is, we have established (a) in the case where $c \neq 0$. Combining the $c = 0$ and $c \neq 0$ cases gives (a).

We now turn to (b). By the alternative definition of $\operatorname{ess\,sup} f$, we may find a sequence $M_1 \geq M_2 \geq \cdots \geq 0$ with $\lim_k M_k = \operatorname{ess\,sup} f$ and $|f| \leq M_k$ μ-a.e. on Ω for every $k \in \mathbb{N}$. For every $k \in \mathbb{N}$, let $A_k = \{\omega \in \Omega : \ |f(\omega)| \leq M_k\}$, so that $\mu(A_k^c) = 0$. Next, let $A = \bigcup_{k=1}^\infty A_k^c$. Then $\mu(A) = 0$, and $\omega \in A^c$ forces $\omega \in \bigcap_{k=1}^\infty A_k$, so that for every $k \in \mathbb{N}$ we have $|f(\omega)| \leq M_k$. It follows for every $\omega \in A^c$ that $|f(\omega)| \leq \lim_k M_k = \operatorname{ess\,sup} f$, completing the proof of (b).

We now turn to (c). From (b), we have that $|f| \leq \operatorname{ess\,sup} f$ μ-a.e. on Ω and $|g| \leq \operatorname{ess\,sup} g$ μ-a.e. on Ω. Then

$$|f + g| \leq |f| + |g| \leq \operatorname{ess\,sup} f + \operatorname{ess\,sup} g \quad \mu\text{-a.e. on } \Omega.$$

Since $\operatorname{ess\,sup} f < +\infty$ and $\operatorname{ess\,sup} g < +\infty$, we have $\operatorname{ess\,sup}(f+g) < +\infty$, that is, $f + g \in L^\infty$. Furthermore, the displayed inequality shows that

$$\operatorname{ess\,sup}(f+g) \leq \operatorname{ess\,sup} f + \operatorname{ess\,sup} g.$$

This completes (c) and the entire proof. ■

We would now like to define $\| \ \|_p$ for when $p = +\infty$.

Definition. We define $\| \ \|_\infty$ on L^∞ by writing $\|f\|_\infty = \operatorname{ess\,sup} f$ whenever $f \in L^\infty$. The function $\| \ \|_\infty$ is referred to as the L^∞ norm.

That $\| \ \|_\infty$ is really a norm on L^∞ is seen by verifying the following four requirements of a norm:

(1) $0 \leq \|f\|_\infty < +\infty$ for each $f \in L^\infty$.

(2) $||f||_\infty = 0$ iff $f = 0$ μ-a.e. on Ω.
(3) $||cf||_\infty = |c| ||f||_\infty$ for every $c \in \mathbb{R}$ and $f \in L^\infty$.
(4) $||f + g||_\infty \leq ||f||_\infty + ||g||_\infty$ for every $f, g \in L^\infty$.

Properties (1) and (2) are trivial. Property (3) is merely part (a) of Claim 1 written in norm notation. Property (4) is merely part (c) of Claim 1 written in norm notation. Note that Property (4) is none other than Minkowski's Inequality $||f + g||_p \leq ||f||_p + ||g||_p$ for the case $p = +\infty$.

We now turn to a definition of the Hölder conjugate p' of p in the case where $p = 1$ or $p = +\infty$. For $0 < p < 1$ and $1 < p < +\infty$, we have written $p' = \frac{p}{p-1}$ for the Hölder conjugate of p. The definition as given doesn't make sense for $p = 1$ or $p = +\infty$, for the expressions $\frac{1}{0}$ and $\frac{+\infty}{+\infty - 1}$ are undefined. We use the following convention.

Convention. We define the Hölder conjugate of $+\infty$ to be $+\infty' = 1$. We define the Hölder conjugate of 1 to be $1' = +\infty$.

By this convention, we have $\frac{1}{p} + \frac{1}{p'} = 1$ whenever $p = 1$ or $p = +\infty$. Therefore, we have $\frac{1}{p} + \frac{1}{p'} = 1$ for every $0 < p \leq +\infty$.

With this convention established, we may now state and prove the form of Hölder's Inequality for the case $p = +\infty$ [and $p = 1$]. Unlike the $1 < p < +\infty$ case, the proof here is very simple due to the above claim, and is left to the reader.

Hölder's Inequality. *Let $f \in L^\infty$ and $g \in L^1$. Then $fg \in L^1$ and the inequality $|\int fg| \leq ||fg||_1 \leq ||f||_\infty ||g||_1$ holds.*

The only real work in this section consists in proving a version of the Riesz-Fischer Theorem for the $p = +\infty$ case. As with the $1 \leq p < +\infty$ case, we view L^∞ as a metric space where the distance between two elements f and g is given by $||f - g||_\infty$. This distance function is easily seen to satisfy the following properties for each $f, g, h \in L^\infty$:

(1) $0 \leq ||f - g||_\infty < +\infty$.

(2) $||f - f||_\infty = 0$, and $||f - g||_\infty = 0$ iff $f = g$ μ-a.e. on Ω.

(3) $||f - g||_\infty = ||g - f||_\infty$.

(4) $||f - h||_\infty \leq ||f - g||_\infty + ||g - h||_\infty$.

That the above definition gives a metric is nearly automatic on the basis of the work that has been done earlier in this section.[9]

As with the $1 \leq p < +\infty$ case, we deal with the notion of convergence relative to the distance function just given.

[9] Again, to have a metric in the strict sense means that $||f - g||_\infty = 0$ forces $f = g$ on Ω. However, we may only infer that $f = g$ μ-a.e. on Ω, which is slightly weaker than saying $f = g$ on Ω. Therefore, in the strictest sense, we don't have a metric. As with the $1 \leq p < +\infty$ case, this slight deficiency won't hurt us a bit, and we still continue to use the word *metric*.

Definition. We will say that a sequence of functions $\{f_n\}_{n=1}^{\infty}$ in L^∞ *converges in* L^∞ to an element $f \in L^\infty$ iff $\lim_n \|f_n - f\|_\infty = 0$. We will write $f_n \overset{\infty}{\to} f$ in this situation.

We now present the Riesz-Fischer Theorem for the case where $p = +\infty$. The proof in this case is much easier than in the case where $1 \le p < +\infty$. In the proof, we shall mainly use the fact that if $f \in L^\infty$, then $|f| \le \|f\|_\infty$ μ-a.e. on Ω. This was proven in part (b) of Claim 1.

Riesz-Fischer Theorem. *L^∞ is a complete metric space. That is, if $\{f_n\}_{n=1}^{\infty}$ is a Cauchy sequence in L^∞, then there is $f \in L^\infty$ such that $f_n \overset{\infty}{\to} f$.*

Proof: Let $\{f_n\}_{n=1}^{\infty}$ denote a Cauchy sequence in L^∞. Then for every $k \in \mathbb{N}$ there is $N_k \in \mathbb{N}$ such that $l, m \ge N_k$ forces $\|f_l - f_m\|_\infty \le \frac{1}{k}$. By part (b) of Claim 1, we may make the following assertion: For every $k \in \mathbb{N}$ there are sets $A(k, l, m) \in \mathcal{F}$ such that

$$l, m \ge N_k \text{ and } \omega \in A(k, l, m) \quad \text{force} \quad |f_l(\omega) - f_m(\omega)| \le \frac{1}{k}$$

and

$$l, m \ge N_k \quad \text{forces} \quad \mu\left(A(k, l, m)^c\right) = 0.$$

Define

$$A = \bigcup_{k=1}^{\infty} \bigcup_{l=N_k}^{\infty} \bigcup_{m=N_k}^{\infty} A(k, l, m)^c.$$

Then $A \in \mathcal{F}$ and $\mu(A) = 0$, for A is a countable union of μ-null sets. Observe that if $\omega \in A^c$, then

$$\text{for every } k \in \mathbb{N} \text{ and every } l, m \ge N_k \text{ we have } |f_l(\omega) - f_m(\omega)| \le \tfrac{1}{k}. \tag{8.8}$$

Therefore, for every $\omega \in A^c$ we have that $\{f_n(\omega)\}_{n=1}^{\infty}$ is a Cauchy sequence in $\mathbb{R}$. Since $\mathbb{R}$ is a complete metric space [under the usual Euclidean distance], $\lim_n f_n(\omega)$ exists in $\mathbb{R}$ for every $\omega \in A^c$. Define f on Ω by writing

$$f(\omega) = \begin{cases} \lim_n f_n(\omega) & \text{if } \omega \in A^c \\ 0 & \text{if } \omega \in A. \end{cases}$$

Then f is $\mathcal{F}/\mathcal{B}^*$-measurable, and letting $m \to \infty$ in (8.8) gives the inequality

$$|f_l(\omega) - f(\omega)| \le \frac{1}{k} \quad \text{for every } k \in \mathbb{N},\ l \ge N_k, \text{ and } \omega \in A^c.$$

This last inequality, the fact that each f_l is in L^∞, and the fact that $f = 0$ on A imply that $f \in L^\infty$. Because $\mu(A) = 0$, we have that $\|f_l - f\|_\infty \le \frac{1}{k}$ for each $k \in \mathbb{N}$ and $l \ge N_k$, so that $\|f_n - f\|_\infty$ can be made arbitrarily small by choosing n to be sufficiently large. This gives $f_n \overset{\infty}{\to} f$ and completes the proof. ∎

Exercises.

1*. In this exercise, we deal with the formal definition of L^∞. We define L^∞ to consist of those equivalence classes $\mathfrak{A} \in \mathcal{E}/\sim$ such that $\operatorname{ess\,sup} f < +\infty$ for some $f \in \mathfrak{A}$. [Recall the discussion in the introduction to this chapter.]
(a) If $f = g$ μ-a.e. on Ω, then $\operatorname{ess\,sup} f = \operatorname{ess\,sup} g$. [Straightforward.] Therefore, we may equivalently say that $\mathfrak{A} \in L^\infty$ iff $\operatorname{ess\,sup} f < +\infty$ for every $f \in \mathfrak{A}$.
(b) If $\mathfrak{A}, \mathfrak{B} \in L^\infty$ and $c \in \mathbb{R}$, then $c\mathfrak{A} \in L^\infty$ and $\mathfrak{A} + \mathfrak{B} \in L^\infty$. Therefore, L^∞ is a vector space.

We define $\| \|_\infty$ on L^∞ by writing $\|\mathfrak{A}\|_\infty = \operatorname{ess\,sup} f$, where $f \in \mathfrak{A}$.

(c) The definition of $\| \|_\infty$ is well defined. [Use (a).]
(d) Let $\mathfrak{A}, \mathfrak{B} \in L^\infty$ and $c \in \mathbb{R}$. Then
(i) $0 \leq \|\mathfrak{A}\|_\infty < +\infty$.
(ii) $\|\mathfrak{A}\|_\infty = 0$ iff $\mathfrak{A} = o$.
(iii) $\|c\mathfrak{A}\|_\infty = |c|\|\mathfrak{A}\|_\infty$.
(iv) $\|\mathfrak{A} + \mathfrak{B}\|_\infty \leq \|\mathfrak{A}\|_\infty + \|\mathfrak{B}\|_\infty$.
Therefore, $\| \|_\infty$ is a norm on L^∞: L^∞ *is a normed linear space.*
(d) For $\mathfrak{A}, \mathfrak{B}, \mathfrak{C} \in L^\infty$, define $\rho(\mathfrak{A}, \mathfrak{B}) = \|\mathfrak{A} - \mathfrak{B}\|_\infty$.
(i) $0 \leq \rho(\mathfrak{A}, \mathfrak{B}) < +\infty$.
(ii) $\rho(\mathfrak{A}, \mathfrak{A}) = 0$, and $\rho(\mathfrak{A}, \mathfrak{B}) = 0$ iff $\mathfrak{A} = \mathfrak{B}$.
(iii) $\rho(\mathfrak{A}, \mathfrak{B}) = \rho(\mathfrak{B}, \mathfrak{A})$.
(iv) $\rho(\mathfrak{A}, \mathfrak{C}) \leq \rho(\mathfrak{A}, \mathfrak{B}) + \rho(\mathfrak{B}, \mathfrak{C})$.
Therefore, ρ is a metric on L^∞.
(e) L^∞ is complete in the metric ρ. Therefore L^∞ is a Banach space.

Comment: All parts of Exercise 1 can be handled quickly by using the results of this section.

2*. Consider the σ-finite measure space $(\Omega, \mathcal{F}, \mu)$.
(a) $f \in L^\infty$ iff there is a bounded $\mathcal{F}/\mathcal{B}^*$-measurable function g on Ω such that $f = g$ μ-a.e. on Ω. [The "if" direction is nearly automatic. For the "only if" part, suppose that $f \in L^\infty$. Define g on Ω by writing $g(\omega) = f(\omega)$ whenever $|f(\omega)| \leq \|f\|_\infty$ and $g(\omega) = \|f\|_\infty$ whenever $|f(\omega)| > \|f\|_\infty$. Then g is bounded, $\mathcal{F}/\mathcal{B}^*$-measurable, and $g = f$ μ-a.e. on Ω.]
(b) If $f \in L^\infty$, then $\|f\|_\infty = \inf\{\sup_{\omega\in\Omega} |g(\omega)| : g \text{ is as in (a)}\}$. [Suppose that there is g as described in part (a) such that $\sup_{\omega\in\Omega} |g(\omega)| < \|f\|_\infty$. Then the set $\{\omega \in \Omega : |f(\omega)| > \sup_{\omega\in\Omega} |g(\omega)|\}$ fails to be μ-null and $f \neq g$ on this set, contradicting the choice of g. This gives $\sup_{\omega\in\Omega} |g(\omega)| \geq \|f\|_\infty$ for all g as in part (a). Next, let $B = \{\omega \in \Omega : |f(\omega)| \leq \|f\|_\infty\}$. We have that fI_B is bounded, $\mathcal{F}/\mathcal{B}^*$-measurable, and equals f except for a μ-null subset of B^c; we also have $\sup_{\omega\in\Omega} |f(\omega)I_B(\omega)| \leq \|f\|_\infty$. This can be used to complete the claim.]

3. Consider the measure space $(\mathbb{R}^k, \mathcal{B}^k, \lambda_k)$. We have that $\operatorname{ess\,sup} f = \sup_{\boldsymbol{x}\in\mathbb{R}^k} |f(\boldsymbol{x})|$ if f is continuous.

4. Quickly prove the $p = +\infty$ version of Hölder's Inequality. [By Claim 1, we have that $|fg| \leq \|f\|_\infty |g|$ μ-a.e. on Ω, so $\int |fg| \leq \int \|f\|_\infty |g| = \|f\|_\infty \int |g| = \|f\|_\infty \|g\|_1$.]

5. Let $(\Omega, \mathcal{F}, \mu)$ be such that $\mu(\Omega) < +\infty$, and let f denote a bounded $\mathcal{F}/\mathcal{B}^*$-measurable function on Ω.
(a) For every $1 \leq p \leq +\infty$ we have $f \in L^p$, hence $\|f\|_p$ exists.
(b) $\lim_{p\to+\infty} \|f\|_p = \|f\|_\infty$.

6. For each $k \in \mathbb{N}$, $L^\infty(\mathbb{R}^k, \mathcal{B}^k, \lambda_k)$ fails to contain an unbounded continuous function. Also, for each $1 \le p < +\infty$, $L^p(\mathbb{R}, \mathcal{B}, \lambda)$ contains an unbounded continuous function. [See Exercise 3.]

7. Let $\Omega = (0,1]$, let $\mathcal{F}$ denote the Borel subsets of $(0,1]$, and let μ denote Lebesgue measure restricted to the Borel subsets of $(0,1]$. Show in this case that L^∞ is not separable. That is, there does not exist a countable dense subset of L^∞.

8.5 CONTAINMENT RELATIONS FOR L^p SPACES

Given a general measure space $(\Omega, \mathcal{F}, \mu)$, it is natural to ask whether there are any containment relations between L^p and L^q, where p and q are distinct positive numbers. It is easy to construct situations where $0 < p < q < +\infty$ but $L^p \not\subseteq L^q$ and $L^q \not\subseteq L^p$. [See Exercise 1.] Therefore, without some additional assumptions, there is not much to be said. However, some fairly interesting results can still be obtained with stronger assumptions.

Claim 1. *If $\mu(\Omega) < +\infty$ and $0 < p < q \le +\infty$, then $L^q \subseteq L^p$. Moreover, if $f \in L^q$ [and hence $f \in L^p$], we have $\|f\|_p \le \|f\|_q \mu(\Omega)^{\frac{1}{p} - \frac{1}{q}}$.*

Proof: Suppose first that $q < +\infty$. Let $r = \frac{q}{p}$, so that $1 < r < +\infty$. Pick $f \in L^q$. Since $\int (|f|^p)^r = \int |f|^q < +\infty$ and $\int |1|^{r'} = \int 1 = \mu(\Omega) < +\infty$, it follows that $|f|^p \in L^r$ and $1 \in L^{r'}$. [Here 1 means the function on Ω that is identically 1.] We now apply Hölder's Inequality to $|f|^p \in L^r$ and $1 \in L^{r'}$ to obtain that $|f|^p = |f|^p \times 1 \in L^1$. This gives $\int |f|^p < +\infty$, whence $f \in L^p$, so that $L^q \subseteq L^p$. For the inequality, we have

$$\begin{aligned} \|f\|_p^p = \int |f|^p = \int |f|^p 1 \quad &\le \quad \||f|^p\|_r \|1\|_{r'} \\ &= \quad \left(\int |f|^{pr}\right)^{\frac{1}{r}} \left(\int 1^{r'}\right)^{\frac{1}{r'}} \\ &= \quad \left(\int |f|^q\right)^{\frac{p}{q}} \mu(\Omega)^{\frac{1}{r'}} \\ &= \quad \|f\|_q^p \mu(\Omega)^{\frac{1}{r'}}, \end{aligned}$$

hence $\|f\|_p \le \|f\|_q \mu(\Omega)^{\frac{1}{r'p}} = \|f\|_q \mu(\Omega)^{\frac{1}{p} - \frac{1}{q}}$. This completes the proof for when $q < +\infty$.

We now deal with the case where $q = +\infty$. Let $f \in L^q = L^\infty$. Then, since $|f| \le \|f\|_\infty$ μ-a.e. on Ω, we have

$$\int |f|^p \le \int \|f\|_\infty^p = \|f\|_\infty^p \int 1 = \|f\|_\infty^p \mu(\Omega) < +\infty,$$

so that $f \in L^p$. This shows that $L^\infty \subseteq L^p$. The above inequality also establishes that $\|f\|_p^p \le \|f\|_\infty^p \mu(\Omega)$, so that $\|f\|_p \le \|f\|_\infty \mu(\Omega)^{\frac{1}{p}}$. This is the desired inequality in the case $q = +\infty$, hence the $q = +\infty$ case stands complete. Combining the $q < +\infty$ and $q = +\infty$ cases gives the result. ∎

In probability theory, where $\mu(\Omega) = 1$, the previous claim is important, and the inequality takes the particularly simple form $\|f\|_p \leq \|f\|_q$.

Our next result is a "sandwich" result. Whereas in the last claim the price to be paid was the assumption that $\mu(\Omega) < +\infty$, here the price to be paid is a stronger hypothesis on the function f [and $\mu(\Omega)$ may be finite or infinite].

Claim 2. *Let $0 < p < r < q < +\infty$ and $f \in L^p \cap L^q$. Then $f \in L^r$. That is, the containment relation $L^p \cap L^q \subseteq L^r$ holds.*

Proof: Since $p < r < q$, there is $0 < c < 1$ with $r = cp + (1-c)q$. Since $\int |f|^{cp\frac{1}{c}} = \int |f|^p < +\infty$ and $\int |f|^{(1-c)q(\frac{1}{c})'} = \int |f|^q < +\infty$, we have that $|f|^{cp} \in L^{\frac{1}{c}}$ and $|f|^{(1-c)q} \in L^{(\frac{1}{c})'}$. [Here, observe that since $0 < c < 1$ we have $1 < \frac{1}{c} < +\infty$ and thus $1 < (\frac{1}{c})' = \frac{1}{1-c} < +\infty$, so it makes sense to discuss $L^{(\frac{1}{c})'}$.] With these two facts, we again use Hölder's Inequality, applying it to the functions $|f|^{cp} \in L^{\frac{1}{c}}$ and $|f|^{(1-c)q} \in L^{(\frac{1}{c})'}$, obtaining that $|f|^r = |f|^{cp}|f|^{(1-c)q} \in L^1$ and hence $\int |f|^r < +\infty$, giving $f \in L^r$. ∎

Let us recall the definition of convexity. Let $0 < p < q < +\infty$, and define an extended real-valued function ϕ on (p,q). The function ϕ is convex iff

$$\phi(cp + (1-c)q) \leq c\phi(p) + (1-c)\phi(q) \quad \text{for every } 0 < c < 1.$$

Also, we shall use the convention that $\log 0 = -\infty$.

Claim 3. *Let $0 < p < q < +\infty$ and let $f \in L^p \cap L^q$, so that $f \in L^r$ for every $r \in (p,q)$ by Claim 2. For $r \in (p,q)$ write $\phi(r) = r \log \|f\|_r$. Then ϕ is convex.*

Proof: If $f = 0$ μ-a.e. on Ω, then $\|f\|_r = 0$ for every $r \in (p,q)$, and hence we have $\phi(r) = -\infty$ for every $r \in (p,q)$, in which case ϕ is trivially convex. We may therefore assume that $|f| > 0$ on a set of positive measure, so that $0 < \|f\|_r < +\infty$ for every $r \in (p,q)$. We now pick $0 < c < 1$. Observe that $p < cp + (1-c)q < q$, hence $f \in L^{cp+(1-c)q}$. We have (i) $|f|^{cp} \in L^{\frac{1}{c}}$ and (ii) $|f|^{(1-c)q} \in L^{(\frac{1}{c})'}$. [These facts were proven in the proof of Claim 2.] We apply Hölder's Inequality to the two functions in (i) and (ii) and note that $(\frac{1}{c})' = \frac{1}{1-c}$, obtaining

$$\begin{aligned}
\int |f|^{cp+(1-c)q} &= \int |f|^{cp}|f|^{(1-c)q} \\
&\leq \left(\int |f|^{cp\frac{1}{c}}\right)^{\frac{1}{\frac{1}{c}}} \left(\int |f|^{(1-c)q(\frac{1}{c})'}\right)^{\frac{1}{(\frac{1}{c})'}} \\
&= \left(\int |f|^p\right)^c \left(\int |f|^q\right)^{1-c} \\
&= \|f\|_p^{cp}\|f\|_q^{(1-c)q}.
\end{aligned}$$

This gives $\|f\|_{cp+(1-c)q}^{cp+(1-c)q} \le \|f\|_p^{cp}\|f\|_q^{(1-c)q}$. Taking logarithms gives

$$[cp + (1-c)q] \log \|f\|_{cp+(1-c)q} \le cp \log \|f\|_p + (1-c)q \log \|f\|_q,$$

and this means that $\phi(cp + (1-c)q) \le c\phi(p) + (1-c)\phi(q)$ for the generic $0 < c < 1$. This gives the result. ∎

Having completed the three claims presented, we specialize somewhat to give an application important in some areas of analysis, although the results given in the section from here on are used nowhere else in this book. We will let Ω denote a nonempty set, and we will take the σ-field $\mathcal{F}$ to be 2^Ω, the collection of all subsets of Ω. We will take μ to be the counting measure on $\mathcal{F}$: $\mu(A)$ denotes the number of points in A if A is finite, and $\mu(A) = +\infty$ when A is infinite. Since $\mathcal{F} = 2^\Omega$, every $f : \Omega \to \bar{\mathbb{R}}$ is $\mathcal{F}/\mathcal{B}^*$-measurable. It has been shown that if $\int f$ exists as a number in $\mathbb{R}$, then

$$\int f = \sup\left\{\sum_{\omega\in F} f(\omega) : F \subseteq \Omega,\, F \text{ finite}\right\} = \sum_{\omega\in\Omega} f(\omega).$$

[The second equality is one of definition, i.e. $\sum_{\omega\in\Omega} f(\omega)$ is defined by the supremum. See Exercise 7 of Section 6.2.] Saying that $f \in L^p(\Omega, \mathcal{F}, \mu)$ means that $\int |f|^p < +\infty$, which here means that

$$\sup\left\{\sum_{\omega\in F} |f(\omega)|^p : F \subseteq \Omega,\, F \text{ finite}\right\} = \sum_{\omega\in\Omega} |f(\omega)|^p$$

is a finite number. It is customary for analysis books to write $l^p(\Omega, \mathcal{F}, \mu)$ for $L^p(\Omega, \mathcal{F}, \mu)$ and l^p for L^p. For a function $f \in l^p$ we have

$$\|f\|_p = \left(\int |f|^p\right)^{\frac{1}{p}} = \left(\sum_{\omega\in\Omega} |f(\omega)|^p\right)^{\frac{1}{p}}.$$

Observe that here we have $|f| \le \operatorname{ess\,sup} f = \sup_{\omega\in\Omega} |f(\omega)|$ for every function f defined on Ω.

We shall prove two results for l^p spaces. These results are due to the very special framework in which we are operating.

Claim 4. *Let $f \in l^p$, where $0 < p < +\infty$. Then there is $0 < M < +\infty$ such that $|f(\omega)| < M$ for every $\omega \in \Omega$, so that $f \in l^\infty$.*

Proof: Pick $0 < p < +\infty$ and $f \in l^p$. We have

$$\sum_{\omega\in\Omega} |f(\omega)|^p = \sup\left\{\sum_{\omega\in F} |f(\omega)|^p : F \text{ finite},\, F \subseteq \Omega\right\} < +\infty. \tag{8.9}$$

Assume to the contrary that for every $0 < M < +\infty$ there is $\omega_M \in \Omega$ such that $|f(\omega_M)| \ge M^{\frac{1}{p}}$. Then, for every $0 < M < +\infty$ and every finite $F \subseteq \Omega$

with $\omega_M \in F$, we have $\sum_{\omega \in F} |f(\omega)|^p \geq |f(\omega_M)|^p \geq M$. Therefore, for every $0 < M < +\infty$ there is a finite subset $F_M \subseteq \Omega$ such that $\sum_{\omega \in F_M} |f(\omega)|^p \geq M$. But then $\sup\{\sum_{\omega \in F} |f(\omega)|^p : F \text{ finite}, F \subseteq \Omega\} = +\infty$, contradicting (8.9). This completes the proof. ■

Claim 5. *For every $0 < p < q < +\infty$ we have $l^p \subseteq l^q \subseteq l^\infty$.*

Proof: Let $f \in l^p$ and let $F \subseteq \Omega$ be finite, so that $\sum_{\omega \in F} |f(\omega)|^p < +\infty$. From Claim 4, find $0 < M < +\infty$ such that $|f(\omega)| < M$ for each $\omega \in \Omega$. Then

$$\sum_{\omega \in F} |f(\omega)|^q = \sum_{\omega \in F} |f(\omega)|^{q-p} |f(\omega)|^p \leq M^{q-p} \sum_{\omega \in F} |f(\omega)|^p < +\infty.$$

Since F is an arbitrary finite subset of Ω, we have

$$\begin{aligned}
\sup\left\{\sum_{\omega \in F} |f(\omega)|^q : F \subseteq \Omega \text{ finite}\right\} &\leq \sup\left\{M^{q-p} \sum_{\omega \in F} |f(\omega)|^p : F \subseteq \Omega \text{ finite}\right\} \\
&= M^{q-p} \sup\left\{\sum_{\omega \in F} |f(\omega)|^p : F \subseteq \Omega \text{ finite}\right\} \\
&< +\infty,
\end{aligned}$$

hence $f \in l^q$. This gives $l^p \subseteq l^q$. The claim that $l^q \subseteq l^\infty$ follows from Claim 4. This completes the proof. ■

We have presented most of the main containment results and inequalities for general L^p spaces [and l^p spaces]. The exercises state a few more containment relations and inequalities.

Exercises.

1. Consider the measure space $(\mathbb{R}, \mathcal{B}, \lambda)$. Let $1 \leq p < q < +\infty$, and let r be such that $\frac{1}{q} < r < \frac{1}{p}$.
(a) Define f on $\mathbb{R}$ by writing $f(x) = x^{-r} I_{(0,1)}(x)$ for each $x \in \mathbb{R}$. Then $f \in L^p$ but $f \notin L^q$. Therefore, we do not in general have $L^p \subseteq L^q$ when $p < q$. [This is straightforward calculus.]
(b) Let $g(x) = x^{-r} I_{(1,+\infty)}(x)$ for every $x \in \mathbb{R}$. Then $g \in L^q$ but $g \notin L^p$. Therefore, we do not in general have $L^q \subseteq L^p$ when $p < q$. [This is again very easy calculus.]

2. Let $0 < p < r < +\infty$. Then $L^p \cap L^\infty \subseteq L^r$, and for any $f \in L^p \cap L^\infty$ we have $\|f\|_r \leq \|f\|_p^{\frac{p}{r}} \|f\|_\infty^{1-\frac{p}{r}}$. This lets us consider the value $+\infty$ in Claim 2. [Let $A = \{\omega \in \Omega : |f(\omega)| \leq \|f\|_\infty\}$, so that $\mu(A^c) = 0$. Then

$$\int |f|^r = \int_A |f|^r = \int_A |f|^p |f|^{r-p} \leq \|f\|_\infty^{r-p} \int_A |f|^p = \|f\|_\infty^{r-p} \|f\|_p^p < +\infty.]$$

3. For any $0 < p < +\infty$ and $0 \leq M < +\infty$ we have

$$\left(\int |f|^p\right)^{\frac{1}{p}} \geq M \mu(\{\omega \in \Omega : |f(\omega)| \geq M\})^{\frac{1}{p}}.$$

This is one form of what is commonly referred to as *Chebyshev's Inequality*. [Break up the integral over Ω into the integrals over $\{\omega \in \Omega : |f(\omega)| \geq M\}$ and its complement.]

4. Let $0 < r < +\infty$ and assume that $f \in L^r \cap L^\infty$, so that $f \in L^p$ for every $r < p < +\infty$ by Exercise 2. We wish to show that $\lim_{p\to\infty} \|f\|_p = \|f\|_\infty$. Follow this outline:

(a) Ignoring the trivial case where $f = 0$ μ-a.e. on Ω, let $f \neq 0$ on a set of positive measure, so that $\|f\|_\infty > 0$. Show that $\underline{\lim}_{p\to\infty} \|f\|_p \geq \|f\|_\infty$. [Let $0 \leq t < \|f\|_\infty$. Use Exercise 3 to obtain $(\int |f|^p)^{\frac{1}{p}} \geq t\mu(\{\omega \in \Omega : |f(\omega)| \geq t\})^{\frac{1}{p}}$. Consider two cases: (i) $\mu(\{\omega \in \Omega : |f(\omega)| \geq t\}) = +\infty$ and (ii) $\mu(\{\omega \in \Omega : |f(\omega)| \geq t\}) < +\infty$. In (i), we have $\int |f|^p = +\infty$ and the claim is trivial. In (ii), we have by elementary calculus that $\lim_{p\to\infty} \mu(\{\omega \in \Omega : |f(\omega)| \geq t\})^{\frac{1}{p}} = 1$, hence $\underline{\lim}_{p\to\infty} \|f\|_p \geq t$. Combine (i) and (ii), using the fact that $0 \leq t < \|f\|_\infty$ is generic.]

(b) Show that $\overline{\lim}_{p\to\infty} \|f\|_p \leq \|f\|_\infty$. [Use the inequality $\|f\|_p \leq \|f\|_r^{\frac{r}{p}} \|f\|_\infty^{1-\frac{r}{p}}$, valid for all $r < p < +\infty$, as given in Exercise 2.]

(c) Put (a) and (b) together to prove the claim.

5. Let $\mu(\Omega) = 1$ and $1 \leq p \leq q \leq +\infty$. Show for arbitrary f that

$$\int |f| \leq \left(\int |f|^p\right)^{\frac{1}{p}} = \left(\int |f|^q\right)^{\frac{1}{q}} \leq \text{ess sup}\, f,$$

so that $\|f\|_1 \leq \|f\|_p \leq \|f\|_q \leq \|f\|_\infty$.

6. We know that $l^p \subseteq l^q \subseteq l^\infty$ when $0 < p < q < +\infty$ by Claim 5. Show further that

$$\text{ess sup}\, f \leq \left(\sum_{\omega\in\Omega} |f(\omega)|^q\right)^{\frac{1}{q}} \leq \left(\sum_{\omega\in\Omega} |f(\omega)|^p\right)^{\frac{1}{p}}$$

for arbitrary $f \in l^p$, so that $\|f\|_\infty \leq \|f\|_q \leq \|f\|_p$ for every $f \in l^p$.

7. Consider $(\mathbb{R}, \mathcal{B}, \lambda)$.

(a) Find f such that $f \in L^p$ but $f \notin L^q$ for every $p < q < +\infty$.

(b) Find f such that $f \in L^q$ for every $0 < q < p$, but $f \notin L^p$.

(c) Find a nonnegative finite-valued function f with $f \notin L^p$ for every $0 < p < +\infty$.

(d) Find f such that $f \in L^p$ but $f \notin L^q$ for every $q \in (0, \infty) - \{p\}$.

8. Let $(\Omega, \mathcal{F}, \mu)$ be such that $\mu(A) > 0$ forces $\mu(A) \geq 1$ for each $A \in \mathcal{F}$. Let $1 \leq p \leq q < +\infty$. Then $L^p \subseteq L^q \subseteq L^\infty$ and $\|f\|_\infty \leq \|f\|_q \leq \|f\|_p$.

8.6 APPROXIMATION

In this section we deal with certain claims regarding collections of functions that are dense in a generic $L^p(\Omega, \mathcal{F}, \mu)$, where $1 \leq p < +\infty$. For this section, we explicitly view L^p as a metric space, where the distance between functions $f, g \in L^p$ is $\|f - g\|_p$.

Recall that $\mathfrak{S}$ [or $\mathfrak{S}(\Omega, \mathcal{F}, \mu)$] denotes the collection of nonnegative finite-valued $\mathcal{F}/\mathcal{B}^*$-measurable simple functions with domain Ω. [See Section 6.1] We also will need the following definition.

Definition: We will write $\mathfrak{S}_0$ for the collection of $\mathcal{F}/\mathcal{B}^*$-measurable finite-valued simple functions on Ω such that $\mu(\{\omega \in \Omega : s(\omega) \neq 0\}) < +\infty$. If confusion is possible, we will write $\mathfrak{S}_0(\Omega, \mathcal{F}, \mu)$ for $\mathfrak{S}_0$.

There are two main differences between $\mathfrak{S}$ and $\mathfrak{S}_0$: (i) functions in $\mathfrak{S}_0$ are not necessarily nonnegative as are the functions in $\mathfrak{S}$, and (ii) whereas it is possible for elements of $\mathfrak{S}$ to be nonzero on a set of infinite measure, elements of $\mathfrak{S}_0$ are nonzero on a set of at most finite measure. With these differences noted, the main containment relation is that nonnegative elements of $\mathfrak{S}_0$ are necessarily in $\mathfrak{S}$.

Our first order of business is to show that $\mathfrak{S}_0$ is a dense subset of L^p, where distance is the usual L^p distance. *Throughout, we assume that* $1 \leq p < +\infty$.

Claim 1. $\mathfrak{S}_0 \subseteq L^p$.

Proof: Pick $s \in \mathfrak{S}_0$; write $s = \sum_{i=1}^n c_i I_{A_i}$, where $c_1, \cdots, c_n \in \mathbb{R}$ and $A_1, \cdots, A_n \in \mathcal{F}$ are disjoint with union Ω. Observing that both quantities $\mu(\{\omega \in \Omega : s(\omega) \neq 0\})$ and $\max\{|c_1|, \cdots, |c_n|\}$ are finite, and using the fact that $|s|^p = \sum_{i=1}^n |c_i|^p I_{A_i}$, we have

$$\begin{aligned} \int |s|^p = \sum_{i=1}^n |c_i|^p \mu(A_i) &= \sum_{\{i:\ c_i \neq 0\}} |c_i|^p \mu(A_i) \\ &\leq (\max\{|c_1|, \cdots, |c_n|\})^p \sum_{\{i:\ c_i \neq 0\}} \mu(A_i) \\ &= (\max\{|c_1|, \cdots, |c_n|\})^p \mu(\{\omega \in \Omega : s(\omega) \neq 0\}) \\ &< +\infty. \end{aligned}$$

This gives $s \in L^p$, proving the claim. ∎

A subcollection A of functions in L^p is dense in L^p iff for every $\epsilon > 0$ and $f \in L^p$ there is $g \in A$ such that $\|f - g\|_p < \epsilon$. This is merely the general metric space definition of denseness adapted to the specifics of L^p space.

Claim 2. *$\mathfrak{S}_0$ is a dense subset of L^p. That is, given $\epsilon > 0$ and $f \in L^p$, there exists $s \in \mathfrak{S}_0$ with $\|s - f\|_p < \epsilon$. Moreover, this $s \in \mathfrak{S}_0$ may be chosen such that $|s| \leq |f|$.*

Proof: Let $f \in L^p$ and $\epsilon > 0$, and first assume that $f \geq 0$. Find a nondecreasing sequence $0 \leq s_1 \leq s_2 \leq \cdots \leq f$ in $\mathfrak{S}$ such that $\lim s_n = f$. Observe that for every $n \in \mathbb{N}$ we have $|f - s_n|^p \leq f^p \in L^1$, hence DCT gives

$$\lim_n \|f - s_n\|_p^p = \lim_n \int |f - s_n|^p = \int \lim_n |f - s_n|^p = \int 0 = 0.$$

Therefore, there is $N \in \mathbb{N}$ such that $\|s_N - f\|_p < \epsilon$. Since $0 \leq s_N \leq f$ and $f \in L^p$, we have $\mu(\{\omega \in \Omega : s_N(\omega) \neq 0\}) < +\infty$. [If s_N is positive on a set of infinite measure, then $\int s_N^p = +\infty$, contradicting the fact that $s_N \in L^p$.] Summarizing, we have $s_N \in \mathfrak{S}_0$ and $\|f - s_N\|_p < \epsilon$. This shows that $\mathfrak{S}_0$ is dense in the collection of nonnegative functions in L^p.

Now reset everything. Let $f \in L^p$ be general, and let $\epsilon > 0$. Write $f = f^+ - f^-$, where both of the nonnegative functions f^+ and f^- are in L^p. We now use the previous paragraph twice. First, we find $s_1 \in \mathfrak{S}_0$ such that $0 \leq s_1 \leq f^+$ and $\|s_1 - f^+\|_p < \frac{\epsilon}{2}$. Next, we find $s_2 \in \mathfrak{S}_0$ such that $0 \leq s_2 \leq f^-$ and $\|s_2 - f^-\|_p < \frac{\epsilon}{2}$. Let $s = s_1 - s_2$. Then s is the difference of two simple functions and hence is simple. Also, since

$$\{\omega \in \Omega : \ s(\omega) \neq 0\} \subseteq \{\omega \in \Omega : \ s_1(\omega) \neq 0\} \cup \{\omega \in \Omega : \ s_2(\omega) \neq 0\},$$

and since $\mu(\{\omega \in \Omega : \ s_j(\omega) \neq 0\}) < +\infty$ for $j = 1$ and $j = 2$, it follows that $\mu(\{\omega \in \Omega : \ s(\omega) \neq 0\}) < +\infty$, hence $s \in \mathfrak{S}_0$. Next, we have

$$\|f - s\|_p = \|(f^+ - s_1) - (f^- - s_2)\|_p \leq \|f^+ - s_1\|_p + \|f^- - s_2\|_p < \frac{\epsilon}{2} + \frac{\epsilon}{2} = \epsilon$$

by Minkowski's Inequality. Next, we have

$$\begin{aligned} |s|^2 = (s_1 - s_2)^2 = s_1^2 + s_2^2 - 2s_1s_2 &= s_1^2 + s_2^2 \\ &\leq (f^+)^2 + (f^-)^2 \\ &= (f^+)^2 + (f^-)^2 + 2f^+f^- \\ &= (f^+ + f^-)^2 \\ &= |f|^2. \end{aligned}$$

[The third equality follows because $0 \leq s_1 \leq f^+$ and $0 \leq s_2 \leq f^-$ forces $0 \leq s_1s_2 \leq f^+f^- = 0$, hence $s_1s_2 = 0$. The fourth equality follows because $f^+f^- = 0$.] This gives $|s| \leq |f|$ and completes the entire proof. ∎

We now deal with the specific measure space $(\mathbb{R}^k, \mathcal{B}^k, \lambda_k)$. We will write $\mathfrak{C}_c$ for the collection of real-valued functions on $\mathbb{R}^k$ that are continuous and have compact support. Since functions in $\mathfrak{C}_c$ are bounded on their support and are zero on the complement of their support, we have that $\mathfrak{C}_c \subseteq L^p(\mathbb{R}^k, \mathcal{B}^k, \lambda_k)$. We will need to recall Lusin's Theorem from Section 5.7. Finally, we still have $1 \leq p < +\infty$.

Claim 3. *Given any bounded $f \in L^p(\mathbb{R}^k, \mathcal{B}^k, \lambda_k)$ and $\epsilon > 0$, we can find $g \in \mathfrak{C}_c$ such that $\|f - g\|_p < \epsilon$ and $\sup_{\omega \in \mathbb{R}^k} |g(\omega)| \leq \sup_{\omega \in \mathbb{R}^k} |f(\omega)|$. In particular, $\mathfrak{C}_c$ is a dense subset of $L^p(\mathbb{R}^k, \mathcal{B}^k, \lambda_k)$.*

Proof: Let $f \in L^p$ and $\epsilon > 0$. By Claims 1 and 2, we may find a function $s \in \mathfrak{S}_0 \subseteq L^p(\mathbb{R}^k, \mathcal{B}^k, \lambda_k)$ with $|s| \leq |f|$ and $\|f - s\|_p < \frac{\epsilon}{2}$. Observe that $\sup_{\omega \in \mathbb{R}^k} |s(\omega)| < +\infty$ since s can assume only a finite number of values, each value being finite. Also, note that $\sup_{\omega \in \mathbb{R}^k} |s(\omega)| \leq \sup_{\omega \in \mathbb{R}^k} |f(\omega)|$ since $|s| \leq |f|$. If s is identically zero, then $s \in \mathfrak{C}_c$ and the proof is complete. We therefore assume that s is not identically zero, so that $\sup_{\omega \in \mathbb{R}^k} |s(\omega)| > 0$.

Note that $\mu(\{\omega \in \Omega : s(\omega) \neq 0\}) < +\infty$ since $s \in \mathfrak{S}_0$. By Lusin's Theorem, there is $g \in \mathfrak{C}_c$ such that $\sup_{\omega \in \mathbb{R}^k} |g(\omega)| \leq \sup_{\omega \in \mathbb{R}^k} |s(\omega)|$ and

$$\lambda_k(\{\omega \in \Omega : g(\omega) \neq s(\omega)\}) < \left(\frac{\epsilon}{4 \sup_{\omega \in \mathbb{R}^k} |s(\omega)|}\right)^p .$$

Write $A = \{\omega \in \Omega : g(\omega) \neq s(\omega)\}$ for brevity. It follows that

$$\begin{aligned}
\|f - g\|_p &= \|(f - s) + (s - g)\|_p \\
&\leq \|f - s\|_p + \|s - g\|_p \\
&< \frac{\epsilon}{2} + \left(\int |s - g|^p\right)^{\frac{1}{p}} \\
&= \frac{\epsilon}{2} + \left(\int_A |s - g|^p\right)^{\frac{1}{p}} \\
&\leq \frac{\epsilon}{2} + \left(\int_A \left(2 \sup_{\eta \in \mathbb{R}^k} |s(\eta)|\right)^p \, d\mu(\omega)\right)^{\frac{1}{p}} \\
&= \frac{\epsilon}{2} + 2 \sup_{\omega \in \mathbb{R}^k} |s(\omega)| \lambda_k(A)^{\frac{1}{p}} \\
&= \frac{\epsilon}{2} + 2 \sup_{\omega \in \mathbb{R}^k} |s(\omega)| \frac{\epsilon}{4 \sup_{\omega \in \mathbb{R}^k} |s(\omega)|} \\
&= \epsilon.
\end{aligned}$$

Since

$$\sup_{\omega \in \mathbb{R}^k} |g(\omega)| \leq \sup_{\omega \in \mathbb{R}^k} |s(\omega)| \leq \sup_{\omega \in \mathbb{R}^k} |f(\omega)|,$$

the entire proof is complete. ■

We next take $k = 1$, working in $(\mathbb{R}, \mathcal{B}, \lambda)$. We will write $\mathfrak{S}_{00}(\mathbb{R}, \mathcal{B}, \lambda)$ for the collection of functions on $\mathbb{R}$ of the form $\sum_{i=1}^n c_i I_{A_i}$, where $c_1, \cdots, c_n \in \mathbb{R}$ and each A_i is a bounded interval of any form. [Such a function is zero on $(A_1 \cup \cdots \cup A_n)^c$.] Observe that Claim 1 gives

$$\mathfrak{S}_{00}(\mathbb{R}, \mathcal{B}, \lambda) \subseteq \mathfrak{S}_0(\mathbb{R}, \mathcal{B}, \lambda) \subseteq L^p(\mathbb{R}, \mathcal{B}, \lambda).$$

Consider $L^p(\mathbb{R}, \mathcal{B}, \lambda)$ for the following claim, where again $1 \leq p < +\infty$.

Claim 4. *$\mathfrak{S}_{00}(\mathbb{R}, \mathcal{B}, \lambda)$ is a dense subset of $L^p(\mathbb{R}, \mathcal{B}, \lambda)$.*

Proof: Since $\mathfrak{S}_0(\mathbb{R}, \mathcal{B}, \lambda)$ is dense in $L^p(\mathbb{R}, \mathcal{B}, \lambda)$ [Claim 2], it is sufficient to show that $\mathfrak{S}_{00}(\mathbb{R}, \mathcal{B}, \lambda)$ is dense in $\mathfrak{S}_0(\mathbb{R}, \mathcal{B}, \lambda)$. To this end, let $s \in \mathfrak{S}_0(\mathbb{R}, \mathcal{B}, \lambda)$ and $\epsilon > 0$. Write $s = \sum_{i=1}^n c_i I_{A_i}$, where $c_1, \cdots, c_n$ are real and nonzero, $A_1, \cdots, A_n \in \mathcal{B}$ are disjoint, $s = 0$ on $(A_1 \cup \cdots \cup A_n)^c$, and

$$\lambda\left(\bigcup_{i=1}^n A_i\right) = \lambda(\{\omega \in \mathbb{R} : s(\omega) \neq 0\}) < +\infty.$$

Observe that $\lambda(A_i) < +\infty$ for $i = 1, \cdots, n$. Let $\delta = \frac{1}{2}(\frac{\epsilon}{1+\sum_{j=1}^{n}|c_j|})^p$, and fix $i \in \{1, \cdots, n\}$ for the time being. Find an open set G_i with $G_i \supseteq A_i$ and $\lambda(G_i - A_i) < \delta$. [This uses the Approximation Theorem from Section 4.2.] Note that $\lambda(G_i) < +\infty$ [since $\lambda(A_i) < +\infty$]. Write $G_i = \bigcup_{j=1}^{\infty} I_{i,j}$, where $I_{i,1}, I_{i,2}, \cdots$ are disjoint bounded open intervals with union G_i. Since $+\infty > \lambda(G_i) = \lambda\left(\bigcup_{j=1}^{\infty} I_{i,j}\right) = \sum_{j=1}^{\infty} \lambda(I_{i,j})$, there is $j_0 \in \mathbb{N}$ [depending on i] such that $\sum_{j=j_0+1}^{\infty} \lambda(I_{i,j}) < \delta$. Observe that we now have

$$\begin{aligned}
\left(\bigcup_{j=1}^{j_0} I_{i,j}\right) \triangle A_i &= \left(\bigcup_{j=1}^{j_0} I_{i,j} - A_i\right) \cup \left(A_i - \bigcup_{j=1}^{j_0} I_{i,j}\right) \\
&\subseteq \left(\bigcup_{j=1}^{\infty} I_{i,j} - A_i\right) \cup \bigcup_{j=j_0+1}^{\infty} I_{i,j} \\
&= (G_i - A_i) \cup \bigcup_{j=j_0+1}^{\infty} I_{i,j}.
\end{aligned}$$

It follows that

$$\begin{aligned}
\lambda\left(\left[\bigcup_{j=1}^{j_0} I_{i,j}\right] \triangle A_i\right) &\leq \lambda\left((G_i - A_i) \cup \bigcup_{j=j_0+1}^{\infty} I_{i,j}\right) \\
&\leq \lambda(G_i - A_i) + \lambda\left(\bigcup_{j=j_0+1}^{\infty} I_{i,j}\right) \\
&= \lambda(G_i - A_i) + \sum_{j=j_0+1}^{\infty} \lambda(I_{i,j}) \\
&< \delta + \delta \\
&= 2\delta.
\end{aligned}$$

Let $B_i = \bigcup_{j=1}^{j_0} I_{i,j}$, so that B_i is a bounded open set and $\lambda(B_i \triangle A_i) < 2\delta$. Therefore,

$$\begin{aligned}
\|I_{A_i} - I_{B_i}\|_p = \left(\int |I_{A_i} - I_{B_i}|^p\right)^{\frac{1}{p}} &= \left(\int I_{A_i \triangle B_i}^p\right)^{\frac{1}{p}} \\
&= \left(\int I_{A_i \triangle B_i}\right)^{\frac{1}{p}} \\
&= \lambda(A_i \triangle B_i)^{\frac{1}{p}} \\
&< (2\delta)^{\frac{1}{p}}.
\end{aligned}$$

We have so far held $i \in \{1, \cdots, n\}$ fixed; we now perform the above construction for each $i \in \{1, \cdots, n\}$, obtaining for each $i \in \{1, \cdots, n\}$ a bounded open set B_i such that $\|I_{A_i} - I_{B_i}\|_p < (2\delta)^{\frac{1}{p}}$.

Having obtained the sets $B_1, \cdots, B_n$, we define $t = \sum_{i=1}^n c_i I_{B_i}$, where $c_1, \cdots, c_n$ are as before and $t = 0$ on $(B_1 \cup \cdots \cup B_n)^c$. Since each B_i is a finite disjoint union of bounded open intervals, we have that $t \in \mathfrak{S}_{00}(\mathbb{R}, \mathcal{B}, \lambda)$. Next, we have

$$\begin{aligned} \|t - s\|_p = \left\| \sum_{i=1}^n c_i (I_{A_i} - I_{B_i}) \right\|_p &\le \sum_{i=1}^n \|c_i (I_{A_i} - I_{B_i})\|_p \\ &= \sum_{i=1}^n |c_i| \|I_{A_i} - I_{B_i}\|_p \\ &< (2\delta)^{\frac{1}{p}} \sum_{i=1}^n |c_i| \\ &= \left[2 \times \frac{1}{2} \left(\frac{\epsilon}{1 + \sum_{i=1}^n |c_i|} \right)^p \right]^{\frac{1}{p}} \sum_{i=1}^n |c_i| \\ &= \frac{\epsilon}{1 + \sum_{i=1}^n |c_i|} \sum_{i=1}^n |c_i| \\ &< \epsilon. \end{aligned}$$

Therefore, given $s \in \mathfrak{S}_0(\mathbb{R}, \mathcal{B}, \lambda)$ and $\epsilon > 0$, there is $t \in \mathfrak{S}_{00}(\mathbb{R}, \mathcal{B}, \lambda)$ with $\|t - s\|_p < \epsilon$. This shows that $\mathfrak{S}_{00}(\mathbb{R}, \mathcal{B}, \lambda)$ is a dense subset of $\mathfrak{S}_0(\mathbb{R}, \mathcal{B}, \lambda)$, and the proof is complete. ∎

We now go from $(\mathbb{R}, \mathcal{B}, \lambda)$ back to $(\mathbb{R}^k, \mathcal{B}^k, \lambda_k)$ for the following result. Recall from Section 7.5 the notion of translation. If $f : \mathbb{R}^k \to \bar{\mathbb{R}}$ is given and if $\boldsymbol{y} \in \mathbb{R}^k$, then $\tau_{\boldsymbol{y}} f$ is the function on $\mathbb{R}^k$ such that $\tau_{\boldsymbol{y}} f(\boldsymbol{x}) = f(\boldsymbol{x} + \boldsymbol{y})$ for every $\boldsymbol{x} \in \mathbb{R}^k$. The results of Claim 2 from Section 7.5 are relevant in this setup and the upcoming claim.

Claim 5. *Let $f \in L^p(\mathbb{R}^k, \mathcal{B}^k, \lambda_k)$, where $1 \le p < +\infty$. Then for every $\boldsymbol{y} \in \mathbb{R}^k$ it follows that $\tau_{\boldsymbol{y}} f \in L^p(\mathbb{R}^k, \mathcal{B}^k, \lambda_k)$. Moreover, $\lim_{\boldsymbol{y} \to \boldsymbol{0}} \|\tau_{\boldsymbol{y}} f - f\|_p = 0$.*

Proof: Let $g = |f|^p$, so that $g \in L^1(\mathbb{R}^k, \mathcal{B}^k, \lambda_k)$. By Claim 2 of Section 7.5, we have $\tau_{\boldsymbol{y}} g \in L^1$ for every $\boldsymbol{y} \in \mathbb{R}^k$. This means that

$$+\infty > \int g(\boldsymbol{x} + \boldsymbol{y})\, d\lambda_k(\boldsymbol{x}) = \int |f(\boldsymbol{x} + \boldsymbol{y})|^p \, d\lambda_k(\boldsymbol{x}) = \int |\tau_{\boldsymbol{y}} f(\boldsymbol{x})|^p \, d\lambda_k(\boldsymbol{x})$$

for every $\boldsymbol{y} \in \mathbb{R}^k$, and hence $\tau_{\boldsymbol{y}} f \in L^p(\mathbb{R}^k, \mathcal{B}^k, \lambda_k)$ for every $\boldsymbol{y} \in \mathbb{R}^k$.

We now turn to establishing the limit relation. Let $\epsilon > 0$. We shall find $\delta > 0$ such that $\|\boldsymbol{y}\| < \delta$ implies $\|\tau_{\boldsymbol{y}} f - f\|_p < \epsilon$. By Claim 3, there is a function $g \in \mathfrak{C}_c$ [and hence $g \in L^p$] such that $\|g - f\|_p < \frac{\epsilon}{3}$. Next, observe that for every $\boldsymbol{y} \in \mathbb{R}^k$ we have $\tau_{\boldsymbol{y}} g \in L^p(\mathbb{R}^k, \mathcal{B}^k, \lambda_k)$ [by the membership claim proved just earlier] and

$$\begin{aligned}\|\tau_{\boldsymbol{y}}g-\tau_{\boldsymbol{y}}f\|_p &= \left(\int |g(\boldsymbol{x}+\boldsymbol{y})-f(\boldsymbol{x}+\boldsymbol{y})|^p\,d\lambda_k(\boldsymbol{x})\right)^{\frac{1}{p}}\\ &= \left(\int |g(\boldsymbol{x})-f(\boldsymbol{x})|^p\,d\lambda_k(\boldsymbol{x})\right)^{\frac{1}{p}}\\ &= \|g-f\|_p\\ &< \frac{\epsilon}{3},\end{aligned}$$

where the second equality follows by applying Claim 2 of Section 7.5 to the $L^1(\mathbb{R}^k,\mathcal{B}^k,\lambda_k)$-function $|g-f|^p$. Next, since g has compact support, there is positive finite r such that $g=0$ on $B(\mathbf{0},r)^c$. Since g is continuous on the compact set $\overline{B(\mathbf{0},r)}$ [and hence uniformly continuous on $\overline{B(\mathbf{0},r)}$], it follows that g is uniformly continuous on $\mathbb{R}^k$. Therefore, there is $0<\delta\le 1$ such that $\|\boldsymbol{y}\|<\delta$ implies

$$|g(\boldsymbol{x}+\boldsymbol{y})-g(\boldsymbol{x})| < \frac{\epsilon}{3\lambda_k(B(\mathbf{0},r+1))^{\frac{1}{p}}} \quad \text{for each } \boldsymbol{x}\in\mathbb{R}^k. \tag{8.10}$$

Therefore, if $\|\boldsymbol{y}\|<\delta$, we have, writing A for $B(\mathbf{0},r+1)$, that

$$\begin{aligned}&\|\tau_{\boldsymbol{y}}g-g\|_p\\ &= \left(\int |g(\boldsymbol{x}+\boldsymbol{y})-g(\boldsymbol{x})|^p\,d\lambda_k(\boldsymbol{x})\right)^{\frac{1}{p}}\\ &= \left(\int_A |g(\boldsymbol{x}+\boldsymbol{y})-g(\boldsymbol{x})|^p\,d\lambda_k(\boldsymbol{x})+\int_{A^c} |g(\boldsymbol{x}+\boldsymbol{y})-g(\boldsymbol{x})|^p\,d\lambda_k(\boldsymbol{x})\right)^{\frac{1}{p}}\\ &< \left(\int_{B(\mathbf{0},r+1)}\left(\frac{\epsilon}{3\lambda_k(B(\mathbf{0},r+1))^{\frac{1}{p}}}\right)^p d\lambda_k(\boldsymbol{x})+\int_{B(\mathbf{0},r+1)^c} 0\,d\lambda_k(\boldsymbol{x})\right)^{\frac{1}{p}}\\ &= \left[\left(\frac{\epsilon}{3\lambda_k(B(\mathbf{0},r+1))^{\frac{1}{p}}}\right)^p \lambda_k(B(\mathbf{0},r+1))\right]^{\frac{1}{p}}\\ &= \frac{\epsilon}{3}.\end{aligned}$$

[The inequality holds for two reasons. The first reason is the inequality of (8.10). Next, we know that $g=0$ on $B(\mathbf{0},r)^c$, and for every $\boldsymbol{x}\in B(\mathbf{0},r+1)^c$ and every $\boldsymbol{y}\in\mathbb{R}^k$ with $\|\boldsymbol{y}\|<\delta\le 1$ we have that $\boldsymbol{x}+\boldsymbol{y}\in B(\mathbf{0},r)^c$, and hence $g(\boldsymbol{x}+\boldsymbol{y})=g(\boldsymbol{x})=0$.] Putting everything together, we see that $\|\boldsymbol{y}\|<\delta$ forces the inequality

$$\begin{aligned}\|\tau_{\boldsymbol{y}}f-f\|_p &= \|(\tau_{\boldsymbol{y}}f-\tau_{\boldsymbol{y}}g)+(\tau_{\boldsymbol{y}}g-g)+(g-f)\|_p\\ &\le \|\tau_{\boldsymbol{y}}f-\tau_{\boldsymbol{y}}g\|_p+\|\tau_{\boldsymbol{y}}g-g\|_p+\|g-f\|_p\\ &< \frac{\epsilon}{3}+\frac{\epsilon}{3}+\frac{\epsilon}{3}\\ &= \epsilon,\end{aligned}$$

whence the proof is complete. ∎

Exercises.

1. The following is an outline of an alternative proof that $\mathfrak{C}_c$ is a dense subset of $L^p(\mathbb{R}^k, \mathcal{B}^k, \lambda_k)$ [where $1 \leq p < +\infty$] that doesn't use Lusin's Theorem [as was used in Claim 3].

(a) Show that $\mathfrak{C}_c$ is dense in $\{I_A : A \in \mathcal{B}^k \text{ is bounded}\}$. [Let $A \in \mathcal{B}^k$ be bounded. Find compact K and open G with $K \subseteq A \subseteq G$ and $\lambda_k(G - K) < \epsilon^p$. Use Urysohn's Lemma to find a $g : \mathbb{R}^k \to \mathbb{R}$ that is continuous with $0 \leq g \leq 1$, $g = 1$ on K, and $g = 0$ on G^c. We have $g \in \mathfrak{C}_c$ and $g(\boldsymbol{x}) - I_A(\boldsymbol{x}) = 0$ for every $\boldsymbol{x} \in K \cup G^c$. Then

$$\|g - I_A\|_p = \left(\int |g - I_A|^p\right)^{\frac{1}{p}} = \left(\int_{G-K} |g - I_A|^p\right)^{\frac{1}{p}} \leq \left(\int_{G-K} 1\right)^{\frac{1}{p}} < \epsilon.]$$

(b) Show that $\mathfrak{C}_c$ is dense in the collection of nonnegative finite-valued $\mathcal{F}/\mathcal{B}^*$-measurable simple functions $\sum_{i=1}^n c_i I_{A_i}$ on Ω such that each A_i is bounded if $c_i > 0$. [Without loss of generality, assume that each $c_i > 0$, so that $A_1, \cdots, A_n$ are bounded. For each $i \in \{1, \cdots, n\}$, find $g_i \in \mathfrak{C}_c$ such that $\|g - I_{A_i}\|_p < \frac{\epsilon}{1+\sum_{i=1}^n |c_i|}$; this requires the result of (a). Let $g = \sum_{i=1}^n c_i g_i$. Then $g \in \mathfrak{C}_c$ and it is easy to show that $\|g - \sum_{i=1}^n c_i I_{A_i}\|_p < \epsilon$.]

(c) Show that $\mathfrak{C}_c$ is dense in the class of nonnegative functions in L^p that are zero outside a bounded set. [Show that the functions in (b) are dense in the collection here in (c) as follows: let $f \geq 0$ be in L^p with $f = 0$ outside some bounded $A \in \mathcal{B}^k$. Find a nondecreasing sequence $\{s_n\}_{n=1}^\infty$ of functions in (b) with $s_n \to f$. Use DCT to show that $\lim_n \|s_n - f\|_p = 0$ by using the fact that $|f - s_n|^p \leq f^p \in L^1$. Since $\mathfrak{C}_c$ is dense in the collection of functions in (b), and since the collection of functions in (b) has been shown by this outline to be dense in the collection of functions given here in (c), the claim of (c) follows.]

(d) Show that $\mathfrak{C}_c$ is dense in the class of functions in L^p that are zero outside a bounded set. [Let $f \in L^p$ be zero outside a bounded set, and pick $\epsilon > 0$. Write $f = f^+ - f^-$. Both f^+ and f^- are nonnegative functions in L^p that are zero outside a bounded set. By (c), there are functions $g_1, g_2 \in \mathfrak{C}_c$ with $\|f^+ - g_1\|_p$ and $\|f^- - g_2\|_p$ both less than $\frac{\epsilon}{2}$. Let $g = g_1 - g_2$. Then $g \in \mathfrak{C}_c$ and $\|f - g\|_p < \epsilon$.]

(e) Show that $\mathfrak{C}_c$ is dense in L^p. [Let $f \in L^p$. Since $|f I_{B(\mathbf{0},n)} - f|^p \leq |f|^p \in L^1$, DCT gives $\lim_n \|f I_{B(\mathbf{0},n)} - f\|_p = 0$. Thus f can be approximated by a function in L^p that vanishes outside a bounded set. Since (d) says that $\mathfrak{C}_c$ is dense in the class of functions in L^p that vanish outside a bounded set, it follows that f can be approximated by a function in $\mathfrak{C}_c$.]

2. We have shown for a general $(\Omega, \mathcal{F}, \mu)$ that $\mathfrak{S}_0$ is a dense subset of L^p for $1 \leq p < +\infty$.

(a) It is also the case that $\mathfrak{S}_0$ is a dense subset of L^∞. Therefore, $\mathfrak{S}_0$ is a dense subset of L^p for $1 \leq p \leq +\infty$. [This should pose no problem.]

(b) What about the case where $0 < p < 1$? [Do Claims 1 and 2 in any way depend on whether or not $p \geq 1$?]

3. Analyze the cases $0 < p < 1$ and $p = +\infty$ for:

(a) Claim 3.

(b) Claim 4.

(c) Claim 5.

4. Let $(\Omega, \mathcal{F}, \mu)$ be such that μ is σ-finite and $\mathcal{F}$ is countably generated. We will show that L^p is separable for $1 \leq p < +\infty$.

Assume that $\mathcal{F}$ is generated by the countable collection $\mathcal{C}$ and Ω is covered by a countable collection $\mathcal{D}$ of $\mathcal{F}$-sets of finite measure. Enumerate the sets in the countable collection $\mathcal{C} \cup \mathcal{D}$ as $E_1, E_2, \cdots$. For each $n \in \mathbb{N}$, write

$$\mathcal{P}_n = \{F_1 \cap \cdots \cap F_n : \ F_i = E_i \text{ or } F_i = E_i^c, \ i = 1, \cdots, n\},$$

and let $\mathcal{F}_n$ denote the collection of finite unions of $\mathcal{P}_n$-sets. Write $\mathcal{F}_0 = \bigcup_{n=1}^{\infty} \mathcal{F}_n$.

(a) Each $\mathcal{F}_n$ is a field on Ω.
(b) $\mathcal{F}_0$ is a countable field that generates $\mathcal{F}$: $\sigma(\mathcal{F}_0) = \mathcal{F}$.
(c) μ is σ-finite on $\mathcal{F}_0$.

Let V denote the collection of functions $\sum_{i=1}^m \alpha_i I_{A_i}$ defined on Ω [the function equals 0 on $(A_1 \cup \cdots \cup A_n)^c$], where the α_i's are rational and the A_i's are $\mathcal{F}_0$-sets of finite measure.

(d) V is a countable collection.
(e) Let $f \in L^p$ and $\epsilon > 0$. There is $g = \sum_{i=1}^m \alpha_i I_{B_i}$ with the α_i's and $\mu(B_i)$'s finite and such that $\|f - g\|_p < \epsilon$; we may assume without loss of generality that each α_i is nonzero and rational. [Use an approximation result in this section.]
(f) For $i = 1, \cdots, m$, there is $A_i \in \mathcal{F}_0$ with $\mu(A_i \Delta B_i) < (\frac{\epsilon}{m|\alpha_i|})^p$. [Use an approximation result from Chapter 3.]
(g) Let $h = \sum_{i=1}^m \alpha_i I_{A_i}$. Then $h \in V$ and $\|g - h\|_p < \epsilon$, hence $\|f - h\|_p < 2\epsilon$. Conclude that L^p is indeed separable.

Comment: The remaining exercises are connected. Some parts are straightforward, while for other parts ample hints are provided. We consider the measure space $(\mathbb{R}^k, \mathcal{B}^k, \lambda_k)$. We define $\mathfrak{C}^\infty$ as the collection of real-valued functions on $\mathbb{R}^k$ having continuous partial derivatives of all orders; $\mathfrak{C}_c^\infty$ will denote those elements of $\mathfrak{C}^\infty$ with compact support.

5. For every $t \in \mathbb{R}$ write $h(t) = e^{-\frac{1}{t}} I_{(0,+\infty)}(t)$. The goal here is to show that h has continuous derivatives of all orders.
(a) Verify for each $n \in \mathbb{N}$ that $h^{(n)}(t) = P_n(\frac{1}{t})e^{-\frac{1}{t}}$ for $t > 0$, where P_n is a polynomial of degree $2n$ with no degree zero terms. [Straightforward; use induction.]
(b) We know that $h^{(n)}(t) = 0$ for every $n \in \mathbb{N}$ and $t < 0$, so all that remains is to discuss the case $t = 0$. Show that $h^{(n)}(0) = 0$ for every $n \in \mathbb{N}$. [Use the definition of the derivative. For $n = 1$, we have $\lim_{\delta \to 0-} \delta^{-1}[h(\delta) - h(0)] = 0$ and $\lim_{\delta \to 0+} \delta^{-1}[h(\delta) - h(0)] = \lim_{x \to \infty} xe^{-x} = 0$, so that $h'(0)$ exists and equals zero. Now assume that $h^{(n)}(0) = 0$. Show that $\lim_{\delta \to 0-} \delta^{-1}[h^{(n)}(\delta) - h(0)] = 0$ and

$$\lim_{\delta \to 0+} \delta^{-1}[h^{(n)}(\delta) - h^{(n)}(0)] = \lim_{\delta \to 0+} \delta^{-1} P_n(\delta^{-1}) e^{-\frac{1}{\delta}} = \lim_{x \to +\infty} x P_n(x) e^{-x} = 0,$$

so that $h^{(n+1)}(0)$ exists and equals 0.]
(c) Put (a) and (b) together to conclude that h indeed possesses continuous derivatives of all orders.

6. Let $g : \mathbb{R}^k \to \mathbb{R}$ be defined by $g(\boldsymbol{x}) = h(1 - \|\boldsymbol{x}\|^2)$. Prove the following easy properties of g:
(a) $g \in \mathfrak{C}^\infty$.

(b) $\|x\| \geq 1$ forces $g(x) = 0$, hence $g \in \mathfrak{C}_c^\infty$.
(c) $0 \leq \|x\| < 1$ forces $g(x) > 0$, hence $\int g = \int_{B(0,1)} g > 0$.
(d) g is bounded.

7. Let $\phi : \mathbb{R}^k \to \mathbb{R}$ be defined by $\phi(x) = \frac{1}{c} g(x)$, where $c = \int g > 0$ [from part (c) in the previous exercise]. We have the following very easy properties of ϕ.
(a) $\phi \in \mathfrak{C}_c^\infty$.
(b) $\phi \geq 0$ and ϕ is bounded.
(c) $\phi(x) > 0$ iff $0 \leq \|x\| < 1$.
(d) $\int \phi = 1$.

8. For each $0 < a < +\infty$ and $x \in \mathbb{R}^k$, write $\phi_a(x) = a^{-n}\phi(\frac{x}{a})$. Prove the following properties of ϕ_a for each $0 < a < +\infty$.
(a) $\phi_a \in \mathfrak{C}_c^\infty$.
(b) ϕ_a is nonnegative and bounded.
(c) $\phi_a(x) > 0$ iff $0 \leq \|x\| < a$. [These first three parts are easy.]
(d) $\int \phi_a = 1$. [Let T denote the $k \times k$ matrix $\frac{1}{a} I$, where I is the identity matrix of order k. We have by Claim 1 of Section 7.5 that

$$\int \phi_a(x)\, d\lambda_k(x) = a^{-n} \int \phi(Tx)\, d\lambda_k(x) = a^{-n} \det(T^{-1}) \int \phi(x)\, d\lambda_k(x) = 1.]$$

9. Let $f \in L^1(\mathbb{R}^k, \mathcal{B}^k, \lambda_k)$. For each $0 < a < +\infty$ and $x \in \mathbb{R}^k$, define

$$\psi_a(x) = \int f(y)\phi_a(x - y)\, d\lambda_k(y).$$

(a) ψ_a is sensibly defined since the integral exists, and ψ_a is bounded. [We know that ϕ_a is bounded, say $\phi_a \leq M_a$, so $|f(y)\phi_a(x-y)| \leq |f(y)|M_a$ for every $x, y \in \mathbb{R}^k$; note that $fM_a \in L^1$.]
(b) For each $i = 1, \cdots, k$ and $x \in \mathbb{R}^k$, we obtain

$$\frac{\partial}{\partial x_i}\psi_a(x) = \int f(y)\frac{\partial}{\partial x_i}\phi_a(x - y)\, d\lambda_k(y),$$

hence ψ_a is continuous. [Use reasoning as given in the hint to (a).]
(c) For every $i, j \in \{1, \cdots, k\}$ and $x \in \mathbb{R}^k$ we have

$$\frac{\partial^2}{\partial x_j \partial x_i}\psi_a(x) = \int f(y)\frac{\partial^2}{\partial x_j \partial x_i}\phi_a(x - y)\, d\lambda_k(y).$$

Conclude that $\frac{\partial}{\partial x_i}\psi_a(x)$ is a continuous function of x. [The same type of argument as used in (b) may be employed here.]
(d) Use (b),(c), and induction to obtain $\psi_a \in \mathfrak{C}^\infty$.
(e) If f has compact support, so does ψ_a. [Say $f = 0$ on $B(0, r)^c$ for some $r > 0$. If $\psi_a(x) \neq 0$, then by definition there is $y \subset \mathbb{R}^k$ such that $f(y)\phi_a(x - y) \neq 0$. If $f(y) \neq 0$, then $y \in B(0, r)$. If $\phi_a(x - y) \neq 0$, then $x - y \in B(0, a)$. Observe that $\|x\| \leq \|y\| + \|x - y\| < r + a$ follows in any case. Thus $\psi_a(x) \neq 0$ forces $x \in B(0, r + a)$, so ψ_a has compact support.]
(f) If f has compact support, then $\psi_a \in \mathfrak{C}_c^\infty$. [Automatic from the previous parts.]

10. Exercises 5-9 lead to the following main result: *$\mathfrak{C}_c^\infty$ is dense in $L^1(\mathbb{R}^k, \mathcal{B}^k, \lambda_k)$.* We shall achieve this by showing that $\mathfrak{C}_c^\infty$ is dense in $\mathfrak{C}_c$. Since $\mathfrak{C}_c$ is dense in L^1, the claim will follow. We shall do this with the given outline.
(a) Let $f \in \mathfrak{C}_c$ and $\epsilon > 0$; find $r > 0$ so that $f = 0$ on $B(\mathbf{0}, r)^c$. Show that there is $0 < a \leq 1$ such that

$$\|\boldsymbol{y} - \boldsymbol{x}\| \leq a \text{ implies } |f(\boldsymbol{y}) - f(\boldsymbol{x})| \leq \frac{\epsilon}{\lambda_k(B(\mathbf{0}, r+1))}.$$

[Note that f is uniformly continuous since f is continuous with compact support.]
(b) We have $f(\boldsymbol{x}) = \int f(\boldsymbol{x})\phi_a(\boldsymbol{x} - \boldsymbol{y})\, d\lambda_k(\boldsymbol{y})$ for every $\boldsymbol{x} \in \mathbb{R}^k$. [Trivial by part (d) of Exercise 8.]
(c) Both f and ψ_a are zero on $B(\mathbf{0}, r+1)^c$. [Recall that $0 < a \leq 1$. The rest is straightforward.]
(d) For every $\boldsymbol{x} \in \mathbb{R}^k$ we have

$$|\psi_a(\boldsymbol{x}) - f(\boldsymbol{x})| \leq \frac{\epsilon}{\lambda_k(B(\mathbf{0}, r+1))}.$$

[We have

$$\begin{aligned}
|\psi_a(\boldsymbol{x}) - f(\boldsymbol{x})| &= \left| \int [f(\boldsymbol{y}) - f(\boldsymbol{x})]\phi_a(\boldsymbol{x} - \boldsymbol{y})\, d\lambda_k(\boldsymbol{y}) \right| \\
&\leq \int |f(\boldsymbol{y}) - f(\boldsymbol{x})|\phi_a(\boldsymbol{x} - \boldsymbol{y})\, d\lambda_k(\boldsymbol{y}) \\
&= \int_{\boldsymbol{y} \in \mathbb{R}^k:\ \|\boldsymbol{x} - \boldsymbol{y}\| \leq a} |f(\boldsymbol{y}) - f(\boldsymbol{x})|\phi_a(\boldsymbol{x} - \boldsymbol{y})\, d\lambda_k(\boldsymbol{y}) \\
&\leq \frac{\epsilon}{\lambda_k(B(\mathbf{0}, r+1))} \int_{\boldsymbol{y} \in \mathbb{R}^k:\ \|\boldsymbol{x} - \boldsymbol{y}\| \leq a} \phi_a(\boldsymbol{x} - \boldsymbol{y})\, d\lambda_k(\boldsymbol{y}) \\
&= \frac{\epsilon}{\lambda_k(B(\mathbf{0}, r+1))}.]
\end{aligned}$$

(e) We have $\|\psi_a - f\|_1 \leq \epsilon$, and hence $\mathfrak{C}_c^\infty$ is dense in $\mathfrak{C}_c$. [We have

$$\begin{aligned}
\|\psi_a - f\|_1 &= \int |\psi_a(\boldsymbol{x}) - f(\boldsymbol{x})|\, d\lambda_k(\boldsymbol{x}) \\
&= \int_{B(\mathbf{0}, r+1)} |\psi_a(\boldsymbol{x}) - f(\boldsymbol{x})|\, d\lambda_k(\boldsymbol{x}) \\
&\leq \left[\frac{\epsilon}{\lambda_k(B(\mathbf{0}, r+1))}\right] \int_{B(\mathbf{0}, r+1)} d\lambda_k(\boldsymbol{x}) \\
&= \frac{\epsilon}{\lambda_k(B(\mathbf{0}, r+1))} \lambda_k(B(\mathbf{0}, r+1)) \\
&= \epsilon.]
\end{aligned}$$

8.7 MORE CONVERGENCE CONCEPTS

In Section 5.6 the notions of convergence μ-a.e. and convergence in measure were discussed relative to an arbitrary measure space $(\Omega, \mathcal{F}, \mu)$, and various connections between the two modes of convergence were discussed. A third mode of convergence has been discussed in this chapter, namely convergence in L^p [also known as *convergence in pth mean*]. In this section, we shall do three things: (i) relate convergence μ-a.e. and convergence in measure to convergence in L^p, (ii) state and prove the *Vitali Convergence Theorem*, and (iii) introduce a wholly new mode of convergence, called *weak convergence in* L^p, connecting this new mode of convergence with the previous modes of convergence. We shall do this in the context of a completely generic measure space $(\Omega, \mathcal{F}, \mu)$.

We now carry out the first of the three steps of the program listed above. We begin with a simple claim.

Claim 1. *If $h \in L^p$ and $\eta > 0$ are such that $\mu(\{\omega \in \Omega : |h(\omega)| \geq \eta^{\frac{1}{2p}}\}) > \sqrt{\eta}$, then $\int |h|^p \geq \eta$. [Here $0 < p < +\infty$.]*

Proof: Let $A = \{\omega \in \Omega : |h(\omega)| \geq \eta^{\frac{1}{2p}}\}$, so that $\mu(A) > \sqrt{\eta}$. Then

$$\begin{aligned} \int |h|^p = \int_A |h|^p + \int_{A^c} |h|^p &\geq \int_A \left(\eta^{\frac{1}{2p}}\right)^p d\mu(\omega) + \int_{A^c} |h|^p \\ &= \sqrt{\eta}\mu(A) + \int_{A^c} |h|^p \\ &> \eta + 0 = \eta, \end{aligned}$$

which completes the proof of this short claim. ∎

The following result says that convergence in L^p implies convergence in measure.

Claim 2. *Let $f, f_1, f_2, \cdots \in L^p$, where $0 < p < +\infty$, and further assume that $f_n \xrightarrow{p} f$. Then $f_n \to f$ in measure.*

Proof: We shall proceed by contraposition, assuming that $f_n \not\to f$ in measure and showing that this forces $f_n \stackrel{p}{\not\to} f$. For each $n \in \mathbb{N}$ and $\theta > 0$, define

$$A(n, \theta) = \{\omega \in \Omega : |f_n(\omega) - f(\omega)| \geq \theta\}.$$

Since $f_n \not\to f$ in measure, there are numbers $\epsilon, \delta > 0$ with the following property:

for every $N \in \mathbb{N}$ there is $n \geq N$ such that $\mu(A(n, \epsilon)) > \delta$.

Therefore, we have $\mu(A(n, \epsilon)) > \delta$ for infinitely many $n \in \mathbb{N}$. Pick η such that $0 < \eta < \min\{\epsilon^{2p}, \delta^2\}$, so that $\eta^{\frac{1}{2p}} < \epsilon$ and $\sqrt{\eta} < \delta$. For every $n \in \mathbb{N}$ we have $A(n, \epsilon) \subseteq A(n, \eta^{\frac{1}{2p}})$, hence for the infinitely many $n \in \mathbb{N}$ mentioned earlier we have

$$\sqrt{\eta} < \delta < \mu(A(n,\epsilon)) \leq \mu(A(n,\eta^{\frac{1}{2p}})).$$

By taking $h = |f_n - f|$ in Claim 1 for the aforementioned infinitely many $n \in \mathbb{N}$, we have $\int |f_n - f|^p \geq \eta > 0$ for the aforementioned infinitely many $n \in \mathbb{N}$. Therefore, it cannot possibly be the case that $\lim_n \int |f_n - f|^p = 0$, which shows that $f_n \stackrel{p}{\nrightarrow} f$. The proof by contraposition is complete. ∎

The converse to Claim 2 is not true in general. In Exercise 2, a simple example will be given where the reader can show that convergence in measure does not necessarily imply convergence in L^p. Thus, we may summarize the basic relationship between convergence in pth mean and convergence in measure: *convergence in pth mean [with $0 < p < +\infty$] implies convergence in measure but not conversely.*

We now turn to discussing the relationship between convergence in L^p and convergence μ-a.e. The main fact is this: *neither convergence in L^p nor convergence μ-a.e. implies the other.* Exercises 3 and 4 will spell out some simple instantiations of this claim. This completes the first of the three parts of this section's program.

We now turn to the second part of this section's program, which is to state and prove the *Vitali Convergence Theorem.* Before doing so, we need a preliminary claim.

Claim 3. *Let $f \in L^1$ and $\epsilon > 0$. There is $E \in \mathcal{F}$ such that $\mu(E) < +\infty$ and $|\int f - \int_E f| < \epsilon$.*

Proof: We first assume that f is a nonnegative function in L^1, picking $\epsilon > 0$. Define $E_n = \{\omega \in \Omega : \ f(\omega) \geq \frac{1}{n}\}$ for each $n \in \mathbb{N}$. Then $\{E_n\}_{n=1}^{\infty}$ is a nondecreasing sequence of $\mathcal{F}$-sets of finite measure[10] with limit

$$\lim_n E_n = \bigcup_{n=1}^{\infty} E_n = \{\omega \in \Omega : \ f(\omega) > 0\}.$$

Therefore, $\{fI_{E_n}\}_{n=1}^{\infty}$ is a nondecreasing sequence of nonnegative functions with limit $fI_{\lim_n E_n}$. Using MCT gives

$$\lim_n \int_{E_n} f = \lim_n \int fI_{E_n} = \int \lim_n fI_{E_n} = \int fI_{\lim_n E_n} = \int_{\lim_n E_n} f = \int f,$$

where the last equality follows since $f = 0$ on $(\lim_n E_n)^c$. It follows that $\{\int_{E_n} f\}_{n=1}^{\infty}$ is a nondecreasing sequence of nonnegative numbers with the finite

[10] If $\mu(E_N) = +\infty$ for some $N \in \mathbb{N}$, then

$$\int_{E_N} f \geq \frac{1}{N}\int_{E_N} 1\, d\mu = \frac{1}{N}\mu(E_N) = +\infty,$$

and hence $\int f \geq \int_{E_N} f = +\infty$, contradicting the fact that $f \in L^1$.

limit $\int f$. We may accordingly find $N \in \mathbb{N}$ so that $0 \le \int f - \int_{E_N} f < \epsilon$, and hence E_N satisfies the requirements of the claim. This proves the claim when $f \in L^1$ is nonnegative.

We now let $f \in L^1$ be arbitrary, again picking $\epsilon > 0$. Write $f = f^+ - f^-$, so that both f^+ and f^- are nonnegative functions in L^1. Using what has just been shown, there is $E^+ \in \mathcal{F}$ with $\mu(E^+) < +\infty$ and $0 \le \int f^+ - \int_{E^+} f^+ < \frac{\epsilon}{2}$. Also, we may find $E^- \in \mathcal{F}$ with $\mu(E^-) < +\infty$ and $0 \le \int f^- - \int_{E^-} f^- < \frac{\epsilon}{2}$. Let $E = E^+ \cup E^-$. Then $\mu(E) < +\infty$, $E^c = E^{+c} \cap E^{-c}$, and

$$\begin{aligned}
\left| \int f - \int_E f \right| &= \left| \int_{E^c} f \right| \\
&= \left| \int_{E^c} f^+ - \int_{E^c} f^- \right| \\
&\le \int_{E^c} f^+ + \int_{E^c} f^- \\
&\le \int_{E^{+c}} f^+ + \int_{E^{-c}} f^- \\
&< \frac{\epsilon}{2} + \frac{\epsilon}{2} \\
&= \epsilon.
\end{aligned}$$

The set E as constructed satisfies the requirements of the claim, and the proof is now complete. ∎

We now state and prove the Vitali Convergence Theorem.

Vitali Convergence Theorem. *Let $(\Omega, \mathcal{F}, \mu)$ denote a measure space with $\mu(\Omega) < +\infty$, and let $1 \le p < +\infty$. Let $\{f_n\}_{n=1}^{\infty}$ denote a sequence of functions in L^p. Let f denote an $\mathcal{F}/\mathcal{B}^*$-measurable function on Ω such that $f_n \to f$ μ-a.e. on Ω and f is finite μ-a.e. on Ω. Then the condition*

(0) $f \in L^p$ and $f_n \xrightarrow{p} f$

is equivalent to the dual conditions

(i) for every $\epsilon > 0$ there is $A \in \mathcal{F}$ [depending on ϵ] with $\mu(A) < +\infty$ and $\int_{A^c} |f_n|^p < \epsilon$ for every $n \in \mathbb{N}$, and

(ii) for every $\epsilon > 0$ there is $\delta > 0$ such that whenever $\mu(E) < \delta$ we have $\int_E |f_n|^p < \epsilon$ for every $n \in \mathbb{N}$. [δ depends only on ϵ, not n.]

Proof: We first assume that $f \in L^p$ with $f_n \xrightarrow{p} f$, and we desire to show that both (i) and (ii) hold. We first attempt to demonstrate (i), picking $\epsilon > 0$. Because $f_n \xrightarrow{p} f$, there is $N \in \mathbb{N}$ such that $\int |f_n - f|^p < \frac{\epsilon}{2}$ for every $n \ge N$. Using Claim 3 and noting that $|f|^p \in L^1$, there is $B \in \mathcal{F}$ with $\mu(B) < +\infty$ and $\int_{B^c} |f|^p < \frac{\epsilon}{2}$. For $j = 1, \cdots, N-1$, we have $|f_j|^p \in L^1$, and we use Claim 3 to obtain $C_j \in \mathcal{F}$ with $\mu(C_j) < +\infty$ and $\int_{C_j^c} |f_j|^p < \frac{\epsilon}{2}$. Let $C = C_1 \cup \cdots \cup C_{N-1}$. Then $\mu(C) < +\infty$ and $\int_{C^c} |f_j|^p \le \int_{C_j^c} |f_j|^p < \frac{\epsilon}{2}$ for $j = 1, \cdots, N-1$. Define

$A = B \cup C$, so that $\mu(A) < +\infty$. If $n \geq N$, we have

$$\begin{aligned}\int_{A^c} |f_n|^p &\leq \int_{B^c \cap C^c} |f_n - f|^p + \int_{B^c \cap C^c} |f|^p \\ &\leq \int |f_n - f|^p + \int_{B^c} |f|^p \\ &< \frac{\epsilon}{2} + \frac{\epsilon}{2} \\ &= \epsilon.\end{aligned}$$

For $n = 1, \cdots, N-1$, we have

$$\int_{A^c} |f_n|^p = \int_{B^c \cap C^c} |f_n|^p \leq \int_{C^c} |f_n|^p < \frac{\epsilon}{2} < \epsilon.$$

We now combine the $n \geq N$ and $n \in \{1, \cdots, N-1\}$ cases to obtain the inequality $\int_{A^c} |f_n|^p < \epsilon$ for every $n \in \mathbb{N}$. This shows that (0) implies (i).

We now prove (ii), still assuming that $f \in L^p$ and $f_n \xrightarrow{p} f$. Let $\epsilon > 0$. Find $\delta_0 > 0$ such that $\mu(E) < \delta_0$ forces $(\int_E |f|^p)^{\frac{1}{p}} < \frac{\epsilon}{2}$. [This is integration property (L15) from Section 6.4.] Find $N \in \mathbb{N}$ such that $n \geq N$ forces $(\int |f_n - f|^p)^{\frac{1}{p}} < \frac{\epsilon}{2}$. Observe for every $E \in \mathcal{F}$ and $n \in \mathbb{N}$ that Minkowski's Inequality gives $||f_n I_E||_p \leq ||f_n I_E - f I_E||_p + ||f I_E||_p$, and hence

$$\left(\int_E |f_n|^p\right)^{\frac{1}{p}} \leq \left(\int_E |f_n - f|^p\right)^{\frac{1}{p}} + \left(\int_E |f|^p\right)^{\frac{1}{p}}.$$

Therefore, if $n \geq N$ and $\mu(E) < \delta_0$, we have

$$\begin{aligned}\left(\int_E |f_n|^p\right)^{\frac{1}{p}} &\leq \left(\int |f_n - f|^p\right)^{\frac{1}{p}} + \left(\int_E |f|^p\right)^{\frac{1}{p}} \\ &< \frac{\epsilon}{2} + \frac{\epsilon}{2} = \epsilon.\end{aligned}$$

For $j = 1, \cdots, N-1$, use (L15) to find $\delta_j > 0$ so that $\mu(E) < \delta_j$ forces $(\int_E |f_n|^p)^{\frac{1}{p}} < \epsilon$. Let $\delta = \min\{\delta_0, \delta_1, \cdots, \delta_{N-1}\} > 0$. Then $(\int_E |f_n|^p)^{\frac{1}{p}} < \epsilon$ if $\mu(E) < \delta$ and $n = 1, \cdots, N-1$. Therefore, $\mu(E) < \delta$ forces $(\int_E |f_n|^p)^{\frac{1}{p}} < \epsilon$ for every $n \in \mathbb{N}$. This shows that (0) implies (ii).

We have shown the following: if $f \in L^p$ and $f_n \xrightarrow{p} f$, then (i) and (ii) hold. We now assume that (i) and (ii) hold, and we desire to show that $f \in L^p$ and $f_n \xrightarrow{p} f$. Let $\epsilon > 0$. By (ii), there exists $\delta > 0$ such that $\mu(E) < \delta$ implies $\int_E |f_n|^p < (\frac{\epsilon}{3})^p$ for each $n \in \mathbb{N}$. Since $f_n \to f$ μ-a.e. on Ω and $\mu(\Omega) < +\infty$, we may invoke Egorov's Theorem [Claim 4 of Section 5.6] to find $B \in \mathcal{F}$ with $\mu(B) < \delta$ and such that $f_n \to f$ uniformly on B^c. Therefore, noting that $\mu(B^c) \leq \mu(\Omega) < +\infty$, we have for large n that

$$\left(\sup_{\omega \in B^c} |f_n(\omega) - f(\omega)|\right)^p \mu(B^c) < \left(\frac{\epsilon}{3}\right)^p.$$

By Fatou's Lemma and the assumption that $f_n \to f$ μ-a.e. on Ω, we have

$$\int_B |f|^p = \int_B \lim_n |f_n|^p = \int_B \underline{\lim}_n |f_n|^p \leq \underline{\lim}_n \int_B |f_n|^p \leq \left(\frac{\epsilon}{3}\right)^p.$$

Next, observe for sufficiently large n that

$$\begin{aligned} &\left(\int |f - f_n|^p\right)^{\frac{1}{p}} \\ \leq\ & \left(\int_B |f - f_n|^p\right)^{\frac{1}{p}} + \left(\int_{B^c} |f - f_n|^p\right)^{\frac{1}{p}} \\ \leq\ & \left(\int_B |f|^p\right)^{\frac{1}{p}} + \left(\int_B |f_n|^p\right)^{\frac{1}{p}} + \left(\int_{B^c} |f - f_n|^p\right)^{\frac{1}{p}} \\ <\ & \frac{\epsilon}{3} + \frac{\epsilon}{3} + \frac{\epsilon}{3} \\ =\ & \epsilon. \end{aligned}$$

[Write $\|f - f_n\|_p \leq \|(f - f_n)I_B\|_p + \|(f - f_n)I_{B^c}\|_p$ for the first inequality. The second inequality comes from writing $\|(f-f_n)I_B\|_p \leq \|fI_B\|_p + \|f_nI_B\|_p$. The third inequality comes from the fact that for large n we have

$$\begin{aligned} \int_{B^c} |f - f_n|^p \leq \int_{B^c} \left[\sup_{\omega \in B^c} |f(\omega) - f_n(\omega)|\right]^p d\mu(\eta) &= \sup_{\omega \in B^c} |f(\omega) - f_n(\omega)|^p \mu(B^c) \\ &< \left(\frac{\epsilon}{3}\right)^p.] \end{aligned}$$

Therefore, we have $\lim_n \int |f - f_n|^p \to 0$. The above shows for large n that $f - f_n \in L^p$. Since each f_n is in L^p, we have for large n that $f = (f - f_n) + f_n$ is the sum of two functions in L^p, and hence $f \in L^p$. We have shown that $f \in L^p$ and $\lim_n \|f_n - f\|_p = 0$, and the entire proof is complete. ■

In Exercise 14, the reader is asked to prove the Vitali Convergence Theorem when $\mu(\Omega) = +\infty$. Some applications of this result are given in Exercises 15 and 16.

With the Vitali Convergence Theorem in tow, the second part of this section's program is now complete. We may now carry out the third and final part of this section's program, defining and discussing a new mode of convergence.

Definition. Consider a generic measure space $(\Omega, \mathcal{F}, \mu)$, and let $1 \leq p < +\infty$. Let $f, f_1, f_2, \cdots \in L^p$. We will say that $f_n \to f$ *weakly [in L^p]* iff

$$\lim_n \int f_n g = \int fg \quad \text{for every } g \in L^{p'}.$$

Some comments pertaining to the above definition should be made. First, recall that $p' = \frac{p}{p-1}$ for $1 < p < +\infty$ and $1' = +\infty$ by the special conventions mentioned in Section 4, so that for any $1 \leq p < +\infty$ we have $\frac{1}{p} + \frac{1}{p'} = 1$. Second, it makes sense to consider the integrals $\int fg, \int f_1 g, \int f_2 g, \cdots$ since Hölder's Inequality implies that $fg, f_1 g, f_2 g, \cdots \in L^1$. Therefore, no problems arise with the definition. Since all quantities involved are finite, the conditions

(1) $\lim_n \int f_n g = \int fg$ for every $g \in L^{p'}$,

(2) $\lim_n |\int f_n g - \int fg| = 0$ for every $g \in L^{p'}$, and

(3) $\lim_n \int (f_n - f)g = 0$ for every $g \in L^{p'}$

are logically equivalent.

For the rest of this section, claims will assume the structure as given in the definition above, namely, that $f, f_1, f_2, \cdots \in L^p$, where $1 \leq p < +\infty$.

Our first result in this direction relates convergence in pth mean and weak convergence. Namely, convergence in pth mean implies weak convergence.

Claim 4. *If $f_n \xrightarrow{p} f$, then $f_n \to f$ weakly in L^p.*

Proof: Let $g \in L^{p'}$. By Hölder's Inequality we have

$$\left|\int f_n g - \int fg\right| = \left|\int (f_n - f)g\right| \leq \int |f_n - f||g| \leq \|f_n - f\|_p \|g\|_{p'} \to 0 \quad \text{as } n \to \infty,$$

where the convergence follows by the fact that $f_n \xrightarrow{p} f$. ∎

The converse of Claim 4 is not true. That is, while convergence in pth mean implies weak convergence in L^p, it is not the case that weak convergence in L^p implies convergence in pth mean. In Exercise 13, the reader will be faced with a challenging example showing that the converse to Claim 4 fails.

The next result relates μ-a.e. convergence to weak convergence. While it is not true that μ-a.e. convergence by itself forces weak convergence [see Exercise 11 for a straightforward example], it is the case that μ-a.e. convergence and an additional assumption will yield weak convergence.

Claim 5. *Let $1 < p < +\infty$ and suppose that $\{\|f_n\|_p\}_{n=1}^{\infty}$ is a bounded sequence. Assume that $f_n \to f$ μ-a.e. on Ω. Then $f_n \to f$ weakly in L^p.*

Proof: Since $\{\|f_n\|_p\}_{n=1}^{\infty}$ is bounded, there is finite $M > 0$ with $\|f_n\|_p \leq M$ for every $n \in \mathbb{N}$. Next, observe by Fatou's Lemma that

$$\|f\|_p^p = \int |f|^p = \int \lim |f_n|^p = \int \underline{\lim} |f_n|^p \leq \underline{\lim} \int |f_n|^p = \underline{\lim} \|f_n\|_p^p \leq M^p.$$

With this, we proceed to establish weak convergence. Let $\epsilon > 0$ and $g \in L^{p'}$. We will show that $|\int (f_n - f)g| \leq \epsilon$ for sufficiently large n. If $\|g\|_{p'} = 0$, then $g = 0$ μ-a.e. on Ω, and we have $|\int (f_n - f)g| = 0 < \epsilon$. We may therefore assume that $\|g\|_{p'} > 0$ in all that follows.

By (L15) applied to the function $|g|^{p'} \in L^1$, there is $\delta > 0$ with the following property:

$$\mu(E) < \delta \quad \text{forces} \quad \left(\int_E |g|^{p'}\right)^{\frac{1}{p'}} < \frac{\epsilon}{6M}. \tag{8.11}$$

By Claim 3 applied to the function $|g|^{p'} \in L^1$, we may find $A \in \mathcal{F}$ with the following property:

$$\mu(A) < +\infty \quad \text{and} \quad \left(\int_{A^c} |g|^{p'}\right)^{\frac{1}{p'}} < \frac{\epsilon}{6M}. \tag{8.12}$$

Using the set A above, we invoke Egorov's Theorem, obtaining a set $B \in \mathcal{F}$ with $B \subseteq A$, $\mu(A - B) < \delta$, and $f_n \to f$ uniformly on B. Using the uniform convergence, there is $N \in \mathbb{N}$ such that

$$|f_n(\omega) - f(\omega)|\mu(B)^{\frac{1}{p}}\|g\|_{p'} < \frac{\epsilon}{3} \quad \text{for every } \omega \in B \text{ and } n \geq N.$$

Integrating this inequality over B yields the following condition:

$$\omega \in B \text{ and } n \geq N \quad \text{imply} \quad \left(\int_B |f_n - f|^p\right)^{\frac{1}{p}} \|g\|_{p'} \leq \frac{\epsilon}{3}. \tag{8.13}$$

[If $\mu(B) = 0$ the last inequality is trivial; if $\mu(B) > 0$ then the inequality preceding (8.13) is equivalent to $\sup_{\omega \in B^c} |f_n(\omega) - f(\omega)| \leq \frac{\epsilon}{3\mu(B)^{\frac{1}{p}}\|g\|_{p'}}$.] Next, breaking Ω into the disjoint sets $A - B$, A^c, and B, we claim that

$$\begin{aligned}
&\left|\int f_n g - \int fg\right| \\
\leq\ & \int |f_n - f||g| \\
=\ & \int_{A-B} |f_n - f||g| + \int_{A^c} |f_n - f||g| + \int_B |f_n - f||g| \\
\leq\ & \|f_n - f\|_p \left(\int_{A-B} |g|^{p'}\right)^{\frac{1}{p'}} + \|f_n - f\|_p \left(\int_{A^c} |g|^{p'}\right)^{\frac{1}{p'}} + \left(\int_B |f_n - f|^p\right)^{\frac{1}{p}} \|g\|_{p'} \\
\leq\ & \frac{\epsilon}{3} + \frac{\epsilon}{3} + \frac{\epsilon}{3} \\
=\ & \epsilon
\end{aligned}$$

for every $n \geq N$. To see this, note that the first inequality and equality are trivial. For the second inequality, observe for the first of the three pieces that

Hölder's Inequality gives

$$\begin{aligned}\int_{A-B} |f_n - f||g| &\le \left(\int_{A-B} |f_n - f|^p\right)^{\frac{1}{p}} \left(\int_{A-B} |g|^{p'}\right)^{\frac{1}{p'}} \\ &\le \left(\int |f_n - f|^p\right)^{\frac{1}{p}} \left(\int_{A-B} |g|^{p'}\right)^{\frac{1}{p'}} \\ &= \|f_n - f\|_p \left(\int_{A-B} |g|^{p'}\right)^{\frac{1}{p'}} ;\end{aligned}$$

the same type of argument yields the other two pieces. We now turn to the third inequality. For the first piece there, we have $\mu(A - B) < \delta$, hence

$$\begin{aligned}\|f_n - f\|_p \left(\int_{A-B} |g|^{p'}\right)^{\frac{1}{p'}} \le \|f_n - f\|_p \frac{\epsilon}{6M} &\le (\|f_n\|_p + \|f\|_p)\frac{\epsilon}{6M} \\ &\le (M + M)\frac{\epsilon}{6M} \\ &= \frac{\epsilon}{3}\end{aligned}$$

by Minkowski's Inequality and (8.11). The second piece is handled by a similar appeal to Minkowski's Inequality and uses (8.12). The third piece is automatic by (8.13). Therefore, $n \ge N$ forces $|\int f_n g - \int fg| \le \epsilon$. This shows that $\lim_n \int f_n g = \int fg$ for every $g \in L^{p'}$ with $\|g\|_{p'} > 0$. Combining the $\|g\|_{p'} = 0$ and $\|g\|_{p'} > 0$ cases shows that $\lim_n \int f_n g = \int fg$ for every $g \in L^{p'}$, whence $f_n \to f$ weakly in L^p. ■

In the above claim, we assumed that $f_n \to f$ μ-a.e. on Ω [among other things]. The next claim says that this particular assumption may be replaced with the weaker assumption that $f_n \to f$ in measure. The main tool will be Claim 1 from Section 5.6.

Claim 6. *Let $1 < p < +\infty$ and suppose that $\{\|f_n\|_p\}_{n=1}^{\infty}$ is a bounded sequence. If $f_n \to f$ in measure, then $f_n \to f$ weakly in L^p.*

Proof: We argue by contradiction, assuming that (i) $f_n \to f$ in measure and (ii) $f_n \not\to f$ weakly in L^p. By the definition of weak convergence, there is g in $L^{p'}$ and $\alpha > 0$ such that $\overline{\lim}_n |\int (f_n - f)g| = \alpha$. By the definition of a superior limit, there exists a subsequence $\{f_{n_k}\}_{k=1}^{\infty}$ of $\{f_n\}_{n=1}^{\infty}$ with the property that $\lim_k |\int (f_{n_k} - f)g| = \alpha$. The subsequence $\{f_{n_k}\}_{k=1}^{\infty}$ converges to f in measure since the original [super]sequence $\{f_n\}_{n=1}^{\infty}$ converges to f in measure. In turn, there exists a subsequence $\{f_{n_{k_j}}\}_{j=1}^{\infty}$ of the subsequence $\{f_{n_k}\}_{k=1}^{\infty}$ such that $f_{n_{k_j}} \to f$ μ-a.e. on Ω as $j \to \infty$ [we use Claim 1 of Section 5.6]. Observe that the sequence $\{\|f_{n_{k_j}}\|_p\}_{j=1}^{\infty}$ is bounded since $\{\|f_n\|_p\}_{n=1}^{\infty}$ is bounded. By Claim 5 applied to $\{f_{n_{k_j}}\}_{j=1}^{\infty}$, we have that $f_{n_{k_j}} \to f$ weakly in L^p as $j \to \infty$, so $\lim_{j\to\infty} |\int(f_{n_{k_j}} - f)g| = 0$. But then

$$\alpha = \lim_{k\to\infty} \left|\int (f_{n_k} - f)g\right| = \lim_{j\to\infty} \left|\int (f_{n_{k_j}} - f)g\right| = 0 < \alpha,$$

a contradiction. We conclude that $f_n \to f$ weakly, completing the proof. ∎

The final claim of this section deals with the case $p = 1$ and relates μ-a.e. convergence to both convergence in L^1 and weak convergence in L^1.

Claim 7. *Let $f, f_1, f_2, \cdots \in L^1$ be such that $\|f_n\|_1 \to \|f\|_1$ and $f_n \to f$ μ-a.e. on Ω. Then*

(a) $\lim_n \int_E |f_n| = \int_E |f|$ for every $E \in \mathcal{F}$,

(b) $f_n \xrightarrow{1} f$ [that is, $f_n \to f$ in L^1], and

(c) $f_n \to f$ weakly in L^1.

Proof: Part (c) follows from part (b) by virtue of Claim 4, hence we need only demonstrate (a) and (b). We do (a) first, picking $E \in \mathcal{F}$. Using Fatou's Lemma multiple times, the hypothesis of μ-a.e. convergence on Ω, and elementary properties of superior limits, we obtain

$$\begin{aligned}
\underline{\lim} \int_E |f_n| \geq \int_E \underline{\lim}\, |f_n| = \int_E \lim |f_n| &= \int_E |f| \\
&= \int |f| - \int_{E^c} |f| \\
&= \int |f| - \int_{E^c} \underline{\lim}\, |f_n| \\
&\geq \int |f| - \underline{\lim} \int_{E^c} |f_n| \\
&= \int \overline{\lim}\, |f_n| + \overline{\lim}\left(-\int_{E^c} |f_n|\right) \\
&\geq \overline{\lim} \int |f_n| + \overline{\lim}\left(-\int_{E^c} |f_n|\right) \\
&\geq \overline{\lim}\left(\int |f_n| - \int_{E^c} |f_n|\right) \\
&= \overline{\lim} \int_E |f_n|.
\end{aligned}$$

[The second inequality above follows because Fatou's Lemma gives

$$\int \overline{\lim}\, |f_n| = -\int \underline{\lim}(-|f_n|) \geq -\underline{\lim} \int -|f_n| = \overline{\lim} \int |f_n|.]$$

The above chain gives

$$\overline{\lim} \int_E |f_n| \leq \int_E |f| \leq \underline{\lim} \int_E |f_n|,$$

and hence $\lim \int_E |f_n|$ exists and equals $\int_E |f|$. This gives (a).

We now complete the proof by showing (b). Let $\epsilon > 0$. Suppose that $\overline{\lim}_n \int |f - f_n| < \epsilon$. From this it will follow that $\lim \int |f - f_n| < \epsilon$, and hence $\|f - f_n\|_1 \to 0$. We therefore desire to prove that $\overline{\lim}_n \int |f - f_n| < \epsilon$. Using Claim 3 for $|f| \in L^1$, there is $A \in \mathcal{F}$ with

$$\mu(A) < +\infty \quad \text{and} \quad \int_{A^c} |f| < \frac{\epsilon}{5}.$$

Using integration property (L15) from Section 6.4, there is $\delta > 0$ such that

$$\mu(E) < \delta \;\text{ implies }\; \int_E |f| < \frac{\epsilon}{5}.$$

Next, we invoke Egorov's Theorem [Claim 4 of Section 5.6] to find $B \in \mathcal{F}$ with $B \subseteq A$, $\mu(A - B) < \delta$, and $f_n \to f$ uniformly on B. Therefore, there is $N \in \mathbb{N}$ such that $|f_n - f|\mu(B) < \frac{\epsilon}{5}$ on B for every $n \geq N$. Putting these things together and using (a), we obtain for $n \geq N$ that

$$\begin{aligned}
&\overline{\lim} \int |f - f_n| \\
= \;& \overline{\lim} \left(\int_{A^c} |f - f_n| + \int_{A-B} |f - f_n| + \int_B |f - f_n| \right) \\
\leq \;& \overline{\lim} \int_{A^c} |f - f_n| + \overline{\lim} \int_{A-B} |f - f_n| + \overline{\lim} \int_B |f - f_n| \\
\leq \;& \overline{\lim} \left(\int_{A^c} |f| + \int_{A^c} |f_n| \right) + \overline{\lim} \left(\int_{A-B} |f| + \int_{A-B} |f_n| \right) + \overline{\lim} \int_B |f - f_n| \\
= \;& \int_{A^c} |f| + \overline{\lim} \int_{A^c} |f_n| + \int_{A-B} |f| + \overline{\lim} \int_{A-B} |f_n| + \overline{\lim} \int_B |f - f_n| \\
\leq \;& 2 \int_{A^c} |f| + 2 \int_{A-B} |f| + \overline{\lim} \sup_{\omega \in B} |f(\omega) - f_n(\omega)| \mu(B) \\
< \;& \frac{2\epsilon}{5} + \frac{2\epsilon}{5} + \frac{\epsilon}{5} \\
= \;& \epsilon.
\end{aligned}$$

This shows that $\overline{\lim} \int |f - f_n| < \epsilon$ for our generic $\epsilon > 0$, and the entire proof is now complete. ■

Exercises.

1. Prove the following simple claims.
(a) Let $\{f_n\}_{n=1}^{\infty}$ denote a Cauchy sequence in L^p, where $0 < p < +\infty$. Show that $\{f_n\}_{n=1}^{\infty}$ is a Cauchy sequence in measure: for every $\epsilon > 0$ and $\delta > 0$ there is $N \in \mathbb{N}$ such that for every $n, m \geq N$ we have $\mu(\{\omega \in \Omega : |f_n(\omega) - f_m(\omega)| > \delta\}) < \epsilon$.
(b) Let $f, f_1, f_2, \cdots \subset L^p$ and suppose that $f_n \xrightarrow{p} f$, where $0 < p < +\infty$. If $g \in L^\infty$, then $fg, f_1g, f_2g, \cdots \in L^p$ and $f_ng \xrightarrow{p} fg$.

2. While convergence in pth mean implies convergence in measure, it is not the case that convergence in measure implies convergence in pth mean. [Consider $(\mathbb{R}, \mathcal{B}, \lambda)$. For each $n \in \mathbb{N}$ and $x \in \mathbb{R}$, let $f_n(x) = n^{\frac{1}{p}} I_{[0,\frac{1}{n}]}(x)$. Then $f_n \to 0$ in measure but not in pth mean.]

3. It is possible for a sequence $\{f_n\}_{n=1}^{\infty}$ in L^p to converge μ-a.e. to some $f \in L^p$ but not in pth mean. That is, convergence μ-a.e. does not force convergence in pth mean. [Consider $(\mathbb{R}, \mathcal{B}, \lambda)$, and let $f_n(\omega) = n^2 I_{(0,\frac{1}{n})}(\omega)$ for every $\omega \in \mathbb{R}$ and $n \in \mathbb{N}$. Then $f_n \to 0$ λ-a.e. on $\mathbb{R}$ but not in pth mean.]

4. It is possible for a sequence $\{f_n\}_{n=1}^{\infty}$ in L^p to convergence in pth mean to zero, but $\{f_n\}_{n=1}^{\infty}$ converges at no point of Ω. [Consider $([0,1], \mathcal{B}, \lambda)$, where here $\mathcal{B}$ denotes the Borel subsets of $[0,1]$ and λ denotes Lebesgue measure restricted to the Borel subsets of $[0,1]$. For every $k \geq 0$ and $0 \leq j \leq 2^k - 1$ write $f_{2^k+j} = kI_{[j2^{-k},(j+1)2^{-k}]}$. For every $\omega \in [0,1]$ we have $\overline{\lim}\, f_n(\omega) = +\infty \neq 0 = \underline{\lim}\, f_n(\omega)$. At the same time $\lim \|f_n\|_1 = 0$.]

5. It is possible to have functions $f, f_1, f_2, \cdots \in L^{p_1} \cap L^{p_2}$ such that $f_n \xrightarrow{p_1} f$ but $f_n \not\xrightarrow{p_2} f$. [Consider $((0,+\infty), \mathcal{B}, \lambda)$, where $\mathcal{B}$ denotes the Borel subsets of $(0,+\infty)$ and λ denotes Lebesgue measure restricted to $\mathcal{B}$ as just given. Let $f_n = n^{-1} I_{(n,2n)}$ for every $n \in \mathbb{N}$. Then $f_n \xrightarrow{p} 0$ for every $1 < p < +\infty$ but $\|f_n\|_1$ fails to converge to zero.]

6. Let $1 < p_0 < +\infty$. There is a sequence $\{f_n\}_{n=1}^{\infty}$ in L^p for all $1 \leq p < p_0$ with $f_n \xrightarrow{p} 0$ for all $1 \leq p < p_0$, but f_n fails to converge in p_0th mean. [Modify one of the hints of the exercises above.]

7. Let $1 \leq p, q < +\infty$, and let $\{f_n\}_{n=1}^{\infty}$ denote a sequence in $L^p \cap L^q$. Suppose that there is $g \in L^p$ and $h \in L^q$ with $f_n \xrightarrow{p} g$ and $f_n \xrightarrow{q} h$. Then $g = h$ μ-a.e. on Ω.

8. Suppose that $\{f_n\}_{n=1}^{\infty}$ is a sequence in L^p, where $1 < p < +\infty$. Suppose that $h \in L^1$ with $f_n \leq h$ μ-a.e. on Ω. If f_n converges in measure to some f, then $f \in L^p$ and $f_n \xrightarrow{p} f$.

9. If $f, f_1, f_2, \cdots \in L^p$ are such that $f_n \to f$ uniformly, then $f_n \xrightarrow{p} f$. [Here $0 < p < +\infty$.]

10. Let $f, f_1, f_2, \cdots \in L^p$ [where $1 \leq p < +\infty$], and suppose that $\|f_n\|_p \to \|f\|_p$. Then either of the conditions $f_n \to f$ μ-a.e. on Ω and $f_n \to f$ in measure is sufficient for the conclusion $f_n \xrightarrow{p} f$.

11. Consider the measure space $([0,1], \mathcal{B}, \lambda)$, where here $\mathcal{B}$ denotes the Borel subsets of $[0,1]$ and λ denotes Lebesgue measure restricted to $\mathcal{B}$. Let $f_n = nI_{[0,n^{-1}]}$ for each $n \in \mathbb{N}$. Then each f_n is in L^p for $1 \leq p < +\infty$, $f_n \to 0$ μ-a.e. on $[0,1]$ and in measure, but $\int f_n g \not\to \int 0g$ when $g = 1$, and hence $f_n \not\to 0$ weakly.

12. Consider the measure space $(\mathbb{R}, \mathcal{B}, \lambda)$. Let $f_n = n^{-1} I_{[1,e^n]}$ for every $n \in \mathbb{N}$. Then each f_n is in L^p for $1 \leq p < +\infty$, $f_n \to 0$ uniformly, but $f_n \not\to 0$ weakly in L^p. [Consider the function $g(\omega) = \omega^{-1} I_{[1,+\infty)}(\omega)$. We have that $g \in L^{p'}$ for each $1 < p < +\infty$, g is bounded, and $\int f_n g = 1$ for every $n \in \mathbb{N}$, while $\int 0g = 0$.]

13. Consider the measure space $([0,2\pi], \mathcal{B}, \lambda)$, where here $\mathcal{B}$ denotes the Borel subsets of $[0,2\pi]$ and λ denotes Lebesgue measure restricted to $\mathcal{B}$. For each $n \in \mathbb{N}$ and $0 \leq \omega \leq 2\pi$, define $f_n(\omega) = \cos n\omega$. Then each f_n is in L^1, and $f_n \to 0$ weakly in L^1. However, $f_n \not\xrightarrow{1} 0$. In fact, we have $f_n \not\to 0$ in measure and $f_n \not\to 0$ μ-a.e. on Ω. [This exercise might be difficult.]

14. Refer to the Vitali Convergence Theorem. Prove that conditions (i) and (ii) imply that $f \in L^p$ and $f_n \to f$ in pth mean when $\mu(\Omega) = +\infty$.

15. Let $(\Omega, \mathcal{F}, \mu)$ denote a measure space with $1 \le p < +\infty$. Let $\{f_n\}_{n=1}^{\infty}$ denote a sequence in L^p, and let f denote an $\mathcal{F}/\mathcal{B}^*$-measurable function that is finite μ-a.e. on Ω with $f_n \to f$ μ-a.e. on Ω. For every $n, k \in \mathbb{N}$, let $B_{n,k} = \{\omega \in \Omega : |f_n(\omega)|^p \ge k\}$. Make the following additional assumption:

(a) For every $\epsilon > 0$ there is a set $A \in \mathcal{F}$ [depending on ϵ] such that $\mu(A) < +\infty$ and $\int_{A^c} |f_n|^p < \epsilon$ for every $n \in \mathbb{N}$.

Then the following four statements are logically equivalent:

(i) $f \in L^p$ and $f_n \xrightarrow{p} f$

(ii) If $\{E_k\}_{k=1}^{\infty}$ is a nonincreasing sequence of $\mathcal{F}$-sets with limit $\varnothing$, then for every $\epsilon > 0$ there is $K \in \mathbb{N}$ such that $k \ge K$ forces $\int_{E_k} |f_n|^p < \epsilon$ for every $n \in \mathbb{N}$.

(iii) For every $\epsilon > 0$ there is $K \in \mathbb{N}$ such that $k \ge K$ forces $\int_{B_{n,k}} |f_n|^p < \epsilon$ for every $n \in \mathbb{N}$.

(iv) For every $\epsilon > 0$ there is $\delta > 0$ such that $\mu(E) < \delta$ forces $\int_E |f_n|^p < \epsilon$ for every $n \in \mathbb{N}$.

[The following outline is useful. By the Vitali Convergence Theorem and Exercise 14, it is immediate that (i) and (iv) are logically equivalent. We show that (i) implies (ii). Let $\epsilon > 0$. Find $N \in \mathbb{N}$ such that $n \ge N$ implies $\|f_n - f\|_p < \epsilon$. For $n \ge N$ and $k \in \mathbb{N}$, Minkowski's Inequality gives

$$\begin{aligned}\left(\int_{E_k} |f_n|^p\right)^{\frac{1}{p}} &\le \left(\int_{E_k} |f|^p\right)^{\frac{1}{p}} + \left(\int_{E_k} |f_n - f|^p\right)^{\frac{1}{p}} \\ &< \left(\int_{E_k} |f|^p\right)^{\frac{1}{p}} + \epsilon.\end{aligned}$$

Apply DCT to $\{|f|^p I_{E_k}\}_{k=1}^{\infty}$ to show that $(\int_{E_k} |f_n|^p)^{\frac{1}{p}} < \epsilon$ for large enough k, say $k \ge K$. Then for every $n \in \mathbb{N}$ and $k \ge K$, we have $(\int_{E_k} |f_n|^p)^{\frac{1}{p}} < 2\epsilon$. For $n \in \{1, \cdots, N-1\}$, observe for every $k \in \mathbb{N}$ that we have

$$\int_{E_k} |f_n|^p \le \int_{E_k} \max\{|f_1|^p, \cdots, |f_{N-1}|^p\} < +\infty,$$

and therefore DCT can be used to obtain (ii).

Now assume (ii), picking $\epsilon > 0$. For each $k \in \mathbb{N}$, define $E_k = \bigcup_{n=k}^{\infty} B_{n,k}$. Then $E_1 \supseteq E_2 \supseteq \cdots$ and $\overline{\lim}_n |f_n(\omega)| = +\infty$ on $\bigcap_{k=1}^{\infty} E_k$; from this we have $\mu(\bigcap_{k=1}^{\infty} E_k) = 0$. Next, write $F_k = E_k - \bigcap_{j=1}^{\infty} E_j$ for every $k \in \mathbb{N}$. We have that $F_1 \supseteq F_2 \supseteq \cdots$ and $\lim_k F_k = \varnothing$. Choose K such that $k \ge K$ implies $\int_{F_k} |f_n|^p < \epsilon$ for every $n \in \mathbb{N}$. For every $n \ge K$ we have $B_{n,K} \subseteq E_K$. Therefore, if $n, k \ge K$, then

$$\int_{B_{n,k}} |f_n|^p \le \int_{B_{n,K}} |f_n|^p \le \int_{E_K} |f_n|^p = \int_{F_K} |f_n|^p < \epsilon.$$

For $n \in \{1, \cdots, K-1\}$, we have $|f_n|^p \le \max\{|f_1|^p, \cdots, |f_{K-1}|^p\} = g$, say, and hence $\int_{B_{n,k}} |f_n|^p \le \int_{B_k} g < +\infty$, where $B_k = \bigcup_{n=1}^{K-1} B_{n,k} = \{\omega \in \Omega : g(\omega) \ge k\}$. Use DCT to derive (iii) in direct fashion.

Now assume (iii). Pick $K \in \mathbb{N}$ such that $k \geq K$ forces $\int_{B_{n,k}} |f_n|^p < \epsilon^p$ for every $n \in \mathbb{N}$. Argue that $\mu(E) < K^{-1}\epsilon^p$ implies

$$\left(\int_E |f_n|^p\right)^{\frac{1}{p}} \leq \left(\int_{E\cap B_{n,K}} |f_n|^p\right)^{\frac{1}{p}} + \left(\int_{E\cap B_{n,K}^c} |f_n|^p\right)^{\frac{1}{p}} < 2\epsilon.$$

Therefore we may deduce (iv) from (iii).]

16. Assume everything in 15, including (a). Assume further that (ii) holds. Argue that (i) and (ii) of the Vitali Convergence Theorem follow as consequences.

Comment: The particular form of the Vitali Convergence Theorem as presented in this text and the previous two exercises [and the outlines of the solution] owe their existence and form to the most excellent presentation of Hewitt and Stromberg, pp. 203-205.

8.8 PRELUDE TO THE RIESZ REPRESENTATION THEOREM

This section serves as a repository of certain preliminary results used for next section's presentation of the famous Riesz Representation Theorem. In particular, this section has a twofold program: (i) a review [or presentation] of linear transformations and some subsidiary concepts, and (ii) the Clarkson Inequalities. The results in this section are used only in the next section, and may be skipped if the next section is being skipped. The presentation in this section as well as the next is explicitly based on that given by Hewitt and Stromberg, where the beauty of their presentation is that (a) it is quite elementary, using elementary calculus considerations in many parts, and (b) it provides a concrete application of the Riesz-Fischer Theorem proven in Section 8.2.

A Brief Discussion of Linear Transformations

Some of the material in this part may be a review of concepts learned before, while some of the material may be new. Either way, there really isn't too much difficulty.

Definition. Let A denote an additive abelian group over a field[11] F. We will say that A *is a linear [vector] space over a field* F iff the following properties hold for every $\alpha, \beta \in F$ and $x, y \in A$:
(1) $\alpha(x + y) = \alpha x + \alpha y$,
(2) $(\alpha + \beta)x = \alpha x + \beta x$,
(3) $\alpha(\beta x) = (\alpha\beta)x$, and
(4) $1x = x$, where 1 denotes the multiplicative identity of F.

[11] Here the word *field* means the structure encountered in algebra, not the definition given in Chapter 1.

If the definition seems rather abstract, consider the following concrete example: *L^p is a linear space over $\mathbb{R}$.* To see this, we clearly have that L^p is an abelian group under function addition, and for $\alpha, \beta \in \mathbb{R}$ and $f, g \in L^p$, we clearly have that
(1) $\alpha(f + g) = \alpha f + \alpha g$,
(2) $(\alpha + \beta)f = \alpha f + \beta f$,
(3) $\alpha(\beta f) = (\alpha\beta)f$, and
(4) $1f = f$ [here 1 denotes the real number 1].
This will be the framework in which the abstract definition will be used.

Definition. Let A and B denote linear spaces over a common field F. A mapping $T : A \to B$ is called a *linear transformation* or *linear operator* iff

$$T(x + y) = T(x) + T(y) \text{ and } T(\alpha x) = \alpha T(x) \quad \text{for every } x, y \in A \text{ and } \alpha \in F.$$

Again, the definition is stated in abstraction. In the next section, we will take $A = L^p(\Omega, \mathcal{F}, \mu)$ [where $1 < p < +\infty$], $F = \mathbb{R}$, and $B = \mathbb{R}$. [We are viewing $B = \mathbb{R}$ as a linear space over itself.] Pick and fix a specific $h \in L^{p'}$. Define $T : L^p \to \mathbb{R}$ by writing $T(f) = \int fh$ for every $f \in L^p$. [That $T(f)$ exists as a real number for every $f \in L^p$ follows immediately from Hölder's Inequality.] Since for every $f, g \in L^p$ and $c \in \mathbb{R}$ we have

$$\int (f + g)h = \int fh + \int gh \quad \text{and} \quad \int (cf)h = c \int fh,$$

we have $T(f + g) = T(f) + T(g)$ and $T(cf) = cT(f)$ for every $f, g \in L^p$ and $c \in \mathbb{R}$, so that T is a linear operator. Since the domain of T is a space of functions, T is often [equivalently] referred to as a *linear functional.*

We note in the abstract definition that $T(0) = T(0 + 0) = T(0) + T(0)$, so that $T(0) = 0$. [The left 0 is the additive identity element of A, and the right 0 is the additive identity element of B.] In the specific example just given, we have that T applied to the zero function gives the number 0: $\int 0h = 0$.

At this point, we must necessarily retreat into formalism. Recall that L^p is actually a collection of equivalence classes of functions, not a collection of functions. Therefore, [using the notation given in the introduction to this chapter], T should in reality be applied to certain equivalence classes $\mathfrak{A}$, not to certain functions f. In the specific example that we have considered, we should in reality write something like the following definition:

$$T(\mathfrak{A}) = \int fh, \quad \text{where } f \text{ is an element of } \mathfrak{A} \text{ and } \mathfrak{A} \in L^p.$$

That the definition is unambiguous follows from the fact that $\int fh = \int gh$ whenever $f = g$ μ-a.e. on Ω, i.e. $f, g \in \mathfrak{A}$.

We claim that T as formally stated is linear. This is quite easy to prove. Pick $\mathfrak{A}, \mathfrak{B} \in L^p$, $h_0 \in \mathfrak{A} + \mathfrak{B}$, $f \in \mathfrak{A}$, and $g \in \mathfrak{B}$. Then $h_0 = f + g$ μ-a.e. on

Ω, and the definition of T in this formal framework gives

$$T(\mathfrak{A}+\mathfrak{B}) = \int h_0 h = \int (f+g)h = \int fh + \int gh = T(\mathfrak{A}) + T(\mathfrak{B}).$$

Now reset everything, picking $c \in \mathbb{R}$, $\mathfrak{A} \in L^p$, $f \in \mathfrak{A}$, and $g \in c\mathfrak{A}$. Then $g = cf$ μ-a.e. on Ω, and hence

$$T(c\mathfrak{A}) = \int gh = \int cfh = c\int fh = cT(\mathfrak{A}).$$

This shows that T as formally stated is linear.

We will be working in the usual informal sense, viewing L^p as a collection of functions [instead of as a collection of equivalence classes] and writing $T(f)$ [instead of $T(\mathfrak{A})$, where $f \in \mathfrak{A}$]. However, we keep the just-given formal discussion in our minds, and we have the following important consequence: *T does not distinguish between functions equal μ-a.e. on Ω.* In terms of our workaday way of stating things, this becomes the following important assertion: *$T(f) = T(g)$ whenever $f = g$ μ-a.e. on Ω.*

We now introduce our third definition for this part of the section.

Definition. Let A and B denote normed linear spaces, with $\|\ \|_A$ and $\|\ \|_B$ denoting the norms on A and B, respectively, and let $T : A \to B$ denote a linear transformation. We will say that T is *bounded* iff there is $0 \leq M < +\infty$ such that $\|T(x)\|_B \leq M\|x\|_A$ for every $x \in A$.

Although $\|\ \|_A$ and $\|\ \|_B$ are formally different entities, we really don't need to subscript the norm symbol $\|\ \|$ by either of A or B, as it should be clear which norm is meant. We therefore omit the subscript. In our specific example under discussion, (i) $A = L^p$, hence $\|\ \|_A$ denotes the L^p norm $\|\ \|_p$, (ii) $B = \mathbb{R}$, so $\|\ \|_B$ denotes absolute value, and (iii) T is bounded iff there is $0 \leq M < +\infty$ with $|\int fh| \leq M\|f\|_p$ for every $f \in L^p$.

For the claims that follow, A and B will denote normed linear spaces, and $T : A \to B$ will denote a linear transformation.

Claim 1. *T is bounded iff T is bounded on the unit sphere of A. That is, T is bounded iff there is $0 \leq M < +\infty$ such that $\|T(x)\| \leq M$ for every $x \in A$ with $\|x\| = 1$.*

Proof: It is automatic that boundedness in general implies boundedness on the unit sphere. For the converse, assume that T is bounded on the unit sphere of A, so that there is $0 \leq M < +\infty$ with $\|T(x)\| \leq M$ for every $x \in A$ with $\|x\| = 1$. Pick $x \in A$. If $\|x\| = 0$, then $x = 0$ [so $T(x) = 0$], hence we trivially have $\|T(x)\| \leq M\|x\|$. If $\|x\| > 0$, then, writing $y = \frac{x}{\|x\|}$, we have $\|y\| = \|\frac{x}{\|x\|}\| = \frac{1}{\|x\|}\|x\| = 1$ and hence $\|T(y)\| \leq M$. Using linearity, we have

$$\|T(y)\| \leq M \text{ iff } \left\|T\left(\frac{x}{\|x\|}\right)\right\| \leq M \text{ iff } \left\|\frac{1}{\|x\|}T(x)\right\| \leq M \text{ iff } \frac{1}{\|x\|}\|T(x)\| \leq M,$$

and hence we have $\|T(x)\| \leq M\|x\|$. Combining the $\|x\| = 0$ and $\|x\| > 0$ cases shows that $\|T(x)\| \leq M\|x\|$ for every $x \in A$, so T is bounded. ∎

The above proof also shows that the set of numbers $M \in [0, +\infty)$ that serve as bounds for T on all of A coincides with the set of numbers $M \in [0, +\infty)$ that serve as bounds for T on the unit sphere of A.

In the specific framework we've been considering, Claim 1 says the following: if there is $0 \leq M < +\infty$ such that $|\int fh| \leq M$ for every $f \in L^p$ with $\|f\|_p = 1$, then $|\int fh| \leq M\|f\|_p$ for every $f \in L^p$ and conversely.

Definition. Let $T : A \to B$ denote a bounded linear transformation. We define the *[operator] norm of* T, denoted by $\|T\|$, by writing

$$\|T\| = \inf\{0 \leq M < +\infty : \|T(x)\| \leq M\|x\| \text{ for every } x \in A\}.$$

By the comments following Claim 1, it is easy to see that

$$\begin{aligned} \|T\| &= \inf\{0 \leq M < +\infty : \|T(x)\| \leq M \text{ for every } x \in A \text{ with } \|x\| = 1\} \\ &= \sup\{\|T(x)\| : x \in A, \|x\| = 1\}. \end{aligned}$$

We now have three uses for the norm notation $\|\ \|$: the norm on A, the norm on B, and the norm of a bounded linear transformation. Again, we may use the same symbol $\|\ \|$ for all three norms, and no confusion will be generated if the context is clear.

In our specific example under consideration, where $T(f) = \int fh$ for every $f \in L^p$ [and $h \in L^{p'}$ is fixed], we have

$$\|T\| = \sup\left\{\left|\int fh\right| : f \in L^p, \|f\|_p = 1\right\}.$$

One order of business in the next section is to show that $\|T\| = \|h\|_{p'}$.

The next claim characterizes the meaning of the norm of T.

Claim 2. *Let $T : A \to B$ denote a bounded linear transformation. Then*

$$\begin{aligned} \|T\| &= \sup\left\{\frac{\|T(x)\|}{\|x\|} : x \in A, x \neq 0\right\} \\ &= \sup\{\|T(x)\| : x \in A, \|x\| = 1\} \\ &= \sup\{\|T(x)\| : x \in A, \|x\| \leq 1\}. \end{aligned}$$

Moreover, we have $\|T(x)\| \leq \|T\|\|x\|$ for every $x \in A$.

Proof: Let C, D, and E denote the sets above, respectively. We wish to show the equation $\|T\| = \sup C = \sup D = \sup E$. On a moment's reflection, we see that $C = D \subseteq E$, and hence $\|T\| = \sup C = \sup D \leq \sup E$. Next, using the equivalent definition of $\|T\|$ given after the definition, we have the inequality $\|T\| \geq \|T(\frac{x}{\|x\|})\| = \frac{\|T(x)\|}{\|x\|} \geq \|T(x)\|$ for every $x \in A$ such that $0 < \|x\| \leq 1$. [We trivially have $\|T(x)\| \leq \|T\|$ when $\|x\| = 0$.] It follows that $\|T\| \geq \sup E$, whence $\|T\| = \sup C = \sup D = \sup E$. Finally, observe for each $x \neq 0$ that $\frac{\|T(x)\|}{\|x\|} \leq \|T\|$ since $\|T\| = \sup C$, and hence we have

$||T(x)|| \leq ||T|| ||x||$ for every nonzero $x \in A$. Since $||T(x)|| \leq ||T|| ||x||$ if $x = 0$, we have $||T(x)|| \leq ||T|| ||x||$ for every $x \in A$. This completes the proof. ∎

Definition. Let A and B denote normed linear spaces and let $T : A \to B$ be a linear transformation. We will say that T *is continuous at* $x \in A$ iff for every $\epsilon > 0$ there is $\delta > 0$ [depending on x and ϵ] such that $||x-y|| < \delta$ implies $||T(x) - T(y)|| < \epsilon$. We will say that T *is continuous on* A iff T is continuous at every $x \in A$. Finally, we will say that T is *uniformly continuous on* A iff for every $\epsilon > 0$ there is $\delta > 0$ [depending *only* on ϵ] such that $||x - y|| < \delta$ implies $||T(x) - T(y)|| < \epsilon$.

The definition of continuity as given is consistent with the usual notions of continuity: if x and y are "close" relative to the metric induced by the norm $|| \, ||_A$, then $T(x)$ and $T(y)$ will be "close" relative to the metric induced by the norm $|| \, ||_B$. Another way of viewing continuity is this: if T is continuous at some $x \in A$, then for every sequence $\{x_n\}_{n=1}^{\infty}$ of elements of A that converges to x in the metric induced by the norm $|| \, ||_A$ [that is, $\lim_n ||x_n - x||_A = 0$], we have $\lim_n T(x_n) = T(\lim_n x_n) = T(x)$. [The left limit is of course the limit relative to distance in B; the right limit is the limit relative to distance in A.] In other words, if $||x_n - x||_A \to 0$, then $||T(x_n) - T(x)||_B \to 0$.

In the specific example we are considering, we have that $h \in L^{p'}$ is fixed and $T : L^p \to \mathbb{R}$ is given for all $f \in L^p$ by the rule $T(f) = \int fh$. Saying that T is continuous at the function $f \in L^p$ means the following: if $\{f_n\}_{n=1}^{\infty}$ is a sequence in L^p with $\lim_n ||f_n - f||_p = 0$, then $\lim_n |\int f_n h - \int fh| = 0$.

The next claim relates boundedness with continuity.

Claim 3. *The following are equivalent:*
(a) T is bounded.
(b) T is uniformly continuous on A.
(c) T is continuous at some $x \in A$.

Proof: If $T(x) = 0$ for every $x \in A$, then (a)-(c) hold. We may therefore assume that T is nonzero, that is, $T(x) \neq 0$ for some $x \in A$. Suppose that (a) holds, so that T is bounded and hence $0 < ||T|| < +\infty$. We show (b), picking $\epsilon > 0$. By Claim 2, we have $||T(x)|| \leq ||T|| ||x||$ for every $x \in A$. Let $\delta = \frac{\epsilon}{||T||} > 0$, so that δ depends only on ϵ. Therefore, if $x, y \in A$ and $||x - y|| < \delta$, we have

$$||T(x) - T(y)|| = ||T(x - y)|| \leq ||T|| ||x - y|| < ||T|| \frac{\epsilon}{||T||} = \epsilon,$$

showing that T is uniformly continuous on A. It follows that (a) implies (b). It is trivial that (b) implies (c), and all that remains is to show that (c) implies (a).

Suppose that T is continuous at some $x_0 \in A$. Then there is $\delta > 0$ such that $||T(x) - T(x_0)|| \leq 1$ whenever $x \in A$ and $||x - x_0|| < \delta$. If $x \in A$ and $||x|| \leq 1$, then $||(\delta x + x_0) - x_0|| = ||\delta x|| = \delta ||x|| \leq \delta$, hence for every $x \in A$

with $||x|| \leq 1$ we have

$$1 \geq ||T(\delta x + x_0) - T(x_0)|| = ||\delta T(x) + T(x_0) - T(x_0)|| = ||\delta T(x)|| = \delta ||T(x)||,$$

so that $||T(x)|| \leq \frac{1}{\delta} < +\infty$ for every $x \in A$ with $||x|| \leq 1$. But then we have $\sup\{||T(x)|| : x \in A, ||x|| \leq 1\} \leq \frac{1}{\delta} < +\infty$, so that T is bounded. This gives (a) and completes the proof. ■

The next claim may seem long, but every part of it is straightforward. The main idea is that we have been calling $||T||$ "the [operator] norm of T" and have used norm notation, but we have not yet shown that we are actually dealing with a bona fide norm. The next proof justifies our usage of the word *norm* and the norm notation. That is, we show that the operator norm is a norm. A preliminary definition follows.

Definition. Let A denote a normed linear space over $\mathbb{R}$, and view $\mathbb{R}$ as a one-dimensional [normed] linear space over itself. We define

$$\mathcal{B}(A, \mathbb{R}) = \{T : A \to \mathbb{R} : T \text{ is a bounded linear transformation}\}.$$

Given elements T and U in $\mathcal{B}(A, \mathbb{R})$ and $\alpha \in \mathbb{R}$, we define $T + U$ and αT by the rules

$$(T + U)(x) = T(x) + U(x) \text{ and } (\alpha T)(x) = \alpha T(x) \quad \text{for every } x \in A.$$

Before doing anything, observe that if T and U are bounded linear transformations from A into $\mathbb{R}$, then $T+U$ and αT [where $\alpha \in \mathbb{R}$] are also bounded linear transformations from A into $\mathbb{R}$. That is, $T + U$ and αT are in $\mathcal{B}(A, \mathbb{R})$ whenever $T, U \in \mathcal{B}(A, \mathbb{R})$ and $\alpha \in \mathbb{R}$. Therefore, $\mathcal{B}(A, \mathbb{R})$ is closed under the above defintion of addition and scalar multiplication. To see this, note that if M_T and M_U are bounds for T and U, respectively, then $||T(x)|| \leq M_T||x||$ and $||U(x)|| \leq M_U||x||$ for every $x \in A$ by Claim 2. It follows that

$$||(T + U)(x)|| = ||T(x) + U(x)|| \leq ||T(x)|| + ||U(x)|| \leq (M_T + M_U)||x||$$

for every $x \in A$, so that $M_T + M_U$ is a bound for $T + U$, and hence $T + U$ is bounded, giving $T + U \in \mathcal{B}(A, \mathbb{R})$. Also, since

$$||(\alpha T)(x)|| = ||\alpha T(x)|| = |\alpha|||T(x)|| \leq |\alpha|M_T||x||$$

for every $x \in A$, we have that $|\alpha|M_T$ is a bound for αT, and hence αT is an element of $\mathcal{B}(A, \mathbb{R})$.

Claim 4. *$\mathcal{B}(A, \mathbb{R})$ is a complete normed linear space over $\mathbb{R}$ relative to the operator norm. That is, $\mathcal{B}(A, \mathbb{R})$ is a Banach space.*

Before proceeding to the proof, we will use the symbol $||\ ||$ in three ways. Given $x \in A$, the symbol $||x||$ denotes $||x||_A$. The norm on $\mathbb{R}$ is the absolute value, so $||T(x)||$ is really $|T(x)|$, though we shall write $||T(x)||$ instead. We shall also use $|| \;||$ in the context of operator norms. No confusion should arise, as the reader can easily distinguish between points of A, elements of $\mathbb{R}$, and linear transformations.

Proof: We break the proof up in various pieces.

(1) *$\mathcal{B}(A, \mathbb{R})$ is a linear space over $\mathbb{R}$.*

Proof: It is clear that $\mathcal{B}(A, \mathbb{R})$ is an additive abelian group under the definition of addition given above. Next, let $\alpha, \beta \in \mathbb{R}$ and let $T, U \in \mathcal{B}(A, \mathbb{R})$. According to the definition, there are four properties whose demonstration is required: (i) $\alpha(T + U) = \alpha T + \alpha U$, (ii) $(\alpha + \beta)T = \alpha T + \beta T$, (iii) $\alpha(\beta T) = (\alpha\beta)T$, and (iv) $1T = T$, where 1 is the multiplicative identity of $\mathbb{R}$, that is, the usual number 1. To see (i), we have for every $x \in A$ that

$$\begin{aligned}(\alpha(T+U))(x) = \alpha(T+U)(x) = \alpha(T(x)+U(x)) &= (\alpha T)(x) + (\alpha U)(x) \\ &= (\alpha T + \alpha U)(x).\end{aligned}$$

The proofs of (ii) and (iii) are similar to the proof of (i). Part (iv) is utterly trivial. This completes the proof of (1).

(2) *The operator norm is truly a norm on $\mathcal{B}(A, \mathbb{R})$.*

Proof: Let $\alpha \in \mathbb{R}$ and $T, S \in \mathcal{B}(A, \mathbb{R})$. We must verify that (i) $||0|| = 0$, where the left 0 is the element of $\mathcal{B}(A, \mathbb{R})$ that assigns each $x \in A$ the value zero, (ii) $||T|| > 0$ if $T \in \mathcal{B}(A, \mathbb{R})$ is such that $T \neq 0$, (iii) $||\alpha T|| = |\alpha|||T||$, and (iv) $||T + S|| \leq ||T|| + ||S||$. Part (i) is trivial. For (ii), suppose that T is such that $T(x) \neq 0$ for at least one $x \in A$. Then $||T(x)|| > 0$, $x \neq 0$, and $||x|| > 0$. Let $y = \frac{x}{||x||}$, so that $y \in A$, $||y|| = 1$, and $||T(y)|| > 0$. Since

$$||T|| = \sup\{||T(z)|| : \; z \in A, \; ||z|| = 1\} \geq ||T(y)|| > 0,$$

we have $||T|| > 0$, giving (ii). For (iii), we have that

$$\begin{aligned}||\alpha T|| &= \sup\left\{\frac{||\alpha T(x)||}{||x||} : \; x \in A, \; x \neq 0\right\} \\ &= \sup\left\{|\alpha|\frac{||T(x)||}{||x||} : \; x \in A, \; x \neq 0\right\} \\ &= |\alpha| \sup\left\{\frac{||T(x)||}{||x||} : \; x \in A, \; x \neq 0\right\} \\ &= |\alpha|||T||.\end{aligned}$$

For (iv), we have that

$$\begin{aligned}&||T + S|| \\ =\;& \sup\{||(T+S)(x)|| : \; x \in A, \; ||x|| = 1\} \\ =\;& \sup\{||T(x) + S(x)|| : \; x \in A, \; ||x|| = 1\} \\ \leq\;& \sup\{||T(x)|| + ||S(x)|| : \; x \in A, \; ||x|| = 1\} \\ \leq\;& \sup\{||T(x)|| : \; x \in A, \; ||x|| = 1\} + \sup\{||S(x)|| : \; x \in A, \; ||x|| = 1\} \\ \leq\;& ||T|| + ||S||.\end{aligned}$$

This shows that the operator norm is truly a norm on $\mathcal{B}(A, \mathbb{R})$, and hence the proof of (2) is complete.

Since the operator norm is truly a norm on $\mathcal{B}(A, \mathbb{R})$, we may now consider the metric induced by the operator norm. Given $T, U \in \mathcal{B}(A, \mathbb{R})$, the distance between T and U is defined as $||T - U||$.

(3) *$\mathcal{B}(A, \mathbb{R})$ is complete in the metric induced by the operator norm.*

Proof: Let $\{T_n\}_{n=1}^{\infty}$ denote a Cauchy sequence of elements of $\mathcal{B}(A, \mathbb{R})$. That is, for every $\epsilon > 0$ there is $N \in \mathbb{N}$ such that $n, m \geq N$ implies $||T_n - T_m|| < \epsilon$. The goal is to show that there is $T \in \mathcal{B}(A, \mathbb{R})$ such that $\lim_n ||T_n - T|| = 0$. We now make some subclaims.

(i) *$\{T_n(x)\}_{n=1}^{\infty}$ is a Cauchy sequence in $\mathbb{R}$ for every $x \in A$.* If $x = 0$, then the subclaim is trivial as $T_n(x) = T_n(0) = 0$ for every $n \in \mathbb{N}$. If $x \neq 0$, then $||x|| > 0$. For given $\epsilon > 0$, there is $N \in \mathbb{N}$ so that $n, m \geq N$ implies $||T_n - T_m|| < \frac{\epsilon}{||x||}$. Then $n, m \geq N$ forces

$$||T_n(x) - T_m(x)|| = ||(T_n - T_m)(x)|| \leq ||T_n - T_m||||x|| < \frac{\epsilon}{||x||}||x|| = \epsilon.$$

This gives the subclaim.

(ii) *For every $x \in A$ there is a real number $T(x)$ with* $\lim_n T_n(x) = T(x)$. This follows immediately from (i) and the fact that $\mathbb{R}$ is complete.

(iii) *The mapping $T : A \to \mathbb{R}$ given in (ii) is linear.* Let $x, y \in A$, $\alpha \in \mathbb{R}$, and $\epsilon > 0$. Find $N_1 \in \mathbb{N}$ such that $n \geq N_1$ forces $||T(x + y) - T_n(x + y)|| < \frac{\epsilon}{3}$. Find $N_2 \in \mathbb{N}$ such that $n \geq N_2$ forces $||T_n(x) - T(x)|| < \frac{\epsilon}{3}$. Find $N_3 \in \mathbb{N}$ so that $n \geq N_3$ forces $||T_n(y) - T(y)|| < \frac{\epsilon}{3}$. Let $N = \max\{N_1, N_2, N_3\}$. Then $n \geq N$ gives

$$\begin{aligned}
& ||T(x+y) - [T(x) + T(y)]|| \\
= \; & ||T(x+y) - T_n(x+y) + T_n(x+y) - T(x) - T(y)|| \\
= \; & ||T(x+y) - T_n(x+y) + T_n(x) - T(x) + T_n(y) - T(y)|| \\
\leq \; & ||T(x+y) - T_n(x+y)|| + ||T_n(x) - T(x)|| + ||T_n(y) - T(y)|| \\
< \; & \frac{\epsilon}{3} + \frac{\epsilon}{3} + \frac{\epsilon}{3} \\
= \; & \epsilon.
\end{aligned}$$

Since $\epsilon > 0$ is arbitrary, we have $||T(x + y) - [T(x) + T(y)]|| = 0$, and hence $T(x + y) = T(x) + T(y)$. We now complete the other part of the program. If $\alpha = 0$, then $||T(\alpha x) - \alpha T(x)|| = 0 < \epsilon$. If $\alpha \neq 0$, then there are elements $N_1, N_2 \in \mathbb{N}$ such that $n \geq N_1$ forces $||T(\alpha x) - T_n(\alpha x)|| < \frac{\epsilon}{2}$ and $n \geq N_2$ forces $||T_n(x) - T(x)|| < \frac{\epsilon}{2|\alpha|}$. Let N denote the maximum of N_1 and N_2.

Then $n \geq N$ forces

$$\begin{aligned}||T(\alpha x) - \alpha T(x)|| &= ||T(\alpha x) - T_n(\alpha x) + \alpha T_n(x) - \alpha T(x)|| \\ &\leq ||T(\alpha x) - T_n(\alpha x)|| + |\alpha|||T_n(x) - T(x)|| \\ &< \frac{\epsilon}{2} + |\alpha|\frac{\epsilon}{2|\alpha|} \\ &= \epsilon.\end{aligned}$$

Since $\epsilon > 0$ is arbitrary, this shows that $||T(\alpha x) - \alpha T(x)|| = 0$, and hence $T(\alpha x) = \alpha T(x)$ for every real α. We have shown that $T(x+y) = T(x) + T(y)$ and $T(\alpha x) = \alpha T(x)$, and hence T is indeed linear.

(iv) *T is a bounded linear transformation from A into $\mathbb{R}$: $T \in \mathcal{B}(A, \mathbb{R})$.*

Proof: Recall that real Cauchy sequences are bounded. Since $\{T_n\}_{n=1}^{\infty}$ is a Cauchy sequence in $\mathcal{B}(A, \mathbb{R})$, there is $N \in \mathbb{N}$ such that $n, m \geq N$ forces the inequality $||T_n - T_m|| < 1$. Therefore, we have $||T_N - T_m|| < 1$ for every $m \geq N$. This forces $||T_m|| < 1 + ||T_N||$ for every $m \geq N$. [If not, there is $m \geq N$ such that $||T_m|| \geq 1 + ||T_N||$, so $||T_m|| - ||T_N|| \geq 1$, hence the inequality $||T_m - T_N|| \geq ||T_m|| - ||T_N|| \geq 1$ obtains, a contradiction.] Let $\beta = \max\{||T_1||, \cdots, ||T_{N-1}||, 1 + ||T_N||\}$. Then for every $n \in \mathbb{N}$ we have $||T_n|| \leq \beta < +\infty$. Let $x \in A$ be such that $||x|| = 1$. Then

$$\begin{aligned}||T(x)|| = ||T(x) - T_n(x) + T_n(x)|| &\leq ||T(x) - T_n(x)|| + ||T_n(x)|| \\ &\leq ||T(x) - T_n(x)|| + \beta \quad \text{for all } n \in \mathbb{N}.\end{aligned}$$

In particular, for sufficiently large n, we have $||T(x) - T_n(x)|| < 1$, so that the inequality $||T(x)|| < 1 + \beta$ obtains. It follows that $||T(x)|| < 1 + \beta < +\infty$ whenever $||x|| = 1$, and hence $\sup\{||T(x)|| : x \in A, ||x|| = 1\} < +\infty$, so that T is bounded by Claim 2. Therefore, $T \in \mathcal{B}(A, \mathbb{R})$.

(v) *$T = \lim_n T_n$ in the metric induced on $\mathcal{B}(A, \mathbb{R})$ by the operator norm.*

Proof: Let $\epsilon > 0$. We want $N \in \mathbb{N}$ so that $n \geq N$ implies $||T_n - T|| \leq \epsilon$. Find $M \in \mathbb{N}$ such that $m, n \geq M$ implies $||T_m - T_n|| < \frac{\epsilon}{2}$. Next, pick $x \in A$ with $||x|| = 1$; choose $m_x \in \mathbb{N}$ with $m_x \geq M$ and $||T(x) - T_{m_x}(x)|| < \frac{\epsilon}{2}$. Then $n \geq M$ implies that

$$\begin{aligned}||T(x) - T_n(x)|| &\leq ||T(x) - T_{m_x}(x)|| + ||T_{m_x}(x) - T_n(x)|| \\ &< \frac{\epsilon}{2} + ||(T_{m_x} - T_n)(x)|| \\ &\leq \frac{\epsilon}{2} + ||T_{m_x} - T_n||||x|| \\ &< \frac{\epsilon}{2} + \frac{\epsilon}{2} \\ &= \epsilon,\end{aligned}$$

so that

$$||T - T_n|| = \sup\{||T(x) - T_n(x)|| : x \in A, \ ||x|| = 1\} \leq \epsilon \quad \text{whenever } n \geq M.$$

This shows that $\lim_n \|T_n - T\| = 0$, completing the proof of (v) and hence of (3).

Putting all of (1), (2), and (3) together completes the entire proof. ■

The real line $\mathbb{R}$ in the previous claim may be replaced with any complete normed linear space B over $\mathbb{R}$. For what we will need, only the case $B = \mathbb{R}$ is required.

The space $\mathcal{B}(A, \mathbb{R})$ [or $\mathcal{B}(A, B)$, where B is a complete normed linear space over $\mathbb{R}$] is called the *dual space* or the *conjugate space* of A. The study of such spaces in abstraction forms a key component of functional analysis. The Riesz Representation Theorem to be found in the next section characterizes the conjugate space of a generic $L^p(\Omega, \mathcal{F}, \mu)$ for $1 < p < +\infty$. That is, the Riesz Representation Theorem characterizes $\mathcal{B}(L^p(\Omega, \mathcal{F}, \mu), \mathbb{R})$ for $1 < p < +\infty$ and a generic measure space $(\Omega, \mathcal{F}, \mu)$.

The Clarkson Inequalities

The Clarkson Inequalities will now be discussed in the context of a generic $L^p(\Omega, \mathcal{F}, \mu)$ for $1 < p < +\infty$. Just as the Hölder Inequality relies on some preliminary calculus, so do the Clarkson Inequalities. These inequalities have different forms as $1 < p < 2$ and $2 \leq p < +\infty$, and the calculus required for these inequalities is different in the two cases. It will be seen that the $2 \leq p < +\infty$ case is more easily handled than the $1 < p < 2$ case.

Claim 5. *Let $2 \leq p < +\infty$. Then for every $0 \leq x \leq 1$ we have*

$$\left(\frac{1+x}{2}\right)^p + \left(\frac{1-x}{2}\right)^p \leq \frac{1}{2}(1 + x^p).$$

Proof: Let $F(x) = (\frac{1+x}{2})^p + (\frac{1-x}{2})^p - \frac{1}{2}(1 + x^p)$ for $0 \leq x \leq 1$. It is sufficient to show that $F(x) \leq 0$ for each $0 \leq x \leq 1$. Observe that $F(0) \leq 0$; here we use the assumption $p \geq 2$. For $0 < x \leq 1$, define

$$\Phi(x) = \frac{2^p}{x^p} F(x) = \left(\frac{1}{x} + 1\right)^p + \left(\frac{1}{x} - 1\right)^p - 2^{p-1}\left(\frac{1}{x^p} + 1\right).$$

Note that $\Phi(1) = 0$. It is sufficient to show that $\Phi'(x) \geq 0$ for each $0 < x < 1$, because if this holds then $\Phi(x) \leq 0$ for each $0 < x \leq 1$, which forces $F(x) \leq 0$ for each $0 \leq x \leq 1$. We have for each $0 < x < 1$ that

$$\Phi'(x) = -\frac{p}{x^{\alpha+2}}\left[(1+x)^\alpha + (1-x)^\alpha - 2^\alpha\right] \quad (\alpha = p - 1 > 1).$$

Define $\Psi(x) = (1+x)^\alpha + (1-x)^\alpha - 2^\alpha$ for every $0 \leq x \leq 1$, so that for every $0 < x < 1$ we have $\Phi'(x) = -\frac{p}{x^{\alpha+2}}\Psi(x)$. To show that $\Phi'(x) \geq 0$ for each $0 < x < 1$, it is therefore sufficient to show that $\Psi(x) \leq 0$ for each $0 < x < 1$. Now $\Psi(1) = 0$, $\Psi(0) < 0$, and

$$\Psi'(x) = \alpha(1+x)^{\alpha-1} - \alpha(1-x)^{\alpha-1} \geq 0 \quad \text{for each } 0 < x < 1.$$

This shows that $\Psi(x) \leq 0$ for each $0 < x < 1$. This completes the proof. ∎

Claim 6. *Let $2 \leq p < +\infty$. Then for every $z, w \in \mathbb{R}$ we have*

$$\left|\frac{z+w}{2}\right|^p + \left|\frac{z-w}{2}\right|^p \leq \frac{|z|^p}{2} + \frac{|w|^p}{2}. \tag{8.14}$$

Proof: Let $w = 0$; the desired inequality is $|\frac{z}{2}|^p + |\frac{z}{2}|^p \leq \frac{|z|^p}{2}$, i.e. $\frac{|z|^p}{2^{p-1}} \leq \frac{|z|^p}{2}$, which holds since $p - 1 \geq 1$. A similar argument shows that the desired inequality holds if $z = 0$. Therefore, if $z = 0$ or $w = 0$, then the desired inequality holds. Assume now that z and w are nonzero, say, $0 < |w| \leq |z|$. By dividing both sides of (8.14) by $|z|^p$, we see that the desired inequality is equivalent to the inequality

$$\left|\frac{1}{2}\left(1+\frac{w}{z}\right)\right|^p + \left|\frac{1}{2}\left(1-\frac{w}{z}\right)\right|^p \leq \frac{1}{2}\left(1+\left|\frac{w}{z}\right|^p\right).$$

However, this inequality is automatic by the previous claim: take $x = \frac{w}{z}$ if w and z have the same algebraic sign, and take $x = -\frac{w}{z}$ if w and z have opposite algebraic sign. This completes the proof. ∎

Clarkson's Inequality. *Let $2 \leq p < +\infty$ and $f, g \in L^p$. Then*

$$\left\|\frac{f+g}{2}\right\|_p^p + \left\|\frac{f-g}{2}\right\|_p^p \leq \frac{\|f\|_p^p}{2} + \frac{\|g\|_p^p}{2}.$$

Proof: Recall our convention of viewing elements of L^p as defined and finite everywhere on Ω. For every $\omega \in \Omega$, the previous claim gives

$$\left|\frac{f(\omega)+g(\omega)}{2}\right|^p + \left|\frac{f(\omega)-g(\omega)}{2}\right|^p \leq \frac{|f(\omega)|^p}{2} + \frac{|g(\omega)|^p}{2},$$

hence integrating over Ω gives

$$\int \left|\frac{f+g}{2}\right|^p + \int \left|\frac{f-g}{2}\right|^p \leq \frac{1}{2}\int |f|^p + \frac{1}{2}\int |g|^p,$$

which gives the desired inequality. ∎

We have considered the case $2 \leq p < +\infty$, and now we turn to the more difficult case $1 < p < 2$. Some notation from advanced calculus must be reviewed first. For $n \in \mathbb{N}$ and $\alpha \in \mathbb{R}$, we write

$$\binom{\alpha}{n} = \frac{\alpha(\alpha-1)\cdots(\alpha-(n-1))}{n!}.$$

We define $\binom{\alpha}{0} = 1$ for each $\alpha \in \mathbb{R}$. From advanced calculus, the series $\sum_{n=0}^{\infty}\binom{\alpha}{n}x^n$ converges uniformly and absolutely to $(1+x)^\alpha$ for each $\alpha > 0$ and $-1 \leq x \leq 1$.

Claim 7. *Let $1 < p \leq 2$. Then for every $0 \leq x \leq 1$ we have*

$$(1+x)^{p'} + (1-x)^{p'} \leq 2(1+x^p)^{\frac{1}{p-1}}. \tag{8.15}$$

Proof: The inequality is automatic for each $0 \leq x \leq 1$ when $p = 2$. We therefore assume that $1 < p < 2$. In this case, the inequality is actually an equality if $x \in \{0, 1\}$, and hence we need only consider $0 < x < 1$ in the case $1 < p < 2$. Observe that as u goes from 0 to 1, the quantity $\frac{1-u}{1+u}$ decreases from 1 to 0. Therefore, we may equivalently demonstrate the inequality

$$\left(1+\frac{1-u}{1+u}\right)^{p'} + \left(1-\frac{1-u}{1+u}\right)^{p'} \leq 2\left[1+\left(\frac{1-u}{1+u}\right)^{p}\right]^{\frac{1}{p-1}} \tag{8.16}$$

for every $0 < u < 1$ [where $1 < p < 2$]. By multiplying both sides of (8.16) by $(1+u)^{p'}$ and then raising both sides of to the $(p-1)$st power, inequality (8.16) after some algebra becomes

$$(1+u^{p'})^{p-1} \leq \frac{1}{2}\left[(1+u)^p + (1-u)^p\right] \tag{8.17}$$

for every $0 < u < 1$. Since (8.17) holds for every $0 < u < 1$ iff (8.16) holds for every $0 < u < 1$, which in turn holds iff (8.15) holds for every $0 < x < 1$, we may complete things by showing that (8.17) holds for every $0 < u < 1$.

We now establish that (8.17) holds for every $0 < u < 1$. We have that

$$\begin{aligned}
&\frac{1}{2}\left[(1+u)^p + (1-u)^p\right] - (1+u^{p'})^{p-1} \\
&= \frac{1}{2}\left[\sum_{k=0}^{\infty}\binom{p}{k}u^k + \sum_{k=0}^{\infty}\binom{p}{k}(-1)^k u^k\right] - \sum_{k=0}^{\infty}\binom{p-1}{k}u^{p'k} \\
&= \frac{1}{2}\left[\sum_{k=0}^{\infty}\binom{p}{k}\left(u^k + (-1)^k u^k\right)\right] - \sum_{k=0}^{\infty}\binom{p-1}{k}u^{p'k} \\
&= \frac{1}{2}\sum_{k \text{ even}}\binom{p}{k}2u^k - \sum_{k=0}^{\infty}\binom{p-1}{k}u^{p'k} \\
&= \sum_{k=0}^{\infty}\binom{p}{2k}u^{2k} - \sum_{k=0}^{\infty}\binom{p-1}{k}u^{p'k} \\
&= \sum_{k=1}^{\infty}\left[\binom{p}{2k}u^{2k} - \binom{p-1}{k}u^{p'k}\right] \\
&= \sum_{k=1}^{\infty}\left[\binom{p}{2k}u^{2k} - \binom{p-1}{2k-1}u^{p'(2k-1)} - \binom{p-1}{2k}u^{p'2k}\right],
\end{aligned} \tag{8.18}$$

where (8.18) follows since

$$\begin{aligned}\sum_{k=1}^{\infty}\binom{p-1}{k}u^{p'k} &= \sum_{k \text{ odd}}\binom{p-1}{k}u^{p'k}+\sum_{k \text{ even}}\binom{p-1}{k}u^{p'k}\\ &= \sum_{k=1}^{\infty}\binom{p-1}{2k-1}u^{p'(2k-1)}+\sum_{k=1}^{\infty}\binom{p-1}{2k}u^{p'2k}.\end{aligned}$$

Now (8.17) holds for every $0 < u < 1$ iff the summation in (8.18) is nonnegative for every $0 < u < 1$. To show that the summation in (8.18) is nonnegative for every $0 < u < 1$, it is sufficient to show that the quantity in square brackets in (8.18) is nonnegative for every $0 < u < 1$ and $k \in \mathbb{N}$. We have, after some straightforward algebra, that

$$\begin{aligned}&\text{the quantity in square brackets in (8.18)}\\ = &\binom{p}{2k}u^{2k}-\binom{p-1}{2k-1}u^{p'(2k-1)}-\binom{p-1}{2k}u^{p'2k}\\ = &u^{2k}\frac{(2-p)(3-p)\cdots(2k-p)}{(2k-1)!}\left\{\frac{1-u^{\frac{2k-p}{p-1}}}{\frac{2k-p}{p-1}}-\frac{1-u^{\frac{2k}{p-1}}}{\frac{2k}{p-1}}\right\} \qquad (8.19)\end{aligned}$$

for every $0 < u < 1$ and $k \in \mathbb{N}$. Now $u^{2k} > 0$, and since $1 < p < 2$ we have that $\frac{(2-p)(3-p)\cdots(2k-p)}{(2k-1)!} > 0$. Since $\frac{1-u^t}{t}$ is decreasing as a function of $0 < t < +\infty$, and since $0 < \frac{2k-p}{p-1} < \frac{2k}{p-1} < +\infty$ for every $k \in \mathbb{N}$, the quantity in curly braces in (8.19) is positive. That is, the quantity in square brackets in (8.18) is positive for every $k \in \mathbb{N}$ and $0 < u < 1$. This shows that (8.17) holds for every $0 < u < 1$, completing the entire proof. ∎

Claim 8. *Let $1 < p \leq 2$. For every $z, w \in \mathbb{R}$ we have the inequality*

$$|z+w|^{p'}+|z-w|^{p'} \leq 2(|z|^p+|w|^p)^{\frac{1}{p-1}}.$$

Proof: If $z = 0$ or $w = 0$, the claim automatically follows. Assume now that z and w are nonzero, say $0 < |z| \leq |w|$. In this case, the desired inequality is equivalent to the inequality

$$\left|1+\frac{z}{w}\right|^{p'}+\left|1-\frac{z}{w}\right|^{p'} \leq 2\left(1+\left|\frac{z}{w}\right|^p\right)^{\frac{1}{p-1}}.$$

Using the previous claim with $x = \frac{z}{w}$ or $x = -\frac{z}{w}$ as z and w have the same or opposite algebraic signs completes the proof. ∎

Clarkson's Inequality. *Let $1 < p < 2$ and $f, g \in L^p$. Then*

$$\left\|\frac{f+g}{2}\right\|_p^{p'}+\left\|\frac{f-g}{2}\right\|_p^{p'} \leq \left[\frac{||f||_p^p}{2}+\frac{||g||_p^p}{2}\right]^{\frac{1}{p-1}}.$$

Proof: We have $0 < p - 1 < 1$ and $\frac{f+g}{2}, \frac{f-g}{2} \in L^p$. The Minkowski-like Inequality from Section 8.4 applied to the elements $|\frac{f+g}{2}|^{p'}$ and $|\frac{f-g}{2}|^{p'}$ in L^{p-1} gives

$$\left\| \left|\frac{f+g}{2}\right|^{p'} \right\|_{p-1} + \left\| \left|\frac{f-g}{2}\right|^{p'} \right\|_{p-1} \leq \left\| \left|\frac{f+g}{2}\right|^{p'} + \left|\frac{f-g}{2}\right|^{p'} \right\|_{p-1}.$$

Using the fact that $\|h^{p'}\|_{p-1} = \|h\|_p^{p'}$ for each $h \in L^p$, we have that

$$\begin{aligned} \left\|\frac{f+g}{2}\right\|_p^{p'} + \left\|\frac{f-g}{2}\right\|_p^{p'} &\leq \left[\int \left(\left|\frac{f}{2} + \frac{g}{2}\right|^{p'} + \left|\frac{f}{2} - \frac{g}{2}\right|^{p'} \right)^{p-1}\right]^{\frac{1}{p-1}} \\ &\leq \left[\int 2^{p-1} \left(\left|\frac{f}{2}\right|^p + \left|\frac{g}{2}\right|^p \right)\right]^{\frac{1}{p-1}} \\ &= \left[\frac{1}{2}\|f\|_p^p + \frac{1}{2}\|g\|_p^p\right]^{\frac{1}{p-1}}, \end{aligned}$$

where the second inequality follows by taking $z = \frac{f}{2}$ and $w = \frac{g}{2}$ in the previous claim. This completes the proof. ■

Exercises.

1. Let T be defined on L^p [where $1 < p < +\infty$] by $T(f) = \int fh$, where $h \in L^{p'}$ is held fixed. Then $\|T\| = \|h\|_{p'}$. [It should be easy to obtain $\|T\| \leq \|h\|_{p'}$. To obtain the reverse inequality, let $f = |h|^{p'-1} \operatorname{sgn} h$. Then $|f|^p = |h|^{p'}$, hence $f \in L^p$. Show next that $\|f\|_p = \|g\|_{p'}^{p'-1}$ and $L(f) = \|f\|_p \|h\|_{p'}$, so that $\|L\| \geq \|h\|_{p'}$.]

2. In the specific example mentioned in this section's discussion on linear transformations, restate Claims 3 and 4 in terms of the specific setup.

3. Use the Clarkson Inequalities to show that L^p $[1 < p < +\infty]$ is *locally uniformly convex*: if $\{f_n\}_{n=1}^{\infty}$ is a sequence in L^p, $f \in L^p$, $\|f\|_p = \|f_1\|_p = \|f_2\|_p = \cdots = 1$, then $\lim_n \|\frac{1}{2}(f_n + f)\|_p = 1$ implies $\lim_n \|f_n - f\|_p = 0$.

4. Suppose that $(\Omega, \mathcal{F}, \mu)$ denotes a measure space, where $\mathcal{F}$ contains at least two disjoint sets of finite positive measure.
(a) There is a positive real number c such that

$$\frac{1}{c} \leq \frac{\|f+g\|_p^2 + \|f-g\|_p^2}{2(\|f\|_p^2 + \|g\|_p^2)} \leq c$$

for every $f, g \in L^p$ such that at least one of $\|f\|_p$ and $\|g\|_p$ is positive.
(b) The unique smallest possible c in the above is $2^{\frac{|2-p|}{p}}$, and the lower and upper bounds in (a) are realized.

8.9 THE RIESZ REPRESENTATION THEOREM

This optional section is dedicated to proving one of the great results of analysis: the Riesz Representation Theorem. This result utilizes much of the theory regarding the abstract Lebesgue integral and the Riesz-Fischer Theorem from Section 8.2, and is one of the cornerstone results of analysis. Throughout this section, we will [as usual] work in a generic measure space $(\Omega, \mathcal{F}, \mu)$, and we will consider the corresponding L^p spaces, where $1 < p < +\infty$.

In the previous section, we dealt with arbitrary bounded linear transformations $T : A \to B$, where A and B are normed linear spaces over $\mathbb{R}$. As stated in that section, we will exclusively take $A = L^p$ and $B = \mathbb{R}$. The norm $\|\ \|_A$ on A is therefore the L^p norm $\|\ \|_p$, and the norm $\|\ \|_B$ on B is nothing more than the absolute value $|\ |$. Therefore, to say that a linear transformation $T : L^p \to \mathbb{R}$ is bounded means that there is $0 \leq M < +\infty$ such that $|T(f)| \leq M\|f\|_p$ for every $f \in L^p$.

The upcoming claim is the first of three claims needed to prove the Riesz Representation Theorem. We take $1 < p < +\infty$ throughout the remainder of this section.

Claim 1. *Let $g \in L^{p'}$, and define $L_g : L^p \to \mathbb{R}$ by writing $L_g(f) = \int fg$ for every $f \in L^p$. Then L_g is a bounded linear functional on L^p and $\|L_g\| = \|g\|_{p'}$.*

Before proving the claim, observe that if $f \in L^p$ and $g \in L^{p'}$, then $fg \in L^1$ by Hölder's Inequality, hence $\int fg$ exists as a real number. Since we have $\int(f_1 + f_2)g = \int f_1 g + \int f_2 g$ and $\int(\alpha f)g = \alpha \int fg$ for every $f, f_1, f_2 \in L^p$, $g \in L^{p'}$, and $\alpha \in \mathbb{R}$, we have $L_g(f_1 + f_2) = L_g(f_1) + L_g(f_2)$ as well as $L_g(\alpha f) = \alpha L_g(f)$ for every $f, f_1, f_2 \in L^p$, $g \in L^{p'}$, and $\alpha \in \mathbb{R}$, hence every mapping L_g as described above is a linear transformation from L^p to $\mathbb{R}$. All that really needs to be shown is that each L_g is bounded and $\|L_g\| = \|g\|_{p'}$.

Proof: Pick and fix $g \in L^{p'}$. For every $f \in L^p$, Hölder's Inequality gives the inequality $|L_g(f)| = |\int fg| \leq \|g\|_{p'}\|f\|_p$, and hence $\|g\|_{p'}$ is a bound for L_g. It follows that L_g is bounded. Since $\|L_g\|$ is the infimum of such bounds, we have $\|g\|_{p'} \geq \|L_g\|$. It is desired to strengthen this to an equality. To do this, consider the function $f = |g|^{p'-1} \operatorname{sgn} g$, so that $|f|^p = |g|^{p'}|\operatorname{sgn} g|^p$. Next, observe that if $g(\omega) = 0$, then $|f(\omega)|^p = 0 = |g(\omega)|^{p'}$. If $g(\omega) \neq 0$, then $|\operatorname{sgn} g(\omega)| = 1$ and we again have $|f(\omega)|^p = |g(\omega)|^{p'}$. Therefore, $|f|^p = |g|^{p'}$ on Ω. It follows that $\int |f|^p = \int |g|^{p'} < +\infty$, and hence $f \in L^p$. Next, observe that $\|g\|_{p'}^{p'-1} = (\int |g|^{p'})^{\frac{p'-1}{p'}} = (\int |f|^p)^{\frac{1}{p}} = \|f\|_p$. We therefore have

$$\begin{aligned}
L_g(f) = \int fg = \int |g|^{p'-1}(\operatorname{sgn} g)g &= \int |g|^{p'} \\
&= \|g\|_{p'}^{p'} \\
&= \|g\|_{p'}^{p'-1}\|g\|_{p'} \\
&= \|f\|_p\|g\|_{p'}.
\end{aligned}$$

That is, for the specific $f \in L^p$ just given, we have $|L_g(f)| = \|g\|_{p'}\|f\|_p$. This shows that any bound for L_g is at least $\|g\|_{p'}$, so that $\|L_g\| \geq \|g\|_{p'}$. It follows that $\|L_g\| = \|g\|_{p'}$ as claimed, and the proof is complete. ∎

Claim 1 shows that one can use integrals to define bounded linear transformations on L^p. The Riesz Representation Theorem provides a sort-of converse to this fact by stating that *every* bounded linear transformation L defined on L^p must have the form L_g as above for some appropriate $g \in L^{p'}$. This very interesting fact is one of the cornerstone results of analysis and functional analysis.

Claim 2. *Let L denote a nonzero bounded linear functional on L^p not identically zero. There is $\phi_0 \in L^p$ such that $\|\phi_0\|_p = 1$ and $L(\phi_0) = \|L\|$.*

Before giving the proof, note that the assumption that L is a nonzero bounded linear functional means that there is $f \in L^p$ such that $L(f) \neq 0$, hence $0 < \|L\| < +\infty$. The conclusion of the claim is that L attains its maximum absolute value on the unit ball of L^p.

Proof: Recall that $\|L\|$ is the infimum of the numbers $0 \leq M < +\infty$ such that $|L(f)| \leq M\|f\|_p$ for every $f \in L^p$. Also, recall that $\|L\|$ is the supremum of the numbers $|L(f)|$, where $f \in L^p$ and $\|f\|_p = 1$. Using these facts, we may find a sequence $\{\phi'_n\}_{n=1}^\infty$ in L^p such that
(i) $\|\phi'_n\|_p = 1$ for every $n \in \mathbb{N}$,
(ii) $|L(\phi'_n)| > \frac{1}{2}\|L\|$ $[> 0]$ for every $n \in \mathbb{N}$, and
(iii) $\lim_n |L(\phi'_n)| = \|L\|$.

We use the sequence $\{\phi'_n\}_{n=1}^\infty$ just given to define a new sequence $\{\phi_n\}_{n=1}^\infty$, where for each $n \in \mathbb{N}$ we define $\phi_n = \phi'_n \operatorname{sgn} L(\phi'_n)$. For each $n \in \mathbb{N}$ we have

$$L(\phi_n) = \begin{cases} L(\phi'_n) & \text{if } L(\phi'_n) > 0 \\ -L(\phi'_n) & \text{if } L(\phi'_n) < 0, \end{cases} \qquad \text{hence} \quad L(\phi_n) = |L(\phi'_n)| > \frac{1}{2}\|L\| > 0.$$

[Recall that we needn't consider the case $L(\phi'_n) = 0$ by (ii).] Next, by considering the cases $L(\phi'_n) > 0$ and $L(\phi'_n) < 0$, we have that $\|\phi_n\|_p = 1$ for every $n \in \mathbb{N}$. We also have that $\lim_n L(\phi_n) = \lim_n |L(\phi'_n)| = \|L\|$ by (iii). Summarizing, we have
(i′) $\|\phi_n\|_p = 1$ for every $n \in \mathbb{N}$,
(ii′) $L(\phi_n) = |L(\phi'_n)| > \frac{1}{2}\|L\| > 0$ for every $n \in \mathbb{N}$, and
(iii′) $\lim_n L(\phi_n) = \|L\|$.
Suppose it is shown that $\{\phi_n\}_{n=1}^\infty$ is a Cauchy sequence in L^p. Then the Riesz-Fischer Theorem implies the existence of $\phi_0 \in L^p$ such that $\phi_n \xrightarrow{p} \phi_0$. Since the boundedness of $\|L\|$ forces L to be continuous on L^p [Claim 3 of the previous section], (iii) will give $\|L\| = \lim_n L(\phi_n) = L(\lim_n \phi_n) = L(\phi_0)$. [The symbol $\lim_n \phi_n$ denotes the L^p limit of ϕ_n, not necessarily the pointwise limit of ϕ_n.] Also, from the conditions $\phi_n \xrightarrow{p} \phi_0$ and (i′), we will have

$$|1 - \|\phi_0\|_p| = |\|\phi_n\|_p - \|\phi_0\|_p| \leq \|\phi_n - \phi_0\|_p \to 0 \quad \text{as } n \to \infty,$$

giving $||\phi_0||_p = 1$. Therefore, the completion of the proof depends on showing that $\{\phi_n\}_{n=1}^{\infty}$ is a Cauchy sequence in L^p.

To show that $\{\phi_n\}_{n=1}^{\infty}$ is a Cauchy sequence in L^p, we proceed by contradiction, assuming that $\{\phi_n\}_{n=1}^{\infty}$ is *not* a Cauchy sequence in L^p. Under this contrary assumption, there is $\alpha > 0$ as well as subsequences $\{\phi_{n_k}\}_{k=1}^{\infty}$ and $\{\phi_{m_k}\}_{k=1}^{\infty}$ of $\{\phi_n\}_{n=1}^{\infty}$ such that $||\phi_{n_k} - \phi_{m_k}||_p > \alpha$ for every $k \in \mathbb{N}$. Note that $0 < \alpha < ||\phi_{n_k} - \phi_{m_k}||_p \leq ||\phi_{n_k}||_p + ||\phi_{m_k}||_p = 1 + 1 = 2$ for each $k \in \mathbb{N}$, so that we have the inequality $0 < \alpha < 2$.

If $2 \leq p < +\infty$, the Clarkson Inequality gives

$$\left\|\frac{\phi_{m_k} + \phi_{n_k}}{2}\right\|_p^p + \left\|\frac{\phi_{m_k} - \phi_{n_k}}{2}\right\|_p^p \leq \frac{1}{2}||\phi_{m_k}||_p^p + \frac{1}{2}||\phi_{n_k}||_p^p = 1$$

for every $k \in \mathbb{N}$, so that for every $k \in \mathbb{N}$ we have

$$\begin{aligned}\left\|\frac{\phi_{m_k} + \phi_{n_k}}{2}\right\|_p^p \leq 1 - \left\|\frac{\phi_{m_k} - \phi_{n_k}}{2}\right\|_p^p &= 1 - \left(\frac{1}{2}\right)^p ||\phi_{m_k} - \phi_{n_k}||_p^p \\ &< 1 - \left(\frac{\alpha}{2}\right)^p \\ &\leq 1 - \left(\frac{\alpha}{2}\right)^2.\end{aligned}$$

Let $\gamma_p^{(1)} = 1 - [1 - (\frac{\alpha}{2})^2]^{\frac{1}{p}}$. Then $0 < \gamma_p^{(1)} < 1$, $\gamma_p^{(1)}$ is free of k, and

$$\left\|\frac{\phi_{m_k} + \phi_{n_k}}{2}\right\|_p \leq 1 - \gamma_p^{(1)} \quad \text{for all } k \in \mathbb{N}, \quad 2 \leq p < +\infty.$$

If $1 < p < 2$ and $k \in \mathbb{N}$, the Clarkson Inequality gives

$$\left\|\frac{\phi_{m_k} + \phi_{n_k}}{2}\right\|_p^{p'} + \left\|\frac{\phi_{m_k} - \phi_{n_k}}{2}\right\|_p^{p'} \leq \left[\frac{||\phi_{m_k}||_p^p}{2} + \frac{||\phi_{n_k}||_p^p}{2}\right]^{\frac{1}{p-1}} = 1,$$

so that

$$\begin{aligned}\left\|\frac{\phi_{m_k} + \phi_{n_k}}{2}\right\|_p^{p'} \leq 1 - \left\|\frac{\phi_{m_k} - \phi_{n_k}}{2}\right\|_p^{p'} &= 1 - \left(\frac{1}{2}\right)^{p'} ||\phi_{m_k} - \phi_{n_k}||_p^{p'} \\ &\leq 1 - \left(\frac{\alpha}{2}\right)^{p'}\end{aligned}$$

for every $k \in \mathbb{N}$. Let $\gamma_p^{(2)} = 1 - [1 - (\frac{\alpha}{2})^{p'}]^{\frac{1}{p'}}$. Then $0 < \gamma_p^{(2)} < 1$ and $\gamma_p^{(2)}$ is free of k. It follows that

$$\left\|\frac{\phi_{m_k} + \phi_{n_k}}{2}\right\|_p \leq 1 - \gamma_p^{(2)} \quad \text{for every } k \in \mathbb{N}, \ 1 < p < 2.$$

Combining the results for $2 \leq p < +\infty$ and $1 < p < 2$, we see that for every $1 < p < +\infty$ there is a number $0 < \gamma_p < 1$ that is free of k and such that

$$\left\|\frac{\phi_{m_k} + \phi_{n_k}}{2}\right\|_p < 1 - \gamma_p \quad \text{for every } k \in \mathbb{N}.$$

Fix $1 < p < +\infty$ for the remainder of the proof.

We now wish to show that $\|\phi_{m_k} + \phi_{n_k}\|_p > 0$ for every $k \in \mathbb{N}$. To see this, observe that if we have $\|\phi_{m_k} + \phi_{n_k}\|_p = 0$ for some k, then $\phi_{m_k} = -\phi_{n_k}$ μ-a.e. on Ω for this k. It follows that $L(\phi_{m_k}) = L(-\phi_{n_k}) = -L(\phi_{n_k})$ for this k, which contradicts (ii′). [Recall the discussion immediately following the definition of a bounded linear transformation given in the previous section.] Therefore, we have that $\|\phi_{m_k} + \phi_{n_k}\|_p > 0$ for every $k \in \mathbb{N}$.

Using the previous paragraph, we may meaningfully define

$$g_k = \frac{\phi_{m_k} + \phi_{n_k}}{\|\phi_{m_k} + \phi_{n_k}\|_p}$$

on Ω for every $k \in \mathbb{N}$, so that $g_k \in L^p$ and $\|g_k\|_p = 1$ for every $k \in \mathbb{N}$. Next, observe for every $k \in \mathbb{N}$ that

$$\begin{aligned} L(g_k) = L\left(\frac{\phi_{m_k} + \phi_{n_k}}{\|\phi_{m_k} + \phi_{n_k}\|_p}\right) &= \frac{1}{\left\|\frac{\phi_{m_k}+\phi_{n_k}}{2}\right\|_p}\left(\frac{1}{2}L(\phi_{m_k}) + \frac{1}{2}L(\phi_{n_k})\right) \\ &> \frac{1}{1-\gamma_p}\left(\frac{1}{2}L(\phi_{m_k}) + \frac{1}{2}L(\phi_{n_k})\right). \end{aligned}$$

From this, we have

$$\begin{aligned} \underline{\lim}_k L(g_k) &\geq \underline{\lim}_k \frac{1}{1-\gamma_p}\left(\frac{1}{2}L(\phi_{m_k}) + \frac{1}{2}L(\phi_{n_k})\right) \\ &= \frac{1}{1-\gamma_p}\left(\frac{1}{2}\underline{\lim}_k L(\phi_{m_k}) + \frac{1}{2}\underline{\lim}_k L(\phi_{n_k})\right) \\ &= \frac{1}{1-\gamma_p}\left(\frac{1}{2}\lim_k L(\phi_{m_k}) + \frac{1}{2}\lim_k L(\phi_{n_k})\right) \\ &= \frac{1}{1-\gamma_p}\left(\frac{1}{2}\|L\| + \frac{1}{2}\|L\|\right) \\ &= \frac{\|L\|}{1-\gamma_p} \\ &> \|L\| \\ &= \|L\|\|g_k\|_p, \end{aligned}$$

which shows that some g_k is such that $L(g_k) > \|L\|\|g_k\|_p$, contradicting the definition of $\|L\|$. The assumption that $\{\phi_n\}_{n=1}^{\infty}$ is not a Cauchy sequence in L^p leads to a contradiction; it must be the case that $\{\phi_n\}_{n=1}^{\infty}$ is in fact a Cauchy sequence in L^p, and thus the proof is finally complete. ■

Claim 3. *Let L denote a nonzero bounded linear functional on L^p. Let $g \in L^p$ be such that $\|g\|_p = 1$ and $L(g) = \|L\|$. [Such a $g \in L^p$ exists by the previous claim.] For every $f \in L^p$ and $t \in \mathbb{R}$, write $\psi_f(t) = \|g + tf\|_p$. If ψ_f is differentiable at $t = 0$, then $\frac{L(f)}{\|L\|} = \psi_f'(0)$.*

Proof: Let $f \in L^p$ be such that ψ_f is differentiable at $t = 0$. We first prove the claim for the case where $\|L\| = 1$, desiring to show the equality $L(f) = \psi_f'(0)$. Observe the following two points:

(i) For every $z \in \mathbb{R}$ we have $g + z(f - L(f)g) \in L^p$ and

$$L(g + z(f - L(f)g)) = L(g) + z[L(f) - L(f)L(g)] = 1 + z[L(f) - L(f)] = 1.$$

(ii) Since $|L(h)| \leq \|L\|\|h\|_p = \|h\|_p$ for every $h \in L^p$, taking $h \in L^p$ to be the function $g + z(f - L(f)g)$ gives $\|g + z(f - L(f)g)\|_p \geq 1$ for all real z.

We now consider the cases $L(f) = 0$ and $L(f) \neq 0$ separately, first assuming that $L(f) \neq 0$. For all real $t \neq -\frac{1}{L(f)}$, write

$$g + tf = (1 + tL(f)) \left[g + \frac{t}{1 + tL(f)}(f - L(f)g)\right].$$

[We needn't worry about when $t = -\frac{1}{L(f)}$ as we are ultimately concerned only with values of t in some small neighborhood of zero.] Observe that

$$\left\|g + \frac{t}{1 + tL(f)}(f - L(f)g)\right\|_p \geq 1$$

for all $t \neq -\frac{1}{L(f)}$ [take $z = \frac{t}{1+tL(f)}$ in (ii)]. If $t = 0$, we have that

$$\left\|g + \frac{t}{1 + tL(f)}(f - L(f)g)\right\|_p = \|g\|_p = 1.$$

It follows for all $t \neq -\frac{1}{L(f)}$ close to zero that

$$\begin{aligned}
& \|g + tf\|_p - \|g\|_p \\
= & \left\|(1 + tL(f)) \left[g + \frac{t}{1 + tL(f)}(f - L(f)g)\right]\right\|_p - 1 \\
= & |1 + tL(f)| \left\|g + \frac{t}{1 + tL(f)}(f - L(f)g)\right\|_p - 1 \\
\geq & |1 + tL(f)| - 1 \\
\geq & tL(f) \quad [\text{since } |1 + a| - 1 \geq a \text{ for all real } a].
\end{aligned}$$

Therefore, for all $t \neq -\frac{1}{L(f)}$ close to zero we have

$$\frac{\psi_f(t) - \psi_f(0)}{t} = \frac{\|g + tf\|_p - \|g\|_p}{t} \begin{cases} \geq L(f) & t > 0 \\ \leq L(f) & t < 0. \end{cases}$$

We have assumed that ψ_f is differentiable at $t = 0$, so the above forces

$$\psi_f'(0) = \lim_{t\to 0} \frac{\psi_f(t) - \psi_f(0)}{t} = \lim_{t\to 0} \frac{\|g + tf\|_p - \|g\|_p}{t} = L(f).$$

This proves the result in the case where $\|L\| = 1$ and $L(f) \neq 0$.

We now keep $\|L\| = 1$ and assume that $L(f) = 0$. By (ii), we have the inequality $\|g + tf\|_p \geq 1$ for every real t [we still have $\|g\|_p = 1$]. Therefore we have the inequality $\|g + tf\|_p - \|g\|_p \geq 0$ for every real t. It follows for every nonzero t that

$$\frac{\|g + tf\|_p - \|g\|_p}{t} \begin{cases} \geq 0 & t > 0 \\ \leq 0 & t < 0. \end{cases}$$

We have assumed that ψ_f is differentiable at $t = 0$, so the above forces

$$\psi_f'(0) = \lim_{t\to 0} \frac{\psi_f(t) - \psi_f(0)}{t} = \lim_{t\to 0} \frac{\|g + tf\|_p - \|g\|_p}{t} = 0 = L(f).$$

This proves the result in the case where $\|L\| = 1$ and $L(f) = 0$, and hence the case where $\|L\| = 1$ has been demonstrated.

We now prove the result for arbitrary nonzero bounded linear functionals $T : L^p \to \mathbb{R}$. Let $T : L^p \to \mathbb{R}$ denote a bounded nonzero linear functional. Define the functional $L = \frac{T}{\|T\|}$. Then L is clearly linear, nonzero, and bounded, with

$$\begin{aligned} \|L\| &= \sup\{|L(f)| : f \in L^p, \|f\|_p = 1\} \\ &= \sup\left\{\frac{|T(f)|}{\|T\|} : f \in L^p, \|f\|_p = 1\right\} \\ &= \frac{1}{\|T\|}\sup\{|T(f)| : f \in L^p, \|f\|_p = 1\} \\ &= \frac{1}{\|T\|}\|T\| \\ &= 1. \end{aligned}$$

By what we have shown, the differentiability of ψ_f at $t = 0$ gives

$$\psi_f'(0) = L(f) = \frac{T(f)}{\|T\|}.$$

This gives the proof for arbitrary nonzero linear functionals T on L^p, and hence the entire proof is complete. ∎

The road has been somewhat long and unavoidably technical, but now we are ready for the main result.

Riesz Representation Theorem. *Let L denote a bounded linear functional on L^p, where $1 < p < +\infty$. Then there is $h \in L^{p'}$ [depending on L] such that $L(f) = \int fh$ for every $f \in L^p$.*

In terms of the notation of Claim 1, the theorem states that when $1 < p < +\infty$, every bounded linear functional on L^p has the form L_h for an appropriately chosen $h \in L^{p'}$. The proof ultimately boils down to the following argument: pick $g \in L^p$ with $\|g\|_p = 1$ and $L(g) = \|L\|$, and show that the desired $h \in L^{p'}$ may be taken as $\|L\||g|^{p-1} \operatorname{sgn} g$. As complicated as the proof below might be, it ultimately revolves around this simple program.

Proof: If $L = 0$ [the functional such that $L(f) = 0$ for every $f \in L^p$], then taking $h = 0$ completes the proof. We next assume that $L \neq 0$, so that $\|L\| > 0$. By Claim 2, there is $g \in L^p$ with $\|g\|_p = 1$ and $L(g) = \|L\|$. Suppose we show for every $f \in L^p$ that ψ_f is differentiable at $t = 0$ and $\psi_f'(0) = \int |g|^{p-2}gf$. [Recall that $\psi_f(t) = \|g + tf\|_p$ for every real t and $f \in L^p$.] Then by Claim 3 we will havefor every $f \in L^p$ that

$$L(f) = \|L\|\psi_f'(0) = \int \|L\||g|^{p-2}gf = \int \|L\||g|^{p-1}(\operatorname{sgn} g)f.$$

If $h = \|L\||g|^{p-1} \operatorname{sgn} g$, then $h \in L^{p'}$ [since $|h|^{p'} \leq \|L\|^{p'}|g|^p \in L^1$], and for every $f \in L^p$ we will have $L(f) = \int fh$. Everything therefore hinges on showing that ψ_f is differentiable at $t = 0$ and $\psi_f'(0) = \int |g|^{p-2}gf$.

In what follows, pick and fix $f \in L^p$, and for each real t define

$$\theta(t) = \psi_f^p(t) = \|tf + g\|_p^p = \int |tf + g|^p.$$

Next, pick and fix $\omega \in \Omega$ and $t_0 \in \mathbb{R}$. If $2 \leq p < +\infty$, or if $1 < p < 2$ and $t_0 f(\omega) + g(\omega) \neq 0$, we have that

$$\begin{aligned}
&\frac{\partial}{\partial t}|tf(\omega) + g(\omega)|^p\Big|_{t=t_0} \\
&= \frac{\partial}{\partial t}\left[(tf(\omega) + g(\omega))^2\right]^{\frac{p}{2}}\Big|_{t=t_0} \\
&= \frac{p}{2}\left[(tf(\omega) + g(\omega))^2\right]^{\frac{p}{2}-1} 2(tf(\omega) + g(\omega))f(\omega)\Big|_{t=t_0} \\
&= p|t_0 f(\omega) + g(\omega)|^{p-2}(t_0 f(\omega) + g(\omega))f(\omega).
\end{aligned}$$

If $1 < p < 2$ and $t_0 f(\omega) + g(\omega) = 0$, then

$$\begin{aligned}
&\frac{\partial}{\partial t}|tf(\omega) + g(\omega)|^p\Big|_{t=t_0} \\
&= \lim_{\delta \to 0} \frac{|(t_0 + \delta)f(\omega) + g(\omega)|^p - |t_0 f(\omega) + g(\omega)|^p}{\delta} \\
&= \lim_{\delta \to 0} \frac{|\delta|^p}{\delta}|f(\omega)|^p \\
&= 0.
\end{aligned}$$

Next, for every real $t_0 \neq 0$ and $1 < p < +\infty$ we have

$$\begin{aligned}
&\frac{\theta(t_0) - \theta(0)}{t_0} \\
&= \int \frac{|t_0 f(\omega) + g(\omega)|^p - |g(\omega)|^p}{t_0}\, d\mu(\omega) \\
&= \int \frac{\partial}{\partial t}|tf(\omega) + g(\omega)|^p \bigg|_{t=t_0^*(\omega)} d\mu(\omega),
\end{aligned}$$

where for each $\omega \in \Omega$ the number $t_0^*(\omega)$ is between 0 and t_0 in accordance with the Mean Value Theorem from calculus. If $2 \leq p < +\infty$, we therefore have

$$\frac{\theta(t_0) - \theta(0)}{t_0} = \int p|t_0^*(\omega)f(\omega) + g(\omega)|^{p-2}(t_0^*(\omega)f(\omega) + g(\omega))f(\omega)\, d\mu(\omega).$$

If $1 < p < 2$, then, writing $A(t_0) = \{\omega \in \Omega : \ t_0^*(\omega)f(\omega) + g(\omega) \neq 0\}$ and $B(t_0) = \{\omega \in \Omega : \ t_0^*(\omega)f(\omega) + g(\omega) = 0\}$ for every real $t_0 \neq 0$, we have

$$\begin{aligned}
&\frac{\theta(t_0) - \theta(0)}{t_0} \\
&= \int \frac{\partial}{\partial t}|tf(\omega) + g(\omega)|^p \bigg|_{t=t_0^*(\omega)} d\mu(\omega) \\
&= \int_{A(t_0)} \frac{\partial}{\partial t}|tf(\omega) + g(\omega)|^p \bigg|_{t=t_0^*(\omega)} d\mu(\omega) + \int_{B(t_0)} \frac{\partial}{\partial t}|tf(\omega) + g(\omega)|^p \bigg|_{t=t_0^*(\omega)} d\mu(\omega) \\
&= \int_{A(t_0)} \frac{\partial}{\partial t}|tf(\omega) + g(\omega)|^p \bigg|_{t=t_0^*(\omega)} d\mu(\omega) + \int_{B(t_0)} 0\, d\mu(\omega) \\
&= \int_{A(t_0)} p|t_0^*(\omega)f(\omega) + g(\omega)|^{p-2}(t_0^*(\omega)f(\omega) + g(\omega))f(\omega)\, d\mu(\omega)
\end{aligned}$$

Next, if $-1 \leq t_0 \leq 1$ [so that $-1 \leq t_0^*(\omega) \leq 1$], and if (a) $2 \leq p < +\infty$ and $\omega \in \Omega$ or (b) $1 < p < 2$ and $\omega \in A(t_0)$, we have

$$\begin{aligned}
&\left| \, p|t_0^*(\omega)f(\omega) + g(\omega)|^{p-2}(t_0^*(\omega)f(\omega) + g(\omega))f(\omega) \, \right| \\
&= p|t_0^*(\omega)f(\omega) + g(\omega)|^{p-1}|f(\omega)| \\
&\leq p\,[|t_0^*(\omega)||f(\omega)| + |g(\omega)|]^{p-1}\, |f(\omega)| \\
&\leq p\,(|f(\omega)| + |g(\omega)|)^{p-1}\, |f(\omega)|.
\end{aligned}$$

Now $f, g \in L^p$, so $|f| + |g| \in L^p$, hence $(|f| + |g|)^{p-1} \in L^{p'}$, whence we have $p(|f| + |g|)^{p-1}|f| \in L^1$ by Hölder's Inequality.

We use these equations to apply DCT. When $2 \le p < +\infty$, we have

$$\begin{aligned}
&\lim_{t_0\to 0} \frac{\theta(t_0)-\theta(0)}{t_0} \\
&= \lim_{t_0\to 0} \int p|t_0^*(\omega)f(\omega)+g(\omega)|^{p-2}(t_0^*(\omega)f(\omega)+g(\omega))f(\omega)\,d\mu(\omega) \\
&= \int \lim_{t_0\to 0} p|t_0^*(\omega)f(\omega)+g(\omega)|^{p-2}(t_0^*(\omega)f(\omega)+g(\omega))f(\omega)\,d\mu(\omega) \\
&= \int p|g(\omega)|^{p-2}g(\omega)f(\omega)\,d\mu(\omega)
\end{aligned}$$

since $t_0 \to 0$ forces $t_0^*(\omega) \to 0$ for every $\omega \in \Omega$. If $1 < p < 2$, then, writing $C = \{\omega \in \Omega : f(\omega) \neq 0\}$, we have

$$\begin{aligned}
&\lim_{t_0\to 0} \frac{\theta(t_0)-\theta(0)}{t_0} \\
&= \lim_{t_0\to 0} \int_{A(t_0)} p|t_0^*(\omega)f(\omega)+g(\omega)|^{p-2}(t_0^*(\omega)f(\omega)+g(\omega))f(\omega)\,d\mu(\omega) \\
&= \lim_{t_0\to 0} \int_{A(t_0)\cap C} p|t_0^*(\omega)f(\omega)+g(\omega)|^{p-2}(t_0^*(\omega)f(\omega)+g(\omega))f(\omega)\,d\mu(\omega) \\
&= \int \lim_{t_0\to 0} p|t_0^*(\omega)f(\omega)+g(\omega)|^{p-2}(t_0^*(\omega)f(\omega)+g(\omega))f(\omega)I_{A(t_0)\cap C}(\omega)d\mu(\omega) \\
&= \int p|g(\omega)|^{p-2}g(\omega)f(\omega)I_{A(0)\cap C}(\omega)\,d\mu(\omega) \\
&= \int p|g(\omega)|^{p-2}g(\omega)f(\omega)\,d\mu(\omega),
\end{aligned}$$

where the fourth equality follows because $\lim_{t_0\to 0} I_{A(t_0)\cap C}(\omega) = I_{A(0)\cap C}(\omega)$ for each $\omega \in \Omega$ [see Exercise 1], and the last equality follows since $A(0) \cap C$ is the set $\{\omega \in \Omega : f(\omega) \neq 0,\ g(\omega) \neq 0\}$. Combining the $2 \le p < +\infty$ and $1 < p < 2$ cases, we may say for every $1 < p < +\infty$ that

$$\theta'(0) = \lim_{t_0\to 0} \frac{\theta(t_0)-\theta(0)}{t_0} = \int p|g|^{p-2}gf.$$

Since $\theta'(0)$ exists and $\theta(t) = \psi_f^p(t)$, we have that $\psi_f(t) = \theta(t)^{\frac{1}{p}}$ and hence ψ_f is differentiable at $t = 0$. In particular, recalling that $||g||_p = 1$, we have

$$\begin{aligned}
\psi_f'(0) = \frac{1}{p}\theta(t)^{\frac{1}{p}-1}\theta'(t)\Big|_{t=0} &= \frac{1}{p}\theta(0)^{\frac{1}{p}-1}\theta'(0) \\
&= \frac{1}{p}\left(\int |g|^p\right)^{\frac{1}{p}-1}\int p|g|^{p-2}gf \\
&= \int |g|^{p-2}gf,
\end{aligned}$$

which is what we wanted to show. This completes the proof. ■

Exercises.

1*. Let the sets $A(t)$ and C be as in the proof of the Riesz Representation Theorem, and let t_n denote a sequence with $t_n \to 0$. We will show that $\lim_n I_{A(t_n)\cap C} = I_{A(0)\cap C}$ on Ω.
(a) It is sufficient to show that $\lim_n [A(t_n) \cap C] = A(0) \cap C$.
(b) $\overline{\lim}_n A(t_n)^c \subseteq A(0)^c$, that is, $A(0) \subseteq \underline{\lim}_n A(t_n)$. [Let $\omega \in \overline{\lim}_n A(t_n)^c$. Then there is an increasing sequence of indices $n_1 < n_2 < \cdots$ with the property that $g(\omega) = -t^*_{n_k}(\omega) f(\omega)$ for each $k \in \mathbb{N}$. If $f(\omega) \neq 0$, then $\frac{g(\omega)}{f(\omega)} = -t^*_{n_k}(\omega)$ for each $k \in \mathbb{N}$. That is, $t^*_{n_1}(\omega) = t^*_{n_2}(\omega) = \cdots$. Since $t_n \to 0$, we have that $t^*_{n_k}(\omega) \to 0$ as $k \to \infty$. It follows that $t^*_{n_k}(\omega) = 0$ for each $k \in \mathbb{N}$. But then $g(\omega) = 0$, so that $\omega \in A(0)^c$. If we instead have $f(\omega) = 0$, then clearly $g(\omega) = 0$, so again we have $\omega \in A(0)^c$.]
(c) $A(0) \cap C \subseteq \underline{\lim}_n [A(t_n) \cap C]$. [Nearly trivial.]
(d) $\overline{\lim}_n [A(t_n) \cap C] \subseteq A(0) \cap C$. [Suppose that there is an increasing sequence of indices $n_1 < n_2 < \cdots$ such that $\omega \in A(t_{n_k}) \cap C$ for each $k \in \mathbb{N}$. Then we have $g(\omega) \neq -t^*_{n_k}(\omega) f(\omega)$ for each $k \in \mathbb{N}$ and $f(\omega) \neq 0$. Therefore, $\frac{g(\omega)}{f(\omega)} = -t^*_{n_k}(\omega)$ for each $k \in \mathbb{N}$, and the same reasoning as in the outline for (b) gives $t^*_{n_k}(\omega) = 0$ for each $k \in \mathbb{N}$. But then $g(\omega) \neq 0$, and since $f(\omega) \neq 0$, we have the desired result.]
(e) By putting (c) and (d) together, show that $\lim_n [A(t_n) \cap C] = A(0) \cap C$.

2. Let $1 < p < +\infty$. Define $T : L^{p'} \to \mathcal{B}(L^p, \mathbb{R})$ [where $\mathcal{B}(L^p, \mathbb{R})$ denotes the collection of bounded linear transformations from L^p to $\mathbb{R}$] by writing $T(g) = L_g$ for every $g \in L^{p'}$. [The notation L_g is as in Claim 1 of this section.]
(a) T is a linear transformation from $L^{p'}$ to $\mathcal{B}(L^p, \mathbb{R})$. [Direct.]
(b) T is norm-preserving: $\|T(g)\| = \|g\|_{p'}$ for every $g \in L^p$. [Use Claim 1.]
(c) T is *onto* $\mathcal{B}(L^p, \mathbb{R})$. [Use the Riesz theorem.]
(d) T is one-to-one. [This follows from some basic properties of the integral.]
(e) $L^{p'}$ and $\mathcal{B}(L^p, \mathbb{R})$ are isomorphic Banach spaces. [Put the previous parts together.]

Comment: By part (e), the spaces $L^{p'}$ and $\mathcal{B}(L^p, \mathbb{R})$, while consisting of formally different entities — the former space consists of functions [or more properly equivalence classes of functions] while the latter space consists of bounded linear functionals on L^p — are identical *when viewed as Banach spaces*. This interesting fact is not used anywhere in this text.

9

The Radon-Nikodym Theorem

This chapter is devoted to proving the classical and foundational theorem of Radon and Nikodym. The introduction here aims to give some background by which the theorem may be appreciated.

Consider a measure space $(\Omega, \mathcal{F}, \mu)$, and let f denote a nonnegative function that is $\mathcal{F}/\mathcal{B}^*$-measurable and defined μ-a.e. on Ω. Define $\nu : \mathcal{F} \to [0, +\infty]$ by writing

$$\nu(A) = \int_A f \, d\mu \quad \text{for every } A \in \mathcal{F}. \tag{9.1}$$

In Section 6.4, property (L14) establishes the following results:
(i) ν is a measure on $\mathcal{F}$,
(ii) for every nonnegative $\mathcal{F}/\mathcal{B}^*$-measurable g defined μ-a.e. on Ω and $E \in \mathcal{F}$ we have $\int_E g \, d\nu = \int_E fg \, d\mu$, and
(iii) for general $\mathcal{F}/\mathcal{B}^*$-measurable g defined μ-a.e. on Ω and $E \in \mathcal{F}$ we have that $\int_E g \, d\nu$ exists iff $\int_E fg \, d\mu$ exists, in which case $\int_E g \, d\nu = \int_E fg \, d\mu$.
We also have the following property, courtesy of (L2):
(iv) $\nu(E) = 0$ for every $E \in \mathcal{F}$ with $\mu(E) = 0$.
Observe that taking $E = \Omega$ and $g = I_A$ [where $A \in \mathcal{F}$] in (ii) gives (9.1).

Therefore, given a nonnegative $\mathcal{F}/\mathcal{B}^*$-measurable function f defined μ-a.e. on Ω that satisfies (9.1), it follows that properties (i)-(iv) hold. The theorem of Radon and Nikodym establishes a sort-of converse. We will ultimately show the following: *if μ is a σ-finite measure on $\mathcal{F}$, if ν is* any *measure with domain $\mathcal{F}$, and if (iv) holds, then there exists a nonnegative $\mathcal{F}/\mathcal{B}^*$-measurable f defined μ-a.e. on Ω such that (9.1) and (i)-(iv) hold.*

Building up the machinery needed to prove this general result is the main order of business of the two upcoming sections. In building up the required machinery, we shall introduce the *Hahn decomposition* and the *Jordan decomposition.* These will be used to give the Radon-Nikodym Theorem. We shall also prove another major result in measure theory, called the *Lebesgue Decomposition Theorem.*

We will exhibit the Radon-Nikodym Theorem in two applications. First, we shall revisit the Riesz Representation Theorem [given in the previous chapter for $1 < p < +\infty$], and we shall rederive the result for $1 \leq p < +\infty$ using the Radon-Nikodym Theorem and avoiding the somewhat tedious analytical manipulations found in the previous chapter. Second, we shall state and prove some results concerning *martingales*, following the work of Andersen and Jessen.

9.1 THE RADON–NIKODYM THEOREM, PART I

Throughout this section, we take Ω to be a generic nonempty set, and $\mathcal{F}$ will denote a generic σ-field on Ω.

Definition. A function $\phi : \mathcal{F} \to \mathbb{R}$ is called an *additive set function [on $\mathcal{F}$]* iff

$$\phi\left(\bigcup_{k=1}^{\infty} A_k\right) = \lim_{n\to\infty} \sum_{k=1}^{n} \phi(A_k) \quad \text{whenever } A_1, A_2, \cdots \in \mathcal{F} \text{ are disjoint.}$$

We discuss some points regarding the definition.

(1) ϕ is real-valued; it cannot assume the values $\pm\infty$.

(2) Given a disjoint sequence $\{A_n\}_{n=1}^{\infty}$ of $\mathcal{F}$-sets, the definition requires that $\lim_{n\to\infty} \sum_{k=1}^{n} \phi(A_k)$ exists as a real number and is equal to the real number $\phi(\bigcup_{k=1}^{\infty} A_k)$. We write $\sum_{k=1}^{\infty} \phi(A_k)$ for $\lim_{n\to\infty} \sum_{k=1}^{n} \phi(A_k)$, so that the definition may be written as follows:

$$\phi\left(\bigcup_{k=1}^{\infty} A_k\right) = \sum_{k=1}^{\infty} \phi(A_k) \quad \text{whenever } A_1, A_2, \cdots \in \mathcal{F} \text{ are disjoint.}$$

(3) Every finite measure is an additive set function; infinite measures are not additive set functions because they assume infinite values.

(4) Taking $A_n = \varnothing$ for each $n \in \mathbb{N}$ in the above definition gives the equation $\phi(\varnothing) = \phi(\varnothing) + \phi(\varnothing) + \cdots$, hence $\phi(\varnothing) = 0$.

(5) ϕ is finitely additive in the sense that $\phi(\bigcup_{k=1}^{n} A_k) = \sum_{k=1}^{n} \phi(A_k)$ whenever $A_1, \cdots, A_n \in \mathcal{F}$ are disjoint. This follows by taking $A_{n+1} = A_{n+2} = \cdots = \varnothing$ in the definition and using the fact that $\phi(\varnothing) = 0$.

(6) We have $\phi(B - A) = \phi(B) - \phi(A)$ whenever $A \subseteq B$ and $A, B \in \mathcal{F}$ by the finiteness and finite additivity of ϕ. To see this, observe that the disjointness

of A and $B - A$ forces $\phi(B) = \phi(A \cup (B - A)) = \phi(A) + \phi(B - A)$, and, since all quantities are finite, we have $\phi(B) - \phi(A) = \phi(B - A)$. In particular, taking $B = \Omega$ gives $\phi(A^c) = \phi(\Omega) - \phi(A)$.

We now present the two main examples of additive set functions.

Example 1. Let μ_1 and μ_2 denote finite measures with domain $\mathcal{F}$. For each $A \in \mathcal{F}$, define $(\mu_1 - \mu_2)(A) = \mu_1(A) - \mu_2(A)$. Observe that $\mu_1 - \mu_2$ is defined and real-valued, and

$$\begin{aligned}(\mu_1 - \mu_2)\left(\bigcup_{k=1}^{\infty} A_k\right) &= \mu_1\left(\bigcup_{k=1}^{\infty} A_k\right) - \mu_2\left(\bigcup_{k=1}^{\infty} A_k\right) \\ &= \sum_{k=1}^{\infty} \mu_1(A_k) - \sum_{k=1}^{\infty} \mu_2(A_k) \\ &= \sum_{k=1}^{\infty} [\mu_1(A_k) - \mu_2(A_k)] \\ &= \sum_{k=1}^{\infty} (\mu_1 - \mu_2)(A_k)\end{aligned}$$

whenever $A_1, A_2, \cdots \in \mathcal{F}$ are disjoint. [The third equality holds because both series $\sum_{k=1}^{\infty} \mu_1(A_k)$ and $\sum_{k=1}^{\infty} \mu_2(A_k)$ converge.] Therefore, $\mu_1 - \mu_2$ is an additive set function on $\mathcal{F}$. More generally, $c_1\mu_1 - c_2\mu_2$ is an additive set function for every $c_1, c_2 \in \mathbb{R}$.

Example 2. Let μ denote a measure with domain $\mathcal{F}$, and let $f \in L^1(\Omega, \mathcal{F}, \mu)$. The mapping ϕ on $\mathcal{F}$ given for all $A \in \mathcal{F}$ by $\phi(A) = \int_A f \, d\mu$ is an additive set function. The verification of this claim is automatic by (L6) in Section 6.4.

Just as we proved certain continuity results for measures in Chapter 2 [properties (M8) and (M9) in particular], there is an analog to the notions of *continuity from below/above* for additive set functions.

Claim 1. *Let ϕ denote an additive set function on $\mathcal{F}$.*
(a) If $\{E_n\}_{n=1}^{\infty}$ is a nondecreasing sequence of $\mathcal{F}$-sets with limit E, then $\lim_n \phi(E_n) = \phi(E)$.
(b) If $\{E_n\}_{n=1}^{\infty}$ is a nonincreasing sequence of $\mathcal{F}$-sets with limit E, then $\lim_n \phi(E_n) = \phi(E)$.

Proof: We do (a) first, assuming that $E_1 \subseteq E_2 \subseteq \cdots$. Then

$$\begin{aligned}\phi(E) = \phi\left(E_1 \cup \bigcup_{n=1}^{\infty}(E_{n+1} - E_n)\right) &= \phi(E_1) + \sum_{n=1}^{\infty} \phi(E_{n+1} - E_n) \\ &= \lim_n \left(\phi(E_1) + \sum_{i=1}^{n} \phi(E_{i+1} - E_i)\right) \\ &= \lim_n \phi\left(E_1 \cup \bigcup_{i=1}^{n}(E_{i+1} - E_i)\right) \\ &= \lim_n \phi(E_n).\end{aligned}$$

[The second equality follows because E_1, $E_2 - E_1$, $E_3 - E_2$, $\cdots$ are disjoint, the third equality follows because the infinite series $\phi(E_1) + \phi(E_2 - E_1) + \cdots$ converges, and the fourth equality follows by the finite additivity of ϕ.] This gives (a).

For (b), assume that $E_1 \supseteq E_2 \supseteq \cdots$ with limit E, so that $E_1^c \subseteq E_2^c \subseteq \cdots$ with limit E^c. [That is, $\lim_n E_n^c = E^c$.] By part (a),

$$\begin{aligned}\phi(E) = \phi(\Omega) - \phi(E^c) &= \phi(\Omega) - \lim_n \phi(E_n^c) \\ &= \lim_n [\phi(\Omega) - \phi(E_n^c)] \\ &= \lim_n \phi(E_n)\end{aligned}$$

This completes the entire proof. ■

Observe that the proof above is very much like the proofs for (M8) and (M9) in Chapter 2. However, the above proof was needed because additive set functions are not necessarily measures.

In all that follows, ϕ denotes an additive set function on $\mathcal{F}$.

Definition. A set $A \in \mathcal{F}$ is *positive [with respect to ϕ]* iff $\phi(E) \geq 0$ for every $E \in \mathcal{F}$ with $E \subseteq A$. A set $A \in \mathcal{F}$ is *negative [with respect to ϕ]* iff $\phi(E) \leq 0$ for every $E \in \mathcal{F}$ with $E \subseteq A$.

In particular, if A is positive [or negative], then $\phi(A) \geq 0$ [or $\phi(A) \leq 0$]. This follows by using the definition with A viewed as a subset of itself.

The first decomposition is as follows.

Definition. We will say that (A, A^c) is a *Hahn decomposition of Ω with respect to ϕ* iff A is a positive set with respect to ϕ and A^c is a negative set with respect to ϕ.

It is in no way obvious that a Hahn decomposition of Ω with respect to a generic additive set function ϕ exists. Claim 3 will show that our definition is not vacuous. Before getting to Claim 3, we will prove the following technical claim.

Claim 2. *There is a set $A^+ \in \mathcal{F}$ such that $\phi(A^+) = \sup\{\phi(A) : A \in \mathcal{F}\}$.*

Proof: Write $\alpha = \sup\{\phi(A) : A \in \mathcal{F}\}$. Choose a sequence $\{A_n\}_{n=1}^\infty$ in $\mathcal{F}$ such that $\phi(A_n) \to \alpha$, and write $A = \bigcup_{n=1}^\infty A_n$. Next, for each $n \in \mathbb{N}$ and $i = 1, \cdots, 2^n$, write

$$B_{ni} = \bigcap_{k=1}^n A'_k, \quad \text{each } A'_k \text{ freely denotes } A_k \text{ or } A - A_k.$$

For each $n \in \mathbb{N}$, we have that $B_{n1}, \cdots, B_{n2^n}$ are disjoint with union A;[1] let $\mathcal{B}_n = \{B_{n1}, \cdots, B_{n2^n}\}$. Each $\mathcal{B}_n$ refines $\mathcal{B}_{n-1}$ in the sense that for every $j \in \{1, \cdots, 2^n\}$ there is a unique $i \in \{1, \cdots, 2^{n-1}\}$ such that $B_{nj} \subseteq B_{n-1,i}$. For each $n \in \mathbb{N}$, define

$$C_n = \bigcup \{B_{ni} \in \mathcal{B}_n : \ \phi(B_{ni}) > 0\} \quad \text{[a disjoint union]}.$$

Since each A_n is a finite disjoint union of certain of the B_{ni}'s, the definition of C_n and the finite additivity of ϕ gives

$$\begin{aligned}
\phi(A_n) &= \phi\left(\bigcup_{i:\ B_{ni} \subseteq A} B_{ni}\right) \\
&= \sum_{i:\ B_{ni} \subseteq A} \phi(B_{ni}) \\
&= \sum_{i:\ B_{ni} \subseteq A,\ \phi(B_{ni})>0} \phi(B_{ni}) + \sum_{i:\ B_{ni} \subseteq A,\ \phi(B_{ni}) \le 0} \phi(B_{ni}) \\
&\le \sum_{i:\ B_{ni} \subseteq A,\ \phi(B_{ni})>0} \phi(B_{ni}) \\
&\le \sum_{i:\ \phi(B_{ni})>0} \phi(B_{ni}) \\
&= \phi\left(\bigcup_{i:\ \phi(B_{ni})>0} B_{ni}\right) \\
&= \phi(C_n)
\end{aligned}$$

for every $n \in \mathbb{N}$. Since the partitions $\mathcal{B}_1, \mathcal{B}_2, \cdots$ are successively finer, the condition $m < n$ forces $(C_m \cup \cdots \cup C_n) - (C_m \cup \cdots \cup C_{n-1})$ to be a subset of C_n that is the finite [possibly empty] disjoint union of certain of the B_{ni}'s; since each such B_{ni} is an element of C_n, we have that $\phi(B_{ni}) > 0$ for each such B_{ni}. It follows that

$$\phi\left((C_m \cup \cdots \cup C_n) - (C_m \cup \cdots \cup C_{n-1})\right) \ge 0$$

by the finite additivity of ϕ. Therefore, we have

$$\phi(C_m \cup \cdots \cup C_{n-1}) \le \phi(C_m \cup \cdots \cup C_n) \quad \text{if } m < n.$$

Therefore, repeatedly using the condition above shows that $m < n$ forces

$$\phi(A_m) \le \phi(C_m) \le \phi(C_m \cup C_{m+1}) \le \cdots \le \phi(C_m \cup \cdots \cup C_n).$$

[1]For $n = 1$, we have $B_{11} = A_1$ and $B_{12} = A - A_1$. For $n = 2$, we have $B_{21} = A_1 \cap A_2$, $B_{22} = A_1 \cap (A - A_2)$, $B_{23} = (A - A_1) \cap A_2$, and $B_{24} = (A - A_1) \cap (A - A_2)$. The pattern for $n \ge 3$ is clear. This should make the preceding comments more understandable.

Next, for all $m \in \mathbb{N}$ define $D_m = \bigcup_{n=m}^{\infty} C_n$, so that D_m is the limit of the nondecreasing sequence $\{C_m \cup \cdots \cup C_{m+k}\}_{k=1}^{\infty}$. We have for every $k, m \in \mathbb{N}$ that $\phi(A_m) \leq \phi(C_m \cup \cdots \cup C_{m+k})$, and Claim 1 gives

$$\phi(D_m) = \lim_{k \to \infty} \phi(C_m \cup \cdots \cup C_{m+k}) \geq \phi(A_m) \quad \text{for every } m \in \mathbb{N}.$$

We now define $A^+ = \bigcap_{m=1}^{\infty} D_m$, so that A^+ is the limit of the nonincreasing sequence $\{D_m\}_{m=1}^{\infty}$, hence

$$\phi(A^+) = \lim_{m \to \infty} \phi(D_m) \geq \lim_{m \to \infty} \phi(A_m) = \alpha$$

by virtue of a second application of Claim 1. Since $\phi(A^+) \leq \alpha$ by the definition of α, we have that $\phi(A^+) = \alpha$. The set A^+ is as promised, and the proof is now complete. ∎

The preceding claim will now be used to show that our definition of a Hahn decomposition of Ω with respect to ϕ is not a vacuous one.

Claim 3. *There exists a Hahn decomposition of Ω with respect to ϕ.*

Proof: As in the previous claim, let $\alpha = \sup\{\phi(A) : A \in \mathcal{F}\}$, and let A^+ be such that $\phi(A^+) = \alpha$. Define A^- to be the complement of A^+. We will show that A^+ is positive and A^- is negative, so that (A^+, A^-) is a Hahn decomposition of Ω with respect to ϕ. We proceed by contradiction in discussing both A^+ and A^-. If A^+ fails to be positive, then there is $E \in \mathcal{F}$ with $E \subseteq A^+$ and $\phi(E) < 0$. Then $\alpha \geq \phi(A^+ - E) = \phi(A^+) - \phi(E) > \phi(A^+) = \alpha$, a contradiction. Therefore, A^+ is positive. Next, if A^- fails to be negative, then there is $E \in \mathcal{F}$ with $E \subseteq A^-$ and $\phi(E) > 0$. Then A^+ and E are disjoint, giving $\phi(A^+ \cup E) = \phi(A^+) + \phi(E) > \phi(A^+) = \alpha$, contradicting the definition of α. Therefore, A^- is negative, and the proof is now complete. ∎

In Exercise 3, we discuss in what sense the Hahn decomposition of Ω with respect to an additive set function ϕ is unique. The result given there is this: *if (A^+, A^-) and (B^+, B^-) are Hahn decompositions of Ω with respect to ϕ, then $\phi(A^+ \triangle B^+) = \phi(A^- \triangle B^-) = 0$.*

We now introduce the second of the three decompositions given in this section. Whereas the Hahn decomposition decomposes Ω into disjoint sets with certain properties, the upcoming decomposition decomposes an additive set function into what will eventually be shown to be the difference of two finite measures.

Definition. Let (A^+, A^-) denote a Hahn decomposition of Ω with respect to the additive set function ϕ. Define $\phi^+ : \mathcal{F} \to \mathbb{R}$ and $\phi^- : \mathcal{F} \to \mathbb{R}$ by writing

$$\phi^+(A) = \phi(A \cap A^+) \quad \text{and} \quad \phi^-(A) = -\phi(A \cap A^-) \quad \text{for every } A \in \mathcal{F}.$$

The functions ϕ^+ and ϕ^- are called the *upper variation* and *lower variation* of ϕ. Observing that $\phi = \phi^+ - \phi^-$, we refer to the representation of ϕ as $\phi^+ - \phi^-$ as the *Jordan decomposition of ϕ*.

Given an additive set function ϕ, the Jordan decomposition of ϕ is unique. An outline of the proof of this fact is given in Exercise 4, and uses the discussion of the uniqueness of Hahn decompositions given in Exercise 3.

Definition. Let μ denote a measure on $\mathcal{F}$. A set $E \in \mathcal{F}$ is called a *support of* μ iff $\mu(E^c) = 0$. We say that E *supports* μ.

Claim 4. *Assume the notation above. The functions ϕ^+ and ϕ^- above are finite measures on $\mathcal{F}$, having supports A^+ and A^-, respectively. [Therefore, ϕ^+ and ϕ^- have disjoint supports.]*

Proof: We deal with ϕ^+ first. For any $A \in \mathcal{F}$ we have $A \cap A^+ \subseteq A^+$, and the positivity of A^+ with respect to ϕ implies that $\phi^+(A) = \phi(A \cap A^+) \geq 0$. Therefore, ϕ^+ is nonnegative. Next, $\phi^+(\varnothing) = \phi(\varnothing \cap A^+) = \phi(\varnothing) = 0$. We now exhibit countable additivity for ϕ^+. Let $\{A_n\}_{n=1}^{\infty}$ denote a disjoint sequence of $\mathcal{F}$-sets. Then

$$
\begin{aligned}
\phi^+\left(\bigcup_{n=1}^{\infty} A_n\right) = \phi\left(\bigcup_{n=1}^{\infty} A_n \cap A^+\right) &= \phi\left(\bigcup_{n=1}^{\infty} (A_n \cap A^+)\right) \\
&= \sum_{n=1}^{\infty} \phi(A_n \cap A^+) \\
&= \sum_{n=1}^{\infty} \phi^+(A_n),
\end{aligned}
$$

where the third equality follows since $\{A_n \cap A^+\}_{n=1}^{\infty}$ is a disjoint sequence of $\mathcal{F}$-sets. This shows that ϕ^+ is a measure. Since $\phi^+(\Omega) = \phi(A^+) \in \mathbb{R}$, ϕ^+ is a finite measure. To see that A^+ is a support of ϕ^+, observe that $\phi^+(A^{+c}) = \phi(A^{+c} \cap A^+) = \phi(\varnothing) = 0$. That ϕ^- is a finite measure with support A^- follows by the same straightforward argument, and is left to the reader in Exercise 2. This completes the proof. ∎

Definition. Measures μ and ν [each with common domain $\mathcal{F}$] are *[mutually] singular* iff μ and ν have disjoint supports. In this case, we write $\mu \perp \nu$, or we may equivalently write $\nu \perp \mu$.

Example 3. The measures ϕ^+ and ϕ^- are singular by virtue of Claim 4: ϕ^+ has support A^+, ϕ^- has support A^-, and $A^+ \cap A^- = \varnothing$.

When $\mu \perp \nu$, there exist sets A and B with $\mu(A^c) = 0$, $\nu(B^c) = 0$, and $A \cap B = \varnothing$. Let $C = \Omega - (A \cup B) = A^c \cap B^c$. We may then state that $B^c = A \cup C$ supports μ since $\mu(B) = \mu((A \cup C)^c) = \mu(A^c \cap C^c) \leq \mu(A^c) = 0$, so that $\mu(B) = 0 = \nu(B^c)$. Therefore, when we say that $\mu \perp \nu$, we may without any loss of generality say that there is a set $D \in \mathcal{F}$ with $\mu(D) = \nu(D^c) = 0$, that is, D supports ν and D^c supports μ.

Definition. Let μ, ν denote measures with common domain $\mathcal{F}$. We will say that ν is *absolutely continuous*[2] *with respect to* μ iff $\nu(A) = 0$ for every $A \in \mathcal{F}$ with $\mu(A) = 0$. We use the notation $\nu \ll \mu$ or $\mu \gg \nu$.

Note that while $\mu \perp \nu$ means the same as $\nu \perp \mu$, it is not necessarily the case that $\nu \ll \mu$ means the same thing as $\mu \ll \nu$.

Claim 5. *Let μ and ν denote finite measures with common domain $\mathcal{F}$. Further assume that $\mu \not\perp \nu$. Then there is a set $A \in \mathcal{F}$ and $\epsilon > 0$ such that $\mu(A) > 0$ and $\epsilon\mu(E) \leq \nu(E)$ for every $E \in \mathcal{F}$ with $E \subseteq A$.*

Proof: For each $n \in \mathbb{N}$, $\nu - \frac{1}{n}\mu$ is an additive set function [see Example 1], and Claim 3 ensures the existence of a Hahn decomposition (A_n^+, A_n^-) of Ω with respect to $\nu - \frac{1}{n}\mu$. Define $M = \bigcup_{n=1}^{\infty} A_n^+$. Then $M^c = \bigcap_{n=1}^{\infty} A_n^-$. For every $n \in \mathbb{N}$ we have $M^c \subseteq A_n^-$, hence $(\nu - \frac{1}{n}\mu)(M^c) \leq 0$, whence $\nu(M^c) \leq \frac{1}{n}\mu(M^c)$. Since μ is finite, this forces $\nu(M^c) = 0$, and thus M supports ν. By the assumption that $\mu \not\perp \nu$, it follows that M^c cannot possibly support μ, which in turn forces $\mu(M) > 0$. Since $M = \bigcup_{n=1}^{\infty} A_n^+$, there is $N \in \mathbb{N}$ such that $\mu(A_N^+) > 0$. Let $\epsilon = \frac{1}{N}$ and $A = A_N^+$ for this aforementioned N. Then $\mu(A) = \mu(A_N) > 0$ and $\epsilon\mu(E) = \frac{1}{N}\mu(E) \leq \nu(E)$ for every $E \in \mathcal{F}$ with $E \subseteq A$. [If $E \subseteq A$, then $E \subseteq A_N^+$; recalling that A_N^+ is a positive set for $\nu - \frac{1}{N}\mu$ gives $(\nu - \frac{1}{N}\mu)(E) \geq 0$, hence $\frac{1}{N}\mu(E) \leq \nu(E)$.] Therefore, the set $A = A_N$ and $\epsilon = \frac{1}{N}$ satisfy the requirements of the claim. ∎

We have talked about the Hahn decomposition of a space Ω into positive and negative sets [relative to a given additive set function]. We have next talked about the Jordan decomposition of an additive set function ϕ into the difference $\phi^+ - \phi^-$ of the finite measures ϕ^+ and ϕ^-. The next and final decomposition, called the *Lebesgue decomposition*, involves [quite roughly] breaking a measure into an "absolutely continuous part" and a "singular part."

Lebesgue Decomposition, Finite Case. *Let μ and ν denote finite measures with domain $\mathcal{F}$. Then there exist finite measures ν_{ac} and ν_s on $\mathcal{F}$ with the following properties:*

(a) $\nu(E) = \nu_{ac}(E) + \nu_s(E)$ for every $E \in \mathcal{F}$: $\nu = \nu_{ac} + \nu_s$.

(b) ν_{ac} is absolutely continuous with respect to μ: $\nu_{ac} \ll \mu$.

(c) ν_s is singular with respect to μ: $\nu_s \perp \mu$.

Proof: Let $\mathcal{G}$ consist of those $\mathcal{F}/\mathcal{B}^*$-measurable $g : \Omega \to [0, +\infty]$ with the property that $\int_E g \, d\mu \leq \nu(E)$ for every $E \in \mathcal{F}$. We have the following three subclaims regarding $\mathcal{G}$:

(i) *$\mathcal{G}$ is closed under finite maxima.*

Proof: Let $g_1, g_2 \in \mathcal{G}$, and let $A = \{\omega \in \Omega : g_1(\omega) \geq g_2(\omega)\}$. Then $A \in \mathcal{F}$ and $\max\{g_1, g_2\}$ is a nonnegative $\mathcal{F}/\mathcal{B}^*$-measurable function on Ω. Next, observe

[2] Other authors will equivalently say that (i) μ *dominates* ν or (ii) ν *is dominated by* μ.

that

$$\begin{aligned}\int_E \max\{g_1, g_2\}\, d\mu &= \int_{E\cap A} \max\{g_1, g_2\}\, d\mu + \int_{E\cap A^c} \max\{g_1, g_2\}\, d\mu \\ &= \int_{E\cap A} g_1\, d\mu + \int_{E\cap A^c} g_2\, d\mu \\ &\le \nu(E\cap A) + \nu(E\cap A^c) \\ &= \nu(E)\end{aligned}$$

for every $E \in \mathcal{F}$, where the inequality follows because $g_1, g_2 \in \mathcal{G}$. This shows that $\max\{g_1, g_2\} \in \mathcal{G}$. By induction, it follows for every $n \in \mathbb{N}$ that $\max\{g_1, \cdots, g_n\} \in \mathcal{G}$ whenever $g_1, \cdots, g_n \in \mathcal{G}$, so that $\mathcal{G}$ is indeed closed under finite maxima.

(ii) *Let $\{g_n\}_{n=1}^{\infty}$ denote a nondecreasing sequence of functions in $\mathcal{G}$. Then $\lim_n g_n \in \mathcal{G}$.*

Proof: Let $g = \lim_n g_n$. Clearly g is a nonnegative $\mathcal{F}/\mathcal{B}^*$-measurable function with domain Ω, and we have $\int_E g_n\, d\mu \le \nu(E)$ for every $E \in \mathcal{F}$ and $n \in \mathbb{N}$. Next, using MCT, observe that

$$\int_E g\, d\mu = \lim_n \int_E g_n\, d\mu \le \nu(E) \quad \text{for every } E \in \mathcal{F}.$$

Therefore, $g \in \mathcal{G}$.

(iii) *Let $\alpha = \sup\{\int g\, d\mu : \ g \in \mathcal{G}\}$. [Note that $\alpha \le \nu(\Omega) < +\infty$ since $\int g\, d\mu \le \nu(\Omega) < +\infty$ for every $g \in \mathcal{G}$.] Then there exists a function $f \in \mathcal{G}$ such that f is finite on Ω and $\int f\, d\mu = \alpha$.*

Proof: By the definition of α as a supremum, we may choose a sequence $\{g_n\}_{n=1}^{\infty}$ of functions in $\mathcal{G}$ such that $\lim_n \int g_n\, d\mu = \alpha$. For every $n \in \mathbb{N}$, define the function $f_n = \max\{g_1, \cdots, g_n\}$. Then $\{f_n\}_{n=1}^{\infty}$ is a nondecreasing sequence of functions in $\mathcal{G}$ by (i), hence $\lim_n f_n \in \mathcal{G}$ by (ii). Using MCT and the fact that $f_n \ge g_n$ for every $n \in \mathbb{N}$, we have

$$\int \lim_n f_n\, d\mu = \lim_n \int f_n\, d\mu \ge \lim_n \int g_n\, d\mu = \alpha.$$

Next, since $\lim_n f_n \in \mathcal{G}$, we have $\int \lim_n f_n\, d\mu \le \alpha$. Therefore, we have $\int \lim_n f_n\, d\mu = \alpha$ and $\lim_n f_n \in \mathcal{G}$. Observe that $\lim_n f_n$ is finite μ-a.e. on Ω since $\int \lim_n f_n\, d\mu = \alpha < +\infty$. Define $f : \Omega \to [0, +\infty)$ by writing

$$f(\omega) = \begin{cases} \lim_n f_n(\omega) & \text{if } \lim_n f_n(\omega) \text{ is finite} \\ 0 & \text{otherwise} \end{cases}$$

for every $\omega \in \Omega$. Note that $f = \lim_n f_n$ μ-a.e. on Ω. We have that f is nonnegative, $\mathcal{F}/\mathcal{B}^*$-measurable, everywhere finite, and for every $E \in \mathcal{F}$ we have $\int_E f\, d\mu = \int_E \lim_n f_n\, d\mu \le \nu(E)$ since $\lim_n f_n \in \mathcal{G}$. It follows that

$f \in \mathcal{G}$. Finally, observe that $\int f\, d\mu = \int \lim_n f_n\, d\mu = \alpha$, so that the function f satisfies the requirements of the subclaim.

With these three subclaims in hand, we now define the promised measures ν_{ac} and ν_s. Define ν_{ac} on $\mathcal{F}$ by writing

$$\nu_{ac}(E) = \int_E f\, d\mu \quad \text{for every } E \in \mathcal{F},$$

where f is as in (iii). [That is, $f \in \mathcal{G}$, $\int f\, d\mu = \alpha < +\infty$, and f is everywhere finite.] Since f is nonnegative, we have that ν_{ac} is nonnegative. Property (L2) from Section 6.4 shows that $\nu_{ac}(\varnothing) = 0$, and property (L6) from the same section shows that ν_{ac} is countably additive. Therefore, ν_{ac} is a measure with domain $\mathcal{F}$. Also, we have

$$\nu_{ac}(\Omega) = \int f\, d\mu = \alpha < +\infty,$$

so ν_{ac} is a finite measure. Also, $\nu_{ac} \ll \mu$ since $\mu(A) = 0$ forces $\int_A f\, d\mu = 0$ by (L2). We have constructed a finite measure ν_{ac} on $\mathcal{F}$ with $\nu_{ac} \ll \mu$, and part (b) is therefore complete.

We next define ν_s on $\mathcal{F}$ by writing

$$\nu_s(E) = \nu(E) - \nu_{ac}(E) = \nu(E) - \int_E f\, d\mu \quad \text{for every } E \in \mathcal{F}.$$

Since ν and ν_{ac} are finite, there are no problems with the preceding definition, and we immediately have $\nu(E) = \nu_{ac}(E) + \nu_s(E)$ for every $E \in \mathcal{F}$, which gives (a). All that remains is to show (c), which means that ν_s must be shown to be a finite measure that is singular with respect to μ.

(iv) *ν_s is a finite measure on $\mathcal{F}$.*

Proof: To see that ν_s is nonnegative, observe for every $E \in \mathcal{F}$ that

$$\nu_s(E) = \nu(E) - \int_E f\, d\mu \geq \nu(E) - \nu(E) = 0,$$

where the inequality follows by the fact that $f \in \mathcal{G}$. Next, it is clear that $\nu_s(\varnothing) = \nu(\varnothing) - \int_\varnothing f\, d\mu = 0 - 0 = 0$. Next, for countable additivity, let $\{E_n\}_{n=1}^\infty$ denote a disjoint sequence of $\mathcal{F}$-sets. Using (L6), we have

$$\begin{aligned}
\nu_s\left(\bigcup_{n=1}^\infty E_n\right) = \nu\left(\bigcup_{n=1}^\infty E_n\right) - \int_{\bigcup_{n=1}^\infty E_n} f\, d\mu &= \sum_{n=1}^\infty \nu(E_n) - \sum_{n=1}^\infty \int_{E_n} f\, d\mu \\
&= \sum_{n=1}^\infty \left[\nu(E_n) - \int_{E_n} f\, d\mu\right] \\
&= \sum_{n=1}^\infty \nu_s(E_n).
\end{aligned}$$

[The third equality follows because $\sum_{n=1}^{\infty} \nu(E_n)$ converges to the finite number $\nu(\bigcup_{n=1}^{\infty} E_n)$ and $\sum_{n=1}^{\infty} \int_{E_n} f\, d\mu$ converges to the finite number $\int_{\bigcup_{n=1}^{\infty} E_n} f\, d\mu$.] This establishes that ν_s is a measure. That ν_s is finite follows from the nearly trivial inequality $\nu_s(\Omega) = \nu(\Omega) - \int f\, d\mu \leq \nu(\Omega) < +\infty$.

(v) $\nu_s \perp \mu$.

Proof: We argue by contradiction, assuming that $\nu_s \not\perp \mu$. Since both μ and ν_s are finite measures [by assumption and (iv)], we use Claim 5, obtaining $\epsilon > 0$ and $A \in \mathcal{F}$ such that $\mu(A) > 0$ and

$$\epsilon\mu(E) \leq \nu_s(E) \quad \text{for every } E \in \mathcal{F} \text{ with } E \subseteq A.$$

Using this, we have for every $E \in \mathcal{F}$ that

$$\begin{aligned}
\int_E (f + \epsilon I_A)\, d\mu &= \int_E f\, d\mu + \epsilon\mu(E \cap A) \\
&\leq \int_E f\, d\mu + \nu_s(E \cap A) \\
&= \int_{E\cap A} f\, d\mu + \nu_s(E \cap A) + \int_{E\cap A^c} f\, d\mu \\
&= \nu(E \cap A) + \int_{E\cap A^c} f\, d\mu \\
&\leq \nu(E \cap A) + \nu(E \cap A^c) \ \text{[since } f \in \mathcal{G}] \\
&= \nu(E).
\end{aligned}$$

Now $f + \epsilon I_A$ is clearly nonnegative, defined on Ω, and $\mathcal{F}/\mathcal{B}^*$-measurable, and for every $E \in \mathcal{F}$ we have $\int_E (f + \epsilon I_A)\, d\mu \leq \nu(E)$. It follows that $f + \epsilon I_A \in \mathcal{G}$. At the same time, $\int (f + \epsilon I_A)\, d\mu = \alpha + \epsilon\mu(A) > \alpha$. This contradicts the fact that $\alpha = \sup\{\int g\, d\mu : g \in \mathcal{G}\}$. The assumption that $\nu_s \not\perp \mu$ is therefore untenable; it follows that $\nu_s \perp \mu$ as claimed.

Summarizing everything, we have that $\nu = \nu_{ac} + \nu_s$, where ν_{ac} and ν_s are finite measures on $\mathcal{F}$. We furthermore have $\nu_{ac} \ll \mu$ and $\nu_s \perp \mu$. Everything is as promised, and the entire proof is now complete. ∎

Using the just-given result, it is a small step forward to prove the finite version of the famous Radon-Nikodym Theorem.

Radon-Nikodym Theorem, Finite Case. *Let μ and ν denote finite measures on $\mathcal{F}$ such that $\nu \ll \mu$. Then there exists a finite-valued nonnegative $\mathcal{F}/\mathcal{B}^*$-measurable function f on Ω with the following properties:*

(a) $\nu(A) = \int_A f\, d\mu$ for every $A \in \mathcal{F}$.

(b) If a nonnegative $\mathcal{F}/\mathcal{B}^*$-measurable function g defined on Ω is such that $\nu(A) = \int_A g\, d\mu$ for every $A \in \mathcal{F}$, then $f = g$ μ-a.e. on Ω. That is, f is unique up to a set of μ-measure zero.

Proof: Return to the previous proof, and recall the finite-valued nonnegative $\mathcal{F}/\mathcal{B}^*$-measurable function f constructed there. We claim that the assumption

$\nu \ll \mu$ gives $\nu(E) = \int_E f\,d\mu$ for every $E \in \mathcal{F}$. To prove this, we use the fact that $\nu_s \perp \mu$ to find a set $S \in \mathcal{F}$ such that $\nu_s(S) = \mu(S^c) = 0$. [That is, S supports μ and S^c supports ν_s.] Since $\mu(S^c) = 0$ and $\nu \ll \mu$, it follows that $\nu(S^c) = 0$. Since $\nu_s \leq \nu$ [recall that $\nu = \nu_{ac} + \nu_s$], we have $\nu_s(S^c) = 0$, so that $\nu_s(\Omega) = 0$. It follows that $\nu_s(E) = 0$ for every $E \in \mathcal{F}$. Therefore, for every $E \in \mathcal{F}$ we have that

$$\nu(E) = \nu_{ac}(E) + \nu_s(E) = \nu_{ac}(E) + 0 = \nu_{ac}(E) = \int_E f\,d\mu,$$

giving (a).

For (b), assume that there is a nonnegative $\mathcal{F}/\mathcal{B}^*$-measurable function g on Ω such that $\nu(E) = \int_E g\,d\mu$ for every $E \in \mathcal{F}$. Then $\int_E f\,d\mu = \int_E g\,d\mu$ for every $E \in \mathcal{F}$. By (L17), we have $f = g$ μ-a.e. on Ω, completing the proof. ∎

We will extend the Radon-Nikodym Theorem and the Lebesgue Decomposition Theorem to the case where μ and ν are σ-finite measures. We do the Lebesgue Decomposition Theorem first, proving a simple preliminary result.

Claim 6. *Let $\rho, \xi_1, \xi_2, \cdots$ denote measures with domain $\mathcal{F}$ such that $\rho \perp \xi_n$ for every $n \in \mathbb{N}$. Then $\rho \perp \sum_{n=1}^{\infty} \xi_n$.*

Proof: For each $n \in \mathbb{N}$, the assumptions imply the existence of a set $E_n \in \mathcal{F}$ such that E_n^c supports ρ and E_n supports ξ_n: $\rho(E_n) = \xi_n(E_n^c) = 0$. Observe that $\bigcap_{n=1}^{\infty} E_n^c$ supports ρ since $\rho(\bigcup_{n=1}^{\infty} E_n) \leq \sum_{n=1}^{\infty} \rho(E_n) = 0$. Next, observe that $\bigcup_{n=1}^{\infty} E_n$ supports $\sum_{n=1}^{\infty} \xi_n$ since

$$\left(\sum_{n=1}^{\infty} \xi_n\right)\left(\bigcap_{j=1}^{\infty} E_j^c\right) = \sum_{n=1}^{\infty} \xi_n\left(\bigcap_{j=1}^{\infty} E_j^c\right) \leq \sum_{n=1}^{\infty} \xi_n(E_n^c) = 0.$$

Therefore, ρ and $\sum_{n=1}^{\infty} \xi_n$ have disjoint supports, completing the proof. ∎

Lebesgue Decomposition, σ-Finite Case. *Let μ and ν denote σ-finite measures with domain $\mathcal{F}$. Then there exist σ-finite measures ν_{ac} and ν_s on $\mathcal{F}$ with the following properties:*

(a) $\nu(E) = \nu_{ac}(E) + \nu_s(E)$ for every $E \in \mathcal{F}$: $\nu = \nu_{ac} + \nu_s$.

(b) ν_{ac} is absolutely continuous with respect to μ: $\nu_{ac} \ll \mu$.

(c) ν_s is singular with respect to μ: $\nu_s \perp \mu$.

Proof: Since μ and ν are σ-finite, we may decompose Ω into a disjoint sequence $\{A_n\}_{n=1}^{\infty}$ of $\mathcal{F}$-sets such that $\mu(A_n), \nu(A_n) < +\infty$ for every $n \in \mathbb{N}$.[3]

[3]Since μ is σ-finite, there is a disjoint sequence $\{B_n\}_{n=1}^{\infty}$ of $\mathcal{F}$-sets of finite μ-measure with $\bigcup_{n=1}^{\infty} B_n = \Omega$. Since ν is σ-finite, there is a disjoint sequence $\{C_n\}_{n=1}^{\infty}$ of $\mathcal{F}$-sets of finite ν-measure with $\bigcup_{n=1}^{\infty} C_n = \Omega$. The collection $\mathcal{C} = \{B_n \cap C_m : n, m \in \mathbb{N}\}$ is disjoint and countably infinite with union Ω. Furthermore, both $\nu(B_n \cap C_m)$ and $\mu(B_n \cap C_m)$ are finite for every $n, m \in \mathbb{N}$. Enumerate the elements of $\mathcal{C}$ as $A_1, A_2, \cdots$ for the result.

For every $n \in \mathbb{N}$, define

$$\mu_n(A) = \mu(A \cap A_n) \quad \text{and} \quad \nu_n(A) = \nu(A \cap A_n) \quad \text{for every } A \in \mathcal{F}.$$

We now make a series of subclaims.

(i) *Each μ_n is a finite measure with support A_n, and $\mu = \sum_{n=1}^{\infty} \mu_n$.*

Proof: Fix $n \in \mathbb{N}$. It is clear that $\mu_n(E) \geq 0$ for every $E \in \mathcal{F}$, and it is equally clear that $\mu_n(\varnothing) = \mu(\varnothing \cap A_n) = \mu(\varnothing) = 0$. If $\{E_j\}_{j=1}^{\infty}$ is a disjoint sequence of $\mathcal{F}$-sets, then $\{E_j \cap A_n\}_{j=1}^{\infty}$ is a disjoint sequence of $\mathcal{F}$-sets and

$$\begin{aligned}\mu_n\left(\bigcup_{j=1}^{\infty} E_j\right) = \mu\left(\bigcup_{j=1}^{\infty} E_j \cap A_n\right) = \mu\left(\bigcup_{j=1}^{\infty}(E_j \cap A_n)\right) &= \sum_{j=1}^{\infty} \mu(E_j \cap A_n) \\ &= \sum_{j=1}^{\infty} \mu_n(E_j).\end{aligned}$$

This shows that μ_n is countably additive, and hence μ_n is a measure. Next, $\mu_n(\Omega) = \mu(\Omega \cap A_n) \leq \mu(A_n) < +\infty$, so μ_n is a finite measure. Next, we have that $\mu_n(A_n^c) = \mu(A_n^c \cap A_n) = \mu(\varnothing) = 0$, so A_n supports μ_n. Finally, for any $E \in \mathcal{F}$ we have that $E \cap A_1, E \cap A_2, \cdots$ are disjoint with union E, hence

$$\begin{aligned}\left(\sum_{n=1}^{\infty} \mu_n\right)(E) = \sum_{n=1}^{\infty} \mu_n(E) = \sum_{n=1}^{\infty} \mu(E \cap A_n) &= \mu\left(\bigcup_{n=1}^{\infty}(E \cap A_n)\right) \\ &= \mu(E).\end{aligned}$$

This completes the proof of (i).

(ii) *Each ν_n is a finite measure with support A_n, and $\nu = \sum_{n=1}^{\infty} \nu_n$.*

Proof: Repeat the proof of (i), replacing μ_n by ν_n and μ by ν.

Using (i) and (ii), we now invoke the finite version of the Lebesgue Decomposition Theorem for each $n \in \mathbb{N}$, obtaining finite measures $\nu_{n,ac}$ and $\nu_{n,s}$ such that

$$\nu_n = \nu_{n,ac} + \nu_{n,s}, \quad \nu_{n,ac} \ll \mu_n, \quad \text{and} \quad \nu_{n,s} \perp \mu_n.$$

With this, we define ν_{ac} and ν_s on $\mathcal{F}$ by writing

$$\nu_{ac} = \sum_{n=1}^{\infty} \nu_{n,ac} \quad \text{and} \quad \nu_s = \sum_{n=1}^{\infty} \nu_{n,s}.$$

On the basis of this definition, (i), and (ii), we have the following subclaims:

(iii) $\nu = \nu_{ac} + \nu_s$, and thus (a) holds.

Proof: For any $E \in \mathcal{F}$, we have

$$\begin{aligned}\nu_{ac}(E) + \nu_s(E) = \sum_{n=1}^{\infty} \nu_{n,ac}(E) + \sum_{n=1}^{\infty} \nu_{n,s}(E) &= \sum_{n=1}^{\infty}[\nu_{n,ac}(E) + \nu_{n,s}(E)] \\ &= \sum_{n=1}^{\infty} \nu_n(E) = \nu(E),\end{aligned}$$

where the second equality follows because all terms are nonnegative and the last equality follows by (ii).

(iv) ν_{ac} *and* ν_s *are* σ*-finite measures.*

Proof: We deal with ν_{ac} only since the proof for ν_s is exactly the same. It is clear that $\nu_{ac} = \sum_{n=1}^{\infty} \nu_{n,ac}$ is nonnegative and $\nu_{ac}(\varnothing) = \sum_{n=1}^{\infty} \nu_{n,ac}(\varnothing) = 0$. If $\{E_j\}_{j=1}^{\infty}$ is a disjoint sequence of $\mathcal{F}$-sets, then

$$\begin{aligned} \nu_{ac}\left(\bigcup_{j=1}^{\infty} E_j\right) = \sum_{n=1}^{\infty} \nu_{n,ac}\left(\bigcup_{j=1}^{\infty} E_j\right) &= \sum_{n=1}^{\infty}\sum_{j=1}^{\infty} \nu_{n,ac}(E_j) \\ &= \sum_{j=1}^{\infty}\sum_{n=1}^{\infty} \nu_{n,ac}(E_j) \\ &= \sum_{j=1}^{\infty} \nu_{ac}(E_j), \end{aligned}$$

where the interchange of summations in the third equality follows because all terms are nonnegative. This shows that ν_{ac} is a measure. Next, fix n. Since A_n supports μ_n, we have that $\mu_n(A_n^c) = 0$, hence $\nu_{n,ac}(A_n^c) = 0$ [since $\nu_{n,ac} \ll \mu_n$], so A_n supports $\nu_{n,ac}$. If $j \neq n$, then $\nu_{j,ac}(A_n) = 0$ since A_j supports $\nu_{j,ac}$ and $A_n \subseteq A_j^c$. It follows that

$$\begin{aligned} \nu_{ac}(A_n) = \sum_{j=1}^{\infty} \nu_{j,ac}(A_n) &= \nu_{n,ac}(A_n) + \sum_{j \neq n} \nu_{j,ac}(A_n) \\ &= \nu_{n,ac}(A_n) \\ &< +\infty \end{aligned}$$

for every $n \in \mathbb{N}$. [Recall that $\nu_{n,ac}$ is a finite measure.] Since $\bigcup_{n=1}^{\infty} A_n = \Omega$, it follows that ν_{ac} is σ-finite, completing the proof of (iv).

The proof will be complete when we show that $\nu_{ac} \ll \mu$ and $\nu_s \perp \mu$.

(v) $\nu_{ac} \ll \mu$.

Proof: If $E \in \mathcal{F}$ is such that $\mu(E) = 0$, then $\sum_{n=1}^{\infty} \mu_n(E) = 0$ by (i), so $\mu_n(E) = 0$ for every $n \in \mathbb{N}$. Since $\nu_{n,ac} \ll \mu_n$ for each $n \in \mathbb{N}$, it follows that $\nu_{n,ac}(E) = 0$ for every $n \in \mathbb{N}$, hence $\nu_{ac}(E) = \sum_{n=1}^{\infty} \nu_{n,ac}(E) = 0$. Therefore, $\mu(E) = 0$ implies $\nu_{ac}(E) = 0$, giving $\nu_{ac} \ll \mu$ as desired.

(vi) $\nu_{n,s} \perp \mu_m$ *for every* $m, n \in \mathbb{N}$.

Proof: Fix $n, m \in \mathbb{N}$. If $m = n$, then the claim reduces to showing that $\nu_{n,s} \perp \mu_n$, which is automatic from our previous invocation of the finite version of the Lebesgue Decomposition Theorem. Assume now that $m \neq n$. By (i), A_m is a support for μ_m, that is, $\mu_m(A_m^c) = 0$. Invoking the fact that ν_n, $\nu_{n,ac}$, and $\nu_{n,s}$ are finite measures and the fact that the sequence $\{A_j\}_{j=1}^{\infty}$

is disjoint, we have

$$\begin{aligned}\nu_{n,s}(A_m) = \nu_n(A_m) - \nu_{n,ac}(A_m) &= \nu(A_m \cap A_n) - \nu_{n,ac}(A_m) \\ &= \nu(\varnothing) - \nu_{n,ac}(A_m) \\ &= -\nu_{n,ac}(A_m),\end{aligned}$$

and since both $\nu_{n,s}$ and $\nu_{n,ac}$ are nonnegative, we must have $\nu_{n,s}(A_m) = 0$. That is, A_m^c is a support for $\nu_{n,s}$. Therefore μ_m and $\nu_{n,s}$ have the disjoint supports A_m and A_m^c, respectively. This gives $\nu_{n,s} \perp \mu_m$ when $m \neq n$. Combining the $m = n$ case with the $m \neq n$ case gives the subclaim.

(vii) $\nu_{n,s} \perp \sum_{m=1}^{\infty} \mu_m$ *for every* $n \in \mathbb{N}$.

Proof: This immediately follows from (vi) and Claim 6.

(viii) $\sum_{n=1}^{\infty} \nu_{n,s} \perp \sum_{m=1}^{\infty} \mu_m$.

Proof: This immediately follows from (vii) and Claim 6.

By (viii), (i), and the definition of ν_s, we immediately have $\nu_s \perp \mu$. Summarizing, we have that $\nu = \nu_{ac} + \nu_s$, where ν_{ac} and ν_s are σ-finite measures with $\nu_{ac} \ll \mu$ and $\nu_s \perp \mu$, and the entire proof is now complete. ∎

In Exercise 15, an outline will be given whereby the reader may show that *the Lebesgue decomposition of* ν *with respect to* μ *is unique.* That is, if $\nu = \nu_1 + \nu_2$ and $\nu = \nu_{ac} + \nu_s$, where ν_1, ν_2, ν_{ac}, and ν_s are σ-finite measures, if $\nu_1, \nu_{ac} \ll \mu$, and if $\nu_2, \nu_s \perp \mu$, then $\nu_1 = \nu_{ac}$ and $\nu_2 = \nu_s$.

We are now in a position to state and prove the final result of this section, which extends the result of the Radon-Nikodym Theorem to the situation where both measures μ and ν are σ-finite on $\mathcal{F}$.

Radon-Nikodym Theorem, σ-Finite Case. *Let* μ *and* ν *denote* σ*-finite measures with domain* $\mathcal{F}$ *and such that* $\nu \ll \mu$. *Then there exists a nonnegative finite-valued* $\mathcal{F}/\mathcal{B}^*$*-measurable function* f *on* Ω *with the following properties:*

(a) $\nu(A) = \int_A f \, d\mu$ *for every* $A \in \mathcal{F}$. *This function* f *is unique up to a set of* μ*-measure zero.*

(b) $\int g \, d\nu = \int fg \, d\mu$ *for every nonnegative* $\mathcal{F}/\mathcal{B}^*$*-measurable* g *defined on* Ω.

(c) For every $g \in L^1(\Omega, \mathcal{F}, \nu)$ *we have* $fg \in L^1(\Omega, \mathcal{F}, \mu)$, *and again the equation* $\int g \, d\nu = \int fg \, d\mu$ *holds.*

Proof: In the preceding proof of the σ-finite version of the Lebesgue Decomposition Theorem, recall the notation, definitions, and results for $\{A_n\}_{n=1}^{\infty}$ and the finite measures $\mu_1, \mu_2, \cdots$ and $\nu_1, \nu_2, \cdots$. The assumption $\nu \ll \mu$ forces $\nu_n \ll \mu_n$ for every $n \in \mathbb{N}$ since $\mu_n(E) = 0$ implies that $\mu(E \cap A_n) = 0$, which [since $\nu \ll \mu$] gives $\nu(E \cap A_n) = 0$, whence $\nu_n(E) = 0$. For each $n \in \mathbb{N}$, apply the finite version of the Radon-Nikodym Theorem to μ_n and ν_n, finding a nonnegative finite-valued $\mathcal{F}/\mathcal{B}^*$-measurable function f_n defined on Ω such that $\nu_n(A) = \int_A f_n \, d\mu_n$ for every $A \in \mathcal{F}$; recall that this function f_n is unique up to a set of μ_n-measure zero. Define $f = \sum_{n=1}^{\infty} f_n I_{A_n}$. [That is, $f = f_n$ on

A_n.] The function f is clearly nonnegative, finite-valued, $\mathcal{F}/\mathcal{B}^*$-measurable, and defined on Ω. Furthermore, recalling that $\nu = \sum_{n=1}^{\infty} \nu_n$ [see (ii) in the previous proof], we have that

$$\begin{aligned} \nu(A) = \sum_{n=1}^{\infty} \nu_n(A) = \sum_{n=1}^{\infty} \int_A f_n \, d\mu_n &= \sum_{n=1}^{\infty} \int_A f_n I_{A_n} \, d\mu \\ &= \int_A \sum_{n=1}^{\infty} f_n I_{A_n} \, d\mu \\ &= \int_A f \, d\mu \end{aligned}$$

for every $A \in \mathcal{F}$, where the third equality directly follows from integration properties (L18) and (L5) found in Section 6.4, and the fourth equality follows by (L11). Property (L17) immediately establishes the uniqueness of f up to a set of μ-measure zero. That is, if g is a nonnegative $\mathcal{F}/\mathcal{B}^*$-measurable function on Ω such that $\nu(A) = \int_A g \, d\mu$ for every $A \in \mathcal{F}$, then $f = g$ μ-a.e. on Ω. This completes part (a). Parts (b) and (c) follow immediately from (L14), hence the entire proof is now complete. ∎

The reader might find it helpful to now reread the introduction to this chapter to see that we have carried through the program described there, at least for the case where both μ and ν are σ-finite. It is actually the case that the results of the Radon-Nikodym Theorem hold if we merely require that μ be σ-finite. That is, instead of needing both μ and ν to be σ-finite, we only need μ to be σ-finite, and ν can be any measure with domain $\mathcal{F}$. This shall be demonstrated in the next section.

We close out the section by discussing some commonly used notation. In the Radon-Nikodym Theorem[s] we have given, the nonnegative finite-valued $\mathcal{F}/\mathcal{B}^*$-measurable f such that $\nu(A) = \int_A f \, d\mu$ for every $A \in \mathcal{F}$ is often referred to as *the Radon-Nikodym derivative of ν with respect to μ*, and one alternatively [but equivalently] may write

$$f = \frac{d\nu}{d\mu} \quad \text{or} \quad d\nu = f \, d\mu$$

to express the fact that f satisfies the conclusions of the Radon-Nikodym Theorem. The notation $d\nu = f \, d\mu$ is merely a formalism; no independent meaning is ascribed to the symbols $d\nu$ and $d\mu$.

Exercises. As usual, Ω is a nonempty set and $\mathcal{F}$ is a σ-field on Ω.

1. In the definition of an additive set function, show that the series $\sum_{n=1}^{\infty} \phi(A_n)$ must converge absolutely. [Observe that $\bigcup_{n=1}^{\infty} A_n = \bigcup_{k=1}^{\infty} A_{n_k}$ for every rearrangement $\{n_k\}_{k=1}^{\infty}$ of the positive integers, hence both $\sum_{n=1}^{\infty} \phi(A_n)$ and $\sum_{k=1}^{\infty} \phi(A_{n_k})$ should be defined and equal. Absolute convergence now follows from the usual calculus considerations.]

2*. In Claim 4, quickly verify that ϕ^- is a finite measure with support A^-.

3. Suppose that (A^+, A^-) and (B^+, B^-) are Hahn decompositions with respect to an additive set function ϕ. Then $\phi(A^+\triangle B^+) = \phi(A^-\triangle B^-) = 0$. [Observe that $A^+ \cap B^-$ is contained in both A^+ and B^-, hence $\phi(A^+ \cap B^-)$ is both nonnegative and nonpositive, and thus equals zero. Similarly, $\phi(A^- \cap B^+) = 0$. Then

$$\phi(A^+\triangle B^+) = \phi(A^+ \cap B^-) + \phi(A^- \cap B^+) = 0.$$

Similarly, $\phi(A^-\triangle B^-) = 0$.]

4. The Jordan decomposition of an additive set function ϕ is unique. [Let (A^+, A^-) and (B^+, B^-) denote Hahn decompositions of Ω with respect to ϕ. Let

$$\phi_A^+(E) = \phi(E \cap A^+) \quad \text{and} \quad \phi_B^+(E) = \phi(E \cap B^+) \quad \text{for every } E \in \mathcal{F};$$

define ϕ_A^- and ϕ_B^- similarly. Then $\phi = \phi_A^+ - \phi_A^-$ is the Jordan decomposition of ϕ relative to the Hahn decomposition (A^+, A^-) and $\phi = \phi_B^+ - \phi_B^-$ is the Jordan decomposition of ϕ relative to the Hahn decomposition (B^+, B^-). Use the result of the previous exercise to show that $\phi_A^+ = \phi_B^+$ and $\phi_A^- = \phi_B^-$.]

5. This problem relates somewhat the notion of absolute continuity with the familiar ϵ-δ concepts.
(a) Let μ and ν denote measures with common domain $\mathcal{F}$ and such that ν is finite. Then $\nu \ll \mu$ iff for every $\epsilon > 0$ there is $\delta > 0$ such that $\mu(A) < \delta$ forces $\nu(A) < \epsilon$. [Suppose first that for every $\epsilon > 0$ there is $\delta > 0$ such that $\mu(A) < \delta$ forces $\nu(A) < \epsilon$. We desire to show that $\nu \ll \mu$. If $\mu(A) = 0$ and $\epsilon > 0$ is given (and the corresponding δ is found), then $\mu(A) < \delta$, hence $\nu(A) < \epsilon$. Since $\epsilon > 0$ is arbitrary, it follows that $\nu(A) = 0$, whence $\nu \ll \mu$. This gives one direction of the claim. Show the other direction of the claim by contraposition: suppose that there is $\epsilon > 0$ such that for every $\delta > 0$ there is a set $A \in \mathcal{F}$ with $\mu(A) < \delta$ and $\nu(A) \geq \epsilon$. In particular, there is $\epsilon > 0$ such that there is a sequence $\{A_n\}_{n=1}^\infty$ of $\mathcal{F}$-sets with $\mu(A_n) < \frac{1}{n^2}$ and $\nu(A_n) \geq \epsilon$ for each $n \in \mathbb{N}$. Let $A = \overline{\lim} A_n = \bigcap_{n=1}^\infty \bigcup_{m=n}^\infty A_m$. For every $n \in \mathbb{N}$ we have

$$\mu(A) \leq \mu\left(\bigcup_{m=n}^\infty A_m\right) \leq \sum_{m=n}^\infty \mu(A_m) = \sum_{m=n}^\infty \frac{1}{m^2},$$

so $\mu(A) = 0$. However, we also have $\nu(A) = \nu(\overline{\lim} A_n) \geq \overline{\lim}\, \nu(A_n) \geq \epsilon > 0$ by property (M10) in Section 2.2. This shows that there is $A \in \mathcal{F}$ such that $\mu(A) = 0$ and $\nu(A) > 0$, so $\nu \not\ll \mu$.]
(b) The claim in (a) is not necessarily true if ν is infinite, since the condition $\nu \ll \mu$ does not imply the ϵ-δ condition. [Let Ω denote the set of integers, let $\mathcal{F} = 2^\Omega$, let ν denote counting measure, so that ν is infinite, and let μ be such that $\mu(\{n\}) = \frac{1}{n^2}$ for each $n \in \Omega$, so that μ is finite.]

6. Let μ, ν, ν_1, and ν_2 denote measures, each having common domain $\mathcal{F}$.
(a) If $\nu_1 \perp \mu$ and $\nu_2 \perp \mu$, then $\nu_1 + \nu_2 \perp \mu$.
(b) If $\nu_1 \ll \mu$ and $\nu_2 \ll \mu$, then $\nu_1 + \nu_2 \ll \mu$.
(c) If $\nu_1 \ll \mu$ and $\nu_2 \perp \mu$, then $\nu_1 \perp \nu_2$.
(d) If $\nu \ll \mu$ and $\nu \perp \mu$, then $\nu = 0$.
(e) If $\mu \perp \mu$, then $\mu = 0$.
(f) If μ and ν are σ-finite with $\nu \ll \mu$, then $\nu(\{\omega \in \Omega : \frac{d\nu}{d\mu}(\omega) = 0\}) = 0$. [All parts for this entire exercise should be straightforward.]

7. Claim 1 may be generalized to the following claim: if $\{E_n\}_{n=1}^{\infty}$ is a sequence of $\mathcal{F}$-sets such that $\lim E_n$ exists [i.e. $\overline{\lim}\, E_n = \underline{\lim}\, E_n$], then $\lim \phi(E_n) = \phi(\lim E_n)$. [Write $\phi = \phi^+ - \phi^-$ for the Jordan decomposition of ϕ, and recall that ϕ^+ and ϕ^- are finite measures. Use the appropriate continuity result[s] from Chapter 2.]

8. Let $f \in L^1(\Omega, \mathcal{F}, \mu)$. Define $\nu(E) = \int_E f\, d\mu$ for every $E \in \mathcal{F}$.
(a) ν is an additive set function such that $\nu^+(E) = \int_E f^+\, d\mu$ and $\nu^-(E) = \int_E f^-\, d\mu$ for every $E \in \mathcal{F}$.
(b) If $A^+ = \{\omega \in \Omega : f(\omega) > 0\}$ and $A^- = A^{+c}$, then (A^+, A^-) is a Hahn decomposition with respect to ν.

9. Let $\mathfrak{M}$ denote the collection of additive set functions ϕ with domain $\mathcal{F}$.
(a) $\mathfrak{M}$ is a linear space over $\mathbb{R}$: for $a, b \in \mathbb{R}$ and $\phi_1, \phi_2 \in \mathfrak{M}$ we have $a\phi_1 + b\phi_2 \in \mathfrak{M}$.
(b) Given $\phi \in \mathfrak{M}$, define $\|\phi\| = \phi^+(\Omega) + \phi^-(\Omega)$, where $\phi = \phi^+ - \phi^-$ is the Jordan decomposition of ϕ. Then $\|\ \|$ is a norm on $\mathfrak{M}$.
(c) Is $\mathfrak{M}$ a Banach space? [In other words, let $\{\phi_n\}_{n=1}^{\infty}$ denote a Cauchy sequence in $\mathfrak{M}$: for every $\epsilon > 0$ there is $N \in \mathbb{N}$ such that $n, m \geq N$ forces $\|\phi_n - \phi_m\| < \epsilon$. Is there a $\phi \in \mathfrak{M}$ such that $\lim \|\phi_n - \phi\| = 0$?]

10. Let (A^+, A^-) denote a Hahn decomposition of the additive set function ϕ, and let $\phi = \phi^+ - \phi^-$ denote the Jordan decomposition. We have

$$\phi^+(A) = \sup\{\phi(E) : E \in \mathcal{F},\ E \subseteq A\} \text{ and } \phi^-(A) = -\inf\{\phi(E) : E \in \mathcal{F},\ E \subseteq A\}$$

for every $A \in \mathcal{F}$. [Let $A, E \in \mathcal{F}$ with $E \subseteq A$. Then

$$\phi(E) = \phi(E \cap A^+) + \phi(E \cap A^-) \leq \phi(E \cap A^+) = \phi^+(E) \leq \phi^+(A).$$

On the other hand, if $E = A \cap A^+$, then $\phi(E) = \phi(A \cap A^+) = \phi^+(A)$. This gives the first equation; the second equation is handled similarly.]

11. The Radon-Nikodym Theorem may fail if μ is not σ-finite, even if ν is finite. [Let Ω be uncountable, let $\mathcal{F} = \{A \subseteq \Omega : A \text{ is amc or } A^c \text{ is amc}\}$, let μ denote counting measure, and let $\nu(A) = 1$ or 0 as A^c is amc or A is amc. In particular, $\nu(\{\omega\}) = 0$ for every $\omega \in \Omega$. If there is an f satisfying the conclusion of the Radon-Nikodym Theorem, then for every $\omega \in \Omega$ we have $0 = \nu(\{\omega\}) = \int_{\{\omega\}} f\, d\mu = f(\omega)$, giving $f = 0$ on Ω. (Recall Exercise 7 from Section 6.2.) Then $1 = \nu(\Omega) = \int f\, d\mu = 0$, a contradiction.]

12. Let μ, ν, ν_1, ν_2, and ρ denote σ-finite measures having domain $\mathcal{F}$. We have the following claims.
(a) If $\nu_1 \ll \mu$ and $\nu_2 \ll \mu$, then $\frac{d(\nu_1 \pm \nu_2)}{d\mu} = \frac{d\nu_1}{d\mu} \pm \frac{d\nu_2}{d\mu}$ μ-a.e. on Ω.
(b) If $\nu \ll \mu$ and $\mu \ll \rho$, then $\nu \ll \rho$ and $\frac{d\nu}{d\rho} = \frac{d\nu}{d\mu}\frac{d\mu}{d\rho}$ μ-a.e. on Ω.
(c) If $\nu \ll \mu$ and $\mu \ll \nu$,[4] then

$$\frac{d\nu}{d\mu} = I_{\{\omega \in \Omega:\ \frac{d\mu}{d\nu}(\omega) > 0\}} \times \frac{1}{\frac{d\mu}{d\nu}} \quad \mu\text{-a.e. on } \Omega.$$

[4] In this case some authors say that μ and ν are *equivalent* and use the notation $\nu \equiv \mu$.

(d) Let $\mu \ll \rho$ and $\nu \ll \rho$. Then $\nu \ll \mu$ iff

$$\rho\left(\left\{\omega \in \Omega : \frac{d\nu}{d\rho}(\omega) > 0,\ \frac{d\mu}{d\rho}(\omega) > 0\right\}\right) = 0,$$

in which case we have

$$\frac{d\nu}{d\mu} = I_{\left\{\omega\in\Omega:\ \frac{d\mu}{d\rho}(\omega)>0\right\}} \times \frac{\frac{d\nu}{d\rho}}{\frac{d\mu}{d\rho}} \quad \mu\text{-a.e. on } \Omega.$$

[This exercise may look harder it really is; it is mostly a consequence of notation and the Radon-Nikodym Theorem. Consider (b). The assumption $\nu \ll \mu$ gives $\nu(E) = \int_E \frac{d\nu}{d\mu}\, d\mu$ for every $E \in \mathcal{F}$; the assumption $\mu \ll \rho$ gives $\mu(E) = \int_E \frac{d\mu}{d\rho}\, d\rho$ for every $E \in \mathcal{F}$. It is nearly obvious that $\nu \ll \rho$, hence $\nu(E) = \int_E \frac{d\nu}{d\rho}\, d\rho$ for every $E \in \mathcal{F}$. Therefore, for every $E \in \mathcal{F}$ we have $\nu(E) = \int_E \frac{d\nu}{d\mu}\, d\mu = \int_E \frac{d\nu}{d\rho}\, d\rho$. Since we have $\mu(E) = \int_E \frac{d\mu}{d\rho}\, d\rho$ for every $E \in \mathcal{F}$, property (L14) from Section 6.4 gives

$$\nu(E) = \int_E \frac{d\nu}{d\mu}\, d\mu = \int_E \frac{d\nu}{d\mu}\frac{d\mu}{d\rho}\, d\rho$$

for every $E \in \mathcal{F}$. Applying the uniqueness claim of the Radon-Nikodym Theorem completes the argument. Parts (a), (c) and (d) are handled similarly.]

13. Let μ and ν denote finite measures with domain $\mathcal{F}$. Let $\mathcal{H} \subseteq \mathcal{F}$ denote a sub-σ-field of $\mathcal{F}$. Let μ^* and ν^* denote the restrictions of μ and ν to $\mathcal{H}$. In accordance with the Lebesgue Decomposition Theorem, let ν_{ac} and ν_s denote the absolutely continuous and singular parts of ν with respect to μ. Similarly, let ν^*_{ac} and ν^*_s denote the absolutely continuous and singular parts of ν^* with respect to μ^*. Then

$$\nu^*_{ac}(E) \geq \nu_{ac}(E) \quad \text{and} \quad \nu^*_s(E) \leq \nu_s(E) \quad \text{for every } E \in \mathcal{H}.$$

[Refer to the proof of the finite version of the Radon-Nikodym Theorem, letting f be nonnegative, finite, and such that $\nu_{ac}(E) = \int_E f\, d\mu$ for every $E \in \mathcal{F}$. Also note that $\nu_s = \nu - \nu_{ac}$. We have $\nu^* \ll \mu^*$, hence there is a nonnegative finite-valued f^* with $\nu^*(E) = \int_E f^*\, d\mu + \nu^*_s(E)$ for every $E \in \mathcal{H}$. Find $S^* \in \mathcal{H}$ such that $\nu^*_s(S^{*c}) = 0$ and $\mu^*(S^*) = \mu(S^*) = 0$, so that ν^*_s and μ have disjoint supports. For $E \in \mathcal{H}$ verify that

$$\begin{aligned}\int_E f^*\, d\mu = \int_{E-S^*} f^*\, d\mu = \int_{E-S^*} f^*\, d\mu^* &= \nu^*(E - S^*)\\ &= \nu(E - S^*)\\ &\geq \int_{E-S^*} f\, d\mu\\ &= \int_E f\, d\mu,\end{aligned}$$

giving $\nu^*_{ac}(E) \geq \nu_{ac}(E)$.]

14. Let μ and ν denote finite measures with domain $\mathcal{F}$. Let $\{\nu_n\}_{n=1}^{\infty}$ denote a sequence of finite measures with domain f such that $\nu = \sum_{n=1}^{\infty} \nu_n$. Assume for each

$n \in \mathbb{N}$ and $A \in \mathcal{F}$ that

$$\nu_n(A) = \int_A f_n \, d\mu + \nu_n'(A),$$

where each f_n is a nonnegative finite-valued function on Ω and each ν_n' is singular with respect to μ. Finally, assume for each $A \in \mathcal{F}$ that

$$\nu(A) = \int_A f \, d\mu + \nu'(A),$$

where f is a nonnegative finite-valued function on Ω and $\nu' \perp \mu$.
(a) $f = \sum_{n=1}^{\infty} f_n$ μ-a.e. on Ω.
(b) $\nu' = \sum_{n=1}^{\infty} \nu_n'$.
(c) $\nu \ll \mu$ iff $\nu_n \ll \mu$ for every $n \in \mathbb{N}$.

15*. Suppose that μ and ν are σ-finite measures on $\mathcal{F}$. The Lebesgue decomposition of ν with respect to μ is unique. That is, if $\nu = \nu_{ac} + \nu_s$ where ν_{ac} and ν_s are σ-finite measures with $\nu_{ac} \ll \mu$ and $\nu_s \perp \mu$, and if in addition $\nu = \nu_{ac}' + \nu_s'$ where ν_{ac}' and ν_s' are σ-finite measures with $\nu_{ac}' \ll \mu$ and $\nu_s' \perp \mu$, then $\nu_{ac} = \nu_{ac}'$ and $\nu_s = \nu_s'$. [Since $\nu_s \perp \mu$, there is $A \in \mathcal{F}$ such that A supports ν_s and A^c supports μ: $\nu_s(A^c) = 0 = \mu(A)$. Since $\nu_s' \perp \mu$, there is $B \in \mathcal{F}$ such that B supports ν_s' and B^c supports μ: $\nu_s'(B^c) = 0 = \mu(B) = 0$. Since $\nu_s(A^c \cap B^c) \le \nu_s(A^c) = 0$ and $\nu_s'(A^c \cap B^c) \le \nu_s'(B^c) = 0$, we have that $A \cup B$ supports both ν_s and ν_s'. Since $\mu(A \cup B) \le \mu(A) + \mu(B) = 0 + 0 = 0$, we have that $(A \cup B)^c$ supports μ. Let $S = A \cup B$, so that $\mu(S) = 0 = \nu_s(S^c) = \nu_s'(S^c)$. It may be verified that

$$\begin{aligned}
\nu_{ac}(E) = \nu_{ac}(E \cap S^c) + \nu_{ac}(E \cap S) &= \nu_{ac}(E \cap S^c) \\
&= \nu_{ac}(E \cap S^c) + \nu_s(E \cap S^c) \\
&= \nu(E \cap S^c) \\
&= \nu_{ac}'(E \cap S^c) + \nu_s'(E \cap S^c) \\
&= \nu_{ac}'(E \cap S^c) \\
&= \nu_{ac}'(E \cap S^c) + \nu_{ac}'(E \cap S) \\
&= \nu_{ac}'(E)
\end{aligned}$$

for every $E \in \mathcal{F}$, giving $\nu_{ac} = \nu_{ac}'$. The other equation $\nu_s = \nu_s'$ automatically follows.]

Comment: Exercises 16-19 show how the Riesz Representation Theorem may be used to prove the finite version of the Radon-Nikodym Theorem. Exercises 16 and 17 are straightforward technical results, while Exercises 18 and 19 proceed to the Radon-Nikodym Theorem.

16. Let μ and ν denote measures with domain $\mathcal{F}$ and such that $\nu(E) \le \mu(E)$ for every $E \in \mathcal{F}$. Let $0 < p < +\infty$ and $f \in L^p(\Omega, \mathcal{F}, \mu)$. Then $\int |f|^p \, d\nu \le \int |f|^p \, d\mu$, so that $f \in L^p(\Omega, \mathcal{F}, \nu)$. [Let $s = \sum_{i=1}^n c_i I_{A_i}$, where $c_1, \cdots, c_n$ are nonnegative real numbers and $A_1, \cdots, A_n$ are disjoint $\mathcal{F}$-sets with union Ω. Then $|s|^p = \sum_{i=1}^n c_i^p I_{A_i}$, and

$$\int |s|^p \, d\nu = \sum_{i=1}^n c_i^p \nu(A_i) \le \sum_{i=1}^n c_i^p \mu(A_i) = \int |s|^p \, d\mu,$$

so the result holds for nonnegative finite-valued $\mathcal{F}/\mathcal{B}^*$-measurable simple functions defined on Ω. Next, let $f \geq 0$ be defined on Ω and $\mathcal{F}/\mathcal{B}^*$-measurable. Let $\{s_n\}_{n=1}^{\infty}$ denote a nondecreasing sequence of nonnegative finite-valued $\mathcal{F}/\mathcal{B}^*$-measurable simple functions defined on Ω with $\lim s_n = f$. By what we have just shown and MCT,

$$\int f^p \, d\nu = \lim \int s_n^p \, d\nu \leq \lim \int s_n^p \, d\mu = \int f^p \, d\mu.$$

For general $f \in L^p(\Omega, \mathcal{F}, \mu)$, write $f = f^+ - f^-$.]

17. This exercise, along with the result in Exercise 16, gives a simple integration result that is used in the next exercise. Let μ and ν denote finite measures with common domain $\mathcal{F}$. Let $g \in L^1(\Omega, \mathcal{F}, \mu + \nu)$. Then g is in both $L^1(\Omega, \mathcal{F}, \mu)$ and $L^1(\Omega, \mathcal{F}, \nu)$ and

$$\int g \, d\nu + \int g \, d\mu = \int g \, d(\mu + \nu).$$

[Observe that $\mu, \nu \leq \mu + \nu$; use Exercise 16 to obtain that g is in both $L^1(\Omega, \mathcal{F}, \mu)$ and $L^1(\Omega, \mathcal{F}, \nu)$. Establish the formula by first considering simple functions. Use MCT and the formula for simple functions to get the result for nonnegative functions. For general functions $L^1(\Omega, \mathcal{F}, \mu + \nu)$, consider positive and negative parts and apply the result for nonnegative functions. This will all be very straightforward.]

18. This exercise relies on Exercise 17, which in turn relies on Exercise 16. Let μ and ν denote finite measures with common domain $\mathcal{F}$ and such that $\nu \ll \mu$. There exists $\mathcal{F}/\mathcal{B}^*$-measurable g defined on Ω with $0 \leq g < 1$ and such that

$$\int f(1-g) \, d\nu = \int fg \, d\mu \quad \text{for every } f \in L^2(\Omega, \mathcal{F}, \mu + \nu).$$

[For $f \in L^2(\Omega, \mathcal{F}, \mu + \nu)$, define $L(f) = \int f \, d\nu$. The finiteness of μ and ν forces the finiteness of $\mu + \nu$, and hence $f \in L^1(\Omega, \mathcal{F}, \mu + \nu)$ by Claim 1 of Section 8.5. Since $\nu \leq \mu + \nu$, the previous exercise gives $f \in L^1(\Omega, \mathcal{F}, \nu)$, hence L is meaningfully defined and finite on $L^2(\Omega, \mathcal{F}, \mu + \nu)$. It is easy to verify that L is linear. Using Hölder's Inequality with $p = p' = 2$ for the functions f and 1, we have for every $f \in L^2(\Omega, \mathcal{F}, \mu + \nu)$ that

$$\begin{aligned} |L(f)| = \left| \int f \, d\nu \right| &\leq \left(\int |f|^2 \, d\nu \right)^{\frac{1}{2}} \left(\int 1^2 \, d\nu \right)^{\frac{1}{2}} \\ &\leq \left(\int |f|^2 \, d(\mu + \nu) \right)^{\frac{1}{2}} \nu(\Omega)^{\frac{1}{2}} \\ &= \|f\|_2 \nu(\Omega)^{\frac{1}{2}} \end{aligned}$$

by virtue of the previous exercise. ($\| \ \|_2$ denotes the norm on $L^2(\Omega, \mathcal{F}, \mu + \nu)$.) It follows that L is a bounded linear functional on $L^2(\Omega, \mathcal{F}, \mu + \nu)$. The Riesz Representation Theorem for $p = 2$ asserts the existence of a function h in $L^2(\Omega, \mathcal{F}, \mu + \nu)$ such that

$$L(f) = \int f \, d\nu = \int fh \, d(\mu + \nu) \quad \text{for every } f \in L^2(\Omega, \mathcal{F}, \mu + \nu).$$

If h is negative on some $B \in \mathcal{F}$ with $(\mu+\nu)(B) > 0$, then $L(I_B) = \nu(B) \geq 0$, but at the same time we have $L(I_B) = \int I_B \, d\nu = \int I_B h \, d(\mu+\nu) = \int_B h \, d(\mu+\nu) < 0$, a contradiction. Therefore, if h is negative on some $B \in \mathcal{F}$, then $(\mu+\nu)(B) = 0$, that is, $h \geq 0$ $(\mu+\nu)$-a.e. on Ω. Next, observe by the previous display and Exercise 17 that

$$\int f(1-h)\, d\nu = \int f \, d\nu - \int fh \, d\nu = \int fh \, d(\mu+\nu) - \int fh \, d\nu = \int fh \, d\mu$$

for every $f \in L^2(\Omega, \mathcal{F}, \mu+\nu)$. Let $E = \{\omega \in \Omega : \ h(\omega) \geq 1\}$. Then we have $I_E \in L^1(\Omega, \mathcal{F}, \mu+\nu)$, and the previous display with $f = I_E$ gives

$$0 \leq \mu(E) = \int I_E \, d\mu \leq \int I_E h \, d\mu = \int I_E(1-h)\, d\nu \leq 0,$$

hence $\mu(E) = 0$. The assumption $\nu \ll \mu$ gives $\nu(E) = 0$. (This is the only time that the assumption $\nu \ll \nu$ is used.) If $g = hI_{E^c}$, then

$$\begin{aligned}
\int f(1-g)\, d\nu = \int f \, d\nu - \int (fI_{E^c})h \, d\nu &= \int fh \, d(\mu+\nu) - \int (fI_{E^c})h \, d\nu \\
&= \int (fI_{E^c})h \, d(\mu+\nu) - \int (fI_{E^c})h \, d\nu \\
&= \int (fI_{E^c})h \, d\mu \quad \text{[Exercise 16]} \\
&= \int fg \, d\mu.
\end{aligned}$$

Therefore, we may take $g = hI_{E^c}$ to complete the proof.]

19. Use Exercise 18 [which relies on Exercises 16 and 17] to prove the finite version of the Radon-Nikodym Theorem: *Let μ and ν denote finite measures with domain $\mathcal{F}$ and such that $\nu \ll \mu$. Then there exists a nonnegative finite-valued $\mathcal{F}/\mathcal{B}^*$-measurable function g^* on Ω such that $\nu(E) = \int_E g^* \, d\mu$ for every $E \in \mathcal{F}$.* [Let $f \geq 0$ be bounded, and let g be as in Exercise 18. Since f and g are bounded and $\mu+\nu$ is finite, it follows for every $n \in \mathbb{N}$ that $(1+g+\cdots+g^{n-1})f \in L^2(\Omega, \mathcal{F}, \mu+\nu)$. Using Exercise 18 we obtain

$$\int (1+g+\cdots+g^{n-1})f(1-g)\, d\nu = \int (1+g+\cdots+g^{n-1})fg \, d\mu \quad \text{for every } n \in \mathbb{N}.$$

Since $0 \leq g < 1$, this gives $\int (1-g^n)f \, d\nu = \int \frac{g}{1-g}(1-g^n)f \, d\mu$ for each $n \in \mathbb{N}$. Using MCT and the fact that $0 \leq g < 1$ gives

$$\begin{aligned}
\int f \, d\nu = \int \lim_n (1-g^n)f \, d\nu &= \lim_n \int (1-g^n)f \, d\nu \\
&= \lim_n \int \frac{g}{1-g}(1-g^n)f \, d\mu \\
&= \int \lim_n \frac{g}{1-g}(1-g^n)f \, d\mu \\
&= \int \frac{g}{1-g} f \, d\mu.
\end{aligned}$$

In particular, if $f = 1$, then $\nu(\Omega) = \int \frac{g}{1-g}\, d\mu$, hence the finiteness of ν forces $\frac{g}{1-g} \in L^1(\Omega, \mathcal{F}, \mu)$, so $\frac{g}{1-g}$ may be taken to be finite on all of Ω. Clearly $\frac{g}{1-g}$ is nonnegative. Write $g^* = \frac{g}{1-g}$ and let $f = I_E$, where $E \in \mathcal{F}$ is arbitrary. We have from the chain of equalities above that $\nu(E) = \int_E g^*\, d\mu$ for every $E \in \mathcal{F}$, so g^* satisfies the conclusion of the finite version of the Radon-Nikodym Theorem. The proof for the uniqueness of g^* as given in the text of course works here — the real work is in showing the existence, not uniqueness.]

Comment: The next two exercises are taken from Section 32 of Halmos. For the next exercise [which is not terribly difficult] recall the discussion on notation found at the end of the section. Exercise 21 requires the solution of Exercise 12(b).

20. Let μ_0, μ_1, and μ_2 denote finite measures. Assume that

$$d\mu_0 = f_1 d(\mu_0 + \mu_1) = f_2 d(\mu_0 + \mu_2) = f d(\mu_0 + \mu_1 + \mu_2).$$

Then we have for $(\mu_0 + \mu_1 + \mu_2)$-a.e. $\omega \in \Omega$ that

$$f(\omega) = \begin{cases} \frac{f_1(\omega)f_2(\omega)}{f_1(\omega)+f_2(\omega)-f_1(\omega)f_2(\omega)} & \text{if } f_1(\omega)f_2(\omega) \neq 0 \\ 0 & \text{if } f_1(\omega) = f_2(\omega) = 0. \end{cases}$$

21. Let $\{\nu_n\}_{n=1}^\infty$ and $\{\mu_n\}_{n=1}^\infty$ denote sequences of finite measures with domain $\mathcal{F}$. For each $n \in \mathbb{N}$ define $\bar{\mu}_n = \sum_{i=1}^n \mu_i$ and $\bar{\nu}_n = \sum_{i=1}^n \nu_i$. Define $\mu = \sum_{n=1}^\infty \mu_n$ and $\nu = \sum_{n=1}^\infty \nu_n$, and further assume that both μ and ν are finite. The goal is to prove the following: *If $\nu_n \ll \mu_n$ for every $n \in \mathbb{N}$, then $\nu \ll \mu$ and $\lim_n \frac{d\bar{\nu}_n}{d\bar{\mu}_n} = \frac{d\nu}{d\mu}$ μ-a.e. on Ω.* Follow Halmos' outline given below.
(a) If $\{E_n\}_{n=1}^\infty$ is a sequence of $\mathcal{F}$-sets with $\bar{\mu}_n(E_n) = 0$ for every $n \in \mathbb{N}$, then $\mu(\overline{\lim}\, E_n) = 0$. [Use the fact that $\bar{\mu}_n(\bigcup_{k=n}^\infty E_k) \leq \sum_{k=n}^\infty \bar{\mu}_k(E_k)$.]
(b) If $\{\phi_n\}_{n=1}^\infty$ and $\{\psi_n\}_{n=1}^\infty$ are sequences of $\mathcal{F}/\mathcal{B}^*$-measurable functions on Ω such that $\phi_n = \psi_n$ μ_n-a.e. on Ω for every $n \in \mathbb{N}$, then

$$\overline{\lim_n}\, \phi_n = \overline{\lim_n}\, \psi_n \quad \text{and} \quad \underline{\lim_n}\, \phi_n = \underline{\lim_n}\, \psi_n \quad \mu\text{-a.e. on } \Omega.$$

[Write $E_n = \{\omega \in \Omega : \phi_n(\omega) \neq \psi_n(\omega)\}$, using part (a).]
(c) From (b), it is sufficient to prove the main claim for any fixed determination of the Radon-Nikodym derivatives $\{\frac{d\bar{\nu}_n}{d\bar{\mu}_n}\}_{n=1}^\infty$. For each $n \in \mathbb{N}$, write $f_n = \frac{d\nu_n}{d\mu}$ and $g_n = \frac{d\mu_n}{d\mu}$. From Exercise 12(b), one such determination is

$$\frac{d\bar{\nu}_n}{d\mu} = \frac{f_1 + \cdots + f_n}{g_1 + \cdots + g_n} \quad \mu_n\text{-a.e. on } \Omega,\ n \in \mathbb{N}.$$

(d) We have $\sum_{n=1}^\infty f_n = \frac{d\nu}{d\mu}$ and $\sum_{n=1}^\infty g_n = 1$, both equalities holding μ-a.e. on Ω. [Use the facts that

$$\sum_{i=1}^n \mu_i(E) = \int_E g_1 + \cdots + g_n\, d\mu \text{ and } \sum_{i=1}^n \nu_i(E) = \int_E (f_1 + \cdots + f_n)\, d\mu$$

for every $n \in \mathbb{N}$. Then use MCT to obtain the result.]
(e) Part (d) directly implies the main claim of this exericse.

9.2 THE RADON–NIKODYM THEOREM, PART II

Let μ and ν denote σ-finite measures on a σ-field $\mathcal{F}$ over Ω. If $\nu \ll \mu$, then the conclusion of the Radon-Nikodym Theorem holds, as proved in the last section. It will turn out that we may relax the requirements on μ and ν. It will ultimately be shown that if μ is a measure satisfying a condition weaker than σ-finiteness, and if ν is *any* measure with $\nu \ll \mu$, then the conclusion of the Radon-Nikodym Theorem holds. The proof of this fact requires discussion of just what this aforementioned weaker condition is, and from this two admittedly uninteresting and quite technical claims are given whereby one may prove the result.

Definition. Let I denote a generic nonempty index set. For each $i \in I$, let $0 \leq x_i \leq +\infty$. We define

$$\sum_{i \in I} x_i = \sup \left\{ \sum_{i \in F} x_i : \; F \text{ is a finite subset of } I \right\}.$$

If I is finite, say $I = \{1, \cdots, n\}$, then $\sum_{i \in I} x_i = \sum_{i=1}^{n} x_i$, so that the above definition reduces to our usual notation in the finite case. If I is countably infinite, say $I = \mathbb{N}$, then $\sum_{i \in I} x_i = \sum_{i=1}^{\infty} x_i$, where $\sum_{i=1}^{\infty} x_i$ is the usual infinite series from calculus, that is, $\sum_{i=1}^{\infty} x_i$ is the limit of the nondecreasing sequence of partial sums $\{\sum_{i=1}^{n} x_i\}_{n=1}^{\infty}$. The main point of the definition is that it gives a definition of a possibly uncountable summation that is consistent with what we have in the at most countably infinite case.

Fact. *Given any nonempty index set I with $0 \leq x_i \leq +\infty$ for each $i \in I$, there is an at most countable subset $I_0 \subseteq I$ such that $\sum_{i \in I_0} x_i = \sum_{i \in I} x_i$.*

Proof: By the definition of $\sum_{i \in I} x_i$ as a supremum, we may find a sequence $\{F_n\}_{n=1}^{\infty}$ of finite subsets of I such that $\lim_n \sum_{i \in F_n} x_i = \sum_{i \in I} x_i$. Let I_0 denote $\bigcup_{n=1}^{\infty} F_n$. Then I_0 is an at most countable subset of I, and hence $\sum_{i \in I_0} x_i \leq \sum_{i \in I} x_i$. At the same time, given any $\epsilon > 0$, there is $N \in \mathbb{N}$ such that $n \geq N$ forces $\sum_{i \in I} x_i - \epsilon < \sum_{i \in F_n} x_i \leq \sum_{i \in I} x_i$. Since each F_n is a subset of I_0, it follows that $\sum_{i \in I_0} x_i = \sum_{i \in I} x_i$. The at most countable subset I_0 of I accordingly satisfies the requirements. ∎

We now introduce the notion of *decomposability*, which will be seen to be a weaker condition than σ-finiteness.

Definition. Let $(\Omega, \mathcal{F}, \mu)$ denote a measure space. Suppose that there is a subcollection $\mathcal{A} \subseteq \mathcal{F}$ with the following four properties:

(i) $\mu(F) < +\infty$ for every $F \in \mathcal{A}$.
(ii) The sets in $\mathcal{A}$ are disjoint with union Ω.
(iii) If $E \in \mathcal{F}$ and $\mu(E) < +\infty$, then $\mu(E) = \sum_{F \in \mathcal{A}} \mu(E \cap F)$.
(iv) If $S \subseteq \Omega$ is such that $S \cap F \in \mathcal{F}$ for every $F \in \mathcal{A}$, then $S \in \mathcal{F}$.

Then we refer to $(\Omega, \mathcal{F}, \mu)$ or μ as *decomposable*, and we equivalently refer to $\mathcal{A}$ as a *decomposition of* $(\Omega, \mathcal{F}, \mu)$.

Example 1. Suppose that $(\Omega, \mathcal{F}, \mu)$ is a σ-finite measure space. By σ-finiteness, there is a disjoint sequence $\{A_n\}_{n=1}^{\infty}$ of $\mathcal{F}$-sets such that $\Omega = \bigcup_{n=1}^{\infty} A_n$ and $\mu(A_n)$ is finite for each $n \in \mathbb{N}$. If $\mathcal{A} = \{A_1, A_2, \cdots\}$, then $\mathcal{A} \subseteq \mathcal{F}$, and $\mathcal{A}$ clearly satisfies requirements (i) and (ii). Next, observe for every $E \in \mathcal{F}$ [in particular for those $E \in \mathcal{F}$ with $\mu(E) < +\infty$] that $\mu(E) = \sum_{n=1}^{\infty} \mu(E \cap A_n) = \sum_{F \in \mathcal{A}} \mu(E \cap F)$, where the last equality follows because $\mathcal{A}$ is countable. This shows that (iii) holds. Next, if $S \subseteq \Omega$ is such that $S \cap F \in \mathcal{F}$ for every $F \in \mathcal{A}$, then $S \cap A_n \in \mathcal{F}$ for every $n \in \mathbb{N}$, and hence $S = \bigcup_{n=1}^{\infty}(S \cap A_n)$ is a countable union of $\mathcal{F}$-sets, giving $S \in \mathcal{F}$. It follows that $\mathcal{A}$ satisfies (iv), and $\mathcal{A} = \{A_1, A_2, \cdots\}$ is accordingly a decomposition of $(\Omega, \mathcal{F}, \mu)$. We have shown that *$\sigma$-finiteness implies decomposability*, so that σ-finiteness is a stronger condition than decomposability, or, equivalently, decomposability is a weaker condition than σ-finiteness.

Example 2. Let Ω denote an uncountable set, let $\mathcal{F} = 2^{\Omega}$, and let μ denote counting measure: for every $A \subseteq \Omega$ we have that $\mu(A)$ is the number points in A if A is finite, while $\mu(A) = +\infty$ if A is infinite. Let $\mathcal{A} = \{\{\omega\} : \omega \in \Omega\}$, the collection of singletons of Ω. Clearly $\mathcal{A}$ satisfies (i) and (ii). Conditions (iii) follows upon a moment's thought, and (iv) trivially holds since $\mathcal{F} = 2^{\Omega}$. Therefore, $\mathcal{A}$ is a decomposition of $(\Omega, \mathcal{F}, \mu)$. At the same time, μ is not σ-finite on $\mathcal{F}$. Therefore, we have an example where decomposability exists in the absence of σ-finiteness.

We now state and prove two admittedly uninteresting and technical claims that are needed in the proof of the more generalized Radon-Nikodym Theorem. These claims are used nowhere else in the text.

Claim 1. *Let μ denote a finite measure with domain $\mathcal{F}$, and let ν denote a measure on $\mathcal{F}$ such that $\nu \ll \mu$. Then there exists $E \in \mathcal{F}$ simultaneously satisfying the following:*

(a) If $A \in \mathcal{F}$ and $A \subseteq E$, then $\nu(A) = 0$ or $\nu(A) = +\infty$.
(b) If $A \in \mathcal{F}$ and $A \subseteq E$, then $\nu(A) = 0$ forces $\mu(A) = 0$.
(c) ν is σ-finite on E^c: E^c is the union of at most countably many $\mathcal{F}$-sets, where each set has finite ν-measure.

Proof: Define the collection

$$\mathcal{D} = \{B \in \mathcal{F} : \nu(C) \in \{0, +\infty\} \text{ for every } C \in \mathcal{F} \text{ with } C \subseteq B\}.$$

Observe that $\varnothing \in \mathcal{D} \subseteq \mathcal{F}$. Define $\alpha = \sup\{\mu(B) : B \in \mathcal{D}\}$. We have

$$\alpha = \sup\{\mu(B) : B \in \mathcal{D}\} \leq \sup\{\mu(B) : B \in \mathcal{F}\} = \mu(\Omega) < +\infty,$$

and hence $\alpha < +\infty$. We now have a series of subclaims.

(1) *$\mathcal{D}$ is closed under finite unions.*

Proof: Let $B_1, B_2 \in \mathcal{D}$. Then every $\mathcal{F}$-set C with $C \subseteq B_1$ is such that $\nu(C) = 0$ or $+\infty$, and every $\mathcal{F}$-set C with $C \subseteq B_2$ is such that $\nu(C) = 0$ or $+\infty$. Next, let $C \in \mathcal{F}$ be such that $C \subseteq B_1 \cup B_2$. Then

$$\nu(C) = \nu(C \cap B_1 \cap B_2^c) + \nu(C \cap B_1^c \cap B_2) + \nu(C \cap B_1 \cap B_2).$$

Now $C \cap B_1 \cap B_2^c$ is an $\mathcal{F}$-subset of B_1, hence $\nu(C \cap B_1 \cap B_2^c) \in \{0, +\infty\}$. Next, $C \cap B_1^c \cap B_2$ is an $\mathcal{F}$-subset of B_2, so $\nu(C \cap B_1^c \cap B_2) \in \{0, +\infty\}$. Similarly, $\nu(C \cap B_1 \cap B_2) \in \{0, +\infty\}$. It follows that $\nu(C)$ is the sum of three numbers, each number being 0 or $+\infty$, and hence $\nu(C) \in \{0, +\infty\}$. Since C is a generic $\mathcal{F}$-subset of $B_1 \cup B_2$, we have $B_1 \cup B_2 \in \mathcal{D}$. Using induction shows that $\mathcal{D}$ is closed under finite unions. This gives (1).

(2) *There exists a nondecreasing sequence* $\{B_n\}_{n=1}^{\infty}$ *of sets in* $\mathcal{D}$ *such that* $\lim_n \mu(B_n) = \alpha$.

Proof: By the definition of α as a supremum, there is a sequence $\{C_n\}_{n=1}^{\infty}$ of $\mathcal{D}$-sets with $\lim_n \mu(C_n) = \alpha$. For each $n \in \mathbb{N}$, let $B_n = C_1 \cup \cdots \cup C_n$. Then each $B_n \in \mathcal{D}$ [by (1)] and $\{B_n\}_{n=1}^{\infty}$ is nondecreasing. Since $\mu(B_n) \geq \mu(C_n)$ for each $n \in \mathbb{N}$, it follows that

$$\varliminf_n \mu(B_n) \geq \varliminf_n \mu(C_n) = \lim_n \mu(C_n) = \alpha.$$

At the same time, $\mu(B_n) \leq \alpha$ for every $n \in \mathbb{N}$ by the definition of α, hence $\varlimsup_n \mu(B_n) \leq \alpha$. Combining inequalities gives $\lim_n \mu(B_n) = \alpha$. This completes the proof of (2).

Before proceeding with more subclaims, we state an additional definition. With the sequence $\{B_n\}_{n=1}^{\infty}$ as given in (2), define

$$D = \lim_n B_n = \bigcup_{n=1}^{\infty} B_n,$$

so that

$$\mu(D) = \mu\left(\lim_n B_n\right) = \lim_n \mu(B_n) = \alpha$$

by measure property (M8) in Section 2.2.

(3) $D \in \mathcal{D}$.

Proof: Let $C \in \mathcal{F}$ be such that $C \subseteq D$. We want to show that $\nu(C)$ is 0 or $+\infty$. Observe that since $B_1 \subseteq B_2 \subseteq \cdots$ and $C \subseteq D = \bigcup_{n=1}^{\infty} B_n$, C may be written as the disjoint union of $C \cap B_1$, $C \cap [B_2 - B_1]$, $C \cap [B_3 - B_2]$, $\cdots$, hence

$$\begin{aligned} \nu(C) &= \nu\left((C \cap B_1) \cup \bigcup_{n=2}^{\infty} (C \cap [B_n - B_{n-1}])\right) \\ &= \nu(C \cap B_1) + \sum_{n=2}^{\infty} \nu(C \cap [B_n - B_{n-1}]). \end{aligned}$$

Since $C \cap B_1$ is an $\mathcal{F}$-subset of the $\mathcal{D}$-set B_1, we have that $\nu(C \cap B_1) \in \{0, +\infty\}$. For each $n \geq 2$, we have that $C \cap (B_n - B_{n-1})$ is an $\mathcal{F}$-subset of the $\mathcal{D}$-set B_n, hence $\nu(C \cap [B_n - B_{n-1}]) \in \{0, +\infty\}$. Therefore, $\nu(C)$ is a countable sum of numbers, each number being 0 or $+\infty$. It follows that $\nu(C)$ is 0 or $+\infty$. Since C is an arbitrary $\mathcal{F}$-subset of D, we have that $D \in \mathcal{D}$, giving (3).

For the next subclaim, we consider the set $D^c = \bigcap_{n=1}^{\infty} B_n^c$.

(4) *If $F \subseteq D^c$ and $\nu(F) > 0$, then there is $F_1 \in \mathcal{F}$ such that $F_1 \subseteq F$ and $0 < \nu(F_1) < +\infty$.*

Proof: Let $F \subseteq D^c$ be such that $\nu(F) > 0$. If in addition we have $\nu(F) < +\infty$, then we may take $F_1 = F$ and be done with things. Therefore, we may assume that $\nu(F) = +\infty$. We argue by contradiction in this case, assuming that every $G \in \mathcal{F}$ with $G \subseteq F$ is such that $\nu(G)$ is 0 or $+\infty$.

Before proceeding further, let $C \in \mathcal{F}$ be such that $C \subseteq F \cup D$. Note that F and D are disjoint since $F \subseteq D^c$. Therefore, $\nu(C) = \nu(C \cap F) + \nu(C \cap D)$. Note that $C \cap F$ is an $\mathcal{F}$-subset of F, so the [contradictory] assumption gives $\nu(C \cap F) \in \{0, +\infty\}$. Next, $C \cap D$ is an $\mathcal{F}$-subset of the $\mathcal{D}$-set D, hence $\nu(C \cap D) \in \{0, +\infty\}$. It follows that $\nu(C) \in \{0, +\infty\}$, whence $F \cup D \in \mathcal{D}$.

Since $\nu \ll \mu$ and $\nu(F) > 0$, we have that $\mu(F) > 0$. By the disjointness of F and D we have

$$\mu(F \cup D) = \mu(F) + \mu(D) > 0 + \alpha = \alpha.$$

Since $F \cup D \in \mathcal{D}$, the definition of α gives $\mu(F \cup D) \leq \alpha$. This is a contradiction, and hence (4) holds in the case where $\nu(F) = +\infty$. Combining the cases where $\nu(F) < +\infty$ and $\nu(F) = +\infty$ gives (4) in all cases.

Before proceeding, we make another definition. Define

$$\mathcal{A} = \{F \in \mathcal{F} : F \subseteq D^c \text{ and } \nu \text{ is } \sigma\text{-finite on } F\}.$$

Note that $\varnothing \in \mathcal{A} \subseteq \mathcal{F}$. Define $\beta = \sup\{\mu(F) : F \in \mathcal{A}\}$, and observe that we have $\beta \leq \sup\{\mu(F) : F \in \mathcal{F}\} = \mu(\Omega) < +\infty$.

(5) *There is a nondecreasing sequence $\{F_n\}_{n=1}^{\infty}$ of $\mathcal{A}$-sets with* $\lim_n \mu(F_n) = \beta$.

Proof: By the definition of β as a supremum, there is a sequence $\{G_n\}_{n=1}^{\infty}$ of $\mathcal{A}$-sets with $\lim_n \mu(G_n) = \beta$. Let $F_n = G_1 \cup \cdots \cup G_n$ for every $n \in \mathbb{N}$. Since $\mathcal{A}$ is easily seen to be closed under finite unions, we have that $\{F_n\}_{n=1}^{\infty}$ is a nondecreasing sequence of $\mathcal{A}$-sets. Since $\mu(F_n) \geq \mu(G_n)$ for each $n \in \mathbb{N}$, it follows that $\underline{\lim}_n \mu(F_n) \geq \underline{\lim}_n \mu(G_n) = \lim_n \mu(G_n) = \beta$. Since $\mu(F_n) \leq \beta$ for each $n \in \mathbb{N}$, we have $\overline{\lim}_n \mu(F_n) \leq \beta$, whence $\lim_n \mu(F_n) = \beta$. This gives (5).

With (5) in tow, we now make yet another definition. Using the sequence $\{F_n\}_{n=1}^{\infty}$ in (5), define

$$F = \lim_n F_n = \bigcup_{n=1}^{\infty} F_n,$$

so that

$$\mu(F) = \mu\left(\lim_n F_n\right) = \lim_n \mu(F_n) = \beta.$$

Since ν is σ-finite on each F_n, ν must be σ-finite on F. Also, noting that each $F_n \subseteq D^c$ gives $F \subseteq D^c$, and hence $F \in \mathcal{A}$.

(6) *ν is σ-finite on D^c.*

Proof: We know that ν is σ-finite on F. Since $F \subseteq D^c$, it is sufficient to show that ν is σ-finite on $D^c - F = D^c \cap F^c$. In turn, it is sufficient to show that $\nu(F^c \cap D^c) = 0$.

To show that $\nu(F^c \cap D^c) = 0$, we argue by contradiction, assuming that $\nu(F^c \cap D^c) > 0$. Using the fact that $F^c \cap D^c \subseteq D^c$ and (4), we may find $H \in \mathcal{F}$ with $H \subseteq F^c \cap D^c$ and $0 < \nu(H) < +\infty$. Observe that $F \cup H \subseteq D^c$ since both F and H are contained in D^c. Since ν is σ-finite on F and $\nu(H) < +\infty$, it follows that ν is σ-finite on $F \cup H$, and thus $F \cup H \in \mathcal{A}$. Next, since $\nu(H) > 0$ and $\nu \ll \mu$, we have that $\mu(H) = 0$. From this and the fact that $F \cap H = \varnothing$, we have $\mu(F \cup H) = \mu(F) + \mu(H) > \beta + 0 = \beta$. However, at the same time we have $\mu(F \cup H) \leq \beta$ by the definition of β and the fact that $F \cup H \in \mathcal{A}$. This is a contradiction. Therefore, it cannot be the case that $\nu(F^c \cap D^c) > 0$; it must be that $\nu(F^c \cap D^c) = 0$. This completes the proof that ν is σ-finite on D^c.

Next, define the collection

$$\mathcal{B} = \{B \in \mathcal{F} : \ B \subseteq D \text{ and } \nu(B) = 0\}.$$

Note that $\varnothing \in \mathcal{B} \subseteq \mathcal{F}$. Next, observe that $\mathcal{B}$ is closed under at most countable unions. Let $\gamma = \sup\{\mu(B) : \ B \in \mathcal{B}\}$, so that

$$\gamma \leq \sup\{\mu(B) : \ B \in \mathcal{F}\} = \mu(\Omega) < +\infty.$$

(7) *There is a nondecreasing sequence* $\{E_n\}_{n=1}^{\infty}$ *of* $\mathcal{B}$*-sets with* $\lim_n \mu(E_n) = \gamma$.

Proof: By the definition of γ as a supremum, there exists a sequence $\{K_n\}_{n=1}^{\infty}$ of $\mathcal{B}$-sets with $\lim_n \mu(K_n) = \gamma$. Let $E_n = K_1 \cup \cdots \cup K_n$ for each $n \in \mathbb{N}$. Then $\{E_n\}_{n=1}^{\infty}$ is a nondecreasing sequence of $\mathcal{B}$-sets. Furthermore, we have $\mu(E_n) \geq \mu(K_n)$ for each $n \in \mathbb{N}$, hence

$$\varliminf_n \mu(E_n) \geq \varliminf_n \mu(K_n) = \lim_n \mu(K_n) = \gamma.$$

We also have $\mu(E_n) \leq \gamma$ for every $n \in \mathbb{N}$, hence $\varlimsup_n \mu(E_n) \leq \gamma$. It follows that $\lim_n \mu(E_n) = \gamma$, and (7) is complete.

Before proceeding further, we give yet another definition. Using the sequence $\{E_n\}_{n=1}^{\infty}$ exhibited in (7), we define

$$G = \lim_n E_n = \bigcup_{n=1}^{\infty} E_n.$$

Observe that $G \subseteq D$ and $\nu(G) = 0$ [since $\nu(E_n) = 0$ for each $n \in \mathbb{N}$], so that $G \in \mathcal{B}$.

With all of this tedious and somewhat labyrinthine work behind us, we define $E = D \cap G^c$, and we claim that E satisfies (a)-(c), that is, $E = D \cap G^c$ is the set promised in the statement of the claim.

To see that E satisfies (a), recall that $D \in \mathcal{D}$ by (3), hence every $A \in \mathcal{F}$ with $A \subseteq D$ is such that $\nu(A) = 0$ or $+\infty$. In particular, $E \subseteq D$, so every $A \in \mathcal{F}$ with $A \subseteq E$ is such that $\nu(A) = 0$ or $+\infty$. Therefore E satisfies (a).

We now take an eye towards showing that $E = D \cap G^c$ satisfies (b), proceeding by contradiction. Suppose that E fails to satisfy (b), so that there is an $\mathcal{F}$-set $A \subseteq E$ with $\nu(A) = 0$ and $\mu(A) > 0$. Then $G \cup A \in \mathcal{B}$[5]. Next, we have $\mu(G \cup A) = \mu(G) + \mu(A) > \gamma + 0 = \gamma$. [Note that G and A are disjoint since $A \subseteq E = D \cap G^c \subseteq G^c$.] However, the definition of γ and the fact that $G \cup A \in \mathcal{B}$ gives $\mu(G \cup A) \leq \gamma$, a contradiction. Therefore, E must satisfy (b).

We now show that $E = D \cap G^c$ satisfies (c), that is, we show that ν is σ-finite on $E^c = D^c \cup G$. Referring to (6), we have that ν is σ-finite on D^c. Since $\nu(G) = 0 < +\infty$, it follows that ν is σ-finite on $D^c \cup G = E^c$.

We have shown that the $\mathcal{F}$-set $E = D \cap G^c$ satisfies (a)-(c), and at last the tedious proof is complete. ∎

Claim 2. *Let $(\Omega, \mathcal{F}, \mu)$ denote a decomposable measure space with $\mathcal{A}$ denoting a decomposition of $(\Omega, \mathcal{F}, \mu)$. Let $A \in \mathcal{F}$ be such that $\mu(A) < +\infty$. There exists an amc subcollection $\mathcal{A}_0 \subseteq \mathcal{A}$ such that*

(a) $\mu(A) = \sum_{F \in \mathcal{A}} \mu(A \cap F) = \sum_{F \in \mathcal{A}_0} \mu(A \cap F)$, *and*

(b) $\mu(A \cap \bigcup_{F \in \mathcal{A} - \mathcal{A}_0} F) = 0$.

Proof: The existence of an at most countable subcollection $\mathcal{A}_0$ of $\mathcal{A}$ that satisfies (a) is a direct consequence of part (iii) of the definition of a decomposition and the earlier fact regarding sums. For this at most countable $\mathcal{A}_0 \subseteq \mathcal{A}$, we have that $A = \left[A \cap \bigcup_{F \in \mathcal{A}_0} F\right] \cup \left[A \cap \bigcup_{F \in \mathcal{A} - \mathcal{A}_0} F\right]$, where the quantities in square brackets are disjoint by part (ii) of the definition of a decomposition. Using part (ii) of the definition a second time gives

$$\begin{aligned}
\mu(A) &= \mu\left(A \cap \bigcup_{F \in \mathcal{A}_0} F\right) + \mu\left(A \cap \bigcup_{F \in \mathcal{A} - \mathcal{A}_0} F\right) \\
&= \mu\left(\bigcup_{F \in \mathcal{A}_0} (A \cap F)\right) + \mu\left(A \cap \bigcup_{F \in \mathcal{A} - \mathcal{A}_0} F\right) \\
&= \sum_{F \in \mathcal{A}_0} \mu(A \cap F) + \mu\left(A \cap \bigcup_{F \in \mathcal{A} - \mathcal{A}_0} F\right).
\end{aligned}$$

Since $\mu(A) < +\infty$, the above display and (a) immediately give part (b), completing the proof. ∎

We are now ready to state and prove our most general form of the Radon-Nikodym Theorem.

[5]Recall that $G \subseteq D$ and $A \subseteq D$, whence $G \cup A \subseteq D$. Since $\nu(G) = 0$ and $\nu(A) = 0$, we have $\nu(G \cap A) = 0$, giving $G \cup A \in \mathcal{B}$.

Radon-Nikodym Theorem, Decomposable Case. *Let $(\Omega, \mathcal{F}, \mu)$ denote a decomposable measure space with decomposition $\mathcal{A}$, and let ν denote any measure with domain $\mathcal{F}$ and $\nu \ll \mu$.*
(a) There exists $\mathcal{F}/\mathcal{B}^*$-measurable $f : \Omega \to [0, +\infty]$ such that $\nu(A) = \int_A f\, d\mu$ for every $A \in \mathcal{F}$ such that μ is σ-finite on A. This function f may be taken to be finite on each $A \in \mathcal{A}$ such that ν is σ-finite on A.
(b) We have $\int g\, d\nu = \int gf\, d\mu$ for every $\mathcal{F}/\mathcal{B}^*$-measurable $g : \Omega \to [0, +\infty]$ such that μ is σ-finite on $\{\omega \in \Omega :\ g(\omega) > 0\}$.
(c) If $g \in L^1(\Omega, \mathcal{F}, \nu)$ and μ is σ-finite on $\{\omega \in \Omega :\ g(\omega) \neq 0\}$, then gf is in $L^1(\Omega, \mathcal{F}, \mu)$ and $\int g\, d\nu = \int gf\, d\mu$.
(d) The function f is unique in the sense that if $h : \Omega \to [0, +\infty]$ is an $\mathcal{F}/\mathcal{B}^*$-measurable function such that $\nu(A) = \int h\, d\mu$ for every $A \in \mathcal{F}$ with $\mu(A) < +\infty$, then $fI_E = hI_E$ μ-a.e. on Ω when μ is σ-finite on E.

Proof: Before doing anything, we define some notation. Given any $A \in \mathcal{F}$, we shall let $\mathcal{F}_A$ denote the σ-field on A consisting of the $\mathcal{F}$-subsets of A: $\mathcal{F}_A = \{B \in \mathcal{F} :\ B \subseteq A\}$. We also write μ_A and ν_A for the restrictions of μ and ν to $\mathcal{F}_A$. Since $\nu \ll \mu$, we have $\nu_A \ll \mu_A$ for every $A \in \mathcal{F}$.

Construction. Let $F \in \mathcal{A}$, so that $\mu(F) < +\infty$ by (ii) of the definition of a decomposition. We may apply Claim 1 in the context of the measure spaces $(F, \mathcal{F}_F, \mu_F)$ and $(F, \mathcal{F}_F, \nu_F)$, obtaining a set $E_F \in \mathcal{F}_F$ with the following properties:
(α) For every $A \in \mathcal{F}_F$ with $A \subseteq E_F$ we have $\nu_F(A) = 0$ or $\nu_F(A) = +\infty$.
(β) For every $A \in \mathcal{F}_F$ with $A \subseteq E_F$, $\nu_F(A) = 0$ forces $\mu_F(A) = 0$.
(γ) $F - E_F$ is σ-finite with respect to ν_F [and hence to ν_{F-E_F}].
[If F is σ-finite with respect to ν_F, we may take $E_F = \varnothing$.]
Observe now that $(F - E_F, \mathcal{F}_{F-E_F}, \nu_{F-E_F})$ is a σ-finite measure space; we also have that $(F - E_F, \mathcal{F}_{F-E_F}, \mu_{F-E_F})$ is a finite measure space. We may apply the σ-finite version of the Radon-Nikodym Theorem given in the previous section to assert the existence of a nonnegative finite-valued $\mathcal{F}_{F-E_F}/\mathcal{B}^*$-measurable function $f^{(F)}$ on $F - E_F$ that is unique up to a set of μ_{F-E_F}-measure zero and such that

$$\nu_{F-E_F}(A) = \int_A f^{(F)}\, d\mu_{F-E_F} \quad \text{for every } A \in \mathcal{F}_{F-E_F}. \tag{9.2}$$

Perform the construction above for every $F \in \mathcal{A}$.

Having performed the preceding construction for every $F \in \mathcal{A}$, we now define a nonnegative function f on Ω by writing

$$f(\omega) = \begin{cases} f^{(F)}(\omega) & \omega \in F - E_F \\ +\infty & \omega \in E_F \end{cases} \qquad \text{whenever } \omega \in F \in \mathcal{A}.$$

By (ii) of the definition of a decomposition, the sets $F \in \mathcal{A}$ are disjoint with union Ω, hence f truly does have domain Ω and no consistency problems arise with how f is defined.

We wish to establish the $\mathcal{F}/\mathcal{B}^*$-measurability of f, using for the first [and only] time requirement (iv) of the definition of a decomposition. To establish

$\mathcal{F}/\mathcal{B}^*$-measurability, we will show that $f^{-1}(\{+\infty\}) \in \mathcal{F}$, $f^{-1}(\{-\infty\}) \in \mathcal{F}$, and $f^{-1}(B) \in \mathcal{F}$ for every $B \in \mathcal{B}$. [See (MF1) in Section 5.1, and note that here $\mathcal{B}$ means the usual Borel σ-field on $\mathbb{R}$.] It is clearly the case that $f^{-1}(\{-\infty\}) = \varnothing \in \mathcal{F}$. Next, since $f^{(F)}$ is finite-valued on $F - E_F$ for every $F \in \mathcal{A}$, we have $f^{-1}(\{+\infty\}) = \bigcup_{F \in \mathcal{A}} E_F$, where the union is disjoint by (ii) of the decomposition definition and the fact that $E_F \subseteq F$ for every $F \in \mathcal{A}$. Since for every $G \in \mathcal{A}$ we have $G \cap \bigcup_{F\in\mathcal{A}} E_F = E_G \in \mathcal{F}$, property (iv) from the definition of a decomposition gives $f^{-1}(\{+\infty\}) \in \mathcal{F}$. Finally, picking $B \in \mathcal{B}$, we recall that $f^{(F)}$ is defined and finite on $F - E_F$ for every $F \in \mathcal{A}$, hence

$$\begin{aligned} f^{-1}(B) = \{\omega \in \Omega :\ f(\omega) \in B\} &= \bigcup_{F\in\mathcal{A}} \{\omega \in F :\ f(\omega) \in B\} \\ &= \bigcup_{F\in\mathcal{A}} \left\{\omega \in F :\ f^{(F)}(\omega) \in B\right\} \\ &= \bigcup_{F\in\mathcal{A}} \left(f^{(F)}\right)^{-1}(B). \end{aligned}$$

Now $f^{(F)}$ is $\mathcal{F}_{F-E_F}/\mathcal{B}^*$-measurable for each $F \in \mathcal{A}$, hence $(f^{(F)})^{-1}(B)$ is in $\mathcal{F}_{F-E_F}$ for every $F \in \mathcal{A}$. Since

$$G \cap f^{-1}(B) = G \cap \bigcup_{F\in\mathcal{A}} \left(f^{(F)}\right)^{-1}(B) = \left(f^{(G)}\right)^{-1}(B) \in \mathcal{F}_{G-E_G} \subseteq \mathcal{F}$$

for every $G \in \mathcal{A}$, (iv) of the decomposition definition gives $f^{-1}(B) \in \mathcal{F}$. This establishes the $\mathcal{F}/\mathcal{B}^*$-measurability of f.

We will now show that $\nu(A) = \int_A f\, d\mu$ for every $A \in \mathcal{F}$ with $\mu(A) < +\infty$. Pick $A \in \mathcal{F}$ with $\mu(A) < +\infty$. By Claim 2, there is an at most countable subcollection $\mathcal{A}_0$ of $\mathcal{A}$ such that

$$\mu(A) = \sum_{F\in\mathcal{A}_0} \mu(A \cap F) \quad \text{and} \quad \mu\left(A \cap \bigcup_{F\in\mathcal{A}-\mathcal{A}_0} F\right) = 0. \tag{9.3}$$

Since $\nu \ll \mu$, we have $\nu(A \cap \bigcup_{F\in\mathcal{A}-\mathcal{A}_0} F) = 0$, and hence

$$\nu(A) = \nu\left(A \cap \bigcup_{F\in\mathcal{A}_0} F\right) = \nu\left(\bigcup_{F\in\mathcal{A}_0} [A \cap F]\right) = \sum_{F\in\mathcal{A}_0} \nu(A \cap F). \tag{9.4}$$

Using (9.2), we have for every $F \in \mathcal{A}$ that

$$\begin{aligned} \nu(A \cap F) &= \nu(A \cap [F - E_F]) + \nu(A \cap E_F) \\ &= \nu_{F-E_F}(A \cap [F - E_F]) + \nu_F(A \cap E_F) \\ &= \int_{A\cap[F-E_F]} f^{(F)}\, d\mu_{F-E_F} + \nu_F(A \cap E_F) \\ &= \int_{A\cap[F-E_F]} f\, d\mu + \nu_F(A \cap E_F). \end{aligned} \tag{9.5}$$

[The last equality follows by noting that $f^{(F)} = f$ on $A \cap [F - E_F]$ for every $F \in \mathcal{A}$ and using integration property (L18) from Section 6.4.] Now $A \cap E_F$ is an $\mathcal{F}$-subset of E_F, so by (α) and (β) we have that $\nu_F(A \cap E_F)$ is 0 or $+\infty$, and if $\nu_F(A \cap E_F) = 0$ then $\mu_F(A \cap E_F) = 0$, so that $\nu_F(A \cap E_F) = 0$ iff $\mu_F(A \cap E_F) = 0$ [recall that $\nu_F \ll \mu_F$]. Therefore, noting that $f = +\infty$ on E_F, we have that

$$\nu_F(A \cap E_F) = \nu(A \cap E_F) = \int_{A \cap E_F} (+\infty)\, d\mu(\omega) = \int_{A \cap E_F} f\, d\mu.$$

[If $\nu_F(A \cap E_F) = 0$, then $\mu(A \cap E_F) = \mu_F(A \cap E_F) = 0$, so that integration property (L2) from Section 6.4 gives $\int_{A \cap E_F} (+\infty)\, d\mu(\omega) = 0$. If $\nu_F(A \cap E_F)$ equals $+\infty$, then $\mu_F(A \cap E_F) = \mu(A \cap E_F) > 0$, so that $\int_{A \cap E_F} (+\infty)\, d\mu(\omega)$ equals $+\infty$.] Going back to (9.5) gives

$$\nu(A \cap F) = \int_{A \cap [F - E_F]} f\, d\mu + \int_{A \cap E_F} f\, d\mu = \int_{A \cap F} f\, d\mu \quad \text{for every } F \in \mathcal{A}. \tag{9.6}$$

Therefore, (9.4) and (9.6) give

$$\begin{aligned} \nu(A) = \sum_{F \in \mathcal{A}_0} \nu(A \cap F) &= \sum_{F \in \mathcal{A}_0} \int_{A \cap F} f\, d\mu \\ &= \int_{A \cap \bigcup_{F \in \mathcal{A}_0} F} f\, d\mu \quad \text{[by (L6)]} \\ &= \int_A f\, d\mu. \end{aligned} \tag{9.7}$$

[The last equality holds because (9.3) and (L2) force $\int_{A \cap \bigcup_{F \in \mathcal{A} - \mathcal{A}_0} F} f\, d\mu = 0$.] This shows that (a) holds when $A \in \mathcal{F}$ is such that $\mu(A) < +\infty$.

We now complete the proof of (a) by assuming that $A \in \mathcal{F}$ is σ-finite with respect to μ. Write $A = \bigcup_{n=1}^{\infty} A_n$, where $A_1, A_2, \cdots$ are disjoint $\mathcal{F}$-sets with finite μ-measure. By (9.7), countable additivity, and integration property (L6), we have

$$\nu(A) = \sum_{n=1}^{\infty} \nu(A_n) = \sum_{n=1}^{\infty} \int_{A_n} f\, d\mu = \int_A f\, d\mu.$$

This at last establishes (a).

We now establish (b), first obtaining the result for nonnegative $\mathcal{F}/\mathcal{B}^*$-measurable simple functions. Suppose that $g = \sum_{i=1}^{n} c_i I_{A_i}$, where $c_1, \cdots, c_n$ are nonnegative and μ is σ-finite on $\{\omega \in \Omega : g(\omega) > 0\}$. Then μ is σ-finite on those sets A_i with $c_i > 0$, and by (a) we have that $\nu(A_i) = \int_{A_i} f\, d\mu$ for those i with $c_i > 0$. It follows that

$$\begin{aligned}
\int g\,d\nu = \sum_{i=1}^{n} c_i\nu(A_i) = \sum_{i:c_i>0} c_i\nu(A_i) &= \sum_{i:c_i>0} c_i \int_{A_i} f\,d\mu \\
&= \sum_{i:c_i>0} c_i \int I_{A_i} f\,d\mu \\
&= \int \sum_{i:c_i>0} c_i I_{A_i} f\,d\mu \\
&= \int \sum_{i=1}^{n} c_i I_{A_i} f\,d\mu \\
&= \int gf\,d\mu.
\end{aligned}$$

Therefore, the claim of (b) holds for simple functions g of the form just stated. Next, assume that $g : \Omega \to [0, +\infty]$ is $\mathcal{F}/\mathcal{B}^*$ measurable and such that μ is σ finite on $\{\omega \in \Omega : g(\omega) > 0\}$. There exists a nondecreasing sequence $\{s_n\}_{n=1}^{\infty}$ of nonnegative $\mathcal{F}/\mathcal{B}^*$-measurable simple functions with $\lim_n s_n = g$. For each $n \in \mathbb{N}$ and $\omega \in \Omega$, we have that $g(\omega) = 0$ forces $s_n(\omega) = 0$, giving

$$\{\omega \in \Omega : s_n(\omega) > 0\} \subseteq \{\omega \in \Omega : g(\omega) > 0\},$$

so that μ is σ-finite on $\{\omega \in \Omega : s_n(\omega) > 0\}$ for every $n \in \mathbb{N}$. Using the fact that (b) thus holds for each s_n, we have $\int s_n\,d\nu = \int s_n f\,d\mu$ for every $n \in \mathbb{N}$, and using MCT on the nonnegative nondecreasing sequence $\{s_n f\}_{n=1}^{\infty}$ gives

$$\begin{aligned}
\int g\,d\nu = \int \lim_n s_n\,d\nu = \lim_n \int s_n\,d\nu &= \lim_n \int s_n f\,d\mu \\
&= \int \lim_n s_n f\,d\mu \\
&= \int gf\,d\mu.
\end{aligned}$$

This completely gives (b).

We now turn to proving (c). Let $g \in L^1(\Omega, \mathcal{F}, \nu)$ be such that μ is σ-finite on the set $\{\omega \in \Omega : g(\omega) \neq 0\}$. Write $g = g^+ - g^-$, where both g^+ and g^- are nonnegative $\mathcal{F}/\mathcal{B}^*$-measurable functions on Ω with $\int g^+\,d\nu < +\infty$ and $\int g^-\,d\nu < +\infty$. Since $g^+(\omega) > 0$ forces $g(\omega) > 0$ and since $g^-(\omega) > 0$ forces $g(\omega) < 0$, we have

$$\{\omega \in \Omega : g^+(\omega) > 0\},\ \{\omega \in \Omega : g^-(\omega) > 0\} \subseteq \{\omega \in \Omega : g(\omega) \neq 0\},$$

and thus μ is σ-finite on both $\{\omega \in \Omega : g^+(\omega) > 0\}$ and $\{\omega \in \Omega : g^-(\omega) > 0\}$. We may therefore apply (b) to g^+ and g^-, obtaining

$$+\infty > \int g^+\,d\nu = \int g^+ f\,d\mu = \int (fg)^+\,d\mu$$

and

$$+\infty > \int g^- \, d\nu = \int g^- f \, d\mu = \int (fg)^- \, d\mu.$$

Therefore, we have that $\int fg \, d\mu$ exists and is finite, that is, $fg \in L^1(\Omega, \mathcal{F}, \mu)$. Putting these equations together, we have

$$\begin{aligned}\int g \, d\nu = \int g^+ \, d\nu - \int g^- \, d\nu &= \int g^+ f \, d\mu - \int g^- f \, d\mu \\ &= \int (g^+ - g^-) f \, d\mu \\ &= \int gf \, d\mu,\end{aligned}$$

which completes the proof of (c).

We now attempt to establish the uniqueness claim in part (d). Assume that $h : \Omega \to [0, +\infty]$ is $\mathcal{F}/\mathcal{B}^*$-measurable and such that $\nu(A) = \int h \, d\mu$ for every $A \in \mathcal{F}$ with $\mu(A) < +\infty$. Before we can establish the claim of (d), we require the following subclaim:

Subclaim: *$f = h$ μ-a.e. on F for every $F \in \mathcal{A}$.*

Proof: Let $F \in \mathcal{A}$ and $A \in \mathcal{F}_{F-E_F}$, so that $\mu(A) \le \mu(F - E_F) \le \mu(F) < +\infty$. It follows that

$$\int_A h \, d\mu = \nu(A) = \int_A f \, d\mu$$

by (a) and the preceding paragraph. By (L18), this gives

$$\int_A h \, d\mu_{F-E_F} = \int_A f \, d\mu_{F-E_F}.$$

Since the set A in $\mathcal{F}_{F-E_F}$ is generic, integration property (L17) [invoked relative to the space $(F - E_F, \mathcal{F}_{F-E_F}, \mu_{F-E_F})$] shows that $f = h$ μ_{F-E_F}-a.e. on $F - E_F$. That is, $f = h$ μ-a.e. on $F - E_F$.

To complete the proof of the subclaim, we must now show that $f = h$ μ-a.e. on E_F. We argue by contradiction, assuming that it is not the case that $f = h$ μ-a.e. on E_F. Under this assumption, there is $A \in \mathcal{F}_{E_F}$ such that $\mu(A) > 0$ and $h(\omega) < +\infty = f(\omega)$ for every $\omega \in A$. [Recall that $f = +\infty$ on E_F, hence $f = +\infty$ on the subset A of E_F.] Since $A \subseteq E_F \subseteq F$ and $\mu(F) < +\infty$, we have $\mu(A) < +\infty$. By the contrapositive of (β), it follows that $\nu(A) > 0$. For every $n \in \mathbb{N}$, let $A_n = \{\omega \in A : h(\omega) < n\}$, so that $\mu(A_n) \le \mu(A) < +\infty$ for every $n \in \mathbb{N}$. Then $\{A_n\}_{n=1}^{\infty}$ is a nondecreasing sequence of $\mathcal{F}$-sets with finite μ-measure, and

$$\lim_n A_n = \bigcup_{n=1}^{\infty} A_n = \{\omega \in A : h(\omega) < +\infty\} = A.$$

We have $0 < \nu(A) = \nu(\lim_n A_n) = \lim_n \nu(A_n)$ by measure property (M8), and hence there is $N \in \mathbb{N}$ such that $\nu(A_N) > 0$. Since $A_N \subseteq A \subseteq E_F$, ($\alpha$) gives $\nu(A_N) = +\infty$.

Therefore, using the fact that $\mu(A_N) < +\infty$ to obtain $\nu(A_N) = \int_{A_N} h\, d\mu$, we have

$$+\infty = \nu(A_N) = \int_{A_N} h\, d\mu \le \int_{A_N} N\, d\mu = N\mu(A_N) < +\infty,$$

which is a contradiction. This shows that $f = h$ μ-a.e. on E_F.

We have shown that $f = h$ μ-a.e. on $F - E_F$ and E_F, and it follows that $f = h$ μ-a.e. on F. Since $F \in \mathcal{A}$ is generic, the proof of the subclaim is complete.

With the subclaim in tow, we may now turn to proving (d). The goal is to show that $fI_G = hI_G$ μ-a.e. on Ω whenever μ is σ-finite on G. Since $fI_G = 0 = hI_G$ on G^c, it suffices to show that $f = h$ μ-a.e. on G whenever μ is σ-finite on G. To this end, let $G \in \mathcal{F}$ be such that μ is σ-finite on G, and write $G = \bigcup_{n=1}^{\infty} G_n$, where $G_1, G_2, \cdots$ are disjoint $\mathcal{F}$-sets of finite μ-measure. If we can show that $f = h$ μ-a.e. on each G_n, then we will have $f = h$ μ-a.e. on G, and (d) will follow. Therefore, we pick and fix $n \in \mathbb{N}$ for the remainder of the proof of (d), desiring to show that $f = h$ μ-a.e. on G_n.

Let $X_n = \{\omega \in G_n : \; h(\omega) \ne f(\omega)\}$. Then $\mu(X_n) \le \mu(G_n) < +\infty$. Invoking (iii) of the decomposition definition gives $\mu(X_n) = \sum_{F\in\mathcal{A}} \mu(X_n \cap F)$. For every $F \in \mathcal{A}$ we have $f = h$ μ-a.e. on F by the subclaim, and hence $\mu(X_n \cap F) = 0$ for every $F \in \mathcal{A}$. But then $\sum_{F\in\mathcal{D}} \mu(X_n \cap F) = 0$ for every finite subcollection $\mathcal{D} \subseteq \mathcal{A}$, whence $\sum_{F\in\mathcal{A}} \mu(X_n \cap F) = 0$, which in turn gives $\mu(X_n) = 0$. This shows that $f = h$ μ-a.e. on G_n, which at last completes (d) and the entire proof itself. ∎

By Example 1 and the version of the Radon-Nikodym Theorem just given, we may state the following:

Radon-Nikodym Theorem. *Let $(\Omega, \mathcal{F}, \mu)$ denote a σ-finite measure space with decomposition $\mathcal{A}$, and let ν denote any measure with domain $\mathcal{F}$ and $\nu \ll \mu$.*

(a) There exists $\mathcal{F}/\mathcal{B}^*$-measurable $f : \Omega \to [0, +\infty]$ such that $\nu(A) = \int_A f\, d\mu$ for every $A \in \mathcal{F}$. This function f may be taken to be finite on each $A \in \mathcal{A}$ that is σ-finite with respect to μ.

(b) We have $\int g\, d\nu = \int gf\, d\mu$ for every $\mathcal{F}/\mathcal{B}^*$-measurable $g : \Omega \to [0, +\infty]$.

(c) If $g \in L^1(\Omega, \mathcal{F}, \nu)$, then gf is in $L^1(\Omega, \mathcal{F}, \mu)$ and $\int g\, d\nu = \int gf\, d\mu$.

(d) The function f is unique in the sense that if $h : \Omega \to [0, +\infty]$ is an $\mathcal{F}/\mathcal{B}^*$-measurable function such that $\nu(A) = \int h\, d\mu$ for every $A \in \mathcal{F}$ with $\mu(A) < +\infty$, then $f = h$ μ-a.e. on Ω.

A pragmatic question is whether all of the extra expended effort in this section is worth having more general results, especially in light of the fact that often measures are finite [as in probability theory] or σ-finite [as in most of analysis]. This is debatable. However, the results of this section are given here for completeness' sake, and the reader may find them of use at some point in the future. Nowhere in the rest of the text are these results used.

Exercises.

1. If $(\Omega, \mathcal{F}, \mu)$ and $(\Omega, \mathcal{F}, \nu)$ are decomposable measure spaces, then $(\Omega, \mathcal{F}, \mu + \nu)$ is a decomposable measure space.

2. Let assumptions be as in Exercise 1. Make the following new definition:[6] μ and ν are *singular* if (i) there is $B \in \mathcal{F}$ such that $\mu(B \cap E) = 0$ for all $E \in \mathcal{F}$ that are σ-finite with respect to μ, and (ii) $\nu(B^c \cap E) = 0$ for all $E \in \mathcal{F}$ that are σ-finite with respect to ν. State and prove an analogous version of the σ-finite version of the Radon-Nikodym Theorem from the previous section, and verify that this version is a bona fide generalization of the decomposable version of the Radon-Nikodym Theorem found in this section.

3. Billingsley gives the following outline for an alternative proof of the final version of the Radon-Nikodym Theorem given in this section. Assume first that μ is finite and $\nu \ll \mu$. [Both measures of course have domain $\mathcal{F}$, and ν is arbitrary.]
(a) Let $\mathcal{H}$ denote the class of $\mathcal{F}$-sets H such that $\mu(E) = 0$ or $\nu(E) = +\infty$ for each $\mathcal{F}$-set $E \subseteq H$. Then $\mathcal{H}$ contains a set H_0 of maximal μ-measure.
(b) Let $\mathcal{C}$ denote the class of sets in $\Omega_0 = H_0^c$ that are countable unions of sets of finite ν-measure. Then $\mathcal{C}$ contains a set C_0 of maximal μ-measure.

Write $D_0 = \Omega_0 - C_0$.

(c) From the maximality of H_0 and C_0, we have $\mu(D_0) = \nu(D_0) = 0$.
(d) Let $\nu_0(A) = \nu(A \cap \Omega_0)$ for every $A \in \mathcal{F}$. Use the Radon-Nikodym Theorem from the previous section on the pair μ, ν_0 to prove the result for the pair μ, ν.
(e) We may now prove the result for the case where μ is σ-finite.
(f) The following additional claim holds: if the Radon-Nikdoym derivative in (e) can be taken everywhere finite, then ν is σ-finite.

9.3 FROM RADON–NIKDOYM TO RIESZ REPRESENTATION

At the end of Chapter 8, we proved the Riesz Representation Theorem:

Riesz Representation Theorem. *Let L denote a bounded linear functional on $L^p(\Omega, \mathcal{F}, \mu)$, where $1 < p < +\infty$. Then there exists $h \in L^{p'}$ such that*

$$L(f) = \int fh \, d\mu \quad \textit{for every } f \in L^p(\Omega, \mathcal{F}, \mu).$$

There are absolutely no restrictions on the nature of the measure space $(\Omega, \mathcal{F}, \mu)$ — μ may be finite, σ-finite, or non-σ-finite. However, some very technical and nonintuitive work was needed for the proof of the theorem. It is possible to prove a slightly less general version of the Riesz Representation Theorem using the Radon-Nikodym Theorem. The price to be paid is that the measure space $(\Omega, \mathcal{F}, \mu)$ must be assumed σ-finite.

The work in this section borrows the definitions and results from the subsection on linear transformations given in Section 8.8. For Section 8.9, we shall be

[6] This definition is only given in this problem and will not be called on later in the text.

using the results up through and including Claim 1 and its proof. The reader will recall that Claim 1 in Section 8.9 is as follows: *Let* $g \in L^{p'}(\Omega, \mathcal{F}, \mu)$, *where* $1 < p < +\infty$, *and define* $L_g : L^p(\Omega, \mathcal{F}, \mu) \to \mathbb{R}$ *by writing* $L_g(f) = \int fg \, d\mu$ *for every* $f \in L^p(\Omega, \mathcal{F}, \mu)$. *Then* L_g *is a bounded linear functional on* $L^p(\Omega, \mathcal{F}, \mu)$ *and* $||L_g|| = ||g||_{p'}$.

With this, we now state and prove the slightly less general form of the Riesz Representation Theorem.

Riesz Representation Theorem. *Let* $(\Omega, \mathcal{F}, \mu)$ *denote a* σ*-finite measure space, and let* $1 < p < +\infty$. *Let* L *denote a bounded linear functional on* $L^p(\Omega, \mathcal{F}, \mu)$. *Then there exists a finite-valued* $g \in L^{p'}(\Omega, \mathcal{F}, \mu)$ *such that*

$$L(f) = \int fg \, d\mu \quad \textit{for every } f \in L^p(\Omega, \mathcal{F}, \mu).$$

This g *is unique up to a* μ*-null set.*

Proof: If $||L|| = 0$, then $L(f) = 0$ for every $f \in L^p(\Omega, \mathcal{F}, \mu)$; here taking $g = 0$ gives the result. We therefore assume that $||L|| > 0$, so that $L(f) \neq 0$ for some $f \in L^p(\Omega, \mathcal{F}, \mu)$. The proof is broken into two major parts: the case where μ is finite, and the case where μ is σ-finite. We prove the finite case first, and we will break the proof of the theorem for this case into a number of convenient subclaims. Until notified otherwise, μ is a finite measure: $\mu(\Omega) < +\infty$.

(i) Let $\phi(A) = L(I_A)$ *for every* $A \in \mathcal{F}$. *Then* ϕ *is an additive set function with domain* $\mathcal{F}$.

Proof: We can clearly consider $L(I_A)$ for every $A \in \mathcal{F}$ since the finiteness of μ ensures that $(\int I_A^p \, d\mu)^{\frac{1}{p}} = (\int I_A \, d\mu)^{\frac{1}{p}} = \mu(A)^{\frac{1}{p}} < +\infty$, so that I_A is in $L^p(\Omega, \mathcal{F}, \mu)$, the domain of L. Since linear functionals are by definition real-valued, so is ϕ. We need only establish that $\phi(\bigcup_{n=1}^{\infty} A_n) = \sum_{n=1}^{\infty} \phi(A_n)$ for every disjoint sequence $\{A_n\}_{n=1}^{\infty}$ of $\mathcal{F}$-sets.

Observe first that for disjoint $\mathcal{F}$-sets A and B we have

$$\phi(A \cup B) = L(I_{A \cup B}) = L(I_A + I_B) = L(I_A) + L(I_B) = \phi(A) + \phi(B);$$

induction therefore gives $\phi(A_1 \cup \cdots \cup A_n) = \sum_{i=1}^{n} \phi(A_i)$ whenever $n \in \mathbb{N}$ and $A_1, \cdots, A_n \in \mathcal{F}$ are disjoint. Next, observe for every $A \in \mathcal{F}$ that

$$|\phi(A)| = |L(I_A)| \leq ||L|| ||I_A||_p = ||L|| \left(\int I_A \, d\mu \right)^{\frac{1}{p}} = ||L|| \mu(A)^{\frac{1}{p}}. \tag{9.8}$$

We may now establish countable additivity. Let $\{A_n\}_{n=1}^{\infty}$ denote a disjoint sequence of $\mathcal{F}$-sets, and pick $\epsilon > 0$. Using the finite additivity given in the previous paragraph, we have for all $N \in \mathbb{N}$ that

$$\begin{aligned} \phi\left(\bigcup_{n=1}^{\infty} A_n\right) = \phi\left(\bigcup_{n=1}^{N} A_n \cup \bigcup_{n=N+1}^{\infty} A_n\right) &= \phi\left(\bigcup_{n=1}^{N} A_n\right) + \phi\left(\bigcup_{n=N+1}^{\infty} A_n\right) \\ &= \sum_{n=1}^{N} \phi(A_n) + \phi\left(\bigcup_{n=N+1}^{\infty} A_n\right). \end{aligned}$$

Using (9.8), we have for every $n \in \mathbb{N}$ that

$$\begin{aligned} \left|\phi\left(\bigcup_{n=1}^{\infty} A_n\right) - \sum_{n=1}^{N}\phi(A_n)\right| &= \left|\phi\left(\bigcup_{n=N+1}^{\infty} A_n\right)\right| \\ &\le \|L\|\mu\left(\bigcup_{n=N+1}^{\infty} A_n\right)^{\frac{1}{p}}. \end{aligned} \tag{9.9}$$

Now μ is finite and $\{\bigcup_{n=N+1}^{\infty} A_n\}_{N=1}^{\infty}$ is nonincreasing with limit $\varnothing$ since the sets $A_1, A_2, \cdots$ are disjoint, hence the sequence $\{\mu(\bigcup_{n=N+1}^{\infty} A_n)\}_{N=1}^{\infty}$ is nonincreasing with limit $\mu(\varnothing) = 0$ by measure property (M9). Therefore, for sufficiently large N, we have that $\mu(\bigcup_{n=N+1}^{\infty} A_n) \le (\frac{\epsilon}{\|L\|})^p$, and in this case (9.9) gives

$$\left|\phi\left(\bigcup_{n=1}^{\infty} A_n\right) - \sum_{n=1}^{N}\phi(A_n)\right| \le \|L\|\mu\left(\bigcup_{n=N+1}^{\infty} A_n\right)^{\frac{1}{p}} \le \|L\|\frac{\epsilon}{\|L\|} = \epsilon.$$

It follows that the sequence $\{\sum_{n=1}^{N}\phi(A_n)\}_{N=1}^{\infty}$ converges to $\phi(\bigcup_{n=1}^{\infty} A_n)$, and hence the countable additivity claim is established. Accordingly, ϕ is an additive set function. This completes the proof of (i).

(ii) There exists finite-valued $g \in L^1(\Omega, \mathcal{F}, \mu)$ such that $\phi(A) = \int_A g\, d\mu$ for every $A \in \mathcal{F}$.

Proof: By Claim 3 of Section 9.1, there exists a Hahn decomposition (A^+, A^-) of Ω with respect to the additive set function ϕ given in (i). Define ϕ^+ and ϕ^- on $\mathcal{F}$ by writing

$$\phi^+(A) = \phi(A \cap A^+) \quad \text{and} \quad \phi^-(A) = -\phi(A \cap A^-) \quad \text{for every } A \in \mathcal{F}.$$

By Claim 4 in Section 9.1, ϕ^+ and ϕ^- are finite measures with domain $\mathcal{F}$ having disjoint supports A^+ and A^-, respectively. Also, if $A \in \mathcal{F}$ is such that $\mu(A) = 0$, then (9.8) gives

$$\phi^+(A) = \phi(A \cap A^+) \le \|L\|\mu(A \cap A^+)^{\frac{1}{p}} \le \|L\|\mu(A)^{\frac{1}{p}} = 0$$

and

$$\phi^-(A) = -\phi(A \cap A^-) = |\phi(A \cap A^-)| \le \|L\|\mu(A \cap A^-)^{\frac{1}{p}} \le \|L\|\mu(A)^{\frac{1}{p}} = 0,$$

so that we have $\phi^+ \ll \mu$ and $\phi^- \ll \mu$. We now invoke the finite version of the Radon-Nikodym Theorem twice, once for the pair of finite measures μ and ϕ^+, and once for the pair of finite measures μ and ϕ^-. Doing this, we may assert the existence of $\mathcal{F}/\mathcal{B}^*$-measurable $g^+, g^- : \Omega \to [0, +\infty)$ such that

$$\phi^+(A) = \int_A g^+\, d\mu \quad \text{and} \quad \phi^-(A) = \int_A g^-\, d\mu \quad \text{for every } A \in \mathcal{F}.$$

Since $\int g^+ \, d\mu = \phi^+(\Omega) < +\infty$ and $\int g^- \, d\mu = \phi^-(\Omega) < +\infty$, we have that $g^+, g^- \in L^1(\Omega, \mathcal{F}, \mu)$. Let $g = g^+ - g^-$. Observe that g is defined on Ω, finite-valued, $\mathcal{F}/\mathcal{B}^*$-measurable, in $L^1(\Omega, \mathcal{F}, \mu)$, and, recalling that $\phi = \phi^+ - \phi^-$, we have for every $A \in \mathcal{F}$ that

$$\phi(A) = \phi^+(A) - \phi^-(A) = \int_A g^+ \, d\mu - \int_A g^- \, d\mu = \int_A g \, d\mu.$$

The function g is as promised, and the proof of (ii) is complete.

(iii) For every finite-valued $\mathcal{F}/\mathcal{B}^*$-measurable simple function $f = \sum_{i=1}^n c_i I_{A_i}$ defined on Ω, we have $L(f) = \int fg \, d\mu$, where g is as in (ii).

Proof: Let f be of the form $\sum_{i=1}^n c_i I_{A_i}$, where $c_1, \cdots, c_n \in \mathbb{R}$ and $A_1, \cdots, A_n$ are disjoint $\mathcal{F}$-sets with union Ω. Since $\mu(A_1), \cdots, \mu(A_n) < +\infty$ by the finiteness of μ, we have

$$\int |f|^p \, d\mu = \int \sum_{i=1}^n |c_i|^p I_{A_i} \, d\mu = \sum_{i=1}^n |c_i|^p \mu(A_i) < +\infty$$

and thus $f \in L^p(\Omega, \mathcal{F}, \mu)$, so that we may meaningfully discuss $L(f)$. We also have that $fg \in L^1(\Omega, \mathcal{F}, \mu)$ since $fg = \sum_{i=1}^n c_i I_{A_i} g$ and $c_i I_{A_i} g \in L^1(\Omega, \mathcal{F}, \mu)$ for $i = 1, \cdots, n$. This means that we may meaningfully discuss $\int fg \, d\mu$. With these preliminary issues established, the proof of the asserted equation is simple:

$$\begin{aligned} L(f) = \sum_{i=1}^n c_i L(I_{A_i}) = \sum_{i=1}^n c_i \int_{A_i} g \, d\mu &= \sum_{i=1}^n c_i \int I_{A_i} g \, d\mu \\ &= \int \sum_{i=1}^n c_i I_{A_i} g \, d\mu \\ &= \int fg \, d\mu. \end{aligned}$$

This completes the proof of (iii).

(iv) The function g as constructed in (ii) is in $L^{p'}(\Omega, \mathcal{F}, \mu)$.

Proof: If $g = 0$ μ-a.e. on Ω, then the claim is trivial. Assume therefore that g is nonzero on a set of positive μ-measure. Find a nondecreasing sequence of nonnegative finite-valued $\mathcal{F}/\mathcal{B}^*$-measurable simple functions $\{s_n\}_{n=1}^\infty$ such that $\lim_n s_n = |g|^{p'}$. Since μ is finite, we have $s_n \in L^1(\Omega, \mathcal{F}, \mu)$ for each $n \in \mathbb{N}$. Since $|g|^{p'}$ is positive on a set of positive μ-measure and $\lim_n s_n = |g|^{p'}$, we have by MCT that

$$0 < \int |g|^{p'} \, d\mu = \int \lim_n s_n \, d\mu = \lim_n \int s_n \, d\mu,$$

so that $\int s_n \, d\mu > 0$ for sufficiently large values of n. Write $h_n = s_n^{\frac{1}{p}} \operatorname{sgn} g$ for each $n \in \mathbb{N}$. Each h_n is simple and in $L^p(\Omega, \mathcal{F}, \mu)$, so $L(h_n) = \int h_n g \, d\mu$ for every $n \in \mathbb{N}$ by (iii). Next, observe that

$$h_n g = s_n^{\frac{1}{p}} (\operatorname{sgn} g) g = s_n^{\frac{1}{p}} |g| \geq s_n^{\frac{1}{p}} s_n^{\frac{1}{p'}} = s_n \quad \text{for every } n \in \mathbb{N},$$

hence for every $n \in \mathbb{N}$ we have

$$\begin{aligned}
\int s_n \, d\mu \leq \int h_n g \, d\mu = L(h_n) &\leq \|L\| \|h_n\|_p \\
&= \|L\| \left(\int \left| s_n^{\frac{1}{p}} \operatorname{sgn} g \right|^p d\mu \right)^{\frac{1}{p}} \\
&= \|L\| \left(\int_{\{\omega \in \Omega: \, g(\omega) \neq 0\}} s_n \, d\mu \right)^{\frac{1}{p}} \\
&\leq \|L\| \left(\int s_n \, d\mu \right)^{\frac{1}{p}}.
\end{aligned}$$

Recalling that $\int s_n \, d\mu > 0$ for sufficiently large n, we may for sufficiently large n divide both sides of the above display by $(\int s_n \, d\mu)^{\frac{1}{p}}$ to obtain the inequality $(\int s_n \, d\mu)^{\frac{1}{p'}} \leq \|L\| < +\infty$, valid for sufficiently large n. By MCT we then have

$$\int |g|^{p'} \, d\mu = \int \lim_n s_n \, d\mu = \lim_n \int s_n \, d\mu \leq \|L\|^{p'} < +\infty,$$

which shows that $g \in L^{p'}(\Omega, \mathcal{F}, \mu)$, completing the proof of (iv).

(v) $L(f) = \int fg \, d\mu$ for every $f \in L^p(\Omega, \mathcal{F}, \mu)$.

Proof: By Hölder's Inequality and (iv) we have $fg \in L^1(\Omega, \mathcal{F}, \mu)$ for every f in $L^p(\Omega, \mathcal{F}, \mu)$, so it makes sense to discuss $\int fg \, d\mu$ whenever $f \in L^p(\Omega, \mathcal{F}, \mu)$. In (iii), we showed that $L(f) = \int fg \, d\mu$ for finite-valued $\mathcal{F}/\mathcal{B}^*$-measurable simple functions f on Ω. We now show that the result holds for general $f \in L^p(\Omega, \mathcal{F}, \mu)$.

Define L_g on $L^p(\Omega, \mathcal{F}, \mu)$ by writing $L_g(f) = \int fg \, d\mu$ for every f in $L^p(\Omega, \mathcal{F}, \mu)$. By Claim 1 of Section 8.9 and (iv), L_g is a bounded linear functional on $L^p(\Omega, \mathcal{F}, \mu)$ with $\|L_g\| = \|g\|_{p'}$. By the boundedness of L_g and Claim 3 of Section 8.8, L_g is continuous [relative to the metric induced by the L^p-norm] on $L^p(\Omega, \mathcal{F}, \mu)$. The assumed boundedness of L and Claim 3 of Section 8.8 force L to be continuous on $L^p(\Omega, \mathcal{F}, \mu)$. Observe that we have $L_g = L$ on the subcollection of $L^p(\Omega, \mathcal{F}, \mu)$ consisting of the simple functions as given in (iii). By Claims 1 and 2 of Section 8.6 and the fact that μ is finite, the collection of simple functions as given in (iii) forms a dense subset of $L^p(\Omega, \mathcal{F}, \mu)$. Since L_g and L are continuous functions on $L^p(\Omega, \mathcal{F}, \mu)$ that coincide on a dense subset of $L^p(\Omega, \mathcal{F}, \mu)$, we have $L_g = L$ on all of $L^p(\Omega, \mathcal{F}, \mu)$. This immediately completes the proof of (v).

(vi) If $h \in L^{p'}(\Omega, \mathcal{F}, \mu)$ is such that $L(f) = \int fh\, d\mu$ for every $f \in L^p(\Omega, \mathcal{F}, \mu)$, then $h = g$ μ-a.e. on Ω.

Proof: For every $f \in L^p(\Omega, \mathcal{F}, \mu)$ we have by (v) that $\int fg\, d\mu = \int fh\, d\mu$. Define L_{g-h} on $L^p(\Omega, \mathcal{F}, \mu)$ by writing $L_{g-h}(f) = \int f(g-h)\, d\mu$ for every f in $L^p(\Omega, \mathcal{F}, \mu)$. It is obvious that $L_{g-h}(f) = 0$ for every $f \in L^p(\Omega, \mathcal{F}, \mu)$, and hence L_{g-h} is a bounded linear functional with bound $\|L_{g-h}\| = 0$. By Claim 1 of Section 8.9, this gives $\|g-h\|_{p'} = 0$, so that $g = h$ μ-a.e. on Ω as claimed. This gives the uniqueness claim of the theorem for the case where μ is finite.

We have now proved the Riesz Representation Theorem for the case where μ is finite. This completes the first of the two major parts of the proof. The second major part of the proof, which we start now, is to show that the theorem holds when μ is σ-finite. Therefore, in all that follows, μ is assumed to be σ-finite. Because of this, there exists a nondecreasing sequence $\{A_n\}_{n=1}^{\infty}$ of $\mathcal{F}$-sets of finite μ-measure such that $\lim_n A_n = \bigcup_{n=1}^{\infty} A_n = \Omega$. We next write

$$\mu_n(A) = \mu(A \cap A_n) \quad \text{for every } A \in \mathcal{F} \text{ and } n \in \mathbb{N},$$

so that each μ_n is a finite measure with domain $\mathcal{F}$. Next, define

$$L_n(f) = L(fI_{A_n}) \quad \text{for each } n \in \mathbb{N} \text{ and } f \in L^p(\Omega, \mathcal{F}, \mu_n).$$

By property (L18) we have that $f \in L^p(\Omega, \mathcal{F}, \mu_n)$ iff $fI_{A_n} \in L^p(\Omega, \mathcal{F}, \mu)$ for every $n \in \mathbb{N}$, so the preceding definition makes sense.

(vii) For every $n \in \mathbb{N}$, L_n is a bounded linear functional on $L^p(\Omega, \mathcal{F}, \mu_n)$ and the inequality $\|L_n\| \leq \|L\|$ obtains.

Proof: Pick $n \in \mathbb{N}$. Using the linearity of L, observe that

$$L_n(f+g) = L((f+g)I_{A_n}) = L(fI_{A_n} + gI_{A_n}) = L(fI_{A_n}) + L(gI_{A_n}) = L_n(f) + L_n(g)$$

and

$$L_n(cf) = L(cfI_{A_n}) = cL(fI_{A_n}) = cL_n(f)$$

for every $f, g \in L^p(\Omega, \mathcal{F}, \mu_n)$ and $c \in \mathbb{R}$, hence L_n is linear. Next, pick $f \in L^p(\Omega, \mathcal{F}, \mu_n)$. Write $\| \ \|_{p,\mu_n}$ for the L^p norm relative to $(\Omega, \mathcal{F}, \mu_n)$, and write $\| \ \|_{p,\mu}$ for the L^p norm relative to $(\Omega, \mathcal{F}, \mu)$. We have

$$\begin{aligned} |L_n(f)| = |L(fI_{A_n})| \leq \|L\| \|fI_{A_n}\|_{p,\mu} &= \|L\| \left(\int |fI_{A_n}|^p\, d\mu \right)^{\frac{1}{p}} \\ &= \|L\| \left(\int |f|^p\, d\mu_n \right)^{\frac{1}{p}} \\ &= \|L\| \|f\|_{p,\mu_n}, \end{aligned}$$

where the third equality follows from integration property (L18) in Section 6.4. It follows that L_n is bounded, and the displayed inequality shows that $\|L_n\| \leq \|L\|$. This completes (vii).

(viii) For every $n \in \mathbb{N}$, there exists finite-valued $g_n \in L^{p'}(\Omega, \mathcal{F}, \mu_n)$ with the following properties:

(1) $g_n = 0$ on A_n^c.

(2) $L_n(f) = \int fg_n \, d\mu_n = \int fI_{A_n}g_n \, d\mu$ for every $f \in L^p(\Omega, \mathcal{F}, \mu_n)$.

(3) $||L_n|| = ||g_n||_{p',\mu_n}$.

(4) g_n is unique up to a set of μ_n-measure zero.

Proof: Let $n \in \mathbb{N}$. The finite case of the Riesz Representation Theorem provides finite-valued $g_n \in L^{p'}(\Omega, \mathcal{F}, \mu_n)$ such that $L_n(f) = \int fg_n \, d\mu_n$ for every $f \in L^p(\Omega, \mathcal{F}, \mu_n)$. Integration property (L18) gives the other equality of (2), and hence g_n satisfies all of (2). By Claim 1 of Section 8.9, g_n satisfies (3). Since for every $f \in L^p(\Omega, \mathcal{F}, \mu_n)$ the value of $L_n(f)$ does not depend on how g_n behaves on A_n^c, we may without loss of generality assume that $g_n = 0$ on A_n^c, so that g_n satisfies (1). Finally, if h_n is a finite-valued element of $L^{p'}(\Omega, \mathcal{F}, \mu_n)$ satisfying (1)-(3), then the same type of argument as in (vi) shows that $h_n = g_n$ μ_n-a.e. on Ω. It follows that g_n satisfies (4), completing the proof of (viii).

Before proceeding further, we need the following simple integration fact.

Fact. *Let $\psi : \Omega \to [0, +\infty]$ be $\mathcal{F}/\mathcal{B}^*$-measurable, and let ξ and ρ denote two measures on $\mathcal{F}$ such that $\xi(A) \le \rho(A)$ for each $A \in \mathcal{F}$. Then $\int \psi \, d\xi \le \int \psi \, d\rho$.*

Proof: If ψ is a nonnegative finite-valued $\mathcal{F}/\mathcal{B}^*$-measurable simple function of the form $\sum_{i=1}^n c_i I_{A_i}$, then

$$\int \psi \, d\xi = \sum_{i=1}^n c_i \xi(A_i) \le \sum_{i=1}^n c_i \rho(A_i) \le \int \psi \, d\rho.$$

If ψ is a nonnegative $\mathcal{F}/\mathcal{B}^*$-measurable function on Ω, then we may find a nondecreasing sequence of nonnegative finite-valued $\mathcal{F}/\mathcal{B}^*$-measurable simple functions $\{s_n\}_{n=1}^\infty$ with $\lim_n s_n = \psi$. Using MCT and what has just been proved for simple functions, we have

$$\int \psi \, d\xi = \lim_n \int s_n \, d\xi \le \lim_n \int s_n \, d\rho = \int \psi \, d\rho,$$

which completes the proof.

The condition $A_1 \subseteq A_2 \subseteq \cdots$ forces $\mu_1 \le \mu_2 \le \cdots$, and the integration fact just mentioned gives

$$\int |f|^p \, d\mu_n \le \int |f|^p \, d\mu_{n+1} < +\infty \quad \text{whenever } f \in L^p(\Omega, \mathcal{F}, \mu_{n+1}).$$

Therefore, we have $L^p(\Omega, \mathcal{F}, \mu_{n+1}) \subseteq L^p(\Omega, \mathcal{F}, \mu_n)$ for every $n \in \mathbb{N}$. Because of the uniqueness of the functions $g_1, g_2, \cdots$ as given in (viii), we may for each $n \in \mathbb{N}$ take $g_{n+1} = g_n$ on A_n.

We now define $g : \Omega \to \mathbb{R}$ by writing $g(\omega) = g_n(\omega)$ whenever $\omega \in A_n$. By the previous paragraph and the fact that $\lim_n A_n = \Omega$, no consistency problems arise from this definition, and clearly g is finite-valued. We will show that g satisfies the remaining conclusions of the theorem.

(ix) For every $n \in \mathbb{N}$, $||I_{A_n} g||_{p',\mu} \leq ||L||$.

Proof: Let $n \in \mathbb{N}$. By the definition of g and the fact that $g_n = 0$ on A_n^c, we have that $I_{A_n} g = g_n$ on Ω, hence

$$\begin{aligned} ||I_{A_n} g||_{p',\mu} = ||g_n||_{p',\mu} = \left(\int |g_n|^{p'} \, d\mu\right)^{\frac{1}{p'}} &= \left(\int |g_n|^{p'} I_{A_n} \, d\mu\right)^{\frac{1}{p'}} \\ &= \left(\int |g_n|^{p'} \, d\mu_n\right)^{\frac{1}{p'}} \\ &= ||g_n||_{p',\mu_n} \\ &= ||L_n|| \\ &\leq ||L||. \end{aligned}$$

[The fourth equality follows by integration property (L18) in Section 6.4, the last equality follows by (viii), and the inequality follows by (vii).] This completes the proof of (ix).

(x) $g \in L^{p'}(\Omega, \mathcal{F}, \mu)$ and $||g||_{p',\mu} \leq ||L||$.

Proof: We have

$$\begin{aligned} \int |g|^{p'} \, d\mu = \int \lim_n I_{A_n} |g|^{p'} \, d\mu = \lim_n \int I_{A_n} |g|^{p'} \, d\mu &= \lim_n ||I_{A_n} g||_{p',\mu}^{p'} \\ &\leq ||L||^{p'} \\ &< +\infty. \end{aligned}$$

[The second equality follows by MCT. The third equality follows because $I_{A_n} |g|^{p'} = |I_{A_n} g|^{p'}$. The first inequality follows by (ix).] This gives both claims immediately and completes the proof of (x).

(xi) For every $f \in L^p(\Omega, \mathcal{F}, \mu)$ we have $L(f) = \int fg \, d\mu$.

Proof: Let $f \in L^p(\Omega, \mathcal{F}, \mu)$. Observe that $|f I_{A_n} g| \leq |f||g| \in L^1$ for every $n \in \mathbb{N}$ by (x) and Hölder's Inequality, so DCT gives

$$\begin{aligned} \int fg \, d\mu = \int \lim_n f I_{A_n} g \, d\mu = \lim_n \int f I_{A_n} g \, d\mu &= \lim_n L_n(f) \\ &= \lim_n L(f I_{A_n}). \end{aligned}$$

[The third equality follows by (viii).] Next, observe that

$$||f - f I_{A_n}||_p^p = ||f I_{A_n^c}||_p^p = \int I_{A_n^c} |f|^p \, d\mu \to \int 0 \, d\mu = 0 \quad \text{as } n \to \infty$$

by the fact that $f \in L^p(\Omega, \mathcal{F}, \mu)$, $\lim_n |f|^p I_{A_n^c} = 0$, and DCT. It follows that $f I_{A_n} \xrightarrow{p} f$. Since L is continuous, we have that $\lim_n L(f I_{A_n}) = L(f)$, completing (xi).

The uniqueness argument in (vi) [where μ was finite] carries over to the σ-finite case without any alterations, so that we may assert the uniqueness of g to a μ-null set. This completes the entire proof. ∎

The preceding proof avoids the "hard analysis" required by the approach in Section 8.9, although the price for doing this is requiring the Radon-Nikodym Theorem and σ-finiteness.

The Riesz Representation Theorem has been proved for $1 < p < +\infty$ and general measure spaces $(\Omega, \mathcal{F}, \mu)$ in Section 8.9. In this section, we have given a proof for the case where $1 < p < +\infty$ and $(\Omega, \mathcal{F}, \mu)$ is σ-finite. We have not yet discussed the cases $p = 1$ or $p = +\infty$, and it is of interest to ask just what [if anything] happens in these cases.

We first discuss the case $p = 1$. Here, the Hölder conjugate of p is $p' = +\infty$, and it will turn out that the Riesz Representation Theorem does *not* hold for general measure spaces; a counterexample is given in Exercise 5. The most general result known to the author is the following: *Let $(\Omega, \mathcal{F}, \mu)$ denote a decomposable measure space, and let L denote a bounded linear functional on $L^1(\Omega, \mathcal{F}, \mu)$. Then there exists $g \in L^\infty(\Omega, \mathcal{F}, \mu)$ such that $L(f) = \int fg\, d\mu$ for every $f \in L^1(\Omega, \mathcal{F}, \mu)$.* Using the work given in the $1 < p < +\infty$ proof above, it is possible to give a fairly short proof of the case $p = 1$ when $(\Omega, \mathcal{F}, \mu)$ is σ-finite, and this proof is sketched in Exercises 7 and 8.

We now discuss the case $p = +\infty$. Here, the Hölder conjugate of p is $p' = 1$. In this situation, the most general result known to the author is that every bounded linear functional L on $L^\infty(\Omega, \mathcal{F}, \mu)$ [where $(\Omega, \mathcal{F}, \mu)$ is arbitrary] may be represented by "an integral with respect to a certain type of finitely additive set function". That is, *if L is a bounded linear functional on $L^\infty(\Omega, \mathcal{F}, \mu)$, then there is $g \in L^1(\Omega, \mathcal{F}, \mu)$ such that $L(f) = \int f\, d\tau$ for every $f \in L^\infty(\Omega, \mathcal{F}, \mu)$, where τ is a function with domain $\mathcal{F}$ such that*
(i) $\sup\{|\tau(A)| : A \in \mathcal{F}\} < +\infty$,
(ii) $\tau(A \cup B) = \tau(A) + \tau(B)$ whenever $A, B \in \mathcal{F}$ are disjoint, and
(iii) $\tau(A) = 0$ if $A \in \mathcal{F}$ is such that $\mu(A \cap E) = 0$ for every $E \in \mathcal{F}$ with $\mu(E) < +\infty$.
Since we have as of yet to define what $\int f\, d\tau$ means for when τ is not a measure but in accordance with (i)-(iii) just given, the statement of the result is somewhat hollow. Exercises 10-12 provide a detailed program by which one may fruitfully define the symbol $\int f\, d\tau$. A program for the proof of the case $p = +\infty$ is given in Exercises 13 and 14.

Exercises. Some of these exercises might seem long. This is more so because of the outlines and hints than of any monumental difficulty.

1. The following is an advanced calculus fact used in the next problem, often called the *diagonal method.* Suppose that each row of the array

$$\begin{array}{cccc} x_{1,1} & x_{1,2} & x_{1,3} & \cdots \\ x_{2,1} & x_{2,2} & x_{2,3} & \cdots \\ \vdots & \vdots & \vdots & \end{array}$$

is a bounded sequence of real numbers. Then there exists a strictly increasing sequence $n_1, n_2, \cdots$ of positive integers such that $\lim_k x_{r,n_k}$ exists for each $r \in \mathbb{N}$. [The first row of the above array is a bounded sequence, and hence there is a convergent subsequence $x_{1,n_{1,1}}, x_{1,n_{1,2}}, x_{1,n_{1,3}}, \cdots$, where the sequence $\{n_{1,k}\}_{k=1}^{\infty}$ is strictly increasing and $\lim_k x_{1,n_{1,k}}$ exists. Return to the original array and look at the second row along the sequence $n_{1,1}, n_{1,2}, \cdots$, so that we are considering the bounded sequence $x_{2,n_{1,1}}, x_{2,n_{1,2}}, x_{2,n_{1,3}}, \cdots$. Select from this bounded sequence a convergent subsequence; write it as $x_{2,n_{2,1}}, x_{2,n_{2,2}}, x_{2,n_{2,3}}, \cdots$. Observe the following: $\{n_{2,k}\}_{k=1}^{\infty}$ is a strictly increasing subsequence of $\{n_{1,k}\}_{k=1}^{\infty}$, and $\lim_k x_{2,n_{2,k}}$ exists. Continue on this process, obtaining a new array

$$\begin{array}{cccc} n_{1,1} & n_{1,2} & n_{1,3} & \cdots \\ n_{2,1} & n_{2,2} & n_{2,3} & \cdots \\ \vdots & \vdots & \vdots & \end{array}$$

such that (i) each row is a strictly increasing sequence of positive integers, (ii) each row is a subsequence of the row above, and (iii) for each $r \in \mathbb{N}$ the limit $\lim_k x_{r,n_{r,k}}$ exists. Therefore, for each $r \in \mathbb{N}$ we have that $x_{r,n_{r,1}}, x_{r,n_{r,2}}, x_{r,n_{r,3}}, \cdots$ is a convergent sequence of the rth row of the original array. Define $n_k = n_{k,k}$ for each $k \in \mathbb{N}$. The n_k's are strictly increasing because each row of the new array is increasing and is contained in the preceding row. Also, $n_r, n_{r+1}, n_{r+2}, \cdots$ is a subsequence of the rth row of the new array for each $r \in \mathbb{N}$. It follows that $x_{r,n_r}, x_{r,n_{r+1}}, x_{r,n_{r+2}}, \cdots$ is a subsequence of $x_{r,n_{r,1}}, x_{r,n_{r,2}}, x_{r,n_{r,3}}, \cdots$ for each r in $\mathbb{N}$, so that $\lim_k x_{r,n_k}$ exists for each $r \in \mathbb{N}$.]

2. Let $(\Omega, \mathcal{F}, \mu)$ denote a σ-finite measure space with $1 < p < +\infty$. Further assume that $\mathcal{F}$ is countably generated, which means that there is a countable class $\mathcal{C}$ with $\sigma(\mathcal{C}) = \mathcal{F}$. Let $B_1^p = \{f \in L^p : \|f\|_p \le 1\}$, so that B_1^p is the unit ball in L^p. We will show that B_1^p *is compact in the topology of weak convergence.* That is, every sequence in B_1^p contains a subsequence converging weakly to an element of B_1^p.

(a) $L^{p'}$ has a countable dense subset, say $\mathcal{G}$. [Recall an approximation result from Chapter 8.]

Append to $\mathcal{G}$ all finite rational linear combinations of its elements. We will refer to this augmented collection as $\mathcal{G}$; $\mathcal{G}$ [the augmented collection] is still countable. Next, let $\{f_n\}_{n=1}^{\infty}$ denote a sequence in B_1^p.

(b) For every $g \in L^{p'}$, $\{\int f_n g \, d\mu\}_{n=1}^{\infty}$ is a bounded sequence of real numbers with bound $\|g\|_{p'}$. From this we may deduce that $\{\int f_n g \, d\mu\}_{n=1}^{\infty}$ is a bounded sequence of real numbers with bound $\|g\|_{p'}$ for every $g \in \mathcal{G}$.

Since G is countable, there is a subsequence $\{n_k\}_{k=1}^{\infty}$ of the sequence $1, 2, \cdots$ such that $\lim_k \int f_{n_k} g \, d\mu$ exists for every $g \in \mathcal{G}$ [Exercise 1]. This is a direct application of the diagonal method from advanced calculus. Write $L_{\mathcal{G}}(g) = \lim_k \int f_{n_k} g \, d\mu$ for every $g \in \mathcal{G}$; Hölder's Inequality gives $|L_{\mathcal{G}}(g)| \le \|g\|_{p'}$ for every $g \in \mathcal{G}$.

(c) $L_{\mathcal{G}}$ is uniformly continuous on its domain $\mathcal{G}$. [We have

$$|L_{\mathcal{G}}(g) - L_{\mathcal{G}}(g')| = \lim_k \left| \int f_{n_k}(g - g') \, d\mu \right| \le \varlimsup_k \|f_{n_k}\|_p \|g - g'\|_{p'} \le \|g - g'\|_{p'}$$

for every $g, g' \in \mathcal{G}$.]

(d) There exists a unique continuous function L on $L^{p'}$ such that $L = L_{\mathcal{G}}$ on $\mathcal{G}$. [From advanced calculus, we know that a uniformly continuous function ψ defined on a dense subset D of a metric space (X, ρ) has a unique continuous extension to all of X.]

(e) For every $g, g' \in \mathcal{G}$ we have $L_{\mathcal{G}}(g + g') = L_{\mathcal{G}}(g) + L_{\mathcal{G}}(g')$. [Direct verification.]

(f) For every $h, h' \in L^{p'}$ we have $L(h + h') = L(h) + L(h')$. [Let $\{g_n\}_{n=1}^{\infty}$ denote a sequence in the dense collection $\mathcal{G}$ with $g_n \xrightarrow{p'} h$; let $\{g_n'\}_{n=1}^{\infty}$ denote a sequence in $\mathcal{G}$ with $g_n' \xrightarrow{p'} h'$. Thus $\{g_n + g_n'\}_{n=1}^{\infty}$ is a sequence in $\mathcal{G}$ with $g_n + g_n' \xrightarrow{p'} h + h'$. Use the continuity of L and (e) to verify the equation

$$\begin{aligned} L(h) + L(h') = \lim_n L(g_n) + \lim_n L(g_n') &= \lim_n [L_{\mathcal{G}}(g_n) + L_{\mathcal{G}}(g_n')] \\ &= \lim_n L_{\mathcal{G}}(g_n + g_n') \\ &= L(h + h').] \end{aligned}$$

(g) Directly verify that $L_{\mathcal{G}}(cg) = cL_{\mathcal{G}}(g)$ for every $g \in \mathcal{G}$ and $c \in \mathbb{R}$. This in turn shows that $L(cg) = cL(g)$ for every $c \in \mathbb{R}$ and $g \in L^{p'}$. [The first part is easy. The second part is obtained in the same spirit as was (f).]

(h) By (g), L is a linear functional on $L^{p'}$. Show in addition that L is bounded with $\|L\| \leq 1$. [We already know that L is continuous from (d); use the relevant claim from Section 8.8 to achieve boundedness. Next, pick $h \in L^{p'}$; let $\{g_n\}_{n=1}^{\infty}$ denote a sequence in $\mathcal{G}$ with $g_n \xrightarrow{p'} h$. Verify that

$$|L(h)| = \lim_n |L(g_n)| = \lim_n |L_{\mathcal{G}}(g_n)| \leq \lim_n \|g_n\|_{p'} = \|h\|_{p'}.]$$

We now invoke the Riesz Representation Theorem for the bounded linear functional L on $L^{p'}$, obtaining a function $f \in L^p$ such that $L(g) = \int fg \, d\mu$ for every $g \in L^{p'}$. Since $\|L\| \leq 1$, we have $\|f\|_p = \|L\| \leq 1$ by Claim 1 of Section 8.9, hence $f \in B_1^p$. Recall the subsequence $\{n_k\}_{k=1}^{\infty}$ given after part (b). Observe that for every $g \in \mathcal{G}$ we have

$$L(g) = \int fg \, d\mu = L_{\mathcal{G}}(g) = \lim_k \int f_{n_k} g \, d\mu.$$

(i) We have that $\{f_{n_k}\}_{k=1}^{\infty}$ converges weakly to f, completing the proof. [Pick g' in $L^{p'}$ and $\epsilon > 0$; find $g \in \mathcal{G}$ with $\|g - g'\|_{p'} < \frac{\epsilon}{3}$. Define $\alpha_k = |\int f_{n_k} g \, d\mu - \int fg \, d\mu|$ for each $k \in \mathbb{N}$, and observe that the above shows $\lim_k \alpha_k = 0$. Verify that

$$\begin{aligned} &\left| \int fg' \, d\mu - \int f_{n_k} g' \, d\mu \right| \\ \leq\ &\left| \int fg' \, d\mu - \int fg \, d\mu \right| + \alpha_k + \left| \int f_{n_k} g \, d\mu - \int f_{n_k} g' \, d\mu \right| \\ \leq\ &\int |f||g' - g| \, d\mu + \alpha_k + \int |f_{n_k}||g' - g| \, d\mu \\ \leq\ &\|f\|_p \|g' - g\|_{p'} + \alpha_k + \|f_{n_k}\|_p \|g' - g\|_{p'} \\ \leq\ &\frac{2\epsilon}{3} + \alpha_k. \end{aligned}$$

For large enough k we thus have $|\int fg'\,d\mu - \int f_{n_k}g'\,d\mu| < \epsilon$. Since $g' \in L^{p'}$ is arbitrary, the entire proof is complete.]

3. Let $f, f', f_1, f_2, \cdots \in L^p$, where the underlying space $(\Omega, \mathcal{F}, \mu)$ is σ-finite and $1 < p < +\infty$. Suppose that $f_n \to f$ weakly in L^p and $f_n \to f'$ weakly in L^p. The goal is to show that $f = f'$ μ-a.e. on Ω, so that *weak limits in* L^p *are unique.*
(a) Quickly show that $\int fg\,d\mu = \int f'g\,d\mu$ for every $g \in L^{p'}$. [Both quantities are equal to $\lim_n \int f_n g\,d\mu$.]
(b) Show that $f = f'$ μ-a.e. on Ω. [Define $L_f(g) = \int fg\,d\mu$ for every $g \in L^{p'}$; define $L_{f'}(g) = \int f'g\,d\mu$ for every $g \in L^{p'}$. Both L_f and $L_{f'}$ are bounded linear functionals on $L^{p'}$ that agree on $L^{p'}$ by (a). Use the uniqueness claim of the Riesz Representation Theorem.]

Comment: The upcoming exercises discuss the Riesz Representation Theorem for the case $p = 1$.

4. Let $(\Omega, \mathcal{F}, \mu)$ denote a σ-finite measure space, and let $g \in L^\infty(\Omega, \mathcal{F}, \mu)$. Define L_g on $L^1(\Omega, \mathcal{F}, \mu)$ by writing $L_g(f) = \int fg\,d\mu$ for every $f \in L^1(\Omega, \mathcal{F}, \mu)$. Then L_g is a bounded linear functional with $||L_g|| = ||g||_\infty$. [It is trivial that L_g is linear. Recalling that $|g| \leq ||g||_\infty$ μ-a.e. on Ω gives $|\int fg\,d\mu| \leq ||g||_\infty ||f||_1$, which shows that L_g is bounded and $||L||_g \leq ||g||_\infty$. If $||g||_\infty = 0$ we trivially have $||L_g|| = ||g||_\infty$. Otherwise, let $||g||_\infty > 0$, $0 < \epsilon < ||g||_\infty$, and let $E = \{\omega \in \Omega : |g(\omega)| > ||g||_\infty - \epsilon\}$. Now $\mu(E) > 0$, and by σ-finiteness there is some subset $A \subseteq E$ with $0 < \mu(A) < +\infty$. Let $f = \frac{1}{\mu(A)} I_A \operatorname{sgn} g$; it is easy to verify that $f \in L^1(\Omega, \mathcal{F}, \mu)$ and $||f||_1 \leq 1$. Then $L_g(f) = \int fg\,d\mu = \frac{1}{\mu(A)} \int_A |g|\,d\mu \geq ||g||_\infty - \epsilon = (||g||_\infty - \epsilon)||f||_1$, so that $||L_g|| \geq ||g||_\infty - \epsilon$. Since $\epsilon > 0$ is arbitrary we have $||L_g|| = ||g||_\infty$.]

Comment. In the text it was stated that the Riesz Representation Theorem does not necessarily hold for arbitrary measure spaces $(\Omega, \mathcal{F}, \mu)$ when $p = 1$. In the two exercises following this next exercise we shall outline a proof of the $p = 1$ case for when $(\Omega, \mathcal{F}, \mu)$ is σ-finite. The next exercise provides a counterexample for a non-σ-finite measure space. This counterexample is not rigorous because certain claims are made without proof. These claims can be verified upon completion of the next chapter, but we state the facts now.

5. Let $\Omega = [0, 1] \times [0, 1]$, and let $\mathcal{F}$ denote the two-dimensional Borel subsets of Ω. For each $E \in \mathcal{F}$, define

$$\nu(E) = \sum_{x \in [0,1]} \lambda(\{y \in [0, 1] : (x, y) \in E\}),$$

where the uncountable sum $\sum_{x \in [0,1]}$ is as defined at the beginning of Section 9.2, and where λ denotes the restriction of one-dimensional Lebesgue measure to the Borel subsets of $[0, 1]$. Next, write

$$\mu(E) = \nu(E) + \sum_{y \in [0,1]} \lambda(\{x \in [0, 1] : (x, y) \in E\})$$

for every $E \in \mathcal{F}$.
(a) Both μ and ν are measures with domain $\mathcal{F}$, and $\nu \leq \mu$. [It should be straightforward to verify that ν is a measure. From this and the interchange of the roles

of x and y one can show that μ is a measure. That $\nu(E) \leq \mu(E)$ for all $E \in \mathcal{F}$ is trivial.]

(b) Define L on $L^1(\Omega, \mathcal{F}, \mu)$ by writing $L(f) = \int f \, d\nu$ for every $f \in L^1(\Omega, \mathcal{F}, \mu)$. Verify that $L^1(\Omega, \mathcal{F}, \mu) \subseteq L^1(\Omega, \mathcal{F}, \nu)$, so that the definition makes sense. Next, verify that L is a bounded linear functional on $L^1(\Omega, \mathcal{F}, \mu)$ with $\|L\| \leq 1$. [Again, this should be straightforward.]

Now assume that the conclusion of the Riesz Representation Theorem holds: there is $g \in L^\infty(\Omega, \mathcal{F}, \mu)$ such that $L = L_g$ for this g, where the notation is as in Exercise 4. That is, assume that $L(f) = \int fg \, d\mu$ for every $f \in L^1(\Omega, \mathcal{F}, \mu)$.

Accept without proof the assertion the existence of some $x_0 \in [0,1]$ such that $\int_{[0,1]} |g(x_0, y)| \, d\lambda(y) = 0$. [The proof of this fact requires the machinery of the next chapter.]

(c) Let $V = \{x_0\} \times [0,1]$ and $f = I_V$. Show that $\nu(V) = 1$ and $f \in L^1(\Omega, \mathcal{F}, \mu)$.

(d) Verify that

$$\begin{aligned} 1 = \nu(V) = \int_\Omega f \, d\nu = L(f) = L_g(f) = \int_\Omega fg \, d\mu &= \int_V g \, d\mu \\ &\leq \int_{[0,1]} g(x_0, y) \, d\lambda(y) \\ &= 0, \end{aligned}$$

so that we achieve a contradiction. [The inequality may be accepted on faith, as it depends on some concepts yet to be defined.] This gives a semirigorous counterexample to the Riesz Representation Theorem in the case $p = 1$. The problem here is that $(\Omega, \mathcal{F}, \mu)$ is not a σ-finite measure space.[7]

6. Let $(\Omega, \mathcal{F}, \mu)$ denote a measure space. Suppose that $\phi(E) = \int f \, d\mu$ for every $E \in \mathcal{F}$, where $f \in L^1(\Omega, \mathcal{F}, \mu)$. ϕ is an additive set function.

(a) Let $P = \{\omega \in \Omega : g(\omega) \geq 0\}$. Then (P, P^c) is a Hahn decomposition of Ω with respect to ϕ. [Obvious.]

(b) Let $\phi = \phi^+ - \phi^-$ denote the Jordan decomposition of ϕ into finite measures ϕ^+ and ϕ^-; recall that Jordan decompositions are unique by Exercise 4 of Section 9.1. Show that $\phi^+(E) = \int_E f^+ \, d\mu$ and $\phi^-(E) = \int_E f^- \, d\mu$ for every $E \in \mathcal{F}$. [We have $\phi^+(E) = \phi(E \cap P) = \int_{E \cap P} f \, d\mu = \int_{E \cap P} f^+ \, d\mu$ for every $E \in \mathcal{F}$, where P is as in part (a).]

(c) Define $|\phi|(E) = \phi^+(E) + \phi^-(E) = \int_E |f| \, d\mu$. $|\phi|$ is a finite measure, and for every $E \in \mathcal{F}$ we have $|\phi|(E) = \sup \sum_{i=1}^n |\phi(E_i)|$, the supremum being over all finite decompositions of E into disjoint $\mathcal{F}$-sets $E_1, \cdots, E_n$. [Let $E \in \mathcal{F}$, and let β denote the supremum. If $E_1, \cdots, E_n$ are disjoint $\mathcal{F}$-sets, then

$$\begin{aligned} \sum_{i=1}^n |\phi(E_i)| = \sum_{i=1}^n |\phi^+(E_i) - \phi^-(E_i)| &\leq \sum_{i=1}^n (\phi^+(E_i) + \phi^-(E_i)) \\ &= \sum_{i=1}^n |\phi|(E_i) \\ &= |\phi|(E), \end{aligned}$$

[7] Actually, $(\Omega, \mathcal{F}, \mu)$ fails even to be decomposable.

which gives $\beta \leq |\phi|(E)$. For the reverse inequality, consider the disjoint $\mathcal{F}$-sets $E \cap P$ and $E \cap P^c$, where P is as in (a). Then

$$\beta \geq |\phi(E \cap P)| + |\phi(E \cap P^c)| = \phi^+(E) + \phi^-(E) = |\phi|(E).$$

This gives $\beta = |\phi|(E)$.]

7. This exercise uses the notation and results of the previous exercise. Let L denote a bounded linear functional on $L^1(\Omega, \mathcal{F}, \mu)$, where $(\Omega, \mathcal{F}, \mu)$ is a finite measure space. We desire to show the following: *There is $g \in L^\infty(\Omega, \mathcal{F}, \mu)$ such that $L(f) = \int fg \, d\mu$ for every $f \in L^1(\Omega, \mathcal{F}, \mu)$.*
(a) Define $\nu(E) = L(I_E)$ for every $E \in \mathcal{F}$. Then ν is an additive set function on $\mathcal{F}$. [Note that part (i) of the proof in this section works for $p = 1$.]
(b) There exists a unique [up to a μ-null set] function $g \in L^1(\Omega, \mathcal{F}, \mu)$ such that $\nu(E) = \int_E g \, d\mu$. [Note that part (ii) of the proof in this section works for $p = 1$.] By the definition of $|\nu|$, we have $|\nu|(E) = \int_E |g| \, d\mu$ for every $E \in \mathcal{F}$.
(c) We may without any loss of generality assume that $|g(\omega)| \leq \|L\|$ for every $\omega \in \Omega$. Therefore, we have $g \in L^\infty(\Omega, \mathcal{F}, \mu)$ as well. [Let $A = \{\omega \in \Omega : \ |g(\omega)| > \|L\|\}$. If $\mu(A) > 0$, then $|\nu|(A) = \int_A |g| \, d\mu > \int_A \|L\| \, d\mu = \|L\|\mu(A)$. By part (c) of the previous exercise, there exists a finite collection $A_1, \cdots, A_n$ of disjoint $\mathcal{F}$-sets with union A such that $\sum_{i=1}^n |\nu(A_i)| > \|L\|\mu(A)$. But then

$$\begin{aligned} \|L\|\mu(A) < \sum_{i=1}^n |\nu(A_i)| = \sum_{i=1}^n |L(I_{A_i})| \ &\leq \ \sum_{i=1}^n \|L\| \|I_{A_i}\|_1 \\ &= \ \|L\| \sum_{i=1}^n \mu(A_i) \\ &= \ \|L\|\mu(A), \end{aligned}$$

a contradiction. Thus $|g| \leq \|L\|$ μ-a.e. on Ω.]
(d) Show that $L(s) = \int sg \, d\mu$ for every finite-valued $\mathcal{F}/\mathcal{B}^*$-measurable simple function s defined on Ω. [Follow the proof of (iii) given in the proof of the section.]
(e) $L(f) = \int fg \, d\mu$ for every $f \in L^1(\Omega, \mathcal{F}, \mu)$. [Follow the proof of (v) given in the proof of the section, or do the usual thing of using MCT to show the result is true for nonnegative functions in $L^1(\Omega, \mathcal{F}, \mu)$ and then using positive and negative parts to give the result in general.]
(f) The function $g \in L^\infty(\Omega, \mathcal{F}, \mu)$ is unique up to a μ-null set. [This should be straightforward. Suppose that $\psi \in L^\infty(\Omega, \mathcal{F}, \mu)$ is such that $\int f\psi \, d\mu = 0$ for every $f \in L^1(\Omega, \mathcal{F}, \mu)$. If $\{\omega \in \Omega : \ \psi(\omega) \neq 0\}$ has positive μ-measure, then, without loss of generality, $A = \{\omega \in \Omega : \psi(\omega) > 0\}$ has positive and finite μ-measure. Let $f = I_A$. Then $\int f\psi \, d\mu = \int_A \psi \, d\mu > 0$, a contradiction. Use this to get the general uniqueness claim.]

8. Extend Exercise 7 to where $(\Omega, \mathcal{F}, \mu)$ is σ-finite. This gives the Riesz Representation Theorem for the case $p = 1$. [Let $\{A_n\}_{n=1}^\infty$ denote a disjoint sequence of $\mathcal{F}$-sets of finite μ-measure with union Ω. For each $n \in \mathbb{N}$ and $E \in \mathcal{F}$, let $\mu_n(E) = \mu(E \cap A_n)$, so that $(\Omega, \mathcal{F}, \mu_n)$ is a finite measure space. Define L_n on $L^1(\Omega, \mathcal{F}, \mu_n)$ by writing $L_n(f) = L(fI_{A_n})$. Then each L_n is a bounded linear functional on $L^1(\Omega, \mathcal{F}, \mu_n)$ with $\|L_n\| \leq \|L\|$. Use the previous exercise now for each $n \in \mathbb{N}$: there is $g_n \in L^\infty(\Omega, \mathcal{F}, \mu_n)$ such that $|g_n(\omega)| \leq \|L_n\| \leq \|L\|$ for every $\omega \in \Omega$

and

$$L(fI_{A_n}) = L_n(f) = \int fg_n \, d\mu_n = \int_{A_n} fg_n \, d\mu \quad \text{for each } f \in L^1(\Omega, \mathcal{F}, \mu_n).$$

Since $L^1(\Omega, \mathcal{F}, \mu) \subseteq L^1(\Omega, \mathcal{F}, \mu_n)$, the above relation holds for each $f \in L^1(\Omega, \mathcal{F}, \mu)$. Now consider the functions $g_1, g_2, \cdots$. For each $n \in \mathbb{N}$ the values of g_n on A_n^c are of no consequence in determining L_n. We may meaningfully define a function g on Ω by writing $g(\omega) = g_n(\omega)$ when $\omega \in A_n$; it is straightforward to verify that $g \in L^\infty(\Omega, \mathcal{F}, \mu)$ and $\|g\|_\infty \leq \|L\|$. At this point, pick $f \in L^1(\Omega, \mathcal{F}, \mu)$. Use the continuity of L, the boundedness of g, and DCT to obtain

$$\begin{aligned} L(f) = \lim_m L\left(\sum_{n=1}^m fI_{A_n}\right) = \lim_m \sum_{n=1}^m L(fI_{A_n}) &= \lim_m \sum_{n=1}^m \int_{A_n} fg \, d\mu \\ &= \lim_m \int \sum_{n=1}^m fI_{A_n} g \, d\mu \\ &= \int fg \, d\mu. \end{aligned}$$

The uniqueness of g up to a μ-null may be established as in the finite case.]

9. Let $\Omega = (0, 1]$, let $\mathcal{F}$ denote the Borel subsets of Ω, and let λ denote Lebesgue measure restricted to $\mathcal{F}$. Then $L^1(\Omega, \mathcal{F}, \mu)$ is not weakly compact. This shows that in Exercise 2 we cannot consider the case $p = 1$, even if the underlying space is finite. [For each $n \in \mathbb{N}$ let $f_n = nI_{(0,n^{-1})}$, and suppose that some subsequence $\{f_{n_k}\}_{k=1}^\infty$ converges weakly as $k \to \infty$ to some $f \in L^1(\Omega, \mathcal{F}, \mu)$. Integrate f_{n_k} against the functions $\operatorname{sgn} fI_{(\epsilon,1)}$ in $L^\infty(\Omega, \mathcal{F}, \mu)$ and show that the hypothesized weak convergence forces $f = 0$ μ-a.e. on Ω. Next, integrate against the constant function 1, (which is clearly in $L^\infty(\Omega, \mathcal{F}, \mu)$), and obtain a contradiction.]

Comment. The next five exercises deal with building up the machinery needed to prove a version of the Riesz Representation Theorem for the case $p = +\infty$. In conjunction with the presentation of this section [or that of Section 8.9] and the discussion of the $p = 1$ case given in the previous exercises, we will have dealt with the Riesz Representation Theorem for $1 \leq p \leq +\infty$.

10. Let $(\Omega, \mathcal{F}, \mu)$ denote a σ-finite measure space. Let $\mathfrak{F}(\Omega, \mathcal{F}, \mu)$ denote the set of functions $\phi : \mathcal{F} \to \mathbb{R}$ such that
(i) $\sup\{|\phi(A)| : A \in \mathcal{F}\} < +\infty$,
(ii) $\phi(A \cup B) = \phi(A) + \phi(B)$ whenever $A, B \in \mathcal{F}$ are disjoint, and
(iii) $\phi(A) = 0$ whenever $A \in \mathcal{F}$ and $\mu(A) = 0$.
Given $\phi \in \mathfrak{F}(\Omega, \mathcal{F}, \mu)$, define $|\phi|$ on $\mathcal{F}$ by writing

$$|\phi|(A) = \sup\left\{\sum_{i=1}^n |\phi(A_i)| : A_1, \cdots, A_n \text{ are disjoint } \mathcal{F}\text{-sets with union } A\right\}$$

for every $A \in \mathcal{F}$. Given $\phi \in \mathfrak{F}(\Omega, \mathcal{F}, \mu)$, define $\|\phi\| = |\phi|(\Omega)$.
(a) Show that $|\phi| \in \mathfrak{F}(\Omega, \mathcal{F}, \mu)$.
(b) Show that $\mathfrak{F}(\Omega, \mathcal{F}, \mu)$ is a normed linear space [with setwise linear operations].

11. Assume the notation of Exercise 10. Let $f = \sum_{j=1}^{m} \alpha_j I_{A_j}$ and $g = \sum_{k=1}^{n} \beta_k I_{B_k}$ denote $\mathcal{F}/\mathcal{B}^*$-measurable simple functions with domain Ω [i.e. $A_1, \cdots, A_m \in \mathcal{F}$ are disjoint with union Ω and $B_1, \cdots, B_n \in \mathcal{F}$ are disjoint with union Ω]. Let $\phi \in \mathfrak{F}(\Omega, \mathcal{F}, \mu)$. Then

$$\left|\sum_{j=1}^{m} \alpha_j \phi(A_j) - \sum_{k=1}^{n} \beta_k \phi(B_k)\right| \leq \|\phi\| \|f - g\|_u,$$

where $\|h\|_u = \sup_{\omega \in \Omega} |h(\omega)|$ for every function h on Ω. [Verify that

$$\begin{aligned}\left|\sum_{j=1}^{m} \alpha_j \phi(A_j) - \sum_{k=1}^{n} \beta_k \phi(B_k)\right| &= \left|\sum_{j=1}^{m}\sum_{k=1}^{n} (\alpha_j - \beta_k)\phi(A_j \cap B_k)\right| \\ &\leq \sum_{j=1}^{m}\sum_{k=1}^{n} |\alpha_j - \beta_k||\phi(A_j \cap B_k)| \\ &\leq \sum_{j=1}^{m}\sum_{k=1}^{n} \|f - g\|_u |\phi(A_j \cap B_k)| \\ &\leq \|f - g\|_u \|\phi\|.]\end{aligned}$$

12. Assume the notation of Exercise 10 and let $\phi \in \mathfrak{F}(\Omega, \mathcal{F}, \mu)$. If $s = \sum_{j=1}^{m} \alpha_j I_{A_j}$ is an $\mathcal{F}/\mathcal{B}^*$-measurable simple function on Ω, define $\int s \, d\phi = \sum_{j=1}^{m} \alpha_j \phi(A_j)$. Next, if f is a bounded $\mathcal{F}/\mathcal{B}^*$-measurable function on Ω, there is a sequence of simple functions $\{s_n\}_{n=1}^{\infty}$ on Ω such that $\|f - s_n\|_u \to 0$.
(a) The real sequence $\{\int s_n \, d\phi\}_{n=1}^{\infty}$ is a Cauchy sequence. [Use Exercise 11.]
Define $\int f \, d\phi = \lim_n \int s_n \, d\phi$.
(b) This definition is well-defined, as $\int f \, d\phi$ does not depend on the particular sequence $\{s_n\}_{n=1}^{\infty}$ of $\mathcal{F}/\mathcal{B}^*$-measurable simple functions on Ω with $\|f - s_n\|_u \to 0$. The definition is also consistent with our definition of $\int \cdots \, d\phi$ for simple functions.
(c) Let f and g denote bounded $\mathcal{F}/\mathcal{B}^*$-measurable functions on Ω, and let $\alpha \in \mathbb{R}$. Show that
(i) $\int \alpha f \, d\phi = \alpha \int f \, d\phi$,
(ii) $\int (f + g) \, d\phi = \int f \, d\phi + \int g \, d\phi$,
(iii) $|\int f \, d\phi| \leq \int |f| \, d|\phi|$, and
(iv) $0 \leq f \leq g$ forces $\int f \, d|\phi| \leq \int g \, d|\phi|$.
(d) Let h denote a bounded $\mathcal{F}/\mathcal{B}^*$-measurable function such that $h = 0$ μ-a.e. on Ω. Then $\int h \, d\phi = 0$. [Let $A = \{\omega \in \Omega : \ h(\omega) \neq 0\}$. Verify that

$$\begin{aligned}\left|\int h \, d\phi\right| \leq \int |h| \, d|\phi| &= \int I_A |h| \, d|\phi| + \int I_{A^c} |h| \, d|\phi| \\ &\leq \|h\|_u |\phi|(A) + 0 \\ &= 0\end{aligned}$$

by appealing to part (c) and to Exercise 10.]
(e) We have defined $\int \cdots \, d\phi$ first for simple functions, then for bounded functions. We wish to extend this to functions in $L^{\infty}(\Omega, \mathcal{F}, \mu)$. To do this, let $g \in L^{\infty}(\Omega, \mathcal{F}, \mu)$, and pick a bounded function $f \in L^{\infty}$ such that $\|f - g\|_{\infty} = 0$. [See Exercise 2 of Section 8.4 to see that such a function f exists.] We define $\int g \, d\phi$ to be $\int f \, d\phi$.

Using (d), show that this definition makes sense in that if g is a bounded function in L^∞ with $\|h-g\|_\infty = 0$, then $\int f\, d\phi = \int h\, d\phi$.

13. Let $(\Omega, \mathcal{F}, \mu)$ denote a σ-finite measure space, and let $\phi \in \mathfrak{F}(\Omega, \mathcal{F}, \mu)$. Define L_ϕ on $L^\infty(\Omega, \mathcal{F}, \mu)$ by writing $L_\phi(g) = \int g\, d\phi$. Then L_ϕ is a bounded linear functional with $\|L_\phi\| = \|\phi\| = |\phi|(\Omega)$. [Linearity follows from Exercise 12. For the other claim, let $g \in L^\infty(\Omega, \mathcal{F}, \mu)$. Pick bounded $f \in L^\infty(\Omega, \mathcal{F}, \mu)$ so that $\|f-g\|_\infty = 0$ and $\|f\|_u = \|g\|_\infty$. (See part (e) of the previous exercise or Exercise 2 of Section 8.4 to see that such a function f exists.) Using the previous exercise gives

$$\begin{aligned} |L_\phi(g)| = \left|\int g\, d\phi\right| = \left|\int f\, d\phi\right| \le \int |f|\, d|\phi| &\le \|f\|_u \|\phi\| \\ &= \|g\|_\infty \|\phi\|. \end{aligned}$$

This gives $\|L_\phi\| \le \|\phi\|$. For the reverse inequality, pick $\epsilon > 0$ and use the definition of $|\phi|$ in Exercise 10 to find a disjoint collection $A_1, \cdots, A_n \in \mathcal{F}$ with union Ω and $\sum_{j=1}^n |\phi(A_j)| > \|\phi\| - \epsilon$. For $j = 1, \cdots, n$, let $\alpha_j = \operatorname{sgn} \phi(A_j)$. Define the function $g = \sum_{j=1}^n \alpha_j I_{A_j}$. We have $g \in L^\infty(\Omega, \mathcal{F}, \mu)$ with $\|g\|_\infty = \|g\|_u \le 1$, and

$$|L_\phi(g)| = \left|\int g\, d\phi\right| = \left|\sum_{j=1}^n \alpha_j \phi(A_j)\right| = \sum_{j=1}^n |\phi(A_j)| \ge \|g\|_\infty(\|\phi\| - \epsilon).$$

Since $\epsilon > 0$ is generic, this gives $\|L_\phi\| \ge \|\phi\|$.]

14. Prove the Riesz Representation Theorem for $p = +\infty$: *Let $(\Omega, \mathcal{F}, \mu)$ denote a σ-finite measure space, and let L denote a bounded linear functional on $L^\infty(\Omega, \mathcal{F}, \mu)$. Then there exists $\phi \in \mathfrak{F}(\Omega, \mathcal{F}, \mu)$ such that $L = L_\phi$, where L_ϕ is as in the previous exercise.* Note the slight difference in the form of the theorem for the $p = +\infty$ case as compared to the case where $1 \le p < +\infty$. Follow this simple outline.

(a) Write $\phi(A) = L(I_A)$ for each $A \in \mathcal{F}$. Then $\phi \in \mathfrak{F}(\Omega, \mathcal{F}, \mu)$. [Observe that

$$\sup\{|\phi(A)| : A \in \mathcal{F}\} \le \sup\{|L(g)| : g \in L^\infty,\ \|g\|_\infty \le 1\} = \|L\| < +\infty.$$

It is straightforward to verify that $\phi(A \cup B) = \phi(A) + \phi(B)$ whenever $A, B \in \mathcal{F}$ are disjoint. Finally, if $A \in \mathcal{F}$ and $\mu(A) = 0$, then $I_A = 0$ μ-a.e. on Ω and $I_A \in L^\infty(\Omega, \mathcal{F}, \mu)$. Note that $\phi(A) = L(I_A) = L(0) = 0$. This shows that (i)-(iii) of the definition in Exercise 10 are satisfied.]

(b) Show that $L(g) = L_\phi(g)$ for every $g \in L^\infty(\Omega, \mathcal{F}, \mu)$. [Let $g \in L^\infty(\Omega, \mathcal{F}, \mu)$. Pick bounded $f \in L^\infty(\Omega, \mathcal{F}, \mu)$ such that $\|g-f\|_\infty = 0$ and $\|f\|_u = \|g\|_\infty$; next choose a sequence $\{s_n\}_{n=1}^\infty$ of $\mathcal{F}/\mathcal{B}^*$-measurable simple functions such that $\|f - s_n\|_u \to 0$. It is straightforward to verify that $L(s_n) = \int s_n\, d\phi$ for each $n \in \mathbb{N}$ by using the linearity of L and Exercise 12. Observe next that $\|f - s_n\|_\infty \to 0$. Using the continuity of L (among other things), verify that

$$L(g) = L(f) = \lim_n L(s_n) = \lim_n \int s_n\, d\phi = \int f\, d\phi = \int g\, d\phi = L_\phi(g).]$$

Comment. The upcoming two exercises are taken from Hewitt and Stromberg [sections (20.37) and (20.38)] and deal with the notions of *filters* and *ultrafilters*. These concepts are not used anywhere else in the book.

15. Suppose that $\phi : 2^{\Omega} \to \{0,1\}$ is such that $\phi(\Omega) = 1$ and $\phi(A \cup B) = \phi(A) + \phi(B)$ whenever A and B are disjoint subsets of Ω. Such a ϕ is referred to as a *finitely additive measure*. Define $\mathcal{U} = \{A \subseteq \Omega : \ \phi(A) = 1\}$.
(a) $\varnothing \notin \mathcal{U}$.
(b) If $A \in \mathcal{U}$ and $A \subseteq B \subseteq \Omega$, then $B \in \mathcal{U}$.
(c) If $A, B \in \mathcal{U}$, then $A \cap B \in \mathcal{U}$.
(d) If $A \subseteq \Omega$, then $A \in \mathcal{U}$ or $A^c \in \mathcal{U}$.

If a collection $\mathcal{V}$ of subsets of Ω satisfies (a)-(c), then $\mathcal{V}$ is called a *filter on* Ω. If $\mathcal{V}$ satisfies (a)-(d), then $\mathcal{V}$ is called an *ultrafilter on* Ω.

Let $\mathcal{V}$ denote an ultrafilter on Ω. Define $\psi : 2^{\Omega} \to \{0,1\}$ by writing $\psi(A) = 1$ if $A \in \mathcal{V}$ and $\psi(A) = 0$ if $A \notin \mathcal{V}$.

(e) ψ is a finitely additive measure.
(f) Let $(\Omega, \mathcal{F}, \mu)$ denote a σ-finite measure space, and let ϕ denote a finitely additive measure in $\mathfrak{F}(\Omega, \mathcal{F}, \mu)$. [See Exercise 10 for the definition of $\mathfrak{F}(\Omega, \mathcal{F}, \mu)$.] Then $\int fg \, d\phi = \int f \, d\phi \int g \, d\phi$ for every $f, g \in L^{\infty}(\Omega, \mathcal{F}, \mu)$ iff $\phi(A)$ is 0 or 1 for every $A \in \mathcal{F}$.

16. Let Ω denote a nonempty set. A filter $\mathcal{U}$ on Ω is called *free* if $\cap\, \mathcal{U} = \varnothing$. [That is, there is no $\omega \in \Omega$ that is common to each of the sets in $\mathcal{U}$.] If a filter is not free, it is referred to as *fixed*.
(a) If $\mathcal{U}$ is a fixed ultrafilter on Ω, then $\mathcal{U} = \{A \subseteq \Omega : \ \omega \in A\}$ for some $\omega \in \Omega$.
(b) If Ω is finite, then every ultrafilter on Ω is fixed.
(c) If Ω is infinite, then there exists a free ultrafilter on Ω. [Define the collection $\mathcal{A} = \{A \subseteq \Omega : \ A^c \text{ is finite}\}$. Then $\mathcal{A}$ is a filter. Use Zorn's Lemma to obtain a maximal filter $\mathcal{V}$ containing $\mathcal{A}$, and then show that $\mathcal{V}$ is a free ultrafilter.]

9.4 MARTINGALE THEOREMS

In this section we present a generalization of Radon-Nikodym derivatives, proving two main theorems regarding such a generalization. We proceed immediately to the main definition of this section.

Definition. Let Ω denote a nonempty set, let $\mathcal{F}$ denote a σ-field of subsets of Ω, and let μ and η denote σ-finite measures with domain $\mathcal{F}$. Using the Lebesgue Decomposition Theorem, we may write $\eta = \eta_{ac} + \eta_s$, where η_{ac} and η_s are σ-finite measures on $\mathcal{F}$ with $\eta_{ac} \ll \mu$ and $\eta_s \perp \mu$. An $\mathcal{F}/\mathcal{B}^*$-measurable function $f : \Omega \to \bar{\mathbb{R}}$ is called a *derivative of* η *with respect to* μ iff (i)-(iv) below hold:

(i) $\eta_{ac}(A) = \int_A f \, d\mu$ for every $A \in \mathcal{F}$.
There are sets $B, P \in \mathcal{F}$ such that
(ii) $\mu(B) = 0$ and $\eta_s(B^c) = 0$,
(iii) $\eta_s(P^c) = 0$, and
(iv) $f(\omega) = +\infty$ if $\omega \in B \cap P$ and $f(\omega) = -\infty$ if $\omega \in B \cap P^c$.

It should be immediately noted that (ii) is a truism when viewed by itself, since we are already assuming that $\eta_s \perp \mu$. Requirement (iii) is also a truism when viewed by itself, as $P = \Omega$ gives $\eta_s(P^c) = \eta_s(\varnothing) = 0$. The connection between the set B in (ii) and the set P in (iii) is given by (iv).

We have the following simple claim regarding derivatives of η with respect to μ.

Claim 1. *Let f denote a derivative of η with respect to μ.*
(a) $f \geq 0$ μ-a.e. on Ω.
(b) If g is a derivative of η with respect to μ, then $f = g$ μ-a.e. on Ω.

Proof: We proceed by contradiction for (a). Let $B = \{\omega \in \Omega : f(\omega) < 0\}$. Assume to the contrary that $\mu(B) > 0$. Since f is a derivative of η with respect to μ, we have [among other things] that $\eta_{ac}(A) = \int_A f\,d\mu$ for every $A \in \mathcal{F}$. In particular, we have $0 \leq \eta_{ac}(B) = \int_B f\,d\mu < 0$, a contradiction. Therefore, it must be the case that $\mu(B) = 0$, so that $f \geq 0$ μ-a.e. on Ω. This gives (a).

We now proceed to (b). Let g denote a derivative of η with respect to μ, so that we have $\int_A g\,d\mu = \eta_{ac}(A) = \int_A f\,d\mu$ for every $A \in \mathcal{F}$. Define f^* and g^* by writing

$$f^*(\omega) = \begin{cases} f(\omega) & f(\omega) \geq 0 \\ 0 & f(\omega) < 0 \end{cases} \qquad \text{and} \qquad g^*(\omega) = \begin{cases} g(\omega) & g(\omega) \geq 0 \\ 0 & g(\omega) < 0 \end{cases}$$

for every $\omega \in \Omega$. Then $f^* \geq 0$ on Ω and $f^* = f$ μ-a.e. on Ω by (a). Similarly, $g^* \geq 0$ on Ω and $g^* = g$ μ-a.e. on Ω. We therefore have that

$$\int_A g^*\,d\mu = \int_A g\,d\mu = \int_A f\,d\mu = \int_A f^*\,d\mu \quad \text{for every } A \in \mathcal{F}.$$

Invoking integration property (L17) from Section 6.4, we obtain that $g^* = f^*$ μ-a.e. on Ω. Since $g^* = g$ μ-a.e. on Ω and $f^* = f$ μ-a.e. on Ω, it follows that $f = g$ μ-a.e. on Ω, completing the proof of (b). This completes the entire proof. ■

We have the separate terms *Radon-Nikodym derivative of η with respect to μ* and *derivative of η with respect to μ*. The next example shows that Radon-Nikodym derivatives are derivatives in the more general sense of this section, and the following example shows that given two σ-finite measures η and μ [on $\mathcal{F}$], a derivative of η with respect to μ always exists.

Example 1. Suppose that η and μ are σ-finite measures with domain $\mathcal{F}$ and such that $\eta \ll \mu$. Then η_s is identically zero, so that $\eta = \eta_{ac}$.[8] The Radon-Nikodym Theorem ensures the existence of a nonnegative $\mathcal{F}/\mathcal{B}^*$-measurable function f that is finite on Ω with $\eta(A) = \eta_{ac}(A) = \int_A f\,d\mu$ for every $A \in \mathcal{F}$. We may take $B = P = \varnothing$

[8]To see this, write $\eta = \eta_{ac} + \eta_s$, where η_{ac} and η_s are σ-finite measures with $\eta_{ac} \ll \mu$ and $\eta_s \perp \mu$. Now η_s and μ have disjoint supports, say $\eta_s(Q^c) = 0 = \mu(Q)$ for some $Q \in \mathcal{F}$. Since $\eta \ll \mu$, we have $\eta(Q) = 0$, whence $\eta_s(Q) = 0$. Since $\eta_s(Q) = \eta_s(Q^c) = 0$, we have $\eta_s(\Omega) = 0$, so $\eta_s(A) = 0$ for every $A \in \mathcal{F}$.

in the definition to see that the Radon-Nikodym derivative f of η with respect to μ is a derivative of η with respect to μ in the sense of our just-given definition. We may therefore say that *a Radon-Nikodym derivative of η with respect to μ is a derivative of η with respect to μ*, so that the definition of a derivative in this section subsumes the definition of a derivative from the Radon-Nikodym framework.

Example 2. Let η and μ denote σ-finite measures on $\mathcal{F}$. Using the Lebesgue Decomposition Theorem, write $\eta = \eta_{ac} + \eta_s$, where η_{ac} and η_s are σ-finite measures on $\mathcal{F}$ with $\eta_{ac} \ll \mu$ and $\eta_s \perp \mu$. Since η_{ac} is a σ-finite measure that is absolutely continuous with respect to μ, the Radon-Nikodym Theorem ensures the existence of a nonnegative finite-valued $\mathcal{F}/\mathcal{B}^*$-measurable function f_0 on Ω with

$$\eta_{ac}(A) = \int_A f_0 \, d\mu \quad \text{for every } A \in \mathcal{F}.$$

Since $\eta_s \perp \mu$, we may find $B \in \mathcal{F}$ with $\mu(B) = \eta_s(B^c) = 0$. Let $P = \Omega$, so that $\eta_s(P^c) = \eta_s(\varnothing) = 0$. Define $f : \Omega \to \bar{\mathbb{R}}$ by writing

$$f(\omega) = \begin{cases} f_0(\omega) & \omega \in B^c \\ +\infty & \omega \in B \cap P = B \\ -\infty & \omega \in B \cap P^c = \varnothing \end{cases} \qquad \text{for every } \omega \in \Omega.$$

Since $f = f_0$ on B^c and $\mu(B) = 0$, we have $f = f_0$ μ-a.e. on Ω, which gives

$$\eta_{ac}(A) = \int_A f_0 \, d\mu = \int_A f \, d\mu \qquad \text{for every } A \in \mathcal{F}.$$

Therefore, f satisfies (i) of the definition. It is immediate that (ii)-(iv) of the definition are satisfied as well, so that f is a derivative of η with respect to μ. We may therefore assert that *the definition of a derivative of η with respect to μ is never vacuous*, as we have just demonstrated that a derivative always exists in the framework of our more general definition.

We now present the main technical claim that characterizes derivatives. The claim is useful in that the equivalent characterization of derivatives does not involve Lebesgue decompositions or Hahn decompositions, as does the definition of a derivative given earlier. We assume the framework and notation of the definition given earlier.

Claim 2. *Let $g : \Omega \to \bar{\mathbb{R}}$ be $\mathcal{F}/\mathcal{B}^*$-measurable. For each $\alpha \in \mathbb{R}$, define*

$$G_\alpha = \{\omega \in \Omega : \; g(\omega) \geq \alpha\} \quad \textit{and} \quad L_\alpha = \{\omega \in \Omega : \; g(\omega) \leq \alpha\}.$$

Then g is a derivative of η with respect to μ iff
(a) $\eta(G_\alpha \cap A) \geq \alpha\mu(G_\alpha \cap A)$ for every $\alpha \in \mathbb{R}$ and $A \in \mathcal{F}$, and
(b) $\eta(L_\alpha \cap A) \leq \alpha\mu(L_\alpha \cap A)$ for every $\alpha \in \mathbb{R}$ and $A \in \mathcal{F}$.

Proof: Suppose that g is a derivative of η with respect to μ. It follows that $\eta_{ac}(A) = \int_A g \, d\mu$ for every $A \in \mathcal{F}$, and there exist B and P as in (ii)-(iv) of the definition: $\mu(B) = 0 = \eta_s(B^c)$, $\eta_s(P^c) = 0$, $g = +\infty$ on $B \cap P$, and $g = -\infty$ on $B \cap P^c$. We first show that (a) holds, picking generic $\alpha \in \mathbb{R}$ and

$A \in \mathcal{F}$. We have:

$$\begin{aligned}
\eta(G_\alpha \cap A) &= \eta_{ac}(G_\alpha \cap A) + \eta_s(G_\alpha \cap A) \\
&= \int_{G_\alpha \cap A} g \, d\mu + \eta_s(G_\alpha \cap A \cap B) + \eta_s(G_\alpha \cap A \cap B^c) \\
&\geq \int_{G_\alpha \cap A} \alpha \, d\mu + \eta_s(G_\alpha \cap A \cap B) + 0 \\
&= \alpha\mu(G_\alpha \cap A) + \eta_s(G_\alpha \cap A \cap B) \\
&\geq \alpha\mu(G_\alpha \cap A).
\end{aligned}$$

This gives (a). For (b), we again pick generic $\alpha \in \mathbb{R}$ and $A \in \mathcal{F}$. We first observe that $L_\alpha \cap A \cap B \subseteq P^c$,[9] and hence $\eta_s(L_\alpha \cap A \cap B) = 0$ by the fact that $\eta_s(P^c) = 0$. We now proceed in the same spirit as before:

$$\begin{aligned}
\eta(L_\alpha \cap A) &= \eta_{ac}(L_\alpha \cap A) + \eta_s(L_\alpha \cap A) \\
&= \int_{L_\alpha \cap A} g \, d\mu + \eta_s(L_\alpha \cap A \cap B) + \eta_s(L_\alpha \cap A \cap B^c) \\
&\leq \int_{L_\alpha \cap A} \alpha \, d\mu + 0 + 0 \quad [L_\alpha \cap A \cap B^c \subseteq B^c \text{ and } \eta_s(B^c) = 0] \\
&= \alpha\mu(L_\alpha \cap A).
\end{aligned}$$

This gives (b), and hence we have shown that derivatives of η with respect to μ satisfy both (a) and (b). This gives one direction of the claim.

We now handle the other direction of the claim. Suppose that g satisfies (a) and (b). We desire to show that g is a derivative of η with respect to μ, that is, we desire to show that (i)-(iv) of the definition are satisfied [relative to g]. Let f denote a derivative of η with respect to μ. [Such an invocation is possible by virtue of Example 2.] Let B and P be as in (ii)-(iv) of the definition: $\mu(B) = 0 = \eta_s(B^c)$, $\eta_s(P^c) = 0$, $f = +\infty$ on $B \cap P$, and $f = -\infty$ on $B \cap P^c$.

We wish to show that $g = f$ μ-a.e. on Ω. To show this claim, we proceed by contradiction, assuming that it is not the case that $g = f$ μ-a.e. on Ω. We make the following intermediate subclaim: if it is not the case that $g = f$ μ-a.e. on Ω, then

(1) There are $\alpha, \beta \in \mathbb{R}$ and $F \in \mathcal{F}$ with $0 < \mu(F) < +\infty$, $\eta_s(F) = 0$, and $f(\omega) \leq \alpha < \beta \leq g(\omega)$ for every $\omega \in F$, or

(2) There are $\alpha, \beta \in \mathbb{R}$ and $F \in \mathcal{F}$ with $0 < \mu(F) < +\infty$, $\eta_s(F) = 0$, and $g(\omega) \leq \alpha < \beta \leq f(\omega)$ for every $\omega \in F$.

[9] If $L_\alpha \cap A \cap B = \varnothing$, this is trivial. If $L_\alpha \cap A \cap B$ is nonempty but is not a subset of P^c, then there is $\omega \in L_\alpha \cap A \cap B \cap P$. Since this ω is in L_α, we have $g(\omega) \leq \alpha < +\infty$. Since this ω is in $B \cap P$, we have $g(\omega) = +\infty$, a contradiction. It follows that $L_\alpha \cap A \cap B \subseteq P^c$ in the nonempty case as well.

If it is not the case that $g = f$ μ-a.e. on Ω, then there is $G \in \mathcal{F}$ with $\mu(G) > 0$ and $g \neq f$ on G. Using the σ-finiteness of μ, we may write Ω as a countable disjoint union $\bigcup_{n=1}^{\infty} A_n$ of $\mathcal{F}$-sets of finite μ-measure. Observe that $G = \bigcup_{n=1}^{\infty}(G \cap A_n)$, and since $\mu(G) > 0$ we have that there is N such that $0 < \mu(G \cap A_N) < +\infty$. Note that $g \neq f$ on $G \cap A_N$. Since $\mu(B) = 0$ forces

$$\mu(G \cap A_N) = \mu(G \cap A_N \cap B) + \mu(G \cap A_N \cap B^c) = \mu(G \cap A_N \cap B^c),$$

we have $0 < \mu(G \cap A_N \cap B^c) < +\infty$. Next, letting $E_1 = \{\omega \in \Omega : \ g(\omega) > f(\omega)\}$ and $E_2 = \{\omega \in \Omega : \ g(\omega) < f(\omega)\}$, observe that

$$\mu(G \cap A_N \cap B^c) = \mu(G \cap A_N \cap B^c \cap E_1) + \mu(G \cap A_N \cap B^c \cap E_2),$$

hence at least one of $\mu(G \cap A_N \cap B^c \cap E_1)$ and $\mu(G \cap A_N \cap B^c \cap E_2)$ is positive and finite. *Without any loss of generality, suppose that* $0 < \mu(G \cap A_N \cap B^c \cap E_1) < +\infty$. Let $F_0 = G \cap A_N \cap B^c \cap E_1$, so that $0 < \mu(F_0) < +\infty$ and $\eta_s(F_0) = 0$ [since $F_0 \subseteq B^c$ and $\eta_s(B^c) = 0$]. Let $F_{\alpha,\beta} = \{\omega \in F_0 : \ f(\omega) \leq \alpha < \beta \leq g(\omega)\}$ for every $\alpha, \beta \in \mathbb{Q}$. Then for every $\alpha, \beta \in \mathbb{Q}$ we have $F_{\alpha,\beta} \subseteq F_0$. Observe that $F_0 = \bigcup_{\alpha,\beta \in \mathbb{Q}} F_{\alpha,\beta}$, a countable union. Since $0 < \mu(F_0) < +\infty$, some F_{α^*,β^*} is such that $0 < \mu(F_{\alpha^*,\beta^*}) < +\infty$. Let $F = F_{\alpha^*,\beta^*}$. Then $0 < \mu(F) < +\infty$, $\eta_s(F) = 0$, and $f \leq \alpha^* < \beta^* \leq g$ on F. This gives (1). If we assume instead that $0 < \mu(G \cap A_N \cap B^c \cap E_2) < +\infty$ in the above italicized sentence and proceed in the same fashion, we see that (2) follows. We have therefore shown the following: *if it is not the case that $f = g$ μ-a.e. on Ω, then (1) or (2) holds.*

Consider option (1) along with the set F and rational numbers α and β given there. By (a) and the fact that $G_\beta^c \cap F = \varnothing$, we have

$$\begin{aligned} \eta(F) = \eta(G_\beta \cap F) + \eta(G_\beta^c \cap F) &= \eta(G_\beta \cap F) \\ &\geq \beta\mu(G_\beta \cap F) \\ &= \beta[\mu(G_\beta \cap F) + \mu(G_\beta^c \cap F)] \\ &= \beta\mu(F). \end{aligned}$$

Since $\eta_s(F) = 0$ and $f \leq \alpha$ on F, we also have

$$\eta(F) = \eta_{ac}(F) + \eta_s(F) = \int_F f \, d\mu + 0 \leq \int_F \alpha \, d\mu = \alpha\mu(F).$$

We therefore have $\beta\mu(F) \leq \eta(F) \leq \alpha\mu(F)$, which is a contradiction since $\alpha < \beta$ and $0 < \mu(F) < +\infty$. Therefore, if it is not the case that $g = f$ μ-a.e. on Ω, then option (1) leads to a contradiction.

Next, consider option (2) along with the set F and the rational numbers α and β given there. Using (b) and the fact that $L_\alpha^c \cap F = \varnothing$, we have

$$\begin{aligned} \eta(F) = \eta(L_\alpha \cap F) + \eta(L_\alpha^c \cap F) &= \eta(L_\alpha \cap F) \\ &\leq \alpha\mu(L_\alpha \cap F) \\ &= \alpha[\mu(L_\alpha \cap F) + \mu(L_\alpha^c \cap F)] \\ &= \alpha\mu(F). \end{aligned}$$

Since $\eta_s(F) = 0$ and $f \geq \beta$ on F, we also have

$$\eta(F) = \eta_{ac}(F) + \eta_s(F) = \int_F f \, d\mu + 0 \geq \int_F \beta \, d\mu = \beta\mu(F).$$

It follows that $\beta\mu(F) \leq \eta(F) \leq \alpha\mu(F)$, which is a contradiction since $\alpha < \beta$ and $0 < \mu(F) < +\infty$. Therefore, if it is not the case that $g = f$ μ-a.e. on Ω, then option (2) also leads to a contradiction. Since both (1) and (2) lead to a contradiction, we must indeed have $g = f$ μ-a.e. on Ω.

We have shown the following: *if g satisfies (a) and (b), and if f is a derivative of η with respect to μ, then $g = f$ μ-a.e. on Ω.*

We still have to show that g itself is a derivative of η with respect to μ. Since $g = f$ μ-a.e. on Ω, we have $\eta_{ac}(A) = \int_A f \, d\mu = \int_A g \, d\mu$ for every $A \in \mathcal{F}$, so that g satisfies (i) of the definition. It remains to show that (ii)-(iv) are satisfied for g. That is, we must exhibit sets B_0 and P_0 such that $\mu(B_0) = \eta_s(B_0^c) = 0$, $\eta_s(P_0^c) = 0$, $g = +\infty$ on $B_0 \cap P_0$, and $g = -\infty$ on $B_0 \cap P_0^c$. Define

$$L = \{\omega \in \Omega : \ g(\omega) = -\infty\}, \ \ G = \{\omega \in \Omega : \ g(\omega) = +\infty\},$$

$$B_0 = B \cap (L \cup G), \ \text{ and } \ P_0 = B \cap G.$$

We now show that g satisfies (ii) with the set B_0. Observe that $\mu(B_0) = 0$ since $B_0 \subseteq B$ and $\mu(B) = 0$. Using the fact that $\eta_{ac} \ll \mu$, we have $\eta_{ac}(B) = 0$, so that $\eta_{ac}(L_\alpha \cap A \cap B) = 0$. By (b), we have for every $A \in \mathcal{F}$ and $\alpha \in \mathbb{R}$ that

$$\eta_s(L_\alpha \cap A \cap B) = \eta(L_\alpha \cap A \cap B) \leq \alpha\mu(L_\alpha \cap A \cap B) \leq \alpha\mu(B) = 0,$$

so that

$$\eta_s(L_\alpha \cap A \cap B) = 0 \ \text{ for every } \alpha \in \mathbb{R} \text{ and } A \in \mathcal{F}. \tag{9.10}$$

Observing that $\{L_n\}_{n=1}^\infty$ is a nondecreasing sequence with limit G^c, we have that $\{L_n \cap A \cap B\}_{n=1}^\infty$ is a nondecreasing sequence with limit $G^c \cap A \cap B$ for every $A \in \mathcal{F}$, hence for every $A \in \mathcal{F}$ we have

$$\eta_s(G^c \cap A \cap B) = \eta_s\left(\lim_n [L_n \cap A \cap B]\right) = \lim_n \eta_s\left(L_n \cap A \cap B\right) = 0$$

by (9.10). In particular, taking $A = L^c$ gives $\eta_s(G^c \cap L^c \cap B) = 0$. Since $B_0^c = B^c \cup [G^c \cap L^c \cap B]$ and $\eta_s(B^c) = 0 = \eta_s(G^c \cap L^c \cap B)$, it follows that $\eta_s(B_0^c) = 0$. We have $\mu(B_0) = 0 = \eta_s(B_0^c)$, so the set B_0 satisfies (ii) of the definition.

We now show that g satisfies (iii) of the definition, that is, $\eta_s(P_0^c) = 0$. Since $\{L_{-n}\}_{n=1}^\infty$ is nonincreasing with limit L, we have for every $A \in \mathcal{F}$ that $\{L_{-n} \cap A \cap B\}_{n=1}^\infty$ is nonincreasing with limit $L \cap A \cap B$. By (9.10), we have

$$\eta_s(L \cap A \cap B) = \eta_s\left(\lim_n [L_{-n} \cap A \cap B]\right) = \lim_n \eta_s(L_{-n} \cap A \cap B) = 0$$

for every $A \in \mathcal{F}$. In particular, taking $A = B$ gives $\eta_s(L \cap B) = 0$. Since we have

$$P_0^c = (B \cap G)^c = (B \cap L^c \cap G^c) \cup (L \cap B) \cup B^c,$$

and since each of $\eta_s(B \cap L^c \cap G^c)$, $\eta_s(L \cap B)$, and $\eta_s(B^c)$ is zero, we indeed have $\eta_s(P_0^c) = 0$. Therefore, P_0 satisfies (iii) of the definition relative to g.

All that remains is (iv) of the definition, but that is easy. Observe that

$$B_0 \cap P_0 = B \cap (L \cup G) \cap G \subseteq G.$$

Since $g = +\infty$ on G, it follows that $g = +\infty$ on $B_0 \cap P_0$. Since $B_0 \cap P_0^c \subseteq L$ and $g = -\infty$ on L, we have $g = -\infty$ on $B_0 \cap P_0^c$. This gives (iv) of the definition.

Summarizing, the function g, which satisfies (a) and (b) by assumption, also satisfies the following four properties:

(i) $\eta_{ac}(A) = \int g\, d\mu$ for every $A \in \mathcal{F}$.

There exist sets B_0, P_0 in $\mathcal{F}$ such that

(ii) $\mu(B_0) = \eta_s(B_0^c) = 0$,

(iii) $\eta_s(P_0^c) = 0$, and

(iv) $g = +\infty$ on $B_0 \cap P_0$ and $g = -\infty$ on $B_0 \cap P_0^c$.

We have accordingly shown that g is a derivative of η with respect to μ, and the proof is therefore complete. ∎

The following two claims are purely technical claims needed for the proof of Claim 5.

Claim 3. *Let $(\Omega, \mathcal{H}, \varphi)$ denote a measure space. Let $\mathcal{H}_0$ denote a sub-σ-field of $\mathcal{H}$, and let φ_0 denote the restriction of φ to $\mathcal{H}_0$. If h is an $\mathcal{H}_0/\mathcal{B}^*$-measurable function on Ω that is nonnegative φ_0-a.e. on Ω, then for every $A \in \mathcal{H}_0$ we have $\int_A h\, d\varphi = \int_A h\, d\varphi_0$.*

Proof: If $h = \sum_{i=1}^n c_i I_{B_i}$, where $B_1, \cdots, B_n \in \mathcal{H}_0$ are disjoint with union Ω and $0 \leq c_1, \cdots, c_n < +\infty$, then for every $A \in \mathcal{H}_0$ we have

$$\int_A h\, d\varphi = \sum_{i=1}^n c_i \varphi(A \cap B_i) = \sum_{i=1}^n c_i \varphi_0(A \cap B_i) = \int_A h\, d\varphi_0,$$

so the claim is true if h is a nonnegative finite-valued $\mathcal{H}_0/\mathcal{B}^*$-measurable simple function on Ω. Next, let h denote a nonnegative $\mathcal{H}_0/\mathcal{B}^*$-measurable function on Ω. Find a nondecreasing sequence $\{s_n\}_{n=1}^\infty$ of nonnegative finite-valued $\mathcal{H}_0/\mathcal{B}^*$-measurable simple functions on Ω with $\lim_n s_n = h$. Since $\int_A s_n\, d\varphi = \int_A s_n\, d\varphi_0$ for every $A \in \mathcal{H}_0$ and $n \in \mathrm{N}$, MCT gives

$$\int_A h\, d\varphi = \lim_n \int_A s_n\, d\varphi = \lim_n \int_A s_n\, d\varphi_0 = \int_A h\, d\varphi_0 \quad \text{for every } A \in \mathcal{H}_0,$$

so the result is true for nonnegative $\mathcal{H}_0/\mathcal{B}^*$-measurable functions on Ω. Finally, let h denote an $\mathcal{H}_0/\mathcal{B}^*$-measurable function on Ω that is nonnegative

φ_0-a.e. on Ω. Define h^* on Ω by stipulating that $h^* = h$ where $h \geq 0$ and $h^* = 0$ where $h < 0$. Then h^* is a nonnegative $\mathcal{H}_0/\mathcal{B}^*$-measurable function on Ω with $h^* = h$ φ_0-a.e. on Ω, and hence

$$\int_A h\, d\varphi = \int_A h^*\, d\varphi = \int_A h^*\, d\varphi_0 = \int_A h\, d\varphi_0 \quad \text{for every } A \in \mathcal{H}_0.$$

This completes the proof. ■

Claim 4. *Let $\mathcal{G}$ denote a field on Ω. Let ν_1 and ν_2 denote measures on $\sigma(\mathcal{G})$ such that $\nu_1 \leq \nu_2$ on $\mathcal{G}$ and ν_2 is σ-finite on $\mathcal{G}$. Then $\nu_1 \leq \nu_2$ on $\sigma(\mathcal{G})$.*

Proof: Pick $B \in \sigma(\mathcal{G})$. If $\nu_2(B) = +\infty$, we trivially have $\nu_1(B) \leq \nu_2(B)$. We next assume that $\nu_2(B) < +\infty$, picking $\epsilon > 0$. By Claim 2 of Section 3.8, there exists a disjoint sequence $\{A_k\}_{k=1}^\infty$ of $\mathcal{G}$-sets such that $B \subseteq \bigcup_{k=1}^\infty A_k$ and $\nu_2(\bigcup_{k=1}^\infty A_k - B) < \epsilon$. Since $\nu_2(B) < +\infty$, this gives $\nu_2(\bigcup_{k=1}^\infty A_k) < \nu_2(B) + \epsilon$, that is, $\sum_{k=1}^\infty \nu_2(A_k) < \nu_2(B) + \epsilon$. We next have that

$$\nu_1(B) \leq \nu_1\left(\bigcup_{k=1}^\infty A_k\right) = \sum_{k=1}^\infty \nu_1(A_k) \leq \sum_{k=1}^\infty \nu_2(A_k) < \nu_2(B) + \epsilon.$$

Since $\epsilon > 0$ is arbitrary, we have $\nu_1(B) \leq \nu_2(B)$. We therefore have the inequality $\nu_1(B) \leq \nu_2(B)$ for every $B \in \sigma(\mathcal{G})$, completing the proof. ■

We now discuss the framework under which we shall operate for the remainder of this section.

Framework. Ω will denote a nonempty set, $\mathcal{F}$ will denote a σ-field of subsets of Ω, and μ and η will denote arbitrary measures on $\mathcal{F}$. We will have a sequence $\{\mathcal{F}_n\}_{n=1}^\infty$ of sub-σ-fields of $\mathcal{F}$. [In particular, $\bigcup_{n=1}^\infty \mathcal{F}_n \subseteq \mathcal{F}$.] For each $n \in \mathbb{N}$, the restrictions of μ and η to $\mathcal{F}_n$ will be denoted by μ_n and η_n, respectively, and we assume that μ_n and η_n are σ-finite on $\mathcal{F}_n$. [This implies that μ and η are σ-finite on $\mathcal{F}$.] For each $n \in \mathbb{N}$, we will let f_n denote a derivative of η_n with respect to μ_n. [By Claim 1, each f_n is nonnegative μ_n-a.e. on Ω, hence each f_n is nonnegative μ-a.e. on Ω.] We will write $\underline{f} = \underline{\lim}_n f_n$ and $\bar{f} = \overline{\lim}_n f_n$, that is,

$$\underline{f} = \lim_n \inf\{f_n, f_{n+1}, \cdots\} \quad \text{and} \quad \bar{f} = \lim_n \sup\{f_n, f_{n+1}, \cdots\}.$$

We will examine two situations under this framework. The first situation is where $\mathcal{F}_1 \subseteq \mathcal{F}_2 \subseteq \cdots$. Here, we will write $\mathcal{F}_\infty = \sigma(\bigcup_{n=1}^\infty \mathcal{F}_n) = \sigma(\lim_n \mathcal{F}_n)$, and we will write μ_∞ and η_∞ for the restrictions of μ and η to $\mathcal{F}_\infty$, respectively. Observe that μ_∞ and η_∞ are σ-finite on $\mathcal{F}_\infty$,[10] and $\mathcal{F}_\infty \subseteq \mathcal{F}$. Also, observe [in this situation] that each f_n is nonnegative μ_∞-a.e. on Ω.

[10]Since μ_1 is σ-finite, we may write Ω as a disjoint union $\bigcup_{n=1}^\infty A_n$, where $A_1, A_2, \cdots \in \mathcal{F}_1$ and each of $\mu(A_1), \mu(A_2), \cdots$ is finite. Since $\mathcal{F}_1 \subseteq \mathcal{F}_\infty$, we have $A_1, A_2, \cdots \in \mathcal{F}_\infty$ and each of $\mu(A_1) = \mu_\infty(A_1)$, $\mu(A_2) = \mu_\infty(A_2)$, $\cdots$ is finite. This shows that μ_∞ is σ-finite on $\mathcal{F}_\infty$; the proof that η_∞ is σ-finite on $\mathcal{F}_\infty$ is exactly the same.

We now prove the first of two theorems, which are due to the work of Andersen and Jessen[11] as presented by Hewitt and Stromberg.[12]

Claim 5. *Suppose that* $\mathcal{F}_1 \subseteq \mathcal{F}_2 \subseteq \cdots$. *Then* $\underline{f}$ *and* $\bar{f}$ *are* $\mathcal{F}_\infty/\mathcal{B}^*$*-measurable, and in addition we have the following:*

(a) $\underline{f}$ *and* $\bar{f}$ *are derivatives of* η_∞ *with respect to* μ_∞.

(b) $\underline{f} = \bar{f}$ μ*-a.e. on* Ω, *so that* $\lim_n f_n$ *exists* μ*-a.e. on* Ω.

(c) $\int_A \lim_n f_n \, d\mu = \eta_{\infty,ac}(A)$ *for every* $A \in \mathcal{F}_\infty$.

Proof: We will show that $\underline{f}$ is $\mathcal{F}_\infty/\mathcal{B}^*$-measurable. Pick $n \in \mathbb{N}$. Observe that f_n is $\mathcal{F}_n/\mathcal{B}^*$-measurable, f_{n+1} is $\mathcal{F}_{n+1}/\mathcal{B}^*$-measurable, f_{n+2} is $\mathcal{F}_{n+2}/\mathcal{B}^*$-measurable, $\cdots$. Because $\mathcal{F}_1 \subseteq \mathcal{F}_2 \subseteq \cdots \subseteq \bigcup_{j=1}^\infty \mathcal{F}_j \subseteq \mathcal{F}_\infty$, we have that each of $f_n, f_{n+1}, f_{n+2}, \cdots$ are $\mathcal{F}_\infty/\mathcal{B}^*$-measurable. Therefore, $\inf\{f_n, f_{n+1}, \cdots\}$ is $\mathcal{F}_\infty/\mathcal{B}^*$-measurable for every $n \in \mathbb{N}$. It follows that $\underline{f} = \underline{\lim}_n f_n = \lim_n \inf\{f_n, f_{n+1}, \cdots\}$ is $\mathcal{F}_\infty/\mathcal{B}^*$-measurable. The proof that $\bar{f} = \overline{\lim}_n f_n$ is $\mathcal{F}_\infty/\mathcal{B}^*$-measurable is almost exactly the same, and hence we have established the measurability claim.

Assume for the moment that (a) holds. By Claim 1, we have that $\underline{f} = \bar{f}$ μ_∞-a.e. on Ω, and hence $\mu_\infty(\{\omega \in \Omega : \underline{f}(\omega) \neq \bar{f}(\omega)\}) = 0$. Since $\mu = \mu_\infty$ on $\mathcal{F}_\infty$, we have $\mu(\{\omega \in \Omega : \underline{f}(\omega) \neq \bar{f}(\omega)\}) = 0$, so that $\underline{f} = \bar{f}$ μ-a.e. on Ω. It follows that $\lim_n f_n$ exists μ-a.e. on Ω, giving (b). Next, observe for every $A \in \mathcal{F}_\infty$ that

$$\int_A \lim_n f_n \, d\mu = \int_A \lim_n f_n \, d\mu_\infty = \int_A \bar{f} \, d\mu_\infty = \eta_{\infty,ac}(A),$$

so that (c) follows. [The first equality follows by noting that $\lim_n f_n$ is nonnegative μ_∞-a.e. on Ω and using Claim 3. The second equality follows because $\lim_n f_n = \bar{f}$ μ_∞-a.e. on Ω. The third equality follows because $\bar{f}$ is a derivative of η_∞ with respect to μ_∞.] Therefore, everything will be complete when (a) is established.

We now turn our attention to proving part (a). For each $\alpha \in \mathbb{R}$, define the sets

$$L_{\alpha,\underline{f}} = \{\omega \in \Omega : \underline{f}(\omega) \leq \alpha\}, \quad L_{\alpha,\bar{f}} = \{\omega \in \Omega : \bar{f}(\omega) \leq \alpha\},$$

$$G_{\alpha,\underline{f}} = \{\omega \in \Omega : \underline{f}(\omega) \geq \alpha\}, \quad G_{\alpha,\bar{f}} = \{\omega \in \Omega : \bar{f}(\omega) \geq \alpha\}.$$

Suppose we show that

$$\eta(L_{\alpha,\underline{f}} \cap A) \leq \alpha\mu(L_{\alpha,\underline{f}} \cap A) \quad \text{for every } A \in \mathcal{F}_\infty \text{ and } \alpha \in \mathbb{R} \tag{9.11}$$

and

$$\eta(G_{\alpha,\bar{f}} \cap A) \geq \alpha\mu(G_{\alpha,\bar{f}} \cap A) \quad \text{for every } A \in \mathcal{F}_\infty \text{ and } \alpha \in \mathbb{R}. \tag{9.12}$$

[11] Danske Vid. Selsk. Mat.-Fys. Medd 25 (1948), Nr. 5

[12] Sections (20.53)-(20.59).

Then, replacing A by $A \cap L_{\alpha,\bar{f}}$ in (9.11), we have

$$\eta(L_{\alpha,\underline{f}} \cap A \cap L_{\alpha,\bar{f}}) \leq \alpha\mu(L_{\alpha,\underline{f}} \cap A \cap L_{\alpha,\bar{f}}) \quad \text{for every } A \in \mathcal{F}_\infty \text{ and } \alpha \in \mathbb{R}.$$

Since $\underline{f} \leq \bar{f}$ gives $L_{\alpha,\bar{f}} \subseteq L_{\alpha,\underline{f}}$ for every $\alpha \in \mathbb{R}$, the above display becomes

$$\eta(L_{\alpha,\bar{f}} \cap A) \leq \alpha\mu(L_{\alpha,\bar{f}} \cap A) \quad \text{for every } A \in \mathcal{F}_\infty \text{ and } \alpha \in \mathbb{R}. \tag{9.13}$$

Next, replacing A by $A \cap G_{\alpha,\underline{f}}$ in (9.12) gives

$$\eta(G_{\alpha,\bar{f}} \cap A \cap G_{\alpha,\underline{f}}) \geq \alpha\mu(G_{\alpha,\bar{f}} \cap A \cap G_{\alpha,\underline{f}}) \quad \text{for every } A \in \mathcal{F}_\infty \text{ and } \alpha \in \mathbb{R}.$$

Since $\underline{f} \leq \bar{f}$ gives $G_{\alpha,\underline{f}} \subseteq G_{\alpha,\bar{f}}$ for every $\alpha \in \mathbb{R}$, the above display becomes

$$\eta(G_{\alpha,\underline{f}} \cap A) \geq \alpha\mu(G_{\alpha,\underline{f}} \cap A) \quad \text{for every } A \in \mathcal{F}_\infty \text{ and } \alpha \in \mathbb{R}. \tag{9.14}$$

Therefore, (9.11) and (9.12) force both (9.13) and (9.14). Observe that (9.11), (9.14), and Claim 2 show that $\underline{f}$ is a derivative of η_∞ with respect to μ_∞. Next, observe that (9.12), (9.13), and Claim 2 show that $\bar{f}$ is a derivative of η_∞ with respect to μ_∞. Therefore, part (a) of the claim [and accordingly the entire proof] will be complete when (9.11) and (9.12) are established.

We now turn to proving (9.11). Pick and fix $\alpha \in \mathbb{R}$; let $\{\alpha_n\}_{n=1}^\infty$ denote a strictly decreasing sequence of real numbers with limit α. For every $n \in \mathbb{N}$ we make the following definitions:

$$H_n = \{\omega \in \Omega : \inf\{f_{n+1}(\omega), f_{n+2}(\omega), \cdots\} < \alpha_n\},$$

$$H_{n,1} = \{\omega \in \Omega : f_{n+1}(\omega) < \alpha_n\},$$

and for each $n \in \mathbb{N}$ and $p \in \{2, 3, \cdots\}$ we define

$$H_{n,p} = \{\omega \in \Omega : \min\{f_{n+1}(\omega), \cdots, f_{n+p-1}(\omega)\} \geq \alpha_n, \ f_{n+p}(\omega) < \alpha_n\}.$$

With these definitions, we make the following subclaims:

(1) $H_{n,p} \in \mathcal{F}_{n+p}$ for every $n, p \in \mathbb{N}$.

Proof: Let $n \in \mathbb{N}$. Since f_{n+1} is $\mathcal{F}_{n+1}/\mathcal{B}^*$-measurable, we have that $H_{n,1} \in \mathcal{F}_{n+1}$. This gives the case where $p = 1$. Now let $p \geq 2$. We have

$$\begin{aligned} & H_{n,p} \\ = \ & \{\omega \in \Omega : \min\{f_{n+1}(\omega), \cdots, f_{n+p-1}(\omega)\} \geq \alpha_n\} \cap \{\omega \in \Omega : f_{n+p}(\omega) < \alpha_n\} \\ = \ & \bigcap_{j=1}^{p-1} \{\omega \in \Omega : f_{n+j}(\omega) \geq \alpha_n\} \ \cap \ \{\omega \in \Omega : f_{n+p}(\omega) < \alpha_n\}. \end{aligned}$$

For $j = 1, \cdots, p-1$, f_{n+j} is $\mathcal{F}_{n+j}/\mathcal{B}^*$-measurable, so for every $j = 1, \cdots, p-1$ we have that $\{\omega \in \Omega : f_{n+j}(\omega) \geq \alpha_n\} \in \mathcal{F}_{n+j}$. Since the σ-fields $\mathcal{F}_k$ are nondecreasing, we have for every $j = 1, \cdots, p-1$ that $\{\omega \in \Omega : f_{n+j}(\omega) \geq \alpha_n\} \in \mathcal{F}_{n+p}$. Next, it is clear that $\{\omega \in \Omega : f_{n+p}(\omega) < \alpha_n\} \in \mathcal{F}_{n+p}$ because f_{n+p} is $\mathcal{F}_{n+p}/\mathcal{B}^*$-measurable. Therefore, $H_{n,p}$ is arrived at by intersecting a finite number of sets in $\mathcal{F}_{n+p}$, and hence $H_{n,p} \in \mathcal{F}_{n+p}$. This gives the case where $p \geq 2$. Combining the $p = 1$ and $p \geq 2$ cases gives the claim.

(2) The sequence $\{H_{n,p}\}_{p=1}^{\infty}$ is disjoint for every $n \in \mathbb{N}$.

Proof: Let $n \in \mathbb{N}$. It is automatic that $H_{n,1} \cap H_{n,p} = \varnothing$ for $p = 2, 3, \cdots$. It is also easy to see that $H_{n,p_1} \cap H_{n,p_2} = \varnothing$ whenever $p_1, p_2 \in \{2, 3, \cdots\}$ with $p_1 < p_2$. Putting these two facts together gives the claim.

(3) For every $n \in \mathbb{N}$, $H_n = \bigcup_{p=1}^{\infty} H_{n,p} \in \bigcup_{p=1}^{\infty} \mathcal{F}_{n+p} \subseteq \bigcup_{k=1}^{\infty} \mathcal{F}_k \subseteq \mathcal{F}_\infty$.

Proof: Let $n \in \mathbb{N}$. We have $H_{n,1}, H_{n,2}, \cdots \subseteq H_n$, and hence $\bigcup_{p=1}^{\infty} H_{n,p} \subseteq H_n$. For the reverse inclusion, observe that if $H_n = \varnothing$ the result trivially follows. Otherwise, pick $\omega \in H_n$. Then $\inf\{f_{n+1}(\omega), f_{n+2}(\omega), \cdots\} < \alpha_n$, so there is $j \in \mathbb{N}$ with $f_{n+j}(\omega) < \alpha_n$; pick the smallest such j. If $j = 1$, then $\omega \in H_{n,1}$. Otherwise, we will have $j \in \{2, 3, \cdots\}$, and because we've picked the smallest j with $f_{n+j}(\omega) < \alpha_n$ we will have each of $f_{n+1}(\omega), \cdots, f_{n+j-1}(\omega)$ at least equal to α_n. This will give $\omega \in H_{n,j}$. Therefore, every $\omega \in H_n$ will also be in some $H_{n,p}$, and hence $H_n \subseteq \bigcup_{p=1}^{\infty} H_{n,p}$. This gives the desired set identity. That $\bigcup_{p=1}^{\infty} H_{n,p} \in \bigcup_{p=1}^{\infty} \mathcal{F}_{n+p}$ follows immediately from the fact that each $H_{n,p}$ is in $\mathcal{F}_{n+p}$ by (1). The other subset relations asserted in this subclaim are obvious.

(4) The sequence $\{H_n\}_{n=1}^{\infty}$ is nonincreasing with limit $L_{\alpha,\underline{f}}$.

Proof: That $H_1 \supseteq H_2 \supseteq \cdots$ follows from the fact that

$$\inf\{f_{n+1}(\omega), f_{n+2}(\omega), \cdots\} \leq \inf\{f_{n+2}(\omega), f_{n+3}(\omega), \cdots\} \text{ for each } \omega \in \Omega \text{ and } n \in \mathbb{N}.$$

Therefore, $\{H_n\}_{n=1}^{\infty}$ is nondecreasing with limit $\bigcap_{n=1}^{\infty} H_n$. We must show that $\bigcap_{n=1}^{\infty} H_n = L_{\alpha,\underline{f}}$. Since $\underline{f} = \underline{\lim}_n f_n = \lim_n \inf\{f_{n+1}, f_{n+2}, \cdots\}$, and since the functions $\inf\{f_{n+1}, f_{n+2}, \cdots\}$ are nondecreasing, the following statements are equivalent:

$\omega \in \bigcap_{n=1}^{\infty} H_n$

$\inf\{f_{n+1}(\omega), f_{n+2}(\omega), \cdots\} < \alpha_n$ for every $n \in \mathbb{N}$

$\lim_n \inf\{f_{n+1}(\omega), f_{n+2}(\omega), \cdots\} \leq \lim_n \alpha_n = \alpha$

$\underline{f}(\omega) \leq \alpha$

$\omega \in L_{\alpha,\underline{f}}$.

This completes the proof of the subclaim.

With subclaims (1)-(4) out of the way, we now proceed back to the main part of the proof, which for the moment means demonstrating (9.11).

Pick $A \in \bigcup_{n=1}^{\infty} \mathcal{F}_n$. There is $N \in \mathbb{N}$ such that $A \in \mathcal{F}_n$ for every $n \geq N$. For every $n \geq N$ and $p \in \mathbb{N}$ we have $H_{n,p} \cap A \in \mathcal{F}_{n+p}$ [by (1)], hence for every

$n \geq N$ we invoke (2) and (3), obtaining

$$\begin{aligned}
\eta(H_n \cap A) = \sum_{p=1}^{\infty} \eta(H_{n,p} \cap A) &= \sum_{p=1}^{\infty} \eta_{n+p}(H_{n,p} \cap A) \\
&\stackrel{(*)}{\leq} \sum_{p=1}^{\infty} \alpha_n \mu_{n+p}(H_{n,p} \cap A) \\
&= \alpha_n \sum_{p=1}^{\infty} \mu(H_{n,p} \cap A) \\
&= \alpha_n \mu(H_n \cap A).
\end{aligned}$$

Proof of Inequality (*): Let $I_{n,p} = \{\omega \in \Omega : \min\{f_{n+1}(\omega), \cdots, f_{n+p-1}(\omega)\} \geq \alpha_n\}$ for $p \geq 2$, and write $I_{n,1} = \Omega$. Define $J_{n,p} = \{\omega \in \Omega : f_{n+p}(\omega) \neq \alpha_n\}$ for each $p \in \mathbb{N}$. Write $L_{\alpha_n, f_{n+p}} = \{\omega \in \Omega : f_{n+p}(\omega) \leq \alpha_n\}$ for every $p \in \mathbb{N}$. Observe that $H_{n,p} = I_{n,p} \cap J_{n,p} \cap L_{\alpha_n, f_{n+p}}$ for each $p \in \mathbb{N}$. Since f_{n+p} is a derivative of η_{n+p} with respect to μ_{n+p} for every $p \in \mathbb{N}$, Claim 2 gives

$$\begin{aligned}
\eta_{n+p}(H_{n,p} \cap A) &= \eta_{n+p}([I_{n,p} \cap J_{n,p} \cap A] \cap L_{\alpha_n, f_{n+p}}) \\
&\leq \alpha_n \mu_{n+p}([I_{n,p} \cap J_{n,p} \cap A] \cap L_{\alpha_n, f_{n+p}}) \\
&= \alpha_n \mu_{n+p}(H_{n,p} \cap A) \quad \text{for every } p \in \mathbb{N}.
\end{aligned}$$

This proves the starred inequality.

Therefore, given $A \in \bigcup_{k=1}^{\infty} \mathcal{F}_k$, we have that

$$\eta(H_n \cap A) \leq \alpha_n \mu(H_n \cap A) \quad \text{for sufficiently large } n. \tag{9.15}$$

Suppose that $A \in \bigcup_{k=1}^{\infty} \mathcal{F}_k$ is such that $\mu(A), \eta(A) < +\infty$. Then, since the sequence $\{H_n \cap A\}_{n=1}^{\infty}$ is nonincreasing with limit $L_{\alpha, \underline{f}} \cap A$ [by (4)], we use measure property (M9) from Section 2.2, obtaining

$$\lim_n \eta(H_n \cap A) = \eta(L_{\alpha, \underline{f}} \cap A) \quad \text{and} \quad \lim_n \mu(H_n \cap A) = \mu(L_{\alpha, \underline{f}} \cap A).$$

Therefore, (9.15) gives $\eta(L_{\alpha, \underline{f}} \cap A) \leq \alpha \mu(L_{\alpha, \underline{f}} \cap A)$ when $A \in \bigcup_{k=1}^{\infty} \mathcal{F}_k$ is such that $\mu(A), \eta(A) < +\infty$.

Next, consider the case where $A \in \bigcup_{k=1}^{\infty} \mathcal{F}_k$ is arbitrary. We use the fact that μ_1 and η_1 are σ-finite on $\mathcal{F}_1$, obtaining a disjoint sequence $\{D_n\}_{n=1}^{\infty}$ of $\mathcal{F}_1$-sets such that $\Omega = \bigcup_{n=1}^{\infty} D_n$, $\mu_1(D_n) = \mu(D_n) < +\infty$ for every $n \in \mathbb{N}$, and $\eta_1(D_n) = \eta(D_n) < +\infty$ for every $n \in \mathbb{N}$. Now $A \cap D_n \in \bigcup_{k=1}^{\infty} \mathcal{F}_k$ and $\mu(A \cap D_n), \eta(A \cap D_n) < +\infty$ for each $n \in \mathbb{N}$, hence for every $n \in \mathbb{N}$ we have $\eta(L_{\alpha, \underline{f}} \cap D_n \cap A) \leq \alpha \mu(L_{\alpha, \underline{f}} \cap D_n \cap A)$ by virtue of the italicized fact at the end of the previous paragraph. Therefore, we have

$$\begin{aligned}
\eta(L_{\alpha, \underline{f}} \cap A) = \sum_{n=1}^{\infty} \eta(L_{\alpha, \underline{f}} \cap D_n \cap A) &\leq \sum_{n=1}^{\infty} \alpha \mu(L_{\alpha, \underline{f}} \cap D_n \cap A) \\
&= \alpha \mu(L_{\alpha, \underline{f}} \cap A).
\end{aligned}$$

We now summarize: in our desire to show (9.11), we have so far shown that

$$\eta(L_{\alpha,\underline{f}} \cap A) \le \alpha\mu(L_{\alpha,\underline{f}} \cap A) \quad \text{for every } A \in \bigcup_{n=1}^{\infty} \mathcal{F}_n.$$

We wish to extend this inequality to every $A \in \mathcal{F}_\infty = \sigma(\bigcup_{n=1}^{\infty} \mathcal{F}_n)$. To do this, define ν_1 and ν_2 on $\mathcal{F}_\infty$ by writing

$$\nu_1(A) = \eta(L_{\alpha,\underline{f}} \cap A) \quad \text{and} \quad \nu_2(A) = \alpha\mu(L_{\alpha,\underline{f}} \cap A) \quad \text{for every } A \in \mathcal{F}_\infty.$$

Because η and μ are σ-finite measures on $\mathcal{F}_\infty$, it is straightforward to show that ν_1 and ν_2 are σ-finite measures on $\mathcal{F}_\infty$. We have that $\nu_1 \le \nu_2$ on the field $\bigcup_{n=1}^{\infty} \mathcal{F}_n$ and ν_2 is σ-finite on $\bigcup_{n=1}^{\infty} \mathcal{F}_n$; it follows by Claim 4 that $\nu_1 \le \nu_2$ on $\sigma(\bigcup_{n=1}^{\infty} \mathcal{F}_n)$. This is precisely the assertion (9.11).

We now turn to prove (9.12), which will complete the entire proof. The method of proof is not too different than that for (9.11), and so we merely give the outline, trusting the competent reader to fill in the details. Pick $\alpha \in \mathbb{R}$, and let $\{\alpha_n\}_{n=1}^{\infty}$ denote a strictly increasing sequence of real numbers with limit α. For every $n \in \mathbb{N}$, define the sets

$$H_n = \{\omega \in \Omega : \sup\{f_{n+1}(\omega), f_{n+2}(\omega), \cdots\} > \alpha_n\},$$

$$H_{n,1} = \{\omega \in \Omega : f_{n+1}(\omega) > \alpha_n\}, \quad \text{and}$$

$$H_{n,p} = \{\omega \in \Omega : \max\{f_{n+1}(\omega), \cdots, f_{n+p-1}(\omega)\} \le \alpha_n,\ f_{n+p} > \alpha_n\},\ p \ge 2.$$

The following may be shown:

(1′) $H_{n,p} \in \mathcal{F}_{n,p}$ for every $n, p \in \mathbb{N}$.

(2′) The sequence $\{H_{n,p}\}_{p=1}^{\infty}$ is disjoint for every $n \in \mathbb{N}$.

(3′) $H_n = \bigcup_{p=1}^{\infty} H_{n,p} \in \bigcup_{k=1}^{\infty} \mathcal{F}_k \subseteq \mathcal{F}_\infty$ for every $n \in \mathbb{N}$.

(4′) $\{H_n\}_{n=1}^{\infty}$ is a nonincreasing sequence with limit $G_{\alpha,\bar{f}}$.

Let $A \in \bigcup_{n=1}^{\infty} \mathcal{F}_n$. There is $N \in \mathbb{N}$ such that $A \in \mathcal{F}_n$ if $n \ge N$. Also, for each $n \ge N$ and $p \in \mathbb{N}$ we have $H_{n,p} \cap A \in \mathcal{F}_{n+p}$ [by (1′)]. Invoking (2′) and (3′), we have

$$\begin{aligned} \eta(H_n \cap A) = \sum_{p=1}^{\infty} \eta(H_{n,p} \cap A) &= \sum_{p=1}^{\infty} \eta_{n+p}(H_{n,p} \cap A) \\ &\ge \sum_{p=1}^{\infty} \alpha_n \mu_{n+p}(H_{n,p} \cap A) \\ &= \alpha_n \sum_{p=1}^{\infty} \mu(H_{n,p} \cap A) \\ &= \alpha_n \mu(H_n \cap A). \end{aligned}$$

[The inequality may be proved by the same type of proof used in proving the earlier starred inequality.] If $\eta(A), \mu(A) < +\infty$, then letting $n \to \infty$ and calling on (4′) gives $\eta(G_{\alpha,\bar{f}} \cap A) \geq \alpha\mu(G_{\alpha,\bar{f}} \cap A)$. For general $A \in \bigcup_{n=1}^{\infty} \mathcal{F}_n$ the same σ-finiteness argument as before gives $\eta(G_{\alpha,\bar{f}} \cap A) \geq \alpha\mu(G_{\alpha,\bar{f}} \cap A)$ for every $A \in \bigcup_{n=1}^{\infty} \mathcal{F}_n$. The same appeal to Claim 4 shows that this inequality holds for every $A \in \mathcal{F}_\infty = \sigma(\bigcup_{n=1}^{\infty} \mathcal{F}_n)$, giving (9.12). This in conjunction with (9.11) gives (a), thereby completing the entire proof. ■

We earlier stated that we wished to discuss the given framework for two situations. The preceding claim discussed the first of the two situations. The second situation is where $\mathcal{F}_1 \supseteq \mathcal{F}_2 \supseteq \cdots$. Here, we will write $\mathcal{F}_0$ for the σ-field $\bigcap_{n=1}^{\infty} \mathcal{F}_n = \lim_n \mathcal{F}_n$, with μ_0 and η_0 denoting the restrictions of μ and η to $\mathcal{F}_0$, respectively.

Claim 6. *Suppose that $\mathcal{F}_1 \supseteq \mathcal{F}_2 \supseteq \cdots$. Then both $\underline{f}$ and $\bar{f}$ are $\mathcal{F}_0/\mathcal{B}^*$-measurable, and we have the following:*

(a) Both $\underline{f}$ and $\bar{f}$ are derivatives of η_0 with respect to μ_0.

(b) $\underline{f} = \bar{f}$ μ-a.e. on Ω, so that $\lim_n f_n$ exists μ-a.e. on Ω.

(c) $\eta_{0,ac}(A) = \int_A \lim_n f_n \, d\mu$ for every $A \in \mathcal{F}_0$.

Proof: We establish the measurability claim regarding $\underline{f}$ and $\bar{f}$. We have that $\underline{f} = \underline{\lim}_n f_n = \lim_n \inf\{f_n, f_{n+1}, \cdots\}$ by definition. Since

$$\inf\{f_n, f_{n+1}, \cdots\} = \lim_k \min\{f_n, \cdots, f_{n+k}\} \quad \text{for every } n \in \mathbb{N},$$

we have that

$$\underline{f} = \lim_n \lim_k \min\{f_n, \cdots, f_{n+k}\} = \lim_n \lim_k \min\{f_{n+m}, \cdots, f_{n+m+k}\}$$

for every $m \in \mathbb{N}$. Now fix $m, n, k \in \mathbb{N}$. We have that f_{n+m} is $\mathcal{F}_{n+m}/\mathcal{B}^*$-measurable, f_{n+m+1} is $\mathcal{F}_{n+m+1}/\mathcal{B}^*$-measurable, $\cdots$, f_{n+m+k} is $\mathcal{F}_{n+m+k}/\mathcal{B}^*$-measurable. Since $\mathcal{F}_{n+m+k} \subseteq \cdots \subseteq \mathcal{F}_{n+m} \subseteq \cdots \subseteq \mathcal{F}_m$, we have that each of $f_{n+m}, \cdots, f_{n+m+k}$ is $\mathcal{F}_m/\mathcal{B}^*$-measurable, so $\min\{f_{n+m}, \cdots, f_{n+m+k}\}$ is $\mathcal{F}_m/\mathcal{B}^*$-measurable. Since k is arbitrary, $\lim_k \min\{f_{n+m}, \cdots, f_{n+m+k}\}$ is $\mathcal{F}_m/\mathcal{B}^*$-measurable. Since n is arbitrary, $\lim_n \lim_k \min\{f_{n+m}, \cdots, f_{n+m+k}\}$ is $\mathcal{F}_m/\mathcal{B}^*$-measurable. It follows that $\underline{f}$ is $\mathcal{F}_m/\mathcal{B}^*$-measurable. Since m is arbitrary, we have that $\underline{f}$ is $\mathcal{F}_m/\mathcal{B}^*$-measurable for every $m \in \mathbb{N}$. Therefore, given arbitrary $B \in \mathcal{B}^*$, we have that $\underline{f}^{-1}(B) \in \mathcal{F}_m$ for every $m \in \mathbb{N}$, and hence $\underline{f}^{-1}(B) \in \bigcap_{m=1}^{\infty} \mathcal{F}_m = \mathcal{F}_0$. It follows that $\underline{f}$ is $\mathcal{F}_0/\mathcal{B}^*$-measurable. The same type of argument shows that $\bar{f}$ is $\mathcal{F}_0/\mathcal{B}^*$-measurable, and the measurability claim is thus established.

Now assume that (a) holds. Then $\mu_0(\{\omega \in \Omega : \underline{f}(\omega) \neq \bar{f}(\omega)\}) = 0$ by Claim 1. Since $\mu = \mu_0$ on $\mathcal{F}_0$, we have $\mu(\{\omega \in \Omega : \underline{f}(\omega) \neq \bar{f}(\omega)\}) = 0$, so that $\underline{f} = \bar{f}$ μ-a.e. on Ω. Therefore, $\lim_n f_n$ exists μ-a.e. on Ω. This gives (b). [Note too that $\lim_n f_n$ is an $\mathcal{F}_0/\mathcal{B}^*$-measurable function defined and nonnegative μ_0-a.e. on Ω.] For (c), we have for every $A \in \mathcal{F}_0$ that

$$\int_A \lim_n f_n \, d\mu = \int_A \lim_n f_n \, d\mu_0 = \int_A \bar{f} \, d\mu_0 = \eta_{0,ac}(A)$$

by using Claim 3, the fact that $\bar{f} = \lim_n f_n$ μ_0-a.e. on Ω, and the fact that $\bar{f}$ is a derivative of η_0 with respect to μ_0. This gives (c). Therefore, all that is really needed is to prove (a). Just as in Claim 5, it suffices to show that

$$\eta(L_{\alpha,\underline{f}} \cap A) \leq \alpha\mu(L_{\alpha,\underline{f}} \cap A) \text{ for every } A \in \mathcal{F}_0 \text{ and } \alpha \in \mathbb{R} \tag{9.16}$$

and

$$\eta(G_{\alpha,\bar{f}} \cap A) \geq \alpha\mu(G_{\alpha,\bar{f}} \cap A) \text{ for every } A \in \mathcal{F}_0 \text{ and } \alpha \in \mathbb{R} \tag{9.17}$$

in order to prove (a). [The sets $L_{\alpha,\underline{f}}$ and $G_{\alpha,\bar{f}}$ are the same as in the proof of Claim 5. Recall in Claim 5 that (9.11) and (9.12) force (9.13) and (9.14). This argument used only the fact that $\underline{f} \leq \bar{f}$.]

For each $\alpha \in \mathbb{R}$, define

$$M_\alpha = \{\omega \in \Omega : \inf\{f_1(\omega), f_2(\omega), \cdots\} < \alpha\}$$

and

$$H_\alpha = \{\omega \in \Omega : \sup\{f_1(\omega), f_2(\omega), \cdots\} > \alpha\}.$$

Since $\mathcal{F}_1 \supseteq \mathcal{F}_2 \supseteq \cdots$ and each f_n is $\mathcal{F}_n/\mathcal{B}^*$-measurable, we have that each f_n is $\mathcal{F}_1/\mathcal{B}^*$-measurable, and hence both $\inf\{f_1, f_2, \cdots\}$ and $\sup\{f_1, f_2, \cdots\}$ are $\mathcal{F}_1/\mathcal{B}^*$-measurable, so $M_\alpha, H_\alpha \in \mathcal{F}_1$ [although they may not be in $\mathcal{F}_0$]. Suppose we show that

$$\eta(M_\alpha \cap A) \leq \alpha\mu(M_\alpha \cap A) \quad \text{for every } A \in \mathcal{F}_0 \text{ and } \alpha \in \mathbb{R} \tag{9.18}$$

and

$$\eta(H_\alpha \cap A) \geq \alpha\mu(H_\alpha \cap A) \quad \text{for every } A \in \mathcal{F}_0 \text{ and } \alpha \in \mathbb{R}. \tag{9.19}$$

We claim that (9.18) forces (9.16); we also claim that (9.19) forces (9.17).

(9.18) forces (9.16): Pick $\alpha \in \mathbb{R}$ and $\epsilon > 0$, and let $A \in \mathcal{F}_0$. Using the containment relation $L_{\alpha,\underline{f}} \subseteq M_{\alpha+\epsilon}$, (9.18) gives

$$\begin{aligned} \eta(L_{\alpha,\underline{f}} \cap A) = \eta(M_{\alpha+\epsilon} \cap L_{\alpha,\underline{f}} \cap A) &\leq (\alpha+\epsilon)\mu(M_{\alpha+\epsilon} \cap L_{\alpha,\underline{f}} \cap A) \\ &= (\alpha+\epsilon)\mu(L_{\alpha,\underline{f}} \cap A). \end{aligned}$$

Therefore, if $A \in \mathcal{F}_0$ is such that both $\eta(A)$ and $\mu(A)$ are finite, the display above forces the inequality $\eta(L_{\alpha,\underline{f}} \cap A) \leq \alpha\mu(L_{\alpha,\underline{f}} \cap A)$. For arbitrary $A \in \mathcal{F}_0$, we use the fact that μ and η are σ-finite on $\mathcal{F}_1$ to write $\Omega = \bigcup_{n=1}^\infty D_n$, where $\{D_n\}_{n=1}^\infty$ is a disjoint sequence of $\mathcal{F}_1$-sets with finite μ- and η-measure. This gives

$$\begin{aligned} \eta(L_{\alpha,\underline{f}} \cap A) = \sum_{n=1}^\infty \eta(L_{\alpha,\underline{f}} \cap A \cap D_n) &\leq \sum_{n=1}^\infty \alpha\mu(L_{\alpha,\underline{f}} \cap A \cap D_n) \\ &= \alpha\mu(L_{\alpha,\underline{f}} \cap A). \end{aligned}$$

This is precisely the statement (9.16).

(9.19) forces (9.17): This proceeds much like the above. Pick $\alpha \in \mathbb{R}$ and $\epsilon > 0$. Using the containment relation $G_{\alpha,\bar{f}} \subseteq H_{\alpha-\epsilon}$, (9.19) gives

$$\begin{aligned} \eta(G_{\alpha,\bar{f}} \cap A) = \eta(H_{\alpha-\epsilon} \cap G_{\alpha,\bar{f}} \cap A) &\geq (\alpha - \epsilon)\mu(H_{\alpha-\epsilon} \cap G_{\alpha,\bar{f}} \cap A) \\ &= (\alpha - \epsilon)\mu(G_{\alpha,\bar{f}} \cap A). \end{aligned}$$

Therefore, if $A \in \mathcal{F}_0$ is such that $\mu(A)$ and $\eta(A)$ are finite, the display above gives the inequality $\eta(G_{\alpha,\bar{f}} \cap A) \geq \alpha\mu(G_{\alpha,\bar{f}} \cap A)$. If $A \in \mathcal{F}_0$ is arbitrary, the same σ-finiteness argument as used before works, and hence (9.17) holds.

Therefore, to complete the proof of (a) [and hence of the entire claim] it suffices to demonstrate both (9.18) and (9.19).

We turn with an eye to establishing (9.18), picking $\alpha \in \mathbb{R}$. Make the following definitions for each $n \in \mathbb{N}$:

$$J_n = \{\omega \in \Omega : \min\{f_1(\omega), \cdots, f_n(\omega)\} < \alpha\},$$

$$J_{n,p} = \{\omega \in \Omega : f_p(\omega) < \alpha, f_{p+1}(\omega) \geq \alpha, \cdots, f_n(\omega) \geq \alpha\},\ p = 1, \cdots, n-1,$$

and

$$J_{n,n} = \{\omega \in \Omega : f_n(\omega) < \alpha\}.$$

The following claims are straightforward to verify:

(1) $J_{n,p} \in \mathcal{F}_p$ for every $n \in \mathbb{N}$ and $p \in \{1, \cdots, n\}$.

(2) For each $n \in \mathbb{N}$ the finite sequence $\{J_{n,p}\}_{p=1}^n$ is disjoint with union J_n.

(3) $J_1 \subseteq J_2 \subseteq \cdots$ and $\lim_n J_n = \bigcup_{n=1}^\infty J_n = M_\alpha$.

For arbitrary $A \in \mathcal{F}_0$ and $n \in \mathbb{N}$, we therefore have

$$\begin{aligned} \eta(J_n \cap A) &= \sum_{p=1}^n \eta(J_{n,p} \cap A) \\ &= \sum_{p=1}^n \eta(\{\omega \in \Omega : f_p(\omega) \leq \alpha\} \cap J_{n,p} \cap A) \\ &\leq \sum_{p=1}^n \alpha\mu(\{\omega \in \Omega : f_p(\omega) \leq \alpha\} \cap J_{n,p} \cap A) \\ &= \alpha \sum_{p=1}^n \mu(J_{n,p} \cap A) \\ &= \alpha\mu(J_n \cap A). \end{aligned}$$

[The second and third equalities hold since $\{\omega \in \Omega : f_p(\omega) \leq \alpha\} \cap J_{n,p} = J_{n,p}$ for each p in $\{1, \cdots, n\}$. The inequality follows by noting that f_p is a derivative of η_p with respect to μ_p for $p = 1, \cdots, n$ and using (b) of Claim 2.] Therefore, for our generic $A \in \mathcal{F}_0$, we have

$$\eta(J_n \cap A) \leq \alpha\mu(J_n \cap A) \quad \text{for every } n \in \mathbb{N}.$$

Next, we let $n \to \infty$ and invoke (3), obtaining

$$\begin{aligned}\eta(M_\alpha \cap A) = \eta\left(\lim_n (J_n \cap A)\right) = \lim_n \eta(J_n \cap A) &\leq \lim_n \alpha\mu(J_n \cap A)\\ &= \alpha\mu\left(\lim_n (J_n \cap A)\right)\\ &= \alpha\mu(M_\alpha \cap A)\end{aligned}$$

for our generic $A \in \mathcal{F}_0$. Since $\alpha \in \mathbb{R}$ is arbitrary, we have established (9.18).

We now turn our attention to establishing (9.19), picking $\alpha \in \mathbb{R}$ in the process. We now make the following new definitions for each $n \in \mathbb{N}$:

$$J_n = \{\omega \in \Omega : \max\{f_1(\omega), \cdots, f_n(\omega)\} > \alpha\},$$

$$J_{n,p} = \{\omega \in \Omega : f_p(\omega) > \alpha,\ f_{p+1}(\omega) \leq \alpha, \cdots,\ f_n(\omega) \leq \alpha\},\ p = 1, \cdots, n-1,$$

and

$$J_{n,n} = \{\omega \in \Omega : f_n(\omega) > \alpha\}.$$

The following claims may be verified in straightforward fashion:

(1′) For every $n \in \mathbb{N}$ and $p = 1, \cdots, n$ we have $J_{n,p} \in \mathcal{F}_p$.

(2′) For every $n \in \mathbb{N}$ the finite sequence $\{J_{n,p}\}_{p=1}^n$ is disjoint with union J_n.

(3′) $J_1 \subseteq J_2 \subseteq \cdots$ and $\lim_n J_n = \bigcup_{n=1}^\infty J_n = H_\alpha$.

Therefore, for arbitrary $A \in \mathcal{F}_0$ and $n \in \mathbb{N}$ we have

$$\begin{aligned}\eta(J_n \cap A) &= \sum_{p=1}^n \eta(J_{n,p} \cap A)\\ &= \sum_{p=1}^n \eta(\{\omega \in \Omega : f_p(\omega) \geq \alpha\} \cap J_{n,p} \cap A)\\ &\geq \sum_{p=1}^n \alpha\mu(\{\omega \in \Omega : f_p(\omega) \geq \alpha\} \cap J_{n,p} \cap A)\\ &= \alpha \sum_{p=1}^n \mu(J_{n,p} \cap A)\\ &= \alpha\mu(J_n \cap A).\end{aligned}$$

[The inequality holds by noting that f_p is a derivative of η_p with respect to μ_p for $p = 1, \cdots, n$ and using (a) of Claim 2.] For our generic $A \in \mathcal{F}_0$, we have shown that

$$\eta(J_n \cap A) \geq \alpha\mu(J_n \cap A) \qquad \text{for every } n \in \mathbb{N}.$$

Letting $n \to \infty$ in the above display and invoking (3′) gives

$$\eta(H_\alpha \cap A) \geq \alpha\mu(H_\alpha \cap A)$$

for our generic $A \in \mathcal{F}_0$. Recalling that $\alpha \in \mathbb{R}$ is arbitrary, we have established (9.19). This completes the entire proof. ■

The proofs of Claims 2, 5, and 6 are quite complex and somewhat forbidding. The reader should not be surprised if the proofs take multiple readings for full comprehension.

The primary use of Claims 5 and 6 [which claims are known as *martingale theorems*] will be for discussing certain limit theorems regarding integrals with respect to measures defined on certain types of product spaces. Claims 5 and 6 will be invoked for certain theorems in that discussion. Therefore, these somewhat abstruse claims may be thought of as nothing more than technical lemmas; they are [for our sakes] not ends in and of themselves. If the reader is curious about some immediate applications of Claims 5 and 6, he may consult (20.61)-(20.66) of Hewitt and Stromberg. As stated, we shall not need such results except for making certain claims in Chapter 11.

Exercises.

1*. Fill in the details for the outline concerning the proof of (9.12). [This should be straightforward if one understands the details concerning the proof of (9.11).]

2*. Prove (1)-(3) and (1′)-(3′) in Claim 6. [Again, this should be easy.]

3. Assume the framework of Claim 5. If $\eta_n \ll \mu_n$ for every $n \in \mathbb{N}$, and if for every E in the field $\bigcup_{k=1}^{\infty} \mathcal{F}_k$ we have $\lim_n \int_E f_n \, d\mu = \int_E f_\infty \, d\mu$, then $\eta_\infty \ll \mu_\infty$. Here f_∞ denotes a derivative of η_∞ with respect to μ_∞. [If $\eta_k \ll \mu_k$ for each $k \in \mathbb{N}$, we have $\eta_k = \eta_{k,ac}$ and $\eta_{k,s} = 0$. Fix $k \in \mathbb{N}$ and $E \in \mathcal{F}_k$; we have

$$\eta(E) = \eta_k(E) = \eta_{k,ac}(E) = \int_E f_k \, d\mu_k = \int_E f_k \, d\mu.$$

For every $n \geq k$, E is also in $\mathcal{F}_n$, so we may also write

$$\eta(E) = \eta_n(E) = \eta_{n,ac}(E) = \int_E f_n \, d\mu_n = \int_E f_n \, d\mu.$$

Let $n \to \infty$ above to obtain

$$\eta(E) = \lim_n \int_E f_n \, d\mu = \int_E f_\infty \, d\mu = \int_E f_\infty \, d\mu_\infty = \eta_{\infty,ac}(E);$$

the last equality follows because f_∞ is a derivative of η_∞ with respect to μ_∞. Therefore $\eta(E) = \eta_\infty(E) = \eta_{\infty,ac}(E)$ for every $E \in \mathcal{F}_k$. Since k is a generic positive integer, it follows that $\eta_\infty = \eta_{\infty,ac}$ on the field $\bigcup_{k=1}^{\infty} \mathcal{F}_k$. Now use the theorem on uniqueness in Section 3.5 to obtain $\eta_\infty = \eta_{\infty,ac}$ on $\mathcal{F}_\infty$.]

10

Products of Two Measure Spaces

In this chapter we are given two σ-finite measure spaces $(\Omega_1, \mathcal{F}_1, \mu_1)$ and $(\Omega_2, \mathcal{F}_2, \mu_2)$. Consider the set $\Omega = \Omega_1 \times \Omega_2$, the ordinary Cartesian product of Ω_1 and Ω_2:

$$\Omega = \Omega_1 \times \Omega_2 = \{(\omega_1, \omega_2) : \omega_1 \in \Omega_1,\ \omega_2 \in \Omega_2\}.$$

We wish to construct an appropriate σ-field $\mathcal{F}$ of subsets of $\Omega_1 \times \Omega_2$ and a measure μ on $\mathcal{F}$ such that

$$\mu(A_1 \times A_2) = \mu_1(A_1)\mu_2(A_2) \quad \text{for every } A_1 \in \mathcal{F}_1 \text{ and } A_2 \in \mathcal{F}_2.$$

Since we want to be able to apply μ to sets of the form $A_1 \times A_2$, where $A_1 \in \mathcal{F}_1$ and $A_2 \in \mathcal{F}_2$, the σ-field $\mathcal{F}$ on $\Omega_1 \times \Omega_2$ should contain $\mathcal{F}_1 \times \mathcal{F}_2$: $\mathcal{F}_1 \times \mathcal{F}_2 \subseteq \mathcal{F}$. We shall accordingly take $\mathcal{F}$ to be the σ-field on $\Omega_1 \times \Omega_2$ generated by $\mathcal{F}_1 \times \mathcal{F}_2$. All that remains is to show that such a μ as above exists. This is not trivial in any sense, but the work is generally straightforward. This construction will be given in detail in the first section. The resulting measure space $(\Omega, \mathcal{F}, \mu)$ will be called the *product* of $(\Omega_1, \mathcal{F}_1, \mu_1)$ and $(\Omega_2, \mathcal{F}_2, \mu_2)$.

The reader has already dealt with a special cases of the discussion above.

Example. Let $(\Omega_1, \mathcal{F}_1, \mu_1) = (\mathbb{R}^k, \mathcal{B}^k, \lambda_k)$, and let $(\Omega_2, \mathcal{F}_2, \mu_2) = (\mathbb{R}^m, \mathcal{B}^m, \lambda_m)$. Here $\mathcal{B}^k$ denotes the σ-field on $\mathbb{R}^k$ consisting of the k-dimensional Borel sets and λ_k denotes k-dimensional Lebesgue measure; $\mathcal{B}^m$ and λ_m denote the m-dimensional Borel sets and m-dimensional Lebesgue measure, respectively. Recall that the symbol $\mathcal{B}^k \otimes \mathcal{B}^m$ was defined to be $\sigma(\mathcal{B}^k \times \mathcal{B}^m)$; $\mathcal{B}^k \otimes \mathcal{B}^m$ is the smallest σ-field on $\mathbb{R}^{k+m}$ that contains all sets of the form $A \times B$, where $A \in \mathcal{B}^k$ and $B \in \mathcal{B}^m$. By

Claim 3 of Section 1.5, we have $\mathcal{B}^k \otimes \mathcal{B}^m = \mathcal{B}^{k+m}$. By Claim 4 of Section 4.2, it was shown that $\lambda_{k+m}(A \times B) = \lambda_k(A)\lambda_m(B)$ for every $A \in \mathcal{B}^k$ and $B \in \mathcal{B}^m$. Therefore, $(\mathbb{R}^{k+m}, \mathcal{B}^{k+m}, \lambda_{k+m})$ is the product of $(\mathbb{R}^k, \mathcal{B}^k, \lambda_k)$ and $(\mathbb{R}^m, \mathcal{B}^m, \lambda_m)$ in the sense described above.

After rigorously defining and constructing the product $(\Omega, \mathcal{F}, \mu)$ of two σ-finite measure spaces $(\Omega_1, \mathcal{F}_1, \mu_1)$ and $(\Omega_2, \mathcal{F}_2, \mu_2)$ as informally defined before, we desire to understand the behavior of integrals $\int_{\Omega_1 \times \Omega_2} f \, d\mu$ of $\mathcal{F}/\mathcal{B}^*$-measurable functions f on $\Omega_1 \times \Omega_2$ in terms of the operations $\int_{\Omega_1} \cdots d\mu_1$ and $\int_{\Omega_2} \cdots d\mu_2$. It will turn out to be the case that we can write something akin to

$$\int_{\Omega_1 \times \Omega_2} f \, d\mu = \int_{\Omega_1} \int_{\Omega_2} f \, d\mu_2 \, d\mu_1 = \int_{\Omega_2} \int_{\Omega_1} f \, d\mu_1 \, d\mu_2$$

under certain conditions on f. At this stage, the meaning of the symbols $\int_{\Omega_1} \int_{\Omega_2} f \, d\mu_2 \, d\mu_1$ and $\int_{\Omega_2} \int_{\Omega_1} f \, d\mu_1 \, d\mu_2$ has yet to be given, but the reader for the time being in this informal discussion may view these "iterated integrals" in the same way as is done with iterated Riemann integrals from calculus: perform the inner integration first, and then perform the outer integration. The specific description of this process is given in Section 2, where we discuss a collection of results that are all referred to by the name *Fubini's Theorem*, and the rigorous result follows fairly easily from the construction of product spaces in Section 1.

This introduction is brief and doesn't do the next section as much justice as desired. The main reason is that there are several little details that must be rigorously stated and demonstrated before we can even begin to carefully talk about the things that are stated quite imprecisely in this introduction. These details will either be interesting on their own or they will be a necessary evil to be endured while plodding to the main results. Either way, they must be handled.

After completion of first two section, some applications of product measures and Fubini's Theorem are presented. One application is to discussing Radon-Nikodym derivatives [see Exercise 15 in Section 2]. Another application involves obtaining a very general form of Minkowski's Inequality that subsumes the version given in Section 8.1. We shall also briefly discuss *convolutions*, using Fubini's Theorem as our main tool. Finally, we shall prove the very classical *Hardy-Littlewood Maximal Inequalities*, which show Fubini's Theorem in action in a Euclidean setting.

10.1 PRODUCT MEASURES

This somewhat technical section lays the groundwork for discussing the famous Fubini Theorems. The section will culminate with a rigorous definition of the *product* of two σ-finite measure spaces $(\Omega_1, \mathcal{F}_1, \mu_1)$ and $(\Omega_2, \mathcal{F}_2, \mu_2)$.

Framework. Ω_1 and Ω_2 will denote nonempty sets, with $\mathcal{F}_1$ and $\mathcal{F}_2$ denoting σ-fields of subsets of Ω_1 and Ω_2, respectively. The ordinary Cartesian product of $\mathcal{F}_1$ and $\mathcal{F}_2$ is denoted by $\mathcal{F}_1 \times \mathcal{F}_2$:

$$\mathcal{F}_1 \times \mathcal{F}_2 = \{E_1 \times E_2 : \ E_1 \in \mathcal{F}_1, \ E_2 \in \mathcal{F}_2\}.$$

We define the *product σ-field of $\mathcal{F}_1$ and $\mathcal{F}_2$* to be the σ-field on $\Omega_1 \times \Omega_2$ generated by $\mathcal{F}_1 \times \mathcal{F}_2$, and we denote this generated σ-field by $\mathcal{F}_1 \otimes \mathcal{F}_2$:

$$\mathcal{F}_1 \otimes \mathcal{F}_2 = \sigma(\mathcal{F}_1 \times \mathcal{F}_2).$$

Given $E \subseteq \Omega_1 \times \Omega_2$ and $\omega_1 \in \Omega_1$, we define the *ω_1-section of E* to be

$$E_{\omega_1} = \{\omega_2 \in \Omega_2 : \ (\omega_1, \omega_2) \in E\}.$$

Given $E \subseteq \Omega_1 \times \Omega_2$ and $\omega_2 \in \Omega_2$, we define the *ω_2-section of E* to be

$$E^{\omega_2} = \{\omega_1 \in \Omega_1 : \ (\omega_1, \omega_2) \in E\}.$$

Given $f : \Omega_1 \times \Omega_2 \to \bar{\mathbb{R}}$ and $\omega_1 \in \Omega_1$, we define $f_{\omega_1} : \Omega_2 \to \bar{\mathbb{R}}$ by writing

$$f_{\omega_1}(\omega_2) = f(\omega_1, \omega_2) \quad \text{for every } \omega_2 \in \Omega_2.$$

Given $f : \Omega_1 \times \Omega_2 \to \bar{\mathbb{R}}$ and $\omega_2 \in \Omega_2$, we define $f^{\omega_2} : \Omega_1 \to \bar{\mathbb{R}}$ by writing

$$f^{\omega_2}(\omega_1) = f(\omega_1, \omega_2) \quad \text{for every } \omega_1 \in \Omega_1.$$

With this notation at hand, we may now begin the fairly long but straightforward development of a product measure. Observe that we have not yet mentioned anything about measures.

Claim 1. *Let $\mathcal{A}$ denote the collection of all finite disjoint unions of $\mathcal{F}_1 \times \mathcal{F}_2$-sets. Then*
(a) $\mathcal{A}$ is a field on $\Omega_1 \times \Omega_2$, and
(b) $\sigma(\mathcal{A}) = \mathcal{F}_1 \otimes \mathcal{F}_2$.

Proof: For (a), we will show that (i) $\Omega_1 \times \Omega_2 \in \mathcal{A}$, (ii) $\mathcal{A}$ is closed under intersections, and (iii) $\mathcal{A}$ is closed under complementation. (i) is trivial. For (ii), let $E, F \in \mathcal{A}$; write $E = \bigcup_{i=1}^{n}(A_i \times B_i)$ and $F = \bigcup_{j=1}^{m}(C_i \times D_i)$, where we have that $A_1, \cdots, A_n, C_1, \cdots, C_m \in \mathcal{F}_1$, $B_1, \cdots, B_n, D_1, \cdots, D_m \in \mathcal{F}_2$, and both unions are disjoint. Then

$$\begin{aligned} E \cap F = \bigcup_{i=1}^{n}(A_i \times B_i) \cap \bigcup_{j=1}^{m}(C_j \times D_j) &= \bigcup_{i=1}^{n}\bigcup_{j=1}^{m}[(A_i \times B_i) \cap (C_j \times D_j)] \\ &= \bigcup_{i=1}^{n}\bigcup_{j=1}^{m}[(A_i \cap C_j) \times (B_i \times D_j)]. \end{aligned}$$

The set $E \cap F$ is clearly a finite union of $\mathcal{F}_1 \times \mathcal{F}_2$-sets. To see that the union is disjoint, pick distinct $(i,j),(i',j') \in \{1,\cdots,n\} \times \{1,\cdots,m\}$, say $i \neq i'$ without any loss of generality. Then

$$\begin{aligned} & [(A_i \cap C_j) \times (B_i \cap D_j)] \cap [(A_{i'} \cap C_{j'}) \times (B_{i'} \cap D_{j'})] \\ = \; & (A_i \cap C_j \cap A_{i'} \cap C_{j'}) \times (B_i \cap D_j \cap B_{i'} \cap D_{j'}) \\ \subseteq \; & (A_i \cap A_{i'}) \times (B_i \cap B_{i'}) \\ = \; & (A_i \times B_i) \cap (A_{i'} \times B_{i'}) \\ = \; & \varnothing \quad \text{since } A_1 \times B_1, \cdots, A_n \times B_n \text{ are disjoint.} \end{aligned}$$

It follows that $E \cap F \in \mathcal{A}$ whenever $E, F \in \mathcal{A}$, and (ii) is established. For (iii), let $E \in \mathcal{A}$, and use the notation from (ii). We have

$$E^c = \bigcap_{i=1}^{n} (A_i \times B_i)^c = \bigcap_{i=1}^{n} [(A_i^c \times B_i) \cup (\Omega_1 \times B_i^c)].$$

For each $i \in \{1,\cdots,n\}$, observe that $(A_i^c \times B_i) \cup (\Omega_1 \times B_i^c)$ is a finite disjoint union of $\mathcal{F}_1 \times \mathcal{F}_2$-sets, and hence is in $\mathcal{A}$. Since E^c is a finite intersection of $\mathcal{A}$-sets, (ii) shows that $E^c \in \mathcal{A}$. This gives (iii), and $\mathcal{A}$ is accordingly a field on $\Omega_1 \times \Omega_2$.

We now turn to (b). Clearly $\mathcal{F}_1 \times \mathcal{F}_2 \subseteq \mathcal{A}$, hence $\sigma(\mathcal{F}_1 \times \mathcal{F}_2) \subseteq \sigma(\mathcal{A})$, so that $\mathcal{F}_1 \otimes \mathcal{F}_2 \subseteq \sigma(\mathcal{A})$. Next, observe that every σ-field containing $\mathcal{F}_1 \times \mathcal{F}_2$ must also contain $\mathcal{A}$. In particular, $\mathcal{F}_1 \otimes \mathcal{F}_2$ is a σ-field that contains $\mathcal{F}_1 \times \mathcal{F}_2$, hence $\mathcal{A} \subseteq \mathcal{F}_1 \otimes \mathcal{F}_2$. This gives $\sigma(\mathcal{A}) \subseteq \sigma(\mathcal{F}_1 \otimes \mathcal{F}_2) = \mathcal{F}_1 \otimes \mathcal{F}_2$, which is the desired reverse inclusion. This completes the entire proof. ■

Claim 2. *Let $E \in \mathcal{F}_1 \otimes \mathcal{F}_2$. Then*
(a) $E_{\omega_1} \in \mathcal{F}_2$ for every $\omega_1 \in \Omega_1$, and
(b) $E^{\omega_2} \in \mathcal{F}_1$ for every $\omega_2 \in \Omega_2$.

Proof: We need only prove (a), as (b) follows by similar reasoning. Define

$$\mathcal{C} = \{E \in \mathcal{F}_1 \otimes \mathcal{F}_2 : \; E_{\omega_1} \in \mathcal{F}_2 \text{ for every } \omega_1 \in \Omega_1\}.$$

First observe that $\Omega_1 \times \Omega_2 \in \mathcal{F}_1 \times \mathcal{F}_2 \subseteq \mathcal{F}_1 \otimes \mathcal{F}_2$ and $(\Omega_1 \times \Omega_2)_{\omega_1} = \Omega_2 \in \mathcal{F}_2$ for every $\omega_1 \in \Omega_1$. Therefore, $\Omega_1 \times \Omega_2 \in \mathcal{C}$. Next, if $E \in \mathcal{C}$, then we have $E^c \in \mathcal{F}_1 \otimes \mathcal{F}_2$ and $(E^c)_{\omega_1} = (E_{\omega_1})^c \in \mathcal{F}_2$ for every $\omega_1 \in \Omega_1$, so that $E^c \in \mathcal{C}$. Next, if $\{E_n\}_{n=1}^{\infty}$ is a sequence of $\mathcal{C}$-sets, then $\bigcup_{n=1}^{\infty} E_n \in \mathcal{F}_1 \otimes \mathcal{F}_2$ and $(\bigcup_{n=1}^{\infty} E_n)_{\omega_1} = \bigcup_{n=1}^{\infty} (E_n)_{\omega_1} \in \mathcal{F}_2$ for every $\omega_1 \in \Omega_1$, whence $\bigcup_{n=1}^{\infty} E_n \in \mathcal{C}$. Therefore, $\mathcal{C}$ is a σ-field on $\Omega_1 \times \Omega_2$ and $\mathcal{C} \subseteq \mathcal{F}_1 \otimes \mathcal{F}_2$. We desire to strengthen this inclusion to an equality. To do this, let $E_1 \in \mathcal{F}_1$ and $E_2 \in \mathcal{F}_2$. Then for every $\omega_1 \in \Omega_1$ we have

$$(E_1 \times E_2)_{\omega_1} = \begin{cases} E_2 & \text{if } \omega_1 \in E_1 \\ \varnothing & \text{if } \omega_1 \notin E_1 \end{cases} \quad \in \mathcal{F}_2.$$

This shows that $\mathcal{F}_1 \times \mathcal{F}_2 \subseteq \mathcal{C}$. Since $\mathcal{C}$ is a σ-field, we have $\mathcal{F}_1 \otimes \mathcal{F}_2 \subseteq \mathcal{C}$. This yields $\mathcal{C} = \mathcal{F}_1 \otimes \mathcal{F}_2$. In other words, every set in $\mathcal{F}_1 \otimes \mathcal{F}_2$ has the property asserted by (a). This completes the entire proof. ■

Claim 3. *Let* $f : \Omega_1 \times \Omega_2 \to \bar{\mathbb{R}}$ *be* $\mathcal{F}_1 \otimes \mathcal{F}_2/\mathcal{B}^*$*-measurable. Then*

(a) f_{ω_1} *is* $\mathcal{F}_2/\mathcal{B}^*$*-measurable for every* $\omega_1 \in \Omega_1$*, and*

(b) f^{ω_2} *is* $\mathcal{F}_1/\mathcal{B}^*$*-measurable for every* $\omega_2 \in \Omega_2$*.*

In other words, the mapping $\omega_2 \mapsto f(\omega_1, \omega_2)$ *on* Ω_2 *is* $\mathcal{F}_2/\mathcal{B}^*$*-measurable for every* $\omega_1 \in \Omega_1$*, and the mapping* $\omega_1 \mapsto f(\omega_1, \omega_2)$ *on* Ω_1 *is* $\mathcal{F}_1/\mathcal{B}^*$*-measurable for every* $\omega_2 \in \Omega_2$*.*

Proof: As with the previous claim, statements (a) and (b) are symmetric and we therefore only need to prove (a). To do this, first consider the case of where $f = I_E$, where $E \in \mathcal{F}_1 \otimes \mathcal{F}_2$. Next, pick $\omega_1 \in \Omega_1$ and $\omega_2 \in \Omega_2$. We have

$$(I_E)_{\omega_1}(\omega_2) = 1 \;\leftrightarrow\; (\omega_1, \omega_2) \in E \;\leftrightarrow\; \omega_2 \in E_{\omega_1} \;\leftrightarrow\; I_{E_{\omega_1}}(\omega_2) = 1,$$

and hence $(I_E)_{\omega_1} = I_{E_{\omega_1}}$. By Claim 2, $E \in \mathcal{F}_1 \otimes \mathcal{F}_2$ forces $E_{\omega_1} \in \mathcal{F}_2$. Therefore, if $f = I_E$ where $E \in \mathcal{F}_1 \otimes \mathcal{F}_2$, we have that f_{ω_1} is the indicator function $I_{E_{\omega_1}}$ of the $\mathcal{F}_2$-set E_{ω_1}, and hence is $\mathcal{F}_2/\mathcal{B}^*$-measurable. This proves (a) when f is an indicator function on $\Omega_1 \times \Omega_2$ of a set in $\mathcal{F}_1 \otimes \mathcal{F}_2$.

Next, let $f = \sum_{i=1}^n c_i I_{E_i}$, where $E_1, \cdots, E_n \in \mathcal{F}_1 \otimes \mathcal{F}_2$ are disjoint with union $\Omega_1 \times \Omega_2$ and $c_1, \cdots, c_n \in \mathbb{R}$, so that f is an $\mathcal{F}_1 \otimes \mathcal{F}_2/\mathcal{B}^*$-measurable simple function on $\Omega_1 \times \Omega_2$. For every $\omega_1 \in \Omega_1$, observe that $f_{\omega_1} = \sum_{i=1}^n c_i I_{(E_i)_{\omega_1}}$, a finite linear combination of the indicator functions $(I_{E_1})_{\omega_1}, \cdots, (I_{E_n})_{\omega_1}$, and each of these is $\mathcal{F}_2/\mathcal{B}^*$-measurable by the previous paragraph. It follows that f_{ω_1} is $\mathcal{F}_2/\mathcal{B}^*$-measurable for every $\omega_1 \in \Omega_1$. This proves the result when f is an $\mathcal{F}_1 \otimes \mathcal{F}_2/\mathcal{B}^*$-measurable simple function on $\Omega_1 \times \Omega_2$.

Next, suppose that f is a nonnegative $\mathcal{F}_1 \otimes \mathcal{F}_2/\mathcal{B}^*$-measurable function on $\Omega_1 \times \Omega_2$. There exists a nondecreasing sequence $\{s_n\}_{n=1}^\infty$ of nonnegative finite-valued $\mathcal{F}_1 \otimes \mathcal{F}_2/\mathcal{B}^*$-measurable simple functions on $\Omega_1 \times \Omega_2$ with $\lim_n s_n = f$. By the previous paragraph, we have that $(s_n)_{\omega_1}$ is $\mathcal{F}_2/\mathcal{B}^*$-measurable for every $\omega_1 \in \Omega_1$ and $n \in \mathbb{N}$. Since $f_{\omega_1} = (\lim_n s_n)_{\omega_1} = \lim_n (s_n)_{\omega_1}$ for every $\omega_1 \in \Omega_1$, we have that f_{ω_1} is the limit of a sequence of $\mathcal{F}_2/\mathcal{B}^*$-measurable functions and hence is itself $\mathcal{F}_2/\mathcal{B}^*$-measurable. This proves the result when f is a nonnegative $\mathcal{F}_1 \otimes \mathcal{F}_2/\mathcal{B}^*$-measurable function on $\Omega_1 \times \Omega_2$.

Finally, if f is a general $\mathcal{F}_1 \otimes \mathcal{F}_2/\mathcal{B}^*$-measurable function on $\Omega_1 \times \Omega_2$, then the functions f^+ and f^- [the usual positive and negative parts of f], both being nonnegative $\mathcal{F}_1 \otimes \mathcal{F}_2/\mathcal{B}^*$-measurable functions on $\Omega_1 \times \Omega_2$, are such that $(f^+)_{\omega_1}$ and $(f^-)_{\omega_1}$ are $\mathcal{F}_2/\mathcal{B}^*$-measurable for every $\omega_1 \in \Omega_1$. Observing that $f_{\omega_1} = (f^+ - f^-)_{\omega_1} = (f^+)_{\omega_1} - (f^-)_{\omega_1}$ for every $\omega_1 \in \Omega_1$, we see for every $\omega_1 \in \Omega_1$ that f_{ω_1} is the difference of two $\mathcal{F}_2/\mathcal{B}^*$-measurable functions on $\Omega_1 \times \Omega_2$ and hence is $\mathcal{F}_2/\mathcal{B}^*$-measurable. This completes the proof. ■

The following definition and claim are reviews of some concepts found in Exercises 1 and 21 of Section 1.3. We are slighly redundant here for the sake of completeness.

Definition. Let Ω denote a nonempty set. A *monotone class [on Ω]* is a collection $\mathcal{M}$ of subsets of Ω with the following two properties:
(a) If $\{E_n\}_{n=1}^{\infty}$ is a nondecreasing sequence of $\mathcal{M}$-sets, then $\bigcup_{n=1}^{\infty} E_n \in \mathcal{M}$.
(b) If $\{E_n\}_{n=1}^{\infty}$ is a nonincreasing sequence of $\mathcal{M}$-sets, then $\bigcap_{n=1}^{\infty} E_n \in \mathcal{M}$.

The point of this definition is that whenever $\{E_n\}_{n=1}^{\infty}$ is a monotone sequence of $\mathcal{M}$-sets, we have $\lim_n E_n \in \mathcal{M}$. The following facts are already known or are straightforward to verify:

(1) Every σ-field is a monotone class, and every field that is a monotone class is a σ-field.
(2) An arbitrary intersection of monotone classes on Ω is itself a monotone class on Ω.
(3) At least two monotone classes on Ω exist: the trivial monotone class $\{\varnothing, \Omega\}$ and the collection 2^{Ω}. Given any collection $\mathcal{F}$ of subsets on Ω, there exists at least one monotone class on Ω that contains $\mathcal{F}$, namely, 2^{Ω}.
(4) Given any collection $\mathcal{F}$ of subsets of Ω, the intersection of all monotone classes on Ω that contain $\mathcal{F}$ is a monotone class that contains $\mathcal{F}$. Furthermore, this constructed monotone class is the smallest monotone class on Ω that contains $\mathcal{F}$.

We next state what is known as *Halmos' Monotone Class Theorem.* It is a strictly set-theoretic fact that we will require in proving Claim 5.

Claim 4. [Halmos' Monotone Class Theorem.] *Let Ω denote a set that is nonempty, and let $\mathcal{F}$ and $\mathcal{M}$ denote, respectively, a field on Ω and a monotone class on Ω. If $\mathcal{F} \subseteq \mathcal{M}$, then $\sigma(\mathcal{F}) \subseteq \mathcal{M}$.*

Proof: Let $m(\mathcal{F})$ denote the minimal monotone class on Ω generated by $\mathcal{F}$ as described in (4) above. We clearly have $\mathcal{F} \subseteq m(\mathcal{F}) \subseteq \mathcal{M}$, hence it is sufficient to show that $\sigma(\mathcal{F}) \subseteq m(\mathcal{F})$. Let us show that $m(\mathcal{F})$ is a field; it will then follow by (1) above that $m(\mathcal{F})$ is a σ-field containing $\mathcal{F}$. Since $\sigma(\mathcal{F})$ is the minimal σ-field that contains $\mathcal{F}$, we will have that $\sigma(\mathcal{F}) \subseteq m(\mathcal{F})$. To complete the proof then, we need only show that $m(\mathcal{F})$ is a field: $\Omega \in m(\mathcal{F})$, $m(\mathcal{F})$ is closed under complementation, and $m(\mathcal{F})$ is closed under finite unions. Since $\Omega \in \mathcal{F} \subseteq m(\mathcal{F})$, we only need to establish the closure of $m(\mathcal{F})$ under complementation and finite unions.

We now turn to complementation. Let $\mathcal{G} = \{A \subseteq \Omega : A^c \in m(\mathcal{F})\}$. Since $A \in \mathcal{F}$ forces $A^c \in \mathcal{F} \subseteq m(\mathcal{F})$, we have $\mathcal{F} \subseteq \mathcal{G}$. Next, let $\{E_n\}_{n=1}^{\infty}$ denote a monotone sequence of $\mathcal{G}$-sets. Then $\{E_n^c\}_{n=1}^{\infty}$ is a monotone sequence of $m(\mathcal{F})$-sets. Since $m(\mathcal{F})$ is a monotone class, we have $(\lim_n E_n)^c = \lim_n E_n^c \in m(\mathcal{F})$, so that $\lim_n E_n \in \mathcal{G}$. This shows that $\mathcal{G}$ is a monotone class on Ω that contains $\mathcal{F}$. By the minimality of $m(\mathcal{F})$, we have $m(\mathcal{F}) \subseteq \mathcal{G}$. Using this, we may now show that $m(\mathcal{F})$ is closed under complementation: $A \in m(\mathcal{F})$ forces $A \in \mathcal{G}$, and hence $A^c \in m(\mathcal{F})$.

For closure under unions, a little more work is required. Define

$$\mathcal{G}_1 = \{A \subseteq \Omega : A \cup B \in m(\mathcal{F}) \text{ for all } B \in \mathcal{F}\}.$$

Observe first that if $A, B \in \mathcal{F}$, then $A \cup B \in \mathcal{F} \subseteq m(\mathcal{F})$, hence $A \in \mathcal{G}_1$, so that $\mathcal{G}_1$ contains $\mathcal{F}$. Next, if $\{E_n\}_{n=1}^{\infty}$ is a monotone sequence of $\mathcal{G}_1$-sets, then $\{E_n \cup B\}_{n=1}^{\infty}$ is a monotone sequence of $m(\mathcal{F})$-sets for every $B \in \mathcal{F}$. It follows that $(\lim_n E_n) \cup B = \lim_n (E_n \cup B) \in m(\mathcal{F})$ since $m(\mathcal{F})$ is a monotone class, which gives $\lim_n E_n \in \mathcal{G}_1$. To summarize, $\mathcal{G}_1$ is a monotone class with $\mathcal{F} \subseteq \mathcal{G}_1$; we invoke the minimality of $m(\mathcal{F})$ once again to obtain $m(\mathcal{F}) \subseteq \mathcal{G}_1$.

Next, define

$$\mathcal{G}_2 = \{B \subseteq \Omega : A \cup B \in m(\mathcal{F}) \text{ for all } A \in m(\mathcal{F})\}.$$

If $B \in \mathcal{F}$ and $A \in m(\mathcal{F}) \subseteq \mathcal{G}_1$, then $A \cup B \in m(\mathcal{F})$; this gives $\mathcal{F} \subseteq \mathcal{G}_2$. Next, if $\{E_n\}_{n=1}^{\infty}$ is a monotone sequence of $\mathcal{G}_2$-sets, then $\{A \cup E_n\}_{n=1}^{\infty}$ is a monotone sequence of $m(\mathcal{F})$-sets for every $A \in m(\mathcal{F})$. Since $m(\mathcal{F})$ is a monotone class, we have that $A \cup \lim_n E_n = \lim_n (A \cup E_n) \in m(\mathcal{F})$ for every $A \in m(\mathcal{F})$, and hence $\lim_n E_n \in \mathcal{G}_2$. This shows that $\mathcal{G}_2$ is a monotone class containing $\mathcal{F}$. We invoke the minimality of $m(\mathcal{F})$ yet again to obtain $m(\mathcal{F}) \subseteq \mathcal{G}_2$.

With this, we may now show that $m(\mathcal{F})$ is closed under finite unions, and hence is a field. Let $A, B \in m(\mathcal{F})$. Then $B \in \mathcal{G}_2$, hence $A \cup B \in m(\mathcal{F})$. This completes the proof. ∎

The upcoming claim is long but straightforward, and is the main key to constructing product measures. It will also be a foundational piece of work in the upcoming Fubini Theorems. We have not as of yet in this section dealt with any measures, as our work has been strictly set-theoretic up to the present. We now bring measures into the mix.

Claim 5. *Let $(\Omega_1, \mathcal{F}_1, \mu_1)$ and $(\Omega_2, \mathcal{F}_2, \mu_2)$ denote σ-finite measure spaces, and pick $E \in \mathcal{F}_1 \otimes \mathcal{F}_2$. Define F on Ω_1 and G on Ω_2 by writing*

$$F(\omega_1) = \mu_2(E_{\omega_1}) \quad \text{and} \quad G(\omega_2) = \mu_1(E^{\omega_2})$$

for every $\omega_1 \in \Omega_1$ and $\omega_2 \in \Omega_2$. Then we have that
(a) F is nonnegative and $\mathcal{F}_1/\mathcal{B}^*$-measurable [hence $\int_{\Omega_1} F\, d\mu_1$ exists],
(b) G is nonnegative and $\mathcal{F}_2/\mathcal{B}^*$-measurable [hence $\int_{\Omega_2} G\, d\mu_2$ exists], and
(c) $\int_{\Omega_1} F\, d\mu_1 = \int_{\Omega_2} G\, d\mu_2$. That is,

$$\int_{\Omega_1} \mu_2(E_{\omega_1})\, d\mu_1(\omega_1) = \int_{\Omega_2} \mu_1(E^{\omega_2})\, d\mu_2(\omega_2).$$

Proof: Before doing anything, observe that Claim 2 ensures that the definitions of F and G make sense. Also, F and G are clearly nonnegative since μ_1 and μ_2 are measures, so the work involves establishing the two measurability claims and the integral identity. The idea of the proof is this: define

$$\mathcal{F} = \{E \in \mathcal{F}_1 \otimes \mathcal{F}_2 : \text{(a), (b), and (c) hold for } E\} \quad [\subseteq \mathcal{F}_1 \otimes \mathcal{F}_2]$$

and

$$\mathcal{A} = \text{ the field of finite disjoint unions of } \mathcal{F}_1 \times \mathcal{F}_2\text{-sets [see Claim 1]},$$

and then show that $\mathcal{F}$ is a monotone class with $\mathcal{A} \subseteq \mathcal{F}$. From Claims 1 and 4, it will then follow that $\mathcal{F}_1 \otimes \mathcal{F}_2 = \sigma(\mathcal{A}) \subseteq \mathcal{F}$, which will give $\mathcal{F}_1 \otimes \mathcal{F}_2 = \mathcal{F}$, completing the proof. Therefore, our proof consists in demonstrating that $\mathcal{A} \subseteq \mathcal{F}$ and $\mathcal{F}$ is a monotone class. These two claims will in turn be broken up into various subclaims.

(i) $\mathcal{F}_1 \times \mathcal{F}_2 \subseteq \mathcal{F}$.

Proof: Let $A_1 \in \mathcal{F}_1$ and $A_2 \in \mathcal{F}_2$; we want $A_1 \times A_2 \in \mathcal{F}$. Observe for every $\omega_1 \in \Omega_1$ and $\omega_2 \in \Omega_2$ that

$$(A_1 \times A_2)_{\omega_1} = \begin{cases} A_2 & \omega_1 \in A_1 \\ \varnothing & \omega_1 \notin A_1 \end{cases} \quad \text{and} \quad (A_1 \times A_2)^{\omega_2} = \begin{cases} A_1 & \omega_2 \in A_2 \\ \varnothing & \omega_2 \notin A_2, \end{cases}$$

so that

$$\mu_2((A_1 \times A_2)_{\omega_1}) = \mu_2(A_2) I_{A_1}(\omega_1) \quad \text{and} \quad \mu_1((A_1 \times A_2)^{\omega_2}) = \mu_1(A_1) I_{A_2}(\omega_2).$$

I_{A_1} is $\mathcal{F}_1/\mathcal{B}^*$-measurable and I_{A_2} is $\mathcal{F}_2/\mathcal{B}^*$-measurable, so $F = \mu_2(A_2)I_{A_1}$ is $\mathcal{F}_1/\mathcal{B}^*$-measurable and $G = \mu_1(A_1)I_{A_2}$ is $\mathcal{F}_2/\mathcal{B}^*$-measurable. This shows that $A_1 \times A_2$ satisfies (a) and (b). Next, we have that

$$\int_{\Omega_1} \mu_2(A_2) I_{A_1}(\omega_1)\, d\mu_1(\omega_1) = \mu_1(A_1)\mu_2(A_2) = \int_{\Omega_2} \mu_1(A_1) I_{A_2}(\omega_2)\, d\mu_2(\omega_2),$$

so that $A_1 \times A_2$ satisfies (c). This gives $A_1 \times A_2 \in \mathcal{F}$, as required.

(ii) $\mathcal{A} \subseteq \mathcal{F}$.

Proof: Let $E \in \mathcal{A}$; write E as the disjoint union of $E_1, \cdots, E_n \in \mathcal{F}_1 \times \mathcal{F}_2$. By (i), we have that the mapping $\omega_1 \mapsto \mu_2((E_i)_{\omega_1})$ is $\mathcal{F}_1/\mathcal{B}^*$-measurable, the mapping $\omega_2 \mapsto \mu_1((E_i)^{\omega_2})$ is $\mathcal{F}_2/\mathcal{B}^*$-measurable, and

$$\int_{\Omega_1} \mu_2((E_i)_{\omega_1})\, d\mu_1(\omega_1) = \int_{\Omega_2} \mu_1((E_i)^{\omega_2})\, d\mu_2(\omega_2), \quad i = 1, \cdots, n.$$

It follows that

$$\omega_1 \mapsto \sum_{i=1}^{n} \mu_2((E_i)_{\omega_1}) = \mu_2\left(\bigcup_{i=1}^{n} (E_i)_{\omega_1}\right) = \mu_2\left(\left(\bigcup_{i=1}^{n} E_i\right)_{\omega_1}\right) = \mu_2(E_{\omega_1}),$$

being a sum of $\mathcal{F}_1/\mathcal{B}^*$-measurable mappings, is itself $\mathcal{F}_1/\mathcal{B}^*$-measurable. By the same reasoning,

$$\omega_2 \mapsto \sum_{i=1}^{n} \mu_1((E_i)^{\omega_2}) = \mu_1\left(\bigcup_{i=1}^{n} (E_i)^{\omega_2}\right) = \mu_1\left(\left(\bigcup_{i=1}^{n} E_i\right)^{\omega_2}\right) = \mu_1(E^{\omega_2})$$

is $\mathcal{F}_2/\mathcal{B}^*$-measurable. The set $E \in \mathcal{A}$ accordingly satisfies (a) and (b). For (c), we must show that E satisfies the equation

$$\int_{\Omega_1} \mu_2(E_{\omega_1})\, d\mu_1(\omega_1) = \int_{\Omega_2} \mu_1(E^{\omega_2})\, d\mu_2(\omega_2).$$

Recalling that

$$\mu_2(E_{\omega_1}) = \sum_{i=1}^{n} \mu_2((E_i)_{\omega_1}) \quad \text{and} \quad \mu_1(E^{\omega_2}) = \sum_{i=1}^{n} \mu_1((E_i)^{\omega_2})$$

for every $\omega_1 \in \Omega_1$ and $\omega_2 \in \Omega_2$, we have

$$\begin{aligned}
\int_{\Omega_1} \mu_2(E_{\omega_1})\, d\mu_1(\omega_1) &= \int_{\Omega_1} \sum_{i=1}^{n} \mu_2((E_i)_{\omega_1})\, d\mu_1(\omega_1) \\
&= \sum_{i=1}^{n} \int_{\Omega_1} \mu_2((E_i)_{\omega_1})\, d\mu_1(\omega_1) \\
&= \sum_{i=1}^{n} \int_{\Omega_2} \mu_1((E_i)^{\omega_2})\, d\mu_2(\omega_2) \\
&= \int_{\Omega_2} \sum_{i=1}^{n} \mu_1((E_i)^{\omega_2})\, d\mu_2(\omega_2) \\
&= \int_{\Omega_2} \mu_1(E^{\omega_2})\, d\mu_2(\omega_2).
\end{aligned}$$

It follows that the set $E \in \mathcal{A}$ satisfies (c), and hence $E \in \mathcal{F}$. Therefore, $\mathcal{A} \subseteq \mathcal{F}$.

The rest of the proof is devoted to showing that $\mathcal{F}$ is a monotone class. That is, we must show that if $\{E_n\}_{n=1}^{\infty}$ is a nondecreasing or nonincreasing sequence of $\mathcal{F}$-sets, then $\lim_n E_n \in \mathcal{F}$. Subclaim (iii) immediately below deals with the nondecreasing situation. Subclaims (iv) and (v) give the buildup for the more difficult nonincreasing situation, which is proved in (vi).

(iii) If $\{E_n\}_{n=1}^{\infty}$ is a nondecreasing sequence of $\mathcal{F}$-sets, then $\lim_n E_n \in \mathcal{F}$.

Proof: Let $\omega_1 \in \Omega_1$, and observe that $\{(E_n)_{\omega_1}\}_{n=1}^{\infty}$ is a nondecreasing sequence of $\mathcal{F}_2$-sets. We have

$$\left(\lim_n E_n\right)_{\omega_1} = \left(\bigcup_{n=1}^{\infty} E_n\right)_{\omega_1} = \bigcup_{n=1}^{\infty} (E_n)_{\omega_1} = \lim_n (E_n)_{\omega_1},$$

and hence the continuity of μ_2 from below [(M8) from Section 2.2] gives

$$F(\omega_1) = \mu_2\left(\left(\lim_n E_n\right)_{\omega_1}\right) = \mu_2\left(\lim_n (E_n)_{\omega_1}\right) = \lim_n \mu_2((E_n)_{\omega_1}).$$

Since each $E_n \in \mathcal{F}$ for each $n \in \mathbb{N}$, the mapping $\omega_1 \mapsto \mu_2((E_n)_{\omega_1})$ is $\mathcal{F}_1/\mathcal{B}^*$-measurable for each $n \in \mathbb{N}$. Since F is the pointwise limit of a sequence of $\mathcal{F}_1/\mathcal{B}^*$-measurable mappings, it follows that F itself is $\mathcal{F}_1/\mathcal{B}^*$-measurable. It follows that $\lim_n E_n$ satisfies (a). By the exact same reasoning, we have for every $\omega_2 \in \Omega_2$ that $(\lim_n E_n)^{\omega_2} = \lim_n (E_n)^{\omega_2}$, $G(\omega_2) = \lim_n \mu_1((E_n)^{\omega_2})$, and

G is $\mathcal{F}_2/\mathcal{B}^*$-measurable, so that $\lim_n E_n$ satisfies (b). For (c), we must show that

$$\int_{\Omega_1} \mu_2\left(\left(\lim_n E_n\right)_{\omega_1}\right) d\mu_1(\omega_1) = \int_{\Omega_2} \mu_1\left(\left(\lim_n E_n\right)^{\omega_2}\right) d\mu_2(\omega_2).$$

Since $E_n \in \mathcal{F}$ for every $n \in \mathbb{N}$, we have

$$\int_{\Omega_1} \mu_2((E_n)_{\omega_1})\, d\mu_1(\omega_1) = \int_{\Omega_2} \mu_1((E_n)^{\omega_2})\, d\mu_2(\omega_2) \quad \text{for every } n \in \mathbb{N}.$$

Twice using MCT, we have

$$\begin{aligned}
\int_{\Omega_1} \mu_2\left(\left(\lim_n E_n\right)_{\omega_1}\right) d\mu_1(\omega_1) &= \int_{\Omega_1} \lim_n \mu_2((E_n)_{\omega_1})\, d\mu_1(\omega_1) \\
&= \lim_n \int_{\Omega_1} \mu_2((E_n)_{\omega_1})\, d\mu_1(\omega_1) \\
&= \lim_n \int_{\Omega_2} \mu_1((E_n)^{\omega_2})\, d\mu_2(\omega_2) \\
&= \int_{\Omega_2} \lim_n \mu_1((E_n)^{\omega_2})\, d\mu_2(\omega_2) \\
&= \int_{\Omega_2} \mu_1\left(\left(\lim_n E_n\right)^{\omega_2}\right) d\mu_2(\omega_2).
\end{aligned}$$

This shows that $\lim_n E_n$ satisfies (c), and hence $\lim_n E_n \in \mathcal{F}$.

To complete matters, we must show that $\mathcal{F}$ is closed under intersections of nonincreasing sequences of $\mathcal{F}$-sets. Unlike the consideration of unions of nondecreasing sequences given just above, this requires some preliminary work.

(iv) Let $\{H_n\}_{n=1}^{\infty}$ denote a nonincreasing sequence of sets in $\mathcal{F}$ such that $H_1 \subseteq A_1 \times A_2$ for some $A_1 \in \mathcal{F}_1$ and $A_2 \in \mathcal{F}_2$ with $\mu_1(A_1), \mu_2(A_2) < +\infty$. Then $\lim_n H_n \in \mathcal{F}$.

Proof: For notational purposes, write H for $\lim_n H_n = \bigcap_{n=1}^{\infty} H_n$. For every $\omega_1 \in \Omega_1$, observe that $\{(H_n)_{\omega_1}\}_{n=1}^{\infty}$ is a nonincreasing sequence in $\mathcal{F}_2$ and

$$H_{\omega_1} = \left(\bigcap_{n=1}^{\infty} H_n\right)_{\omega_1} = \bigcap_{n=1}^{\infty} (H_n)_{\omega_1} = \lim_n (H_n)_{\omega_1}.$$

Since for every $\omega_1 \in \Omega_1$ we have

$$\mu_2((H_1)_{\omega_1}) \le \mu_2((A_1 \times A_2)_{\omega_1}) = \mu_2(A_2) I_{A_1}(\omega_1) \le \mu_2(A_2) < +\infty,$$

we may invoke property (M9) from Section 2.2 to write

$$F(\omega_1) = \mu_2(H_{\omega_1}) = \mu_2\left(\lim_n (H_n)_{\omega_1}\right) = \lim_n \mu_2((H_n)_{\omega_1}).$$

Since $H_n \in \mathcal{F}$ for each $n \in \mathbb{N}$, we have that the mapping $\omega_1 \mapsto \mu_2((H_n)_{\omega_1})$ is $\mathcal{F}_1/\mathcal{B}^*$-measurable for each $n \in \mathbb{N}$. It follows that F, being the pointwise limit of $\mathcal{F}_1/\mathcal{B}^*$-measurable mappings, is itself $\mathcal{F}_1/\mathcal{B}^*$-measurable, so H satisfies (a). By the same type of argument, it may be seen for every $\omega_2 \in \Omega_2$ that

$$H^{\omega_2} = \lim_n (H_n)^{\omega_2} \quad \text{and} \quad \mu_1((H_1)^{\omega_2}) < +\infty,$$

so again (M9) gives

$$G(\omega_2) = \mu_1(H^{\omega_2}) = \mu_1\left(\lim_n (H_n)^{\omega_2}\right) = \lim_n \mu_1((H_n)^{\omega_2}),$$

whence G is $\mathcal{F}_2/\mathcal{B}^*$-measurable, so that H satisfies (b). For (c), the assumption that $H_n \in \mathcal{F}$ for each $n \in \mathbb{N}$ gives

$$\int_{\Omega_1} \mu_2((H_n)_{\omega_1})\, d\mu_1(\omega_1) = \int_{\Omega_2} \mu_1((H_n)^{\omega_2})\, d\mu_2(\omega_2) \quad \text{for each } n \in \mathbb{N},$$

and hence

$$\begin{aligned}
\int_{\Omega_1} F(\omega_1)\, d\mu_1(\omega_1) &= \int_{\Omega_1} \lim_n \mu_2((H_n)_{\omega_1})\, d\mu_1(\omega_1) \\
&= \lim_n \int_{\Omega_1} \mu_2((H_n)_{\omega_1})\, d\mu_1(\omega_1) \\
&= \lim_n \int_{\Omega_2} \mu_1((H_n)^{\omega_2})\, d\mu_2(\omega_2) \\
&= \int_{\Omega_2} \lim_n \mu_1((H_n)^{\omega_2})\, d\mu_2(\omega_2) \\
&= \int_{\Omega_2} G(\omega_2)\, d\mu_2(\omega_2).
\end{aligned}$$

[The second equality holds by DCT:

$$\mu_2((H_n)_{\omega_1}) \le \mu_2((H_1)_{\omega_1}) \le \mu_2((A_1 \times A_2)_{\omega_1}) = \mu_2(A_2) I_{A_1}(\omega_1)$$

for every $n \in \mathbb{N}$ and $\omega_1 \in \Omega_1$, and

$$\int_{\Omega_1} \mu_2(A_2) I_{A_1}(\omega_1)\, d\mu_1(\omega_1) = \mu_1(A_1)\mu_2(A_2) < +\infty.$$

The fourth equality follows by a similar argument employing DCT.] This shows that $H = \lim_n H_n$ satisfies (c), and accordingly $H = \lim_n H_n \in \mathcal{F}$.

In the proof of (iv) above, note the importance of the finiteness assumption. The rest of the proof consists in showing that the finiteness assumption may be removed.

We have not yet used the assumption of σ-finiteness. We do so now, obtaining nondecreasing sequences $\{A_n\}_{n=1}^{\infty}$ and $\{B_n\}_{n=1}^{\infty}$ of $\mathcal{F}_1$-sets and $\mathcal{F}_2$-sets, respectively, such that $\Omega_1 = \bigcup_{n=1}^{\infty} A_n = \lim_n A_n$, $\Omega_2 = \bigcup_{n=1}^{\infty} B_n = \lim_n B_n$, and $\mu_1(A_n), \mu_2(B_n) < +\infty$ for each $n \in \mathbb{N}$. Next, define the collection

$$\mathcal{E} = \{E \in \mathcal{F}_1 \otimes \mathcal{F}_2 : \ E \cap (A_n \times B_n) \in \mathcal{F} \text{ for each } n \in \mathbb{N}\}.$$

This notation shall be used for the remainder of the proof.

(v) $\mathcal{E} = \mathcal{F}_1 \otimes \mathcal{F}_2$.

Proof: It is trivial that $\mathcal{E} \subseteq \mathcal{F}_1 \otimes \mathcal{F}_2$. We desire $\mathcal{F}_1 \otimes \mathcal{F}_2 \subseteq \mathcal{E}$, that is, $\sigma(\mathcal{A}) \subseteq \mathcal{E}$. Utilizing Claim 4, it is sufficient to show that $\mathcal{A} \subseteq \mathcal{E}$ and $\mathcal{E}$ is a monotone class. [Recall that $\mathcal{A}$ is a field by Claim 1.]

We first show that $\mathcal{A} \subseteq \mathcal{E}$. Let $\bigcup_{i=1}^{p}(C_i \times D_i)$ denote an element of $\mathcal{A}$, so that $C_1 \times D_1, \cdots, C_p \times D_p$ are disjoint $\mathcal{F}_1 \times \mathcal{F}_2$-sets. We want to show that $[\bigcup_{i=1}^{p}(C_i \times D_i)] \cap (A_n \times B_n) \in \mathcal{F}$ for every $n \in \mathbb{N}$. This means showing for every $n \in \mathbb{N}$ that $\bigcup_{i=1}^{p}[(C_i \cap A_n) \times (D_i \cap B_n)] \in \mathcal{F}$. Now for every $n \in \mathbb{N}$ the disjointness of $C_1 \times D_1, \cdots, C_p \times D_p$ forces the disjointness of the $\mathcal{F}_1 \times \mathcal{F}_2$-sets $(C_1 \cap A_n) \times (D_1 \cap B_n), \cdots, (C_p \cap A_n) \times (D_p \cap B_n)$. This shows for every $n \in \mathbb{N}$ that the finite union $\bigcup_{i=1}^{p}[(C_i \cap A_n) \times (D_i \cap B_n)]$ of $\mathcal{F}_1 \times \mathcal{F}_2$-sets is disjoint. This gives $\bigcup_{i=1}^{p}[(C_i \cap A_n) \times (D_i \cap B_n)] \in \mathcal{A}$ for each $n \in \mathbb{N}$ [by the definition of $\mathcal{A}$]. By (ii), we have $\mathcal{A} \subseteq \mathcal{F}$, and hence $\bigcup_{i=1}^{p}[(C_i \cap A_n) \times (D_i \cap B_n)] \in \mathcal{F}$ for every $n \in \mathbb{N}$. This gives $\bigcup_{i=1}^{p}(C_i \times D_i) \in \mathcal{E}$, showing that $\mathcal{A} \subseteq \mathcal{E}$.

We now turn to showing that $\mathcal{E}$ is a monotone class. Let $\{E_p\}_{p=1}^{\infty}$ denote a nondecreasing sequence of $\mathcal{E}$-sets; we want $\lim_p E_p = \bigcup_{p=1}^{\infty} E_p \in \mathcal{E}$. To see this, we have for every $n \in \mathbb{N}$ that

$$\bigcup_{p=1}^{\infty} E_p \ \cap \ (A_n \times B_n) = \bigcup_{p=1}^{\infty}[E_p \cap (A_n \times B_n)] = \lim_p [E_p \cap (A_n \times B_n)].$$

Now for each $n \in \mathbb{N}$ we have that $\{E_p \cap (A_n \times B_n)\}_{p=1}^{\infty}$ is a nondecreasing sequence of $\mathcal{F}$-sets, hence $\lim_p[E_p \cap (A_n \times B_n)] \in \mathcal{F}$ by (iii). It follows that $\bigcup_{p=1}^{\infty} E_p \ \cap \ (A_n \times B_n) \in \mathcal{F}$ for each $n \in \mathbb{N}$, and hence $\lim_p E_p \in \mathcal{E}$.

We complete the proof of the fact that $\mathcal{E}$ is a monotone class. Let $\{E_p\}_{p=1}^{\infty}$ denote a nonincreasing sequence of $\mathcal{E}$-sets; we want $\lim_p E_p = \bigcap_{p=1}^{\infty} E_p \in \mathcal{E}$. For every $n \in \mathbb{N}$ we have

$$\bigcap_{p=1}^{\infty} E_p \ \cap \ (A_n \times B_n) = \bigcap_{p=1}^{\infty}[E_p \cap (A_n \times B_n)] = \lim_p [E_p \cap (A_n \times B_n)].$$

Now for each $n \in \mathbb{N}$ we have that $\{E_p \cap (A_n \times B_n)\}_{p=1}^{\infty}$ is a nonincreasing sequence of $\mathcal{F}$-sets with $E_1 \cap (A_n \times B_n) \subseteq A_n \times B_n$ and $\mu_1(A_n), \mu_2(B_n) < +\infty$. By (iv), it follows that $[\bigcap_{p=1}^{\infty} E_p] \cap (A_n \times B_n) = \lim_p[E_p \cap (A_n \times B_n)] \in \mathcal{F}$ for each $n \in \mathbb{N}$. This shows that $\lim_p E_p \in \mathcal{F}$, and completes the proof of (v).

Using (iv) and (v), we may now show that $\mathcal{F}$ is closed under intersections of nonincreasing sequences [of $\mathcal{F}$-sets].

(vi) Let $\{E_p\}_{p=1}^{\infty}$ *denote a nonincreasing sequence in* $\mathcal{F}$. *Then* $\lim_p E_p \in \mathcal{F}$.

Proof: Write E for $\lim_p E_p$. Since $E_p \in \mathcal{F} \subseteq \mathcal{F}_1 \otimes \mathcal{F}_2 = \mathcal{E}$ for every $p \in \mathbb{N}$ [the last equality holds by (v)], we have $E_p \cap (A_n \times B_n) \in \mathcal{F}$ for every $n, p \in \mathbb{N}$. We next have that $E = \lim_n [E \cap (A_n \times B_n)] = \lim_n \lim_p [E_p \cap (A_n \times B_n)]$. Observe for each $n \in \mathbb{N}$ that $\{E_p \cap (A_n \times B_n)\}_{p=1}^{\infty}$ is a nonincreasing sequence of $\mathcal{F}$-sets with $E_1 \cap (A_n \times B_n) \subseteq A_n \times B_n$ and $\mu_1(A_n), \mu_2(B_n) < +\infty$, and hence for each $n \in \mathbb{N}$ we have that $\lim_p [E_p \cap (A_n \times B_n)] \in \mathcal{F}$ by (iv). Therefore, $\{\lim_p [E_p \cap (A_n \times B_n)]\}_{n=1}^{\infty}$ is a nondecreasing sequence of $\mathcal{F}$-sets, hence its limit $\lim_n \lim_p [E_p \cap (A_n \times B_n)]$ is in $\mathcal{F}$ by (iii). This gives $E \in \mathcal{F}$, and the proof of (vi) is complete.

$\mathcal{F}$ is therefore a monotone class, and the proof of this entire claim is now complete. ∎

Definition. Let $(\Omega_1, \mathcal{F}_1, \mu_1)$ and $(\Omega_2, \mathcal{F}_2, \mu_2)$ denote σ-finite measure spaces. Define $\mu_1 \otimes \mu_2$ on $\mathcal{F}_1 \otimes \mathcal{F}_2$ by writing

$$\mu_1 \otimes \mu_2(E) = \int_{\Omega_1} \mu_2(E_{\omega_1})\, d\mu_1(\omega_1) = \int_{\Omega_2} \mu_1(E^{\omega_2})\, d\mu_2(\omega_2)$$

for every $E \in \mathcal{F}_1 \otimes \mathcal{F}_2$.

That there are no logical problems in the definition in terms of integrands being defined, integrals existing, and integrals being equal is exactly the content of Claim 5. The next claim demonstrates some key properties of $\mu_1 \otimes \mu_2$.

Claim 6. *In the framework of the definition above, we have the following:*
(a) $\mu_1 \otimes \mu_2$ is a σ-finite measure on $\mathcal{F}_1 \otimes \mathcal{F}_2$.
(b) $\mu_1 \otimes \mu_2(A \times B) = \mu_1(A)\mu_2(B)$ for every $A \in \mathcal{F}_1$ and $B \in \mathcal{F}_2$.
(c) If ϕ is a measure on $\mathcal{F}_1 \otimes \mathcal{F}_2$ such that $\phi(A \times B) = \mu_1(A)\mu_2(B)$ for every $A \in \mathcal{F}_1$ and $B \in \mathcal{F}_2$, then $\phi = \mu_1 \otimes \mu_2$.

Proof: We do (a). Clearly $\mu_1 \otimes \mu_2$ is nonnegative and $\mu_1 \otimes \mu_2(\varnothing) = 0$. All that remains is countable additivity. Let $\{E_n\}_{n=1}^{\infty}$ denote a disjoint sequence of $\mathcal{F}_1 \otimes \mathcal{F}_2$-sets. Then $\{(E_n)_{\omega_1}\}_{n=1}^{\infty}$ is a disjoint sequence in $\mathcal{F}_2$ and

$$\begin{aligned}
\mu_1 \otimes \mu_2\left(\bigcup_{n=1}^{\infty} E_n\right) &= \int_{\Omega_1} \mu_2\left(\left(\bigcup_{n=1}^{\infty} E_n\right)_{\omega_1}\right) d\mu_1(\omega_1) \\
&= \int_{\Omega_1} \mu_2\left(\bigcup_{n=1}^{\infty} (E_n)_{\omega_1}\right) d\mu_1(\omega_1) \\
&= \int_{\Omega_1} \sum_{n=1}^{\infty} \mu_2((E_n)_{\omega_1})\, d\mu_1(\omega_1) \\
&= \sum_{n=1}^{\infty} \int_{\Omega_1} \mu_2((E_n)_{\omega_1})\, d\mu_1(\omega_1) \\
&= \sum_{n=1}^{\infty} \mu_1 \otimes \mu_2(E_n).
\end{aligned}$$

[The fifth equality follows by (L11) from Section 6.4.] This shows that $\mu_1 \otimes \mu_2$ is a measure.

We now handle σ-finiteness. Write $\Omega_1 = \lim_n A_n$, where $\{A_n\}_{n=1}^{\infty}$ is a nondecreasing sequence in $\mathcal{F}_1$ with $\mu_1(A_n) < +\infty$ for each $n \in \mathbb{N}$. Similarly, write $\Omega_2 = \lim_n B_n$, where $\{B_n\}_{n=1}^{\infty}$ is a nondecreasing sequence in $\mathcal{F}_2$ with $\mu_2(B_n) < +\infty$ for each $n \in \mathbb{N}$. Then $\Omega_1 \times \Omega_2 = \bigcup_{n=1}^{\infty}(A_n \times B_n)$, and

$$\begin{aligned} \mu_1 \otimes \mu_2(A_n \times B_n) &= \int_{\Omega_1} \mu_2((A_n \times B_n)_{\omega_1})\, d\mu_1(\omega_1) \\ &= \int_{\Omega_1} \mu_2(B_n) I_{A_n}(\omega_1)\, d\mu_1(\omega_1) \\ &= \mu_1(A_n)\mu_2(B_n) \\ &< +\infty \end{aligned}$$

for each $n \in \mathbb{N}$. This shows that $\mu_1 \otimes \mu_2$ is σ-finite, giving (a).

For (b), pick $A \in \mathcal{F}_1$ and $B \in \mathcal{F}_2$. Then

$$\begin{aligned} \mu_1 \otimes \mu_2(A \times B) &= \int_{\Omega_1} \mu_2((A \times B)_{\omega_1})\, d\mu_1(\omega_1) \\ &= \int_{\Omega_1} \mu_2(B) I_A(\omega_1)\, d\mu_1(\omega_1) \\ &= \mu_1(A)\mu_2(B), \end{aligned}$$

so that $\mu_1 \otimes \mu_2$ has the stated product property.

For (c), suppose that ϕ is a measure with domain $\mathcal{F}_1 \otimes \mathcal{F}_2$ such that

$$\phi(A \times B) = \mu_1(A)\mu_2(B)\ (= \mu_1 \otimes \mu_2(A \times B)) \quad \text{for each } A \in \mathcal{F}_1 \text{ and } B \in \mathcal{F}_2.$$

Pick $E \in \mathcal{A}$, the field of finite disjoint unions of $\mathcal{F}_1 \times \mathcal{F}_2$-sets. [Recall Claim 1.] Write E as a finite disjoint union $\bigcup_{i=1}^{n}(C_i \times D_i)$ of $\mathcal{F}_1 \times \mathcal{F}_2$ sets. Observe that

$$\phi(E) = \sum_{i=1}^{n} \phi(C_i \times D_i) = \sum_{i=1}^{n} \mu_1 \otimes \mu_2(C_i \times D_i) = \mu_1 \otimes \mu_2(E),$$

and hence $\phi = \mu_1 \otimes \mu_2$ on the field $\mathcal{A}$. Since $\sigma(\mathcal{A}) = \mathcal{F}_1 \otimes \mathcal{F}_2$ and $\mu_1 \otimes \mu_2$ is σ-finite on $\mathcal{F}_1 \times \mathcal{F}_2$ [and hence on $\mathcal{A}$], the Uniqueness Theorem in Section 3.5 shows that $\phi = \mu_1 \otimes \mu_2$ on all of $\mathcal{F}_1 \otimes \mathcal{F}_2$. This completes the entire proof. ∎

At this stage, we have completed the program for this section, and we conclude matters with a definition.

Definition. Let $(\Omega_1, \mathcal{F}_1, \mu_1)$ and $(\Omega_2, \mathcal{F}_2, \mu_2)$ denote σ-finite measure spaces. The σ-field $\mathcal{F}_1 \otimes \mathcal{F}_2$ is called the *product [σ-field] of $\mathcal{F}_1$ and $\mathcal{F}_2$*. The measure $\mu_1 \otimes \mu_2$ on $\mathcal{F}_1 \otimes \mathcal{F}_2$ is called the *product [measure] of μ_1 and μ_2*. The measure space $(\Omega_1 \times \Omega_2, \mathcal{F}_1 \otimes \mathcal{F}_2, \mu_1 \otimes \mu_2)$ is called the *product [measure space] of $(\Omega_1, \mathcal{F}_1, \mu_1)$ and $(\Omega_2, \mathcal{F}_2, \mu_2)$*.

Exercises.

1. Let Ω_1 denote an uncountable set, and let $\mathcal{F}_1$ denote the σ-field of subsets of Ω_1 that are at most countable or have at most countable complement. Let Ω_2 and $\mathcal{F}_2$ be identical to Ω_1 and $\mathcal{F}_1$, respectively. Let $D = \{(\omega_1, \omega_2) \in \Omega_1 \times \Omega_2 : \omega_1 = \omega_2\}$. We have $D_{\omega_1} \in \mathcal{F}_2$ and $D^{\omega_2} \in \mathcal{F}_1$ for every $\omega_1 \in \Omega_1$ and $\omega_2 \in \Omega_2$, but $D \notin \mathcal{F}_1 \otimes \mathcal{F}_2$. This shows that the converse to Claim 2 is not necessarily true. [This should be direct.]

2. Let $A \subseteq \Omega_1$ and $B \subseteq \Omega_2$.
(a) Suppose that $A \times B \neq \varnothing$. Then $A \times B \in \mathcal{F}_1 \otimes \mathcal{F}_2$ iff $A \in \mathcal{F}_1$ and $B \in \mathcal{F}_2$. [The "if" part is almost automatic. For the "only if" part, use Claim 2 and consider $(A \times B)_{\omega_1}$ for $\omega_1 \in A$ and $\omega_1 \notin A$.]
(b) Suppose that $A \times B = \varnothing$. Then obviously $A \times B \in \mathcal{F}_1 \otimes \mathcal{F}_2$, but it is not always the case that $A \in \mathcal{F}_1$ and $B \in \mathcal{F}_2$. [If $A \times B = \varnothing$, then $A = \varnothing$ or $B = \varnothing$. Consider the case where $A = \varnothing$ and B is a nonempty set outside of $\mathcal{F}_2$.]

3. Prove the following set-theoretic facts.
(a) Let $A_1 \times B_1$ and $A_2 \times B_2$ both be nonempty. Then $A_1 \times B_1 \subseteq A_2 \times B_2$ iff $A_1 \subseteq A_2$ and $B_1 \subseteq B_2$.
(b) Let $A_1 \times B_1$ and $A_2 \times B_2$ both be nonempty. Then $A_1 \times B_1 = A_2 \times B_2$ iff $A_1 = A_2$ and $B_1 = B_2$.
(c) Let $A \times B$, $A_1 \times B_1$, and $A_2 \times B_2$ be nonempty. Then $A \times B$ is the disjoint union of $A_1 \times B_1$ and $A_2 \times B_2$ iff either (i) A is the disjoint union of A_1 and A_2 and $B = B_1 = B_2$ or (ii) $A = A_1 = A_2$ and B is the disjoint union of B_1 and B_2.
(d) The "only if" parts of (a) and (b) do not necessarily hold for empty Cartesian products. What about (c)?

4. Let $\mathcal{M}$ denote the collection of all intervals of all forms in $\mathbb{R}$. Consider $\varnothing$ as an interval. Then $\mathcal{M}$ is a monotone class on $\mathbb{R}$, but $\mathcal{M}$ is not a σ-field on $\mathbb{R}$. Therefore, the requirements of a σ-field are more stringent than those for a monotone class. Next, give an even stronger counterexample: find a monotone class $\mathcal{M}$ on $\mathbb{R}$ with $\mathbb{R} \in \mathcal{M}$ and such that $\mathcal{M}$ is closed under complementation, but $\mathcal{M}$ is not a σ-field.

5. Let $\mathcal{F}_1$ and $\mathcal{F}_2$ denote σ-fields on Ω_1 and Ω_2, respectively. It may not be the case that $\mathcal{F}_1 \times \mathcal{F}_2$ is a σ-field on $\Omega_1 \times \Omega_2$. Therefore, our definition of $\mathcal{F}_1 \otimes \mathcal{F}_2$ is not superfluous.

6. Prove Claims 2(b) and 3(b) by mimicking the proofs of Claims 2(a) and 3(a).

7. The product of $(\mathbb{R}^k, \mathcal{B}^k, \lambda_k)$ and $(\mathbb{R}^m, \mathcal{B}^m, \lambda_m)$ is $(\mathbb{R}^{k+m}, \mathcal{B}^{k+m}, \lambda_{k+m})$. In other words, $\mathcal{B}^k \otimes \mathcal{B}^m = \mathcal{B}^{k+m}$ and $\lambda_k \otimes \lambda_m = \lambda_{k+m}$. [Recall Example 1 in the introduction to this chapter.]

8. Let $\Omega_1 = \Omega_2 = [0, 1]$. Let $\mathcal{F}_1 = \mathcal{F}_2$ denote the Borel subsets of $[0, 1]$. Let μ_1 denote Lebesgue measure restricted to $\mathcal{F}_1$, and let μ_2 denote the counting measure on $[0, 1]$. Note that $(\Omega_2, \mathcal{F}_2, \mu_2)$ is not a σ-finite measure space while $(\Omega_1, \mathcal{F}_1, \mu_1)$ is a σ-finite measure space. Let $E = \{(\omega_1, \omega_2) \in \Omega_1 \times \Omega_2 : \omega_1 = \omega_2\}$.
(a) $E \in \mathcal{F}_1 \otimes \mathcal{F}_2$. [Here, $\mathcal{F}_1 \otimes \mathcal{F}_2$ consists of the two-dimensional Borel subsets of $[0, 1] \times [0, 1]$. Show that E is a closed subset of $[0, 1] \times [0, 1]$.]
(b) $\int_{\Omega_1} \mu_2(E_{\omega_1})\, d\mu(\omega_1) = 1$.
(c) $\int_{\Omega_2} \mu_1(E^{\omega_2})\, d\mu(\omega_2) = 0$. In conjunction with (b), this shows that the assumption of σ-finiteness in Claim 5 cannot be dropped.

9. Let $(\Omega_1, \mathcal{F}_1, \mu_1)$ and $(\Omega_2, \mathcal{F}_2, \mu_2)$ denote complete measure spaces.

(a) Suppose that there exists nonempty $A \in \mathcal{F}_1$ with $\mu_1(A) = 0$; suppose also that there is $B \subseteq \Omega_2$ with $B \notin \mathcal{F}_2$ [i.e. $\mathcal{F}_2 \neq 2^{\Omega_2}$]. Then $(\Omega_1 \times \Omega_2, \mathcal{F}_1 \otimes \mathcal{F}_2, \mu_1 \otimes \mu_2)$ fails to be a complete measure space. [Suppose not. We have $A \times B \subseteq A \times \Omega_2$ and $\mu_1 \otimes \mu_2(A \times \Omega_2) = 0$, hence $A \times B \in \mathcal{F}_1 \otimes \mathcal{F}_2$. By Exercise 2(a) this forces $B \in \mathcal{F}_2$, a contradiction.]
(b) As a special case, we may say that although both $(\mathbb{R}^k, \mathcal{L}^k, \bar{\lambda}_k)$ and $(\mathbb{R}^m, \mathcal{L}^m, \bar{\lambda}_m)$ are complete measure spaces, the product measure space $(\mathbb{R}^{k+m}, \mathcal{L}^k \otimes \mathcal{L}^m, \bar{\lambda}_k \otimes \bar{\lambda}_m)$ is not complete. Give another proof that the completion of $(\mathbb{R}^{k+m}, \mathcal{L}^k \otimes \mathcal{L}^m, \bar{\lambda}_k \otimes \bar{\lambda}_m)$ is $(\mathbb{R}^{k+m}, \mathcal{L}^{k+m}, \bar{\lambda}_{k+m})$, namely, $\overline{\mathcal{L}^k \otimes \mathcal{L}^m} = \mathcal{L}^{k+m}$ and $\overline{\bar{\lambda}_k \otimes \bar{\lambda}_m} = \bar{\lambda}_{k+m}$.

10. If $E, F \in \mathcal{F}_1 \otimes \mathcal{F}_2$ are such that $\mu_2(E_{\omega_1}) = \mu_2(F_{\omega_1})$ for every $\omega_1 \in \Omega_1$, then $\mu_1 \otimes \mu_2(E) = \mu_1 \otimes \mu_2(F)$. [Integrate both sides of the given equation over Ω_1 with respect to μ_1.] This is the measure theoretic generalization of what is known as *Cavalieri's Principle.*

11. We know that if $f : \Omega_1 \times \Omega_2 \to \bar{\mathbb{R}}$ is $\mathcal{F}_1 \otimes \mathcal{F}_2/\mathcal{B}^*$-measurable, then f_{ω_1} is $\mathcal{F}_2/\mathcal{B}^*$-measurable for each $\omega_1 \in \Omega_1$ and f^{ω_2} is $\mathcal{F}_1/\mathcal{B}^*$-measurable for each $\omega_2 \in \Omega_2$. We explore some sort-of converses to this result in a special setting.
(a) Suppose that $f : \mathbb{R}^2 \to \mathbb{R}$ is such that f_x is $\mathcal{B}/\mathcal{B}^*$-measurable for each $x \in \mathbb{R}$ and f^y is continuous for each $y \in \mathbb{R}$. Then f is $\mathcal{B}^2/\mathcal{B}^*$-measurable.
(b) Suppose that $f : \mathbb{R}^k \to \mathbb{R}$ is continuous in each of its k arguments separately. [That is, for any $x_2, \cdots, x_k \in \mathbb{R}$ the mapping $x_1 \mapsto f(x_1, \cdots, x_k)$ is continuous; for any $x_1, x_3, \cdots, x_k \in \mathbb{R}$ the mapping $x_2 \mapsto f(x_1, x_3, \cdots, x_k)$ is continuous, and so on.] Then f is $\mathcal{B}^k/\mathcal{B}^*$-measurable.
(c) Let E denote a dense subset of $\mathbb{R}$, and let $f : \mathbb{R}^2 \to \mathbb{R}$ be such that f_x is $\mathcal{L}/\mathcal{B}^*$-measurable for each $x \in E$ and f^y is continuous for $\bar{\lambda}$-a.e. $y \in \mathbb{R}$. Then f is $\mathcal{L}^2/\mathcal{B}^*$-measurable.

12. Consider two σ-finite measure spaces $(\Omega_1, \mathcal{F}_1, \mu_1)$ and $(\Omega_2, \mathcal{F}_2, \mu_2)$.
(a) It is not necessarily the case that $\bar{\mathcal{F}}_1 \otimes \bar{\mathcal{F}}_2$ coincides with $\overline{\mathcal{F}_1 \otimes \mathcal{F}_2}$ [although the former is always contained in the latter], but we do have $\overline{\bar{\mathcal{F}}_1 \otimes \bar{\mathcal{F}}_2} = \overline{\mathcal{F}_1 \otimes \mathcal{F}_2}$.
(b) $\overline{\bar{\mu}_1 \otimes \bar{\mu}_2} = \overline{\mu_1 \otimes \mu_2}$. [It might be interesting to compare these general results with Exercise 9.]

10.2 THE FUBINI THEOREMS

In this section we prove the famous Fubini Theorems. The results of this section are a fairly straightforward continuation of the work done in the previous section. We will work with the σ-finite measure spaces $(\Omega_1, \mathcal{F}_1, \mu_1)$, $(\Omega_2, \mathcal{F}_2, \mu_2)$, and $(\Omega_1 \times \Omega_2, \mathcal{F}_1 \otimes \mathcal{F}_2, \mu_1 \otimes \mu_2)$. We now state an amalgamation of Claim 5 and the definition of product measure from the previous section for easy reference.

Result. *For every $E \in \mathcal{F}_1 \otimes \mathcal{F}_2$, we have that the mapping $\omega_1 \mapsto \mu_2(E_{\omega_1})$ defined on Ω_1 is nonnegative and $\mathcal{F}_1/\mathcal{B}^*$-measurable, the mapping $\omega_2 \mapsto \mu_1(E^{\omega_2})$ defined on Ω_2 is nonnegative and $\mathcal{F}_2/\mathcal{B}^*$-measurable, and*

$$\int_{\Omega_1} \mu_2(E_{\omega_1})\, d\mu_1(\omega_1) = \mu_1 \otimes \mu_2(E) = \int_{\Omega_2} \mu_1(E^{\omega_2})\, d\mu_2(\omega_2).$$

Consider a function $f : \Omega_1 \times \Omega_2 \to \bar{\mathbb{R}}$ that is $\mathcal{F}_1 \otimes \mathcal{F}_2/\mathcal{B}^*$-measurable. In this section we shall consider two situations: (a) f is nonnegative, and (b) $f \in L^1(\Omega_1 \times \Omega_2, \mathcal{F}_1 \otimes \mathcal{F}_2, \mu_1 \otimes \mu_2)$. In both of these situations, the integral $\int_{\Omega_1 \times \Omega_2} f \, d\mu_1 \otimes \mu_2$ is meaningfully defined. Some authors refer to this integral as *the double integral of* f. In (a), this integral exists as a number in $[0, +\infty]$, and in (b), this integral exists as a real number. In either of situations (a) and (b), the notations

$$\int_{\Omega_1 \times \Omega_2} f(\omega_1, \omega_2) \, \mu_1 \otimes \mu_2(d(\omega_1, \omega_2)) \quad \text{and} \quad \int_{\Omega_1 \times \Omega_2} f(\omega_1, \omega_2) \, d\mu_1 \otimes \mu_2(\omega_1, \omega_2)$$

are equivalently employed by various authors. It is sometimes advantageous or convenient to write the arguments ω_1 and ω_2 in the integrand.[1] It should be noted that the subscript $\Omega_1 \times \Omega_2$ is not really necessary, as $\int f \, d\mu_1 \otimes \mu_2$ means the same thing as $\int_{\Omega_1 \times \Omega_2} f \, d\mu_1 \otimes \mu_2$ by our notational conventions given in Section 6.4, but we shall use the subscript anyway.

Continuing with the notational discussion, consider the following symbol

$$\int_{\Omega_1} \int_{\Omega_2} f \, d\mu_2 \, d\mu_1, \quad \text{also written as} \quad \int_{\Omega_1} \int_{\Omega_2} f(\omega_1, \omega_2) \, d\mu_2(\omega_2) \, d\mu_1(\omega_1).^2$$

Here, the integrals are to be understood as they are in multivariable calculus: perform the "inner integral" $\int_{\Omega_2} f \, d\mu_2$ or $\int_{\Omega_2} f(\omega_1, \omega_2) \, d\mu_2(\omega_2)$ first to obtain a function of ω_1 alone, and then perform the "outer integral" $\int_{\Omega_1} \cdots d\mu_1$ or $\int_{\Omega_1} \cdots d\mu_1(\omega_1)$ on this just-obtained function of ω_1 alone. We wish to ask when it makes sense to write something like $\int_{\Omega_1} \int_{\Omega_2} f \, d\mu_2 \, d\mu_1$ or $\int_{\Omega_1} \int_{\Omega_2} f(\omega_1, \omega_2) \, d\mu_2(\omega_2) \, d\mu_1(\omega_1)$. To answer this, first observe [working from "inside-out"] that we need f to be such that $\int_{\Omega_2} f(\omega_1, \omega_2) \, d\mu_2(\omega_2)$ makes sense for μ_1-a.e. $\omega_1 \in \Omega_1$. The quantity $\int_{\Omega_2} f(\omega_1, \omega_2) \, d\mu_2(\omega_2)$ is a function of ω_1 alone, [borrowing calculus terminology, we have "integrated out ω_2"], and we wish to apply $\int_{\Omega_1} \cdots d\mu_1(\omega_1)$ to the function $\int_{\Omega_2} f(\omega_1, \omega_2) \, d\mu_2(\omega_2)$ of ω_1 alone. Now even though $\omega_1 \mapsto \int_{\Omega_2} f(\omega_1, \omega_2) \, d\mu_2(\omega_2)$ may meaningfully exist as an extended real-valued function of ω_1, this extended real-valued function of ω_1 may not be such that applying $\int_{\Omega_1} \cdots d\mu_1$ to it is meaningful. For example, it may turn out that

$$\int_{\Omega_1} \left(\int_{\Omega_2} f(\omega_1, \omega_2) \, d\mu_2(\omega_2) \right)^+ d\mu_1(\omega_1) = +\infty$$

[1] This is the same convention mentioned in Section 6.4, whereby instead of $\int_\Omega f \, d\mu$ one equivalently writes $\int_\Omega f(\omega) d\mu(\omega)$ or $\int_\Omega f(\omega)\mu(d\omega)$.

[2] Some authors also equivalently write $\int_{\Omega_1} d\mu_1(\omega_1) \int_{\Omega_2} f(\omega_1, \omega_2), d\mu_2(\omega)$; the symbol "$d\mu_1(\omega_1)$" is placed on the opposite side of the integral sign $\int_{\Omega_2}$. This breaks with how one is used to seeing iterated integrals in modern treatments of calculus, and hence this notational variant is not used in this book.

and

$$\int_{\Omega_1} \left(\int_{\Omega_2} f(\omega_1, \omega_2) \, d\mu_2(\omega_2) \right)^- d\mu_1(\omega_1) = +\infty,$$

in which case $\int_{\Omega_1} \int_{\Omega_2} f(\omega_1, \omega_2) \, d\mu_2(\omega_2) \, d\mu_1(\omega_1)$ is of the form $(+\infty) - (+\infty)$ and hence is undefined. Another situation for which we may not be able to meaningfully apply the operation $\int_{\Omega_1} \cdots d\mu_1(\omega_1)$ to $\int_{\Omega_2} f(\omega_1, \omega_2) \, d\mu_2(\omega_2)$ is if the mapping $\omega_1 \mapsto \int_{\Omega_2} f(\omega_1, \omega_2) \, d\mu_2(\omega_2)$ fails to be an $\mathcal{F}_1/\mathcal{B}^*$-measurable function of ω_1. Therefore, for the symbol $\int_{\Omega_1} \int_{\Omega_2} f(\omega_1, \omega_2) \, d\mu_2(\omega_2) \, d\mu_1(\omega_1)$ to make sense, we require the function f defined on $\Omega_1 \times \Omega_2$ to be such that:

(1) For every $\omega_1 \in \Omega_1$ the mapping $\omega_2 \mapsto f(\omega_1, \omega_2)$ is $\mathcal{F}_2/\mathcal{B}^*$-measurable and such that $\int_{\Omega_2} f(\omega_1, \omega_2) \, d\mu_2(\omega_2)$ exists for μ_1-a.e. $\omega_1 \in \Omega_1$.
(2) The mapping $\omega_1 \mapsto \int_{\Omega_2} f(\omega_1, \omega_2) \, d\mu_2(\omega_2)$ [defined for the μ_1-a.e. $\omega_1 \in \Omega_1$ in (1)] is $\mathcal{F}_1/\mathcal{B}^*$-measurable and such that we may meaningfully integrate it over Ω_1 with respect to μ_1.

It will turn out that if (a) f is a nonnegative $\mathcal{F}_1 \otimes \mathcal{F}_2/\mathcal{B}^*$-measurable function on $\Omega_1 \times \Omega_2$, or if (b) $f \in L^1(\Omega_1 \times \Omega_2, \mathcal{F}_1 \otimes \mathcal{F}_2, \mu_1 \otimes \mu_2)$, then requirements (1) and (2) just given will be satisfied, and hence both

$$\int_{\Omega_1 \times \Omega_2} f(\omega_1, \omega_2) \, d\mu_1 \otimes \mu_2(\omega_1, \omega_2) \quad \text{and} \quad \int_{\Omega_1} \int_{\Omega_2} f(\omega_1, \omega_2) \, d\mu_2(\omega_2) \, d\mu_1(\omega_1)$$

meaningfully exist *and in fact are equal.* The latter expression is known as the *iterated integral of* f. It will also turn out that either of (a) and (b) force

$$\int_{\Omega_1 \times \Omega_2} f(\omega_1, \omega_2) \, d\mu_1 \otimes \mu_2(\omega_1, \omega_2) \quad \text{and} \quad \int_{\Omega_2} \int_{\Omega_1} f(\omega_1, \omega_2) \, d\mu_1(\omega_1) \, d\mu_2(\omega_2)$$

to both exist and be equal. Therefore, under (a) or (b), we will have that

$$\int_{\Omega_1 \times \Omega_2} f \, d\mu_1 \otimes \mu_2, \quad \int_{\Omega_1} \int_{\Omega_2} f \, d\mu_2 \, d\mu_1, \quad \text{and} \quad \int_{\Omega_2} \int_{\Omega_1} f \, d\mu_1 \, d\mu_2$$

exist and are equal. This is the content of the Fubini Theorems. The Fubini Theorem dealing with (a) is also known as *Tonelli's Theorem.*

With this discussion in tow, we now proceed to one of the major results of analysis and measure theory.

Nonnegative Fubini Theorem. *Let* $f : \Omega_1 \times \Omega_2 \to [0, +\infty]$ *be* $\mathcal{F}_1 \otimes \mathcal{F}_2/\mathcal{B}^*$-*measurable. Then*

(i) For every $\omega_1 \in \Omega_1$ *the mapping* $\omega_2 \mapsto f(\omega_1, \omega_2)$ *on* Ω_2 *is* $\mathcal{F}_2/\mathcal{B}^*$-*measurable and nonnegative.*
(ii) For every $\omega_2 \in \Omega_2$ *the mapping* $\omega_1 \mapsto f(\omega_1, \omega_2)$ *on* Ω_1 *is* $\mathcal{F}_1/\mathcal{B}^*$-*measurable and nonnegative.*
(iii) The mapping $\omega_1 \mapsto \int_{\Omega_2} f(\omega_1, \omega_2) \, d\mu_2(\omega_2)$ *on* Ω_1 *is* $\mathcal{F}_1/\mathcal{B}^*$-*measurable and nonnegative.*

(iv) The mapping $\omega_2 \mapsto \int_{\Omega_1} f(\omega_1,\omega_2)\, d\mu_1(\omega_1)$ on Ω_2 is $\mathcal{F}_2/\mathcal{B}^*$-measurable and nonnegative.
(v) We have

$$\begin{aligned}\int_{\Omega_1\times\Omega_2} f(\omega_1,\omega_2)\, d\mu_1\otimes\mu_2(\omega_1,\omega_2) &= \int_{\Omega_1}\int_{\Omega_2} f(\omega_1,\omega_2)\, d\mu_2(\omega_2)\, d\mu_1(\omega_1)\\ &= \int_{\Omega_2}\int_{\Omega_1} f(\omega_1,\omega_2)\, d\mu_1(\omega_1)\, d\mu_2(\omega_2).\end{aligned}$$

Proof: Statements (i) and (ii) follow from Claim 3 of the previous section, and are restated here in the spirit of completeness. We therefore only need to prove (iii)-(v). Noticing that (iii) and (iv) are symmetric statements and that the two equalities in (v) relating the double integral to the iterated integrals are symmetric, it follows that we may prove (iii)-(v) by merely proving (iii) and the equation

$$\int_{\Omega_1\times\Omega_2} f(\omega_1,\omega_2)\, d\mu_1\otimes\mu_2(\omega_1,\omega_2) = \int_{\Omega_1}\int_{\Omega_2} f(\omega_1,\omega_2)\, d\mu_2(\omega_2) d\mu_1(\omega_1). \tag{10.1}$$

We shall prove (iii) and (10.1) using the usual approach of showing that (iii) and (10.1) hold in turn for $\mathcal{F}_1\otimes\mathcal{F}_2/\mathcal{B}^*$-measurable indicator functions on $\Omega_1\times\Omega_2$, nonnegative $\mathcal{F}_1\otimes\mathcal{F}_2/\mathcal{B}^*$-measurable simple functions on $\Omega_1\times\Omega_2$, and then general nonnegative $\mathcal{F}_1\otimes\mathcal{F}_2/\mathcal{B}^*$-measurable functions on $\Omega_1\times\Omega_2$.

Indicators. Suppose that $f = I_E$, where $E\in\mathcal{F}_1\otimes\mathcal{F}_2$. To verify (iii), we have for each $\omega_1\in\Omega_1$ that

$$\begin{aligned}\int_{\Omega_2} f_{\omega_1}(\omega_2)\, d\mu_2(\omega_2) = \int_{\Omega_2} (I_E)_{\omega_1}(\omega_2)\, d\mu_2(\omega_2) &= \int_{\Omega_2} I_{E_{\omega_1}}(\omega_2)\, d\mu_2(\omega_2)\\ &= \mu_2(E_{\omega_1}).\end{aligned}$$

This shows that the mapping $\omega_1\mapsto\int_{\Omega_2} f(\omega_1,\omega_2)\, d\mu_2(\omega_2)$ on Ω_1 is nothing but the mapping $\omega_1\mapsto\mu_2(E_{\omega_1})$ on Ω_1, which is $\mathcal{F}_1/\mathcal{B}^*$-measurable by the result given at the beginning of this section. Therefore, (iii) holds in this case. For (10.1), we have

$$\begin{aligned}\int_{\Omega_1\times\Omega_2} f(\omega_1,\omega_2)\, d\mu_1\otimes\mu_2(\omega_1,\omega_2) &= \int_{\Omega_1\times\Omega_2} I_E(\omega_1,\omega_2)\, d\mu_1\otimes\mu_2(\omega_1,\omega_2)\\ &= \mu_1\otimes\mu_2(E)\\ &= \int_{\Omega_1}\mu_2(E_{\omega_1})\, d\mu_1(\omega_1)\\ &= \int_{\Omega_1}\int_{\Omega_2} f(\omega_1,\omega_2)\, d\mu_2(\omega_2)\, d\mu_1(\omega_1),\end{aligned}$$

where the third equality follows by the result given at the beginning of this section and the fourth follows by the previous display. This completes the proof for indicator functions.

Nonnegative Simple Functions. Let f denote a simple function of the form $\sum_{i=1}^{n} c_i I_{E_i}$, where $0 \le c_1, \cdots, c_n < +\infty$ and $E_1, \cdots, E_n$ are disjoint $\mathcal{F}_1 \otimes \mathcal{F}_2$-sets with union $\Omega_1 \times \Omega_2$. To verify (iii), first observe that for each $\omega_1 \in \Omega_1$ we have

$$\begin{aligned}\int_{\Omega_2} f(\omega_1, \omega_2)\, d\mu_2(\omega_2) &= \int_{\Omega_2} \sum_{i=1}^{n} c_i I_{E_i}(\omega_1, \omega_2)\, d\mu_2(\omega_2) \\ &= \sum_{i=1}^{n} c_i \int_{\Omega_2} I_{E_i}(\omega_1, \omega_2)\, d\mu_2(\omega_2).\end{aligned}$$

The mapping $\omega_1 \mapsto \int_{\Omega_2} f(\omega_1, \omega_2)\, d\mu_2(\omega_2)$ on Ω_1 is identical to the mapping $\omega_1 \mapsto \sum_{i=1}^{n} c_i \int_{\Omega_2} I_{E_i}(\omega_1, \omega_2)\, d\mu_2(\omega_2)$ on Ω_1, which by the result for indicator functions is a linear combination of the $\mathcal{F}_1/\mathcal{B}^*$-measurable mappings

$$\omega_1 \mapsto \int_{\Omega_2} I_{E_i}(\omega_1, \omega_2)\, d\mu_2(\omega_2) \quad \text{on } \Omega_1 \quad [i = 1, \cdots, n],$$

and hence is $\mathcal{F}_1/\mathcal{B}^*$-measurable. This gives (iii) when f is a nonnegative $\mathcal{F}_1 \otimes \mathcal{F}_2/\mathcal{B}^*$-measurable simple function on $\Omega_1 \times \Omega_2$. To verify (10.1) in this situation, we have, using the result given earlier for simple functions, that

$$\begin{aligned}&\int_{\Omega_1 \times \Omega_2} f(\omega_1, \omega_2)\, d\mu_1 \otimes \mu_2(\omega_1, \omega_2) \\ &= \int_{\Omega_1 \times \Omega_2} \sum_{i=1}^{n} c_i I_{E_i}(\omega_1, \omega_2)\, d\mu_1 \otimes \mu_2(\omega_1, \omega_2) \\ &= \sum_{i=1}^{n} c_i \int_{\Omega_1 \times \Omega_2} I_{E_i}(\omega_1, \omega_2)\, d\mu_1 \otimes \mu_2(\omega_1, \omega_2) \\ &= \sum_{i=1}^{n} c_i \int_{\Omega_1} \int_{\Omega_2} I_{E_i}(\omega_1, \omega_2)\, d\mu_2(\omega_2)\, d\mu_1(\omega_1) \\ &= \int_{\Omega_1} \int_{\Omega_2} \sum_{i=1}^{n} c_i I_{E_i}(\omega_1, \omega_2)\, d\mu_2(\omega_2)\, d\mu_1(\omega_1) \\ &= \int_{\Omega_1} \int_{\Omega_2} f(\omega_1, \omega_2)\, d\mu_2(\omega_2)\, d\mu_1(\omega_1).\end{aligned}$$

This gives (10.1), and the theorem accordingly holds when f is a nonnegative $\mathcal{F}_1 \otimes \mathcal{F}_2/\mathcal{B}^*$-measurable simple function on $\Omega_1 \times \Omega_2$.

Nonnegative Functions. Let f denote a nonnegative $\mathcal{F}_1 \otimes \mathcal{F}_2/\mathcal{B}^*$-measurable function on $\Omega_1 \times \Omega_2$. Let $\{s_n\}_{n=1}^{\infty}$ denote a nondecreasing sequence of nonnegative $\mathcal{F}_1 \otimes \mathcal{F}_2/\mathcal{B}^*$-measurable simple functions with $\lim_n s_n = f$. We have for every $\omega_1 \in \Omega_1$ that $\{(s_n)_{\omega_1}\}_{n=1}^{\infty}$ is a nondecreasing sequence of nonnegative

$\mathcal{F}_2/\mathcal{B}^*$-measurable simple functions with limit f_{ω_1} and

$$\begin{aligned}\int_{\Omega_2} f(\omega_1,\omega_2)\,d\mu_2(\omega_2) &= \int_{\Omega_2} f_{\omega_1}(\omega_2)\,d\mu_2(\omega_2)\\ &= \int_{\Omega_2} \lim_n (s_n)_{\omega_1}(\omega_2)\,d\mu_2(\omega_2)\\ &= \lim_n \int_{\Omega_2} (s_n)_{\omega_1}(\omega_2)\,d\mu_2(\omega_2) \quad \text{[MCT]}\\ &= \lim_n \int_{\Omega_2} s_n(\omega_1,\omega_2)\,d\mu_2(\omega_2).\end{aligned}$$

It therefore follows that the mapping $\omega_1 \mapsto \int_{\Omega_2} f(\omega_1,\omega_2)\,d\mu_2(\omega_2)$ on Ω_1 is identical to the mapping $\omega_1 \mapsto \lim_n \int_{\Omega_2} s_n(\omega_1,\omega_2)\,d\mu_2(\omega_2)$ on Ω_1. From the result for simple functions, we know for each $n \in \mathbb{N}$ that the mapping

$$\omega_1 \mapsto \int_{\Omega_2} s_n(\omega_1,\omega_2)\,d\mu_2(\omega_2) \quad \text{on } \Omega_1$$

is $\mathcal{F}_1/\mathcal{B}^*$-measurable. Since $\omega_1 \mapsto \int_{\Omega_2} f(\omega_1,\omega_2)\,d\mu_2(\omega_2)$ is the pointwise limit of a sequence of $\mathcal{F}_1/\mathcal{B}^*$-measurable mappings on Ω_1, it follows that it too is an $\mathcal{F}_1/\mathcal{B}^*$-measurable mapping on Ω_1. This proves (iii) for nonnegative $\mathcal{F}_1 \otimes \mathcal{F}_2/\mathcal{B}^*$-measurable functions on $\Omega_1 \times \Omega_2$. For the proof of (10.1), we have

$$\begin{aligned}\int_{\Omega_1\times\Omega_2} f(\omega_1,\omega_2)\,d\mu_1\otimes\mu_2(\omega_1,\omega_2) &= \int_{\Omega_1\times\Omega_2} \lim_n s_n(\omega_1,\omega_2)\,d\mu_1\otimes\mu_2(\omega_1,\omega_2)\\ &= \lim_n \int_{\Omega_1\times\Omega_2} s_n(\omega_1,\omega_2)\,d\mu_1\otimes\mu_2(\omega_1,\omega_2)\\ &= \lim_n \int_{\Omega_1}\int_{\Omega_2} s_n(\omega_1,\omega_2)\,d\mu_2(\omega_2)\,d\mu_1(\omega_1)\\ &= \int_{\Omega_1} \lim_n \int_{\Omega_2} s_n(\omega_1,\omega_2)\,d\mu_2(\omega_2)\,d\mu_1(\omega_1)\\ &= \int_{\Omega_1}\int_{\Omega_2} \lim_n s_n(\omega_1,\omega_2)\,d\mu_2(\omega_2)\,d\mu_1(\omega_1)\\ &= \int_{\Omega_1}\int_{\Omega_2} f(\omega_1,\omega_2)\,d\mu_2(\omega_2)\,d\mu_1(\omega_1).\end{aligned}$$

[The second equality follows by MCT. The third equality follows from the result for simple functions just given. The fourth and fifth equalities each follow by applying MCT.] This gives (10.1) for nonnegative $\mathcal{F}_1\otimes\mathcal{F}_2/\mathcal{B}^*$-measurable f on $\Omega_1\times\Omega_2$, and thus the proof of the nonnegative version of Fubini's Theorem is complete. ∎

The previous result is referred to as *Fubini's Theorem*, and the upcoming result is also known as *Fubini's Theorem.* As mentioned before, the nonnegative version of Fubini's Theorem is often called *Tonelli's Theorem*, and the upcoming result is then called *Fubini's Theorem* in contradistinction. We shall merely use the terms *Nonnegative Fubini Theorem* and *Integrable Fubini Theorem* to distinguish between the two results.

Integrable Fubini Theorem. *Let $f : \Omega_1 \times \Omega_2 \to \bar{\mathbb{R}}$ be $\mathcal{F}_1 \otimes \mathcal{F}_2/\mathcal{B}^*$-measurable and such that at least one of*

$$\int_{\Omega_1\times\Omega_2} |f|\, d\mu_1 \otimes \mu_2, \quad \int_{\Omega_1}\int_{\Omega_2} |f|\, d\mu_2\, d\mu_1, \quad \text{or} \quad \int_{\Omega_2}\int_{\Omega_1} |f|\, d\mu_1\, d\mu_2 \tag{10.2}$$

is finite. Then $f \in L^1(\Omega_1 \times \Omega_2, \mathcal{F}_1 \otimes \mathcal{F}_2, \mu_1 \otimes \mu_2)$ and all three integrals are finite and equal. We also have the following:

(i) For μ_1-a.e. $\omega_1 \in \Omega_1$ the mapping $\omega_2 \mapsto f(\omega_1, \omega_2)$ on Ω_2 is in $L^1(\Omega_2, \mathcal{F}_2, \mu_2)$.
(ii) For μ_2-a.e. $\omega_2 \in \Omega_2$ the mapping $\omega_1 \mapsto f(\omega_1, \omega_2)$ on Ω_1 is in $L^1(\Omega_1, \mathcal{F}_1, \mu_1)$.
(iii) The mapping $\omega_1 \mapsto \int_{\Omega_2} f(\omega_1, \omega_2)\, d\mu_2(\omega_2)$ defined μ_1-a.e. on Ω_1 [in accordance with (i)] is in $L^1(\Omega_1, \mathcal{F}_1, \mu_1)$.
(iv) The mapping $\omega_2 \mapsto \int_{\Omega_1} f(\omega_1, \omega_2)\, d\mu_1(\omega_1)$ defined μ_2-a.e. on Ω_2 [in accordance with (ii)] is in $L^1(\Omega_2, \mathcal{F}_2, \mu_2)$.
(v) We have

$$\begin{aligned}\int_{\Omega_1\times\Omega_2} f(\omega_1,\omega_2)\, d\mu_1\otimes\mu_2(\omega_1,\omega_2) &= \int_{\Omega_1}\int_{\Omega_2} f(\omega_1,\omega_2)\, d\mu_2(\omega_2)\, d\mu_1(\omega_1)\\ &= \int_{\Omega_2}\int_{\Omega_1} f(\omega_1,\omega_2)\, d\mu_1(\omega_1)\, d\mu_2(\omega_2).\end{aligned}$$

Proof: If $f : \Omega_1 \times \Omega_2 \to \bar{\mathbb{R}}$ is $\mathcal{F}_1 \otimes \mathcal{F}_2/\mathcal{B}^*$-measurable, then $|f|$ is nonnegative and $\mathcal{F}_1 \otimes \mathcal{F}_2/\mathcal{B}^*$-measurable. By applying the Nonnegative Fubini Theorem to $|f|$, we have

$$\int_{\Omega_1\times\Omega_2} |f|\, d\mu_1 \otimes \mu_2 = \int_{\Omega_1}\int_{\Omega_2} |f|\, d\mu_2\, d\mu_1 = \int_{\Omega_2}\int_{\Omega_1} |f|\, d\mu_1\, d\mu_2,$$

so that all three integrals are finite. In particular, $\int_{\Omega_1\times\Omega_2} |f|\, d\mu_1 \otimes \mu_2$ is finite, which by definition gives $f \in L^1(\Omega_1 \times \Omega_2, \mathcal{F}_1 \otimes \mathcal{F}_2, \mu_1 \otimes \mu_2)$.

Now consider the assertions (i)-(v). Statements (i) and (ii) are symmetric, as are statements (iii) and (iv). The two equations in (v) that relate the double integral to the iterated integral are symmetric, and hence the proof only requires demonstrating (i), (iii), and the identity

$$\int_{\Omega_1\times\Omega_2} f(\omega_1,\omega_2)\, d\mu_1\otimes\mu_2(\omega_1,\omega_2) = \int_{\Omega_1}\int_{\Omega_2} f(\omega_1,\omega_2) d\mu_2(\omega_2)\, d\mu_1(\omega_1). \tag{10.3}$$

Proof of (i): Because of the finiteness assumption in (10.2), we have that $\int_{\Omega_1}\int_{\Omega_2} |f(\omega_1,\omega_2)|\, d\mu_2(\omega_2) d\mu_1(\omega_1) < +\infty$. It follows that

$$\int_{\Omega_2} |f(\omega_1,\omega_2)|\, d\mu_2(\omega_2) < +\infty \quad \text{for } \mu_1\text{-a.e. } \omega_1 \in \Omega_1.$$

In turn, for the aforementioned μ_1-a.e. $\omega_1 \in \Omega_1$, the display above forces the mapping $\omega_2 \mapsto |f(\omega_1, \omega_2)|$ on Ω_2 to be in $L^1(\Omega_2, \mathcal{F}_2, \mu_2)$. Therefore, for the aforementioned μ_1-a.e. $\omega_1 \in \Omega_1$, the mapping $\omega_2 \mapsto f(\omega_1, \omega_2)$ on Ω_2 is in $L^1(\Omega_2, \mathcal{F}_2, \mu_2)$. This gives (i).

Proof of (iii): Define

$$\Omega_{10} = \{\omega_1 \in \Omega_1 : \omega_2 \mapsto f(\omega_1, \omega_2) \text{ is in } L^1(\Omega_2, \mathcal{F}_2, \mu_2)\},$$

so that $\mu_1(\Omega_{10}^c) = 0$ by (i). Also by (i), the integral $\int_{\Omega_2} f(\omega_1, \omega_2)\, d\mu_2(\omega_2)$ exists as a real number for every $\omega_1 \in \Omega_{10}$. Write $f = f^+ - f^-$ as the usual decomposition of f into positive and negative parts. The functions f^+ and f^- are nonnegative and $\mathcal{F}_1 \otimes \mathcal{F}_2/\mathcal{B}^*$-measurable. Applying the Nonnegative Fubini Theorem to f^+ and f^-, the mappings

$$\omega_1 \mapsto \int_{\Omega_2} f^+(\omega_1, \omega_2)\, d\mu_2(\omega_2) \text{ and } \omega_1 \mapsto \int_{\Omega_2} f^-(\omega_1, \omega_2)\, d\mu_2(\omega_2)$$

on Ω_{10} are $\mathcal{F}_1/\mathcal{B}^*$-measurable. It follows that the mapping

$$\omega_1 \mapsto \int_{\Omega_2} f(\omega_1, \omega_2)\, d\mu_2(\omega_2) = \int_{\Omega_2} f^+(\omega_1, \omega_2)\, d\mu_2(\omega_2) - \int_{\Omega_2} f^-(\omega_1, \omega_2)\, d\mu_2(\omega_2)$$

on Ω_{10}, being the difference of $\mathcal{F}_1/\mathcal{B}^*$-measurable mappings on Ω_{10}, is itself an $\mathcal{F}_1/\mathcal{B}^*$-measurable mapping on Ω_{10}. Therefore, $\omega_1 \mapsto |\int_{\Omega_2} f(\omega_1, \omega_2)\, d\mu_2(\omega_2)|$ defined on Ω_{10} is $\mathcal{F}_1/\mathcal{B}^*$-measurable, so we may meaningfully discuss the iterated integral $\int_{\Omega_{10}} \left|\int_{\Omega_2} f(\omega_1, \omega_2)\, d\mu_2(\omega_2)\right| d\mu_1(\omega_1)$. We have

$$\begin{aligned} \int_{\Omega_{10}} \left|\int_{\Omega_2} f(\omega_1, \omega_2)\, d\mu_2(\omega_2)\right| d\mu_1(\omega_1) &\leq \int_{\Omega_{10}} \int_{\Omega_2} |f(\omega_1, \omega_2)|\, d\mu_2(\omega_2)\, d\mu_1(\omega_1) \\ &= \int_{\Omega_1} \int_{\Omega_2} |f(\omega_1, \omega_2)|\, d\mu_2(\omega_2)\, d\mu_1(\omega_1) \\ &< +\infty. \end{aligned}$$

[The equality holds because $\mu_1(\Omega_{10}^c) = 0$, and the last inequality holds because of the assumption of finiteness in (10.2).] Since $\mu_1(\Omega_{10}^c) = 0$, this gives

$$\int_{\Omega_1} \left|\int_{\Omega_2} f(\omega_1, \omega_2)\, d\mu_2(\omega_2)\right| d\mu_1(\omega_1) < +\infty,$$

showing that the mapping $\omega_1 \mapsto \int_{\Omega_2} f(\omega_1, \omega_2)\, d\mu_2(\omega_2)$ defined on Ω_{10} is in $L^1(\Omega_1, \mathcal{F}_1, \mu_1)$, giving (iii).

Proof of (v): This is straightforward. We have

$$\begin{aligned}
\int_{\Omega_1\times\Omega_2} f\,d\mu_1\otimes\mu_2 &= \int_{\Omega_{10}\times\Omega_2} f\,d\mu_1\otimes\mu_2 \\
&= \int_{\Omega_{10}\times\Omega_2} f^+\,d\mu_1\otimes\mu_2 - \int_{\Omega_{10}\times\Omega_2} f^-\,d\mu_1\otimes\mu_2 \\
&= \int_{\Omega_{10}}\int_{\Omega_2} f^+\,d\mu_2\,d\mu_1 - \int_{\Omega_{10}}\int_{\Omega_2} f^-\,d\mu_2\,d\mu_1 \\
&= \int_{\Omega_{10}}\int_{\Omega_2} [f^+ - f^-]\,d\mu_2\,d\mu_1 \\
&= \int_{\Omega_{10}}\int_{\Omega_2} f\,d\mu_2\,d\mu_1 \\
&= \int_{\Omega_1}\int_{\Omega_2} f\,d\mu_2\,d\mu_1.
\end{aligned}$$

[The first equality holds since

$$\begin{aligned}
\mu_1\otimes\mu_2((\Omega_1\times\Omega_2)-(\Omega_{10}\times\Omega_2)) = \mu_1\otimes\mu_2(\Omega_{10}^c\times\Omega_2) &= \mu_1(\Omega_{10}^c)\mu_2(\Omega_2) \\
&= 0\mu_2(\Omega_2) \\
&= 0.
\end{aligned}$$

The third equality follows by applying the Nonnegative Fubini Theorem to f^+ and f^-, and the last equality holds since $\mu_1(\Omega_{10}^c) = 0$.] This completes the entire proof. ■

Stripping away all explanatory material and combining the two Fubini Theorems, we have the following: given an $\mathcal{F}_1\otimes\mathcal{F}_2/\mathcal{B}^*$-measurable f on $\Omega_1\times\Omega_2$ that is nonnegative or in $L^1(\Omega_1\times\Omega_2, \mathcal{F}_1\otimes\mathcal{F}_2, \mu_1\otimes\mu_2)$, we have that all three integrals

$$\int_{\Omega_1\times\Omega_2} f\,d\mu_1\otimes\mu_2,\quad \int_{\Omega_1}\int_{\Omega_2} f\,d\mu_2\,d\mu_1,\quad\text{and}\quad \int_{\Omega_2}\int_{\Omega_1} f\,d\mu_1\,d\mu_2$$

exist and are equal. Recall that all of this takes place relative to σ-finite measure spaces $(\Omega_1, \mathcal{F}_1, \mu_1)$ and $(\Omega_2, \mathcal{F}_2, \mu_2)$.

We now discuss the Fubini Theorem[s] when the σ-finite measure spaces $(\Omega_1, \mathcal{F}_1, \mu_1)$ and $(\Omega_2, \mathcal{F}_2, \mu_2)$ are complete. It will be seen that the essential thrust of the theorems is unchanged, although some of the statements have minor "almost everywhere" modifications. Before proving the complete version of the Nonnegative Fubini Theorem, a very minor result is required.

Claim 1. *Let $(\Omega_1, \mathcal{F}_1, \mu_1)$ and $(\Omega_2, \mathcal{F}_2, \mu_2)$ denote complete σ-finite measure spaces. Let $E\in\mathcal{F}_1\otimes\mathcal{F}_2$ be such that $\mu_1\otimes\mu_2(E) = 0$, and let $F\subseteq E$. Then*

(a) $F^{\omega_2}\in\mathcal{F}_1$ for μ_2-a.e. $\omega_2\in\Omega_2$, and for such ω_2 we have $\mu_1(F^{\omega_2}) = 0$.
(b) $F^{\omega_1}\in\mathcal{F}_2$ for μ_1-a.e. $\omega_1\in\Omega_1$, and for such ω_1 we have $\mu_2(F_{\omega_1}) = 0$.

Proof: Statements (a) and (b) are similar, so we only need to prove (a). Using the definition of product measure and the Nonnegative Fubini Theorem applied to the $\mathcal{F}_1 \otimes \mathcal{F}_2/\mathcal{B}^*$-measurable function I_E on $\Omega_1 \times \Omega_2$, we have

$$\begin{aligned} 0 = \mu_1 \otimes \mu_2(E) &= \int_{\Omega_1 \times \Omega_2} I_E(\omega_1, \omega_2)\, d\mu_1 \otimes \mu_2(\omega_1, \omega_2) \\ &= \int_{\Omega_2} \int_{\Omega_1} I_{E^{\omega_2}}(\omega_1)\, d\mu_1(\omega_1) d\mu_2(\omega_2) \\ &= \int_{\Omega_2} \mu_1(E^{\omega_2})\, d\mu_2(\omega_2). \end{aligned}$$

It follows that $\mu_1(E^{\omega_2}) = 0$ for μ_2-a.e. $\omega_2 \in \Omega_2$. Since the containment relation $F \subseteq E$ forces $F^{\omega_2} \subseteq E^{\omega_2} \in \mathcal{F}_1$ for every $\omega_2 \in \Omega_2$, it follows by the completeness of $(\Omega_1, \mathcal{F}_1, \mu_1)$ that $F^{\omega_2} \in \mathcal{F}_1$ and $\mu_1(F^{\omega_2}) = 0$ for the aforementioned μ_2-a.e. $\omega_2 \in \Omega_2$. This completes the proof. ∎

Nonnegative Complete Fubini Theorem. *Consider the complete and σ-finite measure spaces* $(\Omega_1, \mathcal{F}_1, \mu_1)$ *and* $(\Omega_2, \mathcal{F}_2, \mu_2)$. *Let* $f : \Omega_1 \times \Omega_2 \to [0, +\infty]$ *be* $\overline{\mathcal{F}_1 \otimes \mathcal{F}_2}/\mathcal{B}^*$*-measurable. Then we have the following:*

(i) For μ_1*-a.e.* $\omega_1 \in \Omega_1$ *the mapping* $\omega_2 \mapsto f(\omega_1, \omega_2)$ *is* $\mathcal{F}_2/\mathcal{B}^*$*-measurable.*
(ii) For μ_2*-a.e.* $\omega_2 \in \Omega_2$ *the mapping* $\omega_1 \mapsto f(\omega_1, \omega_2)$ *is* $\mathcal{F}_1/\mathcal{B}^*$*-measurable.*[3].
(iii) The mapping $\omega_1 \mapsto \int_{\Omega_2} f(\omega_1, \omega_2)\, d\mu_2(\omega_2)$ *defined* μ_1*-a.e. on* Ω_1 *[in accordance with (i)] is* $\mathcal{F}_1/\mathcal{B}^*$*-measurable.*
(iv) The mapping $\omega_2 \mapsto \int_{\Omega_1} f(\omega_1, \omega_2)\, d\mu_1(\omega_1)$ *defined* μ_2*-a.e. on* Ω_2 *[in accordance with (ii)] is* $\mathcal{F}_2/\mathcal{B}^*$*-measurable.*
(v) We have that

$$\begin{aligned} \int_{\Omega_1 \times \Omega_2} f(\omega_1, \omega_2)\, d\overline{\mu_1 \otimes \mu_2}(\omega_1, \omega_2) &= \int_{\Omega_1} \int_{\Omega_2} f(\omega_1, \omega_2)\, d\mu_2(\omega_2)\, d\mu_1(\omega_1) \\ &= \int_{\Omega_2} \int_{\Omega_1} f(\omega_1, \omega_2)\, d\mu_1(\omega_1)\, d\mu_2(\omega_2). \end{aligned}$$

Proof: Statements (i) and (ii) are symmetric, as are statements (iii) and (iv). The two equations relating the double integral to the iterated integrals are also symmetric. Therefore, we only need to prove (i), (iii), and the equation

$$\int_{\Omega_1 \times \Omega_2} f(\omega_1, \omega_2)\, d\overline{\mu_1 \otimes \mu_2}(\omega_1, \omega_2) = \int_{\Omega_1} \int_{\Omega_2} f(\omega_1, \omega_2)\, d\mu_2(\omega_2)\, d\mu_1(\omega_1) \tag{10.4}$$

to complete matters. To prove (i), (iii), and (10.4), we proceed in the same fashion as in the proof of the Nonnegative Fubini Theorem, working from indicators to simple functions, and then to nonnegative functions.

[3]If f is $\mathcal{F}_1 \otimes \mathcal{F}_2/\mathcal{B}^*$- measurable, then the Nonnegative Fubini Theorem says that (i) and (ii) hold for all $\omega_1 \in \Omega_1$ and $\omega_2 \in \Omega_2$, respectively.

Indicator Functions. Let $f = I_A$, where $A \in \overline{\mathcal{F}_1 \otimes \mathcal{F}_2}$. Write $A = B \cup C$, where $B \in \mathcal{F}_1 \otimes \mathcal{F}_2$ and C is contained in some $N \in \mathcal{F}_1 \otimes \mathcal{F}_2$ with $\mu_1 \otimes \mu_2(N) = 0$. For every $\omega_1 \in \Omega_1$ we have $A_{\omega_1} = B_{\omega_1} \cup C_{\omega_1}$; the condition $B \in \mathcal{F}_1 \otimes \mathcal{F}_2$ forces $B_{\omega_1} \in \mathcal{F}_2$ for every $\omega_1 \in \Omega_1$. By Claim 1, we have $C_{\omega_1} \in \mathcal{F}_2$ and $\mu_2(C_{\omega_1}) = 0$ for μ_1-a.e. $\omega_1 \in \Omega_1$. Therefore, for the μ_1-a.e. $\omega_1 \in \Omega_1$ such that $C_{\omega_1} \in \mathcal{F}_2$ and $\mu_2(C_{\omega_1}) = 0$, we have that A_{ω_1} is the union of the two $\mathcal{F}_2$-sets B_{ω_1} and C_{ω_1} and hence is in $\mathcal{F}_2$; we also have $\mu_2(A_{\omega_1}) = \mu_2(B_{\omega_1})$ for the aforementioned μ_1-a.e. $\omega_1 \in \Omega_1$. For such $\omega_1 \in \Omega_1$, we have that $f_{\omega_1} = (I_A)_{\omega_1} = I_{A_{\omega_1}}$ is the indicator function of the set $A_{\omega_1} \in \mathcal{F}_2$, and hence f_{ω_1} is $\mathcal{F}_2/\mathcal{B}^*$-measurable for such $\omega_1 \in \Omega_1$. Therefore, (i) is true when $f = I_A$ and $A \in \overline{\mathcal{F}_1 \otimes \mathcal{F}_2}$.

We now prove (iii) for when $f = I_A$ with $A \in \overline{\mathcal{F}_1 \otimes \mathcal{F}_2}$. For the μ_1-a.e. $\omega_1 \in \Omega_1$ mentioned in (i), we have that $\int_{\Omega_2} f(\omega_1, \omega_2)\, d\mu_2(\omega_2)$ exists and

$$\int_{\Omega_2} f(\omega_1, \omega_2)\, d\mu_2(\omega_2) = \int_{\Omega_2} I_{A_{\omega_1}}(\omega_2)\, d\mu_2(\omega_2) = \mu_2(A_{\omega_1}) = \mu_2(B_{\omega_1}).$$

Therefore, the mapping $\omega_1 \mapsto \int_{\Omega_2} f(\omega_1, \omega_2)\, d\mu_2(\omega_2)$ defined for the μ_1-a.e. $\omega_1 \in \Omega_1$ in (i) is identical to the mapping $\omega_1 \mapsto \mu_2(B_{\omega_1})$ defined for the μ_1-a.e. $\omega_1 \in \Omega_1$ mentioned in (i). By the result given at the beginning of this section, this mapping is $\mathcal{F}_1/\mathcal{B}^*$-measurable. Therefore, (iii) holds when $f = I_A$ and $A \in \overline{\mathcal{F}_1 \otimes \mathcal{F}_2}$.

We now prove (10.4) for indicator functions. Observe that

$$\begin{aligned}
\int_{\Omega_1 \times \Omega_2} f(\omega_1, \omega_2)\, d\overline{\mu_1 \otimes \mu_2}(\omega_1, \omega_2) &= \int_{\Omega_1 \times \Omega_2} I_A(\omega_1, \omega_2)\, d\overline{\mu_1 \otimes \mu_2}(\omega_1, \omega_2) \\
&= \overline{\mu_1 \otimes \mu_2}(A) \\
&= \mu_1 \otimes \mu_2(B) \\
&= \int_{\Omega_1} \mu_2(B_{\omega_1})\, d\mu_1(\omega_1) \\
&= \int_{\Omega_1} \int_{\Omega_2} I_B(\omega_1, \omega_2)\, d\mu_2(\omega_2)\, d\mu_1(\omega_1) \\
&= \int_{\Omega_1} \int_{\Omega_2} f(\omega_1, \omega_2)\, d\mu_2(\omega_2)\, d\mu_1(\omega_1).
\end{aligned}$$

[The fourth equality follows by the definition of product measure.] It follows that (10.4) holds when $f = I_A$, where $A \in \overline{\mathcal{F}_1 \otimes \mathcal{F}_2}$. This completes the entire proof when $f = I_A$, with $A \in \overline{\mathcal{F}_1 \otimes \mathcal{F}_2}$.

Nonnegative Simple Functions. Let $f = \sum_{i=1}^n c_i I_{A_i}$, where $c_1, \cdots, c_n \geq 0$ and $A_1, \cdots, A_n$ are disjoint $\overline{\mathcal{F}_1 \otimes \mathcal{F}_2}$-sets with union $\Omega_1 \times \Omega_2$. We apply the result for indicator functions n separate times: for μ_1-a.e. $\omega_1 \in \Omega_1$ the mapping $(I_{A_1})_{\omega_1} = I_{(A_1)_{\omega_1}}$ defined on Ω_2 is $\mathcal{F}_2/\mathcal{B}^*$-measurable; $\cdots$; for μ_1-a.e. $\omega_1 \in \Omega_1$ the mapping $(I_{A_n})_{\omega_1} = I_{(A_n)_{\omega_1}}$ defined on Ω_2 is $\mathcal{F}_2/\mathcal{B}^*$-measurable. Therefore, for μ_1-a.e. $\omega_1 \in \Omega_1$ the mapping $f_{\omega_1} = \sum_{i=1}^n c_i I_{(A_i)_{\omega_1}}$ defined on Ω_2 is a linear combination of $\mathcal{F}_2/\mathcal{B}^*$-measurable functions on Ω_2

and hence is $\mathcal{F}_2/\mathcal{B}^*$-measurable on Ω_2. That is, for μ_1-a.e. $\omega_1 \in \Omega_1$ the mapping $\omega_2 \mapsto \sum_{i=1}^n c_i I_{A_i}(\omega_1, \omega_2)$ on Ω_2 is $\mathcal{F}_2/\mathcal{B}^*$-measurable. Therefore, (i) holds for simple functions as described.

We now handle (iii). For the μ_1-a.e. $\omega_1 \in \Omega_1$ as described in (i), we have that $\int_{\Omega_2} f(\omega_1, \omega_2)\, d\mu_2(\omega_2)$ exists, and

$$\begin{aligned}\int_{\Omega_2} f(\omega_1, \omega_2)\, d\mu_2(\omega_2) &= \int_{\Omega_2} \sum_{i=1}^n c_i I_{A_i}(\omega_1, \omega_2)\, d\mu_2(\omega_2) \\ &= \sum_{i=1}^n c_i \int_{\Omega_2} I_{A_i}(\omega_1, \omega_2)\, d\mu_2(\omega_2).\end{aligned}$$

Using the result given for indicators n separate times, we have that the mapping $\omega_1 \mapsto \int_{\Omega_2} I_{A_1}(\omega_1, \omega_2)\, d\mu_2(\omega_2)$ defined μ_1-a.e. on Ω_1 is $\mathcal{F}_1/\mathcal{B}^*$-measurable; $\cdots$; the mapping $\omega_1 \mapsto \int_{\Omega_2} I_{A_n}(\omega_1, \omega_2)\, d\mu_2(\omega_2)$ defined μ_1-a.e. on Ω_1 is $\mathcal{F}_1/\mathcal{B}^*$-measurable. It follows that the mapping

$$\omega_1 \mapsto \sum_{i=1}^n c_i \int_{\Omega_2} I_{A_i}(\omega_1, \omega_2)\, d\mu_2(\omega_2)$$

defined μ_1-a.e.[4] on Ω_1 is a linear combination of $\mathcal{F}_1/\mathcal{B}^*$-measurable mappings and is thus $\mathcal{F}_1/\mathcal{B}^*$-measurable. The mapping $\omega_1 \mapsto \int_{\Omega_2} f(\omega_1, \omega_2)\, d\mu_2(\omega_2)$ defined μ_1-a.e. on Ω_1 is accordingly $\mathcal{F}_1/\mathcal{B}^*$-measurable, and (iii) thus holds when f is an $\overline{\mathcal{F}_1 \otimes \mathcal{F}_2}/\mathcal{B}^*$-measurable simple function on $\Omega_1 \times \Omega_2$.

We now verify (10.4) for the simple function f as described above. We have

$$\begin{aligned}&\int_{\Omega_1 \times \Omega_2} f(\omega_1, \omega_2)\, d\overline{\mu_1 \otimes \mu_2}(\omega_1, \omega_2) \\ &= \int_{\Omega_1 \times \Omega_2} \sum_{i=1}^n c_i I_{A_i}(\omega_1, \omega_2)\, d\overline{\mu_1 \otimes \mu_2}(\omega_1, \omega_2) \\ &= \sum_{i=1}^n c_i \int_{\Omega_1 \times \Omega_2} I_{A_i}(\omega_1, \omega_2)\, d\overline{\mu_1 \otimes \mu_2}(\omega_1, \omega_2) \\ &= \sum_{i=1}^n c_i \int_{\Omega_1} \int_{\Omega_2} I_{A_i}(\omega_1, \omega_2)\, d\mu_2(\omega_2)\, d\mu_1(\omega_1) \\ &= \int_{\Omega_1} \int_{\Omega_2} \sum_{i=1}^n c_i I_{A_i}(\omega_1, \omega_2)\, d\mu_2(\omega_2)\, d\mu_1(\omega_1) \\ &= \int_{\Omega_1} \int_{\Omega_2} f(\omega_1, \omega_2)\, d\mu_2(\omega_2)\, d\mu_1(\omega_1).\end{aligned}$$

[4]The domain of this mapping is the set of $\omega_1 \in \Omega_1$ such that the previous n measurability conditions are met, hence μ_1-a.e. $\omega_1 \in \Omega_1$ are in the domain.

[We have used the fact that (10.4) holds for indicator functions in the third equality.] This gives (10.4) for nonnegative $\overline{\mathcal{F}_1 \otimes \mathcal{F}_2}/\mathcal{B}^*$-measurable simple functions, and hence the Nonnegative Complete Fubini Theorem holds for nonnegative $\overline{\mathcal{F}_1 \otimes \mathcal{F}_2}/\mathcal{B}^*$-measurable simple functions.

Nonnegative Functions. Let f be an $\overline{\mathcal{F}_1 \otimes \mathcal{F}_2}/\mathcal{B}^*$-measurable and nonnegative function on $\Omega_1 \times \Omega_2$. Find a nondecreasing sequence $\{s_n\}_{n=1}^{\infty}$ of nonnegative $\overline{\mathcal{F}_1 \otimes \mathcal{F}_2}/\mathcal{B}^*$-measurable simple functions on $\Omega_1 \times \Omega_2$ such that $\lim_n s_n = f$. Our first goal is to show that f_{ω_1} is an $\mathcal{F}_2/\mathcal{B}^*$-measurable mapping on Ω_2 for μ_1-a.e. $\omega_1 \in \Omega_1$. Observe first that $f_{\omega_1} = (\lim_n s_n)_{\omega_1} = \lim_n (s_n)_{\omega_1}$ for every $\omega_1 \in \Omega_1$. For every $n \in \mathbb{N}$, the previous work for simple functions shows that for μ_1-a.e. $\omega_1 \in \Omega_1$ the mapping $(s_n)_{\omega_1}$ on Ω_2 is $\mathcal{F}_2/\mathcal{B}^*$-measurable. It follows that for μ_1-a.e. $\omega_1 \in \Omega_1$ the mapping $\lim_n (s_n)_{\omega_1}$ defined on Ω_2 is $\mathcal{F}_2/\mathcal{B}^*$-measurable. This shows that for μ_1-a.e. $\omega_1 \in \Omega_1$ the mapping $\omega_2 \mapsto f(\omega_1, \omega_2)$ defined on Ω_2 is $\mathcal{F}_2/\mathcal{B}^*$-measurable, giving (i) in the general case.

We now handle (iii). For the μ_1-a.e. $\omega_1 \in \Omega_1$ as described in (i), the integral $\int_{\Omega_2} f(\omega_1, \omega_2)\, d\mu_2(\omega_2)$ exists, and thus for the μ_1-a.e. $\omega_1 \in \Omega_1$ as described in (i) we have by MCT that

$$\begin{aligned}\int_{\Omega_2} f(\omega_1, \omega_2)\, d\mu_2(\omega_2) &= \int_{\Omega_2} \lim_n s_n(\omega_1, \omega_2)\, d\mu_2(\omega_2) \\ &= \lim_n \int_{\Omega_2} s_n(\omega_1, \omega_2)\, d\mu_2(\omega_2).\end{aligned}$$

Using the fact that (iii) is true for simple functions shows that for every $n \in \mathbb{N}$ the mapping $\omega_1 \mapsto \int_{\Omega_2} s_n(\omega_1, \omega_2)\, d\mu_2(\omega_2)$ defined μ_1-a.e. on Ω_1 [on D_n, say] is $\mathcal{F}_1/\mathcal{B}^*$-measurable. Therefore, the mapping $\omega_1 \mapsto \int_{\Omega_2} f(\omega_1, \omega_2)\, d\mu_2(\omega_2)$ defined μ_1-a.e. on Ω_1 [more precisely, on $\bigcap_{n=1}^{\infty} D_n$] is the limit of a sequence of $\mathcal{F}_1/\mathcal{B}^*$-measurable mappings each defined μ_1-a.e. on Ω_1 [i.e. the restriction of each mapping $\omega_1 \mapsto \int_{\Omega_2} s_n(\omega_1, \omega_2)\, d\mu_2(\omega_2)$ to $\bigcap_{n=1}^{\infty} D_n$], and hence is itself an $\mathcal{F}_1/\mathcal{B}^*$-measurable mapping defined μ_1-a.e. on Ω_1. Therefore, the mapping $\omega_1 \mapsto \int_{\Omega_2} f(\omega_1, \omega_2)\, d\mu_2(\omega_2)$, defined μ_1-a.e. on Ω_1 is $\mathcal{F}_1/\mathcal{B}^*$-measurable, and so (iii) holds for general nonnegative $\overline{\mathcal{F}_1 \otimes \mathcal{F}_2}/\mathcal{B}^*$-measurable f.

Next, observe that

$$\begin{aligned}\int_{\Omega_1 \times \Omega_2} f(\omega_1, \omega_2)\, d\overline{\mu_1 \otimes \mu_2}(\omega_1, \omega_2) &= \int_{\Omega_1 \times \Omega_2} \lim_n s_n(\omega_1, \omega_2)\, d\overline{\mu_1 \otimes \mu_2}(\omega_1, \omega_2) \\ &= \lim_n \int_{\Omega_1 \times \Omega_2} s_n(\omega_1, \omega_2)\, d\overline{\mu_1 \otimes \mu_2}(\omega_1, \omega_2) \\ &= \lim_n \int_{\Omega_1} \int_{\Omega_2} s_n(\omega_1, \omega_2)\, d\mu_2(\omega_2)\, d\mu_1(\omega_1) \\ &= \int_{\Omega_1} \int_{\Omega_2} \lim_n s_n(\omega_1, \omega_2)\, d\mu_2(\omega_2)\, d\mu_1(\omega_1) \\ &= \int_{\Omega_1} \int_{\Omega_2} f(\omega_1, \omega_2)\, d\mu_2(\omega_2)\, d\mu_1(\omega_1).\end{aligned}$$

[MCT was used once in the second equality and twice in the fourth equality, and the third equality follows because (10.4) holds for each simple function s_n.] This shows that (10.4) holds for general nonnegative $\overline{\mathcal{F}_1 \otimes \mathcal{F}_2}/\mathcal{B}^*$-measurable f, and hence the entire proof of the Nonnegative Complete Fubini Theorem is at a close. ∎

The above proof is somewhat long, but that is more due to the spelling out of details whose spelling out is perhaps not needed at this stage. As stated, the structure of the proof is quite identical to that for the Nonnegative Fubini Theorem given earlier.

Just as there is a "complete" version of the Nonnegative Fubini Theorem, there is a "complete" version of the Integrable Fubini Theorem. With three Fubini Theorems and their proofs spelled out in detail, we shall merely state the fourth Fubini Theorem, as the proof follows wholly standard lines.

Integrable Complete Fubini Theorem. *Let $f : \Omega_1 \times \Omega_2 \to \bar{\mathbb{R}}$ denote a function that is $\overline{\mathcal{F}_1 \otimes \mathcal{F}_2}/\mathcal{B}^*$-measurable and such that at least one of*

$$\int_{\Omega_1 \times \Omega_2} |f| \, d\overline{\mu_1 \otimes \mu_2}, \quad \int_{\Omega_1} \int_{\Omega_2} |f| \, d\mu_2 d\mu_1, \quad or \quad \int_{\Omega_2} \int_{\Omega_1} |f| \, d\mu_1 d\mu_2$$

is finite. Then $f \in L^1(\Omega_1 \times \Omega_2, \overline{\mathcal{F}_1 \otimes \mathcal{F}_2}, \overline{\mu_1 \otimes \mu_2})$ and all three integrals are equal. We also have the following:

(i) For μ_1-a.e. $\omega_1 \in \Omega_1$ the mapping $\omega_2 \mapsto f(\omega_1, \omega_2)$ on Ω_2 is in $L^1(\Omega_2, \mathcal{F}_2, \mu_2)$.

(ii) For μ_2-a.e. $\omega_2 \in \Omega_2$ the mapping $\omega_1 \mapsto f(\omega_1, \omega_2)$ on Ω_1 is in $L^1(\Omega_1, \mathcal{F}_1, \mu_1)$.

(iii) The mapping $\omega_1 \mapsto \int_{\Omega_2} f(\omega_1, \omega_2) \, d\mu_2(\omega_2)$ defined μ_1-a.e. on Ω_1 [in accordance with (i)] is in $L^1(\Omega_1, \mathcal{F}_1, \mu_1)$.

(iv) The mapping $\omega_2 \mapsto \int_{\Omega_1} f(\omega_1, \omega_2) \, d\mu_1(\omega_1)$ defined μ_2-a.e. on Ω_2 [in accordance with (ii)] is in $L^1(\Omega_2, \mathcal{F}_2, \mu_2)$.

(v) We have that

$$\begin{aligned}\int_{\Omega_1 \times \Omega_2} f(\omega_1, \omega_2) \, d\overline{\mu_1 \otimes \mu_2}(\omega_1, \omega_2) &= \int_{\Omega_1} \int_{\Omega_2} f(\omega_1, \omega_2) \, d\mu_2(\omega_2) \, d\mu_1(\omega_1) \\ &= \int_{\Omega_2} \int_{\Omega_1} f(\omega_1, \omega_2) \, d\mu_1(\omega_1) \, d\mu_2(\omega_2).\end{aligned}$$

These four Fubini results constitute the main body of work in this direction. In subsequent sections we shall exhibit the application of the Fubini Theorems to various results in analysis and measure theory. In particular, we will in the next section relate the measure-theoretic results in this section to the special case of Euclidean space.

Exercises.

1. Let $\Omega_1 = \Omega_2 = [0, 1]$, let $\mathcal{F}_1$ and $\mathcal{F}_2$ denote the Borel subsets of $[0, 1]$, let μ_1 denote Lebesgue measure restricted to $\mathcal{F}_1$, and let μ_2 denote counting measure.

Define $E = \{(\omega_1, \omega_2) \in \Omega_1 \times \Omega_2 : \omega_1 = \omega_2\}$. Note that $(\Omega_2, \mathcal{F}_2, \mu_2)$ is not a σ-finite measure space, while $(\Omega_1, \mathcal{F}_1, \mu_1)$ is a σ-finite measure space.
(a) $E \in \mathcal{F}_1 \otimes \mathcal{F}_2$, hence I_E is $\mathcal{F}_1 \otimes \mathcal{F}_2/\mathcal{B}^*$-measurable.
(b) $\int_{\Omega_1} \int_{\Omega_2} I_E \, d\mu_2 \, d\mu_1 = 1$.
(c) $\int_{\Omega_2} \int_{\Omega_1} I_E \, d\mu_1 \, d\mu_2 = 0$. Therefore, the conclusion of the Nonnegative Fubini Theorem fails.
(d) What is $\int_{\Omega_1 \times \Omega_2} I_E \, d\mu_1 \otimes \mu_2 = \mu_1 \otimes \mu_2(E)$?

Comment. The previous exercise shows that the assumption that both $(\Omega_1, \mathcal{F}_1, \mu_1)$ and $(\Omega_2, \mathcal{F}_2, \mu_2)$ be σ-finite is not redundant.

2. Let Ω denote an uncountable set, and let $\mathcal{F}$ denote the collection of $A \subseteq \Omega$ such that A is at most countable or A^c is at most countable. Let μ be such that $\mu(E)$ is 0 or 1 as E is at most countable or E^c is at most countable, respectively.
(a) $(\Omega, \mathcal{F}, \mu)$ is a finite [hence σ-finite] measure space.
(b) Let $(\Omega_1, \mathcal{F}_1, \mu_1)$ and $(\Omega_2, \mathcal{F}_2, \mu_2)$ be identical with $(\Omega, \mathcal{F}, \mu)$. Let E denote a set such that E_{ω_1} is amc for each $\omega_1 \in \Omega_1$ and $(E^{\omega_2})^c$ is amc for each $\omega_2 \in \Omega_2$. Show that

$$\int_{\Omega_1} \int_{\Omega_2} I_E \, d\mu_2 d\mu_1 = 1 \neq 0 = \int_{\Omega_2} \int_{\Omega_1} I_E \, d\mu_1 d\mu_2.$$

(c) Use the Nonnegative Fubini Theorem to deduce that the set E in (b) cannot possibly be an element of $\mathcal{F}_1 \otimes \mathcal{F}_2$. In particular, $\int_{\Omega_1 \times \Omega_2} I_E \, d\mu_1 \otimes \mu_2$ fails to exist since I_E is not $\mathcal{F}_1 \otimes \mathcal{F}_2/\mathcal{B}^*$-measurable.

Comment. The previous exercise shows that the assumption of $\mathcal{F}_1 \otimes \mathcal{F}_2/\mathcal{B}^*$-measurability in the Nonnegative Fubini Theorem is not redundant. It also shows that both iterated integrals can exist while the double integral is undefined.

3. Let $\Omega = [0,1]$, let $\mathcal{L}$ denote the Lebesgue subsets of $[0,1]$, and let $\bar{\lambda}$ denote the restriction of Lebesgue measure to the Lebesgue subsets of $[0,1]$. Let $(\Omega_1, \mathcal{F}_1, \mu_1)$ and $(\Omega_2, \mathcal{F}_2, \mu_2)$ denote copies of the σ-finite measure space $(\Omega, \mathcal{F}, \mu)$.
(a) Suppose that $E \subseteq \Omega_1 \times \Omega_2$ is such that $\mu_2(E_{\omega_1}) = 0 = \mu_1([0,1] - E^{\omega_2})$ for every $\omega_1 \in \Omega_1$ and $\omega_2 \in \Omega_2$. Use the Nonnegative Fubini Theorem to deduce that $E \notin \mathcal{L} \otimes \mathcal{L}$. [Calculate the two iterated integrals of I_E and see that they are unequal; were $E \in \mathcal{L} \otimes \mathcal{L}$ the iterated integrals would agree. This is similar to the previous exercise.]

We desire to show that such a set E actually exists. To do this, assume the Continuum Hypothesis: $\operatorname{card} P_\xi = \aleph_1 = \mathfrak{c}$, where $P_\xi = \{\alpha : \alpha \prec \xi\}$, the set of ordinal numbers preceding ξ, the first uncountable ordinal number.

(b) There exists a one-to-one mapping $\phi : P_\xi \to [0,1]$ that is onto $[0,1]$.
(c) Let $E = \{(\phi(\alpha), \phi(\beta)) : \beta \prec \alpha \prec \xi\}$. Then for every $\omega_1 \in \Omega_1$ we have $\operatorname{card}(E_{\omega_1}) < \mathfrak{c}$ and for every $\omega_2 \in \Omega_2$ we have $\operatorname{card}([0,1] - E^{\omega_2}) < \mathfrak{c}$.
(d) Conclude [using (CH)] that in actuality both E_{ω_1} and $[0,1] - E^{\omega_2}$ are amc for each $\omega_1 \in \Omega_1$ and $\omega_2 \in \Omega_2$, and hence $\bar{\lambda}(E_{\omega_1}) = \bar{\lambda}([0,1] - E^{\omega_2}) = 0$ for every $\omega_1 \in \Omega_1$ and $\omega_2 \in \Omega_2$.

4. Let $\Omega_1 = \Omega_2 = [0,1]$, let $\mathcal{F}_1 = \mathcal{F}_2$ denote the Lebesgue subsets of $[0,1]$, and let $\mu_1 = \mu_2 = \bar{\lambda}$ denote Lebesgue measure restricted to the Lebesgue subsets of $[0,1]$. We will construct, using the Continuum Hypothesis, a set $S \subseteq [0,1] \times [0,1]$ such that $\operatorname{card}(S_{\omega_1}) \leq 1$ and $\operatorname{card}(S^{\omega_2}) \leq 1$ for every $(\omega_1, \omega_2) \in [0,1] \times [0,1]$, yet $S \notin \overline{\mathcal{F}_1 \otimes \mathcal{F}_2}$.

(a) Let $\mathcal{F}$ denote the collection of compact sets $F \subseteq [0,1] \times [0,1]$ such that the inequality $\mu_1 \otimes \mu_2(F) > 0$ obtains. Then $\text{card}(\mathcal{F}) = \mathfrak{c}$, and there is a mapping $\phi : P_\xi \to \mathcal{F}$ that is one-to-one and onto. [We are using the same ordinal notation as in the previous exercise.]
(b) We can choose a point in $\phi(0)$, say $(\omega_{10}, \omega_{20}) \in \phi(0)$.

Write $E_1 = \{(\omega_{10}, \omega_{20})\}$. It is trivial that $(E_1)_{\omega_1}$ and $(E_1)^{\omega_2}$ have cardinal number ≤ 1 for each $\omega_1 \in \Omega_1$ and $\omega_2 \in \Omega_2$.

Suppose that for the given ordinal $\beta \prec \xi$ the set $E_\beta = \{(\omega_{1\alpha}, \omega_{2\alpha}) : \alpha \prec \beta\}$ is such that every ω_1- and ω_2-section of E_β consists of at most one point. Define the set $B_\beta = \{\omega_1 \in \Omega_1 : \bar{\lambda}(\phi(B)_{\omega_1}) > 0\}$.
(c) $\bar{\lambda}(B_\beta) > 0$.
(d) There is a point, call it $\omega_{1\beta}$, in $B_\beta \cap \{\omega_{1\alpha} : \alpha \prec \beta\}^c$.
(e) For this $\omega_{1\beta}$, we have $\bar{\lambda}(\phi(\beta)_{\omega_{1\beta}}) > 0$; from this there exists a point $\omega_{2\beta}$ in $\phi(\beta)_{\omega_{1\beta}} \cap \{\omega_{2\alpha} : \alpha \prec \beta\}^c$.
(f) The set $\{(\omega_{1\alpha}, \omega_{2\alpha}) : \alpha \prec \beta\}$ has ω_1- and ω_2-sections of cardinality at most one.

The ordinal $\beta \prec \xi$ was arbitrary, hence by transfinite induction we may perform (c)-(f) for all ordinals $\alpha \prec \xi$, obtaining a pair $(\omega_{1\alpha}, \omega_{2\alpha})$ as described above for each $\alpha \prec \xi$. Let $S = \{(\omega_{1\alpha}, \omega_{2\alpha}) : \alpha \prec \xi\}$.
(g) $\text{card}(S_{\omega_1}) \leq 1$ and $\text{card}(S^{\omega_2}) \leq 1$ for each $\omega_1 \in \Omega_1$ and $\omega_2 \in \Omega_2$.
(h) $S \cap \phi(\alpha) \neq \varnothing$ for each $\alpha \prec \xi$.
(i) Conclude that $S \notin \overline{\mathcal{F}_1 \otimes \mathcal{F}_2}$. [Every measurable subset of $[0,1] \times [0,1]$ of positive measure contains some $\phi(\alpha)$.]

5. Suppose that $f : \Omega_1 \to [0, +\infty]$ is $\mathcal{F}_1/\mathcal{B}^*$-measurable, and let $g : \Omega_2 \to [0, +\infty]$ be $\mathcal{F}_2/\mathcal{B}^*$-measurable. Define $h(\omega_1, \omega_2) = f(\omega_1)g(\omega_2)$ for each $\omega_1 \in \Omega_1$ and $\omega_2 \in \Omega_2$.
(a) $h : \Omega_1 \times \Omega_2 \to [0, +\infty]$ is $\mathcal{F}_1 \otimes \mathcal{F}_2/\mathcal{B}^*$-measurable.
(b) $\int_{\Omega_1 \times \Omega_2} h \, d\mu_1 \otimes \mu_2 = \int_{\Omega_1} f \, d\mu_1 \int_{\Omega_2} g \, d\mu_2$. [Use the Nonnegative Fubini Theorem, of course.]

6. Assume the Continuum Hypothesis. Let $(\Omega_1, \mathcal{F}_1, \mu_1)$ and $(\Omega_2, \mathcal{F}_2, \mu_2)$ coincide with $([0,1], \mathcal{L}, \bar{\lambda})$, where $\mathcal{L}$ denotes the Lebesgue subsets of $[0,1]$ and $\bar{\lambda}$ denotes Lebesgue measure on the Lebesgue subsets of $[0,1]$.
(a) There is a one-to-one mapping j of $[0,1]$ onto the well-ordered set P_ξ, the set of ordinals preceding the first uncountable ordinal ξ.

Let $E = \{(x,y) \in [0,1] \times [0,1] : j(x) \prec j(y)\}$.
(b) For each $x \in [0,1]$, E_x contains all but countably many points of $[0,1]$.
(c) For each $y \in [0,1]$, E^y contains at most countably many points of $[0,1]$.
(d) Let $f = I_E$. Then f_x and f^y are Borel measurable for each $x, y \in [0,1]$.
(e) We have $\int_{\Omega_2} f_x \, d\mu_2 = 1$ for every $x \in [0,1]$ and $\int_{\Omega_1} f^y \, d\mu_1$ for every $y \in [0,1]$.
(f) $\int_{\Omega_1} \int_{\Omega_2} f \, d\mu_2 \, d\mu_1 = 1 \neq 0 = \int_{\Omega_2} \int_{\Omega_1} f \, d\mu_1 \, d\mu_2$.
(g) Why does (f) not contradict any of the Fubini Theorems?

7. Suppose that $\mu_1(\Omega_1) = \mu_2(\Omega_2) = 1$. Let $A_0 \in \mathcal{F}_1$ be such that $\mu_1(A_0) = \frac{1}{2}$, and let $B_0 \in \mathcal{F}_2$ be such that $\mu_2(B_0) = \frac{1}{2}$. Define $f = 2I_{(A_0 \times \Omega_2) \Delta (\Omega_1 \times B_0)}$. For each $E \in \mathcal{F}_1 \otimes \mathcal{F}_2$, define $\eta(E) = \int_E f \, d\mu_1 \otimes \mu_2$.
(a) η is a finite measure on $\mathcal{F}_1 \otimes \mathcal{F}_2$.
(b) For every $A \in \mathcal{F}_1$ we have $\eta(A \times \Omega_2) = \mu_1(A) = \mu_1 \otimes \mu_2(A \times \Omega_2)$,
(c) For every $B \in \mathcal{F}_2$ we have $\eta(\Omega_1 \times B) = \mu_2(B) = \mu_1 \otimes \mu_2(\Omega_1 \times B)$.

(d) Conclude that the product measure $\mu_1 \otimes \mu_2$ is not uniquely determined by its values on sets of the form $A \times \Omega_2$ and $\Omega_1 \times B$, where $A \in \mathcal{F}_1$ and $B \in \mathcal{F}_2$.

8. Consider the σ-finite measure spaces $(\mathbb{R}, \mathcal{L}, \bar{\lambda})$ and $(\Omega, \mathcal{F}, \mu)$. Let f denote a nonnegative $\mathcal{L}/\mathcal{B}^*$-measurable function on Ω.
(a) The set $\{(\omega, y) \in \Omega \times \mathbb{R} : 0 \le y \le f(\omega)\}$ is in $\mathcal{F} \otimes \mathcal{L}$. [Define $g(\omega, x) = f(\omega)$ for each $(\omega, x) \in \Omega \times \mathbb{R}$. Show that g is $\mathcal{F} \otimes \mathcal{L}/\mathcal{B}^*$-measurable. Write the given set as the inverse image of $[y, +\infty]$ under g.]
(b) $\int_\Omega f \, d\mu = \mu \otimes \bar{\lambda}(\{(\omega, y) \in \Omega \times \mathbb{R} : 0 \le y \le f(\omega)\}.)$
(c) Relate (b) to "the area under the curve".

9. Prove the Integrable Complete Fubini Theorem. [Write $f = f^+ - f^-$ and apply the Nonnegative Complete Fubini Theorem to f^+ and f^-.]

10. Let $\{a_{ij} : i, j \in \mathbb{N}\}$ denote a double sequence of nonnegative numbers.
(a) Deduce the well-known advanced calculus result

$$\sum_{i,j \in \mathbb{N}} a_{ij} = \sum_{i=1}^{\infty} \sum_{j=1}^{\infty} a_{ij} = \sum_{j=1}^{\infty} \sum_{i=1}^{\infty} a_{ij}$$

by relating things to the Fubini framework. [Let $\Omega_1 = \Omega_2 = \mathbb{N}$, let $\mathcal{F}_1 = \mathcal{F}_2 = 2^{\mathbb{N}}$, and let $\mu_1 = \mu_2$ denote counting measure. Show that $\mu_1 \otimes \mu_2$ is counting measure on $\mathcal{F}_1 \otimes \mathcal{F}_2 = 2^{\mathbb{N} \times \mathbb{N}}$. Define $f : \Omega_1 \times \Omega_2 \to [0, +\infty]$ by writing $f(i, j) = a_{ij}$. Then $\int_{\Omega_1 \times \Omega_2} f \, d\mu_1 \otimes \mu_2 = \sum_{i,j \in \mathbb{N}} a_{ij}$ and $\int_{\Omega_1} \int_{\Omega_2} f \, d\mu_2 \, d\mu_1 = \sum_{i=1}^{\infty} \sum_{j=1}^{\infty} a_{ij}$.]
(b) Under what conditions does the preceding equation hold if the a_{ij}'s are not necessarily nonnegative? [Transcribe the Integrable Fubini Theorem conditions into this framework.]

11. The conclusions of the Nonnegative Fubini Theorem holds if instead of assuming that both $(\Omega_1, \mathcal{F}_1, \mu_1)$ and $(\Omega_2, \mathcal{F}_2, \mu_2)$ are σ-finite we merely assume that $\{(\omega_1, \omega_2) \in \Omega_1 \times \Omega_2 : f(\omega_1, \omega_2) \ne 0\}$ is σ-finite with respect to $\mu_1 \otimes \mu_2$.

12. Let $\Omega_1 = \Omega_2 = \mathbb{N}$, let $\mathcal{F}_1 = \mathcal{F}_2 = 2^{\mathbb{N}}$, and let $\mu_1 = \mu_2$ denote counting measure. Define $f : \Omega_1 \times \Omega_2 \to \mathbb{R}$ by writing

$$f(\omega_1, \omega_2) = \begin{cases} 2 - 2^{-\omega_1} & \omega_1 = \omega_2 \\ -2 + 2^{-\omega_1} & \omega_1 = \omega_2 + 1 \\ 0 & \text{otherwise} \end{cases}$$

for each $(\omega_1, \omega_2) \in \Omega_1 \times \Omega_2$.
(a) $\int_{\Omega_1} \int_{\Omega_2} f \, d\mu_2 \, d\mu_1 = 0$ and $\int_{\Omega_2} \int_{\Omega_1} f \, d\mu_1 \, d\mu_2 = -\frac{1}{2}$.
(b) $\int_{\Omega_1 \times \Omega_2} f \, d\mu_1 \otimes \mu_2$ does not exist, as both

$$\int_{\Omega_1 \times \Omega_2} f^+ \, d\mu_1 \otimes \mu_2 = \int_{\Omega_1 \times \Omega_2} f^- \, d\mu_1 \otimes \mu_2 = +\infty.$$

(c) We have that $\int_{\Omega_1 \times \Omega_2} |f| \, d\mu_1 \otimes \mu_2 = +\infty$, hence there is no contradiction with the Integrable Fubini Theorem.
[We have $\int_{\Omega_2} f(\omega_1, \omega_2) \, d\mu_2(\omega_2) = \sum_{\omega_2=1}^{\infty} f(\omega_1, \omega_2) = f(\omega_1, \omega_1) + f(\omega_1, \omega_1 - 1) = 0$ for every $\omega_1 \in \Omega_1$, hence $\int_{\Omega_1} \int_{\Omega_2} f(\omega_1, \omega_2) \, d\mu_2(\omega_2) d\mu_1(\omega_1) = \int_{\Omega_1} 0 \, d\mu_1(\omega_1) = 0$. Next, $\int_{\Omega_1} f(\omega_1, \omega_2) \, d\mu_1(\omega_1) = \sum_{\omega_1=1}^{\infty} f(\omega_1, \omega_2) = f(\omega_2, \omega_2) + f(\omega_2 + 1, \omega_2) = -(\frac{1}{2})^{\omega_2+1}$,

hence $\int_{\Omega_2}\int_{\Omega_1} f(\omega_1,\omega_2)\, d\mu_1(\omega_1)d\mu_2(\omega_2) = \sum_{\omega_2=1}^{\infty} -(\frac{1}{2})^{\omega_2+1} = -\frac{1}{2}$. Next, we have that $\int_{\Omega_1\times\Omega_2} f^+\, d\mu_1 \otimes \mu_2 = f(1,1) + f(2,2) + f(3,3) + \cdots = +\infty$. The double integral of f^- is likewise computed by $f(2,1) + f(3,2) + f(4,3) + \cdots = +\infty$.]

13. There is a measure μ on $\mathcal{F}_1 \otimes \mathcal{F}_2$ such that $\mu(E \times F) = \phi(E, F)$ iff $\phi(E, \cdot)$ is a measure on $\mathcal{F}_2$ for every $E \in \mathcal{F}_1$ and $\phi(\cdot, F)$ is a measure on $\mathcal{F}_1$ for every $F \in \mathcal{F}_2$.

14. Consider a measure space $(\Omega_1, \mathcal{F}_1, \mu_1)$ with $\mu_1(\Omega_1) = 1$. Suppose for every $\omega_1 \in \Omega_1$ that $(\Omega_2, \mathcal{F}_2, \mu_{2,\omega_1})$ is a measure space with $\mu_{2,\omega_1}(\Omega_2) = 1$. Next, suppose for each $B \in \mathcal{F}_2$ that the mapping $\omega_1 \mapsto \mu_{2,\omega_1}(B)$ on Ω_1 is $\mathcal{F}_1/\mathcal{B}^*$-measurable.
(a) If $E \in \mathcal{F}_1 \otimes \mathcal{F}_2$, then the mapping $\omega_1 \mapsto \mu_{2,\omega_1}(E_{\omega_1})$ on Ω_1 is $\mathcal{F}_1/\mathcal{B}^*$-measurable.
(b) For each $E \in \mathcal{F}_1 \otimes \mathcal{F}_2$, define $\pi(E) = \int_{\Omega_1} \mu_{2,\omega_1}(E_{\omega_1})\, d\mu_1(\omega_1)$. Then we have that $(\Omega_1 \times \Omega_2, \mathcal{F}_1 \otimes \mathcal{F}_2, \pi)$ is a measure space with $\pi(\Omega_1 \times \Omega_2) = 1$.
(c) If $f : \Omega_1 \times \Omega_2 \to [0, +\infty]$ is $\mathcal{F}_1 \otimes \mathcal{F}_2/\mathcal{B}^*$-measurable, then the mapping on Ω_1 given by $\omega_1 \mapsto \int_{\Omega_2} f(\omega_1, \omega_2)\, d\mu_{2,\omega_1}(\omega_2)$ is $\mathcal{F}_1/\mathcal{B}^*$-measurable.
(d) For the f in (c) we have

$$\int_{\Omega_1\times\Omega_2} f(\omega_1,\omega_2)\, d\pi(\omega_1,\omega_2) = \int_{\Omega_1}\int_{\Omega_2} f(\omega_1,\omega_2)\, d\mu_{2,\omega_1}(\omega_2)\, d\mu_1(\omega_1).$$

(e) Formulate analogous claims for (c) and (d) if f maps into $\bar{\mathbb{R}}$ instead of $[0, +\infty]$.
(f) Let $\nu(B) = \int_{\Omega_1} \mu_{2,\omega_1}(B)\, d\mu_1(\omega_1)$ for each $B \in \mathcal{F}_2$. Then $\pi(\Omega_1 \times B) = \nu(B)$ for every $B \in \mathcal{F}_2$.
(g) In (f) we have

$$\int_{\Omega_2} f(\omega_2)\, d\nu(\omega_2) = \int_{\Omega_1}\int_{\Omega_2} f(\omega_2)\, d\mu_{2,\omega_1}(\omega_2) d\mu_1(\omega_1)$$

whenever $f : \Omega_2 \to [0, +\infty]$ is $\mathcal{F}_2/\mathcal{B}^*$-measurable.
(h) Redo (g) in the case where f is not necessarily nonnegative.

Comment. The previous exercise is a generalization of the Fubini Theorems. To see this, observe that the equation above reduces to the usual forms if μ_{2,ω_1} does not depend on ω_1.

15. Let Ω_1 denote a set and let $\mathcal{F}_1$ denote a σ-field on Ω_1. Let μ and μ_0 denote measures with domain $\mathcal{F}_1$. Next, let Ω_2 denote a set with $\mathcal{F}_2$ denoting a σ-field on Ω_2. Let ν and ν_0 denote measures with $\mathcal{F}_2$.
(a) If $\mu_0 \ll \mu$ and $\nu_0 \ll \nu$, then $\mu_0 \otimes \nu_0 \ll \mu \otimes \nu$. [Suppose $E \in \mathcal{F}_1 \otimes \mathcal{F}_2$ is such that $\mu \otimes \nu(E) = 0$. Then $0 = \mu \otimes \nu(E) = \int_{\Omega_1} \nu(E_{\omega_1})\, d\mu_1(\omega_1)$, hence $\nu(E_{\omega_1}) = 0$ for μ_1-a.e. $\omega_1 \in \Omega_1$. Thus $\mu(A) = 0$, where $A = \{\omega_1 \in \Omega_1 : \nu(E_{\omega_1}) > 0\}$; the assumption $\mu_0 \ll \mu$ gives $\mu_0(A) = 0$ as well. The assumption $\nu_0 \ll \nu$ shows that for every $\omega_1 \in A^c$ we have $\nu_0(E_{\omega_1}) = 0$. It follows that

$$\begin{aligned} \mu_0 \otimes \nu_0(E) = \int_{\Omega_1} \nu_0(E_{\omega_1})\, d\mu_0(\omega_1) &= \int_A \nu_0(E_{\omega_1})\, d\mu_0(\omega_1) + \int_{A^c} \nu_0(E_{\omega_1})\, d\mu_0(\omega_1) \\ &= 0 + 0 = 0.] \end{aligned}$$

(b) Keep assumptions as in (a). Suppose that f_0 is a Radon-Nikodym derivative of μ_0 with respect to μ, and suppose that g_0 is a Radon-Nikodym derivative of ν_0 with respect to ν. Then $f_0 g_0$ is a Radon-Nikodym derivative of $\mu_0 \otimes \nu_0$ with respect to $\mu \otimes \nu$. In symbols: $\frac{d\mu_0}{d\mu}\frac{d\nu_0}{d\nu} = \frac{d\mu_0\otimes\nu_0}{d\mu\otimes\nu}$ $\mu \otimes \nu$-a.e. on $\Omega_1 \times \Omega_2$. [Let $A \in \mathcal{F}_1 \otimes \mathcal{F}_2$ be such that $\mu_0 \otimes \nu_0(A) < +\infty$. Let $f \in L^1(\Omega_1 \times \Omega_2, \mathcal{F}_1 \otimes \mathcal{F}_2, \mu_0 \otimes \nu_0)$. It is relatively straightforward to verify that

$$\begin{aligned}
\int_{\Omega_1 \times \Omega_2} f(\omega_1, \omega_2)\, d\mu_0 \otimes \nu_0(\omega) &= \int_{\Omega_1} \int_{\Omega_2} f(\omega_1, \omega_2)\, d\nu_0(\omega_2) d\mu_0(\omega_1) \\
&= \int_{\Omega_1} \int_{\Omega_2} f(\omega_1, \omega_2) g_0(\omega_2)\, d\nu(\omega_2) d\mu_0(\omega_1) \\
&= \int_{\Omega_1} \int_{\Omega_2} f(\omega_1, \omega_2) g_0(\omega_2) d\nu(\omega_2) f_0(\omega_1) d\mu(\omega_1) \\
&= \int_{\Omega_1} \int_{\Omega_2} f(\omega_1, \omega_2) f_0(\omega_1) g_0(\omega_2)\, d\nu(\omega_2) d\mu(\omega_1) \\
&= \int_{\Omega_1 \times \Omega_2} f(\omega_1, \omega_2) f_0(\omega_1) g_0(\omega_2)\, d\mu \otimes \nu(\omega_1, \omega_2).
\end{aligned}$$

The last equality requires showing that $(\omega_1, \omega_2) \mapsto f(\omega_1, \omega_2) f_0(\omega_1) g_0(\omega_2)$ on the set $\Omega_1 \times \Omega_2$ is $\mathcal{F}_1 \otimes \mathcal{F}_2 / \mathcal{B}^*$-measurable. For this it is sufficient to show that the mappings $(\omega_1, \omega_2) \mapsto f_0(\omega_1)$ and $(\omega_1, \omega_2) \mapsto g_0(\omega_2)$ on $\Omega_1 \times \Omega_2$ are $\mathcal{F}_1 \otimes \mathcal{F}_2 / \mathcal{B}^*$-measurable; this should be a simple consequence of the facts that f_0 is $\mathcal{F}_1 / \mathcal{B}^*$-measurable and g_0 is $\mathcal{F}_2 / \mathcal{B}^*$-measurable. Taking $f = I_A$ gives

$$\mu_0 \otimes \nu_0(A) = \int_A f_0(\omega_1) g_0(\omega_2)\, d\mu \otimes \nu(\omega_1, \omega_2).$$

The above equation thus holds for every $A \in \mathcal{F}_1 \otimes \mathcal{F}_2$ with $\mu_0 \otimes \nu_0(A) < +\infty$. Using σ-finiteness, it is straightforward to show that the above holds for every $A \in \mathcal{F}_1 \otimes \mathcal{F}_2$.]
(c) Assume now that $\mu_0 \perp \mu$ or $\nu_0 \perp \nu$. Then $\mu_0 \otimes \nu_0 \perp \mu \otimes \nu$. [Assume that $\mu_0 \perp \mu$. Find $B \in \mathcal{F}_1$ with $\mu(B) = 0 = \mu_0(B^c)$. We have $\mu \otimes \nu(B \times \Omega_2) = 0$ and $\mu_0 \otimes \nu_0((B \times \Omega_2)^c) = 0$, so $\mu_0 \otimes \nu_0$ and $\mu \otimes \nu$ have disjoint supports.]
(d) Let η_1 and η_2 be σ-finite measures on $\mathcal{F}_1$, and let η_3 and η_4 denote σ-finite measures on $\mathcal{F}_2$. Then

$$(\eta_1 + \eta_2) \otimes (\eta_3 + \eta_4) = (\eta_1 \otimes \eta_3) + (\eta_1 \otimes \eta_4) + (\eta_2 \otimes \eta_3) + (\eta_2 \otimes \eta_4).$$

[Direct calculation shows that the measure on the left side coincides with the measure on the right on $\mathcal{F}_1 \times \mathcal{F}_2$. Therefore the measures coincide on $\mathcal{A}$, the field of finite disjoint unions of $\mathcal{F}_1 \times \mathcal{F}_2$-sets. Use the Uniqueness Theorem in Section 3.5 to show that the measures on the left and right agree on $\mathcal{F}_1 \otimes \mathcal{F}_2 = \sigma(\mathcal{A})$.]
(e) Write $\mu_0 = \mu_{0,ac} + \mu_{0,s}$ as the [unique] Lebesgue decomposition of μ_0 with respect to μ; write $\nu_0 = \nu_{0,ac} + \nu_{0,s}$ as the [unique] Lebesgue decomposition of ν_0 with respect to ν. Show that

$$\mu_0 \otimes \nu_0 = [\mu_{0,ac} \otimes \nu_{0,ac}] + [(\mu_{0,ac} \otimes \nu_{0,s}) + (\mu_{0,s} \otimes \nu_{0,ac}) + (\mu_{0,s} \otimes \nu_{0,s})]$$

is the [unique] Lebesgue decomposition of $\mu_0 \otimes \nu_0$ with respect to $\mu \otimes \nu$. [We have $\mu_{0,ac} \ll \mu$ and $\nu_{0,ac} \ll \nu$, hence $\mu_{0,ac} \otimes \nu_{0,ac} \ll \mu \otimes \nu$ by (a). Since $\mu_{0,s} \perp \mu$, part (c) gives $\mu_{0,s} \otimes \nu_{0,ac} \perp \mu \otimes \nu$ and $\mu_{0,s} \perp \nu_{0,s} \perp \mu \otimes \nu$. Since $\nu_{0,s} \perp \nu$, we have $\mu_{0,ac} \otimes \nu_{0,ac} \perp \mu \otimes \nu$. It follows that $(\mu_{0,s} \otimes \nu_{0,ac}) + (\mu_{0,ac} \otimes \nu_{0,s}) + (\mu_{0,s} \otimes \nu_{0,s}) \perp \mu \otimes \nu$. Use the uniqueness of the Lebesgue decomposition to finish things.]

10.3 THE FUBINI THEOREMS IN EUCLIDEAN SPACE

We have established the existence of product measures and some results concerning "interchanging the order of integration" for general σ-finite measure spaces $(\Omega_1, \mathcal{F}_1, \mu_1)$ and $(\Omega_2, \mathcal{F}_2, \mu_2)$. In this section we look at the very special but important case where the underlying spaces are Euclidean. We will consider the complete σ-finite measure spaces $(\mathbb{R}^k, \mathcal{L}^k, \bar{\lambda}_k)$ and $(\mathbb{R}^m, \mathcal{L}^m, \bar{\lambda}_m)$, and we shall consider their product, the [incomplete] σ-finite measure space $(\mathbb{R}^{k+m}, \mathcal{L}^k \otimes \mathcal{L}^m, \bar{\lambda}_k \otimes \bar{\lambda}_m)$. We shall actually work with the completion of this measure space, which means that we will be working in $(\mathbb{R}^{k+m}, \mathcal{L}^{k+m}, \bar{\lambda}_{k+m})$. [Recall Claim 7 of Section 4.2.] The two "complete" Fubini Theorems in the previous section assert the following: *If f is a nonnegative $\mathcal{L}^{k+m}$-measurable function on $\mathbb{R}^{k+m}$, or if $f \in L^1(\mathbb{R}^{k+m}, \mathcal{L}^{k+m}, \bar{\lambda}_{k+m})$, then*

$$\begin{aligned} \int_{\mathbb{R}^{k+m}} f(\boldsymbol{x}, \boldsymbol{y})\, d\bar{\lambda}_{k+m}(\boldsymbol{x}, \boldsymbol{y}) &= \int_{\mathbb{R}^k} \int_{\mathbb{R}^m} f(\boldsymbol{x}, \boldsymbol{y})\, d\bar{\lambda}_m(\boldsymbol{y})\, d\bar{\lambda}_k(\boldsymbol{x}) \\ &= \int_{\mathbb{R}^m} \int_{\mathbb{R}^k} f(\boldsymbol{x}, \boldsymbol{y})\, d\bar{\lambda}_k(\boldsymbol{x})\, d\bar{\lambda}_m(\boldsymbol{y}). \end{aligned}$$

[That everything above exists and makes sense is implicit in asserting that the integrals are equal.] We wish to give a few examples of Fubini's Theorem in action in this special situation. Our work depends on the work in Chapter 7, where conditions under which the Riemann and Lebesgue integrals coincide were given. We take $k = m = 1$ in our two examples.

Example 1. Consider calculating the Lebesgue integral

$$\int_B x^2(x+y)\, d\bar{\lambda}_2(x,y), \quad \text{where } B = \{(x,y) \in \mathbb{R}^2 : 0 < x < y < 1\}.$$

The set B is clearly in $\mathcal{L}^2$. Our desired Lebesgue integral is equal to

$$\int_{\mathbb{R}^2} x^2(x+y) I_B(x,y)\, d\bar{\lambda}_2(x,y).$$

This integral exists because the mapping $(x,y) \mapsto x^2(x+y)I_B(x,y)$ on $\mathbb{R}^2$ is nonnegative and $\mathcal{L}^2/\mathcal{B}^*$-measurable. By the Nonnegative Complete Fubini Theorem, we have [writing $A = (0,1)$]

$$\begin{aligned} &\int_{\mathbb{R}^2} x^2(x+y) I_B(x,y)\, d\bar{\lambda}_2(x,y) \\ = &\int_{\mathbb{R}} \int_{\mathbb{R}} x^2(x+y) I_B(x,y)\, d\bar{\lambda}(y)\, d\bar{\lambda}(x) \\ = &\int_A \int_{\mathbb{R}} x^2(x+y) I_B(x,y)\, d\bar{\lambda}(y)\, d\bar{\lambda}(x) + \int_{A^c} \int_{\mathbb{R}} x^2(x+y) I_B(x,y)\, d\bar{\lambda}(y)\, d\bar{\lambda}(x) \\ = &\int_{(0,1)} \int_{\mathbb{R}} x^2(x+y) I_B(x,y)\, d\bar{\lambda}(y)\, d\bar{\lambda}(x), \end{aligned}$$

where the last equality holds because $I_B(x,y) = 0$ if $x \notin (0,1)$. For each $x \in (0,1)$, we have that $I_B(x,y) = 1$ iff $y \in (x,1)$, and hence the last expression becomes

$$\begin{aligned}
&\int_{(0,1)} \int_{\mathbb{R}} x^2(x+y) I_B(x,y)\, d\bar{\lambda}(y)\, d\bar{\lambda}(x) \\
&= \int_{(0,1)} \int_{(x,1)} x^2(x+y) I_B(x,y)\, d\bar{\lambda}(y) d\bar{\lambda}(x) \\
&\quad + \int_{(0,1)} \int_{(x,1)^c} x^2(x+y) I_B(x,y)\, d\bar{\lambda}(y)\, d\bar{\lambda}(x) \\
&= \int_{(0,1)} \int_{(x,1)} x^2(x+y)\, d\bar{\lambda}(y) d\bar{\lambda}(x).
\end{aligned}$$

For every $x \in (0,1)$ the mapping $y \mapsto x^2(x+y)$ on $\mathbb{R}$ is continuous everywhere [hence $\bar{\lambda}$-a.e. on $\mathbb{R}$], and hence for every $x \in (0,1)$ we have

$$\begin{aligned}
\int_{(x,1)} x^2(x+y)\, d\bar{\lambda}(y) &= \int_x^1 x^2(x+y)\, dy \quad \text{[a Riemann integral]} \\
&= -\frac{3}{2}x^4 + x^3 + \frac{1}{2}x^2,
\end{aligned}$$

where the equalities follow by the fact that the Lebesgue and Riemann integrals coincide and by the Second Fundamental Theorem of Calculus. Therefore, since the mapping $x \mapsto -\frac{3}{2}x^4 + x^3 + \frac{1}{2}x^2$ is continuous on $\mathbb{R}$, we have

$$\begin{aligned}
\int_{(0,1)} \int_{(x,1)} x^2(x+y)\, d\bar{\lambda}(y)\, d\bar{\lambda}(x) &= \int_{(0,1)} \left(-\frac{3}{2}x^4 + x^3 + \frac{1}{2}x^2\right) d\bar{\lambda}(x) \\
&= \int_0^1 \left(-\frac{3}{2}x^4 + x^3 + \frac{1}{2}x^2\right) dx \quad \text{[Riemann]} \\
&= \frac{7}{60},
\end{aligned}$$

where again we call on (i) the fact that the Lebesgue and Riemann integrals coincide and (ii) the Second Fundamental Theorem of Calculus. It follows that the Lebesgue integral $\int_B x^2(x+y)\, d\bar{\lambda}_2(x,y)$ is equal to $\frac{7}{60}$. We have also shown that

$$\int_B x^2(x+y)\, d\bar{\lambda}_2(x,y) = \int_0^1 \int_x^1 x^2(x+y)\, dy\, dx \quad \text{[Riemann integral]}.$$

The same type of argument as given earlier shows that we could also write

$$\int_B x^2(x+y)\, d\bar{\lambda}_2(x,y) = \int_0^1 \int_0^y x^2(x+y)\, dx\, dy \quad \text{[Riemann integral]}.$$

The point of this example is not that we really care about the value of the two-dimensional Lebesgue integral of $x^2(x+y)$ over the set B, but that when dealing with a two-dimensional Lebesgue integral of a "nice" function we can resort to our usual multivariable calculus.

Example 2. Suppose that we desire to calculate the Lebesgue integral

$$\int_E y(\sin x)e^{-xy}\, d\bar{\lambda}_2(x,y), \quad \text{where } E = (0,+\infty)\times(0,1).$$

That is, we desire to calculate

$$\int_{\mathbb{R}^2} y(\sin x)e^{-xy} I_E(x,y)\, d\bar{\lambda}_2(x,y).$$

Keep the underlying spaces the same as in the previous example. The integrand is clearly $\mathcal{L}^2/\mathcal{B}^*$-measurable, and we have

$$\begin{aligned}
\int_{\mathbb{R}^2} \left|y(\sin x)e^{-xy} I_E(x,y)\right| d\bar{\lambda}_2(x,y) &\leq \int_{\mathbb{R}^2} |y|e^{-xy} I_E(x,y)\, d\bar{\lambda}_2(x,y) \\
&= \int_{\mathbb{R}}\int_{\mathbb{R}} |y|e^{-xy} I_E(x,y)\, d\bar{\lambda}(x)\, d\bar{\lambda}(y) \\
&= \int_{(0,1)}\int_{(0,+\infty)} ye^{-xy} d\bar{\lambda}(x)\, d\bar{\lambda}(y) \\
&= \int_{(0,1)}\int_0^\infty ye^{-xy}\, dx\, d\bar{\lambda}(y) \\
&= \int_{(0,1)} 1\, d\bar{\lambda}(y) \\
&= \bar{\lambda}((0,1)) \\
&= 1 \quad \text{[a finite number]}.
\end{aligned}$$

[In the first equality, we have applied the Nonnegative Complete Fubini Theorem to the nonnegative $\mathcal{L}^2/\mathcal{B}^*$-measurable mapping $(x,y)\mapsto |y|e^{-xy}I_E(x,y)$ on $\mathbb{R}^2$. The second equality follows by the fact that $I_E(x,y) = I_{(0,+\infty)}(x)I_{(0,1)}(y)$ for each $(x,y)\in\mathbb{R}^2$; we use the same type of argument as given in Example 1. The third equality follows because for each $x\in(0,1)$ the mapping $y\mapsto ye^{-xy}$ is continuous on $\mathbb{R}$, hence for every $x\in(0,1)$ the Lebesgue integral $\int_{(0,+\infty)} ye^{-xy}\, d\bar{\lambda}(y)$ coincides with the Riemann integral $\int_0^\infty ye^{-xy}\, dy$. The fourth inequality follows by the Second Fundamental Theorem of Calculus.] The conditions of the Integrable Complete Fubini Theorem are met, and hence

$$\begin{aligned}
\int_{\mathbb{R}^2} y(\sin x)e^{-xy} I_E(x,y)\, d\bar{\lambda}_2(x,y) &= \int_{\mathbb{R}}\int_{\mathbb{R}} y(\sin x)e^{-xy} I_E(x,y)\, d\bar{\lambda}(x)\, d\bar{\lambda}(y) \\
&= \int_{(0,1)}\int_{(0,+\infty)} y(\sin x)e^{-xy}\, d\bar{\lambda}(x)\, d\bar{\lambda}(y) \\
&= \int_{(0,1)}\int_0^\infty y(\sin x)e^{-xy}\, dx\, d\bar{\lambda}(y) \\
&= \int_{(0,1)} \frac{y}{y^2+1}\, d\bar{\lambda}(y) \\
&= \int_0^1 \frac{y}{y^2+1}\, dy \\
&= \frac{1}{2}\log 2.
\end{aligned}$$

The steps outlined above display are similarly justified as in the previous display. In this and the previous example, we have "shown every step" at the risk of being tedious. In our future work, we will not go through every single step unless the situation warrants it.

If f is a nonnegative $\mathcal{L}^k/\mathcal{B}^*$-measurable function on $\mathbb{R}^k$, we may write

$$\int_{\mathbb{R}^k} f(\boldsymbol{x})\, d\bar{\lambda}_k(\boldsymbol{x})$$

as

$$\int_{\mathbb{R}} \cdots \int_{\mathbb{R}} f(x_1, \cdots, x_k)\, d\bar{\lambda}(x_k) \cdots d\bar{\lambda}(x_1)$$

by repeated application of the Nonnegative Complete Fubini Theorem. There is nothing special about "the order of integration" $d\bar{\lambda}(x_k) \cdots d\bar{\lambda}(x_1)$. We could write [under the same conditions on f]

$$\int_{\mathbb{R}^k} f(\boldsymbol{x})\, d\bar{\lambda}_k(\boldsymbol{x}) = \int_{\mathbb{R}} \cdots \int_{\mathbb{R}} f(x_1, \cdots, x_k)\, d\bar{\lambda}(x_{i_1}) \cdots d\bar{\lambda}(x_{i_k})$$

for any permutation $(i_1, \cdots, i_k)$ of $(1, \cdots, k)$. If f is extended real-valued, and if at least one of $\int_{\mathbb{R}^k} |f(\boldsymbol{x})|\, d\bar{\lambda}_k(\boldsymbol{x})$ or any one of the $k!$ iterated integrals $\int_{\mathbb{R}} \cdots \int_{\mathbb{R}} |f(x_1, \cdots, x_k)|\, d\bar{\lambda}(x_{i_1}) \cdots d\bar{\lambda}(x_{i_k})$ are finite, then we will again have

$$\int_{\mathbb{R}^k} f(\boldsymbol{x})\, d\bar{\lambda}_k(\boldsymbol{x}) = \int_{\mathbb{R}} \cdots \int_{\mathbb{R}} f(x_1, \cdots, x_k)\, d\bar{\lambda}(x_{i_1}) \cdots d\bar{\lambda}(x_{i_k})$$

for any permutation $(i_1, \cdots, i_k)$ of $(1, \cdots, k)$.

One convention that we will follow is to write multiple integrals with respect to Lebesgue measure using Riemann notation. For example, instead of writing $\int_{[a,b]} \int_{[c,d]} \cdots d\bar{\lambda}(x)\, d\bar{\lambda}(y)$ we shall often write $\int_a^b \int_c^d \cdots d\bar{\lambda}(x)\, d\bar{\lambda}(y)$. In fact, for most situations we don't need the symbol $\bar{\lambda}$ in $d\bar{\lambda}(x)$ and can write dx; this is how most authors write a Lebesgue integral with respect to Lebesgue measure. We shall continue to write $d\bar{\lambda}(x)$ for dx in the interest of clarity. At any rate, we shall always be working with Lebesgue integrals unless it is explicitly stated that we are working with Riemann integrals.

Exercises. Here, we switch freely between Riemann integrals and Lebesgue integrals. The reader should be able to verify that the integrals coincide in each case.

1. On $\mathbb{R}$ define the mapping

$$f(x) = \begin{cases} \frac{\sin x}{x} & x > 0 \\ 0 & x \leq 0. \end{cases}$$

(a) We have $\lim_{a\to\infty} \int_0^a f(x)\, dx = \frac{\pi}{2}$. (We mean the limit of a Riemann integral here.) [Show that

$$\int_0^a f(x)\, dx = \frac{\pi}{2} - (\cos a) \int_0^\infty \frac{e^{-ay}}{1+y^2}\, dy - (\sin a) \int_0^\infty \frac{ye^{-ay}}{1+y^2}\, dy$$

by calculating the two-dimensional Riemann integral $\int_{(0,a)\times(0,+\infty)}(\sin x)e^{-xy}\,dA$ in two different ways. Use an appropriate convergence theorem to establish the limit.]
(b) Part (a), in improper Riemann integral notation, says that $\int_0^\infty f(x)\,dx = \frac{\pi}{2}$. However, the Lebesgue integral $\int_{(0,+\infty)} f(x)\,d\bar{\lambda}(x)$ fails to exist. [For $0 < x < +\infty$, observe that $f^+(x) > 0$ iff $f(x) > 0$ iff $2n\pi < x < (2n+1)\pi$ for some nonnegative integer n. Use the following mean value theorem: *If g and h are nonnegative and continuous on (c,d), then there is $\xi \in (c,d)$ with $\int_c^d g(x)h(x)\,dx = g(\xi)\int_c^d h(x)\,dx$.* Take $g(x) = \frac{1}{x}$ and $h(x) = \sin x$, and show for all $n \geq 0$ that

$$\int_{2n\pi}^{(2n+1)\pi} f(x)\,dx = \frac{1}{\xi_n}\int_{2n\pi}^{(2n+1)\pi} \sin x\,dx = \frac{2}{\xi_n}, \quad \text{where } \xi_n \in (2n\pi, (2n+1)\pi).$$

Then $\int_{(0,+\infty)} f^+(x)\,d\bar{\lambda}(x) = \sum_{n=0}^\infty \int_{2n\pi}^{(2n+1)\pi} f(x)\,dx = \sum_{n=0}^\infty \frac{2}{\xi_n} = +\infty$. Similarly, it can be shown that $\int_{(0,+\infty)} f^-(x)\,d\bar{\lambda}(x) = +\infty$.]
(c) We have that

$$\int_0^\infty \frac{\sin ax}{x}\,dx = \begin{cases} \frac{\pi}{2} & a > 0 \\ 0 & a = 0 \\ -\frac{\pi}{2} & a < 0 \end{cases}$$

in the improper Riemann sense, but $\int_{(0,+\infty)} \frac{\sin ax}{x}\,d\bar{\lambda}(x)$ fails to exist as a Lebesgue integral. [Use the result on linear changes of variables.]

2. $\int_{(0,+\infty)} e^{-x^2}\,d\bar{\lambda}(x) = \frac{\sqrt{\pi}}{2}$. [Use Fubini's Theorem to show that

$$\int_0^\infty \int_0^\infty xe^{-(1+y^2)x^2}\,dy\,dx = \int_0^\infty \int_0^\infty xe^{-(1+y^2)x^2}\,dx\,dy.$$

The left integral (by the substitution $u = xy$) is $(\int_0^\infty e^{-x^2}\,dx)^2$. The right integral (by the substitution $u = x^2$) in the inner integral is directly calculated to be $\frac{\pi}{4}$.]

3. This exercise deals with some situations where an appropriate Fubini Theorem conclusion fails to hold.
(a) Define $f : \mathbb{R}^2 \to \mathbb{R}$ by writing

$$f(x,y) = \begin{cases} x^{-2} & 0 < y < x < 1 \\ -y^{-2} & 0 < x < y < 1 \\ 0 & \text{otherwise.} \end{cases}$$

Show that $\int_0^1 \int_0^1 f(x,y)\,d\bar{\lambda}(y)\,d\bar{\lambda}(x) = 1 = -\int_0^1 \int_0^1 f(x,y)\,d\bar{\lambda}(x)\,d\bar{\lambda}(y)$, thus the conclusion of the Integrable Complete Fubini Theorem fails. Also, for this f we have that $\int_{\mathbb{R}^2} |f|\,d\bar{\lambda}_2$, $\int_{\mathbb{R}}\int_{\mathbb{R}} |f(x,y)|\,d\bar{\lambda}(x)\,d\bar{\lambda}(y)$, and $\int_{\mathbb{R}}\int_{\mathbb{R}} |f(x,y)|\,d\bar{\lambda}(y)d\bar{\lambda}(x)$ are $+\infty$. [Use direct calculations for the two iterated integrals of f. Calculate one of the iterated integrals of $|f|$ to obtain $+\infty$, then use the Nonnegative Fubini Theorem to assert that all three integrals of $|f|$ are equal.]
(b) A similar situation occurs when we integrate

$$f(x,y) = \begin{cases} \frac{x^2-y^2}{(x^2+y^2)^2} & 0 < x < 1,\ 0 < y < 1 \\ 0 & \text{otherwise} \end{cases}$$

over $(0,1) \times (0,1)$. [Calculate the two iterated integrals of f to obtain

$$\int_0^1 \int_0^1 f(x,y)\, d\bar{\lambda}(y)\, d\bar{\lambda}(x) = \frac{\pi}{4} = -\int_0^1 \int_0^1 f(x,y)\, d\bar{\lambda}(x)\, d\bar{\lambda}(y).$$

Calculate an iterated integral of $|f|$ to obtain $+\infty$.]

(c) A similar situation occurs when integrating

$$f(x,y) = \begin{cases} \operatorname{sgn}(x-y)e^{-|x-y|} & 0 < x < +\infty,\ 0 < y < +\infty \\ 0 & \text{otherwise} \end{cases}$$

over $(0,+\infty) \times (0,+\infty)$.

(d) $\int_{-1}^1 \int_{-1}^1 f(x,y)\, d\bar{\lambda}(x)\, d\bar{\lambda}(y) = \int_{-1}^1 \int_{-1}^1 f(x,y)\, d\bar{\lambda}(y)\, d\bar{\lambda}(x)$ when

$$f(x,y) = \begin{cases} \frac{xy}{(x^2+y^2)^2} & (x,y) \neq (0,0) \\ 0 & (x,y) = (0,0), \end{cases}$$

but the double integral $\int_{[-1,1]\times[-1,1]} f(x,y)\, d\bar{\lambda}_2(x,y)$ fails to exist.

4. Let $E \in \mathcal{L}^k$ and let $f : E \to [0,+\infty]$ be $\mathcal{L}^k/\mathcal{B}^*$-measurable. Write $\boldsymbol{x}$ for elements of $\mathbb{R}^k$ and y for elements of $\mathbb{R}$.

(a) Let

$$A = \{(\boldsymbol{x},y) \in \mathbb{R}^{k+1} : \boldsymbol{x} \in E,\ 0 \le y < f(x)\}$$

and

$$B = \{(\boldsymbol{x},y) \in \mathbb{R}^{k+1} : \boldsymbol{x} \in E,\ 0 \le y \le f(x)\}.$$

Then A and B are in $\mathcal{L}^k$. [Let $g : \mathbb{R}^{k+1} \to \mathbb{R}$ be given by $g(\boldsymbol{x},y) = f(\boldsymbol{x})$. Then

$$\{(\boldsymbol{x},y) \in E \times \mathbb{R} : g(\boldsymbol{x},y) \le t\} = \{\boldsymbol{x} \in E : f(\boldsymbol{x}) \le t\} \times \mathbb{R} \in \mathcal{L}^{k+1}$$

since $\mathcal{L}^k \times \mathcal{L} \subseteq \mathcal{L}^{k+1}$. This shows that g is $\mathcal{L}^{k+1}/\mathcal{B}^*$-measurable. Now write A and B as inverse g-images of appropriate $\mathcal{B}^*$-sets.]

(b) $\bar{\lambda}_{k+1}(A) = \bar{\lambda}_{k+1}(B)$, the common value being $\int_E f(\boldsymbol{x})\, d\bar{\lambda}_k(\boldsymbol{x})$. [Justify the following:

$$\begin{aligned} \int_E f(\boldsymbol{x})\, d\bar{\lambda}_k(\boldsymbol{x}) &= \int_E \int_0^{f(\boldsymbol{x})} d\bar{\lambda}(y)\, d\bar{\lambda}_k(\boldsymbol{x}) \\ &= \int_{\mathbb{R}^k} \int_{\mathbb{R}} I_{A_{\boldsymbol{x}}}(\boldsymbol{y})\, d\bar{\lambda}(y)\, d\bar{\lambda}_k(\boldsymbol{x}) \\ &= \int_{\mathbb{R}^{k+1}} I_A(\boldsymbol{x},\boldsymbol{y})\, d\bar{\lambda}_{k+1}(\boldsymbol{x},\boldsymbol{y}) \\ &= \bar{\lambda}_{k+1}(A). \end{aligned}$$

Similarly, $\int_E f(\boldsymbol{x})\, d\bar{\lambda}_k(\boldsymbol{x}) = \bar{\lambda}_{k+1}(B)$.]

(c) Relate (b) to the view of the integral of a nonnegative function as the area under the curve.

(d) Use the Nonnegative Fubini Theorem to obtain

$$\begin{aligned}\int_E f(\boldsymbol{x})\, d\bar{\lambda}_k(\boldsymbol{x}) &= \int_0^\infty \bar{\lambda}_k(\{\boldsymbol{x} \in E:\ f(\boldsymbol{x}) > y\})\, d\bar{\lambda}(y) \\ &= \int_0^\infty \bar{\lambda}_k(\{\boldsymbol{x} \in E:\ f(\boldsymbol{x}) \geq y\})\, d\bar{\lambda}(y).\end{aligned}$$

[Verify that $A^y = \{\boldsymbol{x} \in E:\ f(\boldsymbol{x}) > y\}$ for $0 < y < +\infty$ and $A^y = \varnothing$ otherwise. Then verify that

$$\begin{aligned}\int_E f(\boldsymbol{x})\, d\bar{\lambda}_k(\boldsymbol{x}) &= \bar{\lambda}_{k+1}(A) \\ &= \int_{\mathbb{R}^{k+1}} I_A(\boldsymbol{x}, y)\, d\bar{\lambda}_{k+1}(\boldsymbol{x}, y) \\ &= \int_{\mathbb{R}} \int_{\mathbb{R}^k} I_{A^y}(\boldsymbol{x})\, d\bar{\lambda}_k(\boldsymbol{x})\, d\bar{\lambda}(y) \\ &= \int_0^\infty \bar{\lambda}_k(\{\boldsymbol{x} \in E:\ f(\boldsymbol{x}) > y\})\, d\bar{\lambda}(y).]\end{aligned}$$

Comment. In the next exercise, we make the following definitions for positive real α and β:

$$\Gamma(\alpha) = \int_0^\infty x^{\alpha-1} e^{-x}\, d\bar{\lambda}(x) \quad \text{and} \quad B(\alpha, \beta) = \int_0^1 x^{\alpha-1}(1-x)^{\beta-1}\, d\bar{\lambda}(x).$$

These are the *gamma* and *beta* functions, and are well known in applied mathematics.

5. Verify the following facts concerning the gamma function:
(a) $\Gamma(\alpha) = (\alpha - 1)\Gamma(\alpha - 1)$ for all $\alpha > 1$. [Integrate by parts.]
(b) $\Gamma(n) = n!$ for every positive integer n.

Note. Exercises 6-11 form a set of exercises ultimately leading up to the k-dimensional volume of a ball and ellipsoid in $\mathbb{R}^k$.

6. This exercise discusses some more properties of the gamma and beta functions.
(a) Show that $B(\alpha, \beta) = \frac{\Gamma(\alpha)\Gamma(\beta)}{\Gamma(\alpha+\beta)}$ for every $\alpha, \beta > 0$. [Verify that

$$\begin{aligned}&\Gamma(\alpha)\Gamma(\beta) \\ = &\int_0^\infty \int_0^\infty x^{\alpha-1} e^{-x} y^{\beta-1} e^{-y}\, d\bar{\lambda}(x)\, d\bar{\lambda}(y) \\ = &\int_0^\infty \int_x^\infty x^{\alpha-1}(u-x)^{\beta-1} e^{-u}\, d\bar{\lambda}(u)\, d\bar{\lambda}(x) \ \ [y = u - x] \\ = &\int_0^\infty \int_0^u x^{\alpha-1}(u-x)^{\beta-1} e^{-u}\, d\bar{\lambda}(x)\, d\bar{\lambda}(u) \\ = &\int_0^\infty e^{-u} \left[\int_0^u x^{\alpha-1}(u-x)^{\beta-1}\, d\bar{\lambda}(x)\right] d\bar{\lambda}(u) \\ = &\int_0^\infty u^{\alpha+\beta-1} e^{-u}\, d\bar{\lambda}(u) \int_0^1 z^{\alpha-1}(1-z)^{\beta-1}\, d\bar{\lambda}(z) \ \ [z = x/u] \\ = &\ \Gamma(\alpha+\beta) B(\alpha, \beta),\end{aligned}$$

using the Nonnegative Complete Fubini Theorem where appropriate.]
(b) $B(\frac{1}{2}, \frac{1}{2}) = \pi$, hence $\Gamma(\frac{1}{2}) = \sqrt{\pi}$ from (a). [Write $x = \sin^2\theta$ to obtain the identity $B(\alpha, \beta) = 2\int_0^{\pi/2} \sin^{2\alpha-1}\theta \cos^{2\beta-1}\theta \, d\bar{\lambda}(\theta)$. Take $\alpha = \beta = \frac{1}{2}$.]
(c) $\Gamma(\alpha) = 2\int_0^\infty x^{2\alpha-1}e^{-x^2} \, d\bar{\lambda}(x)$, hence (b) may be used to give another proof that $\int_0^\infty e^{-x^2} \, d\bar{\lambda}(x) = \frac{1}{2}\sqrt{\pi}$.

7. Let $f : \mathbb{R} \to [0, +\infty]$ be $\mathcal{L}^k/\mathcal{B}^*$-measurable, and let $c_1, \cdots, c_k$ denote positive real numbers. Writing $\boldsymbol{x} = (x_1, \cdots, x_k)$, show that $\boldsymbol{x} \mapsto f(x_1 + \cdots + x_k)$ on $\mathbb{R}^k$ is $\mathcal{L}^k/\mathcal{B}^*$-measurable and

$$\int_{(0,+\infty)^k} f(x_1 + \cdots + x_k) \prod_{i=1}^k x_i^{c_i - 1} \, d\bar{\lambda}_k(\boldsymbol{x}) = \frac{\prod_{i=1}^k \Gamma(c_i)}{\Gamma(\sum_{i=1}^k c_i)} \int_0^\infty f(t) t^{\sum_{i=1}^k c_i - 1} \, d\bar{\lambda}(t). \tag{10.5}$$

[Use the appropriate form of Fubini's Theorem to obtain

$$\begin{aligned}
&\text{left side of (10.5)} \\
&= \int_{(0,+\infty)^{k-1}} \prod_{i=1}^{k-1} x_i^{c_i-1} \int_0^\infty f(x_1 + \cdots + x_k) x_k^{c_k-1} d\bar{\lambda}(x_k) \, d\bar{\lambda}_{k-1}(x_1, \cdots, x_{k-1}) \\
&= \int_{(0,+\infty)^{k-1}} \prod_{i=1}^{k-1} x_i^{c_i-1} \\
&\quad \times \int_{x_1+\cdots+x_{k-1}}^\infty f(t)(t - x_1 - \cdots - x_{k-1})^{c_k-1} d\bar{\lambda}(t) \, d\bar{\lambda}_{k-1}(x_1, \cdots, x_{k-1}) \\
&= \int_0^\infty f(t) \int_A \left(\prod_{i=1}^{k-1} x_i^{c_i-1}\right) \left(t - \sum_{i=1}^{k-1} x_i\right)^{c_k-1} d\bar{\lambda}_{k-1}(x_1, \cdots, x_{k-1}) \, d\bar{\lambda}(t) \\
&= \int_0^\infty f(t) \int_B t^{\sum_{i=1}^k c_i - 1} \left(\prod_{i=1}^{k-1} y_i^{c_i-1}\right) \\
&\quad \times (1 - y_1 - \cdots - y_{k-1})^{c_k-1} \, d\bar{\lambda}_{k-1}(y_1, \cdots, y_{k-1}) \, d\bar{\lambda}(t) \\
&= \int_0^\infty f(t) t^{\sum_{i=1}^k c_i - 1} \, d\bar{\lambda}(t) \\
&\quad \times \underbrace{\int_B \prod_{i=1}^{k-1} y_i^{c_i-1} (1 - y_1 - \cdots - y_{k-1})^{c_k-1} \, d\bar{\lambda}_{k-1}(y_1, \cdots, y_{k-1})}_{\text{call this } I_2, \text{ free of } f} \\
&= I_1 \times I_2.
\end{aligned}$$

The second equality follows by writing $t = x_1 + \cdots + x_k$. The set A in the third equality is the set of $(x_1, \cdots, x_{k-1})$ with $x_1, \cdots, x_{k-1} > 0$ and $x_1 + \cdots + x_{k-1} < t$. The fourth equality follows by writing $x_i = ty_i$ for $i = 1, \cdots, k-1$; the set B is the set of $(y_1, \cdots, y_{k-1})$ with $y_1, \cdots, y_{k-1} > 0$ and $y_1 + \cdots + y_{k-1} < 1$. If $f(z) = e^{-z}$ the above work and Fubini's Theorem show that the left side of (10.5) is

$$\int_{(0,+\infty)^k} e^{-(x_1+\cdots+x_k)} \prod_{i=1}^k x_i^{c_i-1} \, d\bar{\lambda}_k(\boldsymbol{x}) = \prod_{i=1}^k \Gamma(c_i).$$

But the above derivation shows that in this case the left side of (10.5) is

$$\int_0^\infty e^{-t} t^{\sum_{i=1}^k c_i - 1}\, d\bar{\lambda}(t) \times I_2 = \Gamma\left(\sum_{i=1}^k c_i\right) \times I_2.$$

Therefore, $I_2 = (\prod_{i=1}^k \Gamma(c_i))/\Gamma(\sum_{i=1}^k c_i)$. Going back to general nonnegative f in the derivation above gives the result.]

8. Let f be as in Exercise 7. Let $b_1, \cdots, b_k$ denote positive real numbers, and write $x_i = y_i^{1/b_i}$ for $i = 1, \cdots, k$, obtaining the identity

$$\begin{aligned}&\int_{(0,+\infty)^k} f(x_1^{b_1} + \cdots + x_k^{b_k}) \prod_{i=1}^k x_i^{c_i - 1}\, d\bar{\lambda}_k(\boldsymbol{x}) \\ &= \frac{\prod_{i=1}^k \Gamma(\frac{c_i}{b_i})}{\left(\prod_{i=1}^k b_i\right)\left(\Gamma(\sum_{i=1}^k \frac{c_i}{b_i})\right)} \int_0^\infty f(t) t^{\sum_{i=1}^k \frac{c_i}{b_i} - 1}\, d\bar{\lambda}(t).\end{aligned}$$

9. Let f be as in Exercise 7. Letting $\|\boldsymbol{x}\|$ denote the length of $\boldsymbol{x}$, show that

$$\int_{\mathbb{R}^k} f(\|\boldsymbol{x}\|)\, d\bar{\lambda}_k(\boldsymbol{x}) = \frac{2\pi^{k/2}}{\Gamma(k/2)} \int_0^\infty f(r) r^{k-1}\, d\bar{\lambda}(r).$$

[In Exercise 8, let $b_1 = \cdots = b_k = 2$ and $c_1 = \cdots = c_k = 1$ to obtain

$$\int_{(0,+\infty)^k} f(\|\boldsymbol{x}\|^2)\, d\bar{\lambda}_k(\boldsymbol{x}) = \frac{\pi^{k/2}}{2^k \Gamma(k/2)} \int_0^\infty f(t) t^{\frac{k}{2}-1}\, d\bar{\lambda}(t).$$

If f is the composition of g and the square root, where g is nonnegative and $\mathcal{L}/\mathcal{B}^*$-measurable, we have

$$\begin{aligned}\int_{(0,+\infty)^k} g(\|\boldsymbol{x}\|)\, d\bar{\lambda}_k(\boldsymbol{x}) &= \frac{\pi^{k/2}}{2^k \Gamma(k/2)} \int_0^\infty g(\sqrt{t}) t^{\frac{k}{2}-1}\, d\bar{\lambda}(t) \\ &= \frac{\pi^{k/2}}{2^k \Gamma(k/2)} \int_0^\infty g(r) r^{k-2} 2r\, d\bar{\lambda}(r) \\ &= \frac{\pi^{k/2}}{2^{k-1} \Gamma(k/2)} \int_0^\infty g(r) r^{k-1}\, d\bar{\lambda}(r).\end{aligned}$$

Observe that $\int_{\mathbb{R}^k} g(\|\boldsymbol{x}\|)\, d\bar{\lambda}_k(\boldsymbol{x}) = 2^k \int_{(0,+\infty)^k} g(\|\boldsymbol{x}\|)\, d\bar{\lambda}_k(\boldsymbol{x})$ by the rotational symmetry of the integrand. This gives the result with g replacing f.]

10. The $\bar{\lambda}_k$-measure of a k-dimensional ball with radius r is $\frac{\pi^{k/2} r^k}{\Gamma((k/2)+1)}$. [Use Exercise 9 on the function $I_{B(\mathbf{0},r)}(\boldsymbol{x}) = I_{(0,r)}(\|\boldsymbol{x}\|)$:

$$\bar{\lambda}_k(B(\mathbf{0}, r)) = \int_{\mathbb{R}^k} I_{B(\mathbf{0},r)}(\boldsymbol{x})\, d\bar{\lambda}_k(\boldsymbol{x}) = \frac{2\pi^{k/2}}{\Gamma(k/2)} \int_0^\infty I_{(0,r)}(t) t^{k-1}\, d\bar{\lambda}(t).$$

Here $B(\mathbf{0}, r)$ denotes the ball in $\mathbb{R}^k$ with radius r; we may consider the center to be the origin $\mathbf{0}$ of $\mathbb{R}^k$ by the translation invariance of $\bar{\lambda}_k$.]

11. More generally than Exercise 10, the k-dimensional volume of the ellipsoid given by the equation $\sum_{i=1}^{k} \frac{x_i^2}{c_i^2} < 1$ is $\frac{\pi^{k/2}}{\Gamma(\frac{k}{2}+1)} c_1 \cdots c_n$. [Here $c_1, \cdots, c_n > 0$.]

10.4 THE GENERALIZED MINKOWSKI INEQUALITY

In our discussion of L^p spaces for $1 \leq p \leq +\infty$ we proved the famous and useful Minkowski Inequality: *Let $f, g \in L^p(\Omega, \mathcal{F}, \mu)$. Then $f+g \in L^p(\Omega, \mathcal{F}, \mu)$ and $\|f+g\|_p \leq \|f\|_p + \|g\|_p$.* We shall state and prove a very general version of this inequality, courtesy of the Fubini Theorem[s]. This will provide a good example of "Fubini in action" in a general measure-theoretic context.

Framework. $(\Omega_1, \mathcal{F}_1, \mu_1)$ and $(\Omega_2, \mathcal{F}_2, \mu_2)$ are σ-finite measure spaces, and f denotes a nonnegative $\mathcal{F}_1 \otimes \mathcal{F}_2$-measurable function on $\Omega_1 \times \Omega_2$. We will assume that $1 \leq p < +\infty$. This framework shall be assumed until explicitly mentioned otherwise in this section.

We shall work up to the *Generalized Minkowski Inequality* by virtue of a series of technical claims. We shall use the Fubini Theorems not only to establish identities between double integrals and iterated integrals, but also to establish certain measurability claims. [See (i)-(iv) of the Nonnegative Fubini Theorem.] This is illustrated in our second claim. The first claim that we now present is a mere technical lemma.

Claim 1. *Let $\phi : \Omega_2 \to \bar{\mathbb{R}}$ be $\mathcal{F}_2/\mathcal{B}^*$-measurable. Then the mapping on $\Omega_1 \times \Omega_2$ given by $(\omega_1, \omega_2) \mapsto \phi(\omega_2)$ is $\mathcal{F}_1 \otimes \mathcal{F}_2/\mathcal{B}^*$-measurable. That is, the mapping ϕ, viewed as a function on $\Omega_1 \times \Omega_2$, is $\mathcal{F}_1 \otimes \mathcal{F}_2/\mathcal{B}^*$-measurable.*

Comment. The same type of result holds if $\phi : \Omega_1 \to \bar{\mathbb{R}}$ is $\mathcal{F}_1/\mathcal{B}^*$-measurable: *the mapping $(\omega_1, \omega_2) \mapsto \phi(\omega_1)$ on $\Omega_1 \times \Omega_2$ is $\mathcal{F}_1 \otimes \mathcal{F}_2/\mathcal{B}^*$-measurable.*

Proof: Let $g(\omega_1, \omega_2) = \phi(\omega_2)$ for every $(\omega_1, \omega_2) \in \Omega_1 \times \Omega_2$, and let $B \in \mathcal{B}^*$. Observe that the $\mathcal{F}_2/\mathcal{B}^*$-measurability of ϕ forces $\phi^{-1}(B) \in \mathcal{F}_2$, and hence

$$\begin{aligned} g^{-1}(B) &= \{(\omega_1, \omega_2) \in \Omega_1 \times \Omega_2 : g(\omega_1, \omega_2) \in B\} \\ &= \{(\omega_1, \omega_2) \in \Omega_1 \times \Omega_2 : \phi(\omega_2) \in B\} \\ &= \Omega_1 \times \phi^{-1}(B) \\ &\in \mathcal{F}_1 \times \mathcal{F}_2 \\ &\subseteq \mathcal{F}_1 \otimes \mathcal{F}_2. \end{aligned}$$

Therefore, $g^{-1}(B) \in \mathcal{F}_1 \otimes \mathcal{F}_2$ for our generic $B \in \mathcal{B}^*$. This shows that g is $\mathcal{F}_1 \otimes \mathcal{F}_2/\mathcal{B}^*$-measurable, completing the proof. ■

Claim 2. *The mapping $\omega_2 \mapsto (\int_{\Omega_1} f(\omega_1, \omega_2)^p \, d\mu_1(\omega_1))^{\frac{1}{p}}$ on Ω_2 is $\mathcal{F}_2/\mathcal{B}^*$-measurable and nonnegative, hence we may meaningfully discuss the iterated integral $\int_{\Omega_2} [\int_{\Omega_1} f(\omega_1, \omega_2)^p \, d\mu_1(\omega_1)]^{\frac{1}{p}} \, d\mu_2(\omega_2)$.*

Proof: Since f is a nonnegative $\mathcal{F}_1 \otimes \mathcal{F}_2/\mathcal{B}^*$-measurable mapping with domain $\Omega_1 \times \Omega_2$, the same is true for f^p. Part (iv) of the Nonnegative Fubini

Theorem applied to the nonnegative $\mathcal{F}_1 \otimes \mathcal{F}_2/\mathcal{B}^*$-measurable mapping f^p on $\Omega_1 \times \Omega_2$ immediately asserts that the mapping $\omega_2 \mapsto \int_{\Omega_1} f(\omega_1,\omega_2)^p \, d\mu_1(\omega_1)$ is $\mathcal{F}_2/\mathcal{B}^*$-measurable. Since pth roots of $\mathcal{F}_2/\mathcal{B}^*$-measurable mappings are $\mathcal{F}_2/\mathcal{B}^*$-measurable, we have that the mapping $\omega_2 \mapsto (\int_{\Omega_1} f(\omega_1,\omega_2)^p \, d\mu_1(\omega_1))^{\frac{1}{p}}$ on Ω_2 is $\mathcal{F}_2/\mathcal{B}^*$-measurable and nonnegative. This completes the proof. ∎

Claim 3. *Assume that* $\int_{\Omega_2} [\int_{\Omega_1} f(\omega_1,\omega_2)^p \, d\mu_1(\omega_1)]^{\frac{1}{p}} \, d\mu_2(\omega_2) < +\infty$. *Then for* μ_1*-a.e.* $\omega_1 \in \Omega_1$ *the mapping* $\omega_2 \mapsto f(\omega_1,\omega_2)$ *on* Ω_2 *is in* $L^1(\Omega_2, \mathcal{F}_2, \mu_2)$, *so that* $\int_{\Omega_2} f(\omega_1,\omega_2) \, d\mu_2(\omega_2) < +\infty$ *for* μ_1*-a.e.* $\omega_1 \in \Omega_1$.

Proof: If $p = 1$, the proof is immediate by the Integrable Fubini Theorem. We therefore assume that $1 < p < +\infty$. We first define

$$c = \int_{\Omega_2} \left[\int_{\Omega_1} f(\omega_1,\omega_2)^p \, d\mu_1(\omega_1) \right]^{\frac{1}{p}} d\mu_2(\omega_2) \quad < +\infty,$$

and we write

$$\phi(\omega_2) = \int_{\Omega_1} f(\omega_1,\omega_2)^p \, d\mu_1(\omega_1) \quad \text{for every } \omega_2 \in \Omega_2.$$

Note that (a) $\phi(\omega_2)$ meaningfully exists as a nonnegative number since the mapping $\omega_1 \mapsto f(\omega_1,\omega_2)$ is nonnegative and $\mathcal{F}_1/\mathcal{B}^*$-measurable, and (b) ϕ, which [by Claim 2] is $\mathcal{F}_2/\mathcal{B}^*$-measurable as a function on Ω_2, is $\mathcal{F}_1 \otimes \mathcal{F}_2/\mathcal{B}^*$-measurable when viewed as a function of $(\omega_1,\omega_2) \in \Omega_1 \times \Omega_2$ by virtue of Claim 1. Define the function $g : \Omega_1 \times \Omega_2 \to [0,+\infty]$ for all $(\omega_1,\omega_2) \in \Omega_1 \times \Omega_2$ by the rule

$$g(\omega_1,\omega_2) = \begin{cases} f(\omega_1,\omega_2)\phi(\omega_2)^{-\frac{1}{pp'}} & 0 < \phi(\omega_2) < +\infty \\ 0 & \phi(\omega_2) = 0 \\ +\infty & \phi(\omega_2) = +\infty. \end{cases}$$

Observe that g is $\mathcal{F}_1 \otimes \mathcal{F}_2/\mathcal{B}^*$-measurable since (i) g is a piecewise defined function whose pieces are $\mathcal{F}_1 \otimes \mathcal{F}_2/\mathcal{B}^*$-measurable and (ii) ϕ [viewed as a function on $\Omega_1 \times \Omega_2$] is $\mathcal{F}_1 \otimes \mathcal{F}_2/\mathcal{B}^*$-measurable.

Subclaim 1. For every $\omega_2 \in \Omega_2$, *we have the inequality*

$$f(\omega_1,\omega_2) \le g(\omega_1,\omega_2)\phi(\omega_2)^{\frac{1}{pp'}} \quad \textit{for } \mu_1\textit{-a.e. } \omega_1 \in \Omega_1. \tag{10.6}$$

Proof of Subclaim 1: Pick $\omega_2 \in \Omega_2$. There are three cases to consider regarding ω_2. The first case is where $0 < \phi(\omega_2) < +\infty$. In this case, (10.6) clearly holds as an equality for all $\omega_1 \in \Omega_1$, as this is a mere algebraic rearrangement of the definition of $g(\omega_1,\omega_2)$ in this case.

The next case regarding ω_2 is where $\phi(\omega_2) = +\infty$. In this case, we have by definition that $g(\omega_1,\omega_2) = +\infty$ for every $\omega_1 \in \Omega_1$, and the desired inequality (10.6) takes the trivial form $f(\omega_1,\omega_2) \le (+\infty)(+\infty)$ for every $\omega_1 \in \Omega_1$.

The nontrivial case regarding ω_2 is where $\phi(\omega_2) = 0$, that is, where ω_2 is such that $\int_{\Omega_1} f(\omega_1, \omega_2)^p \, d\mu_1(\omega_1) = 0$. Here, we have for μ_1-a.e. $\omega_1 \in \Omega_1$ that $f^p(\omega_1, \omega_2) = 0$, so that $f(\omega_1, \omega_2) = 0$ for μ_1-a.e. $\omega_1 \in \Omega_1$. Since $g(\omega_1, \omega_2) = 0$ for every $\omega_1 \in \Omega_1$, (10.6) holds for μ_1-a.e. $\omega_1 \in \Omega_1$, taking the form $0 \leq 0 \times 0$. This gives (10.6) in all three cases regarding ω_2, completing the proof of the subclaim.

Subclaim 2. For every $\omega_2 \in \Omega_2$ we have

$$\left(\int_{\Omega_1} g(\omega_1, \omega_2)^p \, d\mu_1(\omega_1) \right)^{\frac{1}{p}} = \left(\int_{\Omega_1} f(\omega_1, \omega_2)^p \, d\mu_1(\omega_1) \right)^{\frac{1}{p^2}} \tag{10.7}$$

Proof of Subclaim 2. Let $\omega_2 \in \Omega_2$. As with the proof of Subclaim 1, we deal with cases regarding ω_2. The first case is where $\phi(\omega_2) = 0$. In this situation, we have $g(\omega_1, \omega_2) = 0$ for every $\omega_1 \in \Omega_1$, and (10.6) gives $f(\omega_1, \omega_2) = 0$ for μ_1-a.e. $\omega_1 \in \Omega_1$. The desired inequality (10.7) here becomes the trivial equation $0 = 0$.

The next case is where $\phi(\omega_2) = +\infty$. Then $g(\omega_1, \omega_2) = +\infty$ for every $\omega_1 \in \Omega_1$, and thus the left side of (10.7) equals $+\infty$. The right side of (10.7) by definition is $\phi(\omega_2)^{\frac{1}{p^2}} = (+\infty)^{\frac{1}{p^2}} = +\infty$, and hence (10.7) takes the form $+\infty = +\infty$ in this case.

The final case is where $0 < \phi(\omega_2) < +\infty$. In this situation we have

$$\begin{aligned}
& \left[\int_{\Omega_1} g(\omega_1, \omega_2)^p \, d\mu_1(\omega_1) \right]^{\frac{1}{p}} \\
= & \left[\int_{\Omega_1} f(\omega_1, \omega_2)^p \phi(\omega_2)^{-\frac{1}{p'}} \, d\mu_1(\omega_1) \right]^{\frac{1}{p}} \\
= & \ \phi(\omega_2)^{-\frac{1}{p'p}} \left[\int_{\Omega_1} f(\omega_1, \omega_2)^p \, d\mu_1(\omega_1) \right]^{\frac{1}{p}} \\
= & \left[\int_{\Omega_1} f(\omega_1, \omega_2)^p \, d\mu_1(\omega_1) \right]^{-\frac{1}{p'p} + \frac{1}{p}} \\
= & \left[\int_{\Omega_1} f(\omega_1, \omega_2)^p \, d\mu_1(\omega_1) \right]^{\frac{1}{p^2}} .
\end{aligned}$$

This proves (10.7) in all three cases and completes the proof of the subclaim.

It follows that

$$\begin{aligned}
& \int_{\Omega_2} f(\omega_1,\omega_2)\, d\mu_2(\omega_2) \\
\leq\ & \int_{\Omega_2} g(\omega_1,\omega_2) \left[\int_{\Omega_1} f(\omega_1,\omega_2)^p\, d\mu_1(\omega_1)\right]^{\frac{1}{pp'}} d\mu_2(\omega_2) \\
\leq\ & \left[\int_{\Omega_2} g(\omega_1,\omega_2)^p\, d\mu_2(\omega_2)\right]^{\frac{1}{p}} \left[\int_{\Omega_2} \left(\int_{\Omega_1} f(\omega_1,\omega_2)^p\, d\mu_1(\omega_1)\right)^{\frac{p'}{pp'}} d\mu_2(\omega_2)\right]^{\frac{1}{p'}} \\
=\ & \left[\int_{\Omega_2} g(\omega_1,\omega_2)^p\, d\mu_2(\omega_2)\right]^{\frac{1}{p}} c^{\frac{1}{p'}} \quad \text{for } \mu_1\text{-a.e. } \omega_1 \in \Omega_1.
\end{aligned}$$

[The first inequality follows by (10.6) and the definition of ϕ. The second inequality follows by fixing ω_1 and applying Hölder's Inequality in Section 8.1 (or the footnote to Hölder's Inequality given in Section 8.1) with the conjugate exponents p and p' to the $\mathcal{F}_2/\mathcal{B}^*$-measurable nonnegative mappings $\omega_2 \mapsto g(\omega_1,\omega_2)$ and $\omega_2 \mapsto [\int_{\Omega_1} f(\omega_1,\omega_2)^p\, d\mu_1(\omega_1)]^{\frac{1}{pp'}}$ on Ω_2, respectively.[5] The equality follows by cancelling the p' terms and invoking the definition of c.] Raising both sides of the above display to the pth power and integrating over Ω_1 with respect to μ_1 gives

$$\begin{aligned}
& \int_{\Omega_1} \left[\int_{\Omega_2} f(\omega_1,\omega_2)\, d\mu_2(\omega_2)\right]^p d\mu_1(\omega_1) \\
\leq\ & \int_{\Omega_1} \int_{\Omega_2} g(\omega_1,\omega_2)^p\, d\mu_2(\omega_2)\, c^{\frac{p}{p'}}\, d\mu_1(\omega_1) \\
=\ & c^{p-1} \int_{\Omega_1} \int_{\Omega_2} g(\omega_1,\omega_2)^p\, d\mu_2(\omega_2)\, d\mu_1(\omega_1) \\
=\ & c^{p-1} \int_{\Omega_2} \int_{\Omega_1} g(\omega_1,\omega_2)^p\, d\mu_1(\omega_1)\, d\mu_2(\omega_2) \\
=\ & c^{p-1} \int_{\Omega_2} \left[\int_{\Omega_1} f(\omega_1,\omega_2)^p\, d\mu_1(\omega_1)\right]^{\frac{1}{p}} d\mu_2(\omega_2) \\
=\ & c^{p-1} c \\
=\ & c^p \quad \text{[a finite number.]}
\end{aligned} \tag{10.8}$$

[The first equality follows because $p' = \frac{p}{p-1}$. The second equality follows by the Nonnegative Fubini Theorem applied to g^p. The third equality follows by

[5] Given ω_1, the $\mathcal{F}_2/\mathcal{B}^*$-measurability of the mapping $\omega_2 \mapsto g(\omega_1,\omega_2)$ on Ω_2 follows directly by Claim 3 of Section 10.1. Next, given ω_1, the $\mathcal{F}_2/\mathcal{B}^*$-measurability of the mapping $\omega_2 \mapsto [\int_{\Omega_1} f(\omega_1,\omega_2)^p\, d\mu_1(\omega_1)]^{\frac{1}{pp'}}$ on Ω_2 follows from Claim 2 of this section.

(10.7). The fourth equality follows by the definition of c.] This gives

$$\int_{\Omega_1}\left[\int_{\Omega_2} f(\omega_1,\omega_2)\, d\mu_2(\omega_2)\right]^p d\mu_1(\omega_1) < +\infty.$$

It follows that

$$\int_{\Omega_2} f(\omega_1,\omega_2)\, d\mu_2(\omega_2) < +\infty \quad \text{for } \mu_1\text{-a.e. } \omega_1 \in \Omega_1.$$

Recalling that f is nonnegative, we have for the μ_1-a.e. $\omega_1 \in \Omega_1$ just mentioned that the mapping $\omega_2 \mapsto f(\omega_1,\omega_2)$ defined on Ω_2 is in $L^1(\Omega_2,\mathcal{F}_2,\mu_2)$. This completes the proof. ∎

Claim 4. *Assume that $\int_{\Omega_2}[\int_{\Omega_1} f(\omega_1,\omega_2)^p\, d\mu_1(\omega_1)]^{\frac{1}{p}}\, d\mu_2(\omega_2) < +\infty$. Then for μ_1-a.e. $\omega_1 \in \Omega_1$ we have that the mapping $\omega_2 \mapsto f(\omega_1,\omega_2)$ on Ω_2 is an element of $L^1(\Omega_2,\mathcal{F}_2,\mu_2)$ and*

$$\left[\int_{\Omega_1}\left(\int_{\Omega_2} f(\omega_1,\omega_2)\, d\mu_2(\omega_2)\right)^p d\mu_1(\omega_1)\right]^{\frac{1}{p}} \leq \int_{\Omega_2}\left[\int_{\Omega_1} f(\omega_1,\omega_2)^p\, d\mu_1(\omega_1)\right]^{\frac{1}{p}} d\mu_2(\omega_2). \tag{10.9}$$

Proof: The first assertion follows immediately from Claim 3. The inequality (10.8) found in the proof of Claim 3 gives (10.9) for $1 < p < +\infty$. When $p = 1$, (10.9) holds [and is actually an equality] by the Nonnegative Fubini Theorem. ∎

We have throughout assumed that f is $\mathcal{F}_1 \otimes \mathcal{F}_2/\mathcal{B}^*$-measurable and nonnegative. We now remove the nonnegativity assumption and restate Claim 4 in this more general framework. This gives the Generalized Minkowski Inequality.

Generalized Minkowski Inequality. *Let $f : \Omega_1 \times \Omega_2 \to \bar{\mathbb{R}}$ be $\mathcal{F}_1 \otimes \mathcal{F}_2/\mathcal{B}^*$-measurable. Then the mapping $\omega_2 \mapsto (\int_{\Omega_1} |f(\omega_1,\omega_2)|^p\, d\mu_1(\omega_1))^{\frac{1}{p}}$ on Ω_2 is nonnegative and $\mathcal{F}_2/\mathcal{B}^*$-measurable, and hence we may meaningfully discuss the nonnegative number $\int_{\Omega_2}\left[\int_{\Omega_1} |f(\omega_1,\omega_2)|^p\, d\mu_1(\omega_1)\right]^{\frac{1}{p}} d\mu_2(\omega_2)$. Furthermore, if*

$$\int_{\Omega_2}\left[\int_{\Omega_1} |f(\omega_1,\omega_2)|^p\, d\mu_1(\omega_1)\right]^{\frac{1}{p}} d\mu_2(\omega_2) < +\infty,$$

then we have as follows:

(a) The mapping $\omega_2 \mapsto f(\omega_1,\omega_2)$ on Ω_2 is in $L^1(\Omega_2,\mathcal{F}_2,\mu_2)$ for μ_1-a.e. ω_1 in Ω_1, so that for μ_1-a.e. $\omega_1 \in \Omega_1$ we may meaningfully discuss the finite number $\int_{\Omega_2} f(\omega_1,\omega_2)\, d\mu_2(\omega_2)$.

(b) The mapping $\omega_1 \mapsto \int_{\Omega_2} f(\omega_1, \omega_2)\, d\mu_2(\omega_2)$ *defined* μ_1*-a.e. on* Ω_1 *[in accordance with (a)] is* $\mathcal{F}_1/\mathcal{B}^*$*-measurable, so that we may discuss the integral* $\int_{\Omega_1} \left| \int_{\Omega_2} f(\omega_1, \omega_2)\, d\mu_2(\omega_2) \right|^p d\mu_1(\omega_1)$.

(c) The inequality

$$\left[\int_{\Omega_1} \left| \int_{\Omega_2} f(\omega_1, \omega_2)\, d\mu_2(\omega_2) \right|^p d\mu_1(\omega_1) \right]^{\frac{1}{p}} \leq \int_{\Omega_2} \left[\int_{\Omega_1} |f(\omega_1, \omega_2)|^p\, d\mu_1(\omega_1) \right]^{\frac{1}{p}} d\mu_2(\omega_2). \tag{10.10}$$

obtains.

Proof: We have that $|f|$ is nonnegative and $\mathcal{F}_1 \otimes \mathcal{F}_2/\mathcal{B}^*$-measurable. Applying Claim 2 to $|f|$ gives the first assertion. For (a), Claim 3 asserts that for μ_1-a.e. $\omega_1 \in \Omega_1$ the mapping $\omega_2 \mapsto |f(\omega_1, \omega_2)|$ on Ω_2 is in $L^1(\Omega_2, \mathcal{F}_2, \mu_2)$. For the same aforementioned μ_1-a.e. $\omega_1 \in \Omega_1$, it follows that the mapping $\omega_2 \mapsto f(\omega_1, \omega_2)$ is in $L^1(\Omega_2, \mathcal{F}_2, \mu_2)$, giving (a).

We do (b). The mappings $\omega_1 \mapsto \int_{\Omega_2} f^+(\omega_1, \omega_2)\, d\mu_2(\omega_2)$ defined μ_1-a.e. on Ω_1 [in accordance with (a)] and $\omega_1 \mapsto \int_{\Omega_2} f^-(\omega_1, \omega_2)\, d\mu_2(\omega_2)$ defined μ_1-a.e. on Ω_1 [in accordance with (a)] are $\mathcal{F}_1/\mathcal{B}^*$-measurable by applying the Nonnegative Fubini Theorem separately to the nonnegative $\mathcal{F}_1 \otimes \mathcal{F}_2/\mathcal{B}^*$-measurable functions f^+ and f^-. Therefore, the mapping

$$\begin{aligned} \omega_1 \quad &\mapsto \quad \int_{\Omega_2} f(\omega_1, \omega_2)\, d\mu_2(\omega_2) \\ &= \quad \int_{\Omega_2} f^+(\omega_1, \omega_2)\, d\mu_2(\omega_2) - \int_{\Omega_2} f^-(\omega_1, \omega_2)\, d\mu_2(\omega_2) \end{aligned}$$

defined μ_1-a.e. on Ω_1 [in accordance with (a)] is $\mathcal{F}_1/\mathcal{B}^*$-measurable. It follows that the mapping $\omega_1 \mapsto |\int_{\Omega_2} f(\omega_1, \omega_2)\, d\mu_2(\omega_2)|^p$ defined μ_1-a.e. on Ω_1 [in accordance with (a)] is $\mathcal{F}_1/\mathcal{B}^*$-measurable and nonnegative, so that we may discuss the iterated integral $\int_{\Omega_1} |\int_{\Omega_2} f(\omega_1, \omega_2)\, d\mu_2(\omega_2)|^p d\mu_1(\omega_1)$. This completes (b).

Part (c) is direct:

$$\begin{aligned} &\left[\int_{\Omega_1} \left| \int_{\Omega_2} f(\omega_1, \omega_2)\, d\mu_2(\omega_2) \right|^p d\mu_1(\omega_1) \right]^{\frac{1}{p}} \\ \leq \; &\left[\int_{\Omega_1} \left(\int_{\Omega_2} |f(\omega_1, \omega_2)|\, d\mu_2(\omega_2) \right)^p d\mu_1(\omega_1) \right]^{\frac{1}{p}} \\ \leq \; &\int_{\Omega_2} \left[\int_{\Omega_1} |f(\omega_1, \omega_2)|^p\, d\mu_1(\omega_1) \right]^{\frac{1}{p}} d\mu_2(\omega_2). \end{aligned}$$

[The first inequality is trivial and the second inequality follows by applying Claim 4 to the nonnegative function $\mathcal{F}_1 \otimes \mathcal{F}_2/\mathcal{B}^*$-measurable function $|f|$ on

$\Omega_1 \times \Omega_2$ and using the assumption that

$$\int_{\Omega_2} \left[\int_{\Omega_1} |f(\omega_1, \omega_2)|^p \, d\mu_1(\omega_1) \right]^{\frac{1}{p}} d\mu_2(\omega_2) < +\infty.]$$

This completes the entire proof. ∎

The statement of the Generalized Minkowski Inequality seems long and pedantic. We could very well just have said that *if* $f : \Omega_1 \times \Omega_2 \to \bar{\mathbb{R}}$ *is* $\mathcal{F}_1 \otimes \mathcal{F}_2/\mathcal{B}^*$*-measurable and if*

$$\int_{\Omega_2} \left[\int_{\Omega_1} |f(\omega_1, \omega_2)|^p \, d\mu_1(\omega_1) \right]^{\frac{1}{p}} d\mu_2(\omega_2) < +\infty,$$

then

$$\begin{aligned} &\left[\int_{\Omega_1} \left| \int_{\Omega_2} f(\omega_1, \omega_2) \, d\mu_2(\omega_2) \right|^p d\mu_1(\omega_1) \right]^{\frac{1}{p}} \\ \leq \; &\int_{\Omega_2} \left[\int_{\Omega_1} |f(\omega_1, \omega_2)|^p \, d\mu_1(\omega_1) \right]^{\frac{1}{p}} d\mu_2(\omega_2). \end{aligned}$$

The length of the statement as originally given is due to the fact that we want to be careful to establish that certain integrals exist. To establish that certain integrals exist, either we must have the integrands nonnegative and measurable, or the integrands must be in the appropriate L^1 space. Besides the result itself, we have tried to be as explicit as possible in showing the use of the Fubini Theorems for establishing certain integration and measurability results.

We now discuss just how the Generalized Minkowski Inequality generalizes the Minkowski Inequality in the case $1 \leq p < +\infty$. Let $(\Omega_1, \mathcal{F}_1, \mu_1)$ denote a σ-finite measure space. Next, let $\Omega_2 = \{\omega_{21}, \omega_{22}\}$, $\mathcal{F}_2 = 2^{\Omega_2}$, and let μ_2 denote counting measure on $\mathcal{F}_2$. The measure space $(\Omega_2, \mathcal{F}_2, \mu_2)$ is trivially σ-finite, and "almost everywhere" statements relative to $(\Omega_2, \mathcal{F}_2, \mu_2)$ are in fact "everywhere" statements. Let $f : \Omega_1 \times \Omega_2 \to \bar{\mathbb{R}}$ be $\mathcal{F}_1 \otimes \mathcal{F}_2/\mathcal{B}^*$-measurable. The condition

$$\int_{\Omega_2} \left[\int_{\Omega_1} |f(\omega_1, \omega_2)|^p \, d\mu_1(\omega_1) \right]^{\frac{1}{p}} d\mu_2(\omega_2) < +\infty$$

in this setting is equivalent to the condition

$$\left[\int_{\Omega_1} |f(\omega_1, \omega_{21})|^p \, d\mu_1(\omega_1) \right]^{\frac{1}{p}} + \left[\int_{\Omega_1} |f(\omega_1, \omega_{22})|^p \, d\mu_1(\omega_1) \right]^{\frac{1}{p}} < +\infty.$$

This is equivalent to saying that both terms of the sum given above are finite: $f^{\omega_{21}}, f^{\omega_{22}} \in L^p(\Omega_1, \mathcal{F}_1, \mu_1)$. The general conclusion

$$\left[\int_{\Omega_1} \left|\int_{\Omega_2} f(\omega_1, \omega_2)\, d\mu_2(\omega_2)\right|^p d\mu_1(\omega_1)\right]^{\frac{1}{p}} \leq \int_{\Omega_2} \left[\int_{\Omega_1} |f(\omega_1, \omega_2)|^p\, d\mu_1(\omega_1)\right]^{\frac{1}{p}} d\mu_2(\omega_2)$$

here becomes

$$\left[\int_{\Omega_1} |f(\omega_1, \omega_{21}) + f(\omega_1, \omega_{22})|^p\, d\mu_1(\omega_1)\right]^{\frac{1}{p}} \leq \left[\int_{\Omega_1} |f(\omega_1, \omega_{21})|^p\, d\mu_1(\omega_1)\right]^{\frac{1}{p}} + \left[\int_{\Omega_1} |f(\omega_1, \omega_{22})|^p\, d\mu_1(\omega_1)\right]^{\frac{1}{p}},$$

that is,

$$\left[\int_{\Omega_1} |f^{\omega_{21}}(\omega_1) + f^{\omega_{22}}(\omega_1)|^p\, d\mu_1(\omega_1)\right]^{\frac{1}{p}} \leq \left[\int_{\Omega_1} |f^{\omega_{21}}(\omega_1)|^p\, d\mu_1(\omega_1)\right]^{\frac{1}{p}} + \left[\int_{\Omega_1} |f^{\omega_{22}}(\omega_1)|^p\, d\mu_1(\omega_1)\right]^{\frac{1}{p}}.$$

This is equivalent to saying that $\|f^{\omega_{21}} + f^{\omega_{22}}\|_p \leq \|f^{\omega_{21}}\|_p + \|f^{\omega_{22}}\|_p$. Summarizing, the Generalized Minkowski Inequality in our special case, stripped of all measurability and existence claims, is this: *Let $f : \Omega_1 \times \Omega_2 \to \bar{\mathbb{R}}$ be such that $f^{\omega_{21}}, f^{\omega_{22}} \in L^p(\Omega_1, \mathcal{F}_1, \mu_1)$. Then $f^{\omega_{21}} + f^{\omega_{22}} \in L^p(\Omega_1, \mathcal{F}_1, \mu_1)$ and $\|f^{\omega_{21}} + f^{\omega_{22}}\|_p \leq \|f^{\omega_{21}}\|_p + \|f^{\omega_{22}}\|_p$.* This is of the same familiar form as found in the original Minkowski Inequality.

Exercises.

1. Consider the σ-finite measure space $(\Omega, \mathcal{F}, \mu)$ and consider the finite measure space $(\Omega_2, \mathcal{F}_2, \mu_2)$, where $\Omega_2 = \{a, b\}$, $\mathcal{F}_2 = 2^{\Omega_2}$, and μ_2 denotes counting measure. Let $f, g \in L^p(\Omega, \mathcal{F}, \mu)$, where $1 \leq p < +\infty$. Define $h : \Omega_1 \times \Omega_2 \to \bar{\mathbb{R}}$ by writing $h(\omega, a) = f(\omega)$ and $h(\omega, b) = g(\omega)$ for each $\omega \in \Omega$. Here, the Generalized Minkowski Inequality becomes as follows: *Let $f, g \in L^p(\Omega, \mathcal{F}, \mu)$. Then $f + g \in L^p(\Omega, \mathcal{F}, \mu)$ and $\|f + g\|_p \leq \|f\|_p + \|g\|_p$.*

2. Let $0 < q \leq p < +\infty$. Without worrying about measurability conditions, obtain the inequality

$$\left[\int_{\Omega_1} \left(\int_{\Omega_2} |f(\omega_1, \omega_2)|^q d\mu_2(\omega_2)\right)^{\frac{p}{q}} d\mu_1(\omega_1)\right]^{\frac{1}{p}} \leq \left[\int_{\Omega_2} \left(\int_{\Omega_1} |f(\omega_1, \omega_2)|^p d\mu_1(\omega_1)\right)^{\frac{q}{p}} d\mu_2(\omega_2)\right]^{\frac{1}{q}}.$$

[We have $1 \le p/q < +\infty$. Play with exponents and use the Generalized Minkowski Inequality.]

3. Let $\mathcal{L}$ denote the Lebesgue subsets of $(0, +\infty)$, and let $\bar{\lambda}$ denote Lebesgue measure on the Lebesgue subsets of $(0, +\infty)$. Consider the σ-finite measure space $((0, +\infty), \mathcal{L}, \bar{\lambda})$. Given a function $f \in L^1((0, +\infty), \mathcal{L}, \bar{\lambda})$, associate the function $F : (0, +\infty) \to [0, +\infty)$ given by $F(x) = \frac{1}{x} \int_0^x f(y)\, d\bar{\lambda}(y)$.
(a) For $1 < p < +\infty$ we have $F \in L^p((0, +\infty), \mathcal{L}, \bar{\lambda})$ and $\|F\|_p \le p'\|f\|_p$. [Replace y by xt, and then use the main result of this section.]
(b) The constant $p' = \frac{p}{p-1}$ is the best [smallest] possible constant for the inequality in (a). [Consider $f(x) = x^{\epsilon - \frac{1}{p}} I_{(0,1)}(x)$ for each $x \in (0, +\infty)$.]
(c) We have excluded the $p = 1$ case so far. Show that if $f \in L^1((0, +\infty), \mathcal{L}, \bar{\lambda})$ is positive, then $F \notin L^1((0, +\infty), \mathcal{L}, \bar{\lambda})$.

Comment: There is a way of doing (a) apart from the Generalized Minkowski Inequality. Write $xf(x) = \int_0^x f(y) y^\alpha y^{-\alpha}\, d\bar{\lambda}(y)$, where $0 < \alpha < \frac{1}{p'}$, and then use Hölder's Inequality to get an upper bound for $F(x)^p$. Integrate to obtain $\int_0^\infty F(x)^p\, d\bar{\lambda}(x) \le (1 - \alpha p')^{1-p} (\alpha p)^{-1} \int_0^\infty f(y)^p\, d\bar{\lambda}(y)$. The inequality in (a) is known as *Hardy's Inequality*. We shall prove a more general version of this result in the final section of this chapter, where the *Hardy-Littlewood Maximal Theorems* are presented as another application of the Fubini Theorems.

4. This exercise is a lemma for the next exercise. Let $(\Omega, \mathcal{F}, \mu)$ denote a measure space with $\mu(\Omega) < +\infty$. Every $\mathcal{F}/\mathcal{B}^*$-measurable function $f : \Omega \to [0, +\infty]$ satisfies

$$\lim_{p \to \infty} \left[\int_\Omega f(\omega)^p\, d\mu(\omega) \right]^{\frac{1}{p}} = \operatorname{ess\,sup}_{\omega \in \Omega} f(\omega).$$

[Suppose f is not essentially bounded. Then $\mu(\{\omega \in \Omega : \ f(\omega) \ge M\}) > 0$ has positive μ-measure for each $M > 0$. Show for each $M > 0$ and $p > 1$ that

$$0 < M\mu(\{\omega \in \Omega : \ f(\omega) \ge M\})^{\frac{1}{p}} \le \left(\int_\Omega |f(\omega)|^p\, d\mu(\omega) \right)^{\frac{1}{p}}$$

by splitting the integral into the integral over the set of ω where $f(\omega) \ge M$ and the set of ω where $f(\omega) < M$. Next, $\lim_{p \to \infty} \mu(\{\omega \in \Omega : \ f(\omega) \ge M\})^{\frac{1}{p}} = 1$ since $\mu(\Omega) < +\infty$. Therefore, given any large M, we may choose p so large that $(\int_\Omega |f(\omega)|^p\, d\mu(\omega))^{\frac{1}{p}}$ exceeds, say, $M/2$. This gives the result when f is not essentially bounded. If on the other hand f is essentially bounded, then $f \in L^p(\Omega, \mathcal{F}, \mu)$ for every $p \ge 1$ by the fact that $\mu(\Omega) < +\infty$. We have shown earlier that $\|f\|_p \to \|f\|_\infty$ in this case; see Exercise 4 in Section 8.5.]

5. We proved the Generalized Minkowski Inequality for $1 \le p < +\infty$. There is a version for the case $p = +\infty$. Let $(\Omega_1, \mathcal{F}_1, \mu_1)$ and $(\Omega_2, \mathcal{F}_2, \mu_2)$ denote σ-finite measure spaces, and let $f : \Omega_1 \times \Omega_2 \to [0, +\infty]$ be $\mathcal{F}_1 \otimes \mathcal{F}_2/\mathcal{B}^*$-measurable.
(a) The mapping $\omega_2 \mapsto \operatorname{ess\,sup}_{\omega_1 \in \Omega_1} f(\omega_1, \omega_2)$ is $\mathcal{F}_2/\mathcal{B}^*$-measurable, hence we may meaningfully discuss $\int_{\Omega_2} \operatorname{ess\,sup}_{\omega_1 \in \Omega_1} f(\omega_1, \omega_2)\, d\mu_2(\omega_2)$.

From now on, assume that $\int_{\Omega_2} \operatorname{ess\,sup}_{\omega_1 \in \Omega_1} f(\omega_1, \omega_2)\, d\mu_2(\omega_2) < +\infty$.

(b) For μ_1-a.e. $\omega_1 \in \Omega_1$ the mapping $\omega_2 \mapsto f(\omega_1, \omega_2)$ is in $L^1(\Omega_2, \mathcal{F}_2, \mu_2)$. That is, for μ_1-a.e. $\omega_1 \in \Omega_1$ we have $\int_{\Omega_2} f(\omega_1, \omega_2)\, d\mu_2(\omega_2) < +\infty$.

(c) The [nonnegative] mapping $\omega_1 \mapsto \int_{\Omega_2} f(\omega_1, \omega_2)\, d\mu_2(\omega_2)$ defined on Ω_1 is $\mathcal{F}_1/\mathcal{B}^*$-measurable, hence we may meaningfully discuss $\operatorname{ess\,sup}_{\omega_1 \in \Omega_1} \int_{\Omega_2} f(\omega_1, \omega_2)\, d\mu_2(\omega_2)$.
(d) We have the inequality

$$\operatorname{ess\,sup}_{\omega_1 \in \Omega_1} \int_{\Omega_2} f(\omega_1, \omega_2)\, d\mu_2(\omega_2) \le \int_{\Omega_2} \operatorname{ess\,sup}_{\omega_1 \in \Omega_1} f(\omega_1, \omega_2)\, d\mu_2(\omega_2).$$

Outline: Consider (a) in the case $\mu_1(\Omega_1) < +\infty$. Exercise 4 shows that the mapping $\omega_2 \mapsto \operatorname{ess\,sup}_{\omega_1 \in \Omega_1} f(\omega_1, \omega_2)$ is the limit as $p \to +\infty$ of the mappings $\omega_2 \mapsto [\int_{\Omega_1} f(\omega_1, \omega_2)^p\, d\mu_1(\omega_1)]^{\frac{1}{p}}$. By our work in the Generalized Minkowski Inequality, the latter mappings are $\mathcal{F}_2/\mathcal{B}^*$-measurable. Use the fact that limits of sequences of measurable mappings are measurable to complete (a) for the case $\mu_1(\Omega_1) < +\infty$. For the general case, use σ-finiteness to write $\Omega_1 = \bigcup_{n=1}^{\infty} A_n$, where $A_n \in \mathcal{F}_1$ and $\mu_1(A_n) < +\infty$ for each $n \in \mathbb{N}$. Show for each $\omega_2 \in \Omega_2$ that

$$\operatorname{ess\,sup}_{\omega_1 \in \Omega_1} f(\omega_1, \omega_2) = \sup_{k \in \mathbb{N}} \operatorname{ess\,sup}_{\omega_1 \in \Omega_1} I_{A_k}(\omega_1) f(\omega_1, \omega_2).$$

The mapping $\omega_2 \mapsto \operatorname{ess\,sup}_{\omega_1 \in \Omega_1} I_{A_k}(\omega_1) f(\omega_1, \omega_2)$ is $\mathcal{F}_2/\mathcal{B}^*$-measurable for each k in $\mathbb{N}$ by the work in the finite case; use the fact that countable suprema of measurable mappings are themselves measurable to complete (a). For (b), the fact that $f(\omega_1, \omega_2) \le \operatorname{ess\,sup}_{\omega_1 \in \Omega_1} f(\omega_1, \omega_2)$ for μ_1-a.e. $\omega_1 \in \Omega_1$ and all $\omega_2 \in \Omega_2$ gives the result. Part (c) follows by the Nonnegative Fubini Theorem. For (d), let c denote the right hand side of the desired inequality. Argue that the set $N = \{(\omega_1, \omega_2) \in \Omega_1 \times \Omega_2 : f(\omega_1, \omega_2) > \operatorname{ess\,sup}_{\eta \in \Omega_1} f(\eta, \omega_2)\}$ is in $\mathcal{F}_1 \otimes \mathcal{F}_2$. For each $\omega_2 \in \Omega_2$, the $\mathcal{F}_1$-set $N^{\omega_2} = \{\omega_1 \in \Omega_1 : f(\omega_1, \omega_2) > \operatorname{ess\,sup}_{\eta \in \Omega_1} f(\eta, \omega_2)\}$ is such that $\mu_1(N^{\omega_2}) = 0$, so $\int_{\Omega_2} \mu_1(N^{\omega_2})\, d\mu_2(\omega_2) = 0$. Using the definition of product measure gives $\int_{\Omega_1} \mu_2(N_{\omega_1})\, d\mu_1(\omega_1) = 0$. It follows for μ_1-a.e. $\omega_1 \in \Omega_1$ that $\mu_2(N_{\omega_1}) = 0$. Argue that for μ_1-a.e. $\omega_1 \in \Omega_1$ we have

$$\begin{aligned}
\int_{\Omega_2} f(\omega_1, \omega_2)\, d\mu_2(\omega_2) &= \int_{(N_{\omega_1})^c} f(\omega_1, \omega_2)\, d\mu_2(\omega_2) \\
&\le \int_{(N_{\omega_1})^c} \operatorname{ess\,sup}_{\eta \in \Omega_1} f(\eta, \omega_2)\, d\mu_2(\omega_2) \\
&= \int_{\Omega_2} \operatorname{ess\,sup}_{\eta \in \Omega_1} f(\eta, \omega_2)\, d\mu_2(\omega_2) \\
&= c.
\end{aligned}$$

This gives the desired inequality.]

10.5 CONVOLUTIONS

In Sections 10.1 and 10.2, we gave the buildup to the Fubini Theorems and the Fubini Theorems themselves. Sections 10.3 and 10.4 presented some applications of the Fubini Theorems. This present section is given in the same spirit, existing to give another illustration of the Fubini Theorems in action.

Our goal this section is to discuss *convolutions* in a measure-theoretic framework. The idea is not to give a comprehensive overview of every major result in this area, but to present a few key results that present Fubini-esque thinking and illustrate measure-theoretic thinking in general. This section exists as a self-contained unit; we do not discuss the results found here in the remainder of the book. While convolutions play a central role in the study of Fourier transforms, Fourier series, and differential equations, there is neither space nor time in this book to present even a representative sample of applications.

Our background space will be $(\mathbb{R}^k, \mathcal{L}^k, \bar{\lambda}_k)$ with k usually if not always being 1 or 2. We consider this space instead of $(\mathbb{R}^k, \mathcal{B}^k, \lambda^k)$ because most analysis applications use the former space instead of the latter.

We begin this section with a preliminary claim.

Claim 1. *Let f and g denote $\mathcal{L}/\mathcal{B}^*$-measurable functions on $\mathbb{R}$. Define $h : \mathbb{R}^2 \to \bar{\mathbb{R}}$ by writing $h(x, y) = f(x - y)g(y)$ for every $(x, y) \in \mathbb{R}^2$. Then h is $\mathcal{L}^2/\mathcal{B}^*$-measurable.*

Proof: Let $T(x, y) = (x - y, y)$ for every $(x, y) \in \mathbb{R}^2$. Then T is a linear and nonsingular mapping. Next, let $h_1(x, y) = f(x)$ for every $(x, y) \in \mathbb{R}^2$. Since f is $\mathcal{L}/\mathcal{B}^*$-measurable, we have that h_1 is $\mathcal{L} \otimes \mathcal{L}/\mathcal{B}^*$-measurable by Claim 1 of Section 10.4. Since $\mathcal{L} \otimes \mathcal{L} \subseteq \mathcal{L}^2$, it follows that h_1 is $\mathcal{L}^2/\mathcal{B}^*$-measurable. By Claim 1 of Section 7.5, it follows that $h_1 \circ T$ is $\mathcal{L}^2/\mathcal{B}^*$-measurable, so the mapping $(x, y) \mapsto h_1(T(x, y)) = h_1(x - y, y) = f(x - y)$ on $\mathbb{R}^2$ is $\mathcal{L}^2/\mathcal{B}^*$-measurable. Next, the $\mathcal{L}/\mathcal{B}^*$-measurability of g implies that the mapping $(x, y) \mapsto g(y)$ on $\mathbb{R}^2$ is $\mathcal{L} \otimes \mathcal{L}/\mathcal{B}^*$-measurable by Claim 1 of Section 10.4. Since $\mathcal{L} \otimes \mathcal{L} \subseteq \mathcal{L}^2$, the mapping $(x, y) \mapsto g(y)$ is $\mathcal{L}^2/\mathcal{B}^*$-measurable. Since the product of $\mathcal{L}^2/\mathcal{B}^*$-measurable mappings is $\mathcal{L}^2/\mathcal{B}^*$-measurable, and since h is the product of the $\mathcal{L}^2/\mathcal{B}^*$-measurable mappings $(x, y) \mapsto f(x - y)$ and $(x, y) \mapsto g(y)$ on $\mathbb{R}^2$, it follows that h is $\mathcal{L}^2/\mathcal{B}^*$-measurable. ∎

Claim 2. *Let $f, g \in L^1(\mathbb{R}, \mathcal{L}, \bar{\lambda})$.*
(a) For $\bar{\lambda}$-a.e. $x \in \mathbb{R}$ the mapping $y \mapsto f(x - y)g(y)$ on $\mathbb{R}$ is in $L^1(\mathbb{R}, \mathcal{L}, \bar{\lambda})$. That is, for $\bar{\lambda}$-a.e. $x \in \mathbb{R}$ we may meaningfully discuss the finite number

$$\int_{\mathbb{R}} f(x - y)g(y) \, d\bar{\lambda}(y).$$

(b) The real-valued mapping $x \mapsto \int_{\mathbb{R}} f(x - y)g(y) \, d\bar{\lambda}(y)$ defined $\bar{\lambda}$-a.e. on $\mathbb{R}$ [in accordance with (a)] is in $L^1(\mathbb{R}, \mathcal{L}, \bar{\lambda})$.

Proof: By Claim 1, the mapping $(x, y) \mapsto f(x - y)g(y)$ on $\mathbb{R}^2$ is $\mathcal{L}^2/\mathcal{B}^*$-measurable, so that the mapping $(x, y) \mapsto |f(x - y)g(y)|$ on $\mathbb{R}^2$ is $\mathcal{L}^2/\mathcal{B}^*$-measurable and nonnegative. We may therefore discuss the double integral

$$\int_{\mathbb{R}^2} |f(x - y)g(y)| \, d\bar{\lambda}_2(x, y).$$

By the Nonnegative [Complete] Fubini Theorem, the iterated integral

$$\int_{\mathbb{R}}\int_{\mathbb{R}} |f(x-y)g(y)|\,d\bar{\lambda}(x)\,d\bar{\lambda}(y)$$

is meaningfully defined and equals $\int_{\mathbb{R}^2} |f(x-y)g(y)|\,d\bar{\lambda}_2(x,y)$. We have

$$\begin{aligned}
\int_{\mathbb{R}}\int_{\mathbb{R}} |f(x-y)g(y)|\,d\bar{\lambda}(x)\,d\bar{\lambda}(y) &= \int_{\mathbb{R}} |g(y)| \int_{\mathbb{R}} |f(x-y)|\,d\bar{\lambda}(x)\,d\bar{\lambda}(y)\\
&= \int_{\mathbb{R}} |g(y)| \int_{\mathbb{R}} |f(x)|\,d\bar{\lambda}(x)\,d\bar{\lambda}(y)\\
&= \int_{\mathbb{R}} |g(y)|\,\|f\|_1\,d\bar{\lambda}(y)\\
&= \|f\|_1 \int_{\mathbb{R}} |g(y)|\,d\bar{\lambda}(y)\\
&= \|f\|_1\|g\|_1 \quad [< +\infty.] \qquad (10.11)
\end{aligned}$$

[The second equality above holds by Claim 2 of Section 7.5, which asserts for every $y \in \mathbb{R}$ that the equality $\int_{\mathbb{R}} |f(x-y)|\,d\bar{\lambda}(x) = \int_{\mathbb{R}} |f(x)|\,d\bar{\lambda}(x)$ holds.] The hypotheses of the Integrable [Complete] Fubini Theorem are satisfied, so we immediately have that (a) the mapping $y \mapsto f(x-y)g(y)$ on $\mathbb{R}$ is in $L^1(\mathbb{R},\mathcal{L},\bar{\lambda})$ for $\bar{\lambda}$-a.e. $x \in \mathbb{R}$, and (b) the mapping $x \mapsto \int_{\mathbb{R}} f(x-y)g(y)\,d\bar{\lambda}(y)$ defined $\bar{\lambda}$-a.e. on $\mathbb{R}$ [in accordance with (a)] is in $L^1(\mathbb{R},\mathcal{L},\bar{\lambda})$. This completes the entire proof. ■

We have not as yet defined the term *convolution.* This is because a rigorous definition of the term involves the work given in the previous two claims. With this work, we may now define the desired concept.

Definition. Let $f, g \in L^1(\mathbb{R},\mathcal{L},\bar{\lambda})$. By Claim 2, we may meaningfully discuss the finite number $\int_{\mathbb{R}} f(x-y)g(y)\,d\bar{\lambda}(y)$ for $\bar{\lambda}$-a.e. $x \in \mathbb{R}$. For all such $x \in \mathbb{R}$ we define

$$f * g(x) = \int_{\mathbb{R}} f(x-y)g(y)\,d\bar{\lambda}(y).$$

The real-valued mapping $f * g$ defined $\bar{\lambda}$-a.e. on $\mathbb{R}$ is called *the convolution of f and g.* The operation $*$ with domain $L^1(\mathbb{R},\mathcal{L},\bar{\lambda}) \times L^1(\mathbb{R},\mathcal{L},\bar{\lambda})$ is called *convolution.*

With the definition in tow, we now investigate some of the rudimentary facts concerning convolutions.

Claim 3. *Let $f, g \in L^1(\mathbb{R},\mathcal{L},\bar{\lambda})$. We have the following:*

(a) $f*g \in L^1(\mathbb{R},\mathcal{L},\bar{\lambda})$, hence we may meaningfully discuss the number $\|f*g\|_1$.

(b) $\|f*g\|_1 \le \|f\|_1\|g\|_1$.

Proof: Part (a) asserts that the mapping $x \mapsto \int_{\mathbb{R}} f(x-y)g(y)\,d\bar{\lambda}(y)$ defined $\bar{\lambda}$-a.e. on $\mathbb{R}$ is in $L^1(\mathbb{R},\mathcal{L},\bar{\lambda})$. This is exactly the content of part (b) of Claim 2.

We may therefore meaningfully discuss $||f*g||_1$. Referring to (10.11) in Claim 2 and using the Nonnegative [Complete] Fubini Theorem in conjunction with the $\mathcal{L}^2/\mathcal{B}^*$-measurability of the mapping $(x,y)\mapsto|f(x-y)g(y)|$ [established in Claim 1], we have that

$$\begin{aligned} ||f*g||_1 = \int_{\mathbb{R}} |f*g(x)|\,d\bar\lambda(x) &= \int_{\mathbb{R}}\left|\int_{\mathbb{R}} f(x-y)g(y)\,d\bar\lambda(y)\right| d\bar\lambda(x) \\ &\le \int_{\mathbb{R}}\int_{\mathbb{R}} |f(x-y)g(y)|\,d\bar\lambda(y)\,d\bar\lambda(x) \\ &= \int_{\mathbb{R}}\int_{\mathbb{R}} |f(x-y)g(y)|\,d\bar\lambda(x)\,d\bar\lambda(y) \\ &\le ||f||_1||g||_1. \end{aligned}$$

This completes the proof. ∎

Claim 4. *Let $\phi:\mathbb{R}\to[0,+\infty]$ be $\mathcal{L}/\mathcal{B}^*$-measurable, and let $x\in\mathbb{R}$. Then the mapping $y\mapsto\phi(x-y)$ on $\mathbb{R}$ is $\mathcal{L}/\mathcal{B}^*$-measurable and*

$$\int_{\mathbb{R}}\phi(y)\,d\bar\lambda(y)=\int_{\mathbb{R}}\phi(x-y)\,d\bar\lambda(y).$$

Furthermore, if $\phi\in L^1(\mathbb{R},\mathcal{L},\bar\lambda)$, then the mapping $y\mapsto\phi(x-y)$ on $\mathbb{R}$ is in $L^1(\mathbb{R},\mathcal{L},\bar\lambda)$ and we again have the equation $\int_{\mathbb{R}}\phi(y)\,d\bar\lambda(y)=\int_{\mathbb{R}}\phi(x-y)\,d\bar\lambda(y)$.

Proof: Let $\phi=\sum_{i=1}^n c_iI_{A_i}$, where $0\le c_1,\cdots,c_n<+\infty$ and $A_1,\cdots,A_n$ are disjoint $\mathcal{L}$-sets with union $\mathbb{R}$. For $i=1,\cdots,n$ we have by the translation invariance of Lebesgue measure and Claim 11 of Section 4.4 that $-A_i\oplus x\in\mathcal{L}$. For each $y\in\mathbb{R}$, we have

$$\phi(x-y)=\sum_{i=1}^n c_iI_{A_i}(x-y)=\sum_{i=1}^n c_iI_{-A_i\oplus x}(y),$$

so that the mapping $y\mapsto\phi(x-y)$ is $\mathcal{L}/\mathcal{B}^*$-measurable. Using this, it follows that

$$\begin{aligned}\int_{\mathbb{R}}\phi(x-y)\,d\bar\lambda(y) = \sum_{i=1}^n c_i\bar\lambda(-A_i\oplus x) &= \sum_{i=1}^n c_i\bar\lambda(A_i) \\ &= \int_{\mathbb{R}}\phi(y)\,d\bar\lambda(y).\end{aligned}$$

[The second equality follows by the translation invariance of Lebesgue measure and Claim 11 of Section 4.4.] This proves the result when ϕ is a nonnegative finite-valued $\mathcal{L}/\mathcal{B}^*$-measurable simple function on $\mathbb{R}$.

Now let ϕ be nonnegative and $\mathcal{L}/\mathcal{B}^*$-measurable. Let $\{s_n\}_{n=1}^\infty$ denote a nondecreasing sequence of nonnegative finite-valued $\mathcal{L}/\mathcal{B}^*$-measurable simple functions on $\mathbb{R}$ with $\lim_n s_n=\phi$. Since the mapping $y\mapsto s_n(x-y)$ on $\mathbb{R}$ is

$\mathcal{L}/\mathcal{B}^*$-measurable for each $n \in \mathbb{N}$, the mapping $y \mapsto \lim_n s_n(x-y) = \phi(x-y)$ is $\mathcal{L}/\mathcal{B}^*$-measurable. By MCT and what we have just shown, we have

$$\begin{aligned} \int_{\mathbb{R}} \phi(x-y)\, d\bar{\lambda}(y) = \lim_n \int_{\mathbb{R}} s_n(x-y)\, d\bar{\lambda}(y) &= \lim_n \int_{\mathbb{R}} s_n(y)\, d\bar{\lambda}(y) \\ &= \int_{\mathbb{R}} \phi(y)\, d\bar{\lambda}(y). \end{aligned}$$

This completes the first part of the proof.

We now assume that $\phi \in L^1(\mathbb{R}, \mathcal{L}, \bar{\lambda})$. Then ϕ^+ and ϕ^- are nonnegative elements of $L^1(\mathbb{R}, \mathcal{L}, \bar{\lambda})$. By what has just been shown, we have

$$\int_{\mathbb{R}} \phi^+(y)\, d\bar{\lambda}(y) = \int_{\mathbb{R}} \phi^+(x-y)\, d\bar{\lambda}(y)$$

and

$$\int_{\mathbb{R}} \phi^-(y)\, d\bar{\lambda}(y) = \int_{\mathbb{R}} \phi^-(x-y)\, d\bar{\lambda}(y),$$

and all integrals are finite. It follows that $\int_{\mathbb{R}} \phi(x-y)\, d\bar{\lambda}(y)$ exists and is finite, and

$$\begin{aligned} \int_{\mathbb{R}} \phi(x-y)\, d\bar{\lambda}(y) &= \int_{\mathbb{R}} \phi^+(x-y)\, d\bar{\lambda}(y) - \int_{\mathbb{R}} \phi^-(x-y)\, d\bar{\lambda}(y) \\ &= \int_{\mathbb{R}} \phi^+(y)\, d\bar{\lambda}(y) - \int_{\mathbb{R}} \phi^-(y)\, d\bar{\lambda}(y) \\ &= \int_{\mathbb{R}} \phi(y)\, d\bar{\lambda}(y). \end{aligned}$$

This completes the entire proof. ■

Claim 5. *Let $f \in L^p(\mathbb{R}, \mathcal{L}, \bar{\lambda})$, where $1 \leq p < +\infty$, and let $h \in \mathbb{R}$. Then $\tau_h f \in L^p(\mathbb{R}, \mathcal{L}, \bar{\lambda})$, and $\|\tau_h f\|_p = \|f\|_p$. [τ_h is the translation operator from Section 7.5.]*

Proof: We have $|f|^p \in L^1(\mathbb{R}, \mathcal{L}, \bar{\lambda})$, hence $\tau_h |f|^p = |\tau_h f|^p \in L^1(\mathbb{R}, \mathcal{L}, \bar{\lambda})$ and

$$\int_{\mathbb{R}} |\tau_h f|^p\, d\bar{\lambda} = \int_{\mathbb{R}} \tau_h |f|^p\, d\bar{\lambda} = \int_{\mathbb{R}} |f|^p\, d\bar{\lambda} \quad [< +\infty]$$

by Claim 2 of Section 7.5. It follows that $\tau_h f \in L^p(\mathbb{R}, \mathcal{L}, \bar{\lambda})$, and the preceding integral identity gives $\|\tau_h f\|_p^p = \|f\|_p^p$, which gives $\|\tau_h f\|_p = \|f\|_p$. This completes the proof. ■

The upcoming result is known as *Young's Theorem.* We will need the following fact: *For $1 < q < +\infty$, there exists an everywhere positive function $h \in L^q(\mathbb{R}, \mathcal{L}, \bar{\lambda})$.* We may, for example, take $h(x) = e^{-x^2}$ for every $x \in \mathbb{R}$. The claim is broken into parts (a)-(e). In the proof, we shall make a few claims that at first glance seem unrelated to the matters at hand, but these claims will allow us to deduce parts (a) and (b). From there, parts (c), (d), and (e) will come at nearly one stroke by invocation of Claim 1 of Section 8.9 and the version of the Riesz Representation Theorem found in Section 8.9, or, alternatively, the version of the Riesz Representation Theorem found in Section 9.3.

Claim 6. *Let $1 < p < +\infty$, and let $f \in L^1(\mathbb{R}, \mathcal{L}, \bar{\lambda})$ with $g \in L^p(\mathbb{R}, \mathcal{L}, \bar{\lambda})$.*

(a) For $\bar{\lambda}$-a.e. $x \in \mathbb{R}$ the mapping $y \mapsto f(x-y)g(y)$ defined on $\mathbb{R}$ is in $L^1(\mathbb{R}, \mathcal{L}, \bar{\lambda})$.

(b) For $\bar{\lambda}$-a.e. $x \in \mathbb{R}$ the mapping $y \mapsto f(y)g(x-y)$ defined on $\mathbb{R}$ is in $L^1(\mathbb{R}, \mathcal{L}, \bar{\lambda})$.

For the $\bar{\lambda}$-a.e. $x \in \mathbb{R}$ simultaneously in both (a) and (b), we abuse notation somewhat[6] *and write*

$$f * g(x) = \int_{\mathbb{R}} f(x-y)g(y)\, d\bar{\lambda}(y)$$

and

$$g * f(x) = \int_{\mathbb{R}} f(y)g(x-y)\, d\bar{\lambda}(y).$$

(c) $f * g \in L^p(\mathbb{R}, \mathcal{L}, \bar{\lambda})$.

(d) $g * f = f * g$ $\bar{\lambda}$-a.e. on $\mathbb{R}$, hence $g * f \in L^p(\mathbb{R}, \mathcal{L}, \bar{\lambda})$ by (c).

(e) $\|f * g\|_p = \|g * f\|_p \le \|f\|_1 \|g\|_p$.

Proof: Let h denote a function with domain $\mathbb{R}$ that is in $L^{p'}(\mathbb{R}, \mathcal{L}, \bar{\lambda})$. The mapping $(x,y) \mapsto f(x-y)g(y)$ on $\mathbb{R}^2$ is $\mathcal{L}^2/\mathcal{B}^*$-measurable by Claim 1, and by Claim 1 of Section 10.4 the mapping $(x,y) \mapsto h(x)$ is $\mathcal{L} \otimes \mathcal{L}/\mathcal{B}^*$-measurable, hence $\mathcal{L}^2/\mathcal{B}^*$-measurable. It follows that the mapping

$$(x,y) \mapsto f(x-y)g(y)h(x) \quad \text{on } \mathbb{R}^2$$

is $\mathcal{L}^2/\mathcal{B}^*$-measurable. Therefore, we may meaningfully discuss the double integral

$$\int_{\mathbb{R}^2} |f(x-y)g(y)h(x)|\, d\bar{\lambda}_2(x,y),$$

which, by the Nonnegative [Complete] Fubini Theorem, allows us to meaningfully discuss the equivalent iterated integral

$$\int_{\mathbb{R}} \int_{\mathbb{R}} |f(x-y)g(y)h(x)|\, d\bar{\lambda}(y) d\bar{\lambda}(x).$$

[6]The slight abuse comes in using the notation $f*g$ and $g*f$ even though it is not necessarily the case that both f and g are in $L^1(\mathbb{R}, \mathcal{L}, \bar{\lambda})$ as the definition requires.

We have that

$$
\begin{aligned}
& \int_{\mathbb{R}} \int_{\mathbb{R}} |f(x-y)g(y)h(x)| \, d\bar{\lambda}(y) \, d\bar{\lambda}(x) \\
= & \int_{\mathbb{R}} |h(x)| \int_{\mathbb{R}} |f(x-y)g(y)| \, d\bar{\lambda}(y) \, d\bar{\lambda}(x) \qquad (10.12) \\
= & \int_{\mathbb{R}} |h(x)| \int_{\mathbb{R}} |f(y)g(x-y)| \, d\bar{\lambda}(y) \, d\bar{\lambda}(x) \qquad (10.13) \\
= & \int_{\mathbb{R}} |f(y)| \int_{\mathbb{R}} |h(x)g(x-y)| \, d\bar{\lambda}(x) \, d\bar{\lambda}(y) \\
\leq & \int_{\mathbb{R}} |f(y)| \|\tau_{-y} g\|_p \|h\|_{p'} \, d\bar{\lambda}(y) \\
= & \int_{\mathbb{R}} |f(y)| \|g\|_p \|h\|_{p'} \, d\bar{\lambda}(y) \\
= & \|f\|_1 \|g\|_p \|h\|_{p'} \quad [< +\infty.] \qquad (10.14)
\end{aligned}
$$

[The first equality is obvious. The second equality follows by fixing x and taking $\phi(y) = |f(x-y)g(y)|$ in Claim 4. The third equality follows by a direct application of the Nonnegative [Complete] Fubini Theorem. The inequality follows by Hölder's Inequality and Claim 5, which asserts for every $y \in \mathbb{R}$ that the mapping $x \mapsto g(x-y)$ on $\mathbb{R}$ is in $L^p(\mathbb{R}, \mathcal{L}, \bar{\lambda})$. The fourth equality follows by Claim 5. The fifth equality is obvious.]

With this preliminary affair finished, we now turn to proving (a) and (b). The previous display holds for all $h \in L^{p'}(\mathbb{R}, \mathcal{L}, \bar{\lambda})$; we will assume in accordance with the remarks immediately preceding this claim that h is everywhere positive. The iterated integral in (10.12) is finite, hence for $\bar{\lambda}$-a.e. $x \in \mathbb{R}$ we have that $|h(x)| \int_{\mathbb{R}} |f(x-y)g(y)| \, d\bar{\lambda}(y)$ is finite. Since h is assumed everywhere positive, it follows that $\int_{\mathbb{R}} |f(x-y)g(y)| \, d\bar{\lambda}(y)$ is finite for $\bar{\lambda}$-a.e. $x \in \mathbb{R}$. In turn, this shows for $\bar{\lambda}$-a.e. $x \in \mathbb{R}$ that the mapping $y \mapsto f(x-y)g(y)$ on $\mathbb{R}$ is in $L^1(\mathbb{R}, \mathcal{L}, \bar{\lambda})$; (a) is accordingly established. For (b), observe that the iterated integral in (10.13) is finite and use the same argument as in (a) [again using the fact that h is everywhere positive] to establish that for $\bar{\lambda}$-a.e. $x \in \mathbb{R}$ the mapping $y \mapsto f(y)g(x-y)$ on $\mathbb{R}$ is in $L^1(\mathbb{R}, \mathcal{L}, \bar{\lambda})$. This gives (b).

Before proceeding further, we make an observation. Because

$$\int_{\mathbb{R}} \int_{\mathbb{R}} |f(x-y)g(y)h(x)| \, d\bar{\lambda}(y) \, d\bar{\lambda}(x) < +\infty$$

and

$$\int_{\mathbb{R}} \int_{\mathbb{R}} |f(y)g(x-y)h(x)| \, d\bar{\lambda}(y) \, d\bar{\lambda}(x) < +\infty$$

for every $h \in L^{p'}(\mathbb{R}, \mathcal{L}, \bar{\lambda})$, we have that

$$\int_{\mathbb{R}} h(x) \int_{\mathbb{R}} f(x-y)g(y) \, d\bar{\lambda}(y) \, d\bar{\lambda}(x) \quad \text{and} \quad \int_{\mathbb{R}} h(x) \int_{\mathbb{R}} f(y)g(x-y) \, d\bar{\lambda}(y) \, d\bar{\lambda}(x)$$

both exist and are equal for every $h \in L^{p'}(\mathbb{R}, \mathcal{L}, \bar{\lambda})$, with the equality following from (a), (b), and Claim 4. [Take $\phi(y) = f(x-y)g(y)$ in Claim 4.]

We now define a functional L on $L^{p'}(\mathbb{R}, \mathcal{L}, \bar{\lambda})$ by writing

$$L(h) = \int_{\mathbb{R}} h(x) f * g(x) \, d\bar{\lambda}(x) = \int_{\mathbb{R}} h(x) \int_{\mathbb{R}} f(x-y)g(y) \, d\bar{\lambda}(y) \, d\bar{\lambda}(x)$$

for every $h \in L^{p'}(\mathbb{R}, \mathcal{L}, \bar{\lambda})$. By the immediately preceding discussion, we also have

$$L(h) = \int_{\mathbb{R}} h(x) g * f(x) \, d\bar{\lambda}(x) = \int_{\mathbb{R}} h(x) \int_{\mathbb{R}} f(y)g(x-y) \, d\bar{\lambda}(y) \, d\bar{\lambda}(x)$$

for every $h \in L^{p'}(\mathbb{R}, \mathcal{L}, \bar{\lambda})$.

L is clearly a linear functional, and the inequality connecting (10.12) and (10.14) gives $|L(h)| \leq ||f||_1 ||g||_p ||h||_{p'}$ for every h in $L^{p'}(\mathbb{R}, \mathcal{L}, \bar{\lambda})$. Therefore, L is in fact a bounded linear functional on $L^{p'}(\mathbb{R}, \mathcal{L}, \bar{\lambda})$ with $||L|| \leq ||f||_1 ||g||_p$. The Riesz Representation Theorem asserts the existence of $\phi \in L^p(\mathbb{R}, \mathcal{L}, \bar{\lambda})$ such that

$$L(h) = \int_{\mathbb{R}} h(x)\phi(x) \, d\bar{\lambda}(x) \quad \text{for every } h \in L^{p'}(\mathbb{R}, \mathcal{L}, \bar{\lambda}).$$

Moreover, we have that $||L|| = ||\phi||_p$ by Claim 1 of Section 8.9, so that the inequality $||\phi||_p \leq ||f||_1 ||g||_p$ obtains. Combining the previous three displays gives

$$\begin{aligned}\int_{\mathbb{R}} h(x) f * g(x) \, d\bar{\lambda}(x) &= \int_{\mathbb{R}} h(x)\phi(x) \, d\bar{\lambda}(x) \\ &= \int_{\mathbb{R}} h(x) g * f(x) \, d\bar{\lambda}(x) \quad \text{for every } h \in L^{p'}(\mathbb{R}, \mathcal{L}, \bar{\lambda}).\end{aligned}$$

Looking at the first equality, this implies[7] that $f * g = \phi$ $\bar{\lambda}$-a.e. on $\mathbb{R}$, and hence $f * g \in L^p(\mathbb{R}, \mathcal{L}, \bar{\lambda})$. Looking at the second equality and using the same reasoning shows that $g * f = \phi$ $\bar{\lambda}$-a.e. on $\mathbb{R}$, so that $g * f \in L^p(\mathbb{R}, \mathcal{L}, \bar{\lambda})$. Since $g * f$ and $f * g$ are both equal to ϕ $\bar{\lambda}$-a.e. on $\mathbb{R}$, it follows that $g * f = f * g$ $\bar{\lambda}$-a.e. on $\mathbb{R}$. Since $||\phi||_p \leq ||f||_1 ||g||_p$ and $f * g = \phi = g * f$ $\bar{\lambda}$-a.e. on $\mathbb{R}$, (e) immediately follows. This gives (c)-(e) at one stroke, completing the proof. ∎

The previous result shows that we may convolve functions in $L^1(\mathbb{R}, \mathcal{L}, \bar{\lambda})$ with functions in $L^p(\mathbb{R}, \mathcal{L}, \bar{\lambda})$, where $1 < p < +\infty$. With Claim 2, we may convolve functions in $L^1(\mathbb{R}, \mathcal{L}, \bar{\lambda})$ functions with functions in $L^1(\mathbb{R}, \mathcal{L}, \bar{\lambda})$. Combining these two cases, we may convolve functions in $L^1(\mathbb{R}, \mathcal{L}, \bar{\lambda})$ with $L^p(\mathbb{R}, \mathcal{L}, \bar{\lambda})$

[7] Let $A = \{x \in \mathbb{R} : f * g(x) > \phi(x)\}$. Using σ-finiteness, write $A = \bigcup_{n=1}^{\infty} A_n$, where $\{A_n\}_{n=1}^{\infty}$ is a disjoint sequence of $\mathcal{L}$-sets with $\bar{\lambda}(A_n) < +\infty$ for each $n \in \mathbb{N}$. Pick $n \in \mathbb{N}$, and consider the function $I_{A_n} \in L^{p'}(\mathbb{R}, \mathcal{L}, \bar{\lambda})$. We have $\int_{A_n} f * g \, d\bar{\lambda} = \int_{A_n} \phi \, d\bar{\lambda}$. Now $f * g > \phi$ on A_n, hence the equality just given forces $\bar{\lambda}(A_n) = 0$. But then $\bar{\lambda}(A) = 0$. If $B = \{x \in \mathbb{R} : f * g(x) < \phi(x)\}$, the same type of argument gives $\bar{\lambda}(B) = 0$. Thus $f * g = \phi$ $\bar{\lambda}$-a.e. on $\mathbb{R}$.

for $1 \le p < +\infty$. Claim 8 will show that we may convolve functions in $L^p(\mathbb{R}, \mathcal{L}, \bar{\lambda})$ with those in $L^{p'}(\mathbb{R}, \mathcal{L}, \bar{\lambda})$ for $1 \le p < +\infty$. A technical result is first required.

Claim 7. *Let $f \in L^p(\mathbb{R}, \mathcal{L}, \bar{\lambda})$, where $1 \le p < +\infty$. Then*

$$\lim_{h \to 0} ||\tau_h f - f||_p = 0.$$

[Here $\tau_h f$ is the function on $\mathbb{R}$ with $\tau_h f(x) = f(x+h)$ for all $x \in \mathbb{R}$.]

Proof: Pick $\epsilon > 0$. By Exercise 2 of Section 5.6, there is a $\mathcal{B}/\mathcal{B}^*$-measurable function g on $\mathbb{R}$ such that $g = f$ $\bar{\lambda}$-a.e. on $\mathbb{R}$. Use Claim 3 of Section 8.6 to find a continuous real-valued function ϕ on $\mathbb{R}$ with compact support [hence $\phi \in L^p(\mathbb{R}, \mathcal{B}, \lambda)$] such that $||\phi - g||_p < \frac{\epsilon}{3}$. [The last norm is relative to $(\mathbb{R}, \mathcal{B}, \lambda)$.] Since $g = f$ $\bar{\lambda}$-a.e. on $\mathbb{R}$, it follows that $||\phi - f||_p < \frac{\epsilon}{3}$, where the norm is with respect to $(\mathbb{R}, \mathcal{L}, \bar{\lambda})$. There exists $\alpha > 0$ such that $\phi = 0$ on $[-\alpha, \alpha]^c$. Since ϕ is continuous on $[-\alpha, \alpha]$, ϕ is uniformly continuous on $\mathbb{R}$. Using this uniform continuity, there exists a number $0 < \delta < 1$ so that for every $t \in \mathbb{R}$ we have

$$|h| < \delta \quad \text{implies} \quad |\phi(t+h) - \phi(t)| < \frac{\epsilon}{3}\left(\frac{1}{2(\alpha+1)}\right)^{\frac{1}{p}}.$$

It follows that if $|h| < \delta$ then

$$\int_{\mathbb{R}} |\phi(t+h) - \phi(t)|^p \, d\bar{\lambda}(t) = \int_{[-\alpha-1, \alpha+1]} |\phi(t+h) - \phi(t)|^p \, d\bar{\lambda}(t) < \left(\frac{\epsilon}{3}\right)^p.$$

In other words, $|h| < \delta$ forces $||\tau_h \phi - \phi||_p < \frac{\epsilon}{3}$. Next, applying Claim 5 to the element $\phi - f$ in $L^p(\mathbb{R}, \mathcal{L}, \bar{\lambda})$, we have

$$||\tau_h \phi - \tau_h f||_p = ||\tau_h(\phi - f)||_p = ||\phi - f||_p < \frac{\epsilon}{3} \quad \text{for all } h \in \mathbb{R}.$$

But then $|h| < \delta$ forces

$$||\tau_h f - f||_p = ||\tau_h f - \tau_h \phi||_p + ||\tau_h \phi - \phi||_p + ||\phi - f||_p < \frac{\epsilon}{3} + \frac{\epsilon}{3} + \frac{\epsilon}{3} = \epsilon.$$

This proves the result. ∎

Claim 8. *Let $1 \le p < +\infty$, with $f \in L^p(\mathbb{R}, \mathcal{L}, \bar{\lambda})$ and $g \in L^{p'}(\mathbb{R}, \mathcal{L}, \bar{\lambda})$. Define [again committing a slight abuse of notation]*

$$f * g(x) = \int_{\mathbb{R}} f(x-y) g(y) \, d\bar{\lambda}(y) \qquad \textit{for each } x \in \mathbb{R}.$$

We have the following:

(a) $f * g$ is meaningfully defined.
(b) $\sup_{x \in \mathbb{R}} |f * g(x)| \le ||f||_p ||g||_{p'}$.
(c) $f * g$ is uniformly continuous on $\mathbb{R}$.

Proof: We first handle (a) in the case $1 < p < +\infty$. For every $x \in \mathbb{R}$, we have the inequality

$$\begin{aligned} &\int_{\mathbb{R}} |f(x-y)g(y)|\, d\bar{\lambda}(y) \\ \leq\ & \left[\int_{\mathbb{R}} |f(x-y)|^p\, d\bar{\lambda}(y)\right]^{\frac{1}{p}} \left[\int_{\mathbb{R}} |g(y)|^{p'}\, d\bar{\lambda}(y)\right]^{\frac{1}{p'}} \\ =\ & \left[\int_{\mathbb{R}} |f(y)|^p\, d\bar{\lambda}(y)\right]^{\frac{1}{p}} \left[\int_{\mathbb{R}} |g(y)|^{p'}\, d\bar{\lambda}(y)\right]^{\frac{1}{p'}} \\ =\ & ||f||_p ||g||_{p'} \quad \text{[a finite number]} \end{aligned}$$

by applying Hölder's Inequality and Claim 4. Therefore, when $1 < p < +\infty$, the expression $\int_{\mathbb{R}} f(x-y)g(y)\, d\bar{\lambda}(y)$ denotes a real number. Accordingly, $f*g(x)$ is a real number for every $x \in \mathbb{R}$ when $1 < p < +\infty$. Now consider the case $p = 1$, so that $p' = +\infty$. Here, saying that $g \in L^{p'}(\mathbb{R}, \mathcal{L}, \bar{\lambda}) = L^\infty(\mathbb{R}, \mathcal{L}, \bar{\lambda})$ means that $g \leq ||g||_\infty = \operatorname{ess\,sup}_{y\in\mathbb{R}} |g(y)|$ $\bar{\lambda}$-a.e. on $\mathbb{R}$ by Claim 1 of Section 8.4. We have for each $x \in \mathbb{R}$ that

$$\begin{aligned} \int_{\mathbb{R}} |f(x-y)g(y)|\, d\bar{\lambda}(y) &\leq \int_{\mathbb{R}} |f(x-y)|||g||_\infty\, d\bar{\lambda}(y) \\ &= ||g||_\infty \int_{\mathbb{R}} |f(x-y)|\, d\bar{\lambda}(y) \\ &= ||g||_\infty \int_{\mathbb{R}} |f(y)|\, d\bar{\lambda}(y) \\ &= ||f||_1 ||g||_\infty \quad \text{[a finite number.]} \end{aligned}$$

[The second equality is justified by Claim 4.] This shows that the expression $\int_{\mathbb{R}} f(x-y)g(y)\, d\bar{\lambda}(y)$ exists as a real number for each $x \in \mathbb{R}$. Accordingly, $f*g$ is defined and finite on $\mathbb{R}$ in the $p = 1$ case as well. This completes (a).

For (b), the work is already done for $1 \leq p < +\infty$:

$$\begin{aligned} |f*g(x)| = \left|\int_{\mathbb{R}} f(x-y)g(y)\, d\bar{\lambda}(y)\right| &\leq \int_{\mathbb{R}} |f(x-y)g(y)|\, d\bar{\lambda}(y) \\ &\leq ||f||_p ||g||_{p'} \quad \text{for each } x \in \mathbb{R}. \end{aligned}$$

Upon taking the supremum of both sides over all real x, (b) follows.

We turn to (c). If $||g||_{p'} = 0$, then $g = 0$ $\bar{\lambda}$-a.e. on $\mathbb{R}$, so that $f*g = 0$ on all of $\mathbb{R}$, showing that $f*g$ is trivially uniformly continuous in this case. We may therefore assume that $||g||_{p'} > 0$. Pick $\epsilon > 0$. For $x \in \mathbb{R}$, let f_x denote the function on $\mathbb{R}$ such that $f_x(y) = f(x-y)$. Observe that $f_x \in L^p(\mathbb{R}, \mathcal{L}, \bar{\lambda})$ since

$$\begin{aligned} \int_{\mathbb{R}} |f_x(y)|^p\, d\bar{\lambda}(y) = \int_{\mathbb{R}} |f(x-y)|^p\, d\bar{\lambda}(y) &= \int_{\mathbb{R}} |f(y)|^p\, d\bar{\lambda}(y) \\ &< +\infty. \end{aligned}$$

[The second equality follows by Claim 4.]

Using Claim 7, there exists $\delta > 0$ such that $||f_x - f_z||_p ||g||_{p'} < \epsilon$ whenever[8] $|x - z| < \delta$. It follows that $|x - z| < \delta$ implies

$$\begin{aligned}
& |f * g(x) - f * g(z)| \\
= & \left| \int_{\mathbb{R}} f(x-y)g(y)\, d\bar{\lambda}(y) - \int_{\mathbb{R}} f(z-y)g(y)\, d\bar{\lambda}(y) \right| \\
\leq & \int_{\mathbb{R}} |f(x-y) - f(z-y)||g(y)|\, d\bar{\lambda}(y) \\
\leq & \left[\int_{\mathbb{R}} |f_x(y) - f_z(y)|^p\, d\bar{\lambda}(y) \right]^{\frac{1}{p}} ||g||_{p'} \\
= & ||f_x - f_z||_p ||g||_{p'} \\
< & \epsilon.
\end{aligned}$$

[The second inequality is Hölder's Inequality; it holds for $1 < p < +\infty$ and $p = 1$.] This completes the proof of (c), finishing the entire proof. ■

The collection of results concerning convolutions given in this section is by no means exhaustive, but it should provide a good illustration of the value of the Fubini Theorems in regards to establishing the existence of certain integrals, the measurability of certain functions defined by integrals, and the equality of iterated integrals. The exercises will give the desirous reader more opportunities to apply the Fubini Theorems [among other things] in the setting of convolutions.

Exercises.

1. Let $f, g, h \in L^1(\mathbb{R}, \mathcal{L}, \bar{\lambda})$.
(a) Convolution is commutative in the sense that $f * g = g * f$ $\bar{\lambda}$-a.e. on $\mathbb{R}$.
(b) Convolution is associative in that $f * (g * h) = (f * g) * h$ $\bar{\lambda}$-a.e. on $\mathbb{R}$.

2. This exercise is a lemma for the next exercise. Let $f \in L^1(\mathbb{R}, \mathcal{L}, \bar{\lambda})$ and $\epsilon > 0$. There exists a closed interval $F = [-\alpha, \alpha] \subseteq \mathbb{R}$ such that $\int_{F^c} f\, d\bar{\lambda} < \epsilon$. [Consider the sequence $\{fI_{[-n,n]}\}_{n=1}^{\infty}$ of functions in $L^1(\mathbb{R}, \mathcal{L}, \bar{\lambda})$. For every $n \in \mathbb{N}$ we have

[8] Pick $x, z \in \mathbb{R}$ and let $h = z - x$. Applying Claim 4 to the function $g(y) = |f(y) - f(h+y)|^p$ [defined for each $y \in \mathbb{R}$] gives

$$\begin{aligned}
||f_x - f_z||_p^p &= \int_{\mathbb{R}} |f(x-y) - f(z-y)|^p\, d\bar{\lambda}(y) \\
&= \int_{\mathbb{R}} |f(x-y) - f(h+x-y)|^p\, d\bar{\lambda}(y) \\
&= \int_{\mathbb{R}} |f(y) - f(h+y)|^p\, d\bar{\lambda}(y) \\
&= ||\tau_h f - f||_p^p.
\end{aligned}$$

By Claim 7, the latter quantity is small for small $|h|$, so that $||f_x - f_z||_p$ is small when $|z - x|$ is small.

$|fI_{[-n,n]}| \leq |f|$ and $\int_{\mathbb{R}} |f| \, d\bar{\lambda} < +\infty$. Therefore $\lim_n \int_{\mathbb{R}} fI_{[-n,n]} \, d\bar{\lambda} = \int_{\mathbb{R}} f \, d\bar{\lambda}$ by DCT. Let $F = [-n, n]$ for sufficiently large n.]

3. In Claim 8, we also have $\lim_{|x|\to+\infty} f * g(x) = 0$. [Let $\epsilon > 0$. Use Exercise 2 to find a closed interval $F = [-\alpha, \alpha] \subseteq \mathbb{R}$ so that

$$\int_{F^c} |f|^p \, d\bar{\lambda} < \left[\frac{\epsilon}{\|f\|_p + \|g\|_{p'}}\right]^p = c^p \text{ and } \int_{F^c} |g|^{p'} \, d\bar{\lambda} < \left[\frac{\epsilon}{\|f\|_p + \|g\|_{p'}}\right]^{p'} = c^{p'}.$$

Verify that $x \in \mathbb{R}$ and $|x| > 2\alpha$ gives $[x - \alpha, x + \alpha] \subseteq F^c$, so that $|x| > 2\alpha$ yields

$$\begin{aligned}
|f * g(x)| &\leq \left|\int_F f(x-y)g(y) \, d\bar{\lambda}(y)\right| + \left|\int_{F^c} f(x-y)g(y) \, d\bar{\lambda}(y)\right| \\
&\leq \left[\int_F |f(x-y)|^p \, d\bar{\lambda}(y)\right]^{\frac{1}{p}} \|g\|_{p'} + \|f\|_p \left[\int_{F^c} |g(y)|^{p'} \, d\bar{\lambda}(y)\right]^{\frac{1}{p'}} \\
&\leq \left[\int_{[x-\alpha,x+\alpha]} |f(y)|^p \, d\bar{\lambda}(y)\right]^{\frac{1}{p}} \|g\|_{p'} + c\|f\|_p \\
&\leq \left[\int_{F^c} |f(y)|^p \, d\bar{\lambda}(y)\right]^{\frac{1}{p}} \|g\|_{p'} + c\|f\|_p \\
&\leq c\|g\|_{p'} + c\|f\|_p \\
&= \epsilon.]
\end{aligned}$$

4. $L^1(\mathbb{R}, \mathcal{L}, \bar{\lambda})$ has no element u such that $u * f = f$ $\bar{\lambda}$-a.e. on $\mathbb{R}$ for every f in $L^1(\mathbb{R}, \mathcal{L}, \bar{\lambda})$. [Suppose such a u exists. Find $\delta > 0$ with $\int_{[-2\delta,2\delta]} |u| \, d\bar{\lambda} < 1$. Let $f = I_{[-\delta,\delta]}$. For $\bar{\lambda}$-a.e. $x \in \mathbb{R}$ we have

$$\begin{aligned}
f(x) = u * f(x) = \int_{\mathbb{R}} u(x-y)f(y) \, d\bar{\lambda}(y) &= \int_{[-\delta,\delta]} u(x-y) \, d\bar{\lambda}(y) \\
&= \int_{[x-\delta,x+\delta]} u(t) \, d\bar{\lambda}(t).
\end{aligned}$$

Since $\bar{\lambda}([-\delta, \delta]) > 0$, there is $x_0 \in [-\delta, \delta]$ with $1 = f(x_0) = \int_{[x_0-\delta,x_0+\delta]} u(t) \, d\bar{\lambda}(t)$. Now $[x_0 - \delta, x_0 + \delta] \subseteq [-2\delta, 2\delta]$, so

$$1 = \left|\int_{x_0-\delta}^{x_0+\delta} u(t) \, d\bar{\lambda}(t)\right| \leq \int_{x_0-\delta}^{x_0+\delta} |u(t)| \, d\bar{\lambda}(t) \leq \int_{-2\delta}^{2\delta} |u(t)| \, d\bar{\lambda}(t) < 1.$$

This gives a contradiction.]

5. A sequence $\{u_n\}_{n=1}^\infty$ of functions in $L^1(\mathbb{R}, \mathcal{L}, \bar{\lambda})$ is called an *approximate unit* if
(i) $u_n \geq 0$ for each $n \in \mathbb{N}$.
(ii) $\|u_n\|_1 = 1$ for each $n \in \mathbb{N}$.
(iii) For every neighborhood V of 0, $\lim_n \int_{V^c} u_n \, d\bar{\lambda} = 0$.
For example, if $u_n = \frac{n}{2} I_{[-\frac{1}{n},\frac{1}{n}]}$ for each $n \in \mathbb{N}$, then $\{u_n\}_{n=1}^\infty$ is an approximate unit. Prove the following: *Let* $\{u_n\}_{n=1}^\infty$ *denote an approximate unit, and let* $1 < p < +\infty$. *For every* $f \in L^p(\mathbb{R}, \mathcal{L}, \bar{\lambda})$, *we have* $\lim_n \|f * u_n - f\|_p = 0$.

[Let $f \in L^p(\mathbb{R}, \mathcal{L}, \bar{\lambda})$ and $\epsilon > 0$. The claim is trivial if $\|f\|_p = 0$, so assume that $\|f\|_p > 0$. Find a neighborhood V of 0 such that $y \in V$ forces $\|\tau_{-y}f - f\|_p < \frac{\epsilon}{2}$. Use (iii) to find $n_0 \in \mathbb{N}$ such that $n \geq n_0$ implies $\int_{V^c} u_n \, d\bar{\lambda} < \frac{\epsilon}{4\|f\|_p}$. Let $n \geq n_0$. We have $f * u_n \in L^p(\mathbb{R}, \mathcal{L}, \bar{\lambda})$ by the results in this section, and for each $h \in L^{p'}(\mathbb{R}, \mathcal{L}, \bar{\lambda})$ we have

$$\begin{aligned}
& \left| \int_{\mathbb{R}} [f * u_n(x) - f(x)] h(x) \, d\bar{\lambda}(x) \right| \\
= & \left| \int_{\mathbb{R}} \left[\int_{\mathbb{R}} f(x-y) u_n(y) \, d\bar{\lambda}(y) - f(x) \right] h(x) \, d\bar{\lambda}(x) \right| \\
= & \left| \int_{\mathbb{R}} \int_{\mathbb{R}} [f(x-y) - f(x)] u_n(y) \, d\bar{\lambda}(y) \, h(x) \, d\bar{\lambda}(x) \right| \\
\leq & \int_{\mathbb{R}} \int_{\mathbb{R}} |f(x-y) - f(x)| |u_n(y)| \, d\bar{\lambda}(y) \, |h(x)| \, d\bar{\lambda}(x) \\
= & \int_{\mathbb{R}} |u_n(y)| \int_{\mathbb{R}} |f(x-y) - f(x)| |h(x)| \, d\bar{\lambda}(x) \, d\bar{\lambda}(y) \\
\leq & \int_{\mathbb{R}} |u_n(y)| \|\tau_{-y} f - f\|_p \|h\|_{p'} \, d\bar{\lambda}(y) \\
\leq & \int_{V} u_n(y) \frac{\epsilon}{2} \|h\|_{p'} \, d\bar{\lambda}(y) + \int_{V^c} 2u_n(y) \|f\|_p \|h\|_{p'} \, d\bar{\lambda}(y) \\
\leq & \left(\frac{\epsilon}{2} \|h\|_{p'} \times 1 \right) + \left(2\|f\|_p \|h\|_{p'} \times \frac{\epsilon}{4\|f\|_p} \right) \\
= & \epsilon \|h\|_{p'}.
\end{aligned}$$

Define L on $L^{p'}(\mathbb{R}, \mathcal{L}, \bar{\lambda})$ by writing $L(h) = \int_{\mathbb{R}} [f * u_n(x) - f(x)] h(x) \, d\bar{\lambda}(x)$. Then L is a bounded linear functional and $\epsilon \geq \|L\| = \|f * u_n - f\|_p$ by Claim 1 of Section 8.9.]

6. Redo Exercise 5 if $p = 1$. [Take $h = 1$ above.]

7. If f and g are elements of $L^1(\mathbb{R}, \mathcal{L}, \bar{\lambda})$ that depend only on $|x|$, then $f * g$ only depends on $|x|$.

8. The following parts give some concrete examples of convolutions courtesy of Jones, pages 278-279.

(a) Let $0 < b \leq a$. Let $f = I_{(-a,a)}$ and $g = I_{(-b,b)}$. Then for each $x \in \mathbb{R}$

$$f * g(x) = \begin{cases} 2b & |x| \leq a - b \\ a + b - |x| & a - b \leq |x| \leq a + b \\ 0 & a + b \leq |x|. \end{cases}$$

(b) Let a and b denote distinct real numbers. Let $f(x) = e^{-ax} I_{[0,+\infty)}(x)$ and $g(x) = e^{-bx} I_{[0,+\infty)}(x)$ for each $x \in \mathbb{R}$. Then

$$f * g(x) = \begin{cases} \frac{e^{-ax} - e^{-bx}}{b-a} & x \geq 0 \\ 0 & x < 0. \end{cases}$$

(c) If in (b) we have $a = b$, then $f * g(x) = xe^{-ax}$ for each $x \in \mathbb{R}$.

(d) If $a+b>0$, $f(x)=e^{-ax^2}$, and $g(x)=e^{-bx^2}$ [for all $x\in\mathbb{R}$], then

$$f*g(x)=\sqrt{\frac{\pi}{a+b}}e^{-\frac{ab}{a+b}x^2} \quad \text{for all } x\in\mathbb{R}.$$

9. This exercise generalizes Young's Theorem [Claim 6] and requires the result of Exercise 3 in Section 8.1. Show the following: *Let* $1\le p,q,r\le+\infty$ *be such that* $\frac{1}{r}=\frac{1}{p}+\frac{1}{q}-1$. *Let* $f\in L^p(\mathbb{R},\mathcal{L},\bar{\lambda})$ *and* $g\in L^q(\mathbb{R},\mathcal{L},\bar{\lambda})$. *Then* $f*g$ *exists* $\bar{\lambda}$*-a.e. on* $\mathbb{R}$ *and is in* $L^r(\mathbb{R},\mathcal{L},\bar{\lambda})$. *Also,* $\|f*g\|_r\le\|f\|_p\|g\|_q$.
(a) If $p=+\infty$, then $q=1$ and $r=+\infty$, and the claim follows easily in this case. The same comment applies to the case where $q=+\infty$ [and hence $p=1$ and $r=+\infty$].
(b) If $r=+\infty$, then p and q are conjugate exponents, and the claim follows by Hölder's Inequality.
(c) Assume that $1\le r<+\infty$ now. This forces $1\le p,q<+\infty$. Show that it is sufficient to prove the claim when both $f,g\ge 0$ and $\|f\|_p=\|g\|_q=1$.
(d) Prove the claim in (c). [Since f,g are nonnegative and $\mathcal{L}/\mathcal{B}^*$-measurable, we may meaningfully discuss $\int_{\mathbb{R}} f(x-y)g(y)\,d\bar{\lambda}(y)=f*g(x)$ for each $x\in\mathbb{R}$. Show that $\frac{1}{r}+\frac{1}{q'}+\frac{1}{p'}=1$, and use Exercise 3 in Section 8.1 to obtain

$$\begin{aligned}
f*g(x) &= \int_{\mathbb{R}} f(x-y)g(y)\,d\bar{\lambda}(y)\\
&= \int_{\mathbb{R}} f(x-y)^{\frac{p}{r}}g(y)^{\frac{q}{r}}f(x-y)^{1-\frac{p}{r}}g(y)^{1-\frac{q}{r}}\,d\bar{\lambda}(y)\\
&\le \left[\int_{\mathbb{R}} f(x-y)^p g(y)^q\,d\bar{\lambda}(y)\right]^{\frac{1}{r}}\left[\int_{\mathbb{R}} f(x-y)^{q'(1-\frac{p}{r})}\,d\bar{\lambda}(y)\right]^{\frac{1}{q'}}\\
&\quad\times\left[\int_{\mathbb{R}} g(y)^{p'(1-\frac{q}{r})}\,d\bar{\lambda}(y)\right]^{\frac{1}{p'}}\\
&= \left[\int_{\mathbb{R}} f(x-y)^p g(y)^q\,d\bar{\lambda}(y)\right]^{\frac{1}{r}}\times 1\times 1
\end{aligned}$$

for each $x\in\mathbb{R}$. This shows that $(f*g)^r\le f^p*g^q\in L^1(\mathbb{R},\mathcal{L},\bar{\lambda})$ by Claim 3. It follows that $f*g\in L^r(\mathbb{R},\mathcal{L},\bar{\lambda})$, and Claim 3 applied to f^p and g^q gives

$$\|f*g\|_r^r=\int_{\mathbb{R}} f*g(x)^r\,d\bar{\lambda}(x)\le\int_{\mathbb{R}} f^p*g^q(x)\,d\bar{\lambda}(x)\le\|f^p\|_1\|g^q\|_1=\|f\|_p^p\|g\|_q^q=1.]$$

10.6 THE HARDY–LITTLEWOOD THEOREMS

In this final section of the chapter we present another application and illustration of the Fubini Theorems. The major result of this section is the *Hardy-Littlewood Maximal Theorem.* This famous and classical theorem concerns itself with Lebesgue measurable functions f in $L^p(\mathbb{R},\mathcal{L},\bar{\lambda})$, where $1\le p<+\infty$.

We will handle the case where $1 < p < +\infty$, and will outline the $p = 1$ case in the exercises.

Framework. $f : \mathbb{R} \to [0, +\infty]$ is $\mathcal{L}/\mathcal{B}^*$-measurable and such that the integral $\int_F f \, d\bar{\lambda}$ is finite for every compact $F \subseteq \mathbb{R}$. We will also make the following definitions for each $x \in \mathbb{R}$:

$$f^{\delta(r)}(x) = \sup_{x<u<+\infty} \frac{1}{u-x} \int_x^u f \, d\bar{\lambda},$$

$$f^{\delta(l)}(x) = \sup_{-\infty<u<x} \frac{1}{x-u} \int_u^x f \, d\bar{\lambda},$$

$$f^{\delta}(x) = \max\{f^{\delta(r)}(x), f^{\delta(l)}(x)\}.$$

The superscripts r and l denote *right* and *left*, respectively. The argument u is to the right of x in the definition of $f^{\delta(r)}(x)$; u is to the left of x in the definition of $f^{\delta(l)}(x)$. Next, we define the following sets for each $t \in \mathbb{R}$:

$$G_t = \{x \in \mathbb{R} : f(x) > t\}, \ \ M_t^{(r)} = \{x \in \mathbb{R} : \ f^{\delta(r)}(x) > t\},$$

$$M_t^{(l)} = \{x \in \mathbb{R} : \ f^{\delta(l)}(x) > t\}, \ \ M_t = \{x \in \mathbb{R} : \ f^{\delta}(x) > t\} = M_t^{(r)} \cup M_t^{(l)}.$$

If $t < 0$, then the nonnegativity of f forces $G_t = M_t^{(r)} = M_t^{(l)} = M_t = \mathbb{R}$. For this reason, our interest shall be for $t \geq 0$.

We now turn to discussing measurability. Given a nonnegative $\mathcal{L}/\mathcal{B}^*$-measurable function f on $\mathbb{R}$ such that $\int_F f \, d\bar{\lambda} < +\infty$ for every compact $F \subseteq \mathbb{R}$, it is far from obvious that the nonnegative functions $f^{\delta(r)}$, $f^{\delta(l)}$, and f^{δ} are $\mathcal{L}/\mathcal{B}^*$-measurable. Also, for $t \geq 0$, it is not obvious that the sets $M_t^{(r)}$, $M_t^{(l)}$, and M_t are in $\mathcal{L}$. It will be first shown that the mappings $s \mapsto \frac{1}{s-x} \int_x^s f \, d\bar{\lambda}$ on $(x, +\infty)$ and $s \mapsto \frac{1}{x-s} \int_s^x f \, d\bar{\lambda}$ on $(-\infty, x)$ are continuous for each $x \in \mathbb{R}$. This fact will then be used to establish that for each $t \geq 0$ the sets $M_t^{(r)}$ and $M_t^{(l)}$ are open [and hence so is M_t, being the union of the two open sets $M_t^{(r)}$ and $M_t^{(l)}$], and hence in $\mathcal{L}$. This will establish the $\mathcal{L}/\mathcal{B}^*$-measurability of the functions $f^{\delta(r)}$, $f^{\delta(l)}$, and f^{δ}. These considerations are the topics of the first two claims.

Claim 1. *Let* $x \in \mathbb{R}$. *Then*

(a) The mapping $s \mapsto \frac{1}{s-x} \int_x^s f \, d\bar{\lambda}$ *on* $(x, +\infty)$ *is continuous.*
(b) The mapping $s \mapsto \frac{1}{x-s} \int_s^x f \, d\bar{\lambda}$ *on* $(-\infty, x)$ *is continuous.*

Proof: Due to the symmetry of claims (a) and (b), we really need to prove (a) only. To prove (a), we note that the mapping $s \mapsto \frac{1}{s-x}$ is trivially continuous on $(x, +\infty)$, and hence (a) will be shown when it is demonstrated that $s \mapsto \int_x^s f \, d\bar{\lambda}$ is continuous on $(x, +\infty)$. To show that $s \mapsto \int_x^s f \, d\bar{\lambda}$ is continuous on $(x, +\infty)$, we will show that $s \mapsto \int_x^s f \, d\bar{\lambda}$ is left-continuous and right-continuous on $(x, +\infty)$.

We first demonstrate left-continuity. Let $s \in (x, +\infty)$. Pick a sequence $\{s_n\}_{n=1}^\infty$ with $x < s_1 \leq s_2 \leq \cdots \leq s$ and $\lim_n s_n = s$. We desire to establish the limit relation

$$\lim_n \int_x^{s_n} f \, d\bar{\lambda} = \int_x^s f \, d\bar{\lambda},$$

that is,

$$\lim_n \int_{\mathbb{R}} f I_{[x,s_n]} \, d\bar{\lambda} = \int_{\mathbb{R}} f I_{[x,s]} \, d\bar{\lambda}.$$

Observe that $\{f I_{[x,s_n]}\}_{n=1}^\infty$ is a nondecreasing sequence of $\mathcal{L}/\mathcal{B}^*$-measurable nonnegative functions on $\mathbb{R}$ with $\lim_n f I_{[x,s_n]} = f I_{[x,s]}$ $\bar{\lambda}$-a.e. on $\mathbb{R}$. By MCT, we have

$$\lim_n \int_{\mathbb{R}} f I_{[x,s_n]} \, d\bar{\lambda} = \int_{\mathbb{R}} \lim_n f I_{[x,s_n]} \, d\bar{\lambda} = \int_{\mathbb{R}} f I_{[x,s]} \, d\bar{\lambda}.$$

This shows that the mapping $s \mapsto \int_x^s f \, d\bar{\lambda}$ on $(x, +\infty)$ is left-continuous.

We now turn to right continuity. Again, let $s \in (x, +\infty)$. This time, let $\{s_n\}_{n=1}^\infty$ denote a nonincreasing sequence in $(s, +\infty)$ with $\lim_n s_n = s$. We again desire to establish the limit relation

$$\lim_n \int_{\mathbb{R}} f I_{[x,s_n]} \, d\bar{\lambda} = \int_{\mathbb{R}} f I_{[x,s]} \, d\bar{\lambda}.$$

Observe that for each $n \in \mathbb{N}$ we have $f I_{[x,s_n]} \leq f I_{[x,s_1]}$. Observe next that $[x, s_1]$ is compact and hence $\int_{[x,s_1]} f \, d\bar{\lambda} < +\infty$. The hypotheses of DCT are satisfied, again giving

$$\lim_n \int_{\mathbb{R}} f I_{[x,s_n]} \, d\bar{\lambda} = \int_{\mathbb{R}} \lim_n f I_{[x,s_n]} \, d\bar{\lambda} = \int_{\mathbb{R}} f I_{[x,s]} \, d\bar{\lambda}.$$

This shows that the mapping $s \mapsto \int_x^s f \, d\bar{\lambda}$ on $(x, +\infty)$ is right-continuous, and thus (a) is established. This completes the proof. ∎

Observe in the above proof that the left continuity claim involves only MCT and uses only the fact that f is nonnegative. However, the right continuity claim requires DCT, which uses the assumption that $\int_F f \, d\bar{\lambda} < +\infty$ for all compact $F \subseteq \mathbb{R}$.

Claim 2. *We have the following:*

(a) For every $t \in \mathbb{R}$, the sets $M_t^{(r)}$ and $M_t^{(l)}$ are open, hence in $\mathcal{L}$.

(b) M_t is open for every $t \in \mathbb{R}$, hence in $\mathcal{L}$.

(c) $f^{\delta(r)}$, $f^{\delta(l)}$, and f^δ are $\mathcal{L}/\mathcal{B}^*$-measurable.

Proof: We do (a) first for $M_t^{(r)}$. If $t < 0$, then $M_t^{(r)} = \mathbb{R}$ by the nonnegativity of f, so $M_t^{(r)}$ is trivially open in this case. To complete the proof that $M_t^{(r)}$ is open, it must be shown that $M_t^{(r)}$ is open when $t \geq 0$. To do this, pick $t \geq 0$. If $M_t^{(r)} = \varnothing$, then $M_t^{(r)}$ is trivially open. The nontrivial case is where $M_t^{(r)} \neq \varnothing$; pick $x_0 \in M_t^{(r)}$. For this x_0, we have $f^{\delta(r)}(x_0) > t$, hence $\sup_{x_0<u<+\infty} \frac{1}{u-x_0} \int_{x_0}^{u} f \, d\bar{\lambda} > t$, whence there is $u_0 \in (x_0, +\infty)$ with $\frac{1}{u_0-x_0} \int_{x_0}^{u_0} f \, d\bar{\lambda} > t$. By part (b) of Claim 1, the mapping $s \mapsto \frac{1}{u_0-s} \int_{s}^{u_0} f \, d\bar{\lambda}$ is continuous on $(-\infty, u_0)$. Since $\frac{1}{u_0-x_0} \int_{x_0}^{u_0} f \, d\bar{\lambda} > t$, there is $\delta > 0$ such that $(x_0-\delta, x_0+\delta) \subseteq (-\infty, u_0)$ and $\frac{1}{u_0-x} \int_{x}^{u_0} f \, d\bar{\lambda} > t$ whenever $x \in (x_0-\delta, x_0+\delta)$. It follows that $x \in (x_0 - \delta, x_0 + \delta)$ implies

$$\sup_{u \in (x,+\infty)} \frac{1}{u-x} \int_{x}^{u} f \, d\bar{\lambda} \geq \frac{1}{u_0 - x} \int_{x}^{u_0} f \, d\bar{\lambda} > t.$$

This shows that $x \in (x_0 - \delta, x_0 + \delta)$ implies $f^{\delta(r)}(x) > t$, so that $x \in M_t^{(r)}$. In other words, given $x_0 \in M_t^{(r)}$, there is $\delta > 0$ such that $(x_0 - \delta, x_0 + \delta) \subseteq M_t^{(r)}$. It follows in the nonempty case with $t \geq 0$ that $M_t^{(r)}$ is open. Therefore, we have shown that $M_t^{(r)}$ is open when (i) $t < 0$, (ii) $t \geq 0$ and $M_t^{(r)} = \varnothing$, and (iii) $t \geq 0$ and $M_t^{(r)} \neq \varnothing$. This covers all of the cases, and $M_t^{(r)}$ is open for all $t \in \mathbb{R}$.

The argument that $M_t^{(l)}$ is open for all $t \in \mathbb{R}$ uses the same three cases (i) $t < 0$, (ii) $t \geq 0$ and $M_t^{(l)} = \varnothing$, and (iii) $t \geq 0$ and $M_t^{(l)} \neq \varnothing$. Cases (i) and (ii) are trivial in the exact same way as before, and the nontrivial case (iii) follows by the same continuity argument as used before, with the exception being that part (a) of Claim 1 is used instead of part (b) of Claim 1. This gives (a).

Part (b) is trivial by (a) and the fact that $M_t = M_t^{(r)} \cup M_t^{(l)}$.

Part (c) is automatic. By (a), the set $M_t^{(r)} = (f^{\delta(r)})^{-1}((t, +\infty])$ is in $\mathcal{L}$ for each $t \in \mathbb{R}$, and hence $f^{\delta(r)}$ is $\mathcal{L}/\mathcal{B}^*$-measurable. Similarly, $f^{\delta(l)}$ is seen to be $\mathcal{L}/\mathcal{B}^*$-measurable. Finally, f^{δ} is $\mathcal{L}/\mathcal{B}^*$-measurable since f^{δ} is the pointwise defined maximum of the $\mathcal{L}/\mathcal{B}^*$-measurable functions $f^{\delta(r)}$ and $f^{\delta(l)}$. This completes (c) and hence the entire proof. ■

Courtesy of the above claims, we may now discuss such things as $\bar{\lambda}(M_t^{(r)})$, $\bar{\lambda}(M_t^{(l)})$, and $\bar{\lambda}(M_t)$ for each real t. We may also discuss the integrals of $f^{\delta(r)}$, $f^{\delta(l)}$, and f^{δ} with respect to $\bar{\lambda}$ over any $\mathcal{L}$-set.

The upcoming claim represents a fair amount of work. The inequalities found therein represent a special case of the Hardy-Littlewood Maximal Theorem.

Claim 3. *Let $0 < t < +\infty$. The following equalities and inequalities obtain:*

(a) $\bar{\lambda}(M_t^{(r)}) = \frac{1}{t} \int_{M_t^{(r)}} f \, d\bar{\lambda}$.

(b) $\bar{\lambda}(M_t^{(l)}) = \frac{1}{t} \int_{M_t^{(l)}} f \, d\bar{\lambda}$.

(c) $\bar{\lambda}(M_t) \leq \frac{2}{t} \int_{M_t} f \, d\bar{\lambda}$.

Proof: The equalities in (a) and (b) are proved in similar fashion; we give a thorough proof of (a). To do (a), pick and fix real $t > 0$. If $M_t^{(r)} = \varnothing$, then (a) reduces to the trivial equation $0 = 0$. We thus assume for the remainder of the proof of (a) that $M_t^{(r)}$ is nonempty. By Claim 2, $M_t^{(r)}$ is open, and hence there is a disjoint collection $\{(\beta_k, \gamma_k)\}_{k=1}^{\infty}$ of [possibly unbounded] open intervals with $M_t^{(r)} = \bigcup_{k=1}^{\infty}(\beta_k, \gamma_k)$. Observe that $\beta_k, \gamma_k \notin M_t^{(r)}$ for each $k \in \mathbb{N}$. *We now fix* $k \in \mathbb{N}$. *Doing this, we fix* $x \in (\beta_k, \gamma_k)$ *and define*

$$N_x = \left\{ s \in (x, \gamma_k] \cap \mathbb{R} : \frac{1}{s-x} \int_x^s f \, d\bar{\lambda} > t \right\}.$$

Subclaim: $N_x \neq \varnothing$.

Proof: There are two cases: (i) $\gamma_k = +\infty$ and (ii) $\gamma_k < +\infty$. We first handle (i). If $\gamma_k = +\infty$, then $N_x = \{s \in (x, +\infty) : \frac{1}{s-x} \int_x^s f \, d\bar{\lambda} > t\}$. Now $x \in M_t^{(r)}$, hence $f^{\delta(r)}(x) = \sup_{u \in (x,+\infty)} \frac{1}{u-x} \int_x^u f \, d\bar{\lambda} > t$. Therefore, there is $s \in (x, +\infty)$ with $\frac{1}{s-x} \int_x^s f \, d\bar{\lambda} > t$, so s is in N_x. It follows that $N_x \neq \varnothing$ when $\gamma_k = +\infty$.

We now assume that $\gamma_k < +\infty$; here $N_x = \{s \in (x, \gamma_k] : \frac{1}{s-x} \int_x^s f \, d\bar{\lambda} > t\}$. We argue by contradiction, assuming that $N_x = \varnothing$. Since $x \in (\beta_k, \gamma_k) \subseteq M_t^{(r)}$, we have

$$f^{\delta(r)}(x) = \sup_{u \in (x,+\infty)} \frac{1}{u-x} \int_x^u f \, d\bar{\lambda} > t,$$

hence there is at least one $w \in (x, +\infty)$ with $\frac{1}{w-x} \int_x^w f \, d\bar{\lambda} > t$. Since N_x is assumed to be empty, it follows that $w \notin (x, \gamma_k]$, so $\gamma_k < w < +\infty$. Next, $\gamma_k \notin N_x$ since N_x is assumed to be empty, hence the definition of N_x forces $\frac{1}{\gamma_k - x} \int_x^{\gamma_k} f \, d\bar{\lambda} \leq t$. We then have

$$\int_{\gamma_k}^w f \, d\bar{\lambda} = \int_x^w f \, d\bar{\lambda} - \int_x^{\gamma_k} f \, d\bar{\lambda} > t(w - x) - t(\gamma_k - x) = t(w - \gamma_k),$$

which means that $\frac{1}{w-\gamma_k} \int_{\gamma_k}^w f \, d\bar{\lambda} > t$, so that

$$\sup_{u \in (\gamma_k,+\infty)} \frac{1}{u-\gamma_k} \int_{\gamma_k}^u f \, d\bar{\lambda} > t,$$

which gives $f^{\delta(r)}(\gamma_k) > t$, forcing $\gamma_k \in M_t^{(r)}$. However, it was stated immediately prior to this subclaim that $\gamma_k \notin M_t^{(r)}$. The assumption that $N_x = \varnothing$ in the case $\gamma_k < +\infty$ leads to the contradictory conclusion that γ_k is both in and not in $M_t^{(r)}$, hence it is the case that $N_x \neq \varnothing$ when $\gamma_k < +\infty$. This completes the proof of the subclaim.

Since $N_x \neq \varnothing$, it has a well-defined supremum; write $s_x = \sup N_x$. Because $x \in (\beta_k, \gamma_k)$ is arbitrary, the above subclaim actually shows that N_y is nonempty for every $y \in (\beta_k, \gamma_k)$. We continue to keep everything fixed.

Subclaim: We have $s_x = \gamma_k$.

Proof: By considering the cases $\gamma_k = +\infty$ and $\gamma_k < +\infty$, it is seen that $s_x \leq \gamma_k$. It is also clear that $x < s_x$. Assume that $s_x < \gamma_k$. From this we employ basic properties of suprema, obtaining

(i) If $x < s < s_x$ and s is sufficiently close to s_x, then $\frac{1}{s-x} \int_x^s f \, d\bar{\lambda} > t$, and

(ii) If $s_x < s < \gamma_k$, then $\frac{1}{s-x}\int_x^s f\,d\bar{\lambda} \leq t$.

Using (i), (ii), and (a) of Claim 1, we have $\frac{1}{s_x - x}\int_x^{s_x} f\,d\bar{\lambda} = t$. Since $x < s_x < \gamma_k$, we have that $s_x \in (\beta_k, \gamma_k)$, so that

$$N_{s_x} = \left\{y \in (s_x, \gamma_k] \cap \mathbb{R} : \frac{1}{y - s_x}\int_{s_x}^{y} f\,d\bar{\lambda} > t\right\} \neq \varnothing.$$

Since $N_{s_x} \neq \varnothing$, we may pick $y \in (s_x, \gamma_k] \cap \mathbb{R}$ such that $\frac{1}{y-s_x}\int_{s_x}^{y} f\,d\bar{\lambda} > t$. Observe next that

$$\int_x^y f\,d\bar{\lambda} = \int_x^{s_x} f\,d\bar{\lambda} + \int_{s_x}^{y} f\,d\bar{\lambda} > t(s_x - x) + t(y - s_x) = t(y - x).$$

Since $y > x$, the above gives $\frac{1}{y-x}\int_x^y f\,d\bar{\lambda} > t$. Since $y \in (s_x, \gamma_k] \cap \mathbb{R} \subseteq (x, \gamma_k] \cap \mathbb{R}$, we actually have that y [which strictly exceeds s_x] is in N_x, contradicting the fact that $s_x = \sup N_x$. The assumption that $s_x < \gamma_k$ yields a contradiction, so it must be the case that $s_x = \gamma_k$.

Subclaim: Given any $x \in (\beta_k, \gamma_k)$, we have that $s_x = \gamma_k$ and $\int_x^{\gamma_k} f\,d\bar{\lambda} \geq t(\gamma_k - x)$.

Proof: Pick $x \in (\beta_k, \gamma_k)$. The previous subclaim gives $s_x = \gamma_k$. For the inequality, we consider the cases where (i) $\gamma_k \in N_x$ and (ii) $\gamma_k \notin N_x$.

We consider the first case, where $\gamma_k \in N_x$. Since by definition $+\infty \notin N_x$, we have $\gamma_k < +\infty$. Here $\frac{1}{\gamma_k - x}\int_x^{\gamma_k} f\,d\bar{\lambda} > t$, so we trivially have $\int_x^{\gamma_k} f\,d\bar{\lambda} \geq t(\gamma_k - x)$ in this case.

For (ii), where $\gamma_k \notin N_x$, we use the facts that $N_x \neq \varnothing$ and $\gamma_k = \sup N_x$ to find a nondecreasing sequence $\{\gamma_{(n)}\}_{n=1}^{\infty}$ of values in (x, γ_k) with $\gamma_{(n)} \in N_x$ for each $n \in \mathbb{N}$ and $\lim_n \gamma_{(n)} = \gamma_k$. [This holds whether $\gamma_k = +\infty$ or $\gamma_k < +\infty$.] For each $n \in \mathbb{N}$ we have $\int_x^{\gamma_{(n)}} f\,d\bar{\lambda} > t(\gamma_{(n)} - x)$. Letting $n \to +\infty$ and using MCT gives

$$\int_x^{\gamma_k} f\,d\bar{\lambda} = \lim_n \int_x^{\gamma_{(n)}} f\,d\bar{\lambda} \geq \lim_n t(\gamma_{(n)} - x) = t(\gamma_k - x),$$

giving the result for case (ii). This completes the proof of the subclaim.

Subclaim: $\int_{\beta_k}^{\gamma_k} f\,d\bar{\lambda} \geq t(\gamma_k - \beta_k)$.

Proof: By the above subclaim, $\int_x^{\gamma_k} f\,d\bar{\lambda} \geq t(\gamma_k - x)$ for each $x \in (\beta_k, \gamma_k)$. Let $\{x_n\}_{n=1}^{\infty}$ denote a nonincreasing sequence in (β_k, γ_k) with limit β_k, so that for every $n \in \mathbb{N}$ we have $\int_{x_n}^{\gamma_k} f\,d\bar{\lambda} \geq t(\gamma_k - x_n)$. Observe that $\{fI_{(x_n,\gamma_k)}\}_{n=1}^{\infty}$ is a nonnegative and nonincreasing sequence; MCT thus gives

$$\begin{aligned}\int_{\beta_k}^{\gamma_k} f\,d\bar{\lambda} = \int_{\mathbb{R}} \lim_n fI_{(x_n,\gamma_k)}\,d\bar{\lambda} = \lim_n \int_{\mathbb{R}} fI_{(x_n,\gamma_k)}\,d\bar{\lambda} &= \lim_n \int_{x_n}^{\gamma_k} f\,d\bar{\lambda} \\ &\geq \lim_n t(\gamma_k - x_n) \\ &= t(\gamma_k - \beta_k).\end{aligned}$$

This gives the subclaim.

Still holding k fixed, we consider the following three cases: (i) $\beta_k = -\infty$ and $\gamma_k < +\infty$, (ii) $\beta_k > -\infty$ and $\gamma_k = +\infty$, and (iii) (β_k, γ_k) is a bounded open interval. For both (i) and (ii), the inequality of the previous subclaim becomes the equality $+\infty = +\infty$, and hence $\int_{\beta_k}^{\gamma_k} f\, d\bar{\lambda} = t(\gamma_k - \beta_k)$ in cases (i) and (ii). For case (iii), we recall that $\beta_k \notin M_t^{(r)}$, so that $f^{\delta(r)}(\beta_k) \leq t$, hence $\sup_{u \in (\beta_k, +\infty)} \frac{1}{u-\beta_k} \int_{\beta_k}^{u} f\, d\bar{\lambda} \leq t$. From this, we have by considering the specific value $\gamma_k \in (\beta_k, +\infty)$ that $\frac{1}{\gamma_k - \beta_k} \int_{\beta_k}^{\gamma_k} f\, d\bar{\lambda} \leq t$. Combined with the reverse inequality of the previous subclaim, we have that $\int_{\beta_k}^{\gamma_k} f\, d\bar{\lambda} = t(\gamma_k - \beta_k)$ in case (iii). Therefore, we have established in general that $\int_{\beta_k}^{\gamma_k} f\, d\bar{\lambda} = t(\gamma_k - \beta_k)$. [That is, the equality holds whether (β_k, γ_k) is bounded or unbounded.] We have held $k \in \mathbb{N}$ and the constituent open subinterval (β_k, γ_k) of $M_t^{(r)}$ fixed. We "unfix" k to obtain

$$\int_{\beta_k}^{\gamma_k} f\, d\bar{\lambda} = t(\gamma_k - \beta_k) \quad \text{for each } k \in \mathbb{N}. \tag{10.15}$$

Since

$$\bar{\lambda}(M_t^{(r)}) = \bar{\lambda}\left(\bigcup_{k=1}^{\infty} (\beta_k, \gamma_k)\right) = \sum_{k=1}^{\infty} \bar{\lambda}((\beta_k, \gamma_k)) = \sum_{k=1}^{\infty} (\gamma_k - \beta_k),$$

it follows upon summing both sides of (10.15) over all $k \in \mathbb{N}$ that

$$\int_{M_t^{(r)}} f\, d\bar{\lambda} = \sum_{k=1}^{\infty} \int_{\beta_k}^{\gamma_k} f\, d\bar{\lambda} = t \sum_{k=1}^{\infty} (\gamma_k - \beta_k) = t\bar{\lambda}(M_t^{(r)}),$$

completing the proof of (a). The proof of (b) follows the same lines.

We now turn to (c); this is a simple consequence of (a) and (b). Using finite subadditivity and the fact that $M_t^{(r)}, M_t^{(l)} \subseteq M_t$, we have

$$\begin{aligned} \bar{\lambda}(M_t) = \bar{\lambda}(M_t^{(r)} \cup M_t^{(l)}) &\leq \bar{\lambda}(M_t^{(r)}) + \bar{\lambda}(M_t^{(l)}) \\ &= \frac{1}{t} \int_{M_t^{(r)}} f\, d\bar{\lambda} + \frac{1}{t} \int_{M_t^{(l)}} f\, d\bar{\lambda} \\ &\leq \frac{1}{t} \int_{M_t} f\, d\bar{\lambda} + \frac{1}{t} \int_{M_t} f\, d\bar{\lambda} \\ &= \frac{2}{t} \int_{M_t} f\, d\bar{\lambda}, \end{aligned}$$

completing the entire proof. ■

With three claims given, there are two more claims to go before stating and proving the main results of this section. For the next claim, recall that $G_t = \{x \in \mathbb{R} : f(x) > t\}$ for each real t.

Claim 4. *Let $0 < k < 1$ and $0 < t < +\infty$. We have the following:*

(a) $\bar{\lambda}(M_t^{(r)}) \leq \frac{1}{(1-k)t} \int_{G_{kt}} f\, d\bar{\lambda}$.

(b) $\bar{\lambda}(M_t^{(l)}) \leq \frac{1}{(1-k)t} \int_{G_{kt}} f\, d\bar{\lambda}$.

(c) $\bar{\lambda}(M_t) \le \frac{2}{(1-k)t}\int_{G_{kt}} f\, d\bar{\lambda}$.

Proof: We prove (a) and (c); the proof of (b) is similar to the proof of (a). Define $g : \mathbb{R} \to [0, +\infty]$ for all $x \in \mathbb{R}$ by the rule

$$g(x) = \begin{cases} f(x) & x \in G_{kt} \\ 0 & x \notin G_{kt}. \end{cases}$$

Since g is a piecewise defined mapping of $\mathcal{L}/\mathcal{B}^*$-measurable mappings, g itself is $\mathcal{L}/\mathcal{B}^*$-measurable. We have for each $x \in \mathbb{R}$ that

$$\begin{aligned}
f^{\delta(r)}(x) &= \sup_{y\in(x,+\infty)} \frac{1}{y-x}\int_x^y f\, d\bar{\lambda} \\
&= \sup_{y\in(x,+\infty)} \frac{1}{y-x}\left[\int_{(x,y)\cap G_{kt}} f\, d\bar{\lambda} + \int_{(x,y)\cap G_{kt}^c} f\, d\bar{\lambda}\right] \\
&\le \sup_{y\in(x,+\infty)} \frac{1}{y-x}\left[\int_{(x,y)\cap G_{kt}} g\, d\bar{\lambda} + \int_{(x,y)\cap G_{kt}^c} kt\, d\bar{\lambda}(z)\right] \\
&\le \sup_{y\in(x,+\infty)} \frac{1}{y-x}\int_{(x,y)\cap G_{kt}} g\, d\bar{\lambda} + \sup_{y\in(x,+\infty)} \frac{1}{y-x}\int_{(x,y)\cap G_{kt}^c} kt\, d\bar{\lambda}(z) \\
&\le \sup_{y\in(x,+\infty)} \frac{1}{y-x}\int_{(x,y)} g\, d\bar{\lambda} + \sup_{y\in(x,+\infty)} \frac{1}{y-x}\int_{(x,y)} kt\, d\bar{\lambda}(z) \\
&= g^{\delta(r)}(x) + kt.
\end{aligned}$$

[The last equality holds because $\int_{(x,y)} kt\, d\bar{\lambda}(z) = kt\bar{\lambda}((x,y)) = kt(y-x)$.] For each real $u > 0$, define $N_u = \{x \in \mathbb{R} : g^{\delta(r)}(x) > u\}$. We claim that $M_t^{(r)} \subseteq N_{(1-k)t}$. To see this, observe that the claim is trivial if $M_t^{(r)}$ is empty. Otherwise, if $x \in M_t^{(r)}$, then $f^{\delta(r)}(x) > t$, hence $g^{\delta(r)}(x) + kt > t$ by the above display, whence $g^{\delta(r)}(x) > (1-k)t$, giving $x \in N_{(1-k)t}$. Using Claim 3(a), there writing $(1-k)t$ for t, g for f, and $N_{(1-k)t}$ for $M_t^{(r)}$, we have

$$\begin{aligned}
\bar{\lambda}(M_t^{(r)}) \le \bar{\lambda}(N_{(1-k)t}) &= \frac{1}{(1-k)t}\int_{N_{(1-k)t}} g\, d\bar{\lambda} \\
&= \frac{1}{(1-k)t}\int_{N_{(1-k)t}\cap G_{kt}} g\, d\bar{\lambda} \\
&= \frac{1}{(1-k)t}\int_{N_{(1-k)t}\cap G_{kt}} f\, d\bar{\lambda} \\
&\le \frac{1}{(1-k)t}\int_{G_{kt}} f\, d\bar{\lambda}.
\end{aligned}$$

[The second equality holds because $g = 0$ on G_{kt}^c. The third equality holds because $g = f$ on G_{kt}.] This gives part (a).

Part (c) is now simple. Using (a) and (b), we have that

$$\begin{aligned}\bar{\lambda}(M_t) = \bar{\lambda}(M_t^{(r)} \cup M_t^{(l)}) &\leq \bar{\lambda}(M_t^{(r)}) + \bar{\lambda}(M_t^{(l)}) \\ &\leq \frac{1}{(1-k)t}\int_{G_{kt}} f\, d\bar{\lambda} + \frac{1}{(1-k)t}\int_{G_{kt}} f\, d\bar{\lambda} \\ &= \frac{2}{(1-k)t}\int_{G_{kt}} f\, d\bar{\lambda}.\end{aligned}$$

This completes the entire proof. ■

This section exists to provide another illustration of the Fubini Theorems, but as of yet no form of the Fubini Theorem has been invoked. That will change, starting with the final technical claim preceding the Hardy-Littlewood Maximal Theorem.

Claim 5. *Let $(\Omega, \mathcal{F}, \mu)$ denote a σ-finite measure space. Let $f : \Omega \to [0, +\infty]$ be $\mathcal{F}/\mathcal{B}^*$-measurable, and let $0 < p < +\infty$. We have the following, where $G_t = \{\omega \in \Omega : f(\omega) > t\}$ for all real $t \geq 0$:*

(a) The nonnegative mapping $t \mapsto pt^{p-1}\mu(G_t)$ on $(0, +\infty)$ is $\mathcal{L}/\mathcal{B}^*$-measurable.

(b) $\int_\Omega f(\omega)^p\, d\mu(\omega) = \int_0^\infty pt^{p-1}\mu(G_t)\, d\bar{\lambda}(t)$.

(c) $\int_E f(\omega)^p\, d\mu(\omega) = \int_0^\infty pt^{p-1}\mu(E \cap G_t)\, d\bar{\lambda}(t)$ for each $E \in \mathcal{F}$.

Proof: Before doing (a), we make some measurability assertions. First, because the mapping $t \mapsto pt^{p-1}$ on $(0, +\infty)$ is $\mathcal{L}/\mathcal{B}^*$-measurable, the mapping $(\omega, t) \mapsto pt^{p-1}$ on $\Omega \times (0, +\infty)$ is $\mathcal{F} \otimes \mathcal{L}$-measurable by Claim 1 of Section 10.4. Next, observe that because f is $\mathcal{F}/\mathcal{B}^*$-measurable, the mapping $(\omega, t) \mapsto f(\omega)$ on $\Omega \times (0, +\infty)$ is $\mathcal{F} \otimes \mathcal{L}/\mathcal{B}^*$-measurable. It follows that

$$A = \{(\omega, t) \in \Omega \times (0, +\infty) : f(\omega) > t\} \in \mathcal{F} \otimes \mathcal{L}$$

since A is the inverse image of $(0, +\infty]$ under the $\mathcal{F} \otimes \mathcal{L}/\mathcal{B}^*$-measurable mapping $(\omega, t) \mapsto f(\omega) - t$ on $\Omega \times (0, +\infty)$. Since $A \in \mathcal{F} \otimes \mathcal{L}$, so is A^c.

Define h on $\Omega \times (0, +\infty)$ by the rule $h(\omega, t) = I_{G_t}(\omega)$. We claim that h is $\mathcal{F} \otimes \mathcal{L}/\mathcal{B}^*$-measurable. To do this, pick $B \in \mathcal{B}^*$; we desire to show that $h^{-1}(B) \in \mathcal{F} \otimes \mathcal{L}$. There are four cases: (1) $1 \in B$, $0 \notin B$, (2) $1 \notin B$, $0 \in B$, (3) $0, 1 \in B$, and (4) $0, 1 \notin B$. For case (1) we have

$$h^{-1}(B) = h^{-1}(\{1\}) = \{(\omega, t) \in \Omega \times (0, +\infty) : h(\omega, t) = 1\} = A \in \mathcal{F} \otimes \mathcal{L}.$$

For case (2), similar reasoning gives $h^{-1}(B) = A^c \in \mathcal{F} \otimes \mathcal{L}$. For case (3), $h^{-1}(B) = \Omega \times (0, +\infty) \in \mathcal{F} \otimes \mathcal{L}$. In case (4) we have $h^{-1}(B) = \varnothing \in \mathcal{F} \otimes \mathcal{L}$.

Therefore, the nonnegative mapping $(\omega, t) \mapsto pt^{p-1}I_{G_t}(\omega)$ on $\Omega \times (0, +\infty)$ is $\mathcal{F} \otimes \mathcal{L}/\mathcal{B}^*$-measurable, as it is a product of the $\mathcal{F} \otimes \mathcal{L}/\mathcal{B}^*$-measurable mappings $(\omega, t) \mapsto pt^{p-1}$ and $(\omega, t) \mapsto I_{G_t}(\omega)$, each on $\Omega \times (0, +\infty)$. We invoke the Nonnegative Fubini Theorem to assert that the mapping

$$t \mapsto \int_\Omega pt^{p-1}I_{G_t}(\omega)\, d\mu(\omega) = pt^{p-1}\int_\Omega I_{G_t}(\omega)\, d\mu(\omega) = pt^{p-1}\mu(G_t)$$

on $(0, +\infty)$ is $\mathcal{L}/\mathcal{B}^*$-measurable. This [at last] gives (a).

We now prove (b). By (a), we may meaningfully discuss the integral $\int_0^\infty pt^{p-1}\mu(G_t)\, d\bar{\lambda}(t)$. Using the Nonnegative Fubini Theorem again, we obtain

$$\begin{aligned}\int_0^\infty pt^{p-1}\mu(G_t)\, d\bar{\lambda}(t) &= \int_0^\infty \int_\Omega pt^{p-1} I_{G_t}(\omega)\, d\mu(\omega)\, d\bar{\lambda}(t)\\ &= \int_\Omega \int_0^\infty pt^{p-1} I_{G_t}(\omega)\, d\bar{\lambda}(t)\, d\mu(\omega)\\ &= \int_\Omega \int_0^{f(\omega)} pt^{p-1}\, d\bar{\lambda}(t)\, d\mu(\omega)\\ &= \int_\Omega f(\omega)^p\, d\mu(\omega).\end{aligned}$$

[The third equality follows because given $\omega \in \Omega$ and $t > 0$ we have that $I_{G_t}(\omega) = I_{(0,f(\omega))}(t)$, hence $\int_0^\infty pt^{p-1} I_{G_t}(\omega) = \int_0^{f(\omega)} pt^{p-1}\, d\bar{\lambda}(t)$. The last equality follows because the continuity of the integrand forces the Lebesgue integral to coincide with the Riemann integral, giving

$$\int_0^{f(\omega)} pt^{p-1}\, d\bar{\lambda}(t) = \int_0^{f(\omega)} pt^{p-1}\, dt = t^p|_0^{f(\omega)} = f(\omega)^p.]$$

This gives (b).

For (c), replace f by fI_E in (b):

$$\begin{aligned}\int_E f^p\, d\mu &= \int_\Omega (fI_E)^p\\ &= \int_0^\infty pt^{p-1}\mu(\{\omega \in \Omega :\ f(\omega)I_E(\omega) > t\})\, d\bar{\lambda}(t)\\ &= \int_0^\infty pt^{p-1}\mu(E \cap G_t)\, d\bar{\lambda}(t).\end{aligned}$$

This completes the entire proof. ∎

Again, note how Fubini's Theorem, besides allowing one to interchange the order of integration, allows certain functions to be demonstrated measurable.

We have now at last come to the famous theorem of Hardy and Littlewood. Everything is stated relative to the framework and definitions given at the beginning of this section. In particular, $f : \mathbb{R} \to [0, +\infty]$ is $\mathcal{L}/\mathcal{B}^*$-measurable and such that $\int_F f\, d\bar{\lambda} < +\infty$ for all compact $F \subseteq \mathbb{R}$.

Hardy-Littlewood Maximal Theorem. *Let $1 < p < +\infty$. We have the following inequalities:*

(a) $[\int_\mathbb{R} [f^{\delta(r)}]^p\, d\bar{\lambda}]^{\frac{1}{p}} \le \frac{p}{p-1}[\int_\mathbb{R} f^p\, d\bar{\lambda}]^{\frac{1}{p}}$. Therefore, $f \in L^p(\mathbb{R}, \mathcal{L}, \bar{\lambda})$ forces $f^{\delta(r)} \in L^p(\mathbb{R}, \mathcal{L}, \bar{\lambda})$ and $||f^{\delta(r)}||_p \le \frac{p}{p-1}||f||_p$.

(b) $[\int_{\mathbb{R}}[f^{\delta(l)}]^p \, d\bar{\lambda}]^{\frac{1}{p}} \leq \frac{p}{p-1}[\int_{\mathbb{R}} f^p \, d\bar{\lambda}]^{\frac{1}{p}}$. *Therefore,* $f \in L^p(\mathbb{R}, \mathcal{L}, \bar{\lambda})$ *forces both* $f^{\delta(l)} \in L^p(\mathbb{R}, \mathcal{L}, \bar{\lambda})$ *and* $\|f^{\delta(l)}\|_p \leq \frac{p}{p-1}\|f\|_p$.

(c) $[\int_{\mathbb{R}}[f^{\delta}]^p \, d\bar{\lambda}]^{\frac{1}{p}} \leq \frac{2p}{p-1}[\int_{\mathbb{R}} f^p \, d\bar{\lambda}]^{\frac{1}{p}}$. *Therefore,* $f \in L^p(\mathbb{R}, \mathcal{L}, \bar{\lambda})$ *forces both* $f^{\delta} \in L^p(\mathbb{R}, \mathcal{L}, \bar{\lambda})$ *and* $\|f^{\delta}\|_p \leq \frac{2p}{p-1}\|f\|_p$.

Proof: Statements (a) and (b) are similar, so we prove only (a). Statement (c) will then follow in straightforward [but somewhat tedious] fashion from (a) and (b).

We turn to (a). If $\int_{\mathbb{R}} f^p \, d\bar{\lambda} = +\infty$, then (a) is trivial. Assume then that $\int_{\mathbb{R}} f^p \, d\bar{\lambda} < +\infty$, that is, $f \in L^p(\mathbb{R}, \mathcal{L}, \bar{\lambda})$. We have

$$\begin{aligned}
&\int_{\mathbb{R}} [f^{\delta(r)}(x)]^p \, d\bar{\lambda}(x) \\
&= \int_0^{\infty} pt^{p-1}\bar{\lambda}(M_t^{(r)}) \, d\bar{\lambda}(t) \\
&= \int_0^{\infty} pt^{p-1}\frac{1}{t}\int_{M_t^{(r)}} f(x) \, d\bar{\lambda}(x) \, d\bar{\lambda}(t) \\
&= \int_0^{\infty} pt^{p-2}\int_{\mathbb{R}} f(x) I_{M_t^{(r)}}(x) \, d\bar{\lambda}(x) \, d\bar{\lambda}(t) \\
&= \int_{\mathbb{R}}\int_0^{\infty} pt^{p-2} f(x) I_{M_t^{(r)}}(x) \, d\bar{\lambda}(t) \, d\bar{\lambda}(x) \\
&= p\int_{\mathbb{R}} f(x)\int_0^{f^{\delta(r)}(x)} t^{p-2} \, d\bar{\lambda}(t) \, d\bar{\lambda}(x) \\
&= p\int_{\mathbb{R}} f(x)\frac{[f^{\delta(r)}(x)]^{p-1}}{p-1} \, d\bar{\lambda}(x) \\
&\leq \frac{p}{p-1}\left[\int_{\mathbb{R}} f(x)^p \, d\bar{\lambda}(x)\right]^{\frac{1}{p}}\left[\int_{\mathbb{R}} [f^{\delta(r)}(x)]^{(p-1)p'}\right]^{\frac{1}{p'}} \\
&= \frac{p}{p-1}\left[\int_{\mathbb{R}} f(x)^p \, d\bar{\lambda}(x)\right]^{\frac{1}{p}}\left[\int_{\mathbb{R}} [f^{\delta(r)}(x)]^{p}\right]^{\frac{1}{p'}}. \qquad (*)
\end{aligned}$$

[The first equality follows by taking $(\Omega, \mathcal{F}, \mu) = (\mathbb{R}, \mathcal{L}, \bar{\lambda})$ in Claim 5, there replacing f with $f^{\delta(r)}$ and G_t with $M_t^{(r)}$. The second equality directly follows from part (a) of Claim 3. The third equality is obvious. The fourth equality follows by the Nonnegative Fubini Theorem.[9] The fifth equality is clear on a moment's thought. The sixth equality is a direct calculation based on the fact

[9]We need the mapping $(x, t) \mapsto pt^{p-2}f(x)I_{M_t^{(r)}}(x)$ on $\mathbb{R} \times (0, +\infty)$ to be $\mathcal{L} \otimes \mathcal{L}/\mathcal{B}^*$-measurable in order to apply the Nonnegative Fubini Theorem. The mapping $t \mapsto pt^{p-2}$ on $(0, +\infty)$ is clearly $\mathcal{L}/\mathcal{B}^*$-measurable, so the mapping $(x, t) \mapsto pt^{p-2}$ on $\mathbb{R} \times (0, +\infty)$ is $\mathcal{L} \otimes \mathcal{L}/\mathcal{B}^*$-measurable. Since $x \mapsto f(x)$ on $\mathbb{R}$ is $\mathcal{L}/\mathcal{B}^*$-measurable, we have that the mapping $(x, t) \mapsto f(x)$ on $\mathbb{R} \times (0, +\infty)$ is $\mathcal{L} \otimes \mathcal{L}/\mathcal{B}^*$-measurable. Next, the inverse image

that the integrand is continuous, hence the Lebesgue integral coincides with the Riemann integral. The inequality follows by Hölder's Inequality applied to the nonnegative functions f and $(f^{\delta(r)})^{p-1}$. The final equality holds since $p' = \frac{p}{p-1}$.]

If $f^{\delta(r)} \in L^p(\mathbb{R}, \mathcal{L}, \bar{\lambda})$, so that $\int_{\mathbb{R}}[f^{\delta(r)}(x)]^p \, d\bar{\lambda}(x)$ is finite [and vice versa], we may go back to $(*)$ and divide both sides of the inequality by the finite number $[\int_{\mathbb{R}}[f^{\delta(r)}(x)]^p \, d\bar{\lambda}(x)]^{\frac{1}{p'}}$. Since $1 - \frac{1}{p'} = \frac{1}{p}$, (a) will immediately follow. Therefore, (a) will be complete when it is shown that $f^{\delta(r)} \in L^p(\mathbb{R}, \mathcal{L}, \bar{\lambda})$. We have

$$\begin{aligned}
\int_{\mathbb{R}}[f^{\delta(r)}(x)]^p \, d\bar{\lambda}(x) &= \int_0^{\infty} pt^{p-1}\bar{\lambda}(M_t^{(r)}) \, d\bar{\lambda}(t) \\
&\leq \int_0^{\infty} pt^{p-1}\frac{1}{(1-k)t}\int_{G_{kt}} f(x) \, d\bar{\lambda}(x) \, d\bar{\lambda}(t) \\
&= \frac{p}{1-k}\int_0^{\infty} t^{p-2}\int_{G_{kt}} f(x) \, d\bar{\lambda}(x) \, d\bar{\lambda}(t) \\
&= \frac{p}{1-k}\int_0^{\infty} t^{p-2}\int_{\mathbb{R}} f(x)I_{G_{kt}}(x) \, d\bar{\lambda}(x) \, d\bar{\lambda}(t) \\
&= \frac{p}{1-k}\int_{\mathbb{R}}\int_0^{\infty} t^{p-2}I_{G_{kt}}(x)f(x) \, d\bar{\lambda}(t) \, d\bar{\lambda}(x) \\
&= \frac{p}{1-k}\int_{\mathbb{R}}\int_0^{f(x)/k} t^{p-2} \, d\bar{\lambda}(t) \, f(x) \, d\bar{\lambda}(x) \\
&= \frac{p}{1-k}\int_{\mathbb{R}}\frac{[f(x)]^{p-1}}{k^{p-1}(p-1)}f(x) \, d\bar{\lambda}(x) \\
&= \frac{p}{(1-k)k^{p-1}(p-1)}\int_{\mathbb{R}} f(x)^p \, d\bar{\lambda}(x) \\
&< +\infty. \quad (**)
\end{aligned}$$

[The first equality follows by Claim 5. The first inequality follows by part (a) of Claim 4. The second and third equalities are obvious. The fourth equality follows by the Nonnegative Fubini Theorem.[10] The fifth equality follows on a moment's thought. The sixth equality follows by direct calculation of the inner integral (as in the display previous to this display). The seventh equality

of $(a, +\infty]$ under the mapping $(x,t) \mapsto I_{M_t^{(r)}}(x)$ on $\mathbb{R} \times (0, +\infty)$ is $\varnothing \in \mathcal{L} \otimes \mathcal{L}$ if $a \geq 1$, $\mathbb{R} \times [0, +\infty) \in \mathcal{L} \otimes \mathcal{L}$ if $a < 0$, and $\{(x,t) \in \mathbb{R} \times (0, +\infty) : f^{\delta(r)}(x) > t\} \in \mathcal{L} \otimes \mathcal{L}$ if $0 \leq a < 1$. [This latter set is the inverse image of $(0, +\infty]$ under the $\mathcal{L} \otimes \mathcal{L}$-measurable mapping $(x,t) \mapsto f^{\delta(r)}(x) - t$; see a nearly identical development in the first two paragraphs of the proof of Claim 5.] Therefore the mapping $(x,t) \mapsto I_{M_t^{(r)}}$ on $\mathbb{R} \times (0, +\infty)$ is $\mathcal{L} \otimes \mathcal{L}/\mathcal{B}^*$-measurable. Putting everything together and using the fact that products of $\mathcal{L} \otimes \mathcal{L}/\mathcal{B}^*$-measurable mappings are $\mathcal{L} \otimes \mathcal{L}/\mathcal{B}^*$-measurable gives the result.

[10] We need the mapping $(x,t) \mapsto t^{p-2}f(x)I_{G_{kt}}(x)$ on $\mathbb{R} \times (0, +\infty)$ to be $\mathcal{L} \otimes \mathcal{L}/\mathcal{B}^*$-measurable in order to apply the Fubini Theorem. The measurability can be established in the same fashion as in the previous footnote or as in the first two paragraphs of the proof of Claim 5. It seems unnecessary to write out another proof of this type.

is obvious, and the final inequality is also obvious.] We have shown that $f^{\delta(r)} \in L^p(\mathbb{R}, \mathcal{L}, \bar{\lambda})$, and the entire proof of (a) is now complete.

We now turn to proving (c). The steps here are very similar as to those in (a). The inequality is trivial if $\int_{\mathbb{R}} f(x)^p \, d\bar{\lambda}(x) = +\infty$, hence we assume that $f \in L^p(\mathbb{R}, \mathcal{L}, \bar{\lambda})$. We have

$$\begin{aligned}
&\int_{\mathbb{R}} [f^{\delta}(x)]^p \, d\bar{\lambda}(x) \\
= \; & \int_0^{\infty} pt^{p-1} \bar{\lambda}(M_t) \, d\bar{\lambda}(t) \\
\leq \; & \int_0^{\infty} pt^{p-1} \frac{2}{t} \int_{M_t} f(x) \, d\bar{\lambda}(x) \, d\bar{\lambda}(t) \\
= \; & 2p \int_0^{\infty} t^{p-2} \int_{\mathbb{R}} I_{M_t}(x) f(x) \, d\bar{\lambda}(x) \, d\bar{\lambda}(t) \\
= \; & 2p \int_{\mathbb{R}} \int_0^{\infty} t^{p-2} I_{M_t}(x) f(x) \, d\bar{\lambda}(t) \, d\bar{\lambda}(x) \\
= \; & 2p \int_{\mathbb{R}} \int_0^{f^{\delta}(x)} t^{p-2} \, d\bar{\lambda}(t) \, f(x) \, d\bar{\lambda}(x) \\
= \; & 2p \int_{\mathbb{R}} \frac{[f^{\delta}(x)]^{p-1}}{p-1} f(x) \, d\bar{\lambda}(x) \\
= \; & \frac{2p}{p-1} \int_{\mathbb{R}} f(x) [f^{\delta}(x)]^{p-1} \, d\bar{\lambda}(x) \\
\leq \; & \frac{2p}{p-1} \left[\int_{\mathbb{R}} f(x)^p \, d\bar{\lambda}(x) \right]^{\frac{1}{p}} \left[\int_{\mathbb{R}} f^{\delta}(x)^p \, d\bar{\lambda}(x) \right]^{\frac{1}{p'}} . \qquad (***)
\end{aligned}$$

[The first equality follows by Claim 5. The inequality follows by part (c) of Claim 3. The second equality is obvious. The third equality follows by the Nonnegative Fubini Theorem.[11] The fourth, fifth, and sixth equalities are all clear on a moment's thought. The final inequality follows by applying Hölder's Inequality to f and f^{δ} and noting that $(p-1)p' = p$. If $\int_{\mathbb{R}} [f^{\delta}(x)]^p \, d\bar{\lambda}(x)$ is finite, that is, if $f^{\delta} \in L^p(\mathbb{R}, \mathcal{L}, \bar{\lambda})$, we may divide both sides of the inequality in $(***)$ by the finite number $[\int_{\mathbb{R}} f^{\delta}(x)^p \, d\bar{\lambda}(x)]^{\frac{1}{p'}}$ and use the identity $1 - \frac{1}{p'} = \frac{1}{p}$ to complete (c). To see that $\int_{\mathbb{R}} f^{\delta}(x)^p \, d\bar{\lambda}(x)$ is finite, we use an argument almost exactly parallel to that given in $(**)$, whereby in the same order we use Claim 5, part (c) of Claim 4, the Nonnegative Fubini Theorem, some routine

[11] As before, one can show that the mapping $(x,t) \mapsto t^{p-2} f(x) I_{M_t}(x)$ on $\mathbb{R} \times (0, +\infty)$ is $\mathcal{L} \otimes \mathcal{L}/\mathcal{B}^*$-measurable.

manipulations, and Hölder's Inequality, obtaining

$$\begin{aligned}
\int_{\mathbb{R}} [f^{\delta}(x)]^p \, d\bar{\lambda}(x) &= \int_0^{\infty} pt^{p-1} \bar{\lambda}(M_t) \, d\bar{\lambda}(t) \\
&\leq \int_0^{\infty} pt^{p-1} \frac{2}{(1-k)t} \int_{G_{kt}} f(x) \, d\bar{\lambda}(x) \, d\bar{\lambda}(t) \\
&= \frac{2p}{1-k} \int_0^{\infty} t^{p-2} \int_{G_{kt}} f(x) \, d\bar{\lambda}(x) \, d\bar{\lambda}(t) \\
&= \frac{2p}{1-k} \int_0^{\infty} t^{p-2} \int_{\mathbb{R}} f(x) I_{G_{kt}}(x) \, d\bar{\lambda}(x) \, d\bar{\lambda}(t) \\
&= \frac{2p}{1-k} \int_{\mathbb{R}} \int_0^{\infty} t^{p-2} I_{G_{kt}}(x) f(x) \, d\bar{\lambda}(t) \, d\bar{\lambda}(x) \\
&= \frac{2p}{1-k} \int_{\mathbb{R}} \int_0^{f(x)/k} t^{p-2} \, d\bar{\lambda}(t) \, f(x) \, d\bar{\lambda}(x) \\
&= \frac{2p}{1-k} \int_{\mathbb{R}} \frac{[f(x)]^{p-1}}{k^{p-1}(p-1)} f(x) \, d\bar{\lambda}(x) \\
&= \frac{2p}{(1-k)k^{p-1}(p-1)} \int_{\mathbb{R}} f(x)^p \, d\bar{\lambda}(x) \\
&< +\infty.
\end{aligned}$$

This completes the proof of (c), and the entire proof is now at a close. ∎

The term *maximal* in the title of the Hardy-Littlewood theorem refers to the inequality regarding f^{δ}. The function f^{δ} is the pointwise maximum of $f^{\delta(r)}$ and $f^{\delta(l)}$, and these two functions in turn are defined as suprema of certain quantities.

There are other versions of the Hardy-Littlewood Maximal Theorem. [A sketch of another version is found in the exercises.] We have not attempted to give [anywhere near] a complete exposition of the family of theorems under the Hardy-Littlewood heading, as the goal has merely been to illustrate the use and application of the Fubini Theorems.

Exercises.

1. Prove Claim 1(b) by mimicking the proof of Claim 1(a).

2. In Claim 2(a), mimic the proof of the fact that $M_t^{(r)}$ is open to show that $M_t^{(l)}$ is open.

3. Prove Claim 3(b) by mimicking the proof of Claim 3(a).

4. Prove Claim 4(b) by mimicking the proof of Claim 4(a).

5. Prove (b) of the Hardy-Littlewood Maximal Theorem for $1 < p < +\infty$ by a fashion similar to that for the proof of part (a) there.

6. In the Hardy-Littlewood Maximal Theorem, we had $1 < p < +\infty$. There is a version of the theorem for the case $p = 1$. The framework is the same as in the $1 < p < +\infty$ case: $f : \mathbb{R} \to [0, +\infty]$ is $\mathcal{L}/\mathcal{B}^*$-measurable and such that $\int_F f \, d\bar{\lambda} < +\infty$ for every compact $F \subseteq \mathbb{R}$. The theorem is this: *Let $E \in \mathcal{L}$ and*

$0 < k < 1$. *Then*

$$\int_E f^{\delta(r)}\, d\bar{\lambda} \leq \frac{1}{k}\bar{\lambda}(E) + \frac{1}{1-k}\int_{\mathbb{R}} f \log^+ f\, d\bar{\lambda}$$

and

$$\int_E f^{\delta(l)}\, d\bar{\lambda} \leq \frac{1}{k}\bar{\lambda}(E) + \frac{1}{1-k}\int_{\mathbb{R}} f \log^+ f\, d\bar{\lambda}.$$

Also, we have

$$\int_E f^{\delta}\, d\bar{\lambda} \leq \frac{1}{k}\bar{\lambda}(E) + \frac{2}{1-k}\int_{\mathbb{R}} f \log^+ f\, d\bar{\lambda}.$$

Here, $\log^+ f$ denotes $\max\{\log f, 0\}$.

[Verify each of the following steps by appealing to the appropriate results of this section:

$$\begin{aligned}
\int_E f^{\delta(r)}\, d\bar{\lambda} &= \int_0^\infty \bar{\lambda}(\{y \in \mathbb{R} :\ I_E(y) f^{\delta(r)}(y) > t\})\, d\bar{\lambda}(t) \\
&= \int_0^\infty \bar{\lambda}(M_t^{(r)} \cap E)\, d\bar{\lambda}(t) \\
&= \int_0^{1/k} \bar{\lambda}(M_t^{(r)} \cap E)\, d\bar{\lambda}(t) + \int_{1/k}^\infty \bar{\lambda}(M_t^{(r)} \cap E)\, d\bar{\lambda}(t) \\
&\overset{(*)}{\leq} \frac{1}{k}\bar{\lambda}(E) + \int_{1/k}^\infty \frac{1}{(1-k)t} \int_{G_{kt}} f(x)\, d\bar{\lambda}(x)\, d\bar{\lambda}(t) \\
&= \frac{1}{k}\bar{\lambda}(E) + \frac{1}{1-k}\int_{1/k}^\infty \frac{1}{t}\int_{\mathbb{R}} I_{G_{kt}}(x) f(x)\, d\bar{\lambda}(x)\, d\bar{\lambda}(t) \\
&= \frac{1}{k}\bar{\lambda}(E) + \frac{1}{1-k}\int_{\mathbb{R}} f(x) \int_{1/k}^\infty \frac{1}{t} I_{G_{kt}}(x)\, d\bar{\lambda}(t)\, d\bar{\lambda}(x).
\end{aligned}$$

Next, show that $\int_{1/k}^\infty \frac{1}{t} I_{G_{kt}}(x)\, d\bar{\lambda}(t)$ is zero or $\log f(x)$ as $0 \leq f(x) \leq 1$ or $f(x) > 1$. That is, show that $\int_{1/k}^\infty \frac{1}{t} I_{G_{kt}}(x)\, d\bar{\lambda}(t) = \log^+ f(x)$ for each $x \in \mathbb{R}$. This gives the $\delta(r)$-inequality, and the $\delta(l)$-inequality is achieved in virtually the same fashion. For the δ-inequality, the same argument as above also works upon replacing $\delta(r)$ by δ, $M_t^{(r)}$ by M_t, and using the inequality $\bar{\lambda}(M_t) \leq \frac{2}{(1-k)t}\int_{G_{kt}} f\, d\bar{\lambda}$ in $(*)$.]

7. Refer to the Hardy-Littlewood Maximal Theorem in the text, where $1 < p < +\infty$. There it was shown that if $f \in L^p(\mathbb{R}, \mathcal{L}, \bar{\lambda})$, then $f^{\delta(r)}, f^{\delta(l)}, f^{\delta} \in L^p(\mathbb{R}, \mathcal{L}, \bar{\lambda})$ and

$$\|f^{\delta(r)}\|_p,\ \|f^{\delta(l)}\|_p \leq \frac{p}{p-1}\|f\|_p \quad \text{and} \quad \|f^{\delta}\|_p \leq \frac{2p}{p-1}\|f\|_p.$$

One question that may arise is whether the constants $\frac{p}{p-1}$ and $\frac{2p}{p-1}$ are the sharpest possible constants. It will turn out that the answer is in the affirmative. Fix $1 < p < +\infty$, and for each $n \in \mathbb{N}$ define f_n on $\mathbb{R}$ by the rule $f_n(x) = x^{(\frac{1}{n}-1)\frac{1}{p}} I_{(0,1)}(x)$. Let $A_{n,p}$ be such that $\|f_n^{\delta(r)}\|_p = A_{n,p}\|f_n\|_p$. Show that $\lim_n A_{n,p} = \frac{p}{p-1}$, and deduce that the constant $\frac{p}{p-1}$ in the Hardy-Littlewood Inequality is the best possible constant. [It is straightforward calculus to compute $\|f_n\|_p$ for each n. To obtain the form of $f_n^{\delta(r)}(x)$ for each n and x, consider the cases $x \geq 1$ and $x < 1$ and use calculus; this won't be hard but it won't be trivial. From this $\|f_n^{\delta(r)}\|_p$ may be obtained for each n, so that the ratio $\|f_n^{\delta(r)}\|_p/\|f_n\|_p$ may be computed for each n. That this quotient tends to $\frac{p}{p-1}$ as $n \to +\infty$ should be seen in straightforward fashion.]

8. For a function f as given in the framework for this section, we defined $f^\delta(x)$ to be the maximum of $f^{\delta(r)}(x)$ and $f^{\delta(l)}(x)$. We may equivalently define $f^\delta(x)$ for each $x \in \mathbb{R}$ by the rule

$$f^\delta(x) = \sup\left\{\frac{1}{u-t}\int_t^u f\,d\bar{\lambda} : \ -\infty < t \le x \le u < +\infty,\ t \ne u\right\}.$$

Comment. Our presentation in the text and in the exercises is based on one form of the Hardy-Littlewood framework. We now present a series of exercises dealing with a more general framework. The measure space will be $(\mathbb{R}^k, \mathcal{L}^k, \bar{\lambda}_k)$, where $k \in \mathbb{N}$ is fixed. The notation $B(\boldsymbol{x}, r)$ denotes the ball of radius r centered at $\boldsymbol{x} \in \mathbb{R}^k$. Given $0 < a < +\infty$, we will write $aB(\boldsymbol{x}, r)$ for $B(\boldsymbol{x}, ar)$. The following exercises give the background leading up to Jones' presentation of an alternative version of the Hardy-Littlewood Maximal Theorem.

9. Prove the *Vitali Covering Theorem: Let $E \subseteq \mathbb{R}^k$ be bounded. Let $\mathcal{F}$ denote a collection of open balls centered at points of E such that every $\boldsymbol{x} \in E$ is the center of some ball in $\mathcal{F}$. Then there exists an at most countable subcollection $B_1, B_2, \cdots$ of balls in $\mathcal{F}$ such that (i) $B_1, B_2, \cdots$ are disjoint, and (ii) $E \subseteq \bigcup_{n\ge 1} 3B_n$.* Follow the given outline.

(a) The theorem is immediate if the radii of the balls are not bounded above. [In this case, we can find a ball B centered at some point of E with radius so large that $E \subseteq B$, trivially giving (i) and (ii).] Therefore, we may assume that the radii of the balls is bounded above. [And of course we may assume that E is nonempty.]

Perform the following inductive scheme: Assume that $B_1, \cdots, B_{n-1}$ have been selected. Let $d_n = \sup\{\operatorname{rad} B : \ B \in \mathcal{F},\ B \cap \bigcup_{m<n} B_m = \varnothing\}$. If there are no $B \in \mathcal{F}$ with $B \cap \bigcup_{m<n} B_m = \varnothing$, the scheme terminates with the sets $B_1, \cdots, B_{n-1}$. Otherwise, choose $B_n \in \mathcal{F}$ such that $\operatorname{rad} B_n > \frac{1}{2} d_n$ and $B_n \cap \bigcup_{m<n} B_m = \varnothing$.

The above scheme is problem-free, and the resulting collection $B_1, B_2, \cdots$ is clearly disjoint. We must show that $E \subseteq \bigcup_{n\ge 1} 3B_n$. Fix $\boldsymbol{x} \in E$ and find $B \in \mathcal{F}$ that is centered at $\boldsymbol{x}$.

(b) There is n such that $B \cap B_n \ne \varnothing$. [Suppose that $B \cap B_n = \varnothing$ for each n. Then the selection process never terminates and $\operatorname{rad} B \le d_n$ for each n. Therefore, we have a countably infinite collection $B_1, B_2, \cdots$ of balls centered at points of E with $\operatorname{rad} B_n > \frac{1}{2} d_n \ge \frac{1}{2} \operatorname{rad} B > 0$. Now $\bigcup_n B_n$ is bounded since E is bounded and the radii of the balls in $\mathcal{F}$ are bounded above, and hence $\bar{\lambda}_k(\bigcup_n B_n) < +\infty$. But at the same time $\bar{\lambda}_k(\bigcup_n B_n) = \sum_n \bar{\lambda}_k(B_n) = +\infty$ by the earlier inequality. This contradiction gives the result.]

By (b) there is n with $B \cap B_n \ne \varnothing$. Let n henceforth denote the smallest such n with $B \cap B_n \ne \varnothing$. Pick $\boldsymbol{y} \in B \cap B_n$.

(c) We have $B \cap \bigcup_{m<n} B_m = \varnothing$, hence $\operatorname{rad} B \le d_n < 2 \operatorname{rad} B_n$.

(d) $\boldsymbol{x} \in 3B_n$, hence $E \subseteq \bigcup_{m\ge 1} 3B_m$, completing the proof. [Let $\boldsymbol{z}$ denote the center of B_n. Then $\|\boldsymbol{x} - \boldsymbol{z}\| \le \|\boldsymbol{x} - \boldsymbol{y}\| + \|\boldsymbol{y} - \boldsymbol{z}\| < \operatorname{rad} B + \operatorname{rad} B_n < 3 \operatorname{rad} B_n$. Therefore, we have that $\boldsymbol{x} \in 3B_n$.]

Definition. Let $f : \mathbb{R}^k \to \bar{\mathbb{R}}$ be such that $\int_F f \, d\bar{\lambda}_k < +\infty$ for each compact $F \subseteq \mathbb{R}^k$. The *Hardy-Littlewood maximal function for* f is the function Mf on $\mathbb{R}^k$ given for all $\boldsymbol{x} \in \mathbb{R}^k$ by

$$Mf(\boldsymbol{x}) = \sup_{0<r<+\infty} \frac{1}{\bar{\lambda}_k(B(\boldsymbol{x}, r))} \int_{B(\boldsymbol{x},r)} |f(\boldsymbol{y})| \, d\bar{\lambda}_k(\boldsymbol{y}).$$

10. Mf is lower semicontinuous, hence is $\mathcal{B}^k/\mathcal{B}^*$-measurable. In particular, Mf is $\mathcal{L}^k/\mathcal{B}^*$-measurable. [Suppose that $Mf(x) > t$. There is $0 < r < +\infty$ such that $t < \frac{1}{\bar{\lambda}_k(B(\boldsymbol{x},r))} \int_{B(\boldsymbol{x},r)} |f(\boldsymbol{y})| \, d\bar{\lambda}_k(\boldsymbol{y})$. We may choose $r' > r$ such that the inequality $t < \frac{1}{\bar{\lambda}_k(B(\boldsymbol{x},r'))} \int_{B(\boldsymbol{x},r)} |f(\boldsymbol{y})| \, d\bar{\lambda}_k(\boldsymbol{y})$ obtains. Now $\|\boldsymbol{x} - \boldsymbol{x}'\| \leq r' - r$ forces $B(\boldsymbol{x}, r) \subseteq B(\boldsymbol{x}', r')$, and the translation invariance of $\bar{\lambda}_k$ gives

$$\begin{aligned} t &< \frac{1}{\bar{\lambda}_k(B(\boldsymbol{x}, r'))} \int_{B(\boldsymbol{x}',r')} |f(\boldsymbol{y})| \, d\bar{\lambda}_k(\boldsymbol{y}) \\ &= \frac{1}{\bar{\lambda}_k(B(\boldsymbol{x}', r'))} \int_{B(\boldsymbol{x}',r')} |f(\boldsymbol{y})| \, d\bar{\lambda}_k(\boldsymbol{y}) \\ &\leq Mf(\boldsymbol{x}'). \end{aligned}$$

Therefore, for all $\boldsymbol{x}'$ sufficiently close to $\boldsymbol{x}$ we have $Mf(\boldsymbol{x}') > t$.]

11. Let $f \in L^1(\mathbb{R}^k, \mathcal{L}^k, \bar{\lambda}_k)$. Then $\bar{\lambda}_k(\{\boldsymbol{x} \in \mathbb{R}^k : Mf(\boldsymbol{x}) > t\}) \leq 3^k t^{-1} \|f\|_1$ for each $0 < t < +\infty$. This is another version of the Hardy-Littlewood Maximal Theorem. [Let E denote $\{\boldsymbol{x} \in \mathbb{R}^k : Mf(\boldsymbol{x}) > t\}$, where $0 < t < +\infty$ is given and fixed. If $E = \varnothing$ the result is trivial. Otherwise, pick $\boldsymbol{x} \in E$, so that $Mf(\boldsymbol{x}) > t$ and there is $0 < r < +\infty$ with $\frac{1}{\bar{\lambda}_k(B(\boldsymbol{x},r))} \int_{B(\boldsymbol{x},r)} |f(\boldsymbol{y})| \, d\bar{\lambda}_k(\boldsymbol{y}) > t$. Therefore, for any $\boldsymbol{x} \in E$ there is a ball B with center $\boldsymbol{x}$ such that $\bar{\lambda}_k(B) < \frac{1}{t} \int_B |f(\boldsymbol{y})| \, d\bar{\lambda}_k(\boldsymbol{y})$. Pick positive real q, and let $\mathcal{F}$ denote the collection of open balls B with centers in $E \cap B(\boldsymbol{0}, q)$ and satisfying the just-mentioned inequality. If $E \cap B(\boldsymbol{0}, q) \neq \varnothing$, there exist disjoint balls $B_1, B_2, \cdots \in \mathcal{F}$ such that $E \cap B(\boldsymbol{0}, q) \subseteq \bigcup_{n \geq 1} 3B_n$. (Vitali Covering Theorem.) Verify that

$$\begin{aligned} \bar{\lambda}_k(E \cap B(\boldsymbol{0}, q)) \leq \sum_{n \geq 1} \bar{\lambda}_k(3B_n) = \sum_{n=1}^{\infty} 3^k \bar{\lambda}_k(B_n) &< \sum_{n \geq 1} \frac{3^k}{t} \int_{B_n} |f(\boldsymbol{y})| \, d\bar{\lambda}_k(\boldsymbol{y}) \\ &= \frac{3^k}{t} \int_{\bigcup_{n \geq 1} B_n} |f(\boldsymbol{y})| \, d\bar{\lambda}_k(\boldsymbol{y}) \\ &\leq 3^k t^{-1} \int_{\mathbb{R}^k} |f(\boldsymbol{y})| \, d\bar{\lambda}_k(\boldsymbol{y}). \end{aligned}$$

Now let $q \to +\infty$ for the result.]

12. Reconsider Exercise 3 of Section 10.4 in the light of this section.

11

Arbitrary Products of Measure Spaces

In the previous chapter, we were given two σ-finite measure spaces $(\Omega_1, \mathcal{F}_1, \mu_1)$ and $(\Omega_2, \mathcal{F}_2, \mu_2)$. We defined and constructed the product measure space $(\Omega_1 \times \Omega_2, \mathcal{F}_1 \otimes \mathcal{F}_2, \mu_1 \otimes \mu_2)$ of $(\Omega_1, \mathcal{F}_1, \mu_1)$ and $(\Omega_2, \mathcal{F}_2, \mu_2)$. The measure $\mu_1 \otimes \mu_2$ is the unique measure with domain $\mathcal{F}_1 \otimes \mathcal{F}_2$ such that

$$\mu_1 \otimes \mu_2(A \times B) = \mu_1(A)\mu_2(B) \quad \text{whenever } A \in \mathcal{F}_1 \text{ and } B \in \mathcal{F}_2.$$

After constructing the above product measure $\mu_1 \otimes \mu_2$, the various forms of the Fubini Theorems were presented; these theorems relate [double] integrals with respect to $\mu_1 \otimes \mu_2$ to [iterated] integrals with respect to μ_1 and μ_2.

This chapter will present a situation that in one sense is more general than the Chapter 10 framework and in one sense is more specific than the Chapter 10 framework: instead of having two σ-finite measure spaces, we will have an arbitrary family $\{(\Omega_i, \mathcal{F}_i, \mu_i)\}_{i \in I}$ of measure spaces indexed by a generic nonempty index set I, and we will have $\mu_i(\Omega_i) = 1$ for each $i \in I$. The previous chapter had $I = \{1, 2\}$; in this chapter we will discuss the situation where I is of any finite size, countably infinite, or even uncountably infinite. We will attempt to define what we mean by the product of the measure spaces $(\Omega_i, \mathcal{F}_i, \mu_i)$ as i ranges over the index set I, and we shall give a generalization of product measure that reduces to the familiar formula in the case where $I = \{1, 2\}$.

The concepts in this chapter are substantial and deep, and matters are not helped by the unavoidable fact that the notation required to accurately express certain concepts is somewhat involved. There is also the fact that many claims require a buildup that can arguably be described as tedious. Our first section defines the requisite notation for this chapter and discusses certain simplifying conventions that will be used. The notation in the first section should be memorized [or made quite familiar to the reader] before proceeding further.

11.1 NOTATION AND CONVENTIONS

In this section, we establish the notation and conventions used throughout this chapter. The reader is urged to be quite familiar with this notation, which is a necessary evil for the results that will follow in subsequent sections.

Index Sets. I will denote a generic nonempty index set consisting of at least two elements. This covers the countably infinite and uncountably infinite cases as well. That is, I may be finite, countably infinite, or uncountably infinite. We will usually use the symbol D to denote a nonempty subset of I, and the symbol F will always denote a finite subset of I.

Measure Spaces. For each $i \in I$ we will have a measure space $(\Omega_i, \mathcal{F}_i, \mu_i)$ with $\mu_i(\Omega_i) = 1$. That is, $\{(\Omega_i, \mathcal{F}_i, \mu_i)\}_{i\in I}$ is a collection of finite measure spaces indexed by the elements of I such that $\mu_i(\Omega_i) = 1$ for each $i \in I$.

Cartesian Products. Let $D \subseteq I$ be nonempty. We define Ω_D to be the Cartesian product $\prod_{i\in D} \Omega_i$.[1] We will denote elements of Ω_D by $\boldsymbol{\omega}_D$ or $(\omega_i)_{i\in D}$. In the special case of where $D = I$, we write Ω for $\Omega_D = \Omega_I = \prod_{i\in I} \Omega_i$.[2] Elements of Ω will be denoted by $\boldsymbol{\omega}$ or $(\omega_i)_{i\in I}$.

The Collections $\mathcal{E}_F$. Given a finite subset $F \subseteq I$, we define $\mathcal{E}_F$ by writing

$$\mathcal{E}_F = \left\{ \prod_{i\in F} A_i : \ A_i \in \mathcal{F}_i \text{ for every } i \in F \right\}.$$

Observe that $\mathcal{E}_F$ is a semiring on Ω_F. [That is, $\varnothing \in \mathcal{E}_F$, $\mathcal{E}_F$ is closed under finite intersections, and $A, B \in \mathcal{E}_F$ with $A \subseteq B$ imply that $B - A$ may be written as a finite disjoint union of $\mathcal{E}_F$-sets.] If $F = \{i_1, \cdots, i_n\}$, we will alternatively write $A_{i_1} \times \cdots \times A_{i_n}$ for $\prod_{i\in F} A_i = \prod_{j=1}^{n} A_{i_j}$.

Fields. Let $D \subseteq I$ be nonempty. We define $\mathcal{G}_D$ to be the field on Ω_D generated by the collection

$$\bigcup_{F\subseteq D} \left\{ \prod_{i\in D} A_i : \ A_i \in \mathcal{F}_i \text{ for each } i \in F; \ A_i = \Omega_i \text{ for each } i \in D - F \right\}.$$

[1]Recall that $\prod_{i\in D} \Omega_i$ [strictly speaking] is the set of functions $\phi : D \to \bigcup_{i\in D} \Omega_i$ such that $\phi(i) \in \Omega_i$ for each $i \in D$. That we can discuss $\prod_{i\in D} \Omega_i$ as a nonempty set for arbitrary nonempty $D \subseteq I$ is a direct consequence of the Axiom of Choice.

[2]We therefore have various uses of the symbol Ω. When Ω is subscripted by a capital letter such as Ω_D or Ω_I, we are talking about a product. When Ω is subscripted by a lowercase letter such as Ω_i, we are talking about one of the component sets. No confusion should arise.

Here, the symbol $\bigcup_{F \subseteq D}$ denotes the union over all finite subsets F of D. The generating collection described above may also be written as

$$\left\{ \prod_{i \in D} A_i : \ A_i \in \mathcal{F}_i \text{ for each } i \in D, \ A_i \neq \Omega_i \text{ for finitely many } i \in D \right\}.$$

If D is in fact a finite nonempty subset of I, then the requirement that $A_i \neq \Omega_i$ for finitely many $i \in D$ is superfluous, and in this case we have

$$\mathcal{G}_D = \sigma\left(\left\{ \prod_{i \in D} A_i : \ A_i \in \mathcal{F}_i \text{ for each } i \in D \right\}\right) = f(\mathcal{E}_D).$$

[Recall that in this context the symbol $f(\mathcal{C})$ denotes the minimal field containing the collection $\mathcal{C}$; see the notation in Section 1.3.] In the special case of where $D = I$, we write $\mathcal{G}$ for $\mathcal{G}_I$. In terms of the notation just given, we have that $\mathcal{G}$ is the field on Ω generated by sets of the form $\prod_{i \in I} A_i$, where $A_i \in \mathcal{F}_i$ for each $i \in I$ and $A_i \neq \Omega_i$ for finitely many $i \in I$:

$$\mathcal{G} = f\left(\left\{ \prod_{i \in I} A_i : \ A_i \in \mathcal{F}_i \text{ for each } i \in I, \ A_i \neq \Omega_i \text{ for finitely many } i \in I. \right\}\right).$$

σ-fields. Let $D \subseteq I$ be nonempty. We define $\mathcal{F}_D$ to be the σ-field on Ω_D generated by the collection

$$\bigcup_{F \subseteq D} \left\{ \prod_{i \in D} A_i : \ A_i \in \mathcal{F}_i \text{ for each } i \in F; \ A_i = \Omega_i \text{ for each } i \in D - F \right\}.$$

Here, the symbol $\bigcup_{F \subseteq D}$ denotes the union over all finite subsets F of D. As in our discussion of $\mathcal{G}_D$, we may equivalently state that $\mathcal{F}_D$ is generated by sets of the form $\prod_{i \in D} A_i$, where $A_i \in \mathcal{F}_i$ for each $i \in D$ and at most finitely many A_i differ from Ω_i:

$$\mathcal{F}_D = \sigma\left(\left\{ \prod_{i \in D} A_i : \ A_i \in \mathcal{F}_i \text{ for each } i \in D, \ A_i \neq \Omega_i \text{ for finitely many } i \in D \right\}\right).$$

If D is in actuality a nonempty finite subset of I, then the requirement that $A_i \neq \Omega_i$ for finitely many $i \in D$ is redundant, so that in this finite case we have

$$\mathcal{F}_D = \sigma\left(\left\{ \prod_{i \in D} A_i : \ A_i \in \mathcal{F}_i \text{ for each } i \in D \right\}\right) = \sigma(\mathcal{E}_D).$$

In the special case of where $D = I$, we write $\mathcal{F}$ for $\mathcal{F}_I$, the σ-field on Ω generated by sets of the form $\prod_{i \in I} A_i$, where $A_i \in \mathcal{F}_i$ for each $i \in I$ and $A_i \neq \Omega_i$ for finitely many $i \in I$:

$$\mathcal{F} = \sigma\left(\left\{ \prod_{i \in I} A_i : \ A_i \in \mathcal{F}_i \text{ for each } i \in I, \ A_i \neq \Omega_i \text{ for finitely many } i \in I \right\}\right).$$

Using the above notation, the goal of this and the next section is to establish the existence of a measure μ with domain $\mathcal{F}$ such that $\mu(\Omega) = 1$ and

$$\mu\left(\prod_{i\in I} A_i\right) = \prod_{j=1}^{n} \mu_{i_j}(A_{i_j})$$

whenever $A_i \in \mathcal{F}_i$ for each $i \in I$ and $A_i \neq \Omega_i$ if and only if $i \in \{i_1, \cdots, i_n\}$. We shall actually construct a measure μ that satisfies a stronger condition, from which the just-given requirement follows as a special case. From this we will establish other results such as various Fubini-esque theorems and the famous Strong Law of Large Numbers.

The preceding notation may understandably seem overwhelming when presented in the abstract; we give illustrations of the notation and its meaning in various situations before proceeding further.

Example 1. Let $I = \{1, \cdots, n\}$, where $n \in \{2, 3, \cdots\}$. Consider a collection of measure spaces $\{(\Omega_i, \mathcal{F}_i, \mu_i)\}_{i\in I}$ with $\mu_i(\Omega_i) = 1$ for each $i \in I$. Observe that every subset D of I will be finite in this situation.

Given a nonempty subset $D = \{i_1, \cdots, i_m\}$ of I, where without loss of generality we have $i_1 < \cdots < i_m$, the set $\Omega_D = \prod_{i\in D} \Omega_i$ is often written as the usual Cartesian product $\Omega_{i_1} \times \cdots \times \Omega_{i_m}$. Regardless of the notation, Ω_D denotes the set of ordered m-tuples $(\omega_{i_1}, \cdots, \omega_{i_m})$, where $\omega_{i_1} \in \Omega_{i_1}, \cdots, \omega_{i_m} \in \Omega_{i_m}$. Here, $\Omega = \Omega_I = \prod_{i\in I} \Omega_i$ is written by some as $\Omega_1 \times \cdots \times \Omega_n$, the set of ordered n-tuples $(\omega_1, \cdots, \omega_n)$ with $\omega_1 \in \Omega_1, \cdots, \omega_n \in \Omega_n$.

Since every subset D of I is finite, we may discuss $\mathcal{E}_D$ for every nonempty $D \subseteq I$. Given a nonempty subset $D = \{i_1, \cdots, i_m\} \subseteq I$ [with $i_1 < \cdots < i_m$ as before], $\mathcal{E}_D$ denotes the collection of subsets $A_{i_1} \times \cdots \times A_{i_m}$ of $\Omega_{i_1} \times \cdots \times \Omega_{i_m}$, where $A_{i_1} \in \mathcal{F}_{i_1}$, $\cdots$, $A_{i_m} \in \mathcal{F}_{i_m}$.

Let us now discuss the field $\mathcal{G}_D$ on Ω_D for arbitrary nonempty $D \subseteq I$, again writing $D = \{i_1, \cdots, i_m\}$ with $i_1 < \cdots < i_m$. Here, the collection

$$\bigcup_{F\subseteq D} \left\{\prod_{i\in D} A_i : A_i \in \mathcal{F}_i \text{ for each } i \in F;\ A_i = \Omega_i \text{ for each } i \in D - F\right\}$$

of subsets of $\Omega_{i_1} \times \cdots \times \Omega_{i_m}$ that by definition generates the field $\mathcal{G}_D$ is seen upon a moment's thought to be none other than $\mathcal{E}_D$: $\mathcal{G}_D = f(\mathcal{E}_D)$. Therefore, in this finite situation we have

$$\mathcal{G}_D = f\left(\{A_{i_1} \times \cdots \times A_{i_m} \subseteq \Omega_{i_1} \times \cdots \times \Omega_{i_m} : A_{i_j} \in \mathcal{F}_{i_j} \text{ for } j = 1, \cdots, m\}\right).$$

Taking $D = I$ gives

$$\mathcal{G} = f(\{A_1 \times \cdots \times A_n \subseteq \Omega_1 \times \cdots \times \Omega_n : A_i \in \mathcal{F}_i \text{ for } i = 1, \cdots, n\}) = f(\mathcal{E}_I).$$

Let us now discuss the σ-field $\mathcal{F}_D$ on Ω_D for arbitrary nonempty $D \subseteq I$, again say $D = \{i_1, \cdots, i_m\}$ with $i_1 < \cdots < i_m$. Just as in the previous paragraph, the finiteness of I gives $\mathcal{F}_D = \sigma(\mathcal{E}_D)$. That is,

$$\mathcal{F}_D = \sigma\left(\{A_{i_1} \times \cdots \times A_{i_m} \subseteq \Omega_{i_1} \times \cdots \times \Omega_{i_m} : A_{i_j} \in \mathcal{F}_{i_j} \text{ for } j = 1, \cdots, m\}\right).$$

Taking $D = I$ gives

$$\mathcal{F} = \sigma\left(\{A_1 \times \cdots \times A_n \subseteq \Omega_1 \times \cdots \times \Omega_n : A_i \in \mathcal{F}_i \text{ for } i = 1, \cdots, n\}\right) = \sigma(\mathcal{E}_I).$$

In this finite situation, let us understand the desired properties regarding the hoped-for measure μ with domain $\mathcal{F}$. The first requirement, namely, the requirement that $\mu(\Omega) = 1$, here means that $\mu(\Omega_1 \times \cdots \times \Omega_n) = 1$. The general requirement that $\mu(\prod_{i \in I} A_i) = \prod_{j=1}^{m} \mu_{i_j}(A_{i_j})$ whenever $A_i \in \mathcal{F}_i$ for each $i \in I$ and $A_i \neq \Omega_i$ iff $i \in \{i_1, \cdots, i_m\}$ here reduces to the assertion that

$$\mu(A_1 \times \cdots \times A_n) = \prod_{i=1}^{n} \mu_i(A_i) \quad \text{whenever} \quad A_1 \in \mathcal{F}_1, \cdots, A_n \in \mathcal{F}_n.$$

In the previous chapter, we have dealt with the case $I = \{1, 2\}$. If we have two measure spaces $(\Omega_1, \mathcal{F}_1, \mu_1)$ and $(\Omega_2, \mathcal{F}_2, \mu_2)$ with $\mu_1(\Omega_1) = \mu_2(\Omega_2) = 1$, then in the notation above the σ-field $\mathcal{F}$ on $\Omega_1 \times \Omega_2$ is none other than what we called $\mathcal{F}_1 \otimes \mathcal{F}_2$ in Chapter 10. The measure μ as desired above is realized in the form of $\mu_1 \otimes \mu_2$ as given in Chapter 10.

Before concluding this example, it should be noted that the seeming excess of notation given in general is not really necessary if our index set I is finite. That is, if I is finite then we can write things in terms of usual Cartesian products and not have to define special symbols. However, the case of interest is where I is infinite.

Example 2. Let I denote a countably infinite set, say $I = \{1, 2, \cdots\}$. Here, the collection $\{(\Omega_i, \mathcal{F}_i, \mu_i)\}_{i \in I}$ of measure spaces with $\mu_i(\Omega_i) = 1$ for each $i \in I$ is nothing but a sequence $\{(\Omega_i, \mathcal{F}_i, \mu_i)\}_{i=1}^{\infty}$ of measure spaces with $\mu_i(\Omega_i) = 1$ for each $i \in \mathbb{N}$. We wish to discuss notation in this countably infinite case.

For any nonempty $D \subseteq I$, say $D = \{i_1, \cdots, i_m\}$ with $i_1 < \cdots < i_m$, we have $\Omega_D = \Omega_{i_1} \times \cdots \times \Omega_{i_m}$. If $D = \{i_1, i_2, \cdots\}$ with $i_1 < i_2 < \cdots$, then

$$\Omega_D = \Omega_{i_1} \times \Omega_{i_2} \times \cdots = \prod_{j=1}^{\infty} \Omega_{i_j}.$$

If $D = I$, then $\Omega = \Omega_I = \Omega_1 \times \Omega_2 \times \cdots = \prod_{i=1}^{\infty} \Omega_i$. The set Ω is nothing other than the set of sequences $(\omega_1, \omega_2, \cdots)$ with $\omega_1 \in \Omega_1$, $\omega_2 \in \Omega_2$, $\cdots$.

We now give some concrete notational examples. The collection $\mathcal{E}_{\{2,7,11,12\}}$ is none other than

$$\{A_2 \times A_7 \times A_{11} \times A_{12} \in \Omega_2 \times \Omega_7 \times \Omega_{11} \times \Omega_{12} : A_j \in \mathcal{F}_j \text{ for } j = 2, 7, 11, 12\}.$$

The collection $\mathcal{E}_{\{1,\cdots,10\}}$ consists of the sets $A_1 \times \cdots \times A_{10} \subseteq \Omega_1 \times \cdots \times \Omega_{10}$ with $A_1 \in \mathcal{F}_1, \cdots, A_{10} \in \mathcal{F}_{10}$.

We now discuss the notation involving $\mathcal{G}_D, \mathcal{F}_D, \mathcal{G}$, and $\mathcal{F}$. Suppose that D is finite, say $D = \{3, 6, 8\}$. Then

$$\mathcal{G}_D = f(\{A_3 \times A_6 \times A_8 \subseteq \Omega_3 \times \Omega_6 \times \Omega_8 : A_j \in \mathcal{F}_j \text{ for } j = 3, 6, 8\}) = f(\mathcal{E}_{\{3,6,8\}})$$

and

$$\mathcal{F}_D = \sigma(\{A_3 \times A_6 \times A_8 \subseteq \Omega_3 \times \Omega_6 \times \Omega_8 : A_j \in \mathcal{F}_j \text{ for } j = 3, 6, 8\}) = \sigma(\mathcal{E}_{\{3,6,8\}}).$$

Suppose that D is an infinite subset of I, say $D = \{5, 6, \cdots\}$. Then $\mathcal{G}_D$ is the field on $\Omega_5 \times \Omega_6 \times \cdots$ generated by the sets of the form $A_5 \times A_6 \times \cdots$, where $A_j \in \mathcal{F}_j$ for $j = 5, 6, \cdots$ and $\{j \in \{5, 6, \cdots\} : A_j \neq \Omega_j\}$ is finite. Next, $\mathcal{F}_D$ is the σ-field on $\Omega_5 \times \Omega_6 \times \cdots$ generated by the sets of the form $A_5 \times A_6 \times \cdots$, where $A_j \in \mathcal{F}_j$ for $j = 5, 6, \cdots$ and $\{j \in \{5, 6, \cdots\} : A_j \neq \Omega_j\}$ is finite. The field $\mathcal{G}$ is the field on $\Omega_1 \times \Omega_2 \times \cdots$ generated by the sets of the form $A_1 \times A_2 \times \cdots$, where $A_i \in \mathcal{F}_i$ for each $i \in \mathbb{N}$ and all but a finite number of A_i's equal Ω_i. The σ-field $\mathcal{F}$ is the σ-field on $\Omega_1 \times \Omega_2 \times \cdots$ generated by the sets of the form $A_1 \times A_2 \times \cdots$, where $A_i \in \mathcal{F}_i$ for each $i \in \mathbb{N}$ and all but a finite number of A_i's equal Ω_i.

In this countably infinite situation, we desire a measure μ on $\mathcal{F}$ with the properties that $\mu(\Omega_1 \times \Omega_2 \times \cdots) = 1$ and

$$\mu\left(\prod_{i=1}^{\infty} A_i\right) = \prod_{j=1}^{n} \mu_{i_j}(A_{i_j})$$

whenever $A_i \in \mathcal{F}_i$ for each $i \in \mathbb{N}$ and $A_i \neq \Omega_i$ iff $i \in \{i_1, \cdots, i_n\}$. This measure μ, among other things, is such that

$$\mu(A_1 \times \Omega_2 \times A_3 \times A_4 \times \Omega_5 \times \Omega_6 \times \cdots) = \mu_1(A_1)\mu_3(A_3)\mu_4(A_4)$$

whenever $A_1 \in \mathcal{F}_1$, $A_3 \in \mathcal{F}_3$, and $A_4 \in \mathcal{F}_4$. Another property of μ is that

$$\mu(A_1 \times \cdots \times A_{300} \times \Omega_{301} \times \Omega_{302} \times \cdots) = \prod_{i=1}^{300} \mu_i(A_i)$$

whenever $A_i \in \mathcal{F}_i$ for $i = 1, \cdots, 300$.

The preceding examples will hopefully make clear the notation given at the beginning of this section. We need the following technical result.

Claim 1. *Let $\Phi : M \to N$ be one-to-one and onto. Let $\mathcal{M}$ denote a σ-field on M that is generated by some collection $\mathcal{M}_0$ of subsets of M. Let $\mathcal{N}$ denote a σ-field on N that is generated by some collection $\mathcal{N}_0$ of subsets of N. Assume that $\Phi(\mathcal{M}_0) \subseteq \mathcal{N}$ and $\Phi^{-1}(\mathcal{N}_0) \subseteq \mathcal{M}$. Then $\Phi(\mathcal{M}) = \mathcal{N}$.*

Proof: Because $\Phi : M \to N$ is one-to-one and onto, the inverse mapping $\Phi^{-1} : N \to M$ exists [and in addition is one-to-one and onto]. The condition $\Phi^{-1}(\mathcal{N}_0) \subseteq \mathcal{M}$ implies that $\sigma(\Phi^{-1}(\mathcal{N}_0)) \subseteq \mathcal{M}$. By Claim 2 of Section 1.3, we have $\sigma(\Phi^{-1}(\mathcal{N}_0)) = \Phi^{-1}(\sigma(\mathcal{N}_0)) = \Phi^{-1}(\mathcal{N})$, hence we have that $\Phi^{-1}(\mathcal{N}) \subseteq \mathcal{M}$. From this we have $\Phi(\Phi^{-1}(\mathcal{N})) \subseteq \Phi(\mathcal{M})$, and using the equality $\Phi(\Phi^{-1}(\mathcal{N})) = \mathcal{N}$ gives $\mathcal{N} \subseteq \Phi(\mathcal{M})$.

We now aim to establish the reverse containment $\Phi(\mathcal{M}) \subseteq \mathcal{N}$. The given condition $\Phi(\mathcal{M}_0) \subseteq \mathcal{N}$ means that $\sigma(\Phi(\mathcal{M}_0)) \subseteq \mathcal{N}$. Writing $\Phi = (\Phi^{-1})^{-1}$ and again using Claim 2 of Section 1.3, we have that

$$\sigma(\Phi(\mathcal{M}_0)) = \sigma((\Phi^{-1})^{-1}(\mathcal{M}_0)) = (\Phi^{-1})^{-1}(\sigma(\mathcal{M}_0)) = (\Phi^{-1})^{-1}(\mathcal{M}) = \Phi(\mathcal{M}),$$

and hence $\Phi(\mathcal{M}) \subseteq \mathcal{N}$. This reverse inclusion completes the proof. ■

The previous claim will now be used to prove the following claim. The content of the claim may seem abstract and overly formal, but from a standpoint of logical consistency it shall be of great importance. We state the claim first and give some examples; this should make the claim's content more understandable. After that, a proof is given.

Claim 2. [Identification Lemma.] *Suppose that $\{D_p\}_{p\in P}$ is a disjoint family of nonempty subsets of I with $\bigcup_{p\in P} D_p = I$. Define $\Phi : \Omega \to \prod_{p\in P} \Omega_{D_p}$ for every $\boldsymbol{\omega} = (\omega_i)_{i\in I} \in \Omega$ by the rule $\Phi(\boldsymbol{\omega}) = ((\omega_i)_{i\in D_p})_{p\in P}$. Then the following results obtain:*

(a) Φ is one-to-one and onto $\prod_{p\in P} \Omega_{D_p}$. Therefore, for every element of Ω there is a unique corresponding element of $\prod_{p\in P} \Omega_{D_p}$, and for every element of $\prod_{p\in P} \Omega_{D_p}$ there is a unique corresponding element of Ω.

Let $\mathcal{F}^*$ denote the σ-field of subsets of $\prod_{p\in P} \Omega_{D_p}$ generated by sets of the form $\prod_{p\in P} A_{D_p}$, where $A_{D_p} \in \mathcal{F}_{D_p}$ for each $p \in P$ and $A_{D_p} \neq \Omega_{D_p}$ for at most finitely many $p \in P$.

(b) For every set in $\mathcal{F}$ there is a unique corresponding set in $\mathcal{F}^*$, and for every set in $\mathcal{F}^*$ there is a unique corresponding set in $\mathcal{F}$. That is, there is a one-to-one correspondence between the sets in $\mathcal{F}$ and the sets in $\mathcal{F}^*$, and vice-versa.

We now give some illustrative examples of this claim and justify the title *Identification Lemma.* We have stated the lemma in full generality. In actuality, we only need the case where P is a two-element set, say $P = \{1, 2\}$, so that the index set I is the disjoint union of D_1 and D_2.

Example 3. Let$I = \{i_1, \cdots, i_m, i_{m+1}\}$, $P = \{1, 2\}$, $D_1 = \{i_1, \cdots, i_m\}$, and $D_2 = \{i_{m+1}\}$. Then D_1 and D_2 are nonempty disjoint subsets of I with union I. Here the mapping Φ is such that

$$\Phi((\omega_{i_1}, \cdots, \omega_{i_m}, \omega_{i_{m+1}})) = ((\omega_{i_1}, \cdots, \omega_{i_m}), \omega_{i_{m+1}})$$

for each $\boldsymbol{\omega} = (\omega_{i_1}, \cdots, \omega_{i_{m+1}}) \in \Omega = \Omega_{i_1} \times \cdots \times \Omega_{i_m} \times \Omega_{i_{m+1}}$. Claim 2 asserts that $\Phi : \Omega_{i_1} \times \cdots \times \Omega_{i_{m+1}} \to (\Omega_{i_1} \times \cdots \times \Omega_{i_m}) \times \Omega_{i_{m+1}}$ is one-to-one and onto. [This should be clear if not obvious.] Put another way, the element $(\omega_{i_1}, \cdots, \omega_{i_m}, \omega_{i_{m+1}})$ in $\Omega_{i_1} \times \cdots \times \Omega_{i_m} \times \Omega_{i_{m+1}}$ uniquely corresponds to the element $((\omega_{i_1}, \cdots, \omega_{i_m}), \omega_{i_{m+1}})$ in $(\Omega_{i_1} \times \cdots \times \Omega_{i_m}) \times \Omega_{i_{m+1}}$, and vice versa. This is the content of (a) of the claim.

The σ-field $\mathcal{F}$ on Ω is the collection of sets $A_{i_1} \times \cdots \times A_{i_{m+1}}$, where $A_{i_j} \in \mathcal{F}_{i_j}$ for each $j \in \{1, \cdots, m+1\}$. [That is, $\mathcal{F} = \sigma(\mathcal{E}_{\{i_1, \cdots, i_{m+1}\}})$.] In the notation of Claim 2, $\mathcal{F}^*$ denotes the σ-field on $(\Omega_{i_1} \times \cdots \times \Omega_{i_m}) \times \Omega_{i_{m+1}}$ generated by sets of the form $A_1 \times A_2$, where $A_1 \in \mathcal{F}_{D_1}$ and $A_2 \in \mathcal{F}_{i_{m+1}}$. That is, $\mathcal{F}^* = \mathcal{F}_{D_1} \otimes \mathcal{F}_{i_{m+1}}$. By (b) of

Claim 2, there is a one-to-one correspondence under Φ between the elements of $\mathcal{F}$ and $\mathcal{F}^*$. That is, there is a one-to-one correspondence under Φ between the sets in $\mathcal{F}_{\{i_1,\cdots,i_m,i_{m+1}\}}$ and the sets in $\mathcal{F}_{\{i_1,\cdots,i_m\}} \otimes \mathcal{F}_{i_{m+1}}$. Furthermore, each $\mathcal{F}^*$-set is the image under Φ of a unique $\mathcal{F}$-set. The correspondence is natural in that whenever $A_{i_j} \in \mathcal{F}_{i_j}$ for $j = 1, \cdots, m+1$ we have

$$\Phi(A_{i_1} \times \cdots \times A_{i_{m+1}}) = (A_{i_1} \times \cdots \times A_{i_m}) \times A_{i_{m+1}}$$

whenever $A_{i_j} \in \mathcal{F}_{i_j}$ for $j = 1, \cdots, m+1$. On the other hand, if a set in $\mathcal{F}^*$ has the special form $(A_{i_1} \times \cdots \times A_{i_m}) \times A_{i_{m+1}}$, for sets $A_{i_j} \in \mathcal{F}_{i_j}$, $j = 1, \cdots, m+1$, there is but one element of $\mathcal{F}$ that is mapped to this set by Φ, namely $A_{i_1} \times \cdots \times A_{i_{m+1}}$.

The elements of $\mathcal{F}$ and $\mathcal{F}^*$ are formally different from a strict set-theoretic perspective, but Claim 2 allows us to uniquely associate a given element of $\mathcal{F}$ with an element of $\mathcal{F}^*$ and vice versa in a natural fashion.

Example 4. Let $I = \{i_1, \cdots, i_n, j_1, \cdots, j_m\}$ be finite, and let $F_1 = \{i_1, \cdots, i_n\}$ with $F_2 = \{j_1, \cdots, j_m\}$. Then F_1 and F_2 are nonempty disjoint subsets of I with union I. Again, let $P = \{1, 2\}$, $D_1 = F_1$, and $D_2 = F_2$. Here

$$\Omega = \Omega_{F_1 \cup F_2} = \Omega_{i_1} \times \cdots \Omega_{i_n} \times \Omega_{j_1} \times \cdots \times \Omega_{j_m}$$

and

$$\Omega_{F_1} \times \Omega_{F_2} = (\Omega_{i_1} \times \cdots \Omega_{i_n}) \times (\Omega_{j_1} \times \cdots \times \Omega_{j_m}).$$

The mapping $\Phi : \Omega \to \prod_{p \in P} \Omega_{D_p}$ is the mapping $\Phi : \Omega_{F_1 \cup F_2} \to \Omega_{F_1} \times \Omega_{F_2}$. For a generic $\boldsymbol{\omega} = (\omega_{i_1}, \cdots, \omega_{i_n}, \omega_{j_1}, \cdots, \omega_{j_m}) \in \Omega_{F_1 \cup F_2}$, we have

$$\Phi(\boldsymbol{\omega}) = ((\omega_{i_1}, \cdots, \omega_{i_n}), (\omega_{j_1}, \cdots, \omega_{j_m}));$$

the mapping Φ is one-to-one and onto $\Omega_{F_1} \times \Omega_{F_2}$. What this means is that although

$$(\omega_{i_1}, \cdots, \omega_{i_n}, \omega_{j_1}, \cdots, \omega_{j_m}) \text{ is different from } ((\omega_{i_1}, \cdots, \omega_{i_n}), (\omega_{j_1}, \cdots, \omega_{j_m}))$$

in a strict set-theoretic sense, we may naturally associate the two objects relative to the one-to-one and onto mapping Φ.

The σ-field $\mathcal{F}$ on $\Omega = \Omega_{F_1 \cup F_2}$ is generated by the sets $A_{i_1} \times \cdots \times A_{i_n} \times A_{j_1} \times \cdots \times A_{j_m}$, where $A_{i_k} \in \mathcal{F}_{i_k}$ for $k = 1, \cdots, n$ and $A_{j_k} \in \mathcal{F}_{j_k}$ for $k = 1, \cdots, m$. That is, $\mathcal{F} = \sigma(\mathcal{E}_{F_1 \cup F_2})$. The σ-field $\mathcal{F}^*$ on $\Omega_{F_1} \times \Omega_{F_2}$ is generated by the sets of the form $A_{F_1} \times A_{F_2}$, where $A_{F_1} \in \mathcal{F}_{F_1}$ and $A_{F_2} \in \mathcal{F}_{F_2}$. That is, $\mathcal{F}^* = \mathcal{F}_{F_1} \otimes \mathcal{F}_{F_2}$. By Claim 2, each element of $\mathcal{F} = \sigma(\mathcal{E}_{F_1 \cup F_2})$ corresponds to one and only one element of $\mathcal{F}^* = \mathcal{F}_{F_1} \otimes \mathcal{F}_{F_2}$, and each element of $\mathcal{F}^* = \mathcal{F}_{F_1} \otimes \mathcal{F}_{F_2}$ corresponds to one and only one element of $\mathcal{F} = \sigma(\mathcal{E}_{F_1 \cup F_2})$. For example, the set

$$A_{i_1} \times \cdots \times A_{i_n} \times A_{j_1} \times \cdots \times A_{j_m} \in \mathcal{F} = \sigma(\mathcal{E}_{F_1 \cup F_2}),$$

where $A_{i_k} \in \mathcal{F}_{i_k}$ for $k = 1, \cdots, n$ and $A_{j_k} \in \mathcal{F}_{j_k}$ for $k = 1, \cdots, m$, uniquely corresponds to the set

$$(A_{i_1} \times \cdots \times A_{i_n}) \times (A_{j_1} \times \cdots \times A_{j_m}) \in \mathcal{F}^* = \mathcal{F}_{F_1} \otimes \mathcal{F}_{F_2},$$

and vice versa. In this sense, the one-to-one and onto correspondence between the elements of $\mathcal{F}$ and $\mathcal{F}^*$ is natural.

Example 5. Let I denote an arbitrary infinite set, and let $D \subseteq I$ be nonempty. We would like to assert that we may naturally associate the sets in $\mathcal{F} = \mathcal{F}_I = \mathcal{F}_{D \cup D^c}$ with those in $\mathcal{F}_D \otimes \mathcal{F}_{D^c}$. This is really a special case of Claim 2 with $P = \{1, 2\}$, $D_1 = D$, and $D_2 = D^c$. The mapping Φ from $\Omega = \Omega_I = \Omega_{D \cup D^c}$ to $\Omega_D \times \Omega_{D^c}$ that sends an element $\boldsymbol{\omega} = (\omega_i)_{i \in D \cup D^c}$ to $((\omega_i)_{i \in D}, (\omega_i)_{i \in D^c})$ is one-to-one and onto. Here, $\mathcal{F}^*$ denotes the σ-field on $\Omega_D \times \Omega_{D^c}$ generated by sets of the form $A_D \times A_{D^c}$, where $A_D \in \mathcal{F}_D$ and $A_{D^c} \in \mathcal{F}_{D^c}$. That is, $\mathcal{F}^* = \mathcal{F}_D \otimes \mathcal{F}_{D^c}$. Under Φ, Claim 2 asserts that each element of $\mathcal{F}$ may be uniquely identified with an element of $\mathcal{F}_D \otimes \mathcal{F}_{D^c}$ and vice versa. Again, this correspondence is natural in that sets of the form $\prod_{i \in I} A_i$, where $A_i \in \mathcal{F}_i$ for each $i \in I$ and at most finitely many A_i are different from Ω_i, correspond to sets of the form $(\prod_{i \in D} A_i) \times (\prod_{i \in D^c} A_i)$ [and vice versa].

We are now ready to give the proof of Lemma 2. In reality, we only need the case where P is a two-element set. In the proof, one claim is made that will be left to the reader for an exercise. We shall point out where this claim is made, and the reader should have no problem filling in the details, especially since a detailed hint is given for proving the claim.

Proof of Lemma 2: We first handle (a). Suppose that $\Phi(\boldsymbol{\omega}^{(1)}) = \Phi(\boldsymbol{\omega}^{(2)})$. Then $((\omega_i^{(1)})_{i \in D_p})_{p \in P} = ((\omega_i^{(2)})_{i \in D_p})_{p \in P}$. It follows for every $p \in P$ that $(\omega_i^{(1)})_{i \in D_p} = (\omega_i^{(2)})_{i \in D_p}$. Therefore, for every $p \in P$ and $i \in D_p$ we have $\omega_i^{(1)} = \omega_i^{(2)}$. Since $\bigcup_{p \in P} D_p = I$, we have that $\omega_i^{(1)} = \omega_i^{(2)}$ for each $i \in I$, which gives $\boldsymbol{\omega}^{(1)} = \boldsymbol{\omega}^{(2)}$. This shows that Φ is one-to-one.

We now show that Φ is onto $\prod_{p \in P} \Omega_{D_p}$. Consider an element $((\xi_i)_{i \in D_p})_{p \in P}$ of $\prod_{p \in P} \Omega_{D_p}$. If $\boldsymbol{\omega} \in \Omega$ is such that $\omega_i = \xi_i$ for every $p \in P$ and $i \in D_p$, then $\Phi(\boldsymbol{\omega}) = ((\omega_i)_{i \in D_p})_{p \in P} = ((\xi_i)_{i \in D_p})_{p \in P}$. This shows that Φ is onto $\prod_{p \in P} \Omega_{D_p}$, and completes the proof of (a).

We now turn to (b), which first requires showing that $\Phi(\mathcal{F}) = \mathcal{F}^*$. We define some notation first. Let $\mathcal{F}^0$ denote the collection of sets $\prod_{i \in I} A_i$ such that $A_i \in \mathcal{F}_i$ for each $i \in I$ and at most finitely many A_i's are different than Ω_i. Then by definition we have $\sigma(\mathcal{F}^0) = \mathcal{F}$. Next, let $\mathcal{N}^0$ denote the collection of sets $\prod_{p \in P} A_{D_p}$, where $A_{D_p} \in \mathcal{F}_{D_p}$ for each $p \in P$ and $A_{D_p} \neq \Omega_{D_p}$ for at most finitely many $p \in P$. We have $\sigma(\mathcal{N}_0) = \mathcal{F}^*$ by definition. Finally, let $\mathcal{N}^1$ denote the collection of sets $\prod_{p \in P} \prod_{i \in D_p} A_i$, where each $A_i \in \mathcal{F}_i$ and at most finitely many of the sets $\prod_{i \in D_p} A_i$ are different from Ω_{D_p}. The σ-field $\mathcal{F}^*$ is also generated by the collection $\mathcal{N}^1$: $\sigma(\mathcal{N}^1) = \mathcal{F}^*$. [See the exercises for the buildup to this claim.] Summarizing the notation, we have $\sigma(\mathcal{F}_0) = \mathcal{F}$ and $\sigma(\mathcal{N}^0) = \sigma(\mathcal{N}^1) = \mathcal{F}^*$.

We ultimately wish to apply Claim 1. There, the set M that is the domain of Φ here is Ω, and the set N that is the image of $M = \Omega$ under Φ is $\prod_{p \in P} \Omega_{D_p}$. The σ-field $\mathcal{M}$ on M there is the σ-field $\mathcal{F}$ on Ω here, and the collection $\mathcal{F}^0$ of the preceding paragraph is the $\mathcal{M}_0$ of Claim 1. The σ-field $\mathcal{N}$ on N in Claim 1 here is the σ-field $\mathcal{F}^*$ on $\prod_{p \in P} \Omega_{D_p}$; the collection $\mathcal{N}_0$ in Claim 1 here is the collection $\mathcal{N}^1$ as given in the preceding paragraph. Claim 1 asserts that if $\Phi(\mathcal{M}_0) \subseteq \mathcal{N}$ and $\Phi^{-1}(\mathcal{N}_0) \subseteq \mathcal{M}$, then $\Phi(\mathcal{M}) = \mathcal{N}$. In the notation of our current claim, we have that if $\Phi(\mathcal{F}^0) \subseteq \mathcal{F}^*$ and $\Phi^{-1}(\mathcal{N}^1) \subseteq \mathcal{F}$, then $\Phi(\mathcal{F}) = \mathcal{F}^*$. Therefore, the proof of the claim that $\Phi(\mathcal{F}) = \mathcal{F}^*$ is complete when the two containment relations $\Phi(\mathcal{F}^0) \subseteq \mathcal{F}^*$ and $\Phi^{-1}(\mathcal{N}^1) \subseteq \mathcal{F}$ are demonstrated.

We establish that $\Phi(\mathcal{F}^0) \subseteq \mathcal{F}^*$. Pick $A \in \mathcal{F}^0$; A has the form $\prod_{i\in I} A_i$ where $A_i \in \mathcal{F}_i$ for each $i \in I$ and at most finitely many of the A_i's are different from Ω_i. It is clear that

$$\Phi(A) = \Phi\left(\prod_{i\in I} A_i\right) = \prod_{p\in P}\prod_{i\in D_p} A_i.$$

Since $A_i \in \mathcal{F}_i$ for each $i \in I$ and at most finitely many A_i's are different from Ω_i, we have for each $p \in P$ that $A_i \neq \Omega_i$ for at most finitely many $i \in D_p$. Therefore, we have (i) $\prod_{i\in D_p} A_i \in \mathcal{F}_{D_p}$ for each $p \in P$, and (ii) $\prod_{i\in D_p} A_i \neq \Omega_{D_p}$ for at most finitely many $p \in P$. From these facts, it follows that $\Phi(A) \in \mathcal{N}^0 \subseteq \mathcal{F}^*$, whence the containment relation $\Phi(\mathcal{F}^0) \subseteq \mathcal{F}^*$ follows.

We now desire to demonstrate the inclusion $\Phi^{-1}(\mathcal{N}^1) \subseteq \mathcal{F}$. Pick a set $B \in \mathcal{N}^1$, so that B has the form $\prod_{p\in P}\prod_{i\in D_p} A_i$ where each $A_i \in \mathcal{F}_i$ and at most finitely many of the A_i's are different from Ω_i. It is clear that

$$\Phi^{-1}(B) = \Phi^{-1}\left(\prod_{p\in P}\prod_{i\in D_p} A_i\right) = \prod_{i\in I} A_i \in \mathcal{F}^0 \subseteq \mathcal{F},$$

so that $\Phi^{-1}(\mathcal{N}^1) \subseteq \mathcal{F}$. This gives the other desired containment relation, and we have shown that $\Phi(\mathcal{F}) = \mathcal{F}^*$.

From the last paragraph, it is immediate that every $A \in \mathcal{F}$ is mapped by Φ to a unique set $\Phi(A) \in \mathcal{F}^*$, and every $B \in \mathcal{F}^*$ is the image under Φ of some $A \in \mathcal{F}$. All that remains is to show that such a set $A \in \mathcal{F}$ is unique. To this end, suppose that $B \in \mathcal{F}^*$, and assume that there are sets A_1 and A_2 in $\mathcal{F}$ such that $\Phi(A_1) = B = \Phi(A_2)$. We wish to show that $A_1 = A_2$. Suppose to the contrary that $A_1 \neq A_2$, where without loss of generality we pick $\omega_1 \in A_1 - A_2$. Since $\omega_1 \in A_1$, we have $\Phi(\omega_1) \in B = \Phi(A_2)$, so that there is $\omega_2 \in A_2$ with $\Phi(\omega_2) = \Phi(\omega_1)$. We have shown that Φ is one-to-one, hence $\omega_1 = \omega_2$, giving $\omega_1 \in A_2$, contradicting the previous fact that $\omega_1 \in A_1 - A_2$. This shows that $A_1 = A_2$, and hence any set in $\mathcal{F}^*$ is the image under Φ of a unique $\mathcal{F}$-set. This completes the entire proof. ∎

It should be mentioned that the proof of Claim 2 is not particularly forbidding. The notation is really the culprit, making things possibly seem more abstruse and complicated than what they really are.

The next section constructs the desired product measure μ as described earlier in this section, and consists of a series of somewhat technical claims and a quite substantive theorem. The reader [once again] is urged to be comfortable with the notation before proceeding further.

Exercises. These exercises are important and should be worked, as they are needed to complete the proof of Lemma 2 and to be called upon in the next section. The reader should have no problems with these exercises, as ample hints are given.

1*. Let Ω_1 denote a nonempty set, and let $\mathcal{A}$ denote a nonempty collection of subsets of Ω_1. Let Ω_2 denote a nonempty set, and let $\mathcal{B}$ denote a nonempty collection of subsets of Ω_2.
(a) Let $B \subseteq \Omega_2$ be nonempty. Then $\sigma_{\Omega_1 \times B}(\mathcal{A} \times \{B\}) = \sigma(\mathcal{A}) \times \{B\}$. [To be sure about notation, the symbol $\sigma_{\Omega_1 \times B}$ denotes the minimal σ-field on $\Omega_1 \times B$, not $\Omega_1 \times \Omega_2$, and the σ on the right side denotes the minimal σ-field on Ω_1. One direction is easy: $\mathcal{A} \times \{B\} \subseteq \sigma(\mathcal{A}) \times \{B\}$, and $\sigma(\mathcal{A}) \times \{B\}$ is a σ-field on $\Omega_1 \times B$. It follows by minimality that $\sigma_{\Omega_1 \times B}(\mathcal{A} \times \{B\}) \subseteq \sigma(\mathcal{A}) \times \{B\}$. We now desire the reverse inclusion $\sigma(\mathcal{A}) \times \{B\} \subseteq \sigma_{\Omega_1 \times B}(\mathcal{A} \times \{B\})$. Define

$$\mathcal{C} = \{A \in \sigma(\mathcal{A}): \ A \times B \in \sigma_{\Omega_1 \times B}(\mathcal{A} \times \{B\})\}.$$

If $A \in \mathcal{A}$, then $A \times B \in \mathcal{A} \times \{B\} \subseteq \sigma_{\Omega_1 \times B}(\mathcal{A} \times \{B\})$, so $A \in \mathcal{C}$. This gives $\mathcal{A} \subseteq \mathcal{C}$. It is straightforward to verify that $\Omega_1 \in \mathcal{C}$, $\mathcal{C}$ is closed under complementation, and $\mathcal{C}$ is closed under countable unions. It follows that $\mathcal{C}$ is a σ-field on Ω_1, so minimality forces $\sigma(\mathcal{A}) \subseteq \mathcal{C}$. Since we trivially have $\mathcal{C} \subseteq \sigma(\mathcal{A})$, it follows that $\mathcal{C} = \sigma(\mathcal{A})$. From this, it is immediate that the desired reverse inclusion $\sigma(\mathcal{A}) \times \{B\} \subseteq \sigma_{\Omega_1 \times B}(\mathcal{A} \times \{B\})$ holds true.]
(b) $\sigma(\mathcal{A} \times \mathcal{B}) = \sigma(\sigma(\mathcal{A}) \times \sigma(\mathcal{B}))$. [To be sure about notation, the left hand σ and the outer right hand σ denote the minimal σ-field on $\Omega_1 \times \Omega_2$. For the proof, it is obvious that $\mathcal{A} \times \mathcal{B} \subseteq \sigma(\mathcal{A}) \times \sigma(\mathcal{B})$, so that $\sigma(\mathcal{A} \times \mathcal{B}) \subseteq \sigma(\sigma(\mathcal{A}) \times \sigma(\mathcal{B}))$. Next, fix $B \in \sigma(\mathcal{B})$. It is clear that $\sigma_{\Omega_1 \times B}(\mathcal{A} \times \{B\}) \subseteq \sigma(\mathcal{A} \times \mathcal{B})$. By part (a), we have the identity $\sigma_{\Omega_1 \times B}(\mathcal{A} \times \{B\}) = \sigma(\mathcal{A}) \times \{B\}$, so that $\sigma(\mathcal{A}) \times \{B\} \subseteq \sigma(\mathcal{A} \times \mathcal{B})$. Since $B \in \sigma(\mathcal{B})$ is arbitrary, we have $\bigcup_{B \in \sigma(\mathcal{B})}[\sigma(\mathcal{A}) \times \{B\}] \subseteq \sigma(\mathcal{A} \times \mathcal{B})$, which is equivalent to asserting that $\sigma(\mathcal{A}) \times \sigma(\mathcal{B}) \subseteq \sigma(\mathcal{A} \times \mathcal{B})$. Therefore, $\sigma(\sigma(\mathcal{A}) \times \sigma(\mathcal{B})) \subseteq \sigma(\mathcal{A} \times \mathcal{B})$, which gives the identity.]

2*. Prove the claim in the Identification Lemma [Lemma 2] for the case where P is a two-element set, which case is really the only one that we use. [For notation's sake, write $P = \{1, 2\}$. Here, $\sigma(\mathcal{N}_0) = \sigma(\mathcal{F}_{D_1} \times \mathcal{F}_{D_2})$. Let $\mathcal{A}_1$ denote the collection of sets of the form $\prod_{i \in D_1} A_i$, where $A_i \in \mathcal{F}_i$ for each $i \in D_1$ and at most finitely many A_i's differ from Ω_i. Then by definition we have $\mathcal{F}_{D_1} = \sigma(\mathcal{A}_1)$. Let $\mathcal{A}_2$ denote the collection of sets of the form $\prod_{i \in D_2} A_i$, where $A_i \in \mathcal{F}_i$ for each $i \in D_2$ and at most finitely many A_i's differ from Ω_i. Then by definition we have $\mathcal{F}_{D_2} = \sigma(\mathcal{A}_2)$ and $\sigma(\mathcal{N}_1) = \sigma(\mathcal{A}_1 \times \mathcal{A}_2)$. Showing that $\sigma(\mathcal{N}_0) = \sigma(\mathcal{N}_1)$ means showing that $\sigma(\mathcal{F}_{D_1} \times \mathcal{F}_{D_2}) = \sigma(\mathcal{A}_1 \times \mathcal{A}_2)$. This is automatic by (b) of the previous exercise.]

3. Prove the Identification Lemma in full generality for the case where P is an arbitrary set. [Here, $\sigma(\mathcal{N}_0) = \sigma(\prod_{p \in P} \mathcal{F}_{D_p})$, and $\sigma(\mathcal{N}_1) = \sigma(\prod_{p \in P} \mathcal{A}_{D_p})$, where for each $p \in P$ $\mathcal{A}_{D_p}$ denotes the collection of sets of the form $\prod_{i \in D_p} A_i$, where $A_i \in \mathcal{F}_i$ for each $i \in D_p$ and $\{i \in D_p : \ A_i \neq \Omega_i\}$ is finite. Observe that $\sigma(\mathcal{A}_{D_p}) = \mathcal{F}_{D_p}$ for each $p \in P$. Try to obtain the identity $\sigma(\prod_{p \in P} \mathcal{A}_{D_p}) = \sigma(\prod_{p \in P} \sigma(\mathcal{A}_{D_p}))$ by proving a generalization of Exercise 1(b).]

4*. Show that $\mathcal{E}_F$ is a semiring on Ω_F.

5*. Let $\mathcal{A}$ denote a semiring on Ω_1, and let $\mathcal{B}$ denote a semiring on Ω_2. Then $\mathcal{A} \times \mathcal{B}$ is a semiring on $\Omega_1 \times \Omega_2$. [Let $A_1 \times B_1 \subseteq A_2 \times B_2$, where $A_1, A_2 \in \mathcal{A}$ and $B_1, B_2 \in \mathcal{B}$. Then $A_1 \subseteq A_2$ and $B_1 \subseteq B_2$, and

$$(A_2 \times B_2) - (A_1 \times B_1) = [A_1 \times (B_2 - B_1)] \cup [(A_2 - A_1) \times B_2],$$

and the union above is disjoint. Now $B_2 - B_1$ may be written as a finite disjoint union $\bigcup_{i=1}^k D_i$ of $\mathcal{B}$-sets, and $A_2 - A_1$ may be written as a finite disjoint union $\bigcup_{j=1}^l C_j$ of $\mathcal{A}$-sets. It follows that

$$\begin{aligned}(A_2 \times B_2) - (A_1 \times B_1) &= \left[A_1 \times \bigcup_{i=1}^k D_i\right] \cup \left[\bigcup_{j=1}^l C_j \times B_2\right] \\ &= \bigcup_{i=1}^k (A_1 \times D_i) \cup \bigcup_{j=1}^l (C_j \times B_2),\end{aligned}$$

so $(A_2 \times B_2) - (A_1 \times B_1)$ is a finite disjoint union of sets in $\mathcal{A} \times \mathcal{B}$.] As a corollary to this result, we have the following result by Exercise 4: *Let F_1 and F_2 denote finite subsets of I. Then $\mathcal{E}_{F_1} \times \mathcal{E}_{F_2}$ is a semiring on $\Omega_{F_1} \times \Omega_{F_2}$.*]

11.2 CONSTRUCTION OF THE PRODUCT MEASURE

Let $\{(\Omega_i, \mathcal{F}_i, \mu_i)\}_{i \in I}$ denote a collection of measure spaces with $\mu_i(\Omega_i) = 1$ for each $i \in I$. The index set I is nonempty, and may be finite, countably infinite, or uncountably infinite. The σ-field $\mathcal{F}$ on $\Omega = \prod_{i \in I} \Omega_i$ as given in the previous section is the σ-field on Ω generated by the sets of the form $\prod_{i \in I} A_i$, where $A_i \in \mathcal{F}_i$ for every $i \in I$ and at most finitely many A_i's differ from Ω_i. Our goal is to prove the existence of a measure μ on $\mathcal{F}$ such that

(i) $\mu(\Omega) = 1$, and

(ii) $\mu(\prod_{i \in I} A_i) = \prod_{j=1}^n \mu_{i_j}(A_{i_j})$ whenever $A_i \in \mathcal{F}_i$ for each $i \in I$ and $A_i = \Omega_i$ for each $i \in I - \{i_1, \cdots, i_n\}$.

If we wanted to use more suggestive but imprecise notation, we could rephrase requirement (ii) as saying that

$$\mu(A_{i_1} \times \cdots \times A_{i_n} \times \Omega_{\{i_1, \cdots, i_n\}^c}) = \prod_{j=1}^n \mu_{i_j}(A_{i_j})$$

whenever $A_{i_1} \in \mathcal{F}_{i_1}, \cdots, A_{i_n} \in \mathcal{F}_{i_n}$. The imprecision comes from the fact that $A_{i_1} \times \cdots \times A_{i_n} \times \Omega_{\{i_1, \cdots, i_n\}^c}$, viewed in the formal set-theoretic sense, is not an element of $\mathcal{F}$, the domain of μ.

We shall construct such a μ by proving a series of technical claims. As stated in the previous section, the hardest part is really the notation, not so much the concepts. Our first claim extends the main result of Section 10.1, namely, the construction of a product measure for the product of two measure spaces.

Claim 1. *Let $m \geq 2$, and let $F = \{i_1, \cdots, i_m\}$ denote a nonempty finite subset of I. There exists a unique measure μ_F on $\mathcal{F}_F = \sigma(\mathcal{E}_F)$ such that*

$$\mu_F\left(\prod_{j=1}^{m} A_{i_j}\right) = \prod_{j=1}^{m} \mu_{i_j}(A_{i_j}) \text{ whenever } A_{i_1} \in \mathcal{F}_{i_1}, \cdots, A_{i_m} \in \mathcal{F}_{i_m}.$$

Proof: If $m = 2$ so that $F = \{i_1, i_2\}$, then the measure $\mu_{i_1} \otimes \mu_{i_2}$ with domain $\mathcal{F}_{i_1} \otimes \mathcal{F}_{i_2}$ constructed in Section 10.1 is the unique measure on $\mathcal{F}_{i_1} \otimes \mathcal{F}_{i_2}$ with the property that $\mu_{i_1} \otimes \mu_{i_2}(A_{i_1} \times A_{i_2}) = \mu_{i_1}(A_{i_1})\mu_{i_2}(A_{i_2})$ for every $A_{i_1} \times A_{i_2} \in \mathcal{F}_{i_1} \times \mathcal{F}_{i_2} = \mathcal{E}_{\{i_1,i_2\}}$. Now by definition we have

$$\mathcal{F}_{i_1} \otimes \mathcal{F}_{i_2} = \sigma(\mathcal{F}_{i_1} \times \mathcal{F}_{i_2}) = \sigma(\mathcal{E}_{\{i_1,i_2\}}) = \mathcal{F}_{\{i_1,i_2\}},$$

so the measure $\mu_{i_1} \otimes \mu_{i_2}$ uniquely serves as our measure μ_F as desired above. Therefore, if F is a two-element subset of I, the claim holds true.

The remainder of the proof is by induction on the number of elements m of F. Suppose that for every m-element subset of I the result holds, and let $F = \{i_1, \cdots, i_m, i_{m+1}\}$, a set of $m+1$ elements. By the induction hypothesis there exists a unique measure η on $\mathcal{F}_{\{i_1,\cdots,i_m\}}$ with the property that

$$\eta(A_{i_1} \times \cdots \times A_{i_m}) = \prod_{j=1}^{m} \mu_{i_j}(A_{i_j})$$

whenever $A_{i_1} \in \mathcal{F}_{i_1}, \cdots, A_{i_m} \in \mathcal{F}_{i_m}$. Next, define the product measure $\eta \otimes \mu_{i_{m+1}}$ on $\mathcal{F}_{\{i_1,\cdots,i_m\}} \otimes \mathcal{F}_{i_{m+1}}$ in accordance with Section 10.1: $\eta \otimes \mu_{i_{m+1}}$ is the unique measure on $\mathcal{F}_{\{i_1,\cdots,i_m\}} \otimes \mathcal{F}_{i_{m+1}}$ such that

$$\eta \otimes \mu_{i_{m+1}}(B \times A_{i_{m+1}}) = \eta(B)\mu_{i_{m+1}}(A_{i_{m+1}})$$

for every $B \in \mathcal{F}_{\{i_1,\cdots,i_m\}}$ and $A_{i_{m+1}} \in \mathcal{F}_{i_{m+1}}$. We now define the promised measure μ_F on $\mathcal{F}_F$, writing

$$\begin{aligned}\mu_F(A) &= \eta \otimes \mu_{i_{m+1}}(\text{unique } \mathcal{F}_{\{i_1,\cdots,i_m\}} \otimes \mathcal{F}_{i_{m+1}}\text{-set corresponding to } A) \\ &= \eta \otimes \mu_{i_{m+1}}(\Phi(A))\end{aligned}$$

for every $A \in \mathcal{F}_F = \mathcal{F}_{\{i_1,\cdots,i_m,i_{m+1}\}}$. [Here Φ is the mapping on $\Omega_{\{i_1,\cdots,i_{m+1}\}}$ such that

$$\Phi((\omega_{i_1}, \cdots, \omega_{i_m}, \omega_{i_{m+1}})) = ((\omega_{i_1}, \cdots, \omega_{i_m}), \omega_{i_{m+1}})$$

whenever $(\omega_{i_1}, \cdots, \omega_{i_{m+1}}) \in \Omega_{i_1} \times \cdots \times \Omega_{i_{m+1}}$, as described in Example 3 of the previous section. Recall that Φ establishes a one-to-one correspondence between the sets in $\mathcal{F}_{\{i_1,\cdots,i_{m+1}\}}$ and $\mathcal{F}_{\{i_1,\cdots,i_m\}} \otimes \mathcal{F}_{i_{m+1}}$.] It is clear that μ_F is nonnegative. Since $\varnothing$ is the unique $\mathcal{F}_{\{i_1,\cdots,i_m\}} \otimes \mathcal{F}_{i_{m+1}}$-set corresponding to $\varnothing$ [viewed as an element of $\mathcal{F}_{\{i_1,\cdots,i_{m+1}\}}$], that is, since $\Phi(\varnothing) = \varnothing$, we have

$$\mu_F(\varnothing) = \eta \otimes \mu_{i_{m+1}}(\varnothing) = 0.$$

Finally, let $\{A_n\}_{n=1}^{\infty}$ denote a disjoint sequence of $\mathcal{F}_{\{i_1,\cdots,i_{m+1}\}}$-sets. Since Φ is one-to-one and onto $(\Omega_{i_1} \times \cdots \times \Omega_{i_m}) \times \Omega_{i_{m+1}}$, it follows that $\{\Phi(A_n)\}_{n=1}^{\infty}$ is a disjoint sequence of $\mathcal{F}_{\{i_1,\cdots,i_m\}} \otimes \mathcal{F}_{i_{m+1}}$-sets and $\Phi(\bigcup_{n=1}^{\infty} A_n) = \bigcup_{n=1}^{\infty} \Phi(A_n)$, so that

$$\begin{aligned}
\mu_F\left(\bigcup_{n=1}^{\infty} A_n\right) = \eta \otimes \mu_{i_{m+1}}\left(\Phi\left(\bigcup_{n=1}^{\infty} A_n\right)\right) &= \eta \otimes \mu_{i_{m+1}}\left(\bigcup_{n=1}^{\infty} \Phi(A_n)\right) \\
&= \sum_{n=1}^{\infty} \eta \otimes \mu_{i_{m+1}}(\Phi(A_n)) \\
&= \sum_{n=1}^{\infty} \mu_F(A_n).
\end{aligned}$$

This shows that μ_F is countably additive and completes the proof that μ_F is a measure on $\mathcal{F}_F = \mathcal{F}_{\{i_1,\cdots,i_m,i_{m+1}\}}$.

We now demonstrate that μ_F has the desired product property. Pick sets $A_{i_1} \in \mathcal{F}_{i_1}, \cdots, A_{i_{m+1}} \in \mathcal{F}_{i_{m+1}}$. Then, since

$$\Phi(A_{i_1} \times \cdots \times A_{i_m} \times A_{i_{m+1}}) = (A_{i_1} \times \cdots \times A_{i_m}) \times A_{i_{m+1}},$$

we have that

$$\begin{aligned}
&\mu_F(A_{i_1} \times \cdots \times A_{i_m} \times A_{i_{m+1}}) \\
= \ & \eta \otimes \mu_{i_{m+1}}((A_{i_1} \times \cdots \times A_{i_m}) \times A_{i_{m+1}}) \\
= \ & \eta(A_{i_1} \times \cdots \times A_{i_m})\mu_{i_{m+1}}(A_{i_{m+1}}) \\
= \ & \mu_{i_1}(A_{i_1}) \cdots \mu_{i_m}(A_{i_m})\mu_{i_{m+1}}(A_{i_{m+1}}).
\end{aligned}$$

This shows that μ_F is a measure on $\mathcal{F}_F$ with the desired product property.

What remains is to establish that μ_F is the *unique* measure on $\mathcal{F}_F$ with the property that $\mu_F(A_{i_1} \times \cdots \times A_{i_{m+1}}) = \prod_{j=1}^{m+1} \mu_{i_j}(A_{i_j})$ whenever $A_{i_1} \in \mathcal{F}_{i_1}$, $\cdots$, $A_{i_{m+1}} \in \mathcal{F}_{i_{m+1}}$. To see this, let ξ denote a measure on $\mathcal{F}_F$ such that $\xi(A_{i_1} \times \cdots \times A_{i_{m+1}}) = \prod_{j=1}^{m+1} \mu_{i_j}(A_{i_j})$ if $A_{i_1} \in \mathcal{F}_{i_1}, \cdots, A_{i_{m+1}} \in \mathcal{F}_{i_{m+1}}$. It follows that $\xi = \mu_F$ on the collection $\mathcal{E}_F$. Observe [or recall from Exercise 4 of the previous section] that $\mathcal{E}_F$ is a semiring on Ω_F and $\sigma(\mathcal{E}_F) = \mathcal{F}_F$. Since $\xi(\Omega_F) = \mu_F(\Omega_F) = 1 < +\infty$, the Uniqueness Theorem in Section 3.5 forces $\xi = \mu_F$ on all of $\mathcal{F}_F$. This establishes the uniqueness claim for μ_F and completes the entire proof. ∎

Definition. In accordance with the previous claim, given any nonempty finite subset $F = \{i_1, \cdots, i_m\}$ of I, we define μ_F to be the unique measure on $\mathcal{F}_F$ such that

$$\mu_F\left(\prod_{j=1}^{m} A_{i_j}\right) = \prod_{j=1}^{m} \mu_{i_j}(A_{i_j}) \text{ whenever } A_{i_1} \in \mathcal{F}_{i_1}, \cdots, A_{i_m} \in \mathcal{F}_{i_m}.$$

The next claim uses Claim 2 and Example 4, both from the previous section. Let $F_1 = \{i_1, \cdots, i_n\}$ and $F_2 = \{j_1, \cdots, j_m\}$ denote disjoint finite nonempty subsets of I. We consider the one-to-one and onto mapping Φ from $\Omega_{F_1 \cup F_2}$ to $\Omega_{F_1} \times \Omega_{F_2}$ given for each $(\omega_{i_1}, \cdots, \omega_{i_n}, \omega_{j_1}, \cdots, \omega_{j_m}) \in \Omega_{F_1 \cup F_2}$ by the rule

$$\Phi((\omega_{i_1}, \cdots, \omega_{i_n}, \omega_{j_1}, \cdots, \omega_{j_m})) = ((\omega_{i_1}, \cdots, \omega_{i_n}), (\omega_{j_1}, \cdots, \omega_{j_m})).$$

Φ establishes a one-to-one correspondence between the sets in $\mathcal{F}_{F_1 \cup F_2}$ and $\mathcal{F}_{F_1} \otimes \mathcal{F}_{F_2}$.

Claim 2. *Let $F_1 = \{i_1, \cdots, i_n\}$ and $F_2 = \{j_1, \cdots, j_m\}$ denote disjoint finite nonempty subsets of I. Consider the measures $\mu_{F_1 \cup F_2}$ and $\mu_{F_1} \otimes \mu_{F_2}$ defined on $\mathcal{F}_{F_1 \cup F_2}$ and $\mathcal{F}_{F_1} \otimes \mathcal{F}_{F_2}$, respectively. Let $B_{F_1} \in \mathcal{F}_{F_1}$ and $B_{F_2} \in \mathcal{F}_{F_2}$. Then*

$$\mu_{F_1 \cup F_2}\left(\Phi^{-1}(B_{F_1} \times B_{F_2})\right) = \mu_{F_1}(B_{F_1})\mu_{F_2}(B_{F_2}) = \mu_{F_1} \otimes \mu_{F_2}(B_{F_1} \times B_{F_2}).$$

Proof: Let $B_{F_1} \in \mathcal{E}_{F_1}$: $B_{F_1} = A_{i_1} \times \cdots \times A_{i_n}$ where $A_{i_1} \in \mathcal{F}_{i_1}, \cdots, A_{i_n} \in \mathcal{F}_{i_n}$. Let $B_{F_2} \in \mathcal{E}_{F_2}$: $B_{F_2} = A_{j_1} \times \cdots \times A_{j_m}$, where $A_{j_1} \in \mathcal{F}_{j_1}, \cdots, A_{j_m} \in \mathcal{F}_{j_m}$. We have that

$$\Phi(A_{i_1} \times \cdots \times A_{i_n} \times A_{j_1} \times \cdots \times A_{j_m}) = (A_{i_1} \times \cdots \times A_{i_n}) \times (A_{j_1} \times \cdots \times A_{j_m}),$$

and hence

$$\begin{aligned}
& \mu_{F_1 \cup F_2}\left(\Phi^{-1}(B_{F_1} \times B_{F_2})\right) \\
= \; & \mu_{F_1 \cup F_2}\left(\Phi^{-1}((A_{i_1} \times \cdots \times A_{i_n}) \times (A_{j_1} \times \cdots \times A_{j_m}))\right) \\
= \; & \mu_{F_1 \cup F_2}(A_{i_1} \times \cdots \times A_{i_n} \times A_{j_1} \times \cdots \times A_{j_m}) \\
= \; & \mu_{i_1}(A_{i_1}) \cdots \mu_{i_n}(A_{i_n})\mu_{j_1}(A_{j_1}) \cdots \mu_{j_m}(A_{j_m}) \\
= \; & \mu_{F_1}(A_{i_1} \times \cdots \times A_{i_n})\mu_{F_2}(A_{j_1} \times \cdots \times A_{j_m}) \\
= \; & \mu_{F_1}(B_{F_1})\mu_{F_2}(B_{F_2}) \\
= \; & \mu_{F_1} \otimes \mu_{F_2}(B_{F_1} \times B_{F_2}).
\end{aligned}$$

We have used the product property for $\mu_{F_1 \cup F_2}$ to obtain the third equality, and the fourth equality follows by using the product property for μ_{F_1} and μ_{F_2}. This shows that the finite measures $\mu_{F_1 \cup F_2} \circ \Phi^{-1}$ and $\mu_{F_1} \otimes \mu_{F_2}$ [both with domain $\mathcal{F}_{F_1} \otimes \mathcal{F}_{F_2}$] agree on $\mathcal{E}_{F_1} \times \mathcal{E}_{F_2}$. By Exercise 5 of the previous section, $\mathcal{E}_{F_1} \times \mathcal{E}_{F_2}$ is a semiring on $\Omega_{F_1} \times \Omega_{F_2}$, and

$$\sigma(\mathcal{E}_{F_1} \times \mathcal{E}_{F_2}) = \sigma(\sigma(\mathcal{E}_{F_1}) \times \sigma(\mathcal{E}_{F_2})) = \sigma(\mathcal{F}_{F_1} \times \mathcal{F}_{F_2}) = \mathcal{F}_{F_1} \otimes \mathcal{F}_{F_2}$$

by Exercise 1(b) of the previous section. The Uniqueness Theorem in Section 3.5 asserts that $\mu_{F_1 \cup F_2} \circ \Phi^{-1}$ and $\mu_{F_1} \otimes \mu_{F_2}$ are the same measure, that is, they coincide on $\mathcal{F}_{F_1} \otimes \mathcal{F}_{F_2}$. In particular, they coincide on the subcollection $\mathcal{F}_{F_1} \times \mathcal{F}_{F_2}$ of $\mathcal{F}_{F_1} \otimes \mathcal{F}_{F_2}$. This is exactly the assertion of the claim, and the proof is therefore complete. ∎

The next claim characterizes the field $\mathcal{G}$ on Ω and is a little technical lemma for the claim that follows. Recall that $\mathcal{G}$ is the field on Ω generated by the sets of the form $\prod_{i\in I} A_i$, where each $A_i \in \mathcal{F}_i$ and all but finitely many of the A_i's are equal to Ω_i. Recall also that given any finite subset $F \subseteq I$, $\mathcal{G}_F$ is the field on Ω_F generated by the sets of the form $\prod_{i\in F} A_i$, where $A_i \in \mathcal{F}_i$ for each $i \in F$, that is, $\mathcal{G}_F$ is the field on Ω_F generated by $\mathcal{E}_F$.

For the following claim, we make a definition. Let F denote a finite subset of I. We define $\Phi_F : \Omega \to \Omega_F \times \Omega_{F^c}$ by writing $\Phi(\omega) = (\omega_F, \omega_{F^c})$ for every $\omega \in \Omega$. Here, $\omega_F = (\omega_i)_{i\in F}$ and $\omega_{F^c} = (\omega_i)_{i\in F^c}$. By Example 5 of the previous section, Φ is one-to-one and onto, establishing a one-to-one correspondence between the elements of $\mathcal{F}$ and $\mathcal{F}_F \otimes \mathcal{F}_{F^c}$.

Claim 3. *For every $A \in \mathcal{G}$ there exists a finite subset $F \subseteq I$ and a set $A_F \in \mathcal{G}_F$ such that $A = \Phi_F^{-1}(A_F \times \Omega_{F^c})$.*

Proof: Let

$$\mathcal{C} = \{A \in \mathcal{G} : \text{ there is finite } F \subseteq I \text{ and } A_F \in \mathcal{G}_F \text{ with } A = \Phi_F^{-1}(A_F \times \Omega_{F^c})\}.$$

Then $\mathcal{C} \subseteq \mathcal{G}$. We desire the reverse inclusion $\mathcal{G} \subseteq \mathcal{C}$. Suppose it is shown that (i) $\mathcal{C}$ contains all sets of the form $\prod_{i\in I} A_i$, where $A_i \in \mathcal{F}_i$ for each $i \in I$ and at most finitely many A_i's differ from Ω_i [recall that these sets form the generating class for $\mathcal{G}$] and (ii) $\mathcal{C}$ is a field on $\mathcal{G}$. Then $\mathcal{C}$ will denote a field on Ω that contains the sets generating the field $\mathcal{G}$. Since $\mathcal{G}$ is the minimal field on Ω containing such sets, we will have $\mathcal{G} \subseteq \mathcal{C}$, giving $\mathcal{G} = \mathcal{C}$, completing the proof. Therefore, we need only show (i) and (ii).

We now show (i). Let $B = \prod_{i\in I} A_i$, where $A_i \in \mathcal{F}_i$ for each $i \in I$ and the set $\{j \in I : A_j \neq \Omega_j\}$ is finite. Let $F = \{j \in I : A_j \neq \Omega_j\}$, and let $A_F = \prod_{i\in F} A_i$. Then F is a finite subset of I and $A_F \in \mathcal{E}_F \subseteq \mathcal{G}_F$; we also have $B = \Phi_F^{-1}(A_F \times \Omega_{F^c})$. This shows that $B \in \mathcal{C}$, completing the proof of (i).

We now show (ii). Observe for any finite subset $F \subseteq I$ that $\Omega_F \in \mathcal{G}_F$ and $\Omega = \Phi_F^{-1}(\Omega_F \times \Omega_{F^c})$, so $\Omega \in \mathcal{C}$. Next, let $A \in \mathcal{C}$, so that A has the form $\Phi_F^{-1}(A_F \times \Omega_{F^c})$ for some finite $F \subseteq I$ and $A_F \in \mathcal{G}_F$. Since

$$A^c = \left(\Phi_F^{-1}(A_F \times \Omega_{F^c})\right)^c = \Phi_F^{-1}\left((A_F \times \Omega_{F^c})^c\right) = \Phi_F^{-1}(A_F^c \times \Omega_{F^c}),$$

and since $A_F^c \in \mathcal{G}_F$, it follows that $A^c \in \mathcal{C}$, so that $\mathcal{C}$ is closed under complementation. We complete (ii) and hence the entire proof by showing that $\mathcal{C}$ is closed under intersection. Let $A, B \in \mathcal{C}$; write $A = \Phi_{F_1}^{-1}(A_{F_1} \times \Omega_{F_1^c})$ for some finite $F_1 \subseteq I$ and $A_{F_1} \in \mathcal{G}_{F_1}$, and write $B = \Phi_{F_2}^{-1}(B_{F_2} \times \Omega_{F_2^c})$ for some finite $F_2 \subseteq I$ and $B_{F_2} \in \mathcal{G}_{F_2}$. Before proceeding further, we must make the following technical subclaim.

Subclaim. *For every $A_{F_1} \in \mathcal{G}_{F_1}$, there is $C_{F_1\cup F_2} \in \mathcal{G}_{F_1\cup F_2}$ such that*

$$\Phi_{F_1}^{-1}(A_{F_1} \times \Omega_{F_1}^c) = \Phi_{F_1\cup F_2}^{-1}(C_{F_1\cup F_2} \times \Omega_{(F_1\cup F_2)^c}). \tag{11.1}$$

Proof of Subclaim: Define the collection

$$\mathcal{D} = \{A_{F_1} \in \mathcal{G}_{F_1} : \text{ there is } C_{F_1 \cup F_2} \in \mathcal{G}_{F_1 \cup F_2} \text{ such that (11.1) holds}\}.$$

We wish to show that $\mathcal{D} = \mathcal{G}_{F_1}$. Since we obviously have $\mathcal{D} \subseteq \mathcal{G}_{F_1}$, we must demonstrate the reverse inclusion $\mathcal{G}_{F_1} \subseteq \mathcal{D}$. We will demonstrate this reverse inclusion by demonstrating that $\mathcal{E}_{F_1} \subseteq \mathcal{D}$ and $\mathcal{D}$ is a field on Ω_{F_1}, for then we will have $\mathcal{G}_{F_1} = f(\mathcal{E}_{F_1}) \subseteq \mathcal{D}$ by the fact that $f(\mathcal{E}_{F_1})$ is the minimal field on Ω_{F_1} that contains $\mathcal{E}_{F_1}$.

We first show that $\mathcal{E}_{F_1} \subseteq \mathcal{D}$. Pick $\prod_{i \in F_1} A_i \in \mathcal{E}_{F_1}$. Then

$$\Phi_{F_1}^{-1}\left(\prod_{i \in F_1} A_i \times \Omega_{F_1^c}\right) = \prod_{i \in I} Q_i,$$

where $Q_i = A_i$ for each $i \in F_1$ and $Q_i = \Omega_i$ for each $i \in F_1^c$. Define the set $C_{F_1 \cup F_2} = \prod_{i \in F_1 \cup F_2} R_i$, where $R_i = A_i$ for each $i \in F_1$ and $R_i = \Omega_i$ for each $i \in F_2$. We have $C_{F_1 \cup F_2} \in \mathcal{E}_{F_1 \cup F_2} \subseteq \mathcal{G}_{F_1 \cup F_2}$, and on a moment's thought we have

$$\Phi_{F_1 \cup F_2}^{-1}\left(C_{F_1 \cup F_2} \times \Omega_{(F_1 \cup F_2)^c}\right) = \prod_{i \in I} Q_i.$$

Comparing the last two displayed equations shows that $\prod_{i \in F_1} A_i \in \mathcal{D}$. That is, we have $\mathcal{E}_{F_1} \subseteq \mathcal{D}$.

We now turn to showing that $\mathcal{D}$ is a field on $\mathcal{G}_{F_1}$: $\Omega_{F_1} \in \mathcal{D}$, $\mathcal{D}$ is closed under complementation, and $\mathcal{D}$ is closed under finite intersections. We first show that $\Omega_{F_1} \in \mathcal{D}$. This is easy if it is observed that

$$\Phi_{F_1}^{-1}\left(\Omega_{F_1} \times \Omega_{F_1}^c\right) = \Omega = \Phi_{F_1 \cup F_2}^{-1}\left(\Omega_{F_1 \cup F_2} \times \Omega_{(F_1 \cup F_2)^c}\right)$$

and $\Omega_{F_1 \cup F_2} \in \mathcal{G}_{F_1 \cup F_2}$. We now discuss closure under complementation. Suppose that $A_{F_1} \in \mathcal{D}$, and let $C_{F_1 \cup F_2} \in \mathcal{G}_{F_1 \cup F_2}$ be such that (11.1) holds. Using straightforward set-theoretic manipulations, we have

$$\begin{aligned}
&\Phi_{F_1}^{-1}(A_{F_1}^c \times \Omega_{F_1^c}) \\
= \;& \Phi_{F_1}^{-1}\left([\Omega_{F_1} \times \Omega_{F_1^c}] - [A_{F_1} \times \Omega_{F_1}^c]\right) \\
= \;& \Phi_{F_1}^{-1}(\Omega_{F_1} \times \Omega_{F_1^c}) - \Phi_{F_1}^{-1}(A_{F_1} \times \Omega_{F_1^c}) \\
= \;& \Phi_{F_1 \cup F_2}^{-1}(\Omega_{F_1 \cup F_2} \times \Omega_{(F_1 \cup F_2)^c}) - \Phi_{F_1 \cup F_2}^{-1}(C_{F_1 \cup F_2} \times \Omega_{(F_1 \cup F_2)^c}) \\
= \;& \Phi_{F_1 \cup F_2}^{-1}\left([\Omega_{F_1 \cup F_2} \times \Omega_{(F_1 \cup F_2)^c}] - [C_{F_1 \cup F_2} \times \Omega_{(F_1 \cup F_2)^c}]\right) \\
= \;& \Phi_{F_1 \cup F_2}^{-1}\left(C_{F_1 \cup F_2}^c \times \Omega_{(F_1 \cup F_2)^c}\right).
\end{aligned}$$

Since $C_{F_1 \cup F_2}^c \in \mathcal{G}_{F_1 \cup F_2}$, it follows that $A_{F_1}^c \in \mathcal{D}$, so that $\mathcal{D}$ is indeed closed under complementation. We now show that $\mathcal{D}$ is closed under finite intersections. Let $A_{F_1}^{(1)}, A_{F_1}^{(2)} \in \mathcal{D}$, and let $C_{F_1 \cup F_2}^{(1)}, C_{F_1 \cup F_2}^{(2)} \in \mathcal{G}_{F_1 \cup F_2}$ denote the corresponding sets for $A_{F_1}^{(1)}$ and $A_{F_1}^{(2)}$ that satisfy (11.1):

$$\Phi_{F_1}^{-1}(A_{F_1}^{(n)} \times \Omega_{F_1^c}) = \Phi_{F_1 \cup F_2}^{-1}\left(C_{F_1 \cup F_2}^{(n)} \times \Omega_{(F_1 \cup F_2)^c}\right) \quad \text{for each } n = 1, 2.$$

We wish to show that $A_{F_1}^{(1)} \cap A_{F_1}^{(2)} \in \mathcal{D}$. Again using standard set-theoretic manipulations, we have

$$\begin{aligned}
\Phi_{F_1}^{-1}\left(\bigcap_{n=1}^{2} A_{F_1}^{(n)} \times \Omega_{F_1^c}\right) &= \Phi_{F_1}^{-1}\left(\bigcap_{n=1}^{2}\left[A_{F_1}^{(n)} \times \Omega_{F_1^c}\right]\right) \\
&= \bigcap_{n=1}^{2} \Phi_{F_1}^{-1}\left(A_{F_1}^{(n)} \times \Omega_{F_1^c}\right) \\
&= \bigcap_{n=1}^{2} \Phi_{F_1 \cup F_2}^{-1}\left(C_{F_1 \cup F_2}^{(n)} \times \Omega_{(F_1 \cup F_2)^c}\right) \\
&= \Phi_{F_1 \cup F_2}^{-1}\left(\bigcap_{n=1}^{2}\left[C_{F_1 \cup F_2}^{(n)} \times \Omega_{(F_1 \cup F_2)^c}\right]\right) \\
&= \Phi_{F_1 \cup F_2}^{-1}\left(\bigcap_{n=1}^{2} C_{F_1 \cup F_2}^{(n)} \times \Omega_{(F_1 \cup F_2)^c}\right).
\end{aligned}$$

Since $C_{F_1 \cup F_2}^{(1)} \cap C_{F_1 \cup F_2}^{(2)} \in \mathcal{F}_{F_1 \cup F_2}$, it follows that $A_{F_1}^{(1)} \cap A_{F_1}^{(2)} \in \mathcal{D}$. This shows that $\mathcal{D}$ is closed under finite intersections, and hence $\mathcal{D}$ is a field on Ω_{F_1}. This completes the proof of the subclaim.

Using the subclaim and the fact that $A = \Phi_{F_1}^{-1}(A_{F_1} \times \Omega_{F_1^c})$ for some finite $F_1 \subseteq I$ and $A_{F_1} \in \mathcal{G}_{F_1}$, it follows that there is $C_{F_1 \cup F_2} \in \mathcal{G}_{F_1 \cup F_2}$ such that

$$\Phi_{F_1}^{-1}(A_{F_1} \times \Omega_{F_1^c}) = \Phi_{F_1 \cup F_2}^{-1}(C_{F_1 \cup F_2} \times \Omega_{F_1 \cup F_2}^c),$$

that is,

$$\Phi_{F_1 \cup F_2}\left(\Phi_{F_1}^{-1}(A_{F_1} \times \Omega_{F_1^c})\right) = C_{F_1 \cup F_2} \times \Omega_{(F_1 \cup F_2)^c}.$$

It follows that

$$A = \Phi_{F_1 \cup F_2}^{-1}\left(\Phi_{F_1 \cup F_2}\left(\Phi_{F_1}^{-1}\left(A_{F_1} \times \Omega_{F_1^c}\right)\right)\right) = \Phi_{F_1 \cup F_2}^{-1}\left(C_{F_1 \cup F_2} \times \Omega_{(F_1 \cup F_2)^c}\right)$$

for the aforementioned $C_{F_1 \cup F_2} \in \mathcal{G}_{F_1 \cup F_2}$. By the same reasoning as given in the subclaim and in the immediately preceding discussion, we next have that

$$B = \Phi_{F_1 \cup F_2}^{-1}\left(\Phi_{F_1 \cup F_2}\left(\Phi_{F_2}^{-1}\left(B_{F_2} \times \Omega_{F_2^c}\right)\right)\right) = \Phi_{F_1 \cup F_2}^{-1}\left(D_{F_1 \cup F_2} \times \Omega_{(F_1 \cup F_2)^c}\right)$$

for an appropriate set $D_{F_1 \cup F_2} \in \mathcal{G}_{F_1 \cup F_2}$. It follows that

$$\begin{aligned}
& A \cap B \\
=\ & \Phi_{F_1 \cup F_2}^{-1}\left(C_{F_1 \cup F_2} \times \Omega_{(F_1 \cup F_2)^c}\right) \cap \Phi_{F_1 \cup F_2}^{-1}\left(D_{F_1 \cup F_2} \times \Omega_{(F_1 \cup F_2)^c}\right) \\
=\ & \Phi_{F_1 \cup F_2}^{-1}\left(\left[C_{F_1 \cup F_2} \cap D_{F_1 \cup F_2}\right] \times \Omega_{(F_1 \cup F_2)^c}\right).
\end{aligned}$$

Since $F_1 \cup F_2$ is a finite subset of I and $C_{F_1 \cup F_2} \cap D_{F_1 \cup F_2} \in \mathcal{G}_{F_1 \cup F_2}$, it follows that $A \cap B \in \mathcal{C}$. The collection $\mathcal{C}$ is closed under finite intersections, and hence is a field. This completes the entire proof. ∎

For the next claim, a bit of terminology is needed. When we say that μ *is a finitely additive measure on the field* $\mathcal{Q}$, we mean that μ is a nonnegative function on $\mathcal{Q}$ with $\mu(\varnothing) = 0$ and $\mu(A \cup B) = \mu(A) + \mu(B)$ whenever $A, B \in \mathcal{Q}$ are disjoint. The reader should also recall the definition of Φ_F [for finite $F \subseteq I$] given immediately before Claim 3.

Claim 4. *Define the collection*

$$\mathcal{H} = \{B \subseteq \Omega : B = \Phi_F^{-1}(A_F \times \Omega_{F^c}) \text{ for some finite } F \subseteq I \text{ and } A_F \in \mathcal{F}_F\}.$$

Then we have the following:

(a) $\mathcal{H}$ *is a field on* Ω *with* $\mathcal{G} \subseteq \mathcal{H} \subseteq \mathcal{F}$.

(b) $\sigma(\mathcal{H}) = \mathcal{F}$.

(c) There exists a unique finitely additive measure μ *on the field* $\mathcal{H}$ *such that*

$$\mu(\Phi_F^{-1}(A_F \times \Omega_{F^c})) = \mu_F(A_F) \quad \text{for every finite } F \subseteq I \text{ and } A_F \in \mathcal{F}_F.$$

Proof: We first handle (a). If $B \in \mathcal{G}$, then $B = \Phi_F^{-1}(A_F \times \Omega_{F^c})$ for some finite $F \subseteq I$ and $A_F \in \mathcal{G}_F$ by Claim 3. Since $\mathcal{G}_F \subseteq \mathcal{F}_F$, it is immediate that $B \in \mathcal{H}$, giving $\mathcal{G} \subseteq \mathcal{H}$.

We now show that $\mathcal{H} \subseteq \mathcal{F}$. Given a set $B = \Phi_F^{-1}(A_F \times \Omega_{F^c})$ in $\mathcal{H}$ where $F \subseteq I$ is finite and $A_F \in \mathcal{F}_F$, then, because Φ_F establishes a one-to-one correspondence between the sets in $\mathcal{F}$ and the sets in $\mathcal{F}_F \otimes \mathcal{F}_{F^c}$, and because $A_F \times \Omega_{F^c} \in \mathcal{F}_F \otimes \mathcal{F}_{F^c}$, it must be the case that B, being the inverse image of the $\mathcal{F}_F \otimes \mathcal{F}_{F^c}$-set $A_F \times \Omega_{F^c}$, is in $\mathcal{F}$. This shows that $\mathcal{H} \subseteq \mathcal{F}$.

To see that $\mathcal{H}$ is a field on Ω, we show that $\Omega \in \mathcal{H}$, $\mathcal{H}$ is closed under complementation, and $\mathcal{H}$ is closed under finite intersections. To see that $\Omega \in \mathcal{H}$, observe that $\Omega_F \in \mathcal{F}_F$ and $\Omega = \Phi_F^{-1}(\Omega_F \times \Omega_{F^c})$ for every finite $F \subseteq I$. Next, let $B \in \mathcal{H}$, so that B has the form $\Phi_F^{-1}(A_F \times \Omega_{F^c})$ for some finite $F \subseteq I$ and $A_F \in \mathcal{F}_F$. Then

$$B^c = (\Phi_F^{-1}(A_F \times \Omega_{F^c}))^c = \Phi_F^{-1}((A_F \times \Omega_{F^c})^c) = \Phi_F^{-1}(A_F^c \times \Omega_{F^c}).$$

Since $A_F^c \in \mathcal{F}_F$, it follows that $B^c \in \mathcal{H}$, and thus $\mathcal{H}$ is closed under complementation. To complete the proof that $\mathcal{H}$ is a field, we must demonstrate closure under finite intersections. This requires the following subclaim, which is quite similar [if not nearly identical] to the subclaim in the proof of Claim 3.

Subclaim 1. Let $A_{F_1} \in \mathcal{F}_{F_1}$. There exists $C_{F_1 \cup F_2} \in \mathcal{F}_{F_1 \cup F_2}$ with the property that

$$\Phi_{F_1}^{-1}(A_{F_1} \times \Omega_{F_1{}^c}) = \Phi_{F_1 \cup F_2}^{-1}(C_{F_1 \cup F_2} \times \Omega_{(F_1 \cup F_2)^c}). \tag{11.2}$$

Proof of Subclaim 1: Define the collection

$$\mathcal{C} = \{A_{F_1} \in \mathcal{F}_{F_1} : \text{ there is } C_{F_1 \cup F_2} \in \mathcal{F}_{F_1 \cup F_2} \text{ such that (11.2) holds}\}.$$

Then $\mathcal{C} \subseteq \mathcal{F}_{F_1}$. The proof will be complete if we demonstrate the reverse inclusion $\mathcal{F}_{F_1} \subseteq \mathcal{C}$. Just as in the proof of the subclaim found in Claim 3, the reader may show that (i) $\mathcal{E}_{F_1} \subseteq \mathcal{C}$ and (ii) $\mathcal{C}$ is a σ-field on Ω_{F_1}. [See Exercise 1.] Upon showing (i) and (ii), it follows that $\mathcal{C}$ is a σ-field on Ω_{F_1} that contains $\mathcal{E}_{F_1}$. Since $\mathcal{F}_{F_1}$ by definition is the minimal σ-field on Ω_{F_1} that contains $\mathcal{C}$, it follows that the reverse inclusion $\mathcal{F}_{F_1} \subseteq \mathcal{C}$ holds, completing the proof of the subclaim.

We now show that $\mathcal{H}$ is closed under finite intersections. Let $B_1, B_2 \in \mathcal{H}$ have the forms $\Phi_{F_1}^{-1}(A_{F_1} \times \Omega_{F_1^c})$ and $\Phi_{F_2}^{-1}(A_{F_2} \times \Omega_{F_2^c})$, respectively, where F_1 and F_2 are finite subsets of I, $A_{F_1} \in \mathcal{F}_{F_1}$, and $A_{F_2} \in \mathcal{F}_{F_2}$. Using the subclaim, we may find sets $C_{F_1 \cup F_2}, D_{F_1 \cup F_2} \in \mathcal{F}_{F_1 \cup F_2}$ such that

$$B_1 = \Phi_{F_1 \cup F_2}^{-1}(C_{F_1 \cup F_2} \times \Omega_{(F_1 \cup F_2)^c}) \quad \text{and} \quad B_2 = \Phi_{F_1 \cup F_2}^{-1}(D_{F_1 \cup F_2} \times \Omega_{(F_1 \cup F_2)^c}),$$

so

$$B_1 \cap B_2 = \Phi_{F_1 \cup F_2}^{-1}([C_{F_1 \cup F_2} \cap D_{F_1 \cup F_2}] \times \Omega_{(F_1 \cup F_2)^c}).$$

Since $C_{F_1 \cup F_2} \cap D_{F_1 \cup F_2} \in \mathcal{F}_{F_1 \cup F_2}$ and $F_1 \cup F_2$ is a finite subset of I, it follows that $B_1 \cap B_2 \in \mathcal{H}$. This shows that $\mathcal{H}$ is closed under finite intersections and completes the proof that $\mathcal{H}$ is a field on Ω. This completes the proof of (a).

For (b), use (a): from $\mathcal{G} \subseteq \mathcal{H} \subseteq \mathcal{F}$ we have $\sigma(\mathcal{G}) \subseteq \sigma(\mathcal{H}) \subseteq \sigma(\mathcal{F})$, hence $\mathcal{F} \subseteq \sigma(\mathcal{H}) \subseteq \mathcal{F}$, giving $\sigma(\mathcal{H}) = \mathcal{F}$.

We now turn to (c). We define μ on $\mathcal{H}$ by the rule

$$\mu(A) = \mu_F(A_F) \quad \text{if } A = \Phi_F^{-1}(A_F \times \Omega_{F^c}) \text{ for some finite } F \subseteq I \text{ and } A_F \in \mathcal{F}_F.$$

Observe that μ clearly has domain $\mathcal{H}$. We must now show that μ is well-defined. That is, a set $A \in \mathcal{H}$ may have various representations

$$\Phi_{F_1}^{-1}(A_{F_1} \times \Omega_{F_1^c}) \quad \text{and} \quad \Phi_{F_2}^{-1}(A_{F_2} \times \Omega_{F_2^c}),$$

where F_1 and F_2 are finite subsets of I with $A_{F_1} \in \mathcal{F}_{F_1}$ and $A_{F_2} \in \mathcal{F}_{F_2}$. In order to show that μ is well-defined, we must show that $\mu(A)$ is invariant with respect to the representation of A. That is, we want $\mu_{F_1}(A_1) = \mu_{F_2}(A_2)$. Write $F_3 = F_1 \cup F_2$, so that F_3 is a finite subset of I. Before proceeding further, we must make some definitions.

We will first define the mapping $\Phi_{F_1,F_3} : \Omega_{F_3} \to \Omega_{F_1} \times \Omega_{F_3 - F_1}$ by writing $\Phi_{F_1,F_3}(\omega_{F_3}) = (\omega_{F_1}, \omega_{F_3 - F_1})$ for every $\omega_{F_3} \in \Omega_{F_3}$. By the Identification Lemma in the previous section, this mapping is one-to-one and onto, establishing a one-to-one correspondence between the sets in $\mathcal{F}_{F_3}$ and $\mathcal{F}_{F_1} \otimes \mathcal{F}_{F_3 - F_1}$ [and vice versa].

We next define the mapping $\Phi_{F_2,F_3} : \Omega_{F_3} \to \Omega_{F_2} \times \Omega_{F_3 - F_2}$ by writing $\Phi_{F_2,F_3}(\omega_{F_3}) = (\omega_{F_2}, \omega_{F_3 - F_2})$ for every $\omega_{F_3} \in \Omega_{F_3}$. As in the previous paragraph, this mapping is one-to-one and onto, establishing a one-to-one correspondence between the sets in $\mathcal{F}_{F_3}$ and those in $\mathcal{F}_{F_2} \otimes \mathcal{F}_{F_3 - F_2}$.

Using this notation, we have the following subclaim.

Subclaim 2. *Let $A_{F_1} \in \mathcal{F}_{F_1}$. Then*

$$\Phi_{F_1}^{-1}(A_{F_1} \times \Omega_{F_1^c}) = \Phi_{F_3}^{-1}\left(\Phi_{F_1,F_3}^{-1}(A_{F_1} \times \Omega_{F_3 - F_1}) \times \Omega_{F_3^c}\right). \tag{11.3}$$

Proof of Subclaim 2: Let $\mathcal{C} = \{A_{F_1} \in \mathcal{F}_{F_1} : \text{(11.3) holds for } A_{F_1}\}$. It is trivial that $\mathcal{C} \subseteq \mathcal{F}_{F_1}$. As in the proof of Subclaim 1, it is straightforward to show that (i) $\mathcal{E}_{F_1} \subseteq \mathcal{C}$ and (ii) $\mathcal{C}$ is a σ-field on Ω_{F_1}. [See Exercise 1.] From this we obtain $\mathcal{F}_{F_1} \subseteq \mathcal{C}$, which gives $\mathcal{C} = \mathcal{F}_{F_1}$, completing the proof.

We now return to the matter at hand, which means showing that $\mu(A)$ is well defined for the given $A \in \mathcal{H}$, that is, we return to the demonstration of the fact that $\mu_{F_1}(A_{F_1}) = \mu_{F_2}(A_{F_2})$. Using Subclaim 2, we have that

$$\Phi_{F_1}^{-1}(A_{F_1} \times \Omega_{F_1^c}) = \Phi_{F_3}^{-1}\left(\Phi_{F_1,F_3}^{-1}(A_{F_1} \times \Omega_{F_3-F_1}) \times \Omega_{F_3^c}\right),$$

and reasoning similar to that given in Subclaim 2 gives

$$\Phi_{F_2}^{-1}(A_{F_2} \times \Omega_{F_2^c}) = \Phi_{F_3}^{-1}\left(\Phi_{F_2,F_3}^{-1}(A_{F_2} \times \Omega_{F_3-F_2}) \times \Omega_{F_3^c}\right).$$

Since $\Phi_{F_1}^{-1}(A_{F_1} \times \Omega_{F_1^c})$ and $\Phi_{F_2}^{-1}(A_{F_2} \times \Omega_{F_2^c})$ are representations of the same set $A \in \mathcal{H}$, we must have that

$$\Phi_{F_3}^{-1}\left(\Phi_{F_1,F_3}^{-1}(A_{F_1} \times \Omega_{F_3-F_1}) \times \Omega_{F_3^c}\right) = \Phi_{F_3}^{-1}\left(\Phi_{F_2,F_3}^{-1}(A_{F_2} \times \Omega_{F_3-F_2}) \times \Omega_{F_3^c}\right).$$

Because Φ_{F_3} establishes a one-to-one correspondence between the sets in $\mathcal{F}$ and those in $\mathcal{F}_{F_3} \otimes \mathcal{F}_{F_3^c}$, the preceding work implies that

$$\Phi_{F_1,F_3}^{-1}(A_{F_1} \times \Omega_{F_3-F_1}) = \Phi_{F_2,F_3}^{-1}(A_{F_2} \times \Omega_{F_3-F_2}).$$

It follows that

$$\mu_{F_3}(\Phi_{F_1,F_3}^{-1}(A_{F_1} \times \Omega_{F_3-F_1})) = \mu_{F_3}(\Phi_{F_2,F_3}^{-1}(A_{F_2} \times \Omega_{F_3-F_2})).$$

By Claim 2, it in turn follows that

$$\mu_{F_1}(A_{F_1})\mu_{F_3-F_1}(\Omega_{F_3-F_1}) = \mu_{F_2}(A_{F_2})\mu_{F_3-F_2}(\Omega_{F_3-F_2}).$$

Since $\mu_{F_3-F_1}(\Omega_{F_3-F_1}) = 1 = \mu_{F_3-F_2}(\Omega_{F_3-F_2})$, the preceding forces

$$\mu_{F_1}(A_{F_1}) = \mu_{F_2}(A_{F_2}),$$

which is the desired equation. We have shown that μ as defined on the field $\mathcal{H}$ is in fact well defined.

We now show that μ is a finitely additive measure on the field $\mathcal{H}$: μ is nonnegative, $\mu(\varnothing) = 0$, and μ is finitely additive. The nonnegativity is immediate. Next, observe that $\varnothing$ [viewed as an $\mathcal{H}$-set] may be represented as $\Phi_F^{-1}(\varnothing \times \Omega_{F^c})$ for any finite subset $F \subseteq I$ [where $\varnothing$ is viewed as an element of $\mathcal{F}_F$], and hence $\mu(\varnothing) = \mu_F(\varnothing) = 0$. All that remains is finite additivity. To this end, let $A_1, A_2 \in \mathcal{H}$ be disjoint. Write $A_1 = \Phi_{F_1}^{-1}(A_{F_1} \times \Omega_{F_1^c})$ for appropriately chosen finite $F_1 \subseteq I$ and $A_{F_1} \in \mathcal{F}_{F_1}$, and write $A_2 = \Phi_{F_2}^{-1}(A_{F_2} \times \Omega_{F_2^c})$

for appropriately chosen finite $F_2 \subseteq I$ and $A_{F_2} \in \mathcal{F}_{F_2}$. Let $F_3 = F_1 \cup F_2$. Using this notation, we have by Claim 2 that

$$A_1 = \Phi_{F_1}^{-1}(A_{F_1} \times \Omega_{F_1^c}) = \Phi_{F_3}^{-1}\left(\Phi_{F_1,F_3}^{-1}(A_{F_1} \times \Omega_{F_3-F_1}) \times \Omega_{F_3^c}\right),$$

and using the same reasoning as in Subclaim 2, we also have that

$$A_2 = \Phi_{F_2}^{-1}(A_{F_2} \times \Omega_{F_2^c}) = \Phi_{F_3}^{-1}\left(\Phi_{F_2,F_3}^{-1}(A_{F_2} \times \Omega_{F_3-F_2}) \times \Omega_{F_3^c}\right).$$

Because A_1 and A_2 are disjoint, it follows that

$$\Phi_{F_1,F_3}^{-1}(A_{F_1} \times \Omega_{F_3-F_1}) \cap \Phi_{F_2,F_3}^{-1}(A_{F_2} \times \Omega_{F_3-F_2}) = \varnothing.$$

Next, we have that

$$A_1 \cup A_2 = \Phi_{F_3}^{-1}\left(\left[\Phi_{F_1,F_3}^{-1}(A_{F_1} \times \Omega_{F_3-F_1}) \cup \Phi_{F_2,F_3}^{-1}(A_{F_2} \times \Omega_{F_3-F_2})\right] \times \Omega_{F_3^c}\right).$$

Putting everything together and using Claim 2 gives

$$\begin{aligned}
& \mu(A_1 \cup A_2) \\
= \;& \mu_{F_3}\left(\Phi_{F_1,F_3}^{-1}(A_{F_1} \times \Omega_{F_3-F_1}) \cup \Phi_{F_2,F_3}^{-1}(A_{F_2} \times \Omega_{F_3-F_2})\right) \\
= \;& \mu_{F_3}\left(\Phi_{F_1,F_3}^{-1}(A_{F_1} \times \Omega_{F_3-F_1})\right) + \mu_{F_3}\left(\Phi_{F_2,F_3}^{-1}(A_{F_2} \times \Omega_{F_3-F_2})\right) \\
= \;& \mu_{F_1}(A_{F_1})\mu_{F_3-F_1}(\Omega_{F_3-F_1}) + \mu_{F_2}(A_{F_2})\mu_{F_3-F_2}(\Omega_{F_3-F_2}) \\
= \;& \mu_{F_1}(A_{F_1}) + \mu_{F_2}(A_{F_2}) \\
= \;& \mu(A_1) + \mu(A_2).
\end{aligned}$$

This shows that μ is finitely additive and completes the proof. ∎

For the next claim, we make a simple assertion. Let Ξ denote a nonempty set and let $\mathcal{Q}$ denote a field on Ξ. *If μ is a finitely additive measure on $\mathcal{Q}$ with $\mu(\Xi) < +\infty$, then*

$$\mu(A) \leq \mu(B) \quad \textit{and} \quad \mu(B - A) = \mu(B) - \mu(A) \quad \textit{if } A \subseteq B,\ A, B \in \mathcal{Q}.$$

To see this, we have $\mu(B) = \mu(A \cup (B - A)) = \mu(A) + \mu(B - A)$ by finite additivity. Since μ is nonnegative, it follows that $\mu(A) \leq \mu(B)$. Also, subtracting $\mu(A)$ from both sides gives the other result.

Claim 5. *Let μ denote a finitely additive measure on a field $\mathcal{Q}$ over Ξ such that $\mu(\Xi) < +\infty$. Suppose that $\mu(A_n)$ decreases to 0 as $n \to \infty$ whenever $\{A_n\}_{n=1}^{\infty}$ is a nonincreasing sequence of $\mathcal{Q}$-sets with limit $\varnothing$. Then μ is a measure on $\mathcal{Q}$.*

Proof: Let $\{B_k\}_{k=1}^{\infty}$ denote a disjoint sequence of $\mathcal{Q}$-sets such that the union $B = \bigcup_{k=1}^{\infty} B_k$ is in $\mathcal{Q}$. Then for each $n \in \mathbb{N}$ we have

$$C_n = \bigcup_{k=n+1}^{\infty} B_k = B - \bigcup_{k=1}^{n} B_k \in \mathcal{Q},$$

and $\{C_n\}_{n=1}^{\infty}$ is nonincreasing with limit $\varnothing$, hence $\mu(C_n)$ decreases to 0 as $n \to \infty$ [by assumption], so that $\lim_n \mu(B - \bigcup_{k=1}^{n} B_k) = 0$. From this, it follows by finite additivity that $\mu(B) - \lim_n \mu(\bigcup_{k=1}^{n} B_k) = 0$. Therefore,

$$\mu(B) = \lim_n \sum_{k=1}^{n} \mu(B_k) = \sum_{k=1}^{\infty} \mu(B_k).$$

This shows that μ is countably additive, demonstrating that μ is a measure on $\mathcal{Q}$. ∎

We now deal with some subsidiary concepts that will be needed as well.

Claim 6. *Let $\mathcal{A}$ denote a semiring on a nonempty set Ξ with $\Xi \in \mathcal{A}$.*

(a) Let $\mathcal{A}^*$ denote the collection of finite disjoint unions of $\mathcal{A}$-sets. Then $\mathcal{A}^*$ is a field on Ξ.

(b) $\mathcal{A}^* = f(\mathcal{A})$.

(c) $\sigma(\mathcal{A}^*) = \sigma(\mathcal{A})$.

Proof: We prove (a). We must show that $\Xi \in \mathcal{A}^*$, $\mathcal{A}^*$ is closed under complementation, and $\mathcal{A}^*$ is closed under finite intersections. Since we have $\Xi \in \mathcal{A} \subseteq \mathcal{A}^*$, it immediately follows that $\Xi \in \mathcal{A}^*$. Next, pick $A \in \mathcal{A}^*$, so that we may write $A = \bigcup_{i=1}^{n} A_i$, a disjoint union of $\mathcal{A}$-sets. Then

$$A^c = \Omega - \bigcup_{i=1}^{n} A_i = \Omega \cap \bigcap_{i=1}^{n} A_i^c.$$

By Claim 1 of Section 2.4, there exists a finite disjoint collection $C_1, \cdots, C_m$ of $\mathcal{A}$-sets with $\Omega \cap \bigcap_{i=1}^{n} A_i^c = \bigcup_{j=1}^{m} C_j$. That is, A^c may be expressed as a finite disjoint union of $\mathcal{A}$-sets, so that $A^c \in \mathcal{A}^*$: $\mathcal{A}^*$ is closed under complementation. Next, pick $A, B \in \mathcal{A}^*$, writing $A = \bigcup_{i=1}^{n} A_i$ and $B = \bigcup_{j=1}^{m} B_j$, where $A_1, \cdots, A_n \in \mathcal{A}$ are disjoint and $B_1, \cdots, B_m \in \mathcal{A}$ are disjoint. Then

$$A \cap B = \bigcup_{i=1}^{n} A_i \cap \bigcup_{j=1}^{m} B_j = \bigcup_{i=1}^{n} \bigcup_{j=1}^{m} (A_i \cap B_j).$$

Since $\mathcal{A}$ is a semiring, we have $A_i \cap B_j \in \mathcal{A}$ for each i and j. Since the A_i's are disjoint and since the B_j's are disjoint, it follows that the collection $\{A_i \cap B_j : i = 1, \cdots, n, \; j = 1, \cdots, m\}$ is a disjoint collection. Therefore, $A \cap B$ is a finite disjoint union of $\mathcal{A}$-sets, whence $A \cap B \in \mathcal{A}^*$. This shows that $\mathcal{A}^*$ is a field on Ξ, proving (a).

We now turn to (b). Since $\mathcal{A} \subseteq \mathcal{A}^*$ and $\mathcal{A}^*$ is a field on Ξ, we have $f(\mathcal{A}) \subseteq \mathcal{A}^*$. We desire the reverse inclusion $\mathcal{A}^* \subseteq f(\mathcal{A})$. Observe first that $f(\mathcal{A})$ is the intersection of all fields on Ξ that contain $\mathcal{A}$. Observe next that every field on Ξ that contains $\mathcal{A}$ will contain the finite unions of $\mathcal{A}$-sets, in particular, the finite disjoint unions of $\mathcal{A}$-sets. That is, every field on Ξ that contains $\mathcal{A}$ will also contain $\mathcal{A}^*$. This forces $\mathcal{A}^* \subseteq f(\mathcal{A})$, proving (b).

For (c), observe that $\sigma(\mathcal{A}^*) = \sigma(f(\mathcal{A}))$ by (b). It is immediate that $\sigma(f(\mathcal{A})) = \sigma(\mathcal{A})$, and hence (c) is complete. This finishes the entire proof. ■

The path has been long and somewhat tedious, but we are ready for the first of the two main results dealing with the construction of the desired product measure on $\mathcal{F}$. The proof of the upcoming claim is quite long and labyrinthine, and the proof is broken up into various parts. For the following proof, the reader is urged to recall the Extension Theorem from Section 3.4 and the Uniqueness Theorem from Section 3.5. Also, recall Halmos' Monotone Class Theorem from Section 10.1: *Let $\mathcal{Q}$ and $\mathcal{M}$ denote a field and a monotone class, respectively, on a nonempty set Ξ. If $\mathcal{Q} \subseteq \mathcal{M}$, then $\sigma(\mathcal{Q}) \subseteq \mathcal{M}$.*

Claim 7. *Let $\{(\Omega_i, \mathcal{F}_i, \mu_i)\}_{i=1}^{\infty}$ denote a sequence of measure spaces with $\mu_i(\Omega_i) = 1$ for each $i \in \mathbb{N}$. There exists a unique measure μ on $\mathcal{F}$ such that*

$$\mu(\Phi_F^{-1}(A_F \times \Omega_{F^c})) = \mu_F(A_F) \text{ whenever } F \subseteq \mathbb{N} \text{ is finite and } A_F \in \mathcal{F}_F.$$

Proof: By Claim 4 there exists a finitely additive measure μ on the field $\mathcal{H}$ such that $\mu(\Phi_F^{-1}(A_F \times \Omega_{F^c})) = \mu_F(A_F)$ whenever $F \subseteq \mathbb{N}$ is finite and $A_F \in \mathcal{F}_F$. Suppose it is shown that μ is actually a measure on the field $\mathcal{H}$. The Extension Theorem in Section 3.4 then asserts that μ may be consistently extended from $\mathcal{H}$ to $\sigma(\mathcal{H}) = \mathcal{F}$. Since $\mu(\Omega) = 1 < +\infty$, the Uniqueness Theorem in Section 3.5 asserts that the extension of μ from $\mathcal{H}$ to $\mathcal{F}$ is the unique such extension. [Equivalently, refer to Claim 3 in Section 3.5.]

To show that μ is actually a measure on $\mathcal{H}$, it is sufficient by Claim 5 to show the following: *if $\{B_n\}_{n=1}^{\infty}$ is a nonincreasing sequence of $\mathcal{H}$-sets with limit $\varnothing$, then* $\lim_n \mu(B_n) = 0$.

To carry out this program, let $\{B_n\}_{n=1}^{\infty}$ denote a sequence of $\mathcal{H}$-sets such that $B_1 \supseteq B_2 \supseteq \cdots$ and $\bigcap_{n=1}^{\infty} B_n = \varnothing$; we want $\lim_n \mu(B_n) = 0$.

We first discuss some notation. Each B_n, being an $\mathcal{H}$-set, has the form $\Phi_{F_n}^{-1}(A_{F_n} \times \Omega_{F_n^c})$ for some finite $F_n \subseteq \mathbb{N}$ and $A_{F_n} \in \mathcal{F}_{F_n}$. For each $n \in \mathbb{N}$, let $k_n = \max F_n$; without any loss of generality we may assume for each $n \in \mathbb{N}$ that $F_n = \{1, 2, \cdots, k_n\}$ and $k_1 < k_2 < \cdots$. We next define the sequence $\{E_m\}_{m=1}^{\infty}$ of $\mathcal{H}$-sets by writing

$$E_m = \begin{cases} \Omega & 1 \leq m < k_1 \\ B_n & k_n \leq m < k_{n+1} \end{cases} \qquad \text{for each } m \in \mathbb{N}.$$

Observe that $E_1 \supseteq E_2 \supseteq \cdots$ and $\bigcap_{m=1}^{\infty} E_m = \bigcap_{n=1}^{\infty} B_n = \varnothing$. Note too that $\lim_m \mu(E_m) = \lim_n \mu(B_n)$, so that $\lim_m \mu(E_m) = 0$ iff $\lim_n \mu(B_n) = 0$. Therefore, in lieu of showing that $\lim_n \mu(B_n) = 0$, we equivalently desire to show that $\lim_m \mu(E_m) = 0$. For each $m \in \mathbb{N}$, define $\mathbb{N}_m = \{1, 2, \cdots, m\}$.

Subclaim 1. *For every $m \in \mathbb{N}$ we may write E_m in the form $\Phi_{\mathbb{N}_m}^{-1}(A_{\mathbb{N}_m} \times \Omega_{\mathbb{N}_m^c})$ for appropriate $A_{\mathbb{N}_m} \in \mathcal{F}_{\mathbb{N}_m}$.*

Proof of Subclaim 1: Let $1 \leq m < k_1$, so that $E_m = \Omega$. Take $A_{\mathbb{N}_m} \in \mathcal{F}_{\mathbb{N}_m}$ to be $\Omega_{\mathbb{N}_m}$; it follows that $E_m = \Omega = \Phi_{\mathbb{N}_m}^{-1}(\Omega_{\mathbb{N}_m} \times \Omega_{\mathbb{N}_m^c}) = \Phi_{\mathbb{N}_m}^{-1}(A_{\mathbb{N}_m} \times \Omega_{\mathbb{N}_m^c})$, proving the result for $1 \le m < k_1$. Next, pick $n \in \mathbb{N}$, and let $k_n \le m < k_{n+1}$. In this situation we have

$$E_m = B_n = \Phi_{F_n}^{-1}(A_{F_n} \times \Omega_{F_n^c})$$

where $F_n = \{1, 2, \cdots, k_n\}$ and A_{F_n} is an appropriately chosen element of $\mathcal{F}_{F_n}$. If $m = k_n$ then we may take $A_{\mathbb{N}_m}$ to be A_{F_n}. In the case where $k_n < m < k_{n+1}$, let $A_{\mathbb{N}_m}$ denote the unique $\mathcal{F}_{\mathbb{N}_m}$-set corresponding to the set $A_{F_n} \times \Omega_{\{k_n+1, k_n+2, \cdots, m\}}$ in $\mathcal{F}_{F_n} \otimes \mathcal{F}_{\{k_n+1, k_n+2, \cdots, m\}}$ under the one-to-one and onto correspondence

$$(\omega_1, \cdots, \omega_{k_n}, \omega_{k_n+1}, \cdots, \omega_m) \leftrightarrow ((\omega_1, \cdots, \omega_{k_n}), (\omega_{k_n+1}, \cdots, \omega_m));$$

we again have $E_m = \Phi_{\mathbb{N}_m}^{-1}(A_{\mathbb{N}_m} \times \Omega_{\mathbb{N}_m^c})$. This gives the result in all cases, completing the proof of the subclaim.

Subclaim 2. *For every $m \in \mathbb{N}$ and $A_{\mathbb{N}_m} \in \mathcal{F}_{\mathbb{N}_m}$, we have that*

$$\mu_{\mathbb{N}_m}(A_{\mathbb{N}_m}) = \int_{\Omega_1} \int_{\Omega_2} \cdots \int_{\Omega_m} I_{A_{\mathbb{N}_m}}(\omega_1, \cdots, \omega_m)\, d\mu_m(\omega_m) \cdots d\mu_1(\omega_1). \qquad (11.4)$$

Proof of Subclaim 2: Fix $m \in \mathbb{N}$. Define

$$\mathcal{C} = \{A_{\mathbb{N}_m} \in \mathcal{F}_{\mathbb{N}_m} : \text{ (11.4) holds for } A_{\mathbb{N}_m}\}.$$

It is trivial that $\mathcal{C} \subseteq \mathcal{F}_{\mathbb{N}_m}$. We desire the reverse inclusion $\mathcal{F}_{\mathbb{N}_m} \subseteq \mathcal{C}$.

Let $A_{\mathbb{N}_m} = \prod_{i=1}^m A_i$, where $A_i \in \mathcal{F}_i$ for each $i \in \{1, \cdots, m\}$. Then [invoking the product property for $\mu_{\mathbb{N}_m}$] we have $\mu_{\mathbb{N}_m}(\prod_{i=1}^m A_i) = \prod_{i=1}^m \mu_i(A_i)$. Since

$$I_{A_{\mathbb{N}_m}}(\omega_1, \cdots, \omega_m) = I_{A_1}(\omega_1) \cdots I_{A_m}(\omega_m)$$

for every $\omega_1 \in \Omega_1, \cdots, \omega_m \in \Omega_m$, we have

$$\begin{aligned} &\int_{\Omega_1} \int_{\Omega_2} \cdots \int_{\Omega_m} I_{A_{\mathbb{N}_m}}(\omega_1, \cdots, \omega_m)\, d\mu_m(\omega_m) \cdots d\mu_1(\omega_1) \\ = &\prod_{i=1}^m \int_{\Omega_i} I_{A_i}(\omega_i)\, d\mu_i(\omega_i) \\ = &\prod_{i=1}^m \mu_i(A_i), \end{aligned}$$

and hence (11.4) holds when $A_{\mathbb{N}_m}$ is of the special form $\prod_{i=1}^m A_i$ with $A_i \in \mathcal{F}_i$ for each $i \in \{1, \cdots, m\}$. That is, we have shown the inclusion $\mathcal{E}_{\mathbb{N}_m} \subseteq \mathcal{C}$.

Next, let $\mathcal{D}_{\mathbb{N}_m}$ denote the collection of finite disjoint unions of $\mathcal{E}_{\mathbb{N}_m}$-sets. Since $\mathcal{E}_{\mathbb{N}_m}$ is a semiring on $\Omega_{\mathbb{N}_m}$ with $\Omega_{\mathbb{N}_m} \in \mathcal{E}_{\mathbb{N}_m}$, part (a) of Claim 6 asserts that $\mathcal{D}_{\mathbb{N}_m}$ is a field on $\Omega_{\mathbb{N}_m}$. Observe by (c) of Claim 6 that $\sigma(\mathcal{D}_{\mathbb{N}_m}) = \sigma(\mathcal{E}_{\mathbb{N}_m}) = \mathcal{F}_{\mathbb{N}_m}$.

We next wish to show that $\mathcal{D}_{\mathbb{N}_m} \subseteq \mathcal{C}$. Let $D \in \mathcal{D}_{\mathbb{N}_m}$, so that D may be written as a finite disjoint union of $A_1, \cdots, A_n \in \mathcal{E}_{\mathbb{N}_m}$. Since

$$\mu_{\mathbb{N}_m}(D) = \sum_{i=1}^n \mu_{\mathbb{N}_m}(A_i) \quad \text{and} \quad I_D = \sum_{i=1}^n I_{A_i},$$

we have that

$$\begin{aligned}
&\int_{\Omega_1}\int_{\Omega_2}\cdots\int_{\Omega_m} I_D(\omega_1,\cdots,\omega_m)\,d\mu_m(\omega_m)\cdots d\mu_1(\omega_1)\\
=\;&\int_{\Omega_1}\int_{\Omega_2}\cdots\int_{\Omega_m}\sum_{i=1}^{n} I_{A_i}(\omega_1,\cdots,\omega_m)\,d\mu_m(\omega_m)\cdots d\mu_1(\omega_1)\\
=\;&\sum_{i=1}^{n}\int_{\Omega_1}\int_{\Omega_2}\cdots\int_{\Omega_m} I_{A_i}(\omega_1,\cdots,\omega_m)\,d\mu_m(\omega_m)\cdots d\mu_1(\omega_1)\\
=\;&\sum_{i=1}^{n}\mu_{\mathbb{N}_m}(A_i),
\end{aligned}$$

and hence $D\in\mathcal{C}$. That is, we have shown the inclusion $\mathcal{D}_{\mathbb{N}_m}\subseteq\mathcal{C}$.

We now wish to show that $\mathcal{C}$ is closed under limits of nondecreasing sequences of $\mathcal{C}$-sets. Let $\{A_n\}_{n=1}^{\infty}$ denote a nondecreasing sequence of $\mathcal{C}$-sets. We wish to show that $\lim_n A_n=\bigcup_{n=1}^{\infty}A_n\in\mathcal{C}$. Since every measure is continuous from below [recall measure property (M8) from Section 2.2], we have

$$\mu_{\mathbb{N}_m}\left(\lim_n A_n\right)=\lim_n \mu_{\mathbb{N}_m}(A_n).$$

By the fact that $I_{\lim_n A_n}=\lim_n I_{A_n}$ and repeated application of MCT, we have that

$$\begin{aligned}
&\int_{\Omega_1}\int_{\Omega_2}\cdots\int_{\Omega_m} I_{\lim_n A_n}(\omega_1,\cdots,\omega_m)\,d\mu_m(\omega_m)\cdots d\mu_1(\omega_1)\\
=\;&\int_{\Omega_1}\int_{\Omega_2}\cdots\int_{\Omega_m}\lim_n I_{A_n}(\omega_1,\cdots,\omega_m)\,d\mu_m(\omega_m)\cdots d\mu_1(\omega_1)\\
=\;&\lim_n\int_{\Omega_1}\int_{\Omega_2}\cdots\int_{\Omega_m} I_{A_n}(\omega_1,\cdots,\omega_m)\,d\mu_m(\omega_m)\cdots d\mu_1(\omega_1)\\
=\;&\lim_n \mu_{\mathbb{N}_m}(A_n).
\end{aligned}$$

This shows that $\lim_n A_n\in\mathcal{C}$, and $\mathcal{C}$ is closed under limits [unions] of nondecreasing sequences of $\mathcal{C}$-sets.

We now show that $\mathcal{C}$ is closed under limits [intersections] of nonincreasing sequences of $\mathcal{C}$-sets. Let $\{A_n\}_{n=1}^{\infty}$ denote a nonincreasing sequence of $\mathcal{C}$-sets. We wish to show that $\lim_n A_n=\bigcap_{n=1}^{\infty}A_n\in\mathcal{C}$. Since $\mu_{\mathbb{N}_m}$ is a finite measure, and since every finite measure is continuous from above [recall measure property (M9) in Section 2.2], it follows that $\mu_{\mathbb{N}_m}(\lim_n A_n)=\lim_n \mu_{\mathbb{N}_m}(A_n)$. Since $I_{A_n}\le 1$ on $\Omega_{\mathbb{N}_m}$ and since

$$\int_{\Omega_1}\cdots\int_{\Omega_m} 1\,d\mu_m(\omega_m)\cdots d\mu_1(\omega_1)=1<+\infty,$$

we may repeatedly apply DCT, obtaining

$$\begin{aligned}
&\int_{\Omega_1}\int_{\Omega_2}\cdots\int_{\Omega_m} I_{\lim_n A_n}(\omega_1,\cdots,\omega_m)\,d\mu_m(\omega_m)\cdots d\mu_1(\omega_1)\\
=&\int_{\Omega_1}\int_{\Omega_2}\cdots\int_{\Omega_m} \lim_n I_{A_n}(\omega_1,\cdots,\omega_m)\,d\mu_m(\omega_m)\cdots d\mu_1(\omega_1)\\
=&\lim_n \int_{\Omega_1}\int_{\Omega_2}\cdots\int_{\Omega_m} I_{A_n}(\omega_1,\cdots,\omega_m)\,d\mu_m(\omega_m)\cdots d\mu_1(\omega_1)\\
=&\lim_n \mu_{\mathbb{N}_m}(A_n).
\end{aligned}$$

This shows that $\lim_n A_n \in \mathcal{C}$, and hence we have shown that $\mathcal{C}$ is closed under limits of nonincreasing sequences of $\mathcal{C}$-sets.

We now summarize. The previous two paragraphs show that $\mathcal{C}$ is a monotone class on $\Omega_{\mathbb{N}_m}$. We have shown that $\mathcal{D}_{\mathbb{N}_m} \subseteq \mathcal{C}$. By Halmos' Monotone Class Theorem in Section 10.1, it follows that $\sigma(\mathcal{D}_{\mathbb{N}_m}) \subseteq \mathcal{C}$, and hence $\mathcal{F}_{\mathbb{N}_m} \subseteq \mathcal{C}$. This is the desired reverse inclusion, and we have shown that $\mathcal{F}_{\mathbb{N}_m} = \mathcal{C}$. This completes the proof of the subclaim.

Fix $m \in \mathbb{N}$. Recalling that $E_m = \Phi_{\mathbb{N}_m}^{-1}(A_{\mathbb{N}_m} \times \Omega_{\mathbb{N}_m^c})$ and using Subclaim 2, we have

$$\begin{aligned}
&\mu(E_m)\\
=&\mu_{\mathbb{N}_m}(A_{\mathbb{N}_m})\\
=&\int_{\Omega_1}\int_{\Omega_2}\cdots\int_{\Omega_m} I_{A_{\mathbb{N}_m}}(\omega_1,\cdots,\omega_m)\,d\mu_m(\omega_m)\cdots d\mu_1(\omega_1). \qquad (11.5)
\end{aligned}$$

Our ultimate desire is to show that $\lim_m \mu(E_m) = 0$. We proceed by contradiction, *assuming to the contrary that $\mu(E_m)$ fails to tend to zero as $m \to \infty$.* [The contradiction will be obtained by constructing an element of the empty set $\bigcap_{m=1}^{\infty} E_m$.]

For each $\omega_1 \in \Omega_1$ and all large $m \in \mathbb{N}$, define

$$f_{1,m}(\omega_1) = \int_{\Omega_2}\cdots\int_{\Omega_m} I_{A_{\mathbb{N}_m}}(\omega_1,\cdots,\omega_m)\,d\mu_m(\omega_m)\cdots d\mu_2(\omega_2).$$

Then, referring to (11.5), we have

$$\mu(E_m) = \int_{\Omega_1} f_{1,m}(\omega_1)\,d\mu_1(\omega_1).$$

Observe that because $\mu_i(\Omega_i) = 1$ for each $i \in \mathbb{N}$ we have $f_{1,m} \leq 1$ on Ω_1 and $\int_{\Omega_1} 1\,d\mu_1(\omega_1) = \mu_1(\Omega_1) = 1 < +\infty$. Observe next that it is not the case that $\lim_m f_{1,m}(\omega_1) = 0$ for each $\omega_1 \in \Omega_1$, for if $\lim_m f_{1,m}(\omega_1) = 0$ for every

$\omega_1 \in \Omega_1$, we would have by DCT that

$$\begin{aligned}\lim_m \int_{\Omega_1} f_{1,m}(\omega_1)\, d\mu_1(\omega_1) &= \int_{\Omega_1} \lim_m f_{1,m}(\omega_1)\, d\mu_1(\omega_1) \\ &= \int_{\Omega_1} 0\, d\mu_1(\omega_1) \\ &= 0,\end{aligned}$$

giving $\lim_m \mu(E_m) = 0$, contradicting the assumption that $\mu(E_m)$ does not tend to zero as $m \to \infty$. Therefore, there is $\xi_1 \in \Omega_1$ such that $f_{1,m}(\xi_1)$ fails to tend to zero as $m \to \infty$. That is, for this $\xi_1 \in \Omega_1$, we have that

$$\int_{\Omega_2} \cdots \int_{\Omega_m} I_{A_{\mathbb{N}_m}}(\xi_1, \omega_2, \cdots, \omega_m)\, d\mu_m(\omega_m) \cdots d\mu_2(\omega_2)$$

does not tend to zero as $m \to \infty$. Next, for each large $m \in \mathbb{N}$ define $f_{2,m}$ on Ω_2 by writing

$$f_{2,m}(\omega_2) = \int_{\Omega_3} \cdots \int_{\Omega_m} I_{A_{\mathbb{N}_m}}(\xi_1, \omega_2, \cdots, \omega_m)\, d\mu_m(\omega_m) \cdots d\mu_3(\omega_3)$$

for each $\omega_2 \in \Omega_2$. It is not the case that $\lim_m f_{2,m}(\omega_2) = 0$ for every $\omega_2 \in \Omega_2$, for otherwise we would have

$$\begin{aligned}0 &= \int_{\Omega_2} 0\, d\mu_2(\omega_2) \\ &= \int_{\Omega_2} \lim_m f_{2,m}(\omega_2)\, d\mu_2(\omega_2) \\ &= \lim_m \int_{\Omega_2} f_{2,m}(\omega_2)\, d\mu_2(\omega_2) \quad \text{[DCT – similar argument as for } f_{1,m}] \\ &= \lim_m \int_{\Omega_2} \int_{\Omega_3} \cdots \int_{\Omega_m} I_{A_{\mathbb{N}_m}}(\xi_1, \omega_2, \cdots, \omega_m)\, d\mu_m(\omega_m) \cdots d\mu_3(\omega_3)\, d\mu_2(\omega_2) \\ &= \lim_m f_{1,m}(\xi_1),\end{aligned}$$

contradicting the fact that $f_{1,m}(\xi_1)$ does not tend to zero as $m \to \infty$. Therefore, we may select a point $\xi_2 \in \Omega_2$ such that $f_{2,m}(\xi_2)$ does not tend to zero as $m \to \infty$. That is,

$$\int_{\Omega_3} \cdots \int_{\Omega_m} I_{A_{\mathbb{N}_m}}(\xi_1, \xi_2, \omega_3, \cdots, \omega_m)\, d\mu_m(\omega_m) \cdots d\mu_3(\omega_3)$$

does not tend to zero as $m \to \infty$. If for all large $m \in \mathbb{N}$ we define $f_{3,m}$ on Ω_3 by writing

$$f_{3,m}(\omega_3) = \int_{\Omega_4} \cdots \int_{\Omega_m} I_{A_{\mathbb{N}_m}}(\xi_1, \xi_2, \omega_3 \cdots, \omega_m)\, d\mu_m(\omega_m) \cdots d\mu_4(\omega_4)$$

for every $\omega_3 \in \Omega_3$, then a similar argument as given before for $f_{1,m}$ and $f_{2,m}$ shows that if it were the case that $\lim_m f_{3,m}(\omega_3) = 0$ for each $\omega_3 \in \Omega_3$, then

$\lim_m f_{2,m}(\xi_2) = 0$, contradicting what has been established for ξ_2. Therefore, there is a point $\xi_3 \in \Omega_3$ such that $f_{3,m}(\xi_3)$ fails to tend to zero as $m \to \infty$, that is,

$$\int_{\Omega_4} \cdots \int_{\Omega_m} I_{A_{\mathbb{N}_m}}(\xi_1, \xi_2, \xi_3, \omega_4, \cdots, \omega_m)\, d\mu_m(\omega_m) \cdots d\mu_4(\omega_4)$$

does not tend to zero as $m \to \infty$. In general, given any $n \in \mathbb{N}$, we define $f_{n,m}$ on Ω_n for all large $m \in \mathbb{N}$ with $m > n$ and $\omega_n \in \Omega_n$ by the rule

$$\begin{aligned} &f_{n,m}(\omega_n) \\ &= \int_{\Omega_{n+1}} \cdots \int_{\Omega_m} I_{A_{\mathbb{N}_m}}(\xi_1, \cdots, \xi_{n-1}, \omega_n, \cdots, \omega_m)\, d\mu_m(\omega_m) \cdots d\mu_{n+1}(\omega_{n+1}). \end{aligned}$$

We then argue that it cannot be the case that $\lim_m f_{n,m}(\omega_n) = 0$ for each $\omega_n \in \Omega_n$ by showing that this condition would force $\lim_m f_{n-1,m}(\xi_{n-1}) = 0$, contradicting what we know concerning the point $\xi_{n-1} \in \Omega_{n-1}$. From this, we deduce that there is $\xi_n \in \Omega_n$ such that $f_{n,m}(\xi_n)$ does not tend to zero as $m \to \infty$. That is,

$$\int_{\Omega_{n+1}} \int_{\Omega_{n+2}} \cdots \int_{\Omega_m} I_{A_{\mathbb{N}_m}}(\xi_1, \cdots, \xi_n, \omega_{n+1}, \cdots, \omega_m)\, d\mu_m(\omega_m) \cdots d\mu_{n+1}(\omega_{n+1})$$

does not tend to zero as $m \to \infty$.

From carrying out the above process indefinitely, we construct an element $\boldsymbol{\xi} = (\xi_1, \xi_2, \cdots, \xi_n, \cdots) \in \Omega$ such that for every $n \in \mathbb{N}$ the numbers

$$\int_{\Omega_{n+1}} \int_{\Omega_{n+2}} \cdots \int_{\Omega_m} I_{A_{\mathbb{N}_m}}(\xi_1, \cdots, \xi_n, \omega_{n+1}, \cdots, \omega_m)\, d\mu_m(\omega_m) \cdots d\mu_{n+1}(\omega_{n+1}) \tag{11.6}$$

[defined for all $m > n$] do not have limit zero as $m \to \infty$.

Subclaim 3. *It cannot be the case that for some $n \in \mathbb{N}$ and some large $m > n$ we have $I_{A_{\mathbb{N}_m}}(\xi_1, \cdots, \xi_n, \omega_{n+1}, \cdots, \omega_m) = 0$ for each $\omega_{n+1} \in \Omega_{n+1}, \cdots, \omega_m \in \Omega_m$.*

Proof of Subclaim 3: Suppose to the contrary that for some $n \in \mathbb{N}$ and some large $m > n$ we have

$$I_{A_{\mathbb{N}_m}}(\xi_1, \cdots, \xi_n, \omega_{n+1}, \cdots, \omega_m) = 0$$

for each $\omega_{n+1} \in \Omega_{n+1}, \cdots, \omega_m \in \Omega_m$. Then for this n and this $m > n$ we have that

$$\int_{\Omega_{n+1}} \int_{\Omega_{n+2}} \cdots \int_{\Omega_m} I_{A_{\mathbb{N}_m}}(\xi_1, \cdots, \xi_n, \omega_{n+1}, \cdots, \omega_m)\, d\mu_m(\omega_m) \cdots d\mu_{n+1}(\omega_{n+1}) = 0.$$

Pick $j \in \mathbb{N}$. Let $\Phi : \Omega_{\mathbb{N}_{m+j}} \to \Omega_{\mathbb{N}_m} \times \Omega_{\{m+1,\cdots,m+j\}}$ be the one-to-one and onto mapping that associates each $(\omega_1, \cdots, \omega_{m+j})$ with $((\omega_1, \cdots, \omega_m), (\omega_{m+1}, \cdots, \omega_{m+j}))$. By the Identification Lemma, Φ establishes a one-to-one correspondence between the sets in $\mathcal{F}_{\mathbb{N}_{m+j}}$ and those in $\mathcal{F}_{\mathbb{N}_m} \times \mathcal{F}_{\{m+1,\cdots,m+j\}}$. Since $E_{m+j} \subseteq E_m$, we have

that

$$\begin{aligned}\Phi^{-1}_{\mathbb{N}_{m+j}}(A_{\mathbb{N}_{m+j}} \times \Omega_{\mathbb{N}^c_{m+j}}) &\subseteq \Phi^{-1}_{\mathbb{N}_m}(A_{\mathbb{N}_m} \times \Omega_{\mathbb{N}^c_m}) \\ &\overset{(*)}{=} \Phi^{-1}_{\mathbb{N}_{m+j}}(\Phi^{-1}(A_{\mathbb{N}_m} \times \Omega_{\{m+1,\cdots,m+j\}}) \times \Omega_{\mathbb{N}^c_{m+j}}),\end{aligned}$$

and hence $A_{\mathbb{N}_{m+j}} \subseteq \Phi^{-1}(A_{\mathbb{N}_m} \times \Omega_{\{m+1,\cdots,m+j\}})$. [To prove the equality $(*)$, let $\mathcal{C}$ denote the set of $A_{\mathbb{N}_m} \in \mathcal{F}_{\mathbb{N}_m}$ for which $(*)$ holds, so that $\mathcal{C} \subseteq \mathcal{F}_{\mathbb{N}_m}$. The reader may verify in Exercise 2 that $\mathcal{E}_{\mathbb{N}_m} \subseteq \mathcal{C}$ and $\mathcal{C}$ is a σ-field on $\mathcal{F}_{\mathbb{N}_m}$. It follows that $\mathcal{F}_{\mathbb{N}_m} \subseteq \mathcal{C}$, whence the equality $(*)$ holds.] It follows that

$$\begin{aligned} & I_{A_{\mathbb{N}_{m+j}}}(\xi_1, \cdots, \xi_n, \omega_{n+1}, \cdots, \omega_{m+j}) \\ \leq\ & I_{\Phi^{-1}(A_{\mathbb{N}_m} \times \Omega_{\{m+1,\cdots,m+j\}})}(\xi_1, \cdots, \xi_n, \omega_{n+1}, \cdots, \omega_{m+j}) \\ \overset{(**)}{=}\ & I_{A_{\mathbb{N}_m}}(\xi_1, \cdots, \xi_n, \omega_{n+1}, \cdots, \omega_m) \\ =\ & 0 \text{ for each } \omega_{n+1} \in \Omega_{n+1}, \cdots, \omega_m \in \Omega_m. \end{aligned}$$

[To demonstrate the equality $(**)$, define

$$\mathcal{C} = \{A_{\mathbb{N}_m} \in \mathcal{F}_{\mathbb{N}_m} \colon \text{equality } (**) \text{ holds for each } \omega_{n+1} \in \Omega_{n+1}, \cdots, \omega_m \in \Omega_m\}.$$

It is trivial that $\mathcal{C} \subseteq \mathcal{F}_{\mathbb{N}_m}$. It is straightforward to show that $\mathcal{E}_{\mathbb{N}_m} \subseteq \mathcal{C}$ and $\mathcal{C}$ is a σ-field on $\Omega_{\mathbb{N}_m}$. (See Exercise 2.) From this, it follows that $\mathcal{F}_{\mathbb{N}_m} = \sigma(\mathcal{E}_{\mathbb{N}_m}) \subseteq \mathcal{C}$, whence $\mathcal{F}_{\mathbb{N}_m} = \mathcal{C}$, proving the equality $(**)$.] Repeated application of the Nonnegative Fubini Theorem shows that

$$\int_{\Omega_{n+1}} \cdots \int_{\Omega_{m+j}} I_{A_{\mathbb{N}_{m+j}}}(\xi_1, \cdots, \xi_n, \omega_{n+1}, \cdots, \omega_{m+j}) d\mu_{m+j}(\omega_{m+j}) \cdots d\mu_{n+1}(\omega_{n+1})$$

equals zero. [Write $\int_{\Omega_{n+1}} \cdots \int_{\Omega_{m+j}}$ as $\int_{\Omega_{m+j}} \cdots \int_{\Omega_{m+1}} \int_{\Omega_m} \cdots \int_{\Omega_{n+1}}$.] Since $j \in \mathbb{N}$ is arbitrary, we have that the quantity (11.6) is zero for all M larger than the specific m, and hence tends to zero as $m \to \infty$. This is a contradiction, and establishes the claim.

From Subclaim 3, we have the following: for each $n \in \mathbb{N}$ and every large $m > n$, there exist elements $\omega_{n+1} \in \Omega_{n+1}, \cdots, \omega_m \in \Omega_m$ such that $I_{A_{\mathbb{N}_m}}(\xi_1, \cdots, \xi_n, \omega_{n+1}, \cdots, \omega_m) = 1$. That is, for each $n \in \mathbb{N}$ and every large $m > n$ there exist points $\eta_{n+1} \in \Omega_{n+1}, \cdots, \eta_m \in \Omega_m$ such that $(\xi_1, \cdots, \xi_n, \eta_{n+1}, \cdots, \eta_m) \in A_{\mathbb{N}_m}$. Recall that $E_m = \Phi^{-1}_{\mathbb{N}_m}(A_{\mathbb{N}_m} \times \Omega_{\mathbb{N}^c_m})$ for appropriately chosen $A_{\mathbb{N}_m} \in \mathcal{F}_{\mathbb{N}_m}$; it follows for each $n \in \mathbb{N}$ and all sufficiently large $m > n$ that

$$(\xi_1, \cdots, \xi_n, \eta_{n+1}, \cdots, \eta_m, \omega_{m+1}, \omega_{m+2}, \cdots) \in E_m$$

whenever $\omega_{m+1} \in \Omega_{m+1}, \omega_{m+2} \in \Omega_{m+2}, \cdots$. Recalling that $E_1 \supseteq E_2 \supseteq \cdots$, the display above forces

$$(\xi_1, \cdots, \xi_n, \eta_{n+1}, \cdots, \eta_m, \omega_{m+1}, \omega_{m+2}, \cdots) \in E_n = \Phi^{-1}_{\mathbb{N}_n}(A_{\mathbb{N}_n} \times \Omega_{\mathbb{N}^c_n})$$

for each $n \in \mathbb{N}$, each sufficiently large $m > n$, each $\omega_{m+1} \in \Omega_{m+1}$, each $\omega_{m+2} \in \Omega_{m+2}, \cdots$. From this, we in turn have that $(\xi_1, \cdots, \xi_n) \in A_{\mathbb{N}_n}$ for every $n \in \mathbb{N}$, and hence

$$(\xi_1, \cdots, \xi_n, \omega_{n+1}, \omega_{n+2}, \cdots) \in \Phi_{\mathbb{N}_n}^{-1}(A_{\mathbb{N}_n} \times \Omega_{\mathbb{N}_n^c}) = E_n$$

for every $n \in \mathbb{N}$ and choice of $\omega_{n+1} \in \Omega_{n+1}$, $\omega_{n+2} \in \Omega_{n+2}$, $\cdots$. In particular, $\boldsymbol{\xi} = (\xi_1, \cdots, \xi_n, \xi_{n+1}, \xi_{n+2}, \cdots) \in E_n$ for every $n \in \mathbb{N}$. But then $\boldsymbol{\xi} \in \bigcap_{n=1}^{\infty} E_n$, contradicting the assumption that $\bigcap_{m=1}^{\infty} E_m = \varnothing$. That is, assuming that $\mu(E_m)$ does not tend to zero as $m \to \infty$ yields a contradiction. It is necessarily the case then that $\lim_m \mu(E_m) = 0$. This completes at last the entire proof. ■

What remains is to handle the case where I is uncountably infinite. This may sound imposing, but it isn't, as we can reduce the uncountably infinite case to the countably infinite case without too much effort. The method of proof is the same as is the proof for Claim 7.

Claim 8. *Let $\{(\Omega_i, \mathcal{F}_i, \mu_i)\}_{i \in I}$ denote an uncountable collection of measure spaces with $\mu_i(\Omega_i) = 1$ for each $i \in I$. There exists a unique measure μ on $\mathcal{F}$ such that*

$$\mu(\Phi_F^{-1}(A_F \times \Omega_{F^c})) = \mu_F(A_F) \quad \textit{whenever } F \subseteq I \textit{ is finite and } A_F \in \mathcal{F}_F.$$

Proof: Let μ denote the finitely additive measure on the field $\mathcal{H}$ as described in Claim 4. As in the proof of Claim 7, we will show the following: *if $\{B_n\}_{n=1}^{\infty}$ is a nonincreasing sequence of $\mathcal{H}$-sets with limit $\varnothing$, then* $\lim_n \mu(B_n) = 0$. By Claim 5, it will follow that μ is a measure on $\mathcal{H}$, and both the Extension Theorem and the Uniqueness Theorem take over, giving the result.

Therefore, to start things we let $\{B_n\}_{n=1}^{\infty}$ denote a sequence of $\mathcal{H}$-sets such that $B_1 \supseteq B_2 \supseteq \cdots$ and $\bigcap_{n=1}^{\infty} B_n = \varnothing$; we want $\lim_n \mu(B_n) = 0$. Each B_n has the form $\Phi_{F_n}^{-1}(A_{F_n} \times \Omega_{F_n^c})$, where F_n is a finite subset of I and A_{F_n} is an appropriate element of $\mathcal{F}_{F_n}$. Define $D = \bigcup_{n=1}^{\infty} F_n$; D is either finite or countably infinite. Invoking Claim 1 if D is finite and Claim 7 if D is countably infinite, we may assert the existence of a unique measure μ_D on $\mathcal{F}_D$ with the property that

$$\mu_D(\Phi_{F,D}^{-1}(A_F \times \Omega_{D-F})) = \mu_F(A_F) \quad \text{whenever } F \subseteq D \text{ is finite and } A_F \in \mathcal{F}_F.$$

[For any finite subset F of D, $\Phi_{F,D} : \Omega_D \to \Omega_F \times \Omega_{D-F}$ is the mapping that assigns each $(\omega_i)_{i \in D} \in \Omega_D$ to the element $((\omega_i)_{i \in F}, (\omega_i)_{D-F}) \in \Omega_F \times \Omega_{D-F}$. This mapping is one-to-one and onto, establishing a one-to-one correspondence between the sets in $\mathcal{F}_D$ and the sets in $\mathcal{F}_F \otimes \mathcal{F}_{D-F}$.]

For each $n \in \mathbb{N}$, define $B_n^D = \Phi_{F_n,D}^{-1}(A_{F_n} \times \Omega_{D-F_n}) \subseteq \Omega_D$; we have that $B_n^D \in \mathcal{F}_D$. Because $\{B_n\}_{n=1}^{\infty}$ is nonincreasing to $\varnothing$, we have that $\{B_n^D\}_{n=1}^{\infty}$ is a nonincreasing sequence of $\mathcal{F}_D$-sets with limit $\varnothing$. Since μ_D is a finite measure on $\mathcal{F}_D$, it follows by (M9) from Section 2.2 that $\lim_n \mu_D(B_n^D) = 0$. Invoking the product property for μ_D in the display above and using the product property of μ as given in Claim 4, we have

$$\begin{aligned} 0 = \lim_n \mu_D(B_n^D) &= \lim_n \mu_D(\Phi_{F_n,D}^{-1}(A_{F_n} \times \Omega_{D-F_n})) \\ &= \lim_n \mu_{F_n}(A_{F_n}) \\ &= \lim_n \mu(\Phi_{F_n}^{-1}(A_{F_n} \times \Omega_{F_n^c})) \\ &= \lim_n \mu(B_n). \end{aligned}$$

This completes the proof. ∎

With a great deal of work behind us, let us summarize affairs. The main result is this: given an arbitrary collection $\{(\Omega_i, \mathcal{F}_i, \mu_i)\}_{i\in I}$ of measure spaces with $\mu_i(\Omega_i) = 1$ for each $i \in I$, there exists a measure μ on $\mathcal{F}$ such that

(1) $\mu(\Omega) = 1$, and

(2) $\mu(\Phi_F^{-1}(A_F \times \Omega_{F^c})) = \mu_F(A_F)$ for every finite subset $F \subseteq I$ and $A_F \in \mathcal{F}_F$, where μ_F is the measure on $\mathcal{F}_F$ as described in Claim 1 [or equivalently the definition following Claim 1].

As a special case of (2), we have the following result. Let $F = \{i_1, \cdots, i_n\}$ denote a finite subset of I, and suppose that $A_F \in \mathcal{F}_F$ has the special form $\prod_{j=1}^n A_{i_j}$, where $A_{i_j} \in \mathcal{F}_{i_j}$ for each $j \in \{1, \cdots, n\}$. [That is, $A_F \in \mathcal{E}_F$.] Then

$$\mu(\Phi_F^{-1}(A_F \times \Omega_{F^c})) = \mu_F(A_F) = \prod_{j=1}^n \mu_{i_j}(A_{i_j}).$$

If we define $A_i = \Omega_i$ for each $i \in F^c$, then $\Phi_F^{-1}(A_F \times \Omega_{F^c}) = \prod_{i\in I} A_i$, and the equation above says that

$$\mu\left(\prod_{i\in I} A_i\right) = \prod_{j=1}^n \mu_{i_j}(A_{i_j}).$$

If $A_i = \Omega_i$ for those $i \in F$ as well, then we obtain $\mu(\Omega) = 1$.

Conventions. We have deliberately been formal in our construction process. For example, we have viewed Ω as different from $\Omega_F \times \Omega_{F^c}$, and we have viewed $\mathcal{F}_F \otimes \mathcal{F}_{F^c}$ as different from $\mathcal{F}$. More examples could be given, but the idea is clear. The desire to carefully treat certain objects as distinct is what creates a fair amount of extra work. It is debatable whether treating, say, Ω as something different from $\Omega_F \times \Omega_{F^c}$, or, say, $\mathcal{F}$ as something different from $\mathcal{F}_F \otimes \mathcal{F}_{F^c}$ is worth the extra work entailed. Many mathematicians would say that the extra rigor is not worth the extra work. We have geared our presentation toward obtaining as rigorous a result as possible with the idea of letting the reader decide for himself if the extra care and worry are worth it. Regardless, the hard work has been done now in a careful and somewhat tedious fashion, and for the rest of the chapter we shall adopt certain simplifying conventions.

Cartesian Products. We shall no longer view $\prod_{i\in I} A_i$ as distinct from the product $\prod_{i\in F} A_i \times \prod_{i\in F^c} A_i$, even though in a strict set-theoretic sense the two objects are different. The Identification Lemma asserts the existence of a natural one-to-one correspondence between the elements of the two objects, and hence we will not run aground in treating the element ω of $\prod_{i\in I} A_i$ as identical with the element

$$(\omega_F, \omega_{F^c}) \in \prod_{i\in F} A_i \times \prod_{i\in F^c} A_i.$$

As a special case, we identify Ω with $\Omega_F \times \Omega_{F^c}$. Other situations not mentioned here may arise, and the reader is trusted to not have any problems in recognizing when an identification of objects [that are formally different according to strict set-theoretic standards] is made.

σ-fields. The Identification Lemma asserts that there is a one-to-one correspondence between the sets in $\mathcal{F}$ and $\mathcal{F}_F \otimes \mathcal{F}_{F^c}$. We shall in the future treat a set in $\mathcal{F}$ as identical with its corresponding set in $\mathcal{F}_F \otimes \mathcal{F}_{F^c}$. As a consequence, we will write $\mu(A_F \times \Omega_{F^c})$, whereas if we wanted to truly be strict [as was done in this section] we would write $\mu(\Phi_F^{-1}(A_F \times \Omega_{F^c}))$. We will write

$$\mu(A_{i_1} \times \cdots \times A_{i_n} \times \Omega_{\{i_1,\cdots,i_n\}^c}) = \prod_{j=1}^{n} \mu_{i_j}(A_{i_j}),$$

where the left side in the strictest sense denotes μ applied to the unique $\mathcal{F}$-set corresponding to the element

$$A_{i_1} \times \cdots \times A_{i_n} \times \Omega_{\{i_1,\cdots,i_n\}^c} \in \mathcal{F}_{i_1} \times \cdots \times \mathcal{F}_{i_n} \times \mathcal{F}_{\{i_1,\cdots,i_n\}^c}.$$

Other similar situations may arise, and the competent reader is trusted to know when an identification of objects that are different in a strict set-theoretic sense is being made, and to know exactly what the formal statements would require.

Measurable Functions. As a consequence of the previous paragraphs, we will view [say] $\mathcal{F}_F \otimes \mathcal{F}_{F^c}/\mathcal{B}^*$-measurable functions as $\mathcal{F}/\mathcal{B}^*$-measurable.

We close this long section with the following consequence of our identification conventions given just above.

Claim 9. *Let I denote an arbitrary index set, and let D denote a nonempty proper subset of I. Let μ_D and μ_{D^c} denote the product measures on the σ-fields $\mathcal{F}_D$ and $\mathcal{F}_{D^c}$ as constructed in Claims 1, 7, and 8. Under the identification of $\mathcal{F}$ with $\mathcal{F}_D \otimes \mathcal{F}_{D^c}$, we have that $\mu = \mu_D \otimes \mu_{D^c}$. That is, we may identify $(\Omega, \mathcal{F}, \mu)$ with $(\Omega_D \times \Omega_{D^c}, \mathcal{F}_D \otimes \mathcal{F}_{D^c}, \mu_D \otimes \mu_{D^c})$.*

Proof: Let $F \subseteq I$ be finite with $A_F \in \mathcal{E}_F$. Write $A_F = \prod_{i\in F} A_i$, where $A_i \in \mathcal{F}_i$ for each $i \in F$. Define

$$A_{F\cap D} = \prod_{i\in F\cap D} A_i \in \mathcal{E}_{F\cap D} \quad \text{and} \quad A_{F\cap D^c} = \prod_{i\in F\cap D^c} A_i \in \mathcal{E}_{F\cap D^c}.$$

Identifying $\mathcal{E}_F$ with $\mathcal{E}_{F\cap D} \times \mathcal{E}_{F\cap D^c}$, and writing $A_F = A_{F\cap D} \times A_{F\cap D^c}$, we have

$$\begin{aligned}
& \mu(A_F \times \Omega_{F^c}) \\
= \;& \mu_F(A_F) \;\; \text{[product property of } \mu] \\
= \;& \mu_{(F\cap D)\cup(F\cap D^c)}(A_{F\cap D} \times A_{F\cap D^c}) \\
= \;& \mu_{F\cap D}(A_{F\cap D})\mu_{F\cap D^c}(A_{F\cap D^c}) \;\; \text{[Claim 2]} \\
= \;& \mu_D(A_{F\cap D} \times \Omega_{F^c\cap D})\mu_{D^c}(A_{F\cap D^c} \times \Omega_{F^c\cap D^c}) \;\; \text{[product property of } \mu_D, \mu_{D^c}] \\
= \;& \mu_D \otimes \mu_{D^c}(A_{F\cap D} \times \Omega_{F^c\cap D} \times A_{F\cap D^c} \times \Omega_{F^c\cap D^c}) \;\; \text{[definition of } \mu_D \otimes \mu_{D^c}] \\
= \;& \mu_D \otimes \mu_{D^c}(A_F \times \Omega_{F^c}) \;\; \text{[identification convention]}.
\end{aligned}$$

This shows that $\mu = \mu_D \otimes \mu_{D^c}$ on the collection

$$\{E \subseteq \Omega : \text{ there is finite } F \subseteq I \text{ and } A_F \in \mathcal{F}_F \text{ with } E = A_F \times \Omega_{F^c}\}.$$

By our identification conventions, the collection above generates the σ-field $\mathcal{F}$, and so the Uniqueness Theorem from Section 3.5 shows that $\mu = \mu_D \otimes \mu_{D^c}$ on $\mathcal{F}$. This completes the proof. ■

Exercises.

1*. Refer to the proof of Claim 4.
(a) Prove Subclaim 1.
(b) Prove Subclaim 2. [Both subclaims are demonstrated in the same fashion as was the subclaim found in the proof of Claim 3.]
(c) Why can't we use the same type of proof as used to demonstrate the finite additivity of μ to show that μ as defined on $\mathcal{H}$ is countably additive? [Hint: given a sequence $\{F_n\}_{n=1}^{\infty}$ of finite subsets of I, it is not necessarily the case that $\bigcup_{n=1}^{\infty} F_n$ is a finite subset of I.]

2*. Prove equalities (*) and (**) given in the proof of Subclaim 3 of Claim 7, following the outlines given there. [The proof of the equality (*) is direct and should pose no problem. For the equality (**), it should be straightforward by now to show that $\mathcal{E}_{\mathbb{N}_m} \subseteq \mathcal{C}$ and $\Omega_{\mathbb{N}_m} \in \mathcal{C}$. Use the identity $I_{A^c} = 1 - I_A$ to show that $\mathcal{C}$ is closed under complementation. To see that $\mathcal{C}$ is closed under countable unions, use the identity $I_{\bigcup_{n=1}^{\infty} A_n} = \sup_n I_{A_n}$.]

3. Let $\{(\Omega_i, \mathcal{F}_i, \mu_i)\}_{i=1}^{\infty}$ denote a sequence of measure spaces such that $\mu_i(\Omega_i) = 1$ for all but finitely many $i \in \mathbb{N}$ and the remaining [finitely many] measure spaces are σ-finite. There exists a measure μ on $\mathcal{F}$ such that $\mu(A_F \times \Omega_{F^c}) = \mu_F(A_F)$ for every nonempty finite $F \subseteq I$ and $A_F \in \mathcal{F}_F$. [Note that Claim 1 of this section does not require the individual component spaces to have measure one, and hence we may construct a measure μ_F as given in the definition following Claim 1 for this slightly more general situation. From here, carry on the program of construction as given in the section.]

4*. Let $\{(\Omega_i, \mathcal{F}_i, \mu_i)\}_{i=1}^{\infty}$ denote a sequence of measure spaces with $\mu_i(\Omega_i) = 1$ for each $i \in \mathbb{N}$. If E has the form $\prod_{i=1}^{\infty} A_i$ where $A_i \in \mathcal{F}_i$ for each $i \in \mathbb{N}$, then $E \in \mathcal{F}$ and

$$\mu(E) = \prod_{i=1}^{\infty} \mu_i(A_i) = \lim_n \prod_{i=1}^{n} \mu_i(A_i).$$

[Write E as the limit of the sequence $\{A_1 \times \cdots \times A_n \times \Omega_{n+1} \times \Omega_{n+2} \times \cdots\}_{n=1}^{\infty}$. This sequence is in $\mathcal{F}$ and is decreasing to $\prod_{i=1}^{\infty} A_i$. Use the fact that measures are continuous from above and invoke the product property of μ.]

5. Let $I = \mathbb{N}$, let $\Omega_i = \{0, 1\}$ for each $i \in \mathbb{N}$, and let $\mathcal{F}_i = 2^{\Omega_i}$ for each $i \in \mathbb{N}$. For each $i \in \mathbb{N}$, let μ_i be defined on $\mathcal{F}_i$ by writing $\mu_i(\{0\}) = \mu_i(\{1\}) = \frac{1}{2}$. Let $\Omega = \prod_{i=1}^{\infty} \Omega_i$, let $\mathcal{F}$ denote the usual σ-field on Ω, and let μ denote the measure of Claim 7.

(a) For any finite subset $\{i_1, \cdots, i_k\}$ of I and sequence $(a_{i_1}, \cdots, a_{i_k})$ of zeros and ones, define

$$A(a_{i_1}, \cdots, a_{i_k}) = \{\boldsymbol{\omega} \in \Omega : \omega_{i_1} = a_{i_1}, \cdots, \omega_{i_k} = a_{i_k}\}.$$

Show that $\mu(A(a_{i_1}, \cdots, a_{i_k})) = 2^{-k}$.

(b) For each $i \in \mathbb{N}$, define f_i on Ω by writing $f_i(\boldsymbol{\omega}) = \omega_i - \frac{1}{2}$. Next, for each $n \in \mathbb{N}$, define $h_n = \frac{1}{n}(f_1 + \cdots + f_n)$.

(i) Show that $f_j = I_{A(1_j)} - \frac{1}{2}$ for each $j \in \mathbb{N}$.

(ii) Show that $f_j f_k = I_{E(1_j, 1_k)} - \frac{1}{2} I_{E(1_j)} - \frac{1}{2} I_{E(1_k)} + \frac{1}{4}$ for all $j, k \in \mathbb{N}$ with $j \neq k$.

(iii) Show that $f_j^2 = \frac{1}{4}$ for each $j \in \mathbb{N}$.

(c) Show that $\|h_n\|_2^2 = \frac{1}{n^2} \sum_{j=1}^{n} \sum_{k=1}^{n} \int_{\Omega} f_j f_k \, d\mu = \frac{1}{4n^2} \sum_{j=1}^{n} \sum_{k=1}^{n} \delta_{jk} = \frac{1}{4n}$, where δ_{jk} is 1 or 0 as $j = k$ or $j \neq k$. Therefore, h_n converges to 0 in the L^2 norm.

(d) Show that h_n converges in measure to 0. [Recall Claim 2 of Section 8.7.] That is, for every $\epsilon > 0$ we have

$$\lim_n \mu\left(\left\{\boldsymbol{\omega} \in \Omega : \left|\frac{\omega_1 + \cdots + \omega_n}{n} - \frac{1}{2}\right| > \epsilon\right\}\right) = 0.$$

(e) Let 0 and 1 correspond to obtaining heads or tails, respectively, upon flipping a fair coin. Then $\frac{1}{n}(\omega_1 + \cdots + \omega_n)$ may in a certain sense be thought of as the proportion of tails obtained in the first n flips. Interpret (d) in this coin-toss setting.

6. Halmos gives the following application of the theory of this section to give a completely nontopological construction of Lebesgue measure on the real line. For each $n \in \mathbb{N}$, let $\Omega_n = \{0, 1\}$, $\mathcal{F}_n = 2^{\Omega_n}$, and let $\mu_n(\{0\}) = \mu_n(\{1\}) = \frac{1}{2}$. Let $(\Omega, \mathcal{F}, \mu)$ denote the product measure space [for the sequence $\{(\Omega_n, \mathcal{F}_n, \mu_n)\}_{n=1}^{\infty}$] as constructed in this section.

(a) For every $\boldsymbol{\omega} = (\omega_1, \omega_2, \cdots) \in \Omega$, we have $\{\boldsymbol{\omega}\} \in \mathcal{F}$ and $\mu(\{\boldsymbol{\omega}\}) = 0$. [Refer to Exercise 4.]

(b) The set $\bar{E}$ of all points $\boldsymbol{\omega} \in \Omega$ for which $\omega_i = 1$ for all but a finite number of i is in $\mathcal{F}$ and has μ-measure zero. [Show that $\bar{E}$ is countably infinite.]

Consider now the measure space $(\bar{\Omega}, \bar{\mathcal{F}}, \bar{\mu})$, where $\bar{\Omega} = \Omega - \bar{E}$, $\bar{\mathcal{F}}$ denotes the σ-field consisting of the sets of the form $A \cap \bar{\Omega}$ as A ranges over $\mathcal{F}$, and $\bar{\mu}(E \cap \bar{\Omega}) = \mu(E)$ for every $E \in \mathcal{F}$. [It is easy to see that $\bar{\mu}$ is consistently defined.]

(c) Given $\boldsymbol{\omega} \in \bar{\Omega}$, write $z(\boldsymbol{\omega}) = \sum_{i=1}^{\infty} \frac{\omega_i}{2^i}$. Show that z establishes a one-to-one correspondence between $\bar{\Omega}$ and $[0, 1)$. [Consider the binary expansion of each element in $[0, 1)$ with the agreement that, if the expansion is not unique, the terminating expansion is to be preferred to the infinite expansion.]

(d) Let $0 \leq a < b \leq 1$. Let $A = \{x : a \leq x < b\}$ and $E = \{\omega \in \bar{\Omega} : z(\omega) \in A\}$. Then $E \in \bar{\mathcal{F}}$ and $\bar{\mu}(E) = b - a$. [Hint: it is sufficient to consider the case where a and b are binary rational numbers.]
(e) If A is any Borel subset of $[0,1)$ and $E = \{\omega \in \bar{\Omega} : z(\omega) \in A\}$, then $E \in \bar{\mathcal{F}}$ and $\bar{\mu}(E)$ equals the Lebesgue measure of A. [Use the uniqueness results from Section 3.5.]
(f) Lebesgue measure on $[0,1)$ has been constructed. By considering the real line as a countable disjoint union of the intervals $[n, n+1)$ as n ranges over $\mathbb{Z}$, construct Lebesgue measure on the real line. [This part doesn't use the product theory of the section.]

11.3 CONVERGENCE THEOREMS IN PRODUCT SPACE

With the construction of product measure finally behind us, we may now turn to discussing some results that follow from the existence of such a product measure. In this section, we prove two convergence theorems due to Jessen and the famous Zero-One Law due to Kolmogorov. These facts will in turn be used in the following section to prove the famous L^2 version of the Strong Law of Large Numbers.

Framework. Let $(\Omega_i, \mathcal{F}_i, \mu_i)$ denote a measure space with $\mu_i(\Omega_i) = 1$ for each i in I. We will take $I = \mathbb{N}$ in this section. Let $\mathbb{N}_m = \{1, \cdots, m\}$ for each $m \in \mathbb{N}$. Let $f \in L^1(\Omega, \mathcal{F}, \mu)$, where $\mathcal{F}$ and μ are the [product] σ-field and measure discussed and constructed in the previous two sections of this chapter. For each $\omega \in \Omega$ and $n \in \mathbb{N}$, we write

$$\begin{aligned} f_n(\omega) &= f_n(\omega_1, \cdots, \omega_n, \omega_{n+1}, \cdots) \\ &= \int_{\Omega_{\mathbb{N}_n^c}} f(\omega_1, \cdots, \omega_n, \xi_{n+1}, \cdots)\, d\mu_{\mathbb{N}_n^c}(\xi_{n+1}, \cdots). \end{aligned}$$

[Note for each $\omega \in \Omega$ that $f_n(\omega)$ does not depend on $\omega_{n+1}, \omega_{n+2}, \cdots$.] For each $\omega \in \Omega$ and $n \in \mathbb{N}$, we write

$$\begin{aligned} f_n'(\omega) &= f_n'(\omega_1, \cdots, \omega_n, \omega_{n+1}, \cdots) \\ &= \int_{\Omega_{\mathbb{N}_n}} f(\xi_1, \cdots, \xi_n, \omega_{n+1}, \cdots)\, d\mu_{\mathbb{N}_n}(\xi_1, \cdots, \xi_n). \end{aligned}$$

[Note for each $\omega \in \Omega$ that $f_n'(\omega)$ does not depend on $\omega_1, \cdots, \omega_n$.]

The fact that the definitions of f_n and f_n' make sense for each $n \in \mathbb{N}$ and $\omega \in \Omega$ immediately follows by the Nonnegative and Integrable Fubini Theorems. [See Exercise 1.]

We now turn to the first of the three main results in this section. The reader is encouraged to review the results of Section 9.4. We shall require Claim 1, Example 1, Claim 3, and Claim 5 from that section.

Claim 1. *$\lim_n f_n(\omega) = f(\omega)$ for μ-a.e. $\omega \in \Omega$.*

Proof: First consider the case of where f is nonnegative. Let $\mathcal{F}^{(n)}$ denote the collection $\{A_{\mathbb{N}_n} \times \Omega_{\mathbb{N}_n^c} : A_{\mathbb{N}_n} \in \mathcal{F}_{\mathbb{N}_n}\}$ for each $n \in \mathbb{N}$. Then $\mathcal{F}^{(n)}$ is a σ-field on Ω for each $n \in \mathbb{N}$.

Subclaim 1. *$\{\mathcal{F}^{(n)}\}_{n=1}^{\infty}$ is a nondecreasing sequence of σ-fields on Ω such that $\sigma(\bigcup_{n=1}^{\infty} \mathcal{F}^{(n)}) = \mathcal{F}$.*

Proof of Subclaim 1: Pick $n \in \mathbb{N}$, and let $A \in \mathcal{F}^{(n)}$, so that A has the form $A_{\mathbb{N}_n} \times \Omega_{\mathbb{N}_n^c}$ for some $A_{\mathbb{N}_n} \in \mathcal{F}_{\mathbb{N}_n}$. Note that

$$A_{\mathbb{N}_n} \times \Omega_{n+1} \in \mathcal{F}_{\mathbb{N}_n} \times \mathcal{F}_{n+1} \subseteq \mathcal{F}_{\mathbb{N}_n} \otimes \mathcal{F}_{n+1} = \mathcal{F}_{\mathbb{N}_{n+1}};$$

the last equality follows by our identification conventions. Since

$$A = A_{\mathbb{N}_n} \times \Omega_{\mathbb{N}_n^c} = (A_{\mathbb{N}_n} \times \Omega_{n+1}) \times \Omega_{\mathbb{N}_{n+1}^c},$$

it follows that $A \in \mathcal{F}^{(n+1)}$. This shows that $\mathcal{F}^{(1)} \subseteq \mathcal{F}^{(2)} \subseteq \cdots$. The fact that $\sigma(\bigcup_{n=1}^{\infty} \mathcal{F}^{(n)}) = \mathcal{F}$ should be clear on a moment's thought. This completes the proof of the subclaim.

In the notation of the framework given in Section 9.4, the set Ω there is the set $\Omega = \prod_{i=1}^{\infty} \Omega_i$ here. The σ-field $\mathcal{F}$ given there is the product σ-field $\mathcal{F}$ here. The σ-fields $\mathcal{F}_n$ given there are the σ-fields $\mathcal{F}^{(n)}$ here. The σ-field $\mathcal{F}_\infty = \sigma(\bigcup_{n=1}^{\infty} \mathcal{F}_n)$ given there is $\mathcal{F}$ here [by Subclaim 1]. The measure μ there is the finite product measure μ here. The measure η there will here be defined by writing

$$\eta(A) = \int_A f \, d\mu \quad \text{for every } A \in \mathcal{F}.$$

Then η is a finite measure on $\mathcal{F}$ and $\eta \ll \mu$ [so that $\eta_{ac} = \eta$], and by definition we have that f is a Radon-Nikodym derivative of η with respect to μ. By Example 1 of Section 9.4, it immediately follows that f is a derivative of η with respect to μ. Let $\eta^{(n)}$ and $\mu^{(n)}$ denote the restrictions of η and μ to $\mathcal{F}^{(n)}$ [respectively] for each $n \in \mathbb{N}$:

$$\eta^{(n)}(A) = \int_A f \, d\mu \quad \text{for every } A \in \mathcal{F}^{(n)}.$$

Subclaim 2. *Let $n \in \mathbb{N}$. We have the following:*

(a) f_n is $\mathcal{F}^{(n)}/\mathcal{B}^*$-measurable [and nonnegative since f is nonnegative], hence we may meaningfully discuss $\int_A f_n \, d\mu^{(n)}$ for every $A \in \mathcal{F}^{(n)}$.

(b) f_n is a Radon-Nikodym derivative of $\eta^{(n)}$ with respect to $\mu^{(n)}$, hence f_n is a derivative of $\eta^{(n)}$ with respect to $\mu^{(n)}$.

Proof of Subclaim 2: The nonnegative function f is $\mathcal{F}/\mathcal{B}^*$-measurable by virtue of being in $L^1(\Omega, \mathcal{F}, \mu)$. Identifying $\mathcal{F}$ with $\mathcal{F}_{\mathbb{N}_n} \otimes \mathcal{F}_{\mathbb{N}_n^c}$, it follows from the Nonnegative Fubini Theorem that the mapping

$$(\omega_1, \cdots, \omega_n) \mapsto \int_{\Omega_{\mathbb{N}_n^c}} f(\omega_1, \cdots, \omega_n, \xi_{n+1}, \cdots) \, d\mu_{\mathbb{N}_n^c}(\xi_{n+1}, \cdots)$$

is $\mathcal{F}_{\mathbb{N}_n}/\mathcal{B}^*$-measurable. Given any $B \in \mathcal{B}^*$, the inverse image of B under the mapping

$$(\omega_1, \cdots, \omega_n, \omega_{n+1}, \cdots) \mapsto \int_{\Omega_{\mathbb{N}_n^c}} f(\omega_1, \cdots, \omega_n, \xi_{n+1}, \cdots) \, d\mu_{\mathbb{N}_n^c}(\xi_{n+1}, \cdots)$$

is the set

$$\begin{aligned} &\left\{ \boldsymbol{\omega} \in \Omega : \int_{\Omega_{\mathbb{N}_n^c}} f(\omega_1, \cdots, \omega_n, \xi_{n+1}, \cdots) \, d\mu_{\mathbb{N}_n^c}(\xi_{n+1}, \cdots) \in B \right\} \\ = &\left\{ (\omega_1, \cdots, \omega_n) \in \Omega_{\mathbb{N}_n} : \int_{\Omega_{\mathbb{N}_n^c}} f(\omega_1, \cdots, \omega_n, \xi_{n+1}, \cdots) \, d\mu_{\mathbb{N}_n^c}(\xi_{n+1}, \cdots) \in B \right\} \\ &\times \Omega_{\mathbb{N}_n^c}. \end{aligned}$$

The first set in the product above is in $\mathcal{F}_{\mathbb{N}_n}$ by the $\mathcal{F}_{\mathbb{N}_n}/\mathcal{B}^*$-measurability of the mapping

$$(\omega_1, \cdots, \omega_n) \mapsto \int_{\Omega_{\mathbb{N}_n^c}} f(\omega_1, \cdots, \omega_n, \xi_{n+1}, \cdots) \, d\mu_{\mathbb{N}_n^c}(\xi_{n+1}, \cdots).$$

Therefore, the product displayed above is an element of $\mathcal{F}^{(n)}$. It follows that the mapping

$$(\omega_1, \cdots, \omega_n, \omega_{n+1}, \cdots) \mapsto \int_{\Omega_{\mathbb{N}_n^c}} f(\omega_1, \cdots, \omega_n, \xi_{n+1}, \cdots) \, d\mu_{\mathbb{N}_n^c}(\xi_{n+1}, \cdots)$$

is $\mathcal{F}^{(n)}/\mathcal{B}^*$-measurable. That is, f_n is $\mathcal{F}^{(n)}/\mathcal{B}^*$-measurable. This proves (a).

We now turn to (b). Pick $A \in \mathcal{F}^{(n)}$; write $A = A_{\mathbb{N}_n} \times \Omega_{\mathbb{N}_n^c}$ for an appropriate set $A_{\mathbb{N}_n}$ in $\mathcal{F}_{\mathbb{N}_n}$. We wish to demonstrate the equation

$$\eta^{(n)}(A) = \int_A f_n \, d\mu^{(n)}, \quad \text{that is,} \quad \eta(A) = \int_A f_n \, d\mu.$$

Observe that

$$\begin{aligned} \eta(A) &= \int_\Omega I_A(\boldsymbol{\omega}) f(\boldsymbol{\omega}) \, d\mu(\boldsymbol{\omega}) \\ &= \int_{\Omega_{\mathbb{N}_n} \times \Omega_{\mathbb{N}_n^c}} I_{A_{\mathbb{N}_n} \times \Omega_{\mathbb{N}_n^c}}(\boldsymbol{\omega}_{\mathbb{N}_n}, \boldsymbol{\omega}_{\mathbb{N}_n^c}) f(\boldsymbol{\omega}_{\mathbb{N}_n}, \boldsymbol{\omega}_{\mathbb{N}_n^c}) \, d\mu_{\mathbb{N}_n} \otimes \mu_{\mathbb{N}_n^c}(\boldsymbol{\omega}_{\mathbb{N}_n}, \boldsymbol{\omega}_{\mathbb{N}_n^c}) \\ &= \int_{\Omega_{\mathbb{N}_n}} \int_{\Omega_{\mathbb{N}_n^c}} I_{A_{\mathbb{N}_n} \times \Omega_{\mathbb{N}_n^c}}(\boldsymbol{\omega}_{\mathbb{N}_n}, \boldsymbol{\omega}_{\mathbb{N}_n^c}) f(\boldsymbol{\omega}_{\mathbb{N}_n}, \boldsymbol{\omega}_{\mathbb{N}_n^c}) \, d\mu_{\mathbb{N}_n^c}(\boldsymbol{\omega}_{\mathbb{N}_n^c}) \, d\mu_{\mathbb{N}_n}(\boldsymbol{\omega}_{\mathbb{N}_n}) \\ &= \int_{\Omega_{\mathbb{N}_n}} I_{A_{\mathbb{N}_n}}(\boldsymbol{\omega}_{\mathbb{N}_n}) \int_{\Omega_{\mathbb{N}_n^c}} f(\boldsymbol{\omega}_{\mathbb{N}_n}, \boldsymbol{\omega}_{\mathbb{N}_n^c}) \, d\mu_{\mathbb{N}_n^c}(\boldsymbol{\omega}_{\mathbb{N}_n^c}) \, d\mu_{\mathbb{N}_n}(\boldsymbol{\omega}_{\mathbb{N}_n}) \\ &= \int_{A_{\mathbb{N}_n}} f_n(\boldsymbol{\omega}_{\mathbb{N}_n}, \boldsymbol{\xi}_{\mathbb{N}_n^c}) \, d\mu_{\mathbb{N}_n}(\boldsymbol{\omega}_{\mathbb{N}_n}). \qquad (*) \end{aligned}$$

The first equality is clear. The second equality follows by Claim 9 from the previous section. The third equality follows by the Nonnegative Fubini Theorem. The fourth equality holds because $I_{A_{\mathbb{N}_n} \times \Omega_{\mathbb{N}_n^c}}(\boldsymbol{\omega}_{\mathbb{N}_n}, \boldsymbol{\omega}_{\mathbb{N}_n^c}) = I_{A_{\mathbb{N}_n}}(\boldsymbol{\omega}_{\mathbb{N}_n})$ for every $\boldsymbol{\omega}_{\mathbb{N}_n} \in \Omega_{\mathbb{N}_n}$

and $\omega_{\mathbb{N}_n^c} \in \Omega_{\mathbb{N}_n^c}$. The fifth equality holds by the definition of f_n; note for every $\omega_{\mathbb{N}_n} \in \Omega_{\mathbb{N}_n}$ that $f_n(\omega_{\mathbb{N}_n}, \xi_{\mathbb{N}_n^c})$ does not depend on $\xi_{\mathbb{N}_n^c}$.

By similar reasoning, we next have that

$$\begin{aligned}
&\int_A f_n(\omega)\, d\mu(\omega) \\
&= \int_\Omega I_A(\omega) f_n(\omega)\, d\mu(\omega) \\
&= \int_{\Omega_{\mathbb{N}_n} \times \Omega_{\mathbb{N}_n^c}} I_{A_{\mathbb{N}_n} \times \Omega_{\mathbb{N}_n^c}}(\omega_{\mathbb{N}_n}, \omega_{\mathbb{N}_n^c}) f_n(\omega_{\mathbb{N}_n}, \omega_{\mathbb{N}_n^c})\, d\mu_{\mathbb{N}_n} \otimes \mu_{\mathbb{N}_n^c}(\omega_{\mathbb{N}_n}, \omega_{\mathbb{N}_n^c}) \\
&= \int_{\Omega_{\mathbb{N}_n}} I_{A_{\mathbb{N}_n}}(\omega_{\mathbb{N}_n}) f_n(\omega_{\mathbb{N}_n}, \xi_{\mathbb{N}_n^c}) \int_{\Omega_{\mathbb{N}_n^c}} d\mu_{\mathbb{N}_n^c}(\omega_{\mathbb{N}_n^c})\, d\mu_{\mathbb{N}_n}(\omega_{\mathbb{N}_n}) \\
&= \int_{A_{\mathbb{N}_n}} f_n(\omega_{\mathbb{N}_n}, \xi_{\mathbb{N}_n^c})\, d\mu_{\mathbb{N}_n}(\omega_{\mathbb{N}_n}) \qquad (**).
\end{aligned}$$

[The fourth equality holds by the Nonnegative Fubini Theorem and the fact that $f_n(\omega_{\mathbb{N}_n}, \omega_{\mathbb{N}_n^c})$ does not depend on $\omega_{\mathbb{N}_n^c}$.] Comparing the expressions for (*) and (**) shows that

$$\eta(A) = \int_A f_n(\omega)\, d\mu(\omega).$$

It follows that f_n is a Radon-Nikodym derivative of $\eta^{(n)}$ with respect to $\mu^{(n)}$, proving (b) and completing the proof of the subclaim.

The hypotheses of Claim 5 of Section 9.4 are satisfied: part (b) there asserts that $\lim_n f_n$ exists μ-a.e. on Ω, and part (c) there gives the equation $\eta(A) = \int_A \lim_n f_n\, d\mu$ for every $A \in \mathcal{F}$. At the same time, we recall the definition of η to obtain the equation $\eta(A) = \int_A f\, d\mu$, valid for each $A \in \mathcal{F}$. That is, both $\lim_n f_n$ and f are Radon-Nikodym derivatives of η with respect to μ. This means [in the terminology of Section 9.4] that both $\lim_n f_n$ and f are derivatives of η with respect to μ. Since derivatives are unique up to sets of μ-measure zero [Claim 1 of Section 9.4], it follows that $\lim_n f_n = f$ μ-a.e. on Ω, completing the proof for the case where f is nonnegative. For the situation where f is a general element of $L^1(\Omega, \mathcal{F}, \mu)$, write $f = f^+ - f^-$ and apply the result to the nonnegative functions f^+ and f^- separately. [See Exercise 2.] ∎

The following claim is due to Kolmogorov, and, besides being needed for our development, is of independent interest. A key component of the proof is the result of Claim 1.

Definition. Let $\boldsymbol{\omega}$ and $\boldsymbol{\xi}$ denote elements of Ω. We will say that $\boldsymbol{\omega}$ and $\boldsymbol{\xi}$ are *ultimately equal* if and only if there is N such that $\omega_n = \xi_n$ for all $n \geq N$. We write $\boldsymbol{\omega} =_u \boldsymbol{\xi}$ in this situation.

Claim 2. Kolmogorov's Zero-One Law. *Let $U \in \mathcal{F}$ have the following property: for every $\boldsymbol{\omega} \in \Omega$, $\boldsymbol{\omega} \in U$ if and only if every $\boldsymbol{\xi} \in \Omega$ with $\boldsymbol{\xi} =_u \boldsymbol{\omega}$ is in U. Then $\mu(U) \in \{0, 1\}$.*

Proof: Define f on Ω to be I_U. For every $n \in \mathbb{N}$ and $\boldsymbol{\omega}, \boldsymbol{\xi} \in \Omega$, adopt the notation established for this section, writing

$$f_n(\boldsymbol{\omega}) = \int_{\Omega_{\mathbb{N}_n^c}} I_U(\omega_1, \cdots, \omega_n, \rho_{n+1}, \cdots)\, d\mu_{\mathbb{N}_n^c}(\rho_{n+1}, \cdots)$$

and

$$f_n(\boldsymbol{\xi}) = \int_{\Omega_{\mathbb{N}_n^c}} I_U(\xi_1, \cdots, \xi_n, \rho_{n+1}, \cdots)\, d\mu_{\mathbb{N}_n^c}(\rho_{n+1}, \cdots).$$

Fix $n \in \mathbb{N}$ and $\boldsymbol{\omega}, \boldsymbol{\xi} \in \Omega$. For any $(\rho_{n+1}, \rho_{n+2}, \cdots) \in \Omega_{\mathbb{N}_n^c}$, we have

$$(\omega_1, \cdots, \omega_n, \rho_{n+1}, \rho_{n+2}, \cdots) =_u (\xi_1, \cdots, \xi_n, \rho_{n+1}, \rho_{n+2}, \cdots).$$

By the assumption on U, there are exactly two possibilities for each element $(\rho_{n+1}, \rho_{n+2}, \cdots) \in \Omega_{\mathbb{N}_n^c}$:

$$(\omega_1, \cdots, \omega_n, \rho_{n+1}, \rho_{n+2}, \cdots),\ (\xi_1, \cdots, \xi_n, \rho_{n+1}, \rho_{n+2}, \cdots) \in U,$$

or

$$(\omega_1, \cdots, \omega_n, \rho_{n+1}, \rho_{n+2}, \cdots),\ (\xi_1, \cdots, \xi_n, \rho_{n+1}, \rho_{n+2}, \cdots) \notin U.$$

Let $A \subseteq \Omega_{\mathbb{N}_n^c}$ denote the set of $(\rho_{n+1}, \rho_{n+2}, \cdots)$ such that the first possibility holds, and let $A^c \subseteq \Omega_{\mathbb{N}_n^c}$ denote the set of $(\rho_{n+1}, \rho_{n+2}, \cdots)$ such that the second possibility holds. We have

$$\begin{aligned}
& f_n(\boldsymbol{\omega}) \\
= & \int_A I_U(\omega_1, \cdots, \omega_n, \rho_{n+1}, \cdots)\, d\mu_{\mathbb{N}_n^c}(\rho_{n+1}, \cdots) \\
& + \int_{A^c} I_U(\omega_1, \cdots, \omega_n, \rho_{n+1}, \cdots)\, d\mu_{\mathbb{N}_n^c}(\rho_{n+1}, \cdots) \\
= & \int_A 1\, d\mu_{\mathbb{N}_n^c}(\rho_{n+1}, \cdots) + \int_{A^c} 0\, d\mu_{\mathbb{N}_n^c}(\rho_{n+1}, \cdots) \\
= & \int_A I_U(\xi_1, \cdots, \xi_n, \rho_{n+1}, \cdots)\, d\mu_{\mathbb{N}_n^c}(\rho_{n+1}, \cdots) \\
& + \int_{A^c} I_U(\xi_1, \cdots, \xi_n, \rho_{n+1}, \cdots)\, d\mu_{\mathbb{N}_n^c}(\rho_{n+1}, \cdots) \\
= & f_n(\boldsymbol{\xi}).
\end{aligned}$$

Since $\boldsymbol{\omega}$ and $\boldsymbol{\xi}$ are arbitrary points in Ω, it follows that f_n is constant on Ω. Since $n \in \mathbb{N}$ is arbitrary, it follows that $\{f_n\}_{n=1}^{\infty}$ is a sequence of constant functions on Ω. By Claim 1, we have that $I_U = f = \lim_n f_n$ μ-a.e. on Ω, and hence I_U, being the μ-a.e. limit of the sequence of constant functions $\{f_n\}_{n=1}^{\infty}$, must itself be constant μ-a.e. on Ω. Since I_U can assume only the values 0 and 1, it follows that $I_U = 0$ μ-a.e. on Ω or $I_U = 1$ μ-a.e. on Ω. In the former case, it follows that $\mu(U) = \int_\Omega I_U\, d\mu = \int_\Omega 0\, d\mu = 0$; in the latter case, it same argument gives $\mu(U) = 1$. This completes the proof. ∎

The following claim is a necessary fact for Claim 4. The proof is simple, and the details are left to the reader. [See Exercise 4.]

Claim 3. *Let $(\Xi, \mathcal{Q}, \mu)$ denote an arbitrary measure space, and assume that every $A \in \mathcal{Q}$ is such that $\mu(A) \in \{0,1\}$. Let $f : \Xi \to \bar{\mathbb{R}}$ denote a $\mathcal{Q}/\mathcal{B}^*$-measurable function. Then there is α so that $f = \alpha$ μ-a.e. on Ξ.*

Proof: The claim holds if f is a nonnegative finite-valued $\mathcal{Q}/\mathcal{B}^*$-measurable simple function on Ξ. If f is nonnegative, then writing f as the limit of a nondecreasing sequence of nonnegative finite-valued $\mathcal{Q}/\mathcal{B}^*$-measurable simple functions shows that the result holds for f. For general f, write $f = f^+ - f^-$ and apply the result to the nonnegative functions f^+ and f^-. ∎

We have shown that $\lim_n f_n = f$ μ-a.e. on Ω in Claim 1. Claims 2 and 3 will now be used to prove an analogous result concerning the limit of the sequence $\{f_n'\}_{n=1}^{\infty}$ as defined at the beginning of this section. As with the proof of Claim 1, we lean heavily on the work from Section 9.4. In particular, we require the result of Claim 6 in that section.

Claim 4. $\lim_n f_n'(\omega)$ *exists and equals the number* $\int_\Omega f\, d\mu$ *for* μ*-a.e.* $\omega \in \Omega$.

Proof: We first assume that f is nonnegative. For each $n \in \mathbb{N}$, write

$$\mathcal{F}^{(n)} = \{\Omega_{\mathbb{N}_n} \times A_{\mathbb{N}_n^c} : A_{\mathbb{N}_n^c} \in \mathcal{F}_{\mathbb{N}_n^c}\}.$$

Then $\mathcal{F}^{(n)}$ is a σ-field on Ω for every $n \in \mathbb{N}$.

Subclaim 1. $\mathcal{F} \supseteq \mathcal{F}^{(1)} \supseteq \mathcal{F}^{(2)} \supseteq \cdots$.

Proof of Subclaim 1: Pick $n \in \mathbb{N}$. Our first goal is to show that $\mathcal{F}^{(n)} \subseteq \mathcal{F}$. Pick $A \in \mathcal{F}^{(n)}$. Then we may write $A = \Omega_{\mathbb{N}_n} \times A_{\mathbb{N}_n^c}$ for an appropriate set $A_{\mathbb{N}_n^c}$ in $\mathcal{F}_{\mathbb{N}_n^c}$. Since we are identifying $\mathcal{F}_{\mathbb{N}_n} \otimes \mathcal{F}_{\mathbb{N}_n^c}$ with $\mathcal{F}$, it follows that

$$A \in \mathcal{F}_{\mathbb{N}_n} \times \mathcal{F}_{\mathbb{N}_n^c} \subseteq \mathcal{F}_{\mathbb{N}_n} \otimes \mathcal{F}_{\mathbb{N}_n^c} = \mathcal{F}.$$

This shows that $\mathcal{F}^{(n)} \subseteq \mathcal{F}$.

We will complete matters by demonstrating that $\mathcal{F}^{(n+1)} \subseteq \mathcal{F}^{(n)}$. Pick A in $\mathcal{F}^{(n+1)}$; write $A = \Omega_{\mathbb{N}_{n+1}} \times A_{\mathbb{N}_{n+1}^c}$ for an appropriate set $A_{\mathbb{N}_{n+1}^c} \in \mathcal{F}_{\mathbb{N}_{n+1}^c}$. Using our identification conventions, we may write

$$A = \Omega_{\mathbb{N}_n} \times \left[\Omega_{n+1} \times A_{\mathbb{N}_{n+1}^c}\right].$$

Observe that $\Omega_{n+1} \times A_{\mathbb{N}_{n+1}^c} \in \mathcal{F}_{n+1} \times \mathcal{F}_{\mathbb{N}_{n+1}^c} \subseteq \mathcal{F}_{n+1} \otimes \mathcal{F}_{\mathbb{N}_{n+1}^c} = \mathcal{F}_{\mathbb{N}_n^c}$. From this, it follows by definition that $A \in \mathcal{F}^{(n)}$. This shows that $\mathcal{F}^{(n+1)} \subseteq \mathcal{F}^{(n)}$, and completes the proof of the subclaim.

Define $\mathcal{F}_0$ to be $\bigcap_{n=1}^{\infty} \mathcal{F}^{(n)}$. Since $\mathcal{F}_0$ is a countable intersection of σ-fields on Ω, $\mathcal{F}_0$ is itself a σ-field on Ω. We now state a very special property enjoyed by sets in $\mathcal{F}_0$.

Subclaim 2. *If $U \in \mathcal{F}_0$, then $\mu(U) \in \{0,1\}$.*

Proof of Subclaim 2: Let $U \in \mathcal{F}_0$. If $U = \varnothing$ then $\mu(U) = 0$, so we may assume that U is nonempty. We ultimately wish to invoke the Zero-One Law. To do this, we must show the following: *for every $\boldsymbol{\omega} \in \Omega$, the condition $\boldsymbol{\omega} \in U$ is equivalent to the assertion that $\boldsymbol{\xi} \in U$ for every $\boldsymbol{\xi} \in \Omega$ that is ultimately equal to $\boldsymbol{\omega}$.*

Pick $\boldsymbol{\omega} \in \Omega$, and assume that every $\boldsymbol{\xi}$ with $\boldsymbol{\xi} =_u \boldsymbol{\omega}$ is in U. In particular, $\boldsymbol{\omega} =_u \boldsymbol{\omega}$, so $\boldsymbol{\omega} \in U$. This gives one direction of the desired italicized assertion.

We now turn to the more difficult direction. Let $\boldsymbol{\omega} \in \Omega$ be such that $\boldsymbol{\omega} \in U$, and let $\boldsymbol{\xi} \in \Omega$ be ultimately equal to $\boldsymbol{\omega}$. Then we have that

$$\boldsymbol{\omega} = (\omega_1, \cdots, \omega_n, \rho_{n+1}, \rho_{n+2}, \cdots) \text{ and } \boldsymbol{\xi} = (\xi_1, \cdots, \xi_n, \rho_{n+1}, \rho_{n+2}, \cdots)$$

for some $n \in \mathbb{N}$ and choice of $(\rho_{n+1}, \rho_{n+2}, \cdots) \in \Omega_{\mathbb{N}_n^c}$. Since

$$U \in \mathcal{F}_0 = \bigcap_{m=1}^{\infty} \mathcal{F}^{(m)} \subseteq \mathcal{F}^{(n)},$$

we may write $U = \Omega_{\mathbb{N}_n} \times A_{\mathbb{N}_n^c}$ for some $A_{\mathbb{N}_n^c} \in \mathcal{F}_{\mathbb{N}_n^c}$. Since $(\omega_1, \cdots, \omega_n, \rho_{n+1}, \cdots)$ is in U, it follows that $(\rho_{n+1}, \rho_{n+2}, \cdots) \in A_{\mathbb{N}_n^c}$. But then $\boldsymbol{\xi}$ is an element of $\Omega_{\mathbb{N}_n} \times A_{\mathbb{N}_n^c}$, that is, $\boldsymbol{\xi} \in U$. This gives the more difficult direction of the italicized assertion, and the italicized assertion stands.

Summarizing, we see that the set $U \in \mathcal{F}_0$ satisfies the requirements of the Zero-One Law. Accordingly, we have that $\mu(U) \in \{0, 1\}$. This gives the proof of the subclaim.

We ultimately wish to apply Claim 6 of Section 9.4 to the matters at hand. To do so, we will need the following subclaim. For the subclaim, we let $\mu^{(n)}$ denote the restriction of μ to $\mathcal{F}^{(n)}$ for each $n \in \mathbb{N}$. We will write μ_0 for the restriction of μ to $\mathcal{F}_0$. Next, we define

$$\eta(A) = \int_A f \, d\mu \quad \text{for every } A \in \mathcal{F},$$

so that η is a finite measure, $\eta \ll \mu$, and f is a Radon-Nikodym derivative of η with respect to μ. [Therefore, in the parlance of Section 9.4, f is a derivative of η with respect to μ.] We will write η_0 for the restriction of η from $\mathcal{F}$ to the sub-σ-field $\mathcal{F}_0$.

Subclaim 3. *Let $n \in \mathbb{N}$. We have the following:*

(a) f_n' is $\mathcal{F}^{(n)}/\mathcal{B}^*$-measurable.

(b) Let $\eta^{(n)}$ denote the restriction of η to $\mathcal{F}^{(n)}$. Then f_n' is a Radon-Nikodym derivative of $\eta^{(n)}$ with respect to $\mu^{(n)}$, so that f_n' is a derivative of $\eta^{(n)}$ with respect to $\mu^{(n)}$.

Proof of Subclaim 3: Pick $n \in \mathbb{N}$. We view $(\Omega_{\mathbb{N}_n} \times \Omega_{\mathbb{N}_n^c}, \mathcal{F}_{\mathbb{N}_n} \otimes \mathcal{F}_{\mathbb{N}_n^c}, \mu_{\mathbb{N}_n} \otimes \mu_{\mathbb{N}_n^c})$ as identical to $(\Omega, \mathcal{F}, \mu)$ in accordance with Claim 9 of the previous section. For generic $B \in \mathcal{B}^*$, the inverse image of B under the mapping

$$\boldsymbol{\omega} \mapsto \int_{\Omega_{\mathbb{N}_n}} f(\xi_1, \cdots, \xi_n, \omega_{n+1}, \cdots) \, d\mu_{\mathbb{N}_n}(\xi_1, \cdots, \xi_n)$$

on Ω is

$$\Omega_{\mathbb{N}_n} \times \left\{ (\omega_{n+1}, \omega_{n+2}, \cdots) \in \Omega_{\mathbb{N}_n^c} : \int_{\Omega_{\mathbb{N}_n}} f(\xi_1, \cdots, \xi_n, \omega_{n+1}, \cdots) \, d\mu_{\mathbb{N}_n}(\xi_1, \cdots, \xi_n) \in B \right\}$$

Since the mapping

$$(\omega_{n+1}, \omega_{n+2}, \cdots) \mapsto \int_{\Omega_{\mathbb{N}_n}} f(\xi_1, \cdots, \xi_n, \omega_{n+1}, \cdots) \, d\mu_{\mathbb{N}_n}(\xi_1, \cdots, \xi_n)$$

on $\Omega_{\mathbb{N}_n^c}$ is $\mathcal{F}_{\mathbb{N}_n^c}/\mathcal{B}^*$-measurable by the Nonnegative Fubini Theorem, the second set in the previous displayed product is therefore an element of $\mathcal{F}_{\mathbb{N}_n^c}$. It follows by the definition of $\mathcal{F}^{(n)}$ that the inverse image of B under the mapping

$$\boldsymbol{\omega} \mapsto \int_{\Omega_{\mathbb{N}_n}} f(\xi_1, \cdots, \xi_n, \omega_{n+1}, \cdots) \, d\mu_{\mathbb{N}_n}(\xi_1, \cdots, \xi_n)$$

on Ω is in $\mathcal{F}^{(n)}$. This proves that $\mathcal{F}^{(n)}/\mathcal{B}^*$-measurability of f'_n.

We now turn to (b), which requires showing that

$$\eta^{(n)}(A) = \int_A f'_n \, d\mu^{(n)} \quad \text{for every } A \in \mathcal{F}^{(n)}.$$

By Claim 3 of Section 9.4 and the fact that $\eta^{(n)} = \eta$ on $\mathcal{F}^{(n)}$, it is sufficient to show that

$$\eta(A) = \int_A f'_n \, d\mu \quad \text{for every } A \in \mathcal{F}^{(n)}.$$

To this end, let $A \in \mathcal{F}^{(n)}$; write $A = \Omega_{\mathbb{N}_n} \times A_{\mathbb{N}_n^c}$ for an appropriately chosen set $A_{\mathbb{N}_n^c}$ in $\mathcal{F}_{\mathbb{N}_n^c}$. Freely using the Nonnegative Fubini Theorem and our identification conventions, we have

$$\begin{aligned}
&\int_A f'_n(\boldsymbol{\omega}) \, d\mu(\boldsymbol{\omega}) \\
&= \int_\Omega I_{\Omega_{\mathbb{N}_n} \times A_{\mathbb{N}_n^c}}(\boldsymbol{\omega}_{\mathbb{N}_n}, \boldsymbol{\omega}_{\mathbb{N}_n^c}) \int_{\Omega_{\mathbb{N}_n}} f(\boldsymbol{\xi}_{\mathbb{N}_n}, \boldsymbol{\omega}_{\mathbb{N}_n^c}) \, d\mu_{\mathbb{N}_n}(\boldsymbol{\xi}_{\mathbb{N}_n}) \, d\mu(\boldsymbol{\omega}) \\
&= \int_{\Omega_{\mathbb{N}_n^c}} \int_{\Omega_{\mathbb{N}_n}} I_{A_{\mathbb{N}_n^c}}(\boldsymbol{\omega}_{\mathbb{N}_n^c}) \int_{\Omega_{\mathbb{N}_n}} f(\boldsymbol{\xi}_{\mathbb{N}_n}, \boldsymbol{\omega}_{\mathbb{N}_n^c}) \, d\mu_{\mathbb{N}_n}(\boldsymbol{\xi}_{\mathbb{N}_n}) \, d\mu_{\mathbb{N}_n}(\boldsymbol{\omega}_{\mathbb{N}_n}) \, d\mu_{\mathbb{N}_n^c}(\boldsymbol{\omega}_{\mathbb{N}_n^c}).
\end{aligned}$$

The quantity

$$I_{A_{\mathbb{N}_n^c}}(\boldsymbol{\omega}_{\mathbb{N}_n^c}) \int_{\Omega_{\mathbb{N}_n}} f(\boldsymbol{\xi}_{\mathbb{N}_n}, \boldsymbol{\omega}_{\mathbb{N}_n^c}) \, d\mu_{\mathbb{N}_n}(\boldsymbol{\xi}_{\mathbb{N}_n})$$

is constant with respect to $\boldsymbol{\omega}_{\mathbb{N}_n}$, and hence we have that

$$\begin{aligned}
&\int_A f'_n(\boldsymbol{\omega}) \, d\mu(\boldsymbol{\omega}) \\
&= \int_{\Omega_{\mathbb{N}_n^c}} I_{A_{\mathbb{N}_n^c}}(\boldsymbol{\omega}_{\mathbb{N}_n^c}) \int_{\Omega_{\mathbb{N}_n}} f(\boldsymbol{\xi}_{\mathbb{N}_n}, \boldsymbol{\omega}_{\mathbb{N}_n^c}) \, d\mu_{\mathbb{N}_n}(\boldsymbol{\xi}_{\mathbb{N}_n}) \, d\mu_{\mathbb{N}_n^c}(\boldsymbol{\omega}_{\mathbb{N}_n^c}).
\end{aligned}$$

By the Nonnegative Fubini Theorem, the last quantity is equal to

$$\begin{aligned}
&\int_{\Omega_{\mathbf{N}_n}\times\Omega_{\mathbf{N}_n^c}} I_{\Omega_{\mathbf{N}_n}\times A_{\mathbf{N}_n^c}}(\omega_{\mathbf{N}_n},\omega_{\mathbf{N}_n^c})f(\omega_{\mathbf{N}_n},\omega_{\mathbf{N}_n^c})\,d\mu_{\mathbf{N}_n}\otimes\mu_{\mathbf{N}_n^c}(\omega_{\mathbf{N}_n},\omega_{\mathbf{N}_n^c})\\
&= \int_\Omega I_A(\omega)f(\omega)\,d\mu(\omega)\\
&= \int_A f(\omega)\,d\mu(\omega)\\
&= \eta(A).
\end{aligned}$$

Summarizing, we have shown that $\eta(A) = \int_A f_n'(\omega)\,d\mu(\omega)$ for our generic A in $\mathcal{F}^{(n)}$. This shows that f_n' is a Radon-Nikodym derivative of $\eta^{(n)}$ with respect to $\mu^{(n)}$, and completes the proof of this subclaim.

We now invoke the conclusions of Claim 6 of Section 9.4 to assert that (i) $\lim_n f_n'(\omega)$ exists for μ-a.e. $\omega \in \Omega$, (ii) $\lim_n f_n'$ is $\mathcal{F}_0/\mathcal{B}^*$-measurable, and (iii) $\eta_{0,ac}(A) = \int_A \lim_n f_n'\,d\mu$ for every $A \in \mathcal{F}_0$, where $\eta_{0,ac}$ denotes the absolutely continuous part [with respect to μ_0] of η_0. Since $\eta \ll \mu$ by how η was defined, and since η_0 and μ_0 are the restrictions of η and μ to the sub-σ-field $\mathcal{F}_0$ of $\mathcal{F}$, it follows that $\eta_0 \ll \mu_0$, so that $\eta_{0,ac} = \eta_0$. Conclusion (iii) just stated becomes the assertion that $\eta_0(A) = \int_A \lim_n f_n'\,d\mu$ for every $A \in \mathcal{F}_0$. That is, $\lim_n f_n'$ is a Radon-Nikodym derivative of η_0 with respect to μ_0. In particular, we have by Claim 3 of Section 9.4 that

$$\eta(\Omega) = \eta_0(\Omega) = \int_\Omega \lim_n f_n'\,d\mu_0 = \int_\Omega \lim_n f_n'\,d\mu.$$

At the same time, the original definition of η gives $\eta(\Omega) = \int_\Omega f\,d\mu$, so that

$$\int_\Omega f\,d\mu = \int_\Omega \lim_n f_n'\,d\mu.$$

Since $\lim_n f_n'$ is $\mathcal{F}_0/\mathcal{B}^*$-measurable, we may combine Subclaim 2 and Claim 3 to assert that there is a constant α such that $\lim_n f_n'(\omega) = \alpha$ for μ_0-a.e. $\omega \in \Omega$. Therefore, $\lim_n f_n'(\omega) = \alpha$ for μ-a.e. $\omega \in \Omega$. It follows that

$$\int_\Omega \lim_n f_n'\,d\mu = \int_\Omega \alpha\,d\mu = \alpha\mu(\Omega) = \alpha = \lim_n f_n' \quad \mu\text{-a.e. on } \Omega.$$

Combining the previous two displays gives $\int_\Omega f\,d\mu = \lim_n f_n'$ μ-a.e. on Ω, which is precisely the desired result. This completes the proof for when f is nonnegative. The proof for general $f \in L^1(\Omega, \mathcal{F}, \mu)$ follows by writing $f = f^+ - f^-$ and applying the result separately to f^+ and f^-. [See Exercise 5.] ∎

The convergence results found in Claims 1 and 4 are due to the work of B. Jessen. For our purposes, Claims 1, 2, and 3 are subsidiary to Claim 4. That is, Claim 4 represents the main result of this section; we shall use this result to prove the famous Strong Law of Large Numbers in the upcoming section.

Exercises.

1*. Quickly verify that (i) the definitions of f_n and f'_n make sense for each $\omega \in \Omega$ when f is nonnegative, and (ii) the definitions make sense for μ-a.e. $\omega \in \Omega$ when f in $L^1(\Omega, \mathcal{F}, \mu)$ is general. [Let f be nonnegative. View

$$(\Omega, \mathcal{F}, \mu) \quad \text{as} \quad (\Omega_{\mathbf{N}_n} \times \Omega_{\mathbf{N}_n^c}, \mathcal{F}_{\mathbf{N}_n} \otimes \mathcal{F}_{\mathbf{N}_n^c}, \mu_{\mathbf{N}_n} \otimes \mu_{\mathbf{N}_n^c}).$$

From Section 10.1, we have for each $\omega_{\mathbf{N}_n} \in \Omega_{\mathbf{N}_n}$ that the mapping $\xi_{\mathbf{N}_n^c} \mapsto f(\omega_{\mathbf{N}_n}, \xi_{\mathbf{N}_n^c})$ on $\Omega_{\mathbf{N}_n^c}$ is nonnegative and $\mathcal{F}_{\mathbf{N}_n^c}/\mathcal{B}^*$-measurable. Therefore, the expression

$$\int_{\Omega_{\mathbf{N}_n^c}} f(\omega_{\mathbf{N}_n}, \xi_{\mathbf{N}_n^c})\, d\mu_{\mathbf{N}_n^c}(\xi_{\mathbf{N}_n^c})$$

is defined for each $\omega_{\mathbf{N}_n} \in \Omega_{\mathbf{N}_n}$. Since the above expression depends only on $\omega_{\mathbf{N}_n}$, we may consider without fear the mapping that assigns every $\omega \in \Omega$ to the above expression. That is, $f_n(\omega)$ makes sense for each $\omega \in \Omega$ when f is nonnegative. Now let $f \in L^1(\Omega, \mathcal{F}, \mu)$ be general. For $\mu_{\mathbf{N}_n}$-a.e. $\omega_{\mathbf{N}_n} \in \Omega_{\mathbf{N}_n}$ the mapping given by $\xi_{\mathbf{N}_n^c} \mapsto f(\omega_{\mathbf{N}_n}, \xi_{\mathbf{N}_n^c})$ on $\Omega_{\mathbf{N}_n^c}$ is in $L^1(\Omega_{\mathbf{N}_n^c}, \mathcal{F}_{\mathbf{N}_n^c}, \mu_{\mathbf{N}_n^c})$, and hence for $\mu_{\mathbf{N}_n}$-a.e. $\omega_{\mathbf{N}_n} \in \Omega_{\mathbf{N}_n}$ the expression

$$\int_{\Omega_{\mathbf{N}_n^c}} f(\omega_{\mathbf{N}_n}, \xi_{\mathbf{N}_n^c})\, d\mu_{\mathbf{N}_n^c}(\xi_{\mathbf{N}_n^c})$$

makes sense. Again noting that the above expression depends only on $\omega_{\mathbf{N}_n}$, we have that we may consider without fear the mapping defined μ-a.e. on Ω that assigns ω to the above expression. That is, $f_n(\omega)$ is meaningfully defined for μ-a.e. $\omega \in \Omega$. This gives the result for f_n; the result for f'_n is similar.

As a side note, since each f_n is in reality defined only μ-a.e. on Ω, we may redefine f_n to be [say] zero at those $\omega \in \Omega$ for which $f_n(\omega)$ is undefined. The modified f_n's now exist on all of Ω, and we may proceed as before. Analogous comments hold for the f'_n's.]

2*. From the fact that Claim 1 holds for nonnegative f, show that Claim 1 holds for general f in $L^1(\Omega, \mathcal{F}, \mu)$. [We have that $f = f^+ - f^-$, where f^+ and f^- are nonnegative functions in L^1. By Claim 1, we have that $\lim_n (f^+)_n = f^+$ μ-a.e. on Ω and $\lim_n (f^-)_n = f^-$ μ-a.e. on Ω. Therefore

$$\begin{aligned}
f(\omega) &= f^+(\omega) - f^-(\omega) \\
&= \lim_n \left[(f^+)_n(\omega) - (f^-)_n(\omega)\right] \\
&= \lim_n \left[\int_{\Omega_{\mathbf{N}_n^c}} f^+(\omega_{\mathbf{N}_n}, \xi_{\mathbf{N}_n^c})\, d\mu_{\mathbf{N}_n^c}(\xi_{\mathbf{N}_n^c}) - \int_{\Omega_{\mathbf{N}_n^c}} f^-(\omega_{\mathbf{N}_n}, \xi_{\mathbf{N}_n^c})\, d\mu_{\mathbf{N}_n^c}(\xi_{\mathbf{N}_n^c})\right] \\
&= \lim_n \left[\int_{\Omega_{\mathbf{N}_n^c}} \left(f^+(\omega_{\mathbf{N}_n}, \xi_{\mathbf{N}_n^c}) - f^-(\omega_{\mathbf{N}_n}, \xi_{\mathbf{N}_n^c})\right) d\mu_{\mathbf{N}_n^c}(\xi_{\mathbf{N}_n^c})\right] \\
&= \lim_n \int_{\Omega_{\mathbf{N}_n^c}} f(\omega_{\mathbf{N}_n}, \xi_{\mathbf{N}_n^c})\, d\mu_{\mathbf{N}_n^c}(\xi_{\mathbf{N}_n^c}) \\
&= \lim_n f_n(\omega) \quad \text{for } \mu\text{-a.e. } \omega \in \Omega.
\end{aligned}$$

This gives Claim 1 in general.]

3. Consider the framework of this section. For each $n \in \mathbb{N}$, let $E^{(n)} \in \mathcal{F}_n$, and write $E_n = E^{(n)} \times \Omega_{\{n\}^c}$. Define $F = \bigcap_{n=1}^{\infty} \bigcup_{k=n}^{\infty} E_k$. Use Kolmogorov's Zero-One Law to show that $\mu(F)$ must be zero or one. [If $F = \varnothing$, the result is trivial, so we may assume that F is nonempty. Let $\boldsymbol{\omega} \in \Omega$ denote an element of F. This is equivalent to the assertion that for each $n \in \mathbb{N}$ there is $k \geq n$ so that $\boldsymbol{\omega} \in E_k = E^{(k)} \times \Omega_{\{k\}^c}$. This is equivalent to the assertion that $\boldsymbol{\omega}$ is in infinitely many of the sets $E^{(k)} \times \Omega_{\{k\}^c}$. This in turn is logically equivalent to the assertion that $\{i \in \mathbb{N} : \ \omega_i \in E^{(i)}\}$ is infinite. If $\boldsymbol{\xi}$ is ultimately equal to $\boldsymbol{\omega}$, then there is N such that $\xi_n = \omega_n$ for every $n \geq N$. From this, the set $\{i \in \mathbb{N} : \ \xi_i \in E^{(i)}\}$ is infinite. By what we have just stated, this forces $\boldsymbol{\xi} \in F$. The condition required for application of the Zero-One Law is accordingly satisfied.]

4*. Prove Claim 3. [If $f = \sum_{i=1}^{n} c_i I_{A_i}$ where $A_1, \cdots, A_n$ are disjoint $\mathcal{Q}$-sets with union Ξ and $0 \leq c_1, \cdots, c_n < +\infty$. Since $\mu(\Xi) = 1$ and since $\mu(A_i) \in \{0, 1\}$ for each $i \in \{1, \cdots, n\}$, it must be the case that some A_i is such that $\mu(A_i) = 1$. For this i, we have that $f = c_i$ μ-a.e. on Ξ. This gives the claim when f is simple. If f is nonnegative, let $\{s_n\}_{n=1}^{\infty}$ denote a nondecreasing sequence of nonnegative $\mathcal{Q}/\mathcal{B}^*$-measurable simple functions with limit f. Each s_n is constant μ-a.e. on Ξ. Since $\lim_n s_n = f$, it follows that f is constant μ-a.e. on Ξ. If f is general, then $f = f^+ - f^-$, where f^+ and f^- are nonnegative $\mathcal{Q}/\mathcal{B}^*$-measurable functions on Ξ. Since f^+ is constant μ-a.e. on Ξ and since f^- is constant μ-a.e. on Ξ, it follows that f is constant μ-a.e. on Ξ.]

5*. Using the fact that Claim 4 holds for nonnegative f in $L^1(\Omega, \mathcal{F}, \mu)$, quickly show that Claim 4 holds for general f in $L^1(\Omega, \mathcal{F}, \mu)$. [Use the same argument as given in the hint for Exercise 2.]

6. Return to the framework of Exercise 5 in the previous section for the following exercise, which is taken from (22.19) of Hewitt and Stromberg.
(a) Let D denote an infinite subset of $\mathbb{N}$, and fix $\boldsymbol{\omega} \in \Omega$. Define the $\mathcal{F}$-set B by writing $B = \{\boldsymbol{x} \in \Omega : \ \omega_n = x_n \text{ for every } n \in D\}$. Then $\mu(B) = 0$. [This should be straightforward.]
(b) Using the notation of Exercise 5, we have for each $n \in \mathbb{N}$ that

$$A(0_1, 0_3, 0_5, \cdots, 0_{2n-1}) = \{\boldsymbol{x} \in \Omega : \ x_1 = x_3 = \cdots = x_{2n-1} = 0\}.$$

Let the above set be denoted by S_n. Then $\lim_n 2^n I_{S_n}(\boldsymbol{\omega}) = 0$ for μ-a.e. $\boldsymbol{\omega} \in \Omega$. [Let η denote the product measure on $\mathcal{F}$ defined from the measures η_n as follows: if n is even, then $\eta_n(\{0\}) = \eta_n(\{1\}) = \frac{1}{2}$, and if n is odd, then $\eta_n(\{0\}) = 1$ and $\eta_n(\{1\}) = 0$. Show that $\eta \perp \mu$. Let $\mathcal{F}^{(n)}$ be as defined in the proof of Claim 1. Show next that the restriction of η to $\mathcal{F}^{(n)}$ is absolutely continuous with respect to the restriction of μ to $\mathcal{F}^{(n)}$, and demonstrate that $2^n I_{S_n}$ is a Radon-Nikodym derivative of η (restricted to $\mathcal{F}^{(n)}$) with respect to μ (restricted to $\mathcal{F}^{(n)}$). Apply Claim 5 from Section 9.4 to achieve the conclusion.]

7. If f is an $\mathcal{F}/\mathcal{B}^*$-measurable function on Ω such that

$$f(\omega_1, \cdots, \omega_n, \rho_{n+1}, \rho_{n+2}, \cdots) = f(\xi_1, \cdots, \xi_n, \rho_{n+1}, \rho_{n+2}, \cdots)$$

for every $n \in \mathbb{N}$ and choice of $(\omega_1, \cdots, \omega_n)$ and $(\xi_1, \cdots, \xi_n)$ in $\Omega_{\mathbb{N}_n}$, then f is constant μ-a.e. on Ω. [Use either the Zero-One Law or Claim 4.]

Comment. The remaining exercises build up to a general convergence result due to B. Jessen, as given in (22.11) and (22.14) of Hewitt and Stromberg. The next exercise is a straightforward technical lemma. After that, we give the general framework.

8. Let $(\Omega_1, \mathcal{F}_1, \mu_1)$ and $(\Omega_2, \mathcal{F}_2, \mu_2)$ denote measures spaces with $\mu_1(\Omega_1) = 1$ and $\mu_2(\Omega_2) = 1$. Write

$$(\Omega, \mathcal{F}, \mu) = (\Omega_1 \times \Omega_2, \mathcal{F}_1 \otimes \mathcal{F}_2, \mu_1 \otimes \mu_2).$$

Let $1 \le p < +\infty$. For $f \in L^p(\Omega, \mathcal{F}, \mu)$, let Sf denote the function on Ω such that

$$Sf(\omega_1, \omega_2) = \int_{\Omega_2} f(\omega_1, \xi_2)\, d\mu_2(\xi_2)$$

for every $(\omega_1, \omega_2) \in \Omega$. Then $Sf \in L^p(\Omega, \mathcal{F}, \mu)$ and $\|Sf\|_p \le \|f\|_p$. [Since $\mu(\Omega) = 1$, we have $L^p(\Omega, \mathcal{F}, \mu) \subseteq L^1(\Omega, \mathcal{F}, \mu)$ by Claim 1 of Section 8.5, so $f \in L^1(\Omega, \mathcal{F}, \mu)$. Using the Fubini Theorem, the mapping

$$\omega_1 \mapsto \int_{\Omega_2} f(\omega_1, \xi_2)\, d\mu_2(\xi_2)$$

is in $L^1(\Omega_1, \mathcal{F}_1, \mu_1)$. This function is $\mathcal{F}_1/\mathcal{B}^*$-measurable, and since the mapping

$$(\omega_1, \omega_2) \mapsto \int_{\Omega_2} f(\omega_1, \xi_2)\, d\mu_2(\xi_2)$$

does not depend on ω_2, it is $\mathcal{F}/\mathcal{B}^*$-measurable. Next, verify that

$$\begin{aligned}
\|Sf\|_p^p &= \int_{\Omega_1 \times \Omega_2} \left| \int_{\Omega_2} f(\omega_1, \xi_2)\, d\mu_2(\xi_2) \right|^p d\mu_1 \otimes \mu_2(\omega_1, \omega_2) \\
&\le \int_{\Omega_1 \times \Omega_2} \left[\int_{\Omega_2} |f(\omega_1, \xi_2)|\, d\mu_2(\xi_2) \right]^p d\mu_1 \otimes \mu_2(\omega_1, \omega_2) \\
&\le \int_{\Omega_1 \times \Omega_2} \int_{\Omega_2} |f(\omega_1, \xi_2)|^p\, d\mu_2(\xi_2)\, d\mu_1 \otimes \mu_2(\omega_1, \omega_2) \\
&= \int_{\Omega_2} \left(\int_{\Omega_1} \int_{\Omega_2} |f(\omega_1, \xi_2)|^p\, d\mu_2(\xi_2)\, d\mu_1(\omega_1) \right) d\mu_2(\omega_2) \\
&= \int_{\Omega_2} \left[\int_{\Omega_1 \times \Omega_2} |f(\omega_1, \xi_2)|^p\, d\mu_1 \otimes \mu_2(\omega_1, \xi_2) \right] d\mu_2(\omega_2) \\
&= \int_{\Omega_2} \|f\|_p^p\, d\mu_2(\omega_2) \\
&= \|f\|_p^p.
\end{aligned}$$

(The second inequality follows by Claim 1 of Section 8.5.) This shows that Sf is in $L^p(\Omega, \mathcal{F}, \mu)$ and $\|Sf\|_p \le \|f\|_p$.]

Framework. Let $(\Omega_i, \mathcal{F}_i, \mu_i)$ denote an infinite collection of measure spaces with $\mu_i(\Omega_i) = 1$ for each $i \in I$. For every finite subset $F \subseteq I$, we shall use our identification conventions as given in Claim 9 of the previous section to identify the infinite product space $(\Omega, \mathcal{F}, \mu)$ with $(\Omega_F \times \Omega_{F^c}, \mathcal{F}_F \otimes \mathcal{F}_{F^c}, \mu_F \otimes \mu_{F^c})$. We will take

$1 \leq p < +\infty$ and $f \in L^p(\Omega, \mathcal{F}, \mu)$. We will define f_{F^c} on Ω by writing

$$f_{F^c}(\boldsymbol{\omega}_F, \boldsymbol{\omega}_{F^c}) = \int_{\Omega_F} f(\boldsymbol{\xi}_F, \boldsymbol{\omega}_{F^c})\, d\mu_F(\boldsymbol{\xi}_F).$$

Therefore, f_{F^c} is a function of the form Sf as given in the previous exercise. We next define f_F on Ω by writing

$$f_F(\boldsymbol{\omega}_F, \boldsymbol{\omega}_{F^c}) = \int_{\Omega_{F^c}} f(\boldsymbol{\omega}_F, \boldsymbol{\xi}_{F^c})\, d\mu_{F^c}(\boldsymbol{\xi}_{F^c}).$$

The following exercise uses the framework given above and the result of Exercise 8. Sets listed as F [with or without a subscript] are finite sets.

9. The following results obtain:
(a) $f_{F^c} \in L^p(\Omega, \mathcal{F}, \mu)$, and for every $\epsilon > 0$ there is finite $F_0 \subseteq I$ such that $F \supseteq F_0$ forces $\|f_{F^c} - \int_\Omega f\, d\mu\|_p < \epsilon$.
(b) $f_F \in L^p(\Omega, \mathcal{F}, \mu)$, and for every $\epsilon > 0$ there is finite $F_0 \subseteq I$ such that $F \supseteq F_0$ forces $\|f_F - f\|_p = 0$.

Outline: Let $f = I_{A_{F_0} \times \Omega_{F_0^c}}$, where $A_{F_0} \in \mathcal{F}_{F_0}$. Given any finite $F \supseteq F_0$, we have

$$\begin{aligned} f_{F^c}(\boldsymbol{\omega}_F, \boldsymbol{\omega}_{F^c}) &= \int_{\Omega_F} I_{A_{F_0} \times \Omega_{F_0^c}}(\boldsymbol{\xi}_F, \boldsymbol{\omega}_{F^c})\, d\mu_F(\boldsymbol{\xi}_F) \\ &= \int_{\Omega_F} I_{A_{F_0} \times \Omega_{F \cap F_0^c} \times \Omega_{F^c}}(\boldsymbol{\xi}_F, \boldsymbol{\omega}_{F^c})\, d\mu_F(\boldsymbol{\xi}_F) \end{aligned}$$

As a function on Ω_F, $I_{A_{F_0} \times \Omega_{F \cap F_0^c} \times \Omega_{F^c}}$ is merely the indicator function of the set $A_{F_0} \times \Omega_{F \cap F_0^c}$. It follows that the above integral is

$$\mu_F(A_{F_0} \times \Omega_{F \cap F_0^c}) = \mu_{F_0}(A_{F_0}) = \mu(A_{F_0} \times \Omega_{F_0^c}) = \int_\Omega f\, d\mu,$$

and thus

$$f_{F^c}(\boldsymbol{\omega}_F, \boldsymbol{\omega}_{F^c}) = \int_\Omega f\, d\mu.$$

Therefore, (a) is trivially satisfied in the situation where f has the special form as given. To establish (b) for the situation where f has the special form as given, let $F \supseteq F_0$. We have that the integrand in

$$f_F(\boldsymbol{\omega}_F, \boldsymbol{\omega}_{F^c}) = \int_{\Omega_{F^c}} I_{A_{F_0} \times \Omega_{F \cap F_0^c} \times \Omega_{F^c}}(\boldsymbol{\omega}_F, \boldsymbol{\xi}_{F^c})\, d\mu_{F^c}(\boldsymbol{\xi}_{F^c})$$

equals 1 for every $\boldsymbol{\xi}_{F^c}$ if $\boldsymbol{\omega}_{F|F_0} \in A_{F_0}$, where the symbol $\boldsymbol{\omega}_{F|F_0}$ denotes the coordinates of $\boldsymbol{\omega}_F$ whose indices are also in F_0 [that is, the restriction of $\boldsymbol{\omega}_F$ to Ω_{F_0}]. In this case we have $f_F = f = 1$. In the case where $\boldsymbol{\omega}_{F|F_0} \notin A_{F_0}$, the integrand is zero and hence $f_F(\boldsymbol{\omega}_F, \boldsymbol{\omega}_{F^c}) = 0$. It follows that

$$f_F(\boldsymbol{\omega}_F, \boldsymbol{\omega}_{F^c}) = I_{A_{F_0} \times \Omega_{F_0^c}}(\boldsymbol{\omega}_F, \boldsymbol{\omega}_{F^c}),$$

and hence (b) is established for when f has the special form as given.

We now desire to prove (a) and (b) for general $f \in L^p(\Omega, \mathcal{F}, \mu)$. Let $\mathcal{Z}$ denote the subset of $L^p(\Omega, \mathcal{F}, \mu)$ for which (a) and (b) hold. The first goal is to show that $\mathcal{Z}$ is a closed linear subspace of $L^p(\Omega, \mathcal{F}, \mu)$. The linearity of $\mathcal{Z}$ is straightforward to demonstrate, and the real work comes in showing that $\mathcal{Z}$ is closed. To this end, let $f^{(n)} \in \mathcal{Z}$ for each $n \in \mathbb{N}$, and suppose that $\lim_n \|f^{(n)} - f\|_p = 0$. Given arbitrary $\epsilon > 0$ and $n \in \mathbb{N}$, we may find a set F_n such that $\|f^{(n)}_{F_n^c} - \int_\Omega f^{(n)}\, d\mu\|_p < \epsilon$ and the same inequality holds true if F_n is replaced by any finite superset of F_n. Using Exercise 8, the inequality $\|f^{(n)} - f\|_p < \delta$ forces the inequalities

$$\|f^{(n)}_{F_n^c} - f_{F^c}\|_p < \delta \quad \text{and} \quad \|f^{(n)}_{F^c} - f_{F^c}\|_p < \delta$$

for each $n \in \mathbb{N}$, $\delta > 0$, and finite $F \subseteq I$. Choose n sufficiently large so that $\|f^{(n)} - f\|_p < \epsilon$, and pick $F \supseteq F_n$. Verify next that

$$\begin{aligned}
\left\| f_{F^c} - \int_\Omega f\, d\mu \right\|_p &\leq \|f_{F^c} - f^{(n)}_{F^c}\|_p + \left\| f^{(n)}_{F^c} - \int_\Omega f^{(n)}\, d\mu \right\|_p \\
&\quad + \left\| \int_\Omega f^{(n)}\, d\mu - \int_\Omega f\, d\mu \right\|_p \\
&< \epsilon + \epsilon + \left| \int_\Omega f^{(n)}\, d\mu - \int_\Omega f\, d\mu \right| \\
&\leq 2\epsilon + \int_\Omega |f^{(n)} - f|\, d\mu \\
&\leq 2\epsilon + \|f^{(n)} - f\|_p \\
&< 3\epsilon.
\end{aligned}$$

[We have used Claim 1 from Section 8.5 and the fact that $\|c\|_p = |c|$ for any constant c.] Therefore, if $F \supseteq F_n$, then $\|f_{F^c} - \int_\Omega f\, d\mu\|_p < 3\epsilon$, and hence (a) follows for our generic f. One can use similar reasoning to show that (b) holds for our generic f. This shows that $\mathcal{Z}$ is closed with respect to the L^p metric.

Now return to the first part of this outline to see that $\mathcal{Z}$ contains all functions of the form I_A for $A \in \mathcal{H}$, where $\mathcal{H}$ is the field as defined in Section 2 of this chapter. By Exercise 8, the fact that $\mathcal{Z}$ is closed, the identity $\|I_E\|_p = \mu(E)^{\frac{1}{p}}$, and Claim 2 of Section 3.8, it may be shown that $\mathcal{Z}$ contains all functions of the form I_B for $B \in \mathcal{F}$. From this, it follows that $\mathcal{Z}$ contains all $\mathcal{F}/\mathcal{B}^*$-measurable simple functions on Ω. Since these simple functions are dense in $L^p(\Omega, \mathcal{F}, \mu)$ by Claims 1 and 2 of Section 8.6, we may assert that $\mathcal{Z}$ contains all functions in $L^p(\Omega, \mathcal{F}, \mu)$.

11.4 THE L^2 STRONG LAW

We will apply the results of the previous section [Claim 4 in particular] to obtain one of the truly great results of measure theory: the *Strong Law of Large Numbers.* This result is one of the linchpins of probability theory and statistical inference, and represents a grand culmination of the work in this book.

There are two versions of the Strong Law that we shall consider. One version is called the L^2 *Strong Law*, and the other version is called the L^1 *Strong Law*. The L^2 version is usually credited to A. N. Kolmogorov, and the L^1 version that we shall prove has a proof that is usually credited to Etemadi.

Before proceeding, we will require two elementary lemmas from advanced calculus.

Claim 1. *Let $\{x_n\}_{n=1}^{\infty}$ denote a sequence of real numbers having the real limit x. Then $\lim_n \frac{1}{n}\sum_{i=1}^{n} x_i = x$.*

Proof: Let $\epsilon > 0$, and choose N so that $k > N$ forces $|x_k - x| < \frac{\epsilon}{2}$. For any $n > N$, observe that

$$\begin{aligned}
\left|\frac{1}{n}\sum_{i=1}^{n} x_i - x\right| &= \left|\frac{1}{n}\sum_{i=1}^{n} x_i - \frac{1}{n}\sum_{i=1}^{n} x\right| \\
&\leq \frac{1}{n}\sum_{i=1}^{n} |x_i - x| \\
&= \frac{1}{n}\sum_{i=1}^{N} |x_i - x| + \frac{1}{n}\sum_{i=N+1}^{n} |x_i - x| \\
&< \frac{1}{n}\sum_{i=1}^{N} |x_i - x| + \frac{n-N}{n}\frac{\epsilon}{2} \\
&< \frac{1}{n}\sum_{i=1}^{N} |x_i - x| + \frac{\epsilon}{2}.
\end{aligned}$$

For sufficiently large $n > N$ we have that $\frac{1}{n}\sum_{i=1}^{N} |x_i - x| < \frac{\epsilon}{2}$. Therefore, we have that $|\frac{1}{n}\sum_{i=1}^{n} x_i - x| < \frac{\epsilon}{2} + \frac{\epsilon}{2} = \epsilon$ for all sufficiently large $n > N$. Since $\epsilon > 0$ is arbitrary, the proof is complete. ∎

Claim 2. *Let $\{x_n\}_{n=1}^{\infty}$ denote a sequence of real numbers such that $\sum_{i=1}^{\infty} \frac{x_i}{i}$ converges. Then $\lim_n \frac{1}{n}\sum_{i=1}^{n} x_i = 0$.*

Proof: We define some preliminary notation. Let $s_0 = 0$ and $s_n = \sum_{i=1}^{n} \frac{x_i}{i}$ for each $n \in \mathbb{N}$. Next, write $t_n = \sum_{i=1}^{n} x_i$ for each $n \in \mathbb{N}$, so that we have $x_k = k(s_k - s_{k-1})$ for each $k \in \mathbb{N}$. Using simple algebra, we have

$$\begin{aligned}
t_{n+1} = \sum_{k=1}^{n+1} k(s_k - s_{k-1}) &= (n+1)s_{n+1} + \sum_{k=1}^{n} ks_k - \sum_{k=2}^{n+1} ks_{k-1} \\
&= (n+1)s_{n+1} + \sum_{k=1}^{n} ks_k - \sum_{k=1}^{n} (k+1)s_k \\
&= (n+1)s_{n+1} - \sum_{k=1}^{n} s_k \quad \text{for } n = 1, 2, \cdots.
\end{aligned}$$

Dividing both sides by $n+1$ gives

$$\frac{t_{n+1}}{n+1} = s_{n+1} - \left(\frac{n}{n+1}\right)\frac{1}{n}\sum_{k=1}^{n} s_k$$

for $n = 1, 2, \cdots$. By assumption and the definition of the numbers s_n, we have that $\{s_n\}_{n=1}^{\infty}$ converges to some real x. By Claim 1, it in turn follows that $\lim_n \frac{1}{n}\sum_{k=1}^{n} s_k$ converges to x as $n \to \infty$. Therefore the right side of the above display tends to $x - x = 0$ as $n \to \infty$, giving $\lim_n \frac{t_{n+1}}{n+1} = 0$. This is the desired limit relation. ∎

The next claim is also of an elementary nature, and will help streamline the proof of the L^2 Strong Law. We will consider a sequence $\{(\Omega_i, \mathcal{F}_i, \mu_i)\}_{i=1}^{\infty}$ of measure spaces with $\mu_i(\Omega_i) = 1$ for each $i \in \mathbb{N}$. Let $(\Omega, \mathcal{F}, \mu)$ denote the usual product measure space as constructed in Sections 1 and 2.

Claim 3. *Let $F \subseteq \mathbb{N}$ be nonempty and finite, and for each $i \in F$ let f_i denote a function in $L^1(\Omega_i, \mathcal{F}_i, \mu_i)$. Let g denote the mapping on Ω given for every $\boldsymbol{\omega} \in \Omega$ by the rule $\boldsymbol{\omega} \mapsto \prod_{i\in F} f_i(\omega_i)$. Then*

$$\int_{\Omega} g(\boldsymbol{\omega})\, d\mu(\boldsymbol{\omega}) = \prod_{i\in F}\int_{\Omega_i} f_i(\omega_i)\, d\mu_i(\omega_i). \tag{11.7}$$

Proof: We proceed by mathematical induction. Let F denote a singleton, say $F = \{i_1\}$. Then

$$\begin{aligned}
\int_{\Omega} g(\boldsymbol{\omega})\, d\mu(\boldsymbol{\omega}) &= \int_{\Omega} f_{i_1}(\omega_{i_1})\, d\mu(\boldsymbol{\omega}) \\
&= \int_{\Omega_{i_1}\times\Omega_{\{i_1\}^c}} f_{i_1}(\omega_{i_1})\, d\mu_{i_1}\otimes\mu_{\{i_1\}^c}(\omega_{i_1}, \boldsymbol{\omega}_{\{i_1\}^c}) \\
&= \int_{\Omega_{i_1}}\int_{\Omega_{\{i_1\}^c}} f_{i_1}(\omega_{i_1})\, d\mu_{\{i_1\}^c}(\boldsymbol{\omega}_{\{i_1\}^c})\, d\mu_{i_1}(\omega_{i_1}) \\
&= \int_{\Omega_{i_1}} f_{i_1}(\omega_{i_1})\int_{\Omega_{\{i_1\}^c}} d\mu_{\{i_1\}^c}(\boldsymbol{\omega}_{\{i_1\}^c})\, d\mu_{i_1}(\omega_{i_1}) \\
&= \int_{\Omega_{i_1}} f_{i_1}(\omega_{i_1})\, d\mu_{i_1}(\omega_{i_1}).
\end{aligned}$$

The second equality follows by our usual identification conventions, and the third equality follows by the Integrable Fubini Theorem. The fifth equality follows because $\mu_{\{i_1\}^c}(\Omega_{\{i_1\}^c}) = 1$. This shows that (11.7) holds when F is a one-element subset of $\mathbb{N}$.

We now perform the requisite induction step. Assume that (11.7) holds for every n-element subset of $\mathbb{N}$. Let $F = \{i_1, \cdots, i_n, i_{n+1}\}$ denote an $(n+1)$-element subset of $\mathbb{N}$. Let $F_1 = \{i_1, \cdots, i_n\}$, an n-element subset of F. Making free use of the Integrable Fubini Theorem, we have

$$\begin{aligned}
&\int_\Omega g(\boldsymbol{\omega})\,d\mu(\boldsymbol{\omega}) \\
&= \int_{\Omega_F \times \Omega_{F^c}} \prod_{j=1}^{n+1} f_{i_j}(\omega_{i_j})\,d\mu_F \otimes \mu_{F^c}(\boldsymbol{\omega}_F, \boldsymbol{\omega}_{F^c}) \\
&= \int_{\Omega_F} \int_{\Omega_{F^c}} \prod_{j=1}^{n+1} f_{i_j}(\omega_{i_j})\,d\mu_{F^c}(\boldsymbol{\omega}_{F^c})\,d\mu_F(\boldsymbol{\omega}_F) \\
&= \int_{\Omega_F} \prod_{j=1}^{n+1} f_{i_j}(\omega_{i_j}) \int_{\Omega_{F^c}} d\mu_{F^c}(\boldsymbol{\omega}_{F^c})\,d\mu_F(\boldsymbol{\omega}_F) \\
&= \int_{\Omega_F} \prod_{j=1}^{n+1} f_{i_j}(\omega_{i_j})\,d\mu_F(\boldsymbol{\omega}_F) \\
&= \int_{\Omega_{F_1} \times \Omega_{i_{n+1}}} \prod_{j=1}^{n+1} f_{i_j}(\omega_{i_j})\,d\mu_{F_1} \otimes \mu_{i_{n+1}}(\boldsymbol{\omega}_{F_1}, \omega_{i_{n+1}}) \\
&= \int_{\Omega_{F_1}} \prod_{j=1}^{n} f_{i_j}(\omega_{i_j}) \int_{\Omega_{i_{n+1}}} f_{i_{n+1}}(\omega_{i_{n+1}})\,d\mu_{i_{n+1}}(\omega_{i_{n+1}})\,d\mu_{F_1}(\boldsymbol{\omega}_{F_1}) \\
&= \int_{\Omega_{i_{n+1}}} f_{i_{n+1}}(\omega_{i_{n+1}})\,d\mu_{i_{n+1}}(\omega_{i_{n+1}}) \times \int_{\Omega_{F_1}} \prod_{j=1}^{n} f_{i_j}(\omega_{i_j})\,d\mu_{F_1}(\boldsymbol{\omega}_{F_1}).
\end{aligned}$$

Now consider the second factor above. We have

$$\begin{aligned}
&\int_{\Omega_{F_1}} \prod_{j=1}^{n} f_{i_j}(\omega_{i_j})\,d\mu_{F_1}(\boldsymbol{\omega}_{F_1}) \\
&= \int_{\Omega_{F_1}} \prod_{j=1}^{n} f_{i_j}(\omega_{i_j}) \int_{\Omega_{F_1^c}} d\mu_{F_1^c}(\boldsymbol{\omega}_{F_1^c})\,d\mu_{F_1}(\boldsymbol{\omega}_{F_1}) \\
&= \int_{\Omega_{F_1} \times \Omega_{F_1^c}} \prod_{j=1}^{n} f_{i_j}(\omega_{i_j})\,d\mu_{F_1} \otimes \mu_{F_1^c}(\boldsymbol{\omega}_{F_1}, \boldsymbol{\omega}_{F_1^c}) \\
&= \int_\Omega \prod_{j=1}^{n} f_{i_j}(\omega_{i_j})\,d\mu(\boldsymbol{\omega}) \\
&= \prod_{j=1}^{n} \int_{\Omega_{i_j}} f_{i_j}(\omega_{i_j})\,d\mu_{i_j}(\omega_{i_j})
\end{aligned}$$

by the induction hypothesis. Substitution of this last display into the previous display yields (11.7) for our generic $(n+1)$-element subset F of $\mathbb{N}$. This completes the induction step, giving the proof. ∎

For the upcoming proof of the L^2 Strong Law, we recall a few results. The first result to recall for an arbitrary finite measure space $(\Omega, \mathcal{F}, \mu)$ with $\mu(\Omega) = 1$ is that

$$L^2(\Omega, \mathcal{F}, \mu) \subseteq L^1(\Omega, \mathcal{F}, \mu) \text{ and } \|f\|_1 \leq \|f\|_2 \text{ if } f \in L^2(\Omega, \mathcal{F}, \mu).$$

[This is Claim 1 of Section 8.5.] In particular, it follows that if $\{f_n\}_{n=1}^{\infty}$ is a sequence in $L^2(\Omega, \mathcal{F}, \mu)$ that converges in L^2 to a function $f \in L^2(\Omega, \mathcal{F}, \mu)$, [that is, $f_n \xrightarrow{2} f$], then f and each f_n are in $L^1(\Omega, \mathcal{F}, \mu)$ and

$$\|f_n - f\|_1 \leq \|f_n - f\|_2 \to 0 \text{ as } n \to \infty,$$

so that $f_n \xrightarrow{1} f$.

The next result to recall is as follows: if $\{f_n\}_{n=1}^{\infty}$ is a sequence of functions in $L^1(\Omega, \mathcal{F}, \mu)$ that converges in L^1 to functions f and g in $L^1(\Omega, \mathcal{F}, \mu)$, then $f = g$ μ-a.e. on Ω. To see this, we have that

$$\|f - g\|_1 = \|(f - f_n) + (f_n - g)\|_1 \leq \|f_n - f\|_1 + \|f_n - g\|_1 \to 0 \text{ as } n \to \infty,$$

so that $\|f - g\|_1 = 0$. From this, we deduce that $f = g$ μ-a.e. on Ω.

We next recall the Riesz-Fischer Theorem regarding the completeness of $L^p(\Omega, \mathcal{F}, \mu)$ [with respect to distance as defined by the L^p norm] for $p \geq 1$. [See Section 8.2.] This means that if $\{f_n\}_{n=1}^{\infty}$ is a Cauchy sequence in $L^p(\Omega, \mathcal{F}, \mu)$, then there is some $f \in L^p(\Omega, \mathcal{F}, \mu)$ such that $f_n \xrightarrow{p} f$ as $n \to \infty$. That is, $\{f_n\}_{n=1}^{\infty}$ is a convergent sequence in $L^p(\Omega, \mathcal{F}, \mu)$.

We are now ready to state the L^2 Strong Law. The framework is as follows: $\{(\Omega_i, \mathcal{F}_i, \mu_i)\}_{i=1}^{\infty}$ is a sequence of measure spaces with $\mu_i(\Omega_i) = 1$ for each $i \in \mathbb{N}$. We will write $(\Omega, \mathcal{F}, \mu)$ for the usual product measure space.

L^2 Strong Law of Large Numbers. *For each $k \in \mathbb{N}$, let g_k denote a function in $L^2(\Omega_k, \mathcal{F}_k, \mu_k)$ such that $\int_{\Omega_k} g_k \, d\mu_k = 0$. Let f_k denote the function on Ω that assigns each $\omega \in \Omega$ to $g_k(\omega_k)$. Assume that $\sum_{k=1}^{\infty} \frac{\|g_k\|_2^2}{k^2}$ converges. Then*

$$\lim_n \frac{1}{n} \sum_{k=1}^{n} f_k(\omega) = 0 \text{ for } \mu\text{-a.e. } \omega \in \Omega.$$

Proof: We first observe that

$$\int_{\Omega} |f_k(\omega)|^2 \, d\mu(\omega) = \int_{\Omega} |g_k(\omega_k)|^2 \, d\mu(\omega) = \int_{\Omega_k} |g_k(\omega_k)|^2 \, d\mu_k(\omega_k) < +\infty$$

for each $k \in \mathbb{N}$ by definition and Claim 3, and hence each f_k is in $L^2(\Omega, \mathcal{F}, \mu)$. Therefore, we have that $\sum_{k=1}^{n} \frac{f_k}{k} \in L^2(\Omega, \mathcal{F}, \mu)$ for every $n \in \mathbb{N}$. By Claim 1 of Section 8.5, we have that $\sum_{k=1}^{n} \frac{f_k}{k} \in L^1(\Omega, \mathcal{F}, \mu)$ for every $n \in \mathbb{N}$.

We now claim that $\{\sum_{k=1}^{n} \frac{f_k}{k}\}_{n=1}^{\infty}$ is a Cauchy sequence in $L^2(\Omega, \mathcal{F}, \mu)$. For every $m, n \in \mathbb{N}$ with $m < n$, we have

$$\begin{aligned}
&\left\| \sum_{k=1}^{n} \frac{f_k}{k} - \sum_{k=1}^{m} \frac{f_k}{k} \right\|_2^2 \quad \text{[norm with respect to } L^2(\Omega, \mathcal{F}, \mu)] \\
&= \int_\Omega \left| \sum_{k=m+1}^{n} \frac{f_k}{k} \right|^2 d\mu \quad \text{[definition of norm]} \\
&= \sum_{j,k=m+1}^{n} \frac{1}{jk} \int_\Omega |f_j f_k| \, d\mu \quad \text{[algebra and linearity of } \textstyle\int_\Omega] \\
&= \sum_{j \neq k} \frac{1}{jk} \int_{\Omega_j} g_j \, d\mu_j \int_{\Omega_k} g_k \, d\mu_k + \sum_{k=m+1}^{n} \frac{1}{k^2} \int_{\Omega_k} |g_k|^2 \, d\mu_k \quad \text{[Claim 3]} \\
&= \sum_{k=m+1}^{n} \frac{\|g_k\|_2^2}{k^2} \quad \text{[norm with respect to } L^2(\Omega_k, \mathcal{F}_k, \mu_k).]
\end{aligned}$$

[The last equality holds since $\int_{\Omega_j} g_j \, d\mu_j = 0$ for each $j \in \mathbb{N}$.] It follows that

$$\left\| \sum_{k=1}^{n} \frac{f_k}{k} - \sum_{k=1}^{m} \frac{f_k}{k} \right\|_2^2 = \sum_{k=m+1}^{n} \frac{\|g_k\|_2^2}{k^2} \to 0 \quad \text{as } m, n \to \infty$$

since $\sum_{k=1}^{\infty} \frac{\|g_k\|_2^2}{k^2}$ is assumed to converge. This shows that $\{\sum_{k=1}^{n} \frac{f_k}{k}\}_{n=1}^{\infty}$ is a Cauchy sequence in $L^2(\Omega, \mathcal{F}, \mu)$.

We now invoke the Riesz-Fischer Theorem from Section 8.2 to assert that $\{\sum_{k=1}^{n} \frac{f_k}{k}\}_{n=1}^{\infty}$ converges in $L^2(\Omega, \mathcal{F}, \mu)$ to some function $h \in L^2(\Omega, \mathcal{F}, \mu)$. Since $L^2(\Omega, \mathcal{F}, \mu) \subseteq L^1(\Omega, \mathcal{F}, \mu)$ [Claim 1, Section 8.5], we have that (i) h is in $L^1(\Omega, \mathcal{F}, \mu)$ and (ii) $\{\sum_{k=1}^{n} \frac{f_k}{k}\}_{n=1}^{\infty}$ converges in $L^1(\Omega, \mathcal{F}, \mu)$ to h. The previous display [again in conjunction with Claim 1 of Section 8.5] also shows that $\{\sum_{k=1}^{n} \frac{f_k}{k}\}_{n=1}^{\infty}$ is a Cauchy sequence in $L^1(\Omega, \mathcal{F}, \mu)$. It follows that $\{\sum_{k=1}^{n} \frac{f_k}{k}\}_{n=1}^{\infty}$ converges in $L^1(\Omega, \mathcal{F}, \mu)$ to the limit [as $n \to \infty$] of the partial sums $\sum_{k=1}^{n} \frac{f_k}{k}$. That is, $\sum_{k=1}^{\infty} \frac{f_k}{k}$ exists as a function in $L^1(\Omega, \mathcal{F}, \mu)$, and $\{\sum_{k=1}^{n} \frac{f_k}{k}\}_{n=1}^{\infty}$ converges in $L^1(\Omega, \mathcal{F}, \mu)$ to $\sum_{k=1}^{\infty} \frac{f_k}{k}$. From the comments preceding the claim, it follows that $h = \sum_{k=1}^{\infty} \frac{f_k}{k}$ μ-a.e. on Ω.

We now claim that $\int_\Omega f_k \, d\mu = 0$ for every $k \in \mathbb{N}$. To see this, use Claim 3 to observe that

$$\int_\Omega f_k \, d\mu = \int_\Omega g_k(\omega_k) \, d\mu(\omega) = \int_{\Omega_k} g_k(\omega_k) \, d\mu_k(\omega_k) = 0.$$

We next claim that $\int_\Omega h\,d\mu = 0$. To see this, recall that we have just shown that $\int_\Omega f_k\,d\mu = 0$ for each $k \in \mathbb{N}$, and for each $n \in \mathbb{N}$ write

$$\begin{aligned}\left|\int_\Omega h\,d\mu\right| &= \left|\sum_{k=1}^n \int_\Omega \frac{f_k}{k}\,d\mu + \int_\Omega \left(h - \sum_{k=1}^n \frac{f_k}{k}\right) d\mu\right| \\ &\le \int_\Omega \left|h - \sum_{k=1}^n \frac{f_k}{k}\right| d\mu \\ &= \left\|h - \sum_{k=1}^n \frac{f_k}{k}\right\|_1,\end{aligned}$$

which goes to zero as $n \to \infty$. This shows that $\int_\Omega h\,d\mu = 0$.

We now turn to Claim 4 of the previous section to write

$$0 = \int_\Omega h(\omega)\,d\mu(\omega) = \lim_n \int_{\Omega_{\mathbb{N}_n}} h(\boldsymbol{\xi}_{\mathbb{N}_n}, \boldsymbol{\omega}_{\mathbb{N}_n^c})\,d\mu_{\mathbb{N}_n}(\boldsymbol{\xi}_{\mathbb{N}_n}) \text{ for } \mu\text{-a.e. } \boldsymbol{\omega} \in \Omega.$$

Adopting the notation from the previous section, the last integral above is $h_n'(\boldsymbol{\omega})$. Observe that $h_n'(\boldsymbol{\omega})$ does not depend on $(\omega_1, \cdots, \omega_n)$, that is, h_n', although defined on Ω, is independent of the first n coordinates.

Observe next for each $n \in \mathbb{N}$ and $\boldsymbol{\omega} \in \Omega$ that

$$\begin{aligned}&h_n'(\boldsymbol{\omega}) \\ &= \int_{\Omega_{\mathbb{N}_n}} h(\boldsymbol{\xi}_{\mathbb{N}_n}, \boldsymbol{\omega}_{\mathbb{N}_n^c})\,d\mu_{\mathbb{N}_n}(\boldsymbol{\xi}_{\mathbb{N}_n}) \\ &= \int_{\Omega_{\mathbb{N}_n}} \sum_{k=1}^\infty \frac{f_k(\boldsymbol{\xi}_{\mathbb{N}_n}, \boldsymbol{\omega}_{\mathbb{N}_n^c})}{k}\,d\mu_{\mathbb{N}_n}(\boldsymbol{\xi}_{\mathbb{N}_n}) \\ &= \int_{\Omega_{\mathbb{N}_n}} \sum_{k=1}^n \frac{f_k(\boldsymbol{\xi}_{\mathbb{N}_n}, \boldsymbol{\omega}_{\mathbb{N}_n^c})}{k}\,d\mu_{\mathbb{N}_n}(\boldsymbol{\xi}_{\mathbb{N}_n}) + \int_{\Omega_{\mathbb{N}_n}} \sum_{k=n+1}^\infty \frac{f_k(\boldsymbol{\xi}_{\mathbb{N}_n}, \boldsymbol{\omega}_{\mathbb{N}_n^c})}{k}\,d\mu_{\mathbb{N}_n}(\boldsymbol{\xi}_{\mathbb{N}_n}).\end{aligned}$$

Next, observe for each $n \in \mathbb{N}$, $k \in \{1, \cdots, n\}$, and $\boldsymbol{\omega} \in \Omega$ that

$$\begin{aligned}\int_{\Omega_{\mathbb{N}_n}} f_k(\boldsymbol{\xi}_{\mathbb{N}_n}, \boldsymbol{\omega}_{\mathbb{N}_n^c})\,d\mu_{\mathbb{N}_n}(\boldsymbol{\xi}_{\mathbb{N}_n}) &= \int_{\Omega_{\mathbb{N}_n}} g_k(\xi_k)\,d\mu_{\mathbb{N}_n}(\boldsymbol{\xi}_{\mathbb{N}_n}) \\ &= \int_{\Omega_k} g_k(\xi_k)\,d\mu_k(\xi_k) \\ &= 0\end{aligned}$$

by assumption and Claim 3, and hence we have that

$$\begin{aligned} h_n'(\omega) &= \int_{\Omega_{\mathbb{N}_n}} \sum_{k=n+1}^{\infty} \frac{f_k(\xi_{\mathbb{N}_n}, \omega_{\mathbb{N}_n^c})}{k} \, d\mu_{\mathbb{N}_n}(\xi_{\mathbb{N}_n}) \\ &= \int_{\Omega_{\mathbb{N}_n}} \sum_{k=n+1}^{\infty} \frac{g_k(\omega_k)}{k} \, d\mu_{\mathbb{N}_n}(\xi_{\mathbb{N}_n}) \\ &= \sum_{k=n+1}^{\infty} \frac{g_k(\omega_k)}{k} \\ &= \sum_{k=n+1}^{\infty} \frac{f_k(\omega)}{k} \quad \text{for } \mu\text{-a.e. } \omega \in \Omega. \end{aligned}$$

It follows that

$$0 = \lim_n h_n'(\omega) = \lim_n \sum_{k=n+1}^{\infty} \frac{f_k(\omega)}{k} \quad \text{for } \mu\text{-a.e. } \omega \in \Omega.$$

This means that

$$\sum_{k=1}^{\infty} \frac{f_k(\omega)}{k} \text{ converges for } \mu\text{-a.e. } \omega \in \Omega.$$

By Claim 2, this means that

$$\lim_n \frac{1}{n} \sum_{k=1}^{n} f_k(\omega) = 0 \quad \text{for } \mu\text{-a.e. } \omega \in \Omega.$$

This is precisely the desired limit relation, and the proof of the L^2 version of Strong Law of Large Numbers is now complete. ■

The reason why the previous version is referred to as the L^2 version should be obvious. In the next section, we shall as our final piece of work in this book show that we can merely require that the functions g_k be in $L^1(\Omega_k, \mathcal{F}_k, \mu_k)$. That is, we shall obtain the conclusion of the Strong Law as given in this section with weaker assumptions.

Exercises.

1. In Claim 3, show that the function g as defined there is an element of $L^1(\Omega, \mathcal{F}, \mu)$, so that we may meaningfully discuss the integral $\int_\Omega g \, d\mu$.

2. In the statement of the L^2 Strong Law, replace the assumption that $\sum_{k=1}^{\infty} \frac{\|g_k\|_2^2}{k^2}$ converges with the assumption that $\sum_{k=1}^{\infty} c_k \|g_k\|_2^2$ converges, where $\{c_k\}_{k=1}^{\infty}$ is a sequence of positive numbers such that $\sum_{k=1}^{\infty} c_k < +\infty$. Then $\sum_{k=1}^{\infty} c_k f_k(\omega)$ converges for μ-a.e. $\omega \in \Omega$.

3. In the statement of the L^2 Strong Law, replace the assumption that $\sum_{k=1}^{\infty} \frac{\|g_k\|_2^2}{k^2}$ converges with the assumption that $\lim_n \frac{1}{n^2} \sum_{k=1}^{n} \|f_k\|_2^2 = 0$. Deduce for every $\epsilon > 0$

that

$$\lim_n \mu\left(\left\{\omega \in \Omega : \left|\frac{1}{n}\sum_{k=1}^{n} f_k(\omega)\right| > \epsilon\right\}\right) = 0,$$

which shows that $\{\frac{1}{n}\sum_{k=1}^{n} f_k\}_{n=1}^{\infty}$ converges in μ-measure to zero. This result is known as the *Weak Law of Large Numbers.*

4. Reconsider the framework of Exercise 5 in Section 2 of this chapter, and recall the definitions of f_n and h_n for each $n \in \mathbb{N}$.
(a) The hypotheses of the Strong Law are satisfied since $\|f_j\|_2^2 = \frac{1}{4}$ for each $j \in \mathbb{N}$, and hence $\lim_n h_n = \lim_n \frac{1}{n}(f_1 + \cdots + f_n) = 0$ μ-a.e. on Ω, that is h_n converges to zero μ-a.e. on Ω.
(b) Show that the result of (a) implies the result of part (d) of the aforementioned exercise. [That is, the Strong Law provides a much stronger conclusion than that of (d). Here, we have $h_n \to 0$ μ-a.e. on Ω, while there we have $h_n \to 0$ in μ-measure. Recall the relationship between convergence μ-a.e. and convergence in μ-measure when the measure of the space is finite.]
(c) As with part (e) of Exercise 5 in Section 2 of this chapter, discuss the result of (a) in terms of tossing a fair coin.

5. For each $n \in \mathbb{N}$, make the following definitions: $\Omega_n = \{0, 1, \cdots, k-1\}$, $\mathcal{F}_n = 2^{\Omega_n}$, and μ_n is the counting measure. Consider the usual product space $(\Omega, \mathcal{F}, \mu)$ as constructed in Sections 1 and 2. Fix $m \in \{0, 1, \cdots, k-1\}$, and for each $n \in \mathbb{N}$ define the mapping g_n on Ω by writing $g_n(\omega) = 1$ if $\omega_n = m$, and $g_n(\omega) = 0$ otherwise. Show that $\frac{1}{n}\sum_{i=1}^{n} g_i \to \frac{1}{k}$ in μ-measure. That is, show for every $\epsilon > 0$ that

$$\lim_n \mu\left(\left\{\omega \in \Omega : \left|\frac{1}{n}\sum_{i=1}^{n} g_i(\omega) - \frac{1}{k}\right| > \epsilon\right\}\right) = 0.$$

11.5 PRELUDE TO THE L^1 STRONG LAW

This section is dedicated to proving some background results needed for the proof of the L^1 Strong Law. As in the previous section, this result — one of the truly great results in mathematics — is of foundational importance in statistical inference and probability. In the statement of the L^2 Strong Law given last section, we required the functions g_k to be in $L^2(\Omega_k, \mathcal{F}_k, \mu_k)$ and such that the series $\sum_{k=1}^{\infty} \frac{\|g_k\|_2^2}{k^2}$ converges. The L^1 version of the Strong Law replaces these requirements with the requirement that g_k is in $L^1(\Omega_k, \mathcal{F}_k, \mu_k)$ for each $k \in \mathbb{N}$. The proof of the L^1 Strong Law comes with a price, namely, it will require a fair number of preliminary results, and we shall require the L^2 Strong Law in the course of its proof. The notation and framework remain unchanged from the previous section.

Claim 1. *Let $0 \leq x_k < 1$ for each $k \in \mathbb{N}$. Then $\sum_{k=1}^{n} x_k \leq \prod_{k=1}^{n}(1 + x_k)$ for every $n \in \mathbb{N}$.*

Proof: We use an elementary induction proof. The result is trivial if $n = 1$. Next, assume that $\sum_{k=1}^{n} x_k \le \prod_{k=1}^{n}(1 + x_k)$ for a given n. We have that

$$\begin{aligned}\sum_{k=1}^{n+1} x_k = \sum_{k=1}^{n} x_k + x_{n+1} &\le \prod_{k=1}^{n}(1+x_k) + x_{n+1} \\ &= \left[\prod_{k=1}^{n}(1+x_k)\right]\left[1 + \frac{x_{n+1}}{\prod_{k=1}^{n}(1+x_k)}\right] \\ &\le \left[\prod_{k=1}^{n}(1+x_k)\right](1+x_{n+1}) \\ &= \prod_{k=1}^{n+1}(1+x_k).\end{aligned}$$

This completes the proof by induction. ■

Claim 2. *We have* $1 - x \ge \frac{1}{1+2x}$ *for every* $0 \le x \le \frac{1}{2}$.

Proof: Let $f(x) = 1 - x$ and $g(x) = \frac{1}{1+2x}$ for each $0 \le x \le \frac{1}{2}$. We have that $f(0) = g(0) = 1$ and $f(\frac{1}{2}) = g(\frac{1}{2}) = \frac{1}{2}$. Observe that f is a decreasing linear function on $[0, \frac{1}{2}]$. Next, observe that g has a negative first derivative and a positive second derivative on $(0, \frac{1}{2})$. That is, g is concave upward on $(0, \frac{1}{2})$. It follows that $g \le f$ on $[0, \frac{1}{2}]$. ■

The following result is a classical and well-known inequality from advanced calculus.

Claim 3. $1 + x \le e^x$ *for every* $x \ge 0$.

Proof: For positive x, we have $1 + x = \sum_{n=0}^{1} \frac{x^n}{n!} < \sum_{n=0}^{\infty} \frac{x^n}{n!} = e^x$. If $x = 0$ the inequality is trivially valid. ■

Claim 4. *Let* $0 \le x_k < 1$ *for every* $k \in \mathbb{N}$. *We have the following:*

(a) If $\sum_{k=1}^{\infty} x_k = +\infty$, *then* $\prod_{k=1}^{\infty}(1 - x_k) = 0$.

(b) If $\sum_{k=1}^{\infty} x_k < +\infty$, *then* $\prod_{k=1}^{\infty}(1 - x_k) > 0$.

Proof: Before doing anything, observe that since $0 < 1 - x_k \le 1$ for each $k \in \mathbb{N}$ we may talk about the limit $\prod_{k=1}^{\infty}(1-x_k)$ of the nonincreasing sequence $\{\prod_{k=1}^{n}(1 - x_k)\}_{n=1}^{\infty}$. We now proceed to the proof itself. For every $k \in \mathbb{N}$ it is obvious that $(1 - x_k)(1 + x_k) = 1 - x_k^2 \le 1$, and hence $1 + x_k \le \frac{1}{1-x_k}$. Using Claim 1 and what has just been shown, we have for every $n \in \mathbb{N}$ that

$$\sum_{k=1}^{n} x_k \le \prod_{k=1}^{n}(1 + x_k) \le \frac{1}{\prod_{k=1}^{n}(1 - x_k)}.$$

If $\sum_{k=1}^{\infty} x_k = +\infty$, then letting $n \to \infty$ in both sides of the above display gives $[\lim_n \prod_{k=1}^{n}(1 - x_k)]^{-1} = +\infty$, which means that $\{\prod_{k=1}^{n}(1 - x_k)\}_{n=1}^{\infty}$ is nonincreasing to zero, and hence $\prod_{k=1}^{\infty}(1 - x_k) = 0$. This proves (a).

We now turn to (b), assuming that $\sum_{k=1}^{\infty} x_k < +\infty$. Because the series $\sum_{k=1}^{\infty} x_k$ converges, we have that $x_k \leq \frac{1}{2}$ for all sufficiently large k, say, if $k \geq m$. Next, if $k \geq m$, Claims 2 and 3 together give

$$\prod_{k=m}^{n}(1-x_k) \geq \frac{1}{\prod_{k=m}^{n}(1+2x_k)} \geq \exp\left(-2\sum_{k=m}^{n} x_k\right) \geq \exp\left(-2\sum_{k=1}^{\infty} x_k\right)$$

for every $n \geq m$. Therefore, there exists $m \in \mathbb{N}$ such that

$$\prod_{k=m}^{n}(1-x_k) \geq \exp\left(-2\sum_{k=1}^{\infty} x_k\right) \quad \text{for every } n \geq m.$$

From this, it follows that $\prod_{k=m}^{\infty}(1-x_k) \geq \exp\left(-2\sum_{k=1}^{\infty} x_k\right)$, giving

$$\begin{aligned}\prod_{k=1}^{\infty}(1-x_k) &= \prod_{k=1}^{m-1}(1-x_k) \times \prod_{k=m}^{\infty}(1-x_k) \\ &\geq \prod_{k=1}^{m-1}(1-x_k) \times \exp\left(-2\sum_{k=1}^{\infty} x_k\right) \\ &> 0.\end{aligned}$$

This establishes (b) and completes the proof. ■

The following claim establishes an elementary fact about product measures that will prove useful.

Claim 5. *Suppose that $A_n \in \mathcal{F}_n$ for each $n \in \mathbb{N}$. Then $\prod_{k=1}^{\infty} A_k \in \mathcal{F}$ and*

$$\mu\left(\prod_{k=1}^{\infty} A_k\right) = \prod_{k=1}^{\infty} \mu_k(A_k).$$

Proof: For each $n \in \mathbb{N}$, define $B_n = A_1 \times \cdots \times A_n \times \Omega_{n+1} \times \Omega_{n+2} \times \cdots$. The sequence $\{B_n\}_{n=1}^{\infty}$ is a nonincreasing sequence of $\mathcal{F}$-sets with limit $\prod_{k=1}^{\infty} A_k$. Since σ-fields are closed under limits of nonincreasing sequences, we have that $\prod_{k=1}^{\infty} A_k \in \mathcal{F}$, whence the quantity $\mu(\prod_{k=1}^{\infty} A_k)$ may be meaningfully discussed. Using the fact that measures are continuous from above [measure property (M9) in Section 2.2] and the product property of our constructed product measures in Section 2, we have

$$\begin{aligned}\mu\left(\prod_{k=1}^{\infty} A_k\right) = \mu\left(\lim_n B_n\right) = \lim_n \mu(B_n) &= \lim_n \mu_{\mathbb{N}_n}\left(\prod_{k=1}^{n} A_k\right) \\ &= \lim_n \prod_{k=1}^{n} \mu_k(A_k) \\ &= \prod_{k=1}^{\infty} \mu_k(A_k).\end{aligned}$$

This proves the claim. ∎

Claim 6. *For every $p \in \mathbb{N}$ we have $p\sum_{k=p}^{\infty} \frac{1}{k^2} \leq 2$.*

Proof: This is mere calculus. Viewing an integral as a limit of Riemann sums, we have for each $p \in \mathbb{N}$ that

$$\sum_{k=p+1}^{\infty} \frac{1}{k^2} \leq \int_p^{\infty} \frac{1}{x^2}\, dx = \frac{1}{p},$$

and hence

$$p\sum_{k=p}^{\infty} \frac{1}{k^2} = p\left[\frac{1}{p^2} + \sum_{k=p+1}^{\infty} \frac{1}{k^2}\right] \leq p\left[\frac{1}{p^2} + \frac{1}{p}\right] = \frac{1}{p} + 1 \leq 2$$

for each $p \in \mathbb{N}$. This proves the claim. ∎

The following claim is one part of what is known as the *Borel-Cantelli Lemma*, and is of independent interest in probability theory. For our purposes, it is one of the ingredients to be used in the proof of the L^1 Strong Law.

Claim 7. *For each $n \in \mathbb{N}$, let $E^{(n)} \in \mathcal{F}_n$, and write $E_n = E^{(n)} \times \Omega_{\{n\}^c}$. Next, define $F = \bigcap_{n=1}^{\infty} \bigcup_{k=n}^{\infty} E_k$.*

(a) $\mu(F) \in \{0, 1\}$.

(b) If $\sum_{n=1}^{\infty} \mu(E_n) = +\infty$ [that is, if $\sum_{n=1}^{\infty} \mu_n(E^{(n)})$ diverges], then $\mu(F) = 1$.

(c) If $\sum_{n=1}^{\infty} \mu(E_n) < +\infty$ [if $\sum_{n=1}^{\infty} \mu_n(E^{(n)})$ converges], then $\mu(F) = 0$.

Proof: We first handle part (a). If $F = \varnothing$, then $\mu(F) = 0$, so we may assume that $F \neq \varnothing$. Observe that the following statements are logically equivalent: (i) $\boldsymbol{\omega} \in F$, (ii) $\boldsymbol{\omega} \in E_n$ for infinitely many n, (iii) $\boldsymbol{\omega} \in E^{(n)} \times \Omega_{\{n\}^c}$ for infinitely many n, (iv) the set $\{i \in \mathbb{N} : \omega_i \in E^{(i)}\}$ is infinite.

We now complete (a). Let $\boldsymbol{\omega} \in F$. If $\boldsymbol{\xi} \in \Omega$ is such that $\boldsymbol{\xi}$ is ultimately equal to $\boldsymbol{\omega}$, then $\xi_n = \omega_n$ for all sufficiently large n, and hence $\{i \in \mathbb{N} : \xi_i \in E^{(i)}\}$ is infinite. This means that $\boldsymbol{\xi} \in F$. That is, every $\boldsymbol{\xi}$ that is ultimately equal to $\boldsymbol{\omega}$ is in F, and so F satisfies the requirements of the Zero-One Law [Claim 2 of Section 11.3], that is, $\mu(F) \in \{0, 1\}$. This gives (a).

We now turn to (b). Assume that $\sum_{n=1}^{\infty} \mu(E_n) = +\infty$, that is, assume that $\sum_{n=1}^{\infty} \mu_n(E^{(n)}) = +\infty$. First observe that

$$\begin{aligned} F^c = \bigcup_{n=1}^{\infty} \bigcap_{k=n}^{\infty} E_k^c &= \bigcup_{n=1}^{\infty} \bigcap_{k=n}^{\infty} \left[E^{(k)c} \times \Omega_{\{k\}^c}\right] \\ &= \bigcup_{n=1}^{\infty} \left[\prod_{k=n}^{\infty} E^{(k)c} \times \Omega_{\mathbb{N}_{n-1}}\right]. \end{aligned}$$

Next, observe that $\{\prod_{k=n}^{\infty} E^{(k)c} \times \Omega_{\mathbb{N}_{n-1}}\}_{n=1}^{\infty}$ is nondecreasing to F^c. Using (M8) from Section 2.2, we have that

$$\begin{aligned} \mu(F^c) &= \mu\left(\lim_n \left[\prod_{k=n}^{\infty} E^{(k)c} \times \Omega_{\mathbb{N}_{n-1}}\right]\right) \\ &= \lim_n \mu\left(\prod_{k=n}^{\infty} E^{(k)c} \times \Omega_{\mathbb{N}_{n-1}}\right) \\ &= \lim_n \prod_{k=n}^{\infty} \mu_k(E^{(k)c}) \quad \text{[Claim 5]} \\ &= \lim_n \prod_{k=n}^{\infty} \left(1 - \mu_k(E^{(k)})\right). \end{aligned}$$

There are now two cases to consider: (i) for every $n \in \mathbb{N}$ there is $k \geq n$ with $\mu_k(E^{(k)}) = 1$, and (ii) $0 \leq \mu_k(E^{(k)}) < 1$ for all large k. In (i), we clearly have $\mu(F^c) = 0$. In (ii), we recall that $\sum_{k=n}^{\infty} \mu_k(E^{(k)}) = +\infty$ for every $n \in \mathbb{N}$ and appeal to Claim 4 to obtain the equation $\prod_{k=n}^{\infty}(1 - \mu_k(E^{(k)})) = 0$ for every $n \in \mathbb{N}$, and hence $\mu(F^c) = 0$ in this case as well. Therefore, in either case we have $\mu(F^c) = 0$, and hence $\mu(F) = 1$. This proves (b).

We now turn to (c), assuming that $\sum_{n=1}^{\infty} \mu_n(E^{(n)}) < +\infty$. For every $n \in \mathbb{N}$, we have that

$$\mu(F) \leq \mu\left(\bigcup_{k=n}^{\infty} E_k\right) \leq \sum_{k=n}^{\infty} \mu(E_k) = \sum_{k=n}^{\infty} \mu_k(E^{(k)}).$$

Letting $n \to \infty$ in both sides of the above display gives $\mu(F) = 0$. This gives (c) and completes the proof. ■

We have proven a fair number of claims, and a few more remain for discussion before stating and proving the L^1 Strong Law. The reader is urged to review Claim 5 of Section 10.6.

Framework for the Remainder of this Section: $\{(\Omega_k, \mathcal{F}_k, \mu_k)\}_{k=1}^{\infty}$ will denote a sequence of measure spaces with $\mu_k(\Omega_k) = 1$ for every $k \in \mathbb{N}$. We will let $(\Omega, \mathcal{F}, \mu)$ denote the usual product measure space as constructed in Sections 1 and 2 of this chapter. For each $k \in \mathbb{N}$, we will let $g_k \in L^1(\Omega_k, \mathcal{F}_k, \mu_k)$. Furthermore, for every $\alpha \in \mathbb{R}$ the numbers $\mu_k(\{\omega_k \in \Omega_k : g_k(\omega_k) > \alpha\})$ are assumed to not depend on k; they depend only on α.

Claim 8. *The following results hold:*
(a) The numbers $\mu_k(\{\omega_k \in \Omega_k : |g_k(\omega_k)| > \alpha\})$ are independent of k for each $\alpha \in \mathbb{R}$.

(b) The numbers $\mu_k(\{\omega_k \in \Omega_k : \alpha < |g_k(\omega_k)| \leq \beta\})$ are independent of k for each α and β in $\mathbb{R}$.[This also holds if g_k replaces $|g_k|$.]

Proof: We consider (a), picking $\alpha \in \mathbb{R}$ and $k \in \mathbb{N}$. Observe that

$$\begin{aligned}
& \mu_k(\{\omega_k \in \Omega_k : |g_k(\omega_k)| > \alpha\}) \\
= \; & \mu_k(\{\omega_k \in \Omega_k : g_k(\omega_k) > \alpha\}) + \mu_k(\{\omega_k \in \Omega_k : g_k(\omega_k) < -\alpha\}) \\
= \; & \mu_k(\{\omega_k \in \Omega_k : g_k(\omega_k) > \alpha\}) + 1 - \mu_k(\{\omega_k \in \Omega_k : g_k(\omega_k) \geq -\alpha\}) \\
= \; & \mu_k(\{\omega_k \in \Omega_k : g_k(\omega_k) > \alpha\}) + 1 - \mu_k\left(\lim_n \{\omega_k \in \Omega_k : g_k(\omega_k) > -\alpha - n^{-1}\}\right) \\
= \; & \mu_k(\{\omega_k \in \Omega_k : g_k(\omega_k) > \alpha\}) + 1 - \lim_n \mu_k\left(\{\omega_k \in \Omega_k : g_k(\omega_k) > -\alpha - n^{-1}\}\right),
\end{aligned}$$

which by assumption does not depend on k. This proves (a).

For (b), we have for given $\alpha < \beta$ that

$$\begin{aligned}
& \mu_k(\{\omega_k \in \Omega_k : \alpha < |g_k(\omega_k)| \leq \beta\}) \\
= \; & \mu(\{\omega_k \in \Omega_k : |g_k(\omega_k)| > \alpha\}) - \mu_k(\{\omega_k \in \Omega_k : |g_k(\omega_k)| > \beta\}),
\end{aligned}$$

which by (a) is a difference of two numbers that are free of k. It follows that the number $\mu_k(\{\omega_k \in \Omega_k : \alpha < |g_k(\omega_k)| \leq \beta\})$ does not depend on k, proving (b). This completes the entire proof. ■

Claim 9. *We have the following results:*

(a) Let $k \in \mathbb{N}$ and $0 \leq t < +\infty$. The number

$$\mu_k\left(\{\omega_k \in \Omega_k : |g_k(\omega_k)| I_{\{\eta_k \in \Omega_k :\, |g_k(\eta_k)| \leq k\}}(\omega_k) > t\}\right)$$

equals

$$\mu_1\left(\{\omega_1 \in \Omega_1 : |g_1(\omega_1)| I_{\{\eta_1 \in \Omega_1 :\, |g_1(\eta_1)| \leq k\}}(\omega_1) > t\}\right).$$

(b) Let $k \in \mathbb{N}$. The integral

$$\int_{\Omega_k} \left[g_k(\omega_k) I_{\{\eta_k \in \Omega_k :\, |g_k(\eta_k)| \leq k\}}(\omega_k)\right]^2 d\mu_k(\omega_k)$$

equals the integral

$$\int_{\Omega_1} \left[g_1(\omega_1) I_{\{\eta_1 \in \Omega_1 :\, |g_1(\eta_1)| \leq k\}}(\omega_1)\right]^2 d\mu_1(\omega_1).$$

Proof: First consider (a). We have by Claim 8 that

$$\begin{aligned}
& \mu_k\left(\{\omega_k \in \Omega_k : |g_k(\omega_k)| I_{\{\eta_k \in \Omega_k :\, |g_k(\eta_k)| \leq k\}}(\omega_k) > t\}\right) \\
= \; & \mu_k(\{\omega_k \in \Omega_k : t < |g_k(\omega_k)| \leq k\}) \\
= \; & \mu_1(\{\omega_1 \in \Omega_1 : t < |g_1(\omega_1)| \leq k\}) \\
= \; & \mu_1\left(\{\omega_1 \in \Omega_1 : |g_1(\omega_1)| I_{\{\eta_1 \in \Omega_1 :\, |g_1(\eta_1)| \leq k\}}(\omega_1) > t\}\right),
\end{aligned}$$

giving the first result.

For (b), we require Claim 5(b) of Section 10.6, there taking $p = 2$. Using this result and (a) just given, we have that

$$\begin{aligned}
&\int_{\Omega_k} \left[g_k(\omega_k) I_{\{\eta_k \in \Omega_k:\ |g_k(\eta_k)| \le k\}}(\omega_k)\right]^2 \, d\mu_k(\omega_k) \\
&= \int_0^\infty 2t\mu_k\left(\left\{\omega_k \in \Omega_k :\ |g_k(\omega_k)| I_{\{\eta_k \in \Omega_k:\ |g_k(\eta_k)| \le k\}}(\omega_k) > t\right\}\right) \, d\bar{\lambda}(t) \\
&= \int_0^\infty 2t\mu_1\left(\left\{\omega_1 \in \Omega_1 :\ |g_1(\omega_1)| I_{\{\eta_1 \in \Omega_1:\ |g_1(\eta_1)| \le k\}}(\omega_1) > t\right\}\right) \, d\bar{\lambda}(t) \\
&= \int_{\Omega_1} \left[g_1(\omega_1) I_{\{\eta_1 \in \Omega_1:\ |g_1(\eta_1)| \le k\}}(\omega_1)\right]^2 \, d\mu_1(\omega_1),
\end{aligned}$$

which completes the proof. ■

The next result is similar to that of the previous claim. We will again require Claim 5 of Section 10.6, there taking $p = 1$. We require the following notation:

$$f'_k(\boldsymbol{\omega}) = g_k(\omega_k) I_{\{\eta_k \in \Omega_k:\ |g_k(\eta_k)| \le k\}}(\omega_k), \quad \boldsymbol{\omega} \in \Omega,\ k \in \mathbb{N},$$

and

$$h_k(\omega_1) = g_1(\omega_1) I_{\{\eta_1 \in \Omega_1:\ |g_1(\eta_1)| \le k\}}(\omega_1), \quad \omega_1 \in \Omega_1, \quad k \in \mathbb{N}.$$

Claim 10. *For every $k \in \mathbb{N}$, we have*

$$\int_\Omega f'_k(\boldsymbol{\omega}) \, d\mu(\boldsymbol{\omega}) = \int_{\Omega_1} h_k(\omega_1) \, d\mu_1(\omega_1).$$

Proof: Given $k \in \mathbb{N}$, we have that

$$\begin{aligned}
&\int_\Omega f'_k(\boldsymbol{\omega}) \, d\mu(\boldsymbol{\omega}) \\
&= \int_\Omega g_k(\omega_k) I_{\{\eta_k \in \Omega_k:\ |g_k(\eta_k)| \le k\}}(\omega_k) \, d\mu(\boldsymbol{\omega}) \\
&= \int_{\Omega_k} g_k(\omega_k) I_{\{\eta_k \in \Omega_k:\ |g_k(\eta_k)| \le k\}}(\omega_k) \, d\mu_k(\omega_k) \quad \text{[Claim 3, Section 4]} \\
&= \int_{\Omega_k} \left[g_k(\omega_k) I_{\{\eta_k \in \Omega_k:\ |g_k(\eta_k)| \le k\}}(\omega_k) + k\right] \, d\mu_k(\omega_k) - k \\
&= \int_0^\infty \mu_k(\{\omega_k \in \Omega_k :\ g_k(\omega_k) I_{\{\eta_k \in \Omega_k:\ |g_k(\eta_k)| \le k\}}(\omega_k) > t - k\}) \, d\bar{\lambda}(t) \ - k,
\end{aligned}$$

where the last equality follows by applying Claim 5 of Section 10.6 to the nonnegative function

$$\omega_k \mapsto g_k(\omega_k) I_{\{\eta_k \in \Omega_k:\ |g_k(\eta_k)| \le k\}}(\omega_k) + k \quad \text{on } \Omega_k.$$

For the integrand of the just-mentioned integral, consider two cases for the value t: (i) $t \geq k$ and (ii) $t < k$. In case (i), Claim 9 gives

$$\begin{aligned} & \mu_k(\{\omega_k \in \Omega_k : g_k(\omega_k) I_{\{\eta_k \in \Omega_k :\, |g_k(\eta_k)| \leq k\}}(\omega_k) > t - k\}) \\ = \;& \mu_k(\{\omega_k \in \Omega_k : t - k < g_k(\omega_k) \leq k\}) \\ = \;& \mu_1(\{\omega_1 \in \Omega_1 : t - k < g_1(\omega_1) \leq k\}) \\ = \;& \mu_1(\{\omega_1 \in \Omega_1 : g_1(\omega_1) I_{\{\eta_1 \in \Omega_1 :\, |g_1(\eta_1)| \leq k\}}(\omega_1) > t - k\}). \end{aligned}$$

For case (ii), where $t - k < 0$, the reader may verify that

$$g_k(\omega_k) I_{\{\eta_k \in \Omega_k :\, |g_k(\eta_k)| \leq k\}}(\omega_k) > t - k$$

if and only if

$$|g_k(\omega_k)| > k \quad \text{or} \quad t - k < g_k(\omega_k) \leq k.$$

Therefore, we have that

$$\begin{aligned} & \mu_k(\{\omega_k \in \Omega_k : g_k(\omega_k) I_{\{\eta_k \in \Omega_k :\, |g_k(\eta_k)| \leq k\}}(\omega_k) > t - k\}) \\ = \;& \mu_k(\{\omega_k \in \Omega_k : |g_k(\omega_k)| > k\}) + \mu_k(\{\omega_k \in \Omega_k : t - k < g_k(\omega_k) \leq k\}) \\ = \;& \mu_1(\{\omega_1 \in \Omega_1 : |g_1(\omega_1)| > k\}) + \mu_1(\{\omega_1 \in \Omega_1 : t - k < g_1(\omega_1) \leq k\}) \\ = \;& \mu_1(\{\omega_1 \in \Omega_1 : g_1(\omega_1) I_{\{\eta_1 \in \Omega_1 :\, |g_1(\eta_1)| \leq k\}}(\omega_1) > t - k\}). \end{aligned}$$

[The second equality follows by invoking (a) and (b) of Claim 8. The last equality follows by the same logic that upholds the first equality.]

Combining cases (i) and (ii), we have that

$$\begin{aligned} & \int_0^\infty \mu_k(\{\omega_k \in \Omega_k : g_k(\omega_k) I_{\{\eta_k \in \Omega_k :\, |g_k(\eta_k)| \leq k\}}(\omega_k) > t - k\})\, d\bar{\lambda}(t) \; - k \\ = \;& \int_0^\infty \mu_1(\{\omega_1 \in \Omega_1 : g_1(\omega_1) I_{\{\eta_1 \in \Omega_1 :\, |g_1(\eta_1)| \leq k\}}(\omega_1) > t - k\})\, d\bar{\lambda}(t) \; - k \\ = \;& \int_{\Omega_1} \left[g_1(\omega_1) I_{\{\eta_1 \in \Omega_1 :\, |g_1(\eta_1)| \leq k\}}(\omega_1) + k\right] \, d\mu_1(\omega_1) - k \\ = \;& \int_{\Omega_1} g_1(\omega_1) I_{\{\eta_1 \in \Omega_1 :\, |g_1(\eta_1)| \leq k\}}(\omega_1)\, d\mu_1(\omega_1) \\ = \;& \int_{\Omega_1} h_k(\omega_1)\, d\mu_1(\omega_1). \end{aligned}$$

[We have again used Claim 5 of Section 10.6 in the second equality above.] This completes the proof. ■

By presenting the many uninteresting technical claims for the L^1 Strong Law in this section, we are able to prove the L^1 Strong Law without having too many exterior details detract from the main structure of the argument. In fact, the proof of the L^1 Strong Law will be relatively easy now that we have paid the requisite price.

11.6 THE L^1 STRONG LAW

In this section, we use the same framework of the previous section. That is, for each $k \in \mathbb{N}$ we are given a measure space $(\Omega_k, \mathcal{F}_k, \mu_k)$ with $\mu_k(\Omega_k) = 1$, and $(\Omega, \mathcal{F}, \mu)$ will denote the usual product measure space as constructed in the first two sections of this chapter. With this framework and the work given in the previous section, we are now in a position to give a relatively succinct proof of the L^1 version of the Strong Law of Large Numbers.

L^1 **Strong Law.** *For each $k \in \mathbb{N}$, let g_k denote a function in $L^1(\Omega_k, \mathcal{F}_k, \mu_k)$. Suppose that the numbers $\mu_k(\{\omega_k \in \Omega_k : g_k(\omega_k) > \alpha\})$ are independent of k for each $\alpha \in \mathbb{R}$. For each $k \in \mathbb{N}$, let f_k denote the mapping on Ω given by the rule $\boldsymbol{\omega} \to g_k(\omega_k)$. Then*

$$\lim_n \frac{1}{n} \sum_{k=1}^{n} f_k(\boldsymbol{\omega}) = \int_\Omega f_1 \, d\mu \quad \textit{for } \mu\textit{-a.e. } \boldsymbol{\omega} \in \Omega.$$

Proof: For each $k \in \mathbb{N}$, $\boldsymbol{\omega} \in \Omega$, and $\omega_1 \in \Omega_1$, we write

$$f'_k(\boldsymbol{\omega}) = \begin{cases} f_k(\boldsymbol{\omega}) & |f_k(\boldsymbol{\omega})| \le k \\ 0 & \text{otherwise} \end{cases} = \begin{cases} g_k(\omega_k) & |g_k(\omega_k)| \le k \\ 0 & \text{otherwise} \end{cases}$$

and

$$h_k(\omega_1) = \begin{cases} g_1(\omega_1) & |g_1(\omega_1)| \le k \\ 0 & \text{otherwise.} \end{cases}$$

[Therefore, each f'_k is defined on Ω while each h_k is defined on Ω_1.]

We now break the proof up into a number of parts.

(i) $\lim_n[\frac{1}{n}\sum_{k=1}^n f_k(\boldsymbol{\omega}) - \frac{1}{n}\sum_{k=1}^n f'_k(\boldsymbol{\omega})] = 0$ *for μ-a.e. $\boldsymbol{\omega} \in \Omega$.*

Proof of (i): For each $k \in \mathbb{N}$, let $E^{(k)} = \{\omega_k \in \Omega_k : |g_k(\omega_k)| > k\}$ and let $E_k = E^{(k)} \times \Omega_{\{k\}^c}$. Then

$$\begin{aligned}
\sum_{k=1}^{\infty} \mu_k(E^{(k)}) &= \sum_{k=1}^{\infty} \mu_k(\{\omega_k \in \Omega_k : |g_k(\omega_k)| > k\}) \\
&= \sum_{k=1}^{\infty}\sum_{n=k}^{\infty} \mu_k(\{\omega_k \in \Omega_k : n < |g_k(\omega_k)| \le n+1\}) \\
&= \sum_{k=1}^{\infty}\sum_{n=k}^{\infty} \mu_1(\{\omega_1 \in \Omega_1 : n < |g_1(\omega_1)| \le n+1\}) \\
&= \sum_{k=1}^{\infty} k\mu_1(\{\omega_1 \in \Omega_1 : k < |g_1(\omega_1)| \le k+1\}) \\
&\le \int_{\Omega_1} |g_1(\omega_1)| \, d\mu_1(\omega_1) \\
&< +\infty.
\end{aligned}$$

[The third equality follows by assumption and Claim 8(b) from the previous section. The fourth equality follows since $\sum_{k=1}^{\infty}\sum_{n=k}^{\infty} a_n = \sum_{k=1}^{\infty} ka_k$ when both series exist. The first inequality should be clear, and the final inequality follows because $g_1 \in L^1(\Omega_1, \mathcal{F}_1, \mu_1)$.]

We have shown that $\sum_{k=1}^{\infty} \mu_k(E^{(k)})$ converges, and hence the Borel-Cantelli Lemma [Claim 7(c)] implies that

$$\mu\left(\bigcap_{n=1}^{\infty}\bigcup_{k=n}^{\infty} E_k\right) = 0.$$

This in turn forces

$$\mu\left(\bigcup_{n=1}^{\infty}\bigcap_{k=n}^{\infty} E_k^c\right) = 1.$$

Therefore, μ-a.e. $\omega \in \Omega$ have the property that ω is in $\bigcap_{k=n}^{\infty} E_k^c$ for some n [which depends on ω]. By the definition of the sets E_k, this implies the following: *for almost all $\omega \in \Omega$, there is a positive integer m_0 depending on ω such that*

$$f_m(\omega) = f'_m(\omega) \quad \textit{if } m \geq m_0.$$

By Claim 1 of Section 4, this forces

$$\lim_n \left[\frac{1}{n}\sum_{k=1}^{n} f_k(\omega) - \frac{1}{n}\sum_{k=1}^{n} f'_k(\omega)\right] = 0 \quad \text{for } \mu\text{-a.e. } \omega \in \Omega$$

as claimed.

(ii) $\sum_{k=1}^{\infty} \frac{1}{k^2}\|f'_k - \int_\Omega f'_k\, d\mu\|_2^2 < +\infty$.

Proof of (ii): We make some direct calculations and use Claim 9(b) from the previous section. We have

$$\begin{aligned}
&\sum_{k=1}^{\infty} \frac{1}{k^2}\left\|f'_k - \int_\Omega f'_k\, d\mu\right\|_2^2 \\
&= \sum_{k=1}^{\infty} \frac{1}{k^2}\int_\Omega \left(f'_k - \int_\Omega f'_k\, d\mu\right)^2 d\mu \\
&= \sum_{k=1}^{\infty} \frac{1}{k^2}\int_\Omega \left(f_k'^2 - 2f'_k\int_\Omega f'_k\, d\mu + \left(\int_\Omega f'_k\, d\mu\right)^2\right) d\mu \\
&= \sum_{k=1}^{\infty} \frac{1}{k^2}\left[\int_\Omega f_k'^2\, d\mu - 2\int_\Omega f'_k\, d\mu\int_\Omega f'_k\, d\mu + \left(\int_\Omega f'_k\, d\mu\right)^2\right] \\
&= \sum_{k=1}^{\infty} \frac{1}{k^2}\left[\int_\Omega f_k'^2\, d\mu - \left(\int_\Omega f'_k\, d\mu\right)^2\right] \\
&\leq \sum_{k=1}^{\infty} \frac{1}{k^2}\int_\Omega f_k'^2\, d\mu.
\end{aligned}$$

We next have, using Claim 9(b) and Claim 3 of Section 4, that

$$\begin{aligned}
&\sum_{k=1}^{\infty} \frac{1}{k^2} \int_{\Omega} f_k'^2 \, d\mu \\
=\ &\sum_{k=1}^{\infty} \frac{1}{k^2} \int_{\Omega} g_k(\omega_k)^2 I_{\{\eta_k \in \Omega_k:\ |g_k(\eta_k)| \le k\}}(\omega_k) \, d\mu(\boldsymbol{\omega}) \\
=\ &\sum_{k=1}^{\infty} \frac{1}{k^2} \int_{\Omega_k} g_k(\omega_k)^2 I_{\{\eta_k \in \Omega_k:\ |g_k(\eta_k)| \le k\}}(\omega_k) \, d\mu_k(\omega_k) \\
=\ &\sum_{k=1}^{\infty} \frac{1}{k^2} \int_{\Omega_1} g_1(\omega_1)^2 I_{\{\eta_1 \in \Omega_1:\ |g_1(\eta_1)| \le k\}}(\omega_1) \, d\mu_1(\omega_1) \\
=\ &\sum_{k=1}^{\infty} \frac{1}{k^2} \int_{\Omega_1} h_k(\omega_1)^2 \, d\mu_1(\omega_1) \\
=\ &\int_{\Omega_1} \left(\sum_{k=1}^{\infty} \frac{1}{k^2} h_k^2 \right) d\mu_1.
\end{aligned}$$

Now pick an element $\omega_1 \in \Omega_1$. If $g_1(\omega_1) = 0$, then $h_k(\omega_1) = g_1(\omega_1) = 0$ for all $k \in \mathbb{N}$, and hence $\sum_{k=1}^{\infty} \frac{1}{k^2} h_k(\omega_1)^2 = 0$. On the other hand, if $g_1(\omega_1) \neq 0$, then there exists a unique $p \in \mathbb{N}$ such that $p-1 < |g_1(\omega_1)| \le p$. For this p, we have that $h_1(\omega_1) = \cdots = h_{p-1}(\omega_1) = 0$ and $h_p(\omega_1) = h_{p+1}(\omega_1) = \cdots = g_1(\omega_1)$. Therefore, using Claim 6 of the previous section, we have

$$\sum_{k=1}^{\infty} \frac{1}{k^2} h_k^2(\omega_1) = \sum_{k=p}^{\infty} \frac{1}{k^2} g_1(\omega_1)^2 \le |g_1(\omega_1)| p \sum_{k=p}^{\infty} \frac{1}{k^2} \le 2|g_1(\omega_1)|.$$

Putting the previous displays together and summarizing, we have that

$$\begin{aligned}
\sum_{k=1}^{\infty} \frac{1}{k^2} \left\| f_k' - \int_{\Omega} f_k' \, d\mu \right\|_2^2 &\le \sum_{k=1}^{\infty} \frac{1}{k^2} \int_{\Omega} f_k'^2 \, d\mu \\
&= \int_{\Omega_1} \left(\sum_{k=1}^{\infty} \frac{1}{k^2} h_k^2 \right) d\mu_1 \\
&\le 2 \int_{\Omega_1} |g_1(\omega_1)| \, d\mu_1(\omega_1) \\
&< +\infty.
\end{aligned}$$

This completes the proof of (ii).

Before going on to the next part, note that the functions $f_k' - \int_{\Omega} f_k' \, d\mu$ defined on Ω are in $L^1(\Omega, \mathcal{F}, \mu)$ and hence are in $L^2(\Omega, \mathcal{F}, \mu)$. Also note that

$$\int_{\Omega} \left(f_k'(\boldsymbol{\omega}) - \int_{\Omega} f_k' \, d\mu \right) d\mu(\boldsymbol{\omega}) = 0.$$

Utilizing the L^2 version of the Strong Law, we deduce that

$$\lim_n \frac{1}{n} \sum_{k=1}^{n} \left[f'_k(\omega) - \int_\Omega f'_k \, d\mu \right] = 0 \quad \text{for } \mu\text{-a.e. } \omega \in \Omega.$$

In other words, we have that

$$\lim_n \left[\frac{1}{n} \sum_{k=1}^{n} f'_k(\omega) - \frac{1}{n} \sum_{k=1}^{n} \int_\Omega f'_k \, d\mu \right] = 0 \quad \text{for } \mu\text{-a.e. } \omega \in \Omega.$$

(iii) $\lim_n \frac{1}{n} \sum_{k=1}^{n} \int_\Omega f'_k \, d\mu = \int_\Omega f_1 \, d\mu$.

Proof of (iii): Recalling Claim 10 from the previous section, we have that

$$\int_\Omega f'_k \, d\mu = \int_{\Omega_1} h_k \, d\mu_1 \quad \text{for every } k \in \mathbb{N}.$$

Using the Dominated Convergence Theorem, we have that

$$\begin{aligned} \lim_k \int_{\Omega_1} h_k \, d\mu_1 = \int_{\Omega_1} \lim_k h_k \, d\mu_1 &= \int_{\Omega_1} g_1 \, d\mu_1 \\ &= \int_\Omega f_1 \, d\mu. \end{aligned}$$

Therefore, using Claim 1 of Section 4, we have that

$$\lim_n \frac{1}{n} \sum_{k=1}^{n} \int_\Omega f'_k \, d\mu = \lim_n \frac{1}{n} \sum_{k=1}^{n} \int_{\Omega_1} h_k \, d\mu_1 = \lim_k \int_{\Omega_1} h_k \, d\mu_1 = \int_\Omega f_1 \, d\mu.$$

This gives (iii).

We now complete the entire proof. Recall that

$$\lim_n \left[\frac{1}{n} \sum_{k=1}^{n} f_k(\omega) - \frac{1}{n} \sum_{k=1}^{n} f'_k(\omega) \right] = 0 \quad \text{for } \mu\text{-a.e. } \omega \in \Omega.$$

By the comments immediately preceding (iii), we have that

$$\lim_n \left[\frac{1}{n} \sum_{k=1}^{n} f'_k(\omega) - \frac{1}{n} \sum_{k=1}^{n} \int_\Omega f'_k \, d\mu \right] = 0 \quad \text{for } \mu\text{-a.e. } \omega \in \Omega.$$

By (iii), the above becomes

$$\lim_n \left[\frac{1}{n} \sum_{k=1}^{n} f'_k(\omega) - \int_\Omega f_1 \, d\mu \right] = 0 \quad \text{for } \mu\text{-a.e. } \omega \in \Omega.$$

It follows that

$$\lim_n \left[\frac{1}{n} \sum_{k=1}^{n} f_k(\omega) - \int_\Omega f_1 \, d\mu \right] = 0 \quad \text{for } \mu\text{-a.e. } \omega \in \Omega.$$

This completes the entire proof.

References

1. Edwin Hewitt and Karl Stromberg, *Real and Abstract Analysis: A Modern Treatment of the Theory of Functions of a Real Variable*, Springer-Verlag, New York, 1969.

This book is, in the author's opinion, the ultimate graduate-level analysis textbook. The discussion of the cardinality of $\mathcal{B}^k$ in Section 1.6, the treatment of linear functionals and the Riesz Representation Theorem in Sections 8.8-9, the generalized version of the Radon-Nikodym Theorem in Section 9.2, the discussion of general derivatives in Section 9.4, the form of the Hardy-Littlewood Maximal Theorem given in Section 10.6, and the entire Chapter 11 discussion of infinite-dimensional product spaces and measures are based upon the presentations of those same topics found in this magisterial work. More generally, the influence of Hewitt and Stromberg may be seen indirectly in everything that is done in this book, from using functional notation to describe integrals to presenting constructions based on ordinal numbers, the Continuum Hypothesis, and the Axiom of Choice.

2. Patrick Billingsley, *Probability and Measure, 3rd Edition*, John Wiley and Sons, New York, 1995.

Billingsley's famous text discusses measure theory in conjunction with probability theory, using the former theory to prove results in the latter theory, and using the latter theory to motivate problems and concepts in the former theory. This was the first book from which the author learned the basics of measure theory. The influence of this book is in the following: (i) the construction of the ethereal Borel sets given in Section 1.7, (ii) the general framework of Sections 2.3 and 2.4 [and the proof of the theorem in Section 2.4], where measures are associated with certain types of point functions, (iii) the entire Chapter 3 framework for demonstrating the existence of measure extensions, (iv) the order of development in Section 9.1, where we use additive set functions instead of signed measures, and (v) the Radon-Nikodym-based development of the Riesz Representation Theorem in Section 9.3. The impossibility results found in Section 4.8 are explicitly taken from Billingsley's presentation. Like Hewitt and Stromberg's text mentioned above, where Billingsley's influence is not explicit, it is certainly implicit.

3. Paul Halmos, *Measure Theory*, Springer-Verlag, New York, 1974.

This is the classic text on measure theory, first appearing in 1950. Halmos' book uses σ-rings and rings in place of σ-fields and fields, and achieves results in slightly greater generality. The discussions and treatments of measure theory in Halmos' book are implicitly behind various topics in this book, such as infinite-dimensional product spaces and the Radon-Nikodym Theorem in Section 9.1. The desire for a more modern and more expository version of Halmos' text is one of the major reasons why the author undertook the effort to write the book. It should also be noted that Halmos' text, besides covering much of the material found in this book, covers a few other general topics that this book cannot cover due to limitations of space, such as locally compact spaces and Haar measure.

4. Frank Jones, *Lebesgue Integration in Euclidean Space*, Jones & Bartlett, Boston 1993.

This is a very readable text that is noteworthy for several things: (i) the detailed "analysis-based" development of Lebesgue measure found within, (ii) the very readable discussion of the translation invariance of Lebesgue measure and of the effects of linear transformations on Lebesgue measure, (iii) his explicit construction of a non-Borel Lebesgue set, (iv) a very easy-to-read but detailed chapter on L^p spaces, and (v) a patient discussion of Lebesgue integrals with respect to Lebesgue measure. Jones' text applies Lebesgue measure and the Fubini Theorems to [among other things] trigonometric series, Fourier transforms, a discussion of the relations between derivatives and integrals, and the general change of variables theorem for a k-dimensional Lebesgue integral. These topics are not found in this book. On the other hand, this book covers some measure-theoretic items not found in Jones' book: the Radon-Nikodym Theorem, the Extension Theorem for measures on general measure spaces, product measures for infinite dimensional product spaces, and the Riesz Representation Theorem. This text follows Jones' most excellent presentation in the discussion of L^p spaces for all three of the cases $0 < p < 1$, $1 \leq p < +\infty$, and $p = +\infty$, the Riesz-Fischer Theorem in Section 8.2, the construction of a non-Lebesgue set in Section 4.5, and the construction of a non-Borel Lebesgue set in Section 4.7. Sections 7.1-7.4 derive their impetus from Jones' discussion. Our discussions of the Generalized Minkowski Inequality and convolutions in Sections 10.4 and 10.5 also follow Jones' presentation. This text is a valued member of the author's library.

5. Walter Rudin, *Real and Complex Analysis, 3rd Edition*, McGraw-Hill, New York, 1987.

   This books stands on its own as one of the time-tested gold standards for analysis books. Whereas we have discussed measures that take values in $[0, +\infty]$, Rudin discusses measures that take on complex values. Accordingly, his presentation of the Radon-Nikodym Theorem and the various decompositions mentioned in Chapter 9 of this book are more general. In Section 5.7, the author has used Rudin's form of Lusin's Theorem. In Sections 8.8 and 8.9, the author has implicitly used some facets of Rudin's presentation in the discussion of the Riesz Representation Theorem. The same comments hold for the discussions of product measures in Section 10.1 and some convolution results in Section 10.5.

6. Kai Lai Chung, *A Course in Probability Theory, 2nd Edition*, Academic Press, New York, 1974.

   This is a book on probability theory and not measure theory [as the title indicates], but the author has found Chung's brief presentation of some measure-theoretic terms to be quite useful. The discussion of the various modes of convergence follows in part Chung's presentation. The discussion of the correspondence between measures and certain types of functions found in Sections 2.3 and 2.4 of this book also borrows some elements of Chung's discussion. Chung's book contains a wealth of applications of measure theory to situations in probability theory, and remains one of the standard books on the topic.

7. M. M. Rao, *Measure Theory and Integration*, John Wiley and Sons, New York, 1987.

   This book is a modern treatise on abstract measure theory for measure theory's sake that is almost encyclopedic in scope. While it is difficult to say exactly where features of Rao's presentation are explicitly used in this text, this book is considered a major reference because the author learned much about measure theory from this book after studying Billingsley's book mentioned above. Rao derives results concerning extensions of measures in greater generality than in this book. He also discusses many advanced concepts of measure theory that do not appear in this book. Rao's book also serves as an excellent reference on any mainstream topic in measure theory.

8. Herbert Federer, *Geometric Measure Theory*, Springer-Verlag, New York, 1969.

   This book is quite large, and the author is qualified to comment only on Chapter 2 of his work, which deals with general measure theory. Nothing from this book is explicitly used, although again the work done in this chapter of Federer's work helped the present author think certain problems through with regards to order and style of presentation. Chapter 2 of Federer [and indeed the entire book] is very densely written, and the competent reader may find that getting through a single page takes hours. The author believes that an entire year-long graduate course in measure theory or advanced analysis could be formed by a thorough study of this chapter.

Minor References

The texts listed below have been useful in helping the author clarify his thought or presentation of certain topics at a few places in the book. There is no significance to be found in the order of their listing below.

9. J. F. C. Kingman and S. J. Taylor, *Introduction to Measure and Probability*, Cambridge University Press, 1966.
10. George G. Roussas, *Statistics 235A Lecture Notes*, unpublished.
11. Joseph L. Doob, *Measure Theory*, Springer-Verlag, New York, 1994.
12. Bernard Gelbaum and John Olmstead, *Counterexamples in Analysis*, Holden-Day, San Francisco, 1964.
13. Walter Rudin, *Principles of Mathematical Analysis, 3rd Edition*, McGraw-Hill, New York, 1976.
14. H. S. Bear, *A Primer of Lebesgue Integration*, Academic Press, New York, 1995.
15. Patrick Billingsley, *Convergence of Probability Measures, 2nd Edition*, John Wiley and Sons, New York, 1999.
16. N. L. Carothers, *Real Analysis*, Cambridge University Press, 2000.
17. Waclaw Sierpiński, *General Topology*, [Translated by C. Cecelia Krieger], Dover Publications, Mineola, New York, 2000 [originally published in 1956 by the University of Toronto Press].

18. Serge Lang, *Real Analysis, Second Edition*, Addison-Wesley, Reading, Massachusetts, 1983.

19. A. N. Kolmogorov and S. V. Fomin, *Introductory Real Analysis*, [Translated by Richard Silverman], Dover Publications, Mineola, New York, 1975 [originally published by the Moscow University Press in 1968].

20. Jeffrey Rosenthal, *A First Look at Rigorous Probability Theory*, World Scientific, Singapore, 2000.

21. Richard Durrett, *Probability: Theory and Examples, 2nd Edition*, Duxbury Press, Wadsworth Press, Belmont, New York, 1996.

22. Daniel Stroock, *Probability Theory: An Analytic View, Revised Edition*, Cambridge University Press, 1993.

23. Robert V. Ash, *Real Analysis and Probability*, Academic Press, New York, 1972.

24. Howard Tucker, *A Graduate Course in Probability*, Academic Press, New York, 1967.

25. H. L. Royden, *Real Analysis, 3rd Edition*, Prentice Hall, New York, 1988.

26. J. Galambos, *Advanced Probability Theory*, Marcel Dekker Publishing, 1988.

Index

WILEY SERIES IN PROBABILITY AND STATISTICS
ESTABLISHED BY WALTER A. SHEWHART AND SAMUEL S. WILKS

BENDAT and PIERSOL · Random Data: Analysis and Measurement Procedures, *Third Edition*
BERRY, CHALONER, and GEWEKE · Bayesian Analysis in Statistics and Econometrics: Essays in Honor of Arnold Zellner
BERNARDO and SMITH · Bayesian Theory
BHAT and MILLER · Elements of Applied Stochastic Processes, *Third Edition*
BHATTACHARYA and JOHNSON · Statistical Concepts and Methods
BHATTACHARYA and WAYMIRE · Stochastic Processes with Applications
BILLINGSLEY · Convergence of Probability Measures, *Second Edition*
BILLINGSLEY · Probability and Measure, *Third Edition*
BIRKES and DODGE · Alternative Methods of Regression
BLISCHKE AND MURTHY (editors) · Case Studies in Reliability and Maintenance
BLISCHKE AND MURTHY · Reliability: Modeling, Prediction, and Optimization
BLOOMFIELD · Fourier Analysis of Time Series: An Introduction, *Second Edition*
BOLLEN · Structural Equations with Latent Variables
BOROVKOV · Ergodicity and Stability of Stochastic Processes
BOULEAU · Numerical Methods for Stochastic Processes
BOX · Bayesian Inference in Statistical Analysis
BOX · R. A. Fisher, the Life of a Scientist
BOX and DRAPER · Empirical Model-Building and Response Surfaces
*BOX and DRAPER · Evolutionary Operation: A Statistical Method for Process Improvement
BOX, HUNTER, and HUNTER · Statistics for Experimenters: An Introduction to Design, Data Analysis, and Model Building
BOX and LUCEÑO · Statistical Control by Monitoring and Feedback Adjustment
BRANDIMARTE · Numerical Methods in Finance: A MATLAB-Based Introduction
BROWN and HOLLANDER · Statistics: A Biomedical Introduction
BRUNNER, DOMHOF, and LANGER · Nonparametric Analysis of Longitudinal Data in Factorial Experiments
BUCKLEW · Large Deviation Techniques in Decision, Simulation, and Estimation
CAIROLI and DALANG · Sequential Stochastic Optimization
CHAN · Time Series: Applications to Finance
CHATTERJEE and HADI · Sensitivity Analysis in Linear Regression
CHATTERJEE and PRICE · Regression Analysis by Example, *Third Edition*
CHERNICK · Bootstrap Methods: A Practitioner's Guide
CHERNICK and FRIIS · Introductory Biostatistics for the Health Sciences
CHILÈS and DELFINER · Geostatistics: Modeling Spatial Uncertainty
CHOW and LIU · Design and Analysis of Clinical Trials: Concepts and Methodologies
CLARKE and DISNEY · Probability and Random Processes: A First Course with Applications, *Second Edition*
*COCHRAN and COX · Experimental Designs, *Second Edition*
CONGDON · Bayesian Statistical Modelling
CONOVER · Practical Nonparametric Statistics, *Second Edition*
COOK · Regression Graphics
COOK and WEISBERG · Applied Regression Including Computing and Graphics
COOK and WEISBERG · An Introduction to Regression Graphics
CORNELL · Experiments with Mixtures, Designs, Models, and the Analysis of Mixture Data, *Third Edition*
COVER and THOMAS · Elements of Information Theory
COX · A Handbook of Introductory Statistical Methods
*COX · Planning of Experiments
CRESSIE · Statistics for Spatial Data, *Revised Edition*
CSÖRGŐ and HORVÁTH · Limit Theorems in Change Point Analysis

*Now available in a lower priced paperback edition in the Wiley Classics Library.

DANIEL · Applications of Statistics to Industrial Experimentation
DANIEL · Biostatistics: A Foundation for Analysis in the Health Sciences, *Sixth Edition*
*DANIEL · Fitting Equations to Data: Computer Analysis of Multifactor Data, *Second Edition*
DASU and JOHNSON · Exploratory Data Mining and Data Cleaning
DAVID and NAGARAJA · Order Statistics, *Third Edition*
*DEGROOT, FIENBERG, and KADANE · Statistics and the Law
DEL CASTILLO · Statistical Process Adjustment for Quality Control
DETTE and STUDDEN · The Theory of Canonical Moments with Applications in Statistics, Probability, and Analysis
DEY and MUKERJEE · Fractional Factorial Plans
DILLON and GOLDSTEIN · Multivariate Analysis: Methods and Applications
DODGE · Alternative Methods of Regression
*DODGE and ROMIG · Sampling Inspection Tables, *Second Edition*
*DOOB · Stochastic Processes
DOWDY and WEARDEN · Statistics for Research, *Second Edition*
DRAPER and SMITH · Applied Regression Analysis, *Third Edition*
DRYDEN and MARDIA · Statistical Shape Analysis
DUDEWICZ and MISHRA · Modern Mathematical Statistics
DUNN and CLARK · Applied Statistics: Analysis of Variance and Regression, *Second Edition*
DUNN and CLARK · Basic Statistics: A Primer for the Biomedical Sciences, *Third Edition*
DUPUIS and ELLIS · A Weak Convergence Approach to the Theory of Large Deviations
*ELANDT-JOHNSON and JOHNSON · Survival Models and Data Analysis
ENDERS · Applied Econometric Time Series
ETHIER and KURTZ · Markov Processes: Characterization and Convergence
EVANS, HASTINGS, and PEACOCK · Statistical Distributions, *Third Edition*
FELLER · An Introduction to Probability Theory and Its Applications, Volume I, *Third Edition,* Revised; Volume II, *Second Edition*
FISHER and VAN BELLE · Biostatistics: A Methodology for the Health Sciences
*FLEISS · The Design and Analysis of Clinical Experiments
FLEISS · Statistical Methods for Rates and Proportions, *Third Edition*
FLEMING and HARRINGTON · Counting Processes and Survival Analysis
FULLER · Introduction to Statistical Time Series, *Second Edition*
FULLER · Measurement Error Models
GALLANT · Nonlinear Statistical Models
GHOSH, MUKHOPADHYAY, and SEN · Sequential Estimation
GIFI · Nonlinear Multivariate Analysis
GLASSERMAN and YAO · Monotone Structure in Discrete-Event Systems
GNANADESIKAN · Methods for Statistical Data Analysis of Multivariate Observations, *Second Edition*
GOLDSTEIN and LEWIS · Assessment: Problems, Development, and Statistical Issues
GREENWOOD and NIKULIN · A Guide to Chi-Squared Testing
GROSS and HARRIS · Fundamentals of Queueing Theory, *Third Edition*
*HAHN and SHAPIRO · Statistical Models in Engineering
HAHN and MEEKER · Statistical Intervals: A Guide for Practitioners
HALD · A History of Probability and Statistics and their Applications Before 1750
HALD · A History of Mathematical Statistics from 1750 to 1930
HAMPEL · Robust Statistics: The Approach Based on Influence Functions
HANNAN and DEISTLER · The Statistical Theory of Linear Systems
HEIBERGER · Computation for the Analysis of Designed Experiments
HEDAYAT and SINHA · Design and Inference in Finite Population Sampling
HELLER · MACSYMA for Statisticians

*Now available in a lower priced paperback edition in the Wiley Classics Library.

HINKELMAN and KEMPTHORNE: · Design and Analysis of Experiments, Volume 1: Introduction to Experimental Design
HOAGLIN, MOSTELLER, and TUKEY · Exploratory Approach to Analysis of Variance
HOAGLIN, MOSTELLER, and TUKEY · Exploring Data Tables, Trends and Shapes
*HOAGLIN, MOSTELLER, and TUKEY · Understanding Robust and Exploratory Data Analysis
HOCHBERG and TAMHANE · Multiple Comparison Procedures
HOCKING · Methods and Applications of Linear Models: Regression and the Analysis of Variance, *Second Edition*
HOEL · Introduction to Mathematical Statistics, *Fifth Edition*
HOGG and KLUGMAN · Loss Distributions
HOLLANDER and WOLFE · Nonparametric Statistical Methods, *Second Edition*
HOSMER and LEMESHOW · Applied Logistic Regression, *Second Edition*
HOSMER and LEMESHOW · Applied Survival Analysis: Regression Modeling of Time to Event Data
HØYLAND and RAUSAND · System Reliability Theory: Models and Statistical Methods
HUBER · Robust Statistics
HUBERTY · Applied Discriminant Analysis
HUNT and KENNEDY · Financial Derivatives in Theory and Practice
HUSKOVA, BERAN, and DUPAC · Collected Works of Jaroslav Hajek—with Commentary
IMAN and CONOVER · A Modern Approach to Statistics
JACKSON · A User's Guide to Principle Components
JOHN · Statistical Methods in Engineering and Quality Assurance
JOHNSON · Multivariate Statistical Simulation
JOHNSON and BALAKRISHNAN · Advances in the Theory and Practice of Statistics: A Volume in Honor of Samuel Kotz
JUDGE, GRIFFITHS, HILL, LÜTKEPOHL, and LEE · The Theory and Practice of Econometrics, *Second Edition*
JOHNSON and KOTZ · Distributions in Statistics
JOHNSON and KOTZ (editors) · Leading Personalities in Statistical Sciences: From the Seventeenth Century to the Present
JOHNSON, KOTZ, and BALAKRISHNAN · Continuous Univariate Distributions, Volume 1, *Second Edition*
JOHNSON, KOTZ, and BALAKRISHNAN · Continuous Univariate Distributions, Volume 2, *Second Edition*
JOHNSON, KOTZ, and BALAKRISHNAN · Discrete Multivariate Distributions
JOHNSON, KOTZ, and KEMP · Univariate Discrete Distributions, *Second Edition*
JUREČKOVÁ and SEN · Robust Statistical Procedures: Aymptotics and Interrelations
JUREK and MASON · Operator-Limit Distributions in Probability Theory
KADANE · Bayesian Methods and Ethics in a Clinical Trial Design
KADANE AND SCHUM · A Probabilistic Analysis of the Sacco and Vanzetti Evidence
KALBFLEISCH and PRENTICE · The Statistical Analysis of Failure Time Data, *Second Edition*
KASS and VOS · Geometrical Foundations of Asymptotic Inference
KAUFMAN and ROUSSEEUW · Finding Groups in Data: An Introduction to Cluster Analysis
KEDEM and FOKIANOS · Regression Models for Time Series Analysis
KENDALL, BARDEN, CARNE, and LE · Shape and Shape Theory
KHURI · Advanced Calculus with Applications in Statistics, *Second Edition*
KHURI, MATHEW, and SINHA · Statistical Tests for Mixed Linear Models
KLEIBER and KOTZ · Statistical Size Distributions in Economics and Actuarial Sciences
KLUGMAN, PANJER, and WILLMOT · Loss Models: From Data to Decisions

*Now available in a lower priced paperback edition in the Wiley Classics Library.

KLUGMAN, PANJER, and WILLMOT · Solutions Manual to Accompany Loss Models: From Data to Decisions
KOTZ, BALAKRISHNAN, and JOHNSON · Continuous Multivariate Distributions, Volume 1, *Second Edition*
KOTZ and JOHNSON (editors) · Encyclopedia of Statistical Sciences: Volumes 1 to 9 with Index
KOTZ and JOHNSON (editors) · Encyclopedia of Statistical Sciences: Supplement Volume
KOTZ, READ, and BANKS (editors) · Encyclopedia of Statistical Sciences: Update Volume 1
KOTZ, READ, and BANKS (editors) · Encyclopedia of Statistical Sciences: Update Volume 2
KOVALENKO, KUZNETZOV, and PEGG · Mathematical Theory of Reliability of Time-Dependent Systems with Practical Applications
LACHIN · Biostatistical Methods: The Assessment of Relative Risks
LAD · Operational Subjective Statistical Methods: A Mathematical, Philosophical, and Historical Introduction
LAMPERTI · Probability: A Survey of the Mathematical Theory, *Second Edition*
LANGE, RYAN, BILLARD, BRILLINGER, CONQUEST, and GREENHOUSE · Case Studies in Biometry
LARSON · Introduction to Probability Theory and Statistical Inference, *Third Edition*
LAWLESS · Statistical Models and Methods for Lifetime Data, *Second Edition*
LAWSON · Statistical Methods in Spatial Epidemiology
LE · Applied Categorical Data Analysis
LE · Applied Survival Analysis
LEE and WANG · Statistical Methods for Survival Data Analysis, *Third Edition*
LePAGE and BILLARD · Exploring the Limits of Bootstrap
LEYLAND and GOLDSTEIN (editors) · Multilevel Modelling of Health Statistics
LIAO · Statistical Group Comparison
LINDVALL · Lectures on the Coupling Method
LINHART and ZUCCHINI · Model Selection
LITTLE and RUBIN · Statistical Analysis with Missing Data, *Second Edition*
LLOYD · The Statistical Analysis of Categorical Data
MAGNUS and NEUDECKER · Matrix Differential Calculus with Applications in Statistics and Econometrics, *Revised Edition*
MALLER and ZHOU · Survival Analysis with Long Term Survivors
MALLOWS · Design, Data, and Analysis by Some Friends of Cuthbert Daniel
MANN, SCHAFER, and SINGPURWALLA · Methods for Statistical Analysis of Reliability and Life Data
MANTON, WOODBURY, and TOLLEY · Statistical Applications Using Fuzzy Sets
MARDIA and JUPP · Directional Statistics
MASON, GUNST, and HESS · Statistical Design and Analysis of Experiments with Applications to Engineering and Science, *Second Edition*
McCULLOCH and SEARLE · Generalized, Linear, and Mixed Models
McFADDEN · Management of Data in Clinical Trials
McLACHLAN · Discriminant Analysis and Statistical Pattern Recognition
McLACHLAN and KRISHNAN · The EM Algorithm and Extensions
McLACHLAN and PEEL · Finite Mixture Models
McNEIL · Epidemiological Research Methods
MEEKER and ESCOBAR · Statistical Methods for Reliability Data
MEERSCHAERT and SCHEFFLER · Limit Distributions for Sums of Independent Random Vectors: Heavy Tails in Theory and Practice
*MILLER · Survival Analysis, *Second Edition*
MONTGOMERY, PECK, and VINING · Introduction to Linear Regression Analysis, *Third Edition*

*Now available in a lower priced paperback edition in the Wiley Classics Library.

MORGENTHALER and TUKEY · Configural Polysampling: A Route to Practical Robustness
MUIRHEAD · Aspects of Multivariate Statistical Theory
MURRAY · X-STAT 2.0 Statistical Experimentation, Design Data Analysis, and Nonlinear Optimization
MURTHY, XIE, and JIANG · Weibull Models
MYERS and MONTGOMERY · Response Surface Methodology: Process and Product Optimization Using Designed Experiments, *Second Edition*
MYERS, MONTGOMERY, and VINING · Generalized Linear Models. With Applications in Engineering and the Sciences
NELSON · Accelerated Testing, Statistical Models, Test Plans, and Data Analyses
NELSON · Applied Life Data Analysis
NEWMAN · Biostatistical Methods in Epidemiology
OCHI · Applied Probability and Stochastic Processes in Engineering and Physical Sciences
OKABE, BOOTS, SUGIHARA, and CHIU · Spatial Tesselations: Concepts and Applications of Voronoi Diagrams, *Second Edition*
OLIVER and SMITH · Influence Diagrams, Belief Nets and Decision Analysis
PALTA · Quantitative Methods in Population Health: Extensions of Ordinary Regressions
PANKRATZ · Forecasting with Dynamic Regression Models
PANKRATZ · Forecasting with Univariate Box-Jenkins Models: Concepts and Cases
*PARZEN · Modern Probability Theory and Its Applications
PEÑA, TIAO, and TSAY · A Course in Time Series Analysis
PIANTADOSI · Clinical Trials: A Methodologic Perspective
PORT · Theoretical Probability for Applications
POURAHMADI · Foundations of Time Series Analysis and Prediction Theory
PRESS · Bayesian Statistics: Principles, Models, and Applications
PRESS · Subjective and Objective Bayesian Statistics, *Second Edition*
PRESS and TANUR · The Subjectivity of Scientists and the Bayesian Approach
PUKELSHEIM · Optimal Experimental Design
PURI, VILAPLANA, and WERTZ · New Perspectives in Theoretical and Applied Statistics
PUTERMAN · Markov Decision Processes: Discrete Stochastic Dynamic Programming
*RAO · Linear Statistical Inference and Its Applications, *Second Edition*
RENCHER · Linear Models in Statistics
RENCHER · Methods of Multivariate Analysis, *Second Edition*
RENCHER · Multivariate Statistical Inference with Applications
RIPLEY · Spatial Statistics
RIPLEY · Stochastic Simulation
ROBINSON · Practical Strategies for Experimenting
ROHATGI and SALEH · An Introduction to Probability and Statistics, *Second Edition*
ROLSKI, SCHMIDLI, SCHMIDT, and TEUGELS · Stochastic Processes for Insurance and Finance
ROSENBERGER and LACHIN · Randomization in Clinical Trials: Theory and Practice
ROSS · Introduction to Probability and Statistics for Engineers and Scientists
ROUSSEEUW and LEROY · Robust Regression and Outlier Detection
RUBIN · Multiple Imputation for Nonresponse in Surveys
RUBINSTEIN · Simulation and the Monte Carlo Method
RUBINSTEIN and MELAMED · Modern Simulation and Modeling
RYAN · Modern Regression Methods
RYAN · Statistical Methods for Quality Improvement, *Second Edition*
SALTELLI, CHAN, and SCOTT (editors) · Sensitivity Analysis
*SCHEFFE · The Analysis of Variance
SCHIMEK · Smoothing and Regression: Approaches, Computation, and Application
SCHOTT · Matrix Analysis for Statistics

*Now available in a lower priced paperback edition in the Wiley Classics Library.

SCHUSS · Theory and Applications of Stochastic Differential Equations
SCOTT · Multivariate Density Estimation: Theory, Practice, and Visualization
*SEARLE · Linear Models
SEARLE · Linear Models for Unbalanced Data
SEARLE · Matrix Algebra Useful for Statistics
SEARLE, CASELLA, and McCULLOCH · Variance Components
SEARLE and WILLETT · Matrix Algebra for Applied Economics
SEBER and LEE · Linear Regression Analysis, *Second Edition*
SEBER · Multivariate Observations
SEBER and WILD · Nonlinear Regression
SENNOTT · Stochastic Dynamic Programming and the Control of Queueing Systems
*SERFLING · Approximation Theorems of Mathematical Statistics
SHAFER and VOVK · Probability and Finance: It's Only a Game!
SMALL and McLEISH · Hilbert Space Methods in Probability and Statistical Inference
SRIVASTAVA · Methods of Multivariate Statistics
STAPLETON · Linear Statistical Models
STAUDTE and SHEATHER · Robust Estimation and Testing
STOYAN, KENDALL, and MECKE · Stochastic Geometry and Its Applications, *Second Edition*
STOYAN and STOYAN · Fractals, Random Shapes and Point Fields: Methods of Geometrical Statistics
STYAN · The Collected Papers of T. W. Anderson: 1943–1985
SUTTON, ABRAMS, JONES, SHELDON, and SONG · Methods for Meta-Analysis in Medical Research
TANAKA · Time Series Analysis: Nonstationary and Noninvertible Distribution Theory
THOMPSON · Empirical Model Building
THOMPSON · Sampling, *Second Edition*
THOMPSON · Simulation: A Modeler's Approach
THOMPSON and SEBER · Adaptive Sampling
THOMPSON, WILLIAMS, and FINDLAY · Models for Investors in Real World Markets
TIAO, BISGAARD, HILL, PEÑA, and STIGLER (editors) · Box on Quality and Discovery: with Design, Control, and Robustness
TIERNEY · LISP-STAT: An Object-Oriented Environment for Statistical Computing and Dynamic Graphics
TSAY · Analysis of Financial Time Series
UPTON and FINGLETON · Spatial Data Analysis by Example, Volume II: Categorical and Directional Data
VAN BELLE · Statistical Rules of Thumb
VESTRUP · The Theory of Measures and Integration
VIDAKOVIC · Statistical Modeling by Wavelets
WEISBERG · Applied Linear Regression, *Second Edition*
WELSH · Aspects of Statistical Inference
WESTFALL and YOUNG · Resampling-Based Multiple Testing: Examples and Methods for *p*-Value Adjustment
WHITTAKER · Graphical Models in Applied Multivariate Statistics
WINKER · Optimization Heuristics in Economics: Applications of Threshold Accepting
WONNACOTT and WONNACOTT · Econometrics, *Second Edition*
WOODING · Planning Pharmaceutical Clinical Trials: Basic Statistical Principles
WOOLSON and CLARKE · Statistical Methods for the Analysis of Biomedical Data, *Second Edition*
WU and HAMADA · Experiments: Planning, Analysis, and Parameter Design Optimization
YANG · The Construction Theory of Denumerable Markov Processes
*ZELLNER · An Introduction to Bayesian Inference in Econometrics
ZHOU, OBUCHOWSKI, and McCLISH · Statistical Methods in Diagnostic Medicine

*Now available in a lower priced paperback edition in the Wiley Classics Library.